Lecture Notes in Computer Science 3322

Commenced Publication in 1973
Founding and Former Series Editors:
Gerhard Goos, Juris Hartmanis, and Jan van Leeuwen

Reinhard Klette   Jovisa Žunić (Eds.)

Combinatorial Image Analysis

10th International Workshop, IWCIA 2004
Auckland, New Zealand, December 1-3, 2004
Proceedings

Volume Editors

Reinhard Klette
University of Auckland
Tamaki Campus, CITR
Glen Innes, Morrin Road, Building 731, Auckland 1005, New Zealand
E-mail: r.klette@cs.auckland.ac.nz

Jovisa Žunić
Exeter University
Computer Science Department
Harrison Building, Exeter EX4 4QF, U.K.
E-mail: J.Zunic@ex.ac.uk

Library of Congress Control Number: 2004115523

CR Subject Classification (1998): I.4, I.5, I.3.5, F.2.2, G.2.1, G.1.6

ISSN 0302-9743
ISBN 3-540-23942-1 Springer Berlin Heidelberg New York

Springer is a part of Springer Science+Business Media

springeronline.com

Printed in Germany

Typesetting: Camera-ready by author, data conversion by Scientific Publishing Services, Chennai, India
Printed on acid-free paper SPIN: 11360971 06/3142 5 4 3 2 1 0

Preface

This volume presents the proceedings of the 10th International Workshop on Combinatorial Image Analysis, held December 1–3, 2004, in Auckland, New Zealand. Prior meetings took place in Paris (France, 1991), Ube (Japan, 1992), Washington DC (USA, 1994), Lyon (France, 1995), Hiroshima (Japan, 1997), Madras (India, 1999), Caen (France, 2000), Philadelphia (USA, 2001), and Palermo (Italy, 2003). For this workshop we received 86 submitted papers from 23 countries. Each paper was evaluated by at least two independent referees. We selected 55 papers for the conference. Three invited lectures by Vladimir Kovalevsky (Berlin), Akira Nakamura (Hiroshima), and Maurice Nivat (Paris) completed the program.

Conference papers are presented in this volume under the following topical part titles: discrete tomography (3 papers), combinatorics and computational models (6), combinatorial algorithms (6), combinatorial mathematics (4), digital topology (7), digital geometry (7), approximation of digital sets by curves and surfaces (5), algebraic approaches (5), fuzzy image analysis (2), image segmentation (6), and matching and recognition (7). These subjects are dealt with in the context of digital image analysis or computer vision.

The editors thank all the referees for their big effort in reading and evaluating the submissions and maintaining the high standard of IWCIA conferences. We are also thankful to the sponsors of IWCIA 2004: the University of Auckland, in particular its Tamaki campus, for hosting the workshop, IAPR (the International Association for Pattern Recognition) for advertising the event, the Royal Society of New Zealand for its financial support, and CITR (the Centre for Image Technology and Robotics at Tamaki campus) and the Computer Science Department of the University of Auckland for providing the day-by-day support during the organizing the event. Also, many thanks to the members of the organizing and scientific committees, which made this conference possible.

September 2004 Reinhard Klette and Joviša Žunić

Organization

IWCIA 2004 was organized by the CITR—Centre for Image Technology and Robotics at Tamaki Campus—and the Computer Science Department, of the University of Auckland, New Zealand.

Executive Committee

Conference Co-chairs	Reinhard Klette (University of Auckland)
	Joviša Žunić (Exeter University)
Scientific Secretariat	Patrice Delmas
	Gisela Klette
Organizing Committee	Penny Barry
	Cliff Hawkis
	Reinhard Klette (Chair)
	Cecilia Lourdes

Referees

E. Andres
J. Baltes
R. Barneva
G. Bertrand
G. Borgefors
V. Brimkov
T. Buelow
C. Calude
J.-M. Chassery
C.-Y. Chen
D. Coeurjolly
M. Conder
I. Debled-Rennesson
P. Delmas
U. Eckhardt
V. di Gesu
A. Hanbury
G. Herman
A. Imiya
K. Inoue
K. Kawamoto
N. Kiryati
C. Kiselman
R. Klette
T.Y. Kong
V. Kovalevski
R. Kozera
W. Kropatsch
A. Kuba
L.J. Latecki
B. MacDonald
M. Moell
K. Morita
I. Nystrom
R. Reulke
J.B.T.M. Roerdink
C. Ronse
B. Rosenhahn
R. Strand
M. Tajine
G. Tee
K. Voss
T. Wei
J. Weickert
G. Woeginger
Q. Zang
J. Žunić

Sponsoring Institutions

University of Auckland, New Zealand
IAPR, International Association for Pattern Recognition
Royal Society of New Zealand

Table of Contents

Discrete Tomography

Combinatorics and Computational Models

Combinatorial Algorithms

Combinatorial Mathematics

Digital Topology

Digital Geometry

Approximation of Digital Sets by Curves and Surfaces

Algebraic Approaches

Fuzzy Image Analysis

Image Segmentation

Matching and Recognition

Binary Matrices Under the Microscope: A Tomographical Problem

Andrea Frosini[1] and Maurice Nivat[2]

[1] Dipartimento di Scienze Matematiche ed Informatiche "Roberto Magari", Università degli Studi di Siena, Pian dei Mantellini 44, 53100, Siena, Italy
frosini@unisi.it

[2] Laboratoire d'Informatique, Algorithmique, Fondements et Applications (LIAFA), Université Denis Diderot 2, place Jussieu 75251 Paris 5Cedex 05, France
Maurice.Nivat@liafa.jussieu.fr

Abstract. A binary matrix can be scanned by moving a fixed rectangular window (sub-matrix) across it, rather like examining it closely under a microscope. With each viewing, a convenient measurement is the number of 1s visible in the window, which might be thought of as the *luminosity* of the window. The *rectangular scan* of the binary matrix is then the collection of these luminosities presented in matrix form. We show that, at least in the technical case of a *smooth* $m \times n$ binary matrix, it can be reconstructed from its rectangular scan in polynomial time in the parameters m and n, where the degree of the polynomial depends on the size of the window of inspection. For an arbitrary binary matrix, we then extend this result by determining the entries in its rectangular scan that preclude the smoothness of the matrix.

Keywords: Discrete Tomography, Reconstruction algorithm, Computational complexity, Projection, Rectangular scan.

1 Introduction and Definitions

The aim of *discrete tomography* is the retrieval of geometrical information about a physical structure, regarded as a finite set of points in the integer square lattice $\mathbb{Z} \times \mathbb{Z}$, from measurements, generically known as *projections*, of the number of atoms in the structure that lie on lines with fixed scopes. A common simplification is to represent a finite physical structure as a binary matrix, where an entry is 1 or 0 according as an atom is present or absent in the structure at the corresponding point of the lattice. The challenge is then to reconstruct key features of the structure from some scan of projections.

Our interest here, following [2], is to probe the structure, not with lines of fixed scope, but with their natural two dimensional analogue, rectangles of fixed scope, much as we might examine a specimen under a microscope or magnifying glass. For each position of our rectangular probe, we count the number of visible atoms, or, in the simplified binary matrix version of the problem, the number of 1s in the prescribed rectangular window, which we term its *luminosity*. In the

R. Klette and J. Žunić (Eds.): IWCIA 2004, LNCS 3322, pp. 1–22, 2004.

matrix version of the problem, these measurements can themselves be organized in matrix form, called the *rectangular scan* of the original matrix. Our first objective is then to furnish a strategy to reconstruct the original matrix from its rectangular scan. As we also note, our investigation is closely related to results on tiling by translation in the integer square lattice discussed in [2].

To be more precise, let M be an $m \times n$ binary matrix, and, for fixed p and q, with $1 \leq p \leq m, 1 \leq q \leq n$, consider a $p \times q$ window $R_{p,q}$ allowing us to view the intersection of any p consecutive rows and q consecutive columns of M. Then, the number $R_{p,q}(M)[i,j]$ of 1s in M on view when the top left hand corner of $R_{p,q}$ is positioned over the (i,j)-entry, $M[i,j]$, of M, is given by summing all the entries on view:

$$R_{p,q}(M)[i,j] = \sum_{r=0}^{p-1}\sum_{c=0}^{q-1} M[i+r, j+c], \quad 1 \leq i \leq m-p+1, \quad 1 \leq j \leq n-q+1.$$

Thus, we obtain an $(m-p+1) \times (n-q+1)$ matrix $R_{p,q}(M)$ with non-negative integer entries $R_{p,q}(M)[i,j]$, as illustrated in Fig. 1. We call $R_{p,q}(M)$ the (p,q)-*rectangular scan* of M; when p and q are understood, we write $R(M) = R_{p,q}(M)$, and speak more simply of the *rectangular scan*. (This terminology is a slight departure from that found in [2].) In the special case when $R(M)$ has all entries equal, say k, we say that the matrix M is *homogeneous* of *degree* k.

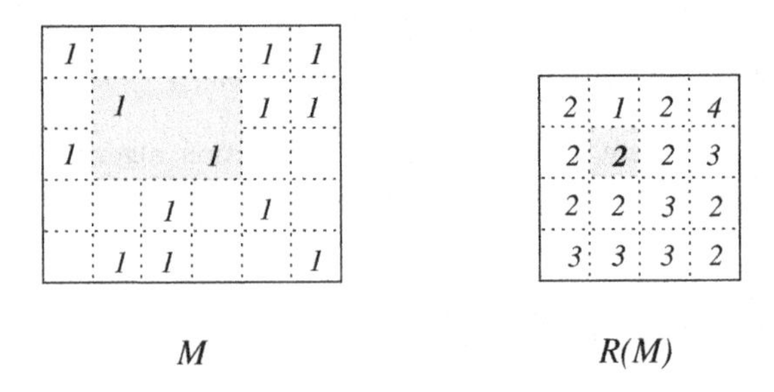

Fig. 1. *A binary matrix M (entries 0 are omitted) and its $(2,3)$-rectangular scan. The sum of the elements of M inside the highlighted rectangle determines the highlighted element of $R(M)$*

Given any $m \times n$ matrix A and integers p and q with $1 \leq p \leq m$ and $1 \leq q \leq n$, we define an $(m-p) \times (n-q)$ matrix $\chi_{p,q}(A) = (\chi_{p,q}(A)[i,j])$ by setting, for $1 \leq i \leq m-p$, $1 \leq j \leq n-q$:

$$\chi_{p,q}(A)[i,j] = A[i,j] + A[i+p, j+q] - A[i+p, j] - A[i, j+q];$$

note that, in the case where A is a *binary* matrix, these entries take only the values -2, -1, 0, 1, or 2 (see Fig. 4). As usual, when p and q can be understood without ambiguity, we suppress them as subscripts. In the event that the matrix

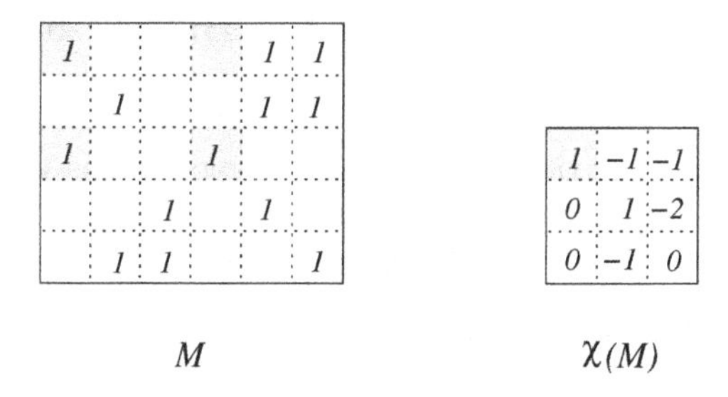

Fig. 2. *A matrix M and its corresponding matrix $\chi_{3,4}(M)$*

$\chi(A)$ is a zero matrix, the matrix A is said to be *smooth*. Notice that the homogeneous matrices are *properly* included in the smooth matrices; an example of a smooth matrix that is not homogeneous is shown in Fig. 3.

We conclude this introductory section with three observations that are direct consequences of our definitions and to which we shall have frequent recourse in what follows. Since their proofs are a matter of simple computation, they are omitted.

Lemma 1. *If M_1 and M_2 are two $m \times n$ binary matrices, then*

$$R(M_1 + M_2) = R(M_1) + R(M_2) \quad \textit{and} \quad \chi(M_1 + M_2) = \chi(M_1) + \chi(M_2).$$

Lemma 2. *If M is a binary matrix, then*

$$\chi_{1,1}(R_{p,q}(M)) = \chi_{p,q}(M).$$

Thus the rectangular scan $R(M)$ of a binary matrix M already contains sufficient information to compute $\chi(M)$ and so to tell whether M is smooth. Notice that, with a certain terminological inexactitude, we can also say, in the case where M is smooth, that $R(M)$ is smooth (more precisely, $R(M)$ is $(1,1)$-smooth, while M itself is (p,q)-smooth, as our more careful statement of the Lemma makes clear).

An appeal to symmetry and induction yields the following generalization of [2, Lemma 2.2].

Lemma 3. *If M be a smooth matrix then, for any integers α and β such that $1 \leq i + \alpha p \leq m$ and $1 \leq j + \beta q \leq n$,*

$$M[i,j] + M[i + \alpha p, j + \beta q] = M[i + \alpha p, j] + M[i, j + \beta q].$$

Finally, we say that an entry $M[i,j]$ of a matrix M is *(p,q)-invariant* if, for any integer α such that $1 \leq i + \alpha p \leq m$ and $1 \leq j + \beta q \leq n$,

$$M[i + \alpha p, j + \alpha q] = M[i,j].$$

If all the entries of M are (p,q)-invariant, then M is said to be (p,q)-invariant.

2 A Decomposition Theorem for Binary Smooth Matrices

In this section we extend the studies about homogeneous matrices started in [2] to the class of smooth matrices, and we prove a decomposition theorem by looking at the invariance of their elements.

Lemma 4. *If M is a smooth matrix, then each of its elements is $(p,0)$-invariant or $(0,q)$-invariant.*

Proof. Since M is smooth, for each $1 \leq i \leq m-p$ and $1 \leq j \leq n-q$, it holds

$$M[i,j] + M[i+p, j+q] = M[i+p, j] + M[i, j+q].$$

Let us consider the following four possibilities for the element $M[i,j]$:

- $M[i,j] \neq M[i+p,j]$: by Lemma 3 for $\alpha = 1$ and for all $\beta \in \mathbb{Z}$ such that $1 \leq j+\beta q \leq n$, it holds $M[i,j+\beta q] = M[i,j]$ and $M[i+p,j+\beta q] = M[i+p,j]$, so $M[i,j]$ is $(0,q)$-invariant.
- $M[i,j] \neq M[i,j+q]$: by reasoning similarly to the previous point, we obtain that $M[i,j]$ is $(p,0)$-invariant.
- $M[i,j] = M[i,j+q] = M[i+p,j]$: if there exists $\alpha_0 \in \mathbb{Z}$ such that $M[i+\alpha_0 p, j] \neq M[i,j]$, again reasoning as in the first case we obtain that $M[i,j]$ is $(0,q)$-invariant. On the other hand, if for all $1 \leq i+\alpha p \leq m$ it holds that $M[i+\alpha p, j] = M[i,j]$, then $M[i,j]$ is $(p,0)$-invariant.

Finally, if $m-p+1 \leq i \leq m$ and $n-q+1 \leq j \leq n$, a similar reasoning leads again to the thesis. □

The reader can check that each entry of the smooth matrix M in Fig. 3 is $(2,0)$-invariant (the highlighted ones) or $(0,3)$-invariant.

Theorem 1. *A matrix M is smooth if and only if it can be obtained by summing up a $(p,0)$-invariant matrix M_1 and a $(0,q)$-invariant matrix M_2 such that they do not have two entries 1 in the same position.*

Proof. ($\Rightarrow$) Let M_1 and M_2 contain the $(p,0)$-invariant and the $(0,q)$-invariant elements of M, respectively. By Lemma 4, the thesis is achieved.

($\Leftarrow$) Since M_1 is $(p,0)$-invariant, then for each $1 \leq i \leq m-p$, $1 \leq j \leq n-q$ it holds

$$\begin{aligned}\chi(M_1)[i,j] &= M_1[i,j] + M_1[i+p,j+q] - M_1[i+p,j] - M_1[i,j+q] = \\ &= M_1[i,j] + M_1[i,j+q] - M_1[i,j] - M_1[i,j+q] = 0\end{aligned}$$

so, by definition, M_1 is smooth. The same result holds for M_2 and, consequently, for $M = M_1 + M_2$. □

Obviously, the converse of Lemma 4 holds. Now we furnish a series of results which lead to the formulation of a reconstruction algorithm for smooth matrices from their rectangular scan:

Lemma 5. *The following statements hold:*
1) *if M is $(0,q)$-invariant, then $R(M)$ has constant rows;*
2) *if M is $(p,0)$-invariant, then $R(M)$ has constant columns.*

Proof. 1) For each $1 \leq i \leq m-p+1$ and $1 \leq j \leq n-q$, we prove that $R(M)[i,j] = R(M)[i,j+1]$:

$$R(M)[i,j+1] = \sum_{r=0}^{p-1}\sum_{c=0}^{q-1} M[i+r,j+1+c] =$$
$$\sum_{r=0}^{p-1}\sum_{c=1}^{q-1} M[i+r,j+c] + \sum_{r=0}^{p-1} M[i+r,j+q] =$$

since M is $(0,q)$-invariant

$$\sum_{r=0}^{p-1}\sum_{c=1}^{q-1} M[i+r,j+c] + \sum_{r=0}^{p-1} M[i+r,j] = R(M)[i,j].$$

2) The proof is similar to 1). □

As a direct consequence of the above results we have:

Theorem 2. *A binary matrix M is smooth if and only if $R(M)$ can be decomposed into two matrices R_r and R_c having constant rows and columns, respectively.*

Fig. 3 shows that the converse of the two statements of Lemma 5 does not hold in general. However, we can prove the following weaker version:

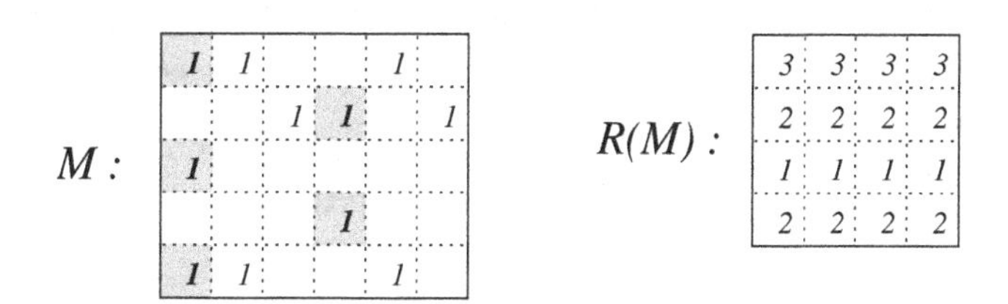

Fig. 3. *A non invariant matrix M whose $(2,3)$-rectangular scan has constant rows*

Lemma 6. *Let M be a binary matrix. The following statements hold:*
1) *if $R(M)$ has constant rows, then there exists a $(0,q)$-invariant matrix M' such that $R(M) = R(M')$;*
2) *if $R(M)$ has constant columns, then there exists a $(p,0)$-invariant matrix M'' such that $R(M) = R(M'')$.*

Proof. 1) We define the matrix M' as follows: the first p columns of M' are equal to those of M, while the other entries of M' are set according to the desired $(0,q)$-invariance. It is easy to verify that $R(M') = R(M)$.
2) The proof is similar to 1). □

2.1 A Reconstruction Algorithm for Smooth Matrices

Lemma 6 assures the correctness of the following reconstruction algorithm:

RecConstRows(A, p, q)

Input: an integer matrix A having constant rows and two integers p and q.

Output: a $(0, q)$-invariant matrix M, having A as (p, q) rectangular scan, if it exists, else return FAILURE.

Procedure:

Step 1: let $e_1, \ldots, e_k$ be the sequence of all the possible configurations of the elements $M[i, j]$, with $1 \leq i \leq p$ and $1 \leq j \leq q$, whose sum equals $A[1, 1]$;

Step 2: for $1 \leq t \leq k$,

Step 2.1: initialize matrix M with the configuration e_t;

Step 2.2: complete the first q columns of M according with the entries of the first column of A, if possible, and set t to the value $k+2$, else increase t by one;

Step 3: if t is equal to $k + 1$ then FAILURE, else complete the entries of M according to the $(0, q)$-invariance constraint, and return M as OUTPUT.

A simple remark about Step 2.2 is needed: the strategy which allows to complete the first q columns of M, i.e. from row $p+1$ to row m, can be a greedy one. More precisely, for $2 \leq i \leq m-p+1$, one can set in the leftmost positions of row $p+i-1$ of M as many entries 1 as needed in order to reach the value $A[i, 1]$. A similar algorithm, say RecConstCols(A, p, q), can be defined to reconstruct a smooth matrix whose rectangular scan A has constant columns.

Theorem 3. *The computational complexity of* RecConstRows(A, p, q) *is* $O(k'(m\,n))$, *where* m *and* n *are the dimensions of the reconstructed matrix* M, *and* k' *is exponential in* p *and* q.

The result immediately follows after observing that k' grows as fast as

$$k = \binom{p\,q}{A[1,1]}.$$

Example 1. Let us follow the computation RecConstRows($A, 3, 4$), with A as depicted in Fig. 4.

A :

5	5	5	5	5	5
7	7	7	7	7	7
6	6	6	6	6	6
8	8	8	8	8	8
5	5	5	5	5	5

Fig. 4. The matrix A of Example 1

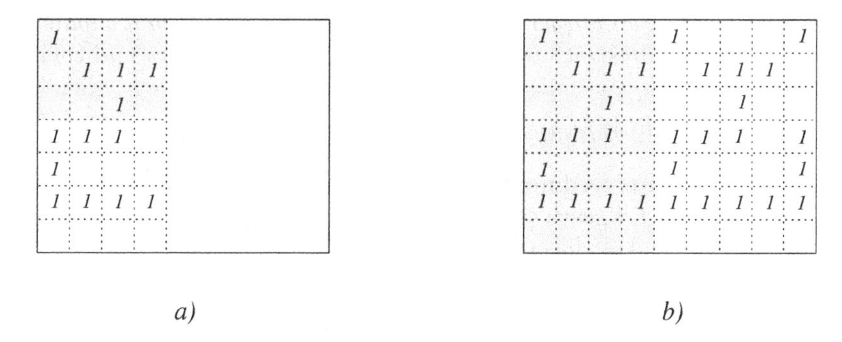

a) b)

Fig. 5. *Two steps of* RECCONSTROWS$(A, 3, 4)$

In Step 1, all the possible $\binom{k=12}{5}$ configurations of a 3×4 rectangle containing exactly five entries 1, are listed.

In Step 2, the matrix M is created and its upper-left 3×4 submatrix is updated with each of the k configurations, until one is found which can be extended to complete the first four columns of M. In Fig. 5 a), it is highlighted one of the configurations, and a way of filling the first four columns of M starting from it, by means of a greedy strategy.

Finally, in Step 3, the placed entries are propagated in order to make M a $(0, 4)$-invariance matrix (Fig. 5, b)).

Obviously, a smarter strategy for lowering the complexity of the part of the procedure which is exponential in p and q, can be found, but it could compromise the easiness of its formulation in order to get a non significant improvement from a theoretical point of view.

Two simple lemmas are needed, if we want to keep the complexity of the general reconstruction procedure still polynomial in m and n:

Lemma 7. *Let M be a smooth matrix. If there exist two columns, say j and j', such that $R(M)[1, j'] \leq R(M)[1, j]$, then, for each $1 \leq i \leq m - p + 1$ we have $R(M)[i, j'] \leq R(M)[i, j]$.*

Proof. Let us consider a decomposition of $R(M)$ into matrices R_r and R_c having constant rows and columns, respectively, as stated in Theorem 2; moreover it is convenient to denote by r_i (resp. c_j) the common value of the elements of the i-th row (resp. j-th column) of the matrix R_r (resp. R_c).

By hypothesis, we have that $R(M)[1, j'] \leq R(M)[1, j]$, which means that $r_1 + c_{j'} \leq r_1 + c_j$, and then $c_{j'} \leq c_j$. Therefore, for any $1 \leq i \leq m - p + 1$ we have $r_i + c_{j'} \leq r_i + c_j$, so the thesis. □

As a neat consequence of Lemma 7, if there exists a row i_0, and two columns j and j' of $R(M)$ such that $R(M)[i_0, j'] = R(M)[i_0, j]$, then, for all $1 \leq i \leq m - p + 1$, it holds $R(M)[i, j'] = R(M)[i, j]$.

Lemma 8. *Let A be an integer matrix. If A is $(1,1)$-smooth, then it admits $k+1$ different decompositions into two matrices having constant rows and columns, where*

$$k = min_{i,j}\{R(M)[i,j] \,:\, 1 \leq i \leq m-p+1, 1 \leq j \leq n-q+1\}.$$

Proof. Let us give a procedure which lists all the possible couples of matrices (A_r^t, A_c^t), with $0 \leq t \leq k$, each of them being a decomposition of A such that A_r^t has constant rows and A_c^t has constant columns:

DECOMPOSE (A)

Input: an integer matrix A of dimension $m \times n$.

Output: a sequence of couples of matrices $(A_r^0, A_c^0), \ldots, (A_r^k, A_c^k)$, with k being the minimum value of A, such that, for each $1 \leq t \leq k$, A_r^t has constant rows, A_c^t has constant columns, and $A_r^t + A_c^t = A$. If such a sequence does not exist, then return FAILURE.

Procedure:

Step 1: from each element of A, subtract the value k and store the result in the matrix A_c;

Step 2: for each $1 \leq i \leq m$

Step 2.1: compute

$$k_i = min_j\{A_c[i,j] \,:\, 1 \leq j \leq n\};$$

Step 2.2: subtract the value k_i from each element of A_c;

Step 2.3: set all the elements of column i of matrix A_r to the value k_i;

Step 3: if the matrix A_c has not constant columns, then give FAILURE as output, else, for each $0 \leq t \leq k$, create matrices A_r^t and A_c^t such that

$$A_r^t[i,j] = A_r[i,j] + t \qquad \text{and} \qquad A_c^t[i,j] = A_c[i,j] + k - t,$$

with $1 \leq i \leq m$ and $1 \leq j \leq n$. Give (A_r^t, A_c^t) as output.

Example 2 shows a run of the algorithm. By construction, each couple (A_r^t, A_c^t) is a decomposition of A, and furthermore, A_r^t has constant rows.

What remains to prove is that the matrix A_c used in Step 3 has constant columns (and, consequently, the same hold for all the matrices $A_c^t = A_c + k - t$). As usual, let us denote by r_i the common value of the elements of the i-th row of A_r, and let us proceed by contradiction, assuming that A_c has not constant columns. Since A is the sum of a column constant and a row constant matrix, and for all $1 \leq i \leq m-p+1$ and $1 \leq j \leq n-q+1$, it holds $A[i,j] = r_i + A_c[i,j] + k$, then A_c is also the sum of a column constant matrix, say A_{cc}, and a row constant matrix, say A_{cr}, which has at least one row, say i_0, whose elements have value $k'_{i_0} \neq 0$.

This situation generates an absurdity, since k_{i_0} computed in Step 2.1 turns out no longer to be the minimum of row i_0 in A_c, updated to that step.

Since a matrix having constant rows (resp. columns) cannot be obtained as sum of a matrix having constant rows and a matrix having constant columns unless the latter is a constant matrix, then the $k+1$ decompositions listed by the algorithm are all the possible ones. □

Example 2. Let us follow the steps of the procedure DECOMPOSE(A), with matrix A depicted in Fig.6, as described in the proof of Lemma 8.

A :

6	5	5	4	5
5	4	4	3	4
4	3	3	2	3
5	4	4	3	4

Fig. 6. *The* $(1,1)$*-smooth matrix* A *of Example 2*

Step 1: we subtract from all the elements of A, the value $k = 2$, i.e. its minimum element, and we store the obtained result in the matrix A_c.

Step 2: for each $1 \leq i \leq m-p+1$, we find the minimum value k_i among the elements of row i of A_c (Step 2.1), we subtract it from all these elements (Step 2.2), and finally, we set the elements in row i of A_r to the value k_i (Step 2.3). In our case, the minimums are $k_1 = 2$, $k_2 = 1$, $k_3 = 0$ and $k_4 = 1$.

A_r^0 :

2	2	2	2	2
1	1	1	1	1
0	0	0	0	0
1	1	1	1	1

A_r^1 :

3	3	3	3	3
2	2	2	2	2
1	1	1	1	1
2	2	2	2	2

A_r^2 :

4	4	4	4	4
3	3	3	3	3
2	2	2	2	2
3	3	3	3	3

A_c^0 :

4	3	3	2	3
4	3	3	2	3
4	3	3	2	3
4	3	3	2	3

A_c^1 :

3	2	2	1	2
3	2	2	1	2
3	2	2	1	2
3	2	2	1	2

A_c^2 :

2	1	1	0	1
2	1	1	0	1
2	1	1	0	1
2	1	1	0	1

Fig. 7. *The three decompositions of A*

Step 3: the matrix A_c updated at the end of Step 2 has constant columns, so the three different decompositions of A can be computed and listed.
The output is depicted in Fig. 7.

All the previous results are useful for getting the main one of this section, i.e. the following general procedure which reconstructs a smooth matrix from its rectangular scan:

RecSmooth(A, p, q)

Input: a $(1,1)$-smooth matrix A and two integers p and q.

Output: a binary (smooth) matrix M such that $R_{p,q}(M) = A$, if it exists, else return FAILURE.

Procedure:

Step 1: let $m - p + 1 \times n - q + 1$ be the dimension of A (this choice allows to obtain the matrix M, if it exists, of dimension $m \times n$), and let

$$k = min_{i,j}\{A[i,j] \; : \; 1 \leq i \leq m - p + 1, 1 \leq j \leq n - q + 1\}.$$

For each $0 \leq t \leq k$,

Step 1.1: let A_r^t and A_c^t be the t-th decomposition of A into matrices obtained by the call Decompose(A);

Step 1.2: use procedure RecConstRows to obtain a sequence s_1 of *all* the matrices having constant rows and whose rectangular scan is A_r^t. Use procedure RecConstCols to obtain a sequence s_2 of *all* the matrices having constant columns and whose rectangular scan is A_c^t;

Step 1.3: make all the possible sums of an element of s_1 with an element of s_2. If an $m \times n$ binary matrix M is obtained, then goto Step 2, else increase t by one (and go back to Step 1);

Step 2: if $t \neq k + 1$ then gives M as output, else return FAILURE.

The correctness of this procedure immediately follows from the correctness of RecConstRows and RecConstCols, and from Lemma 8.

Example 3. Let us consider the decomposition of a rectangular scan into the matrices depicted in Fig. 8, and let $p = 3$ and $q = 3$.

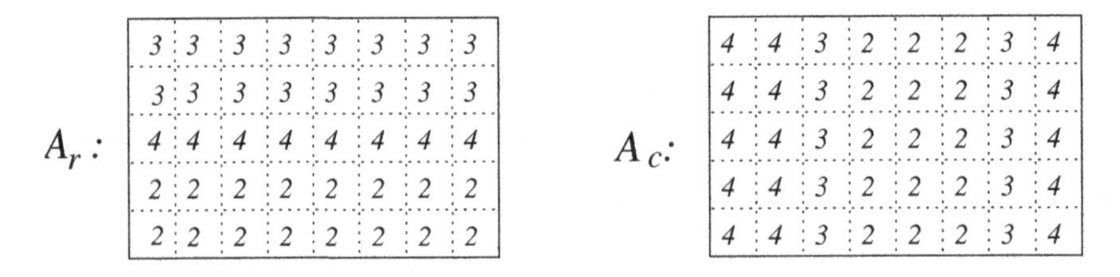

Fig. 8. *The decomposition of a rectangular scan used in Example 3*

We reconstruct a matrix $M = M_1 + M_2$ such that $R(M) = A$, $R(M_1) = A_r$ and $R(M_2) = A_c$.

We start to compute and list all the solutions of RecConstRows($A_r, 3, 3$) and RecConstCols($A_c, 3, 3$), and we arrange them in two sequences s_1 and s_2 as requested in Step 1.2. Then, following Step 1.3, the elements of s_1 and s_2 are summed up in all possible ways until the binary matrix M is obtained.

We are aware that the search for M can be carried on in a smarter way, but this will bring no effective contribution to the decreasing of the computational

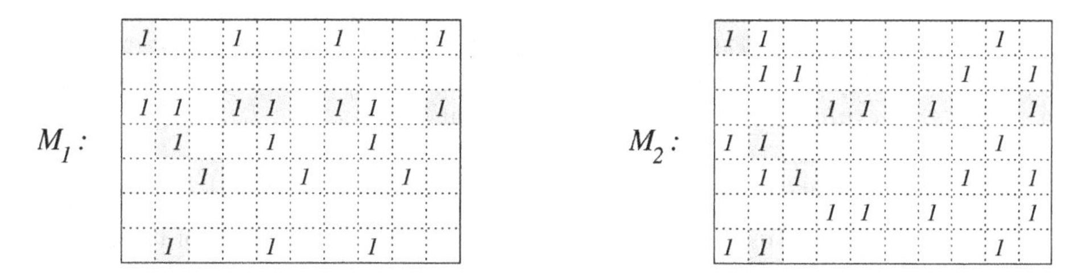

Fig. 9. *Two non-disjoint solutions M_1 and M_2 which are $(0,3)$-invariant and $(3,0)$-invariant, as stated in Lemma 6*

M :

1	1	1	1			1		1	1
1	1			1			1		1
	1	1	1	1	1	1	1	1	1
	1	1						1	
1	1	1		1	1		1	1	1
			1			1			1
1	1	1	1			1		1	1

Fig. 10. *One of the final solutions of the instance in Example 3*

complexity of the reconstruction, and, on the other hand, it will add new lemmas and proofs.

Fig. 9 shows a wrong choice for the matrices M_1 and M_2, since their sum is not a binary matrix. In Fig. 10, it is depicted one of the final solutions M, obtained by a right choice of M_1, whose entries 1 are exactly the highlighted ones of M, and M_2.

The following theorem holds:

Theorem 4. *The computational complexity of* RECSMOOTH(A, p, q) *is polynomial in m and n.*

Proof. The complexity of the algorithm can be computed as the sum of the complexities of all its steps, in particular:

Step 1 finds the minimum element k of A in $O(m\,n)$ (we point out that $k \leq p\,q$ is an immediate consequence of the definition of rectangular scan);
Step 1.1 lists $k+1$ couples of matrices in $O(m\,n)$, and for each of them,
Step 1.2 constructs two sequences of matrices in $O(k'(m\,n))$, with k' independent from m and n, and exponential in p and q, as stated in Theorem 3.
Step 1.3 compares the elements of the two sequences of matrices in $O(m\,n)$.
Step 2 gives matrix M as output in $O(m\,n)$.

The total complexity of the reconstruction process is then $O(m\,n)$. □

Let us call RECSMOOTHALL(A, p, q) the algorithm which naturally extends RECSMOOTH(A, p, q) and which lists all the matrices whose rectangular scan is A. Using the results of this section, we point out that:

Remark 1. If M is a smooth matrix, then RECSMOOTHALL$(R(M), p, q)$ lists it.

This result, widely used in next section, completes the first part of the paper related to the analysis and reconstruction of smooth matrices.

3 The Reconstruction of Binary Matrices from Their Rectangular Scan

In this section, we concentrate on the entries of a rectangular scan which prevent it from being smooth. In particular we define a polynomial time algorithm which lists all the possible matrices which are consistent with them, then we integrate it with the algorithm for reconstructing a smooth matrix defined in the previous section in order to solve the reconstruction problem of a binary matrix from its rectangular scan.

Let $1 \leq a \leq p$, $1 \leq b \leq q$, and A be an integer $m \times n$ matrix. We define (a, b) *subgrid* of A to be the submatrix

$$S(A)_{a,b}[i,j] = A[a + (i-1)\,p, b + (j-1)\,q]$$

where $a + (i-1)\,p \leq m$, and $b + (j-1)\,q \leq n$ (so $S(A)_{a,b}$ turns out to have dimension $\lceil (m-a+1)/p \rceil \times \lceil (n-b+1)/q \rceil$). If we consider again the binary matrix M, by definition it holds that

$$\chi(M)[a + (i-1)\,p, b + (j-1)\,q] = S(\chi(M))_{a,b}[i,j] =$$
$$= S(M)_{a,b}[i,j] + S(M)_{a,b}[i+1,j+1] - S(M)_{a,b}[i+1,j] - S(M)_{a,b}[i,j+1].$$

The matrix V of dimension $m \times n$ is said to be a *valuation* of $S(\chi(M))_{a,b}$ if, for each $1 \leq i \leq m$, $1 \leq j \leq n$,

- if $i \neq (a) mod_p$ and $j \neq (b) mod_q$ then $V[i,j] = 0$;
- $S(\chi(M))_{a,b} = S(\chi(V))_{a,b}$ (see Fig. 11).

This definition immediately leads to the following lemmas:

Lemma 9. *Let $S(\chi(M))_{a,b}$ and $S(\chi(M))_{a',b'}$ be two subgrids whose valuations are V and V', respectively. If $a \neq a'$ or $b \neq b'$, then for each $1 \leq i \leq m$ and $1 \leq j \leq n$, $V[i,j] = 1$ implies $V'[i,j] = 0$.*

This lemma states that the positions having value 1 of two valuations of different subgrids are disjoint.

Lemma 10. *Let V be a valuation of $S(\chi(M))_{a,b}$, and let i_0 be a row of $S(V)_{a,b}$ having all the elements equal to 1. The matrix V' which is equal to V, except in the elements of the row i_0 of $S(V')_{a,b}$ which are set to 0, is again a valuation of $S(\chi(M))_{a,b}$.*

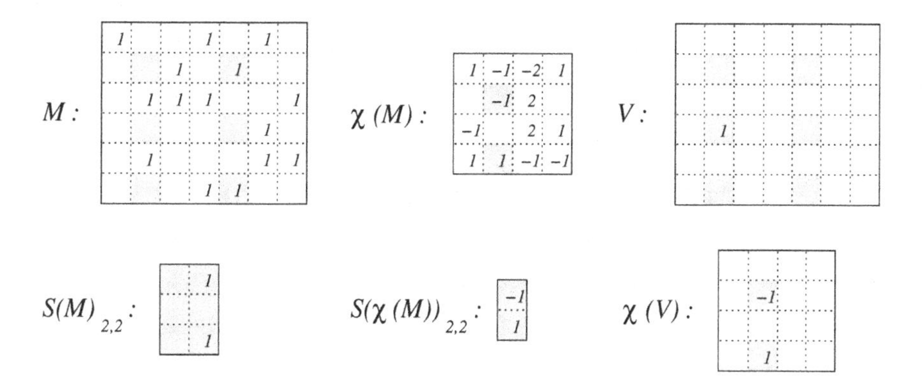

Fig. 11. *The subgrids of the matrices* M *and* $\chi(M)$ *with respect to the position* $(2, 2)$. *Matrix* V *is one of the possible valuations of* $S(\chi(M))_{2,2}$

Roughly speaking, homogeneous changes to the entries of a row of $S(V)_{a,b}$ do not change the rectangular scan of V. A result similar to Lemma 10 obviously holds when we act on a column of $S(V)_{a,b}$.

If V and V' are two valuations as in Lemma 10, then we say that the valuation V is *greater than* the valuation V'. This relation can be easily extended to a finite partial order on the valuations of the subgrids of $\chi(M)$.

Remark 2. Let $1 \leq i \leq m - p$ and $1 \leq j \leq n - q$. If $\chi(M)[i, j] = 2$, then $M[i, j] = M[i + p, j + q] = 1$, and $M[i + p, j] = M[i, j + q] = 0$.

The proof is immediate. A symmetric result holds if $\chi(M)[i, j]$ has value -2.

Now we are ready to state the following crucial lemma which immediately leads to the definition of the procedure RECONSTRUCTION(A, p, q), which reconstructs a binary matrix M from its (p, q) rectangular scan A:

Lemma 11. *Given a binary matrix* M, *for each integers* $1 \leq a \leq p$, $1 \leq b \leq q$, *the number of minimal elements in the partial ordering of the valuations of* $S(\chi(M))_{a,b}$ *is linear with respect to the dimension* $m \times n$ *of* M. *Furthermore, each one of these minimal elements can be reconstructed in polynomial time with respect to* m *and* n.

The proof of this lemma is quite long and tedious, so the authors decided to sketch it separately, in Appendix A. The motivation of this choice is the intent of not moving the attention of the reader too far from the path which leads to the reconstruction of a matrix from its rectangular scan. In addition, we furnish the following example where all the valuations of a matrix $S(\chi(M))_{a,b}$ are listed. The chosen matrix constitutes a sort of worst case configuration.

Example 4. Let us find all the possible minimal valuations of the matrix $S(\chi(M))_{a,b}$ of Fig. 12.

We proceed from the leftmost entry of $S(\chi(M))_{a,b}$ different from 0, till the rightmost one, by examining, one by one, all its possible minimal valuations.

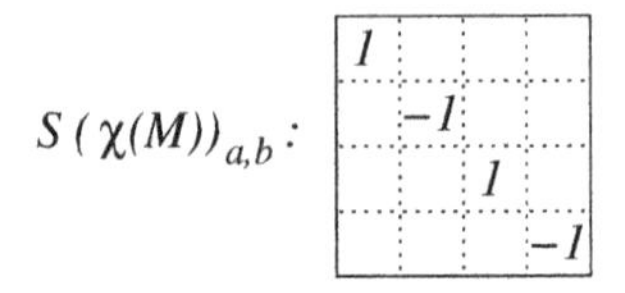

Fig. 12. *The matrix $S(\chi(M))_{a,b}$ of Example 4*

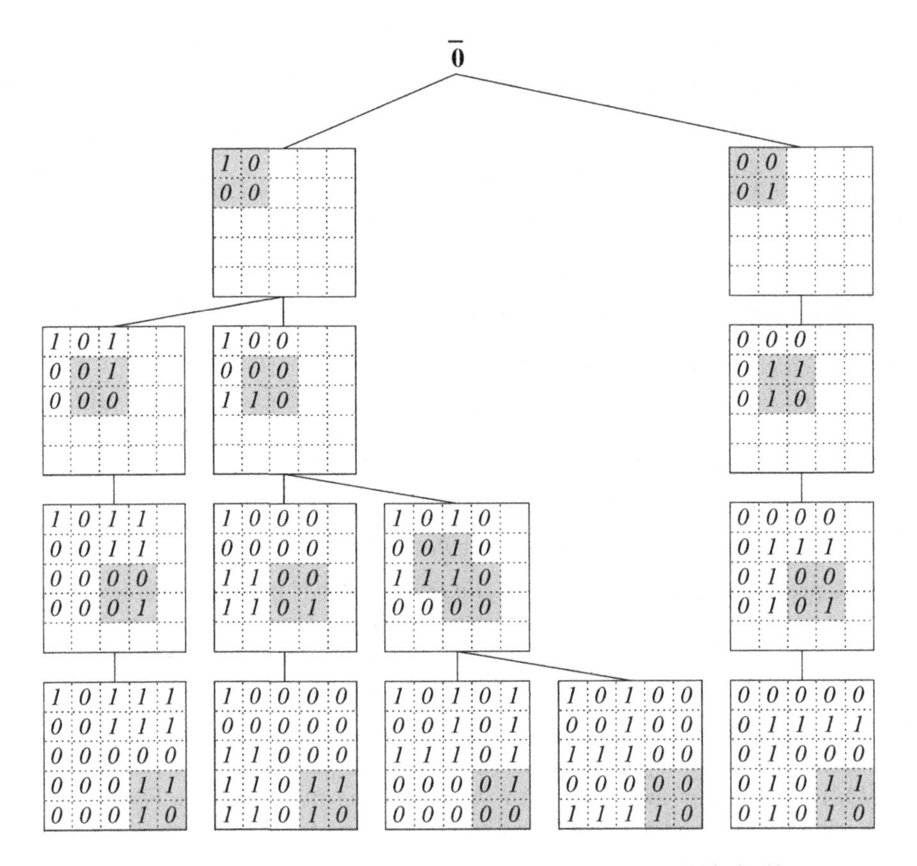

Fig. 13. *The computation of the minimal valuations of $S(\chi(M))_{a,b}$*

The computation is represented in Fig. 13, by using a tree whose root is the (a, b) subgrid of the valuation whose entries are all 0, and whose nodes at level k are all the possible (a, b) subgrids of the valuations of the first k entries different from 0 of $S(\chi(M))_{a,b}$.

On each matrix, the highlighted cells refer the correspondent entry 1 or -1 of $S(\chi(M))_{a,b}$ which is being considered.

It is easy to check that any further addition of 1 or -1 entries in $S(\chi(M))_{a,b}$ does not increase the number of its possible minimal valuations. On the contrary, there often appear non consistent configurations.

RECONSTRUCTION(A, p, q)

Input: an integer matrix A and two integers p and q.

Output: a binary matrix M having A as (p, q) rectangular scan, if it exists, else return FAILURE.

Procedure:

Step 1: for each $1 \leq a \leq p$ and $1 \leq b \leq q$ compute the sequence of valuations $V_{a,b}^{(1)}, \ldots, V_{a,b}^{(k)}$ of $S(\chi(A))_{a,b}$, as shown in Example 4;

Step 2: sum in all possible ways an element from each sequence of valuations (computed in Step 1), and let $M_1, \ldots M_{k'}$ be the obtained sequence of binary matrices;

Step 3: compute the sequence of smooth matrices $A_1, \ldots, A_{k'}$ such that $A_i = A - R(M_i)$, for each $1 \leq i \leq k'$;

Step 4: for each $1 \leq i \leq k'$

Step 4.1 compute RECSMOOTHALL(A_i, p, q) and let $M_i^{(1)}, \ldots, M_i^{(k'')}$ be the output sequence;

Step 4.2: for each $1 \leq j \leq k''$ compute $A_i + M_i^{(j)} = M$. If M is a binary matrix, then return M;

Step 5: return FAILURE.

Lemma 9 assures that each sum performed in Step 2 always returns a binary matrix. Furthermore, since the reconstruction algorithm exhaustively searches all the possible solutions consistent with a given rectangular scan, then its correctness is assured.

However, one can ask whether such a search always produces an output in an amount of time which is polynomial in the dimensions of the solution. The answer is given by the following.

Theorem 5. *The computational complexity of* RECONSTRUCTION(A, p, q) *is polynomial in the dimension* $m \times n$ *of each solution.*

Proof. The complexity of the algorithm can be computed from the complexities of each one of its steps, in particular:

Step 1: for each $1 \leq a \leq p$ and $1 \leq b \leq q$, $S(\chi(A))_{a,b}$ can be computed in $O(m\, n)$. Lemma 11 assures that the number of valuations of $S(\chi(A))_{a,b}$ is $O(m\, n)$, and that each of them can be reconstructed in $O(m\, n)$, so the total amount of time required by Step 1 is $O((m\, n)^2)$.

Step 2: an increasing in the complexity comes from this step: we start from the sequences of valuations created in Step 1, whose number is $p\, q$ at most, and having $O(m\, n)$ elements, and we proceed in summing an element from each of them, in all the possible ways. Since each sum is performed in $O(m\, n)$, and the total number of sums is $O((m\, n)^{p\, q})$, then the complexity increases to $O((m\, n)^{p\, q+1})$. The same amount of time is obviously required in Step 3.

Step 4: by Theorem 4, each call of RECSMOOTHALL takes $O(mn)$ time. Since $O((m\, n)^{p\, q})$ calls are performed (one for each matrix obtained in Step 3), then the total amount of time needed in Step 4 is again $O((m\, n)^{p\, q+1})$, which is also the complexity of the procedure. □

A final example tries to clarify the reconstruction process:

Example 5. Let us follow the computation of RECONSTRUCTION$(A, 3, 3)$, where the rectangular scan A is depicted in Fig. 14.

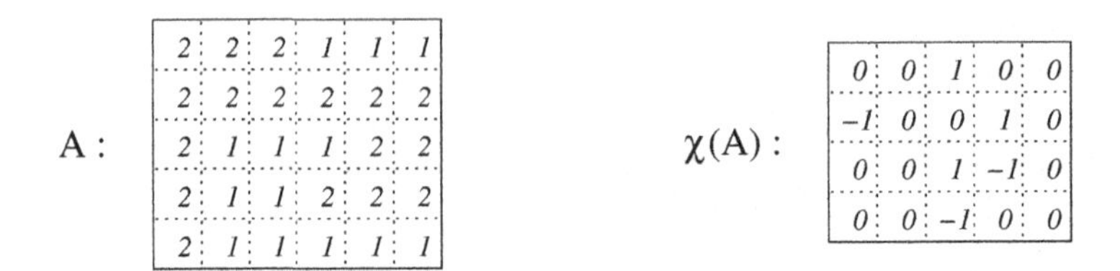

Fig. 14. *The matrix A and the computation of $\chi(A)$*

Step 1: since $\chi(A)$ has some elements different from 0, then A is not smooth. All the valuations of $S(\chi(A))_{a,b}$ are computed:

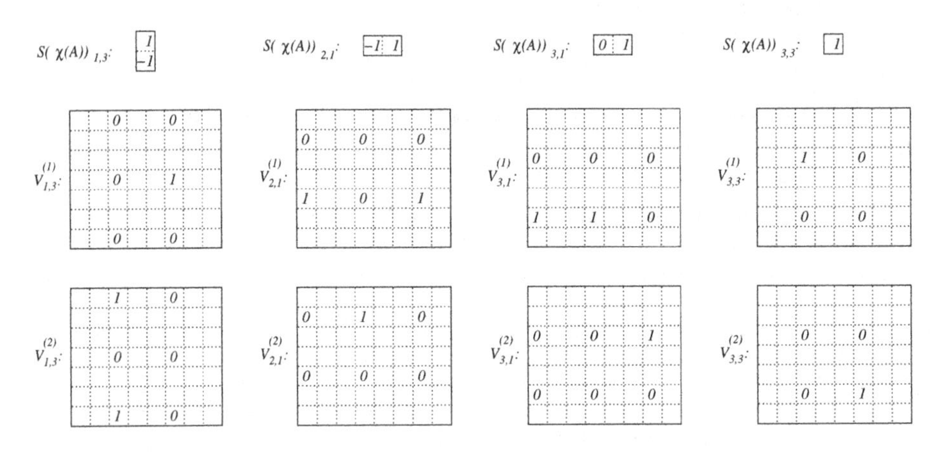

Fig. 15. *All the valuations of $S(\chi(A))_{a,b}$ which are different from the zero matrix*

If $(a, b) = (1, 1)$: the 2×2 matrix $S(\chi(A))_{1,1}$ has all its elements equal to 0, so its minimal valuation is the 7×8 matrix having all the elements equal to 0 (such matrix can be ignored since it gives no contribution to the final sum). The same valuation can be found when (a, b) is set to $(1, 2)$, $(2, 2)$, $(2, 3)$ and $(3, 2)$.

On the other hand, for each $(a, b) \in \{(1, 3), (2, 1), (3, 1), (3, 3)\}$, the matrix $S(\chi(A))_{a,b}$ and its minimal valuations (the 7×8 matrices below it) are depicted in Fig. 15.

Steps 2 and 3: all the possible sums containing exactly an element from each minimal valuations of (a, b) obtained in Step 1 are performed, and then their rectangular scans are subtracted from A, obtaining a new sequence of smooth matrices.

The elements of such a sequence which have positive entries, and so which can be used as input for the algorithm RecAllSmooth, are listed in Fig. 16.

Step 4: each matrix depicted in Fig. 16 is used as input of RecAllSmooth, then Step 4.2 is performed until a solution is obtained. Fig. 17 shows an element of the output of RecAllSmooth$(A - R(V_{1,3}^{(1)} + V_{2,1}^{(2)} + V_{3,1}^{(2)} + V_{3,3}^{(2)}), 3, 3)$, and the correspondent matrix M, solution of the starting reconstruction problem.

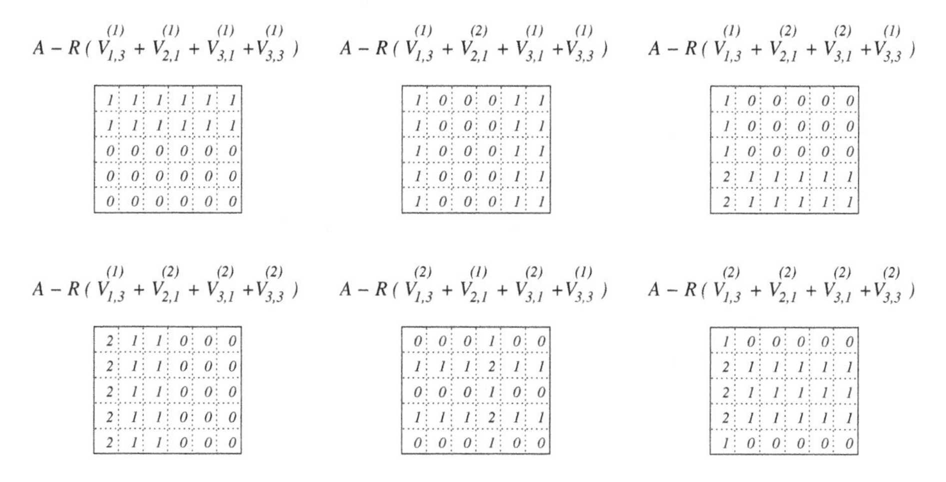

Fig. 16. *The matrices having positive elements computed in Step* 3

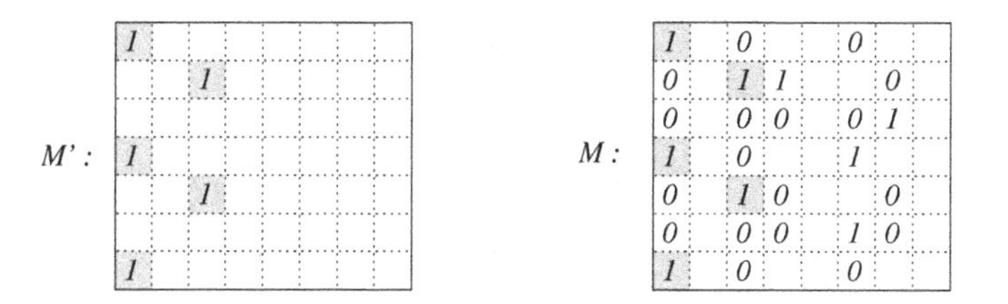

Fig. 17. *The matrices M' and M whose rectangular scans are $A - R(V_{1,3}^{(1)} + V_{2,1}^{(2)} + V_{3,1}^{(2)} + V_{3,3}^{(2)})$ and A, respectively*

References

[1] Herman, G.T., Kuba, A., (eds.): Discrete Tomography: Foundations Algorithms and Applications, Birkhauser Boston, Cambridge, MA (1999)

[2] Nivat, M.: Sous-ensembles homogénes de $\mathbb{Z}^2$ et pavages du plan, C. R. Acad. Sci. Paris, **Ser. I 335** (2002) 83–86

[3] Ryser, H.: Combinatorial properties of matrices of zeros and ones, Canad. J. Math. **9** (1957) 371–377

[4] Tijdeman, R., Hadju, L.: An algorithm for discrete tomography, Linear Algebra and Its Applications **339** (2001) 147–169.

Appendix A

A sketched proof of Lemma 11:

In this proof we assume that $S(\chi(M))_{a,b}$ has dimension $m \times n$, just to simplify the notation. We order the (positions of the) non zero elements of $S(\chi(M))_{a,b}$ according to the numbering of its columns, i.e. from left to right, and, in the same column, according to the numbering of its rows, i.e. from up to bottom, and let $p_1, \ldots, p_t$ be the obtained sequence. We prove the thesis by induction on the number t of elements of the sequence, i.e. we prove that the addition of new nonzero elements in $S(\chi(M))_{a,b}$ does not increase "too much" the number of its possible valuations.

Since Remark 2 states that the presence of entries 2 or -2 in $S(\chi(M))_{a,b}$ does not increase the number of its possible valuations, then we are allowed to focus our attention exactly on the elements having value 1 or -1.

Base $t = 1$: if $p_1 = (i, j)$ and $S(\chi(M))_{a,b}[i, j] = 1$, then the four possible valuations are depicted in Fig. 18. Among them, only V_1 and V_2 are minimal: they can be reached both from V_3 and from V_4 by deleting two rows or two columns entirely filled with entries 1 (see Lemma 10).

If $S(\chi(M))_{a,b}[i, j] = -1$ a symmetrical result holds.

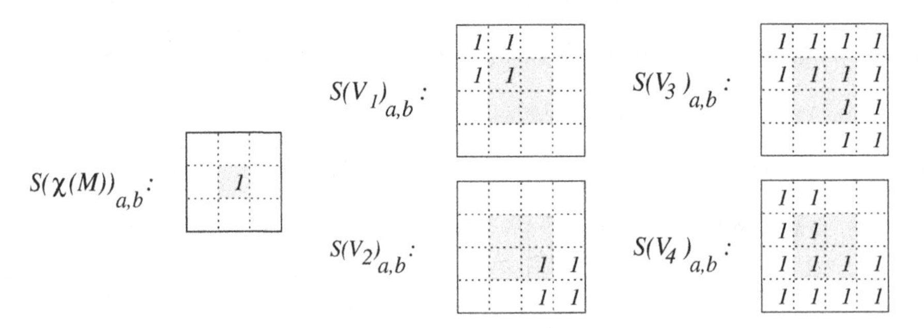

Fig. 18. *The four valuations when $t = 1$, $p_1 = (2, 2)$, and $S(\chi(M))_{a,b}[2, 2] = 1$*

Step $t \to t + 1$: let $S(\chi(M))_{a,b}$ have a sequence $p_1, \ldots, p_{t+1}$ of nonzero points, and let n_t be the number of different minimal valuations of $p_1, \ldots, p_t$, where n_t is linear in t, by inductive hypothesis.

In the sequel, we will show all the possible ways of extending V to the valuation V' which includes the point p_{t+1}.

The thesis will be achieved after proving that only a "small number" of different configurations of V can be extended in more than one way, and then, that the increment in the number of valuations n_t, after adding a single point is bounded by a constant, i.e. $n_t + k \leq n_{t+1}$.

Furthermore, we assume that $p_{t+1} = (i, j)$ and $S(\chi(M))_{a,b}[i, j] = 1$ (if $S(\chi(M))_{a,b}[i, j] = 1$ then symmetrical results hold).

Some pictures are supplied in order to make the different cases transparent. A last remark is needed: since all the elements of $S(\chi(M))_{a,b}$ whose position is greater than $p_{t+1} = (i, j)$ have value 0, then, for all $1 \leq i' \leq m+1$ and $j < j_1 \leq n+1$, it holds $S(V)_{a,b}[i', j+1] = S(V)_{a,b}[i', j_1]$.

Let us call

0-*row*: a row of $S(V)_{a,b}$ whose elements have all value 0;
1-*row*: a row of $S(V)_{a,b}$ whose element in column $j+1$ has value 1;
10-*row*: a row of $S(V)_{a,b}$ which is neither 0-row nor 1-row.

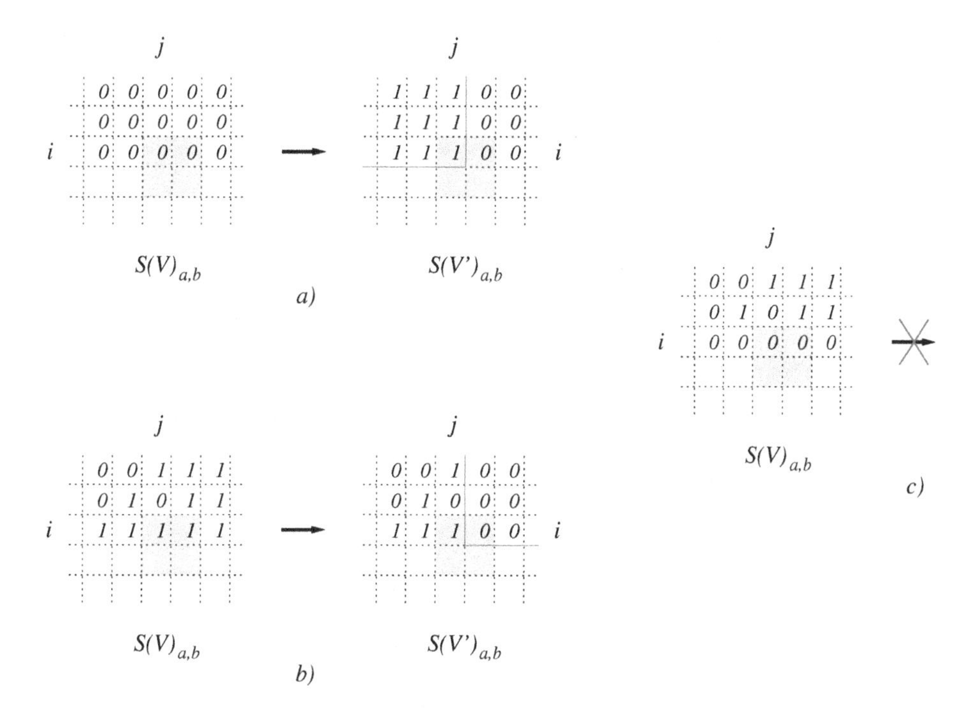

Fig. 19. *Two ways of extending $S(V)_{a,b}$ by adding one single point*

So, let us examine some configurations of $S(V)_{a,b}$:

i) if there exists a valuation V' such that $S(V')_{a,b}$ coincides with $S(V)_{a,b}$ in the rows from $i+1$ to m, and the positions in its first i rows are computed in accordance with the imposed value $S(V)_{a,b}[i, j] = 1$, then V' extends V with the addition of p_{t+1}, as depicted in Fig. 19 *a*);

ii) if, for all $1 \leq i' \leq i$, row i' is a 1-row of $S(V)_{a,b}$, then the valuation V can be extended to V' by modifying its first i rows, as shown in Fig. 19 *b*);

iii) if the configuration of the first i rows of $S(V)_{a,b}$ is different from those of *i*) and *ii*), then the valuation V can not be extended by modifying the first i rows, as one can deduce from Fig. 19 *c*);

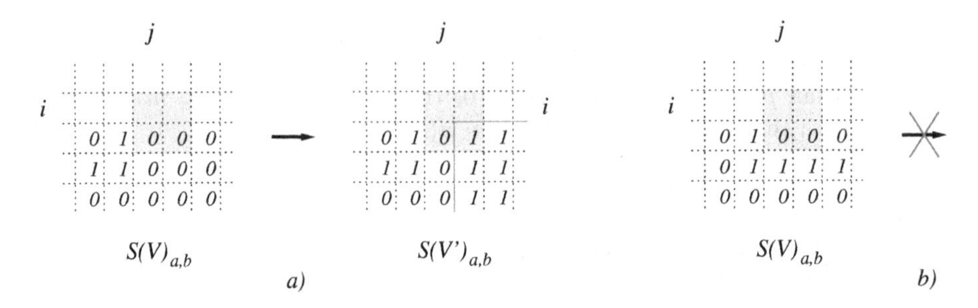

Fig. 20. *The way of extending a valuation $S(V)_{a,b}$ by modifying the rows greater than i*

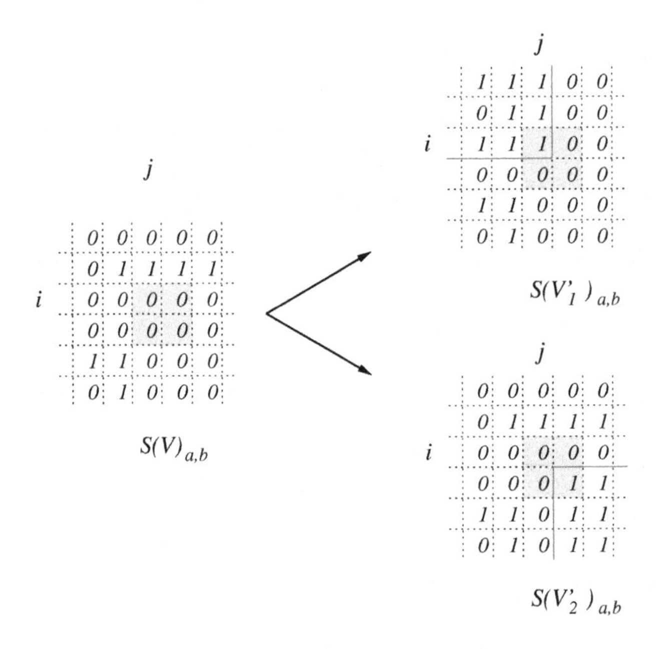

Fig. 21. *A configuration which admits two different valuations*

Comment: let the valuation $S(V)_{a,b}$ be such that, for all $1 \leq i' \leq i$, row i' is a 1-row, and, for all $1 \leq j' \leq j$, $S(V)_{a,b}[i', j'] = 0$. Acting as in case i), one can extend V to V', but an immediate check reveals that such a V' is not minimal.

iv) if, for all $i < i' \leq m$, row i' is a 0-row or a 10-row, then V can be extended to V' by modifying the rows from $i+1$ to m, as shown in Fig. 20 *a*);

v) finally, if, for all $i < i' \leq m$, row i' is a 1-row, then the valuation V can not be extended by modifying the rows from $i+1$ to m (Fig. 20 *b*)).

Furthermore, there exist $S(V)_{a,b}$ whose configuration of the first $j+1$ columns fit both in i) and in iv) (see Fig. 21), or both in ii) and in iv) at the same time (see Fig. 22).

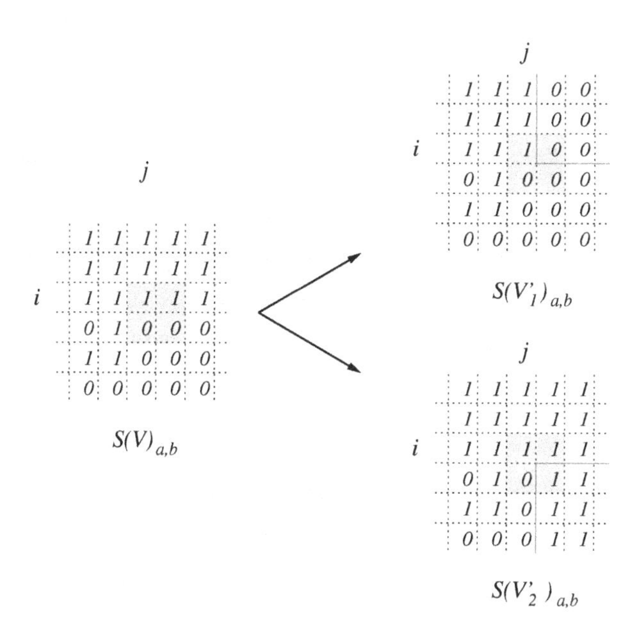

Fig. 22. *A configuration which admits two different valuations. One of them is not minimal*

Since in both cases two new valuations extend V, it may arise the possibility of the existence of some *ad hoc* configurations having a number of valuation which is exponential in t.

This doubt vanishes as soon as we observe that

- in Fig. 21 only one valuation at a time admits again two different extensions in a position greater than (i, j);
- in Fig. 22, valuation $S(V_2')_{a,b}$ is not minimal.

As a matter of fact, we have proved that, if n_t is the number of different configurations for a sequence of t nonzero points of $S(\chi(M))_{a,b}$, then it holds $n_t + 1 \geq n_{t+1}$. However, the proof is still uncomplete, since there are some configurations not yet considered: they appear when we choose to extend V with two points p_{t+1} and p_{t+2} at the same time.

Step $t \rightarrow t+2$: let us assume that $p_{t+1} = (i, j)$, $p_{t+2} = (i', j')$, $S(V)_{a,b}[i, j] = 1$, and $S(V)_{a,b}[i', j'] = -1$, and let us consider the following possible extension of V which can not be obtained by adding one point at a time:

vi) if $j = j'$, for all $i < i_0 \leq i'$, row i_0 is not a 1-row of $S(V)_{a,b}$, and there exists a 1-row below row i' (this condition assures that this configuration can not be achieved by adding the two points separately), then V can be extended to the valuation V' which comprehends p_{t+1} and p_{t+2}, by modifying the rows from $i+1$ to i', as shown in the matrix $S(V')_{a,b}$ of Fig. 23, i.e. it is created a strip of entries 1 starting in column j and including the rows from $i+1$ to i';

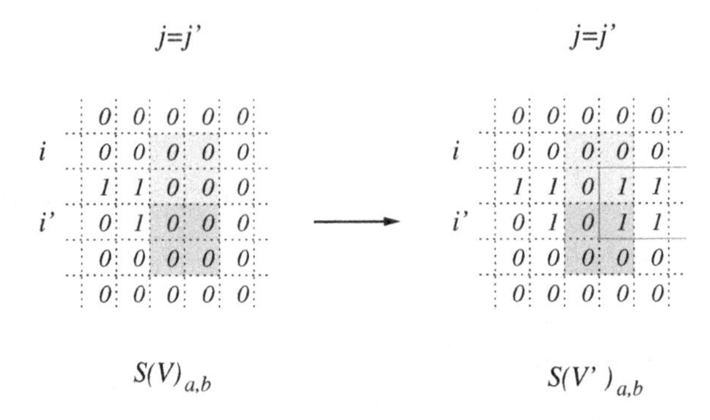

Fig. 23. *A particular configuration described by case vi)*

If we assume $S(V)_{a,b}[i,j] = -1$ and $S(V)_{a,b}[i',j'] = 1$ a result symmetrical to $vi)$ holds. Furthermore, in that case, there exists a second valuation which allows a strip of entries 0 to be created when all the rows from $i+1$ to i' are 1-rows.

Our search for all the possible extensions of V is now complete: some of them are showed with details, others are just left to the reader.

We observe that there exists $S(V)_{a,b}$ which fits in $vi)$, and which allows a second extension, achieved by adding p_{t+1} and p_{t+2} one by one, as shown in Fig. 23. However, as one can immediately deduce from the same figure, only one valuations at a time can present a configuration which admits again two different extensions in a position greater than (i',j').

So, we still have the bound $n_t \leq n_{t+2}+1$ to the number of different valuations of $p_1,\ldots,p_{t+2}$, and consequently the thesis. □

On the Reconstruction of Crystals Through Discrete Tomography

K.J. Batenburg[1,2] and W.J. Palenstijn[1]

[1] Leiden University, P.O. Box 9512, 2300 RA Leiden, The Netherlands
[2] CWI, P.O. Box 94079, 1090 GB Amsterdam, The Netherlands
{kbatenbu, wpalenst}@math.leidenuniv.nl

Abstract. We consider the application of discrete tomography to the reconstruction of crystal lattices from electron microscopy images. The model that is commonly used in the literature to describe this problem assumes that the atoms lie in a strictly regular grid. In practice, this is often not the case. We propose a model that allows for nonregular atom positions. We describe a two-step method for reconstructing the atom positions and types. For the first step, we give an algorithm and evaluate its performance.

1 Introduction

Over the past ten years, the research field of discrete tomography has received considerable attention [3]. Among the principal motivations for studying the tomographic reconstruction of images which have a small discrete set of pixel values is a demand from materials science. Advances in the field of electron microscopy have made it possible to count the number of atoms in each column of a crystal lattice along a small number of directions [2] (see Figure 1). For this application the reconstruction problem consists of retrieving the individual atom positions from a small number of projections.

In the mathematical model that is commonly used in the literature to describe the reconstruction problem, it is assumed that the atoms lie in a strictly regular grid. Each column of atoms corresponds to a single column of pixels in the resulting image.

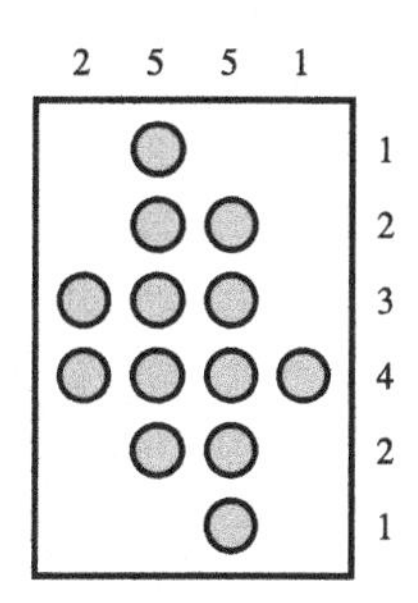

Fig. 1. Atom grid with hor. and vert. projections

The number of atoms in each column is not measured directly by the detector array of the microscope. Processing of the raw measured data results in a *reconstructed exit wave* of the crystal. The phase of the exit wave (see Figure 2a) can be regarded as a projected image of the crystal. Figure 2b shows the measured phase along a line of projected atom columns. In the exit wave phase image each atom column has a width of several pixels, typically around 10.

R. Klette and J. Žunić (Eds.): IWCIA 2004, LNCS 3322, pp. 23–37, 2004.

The atom columns appear as spikes in the measured projection. For crystals that consist of only one type of atom, the height of a peak depends (almost) linearly on the number of atoms in the corresponding column. Therefore, the number of atoms in a column can be found by dividing the peak height by the projection height of a single atom and rounding to the nearest integer.

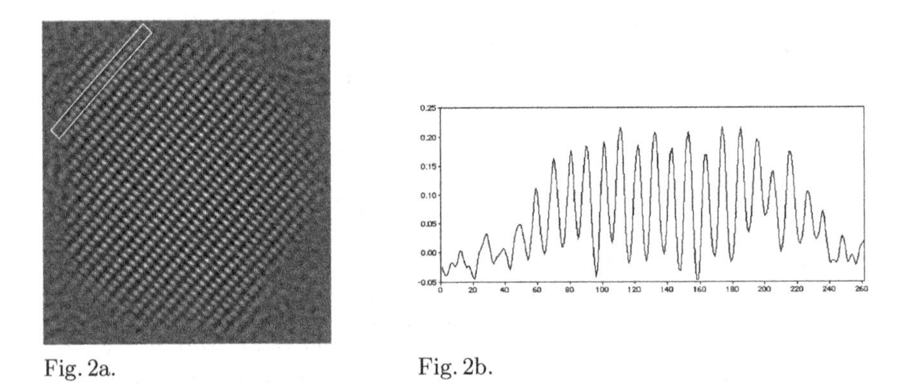

Fig. 2a. Fig. 2b.

Fig. 2. a) Exit wave reconstruction of a CdSe nanocrystal b) Measured phase along a line in the crystal projection (marked in the left figure); courtesy of dr. Ch. Kisielowski, NCEM

For a solid crystal without any irregularities the grid model corresponds well to physical reality. Unfortunately, many crystals that are of interest in materials science contain one or more irregularities, known as *defects*, in the regular grid structure. Figure 3 shows examples of the various types of defects that may occur in a crystal that contains two different atom types, indicated by light and dark gray circles. The defects marked by a, c and d are examples of *point defects*. Defect b is an example of an *edge dislocation*. Note that in most cases the surrounding atoms have moved away from their grid positions.

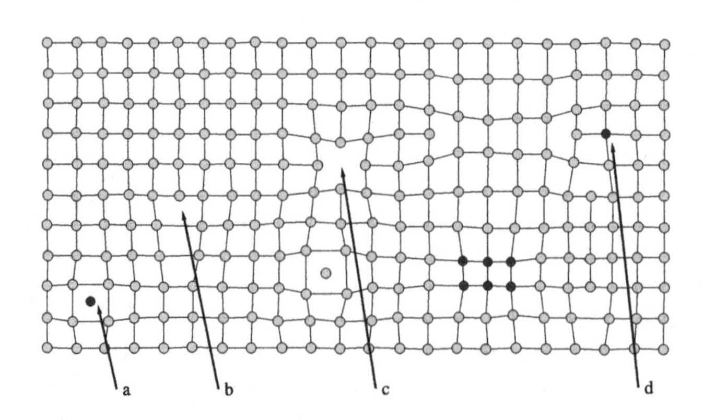

Fig. 3. Overview of various crystal defects; courtesy of prof. dr. H. Föll

If a crystal contains defects, the atoms do not lie in straight columns, which makes the method of counting atoms by looking at the peak heights inapplicable.

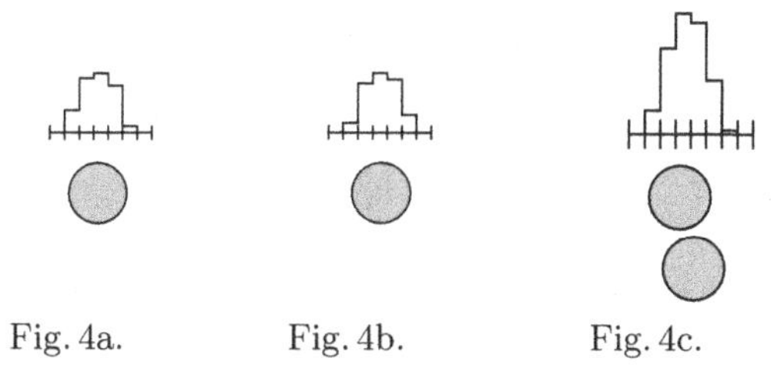

Fig. 4. a) A single atom and its measured projection. b) The same atom shifted slightly to the right. c) Two atoms for which the projection shows only one peak

Figure 4a shows a single atom and its projection as measured by the detector array. Figure 4b shows the same atom shifted slightly to the right. Although the displacement is less than the width of a detector cell, we can tell that the atom has moved from the observed pattern. Clearly, when measuring a single atom, assuming perfect, noiseless measurements, it is possible to determine its position with greater accuracy than the actual detector width. Figure 4c shows two atoms which are not vertically aligned. The measured data shows only a single peak. Yet, by using the a priori knowledge that the peak is a superposition of two atom projections, it is possible to recover the horizontal positions of both atoms exactly by solving a least squares problem (see Section 3.2).

This raises the question if recovering the positions of the atoms is also possible for larger atom configurations, consisting of many more atoms. If the configuration contains several different atom types, it is not even clear if the number of atoms of each type can be computed from the measured projection data.

When performing a tomographic reconstruction, projection data from several directions is available. We want to reconstruct the measured set of atoms (i.e., the number of atoms, their horizontal and vertical coordinates and their types) from the projections.

We propose a two-step method which first performs a reconstruction on each projection separately. After the atom positions in the direction orthogonal to the projection direction and the atom types have been determined for all separate projections, the resulting discrete tomography problem is much simpler than the original one. In this paper we focus on the first step.

Figure 5 shows the basic idea of the two-step approach. First the x-coordinates of the atoms and their types are computed from the vertical projection. Then the y-coordinates of the atoms, along with their types, are computed from the horizontal projections. The process may be repeated if more projections are available. Subsequently the resulting data is combined into a discrete tomography problem. As the computation for each separate projection also yields the types of the projected atoms, it may be possible to solve the tomography problem for each atom type independently if the sample contains several atom types.

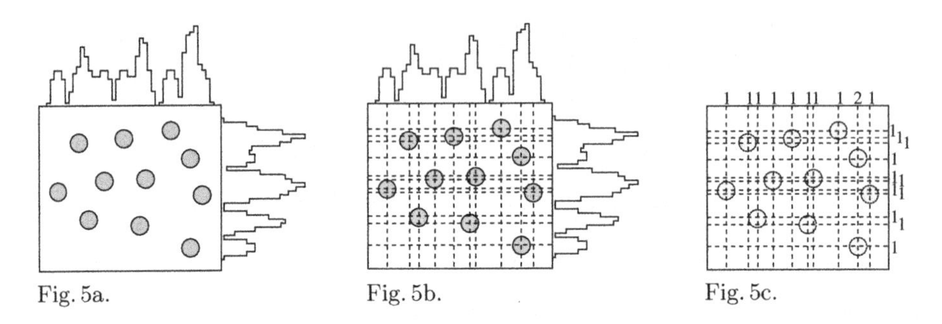

Fig. 5. a) A set of atoms and two of its projections. b) Coordinates of the atoms, reconstructed from the projections. c) The resulting discrete tomography problem

In this paper we explore to what extent the atom positions and their types can be recovered from a single projection. We restrict ourselves to the processing of 1-dimensional projections of 2-dimensional images. We present an algorithm that computes the atom positions and their types. Our experimental results demonstrate that even when many atoms are projected on top of each other it is still possible to recover the positions and types of the individual atoms that constitute the projection. The algorithm does not perform an exhaustive search and is not guaranteed to find a solution. We evaluate its performance and show that it is capable of finding accurate reconstructions for a varied set of test data. The results show that even in the presence of a limited amount of noise the algorithm performs well.

2 Preliminaries

Although the model that we study in this paper is quite flexible, we have to make several simplifications in comparison with physical reality. In this section we define our model and describe in what ways the model may deviate from actual physical experiments.

We restrict ourselves to 2-dimensional images. Although generalization of our methods to 3-dimensional images is not straightforward, we consider the development of algorithms for 2D images as a necessary first step towards the development of more complicated algorithms for the 3D case.

Figure 6a shows a schematic representation of the experimental setup that we will study. Using an electron microscope, one can measure the effect of an atom on a passing electron beam. The figure shows a single atom a, centered in (x_a, y_a), and its effect on a vertical electron beam. The effect of the atom can be described by the *projection function* $f_a(x)$, which gives the magnitude of the effect for any given x. We assume the following properties of f_a:

- f_a is independent of the y-coordinate of the atom;
- f_a is not the zero-function;

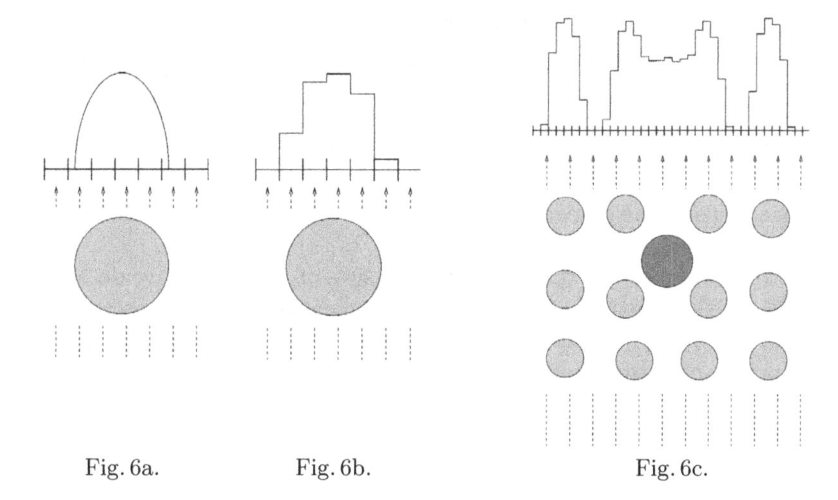

Fig. 6a. Fig. 6b. Fig. 6c.

Fig. 6. a) A single atom and its projection function. b) The projection of an atom as measured by the detector array. c) A set of atoms and its measured projection

- there exists an r such that $f_a(x) = 0$ if $|x - x_a| > r$. The smallest such r is the *atom radius* r_a;
- f_a is continuous;
- $f_a(x_a + x)$ is independent of the atom position x_a. In other words: when we move the atom horizontally, its projection function shifts along with it;
- f_a is nondecreasing for $x \leq x_a$;
- f_a is symmetric, i.e., $f_a(x_a - x) = f_a(x_a + x)$ for all x.

The first five properties are essential to the algorithm that we propose. The remaining properties are assumed for computational convenience and efficiency and are physically realistic. It is possible to remove these assumptions by adapting our algorithm if necessary.

We use the convention of representing atoms by circles in our figures. We assume that the projection function of an atom is independent of the orientation of the atom. In other words, an atom looks the same from all directions.

The detector array does not measure the projection function directly: each detector cell measures the average value of the projection function on a certain horizontal interval (see Figure 6b).

In practice, it is not possible to isolate a single atom. When the atom is part of a larger crystal (see Figure 6c), it is only possible to measure the projection function f_{tot} of the whole crystal. We assume that f_{tot} is the sum of the projection functions of all individual atoms that make up the crystal:

$$f_{\text{tot}}(x) = \sum_{\text{atoms } a} f_a(x).$$

This assumption is not completely valid in practice. The projection functions of the individual atoms do not add up linearly although the nonlinear effects are small. The method that we describe in this paper can also be used in the nonlinear case, as long as adding an atom a will result in a sufficiently large increase of $f_{\text{tot}}(x)$. We assume that the measurable effect of the atoms on the electron beam is limited to the interval $[0, w]$ measured by the microscope where w is the width of the detector array. In other words, $f_{\text{tot}}(x) = 0$ for $x < 0$ or $x > w$. It is possible to drop this assumption, i.e., to allow for a sample which stretches beyond the detector array, but this will result in a severe loss of accuracy near the edges.

The detector that measures the projection consists of n_d consecutive detector cells that have a fixed width w_{det}. We denote the measurement of detector cell $i = 0, 1, \ldots, n_d - 1$ by m_i. For noiseless measurements, m_i is the integral of the projection function over the interval $[x_i, x_i + w_{\text{det}}]$:

$$m_i = \int_{x_i}^{x_i + w_{\text{det}}} f_{\text{tot}}(x)\, dx$$

Note that since the detector cells are equally spaced, $x_i = i w_{\text{det}}$.

We assume that the crystal consists of a small number of *atom types*, e.g., Cd, Se, Au, etc. Atoms of the same type are indistinguishable. Therefore, each atom type has a single projection function modulo shifts. The atom types that make up the crystal and their projection functions are known in advance. The main problem that we study in this paper concerns the recovery of the individual atoms from the collective projection:

Problem 1. (Reconstruction Problem) Given the detector measurements and a set of atom types with their projection functions, determine a set A of atoms (their types and x-coordinates), such that the euclidean distance between the projection of A and the measured projection data is minimal.

It is possible that a solution to the reconstruction problem exists which is quite different from the actual atom configuration. The simplest example of this would occur if two atom types share the same projection function. If certain projection functions are integral linear combinations of other projection functions, similar problems occur. The samples that we consider are typically very thin, around 10 atoms thick, so only linear combinations with equally small coefficients can lead to such problems.

The detector measurements always contain a certain amount of noise. A high noise level may also result in a solution to the reconstruction problem that is different from the actual measured atom configuration.

We now choose a particular set of atom projection functions for demonstrating the basic concepts of our model and our algorithm. We assume that the atom projection functions are projections of circles, multiplied by a *density*. Each atom type t has an associated pair (r_t, ρ_t), the radius and density of the atom. The projection function of an atom a of type t, centered in x_a, is given by

$$f_a(x) = \begin{cases} 2\rho_t\sqrt{r_t^2 - (x - x_a)^2} & \text{if } x \in [x_a - r_t, x_a + r_t] \\ 0 & \text{otherwise.} \end{cases}$$

Put $u = \sqrt{r_t^2 - (x - x_a)^2}$. Then the function F_a, defined by:

$$F_a(x) = \begin{cases} 0 & \text{if } x \leq x_a - r_t \\ \rho_t((x - x_a)u + r_t^2(\arctan(\frac{x - x_a}{u}) + \frac{\pi}{2})) & \text{if } x \in (x_a - r_t, x_a + r_t) \\ \rho_t \pi r_t^2 & \text{if } x \geq x_a + r_t \end{cases}$$

is a primitive function of the projection function. Hence, the contribution $m_i(a)$ of a to the value measured by detector cell i is given by

$$m_i(a) = \int_{x_i}^{x_i + w_{\text{det}}} f_a(x)\, dx \quad = \quad F_a(x_i + w_{\text{det}}) - F_a(x_i).$$

Our algorithm for solving the reconstruction problem searches for the atom types and their x-coordinates from left to right. The interval $[0, w]$ is split into grid cells and each of the cells is assigned a number of atoms, possibly of different types. We call such an assignment of atoms to grid cells a *configuration*. An important step of the algorithm is to determine the minimum and maximum contribution of an atom a to the measured value m_i in detector i if a is assigned to grid cell $c = [l_c, r_c]$. Figure 7 shows the maximum (bold) and minimum (dashed) values that could be measured in each detector cell if a single atom lies somewhere within the indicated interval from the left circle to the right circle.

Because of the restrictions on f_a (it is symmetric around x_a and nondecreasing for $x \leq x_a$), these values can be easily determined. We denote the minimal contribution of a to detector cell i by $m_{i,\min}(a, c)$ and the maximal contribution by $m_{i,\max}(a, c)$.

For a given configuration s, let c_a denote the cell in which atom a lies. Then

$$m_{i,\min}(s) = \sum_{\text{atoms } a \in s} m_{i,\min}(a, c_a)$$

is a lower bound on the value measured in detector i and

$$m_{i,\max}(s) = \sum_{\text{atoms } a \in s} m_{i,\max}(a, c_a)$$

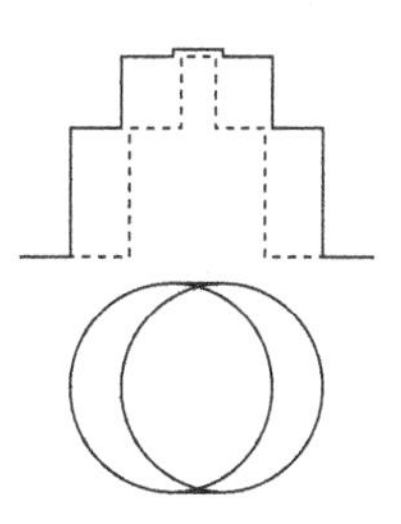

Fig. 7. max. (bold) and min. (dashed) values measured in each detector cell, for a single atom

is an upper bound on the value measured in detector i. We use these bounds extensively in the *block phase* of our algorithm (see Section 3.1). We say that a configuration s is k-*admissible* if

$$\begin{array}{ll} m_{i,\min}(s) \leq m_i \leq m_{i,\max}(s) & \text{for } 0 \leq i \leq k \text{ and} \\ m_{i,\min}(s) \leq m_i & \text{for } i > k. \end{array}$$

Suppose that we try to add atoms to s in order to obtain a configuration which matches the measured data in all detector cells. If we require that the additional atoms only affect the values measured in detector cells to the right of cell k, then k-admissibility is a necessary condition for s to be extendable to a fitting configuration.

In a given configuration s, the position of each atom is determined up to the width of a grid cell. During the fitting phase of our algorithm (see Section 3.2), the coordinates of the atoms are fixed at real values that lie within their respective cells. We call the resulting assignment $\tilde{s}$ of coordinates to atoms an *exact configuration* for s. For an exact configuration $\tilde{s}$, the *simulated measurement* (the value that would be measured by the detector array for this configuration) is given by

$$m_i(\tilde{s}) = \int_{x_i}^{x_i + w_{\text{det}}} \sum_{a \in \tilde{s}} f_a(x)\, dx.$$

We define the *k-distance* of an exact configuration $\tilde{s}$ to the projection data as the square root of $\sum_{i=1}^{k} (m_i - m_i(\tilde{s}))^2$.

3 Algorithm

In this section we describe an algorithm for solving the reconstruction problem. Our main goal is to demonstrate that it is indeed possible to recover the atom types and their approximate positions from the projection data in many cases. We use heuristics to limit the search space. It may happen that the algorithm fails to find an optimal solution or even that it fails to find an approximate solution. However, we demonstrate in Section 4 that our algorithm is capable of finding high quality solutions for a diverse set of test cases.

The reconstruction problem faces us with the task of recovering both the number of atoms of each type and the x-coordinates of all these atoms. The x-coordinates are real values, so they can take an infinite number of different values. In order to obtain a finite search space, we impose a (1-dimensional) grid on the interval $[0, w]$ and determine for each atom in which grid cell it lies, instead of determining the exact atom positions. The imposed grid is typically finer than the detector grid, e.g., twice as fine. We call this grid the *fine grid.* It is relatively easy to calculate lower and upper bounds on the number of atoms present in each grid cell.

We say that a grid cell c in the fine grid *contains* an atom a if the *left side* of a, defined as $x_a - r_a$ is in c. In this way, the projection of a has no effect on the values measured by detector cells to the left of c. Our algorithm constructs atom configurations incrementally from left to right. Once a configuration has been constructed up to grid cell c, we can check if the configuration is k-admissible, where k is the index of the last detector cell that is entirely left of the right boundary of c.

The main loop of our algorithm iterates over the fine grid, from left to right, in steps that comprise a number of fine grid cells. We call a group of fine cells that constitute such a step a *block.* The size of a block is always a power of two (see Section 3.1). We denote the number of blocks that cover the interval $[0, w]$ by n_b.

The algorithm maintains a set S of configurations that partially (up to the current block of fine grid cells) fit the measured data well. At the start of iteration

b, a configuration $s \in S$ contains atoms up to block $b-1$. New atoms are then added to s in block b, forming a new configuration t, which is called a *b-extension* of s.

When the end of the projection data has been reached, the set S contains configurations that fit the measured data well on the interval $[0, w]$. Each configuration provides an estimate of the atom positions and their types. We remark that the atom coordinates are not determined exactly; they are determined up to grid cells in the fine grid. Figure 8 shows an outline of our algorithm. In the next sections the various steps of the algorithm are described in more detail.

```
S_{-1} := {empty configuration};
for b := 0 to n_b do
begin
  S_b := ∅;
  k_b := index of last detector cell covered by blocks 0,...,b;
  foreach s ∈ S_{b-1} do
  begin
    foreach b-extension t of s which is k_b-admissible (see Section 3.1) do
    begin
      fit t to the measured data (see Section 3.2);
      if t fits the data well enough then
        S_b := S_b ∪ {t};
    end
  end
  cull S_b (see Section 3.3);
end
```

Fig. 8. Outline of the algorithm

3.1 Block Phase

When searching for b-extensions of a configuration s, we want to determine the possible sets of atoms that the fine grid cells in b can contain such that the corresponding b-extension still fits the measured data. We first determine all such sets for the entire block, where we require the atoms to lie within the left and right boundary of the block without restricting them to a single fine cell. To this end, we extend the admissibility concept to the "coarse" grid of blocks, i.e., we determine the minimal and maximal contribution of atoms contained in the current block to the measured values m_i.

We determine all sets of atoms that the current block can contain which satisfy this "extended k_b-admissibility". Subsequently, the block grid is refined by splitting the block into its left and right halves. As each atom must lie on one of both sides, we form all partitions of the atom set into a left and a right half. For each partition, we adjust the upper and lower bounds on the projection. If these (narrower) bounds still satisfy the extended k_b-admissibility, we recursively split it into a new grid that is again twice as fine. We repeat this procedure until we have reached the fine grid level. Note that we always choose a power of two as the block size.

As an example, we consider a single block of four fine grid cells in the reconstruction of a crystal that consists of only one atom type (see Figure 9). The root node of the tree represents the current block b. By using the lower and upper bounds on the contribution of atoms in b to the measured data it has been determined that block b must contain 7 atoms. These atoms can be split in various ways among the left and right halves of b. Only two partitions satisfy the finer admissibility bounds: 5 atoms left and 2 atoms right or 4 atoms left and 3 atoms right. By repeating this procedure we end up at the leaf nodes, which represent assignments of the atoms to the four fine grid cells in the block.

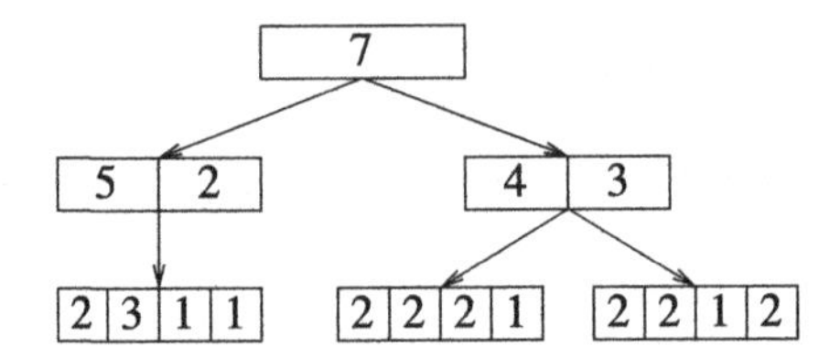

Fig. 9. Admissible configurations are determined recursively, forming a tree of refinements

3.2 Fitting Phase

In the block phase, candidate atom configurations are selected which satisfy the admissibility criterion. However, admissibility is not a sufficient condition that a configuration must satisfy to fit the measured data. The main problem with the admissibility condition is that it considers all detector values separately. For example, the measured effect of an atom can never be equal to its lower bound for all detectors simultaneously.

For a given configuration s which has been constructed up to block b, the fitting procedure constructs an exact configuration $\tilde{s}$ for which the atom positions lie in the prescribed cells such that the k_b-distance of $\tilde{s}$ to the measured data is minimal:

$$\textbf{minimize} \quad \sum_{i=1}^{k_b} (m_i - m_i(\tilde{s}))^2.$$

If the k_b-distance of the resulting exact configuration is larger than a constant D (the *fitting cutoff*), we can conclude that the configuration s is most likely not a good partial solution to the reconstruction problem and it is discarded.

For solving this least squares problem, we use the Levenberg-Marquardt (LM) algorithm [1, §4.7.3]. The LM algorithm requires an initial approximation to the solution of the least squares problem, from which it iteratively moves to a local optimum. For a configuration t that is a b-extension of a configuration s, the x-coordinates of atoms that have been added in the current block b are initialized randomly in a small interval around the center of their respective fine cells. Atoms from previous blocks are initialized at the values found by the LM algorithm when applied to s.

As the algorithm progresses, the number of atoms up to the current block – and consequently the number of variables in the least squares problem – will become increasingly large. It is very unlikely that the addition of a new block will significantly affect the positions of atoms that are far to the left of the current block. The positions of such atoms have already been determined by many prior LM steps. In order to limit the number of variables in the least squares problem, we fix the positions of these atoms. To be precise, all atoms in cells that are at least H fine cells to the left of the start of the current block are fixed, where H is a positive integer constant. Consequently, the terms of the least squares problem that correspond to detectors that are solely affected by the fixed atoms are also removed.

The LM algorithm uses the partial derivatives of $m_i(\tilde{s})$ with respect to the atom positions x_a. These derivatives can be expressed easily in terms of the projection function:

$$\frac{\partial}{\partial x_a} m_i(\tilde{s}) = f_a(x_i) - f_a(x_i + w_{\text{det}})$$

where x_i is the left bound of detector cell i.

Although the LM algorithm will find a locally optimal solution of the least squares problem, it will not necessarily find a global optimum. In its basic form, the LM algorithm does not allow boundary constraints. We implemented the boundary constraints by adding a *penalty function* for each atom to the sum of squared differences. For an atom a that must lie within the interval $[l_a, r_a]$, the penalty is defined as:

$$p_a = \begin{cases} 1000(l_a - x_a)^{1.1} & \text{if } x_a < l_a \\ 0 & \text{if } l_a \leq x_a \leq r_a \\ 1000(x_a - r_a)^{1.1} & \text{if } x_a > r_a. \end{cases}$$

Note that the penalty function is continuously differentiable, which is a requirement for the LM algorithm. The reason for using the penalty method over other methods that use hard boundary constraints is that it provides more information. When an atom a is slightly outside its bounds in the optimal solution, this is a strong indication that a better configuration exists, for which all variables are within their bounds.

Strictly speaking, the formulation of the minimization problem that is solved by the LM algorithm does not depend on the concept of configurations at all (not taking the penalty functions into account). Without a proper start solution and penalty functions, however, the problem will be extremely difficult to solve analytically, as it has a huge number of local optima.

3.3 Culling Phase

As the algorithm proceeds from left to right, the number of states in S_b will typically become larger and larger, unless we resort to *culling* the set after each block.

Although we do not make rigid assumptions on the horizontal positions of the atoms, we expect that for crystals the atoms will tend to lie in columns, resulting in separated peaks. It may happen that a single peak in the projection data can be approximated well by several different atom configurations. Suppose that such a peak is followed by one or more zero measurements. If we can extend any of the partial atom configurations (up to the peak) to a configuration that fits the whole projection, we can extend all the other configurations in exactly the same way. Therefore, we delete almost all configurations $s \in S_b$ for which

$$m_{i,\min}(s) = m_{i,\max}(s) = 0 \quad \text{for all detectors } i \text{ to the right of } b.$$

We keep only those configurations s for which the k_b-distance of the corresponding exact configuration $\tilde{s}$ to the measured data is minimal. Note that we can store the deleted configurations so that we can retrieve alternative partial solutions later if desired. This form of culling reduces the number of configurations enormously between consecutive peaks.

Different configurations always result in different sets of boundary constraints for the LM algorithm. This does not mean, however, that the corresponding exact configurations cannot be very similar. For example, suppose that in the measured atom configuration, an atom is near a fine cell boundary. There will be two configurations that approximate the actual configuration very well: one that places the atom in the cell to the left of the boundary and one that places the atom to the right. After the fitting phase, the resulting exact configurations will typically be almost identical.

Note that the penalty approach in the fitting phase allows atoms to move slightly out of their boundaries. We define the *boundary violation* of an atom a as the squared distance to its nearest cell bound if a lies outside its assigned cell and zero otherwise.

To prevent the superfluous processing of nearly identical configurations, we delete configurations for which the corresponding exact configuration is *almost identical* to another one, retaining only the configuration for which the exact configuration adheres best to its cell boundaries, i.e., for which the sum of the boundary violations is minimal. We say that two exact configurations $\tilde{s}$ and $\tilde{s}'$ are *almost identical* if

- the number of atoms of each type is the same in $\tilde{s}$ and $\tilde{s}'$;
- for each pair of corresponding atoms between $\tilde{s}$ and $\tilde{s}'$ (when sorted by x-coordinate) the distance between their respective x-coordinates is smaller than a positive constant C.

This form of culling reduces the number of configurations significantly while processing peaks in the projection data.

3.4 Noise

The algorithm that we described does not take noise into account. When working with practical data, however, noise will always be present. We assume that we have a good indication of the noise level in advance. It is not difficult to adapt our algorithm to find solutions in the case of noise. Comparisons of $m_{i,\min}$ or $m_{i,\max}$ with the projection data are made less strict, allowing a margin that depends on the noise level. Additionally, the fitting cutoff D has to be increased since solutions will fit the data less accurately.

4 Experimental Results

In this section we report reconstruction results of our algorithm for a set of characteristic test images. As the purpose of these experiments is mainly to show the feasibility of the approach, we do not provide large-scale statistical data on the performance.

We implemented the algorithm in C++. For solving the nonlinear least squares problems, we used the MINPACK implementation of the LM-algorithm. We used a 1.4GHz Opteron machine with 2Gb of RAM.

In some of the tests we added noise to the projection data to simulate noisy measurements. For each detector i a random sample r from a normal distribution with average $\mu = 1$ and variance σ^2 is generated. The original measurement m_i for that detector is replaced by rm_i. For each test the value of σ is indicated in the table. If $\sigma = 0$ no noise was added.

For all tests, we set $w_{\text{det}} = 1$ and scaled all other quantities accordingly. We used the circle projections, described in Section 2. We set the value of the constant H (see Section 3.2) to 40 times the number of fine cells per detector. The constant C (see Section 3.3) was set to 0.2 times the width of a fine cell. The fitting cutoff D was typically set to 0.01 for test cases without noise, to 3 for test cases with noise level $\sigma = 0.01$, to 5 for $\sigma = 0.03$, and to 10 for $\sigma = 0.05$. In cases where this value of D did not result in a solution, we increased the fitting cutoff slightly.

The first test set is the atom configuration in Figure 3. Table 1 shows the reconstruction results for two choices of atom radii. The atom densities are the same for both tests and the data contains no noise. Each block of fine grid cells has a size of two detector cells. The number of fine grid cells per detector cell is indicated in the table. In the column "atoms(type)" the number of atoms of types 0 and 1 is listed. The column "type errors" indicates the number of atoms that were classified as the wrong atom type in the reconstruction. The column "cell errors" indicates the number of atoms that the algorithm placed in the wrong cell. We call a cell error an "off by one error" if the cell in the reconstruction is directly adjacent to the actual cell and an "off by >1 error" if this is not the case.

For the next set of tests, we decomposed Figure 3 into several slices and modified some slices to create additional test cases (see Figure 10). The slices each have their own characteristics. Some contain two atom types, others only one. Yet, we assume for the cases a, b, c, d, e, f and g that the slices contain two atom types, so that the algorithm must find out by itself if only one atom type occurs. The results are shown in Table 2. For the test cases c* and h we use 1 and 3 atom types respectively. For the values (r, ρ) of the three atom types we used $(5, 1)$, $(4.5, 1.41)$ and $(5.4, 1.37)$ respectively.

Table 1. Reconstruction results for the atom configuration in Figure 3, using different pairs of atom radii

	atoms(type)	(r_0, ρ_0) (r_1, ρ_1)	fine cells per det.	runtime (min)	type errors	cell errors	
						off by one	off by > 1
Fig. 3	241(0) 8(1)	(5, 1) (4.5, 1.41)	2	26	0	6	0
		(7, 1) (6, 1.41)	1	78	0	6	0

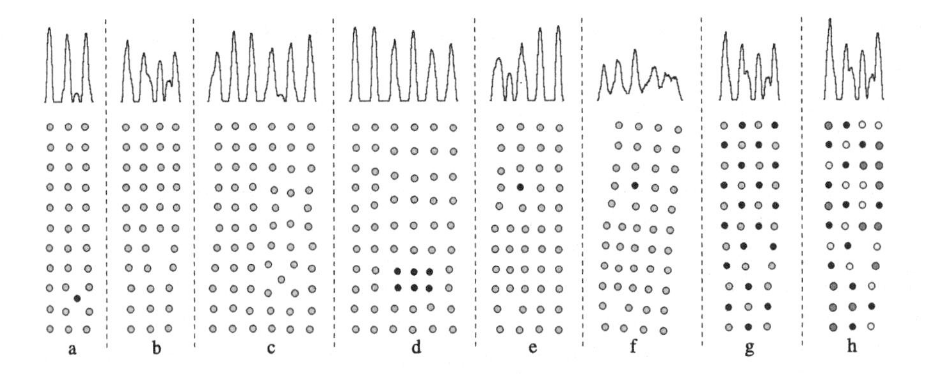

Fig. 10. Test configurations with their projections

Table 2. Reconstruction results for the slices in Figure 10 using different noise levels

	atoms(type)	σ	fine cells per det.	runtime (min)	type errors	cell errors	
						off by one	off by > 1
Fig. 10a	33(0) 1(1)	0	2	8	0	1	0
		0.01	1	15	0	5	0
Fig. 10b	39(0) 0(1)	0	2	17s	0	0	0
		0.01	1	13	0	6	0
Fig. 10c	66(0) 0(1)	0	2	2	0	0	0
		0.01	1	27	0	9	0
Fig. 10c*	66	0.03	1	26s	N/A	17	0
		0.05	1	2		21	2
Fig. 10d	56(0) 6(1)	0	2	15	0	4	0
		0.01	1	19	0	7	0
Fig. 10e	47(0) 1(1)	0	2	5	0	3	0
		0.01	1	172	5	9	1
Fig. 10f	47(0) 1(1)	0	2	17s	0	0	0
		0.01	1	91	0	4	0
Fig. 10g	20(0) 19(1)	0	2	1	0	0	0
		0.01	1	35	0	7	0
Fig. 10h	13(0) 13(1) 13(2)	0	1	187	0	0	0

5 Discussion

The experimental results show that our algorithm is able to reconstruct all test sets accurately when there is no noise. Noise is clearly a problem for our algorithm. When reconstructing atom configurations for which we know in advance that they contain only a single atom type, the tolerance for noise is much higher than for the case of multiple atom types. Even for a noise level of $\sigma = 0.05$ the reconstruction is still quite accurate, considering that the size of a fine grid cell is 1/10th the size of an atom. When there is more than one atom type the runtime becomes prohibitively large for noisy data. For three atom types a noise level of $\sigma = 0.01$ already resulted in a runtime that was unacceptably large (not shown in the table). Our experiments suggest that for two atom types a noise level around $\sigma = 0.01$ still allows accurate reconstruction in reasonable time. We performed some additional experiments with projection data of thicker samples, containing longer atom columns. The results suggest that the runtime increases strongly when the column height increases.

6 Conclusions

In this paper we demonstrated that it is indeed possible to reconstruct the atom types and their approximate x-coordinates from the measured projection data, even in the presence of limited noise. For a noise level of $\sigma = 0.01$ we are able to obtain quite accurate reconstructions even when the sample contains two atom types. Our algorithm is not guaranteed to find the optimal solution of the reconstruction problem, yet it provides good reconstruction results on our set of characteristic test images. In future research we will address the second step in the reconstruction procedure: solving the 2D tomography problem that results after the individual projections have been processed by our algorithm.

References

1. Gill, P.E., Murray, W., Wright, M.H.: Practical Optimization, Academic Press, London and New York, (1981)
2. Jinschek, J.R., Batenburg, K.J., Calderon, H., Van Dyck, D., Chen, F.-R., Kisielowski, Ch.: Prospects for bright field and dark field electron tomography on a discrete grid, Microscopy and Microanalysis, Vol. 10, Supplement 3, Cambridge Journals Online (2004)
3. Herman, G.T., Kuba, A. (ed.): Discrete Tomography: Foundations, Algorithms and Applications, Birkhäuser Boston (1999)

Binary Tomography by Iterating Linear Programs from Noisy Projections

Stefan Weber[1], Thomas Schüle[1,3],
Joachim Hornegger[2], and Christoph Schnörr[1]

[1] University of Mannheim, Dept. M&CS, CVGPR-Group,
D-68131 Mannheim, Germany
www.cvgpr.uni-mannheim.de
{wstefan, schuele, schnoerr}@uni-mannheim.de
[2] Friedrich-Alexander University,
Erlangen-Nürnberg Dept. CS, D-91058 Erlangen, Germany
www5.informatik.uni-erlangen.de
joachim@hornegger.de
[3] Siemens Medical Solutions, D-91301 Forchheim, Germany
www.siemensmedical.com

Abstract. In this paper we improve the behavior of a reconstruction algorithm for binary tomography in the presence of noise. This algorithm which has recently been published is derived from a primal-dual subgradient method leading to a sequence of linear programs. The objective function contains a smoothness prior that favors spatially homogeneous solutions and a concave functional gradually enforcing binary solutions. We complement the objective function with a term to cope with noisy projections and evaluate its performance.

Keywords: Discrete Tomography, Combinatorial Optimization, Linear Programming, D.C. Programming, Noise Suppression.

1 Introduction

Discrete Tomography is concerned with the reconstruction of discrete-valued functions from projections. Historically, the field originated from several branches of mathematics like, for example, the combinatorial problem to determine binary matrices from its row and column sums (see the survey [1]). Meanwhile, however, progress is not only driven by challenging theoretical problems [2, 3] but also by real-world applications where discrete tomography might play an essential role (cf. [4, chapters 15–21]).

The work presented in this paper is motivated by the reconstruction of volumes from *few* projection directions within a *limited* range of angles. From the viewpoint of established mathematical models [5], this is a severely ill-posed problem. The motivation for considering this difficult problem relates to the observation that in some specific medical scenarios, it is reasonable to assume that

R. Klette and J. Žunić (Eds.): IWCIA 2004, LNCS 3322, pp. 38–51, 2004.

the function f to be reconstructed is *binary-valued*. This poses one of the essential questions of discrete tomography: how can knowledge of the discrete range of f be exploited in order to regularize and solve the reconstruction problem?

1.1 Motivation

Consider the 32×32 image on the left side of figure 1 which shows a black rectangle. Given the horizontal and the vertical projection, see figure 2, it is obviously easy to recover the original object from these projections.

Now let us assume that for some reason in each projection the ray in the middle does not measure the correct value, in fact it measures a longer value in the first (figure 3) and a smaller one in the second case (figure 4). The question arises how does a reconstruction algorithm based on linear programming (see section 3) behave on such disturbed data? In the first case (figure 3) there is

Fig. 1. Consider the following binary reconstruction problem: The horizontal and the vertical projection of the left image, 32×32, are given, see figure 2. For some reason one ray in both projections does not measure the correct value, but a higher in the first (figure 3) and a smaller one in the second case (figure 4). The higher measurement does not bother the reconstruction algorithm at all since there are other constraints which are previously met. However, in the second case the constraint with the smaller value is fulfilled before all others and hence the algorithm reacts sensitive to this kind of error, as can be seen in the right image

no problem at all since the constraints of other rays are met first. Only the constraint of the wrong projection ray is not fulfilled entirely, means the inequality constraint, see equation (5), is "less than" for a given solution. Anyhow, the reconstruction algorithm will deliver the correct solution. Unfortunately, in the second case (figure 4) the opposite is true. The constraint of the wrong measurement is met first and hinders the other constraints from being fulfilled entirely. This is shown in the right image of figure 1 where the reconstruction problem was solved with (ILP) (one iteration; $\alpha = 0.0$), see section 3.3. Even for $\alpha > 0$ which enforces more homogeneous reconstructions the gap is not filled up due to the hard constraints.

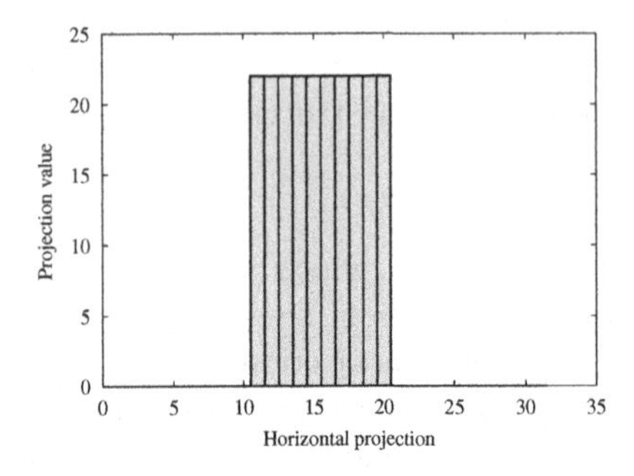

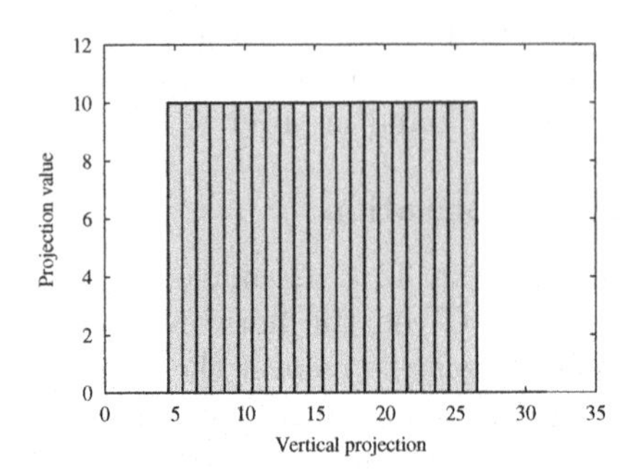

Fig. 2. Correct horizontal (left) and vertical projection (right) of the image shown on the left side of figure 1

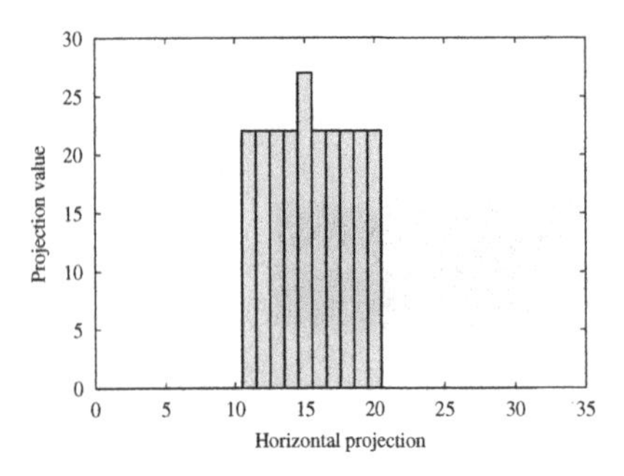

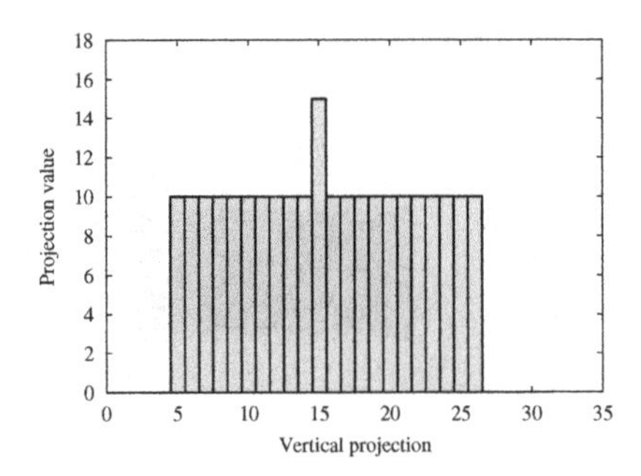

Fig. 3. First error case: The detector at position 15 measures a longer value in both projections

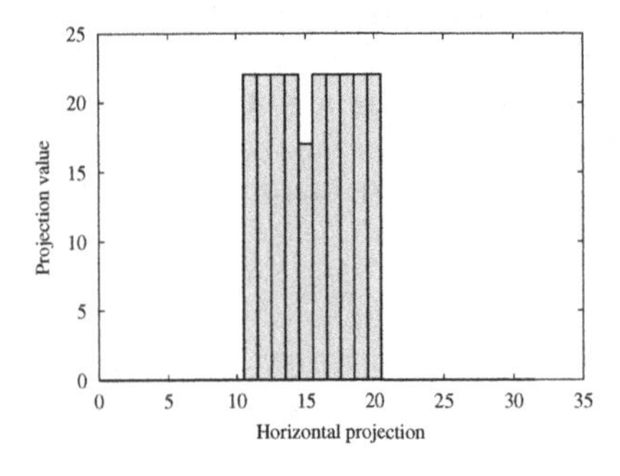

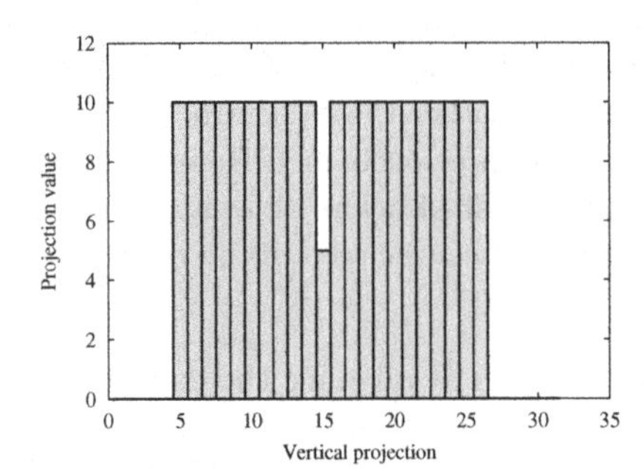

Fig. 4. Second error case: The detector at position 15 measures a lower value in both projections

The motivation of this paper is to overcome this systematic drawback that occurs in case of *noisy* projection data. This is done by the modification of our (ILP) algorithm which we will describe in section 4.1.

2 Problem Statement

The reconstruction problem we consider here is represented by a linear system of equations $Ax = b$. Each projection ray corresponds to a row of matrix A, and its projection value is the corresponding component of b. The entries of A are given as the length of the intersection of a particular pixel (voxel in the 3D case) and the corresponding projection ray (see Fig. 5). Each component $x_i \in \{0, 1\}$ indicates whether the corresponding pixel (belongs to the reconstructed object, $x_i = 1$, or not, $x_i = 0$ (see Fig. 5). The reconstruction problem is to compute the binary indicator vector x from the *under*-determined linear system of projection equations:

$$Ax = b, \quad x = (x_1, ..., x_n)^\top \in \{0, 1\}^n \tag{1}$$

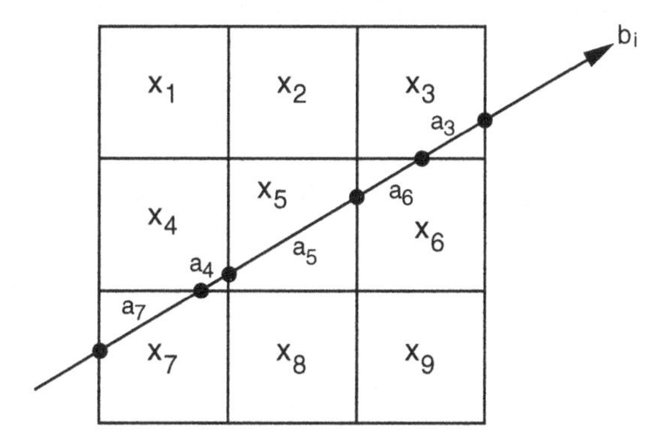

Fig. 5. Discretization model leading to the algebraic representation of the reconstruction problem: $Ax = b$, $x \in \{0, 1\}^n$

3 Related and Prior Work

In order to take advantage of a continuous problem formulation and numerical interior point methods, Fishburn et al. [6] considered the relaxation $x_i \in [0, 1]$, $i = 1, \ldots, n$, and investigated the following linear programming approach for computing a feasible point:

$$\min_{x \in [0,1]^n} \langle 0, x \rangle, \quad Ax = b \tag{2}$$

In particular, the information provided by feasible solutions in terms of additivity and uniqueness of subsets $\mathcal{S} \subset \mathbb{Z}^n$ is studied in [6].

3.1 Best Inner Fit (BIF)

Gritzmann et al. [7] introduced the following linear *integer* programming problem for binary tomography:

$$\max_{x \in \{0,1\}^n} \langle e, x \rangle, \quad e := (1, \ldots, 1)^\top, \quad Ax \leq b\,, \tag{3}$$

and suggested a range of greedy approaches within a general framework for local search. Compared to (2), the objective function (3), called *best-inner-fit (BIF)* in [7], looks for the maximal set compatible with the measurements. Furthermore, the formulation of the projection constraints is better suited to cope with measurement errors and noise.

3.2 Regularized Best Inner Fit (BIF2)

In [8, 9], we studied the relaxation of (3) $x_i \in [0,1], \forall i$, supplemented with a standard smoothness prior enforcing spatial coherency of solutions

$$\sum_{\langle i,j \rangle} (x_i - x_j)^2 \tag{4}$$

Here, the sum runs over all 4 nearest neighbors of the pixel grid (6 neighbors in the 3D case). In order to incorporate this prior into the linear programming approach (3), we used the following approximation by means of auxiliary variables $\{z_{\langle i,j \rangle}\}$:

$$\min_{x \in [0,1]^n, \{z_{\langle i,j \rangle}\}} -\langle e, x \rangle + \frac{\alpha}{2} \sum_{\langle i,j \rangle} z_{\langle i,j \rangle} \tag{5}$$

$$\text{subject to} \quad Ax \leq b\,, \quad z_{\langle i,j \rangle} \geq x_i - x_j\,,\; z_{\langle i,j \rangle} \geq x_j - x_i$$

3.3 Iterated Linear Programming (ILP)

In [10], we added to the relaxation in (5) a concave functional which is minimal at the vertices of the domain $[0,1]^n$ enforcing binary solutions.

$$\frac{\mu}{2} \langle x, e - x \rangle = \frac{\mu}{2} \sum_i x_i - x_i^2\,, \tag{6}$$

The strategy is to choose an increasing sequence of values for μ and to minimize for each of them (7).

$$\min_{x \in [0,1]^n, \{z_{\langle i,j \rangle}\}} -\langle e, x \rangle + \frac{\alpha}{2} \sum_{\langle i,j \rangle} z_{\langle i,j \rangle} + \frac{\mu}{2} \langle x, e - x \rangle \tag{7}$$

$$\text{subject to} \quad Ax \leq b\,, \quad z_{\langle i,j \rangle} \geq x_i - x_j\,,\; z_{\langle i,j \rangle} \geq x_j - x_i$$

Problem (7) is no longer convex, of course, but can be reliably minimized with a sequence of linear programs. This will be explained in section 4.1.

4 Noise Suppression

In case of noisy projection information we cannot consider the entries of the right-hand side vector b as fixed anymore, see section 1.1. Instead, the algorithm should take errors into account and suppress effects on the reconstruction as much as possible.

4.1 Iterated Linear Programming with Soft Bounds (ILPSB)

According to the chosen discretization scheme, section 2 and equation (3.1), each ray is represented by an equation of the form $a_i^\top x \leq b_i$, where a_i is the i-th row of matrix A. In order to handle false projections, we introduce the error variables γ_i leading to the modified equations $a_i^\top x + \gamma_i = b_i, \quad \gamma_i \in \mathbb{R}$. Since we do not wish to perturb the projection equations arbitrarily, we include the term $\sum_i \lambda_i$ into the objective function, where:

$$\lambda_i := \begin{cases} \tau_0 \gamma_i & \text{if} \quad \gamma_i \geq 0 \\ -\tau_1 \gamma_i & \text{else} \end{cases}, \quad \tau_0 > 0, \ \tau_1 > 0 \tag{8}$$

The parameters τ_0 and τ_1 allow to assign different weights to positive and negative deviations from the measurement b_i. Choosing $\tau_0 > \tau_1$ prefers an approximation of the best inner fit constraints, $Ax \leq b$. Consider again $a_i^\top x + \gamma_i = b_i$, in order to met equality it is favorable to set more x_i instead of compensating with the expensive γ_i. Conversely, the choice of $\tau_0 < \tau_1$ approximates the best outer fit constraints. Finally, if $\tau_0 = \tau_1 = \tau$ the term $\sum_i \lambda_i$ results in $\tau \sum_i |\gamma_i| = \tau ||Ax - b||_1$. Hence, instead of (7), we consider the following optimization problem:

$$\min_{x \in [0,1]^n, \{z_{\langle i,j \rangle}\}} \frac{\alpha}{2} \sum_{\langle i,j \rangle} z_{\langle i,j \rangle} + \frac{\mu}{2} \langle x, e - x \rangle + \beta \sum_{i=1}^{m} \lambda_i \tag{9}$$

$$\text{subject to} \quad \begin{aligned} & \tilde{A} \begin{pmatrix} x \\ \gamma \end{pmatrix} = b, \quad \tilde{A} := \begin{pmatrix} a_{11} & \dots & a_{1,n} & 1 & & \\ \vdots & \ddots & \vdots & & \ddots & \\ a_{m,1} & \dots & a_{m,n} & & & 1 \end{pmatrix} \\ & 0 \leq x_i \leq 1, \quad \gamma_i \in \mathbb{R}, \\ & \lambda_i \geq \tau_0 \gamma_i, \quad \lambda_i \geq -\tau_1 \gamma_i, \\ & z_{\langle i,j \rangle} \geq x_i - x_j, \quad z_{\langle i,j \rangle} \geq x_j - x_i \end{aligned}$$

Compared to (ILP), equation (7), we can skip the term $-\langle e^\top x \rangle$ in the objective function of equation (9) since minimizing λ_i forces x to satisfy the projection equations.

Further, the regularization parameter β controls the error tolerance.

4.2 Optimization

As the original (ILP) approach (section 3.3), this problem is not convex. To explain our approach for computing a minimizer, we put

$$z := (x^\top, \dots, z_{\langle i,j \rangle}, \dots, \lambda^\top)^\top \tag{10}$$

and rewrite all constraints from equation (9), in the form

$$\hat{A} z \leq \hat{b}\,, \tag{11}$$

Using the notation

$$\delta_C(z) = \begin{cases} 0 & , z \in C \\ +\infty & , z \notin C \end{cases}$$

for the indicator functions of a convex set C, problem (9) then reads:

$$\min_z f(z) \,,$$

where (cf. definition (10))

$$f(z) = \frac{\alpha}{2} \sum_{\langle i,j \rangle} z_{\langle i,j \rangle} + \beta \sum_{i=1}^{m} \lambda_i + \frac{\mu}{2} \langle x, e - x \rangle + \delta_K(\hat{b} - \hat{A}z) \,, \tag{12}$$

$$= g(z) - h(z) \,, \tag{13}$$

$K = \mathbb{R}^n_+$ is the standard cone of nonnegative vectors, and

$$g(z) = \frac{\alpha}{2} \sum_{\langle i,j \rangle} z_{\langle i,j \rangle} + \beta \sum_{i=1}^{m} \lambda_i + \delta_K(\hat{b} - \hat{A}z) \,, \tag{14}$$

$$h(z) = \frac{\mu}{2} \langle x, x - e \rangle \,. \tag{15}$$

Note that both functions $g(z)$ and $h(z)$ are convex, and that $g(z)$ is non-smooth due to the linear constraints.

To proceed, we need the following basic concepts [11] defined for a function $f : \mathbb{R}^n \to \overline{R}$ and a set $C \subset \mathbb{R}^n$:

$$\operatorname{dom} f = \{x \in \mathbb{R}^n \mid f(x) < +\infty\} \qquad \text{effective domain of } f$$

$$f^*(y) = \sup_{x \in \mathbb{R}^n} \{\langle x, y \rangle - f(x)\} \qquad \text{(conjugate function)}$$

$$\partial f(\overline{x}) = \{v \mid f(x) \geq f(\overline{x}) + \langle v, x - \overline{x} \rangle \,,\ \forall x\} \qquad \text{subdifferential of } f \text{ at } \overline{x}$$

We adopt from [12, 13] the following two-step subgradient algorithm for minimizing (13):

Subgradient Algorithm:

Choose $z^0 \in \operatorname{dom} g$ arbitrary.

For $k = 0, 1, \ldots$ compute:

$$y^k \in \partial h(z^k) \tag{16}$$

$$z^{k+1} \in \partial g^*(y^k) \tag{17}$$

The investigation of this algorithm in [13] includes the following results:

Proposition 1 ([13]). *Assume $g, h : \mathbb{R}^n \to \overline{\mathbb{R}}$ be proper, lower-semicontinuous and convex, and*

$$\operatorname{dom} g \subset \operatorname{dom} h \,, \quad \operatorname{dom} h^* \subset \operatorname{dom} g^* \,. \tag{18}$$

Then

(i) the sequences $\{z^k\}, \{y^k\}$ *according to (16), (17) are well-defined,*
(ii) $\{g(z^k) - h(z^k)\}$ *is decreasing,*
(iii) every limit point z^* *of* $\{z^k\}$ *is a critical point of* $g - h$.

Reconstruction Algorithm.
We apply (16), (17) to problem (9). Condition (18) holds, because obviously $\operatorname{dom} g \subset \operatorname{dom} h$, and $g^*(y) = \sup_z \{\langle z, y\rangle - g(z)\} < \infty$ for any finite vector y.

(16) reads

$$y^k = \nabla h(z^k) = \mu(x^k - \frac{1}{2}e) \tag{19}$$

since

$$\partial h(\overline{z}) = \{\nabla h(\overline{z})\}$$

if h is differentiable [11]. To compute (17), we note that g is proper, lower-semicontinuous, and convex. It follows [11] that

$$\partial g^*(\overline{y}) = \{z \mid g^*(y) \geq g^*(\overline{y}) + \langle z, y - \overline{y}\rangle,\ \forall y\} \tag{20}$$

$$= \operatorname{argmax}_z\{\langle \overline{y}, z\rangle - g(z)\}\ , \tag{21}$$

which is a *convex* optimization problem. Hence, (17) reads:

$$z^{k+1} \in \operatorname{argmin}_z\{g(z) - \langle y^k, z\rangle\}$$

Inserting y^k from (19), we finally obtain by virtue of (14), (11), and (10):

Reconstruction Algorithm (μ Fixed).
Choose $z^0 \in \operatorname{dom} g$ arbitrary.

For $k = 0, 1, \ldots$, compute z^{k+1} as minimizer of the linear program:

$$\min_{x \in [0,1]^n, \{z_{\langle i,j\rangle}\}, \lambda \in \mathbb{R}^m_{\geq 0}} -\left\langle \mu(x^k - \frac{1}{2}e), x\right\rangle + \frac{\alpha}{2}\sum_{\langle i,j\rangle} z_{\langle i,j\rangle} + \beta \sum_{i=1}^{m} \lambda_i \tag{22}$$

$$\text{subject to} \quad \begin{aligned} &\tilde{A}\begin{pmatrix} x \\ \gamma \end{pmatrix} = b \\ &0 \leq x_i \leq 1, \quad \gamma_i \in \mathbb{R}, \\ &\lambda_i \geq \tau_0 \gamma_i, \quad \lambda_i \geq -\tau_1 \gamma_i, \\ &z_{\langle i,j\rangle} \geq x_i - x_j, \quad z_{\langle i,j\rangle} \geq x_j - x_i \end{aligned}$$

In practice, we start with $\mu = 0$ and repeat the reconstruction algorithm for increasing values of μ, starting each iteration with the previous reconstruction z^k. This outer iteration loop terminates when $\forall i,\ \min\{x_i, 1 - x_i\} < \varepsilon$. Throughout all experiments in section 5, (ILP) or $(ILPSB)$, μ was increased by 0.1.

Note that for $\mu = 0$, we minimize (5), whereas for $\mu > 0$ it pays to shift in (22) the current iterate in the direction of the negative gradient of the "binarization" functional (6). While this is an intuitively clear modification of (5), convergence of the sequence of minimizers of (22) due to proposition 1 is not obvious.

5 Experimental Evaluation

For evaluation purposes, we took three parallel projections, 0°, 45°, and 90°, of the 64 × 64 image shown in figure 6(a). In case of noiseless projections (ILP) and ($ILPSB$) are able to find the correct reconstruction within 10 iterations, figure 6(b)-(d).

We independently added for each projection direction a value $\delta b_i \sim \mathcal{N}(0, \sigma)$ to the respective measurement b_i in order to simulate the presence of noise. Roughly speaking, in the experiments with $\sigma = 1.0$ a projection value can differ between ±2 from its correct value and in case of $\sigma = 2.0$ even between ±4. Relative to the image size, 64 × 64, the choice of σ seems to be reasonable for real application scenarios.

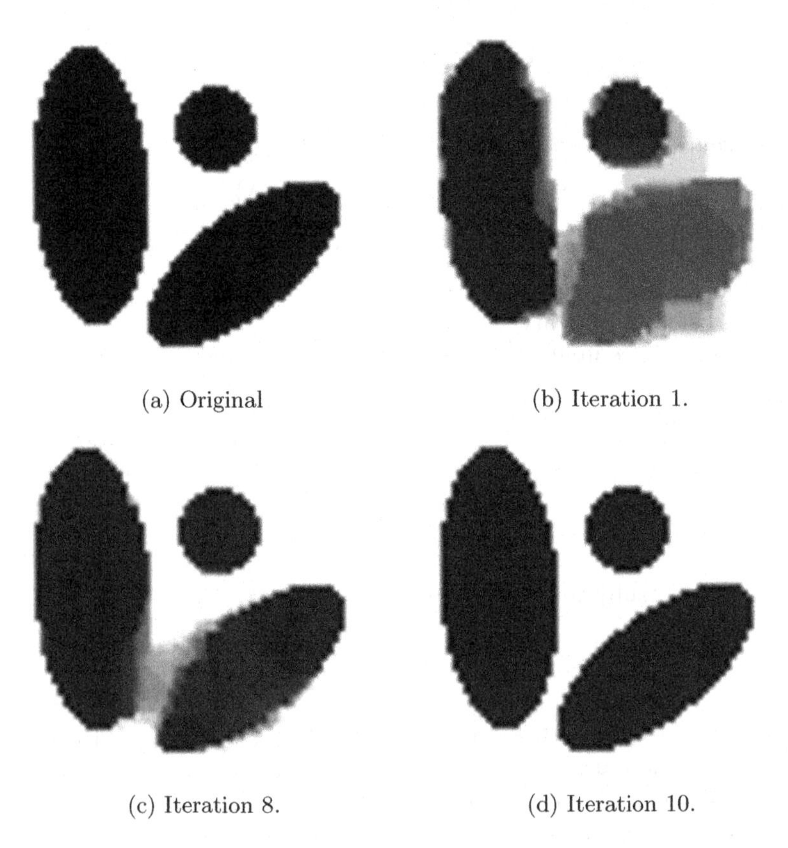

(a) Original (b) Iteration 1.

(c) Iteration 8. (d) Iteration 10.

Fig. 6. (a) Shows the original image, 64 × 64, from which we have taken three parallel projections, 0°, 45°, and 90°. (b)-(d) In case of noiseless projections (ILP), $\alpha = 0.25$, and ($ILPSB$), $\alpha = 0.25$ and $\beta = 1.0$, are able to find the correct solution within 10 iterations

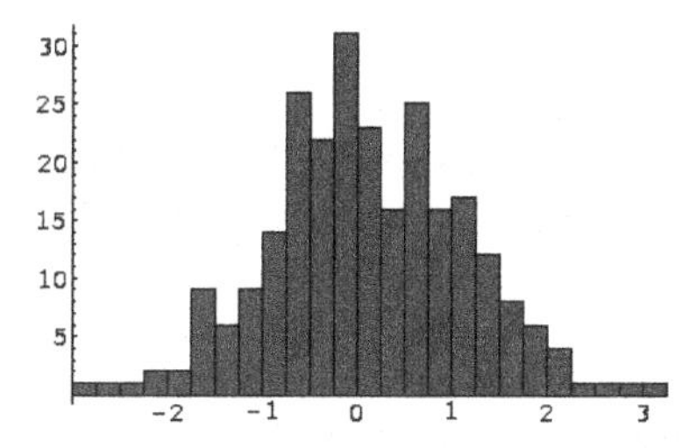

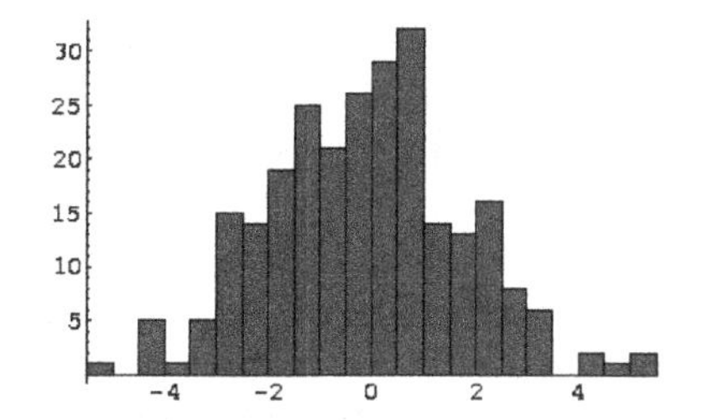

Fig. 7. Each histogram was created from 255 (64 (horizontal rays) + 64 (vertical rays) + (127 (diagonal rays))) samples of different normal distributions, $\mu = 0.0$ (in both cases), $\sigma = 1.0$ (left) and $\sigma = 2.0$ (right). In order to simulate noise, we added independently for each projection direction a value $\delta b_i \sim \mathcal{N}(0, \sigma)$ to the respective measurement b_i

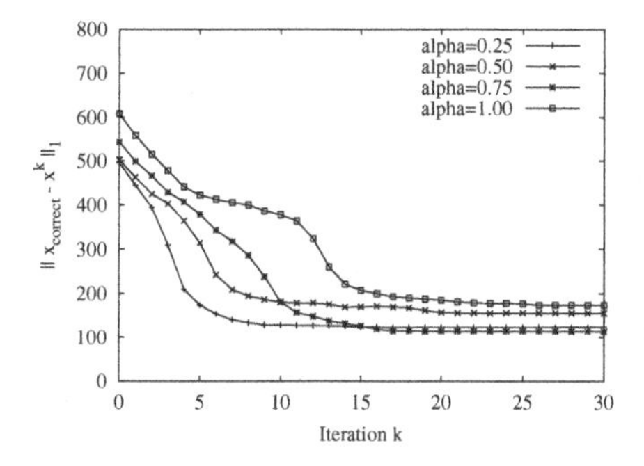

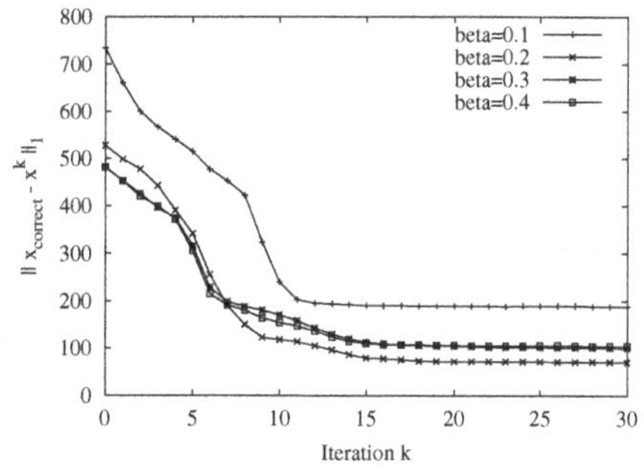

Fig. 8. Experiments with $\sigma = 1.0$: Plots the difference between the original image and the solution at iteration k for (ILP)(left) and $(ILPSB)$(right). The tables 1 and 2 give the final numerical values of these experiments

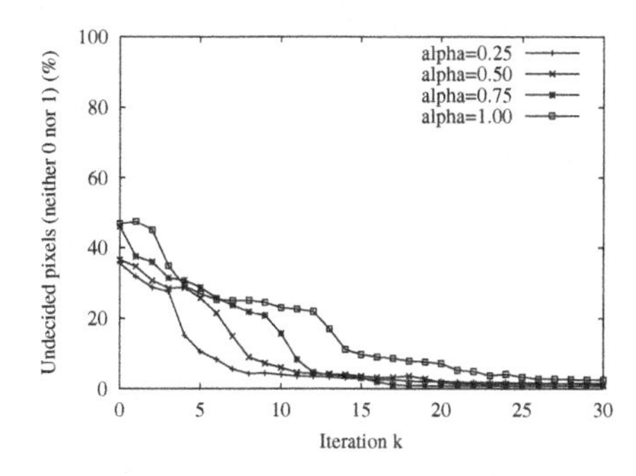

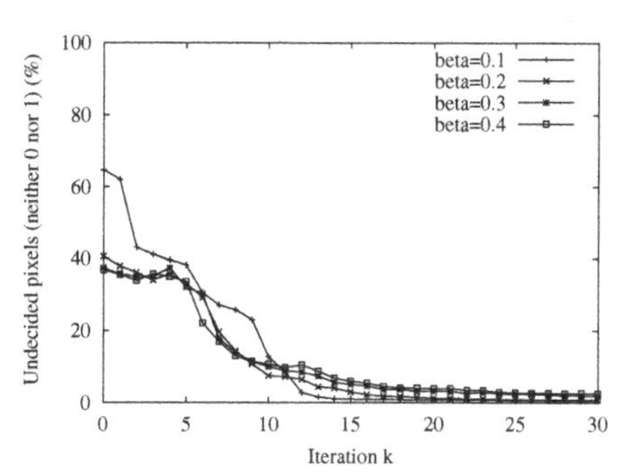

Fig. 9. Experiments with $\sigma = 1.0$: Plots the number of undecided pixels, i.e. pixels that are neither 0 nor 1, at iteration k for (ILP)(left) and $(ILPSB)$(right). For the final numerical values see the tables 1 and 2

In order to find a suitable choice of α we decided to check (ILP) with $\alpha \in \{0.25, 0.5, 0.75, 1.0\}$. In case of noiseless projections, $\alpha = 0.25$ is a good choice. However, in combination with noisy projections our experiments show that α should be set higher ($\alpha \in [0.5, 0.75]$). The (ILP) approach achieved best results with $\alpha = 0.75$ ($\sigma = 1.0$) and $\alpha = 0.5$ ($\sigma = 2.0$).

We checked $(ILPSB)$ for different choices of β. In case of $\sigma = 1.0$ we set the parameters to $\tau_1 = 1.0$, $\tau_0 = 3.0$, $\alpha = 0.5$ and for $\sigma = 2.0$ to $\tau_1 = 1.0$, $\tau_0 = 5.0$, $\alpha = 1.0$. In our experiments best performance was achieved with $\beta = 0.2$. In both cases $(ILPSB)$ reached better final results than (ILP).

Numerical results of our experiments are given in table 1 and plots are shown in the figures 8, 9, 10, and 11. Images of intermediate and the final reconstruction are presented in figure 12.

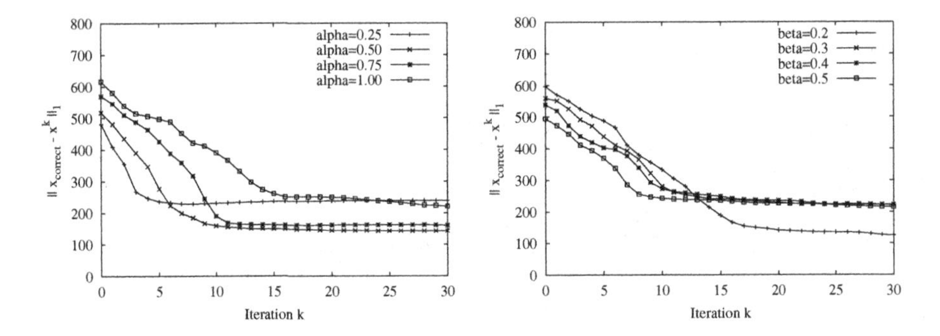

Fig. 10. Experiments with $\sigma = 2.0$: Difference between the original image and the solution at iteration k for (ILP)(left) and $(ILPSB)$(right). For the final numerical values see the tables 1 and 2

Table 1. Summary of the (ILP) results for different α and σ. The quality of the reconstruction (third column) was simply measured by the difference between the original and the solution, i.e. $||x_{correct} - x_{solution}||_1$. Further, we measured the number of pixels that have not been decided, i.e. that are neither 0 nor 1 (fourth column). The best result of (ILP) was obtained with $\alpha = 0.75$ in case of $\sigma = 1.0$ and $\alpha = 0.5$ for $\sigma = 2.0$. Plots of this experiments are shown in the figures 8(left), 9(left), 10(left), and 11(left)

α	σ	difference	undecided
0.25	1.0	124.45	1.00 %
0.50	1.0	156.75	0.63 %
0.75	1.0	112.24	0.49 %
1.00	1.0	172.59	1.03 %
0.25	2.0	240.83	0.98 %
0.50	2.0	142.73	0.90 %
0.75	2.0	159.67	1.25 %
1.00	2.0	215.96	0.93 %

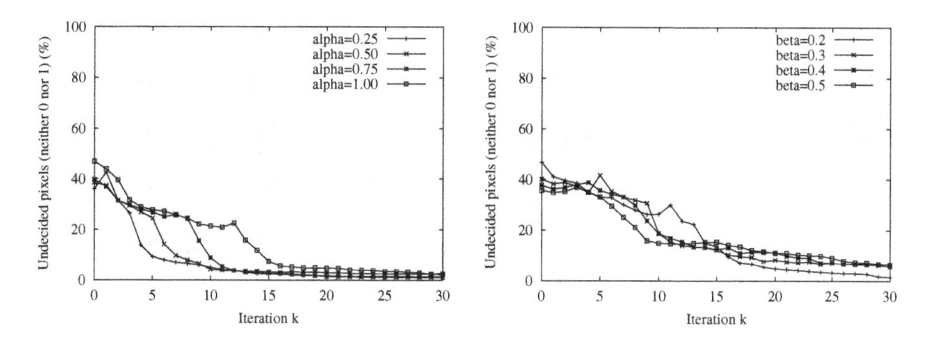

Fig. 11. Experiments with $\sigma = 2.0$: Number of undecided pixels at iteration k for (ILP)(left) and $(ILPSB)$(right). For the final numerical values see the tables 1 and 2

Table 2. $(ILPSB)$ results for different α and σ. The third column shows the difference between original and solution, $||x_{correct} - x_{solution}||_1$, and the fourth column the number of undecided pixels. The $(ILPSB)$ approach yields best results for $\beta = 0.2$. In both cases these results were better than the best results achieved by (ILP), see table 1. Plots of this experiments are shown in the figures 8(right), 9(right), 10(right), and 11(right)

β	σ	difference	undecided
0.1	1.0	191.00	0.00 %
0.2	1.0	68.04	0.05 %
0.3	1.0	93.02	0.05 %
0.4	1.0	104.87	0.10 %
0.2	2.0	119.51	0.17 %
0.3	2.0	232.75	0.20 %
0.4	2.0	220.80	0.20 %
0.5	2.0	194.44	0.59 %

6 Conclusion

In this paper we presented the $(ILPSB)$ approach which is a modification of (ILP) with noise suppression. For evaluation purposes, noise was simulated by sampling normal distributions with $\mu = 0.0$ and $\sigma \in \{1.0, 2.0\}$. In order to compare both approaches we measured the difference between the solution and the original image. Further, we considered the number of pixels that were not decided, i.e. neither 0 nor 1. In our experiments $(ILPSB)$ achieved better results than (ILP) under both criteria.

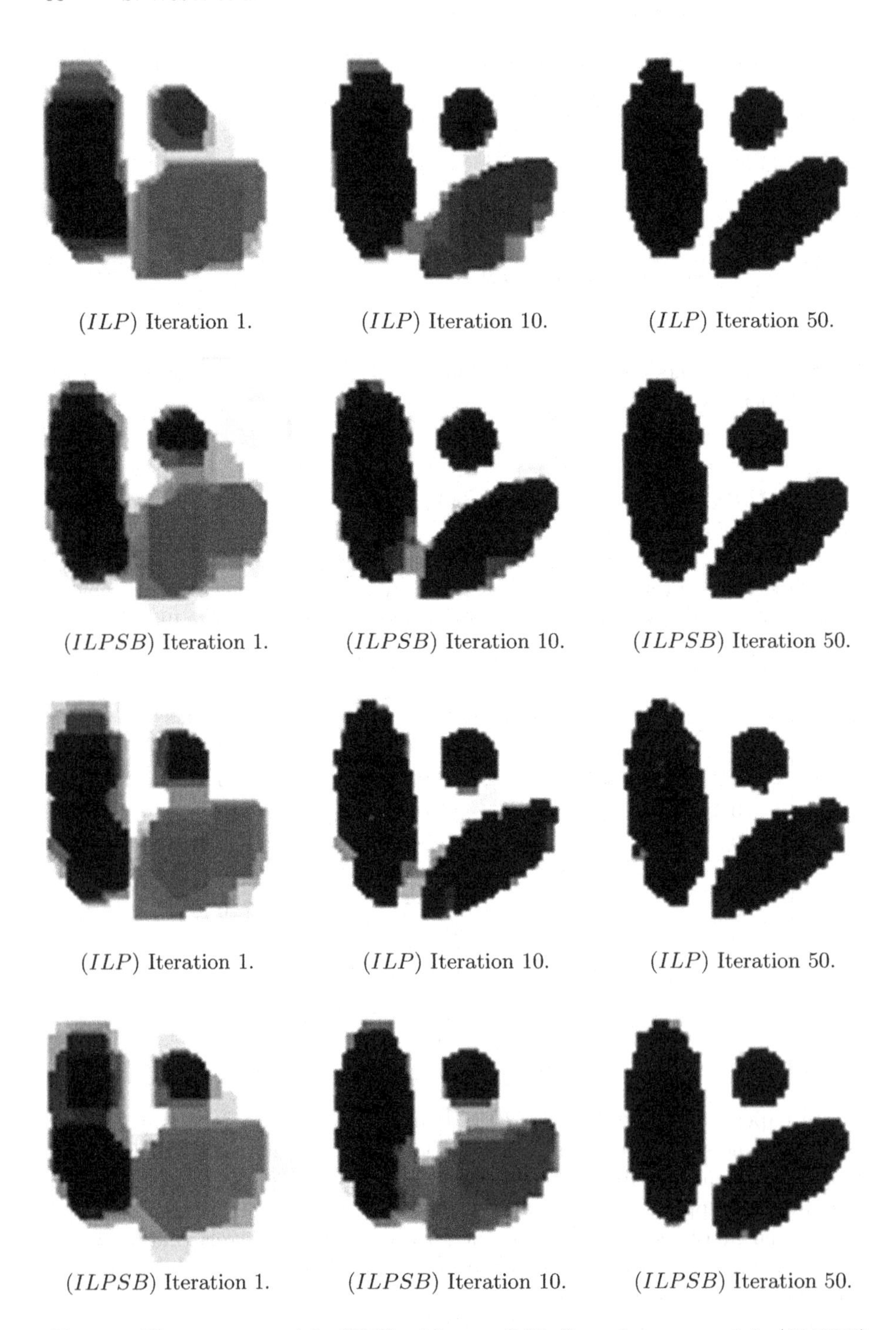

(ILP) Iteration 1. (ILP) Iteration 10. (ILP) Iteration 50.

($ILPSB$) Iteration 1. ($ILPSB$) Iteration 10. ($ILPSB$) Iteration 50.

(ILP) Iteration 1. (ILP) Iteration 10. (ILP) Iteration 50.

($ILPSB$) Iteration 1. ($ILPSB$) Iteration 10. ($ILPSB$) Iteration 50.

Fig. 12. First row: $\sigma = 1.0$, (ILP) with $\alpha = 0.75$. Second row: $\sigma = 1.0$, ($ILPSB$) with $\beta = 0.2$. Third row: $\sigma = 2.0$, (ILP) with $\alpha = 0.5$. Fourth row: $\sigma = 2.0$, ($ILPSB$) with $\beta = 0.2$

References

1. Kuba, A., Herman, G.: Discrete tomography: A historical overview. In Herman, G.T., Kuba, A., eds.: Discrete Tomography: Foundations, Algorithms, and Applications. Birkhäuser (1999) 3–34
2. Gardner, R., Gritzmann, P.: Discrete tomography: Determination of finite sets by x-rays. Trans. Amer. Math. Soc. **349** (1997) 2271–2295
3. Gritzmann, P., Prangenberg, D., de Vries, S., Wiegelmann, M.: Success and failure of certain reconstruction and uniqueness algorithms in discrete tomography. Int. J. Imag. Syst. Technol. **9** (1998) 101–109
4. Herman, G., Kuba, A., eds.: Discrete Tomography: Foundations, Algorithms, and Applications. Birkhäuser Boston (1999)
5. Natterer, F., Wübbeling, F.: Mathematical Methods in Image Reconstruction. SIAM, Philadelphia (2001)
6. Fishburn, P., Schwander, P., Shepp, L., Vanderbei, R.: The discrete radon transform and its approximate inversion via linear programming. Discr. Appl. Math. **75** (1997) 39–61
7. Gritzmann, P., de Vries, S., Wiegelmann, M.: Approximating binary images from discrete x-rays. SIAM J. Optimization **11** (2000) 522–546
8. Weber, S., Schnörr, C., Hornegger, J.: A linear programming relaxation for binary tomography with smoothness priors. In: Proc. Int. Workshop on Combinatorial Image Analysis (IWCIA'03). (2003) Palermo, Italy, May 14-16/2003.
9. Weber, S., Schüle, T., Schnörr, C., Hornegger, J.: A linear programming approach to limited angle 3d reconstruction from dsa projections. Special Issue of Methods of Information in Medicine **4** (2004) (in press).
10. Weber, S., Schnörr, C., Schüle, T., Hornegger, J.: Binary tomography by iterating linear programs. Technical report 5/2004, University of Mannheim (2004)
11. Rockafellar, R.: Convex analysis. 2 edn. Princeton Univ. Press, Princeton, NJ (1972)
12. Dinh, T.P., Elbernoussi, S.: Duality in d.c. (difference of convex functions) optimization subgradient methods. In: Trends in Mathematical Optimization, Int. Series of Numer. Math. Volume 84. Birkhäuser Verlag, Basel (1988) 277–293
13. Dinh, T.P., An, L.H.: A d.c. optimization algorithm for solving the trust-region subproblem. SIAM J. Optim. **8** (1998) 476–505

Hexagonal Pattern Languages

K.S. Dersanambika[1,*], K. Krithivasan[1], C. Martin-Vide[2], and K.G. Subramanian[3,**]

[1] Department of Computer Science and Engineering,
Indian Institute of Technology, Madras
Chennai - 600 036, India
dersanapdf@yahoo.com
kamala@iitm.ernet.in
[2] Rovira I Virgili University,
Pl. Imperial Tarraco 1, 43005 Tarragona, Spain
cmv@correu.urv.es
[3] Department of Mathematics,
Madras Christian College, Chennai - 600 059, India
kgsmani@vsnl.net

Abstract. Hexagonal tiling systems, hexagonal local picture languages and hexagonal recognizable picture languages are defined in this paper. Hexagonal Wang tiles and systems are also introduced. It is noticed that the family of hexagonal picture languages defined by hexagonal Wang systems coincides with the family of hexagonal picture languages recognized by hexagonal tiling system. Similar to hv-domino systems describing rectangular arrays, we define xyz-domino systems and characterize hexagonal picture languages using this. Unary hexagonal picture languages are also considered and we analyze some of their properties.

1 Introduction

Recently, searching for a sound definition of finite state recognizability for picture languages, so that the new definition of the recognizable picture languages inherits many properties from the existing cases, in [3, 4], local and recognizable picture languages in terms of tiling systems were introduced and studied. Subsequently hv-local picture languages via domino systems were defined in [6] and recognizability of a picture language, defined in terms of domino systems is proved to be equivalent to recognizability defined by tiling systems. Informally a picture language is defined as local by sensing the presence of a specified set of square tiles in each picture of the language and requiring no other square tile. Recognizability is defined by projection of local properties. The chapter on two dimensional languages in [4] gives account of these details. On the other hand,

* This work is partially supported by University Grants Commission, India.
** Work partially supported by Spanish Ministry for Education, Culture and Sport, grant No.SAB 2001-0007.

R. Klette and J. Žunić (Eds.): IWCIA 2004, LNCS 3322, pp. 52–64, 2004.

in [8] Wang systems, that are labelled Wang tiles, are introduced as a new formalism to recognize rectangular picture languages and are proved to be again equivalent to recognizability defined by tiling systems.

It is very natural to consider hexagonal tiles on triangular grids that correspond to rectangular tiles on rectangular grids. In fact motivated by the studies of [3, 4, 6, 8], in this paper we define hexagonal local picture languages and hexagonal recognizable picture languages. In fact we require certain hexagonal tiles only to be present in each hexagonal picture of a hexagonal local picture language. This leads on to the notion of hexagonal tiling system defining hexagonal recognizable picture languages.

We also define hexagonal Wang tiles and hexagonal Wang systems and show the equivalence between hexagonal tiling system and hexagonal Wang system. Similar to hv-domino systems[6], we define xyz-domino systems and characterize hexagonal picture languages using these systems. Unary hexagonal picture languages are also considered and we analyze some properties of hexagonal picture languages over one letter alphabet.

2 Preliminaries

We review here the notions of hexagonal pictures and hexagonal picture languages [10]. For notions relating to formal language theory we refer to [9]. Let Σ be a finite alphabet of symbols.

A hexagonal picture p over Σ is a hexagonal array of symbols of Σ. The set of all hexagonal arrays of the alphabet Γ is denoted by Γ^{**H}. A hexagonal picture language L over Γ is a subset of Γ^{**H}.

Example 1. A hexagonal picture over the alphabet {a,b,c,d} is shown in Figure 1.

```
    a   a   a
  a   b   d   c
b   c   c   a   c
  a   b   d   a
    a   a   d
```

Fig. 1. A hexagonal picture

Definition 1. *If $x \in \Gamma^{**H}$, then $\hat{x}$ is the hexagonal array obtained by surrounding x with a special boundary symbol $\# \notin \Gamma$.*

Example 2. A hexagonal picture over the alphabet {a,b,c,d,e,f} surrounded by #, is shown in Figure 2.

With respect to a triad of triangular axes x,y,z the coordinates of each element of the hexagonal picture in Fig 1 is shown in Figure 3.

We now define the notion of projection of a hexagonal picture and of a hexagonal picture language. Let Γ and Σ be two finite alphabets and $\Pi : \Gamma \to \Sigma$ be a mapping which we call, a projection.

```
      #   #   #   #
    #   a   b   c   #
  #   a   d   b   a   #
#   b   c   d   c   b   #
  #   c   d   e   f   #
    #   a   b   c   #
      #   #   #   #
```

Fig. 2. A hexagonal picture surrounded by #

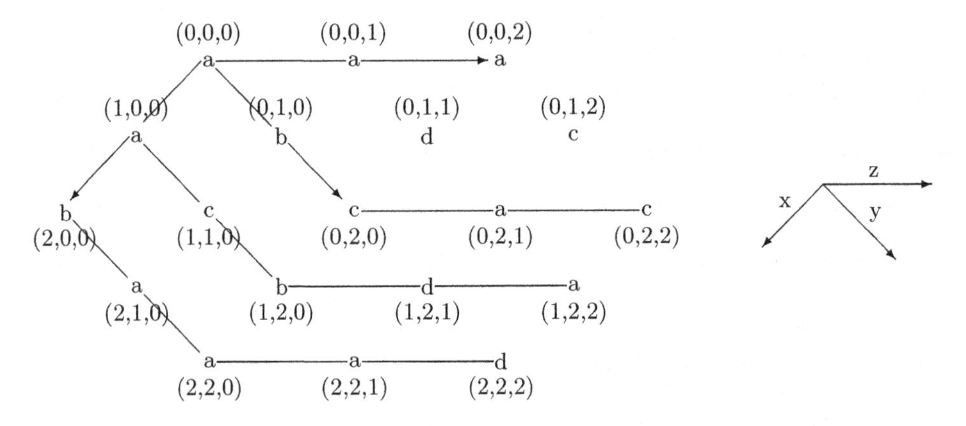

Fig. 3. Coordinates of elements of hexagonal picture of Fig. 1

Definition 2. *Let $p \in \Gamma^{**H}$ be a hexagonal picture. The projection by mapping Π of picture p is the picture $p' \in \Sigma^{**H}$ such that $p'(i,j,k) = \Pi(p(i,j,k))$ for all $1 \leq i \leq l-1$, $1 \leq j \leq m-1$ and $1 \leq k \leq n-1$, where (l,m,n) is called the size of the hexagonal picture.*

Definition 3. *Let $L \subset \Gamma^{**H}$ be a hexagonal picture language. The projection by mapping Π of L is the language $L_1 = \{p' \mid p' = \Pi(p), \forall p \in L\} \subseteq \Sigma^{**H}$.*

Remark 1. As in the case of a hexagonal picture we will denote by $\Pi(L)$ the projection by mapping Π of a hexagonal picture language L. That is $L_1 = \Pi(L)$.

3 Recognizability of Hexagonal Pictures

In this section, the main notions of hexagonal local and hexagonal recognizable picture languages are introduced. For this purpose, we first define a hexagonal tile.

A hexagonal picture of the form shown in Figure 4 is called a hexagonal tile over an alphabet{a,...,g}.

Given a hexagonal picture p of size (l,m,n), for $g \leq l$, $h \leq m$ and $k \leq n$. we denote by $B_{g,h,k}(p)$ the set of all hexagonal blocks (or hexagonal subpictures) of p of size (g,h,k). $B_{2,2,2}$ is in fact a set of hexagonal tiles.

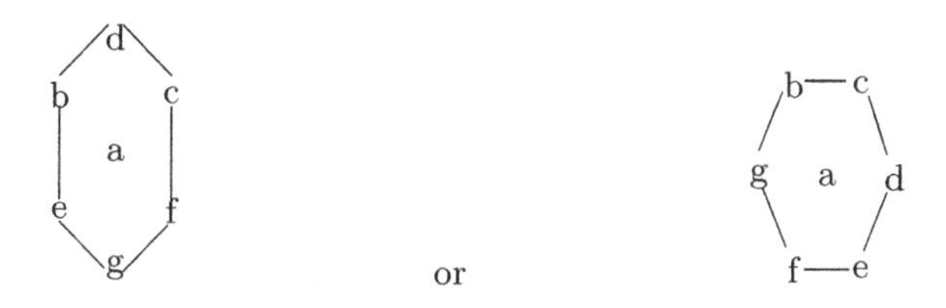

Fig. 4. Two hexagonal tiles

3.1 The Family HLOC

Definition 4. *Let Γ be a finite alphabet. A hexagonal picture language $L \subseteq \Gamma^{\star\star H}$ is called local if there exists a finite set $\triangle$ of hexagonal tiles over $\Gamma \cup \{\#\}$ such that $L = \{p \in \Gamma^{\star\star H} | \ B_{2,2,2}(\hat{p}) \subseteq \triangle\}$.*

The family of hexagonal local picture languages will be denoted by $HLOC$.

Example 3. Let $\Gamma = \{a\}$ be an alphabet and let $\triangle$ be a set of hexagonal tiles over Γ. The hexagonal tiles of $\triangle$ are shown in the Figure 5.

The language $L = L(\triangle)$ is the language of hexagonal pictures an element of which is shown in Figure 6 The language L is indeed a hexagonal local picture language. Note that the hexagonal picture language over one letter alphabet with all sides of equal length is not local.

We give another example of a hexagonal local picture language.

Example 4. Consider the hexagonal tiles as in Figure 7.

The language $L = L(\Delta)$ is the language of hexagonal pictures of size $l = m = n$ a member of which is shown in Figure 8.

Applying a coding that replaces 0 and 1 by a, we obtain hexagonal pictures over $\{a\}$ with equal sides.

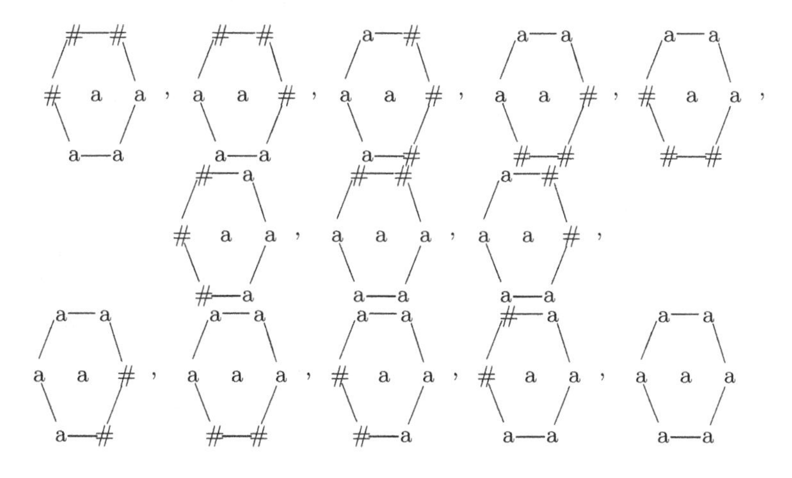

Fig. 5. Hexagonal tiles of Example 3

```
      #   #   #   #   #
    #   a   a   a   a   #
  #   a   a   a   a   a   #
#   a   a   a   a   a   a   #
  #   a   a   a   a   a   #
    #   a   a   a   a   #
      #   #   #   #   #
```

Fig. 6. A hexagonal picture over $\{a\}$

Fig. 7. Hexagonal Tiles of Example 4

```
      #   #   #   #
    #   1   0   0   #
  #   0   0   1   0   #
#   0   1   0   0   1   #
  #   0   0   1   0   #
    #   1   0   0   #
      #   #   #   #
```

Fig. 8. A Hexagonal Picture of L in Example 4

3.2 The Family HREC

We now introduce the family of hexagonal recognizable picture languages using the notion of hexagonal local picture languages introduced in section 3.1 and the notion of projection of a language.

Definition 5. *Let Σ be a finite alphabet. A hexagonal picture language $L \subseteq \Sigma^{**H}$ is called recognizable if there exists a hexagonal local picture language L' (given by a set $\triangle$ of hexagonal tiles) over an alphabet Γ and a mapping $\Pi : \Gamma \to \Sigma$ such that $L = \Pi(L')$.*

The family of hexagonal recognizable picture languages will be denoted by HREC.

Definition 6. *A hexagonal tiling system T is a 4-tuple $(\Sigma, \Gamma, \Pi, \theta)$, where Σ and Γ are two finite sets of symbols, $\Pi : \Gamma \to \Sigma$ is a projection and θ is a set of hexagonal tiles over the alphabet $\Gamma \cup \{\#\}$.*

Definition 7. *The hexagonal picture language $L \subseteq \Sigma^{**H}$ is tiling recognizable if there exists a tiling system $T = (\Sigma, \Gamma, \Pi, \theta)$ such that $L = \Pi(L(\theta))$.*

The language consisting of hexagonal arrays of equal sides over a one letter alphabet {a} is recognizable but not local.

3.3 Hexagonal Wang Systems

Now, we introduce labelled hexagonal Wang tiles. A labelled hexagonal Wang tile is a 7-tuple, consisting of 6 colors chosen from a finite set of colors Q, and a label. The colors are placed at upper left (UL), upper right (UR), Left (L), right (R), lower left (LL), lower right (LR) positions of the label(Figure 9).

Here hexagonal Wang tiles are used to recognize hexagonal picture languages. A hexagonal Wang tile can be represented as in Figure 10.

where p, q, r, s, t and u are colors and a is a label taken from a finite alphabet. Two hexagonal Wang tiles may be adjacent if and only if the adjacent colors are the same.

Definition 8. *A hexagonal Wang system is a triplet $W = (\Sigma, Q, T)$, where Σ is a finite alphabet, Q is a finite set of colors and T is a set of labelled Wang tiles $T \subseteq Q^6 \times \Sigma$.*

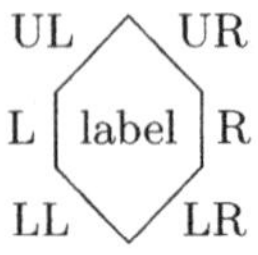

Fig. 9.

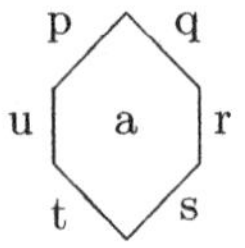

Fig. 10.

We now introduce to the notion of hexagonal tiling in a hexagonal Wang system (HWS).

Definition 9. *Let $W = (\Sigma, Q, T)$ be a hexagonal Wang system. A hexagonal pictures H over T, of size (l,m,n) is a hexagonal tiling if it satisfies the following conditions:*

$H(0,0,n)$ =

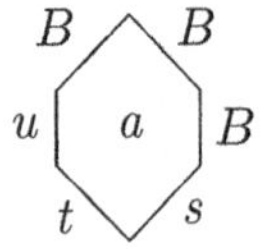

$H(0,m,n)$ =

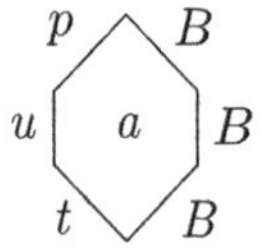

$H(l,m,n)$ =

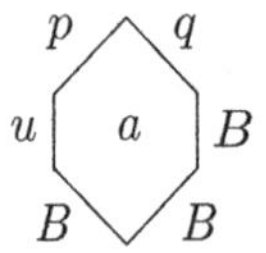

$H(l,m,0)$ =

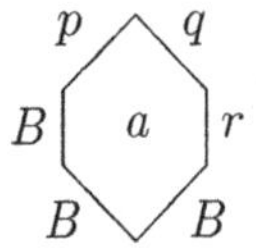

$H(l,0,0)$ =

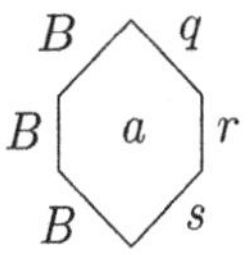

$$H(0,0,0) = \begin{array}{ccc} B & & B \\ B & a & r \\ t & & s \end{array}$$

$$H(0,0,k) = \begin{array}{ccc} B & & B \\ u & a & r \\ t & & s \end{array} \qquad ; k = 1, 2, \ldots, n-1$$

$$H(0,j,n) = \begin{array}{ccc} p & & B \\ u & a & B \\ t & & s \end{array} \qquad ; j = 1, 2, \ldots, m-1$$

$$H(i,m,n) = \begin{array}{ccc} p & & q \\ u & a & B \\ t & & B \end{array} \qquad ; i = 1, 2, \ldots, l-1$$

$$H(l,m,k) = \begin{array}{ccc} p & & q \\ u & a & r \\ B & & B \end{array} \qquad ; k = 1, 2, \ldots, n-1$$

$$H(l,j,0) = \begin{array}{ccc} p & & q \\ B & a & r \\ B & & s \end{array} \qquad ; j = 1, 2, \ldots, m-1$$

$$H(i,0,0) = \begin{array}{ccc} B & & q \\ r & a & B \\ t & & s \end{array} \qquad ; i = 1, 2, \ldots, l-1$$

$$H(i,j,k) = \begin{array}{ccc} p & & q \\ u & a & r \\ t & & s \end{array} \qquad ; \begin{array}{l} i = 0, 1, 2, \ldots, l-1 \\ j = 1, 2, \ldots, m \\ k = 0, 1, 2, \ldots, n-1 \end{array}$$

We obtain the following theorems.

We denote by $L(X)$, the family of hexagonal picture languages recognizable by X systems, $X \in \{HWS, HTS\}$.

Theorem 1. *$L(HWS)$ is closed under projection.*

Theorem 2. *The classes of hexagonal picture languages recognizable by hexagonal tiling and hexagonal Wang systems are equal. That is, $L(HWS) = L(HTS)$.*

3.4 Hexagonal Domino System

Now we consider another formalism to recognize hexagonal pictures which is based on domino systems introduced by Latteux et al [6]. Here we consider dominos of the following types.

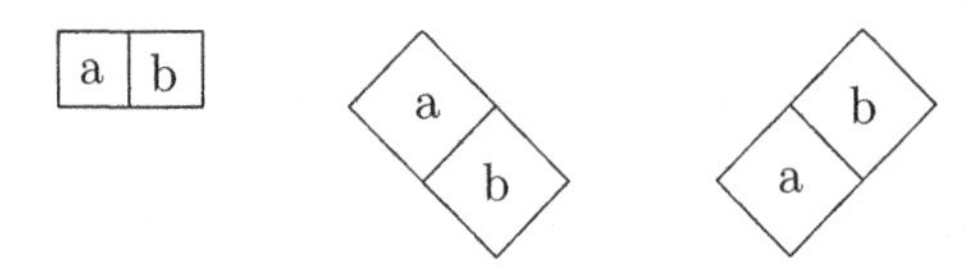

Definition 10. *Let L be a hexagonal picture language. The language L is xyz-local if there exists a set $\triangle$ of dominos as defined above, over the alphabet $\Sigma \cup \{\#\}$ such that*

$$L = \{w \in \Sigma^{\star\star H} \mid \text{All domino tiles relating to } w \text{ are} \subseteq \triangle\}$$

We write $L = L(\triangle)$, if L is xyz-local and $\triangle$ is a set of dominos satisfying the condition in the definition.

Definition 11. *A hexagonal domino system is a 4-tuple $D = (\Gamma, \Sigma, \triangle, \Pi)$, where Σ and Γ are two finite alphabets, $\triangle$ is a set of dominos over Γ and $\Pi : \Gamma \to \Sigma$ is a projection.*

A hexagonal picture language L is xyz-domino recognizable if there exists a domino system D such that $L = \Pi(L(D))$. The class of hexagonal languages recognizable by domino systems is denoted by $L(HDS)$.

Theorem 3. *$HREC = L(HDS)$.*

4 Hexagonal Pictures Over an Alphabet of One Symbol

In this section we analyze some properties of hexagonal picture languages over an alphabet of one symbol. Let $\Sigma = \{0\}$. A hexagonal picture over Σ can be described only by its size (l, m, n) or by a triplet $(a^{l-1}, b^{m-1}, c^{n-1}) \in \{a\}^\star \times \{b\}^\star \times \{c\}^\star$. For example a picture of size $(4, 5, 3)$ in $\{a\}^\star \times \{b\}^\star \times \{c\}^\star$ is an hexagonal array and is denoted by (a^3, b^4, c^2). We use the notation similar to the one given in [8].

Example 5. Let us consider the following hexagonal arrays of one symbol.

```
      0 0 0
     0 0 0 0
    0 0 0 0 0
   0 0 0 0 0 0
    0 0 0 0 0 0
     0 0 0 0 0
      0 0 0 0
       0 0 0
```

The set $\{a\}^\star \times \{b\}^\star \times \{c\}^\star$ is a monoid, so that we can consider the family of rational subsets of $\{a\}^\star \times \{b\}^\star \times \{c\}^\star$.

Infact the operation $\circ$ on $\{a\}^\star \times \{b\}^\star \times \{c\}^\star$ defined as follows. If x and y are two hexagonal arrays of sizes (l, m, n) and (l', m', n') respectively. $x \circ y$ is a hexagonal array of size $(l + l' - 1, m + m' - 1, n + n' - 1)$.

Definition 12. *Let M be a monoid. The class of rational parts $Rat(M)$ is the smallest family of languages $\mathcal{R} \subseteq \mathcal{P}(M)$, where $\mathcal{P}(M)$ denotes the set of all subsets of M.*

1. $\Phi \in \mathcal{R}$,$\{a\} \in \mathcal{R}$ *for all* $a \in M$
2. *If* L_1,$L_2 \in \mathcal{R}$ *then* $L_1 \cup L_2 \in \mathcal{R}$ *and* $L_1 \circ L_2 \in \mathcal{R}$.
3. *If* $L \in \mathcal{R}$ *then* $L^\star = \bigcup_{n \geq 0} L^n \in \mathcal{R}$.

where $\circ$ is the concatenation operator and $\star$ is the star operator [5]. For more details about rational relations, one can refer to [1, 2]. Rational subsets of a monoid can be recognized using automata over monoids.

Definition 13. *Let M be a monoid. An automaton over M, $\mathcal{A} = (Q, M, E, I, F)$ is a direct graph whose edges are labelled by elements of M. Q is a finite set of states, $I \subseteq Q$ is the initial states. $F \subseteq Q$ is the set of final states and $E \subseteq Q \times M \times Q$ is a finite set of labelled edges.*

If $(p, m, q) \in E$, we also write $p \xrightarrow{m} q$. A path Π in $\mathcal{A}$ is a finite sequence of labelled edges

$$\Pi = P_0 \xrightarrow{m_1} p_1 \xrightarrow{m_2} p_2 \xrightarrow{m_3} \cdots \xrightarrow{m_n} p_n$$

A path is successful if $p_0 \in I$ and $p_n \in F$. The set of all the sequences m over M such that there exists a successful path in $\mathcal{A}$ labelled by m is denoted by $\mathcal{L}(\mathcal{A})$.

If $M = \{a\}^\star \times \{b\}^\star \times \{c\}^\star$, the label of edges are triplets of words. Such an automaton can be viewed as an automaton with three tapes and is called 3 tape automaton and is denoted by (3-TA).

Any 3-TA is equivalent to a 3-TA in normal form in which the labels of the edges are $(a, \varepsilon, \varepsilon)$, $(\varepsilon, b, \varepsilon)$, $(\varepsilon, \varepsilon, c)$, i.e. $E \subseteq (Q \times (a, \varepsilon, \varepsilon) \times Q) \cup (Q \times (\varepsilon, b, \varepsilon) \times Q) \cup (Q \times (\varepsilon, \varepsilon, c) \times Q)$. A 3-TA in normal form at any step changes its reading either a letter of the first tape or a letter of second tape or a letter of third tape. A triplet of words is accepted by a 3-TA if starting from an initial state after reading all the symbols in the first, second and third tapes, the automaton reaches a final state.

We call $\mathcal{L}(3-TA)$ the class of rational languages recognized by a 3-TAs, $\mathcal{L}(3-TA) = \{L \subseteq \{a\}^\star \times \{b\}^\star \times \{c\}^\star \mid \exists\, \mathcal{A} \in 3-TA : L = \mathcal{L}(\mathcal{A})\ \ for\ \ some\ 3-TA,\ \ \mathcal{A}\}$.

An example of 3-TA is the following.

Example 6. Let $L = \{(a^n, b^n, c^n) \mid n \geq 0\}$. The language L is recognized by the automaton $A = (Q, \{a\}^\star \times \{b\}^\star \times \{c\}^\star, E, I, T)$, where $Q = \{p_0, p_1, p_2, p_3\}$, $I = \{p_0\}$, $T = \{p_3\}$ and

$$E = \{(p_0, (a, \varepsilon, \varepsilon), p_1), (p_1, (\varepsilon, b, \varepsilon), p_2), (p_2, (\varepsilon, \varepsilon, c), p_3), (p_3, (a, \varepsilon, \varepsilon), p_1)\}$$

We can describe a successful path for $(a^{l-1}, b^{m-1}, c^{n-1})$ using a hexagonal picture A of size (l,m,n) is defined by

1. Case (1)

   $p \xrightarrow{(a,\varepsilon,\varepsilon)} q$ and the $3-TA$ is in a cell with coordinates (x, y, z), it goes to a cell with coordinates (x', y', z') where
   (a) If $z = 0$, then $(x', y', z') = (x + 1, y, 0)$.
   (b) If $z \neq 0$, $x = 0$ and $y \neq m$ then $(x, y', z') = (0, y + 1, z - 1)$.
   (c) If $z \neq 0$, $x = 0$ and $y = m$ then $(x', y', z') = (1, m, z)$.
   (d) If $z \neq 0$ and $x \neq 0$, then $(x', y', z') = (x + 1, y, z)$.

2. Case (2)

   $p \xrightarrow{(\varepsilon,b,\varepsilon)} q$ and the $3-TA$ is in a cell with coordinates (x, y, z), it goes to a cell with coordinates (x', y', z') where
   (a) If $y = m$, then $(x', y', z') = (x + 1, m, z + 1)$.
   (b) If $y \neq m$, then $(x', y', z') = (x, y + 1, z)$.

3. Case (3)

   $p \xrightarrow{(\varepsilon,\varepsilon,c)} q$ and the $3-TA$ is in a cell with coordinates (x, y, z), it goes to a cell with coordinates (x', y', z') where
   (a) If $x = 0$, then $(x', y', z') = (0, y, z + 1)$.
   (b) If $x \neq 0$, $z = 0$, $y \neq m$ then $(x', y', z') = (x - 1, y + 1, 0)$.
   (c) If $x \neq 0$, $z = 0$ and $y = m$ then $(x', y', z') = (x, m, 1)$.
   (d) If $x \neq 0$, and $z \neq 0$ then $(x', y', z') = (x, y, z + 1)$.

For example taking the language of Example 6 a successful path for(a^3, b^3, c^3) is

$$(p_0, (a, \varepsilon, \varepsilon), p_1)\ (p_1, (\varepsilon, b, \varepsilon), p_2)\ (p_2, (\varepsilon, \varepsilon, c), p_3)$$
$$(p_3, (a, \varepsilon, \varepsilon), p_1)\ (p_1, (\varepsilon, b, \varepsilon), p_2)\ (p_2, (a, \varepsilon, \varepsilon), p_3)$$
$$(p_3, (a, \varepsilon, \varepsilon), p_1)\ (p_1, (\varepsilon, b, \varepsilon), p_2)\ (p_2, (a, \varepsilon, \varepsilon), p_3)$$

That is represented by

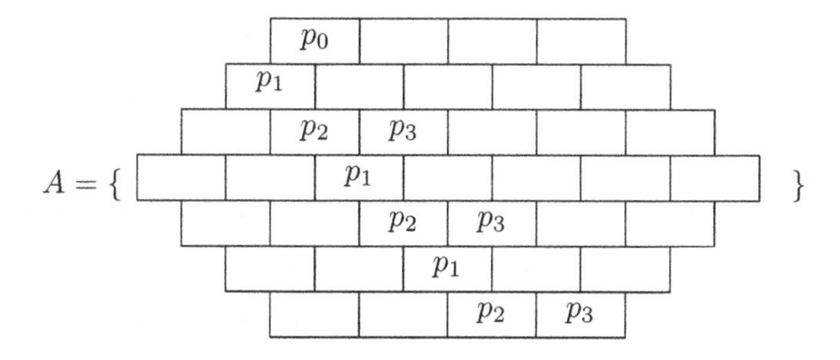

Theorem 4. $\mathcal{L}(3 - TA) = Rat(\{a\}^\star \times \{b\}^\star \times \{c\}^\star)$.

We denote by $HREC$ the family of parts of $\{a\}^\star \times \{b\}^\star \times \{c\}^\star$ that are recognizable as sets of hexagonal pictures over an alphabet of one letter.

Theorem 5. $Rat(\{a\}^\star \times \{b\}^\star \times \{c\}^\star) \subseteq HREC$

To prove this theorem we use 3-TA's (for $Rat(\{a\}^\star \times \{b\}^\star \times \{c\}^\star)$) and Wang system(for HREC).

Theorem 6. $\mathcal{L}(3 - TA) \subseteq \mathcal{L}(HWS)$

5 Conclusion

In this paper, inspired by the work of [3], notions of local and recognizable hexagonal picture languages are introduced. Equivalence of this recognizability to two other kinds of recognizability is brought out. That is , Hexagonal Wang systems and xyz-domino systems are introduced and it it is noticed that they are equivalent to hexagonal tiling systems. Unary hexagonal picture languages are also considered and some of their properties are obtained.

Acknowledgement. The authors are thankful to the referees for their useful comments.

References

[1] J. Berstel. *Transductions and context-free languages.* Teubner, stuttgart, 1979.
[2] S. Eilenberg. *Automata,Languages and Machines*, volume A. Academic Press, 1974.
[3] D. Giammarresi and A. Restivo. Two diamentional finite state recognizability. *fundamenta informaticae*, 25(3,4):399–422, 1966.
[4] D. Giammarresi and A. Restivo, "Two-dimensional Languages". in *Hand book of Formal languages*, eds. A.Salomaa et al., volume 3, Springer-Verlag, Berlin, 1997, pp.215-269.

[5] J.E. Hopcroft and J.D. Ullman. *Introduction to Automata Theory Languages and Computation.* Addision-Wesley, Reading, 1979.
[6] M. Latteux and D. Simplot. Recognizable picture languages and domino tiling. *Theoretical computer science*, 178:275–283, 1997.
[7] M. Mahajan and K. Krithivasan. Hexagonal cellular automata. In R.Narasimhan, editor, *A erspective in Theoretical Computer Science*, volume 16 of *Series in Computer Science*, pages 134–164. World Scientific, Singapur, 1989.
[8] L.D. Prophaetis and S.Varricchio. Recognizability of rectangular pictures by wang systems. *Journel of Automata, Languages and combinatorics*, 2(4):269–288, 1997.
[9] A. Salomaa. *Formal Languages.* Academic Press,Inc, 1973.
[10] G. Siromoney and R. Siromoney. Hexagonal arrays and rectangular blocks. *Computer Graphics and Image Processing*, 5:353–381, 1976.
[11] K.G. Subramaniam. Hexagonal array grammars. *Computer Graphics and Image Processing*, 10(4):388–394, 1979.

A Combinatorial Transparent Surface Modeling from Polarization Images*

Mohamad Ivan Fanany[1], Kiichi Kobayashi[1], and Itsuo Kumazawa[2]

[1] NHK Engineering Service Inc., 1-10-11 Kinuta Setagaya-ku Tokyo, Japan
[2] Imaging Science and Engineering, Tokyo Institute of Technology
fanany@nes.or.jp

Abstract. This paper presents a combinatorial (decision tree induction) technique for transparent surface modeling from polarization images. This technique simultaneously uses the object's symmetry, brewster angle, and degree of polarization to select accurate reference points. The reference points contain information about surface's normals position and direction at near occluding boundary. We reconstruct rotationally symmetric objects by rotating these reference points.

1 Introduction

The simplest way to reconstruct a rotationally symmetric object is by rotating its silhouette [1]. If the object is transparent, however, finding its silhouette is very difficult due to lack of body reflection and inter-reflection effect. On the other hand, current techniques for transparent surface modeling are neither efficient nor effective in dealing with rotationally symmetric transparent objects. Because it relies on complicated light setting aimed to illuminate the whole surface of the object and it suffered much from undesirable inter-reflection effects.

In this study, we pursue a way to obtain accurate reference points that when they are rotated will give accurate surface. To the best of our knowledge, no proposed methods elaborating the extraction of such reference points. The induction of accurate reference points is difficult because it is sensitive to the light wave length, surface's microstructure, bias index, and noise. The key ideas of our method are summarized as follows. First, it is a decision tree induction technique that simultaneously uses object's symmetry, brewster angle, and degree of polarization (DOP) to extract accurate reference points. This technique directly solves the ambiguity problem in determining correct incident angle. Second, it is not necessary to illuminate the whole surface of the object, but only the area near the object's occluding boundary. Third, it gives approximate initial condition for faster iterative relaxation by rotating the normal positions and directions. In this paper, we investigate the effectiveness of this method in reconstructing an

* This work is supported by the National Institute of Information and Communications Technology (NICT) of Japan.

R. Klette and J. Žunić (Eds.): IWCIA 2004, LNCS 3322, pp. 65–76, 2004.

Fig. 1. Transparent objects to be reconstructed

ideal cylindrical acrylic object and a more complicated object such as a plastic coca-cola bottle filled with water shown in Figure 1.

2 Related Works

Transparent surface modeling is a challenging and important problem in computer vision and graphics communities. Despite recent advances in opaque surface modeling, transparent surface modeling relatively has not received much attention. Only recently, some prospective techniques for modeling transparent or specular surface based on polarization images have emerged [2–5]. These techniques, however, commonly face two fundamental difficulties. First, since transparent object has only surface reflection and little body reflection, we have to acquire as much the surface reflection as possible to infer the whole surface area. Second, since the correspondence between the degree of polarization and the obtained incident angle or surface normal is not one to one, we have to solve the ambiguity of selecting the correct value. The first difficulty, namely lack of surface reflection problem, is previously addressed by introducing very complicated light settings such as continuous spherical diffuser illuminated with many point light sources located around the sphere. Such light setting (referred to as photometric sampler firstly proposed by Nayar, et al., [6]) has two limitations: it restricts the object's diameter to be sufficiently small compared to the diffuser's diameter and it suffered much from undesirable inter-reflections. The second difficulty, namely the ambiguity problem, is previously solved by introducing other sources of information such as thermal radiation [3], or new view image [2]. The necessity of such additional information that is not readily available leads to even more impractical and time consuming implementation.

Many transparent objects around us exhibit some form of symmetry. For opaque objects, symmetry is well known of giving a powerful concept which facilitates object characterization and modeling. For instance, the implicit redundancy in symmetric models guides reconstruction process [7, 8], axes of symmetry provide a method for defining a coordinate system for models [9], and symmetries give meaningful hints for shape classification and recognition [10, 11]. For transparent objects, however, the significance of symmetry is largely unknown. Because the symmetry is obscured by highlights, lack of body reflection, and inter-reflections. In fact, these obscuring factors make the methods aimed for opaque surface fail to deal with transparent objects even if the objects are simpler such as those which have symmetrical properties.

Decision trees represent a simple and powerful method of induction from labeled instances [12]. One of the strength of decision tree compares to other methods of induction is that it can be used in situations where considerable uncertainty is present and the representation of the instances is in terms of symbolic or fuzzy attributes [13]. In this paper, we implement a practical decision tree induction technique based on polarization analysis, while at the same time, we avoid the difficulties faced in current transparent surface modeling techniques. Our decision tree directly resolves the ambiguity problem and produces more accurate reference vectors.

3 Polarization Analysis

A more thorough discussion on how to obtain surface normals of transparent surface from polarization analysis of reflected light based on Fresnel equation could be found in [5, 2]. When unpolarized light is incident on dielectric transparent surface with an oblique angle, it will be partially polarized. The total intensity of the reflected light received by camera after passing a polarizer filter is

$$I_s = I_{max} + I_{min}, \tag{1}$$

where

$$I_{max} = \frac{F_\perp}{F_\perp + F_\parallel}, \qquad I_{min} = \frac{F_\parallel}{F_\perp + F_\parallel}. \tag{2}$$

The intensity reflectance $F_\parallel$ and $F_\perp$ are referred to as the Fresnel reflection coefficients. They are defined as

$$\begin{aligned} F_\parallel &= \frac{\tan^2(\phi - \phi^{'})}{\tan^2(\phi + \phi^{'})}, \\ F_\perp &= -\frac{\sin^2(\phi - \phi^{'})}{\sin^2(\phi + \phi^{'})}, \end{aligned} \tag{3}$$

where ϕ and $\phi^{'}$ are respectively incident and refraction angles. There is ϕ that can make $F_\parallel = 0$, that is $\phi = \phi_b$ which is called as Brewster angle. The ϕ_b is given by $\phi + \phi^{'} = \pi/2$ and Snell's law as

$$\phi_b = \arctan(n), \tag{4}$$

where n is the bias index.

The degree of polarization (DOP) is defined as

$$\rho = \frac{I_{max} - I_{min}}{I_{max} + I_{min}}. \tag{5}$$

For unpolarized light $I_{max} = I_{min} = \frac{1}{2} I_s$ hence $\rho = 0$. When $\phi = \phi_b$ (Brewster angle), then $F_{\|} = 0$. Hence $I_{min} = 0$ so $\rho = 1$. Combining Equations (2), (3), and (5), we can rewrite the DOP as

$$\rho = \frac{2 \sin\phi \tan\phi (n^2 - \sin^2\phi)^{1/2}}{n^2 - \sin^2\phi + \sin^2\phi \tan^2\phi}. \tag{6}$$

Thus, theoretically if we know the object's bias index n and ρ, we can estimate ϕ, which in turn will give the surface normal $\mathcal{N}(\alpha, \zeta)$, where α is azimuth and ζ is zenith angles. But practically, it is difficult due to the following factors: ambiguity of estimated (see [5, 2]), relation between light wave length and surface's microstructure, and noise.

4 Light Source Setting

In this study, we use five linear extended light sources (see Figure 2) putted in parallel with respect to the rotational axis of the object to be reconstructed. Such configuration is aimed to extract the boundary normal vectors. A set of reference vectors can later be chosen from these boundary normal vectors. Then we can infer the whole surface seen from the camera by rotating these reference vectors. Theoretically, we can use only one linear extended light source. But practically, there is no guarantee that using only one source would provide adequate number of boundary normal vectors due to noise and complexity of the object shape. So we suggest to use more than one light source. Considering such placement of the camera with respect to the light sources and the object, the only possible surface reflections are occurred in the left half area near occluding boundary. Contrary, the lights coming from the right half area of the surface received no polarization since most of these lights are actually transmitted instead of reflected.

Thus, if we take the simple cylindrical object and put it in this light setting, we perceive two different highlight areas, i.e., A and B, as shown in Figure 3(a). The highlights in A come from the reflection of light by near occluding boundary areas. Whereas, the highlights in B come from the transmission of light by the back surface. Hence, if we analyze the degree of polarization (DOP) image in Figure 3(b), we find that the DOP of area A is relatively high, but in the contrary, the DOP in B is relatively too small to be observed. Realizing such condition, we could expect that correct surface normal extractable in A. Therefore, we can rotate the surface normal obtained in A and override the surface normal in B.

The advantages of this light setting are as follows. First, it is simpler because no diffuser is needed. Second, no restriction on the diameter of the object to be

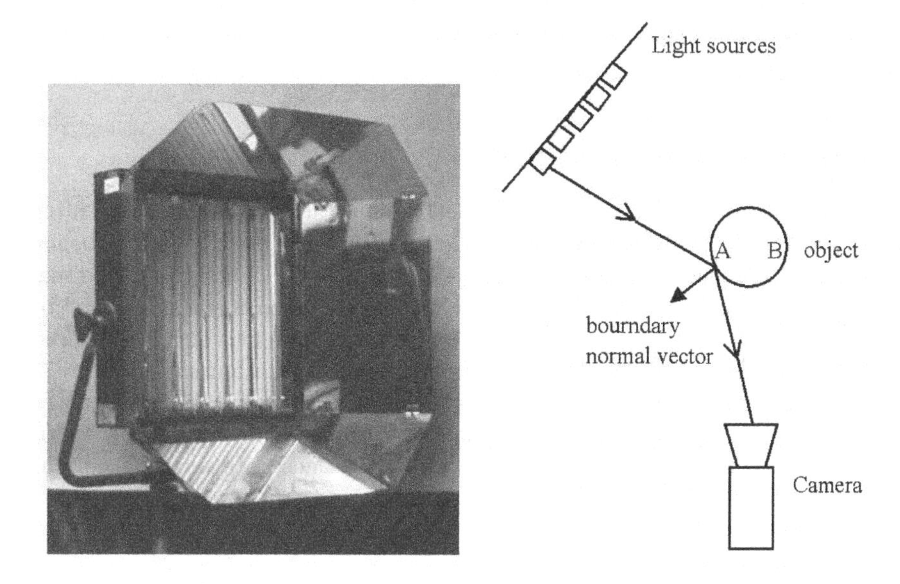

Fig. 2. Light source setting

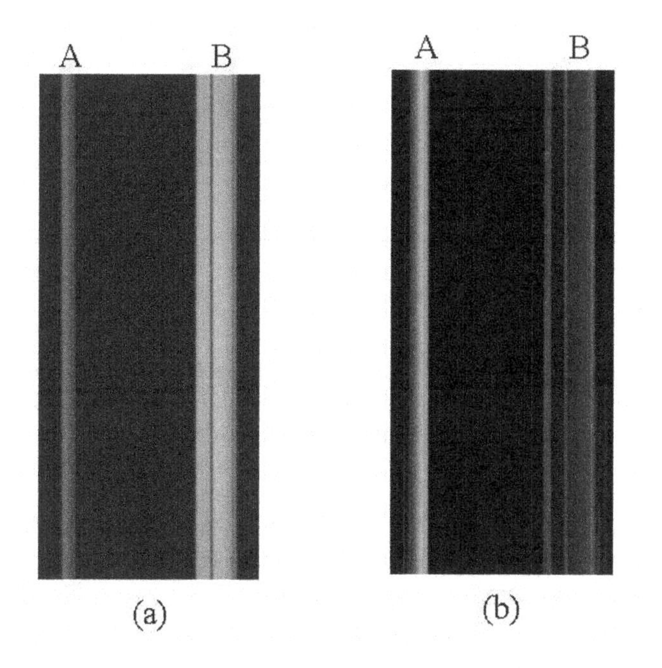

Fig. 3. The polarization image (a) and DOP image (b) of cylinder object put in our light setting

reconstructed relative to the diameter of diffuser. Third, less inter-reflection is incorporated in our light setting. In omni-directional light source using diffuser, such inter-reflection shown to cause inaccurate reconstruction [2].

5 Reference Points Selection

Equation (5) allows us to measure the DOP by rotating a polarization filter placed in front of camera. By rotating the polarization filter, we obtain a sequence of images of an object. We measure from 0 to 180 degrees at 5 intervals. From this process, we obtain 37 images. We observe variance of intensity at each pixel of the 37 images. By using the least-squares minimization, we fit a sinusoidal curve to those intensities and then determine the maximum and minimum intensities, I_{max} and I_{min}. For example we can observe the intensities and its fitted curves (Figure 4) of two pixels in one scanned line. We observe that smaller degree of polarization ρ is more sensitive to noise, so we expect accurate surface's normal cannot be produced in these areas.

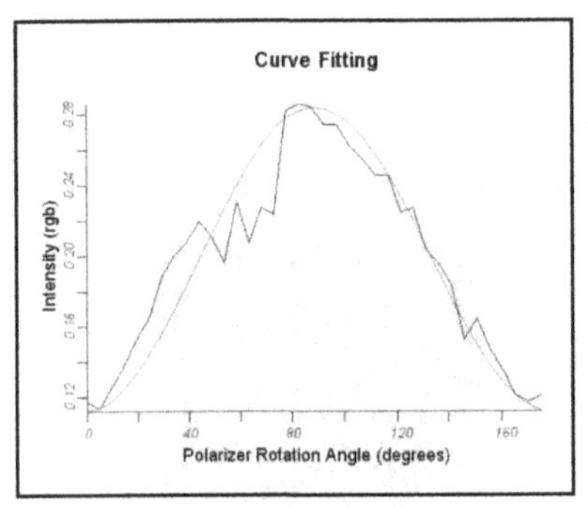

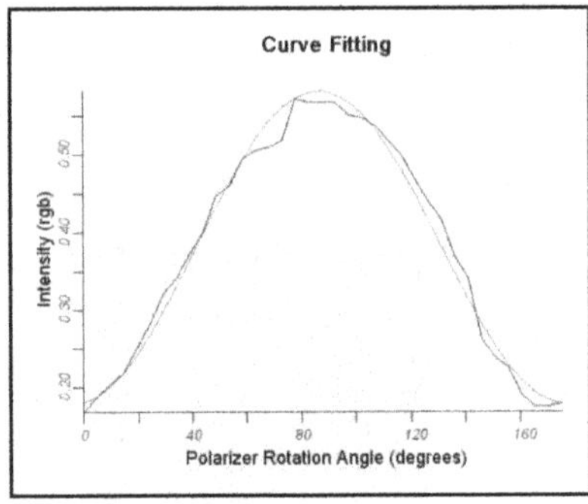

Fig. 4. Curve fitting of intensities of smaller ρ (left) and higher ρ (right)

Using our light source setting mentioned in Section 4 we will observe DOP image as shown in Figure 3. Since the DOP is high on specular area (the area A in Figure 3), we expect that accurate surface's normal $\mathcal{N}(\alpha, \zeta)$ in this area can be extracted. α is extracted by measuring the polarization rotation angle θ that give I_{max}. ζ is extracted from Equation (6) after we know the ρ (DOP) from Equation (5) and bias index n. In this study, we used Hi-Vision camera with long focal length and observe the zoomed in object. Thus we can assume that we observe orthographic projection image where $\zeta = \phi$ (see light reflection from the top of rotationally symmetric object in Figure 5). In addition, we also use normal density filter to reduce undesirable noise.

In a scanned line $\mathcal{L}$ in the DOP image there will be a set of points $P = \{p_1, p_2, \cdots, p_k\}$ where the $\rho(p_1), \rho(p_2), \cdots, \rho(p_k) \geq \rho_{th}$, where ρ_{th} is a given

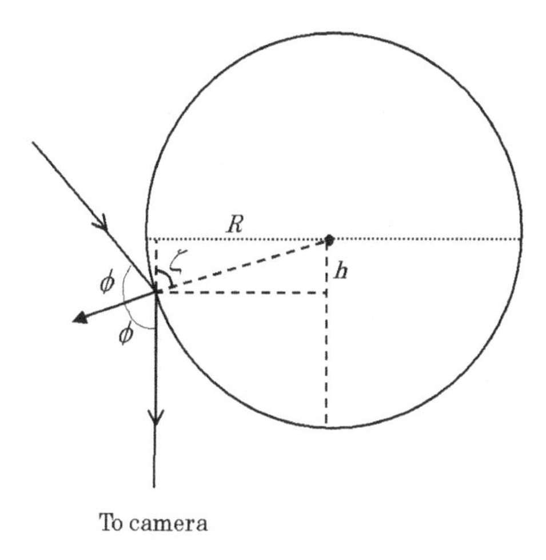

Fig. 5. Inferencing the zenith ζ and height h from orthographic image

DOP threshold. We call P as a set of valid points. Since the $\rho(p_i)$ are great we can expect to find good candidate of reference vectors there. But this does not guarantee that the extracted reference vectors or points are the ones with correct normal direction (in this case it is ζ, since α is measured independently by observing θ) and position (in this case, it is the height $h = R/\cos\zeta$ (see Figure 5).

This study pursue a way to extract such accurate reference vectors for rotation. We investigate three different ways to do this. First, the simplest way is by taking the point p_k where $\rho(p_k) >$ all other points in P. Second, take the representational average of vector position $\bar{h}$ and direction $\bar{\zeta}$ where $\bar{h} = \Sigma h(p_i)/\#P$ ($i = 0, 1, \ldots, \#P$; where $\#P$ is the cardinality of the set P) and $\bar{\zeta} = \Sigma\zeta(p_i/\#P)$. These two simple ways tend to generate reference vectors with wrong h and ζ. In the first way, the ambiguity problem is not addressed and then inconsistent ζs may be generated even though the candidate vector's pixel positions are closed. In the second way, in addition to not resolving the ambiguity problem, it averages these inconsistent candidate vectors. According to our observation the error that can be caused by the second way might be almost $0.5R$.

After observing these two failed methods, we propose the third way, that is a decision tree induction algorithm shown in Figure 6. In this decision tree, the observed specular reflections are obtained on the half left surface area of the rotational symmetric object. When the the observed specular reflection are obtained from the half right area of the object, the second level subtree ($\zeta < \phi_b$ and $\zeta \geq \phi_b$) should be interchanged. This decision tree directly solve the ambiguity problem by incorporating the object symmetry, brewster angle and DOP. The method is easier to understand by direct observation on how this is applied on real data sample (see Subsection 5.1).

Furthermore, the surface obtained by rotating the reference points can be used as approximate initial condition for faster relaxation to recover surface's

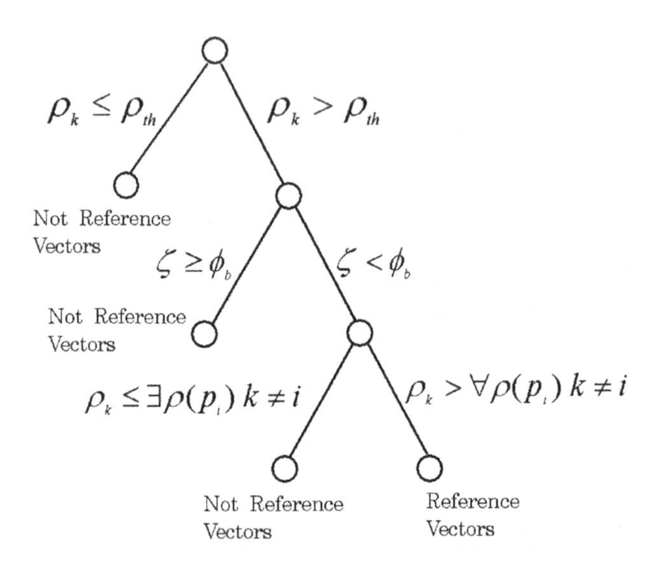

Fig. 6. Decision tree induction of referencing accurate reference points

height from gradient or needle map (Please refer to [14], pages 48–49). We can also further imposes smoothing constraint to the resulted azimuths α and zeniths ζ (see [15]).

5.1 Direct Observation

As an example, we analyze the polarization image of cylindrical acrylic object (Fig. 3. b). The size of the image is 280 pixels width and 540 pixels height. According to a catalogue [16], the refractive index for acrylic object $1.48 \sim 1.50$. If we take the $n = 1.5$ then the Brewster angle is $\phi_b = 0.983$ radian. The DOP histogram of this image is shown in Figure 7. Let us take three arbitrary scanned lines by setting the DOP threshold to $\rho_{th} = 0.68$. We obtain two candidate vectors on a scanned line at $y = 4$, four vectors at $y = 10$, and three vectors at $y = 140$. The azimuth and zenith angles are measured in radian. We list these candidate vectors as follows.

```
y = 4: \\
c = 0, az = 1.5468788, ze = 0.6981317, dop = 0.6918033 \\
c = 1, az = 1.5468788, ze = 1.2566371, dop = 0.6819789 \\
y = 10: \\
c = 0, az = 1.5468788, ze = 1.2566371, dop = 0.6822431 \\
c = 1, az = 1.5468788, ze = 0.7155850, dop = 0.7058824 \\
c = 2, az = 1.6341454, ze = 0.6981317, dop = 0.6917808 \\
c = 3, az = 1.6341454, ze = 1.2566371, dop = 0.6830189 \\
```

```
y = 140: \\
c = 0, az = 1.5468788, ze = 0.6981317, dop = 0.7000000 \\
c = 1, az = 1.5468788, ze = 1.2391838, dop = 0.7213623 \\
c = 2, az = 1.5468788, ze = 0.7155850, dop = 0.7077922 \\
```

Each scanned line is processed independently. At scanned line $y = 4$, we discard the candidate $c = 1$ because its zenith angle is greater that Brewster angle. At $y = 10$, we discard the candidates $c = 0$ and $c = 3$, and select the $c = 1$. At $y = 140$, we discard $c = 1$, even though its DOP is the greatest among the three candidates, and we select $c = 2$. At these three scanned lines, we end up with three reference vectors give two different zenith angles: $0.6981317, 0.7155850$. These two angles give height estimates $h = 0.77R$ and $h = 0.75R$, which are reasonably close.

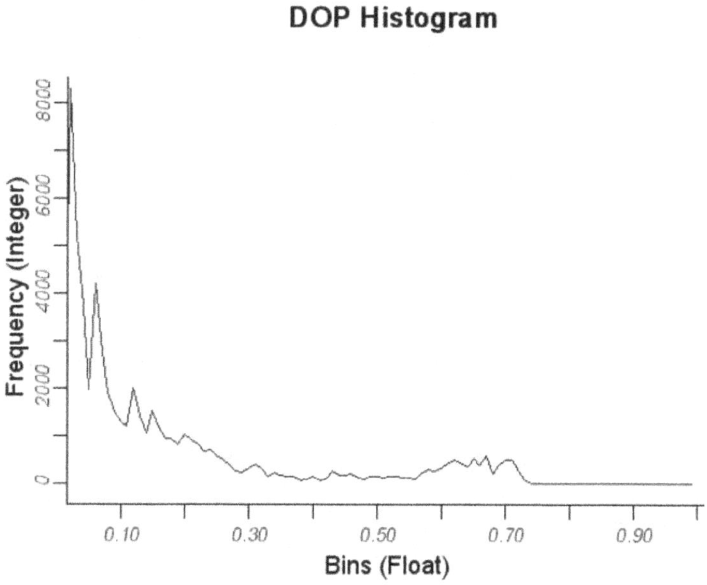

Fig. 7. The DOP histogram of Fig. 3(b)

6 Experiments

In this paper, we investigate the effectiveness of our method in reconstructing a simple cylindrical acrylic object and a more complicated object such as a coca-cola bottle filled with water. Beside more complex, the coca-cola bottle also contains concavities. The normal reconstruction for the two objects are shown in Figure 8. Even though we observe that the generated normals are not perfectly smooth, the estimated height from this normals shown in Figure 9 are acceptable. This reconstructed shapes are the results of relaxation procedure

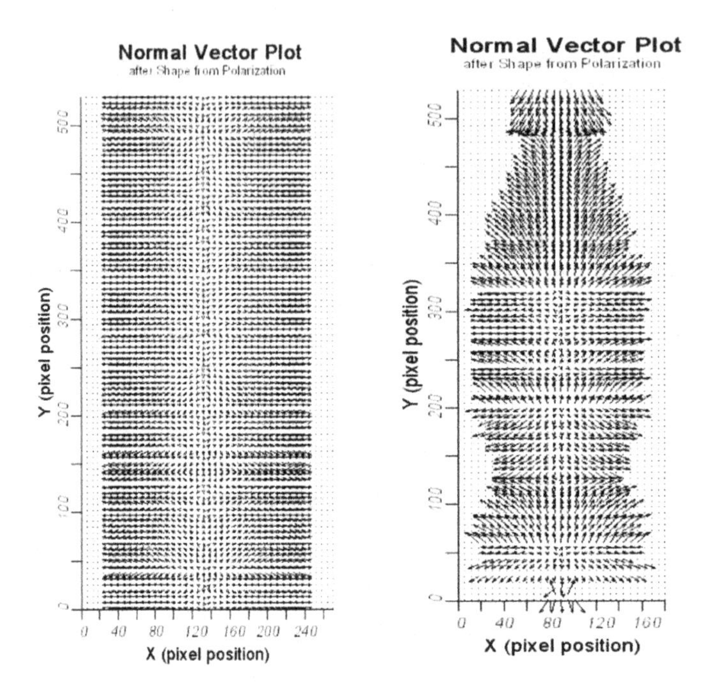

Fig. 8. Reconstructed normals

([14]) by putting the shape and normals from our rotation procedure as initial conditions. By doing so, the relaxation process converges faster (in average, it needs only about 12 iterations) to a smoother and more accurate surface.

We tried to evaluate quantitatively the error generated by our methods. For the acrylic cylindrical object we measured its diameter as 3.0 cm. For the coca-cola bottle we measured several diameters of its parts. The average error for the cylindrical object is 0.038 cm, while for the coca-cola bottle is 0.147 cm.

7 Conclusion

In this paper, we present a simple decision tree induction for transparent surface modeling of rotationally symmetric objects. This decision tree allows practical induction of accurate reference vectors for rotation. The experiment results also show that our light configuration, which allows the decision tree induction is reasonably efficient and effective for reconstructing rotationally symmetric objects. The significance of this study are as follows. First, this will open ways for more practical surface reconstruction based on simple decision tree. Second, the reconstructed object could provide initial estimation that further expanded to deal with concavity and inter-reflection. Even though our method is limited to work on rotationally symmetric objects, it can also further used for non sym-

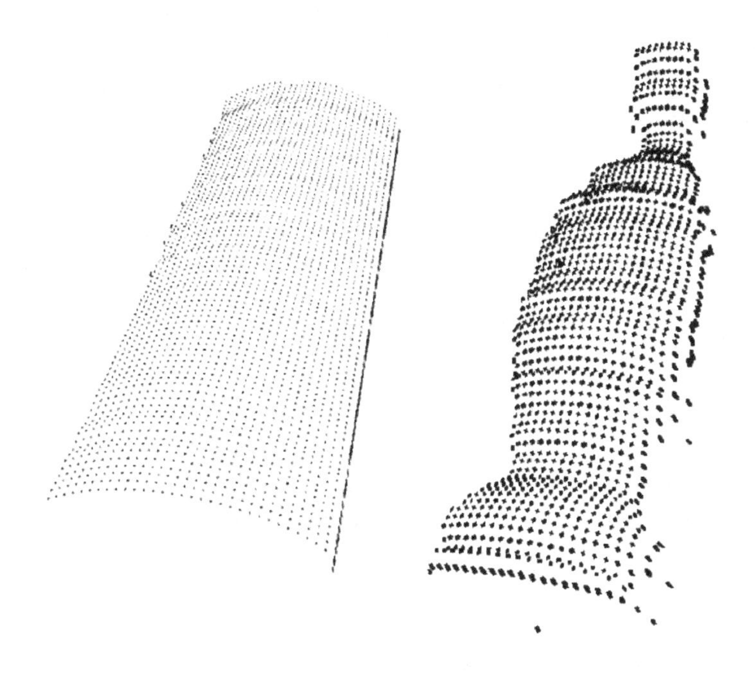

Fig. 9. Reconstructed shapes

metric objects by extracting the symmetries contained in such non symmetric objets. In our view, doing so might be more efficient rather than directly deal with non-symmetric objects.

References

1. Szeliski, R.: From images to models (and beyond): a personal retrospective. In Vision Interface '97, Kelowna, British Columbia. Canadian Image Processing and Pattern Recognition Society, (1997) 126-137.
2. Miyazaki, D., Kagesawa, M., Ikeuchi, K.: Transparent Surface Modeling from a Pair of Polarization Images. IEEE Trans. On PAMI, Vol. 26, No. 1, January (2004) 73–82.
3. Miyazaki, D., Saito, M., Sato, Y., Ikeuchi, I.: Determining Surface Orientations of Transparent Objects Based on Polarization Degrees in Visible and Infrared Wavelength. J. Opt. Soc. Am. A, Vol. 19, No. 4, April (2002) 687–694.
4. Rahmann, S., Centerakis, N.: Reconstruction of Specular Surfaces Using Polarization Imaging, Proc. IEEE Conf. Computer Vision and Pattern Recognition, (2001) 246–253.
5. Saito, M., Sato, Y., Ikeuchi, K., Kashiwagi, H.: Measurement of surface orientation of transparent objects by use of polarization in highlight. J. Opt. Soc. Am. A/Vol. 16, No. 9 September (1999) 2286–2293.

6. Nayar, S. K., Ikeuchi, K., Kanade, T.: Determining Shape and Reflectance of Hybrid Surface by Photometric Sampling. IEEE Trans. Robotics and Automation, Vol. 6, No. 4, August, (1990) 418–431.
7. Mitsumoto, H., Tamura, S., Okazaki., K, Fukui, Y.: Reconstruction using mirror images based on a plane symmetry recovery method. IEEE Trans. on PAMI, Vol. 14, (1992) 941–946.
8. Zabrodsky, H., Peleg, S., Avnir., D.; Symmetry as a continuous feature. IEEE Trans. on PAMI, Vol. 17, (1995) 1154–1156.
9. Liu, Y., Rothfus, W., Kanade, T.: Content-based 3d neororadiologic image retrieval: Preliminary results. IEEE International Workshop on Content-based Access of Image and Video Databases, January, (1998) 91 - 100.
10. Leou, J., Tsai, W.: Automatic rotational symmetry determination for shape analysis. Pattern Recognition 20, (1987) 571–582.
11. Wolfson, H., Reisfeld, D., Yeshurun, Y.: Robust facial feature detection using symmetry. Proc. of Int. Conf. on Pattern Recognition. (1992) 117–120.
12. Quinlan, J.R.: Induction of decision trees. Machine Learning, Vol. 1 (1986) 81–106.
13. Yuan, Y., Shaw, M.J.: Induction of fuzzy decision trees. Fuzzy Sets System, Vol. 69, (1995) 125–139.
14. Horn, B. K. P.: Height and Gradient from Shading. International Journal of Computer Vision, 5:1, (1990) 37–75.
15. Miyazaki, D., Tan, R. T., Hara, K., Ikeuchi, K.: Polarization-based Inverse Rendering from Single View. Proc. International Symposium on CREST Digital Archiving Project, Tokyo (2003) 51–65.
16. http://www.boedeker.com/acryl_p.htm

Integral Trees: Subtree Depth and Diameter

Walter G. Kropatsch[1], Yll Haxhimusa[1], and Zygmunt Pizlo[2]

[1,*] Vienna University of Technology, Institute of Computer Aided Automation,
PRIP 183/2, Favoritenstr, 9, A-1040 Vienna, Austria
{krw, yll}@prip.tuwien.ac.at
[2,**] Department of Psychological Sciences, Purdue University, West Lafayette,
IN 47907-1364
pizlo@psych.purdue.edu

Abstract. Regions in an image graph can be described by their spanning tree. A graph pyramid is a stack of image graphs at different granularities. Integral features capture important properties of these regions and the associated trees. We compute the depth of a rooted tree, its diameter and the center which becomes the root in the top-down decomposition of a region. The integral tree is an intermediate representation labeling each vertex of the tree with the integral feature(s) of the subtree. Parallel algorithms efficiently compute the integral trees for subtree depth and diameter enabling local decisions with global validity in subsequent top-down processes.

1 Introduction

Viola and Jones introduced the 'Integral Image' [1] as an intermediate representation for the image to compute rapidly rectangular features. Each pixel of an integral image stores the sum of values of a window defined by the left upper image corner and the pixel in the lower right corner. The computation of the integral image is linear and the computation of the sum of any rectangular window uses only four pixels of the integral image. Its effectiveness has been demonstrated in people tracking [2]. Rotated windows and articulated movements of arms and legs cause still problems. We follow the strategy to adapt the data structure to the data and compute features on the adapted structures.

On a graph, vertices take the role of pixels in images. Image graphs are embedded in the plane and can take many different forms: the vertices of the 'neighborhood graph' correspond to pixels and are connected by edges if the corresponding pixels are neighbors. In the 'region-adjacency-graph' vertices correspond to regions in the image and edges connect two vertices if the two corresponding regions share a common boundary. Graphs of different granularity can

* Supported by the FWF under grants P14445-MAT, P14662-INF and FSP-S9103-N04.
** Supported by ECVision and by the Air Force Office of Scientific Research.

R. Klette and J. Žunić (Eds.): IWCIA 2004, LNCS 3322, pp. 77–87, 2004.

be related through the concept of dual graph contraction [3] giving rise to graph pyramids representing the regions of an image at multiple resolutions.

We start by further motivating the research by similar problems and solutions in the $k-$traveling salesperson problem and other visual problems. Section 2 transfers the classical parallel algorithm for computing the distance transform of a discrete binary shape from the discrete grid to the plane graph G. We then formulate an algorithm which computes a spanning tree of a given shape by successively removing edges that connect a foreground face with the background (section 3). This is similar to the distance transform and to well-known shrinking and thinning algorithms. However, in contrast to those algorithms, the goal is not to prune the branches of a skeleton of the shape but to determine its 'internal structure'. This internal structure is used in section 4 to determine the diameter and the center of the spanning tree. The diameter of a graph is the longest among the shortest paths between any pair of vertices. Its determination involves the search for the shortest path between any pair of vertices. This is much less complex if the graph is a tree. This is one of the reasons why we first search for a tree spanning the graph and find then the diameter of this tree. Partly as a by-product we compute the maximal path lengths of all branches of the subtrees and the respective diameters (section 5.1). These 'integral features' describe a property of a complete subtree. That is why we chose the name 'integral tree' in analogy to integral image. Integral trees can be used in many ways. We will show first experimental results for a top-down decomposition of the spanning tree into a disjoint set of subtrees with balanced diameters (section 5).

1.1 Further Motivation: TSP and Visual Problem Solving

Let us consider the traveling salesperson problem (TSP) in which n cities must be visited in the shortest time. Suppose that the regulation allows an agent to travel to at most 10 cities. The solution to this problem requires many agents, breaking the original TSP problem into k TSP problems. A simple solution is to cover the vertices of the graph with $k-$tours and to balance the load of the agents, for example by minimizing the maximal tour, or by minimizing the diameter of the subgraph. The TSP is that of finding a shortest tour (minimum length) that visits all the vertices of a given graph with weights on edges. The problem is known to be computationally intractable (NP-hard) [4]. Several heuristics are known to solve practical instances [4]. The TSP has been generalized to *multiple salespersons* ($k-$TSP), allowing each salesperson to visit n/k out of n cities. Another closely related problem is the multiple minimum spanning tree ($k-$MST) problem, where k trees are generated where each tree contains a root, and the size of the largest tree in the forest is minimized. Our goal is to generate a spanning forest that consists of k trees with roots, such that the diameters of the trees are balanced, i.e. none of the diameters of trees in the forest is greatly larger than the other tree diameter.

More recently, pyramid algorithms have been used to model the mental mechanisms involved in solving the visual version of the TSP [5], as well as other types of visual problems [6]. Humans seem to represent states of a problem by clusters (recursively) and determine the sequence of transformations from the start to the goal state by a top-down sequence of approximations. This approach leads to algorithms whose computational complexity is as low as that of the mental processes (i.e. linear), producing solution paths that are close to optimal. It follows that pyramid models may provide the first plausible explanation of the phenomenon of the directedness of thought and reasoning [7].

It is important to emphasize that by "pyramid algorithms" we mean any computational tool that performs image analysis based on multiple representations of the image forming a hierarchy with different scales and resolution, and in which the height (number) of a given level is a logarithmic function of the scale (and resolution) of the operators. Multiresolution pyramids form a subset of the general class of exponential pyramid algorithms. Pyramid algorithms, which incorporate a wider class of operators, are adequate models for the Gestalt rules of perceptual organization such as proximity, good continuation, common fate [8]. They also provide an adequate model of Weber's law and the speed-accuracy tradeoff in size perception, as well as of the phenomenon of mental size transformation [9]. In the case of size processing, modeling visual processes involves both bottom-up (fine to coarse) and top-down (coarse to fine) analyses. The top-down processing seems also critical in solving the image segmentation problem, which is a difficult inverse problem. This problem has received much attention in psychological literature, and is known as figure-ground segregation phenomenon [10].

2 Distance Transform

Let $G(V, E)$ denote a graph embedded in the plane and $\overline{G}(F, \overline{E})$ its dual. Algorithm in Fig. 1 labels each vertex of the graph $G(V, E)$ with the (shortest) distance $d_{\min} : V \mapsto \{0, 1, \ldots, \infty\}$ from the background. Assume that the vertices of the graph describe a binary shape and the edges determine the vertice's neighbors. It is the translation of the parallel algorithm [12] from grids to graphs. Distances of vertices on the boundary to the background are initialized to 1. Edge lengths $l(e) > 0$ in Algorithm Fig. 1 accomodate the fact that lengths other than 1 can appear. On square grids diagonal connections could be weighted by $\sqrt{2}$ or by appropriate chamfer distances [11]. In contracted graphs edges correspond to paths connecting two vertices. In such cases the length of the contracted edge could hold the length of the corresponding path. The integral property resulting

1. Initialize distances $d_{\min}(v) := \begin{cases} 1 & \text{if } v \text{ is on the boundary} \\ \infty & \text{otherwise} \end{cases}$
2. repeat $\forall v \in V$ in parallel:
   $d_{\min}(v) := \min(d_{\min}(v), min\{l(e) + d_{\min}(w) | (v, w) \in E \text{ or } (w, v) \in E\})$

Fig. 1. Algorithm: Parallel distance transform on a graph

from the distance transform is that the boundary of the shape can be reached from any vertex v with a path of length $d_{\min}(v)$ at most.

3 Determine the Spanning Tree

The smallest connected graph covering a given graph is a spanning tree. The diameter of a tree is easier and more efficient to determine than of a graph in general. In addition elongated concave shapes force the diameter to run along the shape's boundary, which is very sensitive to noise.

3.1 Minimal Spanning Tree

The greedy algorithm proceeds as follows: fist it computes distance transform $d_{\min}$; then computes edge weights $w(e) = -d_{\min}(u)d_{\min}(v)$ for all edges $e = (u, v)$; and finally finds minimal spanning tree using Kruskal's greedy algorithm. Skeletons based on morphology or distance transform give usually better results but the subsequent algorithms were able to cope with these deficiences.

3.2 Spanning Skeleton

The construction of the spanning tree is related to the computation of the distance transform and the skeleton of the shape. It operates on the dual graph $\overline{G} = (F, \overline{E})$ consisting of faces F separated by dual edges $\overline{E}$. Let us denote $B \subset F$ the background face(s) and by $\deg_b(f) := |\{(f, b) \in \overline{E}\}|$ the number of edges connecting a face $f \in F$ with the background B. Algorithm in Fig. 2 uses dual graph contraction [3] to successively remove edges connecting the interior of the shape with the background B while simplifying the boundary by removing unnecessary vertices of degree one and two. In our case dual removal of an edge e merges face f with the background face b and corresponds to contracting edge $\overline{e} = (f, b)$ in the dual graph $\overline{G}$. The result is a set of contraction kernels used to build the graph pyramid up to the apex. The searched spanning tree is the equivalent contraction kernel (V, E_{eck}), $E_{eck} \subset E$ [13] of the apex.

1. dually contract vertices of degree 1 and 2 in G; (the connecting edges correspond to self-loops and multi-edges in the dual graph $\overline{G}$.)
2. dually remove all edges $e \in E$ (in parallel) if
   - edge $\overline{e} = (f, b) \in \overline{E}, b \in B$ separates
   - a foreground face $f \in F \setminus B$ from the background
   - in a unique way: $\deg_b(f) = 1$.
3. for all faces $f \in F$ multiply connected with the background, $\deg_b(f) > 1$, do:
   (a) select an edge $\overline{e} = (f, b) \in \overline{E} \subset (F \setminus B) \times B$ and
   (b) dually remove e from E.
4. repeat steps $1 - 3$ until $F = \emptyset$
5. spanning skeleton is the equivalent spanning tree of the surviving vertex of G.

Fig. 2. Algorithm: Spanning Skeleton

3.3 Discussion and Computational Complexity

In Step 1 of the algorithm we distinguish two cases: i) If the vertices of degree less than 3 are adjacent to the background B a complete subtree externally attached to the shape is removed after a number of (sequential) steps corresponding to the length of the longest branch of the tree, and ii) vertices of degrees 1 and 2 may also exist inside the shape if they are not adjacent to the background. They are removed similar to the external tree in the very first step. As before the complexity depends on the longest branch. Since the dual contraction of all trees is independent of each other, the parallel complexity is bound by the longest branch of any tree. Step 2 removes all edges on the boundary of the graph as long as the non-background face is not multiply connected to the background. They are all independent of each other and hence can be removed in one single parallel step. Step 3 removes one of the edges of faces which are multiply connected to the background. Since vertices of degree 2 have been eliminated in step 1 this can only happen at 'thin' parts of the graph (where the removal of 2 or more such edges would disconnect the graph). Only one edge need to be removed to allow the face to merge with the background. Since different faces multiply connected to the background are independent of each other all dual removals can be done in one single parallel step.

The total number of steps needed to complete one iteration of steps 1-3 depends on the longest branch of a tree in step 1 and needs two additional steps. The branches contracted in step one become part of the final spanning tree hence in total, all steps 1 need at most as many steps as the longest path through the tree (i.e. its diameter). The number of iterations is limited by the thickness of the graph since at each iteration one layer of faces adjacent to the background is removed. Hence we conclude that the parallel complexity of the algorithm in the worst case is $\mathcal{O}(diameter(G) + thickness(G))$.

4 Diameter and Integral Tree of Depths

Given a (spanning) tree adapted to the shape we would like to measure distances between any vertices of the tree. Algorithm in Fig. 3a labels each vertex with the length $d_{\max}$ of the longest tree branch away from the center. The result is the same as produced by [14] but it differs by its parallel iterated and local operations. Given the (spanning) tree $T = (V, E_{eck})$ algorithm **Subtree Depth** computes the vertex attribute $d_{\max}$ in $\mathcal{O}(|diameter|/2)$ parallel steps. If the tree is cut at any edge $e = (u, v)$, $d_{\max}(v)$ gives the depth of the remaining tree which includes vertex v. It has the integral property that any leaf of the subtree can be reached along a path not longer than $d_{\max}(v)$. The function $\max2\{M\}$ returns the second largest value of the argument set M, i.e. $\max2(M) := \max(M \setminus \{\max(M)\})$.[1]

[1] If set M has less than two elements, then the function is *not defined* in general. If, however, it appears as a member of a set like in $\min(\cdot)$ or $\max(\cdot)$ then it can simply be ignored or, formally, replaced by the empty set: $\max2(\emptyset) = \max2\{x\} = \emptyset$.

1. Initialize distances $d_{\max}(v) := \begin{cases} 0 & \text{if } v \text{ is a leaf}, \deg_T(v) = 1 \\ \infty & \text{otherwise} \end{cases}$
2. repeat for all vertices $v \in V$ in parallel:
   $d_{\max}(v) := \min(d_{\max}(v), \max2\{l(e) + d_{\max}(w) | (v,w) \in E_{eck} \text{ or } (w,v) \in E_{eck}\})$

a

$$c \to \cdots \to v \not\to w \begin{cases} \to s_1 \cdots \to l_1 \\ \to u \begin{cases} \to s_2 \cdots l_2 \\ \to s_3 \cdots l_3 \\ \to s_4 \cdots l_4 \end{cases} \\ \to s_n \cdots \to l_n \end{cases}$$

b

Fig. 3. a) Algorithm: Subtree Depth and b) subdivision of the Subtree Depth

4.1 Center and Diameter of the Spanning Tree

The sample result is shown in Fig. 4b. Each vertex is labeled with two values, the first being the subtree depth. The **diameter** is the longest path[2] through the tree and consists of the two sub-paths $v_0, v_1, \ldots, v_9$ and $w_9, w_8, \ldots, w_0$ with $d_{\max}(v_i) = d_{\max}(w_i) = i, i = 0 \ldots 9$. Its length is 19. There is one edge (v_9, w_9) of which both ends have (maximal) depth 9. This is the **center of the tree** with the (integral) property that all leafs (i.e. all vertices!) of the tree can be reached in maximally $d_{\max}(v_9) = 9$ steps. The diameter of this tree is obviously 19, an odd number. All trees with an odd diameter have a central edge. Trees with an even diameter have a single maximum $d_{\max}$-value, e.g. a vertex is the center. Similar information is contained in the subtree depth of the other vertices: Given the center of the tree, we can orient the edges such that they either point towards the center or away from the center. Let us assume in the following that all *edges of the tree are oriented towards the center.*

4.2 Computational Complexity of Algorithm Subtree Depth

We consider the number of repetitions of step 2 and the number of steps required to compute max2. First we note that the algorithm stops if the function $d_{\max}(v)$ does not change after updating of step 2. It starts with vertices of subtree depth 0 and increases the distance values at each (parallel) iteration. Hence step 2 need not be repeated more than half the diameter times. To compute the $d_{\max}$-value in step 2 all the neighbors of a vertex need to be considered. Hence this is bounded by the degree of the vertex. In summary the parallel computational complexity is $\mathcal{O}(diameter * maximal\_vertex\_degree)$.

5 Decomposing the Spanning Tree

In [14] we presented an algorithm to decompose a spanning tree into subtrees such that the diameter of each subtree is maximally half the diameter of the

[2] Edge length $l(e) = 1$ is used in all examples.

Table 1. Degrees of the contraction kernels

	'Bister'								
level \ deg	0	1	2	3	4	5	6	8	δ
$0 \to 1$	1653	759							1
$1 \to 2$	2340		24						2
$2 \to 3$	2124		48	24					6
$3 \to 4$	1779		99	8	8	8			10
$4 \to 5$	1111		199	21	22				19
$5 \to 6$	451		244	8	18	8			25
$6 \to 7$	75		174	16	4	8			32
$7 \to 8$	13		48	16			8		43
$8 \to 9$	3		8		2	8			50
$9 \to 10$			1					2	62
$10 \to 11$			1						120

	'Disc'								
level \ deg	0	1	2	3	4	5	6	8	δ
$0 \to 1$	821	380							1
$1 \to 2$	1165		12						2
$2 \to 3$	1057		24	12					6
$3 \to 4$	877		52	4	4	4			10
$4 \to 5$	529		110	8	10				19
$5 \to 6$	229		116	4	8	4			25
$6 \to 7$	37		86	8	2	4			32
$7 \to 8$	5		24	8			4		43
$8 \to 9$			4		1	4			50
$9 \to 10$								1	62

original tree. Recursively continued until the subtrees have a diameter ≤ 2, this strategy creates a hierarchy of log(diameter) height. The only parameter used for this decomposition is the length of the diameter and the center of the tree.

We studied the relation between the shape (two sample examples are shown in Fig. 5a,b, for more examples see [15]) and the resulting graph pyramid. Table 1 lists the observed properties of the contraction kernels used at level k to produce level $k+1$ ($k \to k+1$). For every level the histogram of kernel's degrees is given together with the largest diameter δ of all subtrees at the respective levels. The similarity of the substructure 'Disc' to 'Bister' is obvious and not surprising. The length of the diameter and the center appear to be very robust whereas the fine substructures are sensitive to noise. In particular we observe many spurious branches ($\deg(v) = 0$) and high splitting degrees. This can be avoided to a large extend and optimized using subtree diameters.

5.1 The Integral Tree of Diameters

Subtree depths $d_{\max}$ are upper bounds for reaching any vertex in the outer subtree. Consider the following configuration sketched in Fig. 3b): c denotes the center, l_i are the leafs, v, w, u, s_i are intermediate vertices. $dist(x, y)$ denotes the distance between vertices x and y. The depth of the center c is not shorter than the distance to any leaf[3]: $d_{\max}(c) \geq \text{dist}(c, l_i)$. The actual distance between the center and any vertex v is also bounded: $\text{dist}(c, v) \leq d_{\max}(c) - d_{\max}(v)$. Along the tree's diameter-path the above inequalities are equalities. Assume we cut the tree between vertices v and w. The diameter of the outer subtree of w goes either through w or it connects two subbranches excluding w. If it goes through w its length is the sum of the subtree depth of w and the length of its second longest subbranch. The length of a subbranch is the length of the edge connecting the branch to w plus the subtree depth of the first son in this subbranch: $\delta(w) = d_{\max}(w) + \max 2\{l((w, s)) + d_{\max}(s) | (w, s) \in E_{eck}\}$.

[3] Odd diameters create a central edge splitting the tree in two subtrees for which the above inequalities hold.

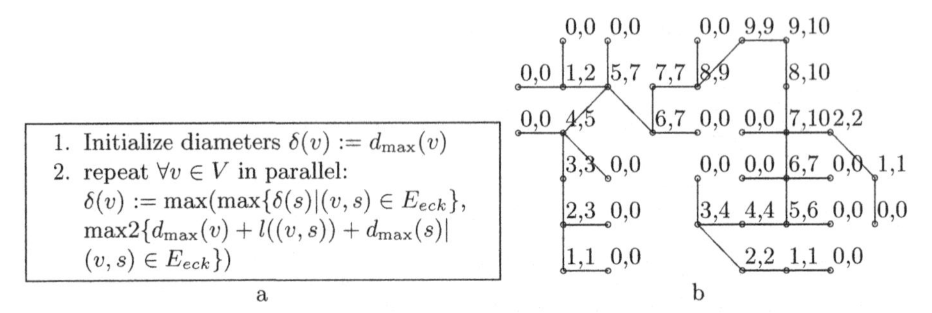

Fig. 4. a) Algorithm: Subtree Diameters δ, and b) Integral trees of depths and diameters $d_{\max}, \delta$

The max2-function is well defined because $d_{\max}(w) > 0$ implies a degree $\deg(w) \geq 2$. If the diameter of subtree w does not go through w it connects two leafs through a vertex, e.g. $u : l_2 \cdots s_2 \leftarrow u \rightarrow s_4 \cdots l_4$. In this case vertex u calculates the diameter as w above and propagates the length of the diameter up to vertex w. The diameters of all subtrees can be computed similar to the Subtree Depth: Algorithm Fig. 4a generates diameters δ (2nd values in Fig. 4b).

5.2 Using Integral Trees for Decomposition

The integral features of depth $d_{\max}$ and diameter δ should enable us to decide locally where it is best to split the spanning tree. Criteria could be a good balance of diameter lengths, a small degree of the top contraction kernels (*"a hand has 5 fingers"*) or more object specific properties that could be known to the system.

Let us consider what happens if we cut the tree at a certain distance from the center by removing the cut-edge. A cut-edge (v, w) is selected if the depth of the outer tree is smaller than a threshold d_T, $d_{\max}(v) < d_T \leq d_{\max}(w)$ (*'cut-edge condition'*). Note that the threshold d_T can depend on the length of the overall diameter $\delta(c)$. After cutting, the longest possible diameter of the outer tree $\delta_{\max}$ is twice the subtree depth of $d_{\max}(v)$ (this was used in [14]). This can be improved using the actual diameters $\delta(v)$ calculated by algorithm subtree-diameters (Fig. 4b). If all edges satisfying the cut-edge condition are rigorously removed the depth of the remaining central tree is reduced by the subtree depth of new leaf $d_{\max}(w) = d_T$. Consequently the diameter of the central tree shrinks by the double amount $\delta_{\text{new}}(c) = \delta_{\text{old}}(c) - 2d_T$. Table in Fig. 5 lists the different diameters and degrees for all possible cut-depths d_T. The decomposition should first split the 'important' components and not be too much influenced by spurious subtrees. The degree of the contraction kernel corresponds exactly to the number of cut-edges. While the 'cut-degree' counts all rigorously created new subtrees including trees with very small depth and diameter (0 in table in Fig. 5), the 'min'-value gives the degree after re-connecting all cut-edges to the central tree which do not increase the largest diameter of all outer and the inner trees. The remaining subtree diameters are bold faced in table in Fig. 5.

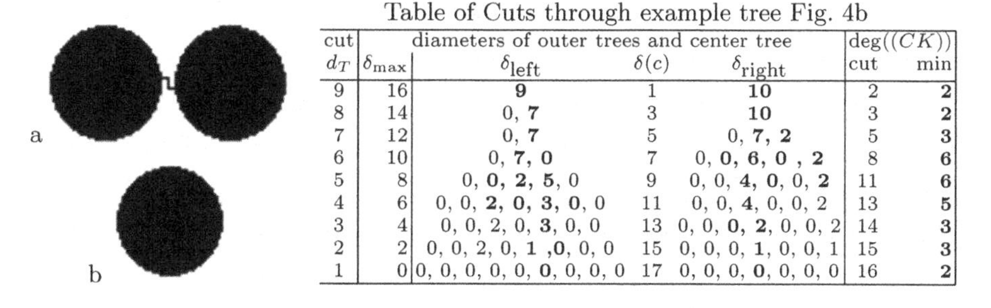

Table of Cuts through example tree Fig. 4b

cut d_T	$\delta_{\max}$	diameters of outer trees and center tree δ_{left}	$\delta(c)$	δ_{right}	deg((CK)) cut	min
9	16	**9**	1	**10**	2	**2**
8	14	0, **7**	3	**10**	3	**2**
7	12	0, **7**	5	0, **7**, **2**	5	**3**
6	10	0, **7**, **0**	7	0, **0**, **6**, **0** , **2**	8	**6**
5	8	0, **0**, **2**, **5**, 0	9	0, 0, **4**, **0**, 0, **2**	11	**6**
4	6	0, 0, **2**, **0**, **3**, **0**, 0	11	0, 0, **4**, 0, 0, 2	13	**5**
3	4	0, 0, 2, 0, **3**, 0, 0	13	0, 0, **0**, **2**, 0, 0, 2	14	**3**
2	2	0, 0, 2, 0, **1** ,**0**, 0, 0	15	0, 0, 0, **1**, 0, 0, 1	15	**3**
1	0	0, 0, 0, 0, 0, **0**, 0, 0, 0	17	0, 0, 0, **0**, 0, 0, 0	16	**2**

Fig. 5. Two example a) Bister($2 \times 1581 + 9$ pixel) and b) Disc (1581 pixel) used in experiments and Table of Cuts

5.3 Experiment: Two Connected Balls (*'Bister'*)

The example of Fig. 5a consists of *two large balls connected by a thin curve.* Bister et.al. [16] used a similar example to demonstrate the shift variance of regular pyramids. The goal of this experiment, refered to as *'Bister'*, is to check whether the simple decomposition expressed by the above description could be derived from the integral tree. Table 2 lists the different subtree depths and diameters in the example *'Bister'* (see subtree depth and diameters of central part in Fig. 6). This shows clearly that the diameters of the two circles (62) propagate up to the center which receives diameter 120. Cutting the path which connects the two large circles produces three subtrees (degree of contraction kernel 2) of which both outer subtrees have diameter 62 from cut-edge with subtree depths (59,60) down to (36,37). With smaller subtree depth the degrees of the contraction kernels start to grow since extra branches of the two circles are cut. We continued the table down to cut-edge (29,30) where the diameter of the center-tree becomes larger than any of the outer trees. We also note that no spurious branches can be integrated in this first level decomposition.

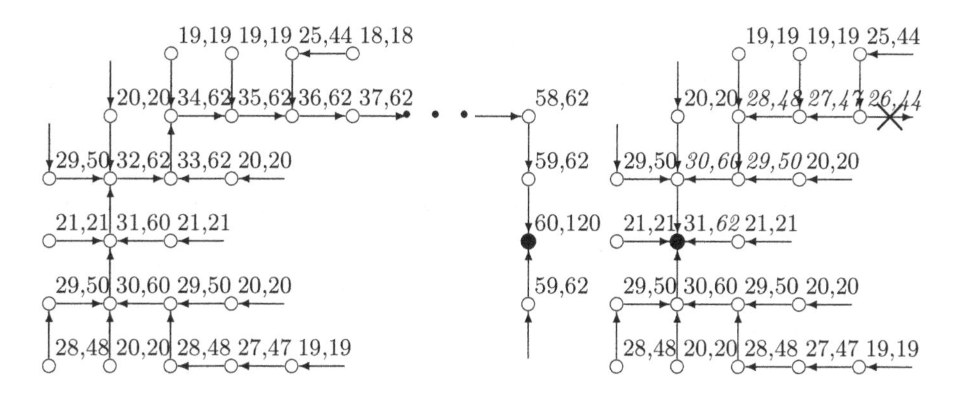

Fig. 6. Center part of left circle of example *Bister before and after cut*

Table 2. Cuts through spanning tree of example *'Bister'*

cut		diameters of outer trees and center tree			deg((CK))
d_T	$\delta_{\max}$	δ_{left}	$\delta(c)$	δ_{right}	cut
60	118	62	0	62	2
59	116	62	2	62	2
			...		
37	72	62	46	62	2
36	70	44, 62	48	44, 62	4
35	68	44,19, 62	50	44,19, 62	6
34	66	44,19,19, 62	52	44,19,19, 62	8
33	64	44,19,19, 62, 20	54	44,19,19, 62, 20	10
32	62	44,19,19,20,50, 60, 20	56	44,19,19,20,50, 60, 20	14
31	60	44,19,19,20,50,21, 60, 21,20	58	44,19,19,20,50,21,60,21,20	18
30	58	44,19,19,20,50,21,50,20,50,21,20	60	44,19,19,20,50,21,50,20,50,21,20	22
			...		

6 Conclusion

We have introduced integral trees that can store integral features or properties. Efficient parallel algorithms have been presented for computing: **i)** the boundary distance $d_{\min}$ of a binary shape; **ii)** the depth of all subtrees $d_{\max}$; and **iii)** the diameter δ of the outer subtrees. These integral features are not just sums over all elements of the subtree but capture properties of the complete substructure. The integral trees have been used to decompose the spanning tree of the shape top-down. The decomposition can use following optimization criteria: i) balance the diameters of the subtrees more efficiently than cutting at a fixed distance from the center or the leafs; unfortunately this often generates contracktion kernels of high degree; ii) set the degree n of the contraction kernel beforehand and find the n subtrees with largest integral feature, e.g. diameter. iii) define the optimization criterium which can be solved using local information provided by the integral tree and some global properties like global size or diameter proportion that are propagated during the top-down process. In future research we plan to apply integral tree for new solutions of the TSP problem as well in tracking.

References

[1] Viola, P., Jones, M.: Robust Real-time Face Detection. International Journal of Computer Vision **57** (2004) 137–154

[2] Beleznai, C., Frühstück, B., Bischof, H., Kropatsch, W.G.: Detecting Humans in Groups using a Fast Mean Shift Procedure. OCG-Schriftenreihe 179, (2004)71–78.

[3] Kropatsch, W.G.: Building Irregular Pyramids by Dual Graph Contraction. IEE-Proc. Vision, Image and Signal Processing **142** (1995) pp. 366–374

[4] Christofides, N.: The Traveling Salesman Problem. John Wiley and Sons (1985)

[5] Graham, S.M., Joshi, A., Pizlo, Z.: The Travelling Salesman Problem: A Hierarchical Model. Memory & Cognition **28** (2000) 1191–1204

[6] Pizlo, Z., Li, Z.: Graph Pyramids as Models of Human Problem Solving. In: Proc. of SPIE-IS&T Electronic Imaging, Computational Imaging, (2004) 5299, 205–215

[7] Humphrey, G.: Directed Thinking. Dodd, Mead, NY (1948)

[8] Pizlo, Z.: Perception Viewed as an Inverse Problem. Vis. Res. **41** (2001) 3145–3161
[9] Pizlo, Z., Rosenfeld, A., Epelboim, J.: An Exponential Pyramid Model of the Time-course of Size Processing. Vision Research **35** (1995) 1089–1107
[10] Koffka, K.: Principles of Gestalt psychology. Harcourt, NY (1935)
[11] Borgefors, G.: Distance Transformation in Arbitrary Dimensions. Computer Vision, Graphics, and Image Processing **27** (1984) 321–145
[12] Borgefors, G.: Distance Transformation in Digital Images. Computer Vision, Graphics, and Image Processing **34** (1986) 344–371
[13] Kropatsch, W.G.: Equivalent Contraction Kernels to Build Dual Irregular Pyramids. Springer-Verlag Advances in Computer Vision (1997) pp. 99–107
[14] Kropatsch, W.G., Saib, M., Schreyer, M.: The Optimal Height of a Graph Pyramid. OCG-Schriftenreihe 160, (2002) 87–94.
[15] Kropatsch, W.G., Haxhimusa, Y., Pizlo, Z.: Integral Trees: Subtree Depth and Diameter. Technical Report No. 92, PRIP, Vienna University of Technology (2004)
[16] Bister, M., Cornelis, J., Rosenfeld, A.: A Critical View of Pyramid Segmentation Algorithms. Pattern Recognition Letters **11** (1990) pp. 605–617

Supercover of Non-square and Non-cubic Grids

Troung Kieu Linh[1], Atsushi Imiya[2], Robin Strand[3], and Gunilla Borgefors[3]

[1] School of Science and Technology, Chiba University,
[2] IMIT, Chiba University, Yayoi-cho 1-33, Inage-ku,
Chiba 263-8522, Japan
[3] Centre for Image Analysis, Lagerhyddsvagen 3,
SE-75237 Uppsala, Sweden

Abstract. We define algebraic discrete geometry of hexagonal- and rhombic-dodecahedral- grids on a plane in a space, respectively. Since, a hexagon and a rhombic-dodecahedron are elements for tilling on a plane and in a space, respectively, a hexagon and a rhombic-dodecahedron are suitable as elements of discrete objects on a plane and in a space, respectively. For the description of linear objects in a discrete space, algebraic discrete geometry provides a unified treatment employing double Diophantus equations. In this paper, we introduce supercove for the hexagonal- and rhombic-dodecahedral- grid-systems on a plane and in a space, respectively.

1 Introduction

In this paper, we deal with discrete linear objects of non-square and non-cubic gris on a plane and in a space, respectively [1–6] and, for these grid systems, we introduce discrete algebraic geometry. Algebraic discrete geometry [7–10] allows us to describe linear manifolds, which are collections of unit elements, in two- and three- dimensional discrete spaces as double Diophantus inequalities.

A hexagon on a plane has both advantages and disadvantages as an elemental cell of discrete objects [1–3]. The area encircled by a hexagon is closer to the area encircled by a circle to compare the area encircled by a square. Although the dual lattice of square grid is the square grid, the dual grid of hexagonal grid is the triangle grid. Therefore, for the multi-resolution analysis we are required to prepare two types of grids.

From the application in omni-directional imaging systems in computer vision and robot vision [11, 12], the spherical camera model is recently given much interests. Though square grid yields a uniform tilling on a plane, it is not suitable as a grid element on the sphere. The hexagonal grid system provides a uniform grid both on the sphere [11, 13, 14] on a plane [1–4]. This property of the hexagonal grid system allows us to use it as a grid system for the numerical integration on the sphere. A sphere is a Riemannian manifold with constant positive curvature. Spherical surface, which is the finite closed manifold with positive constant curvature, and plane, which is the manifold with zero curvature, have geometrically similar properties [15, 16]. The next step of the discrete geometry is the construction of discrete algebraic geometry on the Riemannian manifolds.

R. Klette and J. Žunić (Eds.): IWCIA 2004, LNCS 3322, pp. 88–97, 2004.

2 Hexagonal Grid System on a Plane

We first define hexagonal grids on two-dimensional Euclidean plane (x, y).

Definition 1. *For integers α and β, we call the region*

$$\begin{cases} y_0 - 1 \leq y \leq y_0 + 1 \\ 2x_0 + y_0 - 2 \leq 2x + y \leq 2x_0 + y_0 + 2 \\ 2x_0 - y_0 - 2 \leq 2x - y \leq 2x_0 - y_0 + 2 \\ \{x_0 = 3\alpha, y_0 = 2\beta\} \vee \{x_0 = 3(\alpha + \frac{1}{2}), y_0 = 2(\beta + \frac{1}{2})\} \end{cases} \tag{1}$$

the hexagonal grid centred at (x_0, y_0). Simply, we call it the hexel at $\boldsymbol{x}_0 = (x_0, y_0)$.

The supercover in the hexagonal grid is defined as:

Definition 2. *The supercover in the hexagonal grid system is a collection of all hexagons which cross with a line.*

Figure 1 shows an example of the supercover in the hexagonal grid system on a plane[1].

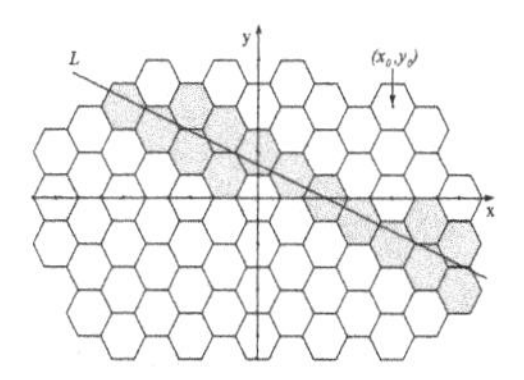

Fig. 1. Supercover in the Hexagonal Grids: Note that these hexagons are not regular

Since the vertices of a hexagon are $(x_0 - 1, y_0)$, $(x_0 + 1, y_0)$, $(x_0 - \frac{1}{2}, y_0 + 1)$, $(x_0 + \frac{1}{2}, y_0 + 1)$, $(x_0 - \frac{1}{2}, y_0 - 1)$, and $(x_0 + \frac{1}{2}, y_0 - 1)$, the distances from these vertices to the centre of the hexel at the point $(x_0, y_0)$are

$$\begin{aligned} D &= \{d_i\}_{i=1}^{6} \\ &= \{r^{-1}(d - a), r^{-1}(d + a), r^{-1}(d + b - \frac{1}{2}a), \\ &\quad r^{-1}(d + b + \frac{1}{2}a), r^{-1}(d - b - \frac{1}{2}a), r^{-1}(d - b + \frac{1}{2}a)\}, \end{aligned} \tag{2}$$

where $d = x_0a + y_0b + \mu$ and $r = \sqrt{a^2 + b^2}$. Therefore, if a line $ax + by + \mu = 0$, for integers a, b, and μ, crosses with a hexagon centred at (x_0, y_0), we have the relation,

$$0 \leq |x_0a + y_0b + \mu| \leq \max\{|a|\ , \frac{1}{2}|a| + |b|\}, \tag{3}$$

[1] Note that our hexels are not regular hexagons. If we select $\beta = \frac{\sqrt{3}}{2}n$, for an integer n, the hexel becomes a regular hexagon.

since the relation

$$\min\{d_i\} \leq 0 \leq \max\{d_i\} \tag{4}$$

implies a double inequality

$$-\max\{|a|\ ,\frac{1}{2}|a|+|b|\} \leq x_0 a + y_0 b + \mu \leq \max\{|a|\ ,\frac{1}{2}|a|+|b|\}. \tag{5}$$

These relations lead to the next theorem.

Theorem 1. *For integers a, b, α, and β, the supercover of a line $ax+by+\mu=0$ on the hexagonal grid system is a collection of hexels whose centres lie in the set*

$$\begin{aligned}&\{(x,y)\mid x=3\alpha, y=2\beta,\ |ax+by+\mu| \leq \max\{|a|\ ,\frac{1}{2}|a|+|b|\}\}\\ &\cup\{(x,y)\mid x=3(\alpha+\frac{1}{2}), y=2(\beta+\frac{1}{2}), |ax+by+\mu| \leq \max\{|a|\ ,\frac{1}{2}|a|+|b|\}\}\end{aligned} \tag{6}$$

Definition 3. *If all elements in a collection of girds P are elements of the supercover of a line, we call that the elements of P are collinear.*

This definition leads to the definitions of recognition and reconstruction of a supercover.

Definition 4. *For a collection of grids P, the process to examine collinearity of elements of a collection of grids is recognition of a supercover. The computation of the parameters of the line from collinear grids is the reconstruction of line.*

These definitions of recognition and reconstruction imply that the computation of parameters of a line from sample hexels achieves both recognition and reconstruction. Therefore, we develop an algorithm for the computation of parameters of a line from a supercover in the hexagonal grid system.

For integers α and β, let

$$\begin{aligned}P = &\{(x_i,y_i)|x_i=3\alpha, y_i=2\beta, i=1,2,\cdots,N\}\\ &\bigcup\{(x_i,y_i)|x_i=3(\alpha+\frac{1}{2}), y_i=2(\beta+\frac{1}{2}), i=1,2,\cdots,N\}\end{aligned} \tag{7}$$

be the centroids of the hexels. Furthermore, let a b and μ be the parameters of the line $ax+by+\mu=0$, which should be reconstructed. If an element x_i in P is an elements of the supercover of line $ax+by+\mu=0$, and, if, for simplicity, we assume that both a and b are positive, the parameters satisfy one of the four system of inequalities,

$$\text{case1}: a \geq 2b > 0, 0 \leq |ax_i+by_i+\mu| \leq a \tag{8}$$

$$\text{case2}: a \geq 2b > 0, 0 \leq |ax_i-by_i+\mu| \leq a \tag{9}$$

$$\text{case3}: 0 < a < 2b, 0 \leq |ax_i+by_i+\mu| \leq \frac{1}{2}a+b \tag{10}$$

$$\text{case4}: 0 < a < 2b, 0 \leq |ax_i-by_i+\mu| \leq \frac{1}{2}a+b. \tag{11}$$

We show the reconstruction algorithm for the case 1. Assuming that, all sample hexels are elements of supercover of line $ax + by + \mu = 0$ for $a \geq 0$ and $b \geq 0$, we have the relations

$$\begin{cases} -(x_i + 1)a - y_i b \leq \mu \leq -(x_j - 1)a - y_j b \\ X_{ij}a + Y_{ij}b \geq 0 \\ a - 2b \geq 0 \\ a, b > 0 \\ i \neq j,\ i, j = 1, 2, \cdots, N. \end{cases}, \tag{12}$$

where $X_{ij} = x_i - x_j + 2$ and $Y_{ij} = y_i - y_j$. This expression allows us to use the algorithm for the reconstruction of Euclidean lines from a collection of pixels.

3 Rhombic-Dodecahedral Grid System in a Space

Setting $\boldsymbol{x}_0 = (x_0, y_0, z_0)$ to be the centroid of a rhombic-dodecahedron in the three-dimensional lattice points $\mathbf{Z}^3$ shown in 8, 14 vertices are

$$\begin{array}{ll} (x_0, y_0, z_0 - 1), & (x_0, y_0, z_0 + 1), \\ (x_0, y_0 - 1, z_0), & (x_0, y_0 + 1, z_0), \\ (x_0 - 1, y_0, z_0), & (x_0 + 1, y_0, z_0), \\ (x_0 + \frac{1}{2}, y_0 - \frac{1}{2}, z_0 - \frac{1}{2}), & (x_0 + \frac{1}{2}, y_0 - \frac{1}{2}, z_0 + \frac{1}{2}), \\ (x_0 + \frac{1}{2}, y_0 + \frac{1}{2}, z_0 - \frac{1}{2}), & (x_0 + \frac{1}{2}, y_0 + \frac{1}{2}, z_0 + \frac{1}{2}), \\ (x_0 - \frac{1}{2}, y_0 - \frac{1}{2}, z_0 - \frac{1}{2}), & (x_0 - \frac{1}{2}, y_0 - \frac{1}{2}, z_0 + \frac{1}{2}), \\ (x_0 - \frac{1}{2}, y_0 + \frac{1}{2}, z_0 - \frac{1}{2}), & (x_0 - \frac{1}{2}, y_0 + \frac{1}{2}, z_0 + \frac{1}{2}), \end{array} \tag{13}$$

and the inside of the polyhedron is defined by the system of inequalities,

$$\begin{cases} x_0 + y_0 - 1 \leq x + y \leq x_0 + y_0 + 1 \\ x_0 - y_0 - 1 \leq x - y \leq x_0 - y_0 + 1 \\ y_0 + z_0 - 1 \leq y + z \leq y_0 + z_0 + 1 \\ y_0 - z_0 - 1 \leq y - z \leq y_0 - z_0 + 1 \\ x_0 + z_0 - 1 \leq x + z \leq x_0 + z_0 + 1 \\ x_0 - z_0 - 1 \leq x - z \leq x_0 - z_0 + 1. \end{cases} \tag{14}$$

This system of inequalities also defines 12 faces of the rhombic-dodecahedron. The grid digitised by rhombic-dodecahedra has the following properties.

Property 1. *The rhombic-dodecahedral grids whose centres satisfy $x_0 + y_0 + z_0 = 2n$ and $x_0 + y_0 + z_0 = 2n + 1$ are elements of a spatial tilling. We call these grids the even grid and odd grid, respectively.*

Property 2. *If these polyhedra share faces and vertices, we call two polyhedra are connected by 12- and 18- connectivities as shown in figure 2 and 3, respectively. The even grid and odd grid induce 12- and 18-connectivities, respectively.*

Property 3. *The even grid and odd grid share faces and vertices, two rhombic-dodecahedra share vertices*

$$\begin{array}{l}(x_A+\frac{1}{2},y_A-\frac{1}{2},z_A-\frac{1}{2}),\ (x_A+\frac{1}{2},y_A-\frac{1}{2},z_A+\frac{1}{2}),\\ (x_A+\frac{1}{2},y_A+\frac{1}{2},z_A-\frac{1}{2}),\ (x_A+\frac{1}{2},y_A+\frac{1}{2},z_A+\frac{1}{2}),\\ (x_A-\frac{1}{2},y_A-\frac{1}{2},z_A-\frac{1}{2}),\ (x_A-\frac{1}{2},y_A-\frac{1}{2},z_A+\frac{1}{2}),\\ (x_A-\frac{1}{2},y_A+\frac{1}{2},z_A-\frac{1}{2}),\ (x_A-\frac{1}{2},y_A+\frac{1}{2},z_A+\frac{1}{2}),\end{array} \tag{15}$$

as shown in Figure 5 or cross each other as shown in Figure 7.

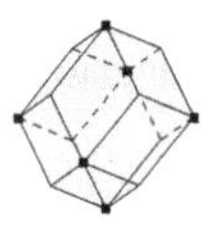

Fig. 2. Adjacent Points of Tow Connected Rhombic-dodecahedron

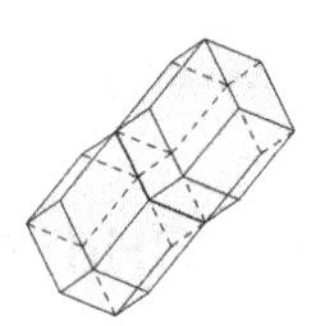

Fig. 3. 12-connectivity

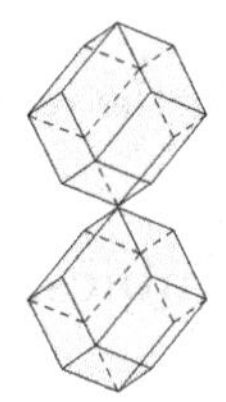

Fig. 4. 18-connectivity

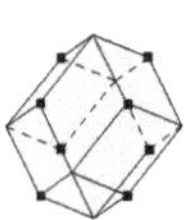

Fig. 5. Vertices of Connected Odd and Even Grids

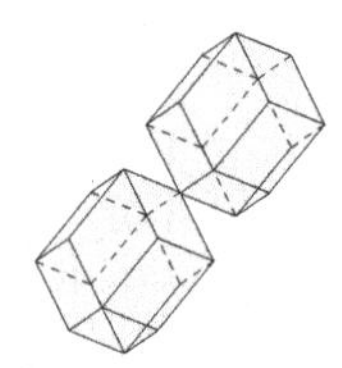

Fig. 6. Connected Odd and Even Rhombic-dodecahedral Grids

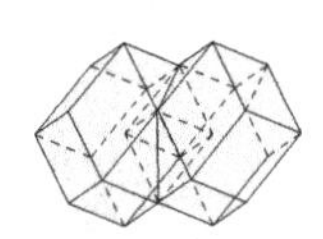

Fig. 7. Crossing Odd and Even Rhombic-dodecahedral Grids

The supercover in the rhombic-dodecahedral grid system is the collection of all rhombic-dodecahedral grids which cross with a plane. For a plane $ax+by+cz+\mu=0$, the distances from the plane to 14 vertices of the rhombic-dodecahedron whose centre is at (x_0,y_0,z_0) are

$$\begin{aligned}D &= \{d_i\}\ ,\ i=1,2,...,14\\ &= \{d_0-\frac{c}{r}, d_0+\frac{c}{r},\ d_0-\frac{b}{r},\\ &\quad d_0+\frac{b}{r}, d_0-\frac{a}{r}, d_0+\frac{a}{r},\\ &\quad d_0+\frac{\frac{1}{2}a-\frac{1}{2}b-\frac{1}{2}c}{r}, d_0+\frac{\frac{1}{2}a-\frac{1}{2}b+\frac{1}{2}c}{r}, d_0+\frac{\frac{1}{2}a+\frac{1}{2}b-\frac{1}{2}c}{r},\\ &\quad d_0+\frac{\frac{1}{2}a+\frac{1}{2}b+\frac{1}{2}c}{r}, d_0+\frac{-\frac{1}{2}a-\frac{1}{2}b-\frac{1}{2}c}{r},\ d_0+\frac{-\frac{1}{2}a-\frac{1}{2}b+\frac{1}{2}c}{r},\\ &\quad d_0+\frac{-\frac{1}{2}a+\frac{1}{2}b-\frac{1}{2}c}{r}, d_0+\frac{-\frac{1}{2}a+\frac{1}{2}b+\frac{1}{2}c}{r}\}\end{aligned} \tag{16}$$

where $d_0 = \frac{ax_0+by_0+cz_0+\mu}{\sqrt{a^2+b^2+c^2}}$. If this rhombic-dodecahedron and the plane cross, we have the relations

$$\min\{d_i\} \leq 0 \leq \max\{d_i\} \quad , \; i = 1, 2, \cdots, 14. \tag{17}$$

Equation (16) implies

$$\begin{aligned} \min\{d_i\} &= d_0 - \frac{1}{r}\max\{|a|, |b|, |c|, \frac{1}{2}(|a|+|b|+|c|)\} \\ \max\{d_i\} &= d_0 + \frac{1}{r}\max\{|a|, |b|, |c|, \frac{1}{2}(|a|+|b|+|c|)\} \end{aligned} \tag{18}$$

Therefore, eq. (17) becomes

$$0 \leq |d_0| \leq \frac{1}{\sqrt{a^2+b^2+c^2}}\max\{|a|, |b|, |c|, \frac{1}{2}(|a|+|b|+|c|)\}. \tag{19}$$

Therefore, we have the equation

$$0 \leq |ax_0 + by_0 + cz_0 + \mu| \leq \max\{|a|, |b|, |c|, \frac{1}{2}(|a|+|b|+|c|)\}. \tag{20}$$

These analysis lead to the theorem.

Theorem 2. *The supercover of plane $ax + by + cz + \mu = 0$ in the rhombic-dodecahedral grid is the collection of the rhombic-dodecahedra which satisfy the condition*

$$|ax + by + cz + \mu| \leq \max\{|a|, |b|, |c|, \frac{1}{2}(|a|+|b|+|c|)\}, \tag{21}$$

for $x + y + z = 2n$ or $x + y + z = 2n + 1$ where a, b, c, μ, and n are integers.

From these inequalities, if a point $\boldsymbol{x}_i = x_i, y_i, z_i)$ is an element of the supercover of plane $ax + by + cx + \mu = 0$, where a, b, c, and μ are integers, the parameters a, b, c, and μ satisfy one of the following four double inequalities,

$$\text{case1} : 0 \leq |ax_i + by_i + cz_i + \mu| \leq |a| \tag{22}$$
$$\text{case2} : 0 \leq |ax_i + by_i + cz_i + \mu| \leq |b| \tag{23}$$
$$\text{case3} : 0 \leq |ax_i + by_i + cz_i + \mu| \leq |c| \tag{24}$$
$$\text{case4} : 0 \leq |ax_i + by_i + cz_i + \mu| \leq \frac{1}{2}(|a|+|b|+|c|), \tag{25}$$

for $x + y + z = 2n$ or $x + y + z = 2n + 1$. These equations can be solved using the same method with the recognition and reconstruction of supercover and Euclidean plane, respectively, in cubic grids.

4 Line Recognition in Rhombic-Dodecahedral Grid System

As shown in Figure 8, the projection of a rhombic-dodecahedron onto the Oxy, Oyz, and Ozx planes are squares whose edges are parallel to $x \pm y = 0$, $y \pm z =$, and $z \pm x = 0$, respectively. We call these squares of projections of a rhombic-dodecahedron the diamond squares. For the reconstruction of Euclidean lines in the rhombic-dodecahedral grid system, we develop an algorithm for the line reconstruction for the diamond grid system whose edges are parallel to $x \pm y = 0$.

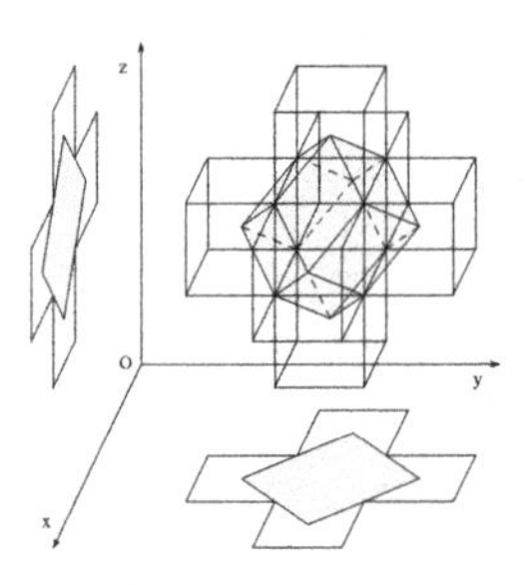

Fig. 8. Projection of a rhombic-dodecahedron to the planes $OxyandOxz$

Definition 5. *For integers x_0 and y_0, the diamond square centred at the point $\boldsymbol{x}_0 = (x_0, y_0)$ is defined by the system of inequalities*

$$\begin{cases} x_0 - y_0 - 1 \leq x - y \leq x_0 - y_0 + 1 \\ x_0 + y_0 - 1 \leq x + y \leq x_0 + y_0 + 1 \end{cases} \tag{26}$$

as shown in Figure 9.

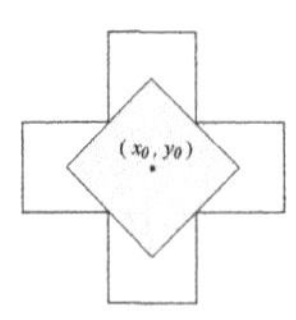

Fig. 9. Diamond Square

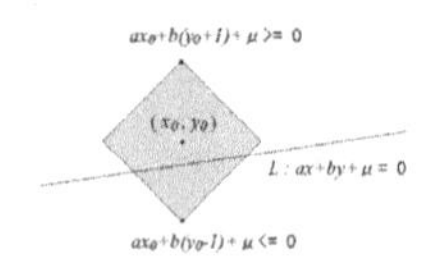

Fig. 10. Line which crosses with Diamond Square

Definition 6. *The supercover in the diamond square grid system is a collection of all diamond squares which cross with a line $ax + by + \mu = 0$, where a, b, and μ are integers.*

For a line $L : ax + by + \mu = 0$, where a, b, and μ are integers as shown in Figure 10, if the condition

$$\begin{aligned} &\{a(x_0 - 1) + by_0 + \mu \leq 0 \leq a(x_0 + 1) + by_0 + \mu\} \\ \vee&\{a(x_0 + 1) + by_0 + \mu \leq 0 \leq a(x_0 - 1) + by_0 + \mu\} \\ \vee&\{ax_0 + b(y_0 - 1) + \mu \leq 0 \leq ax_0 + b(y_0 + 1) + \mu\} \\ \vee&\{ax_0 + b(y_0 + 1) + \mu \leq 0 \leq ax_0 + b(y_0 - 1) + \mu\} \end{aligned} \tag{27}$$

is satisfied a diamond square crosses with line L. Therefore, if the system of double inequalities

$$-\max\{|a|,|b|\} \leq |ax_0 + by_0 + \mu| \leq \max\{|a|,|b|\} \tag{28}$$

is satisfied, the diamond square centred at (x_0, y_0) crosses with line L.

Since there are two classes of diamond squares $x_0 + y_0 = 2n$ and $x_0 + y_0 = 2n + 1$, there exist two types of supercovers for the diamond square grid system as shown in Figures 11 and 12. We call these two types of diamond grids the even diamond grid and the odd diamond grid. These analysis lead to the theorem.

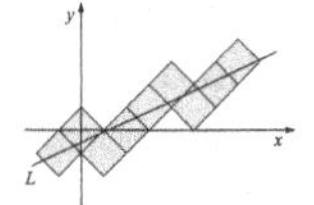

Fig. 11. Supercover on the Even Diamond Grids

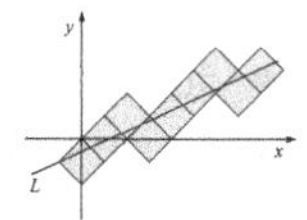

Fig. 12. Supercover on the Odd Diamond Grids

Theorem 3. *The supercover of the line $L : ax + by + \mu = 0$, for integers a, b, and μ, is the collection of the diamond squares*

$$-\max\{|a|,|b|\} \leq |ax + by + \mu| \leq \max\{|a|,|b|\}, \tag{29}$$

for $x + y = 2n$ or $x + y + 2n + 1$.

Next, we develop an algorithm for the reconstruction Euclidean line from the supercover of the diamond square grid system. Let

$$P = \{(x_i, y_i) \in Z^2 | x_i + y_i = 2n,\ i = 1, 2, \cdots, N\} \tag{30}$$

or

$$P = \{(x_i, y_i) \in Z^2 | x_i + y_i = 2n + 1,\ i = 1, 2, \cdots, N\} \tag{31}$$

be sequence of sample points. If the centroid (x_0, y_0) of a diamond square is an element of the supercover of a line $ax + by + \mu = 0$, and for simplicity, if we assume a is positive, the parameters of the line $ax + by + \mu = 0$ satisfy one of four double inequalities,

$$\text{case1} : a \geq b > 0, 0 \leq |ax_i + by_i + \mu| \leq a \tag{32}$$
$$\text{case2} : a \geq b > 0, 0 \leq |ax_i - by_i + \mu| \leq a \tag{33}$$
$$\text{case3} : 0 < a < b, 0 \leq |ax_i + by_i + \mu| \leq b \tag{34}$$
$$\text{case4} : 0 < a < b, 0 \leq |ax_i - by_i + \mu| \leq b \tag{35}$$

For the case 1, if all sample points are centroids of the diamond square in the supercover of $ax + by + \mu = 0$, we have the following inequalities,

$$\begin{cases} -(x_i+1)a - y_i b \leq \mu \leq -(x_j-1)a - y_j b \\ X_{ij}a + Y_{ij}b \geq 0 \\ a - b \geq 0 \\ a, b > 0 \\ i \neq j, i, j = 1, 2, \cdots, N. \end{cases} \quad (36)$$

where Setting $X_{ij} = x_i - x_j + 2$ and $Y_{ij} = y_i - y_j$.

5 Conclusions

We have introduced algebraic discrete geometry for the hexagonal grid system and the rhombic-dodecahedral grid system on a plane and in a space, respectively. Furthermore, we have introduced algebraic discrete geometry of the diamond square grid system on the plane. This diamond square grid is define as the projection of the rhombic-dodecahedral grid system on to the planes perpendicular to the axes of the orthogonal coordinate system.

References

1. Her, I., Geometric transformations on the hexagonal grid IEEE, Trans. Image Processing, **4**, 1213-1222, 1995.
2. Liu, Y.-K, The generation of straight lines on hexagonal grids, Computer Graphic Forum, **12**, 27-31, 1993.
3. Stauton, R. C., An analysis on hexagonal thinning algorithms and skeletal shape representation, Pattern Recognition, **29**, 1131-1146, 1996
4. Middleton, L., Sivaswamy, J., Edge detection in a hexagonal-image processing framework, Image and Vision Computing, **19**, 1071-1081, 2001.
5. McAndrew, A., Osborn, C., The Euler characteristic on the face-centerd cubic lattice, Pattern Recognition Letters, **18**, 229-237, 1997.
6. Saha, P. K., Rosenfeld, A., Strongly normal set of convex polygons or polyhedra, Pattern Recognition Letters, **19**, 1119-1124, 1998.
7. Schramm, J.M., Coplanar tricubes, LNCS, **1347**, 87-98, 1997.
8. Vittone, J., Chassery, J. M., Digital naive planes understanding, Proceedings of SPIE, **3811**, 22-32, 1999.
9. Reveilles, J.-P., Combinatorial pieces in digital lines and planes, Proceedings of SPIE, **2573**, 23-34, 1995.
10. Andres, E., Nehlig, P., Francon, J., Supercover of straight lines, planes, and triangle, LNCS, **1347**, 243-254, 1997.
11. Kimuro, K., Nagata, T., Image processing on an omni-directional view using a hexagonal pyramid, Proc. of JAPAN-USA Symposium on Flexible Automation, **2**, 1215-1218, 1992.
12. Benosman, R., Kang, S.-B. eds., *Panoramic Vision, Sensor, Theory, and Applications*, Springer-Verlag, New York, 2001.
13. Shar, K., White, D., Kimerling, A. J., Geodesic discrete global grid systems, Cartography and Geographic Information Systems, **30**, 121-134, 2003.

14. Randall, D. A., Ringler, T. D., Heikes, R. P., Jones, P., Baumgardner, J., Climate modeling with spherical geodesic grids, IEEE, Computing in Science and Engineering **4**, 32-41, 2002.
15. Morgan, F., *Riemannian Geometry:A beginner's Guide*, Jones and Bartlett Publishers, 1993.
16. Zdunkowski, W., Boot, A., *Dynamics of the Atmosphere,* Cambridge University Press, 2003.
17. Stijnman, M. A., Bisseling, R. H., Barkema, G. T., Partitioning 3D space for parallel many-particle simulations" Computer Physics Communications, **149**, 121-134, 2003.
18. Ibanez,L., Hamitouche, C., Roux C., Ray tracing and 3D object representation in the BCC and FCC grids, LNC, **1347**, 235-241, 1997.

Calculating Distance with Neighborhood Sequences in the Hexagonal Grid

Benedek Nagy

Department of Computer Science, Institute of Informatics,
University of Debrecen,
Debrecen, Hungary
Research Group on Mathematical Linguistics,
Rovira i Virgili University,
Tarragona, Spain
nbenedek@inf.unideb.hu

Abstract. The theory of neighborhood sequences is applicable in many image-processing algorithms. The theory is well examined for the square and the cubic grids. In this paper we consider another regular grid, the hexagonal one, and the distances based on neighborhood sequences are investigated. The points of the hexagonal grid can be embedded into the cubic grid. With this injection we modify the formula which calculates the distances between points in the cubic space to the hexagonal plane. Our result is a theoretical one, which is very helpful. It makes the distances based on neighborhood sequences in the hexagonal grid applicable. Some interesting properties of these distances are presented, such as the non-symmetric distances. It is possible that the distance depends on the ordering of the elements of the initial part of the neighborhood sequence. We show that these two properties are dependent.

Keywords: Digital geometry, Hexagonal grid, Distance, Neighborhood sequences.

1 Introduction

Digital geometry is a part of theoretical image processing. In digital geometry the spaces we work with, consist of points with integer coordinates only. Consequently, we define distance functions, which take integer values. First, mainly the square grid was investigated, since this is the most usual space in image processing. Nowadays other regular and non-regular grids obtain more and more attention. In many applications the hexagonal and the triangular grids fit nicely and give better results. On the square grid one of the first results was the introduction of chessboard and cityblock motions (as neighborhood relations), defined in [14]. In [15, 2] the concept of (periodic) neighborhood sequences were introduced in the n dimensional digital space, which gives us the possibility to mix the possible motions. In [4] the authors extended this theory to infinite sequences, which need not to be periodic. In this paper we use similar generalized

R. Klette and J. Žunić (Eds.): IWCIA 2004, LNCS 3322, pp. 98–109, 2004.

neighborhood sequences. In [2] a formula can be found which give the distance of two points using a given periodic neighborhood sequence. In [13] there is a formula to compute the distance of arbitrary points by arbitrary neighborhood sequences. In these grids, each coordinate value of a point is independent of the others. In 3 dimensions we use 3 coordinates. In [1] the distances in the cubic grid were analyzed using periodic neighborhood sequences.

The neighborhood criteria of the hexagonal grid can be found in [3, 9]. The symmetric coordinate frame we are using in this paper is presented in [11]. The neighborhood sequences have been introduced for this grid in [9, 12].

Note that in this paper the nodes of the hexagonal grid are used. In literature there is a mix of the grid notation. Since the triangular and the hexagonal grids are dual, in image processing the term triangular grid may used for this grid, see for instance [7]. In computer graphics the grids usually defined as we use here, see for example [5].

In this paper, – after explaining the basic concepts of the three-dimensional cubic grid and the hexagonal grid – we show the connection of them. Using the mapping of the hexagonal grid into the cubic one we visualize some concepts of the hexagonal plane. With the help of this injection, we modify the formula from [13] to calculate distances in the hexagonal grid. We present the formula, which give the distance of the points of the hexagonal grid in section 5. After this we note some interesting properties of distances based on neighborhood sequences in the hexagonal grid and we show that there is a strong connection between non-symmetric distances and the property that the distance depends on the ordering of the used elements of neighborhood sequences.

2 Distance Functions in the Cubic Grid

The cubic grid is well-known and often used in image processing, see e.g. [1, 6]. In this space there are 3 kinds of neighborhood criteria. We use some definitions from [2, 4].

Definition 1. *Let p and q be two points in the cubic grid. The i-th coordinate of the point p is indicated by $p(i)$ ($i = 1, 2, 3$), and similarly for q. The points p and q are m-**neighbors** ($m = 1, 2, 3$), if the following two conditions hold:*

- $|p(i) - q(i)| \leq 1$, *for* $1 \leq i \leq 3$,
- $|p(1) - q(1)| + |p(2) - q(2)| + |p(3) - q(3)| \leq m$.

Using these neighborhood criteria the neighborhood sequences are defined in the following way.

*The sequence $B = (b(i))$, $i \in \mathbb{N}$ with $b(i) \in \{1, 2, 3\}$, is called a **neighborhood sequence**. If for some $l \geq 1$, $b(i) = b(i + l)$ holds for every i, then B is called periodic, with a period l. In this case we use the short form $B = (b(1), \ldots, b(l))$.*

*Let p and q be two points. The point sequence $\Pi : p = p_0, p_1, \ldots, p_h = q$, where p_{i-1} and p_i are $b(i)$-neighbors for $1 \leq i \leq h$ – is called a B-**path** from p to q. The length of this path is h. The B-**distance** from p to q is defined as the length of a shortest path, and is denoted by $d(p, q; B)$.*

In digital geometry several shortest paths may exist between two points. The distance functions defined above using neighborhood sequences are not necessarily metrics. In [8] a necessary and sufficient condition is presented for neighborhood seqences to define metrics on $\mathbb{Z}^n$.

In [2] a complex formula was presented which determine distance between two points in the n dimensional digital space using any periodic neighborhood sequence. The formula strongly uses the periodic property. In [13] a general formula can be found which give the result using arbitrary neighborhood sequence. Moreover, in [13] there is a formula especially for the three dimensional case. We recall it:

Proposition 1. *The B-distance of points p and q in the cubic grid is given by*

$$d(p, q; B) = \max\{|w(1)|, d_2, d_3\},$$

where

$$d_2 = \max\left\{ i \,\middle|\, w(1) + w(2) > \sum_{j=1}^{i-1} b^{(2)}(j) \right\} \text{ and}$$

$$d_3 = \max\left\{ i \,\middle|\, w(1) + w(2) + w(3) > \sum_{j=1}^{i-1} b(j) \right\},$$

$w(1)$, $w(2)$, and $w(3)$ are the values $|p(i) - q(i)|$ sorting by non-increasing (i.e. the multiset $\{w(1), w(2), w(3)\}$ contains the same elements as $\{|p(1)-q(1)|, |p(2)-q(2)|, |p(3) - q(3)|\}$, moreover $w(1) \geq w(2) \geq w(3)$) and the values $b^{(2)}(j) = \min\{2, b(j)\}$.

3 The Hexagonal Grid

In this part we recall some important concepts about the hexagonal grid. Usually, we define three types of neighbors on this grid, as Figure 1 shows (see [3, 9, 11]). Each point has three 1-neighbors, nine 2-neighbors (the 1-neighbors, and six more 2-neighbors), and twelve 3-neighbors (nine 2-neighbors, and three more 3-neighbors). We use three coordinate values to represent the points of the grid, see Figure 2.

We would like to give some other definitions and notations. We recall some of them from the literature mentioned earlier. The procedure of assigning coordinate values to a point is given in [11]. We have two types of points according to the sum values of the coordinates. If the sum is 0, then the *parity* of the point is *even*; if the sum is 1, then the point has *odd parity*.

One can see that if two points are 1-neighbors, then their parities are different.

One can check that the formal definition (Definition 1 in Section 2) determines the same neighbors in the hexagonal grid as shown in Fig 1. In the hexagonal grid we use the concepts of neighborhood relations, neighborhood sequences and using a neighborhood sequence B the B-path and B-distance from a point p to

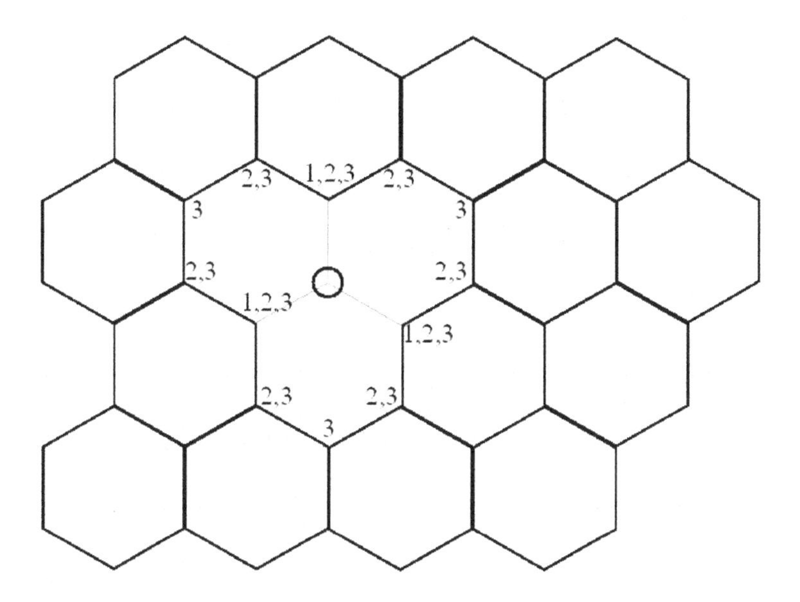

Fig. 1. Types of neighbors on the hexagonal grid

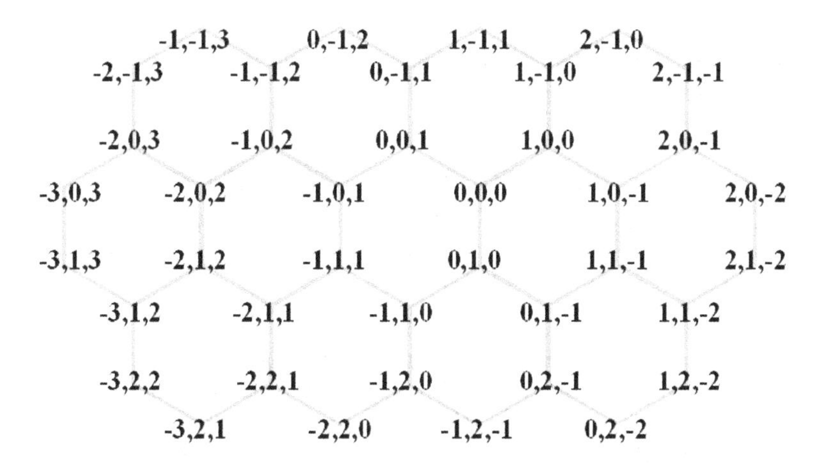

Fig. 2. Coordinate values on the hexagonal grid

a point q in the same way as in the cubic grid (see Definition 1 in the previous section).

As we will see in Section 6, the distance defined by the number of steps of a shortest path can be non-symmetric on the hexagonal grid. It means that there are points and a neighborhood sequence such that the length of the sortest paths depend on the direction.

Definition 2. *The* ***difference*** $w = (w(1), w(2), w(3))$ *is defined for any two points* p *and* q*, in the following way:* $v(i) = p(i) - q(i)$*. Ordering the values* $v(i)$ *by their absolute values non-increasing way, such that* $|w(1)| = \max|v(i)|$, $|w(3)| = \min|v(i)|$ *and* $w(2)$ *has the third value among the values* $v(i)$*. (So, the triplet* w *contains exactly the same values as* v *such that* $|w(1)| \geq |w(2)| \geq |w(3)|$*.)*

If $w(1) + w(2) + w(3) = 0$*, then the* ***parity*** *of* w *is even, else it is odd.*

Note, that the definition above slightly differs on the hexagonal grid and on the cubic grid. On hexagonal grid we use signed values.

We know that the distance from p to q depends on their difference w (including the values $w(i_j) = q(i) - p(i)$, so it is the difference of q and p) and their parities.

The following definition and fact are from [12].

Definition 3. *Let* B *and* C *be two neighborhood sequences. Then* C *is the* ***minimal equivalent neighborhood sequence*** *of* B*, if the following conditions hold:*

- $d(p, q; B) = d(p, q; C)$ *for all points* p*,* q*, and*
- *for each neighborhood sequence* D*, if* $d(p, q; B) = d(p, q; D)$ *for all points* p*,* q*, then* $c(i) \leq d(i)$ *for all* i*.*

Proposition 2. *The minimal equivalent neighborhood sequence* C *of* B *is uniquely determined, and is given by*

- $c(i) = b(i)$*, if* $b(i) < 3$*,*
- $c(i) = 3$*, if* $b(i) = 3$ *and there is no* $j < i$ *such that* $c(j) = 3$*,*
- $c(i) = 3$*, if* $b(i) = 3$ *and there are some* $c(l) = 3$ *where* $l < i$*, and* $\sum_{k=j+1}^{i-1} c(k)$ *is odd, where* $j = \max\{l | l < i, c(l) = 3\}$*,*
- $c(i) = 2$*, otherwise.*

The concept of the minimal equivalent neighborhood sequences shows that it is possible that both theoretically and practically we cannot change all the three coordinate values to step closer to the endpoint even if there is an element 3 in the neighborhood sequence.

4 The Connection Between the Cubic and the Hexagonal Grids

In this section, based on [10], we map the points of the hexagonal plane to the cubic grid. It is natural, because we use 3 coordinate values to represent points in both cases. Let us see what the points of hexagonal grid represent in the cubic grid (Fig. 3). There are 2 parallel planes (according to the parities of points).

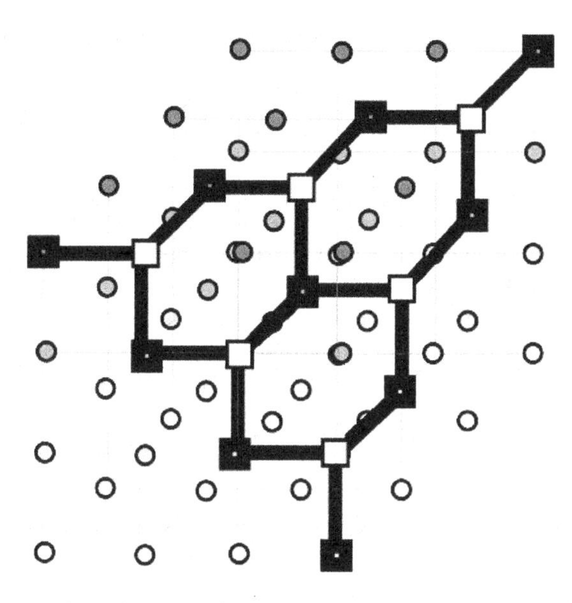

Fig. 3. The points of the hexagonal grid as the points of two parallel planes in the cubic grid

As we can see the considered points of the cubic grid are in the planes in which the sum of coordinate values are 0 (black boxes) or 1 (white boxes).

Therefore we state the following:

Proposition 3. *The hexagonal grid forms two planes in* $\mathbb{Z}^3$.

In Fig. 3 we connect the nodes which are 1-neighbors to obtain the hexagonal grid. In this way the neighborhood relations of the points are the same in these two grids.

In the hexagonal grid, we have some difficulties in changing the coordinate values. Such difficulties do not occur in case of the triangular, square and cubic grids. In the hexagonal grid, when moving from a point to one of its neighbors, we have to care of the parity of these points. Namely, we can change the coordinates of a point in such a way that the sum of the coordinate values must be 0 or 1. The concept of minimal equivalent neighborhood sequence implies that we cannot step out the two possible planes of the cubic grid in our minimal path. We have to face only the step by the first element 3 in the minimal equivalent neighborhood sequence as we detail below.

5 Formula to Calculate Distance in the Hexagonal Grid

In this section we use the previous mapping and modify the formula of Proposition 1 from Section 2 to our case. We will use the minimal equivalent neighborhood sequence C instead of the original neighborhood sequence B.

Now, let us check when it is better to modify all the coordinate values (in a 3-step) than modify only two of them.

At a step by an element 3 of the neighborhood sequence from the point p_{i-1} it is worth modifying all the three coordinate values if and only if one of the following conditions holds

- the parity of p_{i-1} is odd and we need to decrease two coordinates, and increase only one value to go to q: the difference of q and p_{i-1} has two positive and a negative value;
- the parity of p_{i-1} is even and we need to increase two coordinate values, and decrease only one of them: the difference of q and p_{i-1} has two negative and a positive value.

Using these cases, we have the following fact:

Proposition 4. *Let the index of the first occurrence of the element 3 in the neighborhood sequence B be k. Then the previous cases are equivalent to the following possibilities at the first element 3 of B:*

- *the parity of p is even, $\sum_{i=1}^{k-1} b(i)$ is even and we need to decrease two coordinates, and increase only one value to go to q;*
- *the parity of p is odd, $\sum_{i=1}^{k-1} b(i)$ is odd and we need to decrease two coordinates, and increase only one value to go to q;*
- *p is odd, $\sum_{i=1}^{k-1} b(i)$ is even and we need to increase two coordinate values, and decrease only one of them to direction to q;*
- *the parity of p is even, $\sum_{i=1}^{k-1} b(i)$ is odd and we need to increase two coordinate values, and decrease only one of them.*

From the concept of minimal equivalent neighborhood sequences it is obvious that by building a shortest path from the point p to q, it is worth modifying all coordinate values only when there is an element 3 in the minimal equivalent neighborhood sequence of B. Therefore we must use it instead of B, in our calculation. Moreover it is possible (depending on the coordinate values of p and q), that we can use the first element 3 as a value 2 (in the cases that we do not listed above). For this case we will use the concept of reduced minimal equivalent neighborhood sequence, which is defined in the following way:

Definition 4. *The* ***reduced minimal equivalent neighborhood sequence*** *C' of B is given by:*

- *$c'(k) = 2$, where k is the index of the first element 3 of B;*
- *$c'(i) = c(i)$, for all other values of i, where $c(i)$ the correspondent element of the minimal equivalent neighborhood sequence C of B.*

Now we are introducing some notation for our formula.

Let p and q be given points and B be a given neighborhood sequence.

Let $z_1 = 1$ if the parity of p is even and $z_1 = -1$ if the parity of p is odd.

Set $z_2 = 1$ if $\sum_{i=1}^{k-1} b(i)$ is even and $z_2 = -1$ if $\sum_{i=1}^{k-1} b(i)$ is odd.

Let z_3 be the sign of the product $w(1)w(2)w(3)$, i.e. if the difference of q and p has two negative and a positive value then $z_3 = 1$ and if the difference of q and p has two positive and a negative value then $z_3 = -1$; when $w(3) = 0$ then $z_3 = 0$.

Let C and C' be the minimal and the reduced minimal equivalent neighborhood sequences of B.

Finally – using the abbreviations defined above – we can state our formula in its final form:

Theorem 1. *The distance from the point p to q with a given neighborhood sequence B can be calculated as:*

$$d(p, q; B) = \max\{|w(1)|, d_2, d_3'\} \text{ if } z_1 z_2 z_3 = 1;$$

and the distance

$$d(p, q; B) = \max\{|w(1)|, d_2, d_3\} \text{ in other cases;}$$

where

$$d_2 = \max\left(i \;\middle|\; |w(1)| + |w(2)| > \sum_{j=1}^{i-1} c^{(2)}(j)\right),$$

with $c^{(2)}(j) = \min\{2, c(j)\}$, and

$$d_3' = \max\left\{i \;\middle|\; |w(1)| + |w(2)| + |w(3)| > \sum_{j=1}^{i-1} c'(j)\right\},$$

finally

$$d_3 = \max\left(i \;\middle|\; |w(1)| + |w(2)| + |w(3)| > \sum_{j=1}^{i-1} c(j)\right).$$

When the neighborhood sequence does not contain the element 3 the results are the same and both are correct:

$$d(p, q; B) = \max\left\{|w(1)|, d_2, d_3'\right\} = \max\{|w(1)|, d_2, d_3\}.$$

The previous formulae are our main results.

6 On Properties of Distances

In this section, using our formulae we show some interesting properties of these distances. For instance, they may be non-symmetric.

Example:
Let $r = (0,0,0)$ and $s = (-2,1,1)$ be two points, and $B_1 = (3,1,1)$ be a neighborhood sequence.

The minimal equivalent neighborhood sequence of B_1 is:
$C_1 = (3,1,1,2,1,1,2,1,1,...)$ with repeating part $(2,1,1)$.
Let us calculate the value of $d(s,r;B_1)$.
The value of w is $(2,-1,-1)$ in this case. $z_1 z_2 z_3 = 1$, so

$$d(s,r;B_1) = \max\left\{|w(1)|, d_2, d_3'\right\} = 3.$$

Now let us calculate $d(r,s;B_1)$.
In this case w is $(-2,1,1)$. Then

$$d(r,s;B_1) = \max(2,2,2) = 2.$$

Thus this distance function is not symmetric. (Check Fig. 4.)
The non-symmetric distance functions are exotic in the field of digital geometry. In [12] there is a necessary and sufficient condition for a neighborhood sequence to give metric.
Watching the formula for calculating distance we state the following interesting and important property.

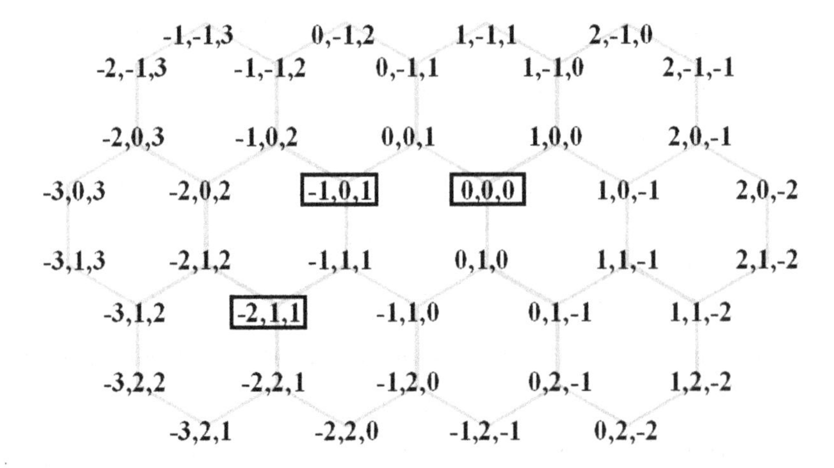

Fig. 4. Non-symmetric distance between the points $r = (0,0,0)$ and $s = (-2,1,1)$ with $B_1 = (3,1,1)$ and a distance not satisfying the triangle inequality using also point $t = (-1,0,1)$ with $B_2 = (2,1,1)$

A distance $d(p, q; B) > k$ depends on the order of the first k elements of B if and only if there is a permutation of these elements such that using it as the initial part of the neighborhood sequence the distance function is not symmetric.

We show how the property above can be used. Let B be a neighborhood sequence for which the distance is not symmetric for all points p and q. So let p and q be two points that $d(p, q; B) \neq d(q, p; B)$. We can assume that $d(p, q; B) = k < d(q, p; B) = l$. If we reorder the first k element of B in reverse order (resulted the neighborhood sequence B_r) then we get that $d(q, p; B_r) = k$. (We have a shortest path from q to p which is the symmetric pair of the original shortest path between p and q.) We note here that in the square and cubic grids the distance does not depend on the permutation of the used element of B.

Now, we are showing an example, when the B-distance fails on the triangular inequality.

Let $r = (0, 0, 0)$, $s = (-2, 1, 1)$ and $t = (-1, 0, 1)$ be three points, and $B_2 = (2, 1, 1)$ be a neighborhood sequence.

It is easy to compute that $d(r, t; B_2) = 1$, $d(t, s; B_2) = 1$ and $d(r, s; B_2) = 3$. (Check on Fig. 4 as well.)

Therefore $d(r, t; B_2) + d(t, s; B_2) < d(r, s; B_2)$.

Now we are presenting interesting examples for the fact, that the B-distances on the hexagonal grid are highly depend on the used neighborhood sequence and the parities and relative direction of the points. They have the following property.

Let $r = (0, 0, 0)$, $s = (-2, 1, 1)$, $p = (-1, 2, -1)$ and $u = (-1, 2, 0)$ be four points, and $B_3 = (3, 1)$, $B_4 = (1, 3)$ be two neighborhood sequences. Then one can compute, that:

$d(r, s; B_3) = 2 < d(r, s; B_4) = 3$,
$d(r, p; B_3) = 3 > d(r, p; B_4) = 2$, and
$d(r, u; B_3) = 2 = d(r, u; B_4)$.

So, one cannot say that the B_3-distances or the B_4-distances are greater than the others, it depends on the points as well.

7 Conclusion

The concept of neighborhood sequences is applicable in image processing (see [6]). In the present paper we used the concept of neighborhood sequences, and with their help, we were able to define distance functions on the hexagonal grid. This method based on the three types of neighborhood relations in the hexagonal grid which are according the neighborhood relations of the cubic grid. Based on the analogy between the points of the cubic grid and the points of hexagonal grid we derived a formula, which give the distance from one point to another. We cannot use the term distance between points because not every distance based on neighborhood sequences is symmetric. We showed that the non-symmetric distances depend on the ordering of elements of the neighborhood sequences.

Our result is a theoretical one; it makes the distances based on neighborhood sequences in the hexagonal grid applicable. It is a future –, but not so hard – work to use this result in real applications. It is a very interesting property that there are non-symmetric distances which may useful in some applications.

Acknowledgements

This research was partly supported by a grant from the Hungarian National Foundation for Scientific Research (OTKA F043090).

References

1. P.E. Danielsson, "3D Octagonal Metrics", *Eighth Scandinavian Conference on Image Processing*, 1993., pp. 727-736.
2. P.P. Das, P.P. Chakrabarti and B.N. Chatterji, "Distance functions in digital geometry", *Information Sciences* **42**, pp. 113-136, 1987.
3. E.S. Deutsch, "Thinning algorithms on rectangular, hexagonal and triangular arrays", *Communications of the ACM*, **15** No.3, pp. 827-837, 1972.
4. A. Fazekas, A. Hajdu and L. Hajdu, "Lattice of generalized neighborhood sequences in nD and ∞D", *Publicationes Mathematicae Debrecen* **60**, pp. 405-427, 2002.
5. H. Freeman, "Algorithm for generating a Digital Straight Line on a Triangular Grid", *IEEE Transactions on Computers* **C-28** pp. 150-152, 1979.
6. A. Hajdu, B. Nagy and Z. Zörgő, "Indexing and segmenting colour images using neighborhood sequences", *IEEE International Conference on Image Processing, ICIP'03*, Barcelona, Sept. 2003. pp. I/957-960.
7. T. Y. Kong and A. Rosenfeld, "Digital Topology: Introduction and Survey", *Computer Vision, Graphics and Image Processing* **48**, pp. 357-393, 1989.
8. B. Nagy, "Distance functions based on neighbourhood sequences", *Publicationes Mathematicae Debrecen* **63**, pp. 483-493, 2003.
9. B. Nagy, "Shortest Path in Triangular Grids with Neighborhood Sequences", *Journal of Computing and Information Technology* **11**, pp. 111-122, 2003.
10. B. Nagy, "A family of triangular grids in digital geometry", *3rd International Symposium on Image and Signal Processing and Analysis (ISPA'03)*, Rome, Italy, Sept. 2003., pp. 101-106, 2003.
11. B. Nagy, "A symmetric coordinate system for the hexagonal networks", *Information Society 2004 – Theoretical Computer Science (IS04-TCS), ACM Slovenija conference*, Ljubljana, Slovenia, Okt. 2004. accepted paper
12. B. Nagy, "Non-metrical distances on the hexagonal plane", *7th International Conference on Pattern Recognition and Image Analysis: New Information Technologies (PRIA-7-2004)*, St. Petersburg, Russian Federation, Okt. 2004. accepted paper.
13. B. Nagy, "Distance with generalised neighborhood sequences in nD and ∞D", *Discrete Applied Mathematics*, submitted.

14. A. Rosenfeld and J.L. Pfaltz, "Distance functions on digital pictures", *Pattern Recognition* **1**, pp. 33-61, 1968.
15. M. Yamashita and T. Ibaraki, "Distances defined by neighborhood sequences", *Pattern Recognition* **19**, pp. 237-246, 1986.

On Correcting the Unevenness of Angle Distributions Arising from Integer Ratios Lying in Restricted Portions of the Farey Plane

Imants Svalbe and Andrew Kingston

Centre for X-ray Physics and Imaging,
School of Physics and Materials Engineering,
Monash University, VIC 3800, AUS
{Imants.Svalbe, Andrew.Kingston}@spme.monash.edu.au

Abstract. In 2D discrete projective transforms, projection angles correspond to lines linking pixels at integer multiples of the x and y image grid spacing. To make the projection angle set non-redundant, the integer ratios are chosen from the set of relatively prime fractions given by the Farey sequence. To sample objects uniformly, the set of projection angles should be uniformly distributed. The unevenness function measures the deviation of an angle distribution from a uniformly increasing sequence of angles. The allowed integer multiples are restricted by the size of the discrete image array or by functional limits imposed on the range of x and y increments for a particular transform. This paper outlines a method to compensate the unevenness function for the geometric effects of different restrictions on the ranges of integers selected to form these ratios. This geometric correction enables a direct comparison to be made of the effective uniformity of an angle set formed over selected portions of the Farey Plane. This result has direct application in comparing the smoothness of digital angle sets.

Keywords: discrete image processing, discrete Radon transforms, Farey sequences and digital angles.

1 Introduction

Discrete Radon transforms [1–3] differ from analogue projections, such as those used in computed tomography (CT). The set of discrete projection angles is explicitly linked to the digital imaging grid on which the projections are made, rather than the free choices in continuous space available for taking measurements in CT. Having the set of projection angles emerge as a property of the transform is an advantage, as it removes the "arbitrary" choice of angles in the continuum. However the distribution of discrete angles that comes with each discrete transform is no longer uniform and some way of comparing angle distributions that arise from different discrete formulations is then needed.

The unevenness function is such a measure; it reflects the sum of the absolute differences of a monotonically arranged set of angles with respect to the same

R. Klette and J. Žunić (Eds.): IWCIA 2004, LNCS 3322, pp. 110–121, 2004.

number of equally spaced angles over the same range. The angle corresponding to an integer ratio is taken as the inverse tangent of the ratio. Thus $0/1, 1/4, 1/2, 3/4$ and $1/1$ is a sequence of ratios with zero unevenness, whilst the sequence of ratios $0/1, 1/3, 1/2, 2/3, 1/1$ has a net unevenness of $1/12 + 0 + 1/12 = 1/6$ for the 3 terms between 0 degrees ($\tan^{-1}(0/1)$) and 45 degrees ($\tan^{-1}(1/1)$). The latter case is shown in Figure 1.

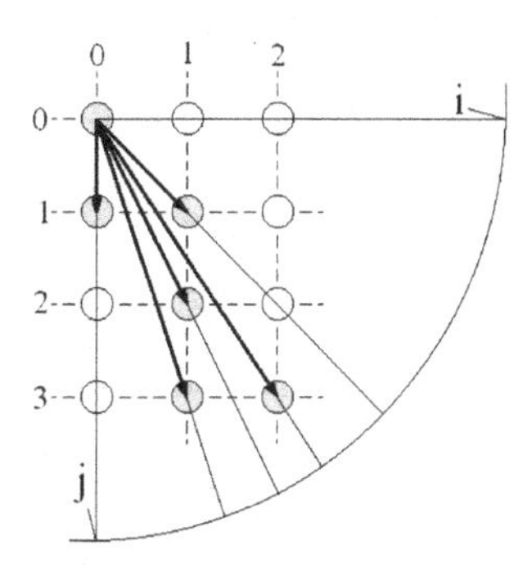

Fig. 1. The set of vectors that represent the angles for ratios 0/1, 1/3, 1/2, 2/3 and 1/1

The use of integer ratios to represent angles occurs frequently in digital transforms. Discrete projective transforms link pixels located at integer multiples i and j of the x and y image grid spacing to form the projection angle $\tan^{-1}(i/j)$. To form non-redundant projection angles, i and j are restricted to the set of relatively prime integers which, as fractions, form members of the Farey sequence [4, 5].

The transforms [1–3] each impose different restrictions on the range of the integers (i, j) that form the discrete projection angle set. These restrictions arise directly from the finite size of the discrete imaging grid, or through some transform dependent functional limit on i and j. If we restrict the portion of the 2D plane from which integer fractions (i, j) can be selected to form discrete projection angles, then the unevenness shifts systematically further from zero as the domain of integers becomes more restricted and further removed from a linear distribution of angles.

To compare the unevenness for angle sets arising from restricted ratios, we need to remove the systematic bias that reflects the restricted domain rather than of the distribution of integers within that domain.

In this paper, we correct the unevenness function for fractions comprised of integers (i, j) that are restricted to specific areas of the 2D plane. We start with the full unrestricted set and then consider the simple restriction $i < m$, $j < n$ (corresponding to the general Farey Sequence [6]). We next consider the length of the segment, $|(i, j)|$, to be less than some radius r, i.e., $|(i, j)| < r$.

The latter case motivated this work, as it corresponds closely to the constraint on the integer ratios that arises in the discrete Radon formalism [2]. We were interested to find if the discrete Radon transform selects Farey fractions in a way that is, in some sense, optimally smooth. The properties of the projective angle sets are of concern in reconstructing images from analogue projections [7] (where digital angles are matched to the nearest uniform continuum angle) or as a means of dynamic image analysis using the projected views of image subsets [8] (where extra angles are added or subtracted as the size of image subset changes).

This paper considers distributions for F_n with $n \lesssim 100$. It takes the simplest approximation of uniform density of angles formed by lines connecting the origin to the points in the Farey plane. We assume this as the density of relatively prime points in the Farey plane is approximately constant. The uniform angle approximation can be shown to be weak. In [9] Boca, Cobeli and Zaharescu derive the limiting distribution for large n for the angles of lines from Farey points to the origin from linear segment or disc bounded sections of the plane. Clearly, more accurate results will be obtained using the distribution based on [9] as the starting point, but for $n < 100$ the accuracy obtained here is sufficient to correct and compare the angle distributions for discrete projection sets.

In Section 2 we review some properties of the Farey sequence and its relation to discrete Radon transforms. Section 3 reviews the unevenness for Farey fractions in F_n where i and $j < n$, whilst Section 4 presents the correction required to "linearise" the unevenness for the general Farey terms for $F_{m,n}$, and Section 5 covers the case where the integers i and j fall inside a circle of radius r. We conclude in Section 6 with some discussion on the relevance of the latter case to the discrete Radon transform of [2] and of the use of geometric corrections to the unevenness in examining the uniformity of relatively prime fractions across arbitrary regions of the 2D plane.

2 Discrete Radon Projections, Relative Primes and the Farey Sequence

The discrete Radon transform sums the contents of image pixels located along straight lines. The lines have slopes that are the ratio of two integers i/j, so that the image is sampled along these lines, but only at grid locations separated by regular intervals of $\sqrt{(i^2 + j^2)}$, as shown in Figure 1. A digital projection is a parallel array of such lines separated by integer translations (usually taken as the unit pixel spacing along the x–axis direction).

For the discrete Radon transform described in [2], the image $I(x, y)$ is chosen to be a square with a side length fixed as a prime number (p) of pixel units. Discrete transforms for arrays that are square but have composite length, L, sides are presented in [10, 11] for $L = 2^n$ and in [12] for $L = p^n$ and [13] for L a general composite number.

The projections are indexed by an integer m, where $0 \leq m \leq p$, by $x = my+t$ for translations $0 \leq t < p$ under modular arithmetic (so that the rays wrap, under periodic boundary conditions, from right to the left image boundary).

Under these constraints, the discrete Radon transform $R(t,m)$ is a 1:1 mapping of the image $I(x,y)$. The image can be reconstructed (using additive operations only), without loss, from the discrete projections, using a back-projection process similar to the projection mechanism.

Linking pixels that are sampled on a given projection ray to their nearest sampled neighbours means each integer m corresponds to a unique ratio of integers, i/j, that gives the slope or angle of the projection. It is shown in [4] that the projection angles for the discrete Radon transform defined in [2] are a subset of the Farey sequence of relatively prime ratios. All Farey fractions i/j that are located at coordinate (i,j) inside a circle of radius $\sqrt{p}$ form part of the projection angle set, with an apparently "random" selection of Farey fractions from radii $\sqrt{p}$ up to $\sqrt{(2p/\sqrt{3})}$ to reach the full complement of the $p+1$ angles that correspond to the discrete projection set for a $p \times p$ image.

3 Unevenness of Farey Terms

A plot of the location (i,j) of the Farey fractions i/j for i and $j < n$ is shown in Figure 2 (the pattern is symmetric about the 45 degree line for the fractions j/i). Relevant recent work on Farey sequences has considered the distribution of Farey points [14] and correlations between Farey fractions [15].

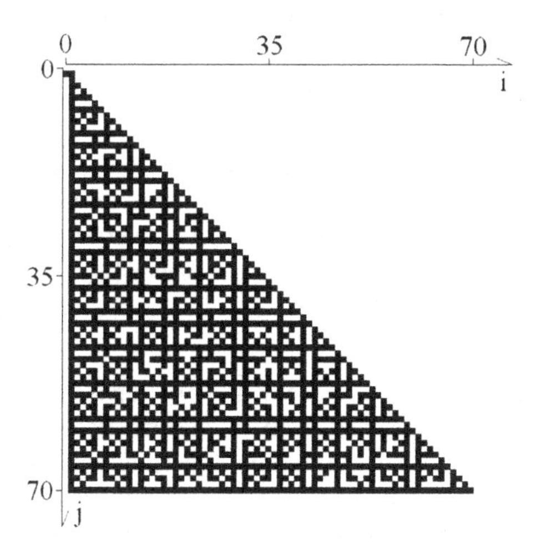

Fig. 2. White points mark the location (i,j) of all positive relatively prime Farey fractions i/j with $i < j$ and $j < n$. For this example, $n = 101$

If there are a total of S Farey points inside the 45 degree triangle formed by the points $i < n$, $j < n$, then we can order these ratios so that their fractional value increases monotonically. This forms the Farey sequence F_n. Let k be the index into this ordered set. We compute the unevenness D_n of the Farey series F_n as $D_n = \sum_{k=0}^{S-1} |d(k)|$ where $d(k) = 1/j - k/S$ and i/j is the kth of S angles between 0/1 and 1/1. Figure 3 shows this process diagrammatically. The number of angles dS enclosed between the dotted lines is proportional to the area dA between the dotted lines, provided that the density of i/j fractions over the 45 degree Farey triangle is uniform. The unevenness, as used in this paper, compares values relative to the linear set of fractions between 0/1 and 1/1. The inverse tangent of these values should be taken to rank the unevenness of terms relative to a uniform distribution of angles.

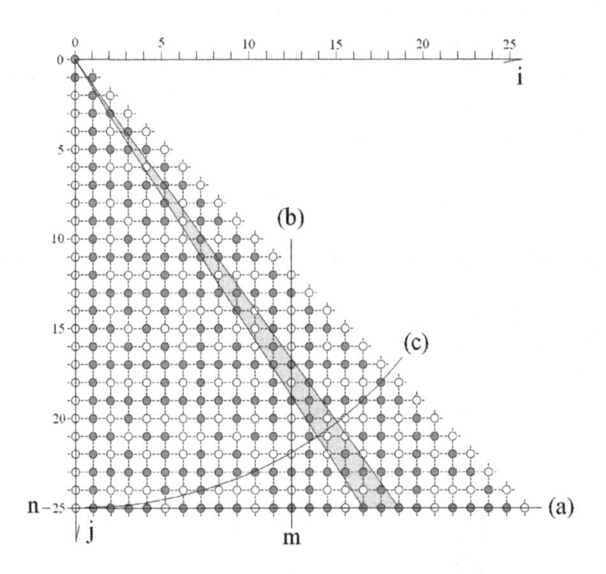

Fig. 3. Distribution of the number of integer ratios as a function of angle. The shaded area encloses part of the angle distribution given by the integer ratios i/j. (a) denotes the full Farey triangle for F_n where $\{i, j\} < n$, (b) illustrates the general Farey case Fm, n where $\{i\} < m$, $\{j\} < n$, and (c) the case where $\{i, j\}$ is restricted to lie inside a circular arc of radius n

All shaded triangles of equal base will here have equal height and hence area. As the angle index k varies from 0 degrees (0/1) to 45 degrees (1/1), dS, which is proportional to dA, remains constant with k. Hence for the full Farey triangle (shown as (a) in Figure 3), the number of angles varies linearly with k in a continuous plane and hence the unevenness would be zero. For discrete points in the plane and if the density of relatively prime ratios was constant, then the unevenness for each F_n would be close to, but not equal to zero. The density of relative primes is approximately constant, as the probability that two numbers are relatively prime is $1/\zeta(2) = 6/\pi^2$ [5]. This relatively poor assumption of an even angle distribution resulting from an "on average" constant density of relative primes is addressed in [9] for bounded sections of the Farey plane that preserve the adjacency property of Farey fractions. In [9] rigorous results are given for the mean density of lattice points visible from the origin as a function of angle. There is great interest in number theory in quantifying the fluctuations in the density of primes (and relative primes). The unevenness function provides a measure of deviations from uniformity and bounds on the deviation of unevenness can be expressed as a variant statement of the Riemann Hypothesis [16].

Here the unevenness function provides a useful comparison of the smoothness with which digitally selected integer ratios fill the range of projection angles. Figure 4 shows $d(k)$, the variation in unevenness as a function of k, for the full Farey sequence F_n with $n = 107$ (for which the total unevenness, $D = 2.426$ over the 3533 terms from 0/1 to 1/1). The work in [9] models the distribution of unevenness values like those shown in Fig. 4.

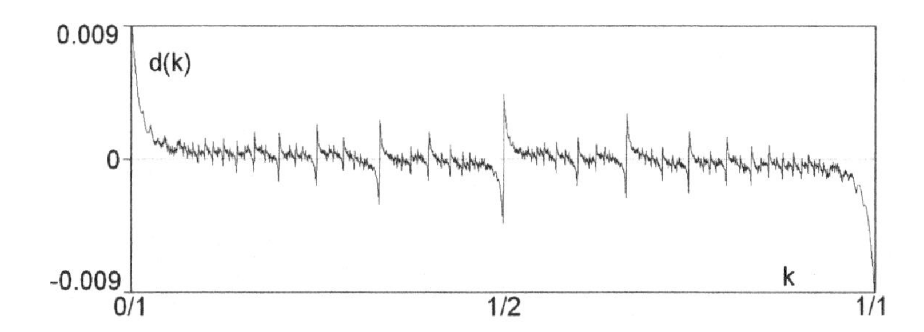

Fig. 4. Unevenness terms $d(k)$, as a function of increasing angle for the Farey sequence with $n = 107$. Note the reflected symmetry about the midpoint, where $i/j = 1/2$ corresponds to $k = 1/2$

4 The Case for $F_{m,n}$

The points chosen to be included as part of Farey plane can be truncated in various ways. As a simple example, we begin by truncating i at an integer $m < n$. This restriction corresponds to formation of the general Farey sequence $F_{m,n}$ [6] where the integer ratios i/j now only come from within $i \leq m$ and $j \leq n$ (see line (b) in Figure 3). [6] also provides efficient algorithms to calculate all terms in such a sequence for any general Farey sequence without the use of recursive generation.

Figure 5 shows components of the observed unevenness as a function of k for $F_{33,66}$. The large unevenness results from the displacement of angles away from the linear trend, due to the selective truncation of the integer ratios within the Farey triangle.

Because the fluctuations in unevenness are approximately equally likely to have either sign, smoothing the curve by some form of local averaging will produce a curve that approximates the underlying distortion due to the geometry. However such an approach will yield only approximate results and has difficulty dealing with the ends of the distribution, where large unevenness contributions of the same size occur.

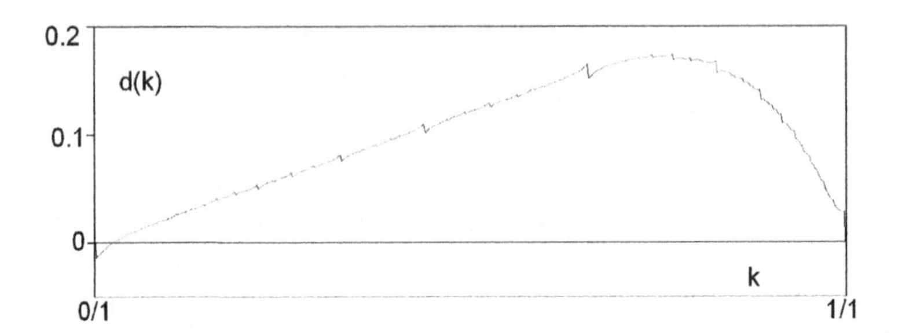

Fig. 5. Unevenness $d(k)$ as a function of k for the general Farey sequence $F_{m,n}$ for $m = 33$, $n = 66$. The total unevenness, D, here is 99.834 over the 1009 terms from 0/1 through to 1/1

This geometric effect should then be compensated for exactly in order to allow a meaningful direct comparison with the corresponding unevenness of the full Farey set, which here would be F_{66}.

If we truncate the allowed values of i at m (as shown in line (b) of Figure 2), we can compute the underlying continuous plane unevenness function as before, based on the incremental changes in area as k increases.

For $0 \leq k \leq m$:

$$A = \frac{nk}{2}. \tag{1}$$

At $k = 0$, $A = 0$ and at $k = m$, $A = nm/2$. This gives the linear variation of angle density with k mentioned in Section 2.

For $m < k \leq n$:

$$A = \frac{nm(k-m)}{2k} + \frac{nm}{2} = \frac{nm(2k-m)}{2k} \tag{2}$$

At $k = m$, $A = nm/2$, at $k = n$, $A = m(2n - m)/2$. The inverse function of (2), expressing k as a function of A, is then

$$k = \frac{nm^2}{2(nm - A)}. \tag{3}$$

The difference between a linear trend and the values of k given by (2) and (3) over 0/1 to 1/1 produces the function required to correct for the truncation of values at m in $F_{m,n}$.

Figure 6 shows the required geometric correction based on (1) and (3) applied to the example of Figure 5. Figure 7 shows Figure 5 corrected by subtracting an appropriately scaled version of the function shown in Figure 6. The total unevenness for Figure 7 is now $D = 1.619$. Note the loss of symmetry about $k = 1/2$ that was previously seen in Figure 4; $i/j = 1/2$ now occurs at $k = 2/3$ in the truncated Farey angle distribution. Figure 8 shows another post-correction example applied to the general Farey sequence $F_{24,80}$.

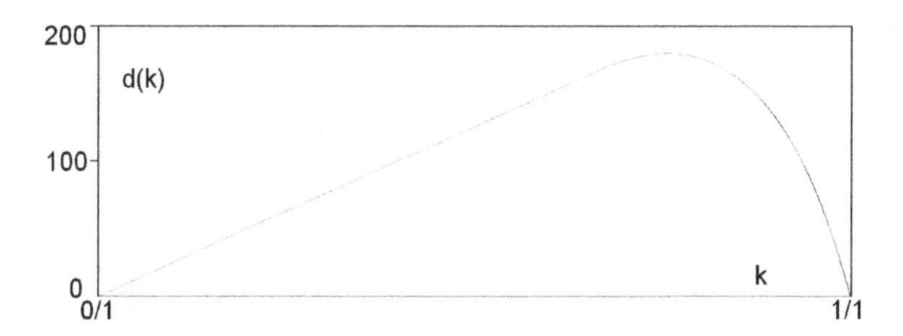

Fig. 6. Geometric correction function for the general Farey sequence $F_{m,n}$ as a function of angle index k, applied to the example $F_{33,66}$

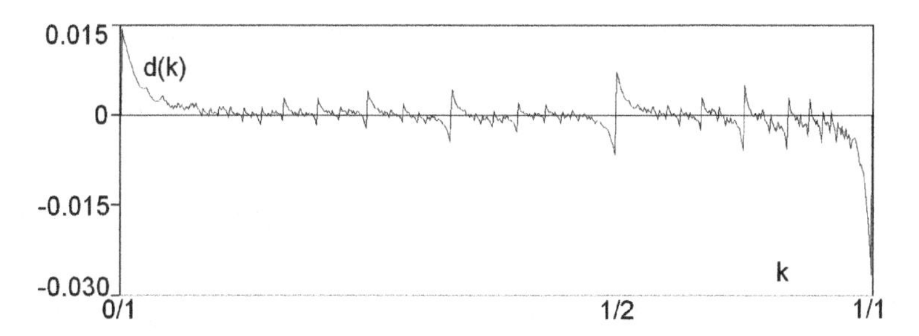

Fig. 7. Corrected unevenness as a function of k for $F_{m,n}$ as a function of angle index k (over 1009 terms) for the example $F_{33,66}$. Here D changes from 99.834 uncorrected to 1.619 corrected

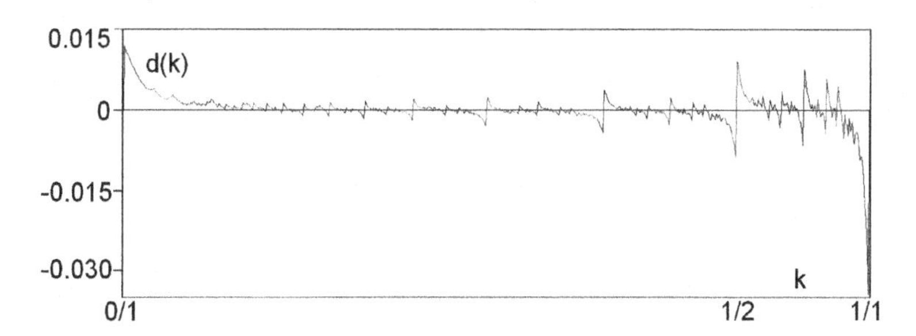

Fig. 8. Corrected unevenness for $F_{m,n}$ as a function of angle index k for $F_{24,80}$ (over 1010 terms). Here D changes from 199.377 uncorrected to 1.479 corrected

5 The Case for F_n Limited by a Disc

The same geometric correction procedure can be applied to the case where we exclude all ratios i/j that lie outside a circle of specified radius, which we take here as $r = n$ (as shown by line (c) in Figure 3). This approximates the discrete Radon transform angle projection selection criteria, as given in Section 2.

We order the sequence of angles by $\theta = \tan^{-1}(k/n)$, so that $k = n\tan(\theta)$. Then $\mathrm{d}k/\mathrm{d}\theta = n\sec^2(\theta)$ and $\mathrm{d}A = n^2\mathrm{d}\theta/2$ so that

$$\frac{\mathrm{d}A}{\mathrm{d}k} = \frac{n}{2\sec^2(\theta)} = \frac{n^3}{2(n^2+k^2)} \tag{4}$$

and

$$A = \frac{n^2\tan^{-1}(k/n)}{2} \tag{5}$$

and

$$k = n\tan(2A/n^2) \tag{6}$$

gives the required correction function for a circular cut-off in the Farey plane when we find the difference between (6) and a linear trend. Figure 9 shows the unevenness as a function of k for the integer ratios i/j of F_{107} limited to $i^2 + j^2 = r^2 < 10,009$.

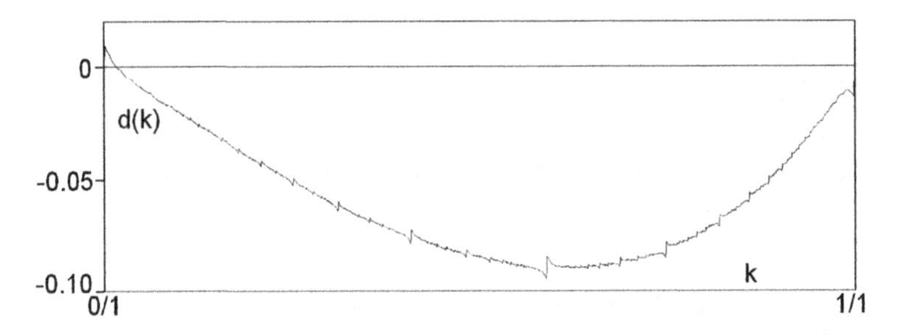

Fig. 9. Unevenness as a function of k for terms in F_n limited to lie inside radius r, here $n = 107$ and $r = \sqrt{(10,009)}$. Here $D = 139.44$. For comparison, D for F_{107} is 2.426 over 2,389 terms

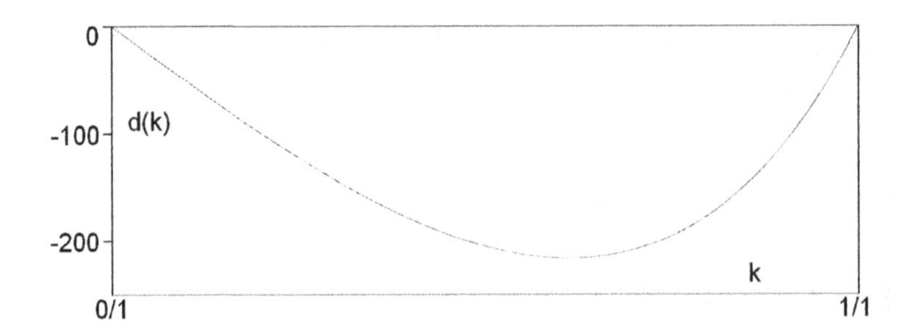

Fig. 10. Geometric correction function given by (6) for the example applied to the case shown in Figure 9

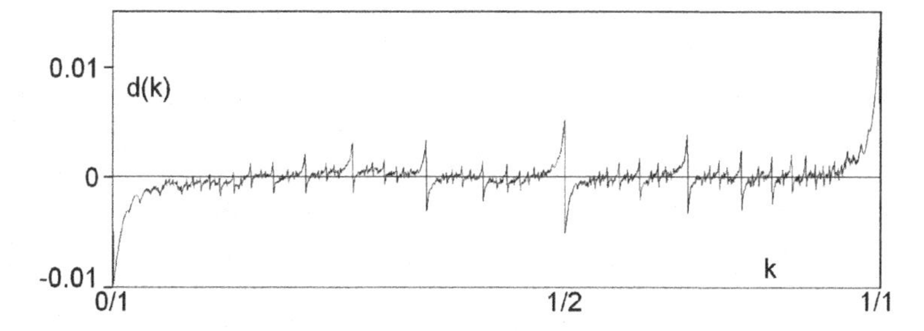

Fig. 11. Figure 9 after geometric correction for the circular truncation in the Farey plane. Here $D = 2.0419$ for the 2,389 angle terms between 0 and 45 degrees and inside radius r, where $r^2 = 10,009$

6 Conclusions

We have developed a simple procedure to enable a direct comparison of the unevenness for angle distributions using differently restricted sections of the Farey plane of relatively prime integer ratios. The procedure involves a geometric correction factor specific to each geometric constraint and assumes a uniform density of points in the Farey plane.

Figure 12 shows the unevenness for a discrete Radon angle set corrected using a hard edged circle with radius equal to the maximum radius for that p. The Radon set shows a slightly greater unevenness than for the full set of relatively prime integer ratios out to the same radius.

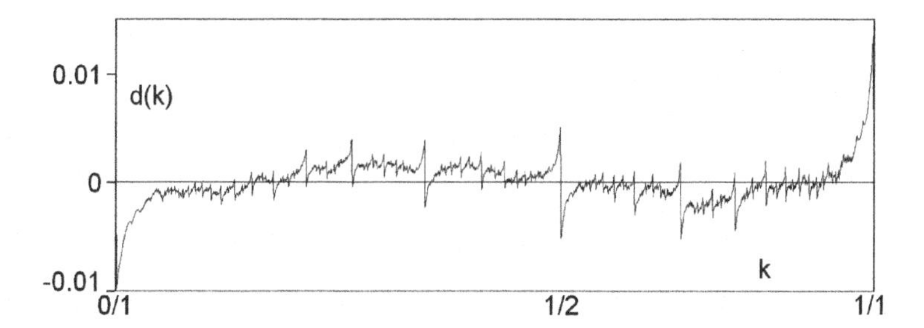

Fig. 12. Compensated unevenness for the discrete radon transform angle distribution at $p = 10,009$, for which $D = 3.274$ (was 143.09 uncorrected)

We are examining the variation of unevenness for the discrete Radon angle sets for primes of similar value, to find which primes have the smoothest digital angle sets, as well as examining the general trends in unevenness as a function of increasing array size p. The same procedure can be used to compare the unevenness for images of the same discrete size but using different transform definitions [1–3] that also lead to different restrictions on the ratio set. These results may also have application in examining the variation in the level of fluctuations in relative primes over sub-regions of the Farey plane. These comparisons need to incorporate the effects of the known limiting distribution of fluctuations described in [9]. This would then provide a level of accuracy needed to compare the unevenness between adjacent prime sized array angle sets. For the DRT Farey points, the adjacency property used in [9] does not hold for points in the DRT Farey set with radius $r \geq \sqrt{p}$, i.e., there are missing Farey fractions, however the theory of [9] is valid for the majority of DRT Farey points which have $r < \sqrt{p}$.

Discrete arrays with sides of composite length [13] also select i/j ratios from the Farey plane, but with an altered distribution of angles to account for the redundancy in sampling due to the non-prime "wrapping" across the array. A similar evaluation of these angle distributions and of those for more general arrays of size $N \times M$ is being undertaken.

Acknowledgments

We thank the referees for drawing our attention to the fundamental work of Boca, Cobeli and Zaharescu [9] and later papers. IS has received research support for this work from the Centre for X-ray Physics and Imaging, the School of Physics and Materials Engineering and the Faculty of Science at Monash University. AK acknowledges support from an Australian Postgraduate Research Scholarship provided through the Australian government, as well as support from the School of Physics and Materials Engineering at Monash University.

References

1. Beylkin, G.: Discrete radon transform. IEEE Transactions on Acoustics, Speech, & Signal Processing **35** (1987) 162–172
2. Matus, F., Flusser, J.: Image representation via a finite Radon transform. IEEE Transactions on Pattern Analysis & Machine Intelligence **15** (1993) 996–1006
3. Guedon, J., Normand, N.: The Mojette transform: applications in image analysis and coding. In: The International Society for Optical Engineering. Volume 3024, pt.2, 1997, p 873-84., SPIE-Int. Soc. Opt. Eng, USA (1997) 873–884
4. Svalbe, I., Kingston, A.: Farey sequences and discrete Radon transform projection angles. In: Electronic Notes in Discrete Mathematics. Volume 12., Elsevier (2003)
5. Hardy, G., Wright, E.: An introduction to the theory of numbers. 4 edn. Clarendon Press, Oxford (1960)
6. Acketa, D., Zunic, J.: On the number of linear partitions of the (m, n)-grid. Information Processing Letters **38** (1991) 163–168
7. Svalbe, I., van der Spek, D.: Reconstruction of tomographic images using analog projections and the digital Radon transform. Linear Algebra and Its Applications **339** (2001) 125–145
8. Kingston, A., Svalbe, I.: Adaptive discrete Radon transforms for grayscale images. In: Electronic Notes in Discrete Mathematics. Volume 12., Elsevier (2003)
9. Boca, F., Cobeli, C., Zaharescu, A.: Distribution of lattice points visible from the origin. Commun. Math. Phys. **213** (2000) 433–470

10. Hsung, T., Lun, D., Siu, W.: The discrete periodic Radon transform. IEEE Transactions on Signal Processing **44** (1996) 2651–2657
11. Lun, D., Hsung, T., Shen, T.: Orthogonal discrete periodic Radon transform. Part I: theory and realization. Signal Processing **83** (2003) 941–955
12. Kingston, A.: Orthogonal discrete Radon transform over p^n. Signal Processing (submitted November 2003)
13. Kingston, A., Svalbe, I.: A discrete Radon transform for square arrays of arbitrary size. submitted to DGCI'05 (2004)
14. Augustin, V., Boca, F., Cobeli, C., Zaharescu, A.: The h-spacing distribution between Farey points. Math. Proc. Cambridge Phil. Soc. **131** (2001) 23–38
15. Boca, F., Zaharescu, A.: The correlations of Farey fractions. preprint http://arxiv.org/ps/math.NT/0404114 (2004)
16. Apostol, T.: Introduction to analytic number theory. Springer-Verlag, New York (1976)

Equivalence Between Regular n-G-Maps and n-Surfaces

Sylvie Alayrangues[1] and Xavier Daragon[2], Jacques-Olivier Lachaud[1], and Pascal Lienhardt[3]

[1] LaBRI - 351 cours de la libération - 33405 Talence Cedex
{alayrang, lachaud}@labri.fr
[2] ESIEE - Laboratoire A²SI, 2 bd Blaise Pascal, cité DESCARTES, BP99, 93162 Noisy le Grand Cedex
daragonx@esiee.fr
[3] SIC - bât SP2MI, Bd M. et P. Curie, BP 30179, 86962 Futuroscope Chasseneuil Cedex
lienhardt@sic.sp2mi.univ-poitiers.fr

Abstract. Many combinatorial structures have been designed to represent the topology of space subdivisions and images. We focus here on two particular models, namely the n-G-maps used in geometric modeling and computational geometry and the n-surfaces used in discrete imagery. We show that a subclass of n-G-maps is equivalent to n-surfaces. We exhibit a local property characterising this subclass, which is easy to check algorithmically. Finally, the proofs being constructive, we show how to switch from one representation to another effectively.

1 Introduction

The representation of space subdivisions and the study of their topological properties are significant topics in various fields of research such as geometric modeling, computational geometry and discrete imagery. A lot of combinatorial structures have already been defined to represent such topologies and specific tools have been developed to handle each of them. Although most of them aim at representing manifold-like underlying spaces they have very variable definitions.

Comparing these structures, and highlighting their similarities or specificities are important for several reasons. It can first create bridges between them and offer the possibility to switch from one framework to another according to the needs of a given application. It may also lead to a more general framework which unify most of these structures. Theoretical results and practical algorithms can also be transferred from one to another. However, these structures are most likely not interchangeable. Indeed, there is yet no complete combinatorial characterisation of manifolds. The structures found in the literature generally propose local combinatorial properties that can only approach the properties of space subdivisions. It is therefore extremely important to know precisely what class of objects is associated to each structure. Several studies have already been carried

R. Klette and J. Žunić (Eds.): IWCIA 2004, LNCS 3322, pp. 122–136, 2004.

out in this direction. Quad-edge, facet-edge and cell-tuples were compared by Brisson in [6]. Lienhardt [15] studied their relations with several structures used in geometric modelling like the n-dimensional (generalized or not) map. The relation between a subclass of orders and cell complexes was also studied in [1]. A similar work was done on dual graphs and maps by Brun and Kropatsch in [7].

We focus here mainly on two structures: the n-surface and the n-dimensional generalized map. The n-surface is a specific subclass of orders defined by Bertrand and Couprie in [4] which is similar to the notion previously defined by Evako *et al.* on graphs in [12]. It is essentially an order relation over a set together with a finite recursive property. It is designed to represent the topology of images and objects within. The generalized map introduced by Lienhardt in [15] is an effective tool in geometric modeling and is also used in computational geometry. It is defined by a set of $n+1$ involutions joigning elements dimension by dimension. Although the definitions of these two structures are very different, we show that a subclass of generalized maps, that we call *regular n-G-maps*, is equivalent to n-surfaces. Furthermore, we provide a simple local characterisation of this subclass. This may have various nice consequences. From a theoretical point of view, some proofs may be simplified by expressing them rather on a model than on the other, some notions can also be extended. Moreover the operators defined on each model may be translated onto the other. A possible application would consist in using the tools defined on orders: homotopic thinning, marching chains using frontier orders [8, 9] to obtain n-surfaces. They can then be transformed into n-G-maps which can easily be handled with their associated construction operators: identification, extrusion, split, merge. To prove the equivalence of these models, we use an intermediary structure defined by Brisson in [6]: the augmented incidence graph. This structure is quite similar to orders although its definition does not involve the same local properties as n-surfaces. Moreover Brisson shows a partial link between n-G-maps and such incidence graphs. He effectively proved that an n-G-map may be built from any augmented incidence graph. In [15], Lienhardt gives a necessary condition to build such an augmented incidence graph from an n-G-map. We show here with a counter-example that it is not sufficient.

The main contributions of these papers are: (i) we prove that n-dimensional augmented incidence graphs and n-surfaces are equivalent structures (Theorem 18), (ii) we complete the works of Brisson and Lienhardt with the characterization of the n-G-map subclass that is equivalent to augmented incidence graphs (Definition 8 of *regular n-G-maps*, Theorem 12 and 14), (iii) we design constructive proofs which allow to effectively switch between the different representation. This result remains very general since any closed n-G-map can be refined into a regular n-G-map with appropriate local subdivisions.

The paper is organized as follows. First, we recall the notions of incidence graphs and orders and show how they are related to each other. We also define precisely the models we wish to compare and give some clues to their equivalence. Then, we give a guideline of the proof before presenting the whole demonstration. We conclude with some perspectives for this work.

2 Models Description

We describe below the models we wish to compare, and we list known results about their relationships. We begin with recalling the general notions related to orders and incidence graphs and we characterize then the appropriate submodels.

2.1 Orders and Incidence Graphs

Orders are used by Bertrand *et al.* [3] to study topological properties of images. The main advantages of this model are its genericity and its simplicity. Orders can be used to represent images of any dimension, whether they are regularly sampled or not.

Definition 1. *An* order *is a pair* $|X| = (X, \alpha)$*, where* X *is a set and* α *a reflexive, antisymmetric, and transitive binary relation. We denote* β *the inverse of* α *and* θ *the union of* α *and* β*.* CF orders *are orders which are* countable*, i.e.* X *is countable, and* locally finite*, i.e.* $\forall x \in X, \theta(x)$ *is finite.*

For any binary relation ρ on a set X, for any element x of X, the set $\rho(x)$ is called the ρ-adherence of x and the set $\rho^\square(x) = \rho(x) \setminus \{x\}$, the strict ρ-adherence of x. A ρ-chain of length n is any sequence x_0, x_1, $\cdots$, x_n such that $x_{k+1} \in \rho^\square(x_k)$. An implicit dimension, $\mathtt{dim}_\alpha(x)$, may be associated to each element of an order [12, 1], as the length of the longest α-chain beginning at it. We choose here to represent orders as simple directed acyclic graphs (see Fig. 2-a), where each node is associated to an element of the order and only direct α-relations[1] are shown. The remaining relations can be deduced by transitivity.

Incidence graphs are used to represent subdivisions of topological spaces. They explicitly deal with the different cells of the subdivision and their incidence relations. They have for example been used by Edelsbrunner [10] to design geometric algorithms.

Definition 2. *An* incidence graph *[6] yielded by a cellular partition of dimension* n *is defined as a directed graph whose nodes correspond to the cells of the partition and where each oriented arc connects an* i*-cell to an* $(i-1)$*-cell to which it is incident. With such a graph is associated a labeling of each node given by the dimension of its associated cell.*

Let us denote by I_i *the index set of the* i*-cells of a cellular partition. The associated* n*-dimensional incidence graph is hence denoted by* $IG_C = (C, \prec)$*, where* $C = \bigcup_{i=0}^{i=n} (\bigcup_{\beta \in I_i} c^i_\beta)$ *is the set of cells and* $\prec$ *is the incidence relation between* $(i-1)$*-cells and* i*-cells,* $i \in \{1, \cdots, n\}$*.*

In the sequel we only consider finite incidence graphs. For convenience, it is also sometimes useful to add two more cells c^{-1} and c^{n+1} to incidence graphs, such that c^{-1} is incident to all 0-cells and all n-cells are incident to c^{n+1}. They

[1] An element x is said to be directly related to x' by α if $x' \in \alpha^\square(x)$ and $\alpha^\square(x) \cap \beta^\square(x') = \emptyset$.

represent no real cells of the subdivision but make easier the definition of generic operators on incidence graphs. An incidence graph with two such cells is called an *extended incidence graph* and is denoted by $IG_C^* = (C^*, \prec)$. Let IG_C be an incidence graph and c an element of C, if c' is linked to c by a chain of cells (eventually empty) related by $\prec$ then we say that c' is a face of c and we denote it by $c' \leq c$. We write $c' < c$ when $c' \neq c$ and $c' \leq c$. An incidence graph is hence represented by a simple directed acyclic graph, where the nodes respectively representing a cell c^{i+1} and a cell c^i are linked by an arc if and only if $c^i \prec c^{i+1}$ (see Fig. 2-b).

There is an obvious relationship between incidence graphs and orders. The incidence graph $IG_C = (C, \prec)$ can indeed be seen as the order $(C, \leq)$ where the dimension associated to each cell is forgotten. $(C, \leq)$ is the order associated to IG_C. Reciprocally, a relation $\prec_{|X|}$ may be defined on the elements of $|X|$, such that $x' \prec_{|X|} x$ is equivalent to $x' \in \alpha(x)$ and $\mathtt{dim}_\alpha(x') = \mathtt{dim}_\alpha(x) - 1$. The incidence graph $(X, \prec_{|X|})$ where each cell of the graph is labeled by its corresponding α-dimension in $|X|$ is the incidence graph associated to $|X|$. However, in the general case, the relation $\leq$ between the cells of an incidence graph built from an order is different from the order relation α on the set X (see [2]). An order $|X| = (X, \alpha)$ is said to be equivalent to an incidence graph $G = (C, \prec)$ if the order $|G| = (C, \leq)$ is isomorphic to $|X|$.

Theorem 3. *An order (X, α) and its associated incidence graph are* equivalent *if and only if:*

1. *each element of x belongs to at least one maximal α-chain,*
2. *$\forall x \in X$, the elements of X directly related to x by α have dimension* $\mathtt{dim}_\alpha(x) - 1$.

When dealing with incidence graphs we hence use the notations defined on orders. An incidence graph or an order are *connected* if and only if any couple of cells can be joined by a sequence of cells related by $\theta^\square$.

We notice here that orders as well as incidence graphs are not able to represent cellular subdivisions with *multi-incidence.* A simple example is the torus made from a single square with opposite borders glued together. They indeed cannot provide information on how the different cells are glued together. They are for example not able to represent differently the two objects of Figs. 2.1 -(a) and 2.1 -(c).

2.2 n-Surfaces, Augmented Incidence Graphs and G-Maps

Orders and incidence graphs can represent a wide range of objects. We concentrate now on subclasses of these models that are used to represent restricted classes of objects. We aim at comparing these structures with n-dimensional generalized maps (n-G-maps) defined by Lienhardt [14–16].

We begin with a subclass of orders defined by Bertrand *et al.* which is close to the notion of manifold proposed by Kovalevsky [4].

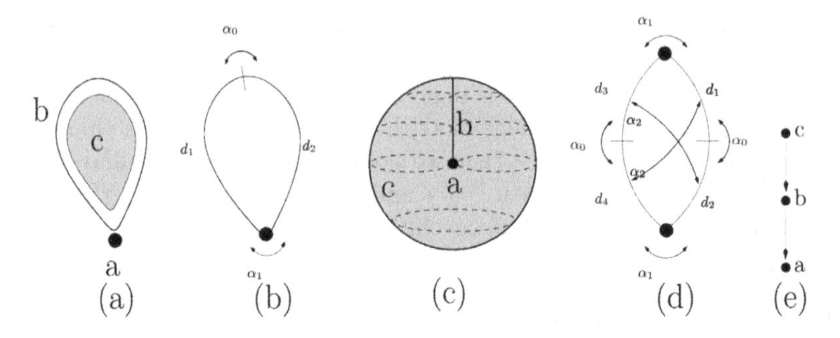

Fig. 1. Both objects (*a*) and (*c*) have the same cells and the same incidence relations, and hence the same incidence graph (*e*). However they are clearly different and are represented by two different 2-*G*-maps, respectively (*b*) and (*d*). The object (*c*) represents the minimal subdivision of the projective plane (see [2])

Definition 4. *Let* $|X| = (X, \alpha)$ *be a non-empty* CF*-order. The order* $|X|$ *is a* 0-surface *if* X *is composed exactly of two points* x *and* y *such that* $y \notin \alpha(x)$ *and* $x \notin \alpha(y)$*. The order* $|X|$ *is an* n-surface, $n > 0$, *if* $|X|$ *is connected and if, for each* $x \in X$*, the order* $|\theta^{\square}(x)|$ *is an* $(n-1)$*-surface.*

It can be recursively proved that Theorem 3 holds for any n-surface (see Fig. 2-a), which is then always isomorphic to its associated incidence graph.

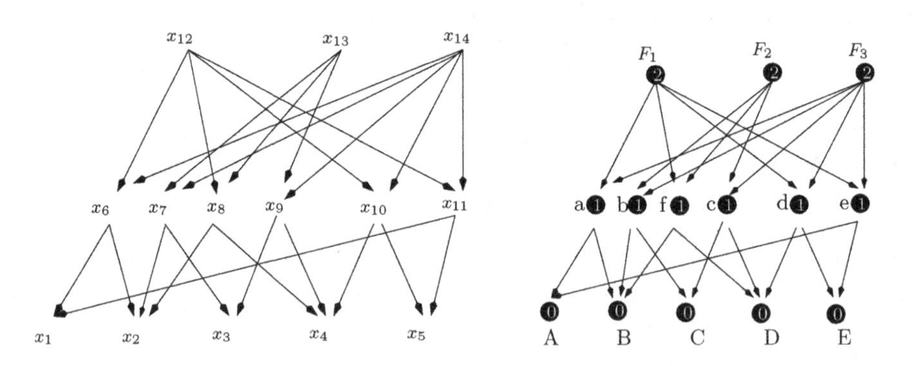

Fig. 2. Example of an order (Left) which is 2-surface and of an incidence graph (Right) which is augmented, and represents the subdivision of Fig. 4-a

We present now a subclass of incidence graphs defined by Brisson [6], to represent CW-complexes whose underlying spaces are d-manifolds. It will form the bridge between n-surfaces and n-G-maps.

Definition 5. *An extended incidence graph* IG^*_C *of dimension* n *is said to be an* augmented incidence graph *when it is connected and:*

1. *each i-cell of C belongs to at least one $(n+1)$-tuple of cells $(c^0, \cdots, c^n)$.*
2. $\forall (c^{i-1}, c^i, c^{i+1}) \in C^* \times C \times C^*$, $c^{i-1} \prec c^i \prec c^{i+1}$, $\exists ! c'^i \in C, c'^i \neq c^i$, $c^{i-1} \prec c'^i \prec c^{i+1}$. (c^{i-1}, c^i, c^{i+1}) *and* (c^{i-1}, c'^i, c^{i+1}) *are called* switch IG^*_C-triplets. *The operator* `switch` *is hence defined by* $\mathtt{switch}(c^{i-1}, c^i, c^{i+1}) = c'^i$. *The cell* c'^i *is then called the* (c^{i-1}, c^{i+1})-twin of c^i in IG^*_C.

The n-G-maps defined by Lienhardt are used to represent the topology of subdivisions of topological spaces. They can only represent quasi-manifolds (see [16]), orientable or not, with or without boundary.

Definition 6. *Let $n \geq 0$, an n-G-*map *is an $(n+2)$-tuple $G = (D, \alpha_0, \cdots, \alpha_n)$ such that:*

- *D is a finite set of darts*
- *α_i, $i \in \{0 \cdots, n\}$ are permutations on D such that:*
  - *$\forall i \in \{0, \cdots, n\}$, α_i is an involution*[2]*.*
  - *$\forall i$, j such that $0 \leq i < i+2 \leq j \leq n$, $\alpha_i \alpha_j$ is an involution (Commutativity property).*

The i-cells of the cellular subdivision represented by an n-G-map are determined by the orbits of the darts of D (see Figs. 4-b, c and d). In the following an orbit of a dart d is denoted by $<\alpha>$ (d) indexed by the indices of the involved permutations.

Definition 7. *Let $G = (D, \alpha_0, \cdots, \alpha_n)$ be an n-G-map. Each i-cell of the corresponding subdivision is given by $<\alpha>_{N-\{i\}}$ (d) where d is a dart incident to this i-cell.*

The set of i-cells is a partition of the darts of the n-G-map, for each i between 0 and n. The incidence relations between the cells is defined by: $c^j = <\alpha>_{N-\{j\}}$ (d) is a face of a cell $c^i = <\alpha>_{N-\{i\}}$ (d) when $j \leq i$ [15]. Two such cells are said to be consecutive when $j = i+1$.

We list below three classical notions related to n-G-maps.

1. *Closeness*: $\forall i \in N = \{0, \cdots, n\}$, α_i is without fixed point: $\forall d \in D, \alpha_i(d) \neq d$
2. *Without Multi-incidence*: $\forall d \in D, \bigcap_{i=0}^{i=n} <\alpha>_{N-\{i\}} (d) = \{d\}$
3. *Connectedness*: $\forall d \in D$, $<\alpha>_N (d) = D$

We note that a subdivision represented by a closed n-G-map has no boundary. A subdivision with multi-incidence is displayed in Fig. 2.1.

The associated incidence graph of an n-G-map is the extended incidence graph corresponding to the cellular subdivision it represents. There is an immediate link between the darts of an n-G-map and the maximal chains, called $(n+1)$ cell-tuples, of its associated incidence graph. A dart d actually defines a unique $(n+1)$ cell-tuple $(c^0, \cdots, c^n)$ with all cells having at least the dart d in

[2] A permutation π on the domain set D is an involution if and only if $\pi \circ \pi =$ *identity on D.*

common (see Definition 7). $(c^0, \cdots, c^n)$ is called the $(n+1)$ cell-tuple associated to d. The condition of non multi-incidence is needed to reciprocally associate a unique dart to each $(n+1)$ cell-tuple. There exists hence a bijection between the set of darts of an n-G-map without multi-incidence and the set of $(n+1)$ cell-tuples in the associated incidence graph. For instance, on Fig. 4-a, the dart 1 is uniquely associated to (A, a, F_1).

However, despite what has been written in [15], the property of non multi-incidence of a generalized map is not sufficient to guarantee that its associated incidence graph is augmented. A counterexample is given in Fig. 3. We introduce hence a more accurate subclass of n-G-maps.

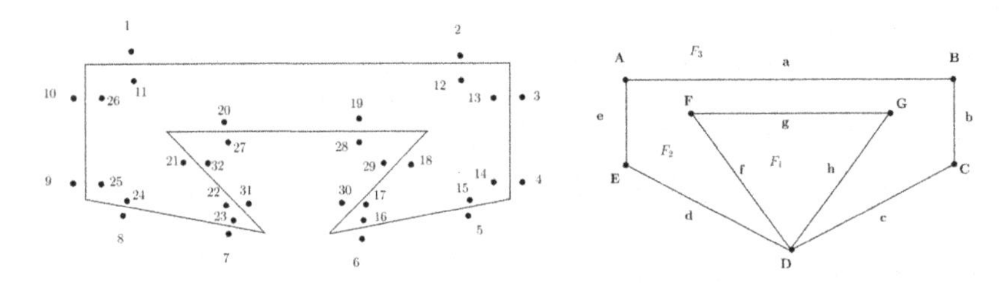

Fig. 3. Example of a closed n-G-map without multi-incidence (Left) and its associated space subdivision (Right) whose associated incidence graph is not augmented: there are four 1-cells (c,d,f,h) between D and F_2. F_1, F_2 and F_3 are respectively defined by the sets of darts $\{27, \ldots, 32\}$, $\{11, \cdots, 26\}$ and $\{1, \cdots, 10\}$

2.3 Regular n-G-Maps

The insufficiency of the non multi-incidence property comes from a subtler kind of multi-incidence. The classical non multi-incidence condition guarantees that there are no multi-incidence on the cellular subdivision associated to the n-G-map. As many other models, each n-G-map may be associated to a "set of simplices" but it has not to be a simplicial complex. It is namely a numbered simplicial set[3] (see Fig. 4-e) in which other kinds of multi-incidence may appear. Such a simplicial set is related to the barycentric triangulation of the corresponding cellular partition. The set of vertices is exactly the set of cells of the incidence graph, each vertex being labeled by the dimension of the corresponding cell. The classical non multi-incidence property implies that two different maximal simplices of the associated numbered simplicial set cannot have the same set of vertices. But it does not force lower dimensional simplices to fullfill the same requirement. We consider hence a restricted subclass of n-G-maps, the *regular n-G-maps*, which avoids more configurations of simplicial multi-incidence.

Definition 8. *A* regular n-G-map *is a connected closed n-G-map without multi-incidence with the additional property,* $\forall i \in \{1, \cdots, n-1\}$ *and* $\forall d \in D$*:*

[3] A numbered simplicial set is a simplicial set in which a positive integer is associated to each 0-simplex, see [16, 17].

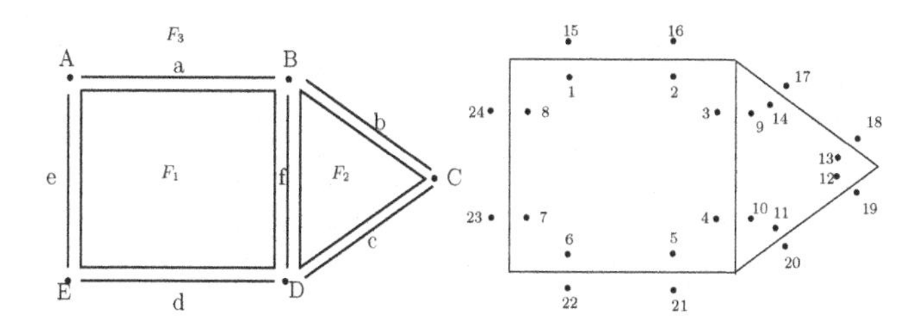

(a) Subdivision of $\mathbb{R}^2$ and its corresponding 2-G-map, $G = (D, \alpha_0, \alpha_1, \alpha_2)$, with $D = \{1, \cdots, 24\}$

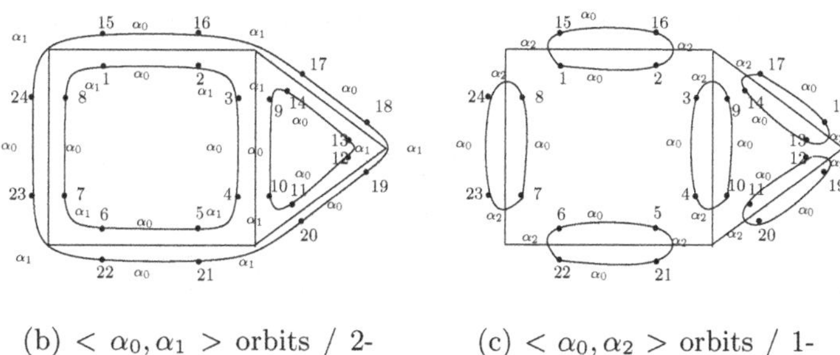

(b) $< \alpha_0, \alpha_1 >$ orbits / 2-cells on a 2-G-map
$< \alpha_0, \alpha_1 > (15) \Leftrightarrow F_3$
$< \alpha_0, \alpha_1 > (9) \Leftrightarrow F_2$
$< \alpha_0, \alpha_1 > (1) \Leftrightarrow F_1$

(c) $< \alpha_0, \alpha_2 >$ orbits / 1-cells on a 2-G-map
$<\alpha_0, \alpha_2> (1) \Leftrightarrow a,$
$<\alpha_0, \alpha_2> (13) \Leftrightarrow b$
$< \alpha_0, \alpha_2 > (11) \Leftrightarrow c,$
$< \alpha_0, \alpha_2 > (5) \Leftrightarrow d$
$< \alpha_0, \alpha_2 > (7) \Leftrightarrow e,$
$< \alpha_0, \alpha_2 > (3) \Leftrightarrow f$

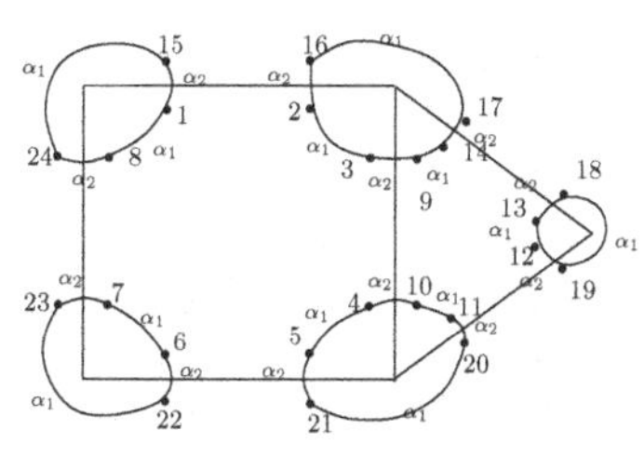

(d) $< \alpha_1 \alpha_2 >$ orbits / 0-cells on a 2-G-map
$< \alpha_1, \alpha_2 > (1) \Leftrightarrow A,$
$< \alpha_1, \alpha_2 > (2) \Leftrightarrow B$
$< \alpha_1, \alpha_2 > (12) \Leftrightarrow C,$
$< \alpha_1, \alpha_2 > (4) \Leftrightarrow D$
$< \alpha_1, \alpha_2 > (6) \Leftrightarrow E$

(e) numbered simplicial set associated to $G' = (D', \alpha_{0_{|D'}}, \alpha_{1_{|D'}}, \alpha_{2_{|D'}})$ with $D' = \{1, \cdots, 14\}$

Fig. 4. A 2-G-map with its associated cellular decomposition and the numbered simplicial set associated to one of its submaps

$$<\alpha>_{N-\{i-1\}}(d) \cap <\alpha>_{N-\{i+1\}}(d) = <\alpha>_{N-\{i-1,i+1\}}(d) \quad (simplicity)$$

The simplicity condition of such an n-G-map impose that the cells of the associated subdivision are more similar to topological disks. It implies that the numbered simplicial set must have a single edge between every two vertices when there is a difference of two between their associated numbers.

This limitation is not too restrictive because of the following property: any closed n-G-map may be refined into a regular n-G-map by appropriate barycentric subdivisions. There is indeed always possible to obtain a simplicial set without multi-incidence from any simplicial set by refining it [13]. Moreover as this process involves barycentric subdivisions [13], we are sure that the resulting simplicial set can be numbered [16]. We also note that the refinement process has not to be done on the whole map but only locally where some multi-incidence appears.

3 Equivalence of Regular n-G-Maps, Augmented Incidence Graph and n-Surfaces

We give first the main ideas and the organisation of the proof. We detail then the whole demonstration.

3.1 Guideline of the Proof

Incidence graphs are used as a bridge between regular n-G-maps and n-surfaces.

Generalized Map and Incidence Graph. An example of a regular n-G-map and an equivalent augmented incidence graph is given in Figs. 4 and 2-b.

We first prove that the incidence graph associated to any regular n-G-map is augmented. It already fullfills a part of the definition since each cell of such an incidence graph belongs to at least one $(n+1)$ cell-tuple. We must then show that the $(n+1)$ involutions of the map induce a switch operator on the incidence graph, which makes it augmented. These involutions are indeed involutions on the darts of the map and thus induce $(n+1)$ involutions on the $(n+1)$ cell-tuples of the associated incidence graph. The regularity of the map allows to prove that these involutions induce a switch property on the $(n+1)$ cell-tuples.

The converse has already partially been proved by Brisson [6]. We begin with proving that the switch operator on an augmented incidence graph of dimension n induces $n+1$ involutions without fixed point on the $(n+1)$ cell-tuples of this graph. We show then that they commute when there is a difference of two between their indices. The $(n+2)$-tuple made of the set containing all $(n+1)$ cell-tuples of the incidence graph and the $(n+1)$ involutions is hence a closed n-G-map without multi-incidence. The switch property allows then to prove that it also verifies the simplicity property.

Incidence Graph and n-Surface. An example of an augmented incidence graph and an equivalent n-surface is displayed on Fig. 2.

The proof is made with an induction over the dimension n. The equivalence is clear for $n = 0$. For $n > 0$, we prove that each subgraph built on the strict θ-adherence of any element of an augmented incidence graph is itself an augmented incidence graph. We also show that an extended incidence graph which is locally everywhere an augmented incidence graph is globally an augmented incidence graph. This means that an n-dimensional augmented incidence graph, $n > 0$, can be recursively defined. It is simply an extended incidence graph such that each subgraph built on the strict θ-adherence of any of its elements is an $(n-1)$-dimensional augmented incidence graph. Now n-dimensional augmented incidence graphs and n-surfaces are equivalent for $n = 0$. Given that they are built with the same recurrence for all $n > 0$, they are hence equivalent for all n.

Organisation of the Proof.

regular n-G-map	$nGIG$-conversion (Theorem 12 p. 132) $\Longrightarrow$ $IGnG$-conversion (Theorem 14 p. 133) $\Longleftarrow$	augmented incidence graph	$IGnS$-conversion (Theorem 18 p. 134) $\Longrightarrow$ $nSIG$-conversion (Theorem 18 p. 134) $\Longleftarrow$	n-surface

3.2 Proof

We first prove the equivalence between regular n-G-maps and augmented incidence graph. We show then that augmented incidence graph and n-surfaces are equivalent. Finally we deduce the link between regular n-G-maps and n-surfaces.

Equivalence Between Regular n-G-Maps and Augmented Incidence Graphs. We first show how to define an $nGIG$-conversion which builds an augmented incidence graph from a regular n-G-map. We then define the $IGnG$-conversion which is the inverse of $nGIG$-conversion up to isomorphism.

$nGIG$-Conversion.
As previously said, there is an n-dimensional incidence graph associated with any regular n-G-map. We are going to prove that this incidence graph is augmented. We need first to state some properties of particular orbits of n-G-maps. The first two lemmas have an interesting interpretation on the numbered simplicial set associated to the n-G-map (Fig. 4-e). The last one is better related to the cellular subdivision. The proofs of the three following lemmas can be found in [2].

The first lemma states that, for any n-G-map without multi-incidence, three 0-simplices with consecutive numbers belong to exactly one 2-simplex.

Lemma 9. *Let $G = (D, \alpha_0, \cdots, \alpha_n)$ be a closed n-G-map without multi-incidence. Let d be any dart of D and $i \in \{1, \cdots, n-1\}$,*

$$<\alpha>_{N-\{i-1\}}(d) \cap <\alpha>_{N-\{i\}}(d) \cap <\alpha>_{N-\{i+1\}}(d) = <\alpha>_{N-\{i-1,i,i+1\}}(d)$$

This second lemma says that, for any n-G-map, a 1-simplex between two 0-simplices numbered $i-1$ and $i+1$ belongs to at most two 2-simplexes.

Lemma 10. *Let $G = (D, \alpha_0, \cdots, \alpha_n)$ be an n-G-map. Let d be any dart of D and $i \in \{1, \cdots, n-1\}$,*

$$<\alpha>_{N-\{i-1,i+1\}}(d) = <\alpha>_{N-\{i-1,i,i+1\}}(d) \cup <\alpha>_{N-\{i-1,i,i+1\}}(d\alpha_i)$$

This third lemma states that every 1-cell of the cellular subdivision associated to an n-G-map has at most two 0-faces and that any $(n-1)$-cell is face of at most two n-cells.

Lemma 11. *Let $G = (D, \alpha_0, \cdots, \alpha_n)$ be an n-G-map and d,d' two darts of D.*

1. $<\alpha>_{N-\{1\}}(d) = (<\alpha>_{N-\{1\}}(d) \cap <\alpha>_{N-\{0\}}(d)) \cup (<\alpha>_{N-\{1\}}(d) \cap <\alpha>_{N-\{0\}}(d\alpha_0))$
2. $<\alpha>_{N-\{n-1\}}(d) = (<\alpha>_{N-\{n-1\}}(d) \cap <\alpha>_{N-\{n\}}(d)) \cup (<\alpha>_{N-\{n-1\}}(d) \cap <\alpha>_{N-\{n\}}(d\alpha_n))$

Theorem 12. *Let $G = (D, \alpha_0, \cdots, \alpha_n)$ be a regular n-G-map. Its associated incidence graph is then augmented. The construction of an augmented incidence graph from a regular n-G-map is called an $nGIG-conversion$.*

Proof. We must prove that the incidence graph associated to the n-G-map has the property needed to build a switch operator. This property can be equivalently expressed on the $(n+1)$ cell-tuples of the graph with the two additional fictive cells c^{-1} and c^{n+1} [6]. For all couple of cells (c^{i-1}, c^{i+1}), there must exist exactly two different cells c^i and c'^i such that all $(n+1)$ cell-tuples containing c^{i-1} and c^{i+1} contains either c^i or c'^i. Moreover since there exists a bijection between the set of darts of a closed n-G-map without multi-incidence and the set of $(n+1)$ cell-tuples of the associated incidence graph, we can equivalently achieve the demonstration with darts or cell-tuples.

Given two cells c^{i-1} and c^{i+1}, let us choose one of the $(n+1)$ cell-tuples containing them and let d be its associated dart. By Definition 7, the dart $d\alpha_i$ also corresponds to an $(n+1)$ cell-tuple containing c^{i-1} and c^{i+1}. But as the map is closed and without multi-incidence, the i-cell associated to d, $<\alpha>_{N-\{i\}}(d)$, is different from the i-cell associated to $d\alpha_i$, $<\alpha>_{N-\{i\}}(d\alpha_i)$. We have then at least two distinct i-cells between c^{i-1} and c^{i+1}. We must prove that there is no other. We translate this condition in terms of orbits of darts.

- If $i \in \{1, \cdots, n-1\}$, $\forall d' \in D$ such that $d' \in <\alpha>_{N-\{i-1\}}(d)$ and $d' \in <\alpha>_{N-\{i+1\}}(d) \Rightarrow$ either $d' \in <\alpha>_{N-\{i\}}(d)$ or $d' \in <\alpha>_{N-\{i\}}(d\alpha_i)$
- $\forall d' \in D$, $d' \in <\alpha>_{N-\{1\}}(d) \Rightarrow d' \in <\alpha>_{N-\{0\}}(d)$ or $d' \in <\alpha>_{N-\{0\}}(d\alpha_0)$,
- $\forall d' \in D$, $d' \in <\alpha>_{N-\{n-1\}}(d) \Rightarrow d' \in <\alpha>_{N-\{n\}}(d)$ or $d' \in <\alpha>_{N-\{n\}}(d\alpha_n)$

The last two points comes directly from Lemma 11. We prove the first point.

$$d' \in <\alpha>_{N-\{i-1\}}(d) \cap <\alpha>_{N-\{i+1\}}(d)$$

$$\overset{simplicity}{\Longleftrightarrow} \quad d' \in <\alpha>_{N-\{i-1,i+1\}}(d)$$

$$\overset{Lemma\ 10}{\Longleftrightarrow} \quad d' \in <\alpha>_{N-\{i-1,i,i+1\}}(d) \cup <\alpha>_{N-\{i-1,i,i+1\}}(d\alpha_i)$$

$$\overset{Lemma\ 9}{\Longleftrightarrow} \quad d' \in (<\alpha>_{N-\{i-1\}}(d) \cap <\alpha>_{N-\{i\}}(d) \cap <\alpha>_{N-\{i+1\}}(d)) \cup (<\alpha>_{N-\{i-1\}}(d\alpha_i) \cap <\alpha>_{N-\{i\}}(d\alpha_i) \cap <\alpha>_{N-\{i+1\}}(d\alpha_i))$$

Otherwise said $d' \in <\alpha>_{N-\{i\}}(d)$ or $d' \in <\alpha>_{N-\{i\}}(d\alpha_i)$. □

IGnG-**Conversion.**
We here show how to build an n-G-map from an augmented incidence graph. The first lemma says that the operator `switch` induces $(n+1)$ involutions on the set of $(n+1)$ cell-tuples of the incidence graph.

Lemma 13. *Let IG_C^* be an augmented incidence graph. Its switch operator induces $(n+1)$ involutions without fixed point α_i, $i \in \{0, \cdots, n\}$ on the set of the $(n+1)$ cell-tuples of IG_C^*, $(c^0, \cdots, c^i, \cdots, c^n)$, defined by:*

$$\alpha_i((c^0, \cdots, c^{i-1}, c^i, c^{i+1}, \cdots, c^n)) = (c^0, \cdots, c^{i-1}, c'^i, c^{i+1}, \cdots, c^n)$$
$$\text{where } c'^i = \texttt{switch}(c^{i-1}, c^i, c^{i+1})$$

Theorem 14. *Let IG_C^* be an augmented incidence graph. Let us define*
- $D = \{(c^0_{\beta_0}, \cdots, c^n_{\beta_n}), c^{-1} \prec c^0_{\beta_0} \prec c^1_{\beta_1} \prec \cdots \prec c^n_{\beta_n} \prec c^{n+1}\}$
- α_i, $i \in \{0, \cdots, n\}$ *such that*

$$(c^0_{\beta_0}, \cdots, c^{i-1}_{\beta_{i-1}}, c^i_{\beta_i}, c^{i+1}_{\beta_{i+1}}, \cdots, c^n_{\beta_n}) \overset{\alpha_i}{\mapsto} (c^0_{\beta_0}, \cdots, c^{i-1}_{\beta_{i-1}}, c^i_{\beta'_i}, c^{i+1}_{\beta_{i+1}}, \cdots, c^n_{\beta_n})$$

with $c^i_{\beta'_i} = \texttt{switch}(c^{i-1}_{\beta_{i-1}}, c^i_{\beta_i}, c^{i+1}_{\beta_{i+1}})$

Then $(D, \alpha_0, \cdots, \alpha_n)$ is a regular n-G-map.
This process is called a IGnG-conversion

Proof. The proof is decomposed in four parts. The *closeness*, *commutativity* and *without multi-incidence* properties have already been proved by Brisson [6] and may also be found in [2]. We just prove here the *simplicity* property. The switch property and the definition of α_i guarantees that if $d' \in <\alpha>_{N-\{i-1\}}(d) \cap <\alpha>_{N-\{i+1\}}(d)$ then either $d' \in <\alpha>_{N-\{i\}}(d)$ or $d' \in <\alpha>_{N-\{i\}}(d\alpha_i)$. Otherwise said,

$$<\alpha>_{N-\{i-1\}}(d) \cap <\alpha>_{N-\{i+1\}}(d)$$

$$\overset{\text{(by Switch prop.)}}{=} \quad (<\alpha>_{N-\{i-1\}}(d) \cap <\alpha>_{N-\{i+1\}}(d) \cap <\alpha>_{N-\{i\}}(d)) \cup (<\alpha>_{N-\{i-1\}}(d) \cap <\alpha>_{N-\{i+1\}}(d) \cap <\alpha>_{N-\{i\}}(d\alpha_i)))$$

$$\overset{\text{(by Lemma 9)}}{=} \quad <\alpha>_{N-\{i-1,i,i+1\}}(d) \cup <\alpha>_{N-\{i-1,i,i+1\}}(d\alpha_i)$$

$$\overset{\text{(by Lemma 10)}}{=} \quad <\alpha>_{N-\{i-1,i+1\}}(d)$$

□

Equivalence Between Augmented Incidence Graphs and n-Surfaces. We state below two lemmas which together provide a recursive definition of n-dimensional augmented incidence graphs. Their proofs can be found in [2].

The first lemma expresses that given an augmented incidence graph IG_C^* all subgraphs of the form $\theta^\square(c)$ with $c \in C$ are augmented incidence graphs too.

Lemma 15. *Let IG_C^* be an augmented incidence graph of dimension $n \geq 1$, then $\forall c \in C, \theta^\square(c)$ is an augmented incidence graph of dimension $n-1$.*

The next lemma shows that an extended incidence graph IG_C^* with dimension at least 1, which is locally everywhere an augmented indicence graph, is also itself an augmented incidence graph.

Lemma 16. *Let IG_C^* be an extended incidence graph of dimension at least 1, such that $\forall c \in C$, $\theta^\square(c)$ is an augmented incidence graph then IG_C^* is also an augmented incidence graph.*

These two lemmas lead to the following theorem which gives a recursive characterization of augmented incidence graphs of dimension n.

Theorem 17. *Let $IG_C^* = (C^*, \prec)$ be an extended incidence graph of dimension n, the two following propositions are equivalent:*

1. *IG_C^* is a non empty augmented incidence graph*
2. *IG_C^* is such that:*
   - *if $n = 0$, C contains exactly two 0-cells c^0 and c'^0, such that c^0 and c'^0 are (c^{-1}, c^1)-twins in IG_C^*.*
   - *if $n > 0$, C is such that for all $c \in C$, $\theta^\square(c)$ is an augmented incidence graph of dimension $n-1$.*

Proof. We are going to prove that (1) $\Leftrightarrow$ (2) for all n. The proof is quite immediate for $n = 0$ [2] (Both models consist in two disconnected points). We show it for $n > 0$:

- $\Rightarrow$ Let IG_C^* be an augmented incidence graph of dimension n. $\forall c \in C_n$, $\theta^\square(c)$ is by Lemma 15 an augmented incidence graph of dimension $n-1$.
- $\Leftarrow$ Let IG_C^* be an extended incidence graph of dimension n fullfilling the conditions of (2). For all $c \in C_n$, $\theta^\square(c)$ is an $(n-1)$-dimensional augmented incidence graph. IG_C^* has dimension strictly greater than 0. It is then by Lemma 16 an augmented incidence graph.

□

This recursive characterization identical to the definition of n-surfaces leads immediately to the following theorem:

Theorem 18. *Let $IG_C = (C, \prec)$ be an incidence graph and $|X| = (X, \alpha)$ an order*

1. *$IG_C^* = (C \cup \{c^{-1}, c^{n+1}\}, \prec)$ is augmented $\Rightarrow$ its associated order is an n-surface*
2. *$|X| = (X, \alpha)$ is an n-surface $\Rightarrow$ its associated incidence graph is augmented*

Equivalence Between Regular n-G-Maps and n-Surfaces. The two preceding results leads to the following equivalence between regular n-G-maps and n-surfaces.

Theorem 19. *Let $G = (D, \alpha_0, \cdots, \alpha_n)$ be an n-G-map and $|X| = (X, \alpha)$ an order such that there exists an isomorphism between their associated incidence graphs. Then the following propositions are equivalent:*

1. *G is a regular n-G-map*
2. *$|X|$ is an n-surface*

If C_G is the set of cells of the subdivision represented by a regular n-G-map G and $\leq_G$ the incidence relation between these cells then $(C_G, \leq_G)$ is the n-surface associated to G by $nGnS$-conversion where the dimension of the cells of C_G are forgotten. Reciprocally, if $D_{|X|}$ is the set of $(n+1)$ α-chains of an n-surface $|X|$, then $(D_{|X|}, \alpha_0, \cdots, \alpha_n)$ is the regular n-G-map associated to $|X|$ by $nSnG$-conversion, where for each $d = (x^0, \cdots, x^{i-1}, x^i, x^{i+1}, \cdots, x^n) \in D_{|X|}$, $d\alpha_i = (x^0, \cdots, x^{i-1}, x'^i, x^{i+1}, \cdots, x^n)$ with $x'^i = (\alpha^\square(x^{i+1}) \cap \beta^\square(x^{i-1})) \backslash \{x^i\}$ if $i \in \{1, \cdots, n-1\}$, $x'^0 = \alpha^\square(x^1) \backslash \{x^0\}$ and $x'^n = \beta^\square(x^{n-1}) \backslash \{x^n\}$.

Theorem 20. *Let $G = (D, \alpha_0, \cdots, \alpha_n)$ be an n-G-map and $|X| = (X, \alpha)$ an order:*

1. *G is a regular n-G-map $\Rightarrow$ its associated order is an n-surface*
2. *$|X|$ is an n-surface $\Rightarrow$ its associated n-G-map is regular*

With the previous construction processes, any n-surface may be built from some regular n-G-map, and any regular n-G-map may be built from some n-surface. We also prove, in [2], that these conversions are inverse to each other up to isomorphism which prove the equivalence of both structures.

4 Conclusion

We have shown that two topological models namely regular n-G-maps and n-surfaces are equivalent structures. This result is important because these models come from various research fields, and are defined very differently. Moreover we have given an explicit way to switch from one representation to another. The equivalence between both models gives us more information on them. It implies for example that the neighbourhood of any cell of a regular n-G-map is a generalized map too.

Future works will be lead into three main directions. It will first be interesting to take advantage of this equivalence by transfering tools, namely operators and properties from one to another or by integrating them in a chain of operations. Besides such models can only represent quasi-manifolds, it would be useful to go on with more general structures such as chains of maps [11] which represent more general subdivisions that are not necessarily quasi-manifolds but such that each cell is a quasi-manifold. We could also focus on subclasses of these models. Finally it could be useful to study more precisely the class of regular n-G-maps, we have introduced here.

References

1. Alayrangues, S., Lachaud, J.-O.: Equivalence Between Order and Cell Complex Representations, Proc. Computer Vision Winter Workshop (CVWW02).
2. Alayrangues, S., Daragon, X., Lachaud, J.-O., Lienhardt, P,: Equivalence between Regular n-G-maps and n-surfaces, Research Report. `http://www.labri.fr/Labri/Publications/Publis-fr.htm`
3. Bertrand, G.: New Notions for Discrete Geometry, Proc. of 8th Discrete Geometry for Computer Imagery (DGCI'99).
4. Bertrand, G.: A Model for Digital Topology, Proc. of 8th Discrete Geometry for Computer Imagery (DGCI'99),
5. Björner, A.: Topological methods, MIT Press, Handbook of combinatorics (vol. 2), 1995.
6. Brisson, E.: Representing Geometric Structures in d Dimensions: Topology and Order, Proceedings of the Fifth Annual Symposium on Computational Geometry, 1989.
7. Brun, L., Kropatsch, W.: Contraction Kernels and Combinatorial Maps, 3rd IAPR-TC15 Workshop on Graph-based Representations in Pattern Recognition, 2001.
8. Daragon, X., Couprie, M., Bertrand, G.: New "marching-cubes-like" algorithm for Alexandroff-Khalimsky spaces, Proc. of SPIE: Vision Geometry XI, 2002.
9. Daragon, X., Couprie, M., Bertrand, G.: Discrete Frontiers, Discrete Geometry for Computer Imagery, Lecture Notes in Computer Science, 2003.
10. Edelsbrunner, H.: Algorithms in combinatorial geometry, Springer-Verlag New York, Inc, 1987.
11. Elter, H.: Etude de structures combinatoires pour la reprsentation de complexes cellulaires, Universit Louis Pasteur, Strasbourg, France, 1994?
12. Evako, A.V., Kopperman R., Mukhin, Y. V.: Dimensional properties of graphs and digital spaces, Journal of Mathematical Imaging and Vision, 1996.
13. Hatcher, A.: Algebraic Topology Cambridge University Press, 2002
14. Lienhardt, P.: Subdivisions of n-dimensional spaces and n-dimensional generalized maps, Proc. 5 th Annual ACM Symp. on Computational Geometry, 1989.
15. Lienhardt, P.: Topological models for boundary representation: a comparison with n-dimensional generalized maps, Computer-Aided Design, 1991.
16. Lienhardt, P.: N-dimensional generalized combinatorial maps and cellular quasi-manifolds, International Journal of Computational Geometry and Applications, 1994.
17. May, P.: Simplicial objects in algebraic topology, von Nostrand, 1967.

$\mathbb{Z}$-Tilings of Polyominoes and Standard Basis

Olivier Bodini[1] and Bertrand Nouvel[2]

[1] LIRMM, 161, rue ADA,
34392 Montpellier Cedex 5, France
[2] LIP, UMR 5668 CNRS-INRIA-ENS Lyon-Univ. Lyon 1,
46 Allée d'Italie, 69364 Lyon Cedex 07, France

Abstract. In this paper, we prove that for every set E of *polyominoes* (for us, a polyomino is a finite union of unit squares of a square lattice), we have an algorithm which decides in polynomial time, for every polyomino P, whether P has or not a $\mathbb{Z}$-tiling (*signed tiling*) by translated copies of elements of E. Moreover, if P is $\mathbb{Z}$-tilable, we can build a $\mathbb{Z}$-tiling of P. We use for this the theory of standard basis on $\mathbb{Z}[X_1, ..., X_n]$. In application, we algorithmically extend results of Conway and Lagarias on $\mathbb{Z}$-tiling problems.

1 Introduction

A *cell* $c(i,j)$ in the square lattice $\mathbb{Z}^2$ denotes the set:

$$c(i,j) := \{(x,y)\,; i \leq x < i+1, j \leq y < j+1\}.$$

So, cells are labelled by their lower left corner and the set of cells S can easily be identified with $\mathbb{Z}^2$. For us, a *polyomino* is a finite -not necessary connected- union of cells. In this paper, we are interested in the study of a variant of the problem of tiling, called the $\mathbb{Z}$*-tiling problem*. Precisely, let P be a polyomino and E a set of polyominoes (*the tiles*); a $\mathbb{Z}$*-tiling of P by E* consists of a finite number of translated tiles placed in the lattice (possibly with overlaps), with each tile be assigned a sign of +1 or -1, such that, for each cell $c(i,j)$ in $\mathbb{Z}^2$, the sum of the signs of the tiles covering $c(i,j)$ is +1 if $c(i,j) \in P$ and 0 if $c(i,j) \notin P$ (Fig. 1). Obviously, a polyomino which is tilable by a set of tiles is also $\mathbb{Z}$-tilable by this set. Consequently, the condition of $\mathbb{Z}$-tilability gives important (and not trivial) necessary conditions of tilability. J.H. Conway and J.C. Lagarias [4] have previously studied this notion. They particularly obtained the following necessary and sufficient condition for the special case where the polyominoes are simply connected:

A simply connected polyomino P has a $\mathbb{Z}$-tiling by a set of simply connected polyominoes E if and only if the combinatorial boundary $[\partial P]$ is included in the tile boundary group $\mathbf{B}(E)$. For these definitions, we can refer to the paper of J.H. Conway and J.C. Lagarias [4].

Nevertheless, this group theoretic theorem presents some drawbacks. Firstly, it only applies to simply connected polyominoes. Secondly, this criterion seems

R. Klette and J. Žunić (Eds.): IWCIA 2004, LNCS 3322, pp. 137–150, 2004.

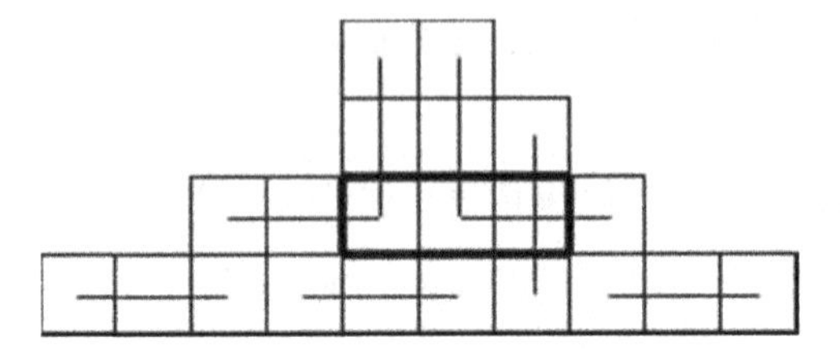

Fig. 1. A (classical) polyomino P which is $\mathbb{Z}$-tilable by bars of length 3. In bold, a negative copy of a bar. The segments indicate positive copies of bars. These positive and negative bars constitute a $\mathbb{Z}$-tiling of P

in general not easier to verify than to solve the original problem. Thirdly, it seems to be impossible to extend theoretic group arguments in higher dimensions. This third open problem has already been mentioned by R. Kenyon in his habilitation [9].

In this paper, we propose another way of solving the problem. We associate for each polyomino P a polynomial in $\mathbb{Z}[X_1, X_2, Y_1, Y_2]$, called *P-polynomial.* We denote it by Q_P. We prove that, given a set E of polyominoes, a polyomino P is $\mathbb{Z}$-tilable by E if and only if $Q_P \in \langle Q_{P'}\ with\ P' \in E, X_1Y_1 - 1, X_2Y_2 - 1\rangle_{\mathbb{Z}}$ (i.e. the ideal of $\mathbb{Z}[X_1, ..., X_n]$ generated by the polynomials $Q_{P'}\ where\ P' \in E$ and by $X_1Y_1 - 1, X_2Y_2 - 1$). This new formulation allows us to use commutative algebraic tools like the standard basis algorithm to solve the problem and we are no more limited to simply connected objects in dimension 2. The reader can find a good introduction to standard basis (for ideals in $\mathbb{K}[X_1, ..., X_n]$ where $\mathbb{K}$ is a field) in [3] or [5]. The algorithm for standard basis is double exponential in the worst case. Nevertheless, for any set of tiles S, we can precompute this basis. After that, for any polyomino we have a polynomial algorithm (division type algorithm) to decide whether P is $\mathbb{Z}$-tilable by S. This leads us to consider coloring arguments. These tools are frequently found in the literature ([7],[8],[10]). This notion gives in general important necessary conditions of tilability. For instance, if a polyomino P is tilable by dominoes, it is easy to prove that P needs to have the same number of black and white squares in a classical chessboard-like coloration. We define in this paper the *generalized coloring* associated to a set of tiles E. This generalized coloring contains all the generalized coloring arguments defined by Conway-Lagarias [4]. Moreover, the generalized coloring of a polyomino P is null if and only if P is $\mathbb{Z}$-tilable by the set E. Finally, we prove that it is possible to determine the generalized coloring of a classical polyomino when we only know the colors of the squares which are adjacent to the boundary of P.

So, if a polyomino P is in a sense "big" (the ratio between perimeter and volume is less than 1/2), then we have a better algorithm to determine the tilability of P.

Now, we are going to introduce the abstract notions that constitute the general framework of this paper. Let us recall that S is the set of all the cells in $\mathbb{Z}^2$, a *$\mathbb{Z}$-weighted polyomino* or simply a *$\mathbb{Z}$-polyomino* is a map P from $S \simeq \mathbb{Z}^2$

into $\mathbb{Z}$ with a finite support. For each cell c, we call *weight of* P in c the number $P(c)$. The space $\mathbb{P}_{\mathbb{Z}}$ of $\mathbb{Z}$-weighted polyominoes has a natural structure of free $\mathbb{Z}$-module. Indeed, the cells of weight 1 constitute a basis of $\mathbb{P}_{\mathbb{Z}}$. We can canonically embed the set of "classical" polyominoes in the $\mathbb{Z}$-module of the $\mathbb{Z}$-polyominoes (assigning 1 to cells covered by the polyomino and 0 to the other cells). We say that a $\mathbb{Z}$-polyomino P is $\mathbb{Z}$-*tilable* by a set of $\mathbb{Z}$-weighted tiles if and only if P is a $\mathbb{Z}$-linear combination of translated elements of this set (the translated of a $\mathbb{Z}$-polyomino $P(.)$ by a vector $v \in \mathbb{Z}^2$ being the $\mathbb{Z}$-polyomino $P(. - v)$).

2 P-Polynomials and Z-Tilings

The notion defined below can easily be extended in a more general framework. For instance, we have similar definitions for $\mathbb{Z}$-polycubes (in this case, the cells are the unit cubes of the regular cubical lattice of $\mathbb{R}^d$) or for $\mathbb{Z}$-polyamands (the cells are the triangles of the regular triangular lattice) or $\mathbb{Z}$-polyhexes (sums of weighted hexagons of the regular hexagonal lattice). In this section, in order to simplify, we only deal with $\mathbb{Z}$-polyominoes and $\mathbb{Z}$-polyhexes. The reader can find in [2] a more general presentation.

Firstly, for $a \in \mathbb{Z}^2$, we define $\mathcal{X}^a$ as:

$$\mathcal{X}^a = X_1^{\frac{a_1+|a_1|}{2}} X_2^{\frac{a_2+|a_2|}{2}} Y_1^{\frac{|a_1|-a_1}{2}} Y_2^{\frac{|a_2|-a_2}{2}}.$$

We encode the square of the plane with 4 parameters (cardinal points) to avoid to work with Laurent polynomials. Let us recall that we denote by $\langle P_1, ..., P_k \rangle_{\mathbb{Z}}$ the ideal of $\mathbb{Z}[X_1, ..., X_n]$ generated by the polynomials $P_1, ..., P_k$. For each $\mathbb{Z}$-polyomino P, we can define its *P-polynomial*

$$Q_P \doteq \sum_{a \in \mathbb{Z}^2} P(c(a))\, \mathcal{X}^a.$$

Lemma 1. *The space* $\mathbb{P}_{\mathbb{Z}}$ *is isomorphic to*

$$\mathbb{Z}[X_1, X_2, Y_1, Y_2] / \langle (X_1Y_1 - 1), (X_2Y_2 - 1) \rangle_{\mathbb{Z}}.$$

Proof. There exists a unique linear map f from $\mathbb{Z}[X_1, X_2, Y_1, Y_2]$ to $\mathbb{P}_{\mathbb{Z}}$ such that $f\left(X_1^{a_1} Y_1^{b_1} X_2^{a_2} Y_2^{b_2}\right)$ is the cell $(a_1 - b_1, a_2 - b_2)$ with weight 1. Now, we must prove that $\ker(f) = \langle (X_1Y_1 - 1), (X_2Y_2 - 1) \rangle_{\mathbb{Z}}$. We proceed by successive divisions by $(X_1Y_1 - 1)$ and $(X_2Y_2 - 1)$ in the successive rings

$$\mathbb{Z}[X_2, Y_1, Y_2][X_1] \quad \text{and} \quad \mathbb{Z}[X_1, Y_1, Y_2][X_2].$$

So, we can write every polynomial Q as follows $Q = R + \sum_{i=1}^{2} Q_i (X_iY_i - 1)$ with R containing only monomials of the form $\mathcal{X}^a$ where $a \in \mathbb{Z}^2$ (i.e. without simultaneously X_i and Y_i). We have the following equivalence: $f(Q)$ is the empty polyomino, denoted by 0 (i.e. the polyomino P with weight 0 on each cell), if and only if $f(R) = 0$ (indeed, $f(Q_i(X_iY_i - 1)) = 0$ because $f(\mathcal{X}^a X_i Y_i) = f(\mathcal{X}^a)$). Moreover, as $\{f(\mathcal{X}^a)\}$ is a basis of the $\mathbb{Z}$-module $\mathbb{P}_\mathbb{Z}$, it is clear that $f(R) = 0 \Leftrightarrow R = 0$ and that $R = 0 \Leftrightarrow Q \in \langle (X_1Y_1 - 1), (X_2Y_2 - 1)\rangle_\mathbb{Z}$. So, $\ker(f) = \langle (X_1Y_1 - 1), (X_2Y_2 - 1)\rangle_\mathbb{Z}$.

Definition 1. *The ideal $Q_P \in \langle Q_{P'}$ with $P' \in E, X_1Y_1 - 1, X_2Y_2 - 1\rangle_\mathbb{Z}$ is called the* tiling ideal *of E. We denote it by $I(E)$.*

Theorem 1. *Let E be a set of $\mathbb{Z}$-polyominoes. A $\mathbb{Z}$-polyomino P is $\mathbb{Z}$-tilable by E if and only if Q_P belongs to the tiling ideal of E.*

Proof. By definition, a $\mathbb{Z}$-polyomino P is $\mathbb{Z}$-tilable by E if and only if

$$P = \sum_{i=1}^{t} \lambda_i P_i(. - a_i)$$

where $\lambda_i \in \mathbb{Z}$, $P_i \in E$ and $a_i \in \mathbb{Z}^2$. Moreover, in

$$\mathbb{Z}[X_1, X_2, Y_1, Y_2]/\langle (X_1Y_1 - 1), (X_2Y_2 - 1)\rangle_\mathbb{Z}$$

we have $Q_{P_i(.-a_i)} = \mathcal{X}^{a_i} Q_{P_i}$ and consequently, $Q_P = \sum_{i=1}^{t} \lambda_i \mathcal{X}^{a_i} Q_{P_i}$. Thus, $Q_P \in I(E)$. Conversely, if $Q_P \in I(E)$, then $Q_P = \sum_{P' \in E} (\sum \lambda_i \mathcal{X}^{a_i}) Q_{P'}$. Indeed, Q_P does not contain monomials where occur simultaneously X_i and Y_i. Now, this implies that $P = \sum_{P' \in E} (\sum \lambda_i P'(. - a_i))$. Finally, P is $\mathbb{Z}$-tilable by E if and only if $Q_P \in I(E)$.

We have the same statement for polyhexes. Indeed, for the hexagonal lattice built in gluing translated copies of the hexagonal convex hull of the points $(0,0), (0,1), \left(\frac{-1}{2}, \frac{\sqrt{3}}{2}\right), \left(\frac{3}{2}, \frac{\sqrt{3}}{2}\right), (0, \sqrt{3}), (1, \sqrt{3})$, we denote by $[a_1, a_2]$ the hexagonal cell whose lower left corner is $\left(\frac{3}{2}(a_1 + a_2), \frac{\sqrt{3}}{2}(-a_1 + a_2)\right)$ where $(a_1, a_2) \in \mathbb{Z}^2$ (Fig.2). For each $\mathbb{Z}$-polyhexe P, we can define the *P-polynomial*

$$Q_P = \sum_{(a_1,a_2) \in \mathbb{Z}^2} P([a_1, a_2]) X^{(a_1,a_2)}.$$

Then, we have the following analogous theorem:

Theorem 2. *Let E be a set of $\mathbb{Z}$-polyhexes. A $\mathbb{Z}$-polyhexe P is $\mathbb{Z}$-tilable by E if and only if*

$$Q_P \in \langle Q_{P'} \textit{ with } P' \in E, X_1Y_1 - 1, X_2Y_2 - 1\rangle_\mathbb{Z}.$$

3 Standard Basis on $\mathbf{Z}[X_1, ..., X_n]$

In this section, we indicate briefly how to solve the problem of membership in an ideal of $\mathbb{Z}[X_1, ..., X_n]$. In fact, we use a non-trivial extended version to

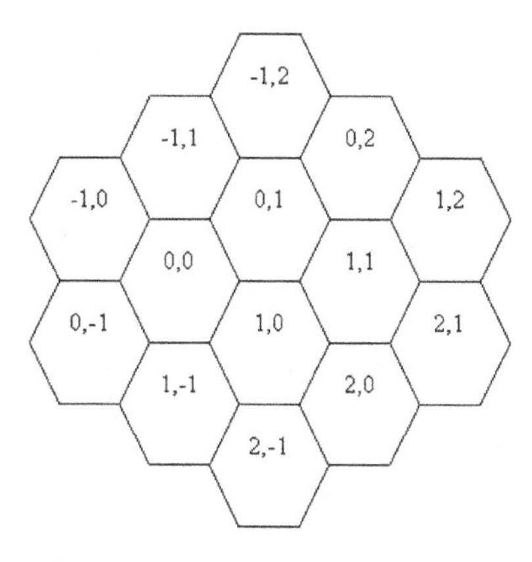

Fig. 2. The hexagonal lattice with its coordinates

$\mathbb{Z}[X_1, ..., X_n]$ of the Buchberger algorithm [2]. The original one only works for an ideal of $\mathbb{K}[X_1, ..., X_n]$ where $\mathbb{K}$ is a field and can be found in [5], [3]. First of all, we have to define a total order on the monomials of $\mathbb{Z}[X_1, ..., X_n]$. Let $\leq^*$ be the lexicographic order on the n-tuples and let $\alpha = (\alpha_1, ..., \alpha_n)$ be in $\mathbb{N}^n$; we denote by X^α the monomial $X_1^{\alpha_1}...X_n^{\alpha_n}$. Then, we define the order $\leq^*$ as $X^\alpha \leq^* X^\beta$ if and only if $\alpha \leq^* \beta$. It is easy to verify that $\leq^*$ is a total order on the monomials of $\mathbb{Z}[X_1, ..., X_n]$ and that we have the following property: For every $\gamma \in \mathbb{N}^n$, if $X^\alpha \leq^* X^\beta$, then $X^{\alpha+\gamma} \leq^* X^{\beta+\gamma}$. This is the *lexicographic order* induced by $X_1 > ... > X_n$.

Let us recall now useful terminology for multivariable polynomials. For each non-empty polynomial $P = \sum_{\alpha \in \mathbb{N}^n} a_\alpha X^\alpha$ in $\mathbb{Z}[X_1, ..., X_n]$:

The *support* of P is $S(P) = \{\alpha \in \mathbb{N}^n \textit{ such that } a_\alpha \neq 0\}$. In particular $S(P)$ is always finite.The *multidegree* of P is $m(P) = \max\{\alpha \in S(P)\}$.
The *leading coefficient* of P is $LC(P) = a_{m(P)}$.
The *leading monomial* of P is $LM(P) = X^{m(P)}$.
The *leading term* of P is $LT(P) = LC(P)LM(P)$.
The *norm* of P is $\|P\|_1 = \sum_{\alpha \in \mathbb{N}^n} |a_\alpha|$

Theorem 3. *Let $F = (P_1, ..., P_s)$ be an s-tuple of polynomials of $\mathbb{Z}[X_1, ..., X_n]$. Then every polynomial P of $\mathbb{Z}[X_1, ..., X_n]$ can be written in the following non-unique form $P = R + \sum_{k=1}^{s} Q_k P_k$ where:*
i) $Q_1, ..., Q_s, R \in \mathbb{Z}[X_1, ..., X_n]$.
ii) $R = \sum_{\alpha \in \mathbb{N}^n} c_\alpha X^\alpha$ and $\forall \alpha \in S(R), c_\alpha X^\alpha$ is not divisible by any of $LT(P_1), ...,$ $LT(P_s)$.

Proof. This proof is an easy consequence of the following generalized division algorithm.

Generalized Division Algorithm.
We denote by $trunc(s)$ the integer part of s.
Input: $(P_1, ..., P_s), P$
Output: $(a_1, ..., a_s), R$
$a_1 := 0, ..., a_s := 0,\ R := 0$
$Q := P$
While $Q \neq 0$ Do
$i := 1$
$division := false$
While ($i \leq s$ and division=false) Do
If $LM(P_i)$ divides $LM(Q)$ and $|LC(P_i)| \leq |LC(Q)|$ Then
$a_i := a_i + trunc(LC(Q)/LC(P_i))LM(Q)/LM(P_i)$
$Q := Q - (trunc(LC(Q)/LC(P_i))LM(Q)/LM(P_i))P_i$
$division := true$
Else
$i := i + 1$
EndIf
EndWhile
If $division = false$ Then
$R := R + LT(Q)$
$Q := Q - LT(Q)$
EndIf
EndWhile
Return $(a_1, ..., a_s), R$

R is the *remainder* of P by $(P_1, ..., P_s)$. We denote it by $\bar{P}^{(P_1,...,P_s)}$.
Roughly speaking, the upper algorithm tries to divide P as much as possible.

Example 1. If we have $P = X_1X_2^2 + X_1X_2 + X_2^2$ and $(P_1 = X_2^2 - 1, P_2 = X_1X_2 - 1)$, then we obtain $P = P_1 \times (X_1 + 1) + P_2 + X_1 + 2$. The remainder is $X_1 + 2$.

Example 2. The remainder of the division of $P = X_1X_2^2 - X_2^2$ by $(P_1 = X_2^2 - 1, P_2 = X_1X_2 - 1)$ is equal to zero. Nevertheless, the division of $P = X_1X_2^2 - X_2^2$ by $(P_1 = X_1X_2 - 1, P_2 = X_2^2 - 1)$ gives $\bar{P}^{(P_1,P_2)} = -X_2^2 + 1$.

So, we point out that the division depends on the ordering in the s-tuple of the polynomials. Actually, the division does not allow us to determine if a polynomial belongs or not to an ideal I of $\mathbb{Z}[X_1, ..., X_n]$.

We recall that in $\mathbb{R}[X]$ a polynomial P belongs to an ideal I if and only if Q divides P where Q is the minimal polynomial of I. We have the following analogous version in $\mathbb{Z}[X_1, ..., X_n]$:

Theorem 4. *For every non-zero ideal I of $\mathbb{Z}[X_1, ..., X_n]$, there exists an s-tuple of polynomials $(P_1, ..., P_s)$ such that P belongs to I if and only if the remainder of P by $(P_1, ..., P_s)$ is equal to zero.*

Such an s-tuple is called a *standard basis* of I. We do not prove here this theorem. The interested reader can find a constructive proof of this latter in the following report [2]. He can also read the paper of J.C. Faugére [6] which presents an efficient algorithm for computing standard basis. To illustrate the interest of this reformulation, we continue this section by an application to a classical problem solved by Conway and Lagarias [4] by using group theoretic arguments. Let T_N denote the triangular array of cells in the hexagonal lattice having $N(N+1)/2$ cells (Fig.3).

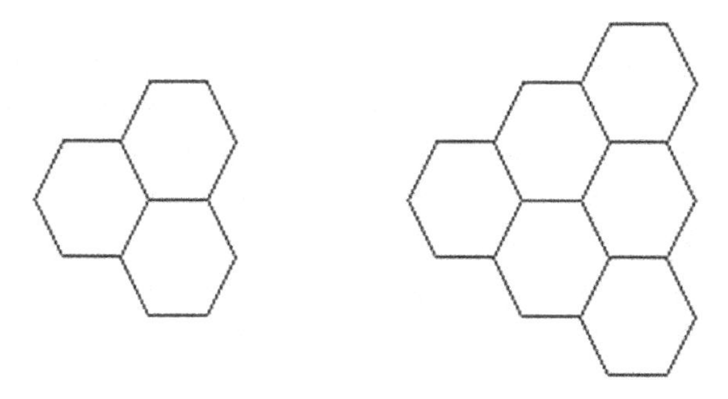

Fig. 3. T_2 and T_3

Theorem 5. *(Conway-Lagarias, Theorems 1.3 and 1.4)*
a) The triangular region T_N in the hexagonal lattice has a $\mathbb{Z}$-tiling by congruent copies of T_2 polyhexes if and only if $N = 0$ or $2 \mod 3$.
b) The triangular region T_N in the hexagonal lattice has a $\mathbb{Z}$-tiling by congruent copies of three-in-line polyhexes if and only if $N = 0$ or $8 \mod 9$.

Proof.
a) Firstly, we put $Y_1 > Y_2 > X_1 > X_2$ and using classical algorithms [2] or [6], we can compute a standard basis of the tiling ideal of T_2:

$$I(E) = \langle X_1 + X_2 + 1, X_1X_2 + X_1 + X_2, Y_1X_1 - 1, Y_2X_2 - 1\rangle_{\mathbb{Z}}.$$

We obtain

$$B = (X_1 + X_2 + 1, X_2^2 + X_2 + 1, Y_2 + X_2 + 1, Y_1 - X_2).$$

Now, as $Q_{T_N} = \sum_{i=0}^{N} X_2^i \left(\sum_{j=0}^{N-i-1} X_1^j \right)$, we can easily compute the remainder of Q_{T_N} by $(X_1 + X_2 + 1, X_2^2 + X_2 + 1, Y_2 + X_2 + 1, Y_1 - X_2)$.

It is equal to $\begin{cases} 0 & \text{if } N = 0 \text{ or } 2 \mod 3. \\ 1 & \text{if } N = 1 \mod 3. \end{cases}$

b) We compute a standard basis of the tiling ideal of the three-in-line polyhexes:

$$I(E) = \langle X_1^2 + X_1 + 1, X_2^2 + X_2 + 1, X_1^2Y_2^2 + X_1Y_2 + 1, Y_1X_1 - 1, Y_2X_2 - 1\rangle_{\mathbb{Z}}.$$

with $Y_1 > Y_2 > X_1 > X_2$. We obtain

$$B = (X_1^2 + X_1 + 1, X_2^2 + X_2 + 1, X_1 + Y_1 + 1, X_2 + Y_2 + 1, 3X_2 + 3X_1 + 3, X_2X_1 - X_1 - X_2 - 2).$$

The remainder of Q_{T_N} by B is equal to
$$\begin{cases} 0 & \text{if } N=0 \text{ or } 8 \mod 9. \\ 1 & \text{if } N=1 \mod 9. \\ X_1 + X_2 + 1 & \text{if } N=2 \text{ or } 3 \mod 9. \\ -2X_1 - 2X_2 - 1 & \text{if } N=4 \mod 9. \\ 2X_1 + 2X_2 + 2 & \text{if } N=5 \text{ or } 6 \mod 9. \\ -X_1 - X_2 & \text{if } N=7 \mod 9. \end{cases}$$

We can notice that this new proof gives a more precise statement than the one obtained by Conway-Lagarias. Moreover, this is an "automatic" proof. Indeed, all the steps can be done computerwise.

4 General Coloring

Let us recall that, given a set E of $\mathbb{Z}$-polyominoes, the ideal

$$I(E) = \langle Q_P; P \in E \text{ and } X_1Y_1 - 1, X_2Y_2 - 1\rangle_{\mathbb{Z}}$$

is the tiling ideal of E. If B is a standard basis of $I(E)$ then B is said to *be associated* to E.

Definition 2. *A* general coloring χ_E *is the map from* $\mathbb{Z}[X_1, X_2, Y_1, Y_2]$ *into* $\mathbb{Z}[X_1, X_2, Y_1, Y_2]$ *such that* $\chi_E(Q) = \bar{Q}^B$.

Theorem 6. *A $\mathbb{Z}$-polyomino P is $\mathbb{Z}$-tilable by a set of $\mathbb{Z}$-polyominoes E if and only if* $\chi_E(P) = 0$.

Proof. By theorem 1, P is $\mathbb{Z}$-tilable by E if and only if Q_P belongs to $I(E)$, and by definition, Q_P belongs $I(E)$ if and only if $\chi_E(Q_P) = 0$.

Remark 1. This definition seems to be tautological. Indeed, this is very useful to have a geometric visualization. Let us observe the following example.

We want to have a coloring argument for the $\mathbb{Z}$-tilability by the set E of classical polyominoes described below (Fig.4) called *T-tetraminoes*. We have

$$I(E) = \langle X_1^2 + X_1X_2 + X_1 + 1, X_1^2X_2 + X_1X_2 + X_1 + X_2, X_1X_2 + X_2^2 + X_2 + 1, X_1X_2^2 + X_1X_2 + X_1 + X_2, X_1Y_1 - 1, X_2Y_2 - 1\rangle.$$

We compute a standard basis for the order $Y_1 > Y_2 > X_1 > X_2$

$$B = (X_1 + 3, X_2 + 3, 8, Y_1 + 3, Y_2 + 3).$$

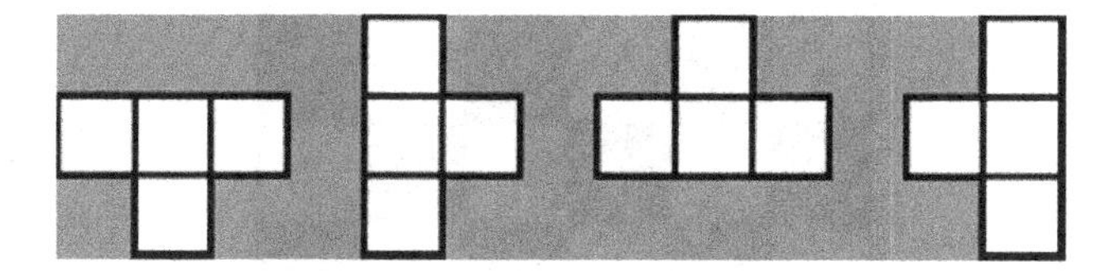

Fig. 4. The T-tetraminoes

So, we have

$$\chi_E(Q_{c(i,j)}) = \chi_E(\mathcal{X}^{(i,j)}) = \begin{cases} 1 & \text{if } i+j=0 \mod 2 \\ 5 & \text{if } i+j=1 \mod 2 \end{cases}.$$

Moreover, we always have the following classical remainder property

$$\chi_E(Q_P) = \chi_E(\sum_{c(i,j)\in P} Q_{c(i,j)}) = \chi_E\left(\sum_{c(i,j)\in P} \chi_E\left(Q_{c(i,j)}\right)\right).$$

Now, as $A = \chi_E\left(\sum_{c(i,j)\in P} Q_{c(i,j)}\right)$ is an integer, $\chi_E(A) = A \mod 8$. So, we obtain the following theorem.

Theorem 7. *Let us suppose that the squares of the plane have a chessboard-like coloration (Fig.5). A polyomino P is $\mathbb{Z}$-tilable by T-tetraminoes if and only if, when assigning 5 on the white squares and 1 on the black ones in P, the sum of values on the squares is a multiple of 8.*

Corollary 1. *A square $n \times n$ is $\mathbb{Z}$-tilable (resp. tilable) by T-tetraminoes if and only if n is a multiple of 4.*

Proof. If n is odd, then the sum of values is odd, and the square $n \times n$ is not $\mathbb{Z}$-tilable by T-tetraminoes. If n is even, an easy computation gives $\chi_E(Q_{n\times n}) = 3n^2 \mod 8$. So, the square $n \times n$ is $\mathbb{Z}$-tilable by T-tetraminoes if and only if n is a multiple of 4. Finally, we observe that the square 4×4 is (classically) tilable by T-tetraminoes. Hence, if n is a multiple of 4, the square $n \times n$ is also tilable by T-tetraminoes.

5 Optimal Quasi-Z-Tilings and Optimal Partial Tilings

Now, to show an other application, we are going to consider the following problem. We want to know the maximum number of translated copies of tiles of type $p \times 1$ and $1 \times q$ (i.e. a horizontal bar of size p and a vertical bar of size q) that we can put in a rectangle $R \times S$. This is what we call an *optimal partial tiling* of $R \times S$ by $p \times 1$ and $1 \times q$. First of all, we notice that

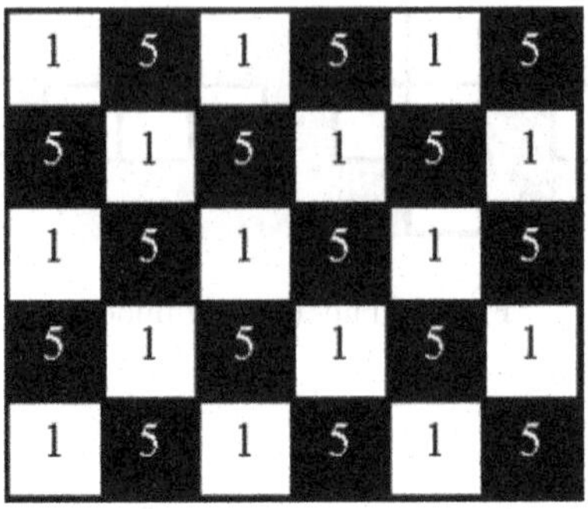

1	5	1	5	1	5
5	1	5	1	5	1
1	5	1	5	1	5
5	1	5	1	5	1
1	5	1	5	1	5

Fig. 5. A general coloring for the T-tetraminoes. The lower left corner square is the square (0,0)

$I(E) = \langle 1 + X_1 + ... + X_1^{(p-1)}, 1 + X_2 + ... + X_2^{(q-1)}, X_1Y_1 - 1, X_2Y_2 - 1\rangle$, $Q_{R\times S} = (1 + X_1 + ... + X_1^{(R-1)})(1 + X_2 + ... + X_2^{(S-1)})$ and that $B = (1 + X_1 + ... + X_1^{(p-1)}, 1 + X_2 + ... + X_2^{(q-1)}, Y_2 + 1 + X_2 + ... + X_2^{(S-2)}, Y_1 + 1 + X_1 + ... + X_1^{(R-2)})$ is a standard basis with respect to $Y_1 > Y_2 > X_1 > X_2$. In the sequel, we denote by $\frac{X^d-1}{X-1}$ the polynomial $1 + ... + X^{d-1}$. Let us continue with a short algebraic lemma:

Lemma 2. *If* $(R \bmod p) < p/2$ *(resp.* $(R \bmod p) > p/2)$ *, the polynomial* $\frac{X^{(R \bmod p)}-1}{X-1}$ *(resp.* $-X^{(R \bmod p)} \times \frac{X^{p-(R \bmod p)}-1}{X-1}$*) is the unique element* Q *of the class of* $\frac{X^R-1}{X-1}$ *in* $\mathbb{Z}[X]/\left\langle \frac{X^p-1}{X-1} \right\rangle$ *with* $\deg Q < p$ *and* $\|Q\|_1$ *minimum among the class.*

If $(R \bmod p) = p/2$*, these polynomials are the unique ones such that* $\deg Q < p$ *and* $\|Q\|_1$ *minimum among the class.*

Proof. Let Q be a polynomial in the class of $\frac{X^R-1}{X-1}$. If Q is minimum for $\| \ \|_1$, so it is for $Q \bmod (X^p - 1)$. Hence, we can suppose that $\deg(Q) < p$. We distinguish two cases:

a) if $\deg(Q) < p - 1$, it is clear that Q is the remainder of P by $\frac{X^p-1}{X-1}$. Thus, $Q = \frac{X^{(R \bmod p)}-1}{X-1}$.
b) if $\deg(Q) = p - 1$, $Q = \frac{X^{(R \bmod p)}-1}{X-1} + k \times \frac{X^p-1}{X-1}$. In $X = 1$, we obtain $Q(1) = (R \bmod p) + kp$. Now, $|Q(1)| \leq \|Q\|_1 \leq (R \bmod p)$. So, $k = -1$ and $Q = -X^{(R \bmod p)} \times \frac{X^{p-(R \bmod p)}-1}{X-1}$. To conclude, it suffices to observe that $\|\frac{X^{(R \bmod p)}-1}{X-1}\|_1 \leq \| - X^{(R \bmod p)} \times \frac{X^{p-(R \bmod p)}-1}{X-1}\|_1$ if and only if $(R \bmod p) \leq p/2$.

Definition 3. *Let* P *be a* $\mathbb{Z}$*-polyomino and* E *a set of tiles. An* optimal quasi-$\mathbb{Z}$-tiling *of* P *by* E*, is a positioning of copies of tiles (that is* $\sum_{i=1}^{t} \lambda_i P_i (. - a_i)$ *where*

$\lambda_i \in \mathbb{Z}$, $P_i \in E$ and $a_i \in \mathbb{Z}^2$) such that $\|Q_P - Q_{\sum_{i=1}^{t} \lambda_i P_i(.-a_i)}\|_1$ is minimum. The value of $\|Q_P - Q_{\sum_{i=1}^{t} \lambda_i P_i(.-a_i)}\|_1$ is called the deficiency *of the optimal quasi-$\mathbb{Z}$-tiling.*

Now, we look at an optimal quasi-$\mathbb{Z}$-tiling of $R \times S$ by $\{p \times 1, 1 \times q\}$. To obtain an element of the class of $Q_{R\times S}$ in $\mathbb{Z}[X_1, X_2, Y_1, Y_2]/I(E)$ minimum for $\| \, \|_1$, it suffices to "reduce" separately $\frac{X_1^R - 1}{X_1 - 1}$ in $\mathbb{Z}[X_1]/\left\langle 1 + ... + X_1^{p-1} \right\rangle$ and $\frac{X_2^S - 1}{X_2 - 1}$ in $\mathbb{Z}[X_2]/\left\langle 1 + ... + X_2^{p-1} \right\rangle$. Indeed, $1 + X_1 + ... + X_1^{(p-1)}$ and $1 + X_2 + ... + X_2^{(q-1)}$ have totally independent variables and the standard basis does not mixed them. So, such a minimal polynomial can be written as $Q_1 \times Q_2$ with:

$$Q_1 \in \left\{ \frac{X_1^{(R \bmod p)} - 1}{X_1 - 1}, -X_1^{(R \bmod p)} \times \frac{X_1^{p-(R \bmod p)} - 1}{X_1 - 1} \right\}$$

and

$$Q_2 \in \left\{ \frac{X_2^{(S \bmod q)} - 1}{X_2 - 1}, -X_2^{(S \bmod q)} \times \frac{X_2^{q-(S \bmod q)} - 1}{X_2 - 1} \right\}.$$

More precisely:

Theorem 8.
- *If $(R \bmod p) \leq p/2$ and $(S \bmod q) \leq q/2$ then the deficiency of an optimum $\mathbb{Z}$-tiling of $R \times S$ by $\{p \times 1, 1 \times q\}$ is $(R \bmod p)(S \bmod q)$.*
- *If $(R \bmod p) \leq p/2$ and $(S \bmod q) > q/2$ then the deficiency of an optimum $\mathbb{Z}$-tiling of $R \times S$ by $\{p \times 1, 1 \times q\}$ is $(R \bmod p)(q - (S \bmod q))$.*
If $(R \bmod p) > p/2$ and $(S \bmod q) \leq q/2$ then the deficiency of an optimum $\mathbb{Z}$-tiling of $R \times S$ by $\{p \times 1, 1 \times q\}$ is $(p - (R \bmod p))(S \bmod q)$.
If $(R \bmod p) > p/2$ and $(S \bmod q) > q/2$ then the deficiency of an optimum $\mathbb{Z}$-tiling of $R \times S$ by $\{p \times 1, 1 \times q\}$ is $(p - (R \bmod p))(q - (S \bmod q))$.

Now, we look for an optimal partial tiling of $R \times S$ by $\{p \times 1, 1 \times q\}$. Let $P_{Q_1 Q_2}$ be the unique $\mathbb{Z}$-polyomino such that $Q_P = Q_1 Q_2$. We notice that $R \times S - P_{Q_1 Q_2}$ is a "true" polyomino if and only if $P(Q_1 Q_2)$ is a "true" sub-polyomino of $R \times S$, (that is to say that all the monomials of $Q_1 Q_2$ are unitary: all the coefficients are $+1$, and belongs to $Q_{R\times S}$). So, in the present case, this arises if $Q_1 Q_2$ is:

$$\frac{X_1^{(R \bmod p)} - 1}{X_1 - 1} \frac{X_2^{(S \bmod q)} - 1}{X_2 - 1}$$

or

$$X_1^{(R \bmod p)} \times \frac{X_1^{p-(R \bmod p)} - 1}{X_1 - 1} \times X_2^{(S \bmod q)} \times \frac{X_2^{q-(S \bmod q)} - 1}{X_2 - 1}.$$

Furthermore, $R \times S - P_{Q_1 Q_2}$ is tilable by $\{p \times 1, 1 \times q\}$. The following drawings explain the tilings (Fig. 6). Thus, the first one is an optimal partial tiling of $R \times S$

by $\{p \times 1, 1 \times q\}$ when $(R \bmod p) \leq p/2$ and $(S \bmod q) \leq q/2$. The second one is an optimal partial tiling of $R \times S$ by $\{p \times 1, 1 \times q\}$ when $(R \bmod p) > p/2$ and $(S \bmod q) > q/2$. Finally, we need to study the cases where:

1) $(R \bmod p) > p/2$ and $(S \bmod q) \leq q/2$,
2) $(R \bmod p) \leq p/2$ and $(S \bmod q) > q/2$.

In fact, these two cases are identical up to rotation. Thus, it suffices to solve the first one. Let us notice in this case that the deficiency of an optimal quasi-$\mathbb{Z}$-tiling of $R \times S$ by $\{p \times 1, 1 \times q\}$ is exactly due to an overlap of some squares. Indeed, all the coefficients of $Q_1 Q_2$ are negative. So, we have an optimal covering. To have an optimal partial tiling of $R \times S$ by $\{p \times 1, 1 \times q\}$ we need to find a polynomial with all its coefficients are positive in the class of $Q_{R \times S}$ in $\mathbb{Z}[X_1, X_2, Y_1, Y_2]/I(E)$ and minimum for $\| \; \|_1$. It is easy to prove that the norm of such a polynomial Q is greater than $\min((R \bmod p) \times (S \bmod q), (p - (R \bmod p)) \times (q - (S \bmod q)))$. Indeed, $\|Q\|_1$ is equal to $Q(1)$. So, we can conclude with the following theorem:

Theorem 9. *The maximum number of unit squares in a rectangle $R \times S$ that we can cover without overlap by translated copies of tiles of type $p \times 1$ and $1 \times q$ is equal to $R \times S - \min((R \bmod p) \times (S \bmod q), (p - (R \bmod p)) \times (q - (S \bmod q)))$.*

In fact, this theorem can easily be extended in higher dimensions. It is a generalization of a theorem of F.W. Barnes [1].

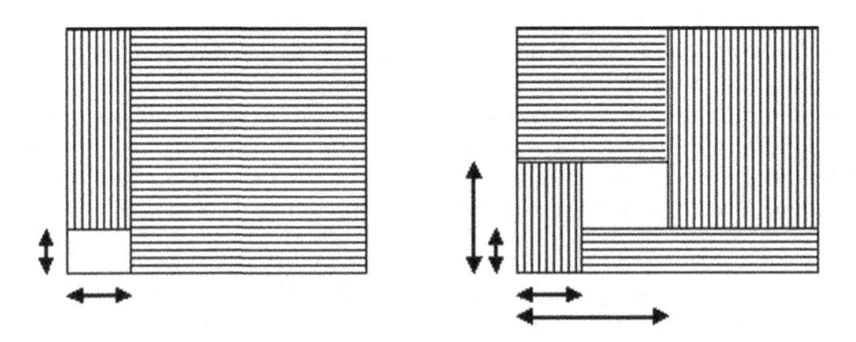

Fig. 6. Left: The optimal tiling when $(R \bmod p) \times (S \bmod q)$ is minimal, the horizontal vector length is $(R \bmod p)$ and the vertical one is $(S \bmod q)$. Right: The optimal tiling when $(p - R \bmod p)) \times (q - (S \bmod q))$ is minimal, the greater horizontal vector length is p and the greater vertical one is q

6 Z-Tilability and Boundary Conditions

In this section, we only deal with classical polyominoes and not with $\mathbb{Z}$-polyominoes. In [4], Conway and Lagarias show that it is possible to know whether a simply connected polyomino P has a $\mathbb{Z}$-tiling by only considering the boundary of P. We prove that we have a similar situation with our characterization. We do not need to compute the remainder of Q_P, but only the remainder of a shorter polynomial associated to the boundary of P.

Theorem 10. *Let P be a polyomino and χ_E be a general coloring associated to E. We suppose that*

$$\chi_E\left(\sum_{i=1}^{2}(X_i+Y_i)\right)-4$$

is not a zero divisor of $\mathbb{Z}[X_1,X_2,Y_1,Y_2]/I(E)$. In this case,

$$\chi_E(Q_P)=0 \text{ if and only if } \sum_{(c_1,c_2)\in S}(\chi_E(Q_{c_1})-\chi_E(Q_{c_2}))=0$$

where the couple (c_1,c_2) belongs to S if it is a domino (union of two adjacent squares) with $c_1\in P$ and $c_2\notin P$.

Proof. Let χ_E be the general coloring, we denote by

$$v_{\chi_E}(\mathcal{X}^a)=\sum_{i=1}^{2}(\chi_E(\mathcal{X}^a)-\chi_E(\mathcal{X}^aX_i))+\sum_{i=1}^{2}(\chi_E(\mathcal{X}^a)-\chi_E(\mathcal{X}^aY_i)).$$

We have $\sum_{(c_1,c_2)\in S}(\chi_E(Q_{c_1})-\chi_E(Q_{c_2})) = \sum_{c\in P}v_{\chi_E}(Q_c)$ because, if the squares c and c' belong to P, the contributions of the couples (c,c') et (c',c) cancel each other out. Now, if we consider that the image of χ_E is in $\mathbb{Z}[X_1,X_2,Y_1,Y_2]/I(E)$, it is obvious that χ_E is a morphism of algebra and that we can rewrite

$$\sum_{c\in P}v_{\chi_E}(Q_c)=\left(4-\chi_E\left(\sum_{i=1}^{2}(X_i+Y_i)\right)\right)\sum_{c\in P}\chi_E(Q_c)$$

in

$\mathbb{Z}[X_1,X_2,Y_1,Y_2]/I(E)$. As $\chi_E\left(\sum_{i=1}^{2}(X_i+Y_i)\right)-4$ is not a zero divisor, $\chi_E(Q_P)=0$ if and only if $\sum_{(c_1,c_2)\in S}(\chi_E(Q_{c_1})-\chi_E(Q_{c_2}))=0$.

To conclude this section, we give an example related to the paper of Thurston [11]. Let us consider that we want to $\mathbb{Z}$-tile a polyomino with dominoes. The associated ideal is $I(E)=\langle X_1+1,X_2+1,X_1Y_1-1,X_2Y_2-1\rangle$. We obtain that $(1+X_1,1+X_2,Y_1+1,Y_2+1)$ is a standard basis of $I(E)$ for the order $Y_1>Y_2>X_1>X_2$. So, $\chi_E(X^{(i,j)})=\begin{cases}1 & \text{if } i+j=0 \mod 2\\ -1 & \text{if } i+j=1 \mod 2\end{cases}$. Hence, a polyomino is $\mathbb{Z}$-tilable by dominoes if and only if it has the same number of black (when $i+j=1 \mod 2$) and white (when $i+j=0 \mod 2$) squares $c_{(i,j)}$. In this case, the polyomino is said *balanced.* Independently, we have

$$\chi_E\left(\sum_{i=1}^{2}(X_i+Y_i)\right)-4=-8$$

which is not a zero divisor. So, we can apply Theorem 10. The values of $\chi_E(Q_{c_1}) - \chi_E(Q_{c_2}) = \begin{cases} 2 & \text{if } c_1 = c_{(i,j)} \text{ and } i+j = 0 \mod 2 \\ -2 & \text{if } c_1 = c_{(i,j)} \text{ and } i+j = 1 \mod 2 \end{cases}$ where $(c_1, c_2) \in S$. Thus, P is balanced if and only if we have the same number of black and white edges on its boundary (a *black* (resp. *white*) *edge* is an edge which borders a black (resp. white) square of P), which is a classical result on domino tilings.

7 Conclusion

We have tried to prove in this paper that the standard basis theory can be very useful (and powerful) to study a lot of tiling problems as:

- $\mathbb{Z}$-tiling problems;
- Coloration argument problems and necessary conditions for "true" tiling;
- Optimal partial tiling problems.

Moreover, this new reformulation does not have the drawbacks of the previous group theoretic representation (simply connectivity, dimension 2, reduction of words in a presentation group). It seems that we have a more practical interpretation to obtain non-trivial results on tilings.

References

1. F.W. Barnes, Best packing of rods into boxes, Discrete Mathematics 142 (1995) 271-275.
2. O. Bodini, Pavage des polyominos et Bases de Grobner, Rapport de recherche N^o RR2001-51, LIP, 2001.
3. B. Buchberger, Introduction to Grobner basis, Logic of computation (Marktoberdorf 95) 35-66, NATO Adv. Sci. Inst. Ser. F Comput. Systems Sci.,157, Springer Berlin (1997).
4. J.H Conway, J.C. Lagarias, Tiling with polyominoes and combinatorial group theory, J.C.T. Series A 53 (1990) 183-208.
5. D. Cox, J. Little, D. O'Shea, Ideals,varieties and algorithms, 2nde edition, Undergraduate Text in Mathematics, Springer Verlag, New York (1997) XV 536 pp.
6. J-C Faugére, A new efficient algorithm for computing Grobner basis, Journal of Pure and Applied Algebra, 139 (1999) 61-88.
7. S. W. Golomb, Tiling with polyominoes, J.C.T. Series A 1 (1966) 280-296.
8. S. W. Golomb, Polyominoes which tile rectangles, J.C.T. Series A 51 (1989) 117-124.
9. R. Kenyon, Sur la dynamique, la combinatoire et la statistique des pavages, Habilitation (1999)
10. D.A. Klarner, Packing a rectangle with congruent n-ominoes, J.C.T. Series A 7 (1969) 107-115.
11. W.P. Thurston, Conway's tiling groups, Amer. Math. Monthly, 97, num. 8 (Oct. 1990) 757-773.

Curve Tracking by Hypothesis Propagation and Voting-Based Verification

Kazuhiko Kawamoto and Kaoru Hirota

Tokyo Institute of Technology, Mail-Box:G3-49, 4259 Nagatsuta,
Midori-ku, Yokohama 226-8502, Japan
kawa@hrt.dis.titech.ac.jp

Abstract. We propose a robust and efficient algorithm for curve tracking in a sequence of binary images. First it verifies the presence of a curve by votes, whose values indicate the number of the points on the curve, thus being able to robustly detect curves against outlier and occlusion. Furthermore, we introduce a procedure for preventing redundant verification by determining equivalence curves in the digital space to reduce the time complexity. Second it propagates the distribution which represents the presence of the curve to the successive image of a given sequence. This temporal propagation enables to focus on the potential region where the curves detected at time $t-1$ are likely to appear at time t. As a result, the time complexity does not depend on the dimension of the curve to be detected. To evaluate the performance, we use three noisy image sequences, consisting of 90 frames with 320×240 pixels. The results shows that the algorithm successfully tracks the target even in noisy or cluttered binary images.

1 Introduction

Object tracking in a sequence of images has various applications, such as motion analysis, object recognition, and video compression, in computer vision. In particular, robust and efficient methods for tracking objects in clutter attracts increasing attention from academic and industry. For a few decades, voting-based algorithms, such as the Hough transform [1] and RANSAC (RANdom SAmple Consensus) [2], have been widely used to detect objects in noisy or cluttered images. However, the voting-based algorithms are, in general, time-consuming, thus being unsuitable for the application to tracking in real or quasi-real time.

We propose a robust and efficient voting-based algorithm for tracking curves in a sequence of noisy binary images. The algorithm roughly consists of two parts. The first part verifies the presence of a curve by votes, whose values indicate the number of the points on the curve, thus being able to robustly detect curves against outlier and occlusion. This voting-based verification is performed in the digital space using the Bresenham algorithm [4]. The digitization enables us to prevent redundant verification by determining equivalence curves in the digital space. The second part propagates the curve detected in the first part to the successive image frame using sequential Monte Carlo (SMC) filtering [15].

R. Klette and J. Žunić (Eds.): IWCIA 2004, LNCS 3322, pp. 151–163, 2004.

This approach, called *hypothesis propagation*, enables us to focus on the potential region where the curve detected at the previous time are likely to appear. The most attractive property is that the computational complexity does not depend on the dimension of the curve to be detected. Furthermore we introduce a procedure for reducing the variance of the state distribution which represents the presence of curves.

To evaluate the tracking performance in clutter, two noisy image sequences, consisting of 90 frames with 320 × 240 pixels, are used. The results shows that the proposed algorithm successfully tracks the circle and the circular arc in the binary images and that the execution time per frame for each image sequence is 23.6 msec and 23.5 msec, respectively, on the average of 10 trials.

This paper is organized as follows. In Sec. 2 we reviews a voting procedure for detecting curves in binary images. In Sec. 3 we present a basic algorithm of the SMC filters, called the bootstrap filter [16]. In Sec 4 we proposes an algorithm for visual tracking with hypothesis propagation and voting-based verification. In Sec. 5 we proposes a procedure for preventing redundant verification by determining equivalence curves in the digital space. In Sec. 6 we report the experimental results.

2 Voting-Based Curve Detection in Binary Image

Let us consider the problem of detecting curves in a binary image. We assume that the curves are analytically expressed with parameter $\boldsymbol{a} = (a_1, a_2, \ldots, a_{n_a})$, i.e., the curves are written by $f(\boldsymbol{r}; \boldsymbol{a}) = 0$, where $\boldsymbol{r} = (x, y)^\top$ denote two-dimensional points in the image. This paper focuses on circle

$$f(\boldsymbol{r}; \boldsymbol{a}) = (x - x_o)^2 + (y - y_o)^2 - r^2 = 0, \tag{1}$$

where $\boldsymbol{a} = (x_o, y_o, r)^\top$. Thus we can reduce the problem to finding the parameter $\boldsymbol{a}$ in the parameter space.

Voting-based algorithms, such as the Hough transform [1] and RANSAC (RANdom SAmple Consensus) [2], can robustly detect the parameter $\boldsymbol{a}$ even in noisy binary images. The algorithms basically consist of two processes: counting (or voting) and thresholding. The counting process evaluates each curve by an additive measure, i.e., the number of the points on the curve is calculated as

$$I(\boldsymbol{a}) = \sum_i h(\boldsymbol{r}_i, \boldsymbol{a}), \quad \text{where } h(\boldsymbol{r}_i, \boldsymbol{a}) = \begin{cases} 1 \text{ if } & f(\boldsymbol{r}_i; \boldsymbol{a}) = 0, \\ 0 \text{ otherwise}, \end{cases} \tag{2}$$

where $\boldsymbol{r}_i$ is the i-th point in the image [3]. Then the thresholding process detects the curve if $I(\boldsymbol{a}) > T$ for a given threshold T.

2.1 Digital Circles by the Bresenham Algorithm

Of course, the definition of the measure in eq. (2) is not appropriate in the digital space, i.e., $f(\boldsymbol{r}; \boldsymbol{a}) \neq 0$ for almost all $\boldsymbol{a} \in \boldsymbol{R}^{n_a}$ and $\boldsymbol{r} \in \boldsymbol{N}^2$. Hence we require the definition of digital curves. Throughout this paper, we use the Bresenham algorithm for drawing circles [4] to produce a sequence of the points on a circle in the digital space. To draw a digital circle by the Bresenham algorithm, we first digitize a real parameter $\boldsymbol{a} \in \boldsymbol{R}^{n_a}$ by

$$\hat{\boldsymbol{a}} = \left\lfloor \boldsymbol{a} + \begin{pmatrix} 0.5 \\ 0.5 \\ 0.5 \end{pmatrix} \right\rfloor = \begin{pmatrix} \lfloor x_o + 0.5 \rfloor \\ \lfloor y_o + 0.5 \rfloor \\ \lfloor r + 0.5 \rfloor \end{pmatrix}, \tag{3}$$

where $\hat{\boldsymbol{a}}$ indicates the digitized vector of a real-number vector $\boldsymbol{a}$, and $\lfloor \cdot \rfloor$ is the floor function, which specifies the largest integer not exceeding a given real number. Then, giving the digital circle $\hat{\boldsymbol{a}}$ to the Bresenham algorithm, we obtain the sequence of the points, $\boldsymbol{r}_1, \boldsymbol{r}_2, \ldots, \boldsymbol{r}_l \in \boldsymbol{N}^2$ on the digital circle. We simply write the Bresenham algorithm by

$$\text{Bresenham}(\hat{\boldsymbol{a}}) = \{\boldsymbol{r}_1, \boldsymbol{r}_2, \ldots, \boldsymbol{r}_l\}. \tag{4}$$

Using this expression, we modify the equation $f(\boldsymbol{r}_i; \boldsymbol{a}) = 0$ in eq. (2) to

$$\boldsymbol{r}_i \in \text{Bresenham}(\hat{\boldsymbol{a}}), \tag{5}$$

thus verifying whether a point $\boldsymbol{r}_i$ is on the circle $\hat{\boldsymbol{a}}$ or not in the digital space.

2.2 Requirements for Voting

The underlying idea of the voting-based algorithm is simple and easy to implement on computers, whereas the algorithms are time-consuming because a lot of iterations are required.

RANSAC requires the number of iterations

$$m = \frac{\log(1-\gamma)}{\log(1-(1-\epsilon)^{n_a})}, \tag{6}$$

where γ is the confidence level, ϵ is the outlier proportion, and n_a is the minimal number of the points that uniquely determines the curve to be detected. Therefore the number of iterations m increases as ϵ increases, which situation happens in the case of noisy images. Figure 1 shows the relationship between m and ϵ, given $\gamma = 0.95$ and $n_a = 3$. In addition, the number of iterations m also increases as n_a increases, i.e., RANSAC requires more computational time when applying to higher dimensional curves.

The Hough transform increments $O(\alpha^{n_a-1})$ bins for each point, where α is the number of bins in each dimension of the parameter space, i.e., a total requirement is $O(M\alpha^{n_a-1})$ if the total number of points is M. Therefore the number of the iterations required exponentially increases as n_a increases. Also the storage required is $O(\alpha^{n_a})$. These complexities make it difficult to apply the Hough transform to higher dimensional curves.

In the literature, numerous techniques for improving the efficiency of the Hough transform and RANSAC have been proposed. For example, randomization [5–8], parallelization [9], and coarse-to-fine search [10–12] are introduced. These techniques are useful for reducing the complexities, but the application to curve detection in real-time remains difficult except for the low dimensional case ($n_a = 2$).

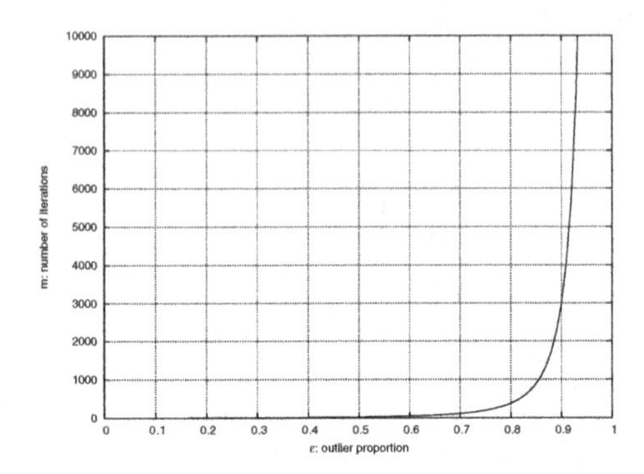

Fig. 1. Number of the iterations required by RANSAC

3 Discrete State Estimation by Bootstrap Filter

The sequential Monte Carlo filters (SMC) [13–15] provide numerical estimates of marginal distribution $p(\boldsymbol{x}_t|\boldsymbol{y}_{1:t})$ up to time t recursively in time as follows:

$$\text{Prediction}: \; p(\boldsymbol{x}_t|\boldsymbol{y}_{1:t-1}) = \int p(\boldsymbol{x}_t|\boldsymbol{x}_{t-1})p(\boldsymbol{x}_{t-1}|\boldsymbol{y}_{1:t-1})d\boldsymbol{x}_{t-1}, \tag{7}$$

$$\text{Filtering}: \; p(\boldsymbol{x}_t|\boldsymbol{y}_{1:t}) = \frac{p(\boldsymbol{y}_t|\boldsymbol{x}_t)p(\boldsymbol{x}_t|\boldsymbol{y}_{1:t-1})}{\int p(\boldsymbol{y}_t|\boldsymbol{x}_t)p(\boldsymbol{x}_t|\boldsymbol{y}_{1:t-1})d\boldsymbol{x}_t}, \tag{8}$$

where $\boldsymbol{x}_{0:t} \stackrel{\text{def}}{=} \{\boldsymbol{x}_0, \boldsymbol{x}_1, \ldots, \boldsymbol{x}_t\}$ and $\boldsymbol{y}_{1:t} \stackrel{\text{def}}{=} \{\boldsymbol{y}_1, \boldsymbol{y}_2, \ldots, \boldsymbol{y}_t\}$ are the signal and the observations up to time t, respectively. The bootstrap filter [16] is the simplest algorithm of the SMC filters. In the computer vision community, an visual tracking algorithm proposed under the framework of the bootstrap filter is known as the conditional density propagation algorithm, namely the Condensation algorithm [17]. The general algorithm, given an initial distribution $p(\boldsymbol{x}_0)$ and a transition equation $p(\boldsymbol{x}_t|\boldsymbol{x}_{t-1})$, is described as follows:

---

1. **Initialization**: $(t = 0)$
   - For $i = 1, 2, \ldots, N$, sample $\boldsymbol{x}_0^{(i)} \sim p(\boldsymbol{x}_0)$ and set $t \leftarrow 1$.
2. **Prediction**:
   - For $i = 1, 2, \ldots, N$, sample $\boldsymbol{x}_t^{(i)} \sim p(\boldsymbol{x}_t|\boldsymbol{x}_{t-1}^{(i)})$
3. **Filtering**:
   - For $i = 1, 2, \ldots, N$, evaluate the likelihoods

$$w_t^{(i)} = p(\boldsymbol{y}_t|\boldsymbol{x}_t^{(i)}).$$

   - For $i = 1, 2, \ldots, N$, normalize the likelihoods

$$w_t^{(i)} \leftarrow \frac{w_t^{(i)}}{\sum_{k=1}^{N} w_t^{(k)}}.$$

- For $i = 1, 2, \ldots, N$, resample $\boldsymbol{x}_t^{(i)}$ according to the likelihoods.
- Set $t \leftarrow t + 1$ and go to step 2.

The bootstrap filter, thus, approximately provides the filtering distribution $p(\boldsymbol{x}_t|\boldsymbol{y}_{1:t})$ with a set of discrete finite sample points $\{\boldsymbol{x}_t^{(i)} \,|\, i = 1, 2, \ldots, N\}$. These sample points are usually called *particles* (this is the reason that the SMC filters are often called the particle filters [18]). Theoretically, the discrete distribution asymptotically converges toward the true one if $N \to \infty$ (however, due to limited computer resource, it is not practical to set the number of sample points, N, to be extremely large).

This simulated-based approach has some attractive properties. First, it handles nonlinear and non-Gaussian state-space models [19], because an analytical expression of the state distribution of interest is not required. Second, the computational complexity does not depend on the dimension of the state; it depends on the number of particles, namely $O(N)$. Third, this discrete representation makes it easy to implement a voting-based algorithm for curve tracking in the framework of the SMC filtering. We will discuss the algorithm in Sec. 4.

4 Voting-Based Curve Tracker with Hypothesis Propagation

We propose an algorithm for curve tracking in a sequence of binary images. We aim to design a tracking algorithm that works in noisy binary images robustly and efficiently. The basic idea is to reuse the curve parameters detected by a voting-based algorithm at the previous time in order to efficiently search the curves at the current time, using the bootstrap filter. We call this approach *hypothesis propagation.* In other words, the algorithm does not waste the information obtained by the voting procedure, which is normally time-consuming, at each time step.

To implement the idea, the algorithm represents the distribution of votes in the parameter space by a set of particles, i.e., each particle $\boldsymbol{x}_t^{(i)}$ represents a curve $\boldsymbol{a}_t^{(i)}$ and its likelihood $w_t^{(i)}$ is evaluated by the received votes $I(\boldsymbol{a}_t^{(i)})$ that are normalized to 1. The most attractive property is that the computational complexity does not depend on the dimension of the curve to be detected. This property is due to the nature of the SMC filtering. Thus the algorithm efficiently deals with higher dimensional curves. The storage requirement is $O(n_a N)$, where n_a is the dimension of the state of interest and N is the number of particles. Therefore the storage size linearly increases as the dimension increases. Note that the Hough transform exponentially increases in size.

In the following, we describe the algorithm for circle tracking in detail, following the order of the general bootstrap filter in Sec. 3.

4.1 Generation of Initial Distribution: Curve Detection

We treat the problem of generating initial distribution $p(\boldsymbol{x}_0)$ as that of detecting a circle in the first frame of a give image sequence. Thus we first detect the circle $\boldsymbol{a}_0 = (x_o(0), y_o(0), r(0))^\top$ using RANSAC or the Hough transform. Then the algorithm generates a set of particles $\{\boldsymbol{x}_0^{(i)} | i = 1, 2, \ldots, N\}$ by adding system noise $\boldsymbol{v} = (v_{x_o}, v_{y_o}, v_r)^\top$ to the parameters:

$$\boldsymbol{x}_0^{(i)} = \boldsymbol{a}_0 + \boldsymbol{v}^{(i)} \text{ with } \boldsymbol{v}^{(i)} \sim \mathcal{N}(\boldsymbol{0}, \boldsymbol{V}), \quad i = 1, 2, \ldots, N, \tag{9}$$

where $\boldsymbol{V} = \mathrm{diag}(\sigma_{x_o}^2, \sigma_{y_o}^2, \sigma_r^2)$, and $\boldsymbol{v}^{(i)}$ is independently chosen from the Gaussian distribution.

4.2 Prediction: Hypothesis Propagation

Let us assume that the parameters of the circle gradually change over time. We model the assumption as the system equation $\boldsymbol{x}_t = \boldsymbol{x}_{t-1} + \boldsymbol{v}$. According to the model, the set of the particles $\{\boldsymbol{x}_t^{(i)} | i = 1, 2, \ldots, N\}$ at time t is generated from $\{\boldsymbol{x}_{t-1}^{(i)} | i = 1, 2, \ldots, N\}$ at time $t-1$, that is,

$$\boldsymbol{x}_t^{(i)} = \boldsymbol{x}_{t-1}^{(i)} + \boldsymbol{v}^{(i)} \text{ with } \boldsymbol{v}^{(i)} \sim \mathcal{N}(\boldsymbol{0}, \boldsymbol{V}), \quad i = 1, 2, \ldots, N. \tag{10}$$

This particle set approximates the prediction distribution $p(\boldsymbol{x}_t | \boldsymbol{x}_{t-1})$, which indicates the set of hypotheses (potential curves) at time t. Note that this approach can deal with more complicated transition models (theoretically, any models). This paper focuses on the simplest linear model to examine the basic performance of the algorithm.

4.3 Filtering: Voting-Based Verification of Curves

The algorithm evaluate the likelihoods of the particles of the prediction distribution by the normalized votes to 1:

$$w_t^{(i)} = \frac{I(\boldsymbol{x}_t^{(i)})}{\sum_{i=1}^{N} I(\boldsymbol{x}_t^{(i)})} = \frac{\sum_k h(\boldsymbol{r}_k, \boldsymbol{x}_t^{(i)})}{\sum_{i=1}^{N} \sum_k h(\boldsymbol{r}_k, \boldsymbol{x}_t^{(i)})}. \tag{11}$$

Note that we redefine the function $h(\cdot, \cdot)$ in Sec. 2 as

$$h(\boldsymbol{r}_i, \boldsymbol{a}) = \begin{cases} 1 \text{ if } \boldsymbol{r}_i \in \mathrm{Bresenham}(\hat{\boldsymbol{a}}), \\ 0 \text{ otherwise}, \end{cases} \tag{12}$$

that is, the algorithm calculates $\sum_k h(\boldsymbol{r}_k, \boldsymbol{x}_t^{(i)})$ in eq. (11) by tracing the sequence of the points on the digital curve

$$\hat{\boldsymbol{x}}_t^{(i)} = (\lfloor x_o(t)^{(i)} + 0.5 \rfloor, \lfloor y_o(t)^{(i)} + 0.5 \rfloor, \lfloor r(t)^{(i)} + 0.5 \rfloor)^\top \tag{13}$$

in the image.

4.4 Resampling and Particle Elimination

The particles $\boldsymbol{x}_t^{(i)}, i = 1, 2, \ldots, N$, are resampled with replacement according to the likelihoods $w_t^{(i)}, i = 1, 2, \ldots, N$, i.e., the particles are drawn with probability $\Pr(X = \boldsymbol{x}_t^{(i)}) = w_t^{(i)}$. This resampling enables us to avoid the degeneracy for which most of the particles have almost zero weight after a few time steps. In other words, the aim of resampling is to concentrate on the particles with large weights [13].

However, resampling cannot prevent the tracker from losing the circle of interest in clutter. For example, consider a tracking problem for noisy binary images, as shown in Fig. 2 (top). Fig. 2 (middle) shows the results obtained by the voting-based tracker with resampling. The circles depicted in the images are the surviving particles after resampling at each time. These results show that the particles lose the circle and widely diffuse over the image as the time increases. This phenomenon arises from the fact that, in noisy images, the relatively high likelihood (votes) is given to the particles which correspond to background noise.

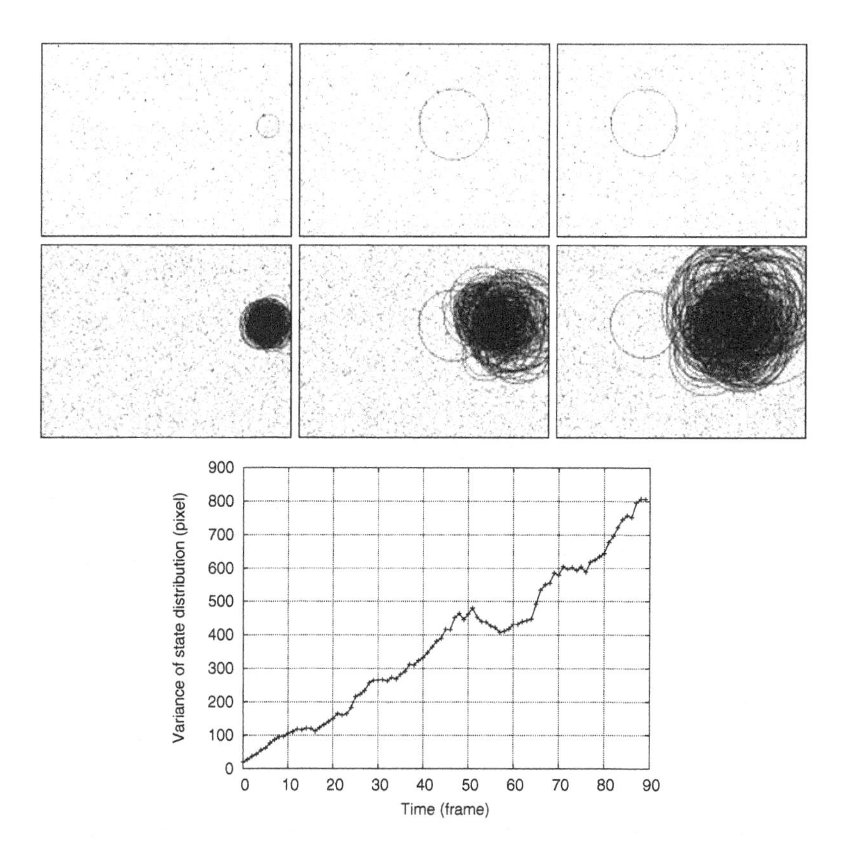

Fig. 2. (Top) examples of a sequence of noisy binary images. (Middle) The tracking results without variance reduction. (Bottom) Temporal change of the variance of the state distribution

From a statistical point of view, this phenomenon causes the variance increase of the state distribution over time. Figure 2 (bottom) shows the trace of the sample covariance matrix over time:

$$\mathrm{tr}V[\boldsymbol{x}] = \sigma_{x_o}^2 + \sigma_{y_o}^2 + \sigma_r^2 \tag{14}$$

$$= \sum_{i=1}^{N} \left((x_o^{(i)} - \mathrm{E}[x_o])^2 + (y^{(i)} - \mathrm{E}[y_o])^2 + (r^{(i)} - \mathrm{E}[r])^2 \right), \tag{15}$$

where $\mathrm{E}[\cdot]$ is the expectation operator. From Fig. 2, one can observe that the variance of the state distribution after resampling increases as the time increase.

Thus, we introduce a procedure for variance reduction. The procedure first selects the mode $\boldsymbol{x}_t^{(m)}$ of the filtering distribution $\{\boldsymbol{x}_t^{(i)} | i = 1, 2, \ldots, N\}$ before resampling, where the mode is selected by $m = \arg\max_i w_t^{(i)}$. Then it assigns the weights of the particles which are distance D away from the mode to 0:

$$w_t^{(i)} = 0 \quad \text{if } \boldsymbol{x}_t^{(i)} \in \left\{ \boldsymbol{x}_t^{(j)} \,|\, j = 1, 2, \ldots, N, d\,(\,\boldsymbol{x}_t^{(j)}, \boldsymbol{x}_t^{(m)}\,) > D \right\}, \tag{16}$$

where $d(\boldsymbol{x}, \boldsymbol{y})$ is the squared Mahalanobis distance defined by

$$d(\boldsymbol{x}, \boldsymbol{y}) = \sqrt{(\boldsymbol{x} - \boldsymbol{y})^\top \boldsymbol{V}^{-1} (\boldsymbol{x} - \boldsymbol{y})}. \tag{17}$$

We illustrate the particle elimination procedure in Fig. 3.

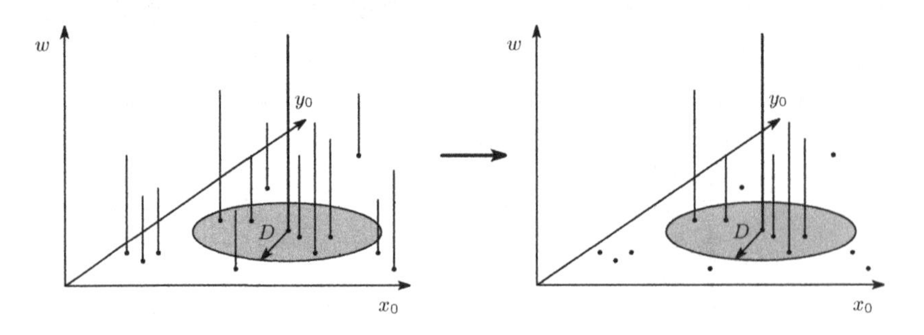

Fig. 3. Eliminating particles away from the mode for variance reduction

Since the particles away from the mode of the distribution have zero weight, they are eliminated after resampling and the other particles around the mode survive over time. Figure 4 (top) shows the result obtained by the voting-based tracker with particle elimination for the input images in Fig. 2(top), and Fig 4 (bottom) shows that the variance does not increase (being almost constant) as the time increase.

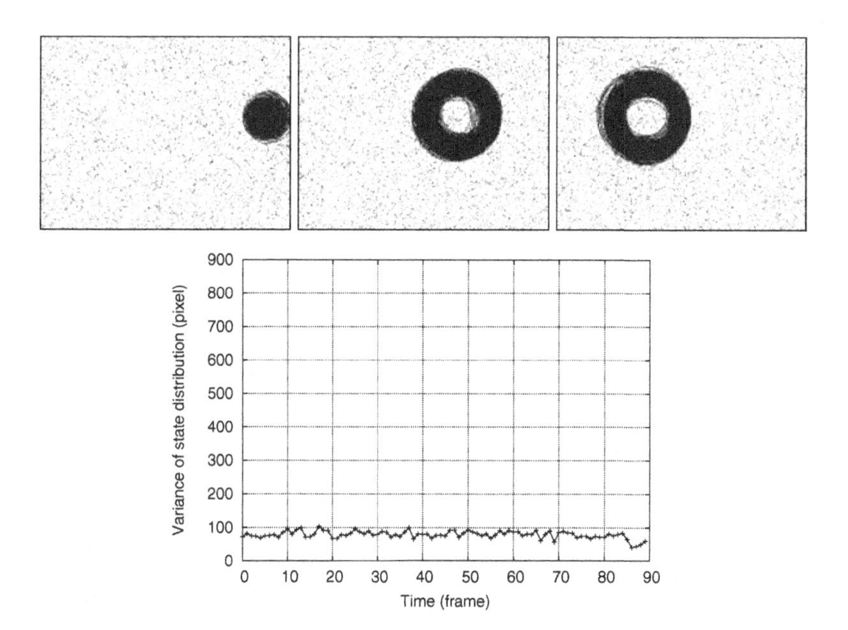

Fig. 4. (Top) The tracking results with particle elimination. (Bottom) Temporal change of the variance of the state distribution

5 Prevention of Redundant Verification

We introduce a procedure for preventing redundant verification by determining equivalence curves in the digital space in order to reduce the time complexity of the algorithm in Sec. 4. The most time-consuming part of the algorithm is the filtering (observation process), i.e., the voting-based verification in Sec. 4.3. This filtering process evaluates each particle $\boldsymbol{x}_t^{(i)}, i = 1, 2, , \ldots, N$, by the summation $\sum_k h(\boldsymbol{r}_k, \boldsymbol{x}_t^{(i)})$. Note that $h(\cdot, \cdot)$ is defined by eq. (12). Therefore, even if the relation $\boldsymbol{x}_t^{(i)} \neq \boldsymbol{x}_t^{(j)}$, for $\boldsymbol{x}_t^{(i)}, \boldsymbol{x}_t^{(j)} \in \boldsymbol{R}^{n_a}$, holds, the relation $\hat{\boldsymbol{x}}_t^{(i)} = \hat{\boldsymbol{x}}_t^{(j)}$, for $\hat{\boldsymbol{x}}_t^{(i)}, \hat{\boldsymbol{x}}_t^{(j)} \in \boldsymbol{N}^{n_a}$, may hold due to digitization. Then the same calculation for the verification of potential curves (hypotheses) repeatedly can be performed for $\boldsymbol{x}_t^{(1)}, \boldsymbol{x}_t^{(2)}, \ldots, \boldsymbol{x}_t^{(N)} \in \boldsymbol{R}^{n_a}$. Figure 5 shows the ratio of the number of "effective" particles to that of the total ones ($N = 500, 1000, 10000$), where any pair of "effective" particles is different from each other in the digital space. In the case of $N = 10000$, only less than 20% of the particles are used in effect, i.e., conversely, the verification for more than 80% of the particles are redundant.

To prevent redundant verification, we determine equivalence curves in the digital space before the filtering process; if $\hat{\boldsymbol{x}}_t^{(i)} = \hat{\boldsymbol{x}}_t^{(j)}$ for $i \neq j$ holds, then we decide that these two vectors are in the same class. This determination for N particles is performed with $\frac{1}{2}N(N+1)$ steps for sorting and N steps for sweeping [20] (Sec. 8,6, pp. 345-346). To maintain each class, we use a singly linked list,

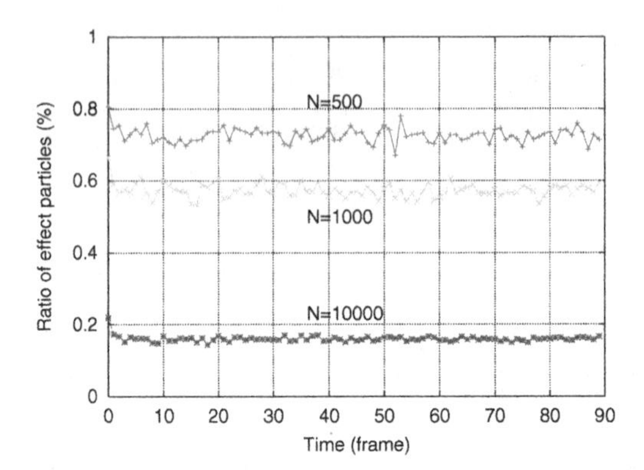

Fig. 5. Ratio of effective particles to total ones

which points to the next node in the list, or to a null value if it is the last node. If a particle is evaluated by its normalized votes in the filtering process, the value is given to the particles in the same class by tracing the elements of the list.

6 Experimental Evaluation in Clutter

To evaluate the basic performance of the proposed algorithm, we use a sequence of noisy binary images, as shown in Fig. 2 (top). This image sequence consists of 90 frames with 320×240 pixels, and a circle moves with evolving time; the center $(x_0(t), y_0(t))^\top$ and the radius $r(t)$ undergo the transition:

$$x_0(t) = 290 - 3t + v, \tag{18}$$

$$y_0(t) = 100 + v, \tag{19}$$

$$r(t) = \begin{cases} r(t-1) + 1 + v, \text{ if } t < \frac{T}{2} \\ r(t-1) - 1 + v, \text{ otherwise} \end{cases}, \tag{20}$$

where v is a white noise generated from the Gaussian distribution with mean 0 and variance 2^2 (pixels), and constant T is the number of the frames. In addition, we add mutually independent 5000 random noises to each image.

First we examine the successful rates of tracking by the algorithm over 90 frames on 10 trials. Table 1 shows the successful rates with $1000, 2000, \ldots, 10000$ particles. For example, the algorithm successfully tracks the circle over the 90 frames at 6 out of 10 times with 1000 particles. We here set the variance of the system noises to be $\sigma_i = 2^2$(pixels)$, i = x_0, y_0, r$. Table 1 shows the algorithm

Table 1. Successful rates of tracking over 90 frames on 10 trials

Number of particles	1000	2000	3000	4000	5000	6000	7000	8000	9000	10000
Successful rate	6/10	9/10	10/10	10/10	10/10	10/10	10/10	10/10	10/10	10/10

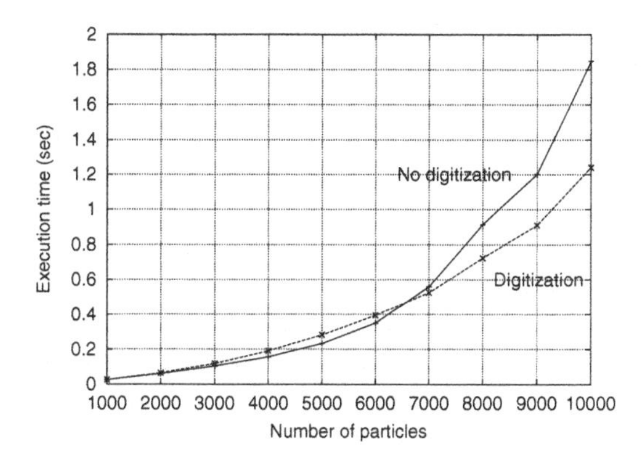

Fig. 6. Execution time

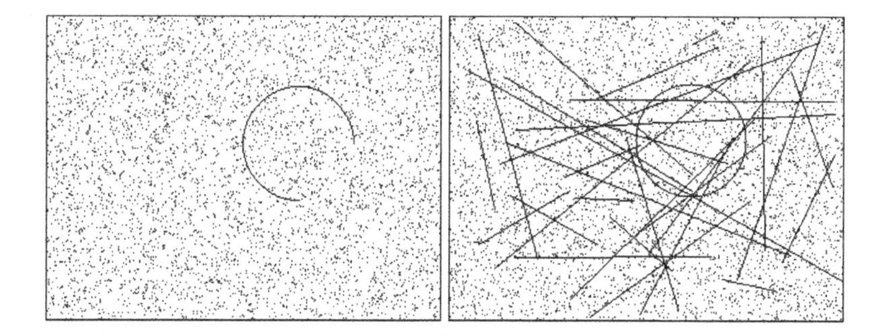

Fig. 7. Examples of input images

tends not to lose the target as the number of particles increases. Hence, although the time and storage complexities increases as the number of particles increases, we should not set particles to be too small in number.

Second we examine the execution time consumed by the tracking algorithm, as shown Fig. 6, on 2.0 GHz Pentium 4 with 256 MB memory. In this figure, "no digitization" and "digitization" indicate the results obtained by the algorithm without and with the digitization, respectively, for preventing redundant verification in Sec. 5. Figure 6 shows the average time of 10 trials (except the case that the algorithm loses the target). From Fig. 6, the procedure for determining equivalence curves in the digital space contributes to the improvement of the efficiency of the algorithm, especially when the particles are large in number. For example, in the case of 10000, the execution time decreases by 32.9%.

We also make other experiments using two binary image sequences, as shown in Fig. 7. In the first case (Fig. 7(left)), we use a circular arc, which lacks part of the circle, as the target to be tracked. In the second case (Fig. 7(right)), we create the images having the more cluttered background than that of the first experiment; we add 20 line segments whose end-points are generated from the uniform distribution over the image plane. In both of the experiments, the other conditions, such as the target transition in eq. (20) and the number of the background noises, are the same as the first experiment. The experimental results show that the proposed algorithm successfully tracks the circle even if the occlusion occurs or the background becomes cluttered.

7 Conclusions

We proposes a visual tracking algorithm based on hypothesis propagation and voting-based verification. The proposed algorithm detects a curve in the image by evaluating the curve by its received votes, and propagates the detected curve to the successive image of a given sequence. Although voting-based algorithms, such as the Hough transform and RANSAC, in general, are time-consuming, this propagation over time provides an efficient search procedure for finding curves. In addition, we introduce a procedure for preventing redundant verification by determining equivalence curves in the digital space. This procedure contributes to the improvement of the efficiency of the algorithm. The experimental results shows the execution time decreases by 32.9% when the particles are 10000 in number. Thus, two combinations of hypothesis propagation and voting-based verification enable us to track curves in clutter robustly and efficiently. Furthermore, we introduce a particle elimination procedure to suppress rapid diffusion of particles over time, which phenomenon is likely to happen in clutter. As a result, the elimination procedure prevents the variance of the filtering distribution from increasing over time.

It might be worth mentioning those which should be developed in the future. The paper assumes that transition model of target objects (e.g., eq. (10)) is known. Then, if the object moves unexpectedly, the algorithm is likely to fail to track it. An adaptive or self-organized process may be useful for treating such a situation.

Acknowledgements

This work was supported in part by the Ministry of Education, Culture, Sports, Science and Technology, Japan, under a Grant-in-Aid for Young Scientists B (No.16700169) and Inamori Foundation.

References

1. R. O. Duda and P. E. Hart, "Use of the Hough transformation to detect lines and curves in pictures," *Comm. ACM*, vol. 15(1), pp. 11–15, Jan. 1972.
2. M. A. Fischer and R. C. Bolles, "Random sample consensus: a paradigm for model fitting with applications to image analysis and automated cartography," *Comm. ACM*, vol. 24(6), pp. 381–395, June 1981.

3. J. P. Princen, J. Illingworth, and J. V. Kittler, "A Formal Definition of the Hough Transform: Properties and Relationships," *J. Mathematical Imaging and Vision*, no. 1, pp. 153–168, 1992.
4. "A Linear Algorithm for Incremental Digital Display of Circular Arcs", *Comm. ACM*, vol. 20(2), pp. 100–106, June 1977.
5. L. Xu and E. Oja. "Randomized Hough Transform: Basic Mechanisms, Algorithms, and Computational Complexities," *CVGIP: Image Understanding*, vol. 57(2), pp. 131–154, 1993.
6. J. R. Bergen and H. Shvaytser, "Probabilistic Algorithm for Computing Hough Transform," *Journal of Algorithms*, vol. 12(4), pp. 639–656, 1991.
7. N. Kiryati, Y. Eldar, and M. Bruckstein, "A Probabilistic Hough Transform," *Pattern Recognition*, vol. 24(4), pp. 303–316, 1991.
8. O. Chum and J. Matas, "Randomized RANSAC with $T_{d,d}$ test," *Proc. the British Machine Vision Conference*, vol. 2, pp. 448-457, Sep. 2002.
9. A. Rosenfeld, J. Jr. Ornelas, and Y. Hung, "Hough Transform Algorithms for Mesh-Connected SIMD Parallel Processors", *Computer Vision, Graphics, and Image Processing*, vol. 41(3), pp. 293–305, 1988.
10. H. Li, M. A. Lavin, and R.L.Master, "Fast Hough transform: a hierarchical approach", *Computer Vision, Graphics, and Image Processing*, vol. 36, pp. 139–161, 1986.
11. J. Illingworth and J. Kittler, "The adaptive Hough transform", *IEEE Trans. Pattern Analysis and Machine Intelligence*, vol. 9(5), pp. 690–698, 1987.
12. P.H.S. Torr, and C. Davidson, "IMPSAC: synthesis of importance sampling and random sample consensus", *IEEE Trans. Pattern Analysis and Machine Intelligence*, vol. 25(3), pp. 354–364.
13. J.S. Liu and R. Chen, "Sequential Monte Carlo methods for dynamic systems", *J. the American Statistical Association*, vol. 93, pp. 1033-1044, 1998.
14. A. Doucet, S. Godill, and C. Andrieu, "On Sequential Monte Carlo Sampling Methods for Bayesian Filtering", *Statistics and Computing*, vol. 10(3), pp. 197–208, 2000.
15. A. Doucet, N. de Freitas, and N. J. Gordon, "*Sequential Monte Carlo Methods in Practice*", Springer-Verlag, May 2001.
16. N. J. Gordon, D. J. Salmond, and A. F. M. Smith, "Novel approach to nonlinear/non-Gaussian Bayesian state estimation", *IEE Proc.-F*, vol. 140(2), pp. 107–113, April 1993.
17. M. Isard and A. Black, "Condensation – Conditional density propagation for visual tracking," *Int. J. Computer Vision*, vol. 29(1), pp.5–28, 1998.
18. M. S. Arulampalam, S. Maskell, N. Gordon, and T. Clapp, "A Tutorial on Particle Filters for Online Nonlinear/Non-Gaussian Bayesian Tracking", *IEEE Trans. Signal Processing*, vol. 50(2), pp. 174–188.
19. G. Kitagawa, "Monte Carlo filter and smoother for non-Gaussian nonlinear state space models", *J. Comput. Graph. Stat.*, vol.5(1), pp. 1–25, 1996.
20. W. H. Press, S. A. Teukolsky, W. T. Vetterling, "Numerical Recipes in C: The Art of Scientific Computing", Cambridge Univ Press, 1993.

3D Topological Thinning by Identifying Non-simple Voxels

Gisela Klette and Mian Pan

CITR, University of Auckland, Tamaki Campus, Building 731,
Auckland, New Zealand

Abstract. Topological thinning includes tests for voxels to be simple or not. A point (pixel or voxel) is simple if the change of its image value does not change the topology of the image. A problem with topology preservation in 3D is that checking voxels to be simple is more complex and time consuming than in 2D. In this paper, we review some characterizations of simple voxels and we propose a new methodology for identifying non-simple points. We implemented our approach by modifying an existing 3D thinning algorithm and achieved an improved running time.

Keywords: simple points, topology, simple deformations, thinning, shape simplification.

1 Introduction and Basic Notions

The basic notion of a simple point is used in topology preserving digital deformations in order to characterize a single element p of a digital image I which can change the value $I(p)$ without destroying the topology of the image. Especially in 3D it is important to find efficient ways of identifying simple voxels as part of algorithms that determine a "central path" [12, 14].

We use common adjacency concepts: 4-,8-(2D), 6-,18-,26-(3D) for the point model and 0-,1-(2D), 0-,1-,2-(3D) for the grid cell model (see Figure 1). Any of these adjacency relations A_α, $\alpha \in \{0, 1, 2, 4, 6, 8, 18, 26\}$, are irreflexive and symmetric. The *α-neighborhood* $N_\alpha(p)$ of a pixel location p includes p and its α-adjacent pixel locations.

In 2D the notions of Crossing numbers or Connectivity numbers are well known (see [10] for a review). Characterizations of simple voxels based on counting numbers of components are complex and time consuming and other concepts have been introduced. We review the notion of attachment sets based on the cell model because we will use this concept in our thinning program.

The *frontier* of a voxel is the union of its six faces. A face of a voxel includes its 4 edges, and each edge includes its 2 vertices. Let p be an n-cell, $0 \leq n \leq 3$. The frontier of a n-cell p is a union of i-cells with $0 \leq i < n$ (i.e., excluding p itself). For example, if p is a voxel (3-cell) then the frontier consists of eight 0-cells, twelve 1-cells and six 2-cells.

R. Klette and J. Žunić (Eds.): IWCIA 2004, LNCS 3322, pp. 164–175, 2004.

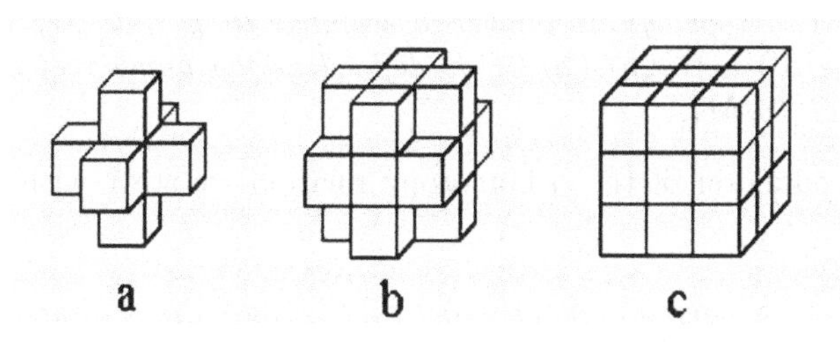

Fig. 1. (a) $N_6(p)$ (b) $N_{18}(p)$ (c) $N_{26}(p)$

Kong [12] defined the I-attachment set of a cell p for the grid cell model as follows, where I is an image:

Definition 1. *Let p and q are grid cells. The I-attachment set of a n-cell p in I is the union of all i-cells, $0 \leq i < n$, on the frontier of p that also lie on frontiers of other grid cells q with $I(p) = I(q), p \neq q$.*

Note that the cardinality of the I-attachment set of a 0-cell is one, and the cardinality of an n-cell is that of the real numbers, for $n = 1, 2, 3$. One example for a 3D I-attachment set is shown in Figure 2.

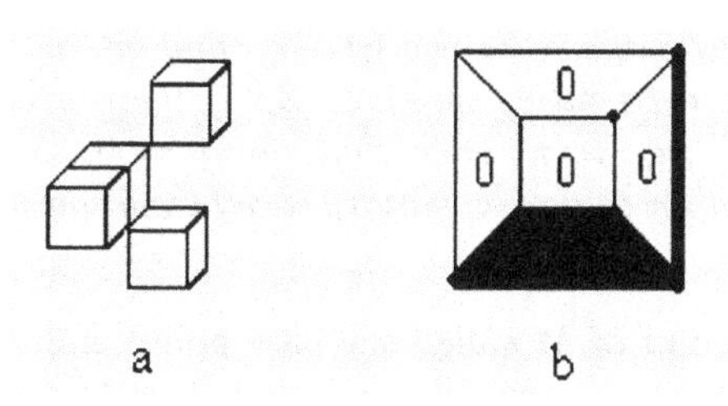

Fig. 2. The voxel in the middle of (a) has the I-attachment set S shown in (b) in form of a Schlegel diagram

To represent the I-attachment set of a voxel we use Schlegel diagrams as Kong proposed in [12].

Geodesic neighborhoods have been introduced by Bertrand [2] in the following way:

Definition 2. *Let $p \in M \subset Z^3$. The geodesic neighborhood of p is defined as follows:*

1. $G_6(p, M) = (A_6(p) \cap M) \cup \{q \in (A_{18}(p) \cap M) \mid q$ *is 6-adjacent to a voxel of* $(A_6(p) \cap M)\}$,
2. $G_{26}(p, M) = A_{26}(p) \cap M$.

Definition 3. *The topological number associated to p and M, denoted by $T_\alpha(p, M)$ for $(\alpha, \alpha') \in (6, 26), (26, 6)$, is defined as the number of α-connected components of $G_\alpha(p, M)$.*

For the introduction of the α-homotopy relation we start with a common definition:

Definition 4. *An α-*path *π with a length l from a point a to a point b in $M \subset Z^3$ is a sequence of voxels $(v_i)_{i=0,\ldots,l}$ such that for $0 \leq i < l$ the voxel v_i is α-adjacent or equal to v_{i+1}, with $v_0 = a$ and $v_l = b$.*

The path π is a *closed path* if $v_0 = v_l$, and it is called a *simple path* if $v_i \neq v_j$ when $i \neq j$ (except for v_0 and v_l if the path is closed). The voxels v_0 and v_l are called *extremities* of π. Given a path $\pi = (v_k)_{k=0,\ldots,l}$, we denote by π^{-1} the sequence $(v'_k)_{k=0,\ldots,l}$ such that $v'_k = v_{l-k}$ for $k \in 0, ..., l$.

Following [4], let $\pi = (v_i)_{i=0,\ldots,l}$ and $\pi' = (v'_k)_{k=0,\ldots,l'}$ be two α-paths and $v_l = v'_0$. We denote by $\pi \bigoplus \pi'$ the path $v_0, ..., v_{i-1}, v'_0, ..., v'_{l'}$ which is the concatenation of the given two paths.

Two closed α-paths π and π' in $M \subset Z^3$ with the same extremities are identical up to an elementary deformation in M if they are of the form $\pi = \pi_1 \bigoplus \gamma \bigoplus \pi_2$ and $\pi' = \pi_1 \bigoplus \gamma' \bigoplus \pi_2$, the α-paths γ and γ' have the same extremities, and they are included in a $2 \times 2 \times 2$ cube if $(\alpha, \alpha') = (26, 6)$, and in a 2×2 square if $(\alpha, \alpha') = (6, 26)$.

Definition 5. *Two α-paths $\pi = (v_i)_{i=0,\ldots,l}$ and $\pi' = (v'_k)_{k=0,\ldots,l'}$ are α-*homotopic *with fixed extremities in $M \subset Z^3$ if there exists a finite sequence of α-paths $\pi = \pi_0, ..., \pi_n = \pi'$ such that, for $i = 0, ...n-1$, the α-path π_i and π_{i+1} are identical up to an elementary α-deformation with fixed extremities ($\pi \simeq_\alpha \pi'$).*

Let B be a fixed point of M called the base point, and $A^\alpha_B(M)$ the set of all closed α-paths $\pi = (v_i)_{i=0,\ldots,l}$ which are included in M and $B = v_0 = v_l$. The α-homotopy relation is an equivalence relation on $A^\alpha_B(M)$, and $E^\alpha(M, B)$ is the set of equivalence classes of this relation. If $\pi \in A^\alpha_B(M)$ then $[\pi]_{E^\alpha(M,B)}$ is the equivalence class of π under this relation.

The concatenation of closed α-paths is compatible with the α-homotopy relation. It defines an operation on the set of equivalence classes $E^\alpha(M, B)$ which associates to the class of π_1 and the class of π_2 the class of $\pi_1 \bigoplus \pi_2$. This operation provides a group structure for the set of all equivalence classes. We call this group the α-fundamental group of M with base point B. If two base points B and B' can be connected by an α-path in M then the α-fundamental group of M with base point B and the α-fundamental group of M with base point B' are isomorphic.

Let $N \subset M$ and $M \subset Z^3$ and let $B \in N$ be a base point. A closed α-path in N is a special α-path in M. If two closed α-paths in N are homotopic then they are also homotopic in M. We define a canonical morphism $i_* : E^\alpha(N, B) \rightarrow$

$E^\alpha(M, B)$ which we call the morphism induced by the inclusion map $i : N \to M$. To the class of a closed α-path $\pi_1 \in A_B^\alpha(N)$ in $E^\alpha(N, B)$, the morphism i_* associates the class of the same α-path in $E^\alpha(M, B)$.

2 Characterizations of Simple Voxels

The literature offers a long list of definitions of simple points in 3D images. The following questions arise: can we find analog characterizations for 3D images based on concepts used for the 2D case; what characterizations are efficient to determine simple voxels in 3D, and are these existing characterizations equivalent. First we review some definitions. In 1994, Bertrand and Malandain [2] proposed the following.

Characterization 1. *A voxel $(p, I(p))$ of an image I is a 26-simple grid point iff it is 26-adjacent to exactly one 26-component of voxels in $A_{26}(p)$ and it is 6-adjacent to exactly one distinct 6-component of voxels in $A_{18}(p)$.*

A voxel $(p, I(p))$ of an image I is a 6-simple grid point iff it is 6-adjacent to exactly one 6-component of voxels in $A_{18}(p)$ and it is 26-adjacent to exactly one distinct 26-component of voxels in $A_{26}(p)$.

The calculation of the numbers of components in the 26-neighborhood of p is time consuming for 3D images. It can be done by using algorithms derived from graph theory. The number of computations depends on the size of the components.

The following characterizations are equivalent for 26-simple 1s [15, 19]:

Characterization 2. *An object voxel p of a 3D image I is 26-simple in I iff*

1. *p is 26-adjacent to another object voxel q, and*
2. *p is 6-adjacent to a background voxel q', and*
3. *the set of object voxels which are 26-adjacent to p is connected, and*
4. *every two background voxels that are 6-adjacent to p are 6-connected by background voxels that share at least one edge with p.*

In terms of topological numbers we can express this characterization as follows [3]:

Characterization 3. *Let $p \in M$ and $(\alpha, \alpha') \in \{(6, 26), (26, 6)\}$. The object voxel p is 26-simple iff $T_\alpha(p, M) = 1$ and $T_{\alpha'}(p, \overline{M}) = 1$.*

The following characterization of simple voxels uses the concept of the I-attachment set of p for 3D images [12]. In Figure 3, the complement of the I-attachment set of p in the frontier of p is not connected and p is not simple.

Characterization 4. *An object voxel p of an image I is 26-simple in I iff the I-attachment set of p, and the complement of that set in the frontier of p, are both non-empty and connected.*

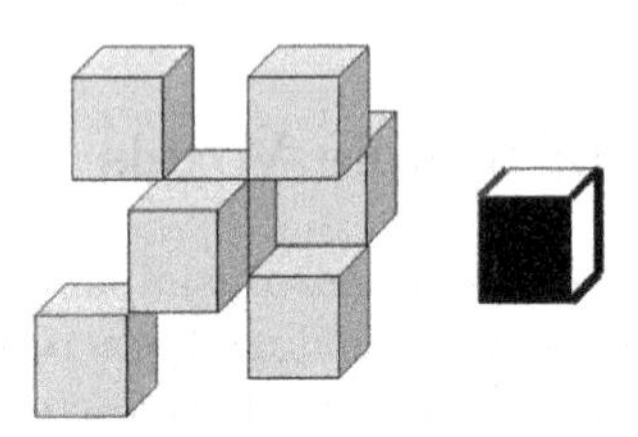

Fig. 3. I-attachment set is not empty, connected and it is not the entire frontier, but p is not simple

This characterization uses the grid cell model; it is equivalent to the previous characterization which is based on the grid point model for 26-simple object voxels [11].

A characterization of 3D simple points based on the α-fundamental group is given in [4, 12].

Characterization 5. *Let $M \subset Z^3$ and $p \in M$. An object voxel p is α-simple ($\alpha \in 6, 26$) iff*

1. *M and $M \backslash p$ have the same number of connected components.*
2. *$\bar{M}$ and $\bar{M} \cup p$ have the same number of connected components.*
3. *For each voxel B in $M \backslash p$, the group morphism $i_* : E^\alpha(M \backslash p, B) \rightarrow E^\alpha(M, B)$ induced by the inclusion map $i : M \backslash p \rightarrow M$ is an isomorphism.*

The main result in [4] is that these three items are sufficient to characterize simple voxels. If a voxel satisfies these three conditions then $T_\alpha(p, M) = 1$ and $T_{\alpha'}(p, \overline{M}) = 1$. Also, if $T_\alpha(p, M) = 1$ and $T_{\alpha'}(p, \overline{M}) = 1$ then these three conditions follow, and even more, the following condition is satisfied as well: for each voxel B' in $\bar{M}$, the group morphism $i'_* : E^{\alpha'}(\bar{M}, B') \rightarrow E^{\alpha'}(\bar{M} \cup p, B')$ induced by the inclusion map $i' : \bar{M} \rightarrow \bar{M} \cup p$ is an isomorphism.

3 Test for Non-simple Voxels

The Euler characteristic of an I-attachment set is easy to compute. We review the definition.

Definition 6. *The* Euler characteristic *$\varepsilon(S(p))$ of the I-attachment set S of a voxel p is equal to the number of 0-cells minus the number of 1-cells plus the number of 2-cells [5].*

Characterization 6. *A voxel p is 26-simple iff the I-attachment set S of p, and the complement of that set in the frontier of p are connected and $\varepsilon(S(p)) = 1$.*

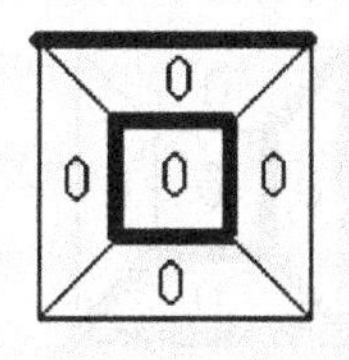

Fig. 4. $\varepsilon(S(p)) = n_0 - n_1 + n_2 = 1$, where $n_0 = 6, n_1 = 5, n_2 = 0$ and p is not simple

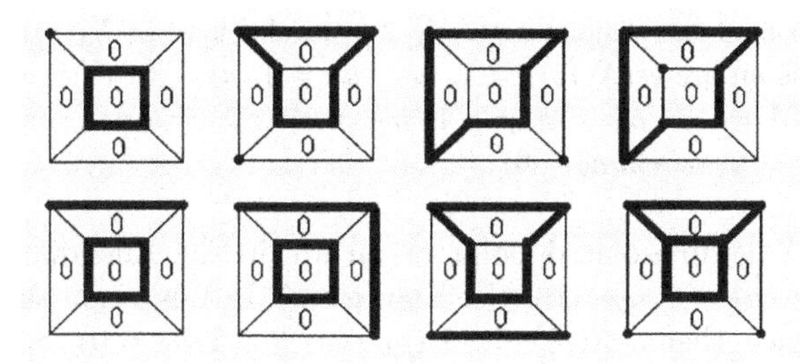

Fig. 5. Voxels that are not simple, for $\varepsilon(S(p)) = 1$ and $N_6 = 6$

Note that a vertex v (0-cell) of a voxel p shares its point with three 1-cells and three 2-cells in the frontier of p. An edge e (1-cell) shares its points with two 0-cells, four other 1-cells and four 2-cells in the frontier of p. A face f (2-cell) shares its points with four 0-cells, eight 1-cells and four other 2-cells. Let $C_x(p)$ be the set of all n-cells ($n \in 0, 1, 2, 3$) of a given n-cell x in the frontier of a voxel p that share points with x. We say that a given n-cell x in the I-attachment set S of p is *isolated* if $C_x(p) \cap S = x$. We say that a given 2-cell y in the complement of the I-attachment set $\bar{S}$ of p is *isolated* if all points in the frontier of y belong to the I-attachment set S of p.

If p is simple then $\varepsilon(S(p)) = 1$. But, if $\varepsilon(S(p)) = 1$ then it does not follow that p is simple, because there are two additional conditions. We investigated the cases where $\varepsilon(S(p)) = 1$ and p is not simple (see figure 4) and we came to the following characterization for non-simple voxels.

Let N_6 be the number of *background voxels* which are *6-adjacent* to p. All configurations for $N_6 = 6$, where $\varepsilon(S(p)) = 1$ and p is not simple are shown in Figure 5.

Proposition 1. *Let $N_6 = 6$. An object voxel p is not simple and $\varepsilon(S(p)) = 1$ iff the I-attachment set S consists of two or three disjoint connected subsets of points. For two sets, one of these sets is a simple curve in the Euclidian space, and the other set is a single point or an arc in the Euclidian space. For three disjoint sets, one of these sets is a non-simple curve in the Euclidian space, and the other sets are isolated points in the Euclidian space.*

Proof. Let p be an *object voxel*. Let X be a subset of S ($X \subseteq S$). If X is a simple curve in the Euclidian space, then $\varepsilon(X(p)) = 0$ ($n_0 = n_1, n_2 = 0$) (See

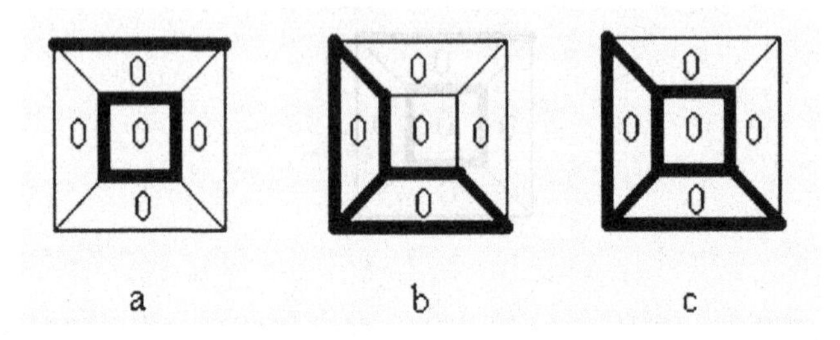

Fig. 6. (a)S consists of two disjoint sets. X_1 is a simple curve ($\varepsilon(X_1(p)) = 0$, $n_0 = n_1$, $n_2 = 0$) and X_2 is an arc ($\varepsilon(X_2(p)) = 1$, $n_0 - n_1 = 1$, $n_2 = 0$) (b)S is a non-simple curve and ($\varepsilon(S(p)) = -1$, $n_0 - n_1 = -1$, $n_2 = 0$) (c)S is a non-simple curve and ($\varepsilon(S(p)) = -2$, $n_0 - n_1 = -2$, $n_2 = 0$)

Figure 6.a). If X is an isolated point or an arc in the Euclidian space, then $\varepsilon(X(p)) = 1$ ($n_0 - n_1 = 1, n_2 = 0$) (See Figure 6.a). If X is a non-simple curve in the Euclidian space, then $\varepsilon(X(p)) \leq -1$ ($n_0 - n_1 \leq -1, n_2 = 0$) (See Figures 6.b and 6.c).

1. We assume p is not simple and $\varepsilon(S(p)) = 1$ and $n_2 = 0$. Based on Theorem 2 we know that the I-attachment set S or the complement of S ($\bar{S}$) are not connected. $\varepsilon(S(p)) = 1$ if $\sum_{i=1}^{3} \varepsilon(X_i(p)) = 1$. If S consists only of one such subset then $\varepsilon(X(p)) = 1$ if $n_0 - n_1 = 1$. But then S and $\bar{S}$ are connected. This is a contradiction to our assumption. For two nonempty subsets X_1 and X_2 we only have the option that $\varepsilon(X_1(p)) = 1$ and $\varepsilon(X_2(p)) = 0$. X_1 can only be a 0-cell or an arc. For X_2 it follows that $n_0 = n_1$ and this is a curve. For three nonempty subsets we have only the option that $n_0 - n_1 \leq -1$ for one subset, and there must be two others with $\varepsilon(X(p)) = 1$. This is only possible if one subset constitutes a non-simple curve and the other two are both isolated points.
2. Now we assume that S consists of two or three disjoint connected subsets of points. For two sets, one of these sets is a simple curve in the Euclidian space, and the other set is a single point or an arc in the Euclidian space. For three disjoint sets, one of these sets is a non-simple curve in the Euclidian space, and the other sets are single points in the Euclidian space. It follows immediately that p is not simple and the $\varepsilon(S(p)) = 1$.

In conclusion, for $N_6 = 6$, all possible cases for a non-simple voxel p and $\varepsilon(S(p)) = 1$ are shown in Figure 5.

For $N_6 = 5$, all cases for a non-simple voxel p and $\varepsilon(S(p)) = 1$ are shown in Figure 7. For $N_6 = 1, 2, 3, 4$, only one case meets this condition, which is shown in Figure 8.

Proposition 2. *Let N_6 be the number of 6-neighbors of p in $\bar{M}$ and $1 < N_6 \leq 6$, $N_{26}(\mathrm{p}) \cap M(\mathrm{p}) > 1$. A voxel p is 26-non-simple iff $\varepsilon(S(p)) \neq 1$, or $\varepsilon(S(p)) = 1$ and S includes an isolated 0-cell or an isolated 1-cell or $\bar{S}$ includes an isolated 2-cell.*

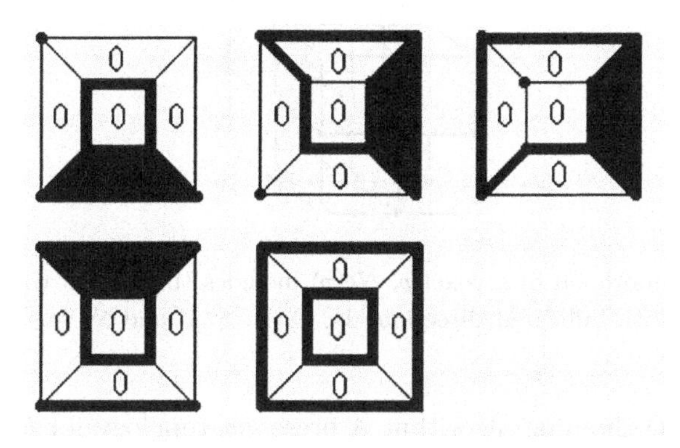

Fig. 7. Voxels are not simple and $\varepsilon(S(p)) = 1$ for $N_6 = 5$

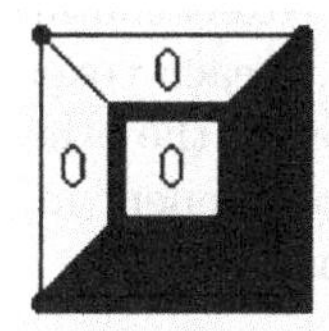

Fig. 8. Voxel is not simple and $\varepsilon(S(p)) = 1$ for $N_6 = 1, 2, 3, 4$

4 Topological Thinning Algorithm

A thinning algorithm is subdivided in a finite number of iterations. In each iteration the object value of voxels satisfying specified deletion conditions are changed into background values. The algorithm stops in one iteration step if no voxel can be found which satisfies the conditions. Thinning algorithms are different with respect to applied local operators. A local operation on a voxel is called sequential if the arguments of the new value of p are the values of the already processed neighbors and the old values of the succeeding neighbors in a defined raster scan sequence. A local operation on a voxel is called parallel if the arguments of the new value of p are the original values in a defined neighborhood of voxels. In all parallel thinning procedures, an efficient test of simplicity of a single voxel is insufficient. We also need to check topology preservation for sets of voxels because sets of voxels are deleted simultaneously in each iteration.

Kong introduced in [12] a method of verifying that a 3D thinning algorithm A preserves topology. The set $D = d_1, d_2, ..., d_k$ is called a *simple set* in I if D can be arranged in a sequence of $d_{l_1}, d_{l_2}, ..., d_{l_k}$ in which each d_{l_j} is simple after $d_{l_1}, d_{l_2}, ..., d_{l_{j-1}}$ is deleted. Let $I_j : 0 \leq j \leq m$ the result of the image after j iterations and let $A_j : 0 < j \leq m$ the application of algorithm A after $(j-1)$ iterations. A component is small if every two elements of the component are 26-adjacent to each other.

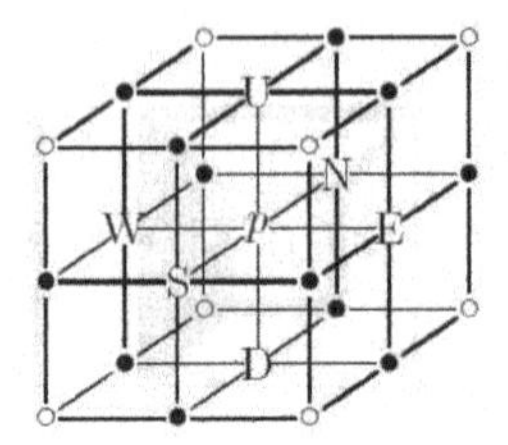

Fig. 9. The neighborhood of a voxel p. $N_6(p)$ includes the central voxel p and six 6-*adjacent* voxels with indicated directions U, D, N, S, E and W for a 6-subiteration algorithm

A parallel 3D thinning algorithm A preserves topology for (26,6)-images if the following conditions hold, for every iteration j and every image I_j.

1. A_j deletes only simple voxels in I_{j-1}.
2. Let T be a set of two or more voxels contained in a block of 1x2x2 or 2x1x2 or 2x2x1 voxels. We have that, if each voxel of T is simple in I_{j-1}, then each voxel is still simple after removing the other voxels in T.
3. In case that there is a small component in I_{j-1} then A_j does not delete at least one voxel of this component.

Three main groups of thinning algorithms have been developed. Examining larger neighborhoods is one possible approach to preserve topology without subiterations. A different approach divides the image into distinct subsets which are successively processed. Many algorithms use different numbers of subiterations (6 or 8 or 12) in order to delete only simple sets of voxels per iteration. A set of border voxels that satisfies specified conditions can be deleted simultaneously in one subiteration.

The modified 3D thinning algorithm in [16, 18] uses 6 subiterations (U, D, N, S, W, E, see Figure 9).

In each subiteration only voxels that have a 6-neighbor in the background in one direction are considered for further tests. Then simple voxels that are not end points are marked. All marked voxels have a second test to secure that they are still simple and not end points after some neighbors have been processed.

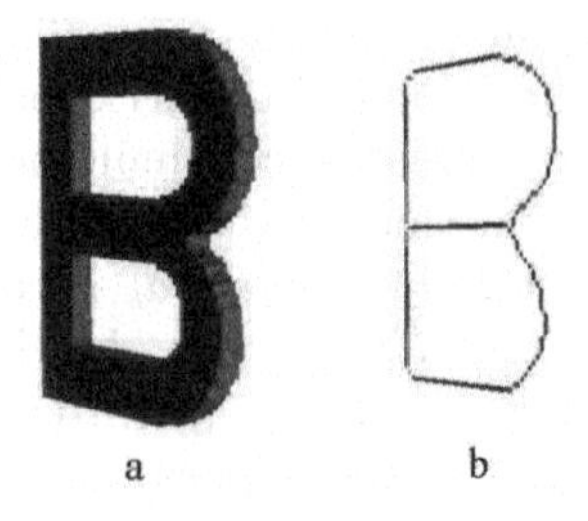

Fig. 10. 3D object and the skeleton of this object

Based on the verification method above we can state that this thinning algorithm preserves topology because of the following:

1. Only simple voxels are deleted per subiteration (condition 1).
2. Simple voxels that are not end points are marked for deletion, and they will be deleted in a second step in case they stay simple after processed neighbors are deleted (condition 2).
3. Condition 3 is verified based on the end point condition.

In [16, 18] the deletion conditions assigned to a subiteration are described by a set of masks. A boolean lookup table indexed by the values of p's 26 neighbors is used. The actual test for simplicity is based on characterization 1.

We applied our approach by replacing the programming code for the test to identify simple voxels in the algorithm provided on the internet.

In case $N_6 = 1$, a voxel is always simple. For $N_6 = 2$, we have only two possibilities: two faces in $\bar{S}$ share an edge or not. For the test of a voxel to be simple or not in a thinning algorithm, the number of 6-neighbors in the background should be counted before a complex procedure starts. For the identification of non-simple voxels can be achieved by the following sequence of tests:

1. If $N_6 = 1$ then voxel p is simple.
2. If $N_6 = 2$ then check whether two of its 6-neighbors are opposite; if not then p is simple.
3. If $N_6 > 2$ then calculate the Euler number $\varepsilon(S(p))$. If $\varepsilon(S(p)) \neq 1$ then p is not simple; otherwise do the following test.
4. If an isolated 0-cell is in S then p is not simple; otherwise do the following test.
5. If an isolated 1-cell is in S then p is not simple; otherwise do the following test.
6. If an isolated 2-cell is in $\bar{S}$ then p is not simple.

5 Conclusions

Simple voxels are of fundamental interest for topology-preserving deformations in 3D binary images. The literature offers characterizations of simple voxels based on different concepts. In this paper we reviewed some characterizations and we characterized non-simple voxels. We used this approach for modifying an existing algorithm and we achieved a significant speed up in running time.

References

1. C. Arcelli and G. Sanniti di Baja: Skeletons of planar patterns. in: *Topological Algorithms for Digital Image Processing* (T. Y. Kong, A. Rosenfeld, eds.), North-Holland, (1996), 99–143.

2. G. Bertrand and G. Maladain: A new characterization of three-dimensional simple points. *Pattern Recognition Letters*, **15**, (1994), 169–175.
3. G. Bertrand: Simple points, topological numbers and geodesic neighborhoods in cubic grids. *Pattern Recognition Letters*, **15**, (1994), 1003–1011.
4. S. Fourey and R. Malgouyres: A concise characterization of 3D simple points. *Discrete Applied Mathematics*, **125**, (2003), 59-80.
5. C. J. Gau and T. Y. Kong: 4D Minimal Non-simple Sets. in: *Discrete Geometry for Computer Imagery*, LNCS 2301, Proc. 10th International Conference, Bordeaux, (2002), 81–91.
6. R. W. Hall: Fast parallel thinning algorithms: parallel speed and connectivity preservation. *Comm. ACM*, **32**(1989), 124–131.
7. R. W. Hall: Parallel connectivity-preserving thinning algorithms. in: *Topological algorithms for Digital Image Processing* (T. Y. Kong, A. Rosenfeld, eds.), North-Holland, (1996), 145–179.
8. C. J. Hilditch: Linear skeletons from square cupboards. in: *Machine Intelligence 4* (B. Meltzer, D. Mitchie, eds.), Edinburgh University Press, (1969), 403–420.
9. G. Klette: Characterizations of simple pixels in binary images. in Proceedings: *Image and Vision Computing New Zealand 2002*, Auckland, (2002), 227–232.
10. G. Klette: A Comparative Discussion of Distance Transformations and Simple Deformations in Digital Image Processing. *Machine Graphics & Vision*,**12**, (2003), 235-256.
11. G. Klette: Simple Points in 2D and 3D Binary Images. in Proceedings of CAIP 2003, LNCS 2756,Springer, Berlin, (2003), 57-64.
12. T. Y. Kong: On topology preservation in 2-D and 3-D thinning. *Int. J. for Pattern Recognition and Artificial Intelligence*, **9**, (1995) 813–844.
13. C. N. Lee, A. Rosenfeld: Simple connectivity is not locally computable for connected 3D images. *Computer Vision, Graphics, and Image Processing*, **51**, (1990) 87–95.
14. C. Lohou and G. Bertrand: A New 3D 6-Subiteration Thinning Algorithm Based on P-Simple Points. in: *Discrete Geometry for Computer Imagery*, LNCS 2301, Proc. 10th International Conference, Bordeaux, (2002), 102–113.
15. G. Maladain and G. Bertrand: Fast characterization of 3D simple points. in: *Proc. 11th IAPR Int. Conf. on Pattern Recognition*, vol. III, The Hague, The Netherlands, (1992), 232–235.
16. K. Palagyi, E. Sorantin, E. Balogh, A. Kuba, C. Halmai, B. Erdohelyi and K. Hausegger: A Sequential 3D Thinning Algorithm and Its Medical Applications, IPMI 2001, LNCS 2082, pages 409-415, Springer Berlin, 2001.
17. K. Palagyi and A. Kuba: Directional 3D Thinning Using 8 Subiterations, in Proceedings: DGCI'99, LNCS 1568, (2003), 325-336.
18. K. Palagyi and A. Kuba: A 3D 6-subiteration thinning algorithm for extracting medial lines, *Pattern Recognition Letters*, **19**: 613-627, 1998.
19. P. K. Saha, B. Chanda, and D. D. Majumder: Principles and algorithms for 2D and 3D shrinking, *Tech. Rep. TR/KBCS/2/91*, NCKBCS Library, Indian Statistical Institute, Calcutta, India, (1991).
20. A. Rosenfeld and J. L. Pfaltz: Sequential operations in digital picture processing. *Comm. ACM*, **13** (1966) 471–494.
21. A. Rosenfeld: Connectivity in digital pictures. *Comm. ACM*, **17** (1970) 146–160.

22. A. Rosenfeld and T. Y. Kong and A. Nakamura: Topology- preserving deformations of two-valued digital pictures. *Graphical Models and Image Processing*, **60**,(1998) 24-34.
23. J. Serra: *Image Analysis and Mathematical Morphology*, vol.2, Academic Press, New York (1982).
24. S. Yokoi and J. I. Toriwaki and T. Fukumura: An analysis of topological properties of digitized binary pictures using local features. *Computer Graphics and Image Processing*, **4**, (1975), 63-73.

Convex Hulls in a 3-Dimensional Space

Vladimir Kovalevsky[1] and Henrik Schulz[2]

[1] Berlin, Germany
kovalev@tfh-berlin.de
[2] Dresden University of Technology, Dresden, Germany
hs24@inf.tu-dresden.de

Abstract. This paper describes a new algorithm of computing the convex hull of a 3-dimensional object. The convex hull generated by this algorithm is an abstract polyhedron being described by a new data structure, the cell list, suggested by one of the authors. The correctness of the algorithm is proved and experimental results are presented.

1 Introduction

The convex hull is a good tool to economically describe a convex object. We present here a new method to compute the convex hull of a three-dimensional digital object using a new data structure, the two-dimensional cell list, suggested by one of the authors [Kov89]. The cell list is well suited to efficiently perform intermediate steps of the algorithm and to economically encode the convex hull.

Since the convex hull is often used in a great number of applications it is an often treated problem in many articles and books on computational geometry. Preparata and Shamos [Pre85] describe a 3D convex hull algorithm of the divide-and-conquer type. The first step in this approach consists in sorting the given points by one of their coordinates. After sorting, the convex hull is computed by a recursive function consisting of two parts: generation of the convex hull of a small subset of points and merging two convex hulls. Since the set of points is sorted, every two subsets are non-intersecting polytopes.

Other approaches such as [Ber00, Cla92] use an incremental method. According to [Ber00] an initial polyhedron of four points is created and then modified by taking the given points into the polyhedron in a random order until it contains the whole set of points. The algorithm connects each point P with the edges of the "horizon", i.e. of the boundary of the subset of faces visible from P. The convex hull is described by a doubly connected edge list. It should be mentioned that the most of these approaches use only simplicial polyhedrons, i.e. the convex hull of the point set is a triangulation.

The algorithm described in this paper can construct the convex hull of any finite set of points which are given by their Cartesian coordinates. However, in the case of a set of voxels, consisting of a few connected components, the algorithm may be essentially accelerated. The idea of this acceleration is based

R. Klette and J. Žunić (Eds.): IWCIA 2004, LNCS 3322, pp. 176–196, 2004.

on the fact, that the convex hull of the whole set is equal to the convex hull of a relatively small subset of "distinguished" voxels or of their "distinguished" vertices.

Let us present some basic definitions necessary for the following sections.

The description of the algorithm and the proof of its correctness are based on the theory of abstract cell complexes (AC complexes) [Kov89]. To remind the reader some basic notions of this theory we have gathered the most important definitions in the Appendix.

Let V be a given set of voxels in a Cartesian three-dimensional space. The voxels of V are specified by their coordinates. In particular the set V may be specified by labeling some elements of a three-dimensional array of bytes or bits. Our aim is to construct the convex hull of the set V while remaining in the frame of abstract cell complexes without using notions from the Euclidean geometry. We consider the convex hull as an abstract polyhedron according to the following definition:

Definition AP: An *abstract polyhedron* is a three-dimensional AC complex containing a single three-dimensional cell whose boundary is a two-dimensional combinatorial manifold without boundary (see Appendix). The two-dimensional cells (2-cells) of the polyhedron are its *faces*, the one-dimensional cells (1-cells) are its *edges* and the zero-dimensional cells (0-cells) are its *vertices* or points.

An abstract polyhedron is called a *geometric* one if coordinates are assigned to each of its vertices. We shall call an abstract geometric polyhedron an AG-polyhedron. Each face of an AG-polyhedron PG must be planar. This means that the coordinates of all 0-cells belonging to the boundary of a face F_i of PG must satisfy a linear equation $H_i(x, y, z) = 0$. If these coordinates are coordinates of some cells of a Cartesian AC complex A then we say that the polyhedron PG is embedded into A or that A contains the polyhedron PG.

Definition CP: An AG-polyhedron PG is called *convex* if the coordinates of each vertex of PG satisfy all the linear inequalities $H_i(x, y, z) \leq 0$ corresponding to all faces F_i of PG. The coefficients of the linear form $H_i(x, y, z)$ are the components of the outer normal of F_i.

A cell c of the complex A containing the convex AG-polyhedron PG is said to *lie in* PG if the coordinates of c satisfy all the inequalities $H_i(x, y, z) \leq 0$ of all faces F_i of PG.

Definition CH: The *convex hull* of a finite set S of points is the smallest convex AG-polyhedron PG containing all points of the set S. "Smallest" means that there exists no convex AG-polyhedron different from PG which contains all points of S and whose all vertices are in PG.

The aim of this paper is to describe the algorithm which constructs the convex hull of an arbitrary set of voxels given by their Cartesian coordinates in a three-dimensional space.

2 The Algorithm

In this section we describe the algorithm of constructing the convex hull of a set V of voxels given as a three-dimensional array in which the voxels of V are labeled. As already mentioned in the Introduction, the algorithm may be essentially accelerated if the set V consists of a few connected components. We shall demonstrate in the next section that the convex hull of the whole set V is equal to the convex hull of a relatively small subset of the so called *local corners.* It is also possible, if desired, to construct the convex hull of the vertices of the voxels, while considering each voxel as a small cube. The latter convex hull is slightly greater than the first: it contains the cubes completely. From this point of view the first convex hull is that of the centers of voxels. Fig. 1 illustrates the idea of these two convex hulls for the 2D case.

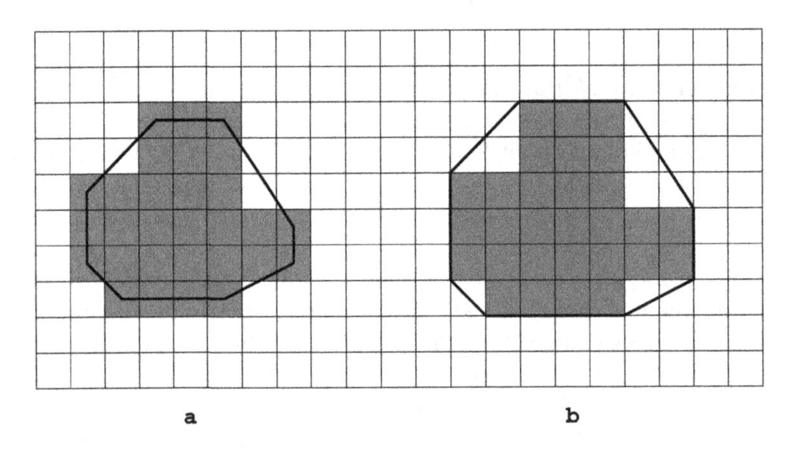

Fig. 1. The convex hull of pixels (a) and that of their vertices (b)

Our algorithm for constructing the convex hull consists of two parts: in the first part a subset of vectors (pointing either to voxels or to the vertices) must be found which are candidates for the vertices of the convex hull. The coordinates of the candidates are saved in an array L. The second part constructs the convex hull of the set L.

A vector $\boldsymbol{v}$ is obviously not suitable as a candidate for a vertex if in the given set V there are two other vectors $\boldsymbol{v}_1$ and $\boldsymbol{v}_2$ such that $\boldsymbol{v}$ lies on the straight line segment connecting $\boldsymbol{v}_1$ and $\boldsymbol{v}_2$, i.e. the vector $\boldsymbol{v}$ is a convex combination of $\boldsymbol{v}_1$ and $\boldsymbol{v}_2$. A vector $\boldsymbol{v}$ is also not suitable as a candidate for a vertex if it may be represented as a convex combination of more than two, say of n voxels $\boldsymbol{v}_i$. Then the vector $\boldsymbol{v}$ may be represented as

$$\boldsymbol{v} = \sum \alpha_i \cdot v_i \quad \text{with } 0 \leq \alpha_i \leq 1 \text{ and } \sum \alpha_i = 1; \quad i = 1, 2, ..., n \tag{1}$$

It is hardly reasonable to test each vector of V whether it is a convex combination of some other vectors, since this problem is equivalent to that of constructing the convex hull. Really, if one knows the convex hull of V, then one also knows the vertices of the convex hull. The vertices are the only vectors which are no convex combinations of other vectors.

However, it is reasonable to test each vector, whether it is a convex combination of vectors *in its neighborhood.* We call a vector which is no convex combination of vectors in its neighborhood a *local corner* of V. The greater the tested neighborhood the less the number of local corners found. Thus, for example, a digital ball B of diameter 32 contains 17077 voxels. The number of local corners of B is 776 when testing 6 neighbors and 360 when testing 26 neighbors of each voxel. We have decided to test 26 neighbors.

When testing a vector $\boldsymbol{v}$ with coordinates (x, y, z) the procedure calculates the coordinates of 13 pairs of vectors from the neighborhood of $\boldsymbol{v}$. The vectors $\boldsymbol{v}_1$ and $\boldsymbol{v}_2$ of each pair are symmetric with respect to $\boldsymbol{v}$, e.g. $\boldsymbol{v}_1 = (x-1, y-1, z-1)$ and $\boldsymbol{v}_2 = (x+1, y+1, z+1)$. If both vectors $\boldsymbol{v}_1$ and $\boldsymbol{v}_2$ are in the set V then the vector $\boldsymbol{v}$ is not a local corner and will be dropped. Only if in each of the 13 pairs the number of vectors belonging to V is 0 or 1, the vector $\boldsymbol{v}$ is recognized as a local corner. The procedure saves the coordinates of each local corner in an array L.

We prefer to consider the problem of recognizing the local corners from the point of view of AC complexes. This gives us the possibility to uniformly examine two different problems: that of constructing the convex hull either of a set of voxels or of the vertices of the voxels.

From the point of view of AC complexes the given set V is the set of three-dimensional cells (3-cells) of a subcomplex M of a three-dimensional Cartesian AC complex A. The complex A represents the topological space in which our procedure is acting. It is reasonable to accept that M is homogeneously three-dimensional (see the Appendix). This means that each cell of M whose dimension is less than 3 is incident to at least one 3-cell of M. With other words, M has no "loose" cells of dimensions less than 3.

Under these assumptions our problem consists in constructing the convex hull either of the interior $Int(M)$ or of the closure $Cl(M)$ of M. The first case corresponds to the convex hull of the set of voxels and the second case to that of the vertices of voxels.

The development of algorithms in digital geometry, to which our problem obviously belongs, and the proof of their correctness becomes easier and more descriptive when using topological coordinates (see Appendix and [Kov01]) rather than standard ones. Using topological coordinates enables one to recognize the dimension of a cell from its coordinates, to easily calculate the coordinates of cells incident to a given cell etc. However, the representation of a Cartesian AC complex in a topological raster [Kov01] demands 8 times more memory space than its representation in a standard raster, where only the voxels get memory elements assigned. The best way to proceed consists in using the standard raster in the computer program, while thinking about the problem in terms of a topological raster.

Consider now the problem of finding the 0-cells which are local corners of the closure $Cl(M)$. When considering only 6 neighbors of a 0-cell then it is easily seen that a 0-cell is a local corner iff it is incident to a *single* 3-cell of M.

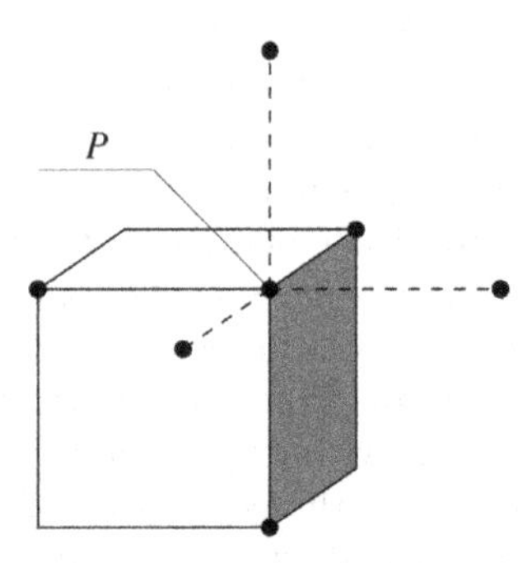

Fig. 2. A 0-cell is a local corner iff it bounds a single 3-cell of M

Really, consider Fig. 2. Here we represent the 0-cells by small dark disks, the 1-cells by line segments, the 2-cells by squares and the 3-cells by cubes. There is only one 3-cell in Fig. 2 and it is in M. If one of the three 1-cells shown in Fig. 2 by dashed lines would be in M then its end points would be also in M since M is closed. In such a case the 0-cell P would be a convex combination of the end points of two 1-cells and thus it would be not a local corner. To each 1-cell of M there must be an incident 3-cell of M since M is homogeneously three-dimensional. Thus in this case there must be at least two 3-cells incident to P. Inversely, if there are more than one 3-cell of M incident to P, then the 0-cells incident to them are also in M, since M is closed. Then there is at least one pair of 0-cells of M such that P is a convex combination of the cells of the pair.

Thus testing the 6-neighborhood of P is rather simple in the topological raster where both the 0-cells and the 3-cells are represented: it is sufficient to test the eight 3-cells incident to P and to count those which are in M. The procedure testing 26 neighbors of P is more complicated.

We have found another solution which needs no topological raster and uses the same procedure which we have described above as the procedure for testing 26 neighbors of a vector. The procedure uses the possibility to interpret the elements of a standard raster as the 0-cells rather than the 3-cells. It is possible to uniquely assign each 0-cell P (except some 0-cells lying in the border of the space) to a 3-cell, e.g. to the farthest one from the coordinate origin among those incident to P. This idea is illustrated by the arrows in the 2D example of Fig. 3a. Thus the given set of voxels may be converted into the set of vertices of these voxels. As the result of the conversion of the given set V some additional elements of the standard raster at the boundary parts of V most remote from the origin become labeled (black squares in the example of Fig. 3b). After this conversion the procedure of testing 26 neighbors of each element of the raster may be applied without any change.

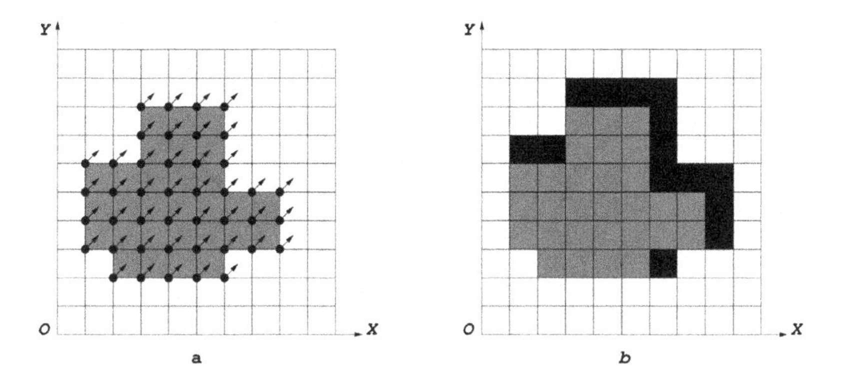

Fig. 3. Converting the coordinates of pixels (gray squares) into those of their vertices; the assignment (a) and the additionally labeled elements (black squares in b)

The second part of our algorithm is that of constructing the convex hull of the set L of the local corners found by the first part.

To build the convex hull of L we first create a simple convex polyhedron spanning four arbitrary non-coplanar local corners of L. It is a tetrahedron. It will be extended step by step until it becomes the convex hull of L. We call it the *current polyhedron* CP.

The surface of the current polyhedron is represented with the data structure called the *two-dimensional cell list* [Kov89] which is now generalized to represent complexes in which a 0-cell may be incident to *any number* of 1-cells. In the original version the list was designed for block complexes [Kov89, Kov01] embedded into a Cartesian complex, where a 0-cell is incident to at most four 1-cells.

The cell list of a two-dimensional complex consists in the general case of three sublists. The kth sublist contains all k-dimensional cells (k-cells), $k = 0, 1, 2$. The 0-cells are the vertices, the 1-cells are the edges, the 2-cells are the faces of the polyhedron. Each entry in the kth sublist corresponds to a k-cell c^k. The entry contains indices of all cells incident to c^k. The entry of a 0-cell contains also its coordinates.

The contents of the cell list is illustrated in Tables 1 to 3 for the case of the surface of a tetrahedron (Fig. 4).

Let us explain the contents of the cell list. The surface is considered as a two-dimensional non-Cartesian complex consisting of vertices (0-cells), edges (1-cells) and faces (2-cells). The sublist of the vertices (Table 1) contains in the first column the identifiers of the vertices, which are simultaneously their indices in the corresponding array.

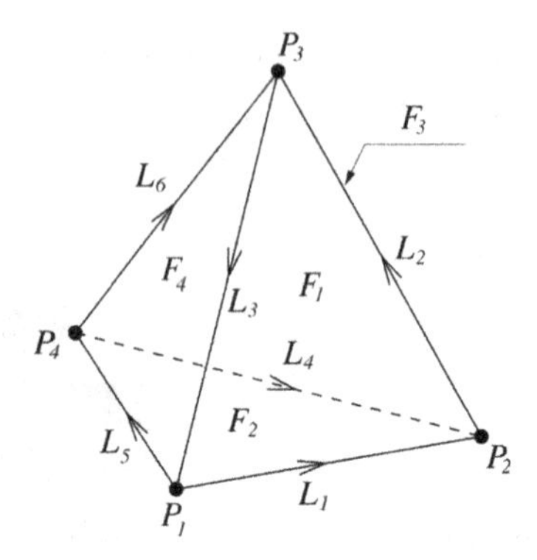

Fig. 4. The tetrahedron and its cells

Table 1. Vertices of the tetrahedron

Vertex	Coordinates	Edges
1	(x_1, y_1, z_1)	-1, 3, -5
2	(x_2, y_2, z_2)	1, -2, 4
3	(x_3, y_3, z_3)	2, -3, 6
4	(x_4, y_4, z_4)	-4, 5, -6

Table 2. Edges of the tetrahedron

Edge	StartP	EndP	LeftF	RightF
1	1	2	1	2
2	2	3	1	3
3	3	1	1	4
4	4	2	2	3
5	1	4	2	4
6	4	3	3	4

Table 3. Faces of the tetrahedron

Face	n	Pairs (P, L)
1	3	$(1, 1), (2, 2), (3, 3)$
2	3	$(1, 5), (4, 4), (2, -1)$
3	3	$(2, -4), (4, 6), (3, -2)$
4	3	$(1, -3), (3, -6), (4, -5)$

The second column contains three integers (x_i, y_i, z_i) for each vertex. These are the coordinates of the vertex. The third column contains in each row the indices of all edges incident to the corresponding vertex. The indices of the edges are signed: the minus sign denotes that the oriented edge points away from the vertex while the plus sign corresponds to an edge pointing to the vertex.

The second sublist (Table 2) is that of edges. Its first column contains the indices. The subsequent columns contain the indices of the starting vertex, of the end vertex, of the left and of the right face, when looking from outside of the polyhedron.

The third sublist (Table 3) is that of faces. The topological relation of a face to other cells is defined by the boundary of the face. The surface of a convex polyhedron is a combinatorial manifold. Therefore the boundary of each face is a 1-manifold, i.e. a simple closed polygon. It is a closed sequence of pairs each of which contains a vertex and an edge. The sequence of pairs stays in a linked list whose content is shown in the third column. The second column contains the number of pairs, which may be different for different faces.

This version of the cell list is redundant because it contains for a pair of two incident cells c^k and c^m both the reference from c^k to c^m and from c^m to c^k. Therefore, for example, the sublist of edges may be computed starting from the sublist of faces. Also the content of the third column of Table 1 may be computed from that data. The redundancy makes the calculation of the convex hull faster because cells incident to each other may be found immediately, without a search. When the calculation of the convex hull is ready, the redundancy of the cell list can be eliminated to save memory space. To exactly reconstruct a convex object from the cell list of its convex hull it suffices to have the coordinates of the vertices and the sublist of the faces where the indices of the edges may be omitted. This is the economical encoding of the convex hull.

The next step in constructing the convex hull is to extend the current polyhedron while adding more and more local corners, some of which become vertices of the convex hull. When the list of the local corners is exhausted the current polyhedron becomes the convex hull of M. The extension procedure is based on the notion of visibility of faces which is defined as follows.

Definition VI. The face F of a convex polyhedron is *visible* from a point P, if P lies in the outer open half-space bounded by the face F, i.e. if the scalar product (N, W) of the outer normal N of the face F and the vector W pointing from a point Q in F to P, is positive. If the scalar product is negative then F is said to be *invisible* from P. If the scalar product is equal to zero then F is said to be *coplanar* with P.

It should be mentioned that the choice of a point Q in F as the starting point of W does not influence the value of the scalar product, since all vectors lying in F are orthogonal to N and therefore their scalar product with N is zero.

To extend the current polyhedron the algorithm takes one local corner after another. For any local corner P it computes the visibility of the faces of the polyhedron from P. Consider first the simpler case when there are no faces of the current polyhedron, which are coplanar with P. The algorithm labels each face of the current polyhedron as being visible from P or not. If the set of visible faces is empty, then the point P is located inside the polyhedron and may be discarded. If one or more faces are visible, then the polyhedron is extended by the point P and some new faces. Each new face connects P with the boundary of the set of visible faces. A new face is a triangle having P as its vertex and one of the edges of the said boundary as its base. All triangles are included into the cell list of the current polyhedron while all visible faces are removed. Also each edge incident to two visible faces and each vertex incident only to visible faces is removed.

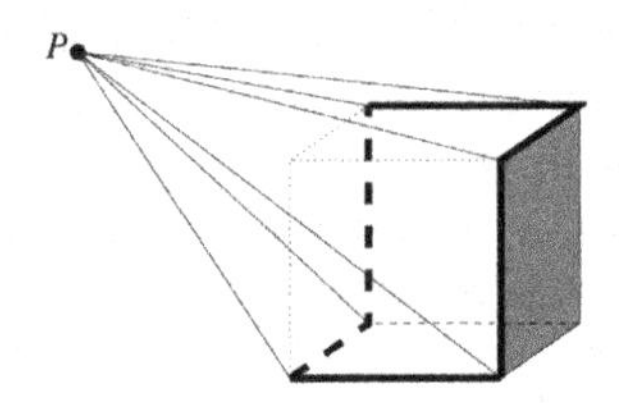

Fig. 5. The current polyhedron (a cube) being extended by the vertex P

In Fig. 5 the boundary of the visible subset is shown by bold lines (solid or dashed). The edges shown by dotted lines must be removed together with the three faces visible from P. The algorithm repeats this procedure for all local corners.

Consider now the problem of coplanar faces. In principle it is possible to treat a face coplanar with the new local corner P in the same way as either a visible or an invisible face. Both variants have its advantages and drawbacks.

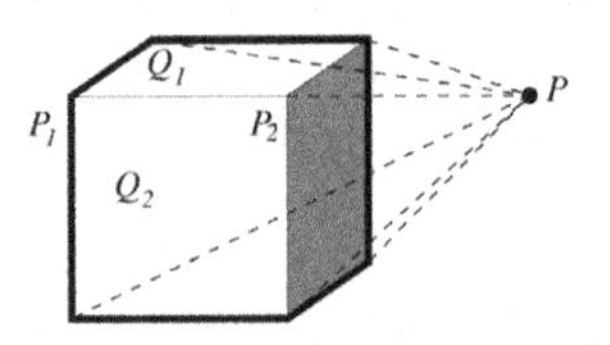

Fig. 6. Treating coplanar faces as visible ones

So when considering faces coplanar with a new vertex P as visible (Fig. 6) then unnecessary many new triangles are constructed some of which are coplanar with each other. They must be merged together later on, and this makes the procedure slower.

When, however, treating a coplanar face as invisible (Fig. 7a) then the number of new faces being constructed may be made smaller: it is possible to connect P only with the end points P_1 and P_2 of the common boundary of the coplanar face Q_2 and of the visible region, rather than with all points of the common boundary. However, in this case some of the faces are no triangles. Their common boundary may consist of more than one edge. The procedure of merging such faces proved

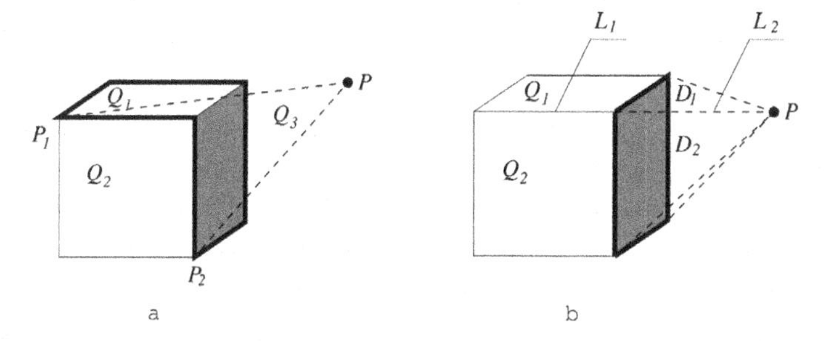

Fig. 7. Treating coplanar faces as invisible; coplanar quadrangles Q_2 and Q_3 (a) and collinear edges L_1 and L_2 (b)

to be much more complicated than that of merging triangles. An example is shown in Fig. 7a: the point P is coplanar with the square Q_2 and with the new quadrangle Q_3. Their common boundary consists of two edges.

When considering coplanar faces as invisible ones some collinear edges may occur (L_1 and L_2 in Fig. 7b). For merging them to a single edge some additional means are necessary.

The best solution seams to consist in considering coplanar faces neither as visible nor as invisible. In this case it is possible to extend a coplanar face towards the point P by a procedure similar to the construction of a two-dimensional convex hull (Fig. 8).

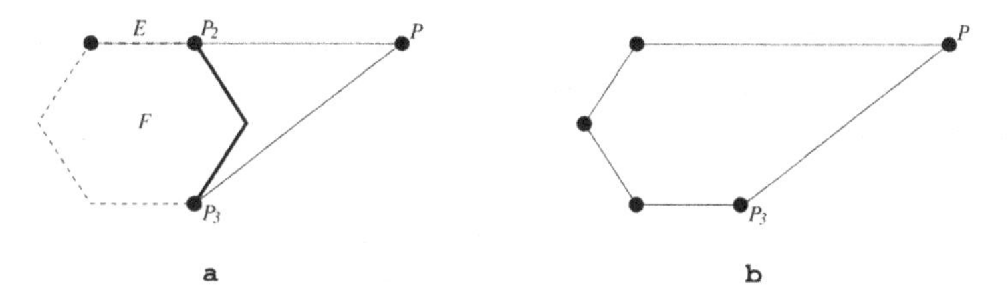

Fig. 8. Constructing a 2D convex hull in the case of a coplanar face: the initial (a) and the resulting (b) face

The number of the new faces being constructed is small. However, the procedure of extending a face towards a coplanar point proved to be still more complicated than that of merging two polygons with more than one common edge. Also some collinear edges may occur in this case (the edges E and (P_2, P) in Fig. 8a). The extension becomes especially complicated in the case when the point P lies on the crossing line of the planes of two adjacent faces (Fig. 6b). Then both of these faces are coplanar with P.

After having tested all three variants we came to the decision that the best solution consists in treating coplanar faces as visible. In this case the program creates sometimes many coplanar triangles which must be merged together. But the procedure of merging triangles is rather simple and fast.

It should be noticed here that merging triangles is not always necessary: many programs of computer graphics work with sets of triangles. Our algorithm gives the possibility to represent the surface of the convex polyhedron as a simplicial complex whose all faces are triangles. To achieve this it suffices to switch off the subroutine of merging. For example, the convex hull of a digital ball of diameter 32 contains 1404 triangles. After merging coplanar triangles the hull contains only 266 faces.

The procedure of adding new faces to the current polyhedron ends after having processed all local corners.

3 Proof of the Correctness

The described algorithm of constructing the convex hull gets a set V of voxels as input and generates an AG-polyhedron K defined by the cell list of its surface. In this section we prove the correctness of the algorithm (Theorem PL). We need the following lemmas LZ, CH and BE.

Lemma LZ. Let V be a set of three-dimensional vectors $\boldsymbol{v} = (v_x, v_y, v_z)$ (pointing to voxels or to vertices of voxels). Let T be a subset of V such that all vectors of T satisfy some given linear inequality, i.e. the inequality cuts T from V. Then T contains at least one vector which is no convex combination of vectors of V.

Proof. Let $H(x, y, z) \geq 0$ be the said inequality. Let us find all vectors of T having $H(x, y, z) = max$. We denote by $TH \subset T$ the subset of all vectors $\mathbf{c} = (c_x, c_y, c_z)$ having the maximum value of H:

$$\mathbf{c} \in TH \longrightarrow H(c_x, c_y, c_z) \geq H(v_x, v_y, v_z) \text{ for each vector } \boldsymbol{v} \in V. \tag{2}$$

The vector $\mathbf{c}$ satisfying (2) can only be a convex combination of vectors from TH since H is linear and its value for a convex combination

$$\mathbf{c} = \alpha \cdot \boldsymbol{v}_1 + (1 - \alpha) \cdot \boldsymbol{v}_2; \text{ with } 0 < \alpha < 1; \ \boldsymbol{v}_1, \boldsymbol{v}_2 \in V. \tag{3}$$

with other vectors would be less than the maximum value of H. If the subset TH contains a single vector $\mathbf{c}$ then we are done: $\mathbf{c}$ is no convex combination of vectors of V. If, however, TH contains more than one vector, then consider the subset $TX \subset TH$ of vectors having the maximum value of the coordinate X. If there is a single such vector then we are done. If not, then all vectors of TX have the same coordinate $x = x_{max}$ and we consider the subset $TY \subset TX$ of vectors having the maximum value of the coordinate Y. This process ends at the latest with the subset $TZ \subset TY$ of vectors having the maximum value of the coordinate Z since there exits only one vector with $x = x_{max}$, $y = y_{max}$, $z = z_{max}$. This vector is no convex combination of vectors of V. □

Lemma CH. Let V be a set of three-dimensional vectors and K an AG-polyhedron satisfying the following three conditions:

1. K is convex;
2. K contains all local corners of V;
3. each vertex of K is identical with one of the local corners of V.

Then K is the convex hull of V according to the Definition CH given in the Introduction.

Proof. First of all let us show that K contains all vectors of V. Suppose there is a subset $T \subset V$ which is not contained in K. This means that the vectors of T do not satisfy at least one of the linear inequalities corresponding to the faces of K. According to Lemma LZ the subset T contains at least one vector which is not a convex combination of vectors of V and thus it is a local corner. This contradicts the second condition of the Lemma. Therefore there is no subset $T \subset V$ outside of K, i.e. K contains all vectors of V.

The polyhedron K is convex and contains all vectors of the set V. To fulfill all conditions of the Definition CH it remains to show that K is the smallest such polyhedron. With other words, we must show that each convex polyhedron K' different from K which is contained in K does not fulfill the conditions of Lemma CH.

If the convex polyhedron K' is contained in K and is different from K then at least one vertex of K' lies in the interior of K. Let us construct the polyhedron K' at first by moving a vertex P of K by a small amount into its interior. Then the original vertex P does not belong to K'. According to the conditions of Lemma CH the vertex P is a local corner of V. Thus there is a local corner of V which is not in K'. Therefore K' does not meet the conditions of Lemma CH.

Each other transformation of K to some polyhedron lying inside of K may be represented as a sequence of movements of vertices of K into its interior. It follows that such a polyhedron does not contain some vertices of K and hence is not in accordance with the conditions of Lemma CH. □

Lemma BE. If an edge E of a polyhedron K lies between a face F_v visible from a point P and another face F_n which is not visible from P then the polyhedron K lies completely in the closed half-space bounded by the plane (P, E) going through P and E.

This means that the coordinates of all vertices of K satisfy the linear inequality $H(x, y, z) \leq 0$ while the linear form $H(x, y, z)$ takes the value zero at the point P and at the endpoints of E. The coefficients of $H(x, y, z)$ are the coordinates of the normal of the plane (P, E), which points to the outside of K.

Proof. If the face F_v is visible from P then the outer normal N_v of F_v and the line segment (Q, P) compose an acute angle $\alpha_v < 90°$ while Q is an arbitrary point in E (Fig. 9). Conversely, the outer normal N_n of the invisible face F_n and the line segment (Q, P) compose an obtuse angle $\alpha_n > 90°$. The angle β_v between the outer normal N_H to the plane (P, E) and N_v is equal to $90° - \alpha_v$

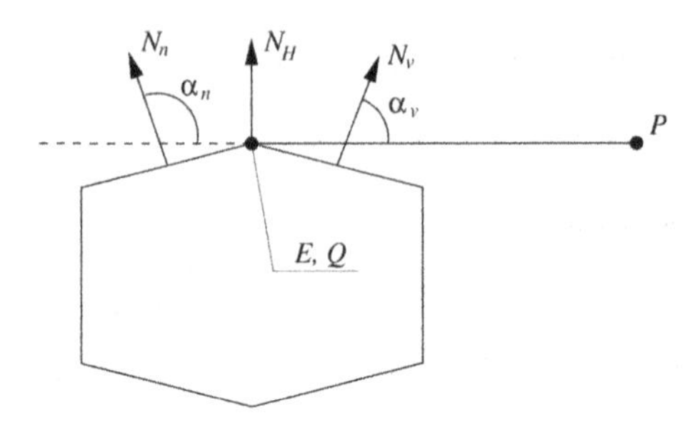

Fig. 9. A section through the polyhedron K and the plane (Q, P)

and is acute. The angle β_n between N_H and N_n is equal to $\beta_n = \alpha_n - 90°$ and is also acute. Consequently the normal N_H to the plane (P, E) is a linear combination of the normals N_v and N_n with positive coefficients (proportional to $cos\beta_v$ and to $cos\beta_n$):

$$N_H = a \cdot N_v + b \cdot N_n; \; a, b > 0 \tag{4}$$

The vector W pointing from Q to an arbitrary point in the polyhedron K has a non-positive scalar product with both N_v and with N_n because K is convex and all its inner point lie on the inner side of each of its faces. According to (4) the scalar product of N_H with the vector W for an arbitrary point of K is non-positive. Thus all points of K lie in the closed half-space bounded by the plane (P, E) which proves the Lemma. □

An angle between two vectors whose end points are cells of a 3D Cartesian AC complex is specified by the scalar and vector product of these vectors.

Theorem PL. The algorithm described in the previous Section constructs the convex hull of the given set of voxels or of vertices of voxels.

Proof. We consider the coordinates of voxels or of vertices of voxels as three-dimensional vectors composing the set V. At its first stage the algorithm saves the local corners of V. Only these vectors are used as candidates for the vertices of the polyhedron being constructed. Thus the third condition of Lemma CH is fulfilled. To fulfill the remaining two conditions it remains to show that the constructed polyhedron is convex and that it contains all local corners of V.

The algorithm starts with a polyhedron K_i, $i = 0$; which is a tetrahedron. It is convex. Then the algorithm takes the next not used local corner P and defines the set of faces which are visible from P (Definition VI in the previous Section). The boundary of the set of visible faces consists of vertices and edges while each edge bounds exactly one visible face. The algorithm constructs a new face for each edge of the boundary. The face spans the edge and the point P. The visible faces become deleted. According to Lemma BE the

Table 4. Comparison of the memory requirements of the cell list and of the triangulation for the example in Fig. 10

	faces	vertices	integers to save
MC-triangulation	3560	~1780	16020
convex hull	63	60	495

Table 5. Experimentally acquired values for half-balls of various diameters

	MC-triangulation			convex hull		
diameter	triangles	integers to save	integers per triangle	faces	inegers in cell list	integers per face
8	504	2268	4.5	30	324	10.80
10	760	3420	4.5	62	454	7.32
12	1072	4824	4.5	50	442	8.84
14	1416	6372	4.5	30	426	14.20
16	1840	8280	4.5	118	1010	8.56
18	2296	10332	4.5	142	1070	7.53
20	2784	12528	4.5	94	938	9.98
22	3488	15696	4.5	118	1142	9.68

polyhedron K_i lies completely in the closed half-space defined by the plane of the new face F. It also lies in the closed half-space corresponding to the inner side of each old not deleted face. Thus the new polyhedron is an intersection of half-spaces and thus it is convex. This is true for any new added local corner P.

Let us show now that K contains all local corners of V. The algorithm processes all local corners of V. Some of them turn out to lie in the current polyhedron. They are not used for the construction of K, but they are nevertheless already contained in K. Other local corners are put into K as their vertices and thus they also become contained in K. All local corners being in K remain in K since the modified polyhedron contains the old one. Thus all local corners are contained in K, all three conditions of Lemma CH are fulfilled and therefore K is the convex hull of V.

□

4 Results of Computer Experiments

We have implemented and tested the described algorithm of constructing the convex hull of a set of voxels or of vertices of voxels. To make a numerical comparison of the memory efficiency we have tested some examples with our algorithm and with the well known Marching Cubes algorithm [Lor87]. It should be mentioned here that we only compare the encoding efficiency of the surface, because the Marching Cubes algotithm does not produce the convex hull of an object.

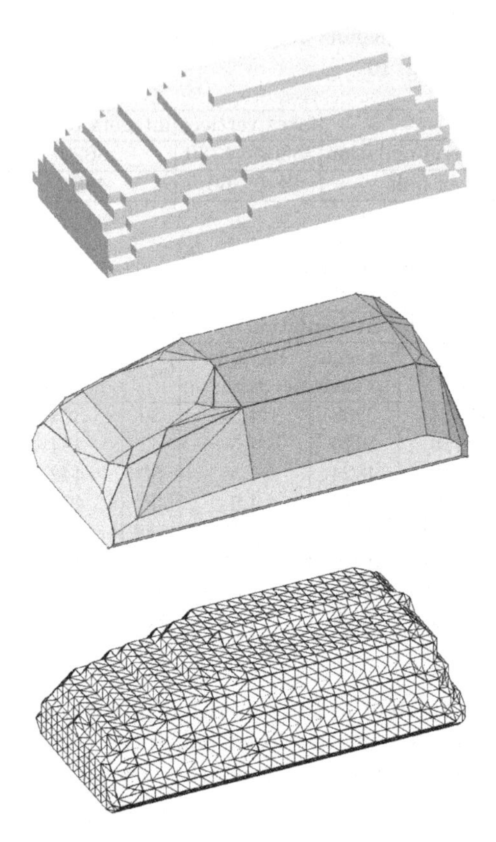

Fig. 10. Example "future car". Top left: The voxel object. It has 3571 voxels and 1782 faces. Top right: Convex hull of this object. Bottom: Triangulation with the Marching Cubes method

We have compared the number of integer values necessary to save a non-redundant cell list with that necessary to save the triangulation of the same object. For the Marching Cubes triangulation method (MC-triangulation) we have assumed the following: one needs to save three coordinates for each vertex and three vertex indices for each triangle.

Since the number N_T of triangles is nearly twice the number N_V of the vertices, we must save on the average

$$size_{tr} = 3 \cdot N_V + 3 \cdot N_T = (3/2 + 3) \cdot N_T \tag{5}$$

integers, i.e. about 4.5 integers for each triangle.

As mentioned above in the Section "The Algorithm" the non-redundant cell list of the convex hull contains the coordinates of the vertices and the sublist of the faces where the indices of the edges are omitted. The later list contains for each face the sequence of the vertex indices in its boundary. In our experiments

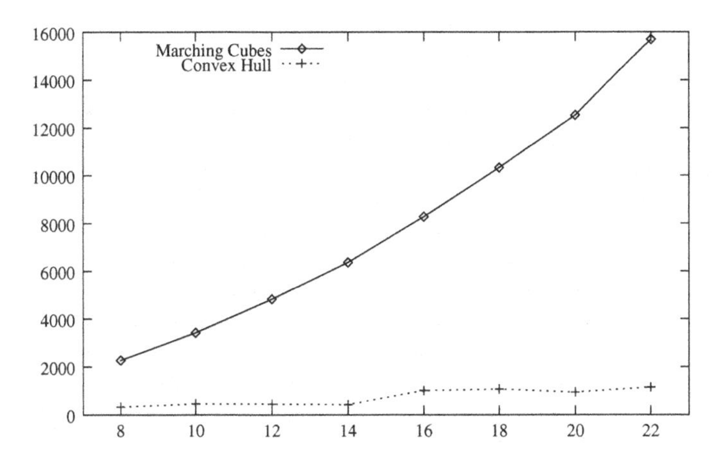

Fig. 11. Number of integers to be saved for half-balls with diameters from 8 to 22 voxels

each face have had on the average about 5 vertices in its boundary. Thus the memory amount necessary to store the non-redundant cell list is approximately equal to

$$size_{cl} = 3 \cdot N_V + 5 \cdot N_F; \tag{6}$$

where N_V and N_F are the numbers of vertices and faces correspondingly.

A simple example is shown in Fig. 10: the object has 3571 voxels and 1782 faces. The convex hull contains only 63 faces.

The following values for the memory requirements have been computed:

The same investigation was made for half-balls with diameters from 8 to 22 voxels (Fig. 12).

The following diagram shows the number of integers necessary to save either the results of the triangulation or the cell list.

As it can be seen from Fig. 12, the convex hull preserves the symmetry of digital objects. The polygons composing the surface are all symmetric. This is an important property of the convex hull, besides its property to be economic, and it is its great advantage as compared with other means representing surfaces of digital objects, e.g. the triangulation or the subdivision into digital plane segments [Kle01].

5 Conclusion

In this paper we present a new algorithm of computing the convex hull of a three-dimensional digitized object represented as a set of voxels. The computed convex hull is an abstract polyhedron which is a particular case of an abstract cell complex. The surface of the polyhedron is encoded by the data structure known as the two-dimensional cell list. This data structure is well suited both to handle the intermediate data during the computation of the convex hull as well as to economically encode the resulting hull. The cell list also provides the possibility to exactly reconstruct the original digitized object. The correctness of the presented algorithm has been proved. Numerous computer experiments demonstrate the memory efficiency of the cell list for convex objects as compared with the well known triangulation method.

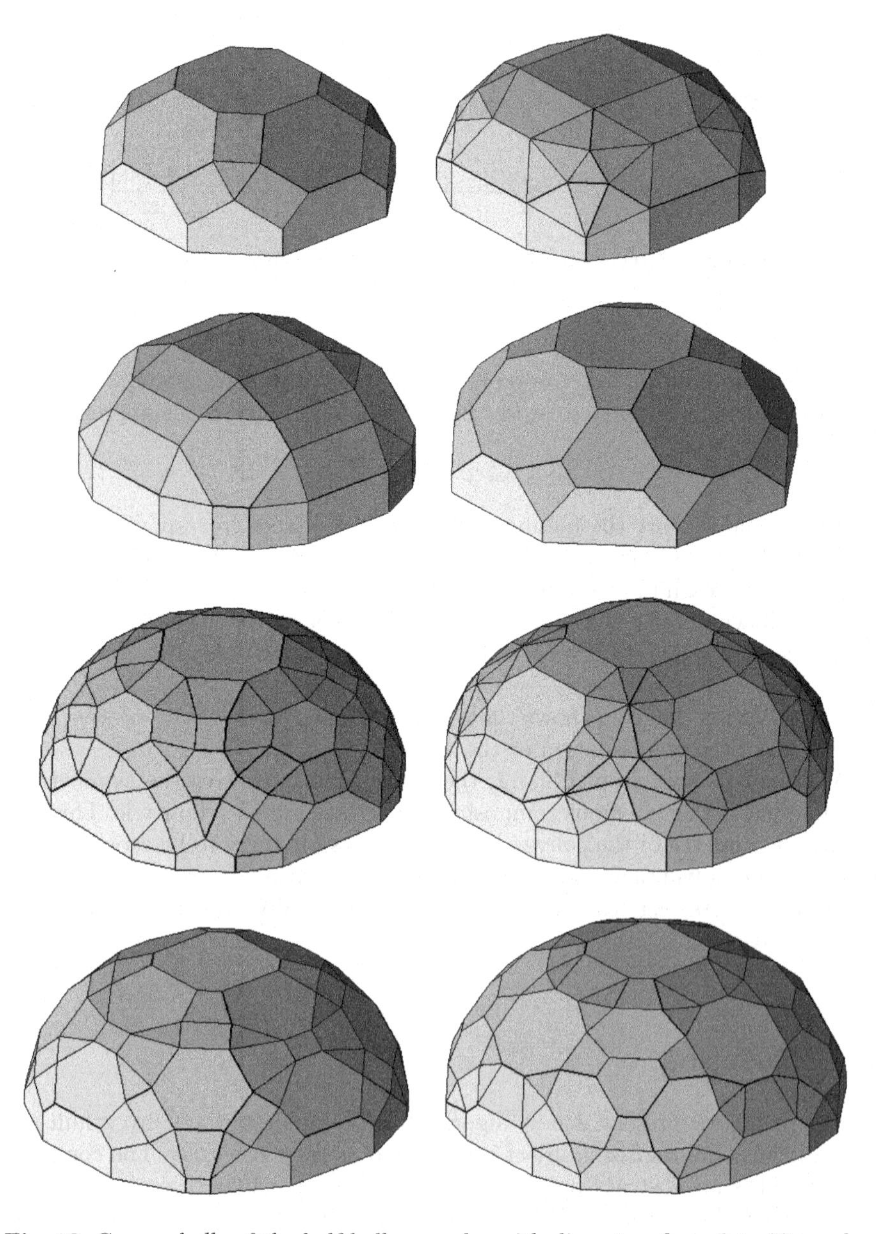

Fig. 12. Convex hulls of the half-ball examples with diameters from 8 to 22 voxels

References

[Ber00] de Berg, M., van Kreveld, M., Overmars, M., Schwarzkopf, O.: *Computational Geometry - Algorithms and Applications.* Springer-Verlag. 2000.

[Cla92] Clarkson, K.L., Mehlhorn, K., Seidel, R.: *Four results on randomized incremental constructions.* Comp. Geom.: Theory and Applications, pages 185-221, 1993. Preliminary version in Proc. Symp. Theor. Aspects of Comp. Sci., 1992.

[Kle01] Klette, R., Sun, H.J.: *A Global Surface Area Estimation Algorithm for Digital Regular Solids.* University of Auckland, CITR-TR-69. 2001.

[Kov89] Kovalevsky, V. A.: *Finite Topology as Applied to Image Analysis.* Computer Vision, Graphics and Image Processing, Vol.45, No.2, pp.141-161. 1989.

[Kov93] Kovalevsky, V. A.: *Digital Geometry based on the Topology of Abstract Cell Complexes.* In Proceedings of the Third International Colloquium "Discrete Geometry for Computer Imagery". University of Strasbourg. pp.259-284. 1993.

[Kov01] Kovalevsky, V. A.: *Algorithms and Data Structures for Computer Topology.* In: Bertrand, G., Imiya, A., Klette, R. (Eds): Digital and Image Geometry. Lecture Notes in Computer Science, Vol.2243, pp.37-58. Springer-Verlag. 2001.

[Kov02] Kovalevsky, V.A.: *Multidimensional Cell Lists for Investigating 3-Manifolds.* Discrete Applied Mathematics, Vol. 125, Issue 1, pp.25-43. 2002.

[Lor87] Lorensen, W.E., Cline, H.E.: *Marching Cubes: A High-Resolution 3D Surface Construction Algorithm.* Computer Graphics, Vol. 21, No. 4, pp.163-169. 1987.

[Pre85] Preparata, F.P., Shamos, M.I.: *Computational Geometry - An Introduction.* Springer-Verlag. 1985.

A Abstract Cell Complexes

In this section we want to remind the reader the basic definitions of the theory of abstract cell complexes [Kov89, Kov93].

Definition ACC: An *abstract cell complex* (AC complex) $C = (E, B, dim)$ is a set E of abstract elements called *cells* provided with an antisymmetric, irreflexive, and transitive binary relation $B \subset E \times E$ called the *bounding relation*, and with a *dimension function* $dim : E \longrightarrow I$ from E into the set I of non-negative integers such that $dim(e') < dim(e'')$ for all pairs $(e', e'') \in B$.

The bounding relation B is a partial order in E. The bounding relation is denoted by $e' < e''$ which means that the cell e' bounds the cell e''.

If a cell e' bounds another cell e'' then e' is called a *side* of e''. The sides of an abstract cell e'' are not parts of e''. The intersection of two distinct abstract cells is always empty, which is different from Euclidean complexes.

If the dimension $dim(e')$ of a cell e' is equal to d then e' is called *d-dimensional* cell or a *d-cell.* An AC complex is called *k-dimensional* or a *k-complex* if the dimensions of all its cells are less or equal to k. Cells of the highest dimension k in an k-complex are called *ground cells.*

Definition SC: A *subcomplex* $S = (E', B', dim')$ of a given AC complex $C = (E, B, dim)$ is an AC complex whose set E' is a subset of E and the relation B' is an intersection of B with $E' \times E'$. The dimension dim' is equal to dim for all cells of E'. A subcomplex of a given complex is uniquely defined by the subset E'. Therefore it is usual to say "subset" instead of "subcomplex".

Definition OP: A subset OS of cells of a subcomplex S of an AC complex C is called *open in* S if OS contains each cell of S which is bounded by a cell of OS.

Definition SON: The smallest subset of a set S which contains a given cell $c \in S$ and is open in S is called *smallest neighborhood* of c relative to S and is denoted by $SON(c, S)$.

Definition CS: A subset CS of cells of an AC complex C is called *closed* if CS contains all cells of C bounding cells of CS.

Definition CL: The smallest subset of a set S which contains a given subset $M \subset S$ and is closed in S is called the *closure* of M relative to S and is denoted by $Cl(M, S)$.

Definition IN: The greatest subset of a set S which is contained in a given subset $M \subset S$ and is open in S is called the *interior* of M relative to S and is denoted by $Int(M, S)$.

Definition BD: The *boundary* ∂S of an n-dimensional subcomplex S of an n-dimensional AC complex C is the closure of the set of all (n-1)-cells of C each of which bounds exactly one n-cell of S.

Definition MA: An *n-dimensional combinatorial manifold* (n-manifold) without boundary is an n-dimensional complex M in which the $SON(P, M)$ of each 0-cell P is homeomorphic to an open n-ball with the cell P lying in the interior of $SON(P, M)$. In a manifold with boundary the $SON(P, M)$ of some 0-cell P may be homeomorphic to a "half-ball", i.e. the 0-cell P lies in the boundary of $SON(P, M)$ rather than in its interior.

Definition IC: Two cells e' and e'' of an AC complex C are called *incident* to each other in C iff either $(e', e'') \in B$, or $(e'', e') \in B$, or $e' = e''$. The incidence relation is symmetric, reflexive and non-transitive.

Definition CN: Two cells e' and e'' of an AC complex C are called *connected* to each other in C iff either e' is incident to e'' or there exists a cell $c \in C$ which is connected to both e' and e''. Because of this recursive definition the connectedness relation is the transitive hull of the incidence relation.

Definition HN: An n-dimensional AC complex C is called *homogeneously n-dimensional* if every k-dimensional cell of C with $k < n$ is incident to at least one n-cell of C.

Definition RG: A *region* is an open connected subset of the space.

Definition SO: A region R of an n-dimensional AC complex C is called *solid* if every cell $c \in C$ which is not in R is incident to an n-cell of the complement $C - R$.

Definition DHS: A *digital half-space* is a solid region of a three-dimensional Cartesian AC complex containing all voxels whose coordinates satisfy a linear inequality.

Definition TL: A connected one-dimensional complex in which all cells, except two of them, are incident to exactly two other cells is called a *topological line*.

Definition CA: By assigning subsequent integer numbers to the cells of a topological line L in such a way that a cell with the number x is incident to cells having the numbers $x - 1$ and $x + 1$, one can define *coordinates* in L which is a one-dimensional space. AC complexes of greater dimensions may be defined as Cartesian products of such one-dimensional AC complexes. A product AC complex is called a *Cartesian complex*.

Definition TR: An n-dimensional array whose each element is assigned to a cell of an n-dimensional Cartesian AC complex while the topological coordinates of the cell serve as the index of the corresponding element of the array is called the *topological raster*.

In a topological raster it is possible to access each cell of any dimension and save a label of any cell. By means of topological coordinates it is easy to find all cells incident to a given cell without a search. It is also possible to specify the dimension of a cell by means of its topological coordinates.

Definition SG: A *standard raster* is an n-dimensional array whose elements represent only the ground cells (i.e. the n-dimensional cells) of an n-dimensional complex, e.g. only the pixels in the 2D case or only the voxels in the 3D case.

A cell c of some lower dimension gets in the standard raster the same coordinates as the ground cell incident to c and lying farther away from the origin of the coordinate system. The dimension of c cannot be specified by means of its coordinates, it must be specified explicitly. To save a label of a cell of some lower dimension (if necessary) some special means are necessary, e.g. it is possible to assign different bits of a byte in the standard raster to cells of different dimension having all the same coordinates.

A Near-linear Time Algorithm for Binarization of Fingerprint Images Using Distance Transform

Xuefeng Liang, Arijit Bishnu, and Tetsuo Asano

JAIST, 1-1, Asahidai, Tatsunokuchi, 9231292, Japan
{xliang, arijit, t-asano}@jaist.ac.jp

Abstract. Automatic Fingerprint Identification Systems (AFIS) have various applications to biometric authentication, forensic decision, and many other areas. Fingerprints are useful for biometric purposes because of their well known properties of distinctiveness and persistence over time. Fingerprint images are characterized by alternating spatial distribution of gray-level intensity values of ridges and ravines/valleys of almost equal width. Most of the fingerprint matching techniques require extraction of minutiae that are the terminations and bifurcations of the ridge lines in a fingerprint image. Crucial to this step, is either detecting ridges from the gray-level image or binarizing the image and then extracting the minutiae. In this work, we focus on binarization of fingerprint images using linear time euclidean distance transform algorithms. We exploit the property of almost equal widths of ridges and valleys for binarization. Computing the width of arbitrary shapes is a non-trivial task. So, we estimate width using distance transform and provide an $O(N^2 \log M)$ time algorithm for binarization where M is the number of gray-level intensity values in the image and the image dimension is $N \times N$. With M for all purposes being a constant, the algorithm runs in near-linear time in the number of pixels in the image.

1 Introduction

Automatic fingerprint identification systems (AFIS) provide widely used biometric techniques for personal identification. Fingerprints have the properties of distinctiveness or individuality, and the fingerprints of a particular person remains almost the same (persistence) over time. These properties make fingerprints suitable for biometric uses. AFISs are usually based on minutiae matching [9, 14, 17, 18]. Minutiae, or Galton's characteristics [11] are local discontinuities in terms of terminations and bifurcations of the ridge flow patterns that constitute a fingerprint. These two types of minutiae have been considered by Federal Bureau of Investigation for identification purposes [29]. A detailed discussion on all the aspects of personal identification using fingerprint as an important biometric technique can be found in Jain et al. [17, 19]. AFIS based on minutiae matching involves different stages (see Figure 1 for an illustration):

1. fingerprint image acquisition;
2. preprocessing of the fingerprint image;

R. Klette and J. Žunić (Eds.): IWCIA 2004, LNCS 3322, pp. 197–208, 2004.

3. feature extraction (e.g. minutiae) from the image;
4. matching of fingerprint images for identification.

The preprocessing phase is known to consume almost 90-95% of the total time of fingerprint identification and verification [3]. That is the reason a considerable amount of research has been focussed on this area.

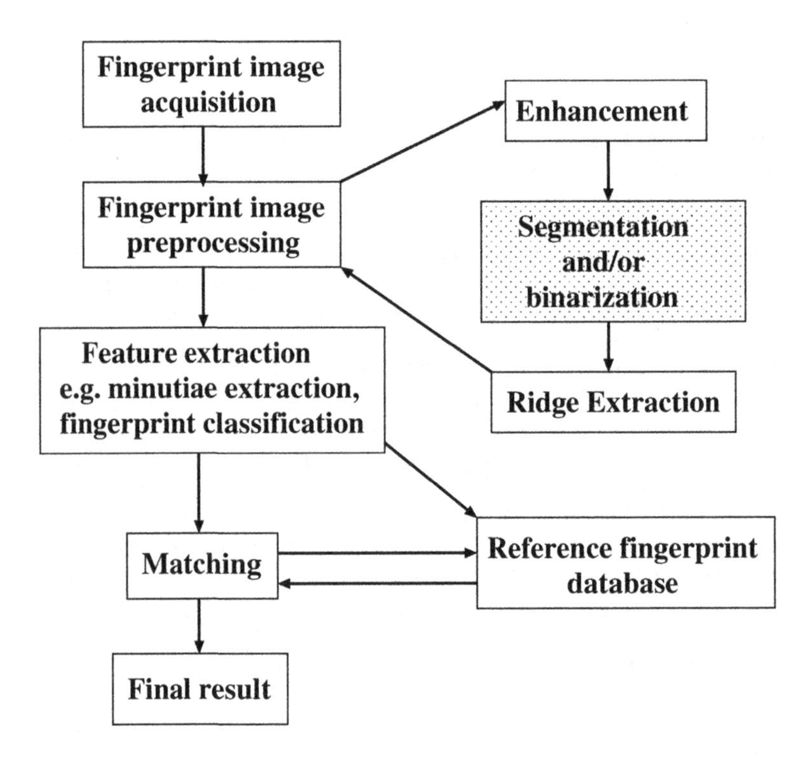

Fig. 1. A flowchart showing different phases of fingerprint analysis. The highlighted module shows the area of our work

Our work proposed in this paper involves binarization of fingerprint images that is to be preceded by an enhancement step. So, below we discuss briefly enhancement. Also, we briefly discuss and review segmentation and binarization methods applied to fingerprint images.

1.1 Enhancement of Fingerprint Images

Fingerprint images require specialised enhancement techniques owing to their inherent characteristics like high noise content, particular structural content of alternating ridges and valleys. Conventional image processing enhancement techniques are not very suitable for a fingerprint image [8]. Fingerprint image enhancement algorithms are available both for binary and gray level images. A binary fingerprint image consists of ridges marked as object (1) pixels and the rest as background pixels (0). Hung [8] designed an algorithm for enhancing a binary fingerprint image based on the structural information of its ridges. Ridge widths are normalized based on some region index. Ridge breaks are corrected using the dual relationship between ridge breaks and valley bridges. However, obtaining a binary fingerprint image from a gray-tone image involves inherent problems of binarization and thinning or ridge extraction procedures [6]. Thus, most of the enhancement algorithms are designed for gray-level fingerprint images. The much widely used PCASYS package [5] uses an enhancement algorithm described earlier [1]. It involves cutting out subregions of the images (a 32×32 block to be specific), taking their FFT and suppression of a band of low and high frequency components followed by some non-linear operations in the frequency domain and transforming it back to the spatial domain. This algorithm was also used by Kovács-Vajna [18]. We have also used this enhancement algorithm in our work owing to its simplicity and elegance.

1.2 Segmentation of Fingerprint Images

In literature concerning fingerprints, some authors have used the term segmentation to mean the process of generating a binary image from a gray-level fingerprint image. But, as suggested in [19], the most widely held view about segmentation of a fingerprint image is the process of separation of fingerprint area (ridge, valley and slope areas in between ridge and valley areas) from the image background. The process of segmentation is useful to extract out meaningful areas from the fingerprint, so that features of the fingerprint are extracted from these areas only. Fingerprint images are characterized by alternating spatial distribution of varying gray-level intensity values of ridges and ravines/valley. This pattern is unique to a fingerprint area compared to the background which does not have this spatial distribution of gray-level values. Also, global thresholding for segmentation does not work as the spatial distribution of gray-level values keeping their alternating structure intact, can vary in the absolute magnitude of their gray-level values. Thus, local thresholding is needed. Exploitation of these property have been the key of most of the segmentation algorithms. O'Gorman and Nickerson [24] used a $k \times k$ spatial filter mask with an appropriate orientation based on user inputs for labeling the pixels as foreground (crest) or background. Mehtre and Chatterjee [21] described a method of segmenting a fingerprint image into ridge zones and background based on some statistics of local orientations of ridges of the original image. A gray-scale variance method is used in the image blocks having uniform gray-level, where the directional method of segmentation fails. Ratha et al. [25] used the fact that noisy regions show no directional dependence, whereas, fingerprint regions exhibit a high variance of their orientation values across the ridge and a low variance along the ridge to design a segmentation algorithm that works on 16×16 block. Maio and Maltoni [20] used the average magnitude of gradient values to discriminate foreground and background regions. The idea behind this is that fingerprint regions are supposed to have more edges than background region and as such would have higher gradient values.

1.3 Binarization of Fingerprint Images

The general problem of image binarization is to obtain a threshold value so that all pixels above or equal to the threshold value are set to object pixel (1) and below the threshold value are set to background (0). Thresholding can be done globally where a single threshold is applied globally or locally where different thresholds are applied to different image regions. Images, in general, have different contrast and intensity, and as such local thresholds work better. The thresholding problem can be viewed as follows. Given an image I with $N \times N$ pixel entries, and gray-level intensity value g ranging from 0, 1, ... to $M-1$, select a value $t \in [0, M-1]$ based on some condition so that a pixel (i, j) is assigned a value of 1 if the gray-level intensity value is greater or equal to t, else assign 0 to the pixel (i, j). The condition mentioned above is decided based on the application at hand. The binarization methods applicable to fingerprint images draw heavily on the special characteristics of a fingerprint image. Moayer and Fu [23] proposed an iterative algorithm using repeated convolution by a Laplacian operator and a pair of dynamic thresholds that are progressively moved towards an unique value. The pair of dynamic thresholds change with each iteration and control the convergence rate to the binary pattern. Xiao and Raafat [30] improved the above method by using a local threshold, to take care of regions with different contrast, and applied after the convolution step. Both of these methods requiring repeated convolution operations are time consuming and the final result depends on the choice of the pair of dynamic thresholds and some other design parameters. Coetzee and Botha [7] proposed an algorithm based on the use of edges in conjunction with the gray-scale image. The resultant binary image is a logical OR of two binary images. One binary image is obtained by a local threshold on the gray scale image and the other binary image is obtained by filling in the area delimited by the edges. The efficiency of this algorithm depends heavily on the efficiency of the edge finding algorithm to find delimiting edges. Ratha et al. [25] proposed a binarization approach based on the peak detection in the gray-level profiles along sections orthogonal to the ridge orientation. The gray-level profiles are obtained by projection of the pixel intensities onto the central section. This heuristic algorithm though working well in practice has a deficiency that it does not retain the full width of the ridges, and as such is not a true binary reflection of the original fingerprint image.

In this work, we propose a combinatorial algorithm for binarization of fingerprint images based on Euclidean distance transform. Most of the previous algorithms discussed here are heuristics in that they do not start with a definition of an optimal threshold. In contrast, we define a condition for an optimal threshold based on equal widths of ridges and valleys. We show how distance transform can be used as a measure for width and then design an algorithm to efficiently compute the threshold for binarization. Using distance transform for binarization has also got another distinct advantage. The next step following binarization is ridge extraction and ridges can be efficiently extracted using distance transform values. As the same feature can be used for both binarization and ridge extraction, a lot of time savings can be obtained in real applications.

The rest of the paper is organised as follows. In Section 2, we briefly review Euclidean Distance Transform algorithm. Section 3 has a discussion on measuring width of shapes using average Distance Transform values. Section 4 discusses the threshold criteria and discusses the algorithm for thresholding and shows results on different fingerprint images. Finally, we finish with some discussions in Section 5.

2 Distance Transform

A two-dimensional binary image I of $N \times N$ pixels is a matrix of size $N \times N$ whose entries are 0 or 1. The pixel in a row i and column j is associated with the Cartesian co-ordinate (i, j). For a given distance function, the *Euclidean distance transform* of a binary image I is defined in [4] as an assignment to each background pixel (i, j) a value equal to the Euclidean distance between (i, j) and the closest feature pixel, i.e. a pixel having a value 1. Breu et al. [4] proposed an optimal $O(N \times N)$ algorithm for computing the *Euclidean distance transform* as defined using Voronoi diagrams. Construction and querying the Voronoi diagrams for each pixel (i, j) take time $\theta(N^2 \log N)$. But, the authors use the fact that both the sites and query points of the Voronoi diagrams are subsets of a two-dimensional pixel array to bring down the complexity to $\theta(N^2)$. In [13], Hirata and Katoh define *Euclidean distance transform* in an almost same way as the assigment to each 1 pixel a value equal to the Euclidean distance to the closest 0 pixel. The authors use a bi-directional scan along rows and columns of the matrix to find out the closest 0. Then, they use an envelope of parabolas whose parameters are obtained from the values of the bi-directional scan. They use the fact that two such parabolas can intersect in at most one point to show that each parabola can occur in the lower envelope at most once to compute the *Euclidean distance transform* in optimal $\theta(N^2)$ time. In keeping with the above, we define two types of *Euclidean distance transform* values. The first one $DT_{1,0}$ is the same as the above. The second one is $DT_{0,1}$ which is the value assigned to a 0 pixel equal to the Euclidean distance to the nearest 1 pixel. Using the results given in [13], we have the following fact:

Fact 1. *Both* $DT_{1,0}$ *and* $DT_{0,1}$ *can be computed in optimal time* $O(N^2)$ *for an* $N \times N$ *binary image. Also, the values of both* $DT_{1,0}$ *and* $DT_{0,1}$ *are greater than or equal to* 1.

3 Distance Transform and Width

The fingerprint images are characterized by almost equal width ridges and valleys as shown in Figure 2. We will use this particular characteristic of the fingerprint image for binarization. Measuring the width for arbitrary shapes is a difficult, non-trivial problem. In this section, we model the problem in a continuous domain to show how distance transform can be used to find equal width ridges and valleys.

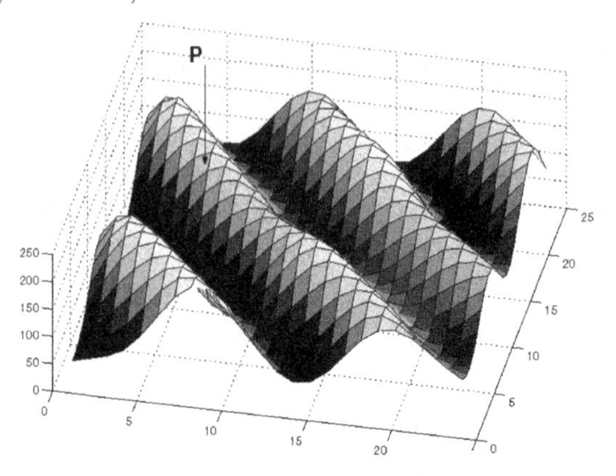

Fig. 2. Magnified view of a part of the gray scale topology of a fingerprint image

3.1 Model in the Continuous Domain

The fingerprint image can be modeled as shown in Figure 3. In the continuous domain, the image is a continuous function $f : (x, y) \rightarrow \mathbb{R}$. A cross section of this function along a direction perpendicular to the ridge increases till it reaches the ridge point which is a maxima, then decreases till it reaches the valley, which is a minima; and this cycle repeats. Let $t \in [0, M]$ be a threshold, such that if f is thresholded at t, and if the value of f is greater than t, it is mapped to 1, else to 0. See Figure 3. The highlighted part shown on the right is the part mapped to 1. After thresholding, the parts would be rectangles as shown in Figure 3. We compute the total distance transform values of the rectangles. Consider a rectangular object $ABCD$ of width w and height h, with $h > w$. The medial axis of this object is given by the line segments $\overline{AE}$, $\overline{BE}$, $\overline{EF}$, $\overline{FD}$, $\overline{FC}$. The medial axis divides the rectangular shape into four regions such that the nearest boundary line from any point in the region is determined. As an example, the region 1 has $\overline{AD}$ as its nearest boundary line and region 3 has $\overline{AB}$ as its nearest boundary line. The total distance transform value for region 1 is $\int_0^{w_i/2}\int_{-y+w_i/2}^{y+(h-w_i/2)}(w_i/2 - y)\,dxdy = (w_i^2 h)/8 - w_i^3/12$. Similarly, the total distance transform value for region 3 is $\int_{x-w_i/2}^{-x+w_i/2}\int_0^{w_i/2} x\,dxdy = w_i^3/24$. So, the total distance transform value $\phi_{dt}(w_i)$ of the rectangle is $w_i^2 h/4 - w_i^3/12$ $= w_i^2/4(h - w_i/3)$. Note that, the total distance transform value increases (decreases) with the increase (decrease) of width because $\phi_{dt}(w_i)' > 0$ and $h > w$. Now, the total distance transform $DT_{1,0}$ is $w_1^2 h/4 - w_1^3/12 + w_3^2 h/4 - w_3^3/12$ and the total distance transform $DT_{0,1}$ is $w_2^2 h/4 - w_2^3/12$. Now, as t increases, both w_1 and w_3 decrease and w_2 increases. This implies that with increase of t, $DT_{1,0}$ decreases and $DT_{0,1}$ increases. So, $DT_{1,0}$ and $DT_{0,1}$ can intersect only once and evidently, $DT_{1,0}$ is equal to $DT_{0,1}$ when $w_1 = w_2$. That is, the optimal value of threshold is reached when $DT_{0,1} = DT_{1,0}$, implying $w_1 = w_2$. This simple analysis shows that total distance transform can be used as a measure of finding a threshold that gives equal width ridges and valleys. Our goal in this work is to find an optimal threshold to binarize the fingerprint image. The optimality

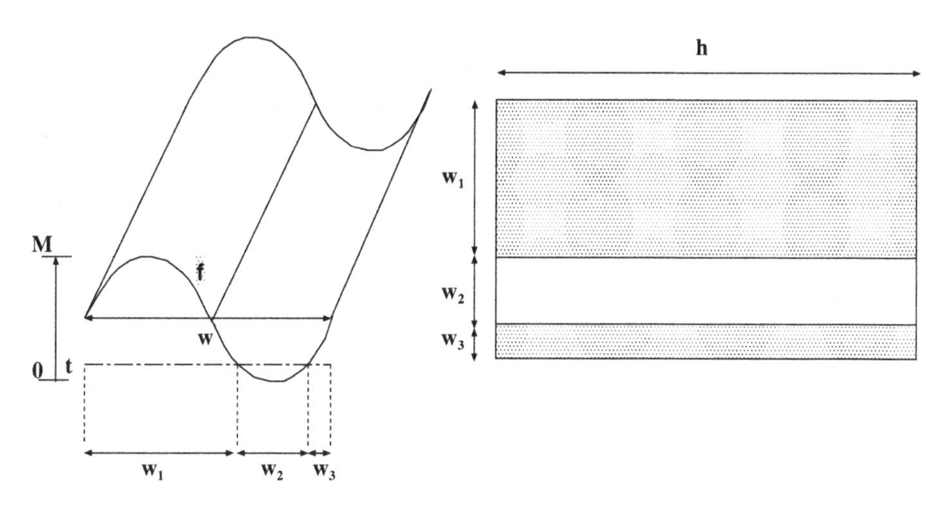

Fig. 3. Diagram of the model

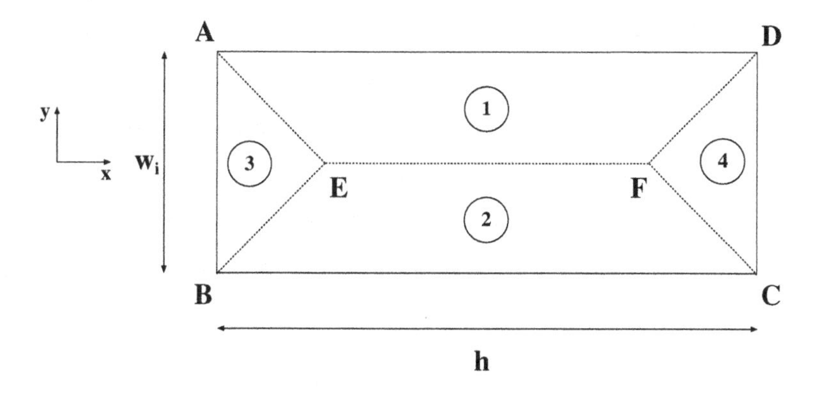

Fig. 4. Diagram for computing total distance transform

criteria is given by the equal width of ridge and valley. So, more formally we have the following definition.

Definition 1. *The optimal threshold is a value* $t \in [0, M]$ *that binarizes the image such that the ridge width is equal to the valley width or sum total of distance transform values are equal.*

3.2 Discrete Image and Distance Transform

In the discrete model, the co-ordinates are discrete given by the pixel locations. The gray-level values g are also discrete taking values from 0 to $M - 1$. So, the observations from the previous subsection do not directly apply. But, the crucial observation from the previous subsection is that sum total of $DT_{1,0}$ values decreases with t and the sum total of $DT_{0,1}$ values increases with t. Then, the optimal threshold t can be obtained as that value of t that makes the width of the 1 region and 0 region equal and can be computed from the intersection of the curves of the sum total of $DT_{0,1}$ and $DT_{1,0}$ values. For, the analysis, we make the following assumption. The pixels take the gray-level intensity values such that all the intermediate gray-level values between the maximum and the minimum are present. With that assumption, we have the following lemma.

Lemma 1. *The sum total of $DT_{1,0}$ values decreases with the threshold t. Similarly, the sum total of $DT_{0,1}$ values increase with the threshold t.*

The proof is easy. We know that each of the Euclidean distance transform values in the discrete domain is greater than or equal to 1 (see Fact 1). So, with the threshold t increasing, pixels in the binary image move from the regions of 1 to 0, thus making $DT_{1,0}$ and $DT_{0,1}$ decreasing and increasing respectively. Also, note that the assumption that the pixels take the gray-level intensity values such that all the intermediate gray-level values between the maximum and the minimum are present, ensures the strictly decreasing and increasing relations of sum total of $DT_{1,0}$ and $DT_{0,1}$ values. Otherwise, it would have been non-increasing and non-decreasing respectively.

Also, in the discrete case, we may not be able to locate a single value, where the functions of sum total of $DT_{1,0}$ and $DT_{0,1}$ meet. So, we modify the definition of the optimal threshold in the discrete case as follows.

Definition 2. *The optimal threshold can be two values t_1 and t_2 such that $t_2 - t_1 = 1$ and the sum total of $DT_{1,0}$ values is greater than the sum total of $DT_{1,0}$ values at t_1 and their relation reverses at t_2.*

With this definition in place, we are in a position to design the algorithm in the next section.

4 Algorithm and Results

4.1 Algorithm for Binarization

To take care of different contrast and intensity across different image regions, we apply local thresholding. We cut out sub-blocks of image region and apply the enhancement algorithm due to [1] followed by our binarization algorithm.

Algorithm for Binarization.

Input: A gray-level fingerprint image I with gray-level intensity varying from 0 to $M-1$, and of size $N \times N$;

Output: A thresholded binary image

1. **do for all** sub-block B_i of the image I;
2. Apply the enhancement algorithm given in [1];

3. $t_1 = 0$, $t_2 = M - 1$; $mid \leftarrow \lceil (t_1 + t_2)/2 \rceil$;
4. **do**
5. $mid \leftarrow \lceil (t_1 + t_2)/2 \rceil$;
6. Compute $SumDT^{mid}_{1,0}$ and $SumDT^{mid}_{0,1}$;
7. **if**($SumDT^{mid}_{1,0} > SumDT^{mid}_{0,1}$) $t_1 \leftarrow mid$;
8. **else** $t_2 \leftarrow mid$;
   while($t_2 - t_1 > 1$)
9. Threshold obtained for binarization is t_1 or t_2;

The loop originating in Step 4 runs $O(\log M)$ times and the dominant computation is the computation of Euclidean Distance Transform and its sum which takes $O(N^2)$ time (see Fact 1). Thus the total time complexity of the binarization process is $O(N^2 \log M)$. With M, the number of gray-levels, being a constant for all practical purposes, the algorithm for binarization runs in time that is linear in the number of pixel entries which is $O(N^2)$.

4.2 Results on Fingerprint Images

We used the fingerprint images from (i) NIST Special Database 4[28], (ii) NIST Special Database 14[5], (iii) Database B1 of FVC2000[10], and (iv) Database B2 of FVC2000[10]. The images of (i) and (ii) are of size 480 × 512. The images of (iii) are of size 300 × 300 and (iv) are of size 364 × 256. All of the images are of 500 dpi resolution. Figures 5-8(a) show the original image, Figures 5-8(b) show the enhanced image due to [1] and Figures 5-8(c) show the resultant binary image obtained by application of our algorithm.

5 Discussions and Conclusions

We have developed a combinatorial algorithm for binarization of fingerprint images expoiting the fingerprint characteristics of equal width ridge and valleys. We used Euclidean Distance Transform as a measure of width as determining width for arbitrary discrete shapes is a non-trivial task. We have reported relevant results from standard image databases widely used. But, the definition 2

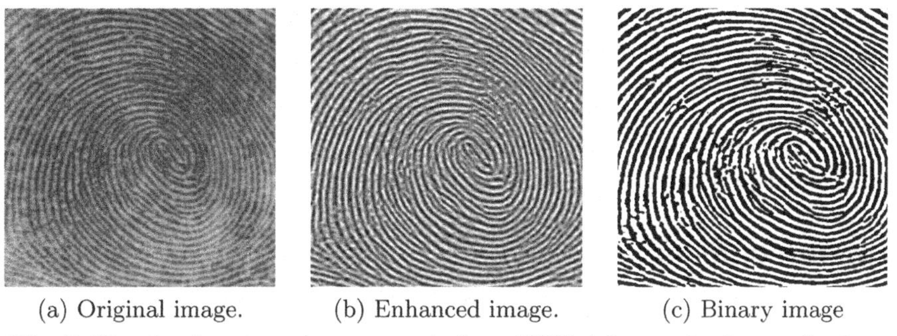

(a) Original image. (b) Enhanced image. (c) Binary image

Fig. 5. Binarization on an image sample from NIST-4 fingerprint image database

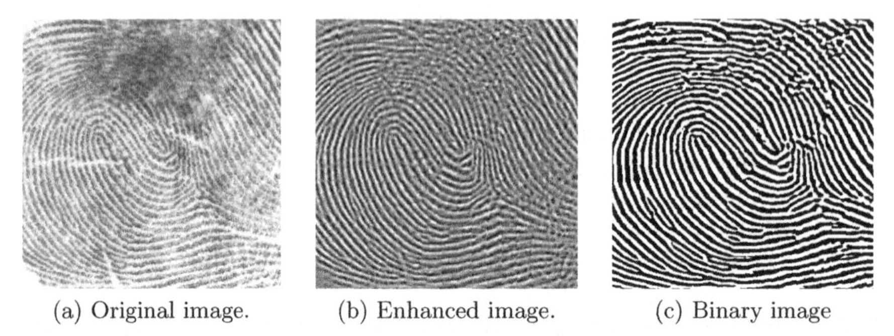

(a) Original image. (b) Enhanced image. (c) Binary image

Fig. 6. Binarization on an image sample from NIST-14 fingerprint image database

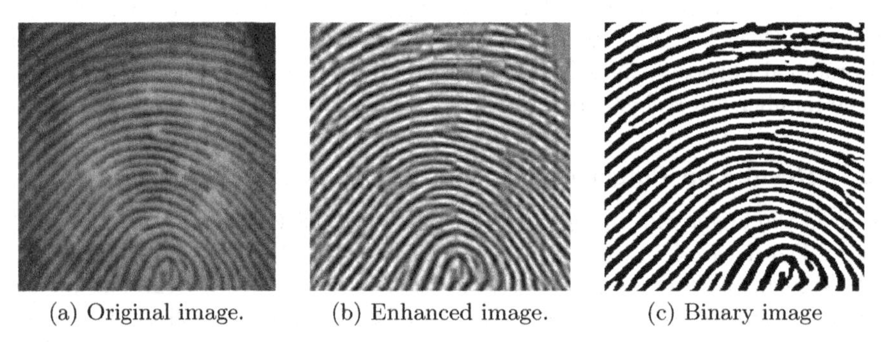

(a) Original image. (b) Enhanced image. (c) Binary image

Fig. 7. Binarization on an image sample from FVC2000 DB1 fingerprint image database

used for our algorithm has a drawback in realistic terms. During the acquisition of fingerprints, ridges, being the elevated structures on the finger, exert more pressure on the device making the acquisition. And as such, the widths of the ridges should be greater than the width of the valley for a more realistic model. But, still the lemma 1 will hold and the algorithm instead of trying to find the crossover point of sum total of $SumDT_{1,0}$ and $SumDT_{0,1}$ will terminate when $SumDT_{1,0}$ is greater than $SumDT_{0,1}$ by a certain ϵ. Determining this ϵ from real fingerprint images is a future problem we would like to address. Also, note that our binarization algorithm using distance transform has a distinct benefit. Please refer to Figure 1. The module following binarization is ridge extraction. Ridge is the skeleton of the thick binary structures obtained from the binarization. Euclidean Distance Transform can be effectively used to find the skeleton [31].

Thus the same feature of distance transform can be used for both binarization and ridge extraction which in real applications can save a lot of time.

Acknowledgment

This research for the first author was conducted as a program for the "Fostering Talent in Emergent Research Fields" in Special Coordination Funds for Promot-

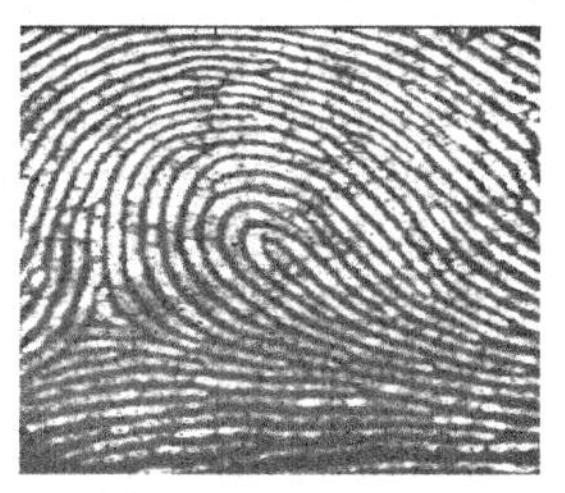

(a) Original image.

(b) Enhanced image.

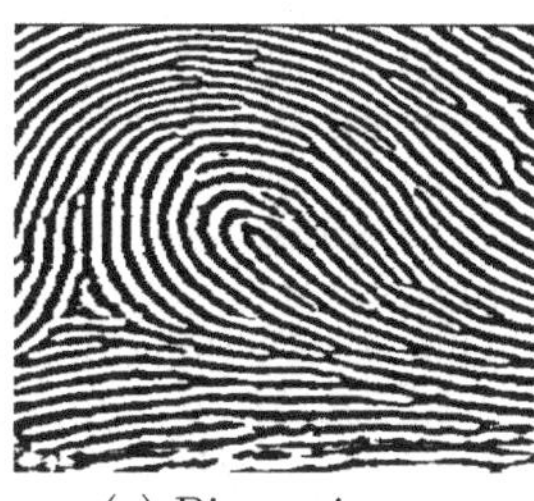

(c) Binary image

Fig. 8. Binarization on an image sample from FVC2000 DB2 fingerprint image database

ing Science and Technology by Ministry of Education, Culture, Sports, Science and Technology. This research for the third author was partially supported by the same Ministry, Grant-in-Aid for Scientific Research (B) and Exploratory Research.

References

1. *Automated Classification System Reader Project (ACS)*, Technical Report, DeLaRue Printrak Inc., Feb., 1985.
2. Bhanu, B. and Tan, X., "Fingerprint Indexing Based on Novel Features of Minutiae Triplets", *IEEE Trans. PAMI*, vol. 25, no. 5, pp. 616-622, 2003.
3. Blue, J. L., Candela G. T., Grother, P. J., Chellappa, R., Wilson, C. L., and Blue, J.D., "Evaluation of Pattern Classifiers for Fingerprint and OCR Application", *Pattern Recognition*, vol. 27, no. 4, pp. 485-501, 1994.
4. Breu, H., Gil, J., Kirkpatrick, D., and Werman, M., "Linear Time Euclidean Distance Transform Algorithms", *IEEE Trans. PAMI*, vol. 17, no. 5, pp. 529-533, 1995.
5. Candela, G. T., Grother, P. J., Watson, C. I., Wilkinson, R. A. and Wilson, C. L., *PCASYS - A Pattern-Level Classification Automation System for Fingerprints*, NISTIR 5647, National Institute of Standards and Technology, August, 1995.
6. Chang, J. -H., Fan, K. -C., "Fingerprint Ridge Allocation in Direct Gray-Scale Domain", *Pattern Recognition*, vol. 34, no. 10, pp. 1907-1925, 2001.
7. Coetzee, L., and Botha, E. C., "Fingerprint Recognition in Low Quality Images", *Pattern Recognition*, vol. 26, no. 10, pp. 1441-1460, 1993.
8. Douglas Hung, D. C., "Enhancement and Feature Purification of Fingerprint Images", *Pattern Recognition*, vol. 26, no. 11, pp. 1661-1771, 1993.
9. Farina, A., Zs. M. Kovács-Vajna, Zs., M. and Leone, A., "Fingerprint Minutiae Extraction from Skeletonized Binary Images", *Pattern Recognition*, vol. 32, pp. 877-889, 1999.
10. Fingerprint Verification Competition, 2000, http://bias.csr.unibo.it/fvc2000/download.asp.
11. Galton, F., "Fingerprints", *London: Macmillan*, 1892.
12. Haralick, R., "Ridges and Valleys on Digital Images", *Computer Vision Graphics Image Processing*, vol. 22, pp. 28-38, 1983.

13. Hirata, T., and Katoh, T., "An Algorithm for Euclidean distance transformation", *SIGAL Technical Report of IPS of Japan*, 94-AL-41-4, pp. 25-31, September, 1994.
14. Hollingum, J., "Automated Fingerprint Analysis Offers Fast Verification", *Sensor Review*, vol. 12, no. 13, pp. 12-15, 1992.
15. Hong, L., Wan, Y., and Jain, A. K., "Fingerprint Image Enhancement: Algorithm and Performance Evaluation" , *IEEE Trans. PAMI*, vol. 20, no. 8, pp. 777-789, 1998.
16. Jain, A. K., Hong, L., and Bolle, R., "On-Line Fingerprint Verification" , *IEEE Trans. PAMI*, vol. 19, no. 4, pp. 302-314, 1997.
17. Jain, A. K., Hong, L., Pankanti, S. and Bolle, R., "An Identity-Authentication System Using Fingerprints", *Proc. of IEEE*, vol. 85, no. 9, pp. 1365-1388, 1997.
18. Kovács-Vajna, Z. M., "A Fingerprint Verification System Based on Triangular Matching and Dynamic Time Warping", *IEEE Trans. PAMI*, vol. 22, no. 11, pp. 1266-1276, 2000.
19. Maltoni, D., Maio, D., Jain, A. K., and Prabhakar, S., *Handbook of Fingerprint Recognition*, Springer-Verlag, New York, 2003.
20. Maio, D. and Maltoni, D., "Direct Gray-Scale Minutiae Detection In Fingerprints", *IEEE Trans. PAMI*, vol. 19, no. 1, pp. 27-39, 1997.
21. Mehtre, B. M. and Chatterjee, B., "Segmentation of Fingerprint Images - A Composite Method", *Pattern Recognition*, vol. 22, pp. 381-385, 1989.
22. Mehtre, B. M., "Fingerprint Image Analysis for Automatic Identification", *Machine Vision and Applications*, vol. 6, no. 2, pp. 124-139, 1993.
23. Moayer, B. and Fu, K., "A Tree System Approach for Fingerprint Pattern Recognition", *IEEE Trans. PAMI*, vol. 8, no. 3, pp. 376-388, 1986.
24. O'Gorman. L. and Nickerson, J. V., "An Approach to Fingerprint Filter Design", *Pattern Recognition*, vol. 22, pp. 29-38, 1989.
25. Ratha N. K., Chen, S. Y., and Jain, A. K., "Adaptive Flow Orientation-Based Feature Extraction in Fingerprint Images", *Pattern Recognition*, vol. 28, no. 11, pp. 1657-1672, 1995.
26. Rosenfeld, A. and Kak, A. C., *Digital Image Processing*, vol. 2, Academic Press Inc., Orlando, Florida, 1982.
27. Senior A., "A Combination Fingerprint Classifier", *IEEE Trans. PAMI*, vol. 23, no. 10, pp. 1165-1174, 2001.
28. Watson, C. I., Wilson, C. L., *Fingerprint Database*, National Institute of Standards and Technology, Special Database 4, FPDB, April, 1992.
29. Wegstein, J. H., "An Automated Fingerprint Identification System", *US Government Publication*, Washington, 1982.
30. Xiao, Q., and Raafat, H., "Fingerprint Image Post-Processing: A Combined Statistical and Structural Approach", *Pattern Recognition*, vol. 24, no. 10, pp. 985-992, 1991.
31. Shih, F. Y. and Pu, C. C.,"A Skeletonization Algorithm by Maxima Tracking on Euclidean Distance Transform", *Pattern Recognition*, vol. 28, no. 3, pp. 331-341, March 1995.

On Recognizable Infinite Array Languages

S. Gnanasekaran[1] and V.R. Dare[2]

[1] Department of Mathematics, Periyar Arts College,
Cuddalore - 607 001, Tamil Nadu, India
sg_vianna@sancharnet.in

[2] Department of Mathematics, Madras Christian College,
Tambaram, Chennai - 600 059, Tamil Nadu, India
rchristian@eth.net

Abstract. A recognizable infinite array language or recognizable $\omega\omega$-language is defined as the image of a local $\omega\omega$-language by an alphabetic morphism. Here, we introduce Wang systems for $\omega\omega$-languages and prove that the class of $\omega\omega$-languages obtained by Wang systems is the same as the class of recognizable $\omega\omega$-languages. We give automata characterization to the recognizable $\omega\omega$-languages. We provide an algorithm for learning recognizable infinite array languages from positive data and restricted superset queries.

Keywords: array, prefix, local language, recognizable array language, on-line tesselation automaton, learning.

1 Introduction

Local sets of words play a considerable role in the theory of recognizable string languages. For example, it is well known that every recognizable string language can be obtained as the image of a local set by an alphabetic morphism [6].

Gimmarresi and Restivo [7] have generalized this notion to picture languages that are sets of rectangular arrays of symbols. Then they have defined the class of recognizable array languages as the set of languages which can be obtained by projection of local array languages. Latteux and Simplot [11] have defined hv-local array languages by replacing in the definition of local array languages the 2×2 tiles by horizontal and vertical dominoes and have proved that every recognizable array language can be obtained as the projection of a hv-local array language. De Prophetics and Varricchio [5] have introduced the notion of labeled Wang tiles. They have defined Wang systems and have shown that the family of array languages recognized by Wang systems coincides with the family of recognizable array languages. Inoue and Nakamura [10] have introduced a model of two-dimensional tape acceptor, called two-dimensional on-line tesselation automaton. Inoue and Takanami have proved that the class of languages accepted by this type of automata is the class of recognizable array languages.

These different notions of recognizability of a set of arrays are extended to infinite array languages or $\omega\omega$-languages. Dare et al [4] have introduced local

R. Klette and J. Žunić (Eds.): IWCIA 2004, LNCS 3322, pp. 209–218, 2004.

$\omega\omega$-languages and have shown that the set of local $\omega\omega$-languages is the set of adherence of local array languages. We [8] have defined hv-local $\omega\omega$-language and proved that recognizable $\omega\omega$-language is a projection of a hv-local $\omega\omega$-language.

In the study of inductive inference of formal languages, Gold [9] has proved that the class of languages containing all finite sets and one infinite set is not identifiable in the limit from postive data only. This implies that even the class of regular languages is not identifiable in the limit from positive data. Angluin [1] has developed several conditions for identifying, a class of languages, in the limit from positive data and presented some examples of these identifiable classes.

In this paper, we introduce Wang systems for $\omega\omega$-languages and show that the class of $\omega\omega$-languages recognized by Wang systems is the same as the class of recognizable $\omega\omega$-languages. We provide automata characterization for the class of recognizable $\omega\omega$-languages. We give an algorithm for learning recognizable infinite array languages from positive data and restricted superset queries [2].

2 Basic Definitions and Results

Let Σ be a finite alphabet. An array or a two-dimensional picture over Σ of size (m, n) is a two-dimensional rectangular arrangement of symbols from Σ in m rows and n columns. We adopt the convention that the bottom-most row is the first row and the left-most column is the first column. For $p \in \Sigma^{**}$, let $l_1(p)$ denote the number of rows of p and $l_2(p)$ denote the number of columns of p. The set of all arrays over Σ is denoted by Σ^{**}. An array language over Σ is a subset of Σ^{**}.

For any array p of size $(m, n), \hat{p}$ is the array of size $(m+2, n+2)$ obtained by surrounding p with a special symbol $\# \notin \Sigma$ and $\hat{}p$ is the array of size $(m+1, n+1)$ obtained by placing a row of #s below the first row of p and placing a column of #s to the left of the first column of p.

An infinite array has infinite number of rows and columns. The set of all infinite arrays over Σ is denoted by $\Sigma^{\omega\omega}$. An $\omega\omega$-language over Σ is a subset of $\Sigma^{\omega\omega}$. For $p \in \Sigma^{\omega\omega}, \hat{p}$ is the infinite array obtained by placing a row of #s below the first row of p and placing a column of #s to the left of the first column of p. For $p \in \Sigma^{**} \cup \Sigma^{\omega\omega}$, we denote by $B_{h,k}(p)$, the set of all blocks of p of size (h, k). A square array of size (2,2) is called a tile.

If an array p has entry $a_{ij} \in \Sigma$ in the $i-th$ row and $j-th$ column, then we write $p = (a_{ij}), i = 1, \ldots, m; j = 1, \ldots, n$ if $p \in \Sigma^{**}$ is of size (m, n) and $p = (a_{ij}), i = 1, 2, \ldots; j = 1, 2, \ldots$ if $p \in \Sigma^{\omega\omega}$. If $p = (a_{ij}) \in \Sigma^{\omega}$, a prefix of p is an array $q = (a_{ij}), i = 1, \ldots, l; j = 1, \ldots, r, 1 \leq l, r < \infty$. The set of all prefixes of p is denoted by $Pref(p)$. If $L \subseteq \Sigma^{**} \cup \Sigma^{\omega\omega}$, then $Pref(L) = \bigcup_{p \in L} Pref(p)$.

A language $L \subseteq \Sigma^{\omega\omega}$ is called $\omega\omega$-local [4] if there exists a finite set θ of tiles over $\Sigma \cup \{\#\}$ such that

$$L = \{p \in \Sigma^{\omega\omega} : B_{2,2}(\hat{p}) \subseteq \theta\}$$

The set of all $\omega\omega$-local languages is denoted by $\omega\omega$-LOC.

Let π be a mapping from Σ to Γ where Σ and Γ are finite alphabets. We call π a projection. If $p = (a_{ij}) \in \Sigma^{\omega\omega}$, then $\pi(p) = (\pi(a_{ij})), i = 1, 2, \ldots; j = 1, 2, \ldots$. We say that $L \subseteq \Sigma^{\omega\omega}$ is recognizable $\omega\omega$-language if there exists a local $\omega\omega$-language L' over Γ and a projection $\pi : \Gamma \to \Sigma$ such that $L = \pi(L')$. The class of all recognizable $\omega\omega$-languages is denoted by $\omega\omega$-REC.

A domino is an array whose size is $(1, 2)$ or $(2, 1)$. $L \subseteq \Sigma^{\omega\omega}$ is called hv-local [8] if there exists a finite set Δ of dominoes over $\Sigma \cup \{\#\}$ such that

$$L = \{p \in \Sigma^{\omega\omega} : B_{1,2}(\hat{p}) \cup B_{2,1}(\hat{p}) \subseteq \Delta\}$$

and we write $L = \mathcal{L}^{\omega\omega}(\Delta)$.

We say that $L \subseteq \Sigma^{\omega\omega}$ is recognized by a domino system if there exists a hv-local $\omega\omega$-language L' over Γ and a projection $\pi : \Gamma \to \Sigma$ such that $L = \pi(L')$. The class of all $\omega\omega$-languages recognized by domino systems is denoted by $\mathcal{L}^{\omega\omega}(DS)$.

Proposition 1. *[8] If $L \subseteq \Sigma^{\omega\omega}$ is hv-local, then L is an $\omega\omega$-local language.*

Proposition 2. *[8] $L \subseteq \Sigma^{\omega\omega}$ is hv-local if and only if $L = L_1 \oplus L_2$ where L_1 and L_2 are local ω-languages.*

Theorem 1. *[8] $\omega\omega$-REC $= \mathcal{L}^{\omega\omega}(DS)$.*

3 Labeled Wang Tiles

In this section we use a different formalism to recognize $\omega\omega$-language. Wang tiles are introduced in [3] for the tiling of Euclidean plane. Then De Prophetics and Varricchio [5] have introduced the notion of labelled Wang tiles simply adding a label, taken in a finite alphabet, to a Wang tile. They have also defined Wang systems and have proved that the family of array languages recognized by Wang systems coincides with the family of recognizable array languages.

We extend the concept of Wang systems to $\omega\omega$-languages in this section and prove that the family of $\omega\omega$-languages recognized by Wang systems is the same as the family of recognizable $\omega\omega$-languages.

A labelled Wang tile is a 5-tuple, consisting of 4 colours, choosen in a finite set of colours Q and a label choosen is a finite alphabet Σ.

Definition 1. *A Wang system is a triplet $W = (\Sigma, Q, T)$ where Σ is a finite alphabet, Q is a finite set of colours and T is a finite set of Wang tiles, $T \subseteq Q^4 \times \Sigma$.*

Definition 2. *Let $W = (\Sigma, Q, T)$ be a Wang system. An $\omega\omega$-array M over T is a tiling, if it satisfies the following conditions:*

1. $M(1,1) = B\,\overset{q}{\underset{B}{\boxed{a}}}\,p$, $M(1,n) = r\,\overset{q}{\underset{B}{\boxed{b}}}\,p, n = 1, 2, \ldots$

   $M(m,1) = B\,\overset{r}{\underset{p}{\boxed{c}}}\,q, m = 1, 2, \ldots$

2. $M(m,n) = \begin{array}{c} r \\ s\boxed{a}q \\ p \end{array}, m,n = 2,3,\ldots$

Here $p,q,r,s \neq B$.

If M is a tiling of W, the label of M, denoted by $|M|$, is an $\omega\omega$-array over Σ, defined by

$$|M|(m,n) = a \Leftrightarrow M(m,n) = \begin{array}{c} r \\ s\boxed{a}q \\ p \end{array}, \text{ for some } p,q,r,s$$

Definition 3. *Let W be a Wang system. An $\omega\omega$-array w is generated by W if there exists a tiling M such that $|M| = w$.*

We denote by $\mathcal{L}^{\omega\omega}(W)$, the language of $\omega\omega$-arrays generated by the Wang system W.

Definition 4. *The $\omega\omega$-array language $L \subseteq \Sigma^{\omega\omega}$ is Wang recognizable if there exists a Wang system W such that $L = \mathcal{L}^{\omega\omega}(W)$. The family of all Wang recognizable $\omega\omega$-languages is denoted by $\mathcal{L}^{\omega\omega}(WS)$.*

Proposition 3. *$\mathcal{L}^{\omega\omega}(WS)$ is closed under projection.*

Proof. Let $W = (\Gamma, Q, T)$ be a Wang system and $\pi : \Gamma \to \Sigma$ be a projection. We have to show that if $L = \mathcal{L}^{\omega\omega}(W)$, then $L' = \pi(L) = \mathcal{L}^{\omega\omega}(W')$ for some Wang system W'.

Let $W' = (\Sigma, Q, T')$ where

$$T' = \left\{ \begin{array}{c} r \\ s\boxed{\pi(a)}q \\ p \end{array} : \begin{array}{c} r \\ s\boxed{a}q \\ p \end{array} \in T \right\}$$

Then $L' = \mathcal{L}^{\omega\omega}(W')$.

Proposition 4. *$\mathcal{L}^{\omega\omega}(WS) \subseteq \omega\omega$-REC.*

Proof. Let $L \in \mathcal{L}^{\omega\omega}(WS)$ and $L = \mathcal{L}^{\omega\omega}(W)$ where $W = (\Sigma, Q, T)$. Let $\Gamma = T$ and $\theta = \theta_1 \cup \theta_2 \cup \theta_3 \cup \theta_4$ where

$$\theta_1 = \left\{ \begin{array}{|c|c|} \hline \# & \begin{array}{c} q \\ B\boxed{a}p \\ B \end{array} \\ \hline \# & \# \\ \hline \end{array} : \begin{array}{c} q \\ B\boxed{a}p \\ B \end{array} \in T \right\}$$

$$\theta_2 = \left\{ \begin{array}{|c|c|} \hline \# & \begin{array}{c} r \\ B\boxed{a}q \\ p \end{array} \\ \hline \# & \begin{array}{c} p \\ B\boxed{b}s \\ t \end{array} \\ \hline \end{array} : \begin{array}{c} r \\ B\boxed{a}q \\ p \end{array}, \begin{array}{c} p \\ B\boxed{b}s \\ t \end{array} \in T \right\}$$

$$\theta_3 = \left\{ \begin{array}{|c|c|} \hline \begin{array}{c} r \\ q\boxed{a}p \\ B \end{array} & \begin{array}{c} s \\ p\boxed{b}t \\ B \end{array} \\ \hline \# & \# \\ \hline \end{array} : \begin{array}{c} q \\ r\boxed{a}p \\ B \end{array}, \begin{array}{c} s \\ p\boxed{b}t \\ B \end{array} \in T \right\}$$

$$\theta_4 = \left\{ \begin{array}{|c|c|} \hline \begin{array}{c} r \\ s\boxed{a}q \\ p \end{array} & \begin{array}{c} v \\ q\boxed{b}u \\ t \end{array} \\ \hline \begin{array}{c} p \\ x\boxed{c}z \\ y \end{array} & \begin{array}{c} t \\ z\boxed{d}f \\ w \end{array} \\ \hline \end{array} : \begin{array}{c} r \\ s\boxed{a}q \\ p \end{array}, \begin{array}{c} v \\ q\boxed{b}u \\ t \end{array}, \begin{array}{c} p \\ x\boxed{c}z \\ y \end{array}, \begin{array}{c} t \\ z\boxed{d}f \\ w \end{array} \in T \right\}$$

Let $L_1 = \mathcal{L}^{\omega\omega}(\theta)$. Define $\pi : \Gamma \to \Sigma$ by $\pi\left(\begin{array}{c} r \\ s\boxed{a}q \\ p \end{array} \right) = a$. Then $L = \pi(L_1)$ and therefore $L \in \omega\omega$-REC.

Proposition 5. *$\omega\omega$-REC $\subseteq \mathcal{L}^{\omega\omega}(WS)$.*

Proof. Let $L \in \omega\omega$-REC. Then there exists a local $\omega\omega$-language L_1 over Γ and a projection $\pi : \Gamma \to \Sigma$ such that $L = \pi(L_1)$. Let $L_1 = \mathcal{L}^{\omega\omega}(\theta)$ where θ is a finite set of tiles over $\Gamma \cup \{\#\}$.

Consider the Wang system $W = (\Gamma, Q, T)$ where

$$Q = (\Gamma \cup \{\#\})^2 \cup \{B\} \text{ where } B \notin \Gamma \cup \{\#\}$$

$$T = T_1 \cup T_2 \cup T_3 \cup T_4 \text{ where}$$

$$T_1 = \left\{ \begin{array}{c} a\# \\ B\boxed{\ a\ }\#a \\ B \end{array} : \begin{array}{|c|c|} \hline \# & a \\ \hline \# & \# \\ \hline \end{array} \in \theta \right\}$$

$$T_2 = \left\{ \begin{array}{c} ba \\ \#a\boxed{\ b\ }\#b \\ B \end{array} : \begin{array}{|c|c|} \hline a & b \\ \hline \# & \# \\ \hline \end{array} \in \theta \right\}$$

$$T_3 = \left\{ \begin{array}{c} b\# \\ B\boxed{\ b\ }ab \\ a\# \end{array} : \begin{array}{|c|c|} \hline \# & b \\ \hline \# & a \\ \hline \end{array} \in \theta \right\}$$

$$T_4 = \left\{ \begin{array}{c} da \\ ba\boxed{\ d\ }cd \\ cb \end{array} : \begin{array}{|c|c|} \hline a & d \\ \hline b & c \\ \hline \end{array} \in \theta \right\}$$

Then $L = \mathcal{L}^{\omega\omega}(W)$. Thus $L \in \mathcal{L}^{\omega\omega}(WS)$ and therefore $\omega\omega$-REC $\subseteq \mathcal{L}^{\omega\omega}(WS)$.

Combining Propositions 4 and 5, we have the result.

Theorem 2. *$\omega\omega$-REC $= \mathcal{L}^{\omega\omega}(WS)$.*

4 Automata Characterization of $\omega\omega$-Recognizable Languages

In this section, we give automata characterization of recognizable $\omega\omega$-languages.

A non-deterministic (deterministic) two-dimensional online tesselation automaton (2-OTA (2-DOTA)) is $M = (Q, \Sigma, \delta, I, F)$ where

Σ is a finite alphabet
Q is a finite set of states
$I \subseteq Q(I = \{q_o\} \subseteq Q)$ is the set of initial states,
$F \subseteq Q$ is the set of final states and
$\delta : Q \times Q \times \Sigma \rightarrow 2^Q(\delta : Q \times Q \times \Sigma \rightarrow Q)$ is the transition function.

Let p be a finite or an infinite array.

A run of M on p consists of associating a state to each position (i, j) of $\hat{p}$. Such state is given by the transition function δ and it depends on the state already associated to positions $(i-1, j)$ and $(i, j-1)$ and on the symbol $p(i, j)$.

At time $t = 0$, an initial state $q_o \in I$ is associated with all positions of $\hat{p}$ holding #. At time $t = 1$, a state from $\delta(q_o, q_o, a_{11})$ is associated with the position $(1, 1)$ holding a_{11}. At time $t = 2$, states are associated simultaneously with positions $(1, 2)$ and $(2, 1)$ respectively holding a_{12} and a_{21}. If q_{11} is the state associated with the position $(1, 1)$, then the states associated with the position $(2, 1)$ is an element of $\delta(q_{11}, q_0, a_{21})$ and to the position $(1, 2)$ is an element of $\delta(q_0, q_{11}, a_{12})$. We then proceed to the next diagonal. The states associated with the position (i, j) by δ depend on the states already associated with the states in the positions $(i-1, j), (i, j-1)$ and the symbol a_{ij}. A 2-OTA M recognizes a finite array p if there exists a run of M on p such that the states associated to position $(l_1(p), l_2(p))$ is a final state. We say that L is recognized by M if L is equal to the set of all arrays recognized by M and we write $L = \mathcal{L}^{**}(M)$. The set of all array languages recognized by a 2-OTA is denoted by $\mathcal{L}^{**}(2\text{-}OTA)$.

A run of an infinite array p is a sequence of states $q_{11}q_{12}q_{21}q_{31}q_{22}q_{13}\ldots$. The run of p is denoted by $r(p)$. We define $inf(r(p))$ as the set of all states which repeat infinitely many times in $r(p)$. We say that $L \subseteq \Sigma^{\omega\omega}$ is recognized by a 2-OTA $M = (Q, \Sigma, \delta, q_0, F)$ if

$$L = \{p \in \Sigma^{\omega\omega} : inf(r(p)) \cap F \neq \phi, \text{ for some run } r(p)\}$$

and we write $L = \mathcal{L}^{\omega\omega}(M)$.

Theorem 3. *$L \in \omega\omega$-REC, if and only if L is recognized by a 2-OTA in which every state is a final state.*

Proof. Let $L \in \mathcal{L}^{\omega\omega}$. Then there exists a finite set θ of tiles such that $L = \mathcal{L}^{\omega\omega}(\theta)$. Consider the 2-OTA $M = (Q, \Sigma, \delta, q_0, Q)$ where $Q = \theta$

$$q_0 = \left\{ \begin{array}{|c|c|}\hline \# & a \\ \hline \# & \# \\ \hline \end{array} : \begin{array}{|c|c|}\hline \# & a \\ \hline \# & \# \\ \hline \end{array} \in \theta \right\}$$ and $\delta : Q \times Q \times \Sigma \to 2^Q$ is defined by

$$\delta\left(\begin{array}{|c|c|}\hline a & b \\ \hline c & d \\ \hline \end{array}, \begin{array}{|c|c|}\hline d & e \\ \hline f & g \\ \hline \end{array}, x \right) = \left\{ \begin{array}{|c|c|}\hline b & x \\ \hline d & e \\ \hline \end{array} : \begin{array}{|c|c|}\hline b & x \\ \hline d & e \\ \hline \end{array} \in \theta \right\}$$

Then $L = \mathcal{L}^{\omega\omega}(M)$. Since the languages recognized by 2-OTA is closed under morphism, every $L \in \omega\omega$-REC is recognized by a 2-OTA with $F = Q$.

Conversely let L be recognized by a 2-OTA $M = (Q, \Sigma, \delta, q_0, Q)$.

Let

$$\Gamma = Q \times (\Sigma \cup \{\#\})$$

$$\theta_1 = \left\{ \begin{array}{|c|c|}\hline (q_0, \#) & (p, a) \\ \hline (q_0, \#) & (q_0, \#) \\ \hline \end{array} : p \in \delta(q_0, q_0, a) \right\}$$

$$\theta_2 = \left\{ \begin{array}{|c|c|}\hline (q_0, \#) & (q, b) \\ \hline (q_0, \#) & (p, a) \\ \hline \end{array} : q \in \delta(q_0, p, b) \right\}$$

$$\theta_3 = \left\{ \begin{array}{|c|c|}\hline (p, a) & (q, b) \\ \hline (q_0, \#) & (q_0, \#) \\ \hline \end{array} : q \in \delta(p, q_0, b) \right\}$$

$$\theta_4 = \left\{ \begin{array}{|c|c|}\hline (p, a) & (q, b) \\ \hline (r, e) & (s, d) \\ \hline \end{array} : q \in \delta(p, s, b) \right\}$$

and $\theta = \theta_1 \cup \theta_2 \cup \theta_3 \cup \theta_4$.

Then $L_1 = \mathcal{L}^{\omega\omega}(\theta)$. Define $\pi : \Gamma \to \Sigma$ by $\pi(p, a) = a$. Then $L = \pi(L_1)$. Therefore $L \in \omega\omega$-REC.

Remark 1. If L is recognized by a 2-OTA, $M = (Q, \Sigma, \delta, q_0, F)$, we show that L is morphic image of a hv-local $\omega\omega$-language.

Using the notation of the previous theorem, let

$$\Gamma_1 = \theta$$

$$\Delta_1 = \left\{ \begin{array}{|cc|cc|}\hline (p, a) & (q, b) & (q, b) & (t, e) \\ (r, c) & (s, d) & (s, d) & (u, f) \\ \hline \end{array} : \begin{array}{|cc|cc|}\hline (p, a) & (q, b) & (q, b) & (t, e) \\ (r, c) & (s, d) & (s, d) & (u, f) \\ \hline \end{array} \in \theta \right\}$$

$$\Delta_2 = \left\{ \begin{array}{|cc|}\hline (t, e) & (u, f) \\ (p, a) & (q, b) \\ \hline (p, a) & (q, b) \\ (r, c) & (s, d) \\ \hline \end{array} : \begin{array}{|cc|}\hline (p, a) & (q, b) \\ (r, c) & (s, d) \\ \hline \end{array}, \begin{array}{|cc|}\hline (p, a) & (q, b) \\ (r, c) & (s, d) \\ \hline \end{array} \in \theta \right\}$$

and $\Delta = \Delta_1 \cup \Delta_2$. Let $L_2 = \mathcal{L}^{\omega\omega}(\Delta)$. Define $\pi_1 : \Gamma_1 \to \Sigma$ by
$\pi\left(\begin{array}{|cc|}\hline (p, a) & (q, b) \\ (r, c) & (s, d) \\ \hline \end{array} \right) = b$. Then L_2 is hv-local and $L = \pi_1(L_2)$.

5 Learning of Recognizable Infinite Array Languages

In this section, we give learning algorithm for recognizable infinite array languages.

We [8] have given algorithm to learn hv-local $\omega\omega$-languages in the limit from positive data that are ultimately periodic arrays.

An infinite array $p \in \Sigma^{\omega\omega}$ is called ultimately periodic if $p = T' \oplus T''$ where T' and T'' are finite sets of ultimately periodic infinite words over Σ.

If $T' = \{u_1 v_1^\omega, \ldots, u_k v_k^\omega\}$ is a finite set of ultimately periodic infinite words, let $l(T') = max\{|u_1 v_1^2|, \ldots, |u_k v_k^2|\}$. Let $p_{(2)}$ be the prefix of p of size $(l(T''), l(T'))$. If $p_{(2)}$ is of size (m, n), the area $A(p)$ of p is mn. The time complexity of the algorithm given in [8] depends on the area of the positive data provided and is bounded by $O(N)$, where N = sum of the areas of the given positive data.

We have proved (Theorem 1) that an $\omega\omega$-language is recognizable if and only if it is a projection of a hv-local $\omega\omega$-language. We will show how to derive a learning algorithm, for recognizable $\omega\omega$-languages, from one that learns hv-local $\omega\omega$-languages.

Let $L \in \omega\omega$-REC. Let L be recognized by a 2-DOTA $M = (Q, \Sigma, \delta, q_0, Q)$. Let $\Gamma = Q \times (\Sigma \cup \{\#\})$ and let π_1 and π_2 be projections on Γ defined by $\pi_1((q,a)) = q$ and $\pi_2((q,a)) = a$. An array p over Γ is called a computation description array if $\pi_1(p)$ is an accepting run of M on $\pi_2(p)$.

Note that

1. The alphabet Γ contains $m(n+1)$ elements, where n is the number of states of minimum 2-OTA for L and $m = |\Sigma|$.
2. For any positive example p of L, let $C(p)$ denote the set of all computation description array for p. Then $C(p)$ has at most $n^{A(p)}$ arrays.
3. If U is a hv-local $\omega\omega$-language over Γ such that $\pi(U) = L$ and E is a characteristic sample for U, then there is a finite set S_L of positive data of L such that $E \subseteq \pi^{-1}(S_L)$.

From the above note, we obtain a learning algorithm for $\omega\omega$-REC.

Algorithm REC

Input: A positive presentation of an unknown recognizable $\omega\omega$-language L,
n=the number of states of the minimum 2-OTA for L.
Output: A finite set Δ for dominoes such that $L = \pi(\mathcal{L}^{\omega\omega}(\Delta))$.
Query: Restricted superset query.
Procedure:
Initialize all parameters:
$E_0 := \phi, \Delta_0 := \phi$, answer := "no"

repeat
  while answer := "no" **do**
    $i := i + 1$;
    read the next positive example p;
    let $C(p) = \{w_1, \ldots, w_k\}$ be the set of all computation descriptions for p;
    let $C(p_{(2)}) = \{w'_1, \ldots, w'_k\}$ where w'_i is the prefix of w_i and size
      of w'_i = size of $p_{(2)}$;
    let $j := 0$;
    while $(j < k)$ and answer := "no" **do**
      $j := j + 1$;
      $E_i := E_i \cup \{w_j\}$;
      scan w'_j to compute $B_{2,2}(w'_j)$;
      let $\Delta_i := \Delta_i \cup \{ \boxed{\alpha|\beta} : \alpha, \beta \in B_{2,2}(w'_j)$ and first column of
        β = second column of $\alpha\} \cup \{ \begin{array}{|c|}\hline \alpha \\ \hline \beta \\ \hline \end{array} : \alpha, \beta \in B_{2,2}(w'_j)$ and first row
        of β = second row of $\alpha\}$ be the new conjecture;
      let answer := Is-superset of $(\pi(\mathcal{L}^{\omega\omega}(\Delta_i), L)$;
      If answer := "yes" **then** $\Delta = \Delta_i$;
    end
  until answer := "yes";
end

To prove that the learning Algorithm REC terminates, we need to prove the following result.

Proposition 6. *Let n be the number of states for the 2-OTA recognizing the unknown recognizable $\omega\omega$-language L. After at most $t(n)$ number of queries, Algorithm REC produces a conjecture Δ_i such that E_i includes a characteristic sample for a hv-local $\omega\omega$-language U with the property that $L = \pi(U)$ where $t(n)$ is a polynomial in n, depending on L.*

Proof. Let L be an unknown recognizable $\omega\omega$-language. Let $M = (Q, \Sigma, \delta, q_0, Q)$ be a 2-OTA which accepts L. Let $|Q| = n$ and $|\Sigma| = n$. Then there is a hv-local $\omega\omega$-language U over $\Gamma = Q \times (\Sigma \cup \{\#\})$ and a projection π such that $\pi(U) = L$. Let E be a characteristic sample for U. In [8], we have proved that $U = U_1 \oplus U_2$ where U_1 and U_2 are local ω-languages and $E = E_1 \oplus E_2$ where E_1 and E_2 are characteristic samples for U_1 and U_2 respectively. If $M_1 = (Q_1, \Sigma, \delta_1, q'_0, Q_1)$ and $M_2 = (Q_2, \Sigma, \delta_2, q''_0, Q_2)$ are B-machines which accept U_1 and U_2 respectively, Saoudi and Yokomori [12] have mentioned (in the proof of Lemma 15) that the lengths of all strings in E_1 and E_2 are not more than $3m_1^2$ and $3m_2^2$ respectively, where $m_1 = |\Sigma||Q_1|^2$ and $m_2 = |\Sigma||Q_2|^2$. Therefore the sizes of the arrays in $E = E_1 \oplus E_2$ are (i, j) where $1 \leq i \leq 3m_2^2$ and $1 \leq j \leq 3m_1^2$. Now find a finite set of positive data S_L of L such that $E \subseteq \pi^{-1}(S_L)$. Since π is area preserving, the areas of all the arrays in S_L are not more that $9m_1^2m_2^2$. Let $S_L = \{w_1, \ldots, w_p\}$ and $l = max\{A(w_1), \ldots, A(w_p)\}$. Then the number of computation description in $\pi^{-1}(S_L)$ is at most $n^{A(w_1)} + \cdots + n^{A(w_p)} \leq pn^l = t(n)$. Now with at most

$t(n)$ number of queries, Algorithm REC finds a finite set E_i of positive data of U with the property that E_i includes a characteristic sample for U and $\pi(U) = L$.

Theorem 4. *Given an unknown recognizable $\omega\omega$-language L, Algorithm REC learns, from positive data and superset queries, a finite set of dominoes Δ such that $L = \pi(\mathcal{L}^{\omega\omega}(\Delta))$.*

References

1. D. Angluin, Inductive inference of formal languages from positive data, Information and Control, **45** (1980), 117-135.
2. D. Angluin, Queries and concept learning, Machine Learning, **2** (1988), 319-342.
3. K. Culik II and J. Kari, An aperiodic set of Wang cubes, Lecture Notes in Computer Science, Vol. **1046**, Springer Verlag, 1996, 137-147.
4. V.R. Dare, K.G. Subramanian, D.G. Thomas and R. Siromoney, Infinite arrays and recognizability, International Journal of Pattern Recognition and Artificial Intelligence, Vol. **14**, No. **4**, (2000), 525-536.
5. L. De Prophetis and S. Varricchio, Recognizability of rectangular pictures by Wang systems, Journal of Automata, Languages and Combinatorics, **2**, (1997), **4**, 269-288.
6. S. Eilenberg, Automata, Languages and Machines, Vol. **A**, Academic Press, New York, (1974).
7. D. Giammerresi and A. Restivo, Recognizable picture languages, Int. J. Pattern Recognition and Artificial Intelligence **6** (1992), 241-256.
8. S. Gnanasekaran and V.R. Dare, Infinite arrays and domino systems, Electronic Notes in Discrete Mathematics, Vol. **12** (2003).
9. E.M. Gold, Language identification in the limit, Information and Control, **10** (1967), 447-474.
10. K. Inoue and A. Nakamura, Some properties of two-dimensional on-line tesselation acceptor, Inf. Sci. **13**, (1977), 95-121.
11. M. Latteux and D. Simplot, Theoretical Computer Science, **178**, (1997), 275-283.
12. A. Saoudi and T. Yokomori, Learning local and recognizable ω-languages and monodic logic programs, Proc. Euro COLT'93.

On the Number of Digitizations of a Disc Depending on Its Position

Martin N. Huxley[1,*] and Joviša Žunić[2,**]

[1] School of Mathematics, Cardiff University,
23 Senghennydd Road, Cardiff CF 24 4YH, U.K.
Huxley@cf.ac.uk

[2] Computer Science Department, Exeter University,
Harrison Building, Exeter EX4 4QF, U.K.
J.Zunic@ex.ac.uk

Abstract. The digitization $\mathbf{D}(R,(a,b))$ of a real disc $D(R,(a,b))$ having radius R and the centre (a,b) consists of all integer points inside of $D(R,(a,b))$, i.e., $\mathbf{D}(R,(a,b)) = D(R,(a,b)) \cap \mathbf{Z}^2$. In this paper we show that that there are

$$3\pi R^2 + \mathcal{O}\left(R^{339/208} \cdot (\log R)^{18627/8320}\right)$$

different (up to translations) digitizations of discs having the radius R. More formally,

$$\#\{\mathbf{D}(R,(a,b)) \mid a \text{ and } b \text{ vary through } [0,1)\}$$

$$= 3\pi R^2 + \mathcal{O}\left(R^{339/208} \cdot (\log R)^{18627/8320}\right).$$

The result is of an interest in the area of digital image processing because it describes (in, let say, a combinatorial way) how big the impact of the object position on its digitization can be.

Keywords: Digital disc, lattice points, enumeration.

1 Introduction

A digital disc is the binary picture (digitization) of a real disc. Formally, it is defined as the set of all integer points inside of a given real disc. It is clear that the digitization of a real disc depends on its size (radius) but also on its position with respect to the integer grid.

The problem of estimating the number of integer points inside a real disc is intensively studied in the literature, particularly in the area of number theory.

---

* This work is a part of INTAS research project on 'Analytical and combinatorial methods in number theory and geometry'.

** J. Žunić is also with the Mathematical Institute, Serbian Academy of Sciences and Arts, Belgrade.

R. Klette and J. Žunić (Eds.): IWCIA 2004, LNCS 3322, pp. 219–231, 2004.

Most attention was given to real discs centred at the origin – for a detailed overview we refer to [7]. A very recent result by Huxley [5], related to the number of lattice points inside a closed convex curve which satisfies some smoothness conditions, says that the number of lattice points inside a disc having the radius R approximates the area of this disc within an $\mathcal{O}\left(R^{\frac{131}{208}} \cdot (\log R)^{\frac{18627}{8320}}\right)$ error. Even though this estimate is related to discs in a general position it is still better than the previously known bounds for discs centred at the origin.

How efficiently a real disc $D(R,(a,b)) : (x-a)^2 + (y-b)^2 \leq R^2$ can be reconstructed from its digitization $D(R,(a,b)) \cap \mathbf{Z}^2$ is the problem considered in [13]. Combining this result with the results from [5] and [6], we have that the approximations

$$R \approx \sqrt{\frac{1}{\pi} \cdot \sum_{(i,j)\in D\cap \mathbf{Z}^2} 1}, \qquad a \approx \frac{\sum\limits_{(i,j)\in D\cap \mathbf{Z}^2} i}{\sum\limits_{(i,j)\in D\cap \mathbf{Z}^2} 1}, \qquad b \approx \frac{\sum\limits_{(i,j)\in D\cap \mathbf{Z}^2} j}{\sum\limits_{(i,j)\in D\cap \mathbf{Z}^2} 1}. \tag{1}$$

are within an $\mathcal{O}\left(R^{-\frac{285}{208}} \cdot (\log R)^{\frac{18627}{8320}}\right)$ error.

Obviously, there are infinitely many real discs (not isometric necessarily) with the same digitization and due to the digitization process there is always an inherent loss of information. Estimates such as those from (1) are commonly used to describe such digitization effects.

On the other side, the position of a real disc has also some impact on the digitization result. Figure 1 illustrates that isometric real discs could have very different digitizations depending on their position. Generally speaking, it could be said that the number of different digitizations of the same object caused by the different positions of it could be (somehow) a measure of the digitization effects, as well. Here, we consider such kind of the problem. Precisely, we study the following question: *How many different (up to translation) digital discs are digitizations of real discs having the same radius?*

We will show that there are

$$3\pi R^2 + \mathcal{O}\left(R^{339/208} \cdot (\log R)^{18627/8320}\right)$$

different (up to translations) digital discs which are digitizations of real discs having the radius R.

2 Definitions and Preliminaries

We will use the following definitions and basic statements.

The set of integers is denoted by $\mathbf{Z}$, while $\mathbf{R}$ means the set of real numbers.

Throughout this paper we consider closed discs of fixed radius R in the Euclidean plane.

For a point $P = (a,b)$ we write:

- $D(R,(a,b))$ (or simply $D(P)$ and $D(a,b)$) for the disc having the radius R and the centre (a,b);

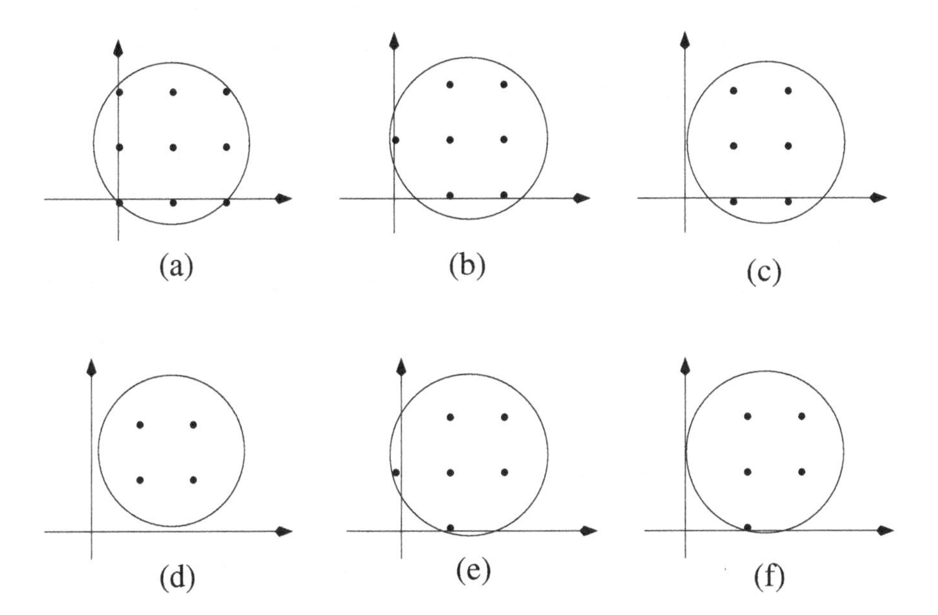

Fig. 1. There are 6 nonisometric digitizations of a disc having the radius $\sqrt{2}$. They can be obtained as the digitization of the following discs: (**a**) $(x-1)^2+(y-1)^2 \le 2$; (**b**) $(x-1.1)^2+(y-1)^2 \le 2$; (**c**) $(x-1.5)^2+(y-1)^2 \le 2$ (**d**) $(x-1.5)^2+(y-1.5)^2 \le 2$; (**e**) $(x-\frac{1+\sqrt{3}}{2})^2+(y-\frac{1+\sqrt{3}}{2})^2 \le 2$ (**f**) $(x-\frac{1+\sqrt{3}}{2}-0.001)^2+(y-\frac{1+\sqrt{3}+0.001}{2})^2 \le 2$

- $\mathbf{D}(R,(a,b))$ (or simply $\mathbf{D}(P)$ and $\mathbf{D}(a,b)$) for the set of integer points in the disc $D(P)$;
  $\mathbf{D}(R,(a,b))$ is the digitization of $D(R,(a,b))$, i.e.,
  $\mathbf{D}(R,(a,b)) = D(R,(a,b)) \cap \mathbf{Z}^2$.
- $N(a,b)$ (i.e., $N(P)$) for the number of integer points in $\mathbf{D}(P)$.

If the point P' is obtained by adding an integer vector (u,v) to P, then the set $\mathbf{D}(P')$ consists of the points of $\mathbf{D}(P)$ translated by the vector (u,v), so that $N(P') = N(P)$. We regard translation by an integer vector as an equivalence relation between the sets $\mathbf{D}(P)$. We call the equivalence classes *digital discs* of radius R, and ask how many different digital discs of radius R there are. Each class contains exactly one representative $\mathbf{D}(P)$ with P in the unit square

$$Q = \{(x,y) | 0 \le x < 1, 0 \le y < 1\}.$$

We consider the sets

$$\mathcal{E}_1 = \bigcap_{P\in Q} D(P), \qquad \mathcal{E}_2 = \bigcup_{P\in Q} D(P), \qquad \text{and} \qquad \mathcal{E} = \mathcal{E}_2 \setminus \mathcal{E}_1.$$

The set $\mathcal{E}_1$ is bounded by four quarter-circles of radius R with centres at the opposite corners of the square Q – see Fig. 2 (a). The set $\mathcal{E}_2$ is bounded by four quarter-circles of radius R with centres at the nearest corners of the square Q,

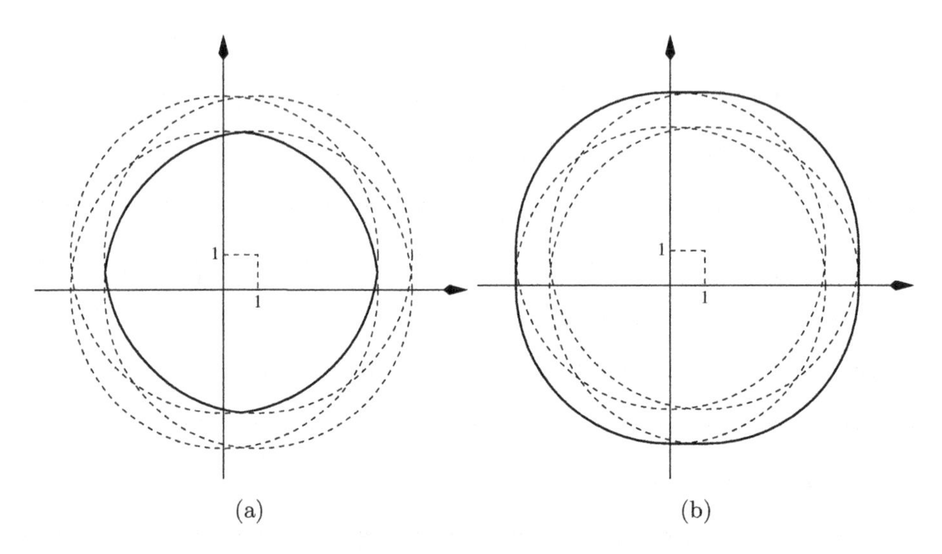

Fig. 2. (a) The bold line is the boundary of $\mathcal{E}_1$, while four circles cetred at $(0,0)$, $(0,1)$, $(1,0)$, and $(1,1)$ are presented by dashed lines; (b) The bold line is the boundary of $\mathcal{E}_2$. The straight line segments on it are: $[(1+R,0),(1+R,1)]$, $[(0,1+R),(1,1+R)]$, $[(-R,0),(-R,1)]$, and $[(0,-R),(1,-R)]$

and by line segments of length one parallel to the sides of the square – see Fig. 2 (b). In each case part of the boundary lies within the set, and part lies outside.

We need the following lemma.

Lemma 1. *The strip*

$$\mathcal{E} = \mathcal{E}_2 \setminus \mathcal{E}_1$$

is $1 + \mathcal{O}(1/R)$ *in width when measured in the direction of either the* x*-axis or the* y*-axis, whichever is closer to being normal to the boundary curves. The area of* $\mathcal{E}$ *is therefore* $4\sqrt{2}R + \mathcal{O}(1)$.

We use a special case of Theorem 5 of [5].

Proposition 1. *Let* S *be a plane region bounded by* c_1 *arcs with the following smoothness property. There is a length scale* $R \geq 2$ *and positive constants* c_2, c_3 *and* c_4 *such that on each arc, when we regard the radius of curvature* ρ *as a function of the tangent angle* ψ, *then*

$$c_2 R \leq \rho \leq c_3 R, \qquad \left| \frac{d\rho}{d\psi} \right| \leq c_4 R.$$

Then the number of integer points in S *is*

$$Area\_of\_S + \mathcal{O}(c_1 R^{\kappa} (\log R)^{\lambda})$$

with $\kappa = 131/208$, $\lambda = 18627/8320$. *The constant implied in the* $\mathcal{O}$*-notation is constructed from* c_2, c_3 *and* c_4.

Taking the set S in Proposition 1 as the disc $D(P)$, we have $N(P) = \pi R^2 + \mathcal{O}(R^\kappa (\log R)^\lambda)$ uniformly in the position of P within the unit square Q.

We call integer points (m, n) in the strip $\mathcal{E}$ *critical points*. If (m, n) is a critical point, then (m, n) lies in the disc $D(P)$ for some positions of the disc $D(P)$, but not for other positions. The straight sections of the boundary of $\mathcal{E}$ can be replaced by circular arcs of radius R without changing the set of integer points in $\mathcal{E}$, so another appeal to Proposition 1 (together with Lemma 1) gives the number of critical points as

$$\#(\mathcal{E} \cap \mathbf{Z}^2) = 4\sqrt{2}R + \mathcal{O}(R^\kappa (\log R)^\lambda). \tag{2}$$

For each critical point (m, n) we draw a circular arc $C(m, n)$ centre (m, n) of radius R.

$C(m, n)$ cuts the boundary of the square Q in two points denoted by $A_1(m, n)$ and $A_2(m, n)$.

If the point P of the square Q is inside the arc $C(m, n)$ then (m, n) lies in the disc $D(P)$. If the point P is outside the arc $C(m, n)$ then (m, n) does not lie in the disc $D(P)$. These arcs divide the square Q into regions. Points P in the same region have the same set $\mathbf{D}(P)$ of integer points, and points in different regions have different sets $\mathbf{D}(P)$. In the next section, we will estimate the number of such regions and consequently, the number of different digitizations of real discs having the same radius.

3 Main Result

To estimate the number of regions of Q made by the arcs $C(m, n)$ (while $(m, n) \in \mathcal{E}$), we will use the well-known Euler's formula, related to planar graphs, which says that the number of faces plus the number of vertices equals the number of edges plus two.

Let us consider the crossing points of arcs $C(m, n)$ and $C(m', n')$. There is a rare case when $C(m, n)$ and $C(m', n')$ cross twice within the square Q, so the points $A_1(m, n)$ and $A_2(m, n)$ both lie outside the arc $C(m', n')$. This can only happen when $m' = -m + \mathcal{O}(1)$, $n' = -n + \mathcal{O}(1)$, so (by (2)) there are $\mathcal{O}(R)$ pairs of arcs with double crossing points. If $C(m, n)$ and $C(m', n')$ cross once inside the square Q, then the points $A_1(m, n)$ and $A_2(m, n)$ lie on opposite sides of the arc $C(m', n')$. Let $E_1(m, n)$ be the subset of the strip $\mathcal{E}$ consisting of those points inside the disc $D(A_1(m, n))$ and outside the disc $D(A_2(m, n))$, and let $E_2(m, n)$ be the subset of the strip $\mathcal{E}$ consisting of those points inside the disc $D(A_2(m, n))$ and outside the disc $D(A_1(m, n))$, i.e.,

$$E_1(m, n) = \mathcal{E} \cap (D(A_1(m, n)) \setminus D(A_2(m, n)))$$

and

$$E_2(m, n) = \mathcal{E} \cap (D(A_2(m, n)) \setminus D(A_1(m, n))).$$

If the arcs $C(m, n)$ and $C(m', n')$ cross, then (m', n') lies in the union $E(m, n)$ of the sets $E_1(m, n)$ and $E_2(m, n)$,

$$E(m,n) = E_1(m,n) \cup E_2(m,n).$$

Lemma 2. *Let L be the number of intersection of arcs $C(m,n)$ and $C(m',n')$ (while $(m,n) \in \mathcal{E}$ and $(m',n') \in \mathcal{E}$) counted accordingly to multiplicity. Then*

$$L = 6\pi R^2 + \mathcal{O}\left(R^{\kappa+1}(\log R)^{\lambda}\right).$$

Proof. To calculate asymptotics, we use a continuous model of the discrete integer lattice. For any point (x,y) in the strip $\mathcal{E}$, not necessarily an integer point, we can form the arc $C(x,y)$, set $E(x,y)$, point $A_1(x,y)$, and point $A_2(x,y)$ by the same construction. Let $d = d(x,y)$ be the distance from $A_1(x,y)$ to $A_2(x,y)$, and let $e(x,y)$ be the area of $E(x,y)$. Then $e(x,y)$ is the sum of the areas of two equal circles radius R with centres d apart, minus twice the area of their intersection. Let $\phi = \phi(x,y)$ be the small angle with

$$\sin\phi = \frac{d}{2R}.$$

The common chord of the two circles subtends an angle $\pi - 2\phi$ at the centre of either circle. The area of the intersection is

$$(\pi - 2\phi)R^2 - 2dR\cos\phi,$$

so we have

$$e(x,y) = 4\phi R^2 + 4dR\cos\phi.$$

As an approximation we have

$$e(x,y) = 6dR + \mathcal{O}(1). \tag{3}$$

We want to add up, for each integer point (m,n) in $\mathcal{E}$, the number of arcs $C(m',n')$ that cross $C(m,n)$ once. By Proposition 1, the sum is

$$\sum_{(m,n)\in\mathcal{E}} \left(e(m,n) + \mathcal{O}(R^{\kappa}(\log\ R)^{\lambda})\right)$$

$$= \sum_{(m,n)\in\mathcal{E}} e(m,n) + \mathcal{O}(R^{\kappa+1}(\log\ R)^{\lambda}). \tag{4}$$

We would like to replace the first term in (4) by

$$\iint_{\mathcal{E}} e(x,y)dxdy.$$

The function $e(x,y)$ is zero on the boundary of the strip $\mathcal{E}$, and has partial derivatives of size R, so the integer lattice has too few grid points to be used for straightforward numerical integration.

Let $S(t)$ be the subset of $\mathcal{E}$ on which $e(x, y) \geq t$, and let T be the maximum of $e(x, y)$. The Riesz interchange principle gives

$$\sum_{(m,n)\in\mathcal{E}} e(m,n) = \int_0^T \left(\sum_{(m,n)\in S(t)} 1 \right) dt. \tag{5}$$

The region $S(t)$ is bounded by contour lines of the function $e(x, y)$. These are the locus of points (x, y) for which the distance $d = d(x, y)$ between $A_1 = A_1(x, y)$ and $A_2 = A_2(x, y)$ takes a fixed value. If we are given A_1 and A_2, then there are two possible points (x, y), both lying on the perpendicular bisector of the line $A_1 A_2$. For example, if A_1 is $(d \cos\theta, 0)$ and A_2 is $(0, d \sin\theta)$, then the two possible points (x, y) are

$$\left(\frac{1}{2} d\cos\theta \pm R\cos\phi\sin\theta, \frac{1}{2} d\sin\theta \pm R\cos\phi\cos\theta\right)$$

$$= (\pm R\sin(\theta \pm \phi), \pm R\cos(\theta \mp \phi)), \tag{6}$$

where the upper signs are taken together. Since ϕ is a small angle, we can see that this part of the contour still has radius of curvature approximately R. There are congruent curves when A_1 and A_2 lie on other adjacent pairs of sides of the square Q, whilst if A_1 and A_2 lie on the same side of the square or on opposite sides of the square, then we get a short straight segment of the contour line, which can be approximated by an arc of a circle radius R as before with negligible corrections to the area and the number of integer points in $S(t)$. Let $f(t)$ be the area of $S(t)$. By Proposition 1 we have

$$\sum_{(m,n)\in S(t)} 1 = f(t) + \mathcal{O}(R^{\kappa}(\log\ R)^{\lambda}),$$

and in (5) since $T = \mathcal{O}(R)$ we have

$$\sum_{(m,n)\in\mathcal{E}} e(m,n) = \int_0^T f(t)dt + \mathcal{O}(R^{\kappa+1}(\log R)^{\lambda}). \tag{7}$$

Since $e(x, y) = 0$ if the point (x, y) lies on the boundary of the strip $\mathcal{E}$, for $t > 0$ the contour lines of $e(x, y)$ bounding $S(t)$ lie entirely within the strip $\mathcal{E}$. For $d \leq 1$ the set $S(t)$ forms a narrower strip within the strip $\mathcal{E}$, bounded by contour lines which are piecewise of the type (6). For $1 < d \leq \sqrt{2}$ the set $S(t)$ is in four disconnected parts, with contour lines of the type (6) ending where $d\cos\theta = \pm 1$ or $d\sin\theta = \pm 1$, so that either A_1 or A_2 becomes a vertex of the square Q. These curves are joined by a short line segment parallel to x- or y-axis as the points A_1 and A_2 move in parallel along opposite sides of the square.

The component of $S(t)$ in the first quadrant is bounded by contour lines parametrised by

$$\left(R\cos\phi\sin\theta + \frac{1}{2} d\cos\theta,\ \ R\cos\phi\cos\theta + \frac{1}{2} d\sin\theta\right)$$

and

$$\left(R\cos\phi\sin\theta - \frac{1}{2}d\cos\theta + 1,\ \ R\cos\phi\cos\theta - \frac{1}{2}d\sin\theta + 1\right),$$

whose polar coordinates are of the form

$$\left(R + \frac{1}{2}d\sin 2\theta + \mathcal{O}\left(\frac{1}{R}\right),\ \ \theta + \mathcal{O}\left(\frac{1}{R}\right)\right),$$

$$\left(R + \sin\theta + \cos\theta - \frac{1}{2}d\sin 2\theta + \mathcal{O}\left(\frac{1}{R}\right),\ \ \theta + \mathcal{O}\left(\frac{1}{R}\right)\right)$$

respectively.

For $d \le 1$ we have

$$\begin{aligned} f(t) &= 4\int_0^{\frac{\pi}{2}} (\sin\theta + \cos\theta - d\sin 2\theta)Rd\theta + \mathcal{O}(1) \\ &= 4(2-d)R + \mathcal{O}(1). \end{aligned}$$

For $1 < d \le \sqrt{2}$ we define an angle δ by $\cos\delta = 1/d$, and then

$$\begin{aligned} f(t) &= 4\int_\delta^{\frac{\pi}{2}-\delta} (\sin\theta + \cos\theta - d\sin 2\theta)Rd\theta + \mathcal{O}(1) \\ &= 4R(2\cos\delta - 2\sin\delta - d\cos 2\delta) + \mathcal{O}(1) \\ &= 4R\left(d - \frac{2\sqrt{d^2-1}}{d}\right) + \mathcal{O}(1). \end{aligned}$$

It is convenient to rescale by $u = t/6R$. Then (3) gives $d = u + \mathcal{O}(1/R)$, so the integral in (7) is

$$\begin{aligned} &6R\int_0^{T/6R} f(6Ru)du \\ &= 6R\int_0^1 4R(2-u)du + 6R\int_1^{\sqrt{2}} 4R\left(u - \frac{2\sqrt{u^2-1}}{u}\right)du + \mathcal{O}(R) \\ &= 24R^2\left(3 - \frac{1}{2} - \frac{1}{2} - 2\int_0^{\pi/4} \tan^2\theta\, d\theta\right) + \mathcal{O}(R) \\ &= 48R^2 - 48R^2\left(1 - \frac{\pi}{4}\right) + \mathcal{O}(R) \\ &= 12\pi R^2 + \mathcal{O}(R). \end{aligned}$$

That completes the proof. ∎

In order to compute the number of vertices of the planar graph made by the arcs $C(m,n)$ and the edges of the unit square Q we have to estimate the maximum number of arcs $C(m,n)$ which intersect in the same point.

Let us call the point P in the square Q a *bad point* if three or more arcs $C(m,n)$ meet at P – i.e, there are critical points $(m_1,n_1),\ldots,(m_k,n_k)$ on the circumference of the disc $D(P)$, with $k \geq 3$. We give the following upper bounds for k and for the number of bad points.

Lemma 3. *Let $P \in Q$. Then the number of bad points is upper bounded by*

$$\mathcal{O}\left(R^{1+\epsilon}\right)$$

while the maximum number of arcs $C(m,n)$ coincident with P is upper bounded by

$$\mathcal{O}\left(R^{\epsilon}\right)$$

for any $\epsilon > 0$.

Proof. Let $(m_1,n_1),\ldots,(m_k,n_k)$ be critical points on the circumference of the disc $D(P)$, with $k \geq 3$. We will prove $k = \mathcal{O}\left(R^{\varepsilon}\right)$.

The point P lies on the perpendicular bisectors

$$2(m_j - m_i)x + 2(n_j - n_i)y = m_i^2 - m_j^2 + n_i^2 - n_j^2,$$

for $1 \leq i < j \leq k$, so P is a rational point $(a/q, b/q)$ with highest common factor $(a,b,q) = 1$, and denominator

$$q \leq 2(2R+1)^2.$$

The critical points (m_j,n_j) lie on the circle

$$\left(x - \frac{a}{q}\right)^2 + \left(y - \frac{b}{q}\right)^2 = R^2.$$

Here R^2 must be a rational number g/h in its lowest terms. Let q_0 be the smallest positive integer with $h|q_0^2$, so that $q_0^2R^2$ is some positive integer G. By unique factorisation q^2R^2 is a positive integer if and only if $q_0|q$, so we can write $q = q_0q'$. The critical points (m_j,n_j) satisfy

$$(q_0q'm_j - a)^2 + (q_0q'n_j - b)^2 = q^2R^2 = q'^2G. \tag{8}$$

We use the unique factorisation of Gaussian integers (see [3]). From (8) there are integers e_j, f_j, r_j, s_j with

$$q_0q'm_j - a + i(q_0q'n_j - b) = (r_j + is_j)^2(e_j + if_j), \tag{9}$$

$$q' = r_j^2 + s_j^2 = (r_j + is_j)(r_j - is_j), \tag{10}$$

$$G = e_j^2 + f_j^2 = (e_j + if_j)(e_j - if_j). \tag{11}$$

Now $(r_j + is_j) | (a + ib)$, and since $(a, b, q') = 1$, the Gaussian highest common factor of $a + ib$ and $a - ib$ is an ideal generated by 1 or by $1 + i$. In both cases the factor $r_j + is_j$ is unique up to multiplication by ± 1 or $\pm i$. We can make a fixed choice of $r + is$, and absorb the remaining factor ± 1 into the factor $e_j + if_j$ in (9). Subtracting the equations (9) corresponding to two critical points (m_j, n_j) and (m_ℓ, n_ℓ), we have

$$q_0 q'(m_j - m_\ell) + i q_0 q'(n_j - n_\ell) = (r + is)^2 (e_j - e_\ell + if_j - if_\ell). \tag{12}$$

The left-hand side of (12) is divisible by $r - is$, so

$$(r + is) | (i + i)(e_j - e_\ell + if_j - if_\ell). \tag{13}$$

We are now ready for a counting argument. We have

$$G = q_0^2 R^2 \leq 4R^2 (2R + 1)^4 = D,$$

say. Let Δ be the maximum, over positive integers $n \leq 8D$, of the number of ways of writing n as a sum of two squares of integers. Then

$$\Delta = \mathcal{O}(R^\varepsilon) \tag{14}$$

for any $\varepsilon > 0$ (see [3]). There are at most Δ possible factors $e_j + if_j$ in (11), so in (13) $r - is$ is a factor of one of at most Δ^2 Gaussian integers α with $|\alpha|^2 \leq 8D$. For each pair j and ℓ, there are at most Δ possibilities for $r - is$, which determines q'. We can identify the bad point $P(a/q, b/q)$ from (11) if we are given one of the critical points (m, n) for which the arc $C(m, n)$ goes through P. There are K critical points, so there can be at most $\Delta^3 K$ bad points $(a/q, b/q)$ in the square Q. Since the number of critical points is $\mathcal{O}(R)$ (see Lemma 1) the first statement of the lemma is proved.

Finally, if we are given a bad point P on k arcs $C(m_j, n_j)$, then each arc corresponds to a factor $e_j + if_j$ in (11), so $k \leq \Delta$. ∎

Now, we give the main result of the paper.

Theorem 1. *The number of different (up to translation) digitizations of real discs having the radius R is*

$$3\pi R^2 + \mathcal{O}\left(R^{\kappa+1} \cdot (\log R)^\lambda\right)$$

with $\kappa = 131/208$ and $\lambda = 18627/8320$.

Proof. Suppose that there are F different sets $\mathbf{D}(P)$, each corresponding to a subregion of the square Q. These subregions are bounded by the boundary of Q and the arcs $C(m, n)$ corresponding to the K critical points. We consider this configuration as a graph whose vertices are the crossing points of arcs $C(m, n)$, and at most $2K$ distinct points where the arcs $C(m, n)$ cut the boundary of Q. Each bad crossing point is counted with multiplicity at most Δ^2 (see (14)) in the sum L (from Lemma 2), and each good crossing point is counted twice, so, by Lemma 1 and Lemma 3, the number of vertices is

$$V = \frac{L}{2} + \mathcal{O}(\Delta^5 K) + \mathcal{O}(K) = \frac{L}{2} + \mathcal{O}(R^{1+\epsilon}) \tag{15}$$

for any $\epsilon > 0$.

The good crossing points have valency 4, and the number of edges is similarly

$$E = L + \mathcal{O}(\Delta^5 K) + \mathcal{O}(K). \tag{16}$$

The exterior of the square Q counts as one face in Euler's formula "faces plus vertices equals edges plus two", so we have

$$F + 1 + V = E + 2, \tag{17}$$

and finally, by (15)-(17) and Lemma 2,

$$F = \frac{L}{2} + \mathcal{O}(R^{1+\epsilon}) = 3\pi R^2 + \mathcal{O}(R^{\kappa+1}(\log\ R)^{\lambda}).$$

Each of the F different regions corresponds to a different digital disc $\mathbf{D}(P)$. ∎

4 Concluding Remarks

In this paper we illustrate how the digitization of a given real disc can depend on its position with respect to the digitization (integer) grid. We prove that if a real disc having the radius R is digitized then the number of its different (up to translation) digitizations is

$$3\pi R^2 + \mathcal{O}\left(R^{339/208} \cdot (\log R)^{18627/8320}\right).$$

There is a variety of digital objects enumeration problems which are already considered in the literature. The problem which comes from the area of digital image analysis is the problem of estimating the number of digital straight line segments that are realizable on a squared integer grid of a size, let say, $n \times n$. This problem can be reformulated noticing that a digital straight line segment corresponds uniquely to a linear dichotomy of an $n \times n$ integer grid. Then, we can use [8] which says that the number P_n of such linear dichotomies (partitions) is

$$P_n = \frac{3 \cdot n^4}{\pi^2} + \mathcal{O}\left(n^3 \cdot \log n\right).$$

It has been done by using the concept of *adjacent pairs* from [9], which enables an efficient characterization of linear dichotomies of an arbitrary planar finite number point set.

Another related problem is studied in [1]. The authors defined a *corner cut* as a set $\mathbf{A} \subset \mathbf{N}_0^2$ of non negative integer points which can be separated from $\mathbf{N}_0^2 \setminus \mathbf{A}$

by a straight line and investigated how many different corner cuts consist of n points. It turns out that such a number C_n has the order of magnitude $n \cdot \log n$, i.e., there are constants $k_1 > 0$ and $k_2 > 0$ such that $k_1 \cdot n \cdot \log n \leq C_n \leq k_2 \cdot n \cdot \log n$. The results is extended ([10]) to an arbitrary d-dimensional space and shows that there are $\mathcal{O}\left(n^{d-1} \cdot (\log n)^{d-1}\right)$ different n point sets $\mathbf{A} \subset \mathbf{N}_0^d$ which can be separated from $\mathbf{N}_0^d \setminus \mathbf{A}$ by a hyperplane.

The number of different n-point sphere corner cuts in d-dimensions is upper bonded by $\mathcal{O}\left(n^{d+1} \cdot (\log n)^{d-1}\right)$ (see [11]). An upper bound for the number of digital discs consisting on a fixed number of points is given in [12]. As another kind of problems related to digital discs, let us mention the *recognition problem* which is solved in [2].

To close, let us mention that the number $\mathcal{S}(n,m)$ of different digital point sets which can be realized as a set intersection of an $m \times n$-integer grid and a real disc (see Fig. 3) is also of interest. $\mathcal{S}(m,n)$ can be understood as the number of

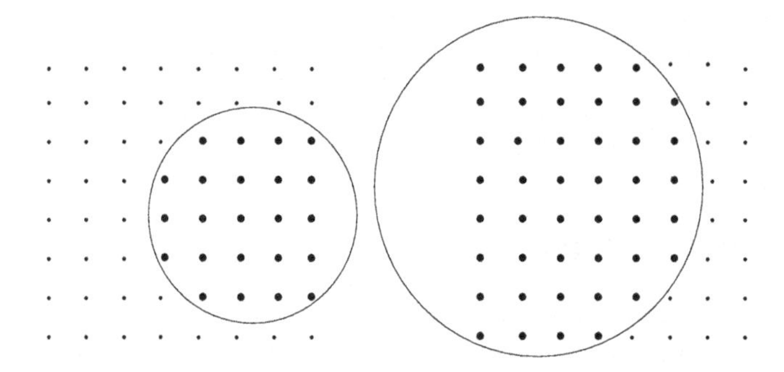

Fig. 3. "Signatures" of two nonisometric discs on a 8×8 integer grid

different "signatures" which real discs can have on an integer grid (i.e., a binary picture) of a given size.

References

1. S. Corteel, G. Rémond, G. Schaeffer, and H. Thomas, "The Number of Plane Corner Cuts," *Advances in Applied Mathematics*, **23** (1999) 49-53.
2. S. Fisk, "Separating point sets by circles, and the recognition of digital discs," *IEEE Trans. on Pattern Analysis and Machine Intelligence,* **8** (1984) 554-556.
3. G. H. Hardy, E. M. Wright, "An Introduction to the Theory of Numbers," 4th edn Oxford University Press, 1959.
4. M. N. Huxley, "Area, Lattice Points, and Exponential Sums," London Math. Soc. Monographs 13, Oxford University Press 1996.
5. M. N. Huxley, "Exponential Sums and Lattice Points III," *Proc. London Math. Soc.*, **87** (2003) 591-609.
6. R. Klette, J. Žunić, "Multigrid convergence of calculated features in image analysis," *Journal of Mathematical Imaging and Vision* **13** (2000) 173-191.

7. E. Krätzel, *Lattice Points* (VEB Deutscher Verlag der Wissenschaften, Berlin, 1988).
8. J. Koplowitz, M. Lindenbaum, A. Bruckstein, "On the number of digital straight lines on a squared grid," *IEEE Trans. Information Theory* **15** (1993) 949-953.
9. M. Lindenbaum, J. Koplowitz, "A new parametrization of digital straight lines," *IEEE Trans. on Pattern Analysis and Machine Intelligence,* **13** (1991) 847-852.
10. U. Wagner, "On the Number of Corner Cuts," *Advances in Applied Mathematics,* **29** (2002) 152-161.
11. J. Žunić, "Cutting Corner with Spheres in *d*-dimensions," *Advances in Applied Mathematics,* **32** (2004) 609-614.
12. J. Žunić, "On the Number of Digital Discs," *Journal of Mathematical Imaging and Vision,* accepted.
13. J. Žunić, N. Sladoje, "Efficiency of Characterizing Ellipses and Ellipsoids by Discrete Moments," *IEEE Trans. on Pattern Analysis and Machine Intelligence,* **22** (2000) 407-414.

On the Language of Standard Discrete Planes and Surfaces

Damien Jamet

LIRMM, Université Montpellier II, 161 rue Ada, 34392 Montpellier Cedex 5 - France
jamet@lirmm.fr

Abstract. A standard discrete plane is a subset of $\mathbb{Z}^3$ verifying the double Diophantine inequality $\mu \leq ax + by + cz < \mu + \omega$, with $(a, b, c) \neq (0, 0, 0)$. In the present paper we introduce a generalization of this notion, namely the $(1, 1, 1)$-discrete surfaces. We first study a combinatorial representation of discrete surfaces as two-dimensional sequences over a three-letter alphabet and show how to use this combinatorial point of view for the recognition problem for these discrete surfaces. We then apply this combinatorial representation to the standard discrete planes and give a first attempt of to generalize the study of the dual space of parameters for the latter [VC00].

1 Introduction

The works related to discrete lines and planes can be roughly divided in two kinds of approaches. In [And93], É. Andrès introduced the arithmetic discrete planes, as a natural generalization of the arithmetic discrete lines introduced by J.P. Réeveillès [Rév91]. Since then, using different approaches, many authors have investigated the recognition problem of discrete planes, that is, « given $\mathcal{V} \subseteq \mathbb{Z}^3$ a set of voxels, does there exist a discrete plane containing $\mathcal{V}$? » (using linear programming [Meg84, PS85, VC00, Buz02], arithmetic structure [DRR96] and Farey series [VC00]). An interesting review of these algorithms can be found in [BCK04].

On the other hand, a wide literature has been devoted to the study of Sturmian words, that is, the infinite words over a binary alphabet which have $n + 1$ factors of length n [Lot02]. These words are also equivalently defined as a discrete approximation of a line with irrational slope. Then, many attempts have been investigated to generalize this class of infinite words to two-dimensional words. For instance, in [Vui98, BV00b, ABS04], it is shown that the orbit of an element $\mu \in [0, 1[$ under the action of two rotations codes a standard discrete plane. Furthermore, the generating problem of one or two-dimensional words characterizing discrete lines or planes is investigated in [BV00b, Lot02, ABS04, BT04].

Let us now introduce some basic notions and notation used in the present paper. Let $\{\vec{e_1}, \vec{e_2}, \vec{e_3}\}$ denote the canonical basis of the Euclidean space $\mathbb{R}^3$. An element of $\mathbb{Z}^3$ is called a *voxel*. It is usual to represent a voxel $(x, y, z) \in \mathbb{Z}^3$ as a unit cube of $\mathbb{R}^3$ centered in (x, y, z). Another equivalent representation is to

R. Klette and J. Žunić (Eds.): IWCIA 2004, LNCS 3322, pp. 232–247, 2004.

consider the unit cube $\{(x+\lambda_1, y+\lambda_2, z+\lambda_3) \mid (\lambda_1,\lambda_2,\lambda_3) \in [0,1]\}$. In the present paper, for clarity issues, we consider the last representation.

Let $(a,b,c,\mu,\omega) \in \mathbb{R}^5$. An *arithmetic discrete plane* with *normal vector* (a,b,c), with *translation parameter* μ, and with *thickness* ω, is the subset of $\mathbb{Z}^3$ defined as follows:

$$\mathfrak{P}(a,b,c,\mu,\omega) = \left\{(x,y,z) \in \mathbb{Z}^3 \mid \mu \leq ax+by+cz < \mu+\omega\right\}. \tag{1}$$

If $\omega = \max\{|a|,|b|,|c|\}$, then $\mathfrak{P}(a,b,c,\mu,\omega)$ is said to be a *naive discrete plane.* If $\omega = |a|+|b|+|c|$, then $\mathfrak{P}(a,b,c,\mu,\omega)$ is said to be a *standard discrete plane.*

Considering the action of the group of isometries on the set of the discrete planes, we can suppose, with no loss of generality, that $0 \leq a \leq b \leq c$ and $c \neq 0$.

It is well known that the naive discrete planes are *functional*, that is, if $0 \leq a \leq b \leq c$, the naive discrete plane $\mathfrak{P}(a,b,c,\mu,\max\{|a|,|b|,|c|\})$ is in bijection with the integral points of the plane $z=0$ by the projection map $\pi_z : \mathbb{R}^3 \rightarrow \{(x,y,z) \in \mathbb{R}^3 \mid z=0\}$ along the vector $(0,0,1)$. In a similar way, in [ABS04], it is shown that, given the affine orthogonal projection along the vector $(1,1,1)$ onto the plane $x+y+z=0$, namely $\pi : \mathbb{R}^3 \longrightarrow \{(x,y,z) \in \mathbb{R}^3 \mid x+y+z=0\}$, and given $\Gamma = \pi\left(\mathbb{Z}^2\right)$, then the restriction $\pi : \mathfrak{P}(a,b,c,\mu,|a|+|b|+|c|) \longrightarrow \Gamma$ is a bijection. In other words, any standard discrete plane can be recoded on a regular lattice (see Section 2).

From now on, let us denote $\mathfrak{P}(a,b,c,\mu)$ the standard discrete plane $\mathfrak{P}(a,b,c,\mu,|a|+|b|+|c|)$. We call *unit cube* any translate of the *fundamental unit cube* with integral vertices, that is, any set $(x,y,z)+\mathcal{C}$ where $(x,y,z) \in \mathbb{Z}^3$ and $\mathcal{C}$ is the fundamental unit cube (see Figure 2(a)):

$$\mathcal{C} = \left\{\lambda_1 e_1 + \lambda_3 e_3 + \lambda_3 e_3 \mid (\lambda_1,\lambda_2,\lambda_3) \in [0,1]^3\right\}.$$

Let us now define the three basic faces (see Figure 1):

$$\begin{aligned}
E_1 &= \{\lambda_2 \overrightarrow{e_2} + \lambda_3 \overrightarrow{e_3} \mid (\lambda_2,\lambda_3) \in [0,1[^2\}, \\
E_2 &= \{-\lambda_1 \overrightarrow{e_1} + \lambda_3 \overrightarrow{e_3} \mid (\lambda_1,\lambda_3) \in [0,1[^2\}, \\
E_3 &= \{-\lambda_1 \overrightarrow{e_1} - \lambda_2 \overrightarrow{e_2} \mid (\lambda_1,\lambda_2) \in [0,1[^2\}.
\end{aligned}$$

Let $(x,y,z) \in \mathbb{Z}^3$. We call *pointed face of type k pointed on (x,y,z)* the set $(x,y,z)+E_k$ with $k \in \{1,2,3\}$. Notice that each face contains exactly one integral point. We call it the *distinguished vertex* of the face.

Let $\mathcal{P}$ be the plane with equation $ax+by+cz=\mu$ with $(a,b,c) \in \mathbb{R}^3$ and $0 \leq a \leq b \leq c$, let $\mathcal{C}_\mathcal{P}$ be the union of the unit cubes intersecting the open half-space $ax+by+cz<\mu$, and let $\mathfrak{P}_\mathcal{P} = \overline{\mathcal{C}_\mathcal{P}} \setminus \mathring{\mathcal{C}}_\mathcal{P}$, where $\overline{\mathcal{C}_\mathcal{P}}$ (resp. $\mathring{\mathcal{C}}_\mathcal{P}$) is the closure (resp. the interior) of the set $\mathcal{C}_\mathcal{P}$ in $\mathbb{R}^3$, provided with its usual topology. In [ABS04], it is proved that the set $\mathfrak{P}_\mathcal{P}$ is partitioned by pointed faces. Moreover, let $\mathcal{V}_\mathcal{P} = \mathfrak{P}_\mathcal{P} \cap \mathbb{Z}^3$ be the set of *vertices* of $\mathfrak{P}_\mathcal{P}$. Then, $\mathcal{V}_\mathcal{P} = \mathfrak{P}(a,b,c,\mu)$ (see (1)). From now on, up to the context and if no confusion is possible, we will call discrete plane indifferently $\mathfrak{P}_\mathcal{P}$ and $\mathfrak{P}(a,b,c,\mu)$.

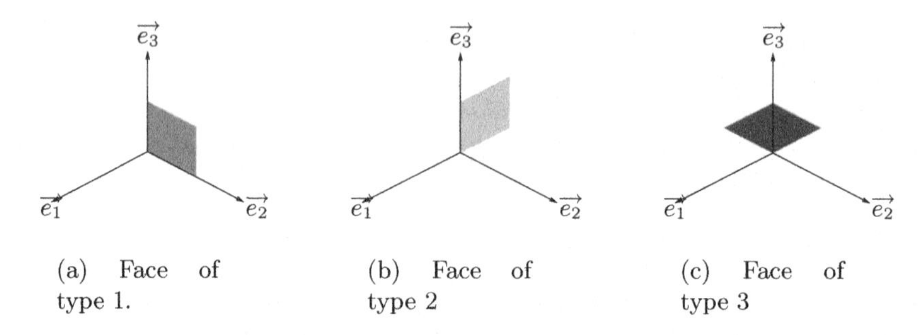

(a) Face of type 1.

(b) Face of type 2

(c) Face of type 3

Fig. 1. The three fundamental faces

In the present paper we introduce a generalization of the concept of standard discrete planes: the $(1,1,1)$-discrete surfaces (see Figure 3). Roughly speaking, a $(1,1,1)$-discrete surface is a subset of $\mathbb{R}^3$, partitionable by the pointed faces and in one-to-one correspondence, by the projection map $\pi : \mathbb{R}^3 \to \{(x,y,z) \in \mathbb{R}^3 \mid x+y+z=0\}$ with the diagonal plane $\{(x,y,z) \in \mathbb{R}^3 \mid x+y+z=0\}$. Then, as performed in the case of the standard discrete planes, given a discrete surface $\mathfrak{S}$, we associate to it a two-dimensional coding depending on the type of the pointed faces partitioning $\mathfrak{S}$. Then, it becomes natural to try to characterize the two-dimensional sequences coding the $(1,1,1)$-discrete surfaces. In other words, given a two-dimensional sequence $U \in \{1,2,3\}^{\mathbb{Z}^2}$, does U code a $(1,1,1)$-discrete surface $\mathfrak{S}$? Is this problem local? that is, does there exist a finite set of two-dimensional finite patterns $\mathcal{E}$ such that: « U codes a $(1,1,1)$-discrete surface if and only if, for all $\omega \in \mathcal{E}$, ω does not belong to the language of U »?

This paper is organized as follows. In Section 2, we define the $(1,1,1)$-discrete surfaces and their two-dimensional codings. In Section 3, after introducing the notions of τ-shape, τ-patterns, τ-complexity and τ-language, we investigate the characterization problem of the sequences $U \in \{1,2,3\}^{\mathbb{Z}^2}$ coding discrete surfaces. Then we give the list $\mathfrak{A}$ of permitted τ-patterns (see Figure 4), and prove:

Theorem 1. *Let $U \in \{1,2,3\}^{\mathbb{Z}^2}$. Then U codes a $(1,1,1)$-discrete surface if and only if $\mathcal{L}_\tau(u) \subseteq \mathfrak{A}$, where $\mathcal{L}_\tau(U)$ is the subset of subwords of U of shape τ.*

In Section 4, we show that the standard discrete planes have a canonical structure of $(1,1,1)$-discrete surface and the language of their two-dimensional codings is completely defined by their normal vector and does not depend on their translation parameter. Next, we prove that the τ-complexity of a standard discrete planes is bounded by 6 and equal to 6 for the standard discrete planes with a $\mathbb{Q}$-free normal vector. Finally, in Section 5, we give a first attempt to generalize the study of the dual space of parameters and its corresponding Farey tessellation [VC00].

2 (1, 1, 1)-Discrete Surfaces and Two-Dimensional Codings

In this section, we introduce the (1, 1, 1)-discrete surfaces and we show how we can recode each discrete surface on a regular lattice.

Let $\pi : \mathbb{R}^3 \longrightarrow \{(x, y, z) \in \mathbb{R}^3 \mid x + y + z = 0\}$ be the affine projection along the vector (1, 1, 1). Then, π is explicitly defined by:

$$\begin{array}{rccl} \pi : & \mathbb{R}^3 & \longrightarrow & \{(x, y, z) \in \mathbb{R}^3 \mid x + y + z = 0\} \\ & (x, y, z) & \mapsto & (x - z)\pi(\vec{e_1}) + (y - z)\pi(\vec{e_2}). \end{array} \tag{2}$$

Let us recall [BV00b, ABS04] that each standard discrete plane is in one-to-one correspondence with the regular lattice $\Gamma = \mathbb{Z}\pi(\vec{e_1}) + \mathbb{Z}\pi(\vec{e_2}) = \pi(\mathbb{Z}^3)$ and is partitioned by integral translates of the three basic faces E_1, E_2 and E_3. Using these properties of standard discrete planes, we define the (1, 1, 1)-discrete surfaces as follows:

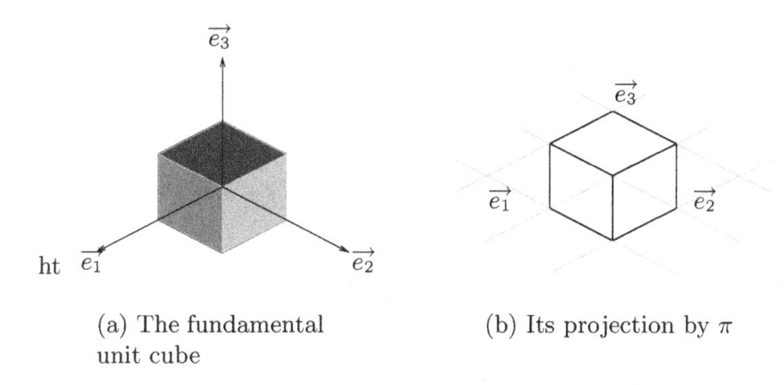

(a) The fundamental unit cube

(b) Its projection by π

Fig. 2. The projection of the fundamental unit cube

Definition 1 (**(1, 1, 1)-Discrete Surface).** *Let $\mathfrak{S} \subseteq \mathbb{R}^3$. Then $\mathfrak{S}$ is said to be a* (1, 1, 1)-discrete surface *(or just a* discrete surface*) if the following conditions hold:*

i) the projection map $\pi : \mathfrak{S} \rightarrow \{(x, y, z) \in \mathbb{R}^3 \mid x + y + z = 0\}$ is a bijection;
ii) $\mathfrak{S}$ is partitioned by pointed faces.

Even if, unfortunately, the terminology can be ambiguous, in particular for the ones who are are accustomed with [Fra95, KI00, Mal97, RKW91], we will use the terminology *discrete surface* instead of (1, 1, 1)-*discrete surface* in the present paper, in order to simplify notations.

Since the plane $x + y + z = 0$ is a disjoint union of a countable set of translates of the tiles $\pi(E_1)$, $\pi(E_2)$ and $\pi(E_3)$, then there exist two sequences $(x_n, y_n, z_n)_{n\in\mathbb{N}} \in (\mathbb{Z}^3)^{\mathbb{N}}$ and $(i_n)_{n\in\mathbb{N}} \in \{1, 2, 3\}^{\mathbb{N}}$ such that $\mathfrak{S} = \bigcup_{n\in\mathbb{N}} (x_n, y_n, z_n) + E_{i_n}$.

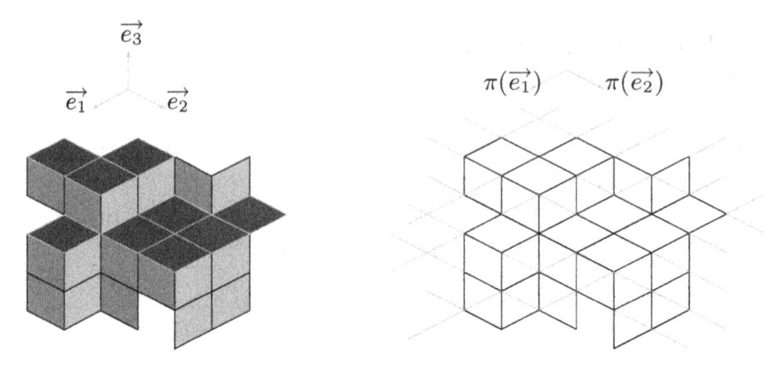

Fig. 3. A piece of a discrete surface and its projection under π

A first property of discrete surfaces is that, given $n \in \mathbb{N}$, the point (x_n, y_n, z_n) cannot have two different types. Moreover, the projection map $\pi : \mathbb{R}^3 \to \{(x, y, z) \in \mathbb{R}^3 \mid x + y + z = 0\}$ provides a one-to-one correspondence between $\{(x_n, y_n, z_n) \mid n \in \mathbb{N}\}$ and Γ. More precisely,

Lemma 1. *Let $\mathfrak{S} = \bigcup_{n \in \mathbb{N}} (x_n, y_n, z_n) + E_{i_n}$ be a discrete surface. Then, the following assertions hold:*

i) $\forall (m, n) \in \mathbb{N}^2$, $(x_m, y_m, z_m) = (x_n, y_n, z_n)$ *implies* $i_n = i_m$*;*
ii) the function $\pi : \{(x_n, y_n, z_n) \mid n \in \mathbb{N}\} \longrightarrow \mathbb{Z}\pi(\vec{e_1}) + \mathbb{Z}\vec{e_2}$ *is a bijection. In other words,* $\{(x_n - z_n, y_n - z_n) \mid n \in \mathbb{N}\} = \mathbb{Z}^2$.

In the present paper, we suppose that the representation of a discrete surface $\mathfrak{S}$ is *reduced*, that is, Lemma 1 i) and ii) are assumed to hold, and we denote by $\mathcal{V}_{\mathfrak{S}} = \mathfrak{S} \cap \mathbb{Z}^3$ the set of *vertices* of $\mathfrak{S}$.

Since every vertex of $\mathfrak{S}$ has a unique type, then, to each $(m, n) \in \mathbb{Z}^2$, we can associate the type of the antecedent $(x, y, z) \in \mathfrak{S}$ of the element $m\pi(\vec{e_1}) + n\vec{e_2} \in \Gamma$. Thus, we obtain the two-dimensional coding of $\mathfrak{S}$ as follows:

Definition 2 (Two-Dimensional Coding). *Let $\mathfrak{S} = \bigcup_{n \in \mathbb{N}} (x_n, y_n, z_n) + E_{i_n}$ be a discrete surface. The two-dimensional coding of $\mathfrak{S}$ is the sequence $U \in \{1, 2, 3\}^{\mathbb{Z}^2}$ defined as follows:*

$$\forall n \in \mathbb{N},\ U_{x_n - z_n, y_n - z_n} = i_n.$$

Since we have a two-dimensional coding over the three-letter alphabet $\{1, 2, 3\}$ of each discrete surface, it becomes natural to investigate the language of these sequences and to study the characterization problem of such a sequence, that is, given a two-dimensional sequence $U \in \{1, 2, 3\}^{\mathbb{Z}^2}$, does it code a discrete surface $\mathfrak{S}$? In the next section, we prove that the language of a discrete surface coding is of finite type and we provide the set of permitted patterns.

For every $(m, n) \in \mathbb{Z}^2$, let $\tau_{m,n} = \{(m, n), (m, n + 1), (m + 1, n + 1)\}$. A *$\tau$-pattern* is a pattern with shape $\overline{\tau}$. Hence, following the definitions above, one can define the τ-language and the τ-complexity of a two-dimensional sequence.

3 Characterization of the Two-Dimensional Coding of the Discrete Surfaces

3.1 Basic Notions on Two-Dimensional Sequences Over a Finite Alphabet

In this section, we recall some basic notions and terminology concerning the two-dimensional sequences over a finite alphabet.

Let $\sim$ be the equivalence relation over the set $\mathfrak{P}(\mathbb{Z}^2)$ of the finite subsets of $\mathbb{Z}^2$, as follows:

$$\forall(\Omega, \Omega') \in \mathfrak{P}(\mathbb{Z}^2)^2, \quad \Omega \sim \Omega' \iff \exists(v_1, v_2) \in \mathbb{Z}^2,\, \Omega = \Omega' + (v_1, v_2).$$

An element $\overline{\Omega}$ of $\mathfrak{P}(\mathbb{Z}^2)/\sim$ is said to be a *shape*.

Let $\mathcal{A}$ be a finite alphabet. Let Ω be a finite subset of $\mathbb{Z}^2$. A function $\omega : \Omega \to \mathcal{A}$ is called a *finite pointed pattern over the alphabet* $\mathcal{A}$. The equivalence relation defined above provides an equivalence relation over the set of the finite pointed patterns over the alphabet $\mathcal{A}$, also denoted $\sim$, as follows: $\forall(\omega, \omega') \in \mathcal{W}_{\mathcal{A}}^2$, $\omega \sim \omega'$ if and only if

$$\exists(v_1, v_2) \in \mathbb{Z}^2,\, \Omega = \Omega' + (v_1, v_2) \text{ and } \forall(m, n) \in \Omega,\, \omega_{m,n} = \omega'_{m+v_1,n+v_2}.$$

Let us notice, that given two finite pointed patterns over the alphabet $\mathcal{A}$, $\omega : \Omega \to \mathcal{A}$ and $\omega' : \Omega' \to \mathcal{A}$, one has $\omega \sim \omega'$ implies that $\Omega \sim \Omega'$. The equivalence class $\overline{\omega}$ of ω is said to be a *pattern of shape* $\overline{\Omega}$. In order to simplify the notation, when no confusion is possible, we will use ω (resp. Ω) instead of $\overline{\omega}$ (resp. $\overline{\Omega}$).

Let $U \in \mathcal{A}^{\mathbb{Z}^2}$ be a two-dimensional sequence and let $\omega : \Omega \to \mathcal{A}$ be a pattern of shape Ω. An *occurrence* of ω in U is an element $(m_0, n_0) \in \mathbb{Z}^2$ such that for all $(m, n) \in \Omega$, $\omega_{m,n} = U_{m_0+m,n_0+n}$. The set of patterns occurring in U is called the *language* of U and is denoted $\mathcal{L}(U)$. Given a shape Ω, the set of patterns with shape Ω occurring in U is called the *Ω-language* of U and is denoted by $\mathcal{L}_\Omega(U)$.

Let Ω be a shape. The *Ω-complexity map* is the function $p_\Omega : \mathcal{A}^{\mathbb{Z}^2} \longrightarrow \mathbb{N} \cup \{\infty\}$ defined as follows:

$$\begin{aligned} p_\Omega : \mathcal{A}^{\mathbb{Z}^2} &\longrightarrow \mathbb{N} \cup \{\infty\} \\ U &\mapsto |\mathcal{L}_\Omega(U)|, \end{aligned}$$

where $|\mathcal{L}_\Omega(U)|$ is the cardinality of the set $\mathcal{L}_\Omega(U)$.

3.2 Characterization of the Two-Dimensional Coding of a Discrete Surface

Let us first reduce the characterization problem to a two-dimensional tiling problem of the plane $\{(x, y, z) \in \mathbb{R}^3 \mid x + y + z = 0\}$. Indeed, a direct consequence of Definitions 1 and 2 is:

Lemma 2. *Let $U \in \{1,2,3\}^{\mathbb{Z}^2}$ be a two-dimensional sequence. The following assertions are equivalent:*

i) the set $\mathfrak{S} = \bigcup_{(m,n)\in\mathbb{Z}^2}\{m\overrightarrow{e_1} + n\overrightarrow{e_2} + E_{U_{m,n}}\}$ is a discrete surface;
ii) the sequence U codes a discrete surface;
iii) the set $\{m\pi(\overrightarrow{e_1}) + n\pi(\overrightarrow{e_2}) + \pi(E_{U_{m,n}})_{|(m,n)\in\mathbb{Z}^2}\}$ is a partition of the plane $x+y+z=0$.

Let $\mathcal{P}_0 = \mathbb{R}\pi(\overrightarrow{e_1}) + \mathbb{R}\pi(\overrightarrow{e_2})$ be the two-dimensional $\mathbb{R}$-vector space of basis $\{\pi(\overrightarrow{e_1}), \pi(\overrightarrow{e_2})\}$. Let $|\cdot|_\infty : \mathcal{P}_0 \longrightarrow \mathbb{R}_+$ be the norm on $\mathcal{P}_0$ defined by:

$$\forall (x,y) \in (\mathbb{R}^2)^2,\ |x\pi(\overrightarrow{e_1}) + y\pi(\overrightarrow{e_2})|_\infty = \max\{|x|, |y|\}.$$

Let d_∞ be the distance on $\mathcal{P}_0$ associated to the norm $|\cdot|_\infty$, that is,

$$\forall (z,z') \in \mathcal{P}_0^2,\ d_\infty(z,z') = |z-z'|_\infty.$$

The following lemma is immediate (see Figure 1):

Lemma 3. *Let $z, z' \in \Gamma = \pi(\mathfrak{S})$, $z'' \in \mathcal{P}_0$ and $(i,i') \in \{1,2,3\}^2$. Then,*

i) $z + \pi(E_i) \cap z' + \pi(E_{i'}) \neq \emptyset \implies d_\infty(z,z') \leq 1$;
ii) $z'' \in z + \pi(E_i) \implies d_\infty(z,z'') < 2$.

An interesting consequence of Lemma 3 is that, given a two-dimensional sequence $U \in \{1,2,3\}^{\mathbb{Z}^2}$, deciding whether U codes a discrete surface is a local problem. Now, it remains to exhibit a set $\mathfrak{A}$ of *permitted* patterns.

Roughly speaking, the characterization problem can be divided in two parts: an « injection problem » and a « surjection problem ». The « injection problem » consists in deciding whether a given union of projections of pointed faces is disjoint. The « surjection problem » consists in deciding whether a given union of projections of pointed faces covers $\mathcal{P}_0$.

Then, let us first investigate the « injection problem ».

Lemma 4. *Let $U \in \{1,2,3\}^{\mathbb{Z}^2}$ be a two-dimensional sequence. The following assertions are equivalent:*

i) The sets $m\pi(\overrightarrow{e_1}) + n\pi(\overrightarrow{e_2}) + \pi(E_{U_{m,n}})$, with $(m,n) \in \mathbb{Z}^2$ are relatively disjoint.
ii) For every $(m,n) \in \mathbb{Z}^2$, the sets $m'\pi(\overrightarrow{e_1}) + n'\pi(\overrightarrow{e_2}) + \pi(E_{U_{m,n}})$, with $(m',n') \in \tau_{m,n}$ are relatively disjoint.

Hence, we have obtained a necessary and sufficient condition to decide whether a union of projections of pointed faces is a disjoint union. It remains to find a similar condition for the « surjection problem ». Since the characterization problem is local, a disjoint union of projections of pointed faces will cover the plane $\mathcal{P}_0 = \{(x,y,z) \in \mathbb{R}^3 \mid x+y+z = 0\}$ if and only if each point z of $\mathcal{P}_0$ will be covered by the projection of a pointed face close to x. This is a direct consequence of Lemma 3. Consequently, given a point $g = m\pi(\overrightarrow{e_1}) + n\pi(\overrightarrow{e_2}) \in \Gamma$, a union

$\bigcup_{(m,n)\in\mathbb{Z}^2} m\pi(\overrightarrow{e_1}) + n\pi(\overrightarrow{e_2}) + \pi(E_{U_{m,n}})$ of projections of pointed faces will cover $\mathcal{P}_0$ if and only if $g + \pi(E_3) \subset \bigcup_{\substack{z=(z_1,z_2)\in\Gamma \\ d_\infty(z,g)<2}} z_1\pi(\overrightarrow{e_1}) + z_2\pi(\overrightarrow{e_2}) + \pi(E_{U_{z_1,z_2}})$. In fact, this problem can be reduced to the study of the τ-patterns.

Lemma 5. *Let $U \in \{1,2,3\}^{\mathbb{Z}^2}$ be a two-dimensional sequence. The following assertions are equivalent:*

i) for every $(m_0, n_0) \in \mathbb{Z}^2$,

$$(m_0+1)\pi(\overrightarrow{e_1})+n_0\pi(\overrightarrow{e_2})+\pi(E_3) \subseteq \bigcup_{(m,n)\in\tau_{m_0,n_0}} m\pi(\overrightarrow{e_1}) + n\pi(\overrightarrow{e_2}) + \pi(E_{U_{m,n}}).$$

ii) $\displaystyle\bigcup_{(m,n)\in\mathbb{Z}^2} m\pi(\overrightarrow{e_1}) + n\pi(\overrightarrow{e_2}) + \pi(E_{U_{m,n}}) = \mathcal{P}_0.$

A simple enumeration gives the permitted τ-patterns (see Figure 4). In fact, we have proved that:

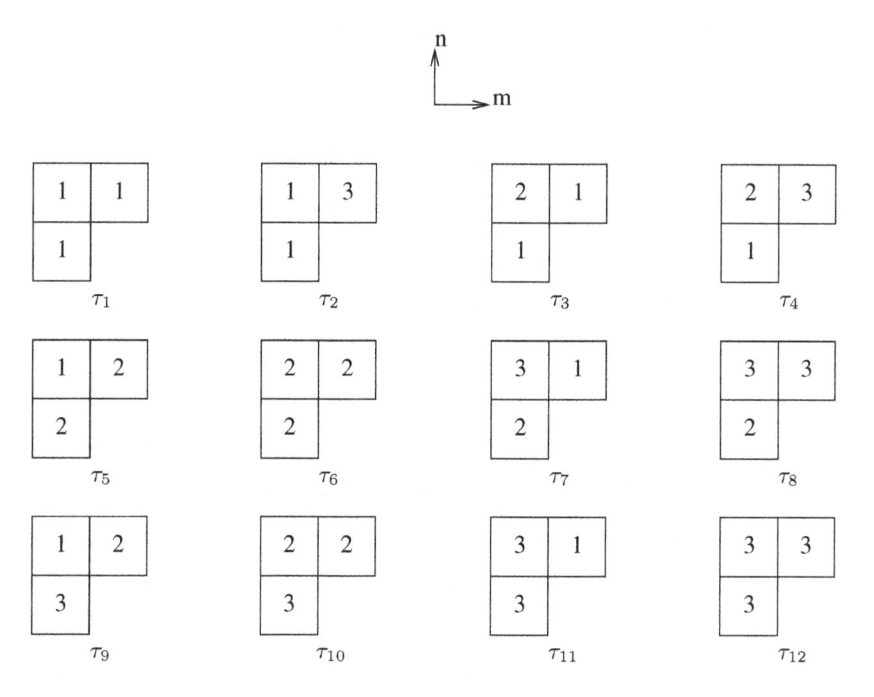

Fig. 4. The set $\mathfrak{A}$ of permitted τ-patterns of a discrete surface

Theorem 2. *Let $\mathfrak{A}$ be the set of allowed τ-patterns (see Figure 4). Let $U \in \{1,2,3\}^{\mathbb{Z}^2}$ be a two-dimensional sequence over the three-letter alphabet $\{1,2,3\}$. Then U codes a discrete surface $\mathfrak{S}$ if and only if $\mathcal{L}_\tau(U) \subseteq \mathfrak{A}$.*

4 A Particular Case of Discrete Surfaces: the Standard Discrete Planes

In this section, we investigate the standard discrete planes with a positive normal vector and show that they admit a canonical structure of discrete surface. From now on, we suppose that $(a,b,c) \in \mathbb{R}^3_+$.

4.1 Preliminaries

For the moment, we have defined the discrete surfaces via a one-to-one condition on the projection map $\pi : \mathfrak{S} \longrightarrow \{(x,y,z) \in \mathbb{R}^3 \mid x+y+z=0\}$. In [BV00b, ABS04], it is proved that a standard discrete plane is in bijection with $\Gamma = \pi(\mathbb{Z}^3)$. Let $\mathcal{P}$ be a plane with equation $ax+by+cz=\mu$. To prove that $\mathfrak{P}_\mathcal{P}$ (see Section 2) is a discrete surface, we have to show that $\pi : \mathfrak{P}_\mathcal{P} \longrightarrow \{(x,y,z) \in \mathbb{R}^3 \mid x+y+z=0\}$ is a bijection, or equivalently, that the coding of $\mathfrak{P}_\mathcal{P}$ codes a discrete surface. Let us recall how to build the two-dimensional coding of $\mathfrak{P}_\mathcal{P}$. It is based on Lemma 6.

Lemma 6. *[ABS04] Let $(x,y,z) \in \mathcal{V}_\mathcal{P}$ and $k \in \{1,2,3\}$. Let $I_1 = [0,a[$, $I_2 = [a,a+b[$ and $I_3 = [a+b,a+b+c[$. Then, the following assertions are equivalent:*

i) the point (x,y,z) is of type k, that is,$(x,y,z)+E_k \subseteq \mathfrak{P}_\mathcal{P}$;
ii) $ax+by+cz-\mu \in I_k$;
iii) $a(x-z)+b(y-z)-\mu \mod (a+b+c) \in I_k$.

The two-dimensional sequence U coding $\mathfrak{P}_\mathcal{P}$ is defined as follows:

$$\forall (m,n) \in \mathbb{Z}^2,\ \forall k \in \{1,2,3\},\ U_{m,n} = k \iff am+bn-\mu \in I_k.$$

The discrete surface structure of $\mathfrak{P}_\mathcal{P}$ follows from:

Theorem 3. *The set $\mathfrak{P}_\mathcal{P}$ is a discrete surface.*

Proof. Let U be the two-dimensional coding of $\mathfrak{P}_\mathcal{P}$. Let us show that U codes a discrete surface. Indeed, since U codes $\mathfrak{P}_\mathcal{P}$, we will deduce that $\mathfrak{P}_\mathcal{P}$ is a discrete surface. Let $k \in \{1,2,3\}$ and let us consider a τ-pattern ω such that $\omega_{0,0} = 1$. Let $(m,n) \in \mathbb{Z}^2$ be an occurrence of ω, that is, $U_{m+i,n+j} = \omega_{k i,j}$ for $(i,j) \in \tau_{0,0}$. Let us first suppose that $(m,n) = (0,0)$. Then, we deduce that $\mu \in [0,a[\mod (a+b+c)$. Hence $\mu+a+b \in [a+b,2a+b[\mod (a+b+c)$. If $a<c$, then $\mu+a+b \in [a+b,a+b+c[\mod (a+b+c)$ and $\omega_{1,1} = 3$. Conversely, if $a>c$, then $2a+b \in [0,a[\mod (a+b+c)$ and $\omega_{1,1} = 1$. In all cases, $\omega_{1,1} \neq 2$. If $(m,n) \neq (0,0)$, we similarly prove that $\omega_{1,1} \in \{1,3\}$. The other forbidden τ-patterns can be excluded in the same way. ■

4.2 Characterization of the Language of a Standard Discrete Plane

In this section, given a standard discrete plane $\mathfrak{P}$, we call *language of a standard discrete plane* the language of the two-dimensional coding of $\mathfrak{P}$.

Let $(\alpha, \beta) \in \mathbb{R}_+^2$. The rotation R_α of angle α modulo β is the function $R_\alpha : [0, \beta[\longrightarrow [0, \beta[$ defined as follows:

$$R_\alpha : \begin{array}{ccc} [0,\beta[& \longrightarrow & [0,\beta[\\ x & \mapsto & x+\alpha \mod \beta. \end{array}$$

From now on, R_a (resp. R_b) denotes the rotation of angle a (resp. of angle b) modulo $a + b + c$.

Lemma 7. *Let $U \in \{1, 2, 3\}^{\mathbb{Z}^2}$ be the two-dimensional coding of the standard discrete plane $\mathfrak{P}(a, b, c, \mu)$. Let $\omega : \Omega \to \{1, 2, 3\}$ be a pattern. Then, the following assertions are equivalent:*

i) $\omega \in \mathcal{L}(U)$, that is, there exists $(k, k') \in \mathbb{Z}^2$ such that:

$$\forall (m, n) \in \Omega,\ \omega_{m,n} = U_{m+k,n+k'}.$$

ii) there exists $(k, k') \in \mathbb{Z}^2$ such that:

$$ak + bk' - \mu \in \bigcap_{(i,j)\in\Omega} R_a^{-i} \circ R_b^{-j} \left(I_{\omega_{i,j}}\right).$$

In [Rév91, And93, VC00], the authors considered standard discrete planes $\mathfrak{P}(a, b, c, \mu)$ with $(a, b, c, \mu) \in \mathbb{Z}^4$ and $\gcd(a, b, c) = 1$. In [BV00b, Lot02, ABS04], the authors investigated the standard discrete lines or planes with a $\mathbb{Q}$-free normal vector. Let us recall that a n-uple $(a_1, \ldots, a_n) \in \mathbb{R}^n$ is said to be $\mathbb{Q}$-free if for every $(x_1, \ldots, x_n) \in \mathbb{Q}^n$, one has:

$$\sum_{i=1}^{n} a_i x_i = 0 \iff \forall i \in [\![1, n]\!],\ x_i = 0.$$

In fact, this two-case division is not necessary to study the language of the two-dimensional coding of a standard discrete plane. More precisely:

Corollary 1. *Let $U \in \{1, 2, 3\}$ be the two-dimensional coding of the standard discrete plane $\mathfrak{P}(a, b, c, \mu)$. Let ω be an Ω-pattern. Then,*

$$\omega \in \mathcal{L}(U) \iff I_\omega = \bigcap_{(i,j)\in\Omega} R_a^{-i} \circ R_b^{-j} \left(I_{\omega_{i,j}}\right) \neq \emptyset.$$

A direct consequence of Corollary 1 is:

Corollary 2. *Let $U \in \{1, 2, 3\}^{\mathbb{Z}^2}$ (resp. $U' \in \{1, 2, 3\}^{\mathbb{Z}^2}$) be the two-dimensional coding of the standard discrete plane $\mathfrak{P}(a, b, c, \mu)$ (resp. $\mathfrak{P}(a', b', c', \mu')$). Let us suppose that $\mathfrak{P}$ is parallel to $\mathfrak{P}'$, that is, there exits $\alpha \in \mathbb{R}$ such that $(a, b, c) = \alpha(a', b', c')$. Then $\mathcal{L}(U) = \mathcal{L}(U')$.*

Since two sequences coding two parallel standard discrete planes have the same language, it becomes natural to investigate the following problem: given two standard discrete planes $\mathfrak{P} = \mathfrak{P}(a,b,c,\mu)$ and $\mathfrak{P}' = \mathfrak{P}(a',b',c',\mu')$ and given $U \in \{1,2,3\}^{\mathbb{Z}^2}$ (resp. $U' \in \{1,2,3\}^{\mathbb{Z}^2}$) the two-dimensional coding of $\mathfrak{P}$ (resp. $\mathfrak{P}'$). Let us suppose that $\mathcal{L}(U) = \mathcal{L}(U')$. Are the standard discrete planes $\mathfrak{P}$ and $\mathfrak{P}'$ parallel? The answer is given by the following theorem:

Theorem 4. *Let $(a,b,c,\mu) \in \mathbb{Z}^4$ (resp. $(a',b',c',\mu') \in \mathbb{Z}^4$). Let $U \in \{1,2,3\}^{\mathbb{Z}^2}$ (resp. $U' \in \{1,2,3\}^{\mathbb{Z}^2}$) be the two-dimensional coding of the standard discrete plane $\mathfrak{P} = \mathfrak{P}(a,b,c,\mu)$ (resp. $\mathfrak{P}' = \mathfrak{P}(a',b',c',\mu')$). Then, the following assertions are equivalent:*

i) the planes $ax + by + cz = \mu$ and $a'x + b'y + c'z = \mu'$ are parallel;
ii) there exists $(m_0, n_0) \in \mathbb{Z}^2$ such that, for every $(m,n) \in \mathbb{Z}^2$, $U_{m,n} = U'_{m+m_0,n+n_0}$;
iii) $\mathcal{L}(U) = \mathcal{L}(U')$.

Proof. Let us first prove that given a square S of edge $a+b+c$, the number of a (resp. b, c) in a subwords $\omega : S \to \{1,2,3\}$ is $a(a+b+c)$ (resp. $b(a+b+c)$, $c(a+b+c)$). In fact, it is sufficient to study the case $S = [\![0, a+b+c-1]\!]^2$. The general case is a direct consequence of Corollary 1.

Let us assume that $\gcd(a,b,c) = 1$. Then, $\gcd(a,b,a+b+c) = 1$. Hence, for every element $k \in [\![0, a-1]\!]$, there exists $(x,y) \in \mathbb{Z}^2$ such that $ax + by - \mu \equiv k \mod a+b+c$. Let $(m,n) \in [\![0, a+b+c-1]\!]^2$ such that $m \equiv x \mod a+b+c$ and $n \equiv y \mod a+b+c$. Then $am + bn - \mu \equiv ax + bx - \mu \equiv x \mod (a+b+c)$ and $U_{m,n} = 1$. Let $k \in [\![0, a+b+c-1]\!]$. Then $U_{m-kb,n+ka} = 1$. Moreover, for all $(m,n) \in [\![0, a+b+c-1]\!]^2$, $(m,n) \equiv (m-kb, n+ka) \mod a+b+c$ if and only if $k = 0$. Indeed, let us suppose that $ka \equiv kb \equiv 0 \mod a+b+c$ and let $(u,v) \in \mathbb{Z}^2$ such that $au + bv \equiv 1 \mod a+b+c$. Then $k \equiv k(au+bv) \equiv kau + kav \equiv 0 \mod a+b+c$. Since $k \in [\![0, a+b+c-1]\!]$, we deduce that $k = 0$. Hence $|U|_a \geq a(a+b+c)$. We similarly prove that $|U|_b \geq b(a+b+c)$ and $|U|_c \geq c(a+b+c)$. Finally, since $|U|_a + |U|_b + |U|_c \geq (a+b+c)^2$, one has the desired result. If $\gcd(a,b,c) = d$, then let us define $a' = a/d$, $b' = b/d$ and $c' = c/d$. Then, let us denote $|\widetilde{U}|_a$ (resp. $|\widetilde{U}|_b$, $|\widetilde{U}|_c$) the number of a (resp. b, c) in the square $S = [\![0, \frac{a+b+c}{d} - 1]\!]^2$. Since $(0, a+b+c)$ and $(a+b+c, 0)$ are two periodic vectors of U, that is, for all $(k,k') \in \mathbb{Z}^2$ and for all $(m,n) \in \mathbb{Z}^2$, we have $U_{m,n} = U_{m+k(a+b+c),n+k'(a+b+c)}$, then $|U|_a = d^2|\widetilde{U}|_a$ (resp. $|U|_b = d^2|\widetilde{U}|_b$, $|U|_c = d^2|\widetilde{U}|_c$). By the same way as above, we obtain that $|\widetilde{U}|_a = a\frac{a+b+c}{d^2}$ (resp. $|\widetilde{U}|_b = b\frac{a+b+c}{d^2}$, $|\widetilde{U}|_c = c\frac{a+b+c}{d^2}$) and the desired result follows.

It is sufficient to prove that ii) $\implies$ iii). Let us suppose that $\mathcal{L}(U) = \mathcal{L}(U')$. Let $k = \operatorname{lcm}(a+b+c, a'+b'+c')/(a+b+c)$ and $k' = \operatorname{lcm}(a+b+c, a'+b'+c')/(a'+b'+c')$. Let $(a_1, b_1, c_1, \mu_1) = k(a,b,c,\mu)$ and $(a'_1, b'_1, c'_1, \mu'_1) = k'(a',b',c',\mu')$. Let $V \in \{1,2,3\}^{\mathbb{Z}^2}$ (resp. $V' \in \{1,2,3\}^{\mathbb{Z}^2}$) be the two-dimensional sequence coding the standard discrete plane $\mathfrak{P}(a_1, b_1, c_1, \mu_1)$ (resp. $\mathfrak{P}(a'_1, b'_1, c'_1, \mu'_1)$). Then, $V = U$ and $V' = U'$. Since $a_1 + b_1 + c_1 = a'_1 + b'_1 + c'_1$ and one has $|V|_a = |V'|_a$ (resp. $|V|_b = |V'|_b$, $|V|_c = |V'|_c$)) (see Lemma 7), we have $a_1(a_1 + b_1 + c_1) = a'_1(a_1 + b_1 + c_1)$ (resp. $b_1(a_1 + b_1 + c_1) = b'_1(a_1 + b_1 + c_1)$, $c_1(a_1 + b_1 + c_1) = c'_1(a_1 + b_1 + c_1)$), that is, $ka = k'a'$, $kb = k'b'$ and $kc = k'c'$. ∎

5 An Analytic Description of the τ-Language of the Standard Discrete Planes

In Theorem 4, we proved that the language of a standard discrete plane with a positive normal vector does not depend on its translation parameter μ and is completely defined by its normal vector (a, b, c).

In this section, we provide an analytic way to describe the τ-language $\mathcal{L}_\tau(\mathfrak{P})$ of a standard discrete plane $\mathfrak{P}$. This kind of investigation can be compared to [Col02, Gér99, VC99, Vui98].

Roughly speaking, to each τ-pattern ω of $\mathfrak{A}$, we associate the subset of the triples (a, b, c) of $\mathbb{R}^3$, such that ω has an occurrence in the two-dimensional coding of any standard discrete planes with normal vector (a, b, c).

Let us recall that we can defined the discrete surface structure of a standard discrete plane if and only if its normal vector (a, b, c) is positive, that is $\min\{a, b, c\} \geq 0$. In the present section, if it is not mentioned, we will suppose a, b, c to be positive and $c \neq 0$.

Since it is easily checked that $\mathfrak{P}(a, b, c, \mu) = \mathfrak{P}(a/c, b/c, 1, \mu/c)$, let us assume that $c = 1$. Hence, to each τ-pattern ω of $\mathfrak{A}$, we will associate the subset of the pairs (a, b) of $\mathbb{R}^2$, such that ω has an occurrence in the two-dimensional coding of any standard discrete planes with normal vector $(a, b, 1)$.

For instance, let us consider the following τ-patterns:

$$\omega = \begin{matrix} 1 & 2 \\ 3 & \end{matrix}$$

Then, following Corollary 1, one has:

$$\begin{aligned} \omega \in \mathcal{L}_\tau(\mathfrak{P}) &\iff I_3 \cap R_b^{-1}(I_1) \cap R_a^{-1} \circ R_b^{-1}(I_2) \neq \emptyset \\ &\iff b < a + c. \end{aligned}$$

Then, assuming that $c = 1$, we associate to ω the set $\{(a, b) \in \mathbb{R}^2 \mid b < a + 1\}$, representing the pairs $(a, b, 1) \in \mathbb{R}^3_+$, such that ω occurs in the two-dimensional coding of any standard discrete plane with normal vector $(a, b, 1)$.

Considering the τ-patterns of Figure 4, we obtain the following graphical representation (see Figure 5):

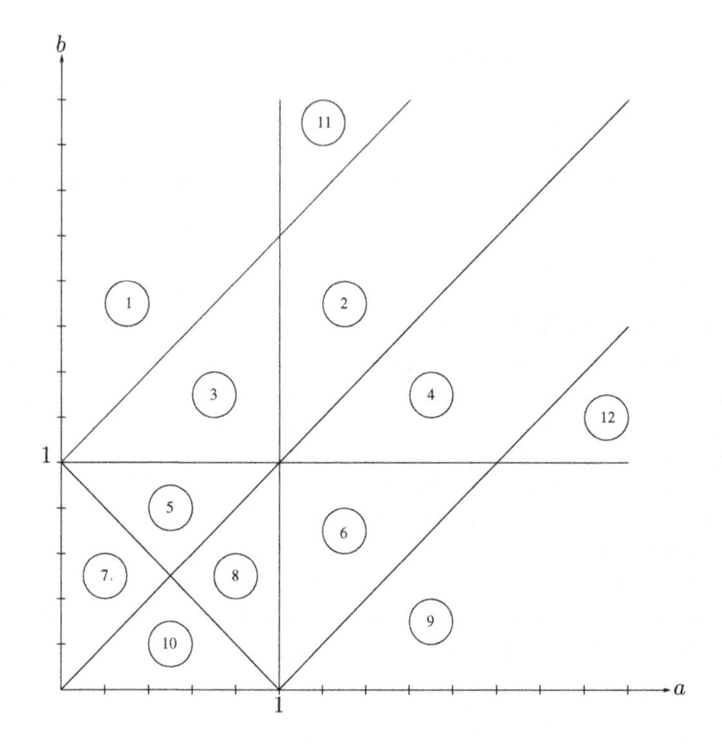

Fig. 5. Graphical representation of the τ-language of the standard discrete planes

zone \ τ-patterns	τ_1	τ_2	τ_3	τ_4	τ_5	τ_6	τ_7	τ_8	τ_9	τ_{10}	τ_{11}	τ_{12}
1				×	×	×	×	×		×		
2			×	×	×		×		×	×		
3				×	×		×	×	×	×		
4		×	×	×	×		×		×			
5				×			×	×	×	×	×	
6		×	×	×			×		×		×	
7				×				×	×	×	×	×
8		×		×			×	×	×		×	
9	×	×	×				×		×		×	
10		×		×				×	×		×	×
11			×	×	×	×	×			×		
12	×	×	×		×		×		×			

For every $k \in [\![1, 12]\!]$, let $\mathcal{L}_k$ be the set of $\mathfrak{A}$-patterns associated to the k-th zone of Figure 5. Then, a direct consequence of Corollary 1 and Figure 5 is:

Theorem 5. *Let* $(a, b) \in \mathbb{N}^2$, $\{n_1, \ldots, n_k\} \subseteq [\![1, 12]\!]$ *be the finite set of all the zones of Figure 5 containing* (a, b) *and* $\mathcal{L}(a, b)$ *be the language of the standard discrete plane with normal vector* $(a, b, 1)$*. Then,*

$$\mathcal{L}(a, b) = \bigcap_{i=1}^{k} \mathcal{L}_i.$$

Let us call *τ-complexity of a standard discrete plane* $\mathfrak{P}$ the τ-complexity of the two-dimensional coding of $\mathfrak{P}$. Then, a direct consequence of Figure 5 and Theorem 5 is:

Corollary 3. *Let* $U \in \{1, 2, 3\}^{\mathbb{Z}^2}$ *be the two-dimensional coding of a standard discrete plane with normal vector* $(a, b, c) \in \mathbb{N}^3$ *with* $c \neq 0$ *and let* $\{n_1, \ldots, n_k\} \subseteq [\![1, 12]\!]$ *be the finite set of all the zones of Figure 5 containing* $(a/c, b/c)$*. Then,*

$$p_{\mathcal{T}}(U) = 6 - k + 1.$$

Remark 1. Let $U \in \{1, 2, 3\}^{\mathbb{Z}^2}$ be the two-dimensional coding of a standard discrete plane $\mathfrak{P}(a, b, c, \mu)$. One can have $p_{\mathcal{T}}(U) = 6$ with while $\{a, b, c\}$ is non-$\mathbb{Q}$-free. For instance, let $a = 1$, $b = 3$ and $c = 5$ (see Figure 6).

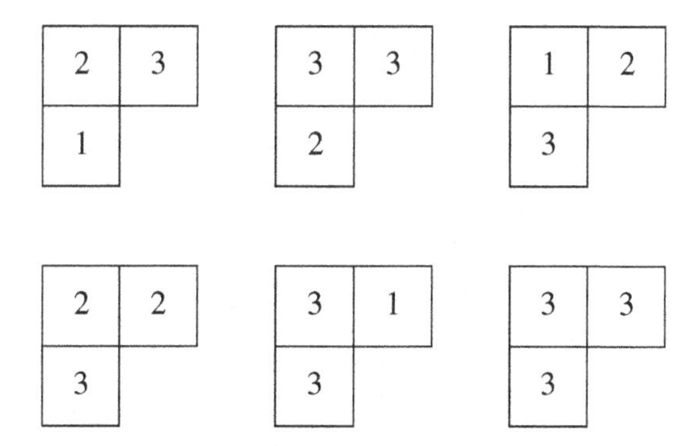

Fig. 6. τ-patterns of the sequence associated to the standard discrete plane $\mathfrak{P}(1, 3, 5, 0)$

Acknowledgments

I would like to thank Pierre Arnoux for having pointed me out the recognition problem of combinatorial codings of discrete surfaces and the referees for the useful suggestions they made.

References

[ABS04] P. Arnoux, V. Berthé, and A. Siegel. Two-dimensional iterated morphisms and discrete planes. *Theoretical Computer Science*, 319:145–176, 2004.

[And93] É. Andrès. Le plan discret. Colloque de géométrie discrète en imagerie : fondements et applications, Septembre 1993.

[BCK04] V. Brimkov, D. Coeurjolly, and R. Klette. Digital planarity - a review. Technical Report RR2004-24, Laboratoire LIRIS - Université Lumière Lyon 2, may 2004.

[Buz02] L. Buzer. An incremental linear time algorithm for digital line and plane recognition using a linear incremental feasibility problem. In *Proceedings of the 10th International Conference on Discrete Geometry for Computer Imagery*, pages 372–381. Springer-Verlag, 2002.

[BT04] V. Berthé and R. Tijdeman. Lattices and multi-dimensional words. *Theoretical Computer Science*, 319:177–202, 2004.

[BV00a] V. Berthé and L. Vuillon. Suites doubles de basse complexité. *Journal de Théorie des Nombres de Bordeaux*, 12:179–208, 2000.

[BV00b] V. Berthé and L. Vuillon. Tilings and rotations on the torus: a two-dimensional generalization of sturmian sequences. *Discrete Mathematics*, 223:27–53, 2000.

[Col02] M.A. Jacob-Da Col. About local configurations in arithmetic planes. *Theor. Comput. Sci.*, 283(1):183–201, 2002.

[DRR96] I. Debled-Rennesson and J.P. Réveillès. Incremental algorithm for recognizing pieces of digital planes. In Robert A. Melter, Angela Y. Wu, and Longin Latecki, editors, *Vision Geometry V*, volume 2826 of *SPIE Proceedings*, pages 140–151, August 1996.

[Fra95] J. Françon. Discrete combinatorial surfaces. *Graph. Models Image Process.*, 57(1):20–26, 1995.

[Gér99] Y. Gérard. Local configurations of digital hyperplanes. In *Proceedings of the 8th International Conference on Discrete Geometry for Computer Imagery*, pages 65–75. Springer-Verlag, 1999.

[KI00] Y. Kenmochi and A. Imiya. Naive planes as discrete combinatorial surfaces. In *Proceedings of the 9th International Conference on Discrete Geometry for Computer Imagery*, pages 249–261. Springer-Verlag, 2000.

[Lot02] Lothaire. *Algebraic Combinatorics on Words*. Cambridge University Press, 2002.

[Mal97] R. Malgouyres. A definition of surfaces of $\mathbb{Z}^3$. a new 3d discrete jordan theorem. *Theor. Comput. Sci.*, 186(1-2):1–41, 1997.

[Meg84] N. Megiddo. Linear programming in linear time when the dimension is fixed. *J. ACM*, 31(1):114–127, 1984.

[PS85] F.P. Preparata and M.I. Shamos. *Computational geometry: an introduction*. Springer-Verlag New York, Inc., 1985.

[Rév91] J.P. Réveillès. *Géométrie discète, calcul en nombres entiers et algorithmique*. PhD thesis, Université Louis Pasteur, Strasbourg, 1991.

[RKW91] A. Rosenfeld, T.Y. Kong, and A.Y. Wu. Digital surfaces. *CVGIP: Graph. Models Image Process.*, 53(4):305–312, 1991.

[VC99] J. Vittone and J.M. Chassery. (n, m)-cubes and farey nets for naive planes understanding. In *Proceedings of the 8th International Conference on Discrete Geometry for Computer Imagery*, pages 76–90. Springer-Verlag, 1999.

[VC00] J. Vittone and J.M. Chassery. Recognition of digital naive planes and polyhedrization. In *DGCI: International Workshop on Discrete Geometry for Computer Imagery*, pages 297–307, 2000.

[Vui98] L. Vuillon. Combinatoire des motifs d'une suite sturmienne bidimensionnelle. *Theor. Comput. Sci.*, 209(1-2):261–285, 1998.

Characterization of Bijective Discretized Rotations

Bertrand Nouvel[1,*] and Eric Rémila[1,2]

[1] Laboratoire de l'Informatique du Parallélisme,
UMR 5668 (CNRS - ENS Lyon - UCB Lyon - INRIA),
Ecole Normale Supérieure de Lyon,
46, Allée d'Italie 69364 Lyon cedex 07 - France
[2] IUT Roanne (Université de Saint-Etienne),
20, avenue de Paris 42334 Roanne Cedex - France
{bertrand.nouvel, eric.remila}@ens-lyon.fr

Abstract. A discretized rotation is the composition of an Euclidean rotation with the rounding operation. For $0 < \alpha < \pi/4$, we prove that the discretized rotation $[r_\alpha]$ is bijective if and only if there exists a positive integer k such as

$$\{\cos\alpha, \sin\alpha\} = \{\frac{2k+1}{2k^2+2k+1}, \frac{2k^2+2k}{2k^2+2k+1}\}$$

The proof uses a particular subgroup of the torus $(\mathbb{R}/\mathbb{Z})^2$.

1 Introduction

In computer graphics, or in physical modeling, most of the time when a rotation has to be done, programmers simply use a discretized rotation, i. e. the composition of a classical rotation with a rounding operation. Unfortunately, this discrete rotation often has regrettable properties in terms of discrete geometry. Actually, for most angles the discretized rotation (restricted to $\mathbb{Z}^2$) is not bijective.

Nevertheless, in [5], Marie-Andrée Jacob and Eric Andrès have proved that for a certain subset of angles (the *integer pythagorean angles*), the discretized rotation is bijective. The proof relies on the classical formalism of discrete geometry, and a particular notion of underlying tile. But the question of the reciprocal was not mentioned and was left open – we did not know if there were some other bijective angles for discretized rotation.

In this paper, we exhibit an alternative proof to the Andres-Jacob result and we prove that the reciprocal is actually true : therefore we obtain a very simple characterization of the bijective angles for the discretized rotation.

In this article, we are going to start out with the minimal definitions we require; particularly those of the angles that are concerned by the Andrès-Jacob

* PhD research supported by TF1 through an ANRT CIFRE convention.

R. Klette and J. Žunić (Eds.): IWCIA 2004, LNCS 3322, pp. 248–259, 2004.

theorem. We continue by giving a characterization of surjective rotations : a discretized rotation is surjective if and only if no integer point has an image by Euclidean rotation and canonical projection to the torus $\mathbb{T}^2 = (\mathbb{R}/\mathbb{Z})^2$ that stands inside a certain frame of the torus $\mathbb{T}^2$. The equivalence in between surjectivity and injectivity for the discretized rotation is then proved.

Afterward, to characterize angles that are bijective, we have examined a particular subgroup of the torus $\mathbb{T}^2$. Naturally, it is then described with great accuracy : more precisely we show that it is possible to identify the smallest vector. This vector can generate the whole studied group.

At the end, all these elements put back together allow us to reprove the Andrès-Jacob theorem and to prove its reciprocal.

2 Pythagorean Angles and Triples

An angle is a real number of the interval $[0...2\pi[$. For sake of simplicity, we will only study (without loss of generality by symmetry arguments) angles which belong to $]0, \frac{\pi}{2}[$.

Definition 1. *An angle α is* pythagorean *if $\cos\alpha$ and $\sin\alpha$ are both rational.*

Notice that α is pythagorean if and only if $\alpha' = \frac{\pi}{2} - \alpha$ is pythagorean.

Proposition 1. *An angle α is pythagorean if and only if there exists a vector $\mathbf{v}$ of $\mathbb{Z}^2 \setminus \{(0,0)\}$ such that $r_\alpha(\mathbf{v})$ also belongs to $\mathbb{Z}^2$.*

Proof. Let α be a pythagorean angle. There exists an integer C such that $C\cos\alpha$ and $C\sin\alpha$ both are integers. This can be interpreted as saying that $r_\alpha(C, 0)$ is in $\mathbb{Z}^2$. Conversely, let $\mathbf{v} = (x, y)$ be an integer vector such that $r_\alpha(\mathbf{v})$ is also in $\mathbb{Z}^2$. We state $(x', y') = r_\alpha(\mathbf{v})$. We have $\cos\alpha = \frac{xx'+yy'}{x^2+y^2}$ and $\sin\alpha = \frac{xy'-yx'}{x^2+y^2}$. This proves that $\cos\alpha$ and $\sin\alpha$ both are rational.

Any pair (p, q) of positive integers such that $q < p$ can generate two pythagorean angles α and α', such that $\alpha + \alpha' = \frac{\pi}{2}$: the first angle transforms (p, q) into (q, p), and the second one transforms (q, p) into $(-q, p)$.

All the pythagorean angles generated as above can be generated with pairs (p, q) such that $gcd(p, q) = 1$ (since, for each positive integer h, the angles generated by the pair (hp, hq) are the same as those generated by (p, q)) and $p - q$ is odd (otherwise the angles generated by the pairs (p, q) are the same as those generated $(\frac{p+q}{2}, \frac{p-q}{2})$). The proposition below claims that the above process generates all the pythagorean angles.

Proposition 2. *An angle α of $]0, \frac{\pi}{2}[$ is pythagorean if and only if there exists a vector (p, q) of $\mathbb{Z}^2$ such that $p > q > 0$, $gcd(p, q) = 1$, $p - q$ is odd and either $r_\alpha(p, q) = (q, p)$ or $r_\alpha(q, p) = (-q, p)$.*

Proof. This is a consequence of classical results related to pythagorean triples (See for example [4] or [10]). For any triple (a, b, c) of positive integers such that

$a^2+b^2=c^2$ and $gcd(a,b,c)=1$, there exists a unique pair (p,q) of positive integers such that $c=p^2+q^2$ and $\{a,b\}=\{p^2-q^2, 2pq\}$. Obviously, we necessarily have : $p>q$, $gcd(p,q)=1$, and $p-q$ odd. The proposition is just an application of this result for (a,b,c) such that $\cos\alpha = a/c$ and $\sin\alpha = b/c$, and c minimal.

Definition 2. *The subset of pythagorean angles consisting in angles generated by pairs $(k+1,k)$ of consecutive integers is called the set of* integer pythagorean angles.

Notice that the pair $(k+1,k)$ leads to the following triple : $(a=2k(k+1)$, $b=2k+1$, $c=2k(k+1)+1)$.

3 Rotation Multiplicities, Holes and Double Points

The *rounding function* is defined so : for each element x of $\mathbb{R}$, $[x]=\lfloor x+\frac{1}{2}\rfloor$ (the function floor which is written $\lfloor x\rfloor$, designates the unique integer such that $\lfloor x\rfloor \le x < \lfloor x\rfloor+1$). On vectors, the discretization is applied component by component : for each vector $\mathbf{v}=(x,y)$ of $\mathbb{R}^2$, we have $[\mathbf{v}]=([x],[y])$. The set of vectors of the real plane that are discretized to a same vector is called a *cell.*

Given an angle α, the *discretized rotation* $[r_\alpha]$ is defined on the set $\mathbb{Z}^2$ as the composition of the Euclidean rotation r_α and of the rounding function $[.]$.

Let $\mathbf{i}_\alpha$ and $\mathbf{j}_\alpha$ be the vectors defined as the rotated images of the vectors of canonical base of the plane : $\mathbf{i}_\alpha = r_\alpha(\mathbf{i}) = (\cos\alpha, \sin\alpha)$ and $\mathbf{j}_\alpha = r_\alpha(\mathbf{j}) = (-\sin\alpha, \cos\alpha)$.

The *multiplicity* $M_\alpha(\mathbf{w})$ maps each vector $\mathbf{w}$ of $\mathbb{Z}^2$ to the number of its antecedents by discretized rotation (notice that M_α is a planar configuration which can be obtained by projection (more exactly by morphism – i. e. cell by cell) of the planar coloration introduced in [7]). Formally, the application M_α is defined by the following equation :

$$M_\alpha(\mathbf{w}) = CARD(\{\mathbf{v}\in\mathbb{Z}^2, [r_\alpha](\mathbf{v})=\mathbf{w}\})$$

Each set of three different points of the grid $r_\alpha(\mathbb{Z}^2)$ contains at least two points that are at a distance of at least $\sqrt{2}$, and the ends of a segment of length of at least $\sqrt{2}$ cannot be in the same cell, therefore there cannot be three different points inside the same cell, thus the multiplicity M_α of any vector will never exceed 2.

A *hole* is a vector characterized by the fact that $M_\alpha(\mathbf{w})=0$: there is a hole in M_α if and only if $[r_\alpha]$ is not surjective. A *double point* is characterized by the fact that $M_\alpha(\mathbf{w})=2$: there is a double point in M_α if and only if $[r_\alpha]$ is not injective. The discretized rotation $[r_\alpha]$ is bijective if and only if for each vector $\mathbf{w}$ of $\mathbb{Z}^2$, $M_\alpha(\mathbf{w})=1$.

Let P_k be the set of vectors of $\mathbb{Z}^2$ such as $M(\mathbf{v})=k$. Normally, P_0 is empty if and only if $[r_\alpha]$ is surjective and P_2 is empty if and only if $[r_\alpha]$ is injective.

We identify the torus $\mathbb{T}^2=(\mathbb{R}/\mathbb{Z})^2$ with the cell $[-\frac{1}{2},+\frac{1}{2}[^2$. A canonical projection from $\mathbb{R}^2$ to $\mathbb{T}^2$ is provided by the operator $\{\mathbf{x}\}=\mathbf{x}-[\mathbf{x}]$. A *frame* is the cartesian product of non-empty half-opened intervals $[a...b[$ on the torus $\mathbb{T}^2 = (\mathbb{R}/\mathbb{Z})^2$.

Theorem 1. *Let $F_{0\downarrow}$ denote the frame :*

$$F_{0\downarrow} = \left[\frac{1}{2} - \sin\alpha, -\frac{1}{2} + \cos\alpha\right[\times \left[-\frac{1}{2}, -\frac{3}{2} + \cos\alpha + \sin\alpha\right[$$

We have : $P_0 \neq \emptyset$ if and only if there exists a vector $\mathbf{v}$ of $\mathbb{Z}^2$ such that $\{r_\alpha\}(\mathbf{v}) \in F_{0\downarrow}$.

More precisely, with the notations above, we have $M_\alpha([r_\alpha](\mathbf{v}) - \mathbf{j}) = 0$.

Proof. First notice that necessarily, for any vector $\mathbf{w}'$ of $\mathbb{Z}^2$, there exists a vector $\mathbf{v}$ of $\mathbb{Z}^2$ such that the distance between $r_\alpha(\mathbf{v})$ and $\mathbf{w}'$ is at most $\sqrt{2}/2$. Let $\mathbf{w}$ be a hole, $\mathcal{H}$ be the discretization cell associated and $\mathcal{D}_\mathbf{w}$ be the closed disk centered in $\mathbf{w}$ of radius $\sqrt{2}/2$. From the remark above, the set $\mathcal{D}_\mathbf{w} \cap r_\alpha(\mathbb{Z}^2)$ is not empty.

On the other hand, since $\mathcal{H}$ is a hole, $r_\alpha(\mathbb{Z}^2)$ does not meet $\mathcal{H}$, and, therefore, $r_\alpha(\mathbb{Z}^2)$ does not meet $\mathcal{H}+\mathbf{t}$, for any vector $\mathbf{t}$ of $r_\alpha(\mathbb{Z}^2)$ (since $r_\alpha(\mathbb{Z}^2)+\mathbf{t} = r_\alpha(\mathbb{Z}^2)$). In particular, we have : $r_\alpha(\mathbb{Z}^2) \cap (\cup_{\mathbf{t}\in\mathcal{T}}(H + \mathbf{t})) = \emptyset$, where $\mathcal{T}$ denotes the set $\mathcal{T} = \{\mathbf{t} \in r_\alpha(\mathbb{Z}^2), \mathbf{t} = x\mathbf{i}_\alpha + y\mathbf{j}_\alpha, -1 \leq x \leq 1, -1 \leq y \leq 1\}$.

Thus $r_\alpha(\mathbb{Z}^2)$ necessarily meets $\mathcal{D}_\mathbf{w} \setminus \cup_{\mathbf{t}\in\mathcal{T}}(H + \mathbf{t})$. This set is contained into the union of four squares (see figure 1) : the square $F_{0\downarrow} + \mathbf{w} + \mathbf{j}$ and its translated copies by vectors $-\mathbf{i}_\alpha$,$-\mathbf{j}_\alpha$ and $-\mathbf{i}_\alpha - \mathbf{j}_\alpha$.

Obviously, the fact that one of these squares encounters $r_\alpha(\mathbb{Z}^2)$ implies that each of them meets $r_\alpha(\mathbb{Z}^2)$. This is especially true for $F_{0\downarrow} + \mathbf{w} + \mathbf{j}$ thus there exists $\mathbf{v}$ of $\mathbb{Z}^2$ such that $\{r_\alpha\}(\mathbf{v}) \in F_{0\downarrow}$.

Conversely, if there exists a vector $\mathbf{v}$ of $\mathbb{Z}^2$ such that $\{r_\alpha\}(\mathbf{v}) \in F_{0\downarrow}$, then there exists a vector $\mathbf{w}$ of $\mathbb{Z}^2$ such that $F_{0\downarrow}+\mathbf{w}+\mathbf{j}$ contains $r_\alpha(\mathbf{v})$. Thus $F_{0\downarrow}+\mathbf{w}+\mathbf{j}$ and its translated copies by vectors $-\mathbf{i}_\alpha$,$-\mathbf{j}_\alpha$ and $-\mathbf{i}_\alpha - \mathbf{j}_\alpha$ meet $r_\alpha(\mathbb{Z}^2)$. This yields that $\mathbf{w}$ is a hole since each vector of its discretization cell is at a distance lower than 1 from an element of $r_\alpha(\mathbb{Z}^2)$.

Theorem 2. *Let F_2 denote the frame :*

$$F_2 = \left[-\frac{1}{2}, \frac{1}{2} - \cos\alpha\right[\times \left[-\frac{1}{2}, \frac{1}{2} - \sin\alpha\right[$$

We have : $P_2 \neq \emptyset$ if and only if there exists a vector $\mathbf{v}$ of $\mathbb{Z}^2$ such that $\{r_\alpha\}(\mathbf{v}) \in F_2$.

More precisely, with the notations above, we have $M_\alpha([r_\alpha](\mathbf{v})) = 2$.

Proof. Since $r_\alpha(\mathbb{Z}^2)$ is invariant by rotation of angle $\pi/2$, $P_2 \neq \emptyset$ if and only if there exists $\mathbf{v}$ such that $[r_\alpha](\mathbf{v}) = [r_\alpha](\mathbf{v} + \mathbf{i})$. This condition trivially gives the result.

3.1 The Non Pythagorean Case

Now, we introduce two groups that have a main importance for the study of discrete rotations :

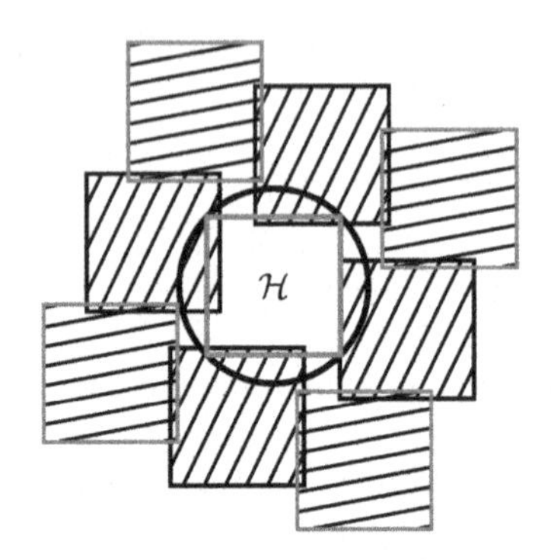

Fig. 1. The main argument for Theorem 1: The set $r_\alpha(\mathbb{Z}^2)$ meets neither the (big) dashed square nor $\mathcal{H}$, but meets the closed disk. Thus $r_\alpha(\mathbb{Z}^2)$ meets one of the four remaining small squares

- the subgroup $\mathcal{G}'_\alpha$ of $\mathbb{T}^2$ defined by : $\mathcal{G}'_\alpha = \{r_\alpha\}(\mathbb{Z}^2)$ which one may also see as : $\mathbb{Z}\{\mathbf{i}_\alpha\} + \mathbb{Z}\{\mathbf{j}_\alpha\}$
- the subgroup $\mathcal{G}_\alpha$ of $\mathbb{R}^2$ defined by : $\mathcal{G}_\alpha = \mathbb{Z}\mathbf{i} + \mathbb{Z}\mathbf{j} + \mathbb{Z}\mathbf{i}_\alpha + \mathbb{Z}\mathbf{j}_\alpha$.

At first glance, we notice that a point $\mathbf{v}$ of $\mathbb{R}^2$ belongs to $\mathcal{G}_\alpha$ if and only if $\{\mathbf{v}\}$ belongs to $\mathcal{G}'_\alpha$ and, moreover $[-1/2, 1/2[^2 \cap \mathcal{G}_\alpha = \mathcal{G}'_\alpha$. We also notice that $\mathcal{G}_\alpha$ and $\mathcal{G}'_\alpha$ are both invariant by rotation of angle $\pi/2$.

We now focus on non pythagorean angles in order to show that in this case, the discretized rotation will be neither injective, nor surjective.

Proposition 3. *Let α denote a non pythagorean angle; For all $\epsilon > 0$, there exists a vector $\mathbf{e}_\epsilon$ of $\mathcal{G}_\alpha$ such that $0 < ||\mathbf{e}_\epsilon|| \leq \epsilon$. Moreover, $\mathcal{G}_\alpha$ contains the group $\mathbb{Z}\mathbf{e}_\epsilon + \mathbb{Z}\mathbf{e}'_\epsilon$, with $\mathbf{e}'_\epsilon = r_{\pi/2}(\mathbf{e}_\epsilon)$.*

Proof. Since α is not pythagorean, the elements of the sequence $(\{n\mathbf{i}_\alpha\})_{n\in\mathbb{N}}$ are pairwise disjoint. Since all of them are in the compact square $[-1/2, 1/2]^2$, there exists a subsequence $(\{n_k\mathbf{i}_\alpha\})_{k\in\mathbb{N}}$ which converges. Thus the sequence $(\{n_{k+1}\mathbf{i}_\alpha\} - \{n_k\mathbf{i}_\alpha\})_{k\in\mathbb{N}}$ is a non ultimately constant sequence which converges to $(0,0)$. Thus, for all $\epsilon > 0$, there exists an integer k such that $0 < ||\{n_{k+1}\mathbf{i}_\alpha\} - \{n_k\mathbf{i}_\alpha\}|| \leq \epsilon$. This element is in $\mathcal{G}_\alpha$, which gives the result (the second part of the proposition is trivial).

Corollary 1. *Let α denote a non pythagorean angle. The associated discretized rotation $[r_\alpha]$ is neither injective nor surjective.*

Proof. Let F be a fixed frame. From the above proposition applied for ϵ sufficiently small, the group $\mathcal{G}'_\alpha$ has a non empty intersection with F. In particular, this is true for the frame of surjectivity $F_{0\downarrow}$ and the frame of injectivity F_2. This gives the result, according to Theorem 1 and Theorem 2.

4 The Pythagorean Case

We fix a pair (p, q) of positive integers such that $p > q$, $gcd(p, q) = 1$ and $p - q$ is odd. Let α be the angle such that $\cos\alpha = a/c$ and $\sin\alpha = b/c$, with $a = p^2 - q^2$, $b = 2pq$ and $c = p^2 + q^2$. We also state : $\alpha' = \frac{\pi}{2} - \alpha$, the angle defined by the other triple $(a' = 2pq, b' = p^2 - q^2, c = p^2 + q^2)$.

For each pair (x, y) of $\mathbb{Z}^2$ we have $[r_\alpha](x, y) = (x', y')$ if and only if $[r_{\alpha'}](y, x) = (y', x')$. Thus $[r_\alpha]$ is bijective (resp. injective/ surjective) if and only if $[r_{\alpha'}]$ is (resp. injective/ surjective).

For the sake of simplicity, we now assume that a is odd (a is the first element of the triple associated with the angle). There is no loss of generality.

4.1 Reduction to Surjectivity

We now prove that, for pythagorean angles, the bijectivity problem is equal to the surjectivity problem.

Lemma 1 (Square Lemma). *Let S be a half-opened square of the plane such that the vectors induced by its edges have integer components. The number of integer vectors contained in S is equal to the area of S.*

Proof. (sketch) The idea (see Figure 2) is to divide the square into three parts, two triangles and another one, and afterward translate the triangles to obtain a polygon with integer sides, vertical or horizontal, which is the disjoint union of two half opened squares. The main arguments used are the facts below :

- the lemma above obviously holds for any half-opened square whose (integer) sides are vertical or horizontal.
- two domains of the plane which can be mutually obtained by integer translation contain the same number of integer vectors

A precise choice can be made for boundaries, in order to get half opened squares at the end.

Theorem 3. *Let α denote a pythagorean angle. The function $[r_\alpha]$ is be one-to-one if and only if it is onto. Thus bijectivity is equivalent to injectivity or surjectivity.*

Proof. We have $r_\alpha((a, -b)) = (c, 0)$ and $r_\alpha((b, a)) = (0, c)$. Thus, for each vector $\mathbf{v}$ of $\mathbb{Z}^2$, we have $[r_\alpha](\mathbf{v} + (a, -b)) = [r_\alpha](\mathbf{v}) + (c, 0)$ and $[r_\alpha](\mathbf{v} + (b, a)) = [r_\alpha](\mathbf{v}) + (0, c)$. This yields that for each vector $\mathbf{w}$ of $\mathbb{Z}^2$, we have $M(\mathbf{w} + (c, 0)) = M(\mathbf{w} + (0, c)) = M(\mathbf{w})$. In other words, the multiplicity is a periodic function.

Consider the real window $[-\frac{1}{2}, -\frac{1}{2} + c[^2$. From the periodicity seen above, $[r_\alpha]$ is injective if and only if there exists no integer vector $\mathbf{w}$ in $[-\frac{1}{2}, -\frac{1}{2} + c[^2$ such that $M(\mathbf{w}) \geq 2$. Similarly, $[r_\alpha]$ is surjective if and only if there exists no integer vector $\mathbf{w}$ in $[-\frac{1}{2}, -\frac{1}{2} + c[^2$, such that $M(\mathbf{w}) = 0$.

On the other hand, for each vector $\mathbf{v}$ of $\mathbb{Z}^2$, $[r_\alpha](\mathbf{v})$ is element of $[-\frac{1}{2}, -\frac{1}{2} + c[^2$ if and only if $\mathbf{v}$ is element of $r_{-\alpha}([-\frac{1}{2}, -\frac{1}{2} + c[^2)$. From the square lemma, the square $r_{-\alpha}([-\frac{1}{2}, -\frac{1}{2}+c[^2)$ contains c^2 integer vectors, as the square $[-\frac{1}{2}, -\frac{1}{2}+c[^2$. Thus there exists an integer vector $\mathbf{w}$ in $[-\frac{1}{2}, -\frac{1}{2}+c[^2$ such that $M(\mathbf{w}) = 0$ if and only if there exists an integer vector $\mathbf{w}$ in $[-\frac{1}{2}, -\frac{1}{2}+c[^2$ such that $M(\mathbf{w}) \geq 2$. This achieves the proof.

Notice, that instead of the square lemma, a corollary of the famous Pick's Theorem may also be used[1]. (Even if the proof of the previous lemma can be a little bit wiser in terms it could require less constraints.)

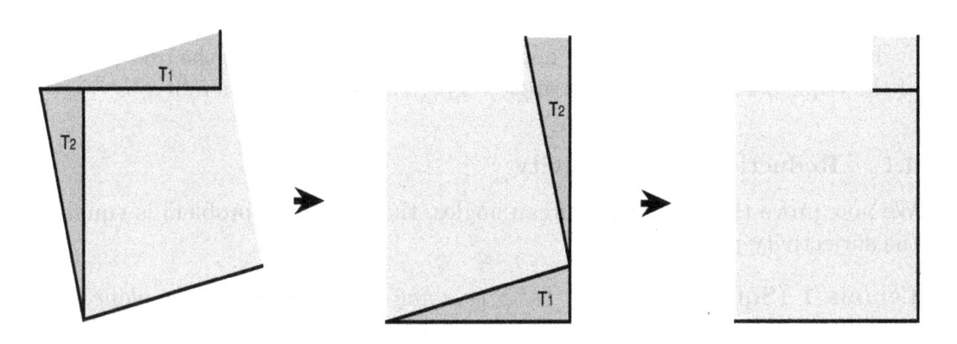

Fig. 2. The scheme of the proof of the square lemma. The triangles T_1 and T_2 are translated, and two squares are obtained. The dark lines point out the boundaries which are inside the domains

4.2 Structural Study of $\mathcal{G}'_\alpha$

The Theorem 1 will be used to characterize surjective (i.e. bijective) rotations. But, it requires to know the precise structure of $\mathcal{G}'_\alpha$ which is $\{r_\alpha\}(\mathbb{Z}^2)$.

Lemma 2 (Membership Criterion). *Let* $\mathbf{v} = (\frac{x}{c}, \frac{y}{c})$, *with* (x, y) *in* $\mathbb{Z}^2$. *There exists an integer* n *such that* $\{n\mathbf{i}_\alpha\} = \{\mathbf{v}\}$ *if and only if* $xb - ya = det \begin{pmatrix} x & a \\ y & b \end{pmatrix} \equiv 0\ [c]$.

Proof. There exists an integer n such that $\{n\mathbf{i}_\alpha\} = \{\mathbf{v}\}$ if and only if there exists a triple (n, n', n'') of integers such that : $(\frac{x}{c}+n', \frac{y}{c}+n'') = n(\frac{a}{c}, \frac{b}{c})$ (notice that $\mathbf{i}_\alpha = (\frac{a}{c}, \frac{b}{c})$).

This is equivalent to the existence of an integer n such that $x \equiv na\ [c]$ and $y \equiv nb\ [c]$. And which is possible if and only if $xa^{-1} \equiv yb^{-1} [c]$ (the inverses are taken in $\mathbb{Z}/c\mathbb{Z}$, the numbers a and b both are invertible since $\frac{a}{c}$ and $\frac{b}{c}$ both are irreducible fractions). The latter equality can be rewritten : $xb - ya \equiv 0\ [c]$.

Proposition 4. *Let* $\mathbf{m}$ *and* $\mathbf{m}'$ *be the vectors defined by* $\mathbf{m} = (\frac{p}{c}, \frac{q}{c})$ *and* $\mathbf{m}' = (\frac{-q}{c}, \frac{p}{c})$. *The group* $\mathcal{G}'_\alpha$ *is cyclic, of order* c, *generated by the vector* $\{\mathbf{m}\}$. *The group* $\mathcal{G}_\alpha$ *is the subgroup of* $\mathbb{R}^2$ *generated by* $\mathbb{Z}\mathbf{m} + \mathbb{Z}\mathbf{m}'$.

[1] See for example [10] or [3].

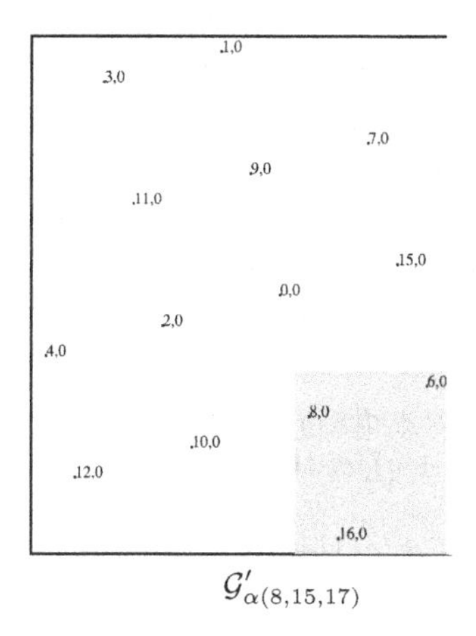

$\mathcal{G}'_{\alpha(8,15,17)}$

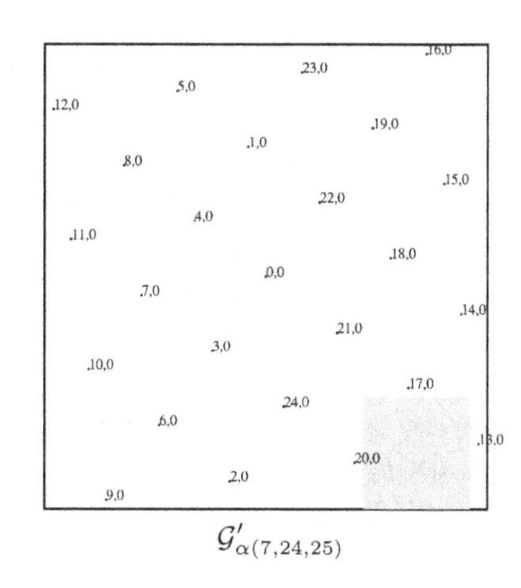

$\mathcal{G}'_{\alpha(7,24,25)}$

Fig. 3. On the left, $\mathcal{G}'_\alpha$ and $F_{0\downarrow}$ for a non integer pythagorean non integer angle. On the right $\mathcal{G}'_\alpha$ and $F_{0\downarrow}$ for a non integer an integer pythagorean angle

Proof. We first notice that the set $\{\{\mathbf{i}_\alpha\}, \{\mathbf{j}_\alpha\}\}$ generates $\mathcal{G}'_\alpha$ and $\mathbf{j}_\alpha = (-\frac{b}{c}, \frac{a}{c})$. Remark that $-b^2 - a^2 = -c^2$, thus, applying the membership criterion, there exists an integer n such that $\{n\mathbf{i}_\alpha\} = \{\mathbf{j}_\alpha\}$. Thus $\{\mathbf{i}_\alpha\}$ generates $\mathcal{G}'_\alpha$, which, therefore, is cyclic. Moreover, since $gcd(a,c) = gcd(b,c) = 1$, it stands that : $\{n\mathbf{i}_\alpha\} = (0,0)$ if and only if $n \equiv 0[c]$. It proves that the order of $\mathcal{G}'_\alpha$ is c.

We have : $pb - qa = p(2pq) - q(p^2 - q^2) = p^2q + q^3 = q(p^2 + q^2) = qc$. Thus, applying the membership criterion, we obtain that $\{\mathbf{m}\}$ is an element of $\mathcal{G}'_\alpha$. Moreover, since $gcd(p,c) = gcd(q,c) = 1$, $\{\mathbf{m}\}$ is of order c and, therefore, generates $\mathcal{G}'_\alpha$.

For the third part of the proposition, remark that $\mathcal{G}_\alpha$ and $\mathbb{Z}\mathbf{m} + \mathbb{Z}\mathbf{m}'$ both are invariant by integer translation, thus each of these groups is defined by its intersection with the cell $[-\frac{1}{2}, \frac{1}{2}[^2$. Moreover, $\mathcal{G}_\alpha$ contains $\mathbb{Z}\mathbf{m} + \mathbb{Z}\mathbf{m}'$ and $\mathcal{G}_\alpha \cap [-\frac{1}{2}, \frac{1}{2}[^2$ (which is $\mathcal{G}'_\alpha$) contains exactly c elements. Thus, we only have to prove that $(\mathbb{Z}\mathbf{m} + \mathbb{Z}\mathbf{m}') \cap [-\frac{1}{2}, \frac{1}{2}[^2$ also contains c elements. In that aim, consider the real plane, seen using the basis $(\mathbf{m}, \mathbf{m}')$: vectors of $\mathbb{Z}\mathbf{m} + \mathbb{Z}\mathbf{m}'$ are seen as the integer vectors. We have $\mathbf{i} = p\mathbf{m} - q\mathbf{m}'$, thus the discretization cell of the origin can be seen as a square on which the square lemma can be applied. Thus, the discretization cell of the origin contains c elements of $\mathbb{Z}\mathbf{m} + \mathbb{Z}\mathbf{m}'$. This achieves the proof.

5 Results

5.1 Proof of the Reciprocal of the Andrès Jacob Theorem

We now prove the reciprocal of Andrès-Jacob Theorem. The outline of the proof is structured as follows : the main idea of the proof is to show that when we are in the non - integer- pythagorean case it is necessity to have a "hole". This necessity is due to the density of $\mathcal{G}'_\alpha$ in $[-\frac{1}{2}, \frac{1}{2}[^2$ (or the density of $\mathcal{G}_\alpha$ in $\mathbb{Z}^2$).

Lemma 3 (Size Lemma). *Let $[x, x+d[\times[y, y+d[$ be a square of the real plane, with $d \geq \frac{p+q}{c}$. This square has a non-empty intersection with $\mathcal{G}_\alpha$.*

Proof. We state : $(x, y) = z\mathbf{m} + t\mathbf{m}'$. Up to a translation of $\lfloor z \rfloor \mathbf{m} + \lfloor t \rfloor \mathbf{m}'$, it can be assumed without loss of generality that $0 \leq z < 1$ and $0 \leq t < 1$, which gives $-q/c \leq x < p/c$ and $0 \leq y < (p+q)/c$. With this hypothesis we have the case by case analysis below (obtained by cutting the square $\{\mathbf{v} \in \mathbb{Z}^2, \mathbf{v} = z\mathbf{m} + t\mathbf{m}', 0 \leq z < 1, 0 \leq t < 1\}$ by vertical and horizontal lines) :

- for $(x, y) = (0, 0)$, the vector $(0, 0)$ is in $[x, x+d[\times[y, y+d[$,
- for $-q/c \leq x < (p-q)/c$ and $0 < y < (p+q)/c$, the vector $\mathbf{m} + \mathbf{m}'$ is in $[x, x+d[\times[y, y+d[$,
- for $(p-q)/c \leq x < (p+q)/c$ and $0 \leq y \leq q/c$, the vector $\mathbf{m}$ is in $[x, x+d[\times[y, y+d[$,
- for $(p-q)/c \leq x < (p+q)/c$ and $q/c < y < (p+q)/c$, the vector $2\mathbf{m} + \mathbf{m}'$ is in $[x, x+d[\times[y, y+d[$.

Theorem 4 (Reciprocal of Andrès Jacob Theorem). *If the angle α is not an integer pythagorean angle then the discretized rotation $[r_\alpha]$ is not bijective.*

Proof. We recall that if the angle α does not belongs to the set of pythagorean angles, then, from Corollary 1 of Proposition 3, $[r_\alpha]$ is not bijective.

For a second time, assume that the angle α is a pythagorean one. With the conventions and notations used above for pythagorean angles, the length $sidef_0$ of the side of the square $F_{0\downarrow}$ is $\cos(\alpha) + \sin(\alpha) - 1 = \frac{a+b-c}{c} = \frac{2q(p-q)}{c}$. In order P_0 to be empty, it is mandatory that $sidef_0 < (p+q)/c$ which gives $2q(p-q) < (p+q)$.

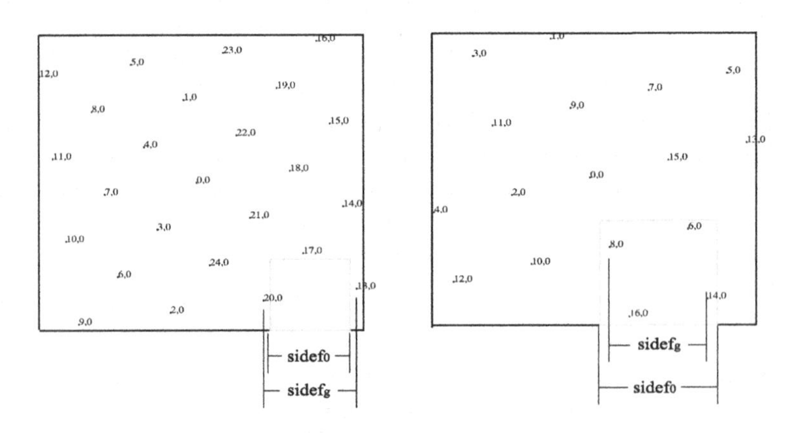

Fig. 4. $sidef_0$ and $\frac{p+q}{c} = sidef_g$ in two samples of Pythagorean angles

We write $p = q + e$, and thus it is obtained $2qe < 2q + e$, which may be rewritten as : $(2q-1)e < 2q$, or $(2q-1)(e-1) < 1$. This is only possible if $e = 1$;

which is equivalent of saying that (p, q) leads to an integer pythagorean triple. Therefore the discretized rotation cannot bijective for pythagorean non integer angles.

5.2 Alternate Proof of the Andrès Jacob Theorem

Theorem 5 (Andrès-Jacob). *If the angle of rotation is integer pythagorean, then the discretized euclidean rotation is bijective.*

The original proof directly proves the injectivity, with arithmetical arguments. We provide here an alternate one that relies on our framework.

Proof. The main idea of the proof aims to show that there is no point of the group $\mathcal{G}'_\alpha$ that stands in the "hole frame" $F_{0\downarrow}$. In order to ensure that the frame is avoided, the position of the points that surround the "hole frame" $F_{0\downarrow}$ has to be stated precisely.

The frame $F_{0\downarrow}$ admits the following coordinates :

$$\begin{aligned} F_{0\downarrow} &= [(F_{0\downarrow})_L, (F_{0\downarrow})_R[\times [(F_{0\downarrow})_D, (F_{0\downarrow})_U[\\ &= \left[\frac{1}{2} - \sin(\alpha), -\frac{1}{2} + \cos(\alpha)\right[\times \left[-\frac{1}{2}, -\frac{3}{2} + \cos(\alpha) + \sin(\alpha)\right[\\ &= \left[\frac{c-2b}{2c}, \frac{2a-c}{2c}\right[\times \left[\frac{-c}{2c}, \frac{2a+2b-3c}{2c}\right[\\ &= \left[\frac{p^2-3q^2}{2c}, \frac{p^2-3q^2}{2c}\right[\times \left[\frac{-c}{2c}, \frac{2a+2b-3c}{2c}\right[\\ &= \left[\frac{-2k^2-2k+1}{2c}, \frac{-2k^2+2k+1}{2c}\right[\times \left[\frac{-2k^2-2k-1}{2c}, \frac{-2k^2+2k-1}{2c}\right[\end{aligned}$$

We have assumed that α is the k-th integer pythagorean angle (i. e. $p = k+1$ and $q = k$); this yields to $a = p^2 - q^2 = 2k+1$, $b = 2pq = 2k(k+1)$ and $c = p^2 + q^2 = 2k^2 + 2k + 1$.

With this hypothesis, we also have : $\mathbf{m} = (\frac{k+1}{c}, \frac{k}{c})$ and $\mathbf{m}' = (\frac{-k}{c}, \frac{k+1}{c})$. Consider the four following vectors:

- $\mathbf{a} = (\frac{-2k^2-2k}{2c}, \frac{-2k^2}{2c}) = (\frac{-k^2-k}{c}, \frac{-k^2}{c})$,
- $\mathbf{b} = \mathbf{a} + \mathbf{m} = (\frac{-2k^2+2}{2c}, \frac{-2k^2+2k}{2c}) = (\frac{-k^2+1}{c}, \frac{-k^2+k}{c})$,
- $\mathbf{c} = \mathbf{a} + \mathbf{m} - \mathbf{m}' = (\frac{-2k^2+2k+2}{2c}, \frac{-2k^2-2}{2c}) = (\frac{-k^2+k+1}{c}, \frac{-k^2-1}{c})$,
- $\mathbf{d} = \mathbf{a} - \mathbf{m}' = (\frac{-2k^2}{2c}, \frac{-2k^2-2k-2}{2c}) = (\frac{-k^2}{c}, \frac{-k^2-k-1}{c})$

According to these definitions, we see that these points surround $F_{0\downarrow}$: $\mathbf{a}_x < (F'_{0\downarrow})_L$, $\mathbf{b}_y > (F'_{0\downarrow})_U$, $\mathbf{c}_x > (F'_{0\downarrow})_R$ and $\mathbf{d}_y < (F'_{0\downarrow})_D$. Therefore we conclude that $F_{0\downarrow}$ is contained in the square $]\mathbf{a}_x, \mathbf{c}_x[\times]\mathbf{d}_y, \mathbf{b}_y[$.

On the other hand, $\mathbf{a}$ belongs to $\mathcal{G}_\alpha$, from the membership criterion:

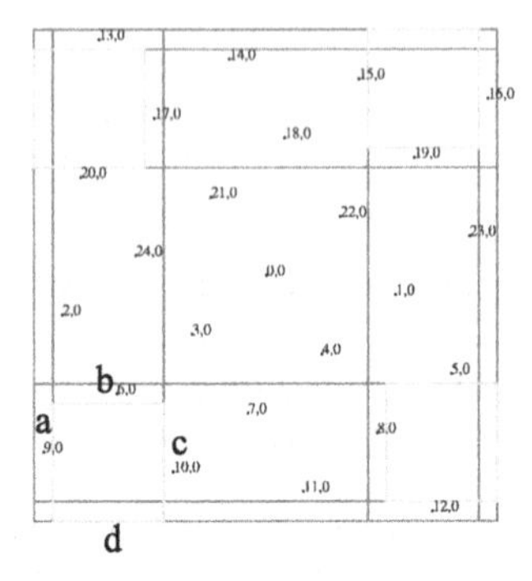

Fig. 5. The position of the points **a**, **b**, **c** and **d** relatively to the hole frame

$$\begin{aligned}\det\begin{pmatrix}-k^2-k & a\\ -k^2 & b\end{pmatrix} &= \det\begin{pmatrix}-k^2-k & 2\,k+1\\ -k^2 & 2\,k\,(k+1)\end{pmatrix}\\ &= -2k^4-2k^3-k^2\\ &= -k^2(2k^2+2k+1)\\ &= -k^2c\\ &\equiv 0[c]\end{aligned}$$

Since $\mathbf{a}$ belongs to the $\mathcal{G}_\alpha$, this implies that $\mathbf{b}$, $\mathbf{c}$ and $\mathbf{d}$ also do. Thus $]\mathbf{a}_x,\mathbf{c}_x[\times]\mathbf{d}_y,\mathbf{b}_y[$ does not meet $\mathcal{G}_\alpha$, since each element of this square is at distance lower than $||\mathbf{m}|| = \frac{\sqrt{c}}{c}$ to one of the vectors $\mathbf{a}$, $\mathbf{b}$, $\mathbf{c}$, or $\mathbf{d}$. Therefore, the set $F_{0\downarrow}\cap\mathcal{G}_\alpha$ is empty, thus $F_{0\downarrow}\cap\mathcal{G}'_\alpha$ is empty, which gives the result from theorem 1[2].

6 Conclusion

The choice of the rounding operator as the discretization function matters for the result : For instance, if we would have taken the floor function as discretization operator, there is no hope to have a bijective rotation : since as long as $\alpha > 0$, $\lfloor r_\alpha(0,0)\rfloor = \lfloor r_\alpha(\mathbf{i})\rfloor$.

While there exists a span of possible discretization functions in between the floor and the rounding, it is the rounding discretization that brings the best point-by-point discretization. The arguments used to prove non-bijectivity hold for any discretization : the only possible bijective rotations are those associated to integer pythagorean angles.

The characterization of angles such that the discretized rotations is bijective, which are somehow the angles for which the discretized rotations has good properties : everything is finite; therefore they are suitable for computations. It has

[2] At the following URL, http://perso.ens-lyon.fr/bertrand.nouvel/work/proofbij.mupad, a mupad session that contains all the proof of our work has been saved, and the interested reader may consult it.

lead to scientific knowledge on the way rotations work : we have got a complete description of G'_α for pythagorean angles. However, there are others classes of angles, such as the hinge angles [8] that should be studied, and these angles seem to be a source of wonderfully challenging problems.

References

1. Eric Andrès. *Discrete Circles, and Discrete Rotations.* PhD thesis, Université Louis Pasteur, 1992.
2. Eric Andres. Habilitation à diriger des recherches : Modélisation analytique discrète d'objets géometriques, 2000.
3. Jean-Marc Chassery and Annick Montanvert *Géométrie discrète en analyse d'images* Hermes mai, 1991.
4. G.H. Hardy and E.M. Wright An Introduction to the Theory of Numbers Oxford University Press, London, 1979.
5. Marie-André Jacob and Eric Andrès. On Discrete Rotations. In *Discrete Geometry for Computer Imagery,* 1995. LLAIC editions (Univ. Clermont-Ferrand 1).
6. Bertrand Nouvel. Action des rotations sur le voisinage dans le plan discret. Master's thesis, ENS-Lyon, 2002.
7. Bertrand Nouvel and Eric Rémila. On Colorations Induced by Discrete Rotations. In *Discrete Geometry for Computer Imagery,* 2003. Lecture Notes in Computer Science, no 2886.
8. Bertrand Nouvel. An Incremental Transitive Algorithm for the Discretized Rotations. Submitted to *Discrete Geometry for Computer Imagery,* 2005.
9. Jean Pierre Réveillès. *Géométrie disrète, Calcul en nombre entiers, et Algorithmique.* PhD thesis, Université Louis Pasteur, 1991.
10. Klaus Voss. *Discrete Images, Objects and Functions in* $\mathbb{Z}^n$ Springer, Berlin, 1993

Magnification in Digital Topology

Akira Nakamura

Hiroshima University
akira668@urban.ne.jp

Dedicated to Professor A. Rosenfeld (1931–2004)

Abstract. When the author was working with Prof. Azriel Rosenfeld on joint research, we proposed a very strong deformation technique in digital topology called "magnification". In this paper, the methods are explained in detail and some applications are given.

1 Introduction

Between 1970 and 1975, Professor Azriel Rosenfeld published a number of mathematically oriented articles [12, 13, 14, 15, 16] and established the filed we now call *digital topology*. It is well-known that [12] is the first journal paper on digital topology. Hence, people call him the "grandfather of digital topology".

Over a long period of time, the author has collaborated with Azriel on various problems in digital pictures. This joint work began from author's first stay at his laboratory called "Computer Vision Laboratory" at University of Maryland. The time was November of 1977. In early years of our collaboration, the author worked on "picture languages" with Azriel. In those days, the author interested in 2D (or 3D) languages and automata. In 1979 Azriel published a pioneering book "Picture Languages" [17] in this field. For the details of his contribution to theory of picture languages, the author reported it in "Foundations of Image Understanding" [10] that is a memorial book for Azriel's career.

At these days, the author worked also on "digital geometry" related to picture languages. Some of the author's papers on this topic are contained in the book "Digital Geometry" [2], a recent publication by Klette and Rosenfeld.

In the 1990's, the author's interest moved into digital topology. In the last decade, we have worked in digital topology [8, 9, 11, 19, 20, 21, 22]. In almost all of these papers, a method called "magnification" plays a key role.

In digital topology, the fundamental concept is a connectivity relation, and various adjacency relations between two lattice points are used. Latecki, Eckhard and Rosenfeld [5] defined a special class of subsets of binary digital images called "well-composed". In this special class in 2D, we have only one connectedness relation, since 4- and 8-connectedness are equivalent, so that sets of this class have very nice topological properties; for example, the Jordan Curve Theorem holds for them, the Euler characteristic is locally computable, and further many topological properties of these pictures can be treated in a continuous analog.

R. Klette and J. Žunić (Eds.): IWCIA 2004, LNCS 3322, pp. 260–275, 2004.

In this paper, we propose a method called "magnification" by which non-well-composed pictures are converted to well-composed ones. Also, a partial magnification is proposed. These deformations are based on the change of simple pixels (of 2D case) or simple voxels (of 3D case) so that it is topology-preserving. We extend the method to fuzzy pictures and also to n-dimensional (nD) pictures. After that, some interesting applications of this method are shown. In the last work with Azriel, the author tried to solve a problem (described in the final section of this paper) by making use of magnification method. The 2D case of this problem is easily shown by Proposition 2.5.4 in [17]. But, this plan has been not yet realized since Azriel passed away Feb.22, 2004 while we were working.

The main parts of this paper have been already published, so that the explanation is roughly given. But, magnification of fuzzy pictures on rectangular array is new, so it is described a little minutely.

2 Magnifications in Binary Pictures

In general, a digital binary picture *P* of n-dimensional space is a mapping from Z^n into {0,1} such that either only finitely many lattice points are mapped into 1's or only finitely many points are mapped into 0's. (The foreground of a segmented image means the objects consisting of finite 1's and the background means all other objects consisting of 0's. Most treatments of the subject allow only the first of these conditions, but we shall find it convenient to allow either.) These 0 and 1 are called *n-xels*, in particular, a 2-xel is called a *pixel* and a 3-xel is a *voxel*. We can treat digital pictures of nD space, For two n-xels p and q, there are many adjacency relations of lattice points, e.g., the 4-adjacency relation (its north, south, east, and west neighbors) and 8-adjacency one (adding its northwest, northeast, southwest, and southeast neighbors to 4-adjacency) in 2D. Similarly, there are the 6-adjacency and 26-adjacency in 3D.

For discussion in continuous analog of digital pictures, it is natural that we put a closed unit hypercube at a lattice point such that a black hypercube is put at 1 and a white hypercube at 0. For example, for 2D we put a unit square and for 3D to a unit cube. We denote a unit hypercube at a lattice point p by [p]. In this case, there are the following two possibilities (1) and (2) on the common boundary (if any) between a black unit hypercube and a white one.

(1) The common boundary is black.
(2) The common boundary is white.

Usually, a digital picture *P* is denoted by *P* = (V, a, b, B). In this notation, V = Z^n and B is the objects of 1's; for Z^2 (a, b) = (8, 4) or (4, 8) and for Z^3 (a, b) = (26, 6) or (6, 26); for Z^n (a, b) = (3^n - 1, 2n) or (2n, 3^n - 1). The (8, 4), (26, 6), (3^n - 1, 2n) correspond the above (1) and the (4, 8), (6, 26), (2n, 3^n - 1) correspond to the above (2). In this paper, we use the case (1).

Here, let us define a simple n-xel (n=2, 3, 4) of B. Let p be a n-xel of B and we consider [p] and the polyhedron [B]. In [3], Kong gave the following definition: p is *simple* iff there is a "deformation retraction" of [p] onto [p]$\cap$([B - p]). This p is a *black simple n-xel.* A white simple n-xel q is dually defined. Let q be an n-xel of the complement of B.

Then, we regard q as 1. If q is a black simple in [B∪q], q is called a *white simple 4-xel.* Let $N_8(p)$ be the set of 8-neighbors of a pixel p, and $N_4(p)$ be the set of 4-neighbors of a pixel p. N(p) is defined as $N_8(p) \cup \{p\}$.

It is well-known that there are the following facts:

- A black pixel p (having value 1) is simple iff (i) p is 8-adjacent to exactly one black component in $N_8(p)$, and (ii) p is 4-adjacent to exactly one white component in $N_8(p)$,
- A white pixel p (having value 0) is simple iff (i) p is 8-adjacent to exactly one black component in $N_8(p)$, and (ii) p is 4-adjacent to exactly one white component in $N_8(p)$.

From the definition, we can show that changing values of simple n-xel preserves topology (in the sense of homotopy). A deformation by a finite (possibly null) series of applications of the change of value of simple 4-xel is called *simple deformation* (abbreviated to SD). If a picture is obtained from another picture by SD, we can define a relation these two pictures by making use of SD. This relation is called *SD-equivalent relation.* It is obvious that "SD-equivalent" is an equivalent relation.

2.1 Magnification in 2D and 3D Cases

In this subsection, we explain the magnification in 2D. Let us consider a digital picture $P = (Z^2, 8, 4, B)$. The magnification means to magnify B to each direction (i.e., x-direction and y-direction) by a factor k (>1). We assume that 1's of P are between y-coordinates h and 1. In other words, the highest level of 1's is h and the lowest level of 1's is 1. This assumption is always satisfied if we re-coordinate the y-coordinate. First, we consider an upward magnification of P (i.e., to the y-direction) by a factor k. After that, we repeat the similar magnification to x-direction.

The upward magnification of P is recursively done, y-level by y-level, from the top y-level to bottom y-level.

(I) Procedure for the Top Level:

Let us consider an arbitrary 1 (say, p) on the top level, and q be a white pixel above p.

We change the value of q to 1, and we repeat this dilation until h x k. We apply the same procedure to an arbitrary black pixel on the top level, and repeat the same thing to every black pixel on the top level.

Then, we repeat the same procedure to each white pixel on the top level. In this case, we do nothing since the pixel above a white pixel on the top level is also white.

(II) Recursive Step:

Assume that the dilation of all pixel at level i+1 has been finished. We dilate every pixel at level i. For this case, we dilate a black pixel at level i before every white pixel at level i. Here, from the assumption of recursive step we have the following situations (a) and (b):

(a) If a pixel $q(x, i+1)$ is black, then for every j such that $i+1<j<k(i+1)+1$ a pixel $p(x, j)$ is black.
(b) If a pixel $q(x, i+1)$ is white, then for every j such that $i+1<j<k(i+1)+1$ a pixel $p(x, j)$ is white.

Let us consider an arbitrary black pixel r on level i. We dilate r until $k \times i$. After that, we dilate another arbitrary black pixel on the level i until $k \times i$. We repeat this procedure. After finishing dilations of all black pixels on level i, we repeat the similar dilation to white pixels on level i.

Then, we dilate every pixel on level 1 after repeating the above procedure, we get a magnified set of B to y-direction Then, we magnify the obtained set to x-direction by the same method.

For 3D pictures, the procedure is almost the same as the 2D case.

Theorem 1: The above magnification of 2D and 3D binary picture is done by SD.

(Proof). It is enough to show that the above-mentioned procedure is done by SD.

For the first dilation of (I), this is immediate since all pixels above the top level are white. For the second dilation of (I), this is also immediate since the conditions (a) and (b) are also satisfied for $i = h$.

For the dilation of (II), the procedure is done by SD since the conditions (a) and (b) have been satisfies.

The magnification procedure in 3D is almost the same as in 2D case. It is enough to SD-dilation first upward (z-direction). After that, repeat it to x-direction then to y-direction. //

We consider a "partial magnification" such that the magnification is performed a limited area of a picture. For the 2D case, this is given in the following proposition:

Proposition 2: We partition a picture into the following regions A, B, C, and D. Let C be an isothetic rectangle whose leftmost and rightmost columns are both constant. Let A and B be isothetic rectangular regions just to the left and right of C that contain all the 1's that lie to the left and right of C, and D be the isothetic rectangular region above A, B, and C that contains all 1's that lie above A, B, and C. Then, A, B, and D can be magnified upward any desired amount using SD.
(Note that C is not magnified.)

(Proof). D can be magnified because SD-magnification works for the rows above any given row that is the top row of A, B and C. Further, A and B can be magnified because the columns of C adjacent to A and B have constant values, and D has been already magnified. //

By considering rectangular parallelepipeds instead of rectangles of Proposition 2, we have the partial magnification of 3D.

Corollary 3: A partial magnification of 3D is done by SD.

(Proof). This is immediate by the same proof method as Proposition 2. //

2.2 Magnification in 4D and nD Cases

In argument of 4D case, it is convenient to regard an n-xel as a closed unit 4D hypercube. As pointed out in [3], a *simple 4-xel* is characterized in the following form:

A 4-xel q(x, y, z, t) of B is simple in B iff the following conditions all hold:

(a) $\cup$Attach(q, B) is nonempty and connected.
(b) $\cup$ Boundary(q) - $\cup$ Attach(q, B) is nonempty and connected.
(c) $\cup$Attach(q, B) is simply connected.

Here, Attach(q, B) is defined as the (possibly empty) xel complex Boundary(q) $\cap$ $\cup${Boundary(x) | x is in (B - {q}) }. In other words, interpreting a 4-xel as a 4D unit hypercube, q is simple in B iff there is a deformation retraction of [q] onto [q] $\cap$ ([B - q]).

Then, we have the following Theorem 4:

Theorem 4: The magnification of 4D binary picture is done by SD.

(Proof). The proof is the same as Theorem 1.

First, we select one (say, t-coordinate) of the coordinations. Then, we apply the same procedure to the t-direction. Then, it is enough to repeat successively the dilation to x-direction, y-direction, and z-direction. //

Kong told the author in his private communication that our magnification procedure will work for nD (n>4) case. In this case, a simple xel is defined as follows [4]:

For any xel-image *I* and $I \ni x$, x is said to be deformationally simple in *I* the polyhedron $\cup$ (*I* -{x}} is strongly deformation retract of the polyhedron $\cup$ *I*.

This concept is obtained as an extension of 4D case. However, it is needed to define exactly "xel" , so that this is an interesting further topic.

3 Magnification in Fuzzy Pictures

3.1 Magnification of Fuzzy Pictures on Hexagonal Grid

A digital binary picture was defined on Z^n, and adjacency relations for 0's and 1's were dual. But, for a fuzzy picture (i.e., gray-scale picture) it is a little difficult to treat types of adjacency since the values are not two sorts. To avoid this trouble, in [1] we use a hexagonal array S. (In the later, we treat also fuzzy pictures on a rectangular array). A fuzzy picture is a mapping s: S $\rightarrow$ [0, 1] such that only finitely many positive values appear in S. (The foreground of a fuzzy picture means the objects consisting of finite positive values and the background means objects of 0's.) In this subsection, we give a magnification method for such pictures. The method was already described in our paper [9]. In fuzzy pictures, each lattice point is denoted by a capital letter.

The neighborhood of a lattice point P in a hexagonal array is illustrated below:

```
  A   B
C   P   D
  E   F
```

The set {A, B, C, D, E, F} is the neighborhood of P and is denoted by $N_h(P)$. This corresponds to the usual notation $N_8(P)$ in a rectangular array. In such a fuzzy picture, we want to define a simple point. Before giving the definition, let us review the concept of a simple (pixel) in nonfuzzy (two-valued) case for a hexagonal array. A black point (having the value 1) is said to be a *black simple point* iff

(b1) P is adjacent to exactly one black(having value 1) component in $N_h(P)$,
(b2) P is adjacent to exactly one white (having value 0) component in $N_h(P)$.

Let S be a subset of S, and P be a simple point of S. Let S^- be a subset of S obtained from S by deleting P from S (changing the value of P from 1 to 0). Then, it is well-known that the numbers of components and holes in S^- are the same as those in S, respectively.

The above definition of simple point was for a black point, but dual to the black case we can also define a white simple point. A white point P (having value 0) is said to be a *white simple point* iff

(w1) P is adjacent to exactly one black component in $N_h(P)$,
(w2) P is adjacent to exactly one white component in $N_h(P)$.

Let S be a subset of S and P be not in S. When we consider the subset $S\cup\{P\}$, the value of P changes 1. If P is a white simple point, we have (b1) and (b2). We denote a subset of S obtained from S by adding a white simple P to S (changing the value of P from 0 to 1) by S^+. Then, it is easily shown that the numbers of components and holes of S are the same those of S^+, respectively.

Based on the above observation, we now define simple points for a fuzzy picture s. Let P be a point of S, and let the "extended" neighborhood $N_h(P)\cup\{P\}$ of P be

```
  A   B
C   P   D
  E   F
```

and membership values of these points are denoted by $s(N_h(P)\cup\{P\})$ (hereafter we denote the membership value of a point X by the small letter x). P is said to be a *negative-simple* iff P is a black simple point of the set of thresholded $s(N_h(P)\cup\{P\})$ by p and P is to be *positive-simple* iff P is a white simple point of thresholded $s(N_h(P)\cup\{P\})$ by p +e.

Let P be a negative-simple point and V be the set $\{x \mid x$ is a value of $N_h(P)$ and $x<p\}$.

Then, maxV is called the *negative value* of P. Similarly, let U be the set $\{x \mid x$ is a value of $N_h(P)$ and $x>p\}$. Then, minU is called the *positive value* of P. If P is a negative-simple point, the replacement of p by the negative value of P is called the *negative-simple point operation.* Similarly, of P is a positive-simple point, the replacement of p by the positive value of P is called the *positive-simple point operation.*

Let s be a fuzzy picture of S. Let s^- be a fuzzy picture which is obtained from s by applying the negative-simple point operation to a point of S, and s^+ be a fuzzy picture which is obtained from s by applying the positive-simple point operation to a point of S.

Example of s, s^-, and s^+:

```
  0.8   0.9          0.8   0.9          0.8   0.9
0.8   0.7   0.3    0.8   0.4   0.3    0.8   0.8   0.3
  0.3   0.4          0.3   0.4          0.3   0.4
       s                  s-                 s+
```

We have the following proposition that has been proved in [9].

Proposition 5: Let us consider the $N_h(P)$ of a point P in a fuzzy picture. If P is a negative-simple point then the degree of local connectedness of pairs of points in $H_h(P)$ does not decrease after applying the negative-simple point operation to P. Also, if P is a positive-simple point then the degree of local connectedness of pairs of points in $N_h(P)$ does not increase after applying the positive-simple point operation to P.

(Proof). This has been proved in [9]. //

The simple deformation in fuzzy pictures is similarly defined as the crisp (two-valued) case. In other words, a deformation by a finite (possibly null) series of applications of the change of value of positive- (or negative-) simple point operation is called *simple deformation* (abbreviated to SD).

Theorem 6: The magnification of fuzzy picture is done by SD.

(Proof). Let the extended neighborhood $N_h(P) \cup \{P\}$ of P be

```
  A   B
C   P   D
  E   F
```

We will describe how to magnify the picture "upward" in the direction PB; magnification in other direction is analogous. With respect to this direction, the picture can be divided into "row". Five of these rows intersect $N_h(P) \cup \{P\}$; the intersections are {E}, {C, F}, {P}, {A, D}, and {D}. Let the nonzero points of the picture be contained in n rows, numbered "upward". Note that if P is in row i, B is in row i+2 and E in row i-2.

As in Section 2.1, our method of magnification is based on "upward" dilation of the constant-value runs of points (i.e., the maximal sequence of points in the PB direction that have the same value) in the PB direction. Let r_i be a run whose uppermost point is in row i. We will dilate each r_i by amount [id/2]; we will do this first for all runs r_n, then for runs r_{n-1}, and so on. Let the run below r_i have its uppermost point in row h so that the length of r_i is (i-h)/2 (note that i-h must be even). When we dilate r_i by [id/2], its length becomes (i-h)/2 + [id/2]. When we later dilate r_h by [hd/2], this erodes the dilated r_i so that its length becomes (i-h)/2 + [id/2] - [hd/2] = (i-h)/2 + (i-h)d/2 = (i-h)(d+1)/2, so that r_i is magnified by factor d+1.

We now show that the dilations can be accomplished by repeated simple point operations. If P is the uppermost point a run r_i, then when we dilate r_i, the runs that contains A, B, and D have already been dilated by greater amount, or infinite runs of 0's. Thus, A, B, and D belong to long runs of a's ,b's, and d's in the PB direction. It is easily verified that , no matter what the values a, b, d , and p are (note that p is not b, since P is the uppermost point of a run), B is either positive-simple or negative-simple (or both).

Suppose $p<b$. If $a \leq p$ and $d \leq p$, p is the negative value of B, so performing the negative simple point operation on B dilates r_i upward by one step. If $p<a<b$ or $p<d<b$ (or both), performing the negative simple point operation on B gives it the value a or d; but B is still simple, and performing the operation again (twice, if we have $p<a<d$ or $p<d<a<b$) gives B the value p, as desired. Similarly, if $p>b$, performing the positive simple point operation on B (one, twice , or three times) gives B the value p. Thus in any case, r_i can be dilated upward by repeatedly performing simple point operations. //

3.2 Magnification of Fuzzy Pictures on Rectangular Array

In 2D rectangular array two-valued pictures, there were two kinds of neighboring relations called 4-adjacency and 8-adjacent. It is standard practice to use different types of connectedness for the 1's and 0's. For fuzzy digital pictures, however, it is impossible to consider such adjacency relations between two pixel, because the fuzzy values are in [0, 1]. To avoid this trouble, in the previous subsection, we have considered the hexagonal array.

Here, let us consider magnifications of fuzzy pictures based on 2D rectangular array and 3D cubical array.

In [18], Rosenfeld discussed topology of fuzzy pictures in the rectangular array. Its central concept is "connectedness" between two fuzzy pixels. In [18], he didn't explicitly mention its adjacency relations. We define a simple pixel of the rectangular array fuzzy pictures. This definition is based on fuzzy topology in [18], especially the notion "thresholdable connected objects" described at page 87 of that paper. Then, we explain a magnification method for such pictures. We will also remark that this magnification method will be extended to 3D fuzzy picture in cubic array.

A fuzzy digital picture on the rectangular array is a mapping of lattice points of Z^2 into [0, 1], such that finitely many lattice points are mapped into positive numbers. Using the same notation as in the hexagonal case, we denote fuzzy pictures by Greek small letters s, t, ...

As the same before, we use small letters p, q, ... to represent fuzzy values of pixels P, Q,..., respectively. For a real number 1 in [0, 1], we consider the following threshold operation L_l:

$L_l(x) = 1$ if x is not smaller than 1, and $L_l(x) = 0$ if $x<1$.

Then, we obtain a two-valued picture $L_l(s)$ by operating L_l to s. Let P be a fuzzy pixel. We consider N(P) of P. Further, we consider pixels in $L_l(N(P))$ that are obtained by threshold operation L_p. In other words, if a fuzzy value t in N(P) is not smaller than p, then we give the value 1 to this pixel T, and if a fuzzy value t in N(P) is smaller than p, we give the value 0 to this pixel T. Thus, we get two-valued picture L_p ((N(P)) from N(P). In this case, P itself changes to L_p(p) (=1). If L_p(p) (=1) is a simple pixel in L_p ((N(P)), then we say that P is a *negative simple fuzzy pixel.*

Also, we consider pixels in L_{p+e} (N(P)) that are obtained by threshold operation L_{p+e}. In other words, if a fuzzy value t in N(P) is larger than p+e then we give the value 1 to this pixel T, and if a fuzzy value t in N(P) is not larger than p+e, we give the value 0 to this pixel T. Thus, we get a two-valued picture L_{p+e}(N(P)) from N(P). In this case, P itself changes to L_{p+e}(p) (=0). If the L_{p+e}(p) (=0) is a white simple pixel in L_{p+e}(N(P)), then we say that P is a *positive simple fuzzy pixel*.

Let P be a negative simple fuzzy pixel. In L_p ((N(P)), we consider a white (=0) 4-component that is 4-adjacent to P. The original fuzzy set of this component is denoted by W(P). Then, the largest fuzzy value in W(P) is called the *negative value* of P.

Let P be a positive simple fuzzy pixel. In L_{p+e}(N(P)), we consider a black (=1) 8-component that is 8-adjacent to P. The original fuzzy set of this component is denoted by B(P). Then, the smallest fuzzy value in B(P) is called the *positive value* of P. If P is a negative simple fuzzy pixel, the replacement of p by a negative value of P is called the *negative simple fuzzy value operation.* Similarly, if P is a positive simple fuzzy pixel, the replacement of p by a positive value of P is called the *positive simple fuzzy pixel operation.*

Let s be a fuzzy picture. Let s^- be a fuzzy picture that is obtained from s by applying the negative fuzzy simple pixel of operation to a pixel in s. Note that s^- means an "eroded" fuzzy picture of s. Let s^+ be a fuzzy picture that is obtained from s by applying the positive simple fuzzy pixel to a pixel in s. Note again that s^+ means a "dilated" fuzzy picture of s.

Let s and t be two fuzzy picture. We say that t differs from s by simple deformation (in short, SD) if t is obtained from s after a finite (possibly null) applications of the negative simple pixel operation and/or the positive simple pixel operation.

Example:

The center 0.4 in N(P) is p:

N(P)

0.8	0.5	0.6
0.1	0.4	0.1
0.1	0.2	0.3

L_p(N(P))

1	1	1
0	1	0
0	0	0

L_{p+e}(N(P))

1	1	1
0	0	0
0	0	0

$N(P)^-$

0.8	0.5	0.6
0.1	0.3	0.1
0.1	0.2	0.3

$N(P)^+$

0.8	0.5	0.6
0.1	0.5	0 1
0.1	0.2	0.3

Proposition 7: Let s and t be two fuzzy pictures such that t is differs from s by SD. Then, for an arbitrary value 1 in [0,1], L_l(s) and L_l(t) are SD-equivalent.

(Proof). To prove this Proposition, it is enough to show that L_l(s) and $L_l(s^-)$ are SD-equivalent and also that L_l(s) and $L_l(s^+)$ are SD-equivalent. Here, we prove that L_l(s) and $L_l(s^-)$ are SD-equivalent. The proof of another case is similar.

Let P be negative simple fuzzy pixel and q be the negative value of P. Since L_l(*s*-N(P)) and $L_l(s^-$ - N(P)) are the same, we consider the region of N(P). We consider the following cases:

(1) 1>p: For this case, L_l(s) and $L_l(s^-)$ are the same, so that Proposition is immediate.
(2) 1=p: For this case, only the value of P changes. From the definition of negative simple fuzzy pixel operation, L_p(N(P)) and $L_p(N(P)^-)$ are SD-equivalent. Hence, Proposition is true.
(3) p>1>q: This case is the same as (2).
(4) 1=q: In this case, L_p(N(P)) and $L_p(N(P)^-)$ are the same, so that Proposition is true.
(5) q>1: In this case, L_l(s) and $L_l(s^-)$ are the same.

Therefore, we have this Proposition. //

SD is not an equivalence relation, since we cannot get N(P) from $N(P)^-$ by SD. See the above example. However, we say that SD of fuzzy pictures is "topology-preserving". Because we have Theorem 7.

Now, let us explain, in detail, a magnification method of 2D rectangular fuzzy pictures. The magnification means the enlarging every pixel to k × k. For example, we can get the following (2) from a given (1) by magnification.

```
(1)
0 0 0 0 0 0
0 a b c d 0
0 0 e f g 0
0 h j k p 0
0 0 0 0 0 0
```

```
(2)
0 0 0 0 0 0 0 0 0 0 0 0 0 0
0 a a a b b b c c c d d d 0
0 a a a b b b c c c d d d 0
0 a a a b b b c c c d d d 0
0 0 0 0 e e e f f f g g g 0
0 0 0 0 e e e f f f g g g 0
0 0 0 0 e e e f f f g g g 0
0 h h h j j j k k k p p p 0
0 h h h j j j k k k p p p 0
0 h h h j j j k k k p p p 0
0 0 0 0 0 0 0 0 0 0 0 0 0 0
```

In this case, letters a, b, c, ... are fuzzy values in (0, 1]. The picture (2) is a magnification of (1) by a factor 3

Theorem 8: The above magnification method is done by SD.

(Proof). Let s be a given fuzzy picture. Let the portion of picture s that contains non-zeros have n rows, which we number 1, ..., n starting from the bottom row; the row of 0's below the bottom row is numbered 0. Each column of s consists of runs of a fuzzy value. We call a pixel of value p *p-pixel.* A p-pixel is upward dilated by a factor of k (where k is a positive integer) as follows:

Assume that a p-pixel is a row h. We dilate this p-pixel until it reaches at a row h×k. In this case, we assume that k is not 1. Because "k=1" means that we do nothing. This dilation is done, row by row, from the top row to bottom row. In other words, only after finishing the dilation of each fuzzy pixel on a row i+1, the next dilation is applied to the row i.

For the top row, there is no problem since the row above the top consists entirely of 0's.

See Fig.1 of a case where k=3.

```
                  0 0 0
                  a b c
                  a b c
                  a b c
                  a b c
                  a b c
0 0 0             a b c
a b c             a b c
d e f             d e f
g h j             g h j
0 0 0             0 0 0
```

Fig. 1

Assume that we can upward SD-dilation until the row h+1. Then, it is enough to show that the above dilation for the row h is done by SD. See Fig.2 of an example for h=2.

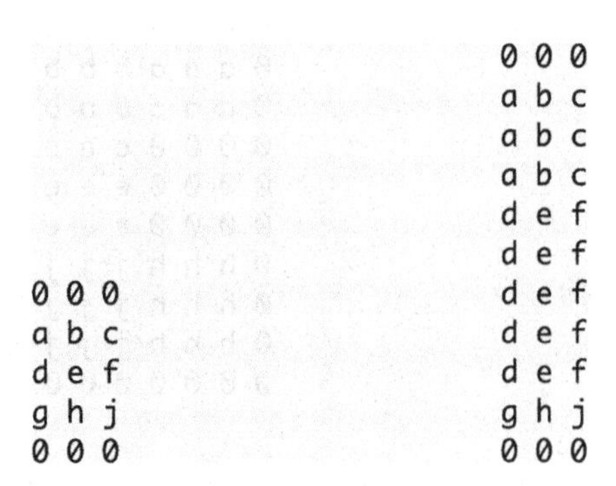

Fig. 2

To prove that this dilation is done by SD, we need a detailed explanation.

From the assumption, we can have the following neighbors N(q) for every pixel q's on the row h+1

```
                 p q r
row h+1          p q r
row h            u v w
```

First, let us consider neighbors N(q)'s satisfying the condition q<v, where v is a pixel below q. This neighbor N(q) is called a *top 0 and bottom 1 neighborhood* (denoted by (0-1)-N(q)). In general, there are many (0-1)-N(q)'s on the row h. As mentioned below, we upward dilate v until h x k. In this case, the application order among (0-1)-N(q)'s is arbitrary.

For a (0-1)-N(q), we can have the following configuration by SD:

```
                 p q r
row h+1          p v r
row h            u v w
```

Because, L_{q+e}(N(q)) for N(q) satisfying the condition is one of the following configurations by threshold.

```
0 0 0    1 0 0    0 0 1    1 0 1
0 0 0    1 0 0    0 0 1    1 0 1
? 1 !    ? 1 !    ? 1 !    ? 1 !
```

In this case, ? is 0 or 1, and also ! is 0 or 1.

Furthermore, by repeating the same argument until the row h x k we have

```
                 p v r
row h+1          p v r
row h            u v w
```

By repeating this procedure for all (0-1)-N(q)'s on the row h, all neighbors N(q)'s on the row h must eventually satisfy that q>v or q=v. But we need not to treat that case q=v.

Because, for this case q=v it is not needed to dilate upward. Hence all N(q)'s that we want apply SD must satisfy q>v. Such a N(q) is called a *top 1 and bottom 0 neighborhood* (denoted by (1-0)-N(q)). Note that there can be many (1-0)-N(q)'s on the row h.

As mentioned below, we can upward dilate v. In this case, the application order among (1-0)-N(q)'s is arbitrary.

Then we have the following configuration by SD:

```
                 p q r
row h+1          p v r
row h            u v w
```

The reason is as follows:

We apply a threshold operation L_q to an initial N(q) satisfying the assumption. Then, we have one of the following configurations:

```
$ 1 %
$ 1 %
? 0 !
```

where $, %, ! are 0 or 1, but $\$ \geq ?$ and $\% \geq !$.

This fact follows from a configuration at the present stage of our dilation. Then, the center 1 is a black simple pixel.

Then, we have:

```
                p q r
row h+1         p v r
row h           u v w
```

Then, we repeat the same dilation until the row h x k. Since the above replacement is done by SD, topology of s doesn't change.

By repeating this procedure for all (1-0)-N(q)'s on the row h, all pixels p's on h are dilated until the row h x k .

```
0 0 0
a b c
a b c
a b c
d e f
d e f
d e f
g h j
g h j
g h j
0 0 0
```

Fig. 3

Therefore, by induction we can upward (to y-direction) magnify s by a factor k. See Fig. 3. Note that (h+1) k - hk = k

After finishing all upward dilation, we apply the same magnification to x-direction.

Therefore we have this theorem. //

4 Applications of Magnification

We used the magnification method in our paper [8, 20]. Adding to those, there are some applications described in the following subsections.

4.1 Well-Composedness in 2D and 3D Pictures

As mentioned in Introduction, Latecki et al. introduced a special class of subsets of binary digital pictures called "well-composed". In the special sets in 2D, we have only one connectedness relation, since 4- and 8-connected are equivalent, so that sets of this class have nice topological properties. Since there is only one connecteness, we can treat digital pictures in a continuous analog.

This is the same for 3D binary pictures.

Proposition 9: Let B be an arbitrary set of 1's in 2D or 3D binary pictures. Then, we can get a well-composed set B* by SD.

(Proof). First, we deform a given B to the magnified set (denoted by m(B)) by SD. Then, it is enough to change a bad pixel (or voxel) that violates well-composedness in m(B). This is possible by SD since m(B) is magnified by a sufficient large amount. //

4.2 Well-Composedness of 4D and nD (n>4)

We can define well-composed sets of 4D rectangular array pictures and construct a well-composed set from a non-well-composed one. For the details, see my draft [11]. Further, Kong told me that this technique works for nD. Also he suggested the following definition of well-composedness in nD pictures.

Let S of n-xels in Euclidean n-space R^n. Here, we stipulate that a real point on the common boundary (if any) between a black n-xel and a white n-xel belongs to both. S is "well-composed" iff the following conditions holds every (real) point p in R^n : The set of n-xels containing p that lie in S, and the set of n-xels containing p that do not lie in S, are both 2n-connected sets.

Based on this definition, we are able to SD-change a non-well-composed picture *P* in 4D to a well-composed one. After magnifying *P*, it is enough to SD-deform each bad 4-xel.

4.3 Multicolor Well-Composed Pictures

In [6], Latecki proposed an interesting concept of multicolor well-composed pictures, and prove some important properties. This is nothing but our fuzzy pictures on rectangular array. Therefore, it is immediate to SD-deform a well-composed fuzzy picture from a non-well composed one. After magnifying, it is enough to SD-deform a bad fuzzy pixel.

5 Further Problems and Concluding Words

There will be various interesting applications of the magnification. It has been known that the converse of Jordan curve theorem (for surfaces in the meaning of Morgenthaler and Rosenfeld) is not true for general 3D pictures. However, for well-composed pictures we are able to have Jordan surface. Also, Latecki [7] shows some interesting theorems on 3D well-composed pictures

Regarding deformation to a well-composed picture of an non-well-composed one in nD pictures, it will be possible but needs further rigorous discussion.

As mentioned in Introduction, this is a memorial talk dedicated to Professor Azriel Rosenfeld. In this talk, the author has reported recent joint works around the magnification that we did. He said often "Magnification is a very strong technique in digital topology". In fact, this magnification method will be applied for the following problem by making use of the proof technique of the Schoenflies theorem (in the conventional topology). Because a magnified set consists of very small unit cubes.

Problem: Let us consider a 3D picture $P = (Z^3, 26, 6, B)$, where B is well-composed and doesn't contain any cavity or any tunnel. Is B SD-equivalent to s single voxel ?

There is another hard problem (called the Animal Problem) similar to this one. Here, animal means the union of lattice cubes homeomorphic to the 3-ball. However, its proof is extremely hard, since animality-preserving is stronger than SD (see the following example). The problem needs another magnification method than that of this paper.

Hence, it will be still a further challenging problem.

Example: An animal {(0, 0, 0), (0, 0, 1), (1, 0 0), (1, 1, 0), (2, 1, 0), (2, 1, 1)} is upward dilated by SD, but it is not done in animality-preserving.

Acknowledgment

The author expresses his appreciation to Prof. T.Y. Kong and Prof. L Latecki for many helpful and stimulating suggestions in preparing this paper. Of course, sincere thanks go to the late Prof. Rosenfeld who was very kind to the author. He was a truly exceptional scientist. It was a great honor to know and to work with him. The author will always be very grateful that he had the good fortune to do so.

The author closes this paper with the last email from Azriel.

To: akira668@urban.ne.jp
Subject: Problem
Cc: ar@vinland.cfar.umd.edu

Dear Akira,
Regrettably, I spent yesterday in the hospital and still have to take
a lot of tests. It may be several weeks before I know how serious the situation is.
When I find out, I'll let you know. Sorry for the interruption to our work.
Azriel

References

1. Aizawa, K. and Nakamura, A.: Grammars on the hexagonal array, *Inter. J. of Pattern Recognition and Artificial Intteligence*, 3 (1989), 469-477.
2. Klette, R. and Rosenfeld, A.: **Digital Geometry**, Morgan Kaufmann, San Francisco, 2004.

3. Kong, T.Y.: Topology-preserving deletion of 1's from 2-, 3-, 4-dimensional binary images, *LNCS*, 1347(1997), 3-18.
4. Kong, T.Y. and Roscoe, A.W.: Simple points in 4-dimensional (and higher-dimensional) binary images, (manuscript) April 2, 2004.
5. Latecki, L. and Eckhard, U., and Rosenfeld, A.: Well-compsed sets, *Comput. Vision Image Understanding*, 61(1995), 70-83.
6. Latecki, L.: Multicolor well-composed pictures, *Pattern Recgnition Letters*, 16(1995), 425-431.
7. Latecki, L.: 3D well-composed pictures, *Graphical Models and Image Processing*, 59(1997), 164-172.
8. Nakamura, A. and Rosenfeld, A.: Digital konts, *Pattern Recognition*, 33(2000), 1541-1553.
9. Nakamura, A. and Rosenfeld, A.: Topology-preserving deformations of fuzzy digital pictures, in **Fuzzy Techniques in Image Processing** edited by E.E. Kerre and M. Nachtegael, Physica-Verlag, Heidelberg, 2000, 394-404.
10. Nakamura, A.: Picture languages, in **Foundations of Image Understanding**, edited by L.S. Davis, Kluwer Academic Publishers, Boston, 2001, 127-155.
11. Nakamura, A.: Magnification method of 4D digital pictures (draft paper).
12. Rosenfeld, A.: Connectivity in digital pictures, *J. ofACM*, 17(1970), 146-160.
13. Rosenfeld, A.: Arcs and curves in digital pictures, *J ofACM*, 20(1973), 81-87.
14. Rosenfeld:, A.: Adjacency in digital pictures, *Information and Control*, 26(1974), 24-33.
15. Rosenfeld, A.: A characterization of parallel thinning algorithms, *Information and Control*, 29(1975), 286-291.
16. Rosenfeld, A.: A converse to the Jordan curve theorem for digital curves, *Information and Control*, 29(1975), 292-293.
17. Rosenfeld, A.: **Picture Languages**, Academic Press, New York, 1979.
18. Rosenfeld, A.: Fuzzy digital topology, *Information and Control*, 40(1979), 76-87.
19. Rosenfeld, A. and Nakamura, A. Local deformation of digital curves, *Pattern Recognition Letters*, 18(1997), 613-620.
20. Rosenfeld, A., Kong, T.Y., and Nakamura, A.: Topology-preserving deformations of two-valued digital pictures, *Graphical Models and Image Processing*, 60(1998), 24-34.
21. Rosenfeld, A., Saha, P.K., and Nakamura, A.: Interchangeable pairs of pixels in two-valued digital images, *Pattern Recognition*, 34(2001), 1853-1865.
22. Rosenfeld, A. and Nakamura, A.: Two simply connected sets that have the same area are IP-equivalent, *Pattern Recognition*, 35(2002), 537-541.

Curves, Hypersurfaces, and Good Pairs of Adjacency Relations

Valentin E. Brimkov[1] and Reinhard Klette[2]

[1] Fairmont State University, 1201 Locust Avenue, Fairmont, West Virginia 26554-2470, USA
vbrimkov@fairmontstate.edu

[2] CITR Tamaki, University of Auckland, Building 731, Auckland, New Zealand
r.klette@auckland.ac.nz

Abstract. In this paper we propose several equivalent definitions of digital curves and hypersurfaces in arbitrary dimension. The definitions involve properties such as one-dimensionality of curves and $(n-1)$-dimensionality of hypersurfaces that make them discrete analogs of corresponding notions in topology. Thus this work appears to be the first one on digital manifolds where the definitions involve the notion of dimension. In particular, a digital hypersurface in nD is an $(n-1)$-dimensional object, as it is in the case of continuous hypersurfaces. Relying on the obtained properties of digital hypersurfaces, we propose a uniform approach for studying good pairs defined by separations and obtain a classification of good pairs in arbitrary dimension.

Keywords: digital geometry, digital topology, digital curve, digital hypersurface, good pair.

1 Introduction

A regular orthogonal grid subdivides $\mathbb{R}^n$ into n-dimensional hypercubes (e.g., unit squares for $n = 2$) defining a class $\mathbb{C}_n^{(n)}$. Let $\mathbb{C}_n^{(k)}$ be the class of all k-dimensional facets of n-dimensional hypercubes, for $0 \leq k < n$. The grid-cell space $\mathbb{C}_n$ is the union of all these classes $\mathbb{C}_n^{(k)}$, for $0 \leq k \leq n$.

In this paper we study digital curves, hypersurfaces, and good pairs of adjacency relations in grid-cell spaces $\mathbb{C}_n$ $(n \geq 2)$, equipped with adjacencies A_α (e.g., $\alpha = 0, 1$ for $n = 2$, and $\alpha = 0, 1, 2$ for $n = 3$)[1]. A *good pair*[2] combines two adjacency relations on $\mathbb{C}_n$. The reason for introducing the first good pairs (α, β) in [8], with (α, β) equal to (1,0) or (0,1), were observations in [28]. (A_α is

[1] In 2D, 0- and 1-adjacency correspond to 8- and 4-adjacency, respectively, while in 3D, 0-, 1-, and 2-adjacency correspond to 26-, 18- and 6-adjacency, respectively. The latter are traditionally used within the grid-point model on $\mathbb{Z}^n$.

[2] The name was created for the oral presentation of [15]. Note that the same term has been used already with different meaning in topology.

R. Klette and J. Žunić (Eds.): IWCIA 2004, LNCS 3322, pp. 276–290, 2004.

the adjacency relation for 1*s*, which are the pixels with value 1, and A_β is the adjacency relation for 0*s*.) The benefit of two alternative adjacencies was then formally shown in [25]: (1,0) or (0,1) define region adjacency graphs for binary pictures which form a rooted tree. This simplifies topological studies of binary pictures.

Good pairs may induce a digital topology[3] on $\mathbb{C}_n^{(n)}$ (and not vice-versa in general). For example, using the good pair (1,0) (or (0,1)) is equivalent to regarding 1-components of 1*s* as open regions and 0-components of 0*s* as closed regions in $\mathbb{C}_n^{(n)}$ (or vice versa). [9] shows that there are two digital topologies on $\mathbb{C}_2$ (where one corresponds to (1,0) or (0,1)), five on $\mathbb{C}_3$, and [16] shows that there are 24 on $\mathbb{C}_4$ (all up to homeomorphisms). This paper provides a complete characterization of good pairs, showing that there are $2n-1$ good pairs on $\mathbb{C}_n$.

The study of good pairs is directed on the understanding of separability properties: which sets defined by one type of adjacency allow to separate sets defined by another type of adjacency. These separating sets can be defined in the form of digital curves in 2D, or as digital surfaces in 3D. In this way, studies of good pairs and of (separating) surfaces are directly related to one-another. Topology of incidence grids is one possible approach: frontiers of closed sets of n-cells define hypersurfaces, consisting of $(n-1)$-cells.

Digital surfaces have been studied under different points of view. The approximation of n-dimensional manifolds by graphs is studied in [29, 30], with a special focus on topological properties of such graphs defined by homotopy and on homology or cohomology groups. [13] defined digital surfaces in $\mathbb{Z}^3$ based on adjacencies of 3-cells. The approximation of boundaries of finite sets of grid points (in n dimensions) based on "continuous analogs" was proposed and studied in [20]. [12] discusses local topologic configurations (stars) for surfaces in incidence grids. Digital surfaces in the context of arithmetic geometry were studied in [4].

A Jordan surface theorem for the Khalimsky topology is proved in [18]. For discrete combinatorial surfaces, see [10]. For obtaining α-surfaces by digitization of surfaces in $\mathbb{R}^3$, see [6]. It is proved in [21] that there is no local characterization of 26-connected subsets S of $\mathbb{Z}^3$ such that its complement $\overline{S}$ consists of two 6-components and every voxel of S is adjacent to both of these components. [21] defines a class of 18-connected surfaces in $\mathbb{Z}^3$, proves a Jordan surface theorem for these surfaces, and studies their relationship to the surfaces defined in [22]. [3] introduces a class of *strong surfaces* and proves that both the 26-connected surfaces of [22] and the 18-connected surfaces of [21] are strong. For 6-surfaces, see [5].

Frontiers in cell complexes (and related topological concepts such as components and fundamental group) were studied in [1]. For characterizations of and algorithms for curves and surfaces in frontier grids, see [11, 19, 27, 31]. G.T. Herman and J.K. Udupa used frontiers in the grid cell model, and V. Kovalevsky general-

[3] A digital topology on $\mathbb{C}_n^{(n)}$ is defined by a family of open subsets that satisfy a number of axioms (see, e.g., Section 6.2 in [14]).

ized these studies using the model of topologic abstract complexes, that can also be modelled by incidence grids. [7] define curves in incidence grids.

In this paper we present alternative definitions of digital hypersurfaces, partially following ideas already published in the cited references above, and prove their equivalence. In short, a digital α-hypersurface is composed by (closed) α-curves; two of such curves are either disjoint and non-adjacent, or disjoint but adjacent, or they have overlapping portions. The main contributions of this paper are as follows ($n \geq 2$):

- We define digital manifolds in arbitrary dimensions, as the definitions involve the notion of *dimension of a digital object* [23]. Thus a digital curve is a one-dimensional digital manifold, while a digital hypersurface in nD is an $(n-1)$-dimensional manifold, in conformity to topology (see, e.g., the topological definitions of curves by Urysohn and Menger, as discussed in [14]). To our knowledge of the available literature, this is the first work involving dimensionality in defining these notions in digital geometry.
- We show that there are two and only two basic types of α-hypersurfaces, one for $\alpha = n-1$ and one for $\alpha = n-2$.
  - For $\alpha = n-2$, the hypersurface S has $(n-2)$-gaps which appear on $(n-2)$-curves that build S and, possibly, between adjacent pairs of such $(n-2)$-curves.
  - For $\alpha = n-1$, the hypersurface S is $(n-2)$-gapfree,[4] but may still have $0, 1, \ldots, (n-4)$ or $(n-3)$-gaps, which may appear between adjacent pairs of $(n-1)$-curves rather than on the curves themselves. The last possibility is when an $(n-1)$-hypersurface is i-gapfree for any $0 \leq i \leq n-2$.
- We investigate combinatorial properties of digital hypersurfaces, showing that any digital hypersurface defines a matroid.
- Relying on the obtained properties of digital hypersurfaces, we define and study good pairs of adjacency relations in arbitrary dimension. We define nD good pairs through separation by digital hypersurfaces and show that there are exactly $2n-1$ good pairs of adjacency relations. We also provide a short review and comments on some other approaches for defining good pairs which have been communicated elsewhere.

Some of the proofs of results reported in this paper follow directly from the definitions, while others are technical and rather lengthy, and cannot be reported in this brief conference submission. Complete proofs will be included in a full-length journal version of this paper.

[4] This was also called "tunnel-free" in earlier publications (e.g., in [2, 24]). The Betti number β_1 defines the number of tunnels in topology. Informally speaking, the location of a tunnel cannot be uniquely identified in general; there is only a unique way to count the number of tunnels. Locations of gaps are identified by defining sets. However, our hypothesis is that tunnel-freeness (i.e., $\beta_1 = 0$) and gap-freeness (in the sense of [2, 24]) are equivalent concepts.

2 Preliminaries

We start with recalling basic definitions; notations follow [14]. In particular, the grid point space $\mathbb{Z}^n$ allows a refined representation by an incidence grid defined on the cellular space $\mathbb{C}_n$ (as defined above).

2.1 Some Definitions

Elements in $\mathbb{C}_n^k$ are k-cells, for $0 \leq k \leq n$. An m-dimensional facet of a k-cell is an m-cell, for $0 \leq m \leq k-1$. Two k-cells are called *m-adjacent* if they share an m-cell. Two k-cells are *properly m-adjacent* if they are m-adjacent but not $(m+1)$-adjacent.

A digital object S is a finite set of n-cells. An *m-path* in S is a sequence of n-cells from S such that every two consecutive n-cells are m-adjacent. The *length* of a path is the number of n-cells it contains. A *proper m-path* is an m-path in which at least two consecutive n-cells are not $(m+1)$-adjacent. Two n-cells of a digital object S are *m-connected* (in S) iff there is an m-path in S between them. A digital object S is *m-connected* iff there is an m-path connecting any two n-cells of S. S is *properly m-connected* iff it contains two n-cells such that all m-paths between them are proper. An *m-component* of S is a maximal (i.e., non-extendable) m-connected subset of S.

Let M be a subset of a digital object S. If $S \setminus M$ is not m-connected then the set M is said to be *m-separating* in S. (In particular, the empty set m-separates any set S which is not m-connected.) Let a digital object M be m-separating but not $(m-1)$-separating in a digital object S. Then M is said to have *k-gaps* for any $k < m$. A digital object without any m-gaps is called *m-gapfree*.

Although the above definition has been used in a number of papers by different authors, one can reasonably argue that it requires further refinement. Consider, for instance, the following example.

Let M_1 and M_2 be two digital objects that are subsets of a superset S, and assume that $M_1 \cap M_2 = \emptyset$ (we may think that M_1 and M_2 are "far away" from each other). In addition, assume that M_1 has a k-gap with respect to an adjacency relation A_α, while M_2 is a closed digital hypersurface that k-separates S. Then it turns out that the digital set $M_1 \cup M_2$, that consists of (at least) two connected components, has no k-gap with respect to A_α.

Despite such kind of phenomena, the above definition is adequate for the studies that follow. Further work by authors will be aimed at contributing to a more restrictive definition which will exclude "counterintuitive" examples as the one above. For this, one can take advantage of some of the results presented in the subsequent sections.

Let M be an m-separating digital object in S such that $S \setminus M$ has exactly two m-components. An *m-simple cell* in M (with respect to S) is an n-cell c such that $M \setminus \{c\}$ is still m-separating in S. An m-separating digital object in S is *m-minimal* (or *m-irreducible*) if it does not contain any m-simple cell (with respect to S).

For a set of n-cells S, by $\overline{S}$ we denote the complement of S to the whole digital space $\mathbb{C}_n^{(n)}$ of all n-cells.

$J^+(A)$ is the outer Jordan digitization (also called *supercover*) of a set $A \subseteq \mathbb{R}^n$, which consists of all n-cells intersected by A.

By $B_\alpha(c)$ we denote the *unit α-ball* with center c consisting of all α-neighbors of c. Furthermore, let $B_\alpha^*(c) = B_\alpha(c) \setminus \{c\}$.

For a given set $M = \{c_1, c_2, \ldots, c_m\} \subseteq \mathbb{C}_n^{(n)}$ of n-cells, we define its α-adjacency graph $G_M^\alpha(V, E)$ with a set of vertices $V = \{v_1, v_2, \ldots, v_m\}$ and a set of edges $E = \{(v_i, v_j) : c_i$ and c_j are $\alpha -$ adjacent$\}$. (In the above definition a graph vertex v_i corresponds to the element $c_i \in M$.)

2.2 Dimension

Mylopoulos and Pavlidis [23] proposed definition of dimension of a (finite or infinite) set of n-cells S with respect to an adjacency relation A_α (for its use see also [14]). Let $B_\alpha^\star(c)$ be the union of $B_\alpha(c)$ with all n-cells c' for which there exist $c_1, c_2 \in B_\alpha(c)$ such that a shortest α-path from c_1 to c_2 not passing through c passes through c'. For example, $B_1^\star(c) = B_0^\star(c) = B_0(c)$ for $n = 2$, and $B_2^\star(c) = B_1^\star(c) = B_1(c)$ and $B_0^\star(c) = B_0(c)$ for $n = 3$.

In what follows we will use the definition of dimension from [23]. Let S be a digital object in $\mathbb{C}_n^{(n)}$ and A_α an adjacency relation on $\mathbb{C}_n^{(n)}$. The *dimension* $\dim_\alpha(S)$ is defined as follows:

(1) $\dim_\alpha(S) = -1$ if $S = \emptyset$,
(2) $\dim_\alpha(S) = 0$ if S is a totally α-disconnected nonempty set (i.e., there is no pair of cells $c, c' \in S$ such that $c \neq c'$ and $\{c, c'\}$ is α-connected),
(3) $\dim_\alpha(S) = 1$ if $\mathrm{card}(B(c) \cap S) \leq 2$ for all $c \in S$, and there is at least one $c \in S$ with $\mathrm{card}(B(c) \cap S) > 0$,
(4) $\dim_\alpha(S) = \max\limits_{c \in S} \dim_\alpha(B^\star(c) \cap S) + 1$ otherwise.

If in the last item of the definition the maximum is reached for an n-cell c, we will also say that S is $\dim_\alpha(S)$-dimensional at c.

An *elementary grid triangle* in $\mathbb{C}_2^{(2)}$ is a set $T = \{(i,j), (i+1,j), (i,j+1)\}$, or a 90, 180, or 270 degree rotation of such a T. A 0-connected set $M \subseteq \mathbb{C}_2^{(2)}$ is two-dimensional with respect to adjacency relation A_0 iff it contains an elementary grid triangle as a proper subset. Similarly, a 1-connected set $M \subseteq \mathbb{C}_2^{(2)}$ is two-dimensional with respect to adjacency relation A_1 iff it contains as a proper subset a 2×2 square of grid points. See [14]. These properties generalize to an arbitrary dimension n, as follows.

Lemma 1. *(a) An α-connected set $M \subseteq \mathbb{C}_n^{(n)}$, $0 \leq \alpha \leq n-2$, is two-dimensional iff it contains as a proper subset an* elementary grid triangle *consisting of three cells c_1, c_2, c_3, such that any two of them are α-adjacent.*

(b) An $(n-1)$-connected set $M \subseteq \mathbb{C}_n^{(n)}$ is two-dimensional iff it contains as a proper subset an elementary grid square *consisting of four cells c_1, c_2, c_3, c_4 with coordinates $c_1 = (i, i, \ldots, i, i)$, $c_2 = (i+1, i, \ldots, i, i)$, $c_3 = (i+1, i+1, \ldots, i, i)$, $c_4 = (i, i+1, \ldots, i, i)$, for some $i \in \mathbb{Z}$.*

3 Digital Curves and Hypersurfaces

In what follows we consider digital analogs of simple closed curves and of hypersurfaces that separate the superspace $\mathbb{C}_n^{(n)}$. We will consider analogs of either bounded closed Jordan hypersurfaces or unbounded hypersurfaces (such as hyperplanes) that separate $\mathbb{R}^n$. (The latter can also be considered as "closed" in the infinite point.) We will not specify whether we consider closed or unbounded hypersurfaces whenever the definitions and results apply to both cases and no confusions arise. We also omit the word "digital" where possible.

The considerations take place in the n-dimensional space $\mathbb{C}_n^{(n)}$ of n-cells. We allow adjacency relations A_α as defined above. We are interested to establish basic definitions for this space that:

- reflect properties which are analogous to the topological connectivity of curves or hypersurfaces in Euclidean topology,
- reflect the one- or $(n-1)$-dimensionality of a curve or hypersurface, respectively, and
- characterize hypersurfaces with respect to gaps.

A digital curve (hypersurface), considered in the context of an adjacency relation A_α, will be called an α-curve (α-hypersurface).

3.1 Digital Curves

A set $\tau \subset \mathbb{C}_n^{(n)}$ is an α-curve iff it is α-connected and one-dimensional with respect to A_α. (Note that Urysohn-Menger curves in $\mathbb{R}^n$ are defined to be one-dimensional continua.) Figure 1 presents examples and counterexamples for $\mathbb{C}_2^{(2)}$.

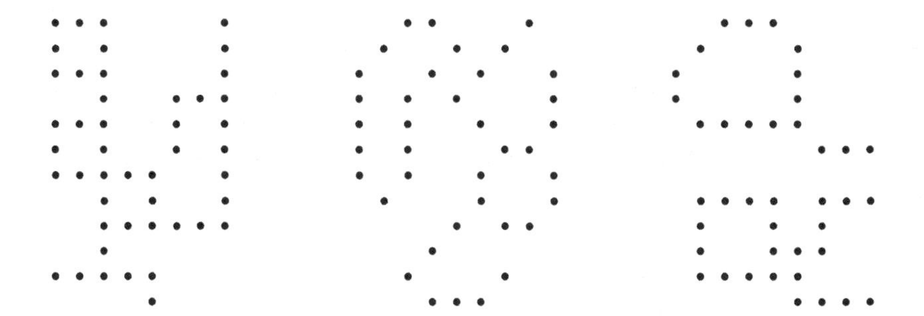

Fig. 1. Examples of a 1-curve (left), 0-curve (middle), and two 0-connected sets in the digital plane that are neither 0- nor 1-curves (right)

In the rest of this section we define and study digital analogs of simple closed curves (i.e., those that have branching index 2 at any point). The following lemma provides necessary and sufficient conditions for a set of n-cells to be connected and a loop with respect to adjacency relation A_α.

Lemma 2. *Let $\rho = \{c_1, c_2, \ldots, c_l\}$ be a set of n-cells. The following properties are equivalent:*

(A1) *c_i is α-adjacent to c_j iff $i = j \pm 1$(modulo l).*
(A2) *ρ is α-connected and $\forall\ c \in \rho$, $\text{card}(B^*_\alpha(c) \cap \rho) = 2$.*
(A3) *The α-adjacency graph $G^\alpha_\rho(V, E)$ is a simple loop.*

The following lemma provides conditions which are equivalent to the one-dimensionality of a set of n-cells.

Lemma 3. *Let $\rho = \{c_1, c_2, \ldots, c_l\}$ be a set of n-cells. The following properties are equivalent:*

(B1) *ρ is one-dimensional with respect to A_α.*
(B2) *If $0 \leq \alpha < n-1$, then ρ does not contain as a proper subset an elementary grid triangle such that any two of its n-cells are α-adjacent; if $\alpha = n-1$, then ρ does not contain as a proper subset an elementary grid square.*
(B3) *$\forall c \in \rho$, the set $B^*_\alpha(c) \cap \rho$ is totally disconnected.*

We list one more condition.
(B4) If $0 \leq \alpha < n-1$, then $l \geq 4$; if $\alpha = n-1$, then $l \geq 8$.

Lemma 4. *Let $\rho = \{c_1, c_2, \ldots, c_l\}$ be a set of n-cells. Then all property pairs ((Ai), (Bj)), for $1 \leq i \leq 3$ and $1 \leq j \leq 4$, are equivalent.*

Thus we are prepared to give the following general definition, summarizing twelve equivalent ways for defining a simple α-curve.

Definition 1. *A* simple α-curve $(0 \leq \alpha \leq n-1)$ *of length l is a set $\rho = \{c_1, c_2, \ldots, c_l\} \subseteq \mathbb{C}^{(n)}_n$, satisfying properties* (Ai) *and* (Bj)*, for some pair of indexes i, j, with $1 \leq i \leq 3$ and $1 \leq j \leq 4$.*

Note that for any $n \leq 2$, four n-cells whose centers form a 1×1 square do not form a digital curve, since such a set of cells would be two-dimensional.

A simple α-curve will also be called a *one-dimensional α-manifold.* In $\mathbb{C}^{(2)}_2$ we have the following:

Proposition 1. *A finite set ρ of pixels is a simple α-curve in $\mathbb{C}^{(2)}_2$ $(\alpha = 0, 1)$ iff it is α-minimal in $\mathbb{C}^{(2)}_2$.*

Note that this last result does not generalize to higher dimensions since a one-dimensional digital object cannot separate $\mathbb{C}^{(n)}_n$ if $n > 2$.

A simple α-curve ρ $(0 \leq \alpha < n-1)$ is a *proper* α-curve (or a proper one-dimensional α-manifold), if it is not an $(\alpha+1)$-curve.

Example 1. A proper 0-curve in $\mathbb{C}^{(2)}_2$ is a 0-curve which is not a 1-curve (see Figure 2, left) It follows that any closed 0-curve is a proper 0-curve.

A *proper* 0-curve in $\mathbb{C}^{(3)}_3$ is a 0-curve which is not a 1- or 2-curve, and a *proper* 1-curve is a 1-curve which is not a 2-curve.

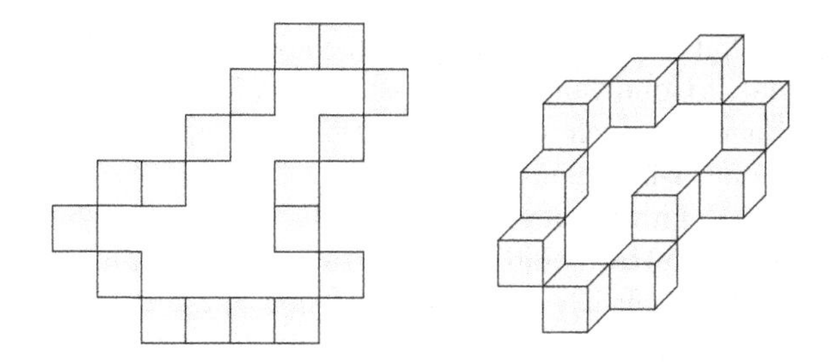

Fig. 2. A proper 0-curve in 2D (left) and an improper 0-curve in 3D (right)

Any 1-curve is a proper 1-curve. This follows from the facts that the curve is closed and one-dimensional with respect to 1-adjacency. If we assume the opposite, we will obtain that the curve is either an infinite sequence of voxels (e.g., of the form $(0,0,1),(0,0,2),(0,0,3),\ldots$) or that it is two-dimensional.

However, a closed 0-curve does not need to be proper (see Figure 2, right).

A *simple α-arc* σ is an α-connected proper subset of a simple α-curve. It contains exactly two n-cells c, c' such that $\mathrm{card}(B^*_\alpha(c)\cap\rho) = \mathrm{card}B^*_\alpha(c')\cap\rho) = 1$.

3.2 Digital Hypersurfaces

We consider digital analogs of hole-free hypersurfaces. Accordingly, we are interested in hypersurfaces without $(n-1)$-gaps, although the theory can be extended to cover this case, as well. However, in the framework of our approach, a hypersurface with $(n-1)$-gaps can be an $(n-2)$-dimensional set of n-cells, while we want a digital hypersurface to be $(n-1)$-dimensional, in conformity with the continuous case.

We give the following recursive definition.

Definition 2. (i) *M is a 1-dimensional* $(n-1)$-manifold *in $\mathbb{C}_n^{(n)}$ if it is an $(n-1)$-curve in $\mathbb{C}_n^{(n)}$.*
M is a k-dimensional $(2 \le k \le n-1)$ $(n-1)$-manifold *in $\mathbb{C}_n^{(n)}$ if*
(1) M is $(n-1)$-connected (or, equivalently, M consists of a single $(n-1)$-component), and
(2) for any $x \in M$ the set $B^*_0(x)\cap M$ is a $(k-1)$-dimensional $(n-1)$-manifold in $\mathbb{C}_n^{(n)}$.
(ii) *M is a k-dimensional* α-manifold $(0 \le \alpha \le n-2)$ *in $\mathbb{C}_n^{(n)}$ if*
(1) M is α-connected (or, equivalently, M consists of a single α-component), and
(2) for any $x \in M$ the set $B^*_0(x) \cap M$ is a $(k-1)$-dimensional α-manifold in $\mathbb{C}_n^{(n)}$ but is not a $(k-1)$-dimensional $(\alpha+1)$-manifold in $\mathbb{C}_n^{(n)}$.
(Such an α-manifold will also be called proper.*)*

In the particular case when S is an $(n-1)$-dimensional α-manifold in $\mathbb{C}_n^{(n)}$ for $\alpha = n-2$ or $n-1$, we say that S is a digital *α-hypersurface*. S is a proper α-hypersurface for $\alpha = n-2$ if it is not an $(n-1)$-hypersurface for $\alpha = n-1$.

It is also clear that any proper one-dimensional α-manifold is an α-curve.

We remark that if Condition (1) is missing, then S may have more than one connected component. In such a case Condition (2) implies that any connected component of S is an α-hypersurface.

Note that in the definition of an α-hypersurface we use the ball $B_0^*(x)$ rather than $B_\alpha^*(x)$, since the latter could cause certain incompatibilities. This can be easily seen in the 3D case: if we use B_2^* to define a 2-surface, $B_2^*(x) \cap S$ may be a 1-curve rather than a 2-curve. Similarly, if we use B_1^* to define a 1-surface, $B_1^*(x) \cap S$ may be a 0-curve rather than a 1-curve. This is avoided by using B_0^* in all cases.

The so defined digital hypersurfaces have the following properties, among others.

Proposition 2. *An α-hypersurface S is $(n-1)$-dimensional at any n-cell in S with respect to adjacency relation A_α.*

Proposition 3. *(a) An $(n-2)$-hypersurface S has $(n-2)$-gaps and is $(n-1)$-gapfree. Moreover, it is $(n-1)$-minimal.*

(b) An $(n-1)$-hypersurface S is $(n-2)$-gapfree. Note that S may have or may not have k-gaps for $0 \leq k \leq n-3$. If S has k-gaps but no $(k+1)$-gaps for $0 \leq k \leq n-3$, then S is $(k+1)$-minimal. If S is k-gapfree for $0 \leq k \leq n-3$, then S is 0-minimal.

Part (b) of this last proposition suggests to distinguish $n-1$ types of hypersurfaces: those that are 0-gapfree, those with 0-gaps but with no 1-gaps, etc., up to those with $(n-2)$-gaps but with no $(n-1)$-gaps.

We label them as $(n-1)_{(0)}$, $(n-1)_{(1)}$,...,$(n-1)_{(n-2)}$-hypersurfaces.

In fact, the concept of minimality itself can provide a complete characterization of a digital hypersurface, as follows.

Definition 3. *A set $S \subset \mathbb{C}_n^{(n)}$ is a k^*-*hypersurface*, for $k = 0, 1, 2, \ldots, n-1$, if S is k-minimal in $\mathbb{C}_n$.*

Theorem 1. *(a) S is an $(n-1)_{(i)}$-hypersurface $(0 \leq i \leq n-2)$ iff it is an i^*-hypersurface.*

(b) S is an $(n-2)$-hypersurface iff it is an $(n-1)^*$-hypersurface.

We remark that a k^*-hypersurface cannot have "singularities," which may appear, e.g., in case of a 3D "pinched sphere" or a "strangled torus." In fact, surfaces of that kind would either be non-simple or three-dimensional or both, so they would not satisfy our definition of surface.

In summary, we have two types of hypersurfaces: $(n-1)$ and $(n-2)$ hypersurfaces, as the $(n-1)$ hypersurfaces can be classified (with respect to their gaps) as $(n-1)$-hypersurfaces of types $0, 1, 2, \ldots, n-2$, respectively.

Indeed, one can consider more general digital hypersurfaces which are not covered by the above definitions. If, for instance, we do not require in Definition

2 the manifold $B_0^*(x) \cap S$ to be proper, we may have a hypersurface where subsets can be of varying hypersurface type. We are interested (see next section) in combinatorial properties of the considered hypersurfaces. More general digital hypersurfaces would be just "mixtures" of patches of hypersurfaces of some of the considered types, and their combinatorial study would lose its focus.

Now let Γ be a closed surface in $\mathbb{R}^n$ and $J^+(\Gamma)$ its outer Jordan digitization.

Definition 4. *Let $\mathcal{D}_k(\Gamma)$ be the family of all subsets of $J^+(\Gamma)$ that are k-minimal, for some $0 \leq k \leq n-1$. We call a set of n-cells $D_k(\Gamma) \in \mathcal{D}_k(\Gamma)$ a* k-digitization *of Γ if the Hausdorff distance $H_d(\Gamma, V(D_k(\Gamma)))$ is minimal, over all the elements of $\mathcal{D}_k(\Gamma)$.*

Proposition 4. *Any k^*-hypersurface is a k-digitization of certain hypersurface Γ.*

Examples of k-digitization (and thus of k^*-hypersurfaces) were actually already provided by the following theorem from [4]:

Theorem 2. *A digital hyperplane P^k defined by $P^k = P^k(b, a_1, a_2, \ldots, a_n, \omega) = \{x \in \mathbb{Z}^n | -\frac{\omega}{2} \leq b + \sum_{i=1}^n a_i x_i < \frac{\omega}{2}\}$, where $\omega = \sum_{i=k+1}^n a_i$, $0 \leq k \leq n-1$, is a k-digitization of the hyperplane $\gamma : b + a_1x_1 + a_2x_2 + \ldots + a_nx_n = 0$.*

The above result is related to a theorem from [2] that characterizes the gaps of analytically defined digital hyperplanes. Specifically, let P^k be a digital hyperplane as in Theorem 2. If $\omega < a_n$, then P^k has $(n-1)$-gaps; for $0 < k < n$, if $\sum_{i=k+1}^n a_i \leq \omega < \sum_{i=k}^n a_i$, then P^k has $(k-1)$-gaps and is k-separating; and if $\omega \geq \sum_{i=1}^n a_i$, P^k is gapfree.

3.3 Hypersurface Digitization Matroid

In this section we briefly investigate the structure of digital hypersurfaces from a combinatorial point of view.

Let E be a finite set and $\mathcal{F}$ a family of subsets of E. Recall that $(E, \mathcal{F})$ is a *matroid*[5] if the following axioms are satisfied:

(1) $\emptyset \in \mathcal{F}$,

(2) if $F_2 \in \mathcal{F}$ and $F_1 \subseteq F_2$, then $F_1 \in \mathcal{F}$,

(3) if $F_1, F_2 \in \mathcal{F}$ and $\text{card}(F_1) < \text{card}(F_2)$, then there is an element $x \in F_2$ such that $F_1 \cup \{x\} \in \mathcal{F}$.

The last condition can be substituted by the following:

(3′) all maximal elements of $\mathcal{F}$ have the same cardinality.

Theorem 3. *For a given k, $0 \leq k \leq n-1$, all k-digitizations of a closed hypersurface Γ and their subsets form a matroid.*

[5] For getting acquainted with matroid theory the reader is referred to the monograph by Welsh [32].

We call it the *hypersurface digitization matroid.* This theorem demonstrates in particular the possibility to generate closed digital hypersurfaces by greedy-type algorithms.

4 Good Pairs

As already mentioned, studies on digital surfaces naturally interfere with studies on good pairs of adjacency relations. An important motivation for studying good pairs is seen in the possibility that some results of digital topology may hold uniformly for several pairs of adjacency relations. Thus one could obtain a proof which is valid for all of them by proving a statement just for a single good pair of adjacencies.

4.1 Variations of the Notion "Good Pair"

Different approaches in the literature lead to diverse proposals of good pairs (note: they may be called differently, but address the same basic concept). It seems to be unrealistic to define good pairs in such a way that this will cover all previous studies. Therefore, instead of looking for a universal definition, it might be more reasonable and useful to propose and study a number of definitions related to the fundamental concepts of digital topology. The rest of this section reviews several possible approaches.

Good pairs in terms of strictly normal digital picture spaces have been considered in [17]. In that framework, it is shown that adjacencies (1,0) and (0,1) in 2D, and (2,0), (0,2), (2,1) and (1,2) in 3D define strictly normal digital picture spaces, while (1,1) and (0,0) in 2D and (2,2), (1,1), (0,0), (1,0) and (0,1) in 3D do not.

In [14] good pairs have been defined for 2D as follows: (β_1, β_2) is called a *good pair* in the 2D grid iff (for $(i, k) \in \{(1, 2), (2, 1)\}$) any simple β_i-curve β_k-separates its (at least one) β_k-holes from the background and any totally β_i-disconnected set cannot β_k-separate any β_k-hole from the background. It follows that (1,0) and (0,1) are good pairs, but (1,1) and (0,0) are not. [14] does not generalize this definition to higher dimensions, but suggests the use of (α, β)-separators for the case $n = 3$. ($M \subseteq \mathbb{Z}^3$ is called an *(α, β)-separator* iff M is α-connected, M divides $\mathbb{Z}^3 \setminus M$ into (exactly) two β-components, and there exists a $p \in M$ such that $\mathbb{Z}^3 \setminus (M \setminus \{p\}) = (\mathbb{Z}^3 \setminus M) \cup \{p\}$ is β-connected.) (α, β)-and (β, α)-separators exist for (α, β) = (0,2),(2,0), (1,2), (2,1), and (1,1). However, there are some difficulties with the case $(\alpha, \beta) = (1, 1)$, as an example from [14] illustrates. Further "strange" examples of separators in $\mathbb{Z}^3$ suggest to refine this notion.

Another approach is based on the following digital variant of the Jordan curve theorem due to A. Rosenfeld [26].

Theorem 4. *If C is the set of points of a simple closed 1-curve (0-curve) and* $\text{card}(C) > 4$ $(\text{card}(C) > 3)$, *then $\overline{C}$ has exactly two 0-components (1-components).*

This theorem defines good pairs of adjacency relations in 2D, as follows. (α, β) *is a* 2D good pair *if for a simple closed α-curve C, $\overline{C}$ has exactly two β-components.* It follows that $(1, 0)$ and $(0, 1)$ are good pairs. It is also easy to see that $(1, 1)$ and $(0, 0)$ are not good pairs.

This above definition can be extended to 3D, as follows: (α, β) *is a* 3D good pair *if for a simple closed α-surface S, $\overline{S}$ has exactly two β-components.* We remark that in view of the definition of an α-surface from Section 3.2, a 0-digital surface would not be a true surface and should not be called "surface" since it would have 2-gaps. In fact, 3D digital surfaces need to be at least 1-connected. Thus (0,2) would not be a good pair in the framework of the Jordan surface theorem approach.

Another approach is based on separation through surfaces (see, e.g., [11, 14]).

Theorem 5. *A simple closed 1-curve (0-curve) γ 0-separates (1-separates) all pixels inside γ from all pixels outside γ. More precisely, we have that a simple closed 1-curve has exactly one 0-hole and a simple closed 0-curve has exactly one 1-hole. A simple closed 1-curve 0-separates its 0-hole from the background and a simple closed 0-curve 1-separates its 1-hole from the background.*

Based on this last theorem, we can give the following definition: (α, β) *is called a* 2D good pair *if any simple closed α-curve β-separates its β-holes from the background.*

Clearly, $(1, 0)$ and $(0, 1)$ are good pairs, while (0,0) is not a good pair. Note that here also (1,1) is a good pair, as distinct from the case of good pairs defined trough Jordan curve theorem.

Let us mention that in a definition from [14] both (α, β) and (β, α) are required to satisfy the conditions of a good pair. To avoid confusion, we suggest to treat this case as a special event: (α, β) *is called a* perfect pair *in 2D if any simple closed α-curve β-separates its β-holes from the background and any simple closed β-curve α-separates its α-holes from the background.*

In what follows we consider good pairs defined by this last approach that seems the most reasonable to the authors.

4.2 Good Pairs for the Space of n-Cells

Definition 5. (α, β) *is called a* good pair *of adjacency relations in $\mathbb{C}_n^{(n)}$ if any closed α-hypersurface β-separates its β-holes from the background.*

Here α is a (possibly composite) label of the hypersurface type in accordance with our hypersurface classification above, while β is an integer representing an adjacency. More precisely, $\alpha = (n-1)_{(i)}$, $0 \leq i \leq n-2$ or $\alpha = n-2$, and $0 \leq \beta \leq n-1$.

Theorem 6. *There are $2n - 1$ good pairs in the n-dimensional digital space: $((n-1)_{(i)}, i)$ for $0 \leq i \leq n-2$, $((n-1)_{(i)}, n-1)$ for $0 \leq i \leq n-2$, and*

$(n-2, n-1)$, where the first component of such a pair labels the type of the hypersurface and the second is an adjacency relation.

Alternatively, the good pairs are (i^*, i) for $0 \leq i \leq n-1$, and $(i^*, n-1)$ for $0 \leq i \leq n-2$.

Note that in a pair of the form $((n-1)_{(i)}, i)$, $0 \leq i \leq n-1$ the first component also specifies an adjacency relation corresponding to the type of the hyperplane.

We illustrate the last theorem for $n = 2$ and $n = 3$. For $n = 2$, the good pairs are $(1_{(0)}, 0)$, $(1_{(0)}, 1)$, and $(0^*, 1)$, which correspond to (1,0), (1,1), and (0,1), respectively. See Figure 3.

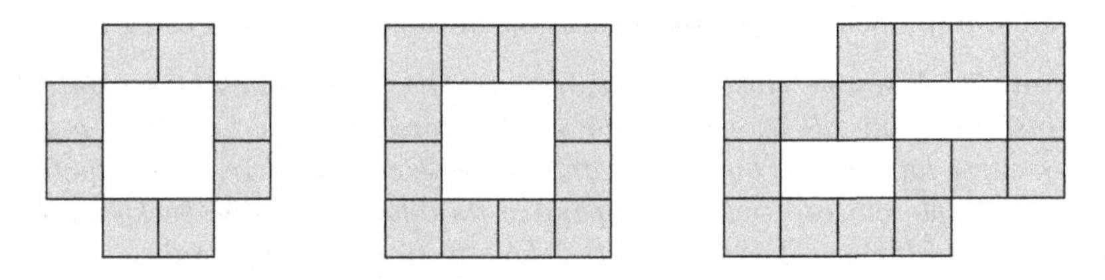

Fig. 3. Illustration to good pairs in 2D: $(0, 1)$ (left), $(1, 1)$ (middle), and $(1, 0)$ (right)

For $n = 3$ the good pairs are $(2_{(0)}, 0)$, $(2_{(1)}, 1)$, $(2_{(0)}, 2)$, $(2_{(1)}, 2)$, and $(1^*, 2)$, which correspond to (2,0), (2,1), (2,2), (2,2), and (1,2), respectively. Note that pair (2,2) is counted twice since there are two different structures corresponding to it.

5 Concluding Remarks

In this paper we proposed several equivalent definitions of digital curves and hypersurfaces in arbitrary dimension. The definitions involve properties (such as one-dimensionality of curves and $(n-1)$-dimensionality of hypersurfaces) that characterize them to be digital analogs of definitions for Euclidean spaces. Further research may pursue designing efficient algorithms for recognizing whether a given set of n-cells is a digital curve or hypersurface.

We also proposed a uniform approach to studying good pairs defined by separation and, in that framework, obtained a classification of good pairs in arbitrary dimension. A future task is seen in extending the obtained results under other reasonable definitions of good pairs.

Acknowledgements

The authors thank the three anonymous referees for their useful remarks and suggestions.

References

1. J.C. Alexander and A.I. Thaler. The boundary count of digital pictures. *J. ACM*, **18**:105–112, 1971.
2. E. Andres, R. Acharya, and C. Sibata. Discrete analytical hyperplanes. *Graphical Models Image Processing*, **59**:302–309, 1997.
3. G. Bertrand and R. Malgouyres. Some topological properties of surfaces in $\mathbb{Z}^3$. *J. Mathematical Imaging Vision*, **11**:207–221, 1999.
4. V.E. Brimkov, E. Andres, and R.P. Barneva. Object discretizations in higher dimensions. *Pattern Recognition Letters*, **23**:623–636, 2002.
5. L. Chen, D.H. Cooley, and J. Zhang. The equivalence between two definitions of digital surfaces. *Information Sciences*, **115**:201–220, 1999.
6. D. Cohen-Or, A. Kaufman, and T.Y. Kong. On the soundness of surface voxelizations. In: T.Y. Kong and A. Rosenfeld, editors. *Topological Algorithms for Digital Image Processing*, 181–204. Elsevier, Amsterdam, The Netherlands, 1996.
7. M. Couprie and G. Bertrand. Tessellations by connection. *Pattern Recognition Letters*, **23**:637–647, 2002.
8. R.O. Duda, P.E. Hart, and J.H. Munson. Graphical-data-processing research study and experimental investigation. TR ECOM-01901-26, Stanford Research Institute, Menlo Park, California, March 1967.
9. U. Eckhardt and L. Latecki. Topologies for the digital spaces $\mathbb{Z}^2$ and $\mathbb{Z}^3$. *Computer Vision Image Understanding*, **90**:295–312, 2003.
10. J. Françon. Discrete combinatorial surfaces. *Graphical Models Image Processing*, **57**:20–26, 1995.
11. G.T. Herman. Boundaries in digital spaces: Basic theory. In: T.Y. Kong and A. Rosenfeld, editors. *Topological Algorithms for Digital Image Processing*, 233–261. Elsevier, Amsterdam, The Netherlands, 1996.
12. Y. Kenmochi, A. Imiya, and A. Ichikawa. Discrete combinatorial geometry. *Pattern Recognition*, **30**:1719–1728, 1997.
13. C.E. Kim. Three-dimensional digital line segments. *IEEE Trans. Pattern Analysis Machine Intelligence*, **5**:231–234, 1983.
14. R. Klette and A. Rosenfeld. *Digital Geometry - Geometric Methods for Digital Picture Analysis*. Morgan Kaufmann, San Francisco, 2004.
15. T.Y. Kong. Digital topology. In: L.S. Davis, editor. *Foundations of Image Understanding*, 33–71. Kluwer, Boston, Massachusetts, 2001.
16. T.Y. Kong. Topological adjacency relations on $\mathbb{Z}^n$. *Theoretical Computer Science*, **283**:3–28, 2002.
17. T.Y. Kong, A.W. Roscoe, and A. Rosenfeld. Concepts of digital topology. *Topology and its Applications*, **46**:219–262, 1992.
18. R. Kopperman, P.R. Meyer, and R. Wilson. A Jordan surface theorem for three-dimensional digital spaces. *Discrete Computational Geometry*, **6**:155–161, 1991.
19. V. Kovalevsky. Multidimensional cell lists for investigating 3-manifolds. *Discrete Applied Mathematics*, **125**:25–44, 2003.
20. J.-O. Lachaud and A. Montanvert. Continuous analogs of digital boundaries: A topological approach to isosurfaces. *Graphical Models*, **62**:129–164, 2000.
21. R. Malgouyres. A definition of surfaces of $\mathbb{Z}^3$: A new 3D discrete Jordan theorem. *Theoretical Computer Science*, **186**:1–41, 1997.
22. D.G. Morgenthaler and A. Rosenfeld. Surfaces in three-dimensional digital images. *Information Control*, **51**:227–247, 1981.

23. J.P. Mylopoulos and T. Pavlidis. On the topological properties of quantized spaces. I. The notion of dimension. *J. ACM*, **18**:239–246, 1971.
24. J.-P. Reveillès. Géométrie discrète, calcul en nombres entiers et algorithmique. Thèse d'état, Université Louis Pasteur, Strasbourg, France, 1991.
25. A. Rosenfeld. Adjacency in digital pictures, *Information and Control* **26**, 24–33, 1974
26. A. Rosenfeld. Compact figures in digital pictures. *IEEE Trans. Systems, Man, Cybernetics,* **4**:221–223, 1974.
27. A. Rosenfeld, T.Y. Kong, and A.Y. Wu. Digital surfaces. *CVGIP: Graphical Models Image Processing*, **53**:305–312, 1991.
28. A. Rosenfeld and J.L. Pfaltz. Sequential operations in digital picture processing. *J. ACM*, **13**:471–494, 1966.
29. G. Tourlakis. Homological methods for the classification of discrete Euclidean structures. *SIAM J. Applied Mathematics*, **33**:51–54, 1977.
30. G. Tourlakis and J. Mylopoulos. Some results in computational topology. *J. ACM*, **20**:430–455, 1973.
31. J.K. Udupa. Connected, oriented, closed boundaries in digital spaces: Theory and algorithms. In: T.Y. Kong and A. Rosenfeld, editors. *Topological Algorithms for Digital Image Processing*, 205–231. Elsevier, Amsterdam, The Netherlands, 1996.
32. D.J.A. Welsh. *Matroid Theory.* Academic Press, London, 1976.

A Maximum Set of (26,6)-Connected Digital Surfaces*

J.C. Ciria[1], A. De Miguel[1], E. Domínguez[1], A.R. Francés[1], and A. Quintero[2]

[1] Dpt. de Informática e Ingeniería de Sistemas, Facultad de Ciencias,
Universidad de Zaragoza, E-50009 – Zaragoza, Spain
{jcciria, admiguel, afrances}@posta.unizar.es
[2] Dpt. de Geometría y Topología, Facultad de Matemáticas, Universidad de Sevilla,
Apto. 1160, E-41080 – Sevilla, Spain
quintero@us.es

Abstract. In the class $\mathcal{H}$ of $(26,6)$–connected homogeneous digital spaces on R^3 we find a digital space E^U with the largest set of digital surfaces in that class. That is, if a digital objet S is a digital surface in any space $E \in \mathcal{H}$ then S is a digital surface in E^U too.

1 Introduction

In the graph-theoretical approach to Digital Topology there is a common agreement about the notion of digital curve: a subset S is a *simple closed curve* if each $p \in S$ is adjacent to exactly two other pixels in S. However, it is not yet well established a general notion of digital surface that naturally extends to higher dimensions and such that these digital objects have properties similar to those held by topological surfaces. Although, in our opinion, this ambitious goal is far to be reached, several relevant contributions can be found in the literature, some of them are quoted next.

In [10], Morgenthaler and Rosenfeld gave for the first time a definition of digital surface in the discrete space $\mathbb{Z}^3$ provided with the usual adjacency pairs $(6,26)$ and $(26,6)$, and showed a discrete version of the Jordan-Brouwer Theorem for these objects. Later on, Kong and Roscoe [7] generalized this result to any adjacency pair $(n,m) \neq (6,6)$, where $\{n,m\} \subseteq \{6,18,26\}$. More recently, Bertrand and Malgouyres [4] and Couprie and Bertrand [6] have defined new families of digital surfaces called *strong n-surfaces*, $n \in \{18,26\}$, and *simplicity n-surfaces*, $n \in \{6,26\}$, respectively. A digital Jordan-Brouwer Theorem for simplycity 26-surfaces can be found in [5].

Despite of these contributions, the most general definition of digital surface is not yet clear, even if we restrict the problem to the $(26,6)$-adjacency. Actually, in [4] and [6] it is shown that the set of Morgenthaler's $(26,6)$-surfaces is a subset

* This work has been partially supported by the projects BFM2001-3195-C03-01 and BFM2001-3195-C03-02 (MCYT Spain).

R. Klette and J. Žunić (Eds.): IWCIA 2004, LNCS 3322, pp. 291–306, 2004.

of the strong 26-surfaces and these are, in turn, strictly contained in the set of simplicity 26-surfaces, and furthermore, in [8], a 26-connected digital object (see Fig. 5(a)) which is not a simplicity 26-surface is still suggested as a new kind of digital surface. On the other hand, these notions do not seem to generalize easily to higher dimensions.

In [3] we propose a different framework for Digital Topology in which a notion of digital surface and, more generally, of digital manifold naturally arises. In this approach a digital space is defined as a pair (K, f), where K is a *device model* used to represent digital images in a discrete setting and f is a *lighting function* that associates to each digital image an Euclidean polyhedron, called its *continuous analogue*. Continuous analogues intend to formalize the idea of "continuous perception" that an observer may take on a digital image. In this way, a digital object S is naturally said to be a digital surface if its "looks like" as a surface; that is if the continuous analogue of S is a (combinatorial) surface. Moreover, for digital spaces satisfying suitable conditions, digital surfaces exhibit properties similar to those of topological surfaces. For example, in [3] and [2] digital versions of the Jordan-Brouwer and Index Theorems are shown for digital manifolds of arbitrary dimension.

This notion of digital surface is closely related to the other definitions quoted previously. Actually, for each one of the families of surfaces mentioned above it is found [1,2,5] a particular digital space (R^3, f) resembling the $(26,6)$-connectedness and whose set of digital surfaces coincides or contains the corresponding family, where the device model R^3 is the standard decomposition of the Euclidean space $\mathbb{R}^3$ by unit cubes, which is canonically identified to $\mathbb{Z}^3$.

The construction of continuous analogues from a lighting function makes use of elements which are not pixels. This way, and despite of its good properties, our notion of digital surface might not be considered completely digital. The relationship quoted above with other definitions of surfaces shows that it is possible to find a purely digital characterization, in terms of adjacency, for our surfaces in some digital spaces. However, to obtain such a characterization for all possible $(26,6)$-connected digital surfaces would require to analyse, case by case, all digital spaces resembling the $(26,6)$-connectedness. In this paper we show that such an enormous effort is not necessary. Instead it will suffice to study the surfaces of just one "universal" digital space. More precisely, we show that there exists a homogeneous $(26,6)$-connected digital space $E^U = (R^3, f^U)$ whose family of digital surfaces is the largest in that class of digital spaces. Moreover, as far as we know, that family strictly contains any set of surfaces defined on $\mathbb{Z}^3$ using the graph-based approach in the literature. We also announce that we have obtained the required characterization in terms of the $(26,6)$-adjacency for this set of surfaces, which will be the subject of a future paper.

This paper is organized as follows. Section 1 reviews the basic notions of our framework for Digital Topology. In Section 2, some values of the lighting function f on arbitrary digital objects in an homogeneous $(26,6)$-connected digital space (R^3, f) are computed and, moreover, the values of f for any digital surface S in (R^3, f) are characterized in Section 3. Finally, in Section 4 we use this characterization to give the universal space $E^U = (R^3, f^U)$ that contains the surfaces of any homogeneous (26, 6)-connected digital space (R^3, f).

2 Preliminaries

In our approach to Digital Topology [3] we propose a multilevel architecture which, using different levels, allows us to represent digital images in various settings. However, for simplicity, we introduce in this section only the levels which are explicitly used in this paper: the *device model* for representing the spatial layout of pixels and the *continuous level* where a continuous interpretation for each digital image is found.

In this paper we only use the device model R^n, termed the *standard cubical decomposition* of the Euclidean n-space $\mathbb{R}^n$, which is the polyhedral complex determined by the collection of unit n-cubes in $\mathbb{R}^n$ whose edges are parallel to the coordinate axes and whose centers are in the set $\mathbb{Z}^n$. Each n-cell in R^n is representing a pixel, and so the digital object displayed in a digital image is a subset of the set $\text{cell}_n(R^n)$ of n-cells in R^n; while the other lower dimensional cells in R^n (actually, k-cubes, $0 \le k < n$) are used to describe how the pixels could be linked to each other.

Remark 1. Each k-cell $\sigma \in R^n$ can be associated to its center $c(\sigma)$ which is a point in the set $\mathcal{Z}^n$, where $\mathcal{Z} = \frac{1}{2}\mathbb{Z} = \{x \in \mathbb{R} \mid x = z/2, z \in \mathbb{Z}\}$. If $\dim \sigma = n$ then $c(\sigma) \in \mathbb{Z}^n$, so that every digital object O in R^n can be naturally identified with a subset of points in $\mathbb{Z}^n$. Henceforth we shall use this identification without further comment.

As it is usual in Polyhedral Topology, given two cells $\gamma, \sigma \in R^n$ we write $\gamma \le \sigma$ if γ is a face of σ, and $\gamma < \sigma$ if in addition $\gamma \ne \sigma$. The interior of a cell σ is the set $\mathring{\sigma} = \sigma - \partial\sigma$, where $\partial\sigma = \cup\{\gamma \mid \gamma < \sigma\}$ stands for the boundary of σ. We refer to [9, 11] for further notions on polyhedral topology.

In order to associate a continuous representation to each digital object we use what is called a lighting function. To introduce this kind of functions we need the following definitions.

Given a cell $\alpha \in R^n$ and a digital object $O \subseteq \text{cell}_n(R^n)$ the *star of* α *in* O is the set $\text{st}_n(\alpha; O) = \{\sigma \in O \mid \alpha \le \sigma\}$ of n-cells (pixels) in O having α as a face. Similarly, the *extended star of* α *in* O is the set $\text{st}^*_n(\alpha; O) = \{\sigma \in O \mid \alpha \cap \sigma \ne \emptyset\}$ of n-cells (pixels) in O intersecting α. Finally, the *support* of O is the set $\text{supp}(O)$ of cells of R^n (not necessarily pixels) that are the intersection of n-cells (pixels) in O; that is, $\alpha \in \text{supp}(O)$ if and only if $\alpha = \cap\{\sigma \mid \sigma \in \text{st}_n(\alpha; O)\}$. In particular, if α is a pixel in O then $\alpha \in \text{supp}(O)$. To ease the writing, we shall use the following notation: $\text{supp}(R^n) = \text{supp}(\text{cell}_n(R^n))$, $\text{st}_n(\alpha; R^n) = \text{st}_n(\alpha; \text{cell}_n(R^n))$ and $\text{st}^*_n(\alpha; R^n) = \text{st}^*_n(\alpha; \text{cell}_n(R^n))$. Finally, we shall write $\mathcal{P}(A)$ for the family of all subsets of a given set A.

A *lighting function* on the device model R^n is a map $f : \mathcal{P}(\text{cell}_n(R^n)) \times R^n \to \{0, 1\}$ satisfying the following five axioms for all $O \in \mathcal{P}(\text{cell}_n(R^n))$ and $\alpha \in R^n$:

(1) *object axiom*: if $\alpha \in O$ then $f(O,\alpha) = 1$;
(2) *support axiom*: if $\alpha \notin \mathrm{supp}(O)$ then $f(O,\alpha) = 0$;
(3) *weak monotone axiom*: $f(O,\alpha) \leq f(\mathrm{cell}_n(R^n),\alpha)$;
(4) *weak local axiom*: $f(O,\alpha) = f(\mathrm{st}_n^*(\alpha;O),\alpha)$; and,
(5) *complement connectivity axiom*: if $O' \subseteq O \subseteq \mathrm{cell}_n(R^n)$ and $\alpha \in K$ are such that $\mathrm{st}_n(\alpha;O) = \mathrm{st}_n(\alpha;O')$, $f(O',\alpha) = 0$ and $f(O,\alpha) = 1$, then the set $\alpha(O',O) = \cup\{\mathring{\omega} \mid \omega < \alpha, f(O',\omega) = 0,\ f(O,\omega) = 1\} \subseteq \partial\alpha$ is non-empty and connected.

If $f(O,\alpha) = 1$ we say that f *lights* the cell α for the object O, otherwise f vanishes on α for O.

Remark 2. a) For any digital object $O \subseteq \mathrm{cell}_n(R^n)$ and any n-cell $\alpha \in R^n$, Axioms 1 and 2 imply that $\alpha \in O$ if and only if $f(O,\alpha) = 1$.

b) Notice that Axiom 5 is equivalent to requiring that if $\mathrm{st}_n(\alpha;O) = \mathrm{st}_n(\alpha;O')$, $f(O,\alpha) = 1$ and $\alpha(O',O)$ is empty or non-connected then necessarily $f(O',\alpha) = 1$. As a consequence, if $O_1 \subseteq O_2 \subseteq \cdots \subseteq O_k$ is a sequence of objects with $\mathrm{st}_n(\alpha;O_1) = \mathrm{st}_n(\alpha;O_k)$, $f(O_k,\alpha) = 1$ and all the sets $\alpha(O_i,O_{i+1})$, $1 \leq i \leq k-1$, are empty or non-connected then $f(O_i,\alpha) = 1$ for all i.

Given a lighting function f on R^n, at the continuous level of our architecture we define the *continuous analogue* of a digital object $O \subseteq \mathrm{cell}_n(R^n)$ as the underlying polyhedron $|\mathcal{A}_O^f|$ of the simplicial complex $\mathcal{A}_O^f$, whose k-simplexes are $\langle c(\alpha_0), c(\alpha_1), \ldots, c(\alpha_k)\rangle$ where $\alpha_0 < \alpha_1 < \cdots < \alpha_k$ are cells in R^n such that $f(O,\alpha_i) = 1$, $0 \leq i \leq k$. The complex $\mathcal{A}_O^f$ is called the *simplicial analogue* of O. Notice that the simplicial analogue $\mathcal{A}_O^f$ is always a full subcomplex of the barycentric subdivision of the standard cubical decomposition complex R^n. Moreover, according to Remark 2(a), the center $c(\sigma)$ of an n-cell $\sigma \in \mathrm{cell}_n(R^n)$ is a 0-simplex of $\mathcal{A}_O^f$ if and only if σ belongs to the digital object O.

For the sake of simplicity, we will usually drop "f" from the notation of the levels of an object. Moreover, for the whole object $\mathrm{cell}_n(R^n)$ we will simply write $\mathcal{A}_{R^n}$ for its simplicial analogue.

A *digital space* is a pair (R^n, f) where f is a lighting function on the device model R^n; that is, a model for representing digital images, R^n, together with a continuous interpretation of each digital image provided by f. See [3] for a definition of digital spaces on more general device models.

As continuous analogues intend to be the "continuous interpretation" of digital images, it is natural to introduce digital notions in terms of the corresponding continuous ones. For example, we will say that an object O is *connected* if its continuous analogue $|\mathcal{A}_O|$ is a connected polyhedron. Similarly, an object O is called a *m-dimensional digital manifold* if $|\mathcal{A}_O|$ is a combinatorial m-manifold without boundary; that is, we call O a *digital surface* (2-manifold) if it looks like as a surface. More precisely, O is a digital surface if for each vertex $v \in \mathcal{A}_O$ its link $\mathrm{lk}(v;\mathcal{A}_O) = \{A \in \mathcal{A}_O \mid v, A < B \in \mathcal{A}_O \text{ and } v \notin A\}$ is a 1-sphere. See [11] for the general definition of combinatorial manifold.

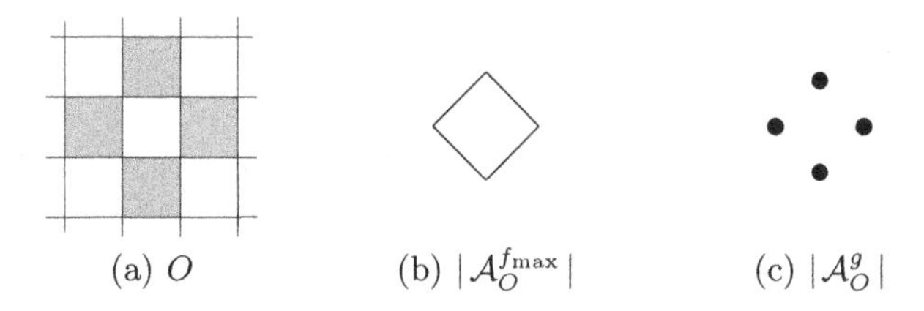

Fig. 1. (a) A digital object $O \subseteq \text{cell}_2(R^2)$ whose continuous analogues in the digital spaces (b) $(R^2, f_{\max})$ and (c) (R^2, g) are a curve and a disconnected set

Example 1. The following are lighting functions on the device model R^2:

(a) $f_{\max}(O, \alpha) = 1$ if and only if $\alpha \in \text{supp}(O)$
(b) $g(O, \alpha) = 1$ if and only if $\alpha \in \text{supp}(O)$ and $\text{st}_2(\alpha; R^2) \subseteq O$
(c) $h(O, \alpha) = f_{\max}(O, \alpha)$ if $c(\alpha) = (x, y) \in \mathcal{Z}^2$ and $x \geq 0$, and $h(O, \alpha) = g(O, \alpha)$ otherwise.

Figures 1(b) and (c) show the continuous analogues that the lighting functions $f_{\max}$ and g, respectively, associate to the object $O \subseteq \text{cell}_2(R^2)$ depicted in Fig. 1(a). According to the previous definitions, O should be considered as a connected curve (1-manifold) in the digital space $(R^2, f_{\max})$ while it is a non connected object in (R^2, g).

It can be readily checked that the functions $f_{\max}$ and g, and hence the corresponding continuous analogues, are invariant under plane motions while the function h does not. This difference suggests the following definition. A digital space (R^n, f) is said to be *homogeneous* if for any spatial motion $\varphi : \mathbb{R}^n \to \mathbb{R}^n$ preserving $\mathbb{Z}^n$ the equality $f(\varphi(O), \varphi(\alpha)) = f(O, \alpha)$ holds for all cells $\alpha \in R^n$ and digital objects $O \subseteq \text{cell}_n(R^n)$. According to Axiom 4, if (R^n, f) is homogeneous, a minimal family of objects, called *canonical patterns*, suffices to determine f. Actually, $f(O, \alpha) = f(\text{patt}(O, \alpha), \alpha_k)$ for any object O and any k-cell $\alpha \in R^n$, $0 \leq k < n$, where α_k is a fixed but arbitrary k-cell and the object $\text{patt}(O, \alpha) \subseteq \text{st}_n^*(\alpha_k; R^n)$, called the *pattern of O in α*, is the unique canonical pattern P for which there exists a spatial motion $\mathbb{R}^3 \to \mathbb{R}^3$ preserving $\mathbb{Z}^3$ that carries $\text{st}_n^*(\alpha; O)$ to P.

Remark 3. Fig. 2 shows the canonical patterns around a vertex α_0 for the device model R^3. Black dots in this figure are actually representing the 3-cells of each canonical pattern $P \subseteq \text{st}_3(\alpha_0; R^3)$, according to the identification in Remark 1, and the vertex α_0 itself corresponds to the center of the cubes. Through this paper we will consider de following subfamilies of canonical patterns:

$\mathfrak{A} = \{A_0, A_1, A_2^a, A_2^b, A_3^a, A_4^a\}$
$\mathfrak{B} = \{B_2^c, B_5^c, B_6^b, B_8\}$
$\mathfrak{C} = \{C_3^c, C_4^b, C_4^e, C_4^f, C_5^b, C_6^c\}$
$\mathfrak{D} = \{D_3^b, D_4^c, D_4^d, D_5^a, D_6^a, D_7\}$

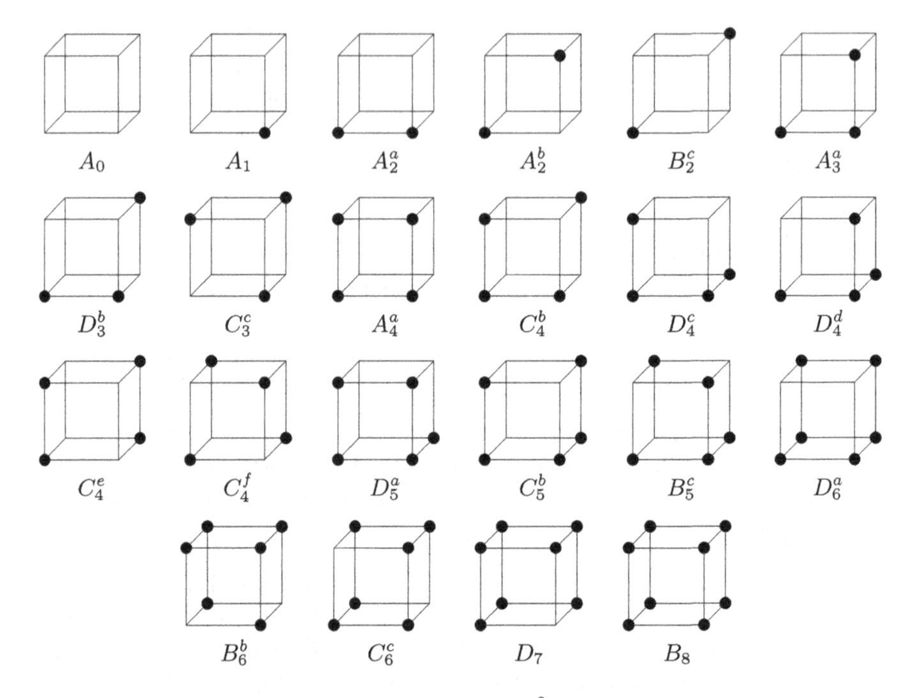

Fig. 2. Canonical patterns of R^3 around a vertex

3 (26, 6)-Connected Digital Spaces

The aim of this section is to find some necessary conditions on the lighting function f of a digital space (R^3, f) providing us with the $(26, 6)$-connectivity usually defined on $\mathbb{Z}^3$ by means of adjacencies. For this, we first recall that a digital object O in (R^3, f) is naturally said to be connected if its continuous analogue $|\mathcal{A}_O|$ is connected. And, in the same way, the complement $\text{cell}_3(R^3) - O$ of O is declared to be connected if $|\mathcal{A}_{R^3}| - |\mathcal{A}_O|$ is connected. These notions of connectedness are characterized by the following notions of adjacency.

Definition 1. *Two cells $\sigma, \tau \in O$ are* $\emptyset$-adjacent *if $f(O, \alpha) = 1$ for some common face $\alpha \leq \sigma \cap \tau$. Moreover, $\sigma, \tau \in \text{cell}_3(R^3) - O$ are* O-adjacent *if $f(\text{cell}_3(R^3), \alpha) = 1$ and $f(O, \alpha) = 0$ for some $\alpha \leq \sigma \cap \tau$.*

More precisely, for $X \in \{\emptyset, O\}$, the notion of X-adjacency directly leads to the notions of X-path, X-component and X-connectedness. Then, it can be proved that $|\mathcal{A}_O|$ is connected if and only if the object O is $\emptyset$-connected and $|\mathcal{A}_{R^3}| - |\mathcal{A}_O|$ is connected if and only if $\text{cell}_3(R^3) - O$ is O-connected. See Section 4 in [3] for a detailed proof of this fact in a much more general context.

By using this characterization, one may get intuitively convinced that the lighting functions $f_{\max}$ and g in Example 1 describe the $(8, 4)$- and $(4, 8)$-adjacencies usually defined on $\mathbb{Z}^2$. Actually the digital spaces $(R^2, f_{\max})$ and (R^2, g) are $(8, 4)$- and $(4, 8)$-connected, respectively, in the sense of the following.

Definition 2. *Given an adjacency pair $(k, \overline{k})$ in $\mathbb{Z}^n$ we say that the digital space (R^n, f) is $(k, \overline{k})$-connected if the two following properties hold for any digital object $O \subseteq \text{cell}_n(R^n)$:*

1. *C is a $\emptyset$-component of O if and only if it is a k-component of O; and,*
2. *C is an O-component of the complement $\text{cell}_n(R^n) - O$ if and only if it is a $\overline{k}$-component.*

Several examples of (homogeneous) $(26, 6)$-connected digital spaces (R^3, f) can be found in [1, 3, 5] as well as examples of $(k, \overline{k})$-connected spaces for $k, \overline{k} \in \{6, 18, 26\}$.

From now on, we assume that (R^3, f) is a given $(26, 6)$-connected digital space for which we collect some necessary conditions on the lighting function f that will be useful in the next section.

Proposition 1. *For all cells $\alpha \in R^3$, $f(\text{cell}_3(R^3), \alpha) = 1$. Thus, $|\mathcal{A}_{R^3}| = \mathbb{R}^3$.*

Proof. By Axiom 2, $f(O_\alpha, \alpha) = 1$ for any 26-connected object $O_\alpha = \{\sigma, \tau\}$, with $\alpha = \sigma \cap \tau$. Hence the result follows by Axiom 3.

Proposition 2. *Let $O \subseteq \text{cell}_3(R^3)$ be a digital object. Given a 2-cell $\gamma \in \text{supp}(O)$, let $\alpha < \gamma$ be a vertex and $\beta_1, \beta_2 < \gamma$ the edges with $\alpha = \beta_1 \cap \beta_2$. Assume that $f(O, \alpha) = 1$ and $f(O, \gamma) = f(O, \beta_1) = f(O, \beta_2) = 0$. Then $f(O, \delta) = 0$ for each face $\delta < \gamma$ such that $\delta \neq \alpha$.*

Proof. As $\gamma \in \text{supp}(O)$, $\text{st}_3(\gamma; O) = \text{st}_3(\gamma; R^3)$. Moreover, by Proposition 1 we know that $f(\text{cell}_3(R^3), \gamma) = 1$. Then, since $f(O, \gamma) = 0$, Axiom 5 yields that $\gamma(O, \text{cell}_3(R^3)) = \cup\{\mathring{\omega} \mid \omega < \gamma, f(O, \omega) = 0, f(\text{cell}_3(R^3), \omega) = 1\}$ is a connected set. Therefore the result follows since this set contains $\mathring{\beta}_1 \cup \mathring{\beta}_2$ and $f(O, \alpha) = 1$.

Proposition 3. *Let $P = \text{patt}(O, \alpha)$ be the pattern of an object O in a vertex $\alpha \in R^3$. Then: (a) $f(O, \alpha) = 0$ if $P \in \mathfrak{A}$ and (b) $f(O, \alpha) = 1$ if $P \in \mathfrak{B} \cup \mathfrak{C}$.*

Proof. Part (a) is an immediate consequence of Axiom 2 since $P \in \mathfrak{A}$ implies $\alpha \notin \text{supp}(O)$.

For $P = B_8$, $\text{st}_3(\alpha; O) = \text{st}_3(\alpha; R^3)$, and (b) follows by Axiom 4 and Proposition 1. Moreover, if $P = B_2^c$, $\text{st}_3(\alpha; O)$ is the object O_α in the proof of Proposition 1 and, by that proof, $f(\text{st}_3(\alpha; O), \alpha) = 1$. Hence $f(O, \alpha) = 1$ by Axiom 4.

For the rest of patterns in $\mathfrak{B} \cup \mathfrak{C}$ we consider the digital object $\overline{O} = (\text{cell}_3(R^3) - \text{st}_3(\alpha; R^3)) \cup \text{st}_3(\alpha; O)$. It is easily checked that its complement $O' = \text{st}_3(\alpha; R^3) - \text{st}_3(\alpha; O)$ is not 6-connected and hence non $\overline{O}$-connected. This shows that $f(\overline{O}, \alpha) = 1$; otherwise $f(\overline{O}, \alpha) = 0$ and $f(\text{cell}_3(R^3), \alpha) = 1$ (Proposition 1) give us the $\overline{O}$-connectivity of O'. Now, Axiom 4 and the equality $\text{st}_3(\alpha; O) = \text{st}_3(\alpha; \overline{O})$ lead to $f(O, \alpha) = 1$.

Remark 4. The behaviour of the lighting function f on the patterns in $\mathfrak{D}$ is not determined by the $(26, 6)$-connectivity. To check this it suffices to consider the homogeneous $(26, 6)$-connected spaces (R^3, f_1) and (R^3, f_2) given by

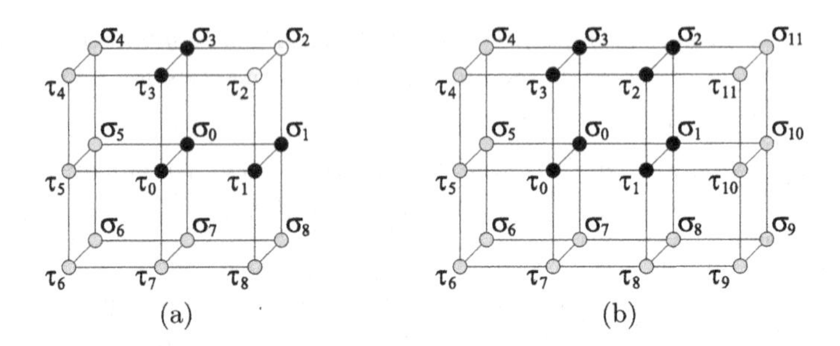

Fig. 3. Auxiliary figures for Propositions 5 and 6

a) If $\dim \alpha \neq 0$ then $f_1(O,\alpha) = 1 = f_2(O,\alpha)$ if and only if $\alpha \in \text{supp}(O)$.
b) If $\dim \alpha = 0$ then $f_1(O,\alpha) = 1$ if and only if $\text{patt}(O,\alpha) \in \mathfrak{B} \cup \mathfrak{C}$ while $f_2(O,\alpha) = 1$ if and only if $\text{patt}(O,\alpha) \in \mathfrak{B} \cup \mathfrak{C} \cup \mathfrak{D}$.

Proposition 4. *Let $O \subseteq \text{cell}_3(R^3)$ be a digital object and $\beta = \langle \alpha_1, \alpha_2 \rangle \in R^3$ an edge such that $f(O,\alpha_1) = f(O,\alpha_2)$. Then $f(O,\beta) = 1$ if either (a) $\text{st}_3(\beta; O) = \text{st}_3(\beta; R^3)$ or (b) $\text{st}_3(\beta; O) = \{\sigma, \tau\}$ and $\beta = \sigma \cap \tau$.*

Proof. For case (b) we consider the digital object $O_1 = (\text{cell}_3(R^3) - \text{st}_3(\beta; R^3)) \cup \{\sigma, \tau\}$. Then $f(O_1, \alpha_i) = 1$ by Proposition 3 since $\text{patt}(O_1, \alpha_i) = B_6^b$, $i = 1, 2$. Moreover, the complement $\text{st}_3(\beta; R^3) - \{\sigma, \tau\}$ of O_1 consists of two non 6-adjacent (equivalently, non O_1-adjacent) 3-cells. Therefore $f(O_1, \alpha) = 1$, and Remark 2(b) yields $f(O, \alpha) = 1$.

In case (a) we have $f(\text{cell}_3(R^3), \beta) = f(\text{cell}_3(R^3), \alpha_i) = 1$ by Proposition 1, and we argue as in case (b) with $O_1 = \text{cell}_3(R^3)$.

Next we will obtain some properties of f related to the pattern D_6^a. For this we use Fig. 3(a) which, according to the identification in Remark 1, depicts $\text{st}_3^*(\gamma; R^3)$ for the 2-cell $\gamma = \sigma_0 \cap \tau_0$. Moreover, let $\alpha_i = \sigma_0 \cap \tau_{2i}$ and $\beta_i = \sigma_0 \cap \tau_{2i-1}$ $(1 \leq i \leq 4)$ denote the vertices and edges of γ respectively. Observe that if the black dots in Fig. 3(a) represent 3-cells contained in a given object O, we have $\text{patt}(O, \alpha_1) = D_6^a$.

Lemma 1. *With the above notation assume that α_1 is the only face of γ for which $f(O, \alpha_1) = 1$. Then $\{\sigma_i, \tau_i\} \not\subseteq O$ for $i = 5, 6, 7$.*

Proof. It is an immediate consequence of the hypothesis and Proposition 4 that $\{\sigma_i, \tau_i\} \not\subseteq O$ for $i = 5, 7$. Therefore $\{\sigma_6, \tau_6\} \subseteq O$ yields a contradiction since then $\text{patt}(O, \alpha_3) \in \mathfrak{B} \cup \mathfrak{C}$ and $f(O, \alpha_3) = 1$ by Proposition 3.

Proposition 5. *With the notation and hypothesis of Lemma 1 assume in addition that (R^3, f) is homogeneous. Then $f(O, \gamma) = 1$ if the following extra condition holds:*

p7) $f(O, \alpha) = 0$ for all digital objects $O \subseteq \text{cell}_3(R^3)$ and vertices $\alpha \in R^3$ with $\text{patt}(O, \alpha) = D_7$.

Proof. By Remark 2(b) it will suffice to show that for the object $O_1 = O \cup \{\sigma_2, \sigma_6, \tau_6\}$, see Fig. 3(a), the sets

$$\gamma(O, O_1) = \cup\{\mathring{\omega} \mid \omega < \gamma, f(O, \omega) = 0, f(O_1, \omega) = 1\}$$

$$\gamma(O_1, \mathrm{cell}_3(R^3)) = \cup\{\mathring{\omega} \mid \omega < \gamma, f(O_1, \omega) = 0, f(\mathrm{cell}_3(R^3), \omega) = 1\}$$

are non-connected since $f(\mathrm{cell}_3(R^3), \gamma) = 1$ by Proposition 1.

Notice that $O_1 \neq O$ by Lemma 1. Next we sate all the values $f(O_1, \alpha_i)$ on the vertices α_i, $1 \leq i \leq 4$, of γ. As $\mathrm{st}_3(\alpha_i; O) = \mathrm{st}_3(\alpha_i; O_1)$ for $i = 2, 4$, Axiom 4 yields $f(O_1, \alpha_i) = 0$. Moreover, $\mathrm{patt}(O_1, \alpha_1) = D_7$ and thus $f(O_1, \alpha_1) = 0$ by the extra condition (p7). In addition, one readily checks that $\mathrm{patt}(O_1, \alpha_3) \in \mathfrak{B} \cup \mathfrak{C}$ and, so, $f(O_1, \alpha_3) = 1$ by Proposition 3.

From the above values and Proposition 4 it follows that $f(O_1, \beta_i) = 1$, $i = 1, 2$. Therefore, a direct checking shows that α_1 and α_2 lie in different components of $\gamma(O_1, \mathrm{cell}_3(R^3))$ and $\mathring{\beta}_1$ and $\mathring{\beta}_2$ lie in different components of $\gamma(O, O_1)$.

Proposition 6. *Assume that (R^3, f) is homogeneous and satisfies condition (p7) in Proposition 5. Let $\gamma_0, \gamma_1 \in R^3$ be two 2-cells which do not lie in the same 3-cell and such that $\gamma_0 \cap \gamma_1 = \langle \alpha, \alpha' \rangle$ is an edge. Then $f(O, \gamma_0) = 1$ for any object $O \subseteq \mathrm{cell}_3(R^3)$ such that* $\mathrm{patt}(O, \alpha) = B_8$ *and, moreover, $f(O, \delta) = 0$ for any face $\delta < \gamma_0$ or $\delta < \gamma_1$ with $\delta \neq \alpha$.*

Proof. We proceed as in the proof of Proposition 5. For this we label the 3-cells in $\mathrm{st}_3^*(\gamma_0; R^3) \cup \mathrm{st}_3^*(\gamma_1; R^3)$ as it is shown in Fig. 3(b), where $\gamma_j = \sigma_j \cap \tau_j$, $j = 0, 1$, and $\alpha_i = \sigma_0 \cap \tau_{2i}$ and $\beta_i = \sigma_0 \cap \tau_{2i-1}$, $1 \leq i \leq 4$, are the vertices and edges, respectively, of γ_0. Notice that $\alpha = \alpha_1$ and $\alpha' = \alpha_4$.

Similarly to Lemma 1 we can show that $\{\sigma_k, \tau_k\} \not\subseteq O$ for $k \in \{5, 7, 8, 10\}$, and then by Proposition 3 we easily derive that $\{\sigma_s, \tau_s\} \not\subseteq O$ for $s = 6, 9$.

Then we consider the new object $O_1 = O \cup \{\sigma_4, \tau_4, \sigma_5, \tau_5, \sigma_7, \tau_7, x_6, x_8\}$ where $x_i = \sigma_i$ if $\sigma_i \in O$ and $x_i = \tau_i$ otherwise. Notice that $O \neq O_1$ by the previous paragraph, moreover we have: $f(O_1, \alpha_1) = f(O_1, \alpha_2) = 1$ by Proposition 3 since $\mathrm{patt}(O_1, \alpha_i) = B_8$, $i = 1, 2$; $f(O_1, \alpha_3) = f(O_1, \alpha_4) = 0$ since $\mathrm{patt}(O_1, \alpha_i) = D_7$, $i = 3, 4$; and, $f(O_1, \beta_2) = f(O_1, \beta_4) = 1$ by Proposition 4. From these values it is not difficult to show that $\mathring{\beta}_2$ and $\mathring{\beta}_4$ belong to different components of $\gamma_0(O, O_1)$ while α_3 and α_4 define different components in $\gamma_0(O_1, \mathrm{cell}_3(R^3))$. Hence these sets are not connected and, by Remark 2(b), we get $f(O, \gamma_0) = 1$.

4 Surfaces on (26, 6)-Connected Digital Spaces

Given a homogeneous (26, 6)-connected digital space (R^3, f), we have already worked out in Section 3 a set of values of f for arbitrary digital objects. Through this section S will stand for a surface in (R^3, f) and we will compute the value $f(S, \delta)$ for each cell $\delta \in R^3$. Indeed, if δ is a 3-cell, we know by Remark 2(a) that $f(S, \delta) = 1$ if and only if $\delta \in S$. So, we proceed with the lower dimensional cells of R^3. For vertices we will prove the following

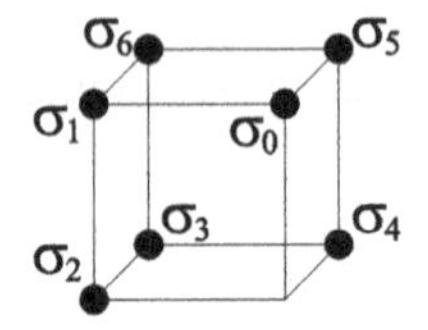

Fig. 4. A labelling for the 3-cells in D_7

Theorem 1. *Let S be a digital surface in a homogeneous $(26,6)$-connected digital space (R^3, f). If $\alpha \in R^3$ is a vertex then $\mathrm{patt}(S,\alpha) \in \mathfrak{A} \cup \mathfrak{C} \cup \mathfrak{D}$. Moreover, $f(S,\alpha) = 0$ if $\mathrm{patt}(S,\alpha) \in \mathfrak{A} \cup \mathfrak{D}$ and $f(S,\alpha) = 1$ if $\mathrm{patt}(S,\alpha) \in \mathfrak{C}$.*

For the proof of Theorem 1 we need the results in Section 3 as well as a series of partial results which follow.

Proposition 7. *Let $\beta \in R^3$ be an edge. Then $f(S,\beta) = 0$ if either $\mathrm{st}_3(\beta; S)$ consists of three 3-cells or if $\mathrm{st}_3(\beta; S) = \mathrm{st}_3(\beta; R^3)$ and $f(S,\alpha_i) = 1$ for some of the two vertices α_1, α_2 of β.*

Proof. Assume on the contrary that $f(S,\beta) = 1$; that is, its center $c(\beta)$ belongs to $\mathcal{A}_S$ and thus the link $L = \mathrm{lk}(c(\beta); \mathcal{A}_S)$ is a 1-sphere.

In case $\mathrm{st}_3(\beta; S) = \{\sigma_1, \sigma_2, \sigma_3\}$, then $c(\sigma_i) \in \mathcal{A}_S$ for $1 \leq i \leq 3$. Moreover, $f(S,\alpha) = 0$ if $\alpha < \beta$ since, otherwise, L would contain the one point union of the three edges $\langle c(\alpha), c(\sigma_i)\rangle$, $1 \leq i \leq 3$. Then, if we assume that $\beta = \sigma_1 \cap \sigma_3$, only the centers of the 2-cells $\sigma_1 \cap \sigma_2$ and $\sigma_2 \cap \sigma_3$ can appear in L since they are the only ones in $\mathrm{supp}(S)$. Hence L is not a 1-sphere. This contradiction shows that $f(S,\beta) = 0$.

If $\mathrm{st}_3(\beta; S) = \{\sigma_1, \sigma_2, \sigma_3, \sigma_4\}$ and $f(S,\beta) = f(S,\alpha_1) = 1$ then L contains the one point union of the four edges $\langle c(\alpha_1), c(\sigma_j)\rangle$, $1 \leq j \leq 4$, and hence it is not a 1-sphere.

Proposition 8. *Assume that (R^3, f) is $(26,6)$-connected. Then $P = \mathrm{patt}(S,\alpha) \notin \{B_2^c, B_5^c, B_6^b\}$ for every vertex $\alpha \in R^3$.*

Proof. Assume on the contrary that P is one of the patterns in the statement. By Proposition 3, $f(S,\alpha) = 1$ and so $c(\alpha) \in \mathcal{A}_S$. However we claim that $L = \mathrm{lk}(c(\alpha); \mathcal{A}_S)$ is not a 1-sphere. Indeed, if $P = B_2^c$, L consists of two points by Axiom 2. Otherwise, if $P \neq B_2^c$, we know by Proposition 7 that $f(S,\beta) = 0$ for each edge $\beta > \alpha$ such that $\mathrm{st}_3(\beta; S)$ contains at least three elements. This condition and the fact that each center $c(\sigma)$, $\sigma \in \mathrm{st}_3(\alpha; S)$, should belong to exactly two edges in the 1-sphere L yields that $f(S,\delta) = 1$ for each other cell $\delta > \alpha$ with $\delta \in \mathrm{supp}(S)$. Then, it is not hard to find a 3-cell $\overline{\sigma} \in \mathrm{st}_3(\alpha; S)$ with $c(\overline{\sigma})$ lying in three edges of L.

Proposition 9. *If $\alpha \in R^3$ is a vertex with $P = \mathrm{patt}(S,\alpha) \in \{D_3^b, D_4^c, D_4^d, D_5^a, D_7\}$ then $f(S,\alpha) = 0$.*

Proof. If $f(S,\alpha) = 1$, Proposition 7 says that $f(S,\beta) = 0$ for all edges $\beta > \alpha$ with $\mathrm{st}_3(\beta;S)$ having at least three elements. Therefore:

a) If $P \neq D_7$ we use Axiom 2 to check that there is at least one 3-cell $\sigma_0 \in \mathrm{st}_3(\alpha;S)$ such that $f(S,\delta) = 0$ for all faces $\alpha < \delta < \sigma_0$ except possibly one. Hence its center $c(\sigma_0)$ is vertex of at most one edge in the link $L = \mathrm{lk}(c(\alpha);\mathcal{A}_S)$ and so L is not a 1-sphere.

b) If $P = D_7$ let us label the 3-cells in $\mathrm{st}_3(\alpha;S)$ as it is shown in Fig. 4. The fact that the centers $c(\sigma_0)$, $c(\sigma_2)$ and $c(\sigma_4)$ are in two edges of L implies that $f(S,\gamma_i) = 1$ for $\gamma_i = \sigma_i \cap \sigma_{i+1 \bmod 6}$ $(0 \leq i \leq 5)$. But then the centers $c(\gamma_i)$ and $c(\sigma_i)$ generate a cycle in L leaving $c(\sigma_6)$ out. Therefore L is not a 1-sphere.

Proposition 10. *Assume that* (R^3, f) *is homogeneous and* (26, 6)*-connected. Then* $f(S,\alpha) = 0$ *for all vertices* $\alpha \in R^3$ *with* $\mathrm{patt}(S,\alpha) = D_6^a$.

Proof. Let us assume that $f(S,\alpha) = 1$, and hence $L = \mathrm{lk}(c(\alpha),\mathcal{A}_S)$ is a 1-sphere. Moreover, by Proposition 7 and Axiom 2 we know that $f(S,\beta) = 0$ for all the six edges $\beta > \alpha$. From these facts we infer that $f(S,\sigma\cap\tau) = 1$ for all the 2-cells $\sigma\cap\tau$, where $\sigma,\tau \in \mathrm{st}_3(\alpha;S)$, except for $\gamma = \sigma_0 \cap \tau_0$ where σ_0, τ_0 are the only two 3-cells which are 6-adjacent to exactly three other elements in $\mathrm{st}_3(\alpha;S)$. Furthermore, Proposition 2 yields that $f(S,\delta) = 0$ for any cell $\alpha \neq \delta < \gamma$. This leads to a contradiction with Proposition 5 where condition (p7) is guaranteed by Proposition 9.

The proof of the next result use Proposition 3 to show that $f(S,\alpha) = 1$ and then follows the same steps as the proof of Proposition 10 by using Proposition 6 instead of Proposition 5 in the argument.

Proposition 11. *Assume that* (R^3, f) *is homogeneous and* (26, 6)*-connected. Then* $\mathrm{patt}(S,\alpha) \neq B_8$ *for any vertex* $\alpha \in R^3$.

We are now ready to prove Theorem 1.

Proof (of Theorem 1). The patterns in $\mathfrak{B}$ are ruled out by Propositions 8 and 11. Moreover, the vanishing of f on vertices with patterns in $\mathfrak{A}$ is given by Proposition 3, while for vertices with patterns in $\mathfrak{D}$ it follows from Propositions 9 and 10. Finally, the lighting of vertices with patterns in $\mathfrak{C}$ is given by Proposition 3.

Remark 5. Theorem 1 can be used to determine the value of the lighting function f on the cells $\beta \in \mathrm{supp}(\mathrm{st}_3(\alpha;S))$ where $\alpha \in R^3$ is a vertex such that $P = \mathrm{patt}(S,\alpha) \in \mathfrak{C}$. More precisely, by Theorem 1 $f(S,\alpha) = 1$, and so $L = \mathrm{lk}(c(\alpha);\mathcal{A}_S)$ is a 1-sphere containing the center of each 3-cell in $\mathrm{st}_3(\alpha;S)$. If $P \neq C_4^f$ this implies that $f(S,\beta) = 1$ for each cell $\beta \in \mathrm{supp}(\mathrm{st}_3(\alpha;S))$ except for those on which f vanishes by Proposition 7. For $P = C_4^f$, the fact that L is a cycle yields that exactly four of the six edges $\beta \in \mathrm{supp}(\mathrm{st}_3(\alpha;S))$ are lighted by f and, moreover, the two remaining edges are contained in a straight-line in $\mathbb{R}^3$.

The values of f on edges are characterized as follows.

Theorem 2. *Assume that (R^3, f) is homogeneous and $(26,6)$-connected. Then for any edge $\beta = \langle\alpha_1, \alpha_2\rangle \in \mathrm{supp}(S)$: (a) if $\mathrm{st}_3(\beta; S) = \mathrm{st}_3(\beta; R^3)$ then $f(S,\beta) = 1$; (b) if $\mathrm{st}_3(\beta; S)$ contains three cells then $f(S,\beta) = 0$; and, (c) if $\mathrm{st}_3(\beta; S)$ contains two elements then $f(S,\beta) = 1$ if and only if both vertices of β are lighted by f.*

Proof. In case (a) $\mathrm{patt}(S, \alpha_i) \in \mathfrak{A} \cup \mathfrak{D}$ and, moreover, $f(S, \alpha_i) = 0$, $i = 1, 2$, by Theorem 1. Therefore $f(S, \beta) = 1$ by Proposition 4. Case (b) follows directly from Proposition 7. Finally, in case (c), as $\mathrm{st}_3(\beta; S)$ contains only two elements and $\beta \in \mathrm{supp}(S)$ it follows that $\gamma \notin \mathrm{supp}(S)$ for all 2-cells $\gamma > \alpha$. Moreover, since $L = \mathrm{lk}(c(\beta); \mathcal{A}_S)$ is a 1-sphere then necessarily $c(\alpha_1), c(\alpha_2) \in L$ and so $f(S, \alpha_1) = f(S, \alpha_2) = 1$.

Conversely, if $f(S, \alpha_i) = 1$, $i = 1, 2$, we get $f(S, \beta) = 1$ by Proposition 4.

Corollary 1. *Let $\beta = \langle\alpha_1, \alpha_2\rangle \in R^3$ be an edge with $\mathrm{st}_3(\beta; S) = \mathrm{st}_3(\beta; R^3)$. Then $f(S, \gamma) = 1$ for each 2-cell $\gamma > \beta$.*

Proof. By Theorem 2 $f(S, \beta) = 1$ and, thus, $f(S, \alpha_i) = 0$, $i = 1, 2$, by Proposition 7. Then, the requirement that $\mathrm{lk}(c(\beta); \mathcal{A}_S)$ is a 1-sphere implies the result.

We finish this section with the characterization of f on 2-cells. Namely

Theorem 3. *Assume that (R^3, f) is homogeneous and $(26,6)$-connected. Then, for any 2-cell $\gamma \in R^3$, $f(S, \gamma) = 1$ if and only if $\gamma \in \mathrm{supp}(S)$.*

Proof. The "only if" part is just Axiom 2. For the converse let us assume $\gamma \in \mathrm{supp}(S)$; that is, $\gamma = \sigma \cap \tau$ with $\sigma, \tau \in S$.

If $f(S, \alpha) = 1$ for some vertex $\alpha < \gamma$, Theorem 1 yields $\mathrm{patt}(S, \alpha) \in \mathfrak{C}$, and the fact that σ is 6-adjacent to τ implies that necessarily $\mathrm{patt}(S, \alpha) \neq C_4^f$. Hence the result follows by Remark 5. Also, if $f(S, \beta) = 1$ for some edge $\beta < \gamma$, then $\mathrm{st}_3(\beta; S) = \mathrm{st}_3(\beta; R^3)$ by Theorem 2, and the result follows by Corollary 1. Finally we study the case $f(S, \delta) = 0$ for each proper face $\delta < \gamma$. In particular, if δ is a vertex Theorem 1 yields that $\mathrm{patt}(S, \delta) \in \mathfrak{A} \cup \mathfrak{D}$. Moreover, Theorem 2 and $\sigma, \tau \in \mathrm{st}_3(\delta; S)$ implies that $\mathrm{patt}(S, \delta) \in X = \{A_2^a, A_3^a, D_3^b, D_4^c, D_4^d, D_5^a\}$. In addition, notice that $\mathrm{st}_3^*(\gamma; S)$ is a 26-connected digital object and, thus, there must exist a $\emptyset$-path in this set from σ to τ. However we will show by induction that any $\emptyset$-path $\Sigma_k = (\sigma_i)_{i=0}^k$ starting at $\sigma_0 = \sigma$ satisfies the condition

$$\sigma_i \cap \sigma \not< \tau \text{ for } i \leq k \ . \quad (1)$$

and so Σ_k cannot finish at τ. This contradiction completes the proof.

Firstly we notice that $\sigma_i \cap \sigma \neq \emptyset$ for $i \leq k$. Then, the inductive process works as follows. For $k = 1$, if $\sigma \cap \sigma_1 < \tau$ then $f(S, \varepsilon) = 0$ for each $\varepsilon \leq \sigma \cap \sigma_1 = \sigma_0 \cap \sigma_1$, which is not possible since σ_0 and σ_1 are $\emptyset$-adjacent in $\mathrm{st}_3(\alpha; S)$. Now assume that Condition 1 holds for $k - 1$ and let $\varepsilon \leq \sigma_{k-1} \cap \sigma_k$ a face with $f(S, \varepsilon) = 1$ joining σ_{k-1} to σ_k in the $\emptyset$-path Σ_k. Notice that $\varepsilon \not< \tau$, otherwise Lemma 2 below shows that $\varepsilon < \sigma$ and hence $f(S, \varepsilon) = 0$. Moreover, if $\varepsilon < \sigma$ then Condition 1 holds for k by arguing as in the case $k = 1$. Finally, if $\varepsilon \not< \tau$ and $\varepsilon \not< \sigma$, the set X above is reduced to the patterns D_4^c, D_4^d and D_5^a, which are studied individually to derive the result.

Lemma 2. *Let $\sigma, \tau \in \text{cell}_3(R^3)$ two 6-adjacent 3-cells and $\gamma = \sigma \cap \tau$. If $\rho \in \text{st}_3^*(\gamma; R^3)$ and $\rho \cap \sigma \not< \tau$ then $\rho \cap \tau < \sigma$.*

Proof. From the hypothesis the three cells ρ, σ and τ share a vertex α. Moreover, if we consider the digital object $O = \{\sigma, \tau, \rho\}$ the pattern $\text{patt}(O, \alpha)$ is either A_3^a or D_3^b. Then a simple analysis of these cases yields the result.

5 A Universal Space for the (26, 6)-Connected Surfaces

In this section we define a new lighting function on the device model R^3 which gives us a homogeneous (26, 6)-connected digital space whose family of surfaces is the largest among the class of such digital spaces. Namely, we prove the following

Theorem 4. *There exists a homogeneous (26, 6)-connected digital space $E^U = (R^3, f^U)$ which is universal in the following sense: any digital surface S in an arbitrary homogeneous (26, 6)-connected digital space (R^3, f) is also a digital surface in E^U.*

The lighting function $f^U : \mathcal{P}(\text{cell}_3(R^3)) \times R^3 \to \{0, 1\}$ is defined as follows. Given a digital object $O \subseteq \text{cell}_3(R^3)$ and a cell $\delta \in R^3$, $f^U(O, \delta) = 1$ if and only if one of the following conditions holds:

1. $\dim \delta = 0$ and $\text{patt}(O, \delta) \in \mathfrak{B} \cup \mathfrak{C}$.
2. $\dim \delta > 1$ and $\delta \in \text{supp}(O)$.
3. $\dim \delta = 1$ and either $\text{st}_3(\delta; R^3) \subseteq O$ or $\text{st}_3(\delta; O) = \{\sigma, \tau\}$, with $\delta = \sigma \cap \tau$, and $f^U(O, \alpha_1) = f^U(O, \alpha_2)$ for the vertices α_1, α_2 of δ.

It is not difficult, but a tedious task, to check that f^U is a lighting function and to prove that (R^3, f^U) is actually a homogeneous (26, 6)-connected digital space. Any way, the following property, which will be used later, is immediate from the definition of f^U.

Proposition 12. *If $O \subseteq \text{cell}_3(R^3)$ is a digital object and $\alpha \in R^3$ a cell for which $\text{st}_3(\alpha; R^3) \subseteq O$ then $f^U(O, \alpha) = 1$; that is, (R^3, f^U) is a* solid digital space *as defined in [2]. In particular, $f^U(\text{cell}_3(R^3), \alpha) = 1$ for any cell $\alpha \in R^3$, and hence $| \mathcal{A}_{R^3} | = \mathbb{R}^3$.*

Proof (Theorem4: Universality of E^U). The result is an immediate consequence of Proposition 13 below which provides us with the equality $f^U(S, \delta) = f(S, \delta)$ for all cells $\delta \in R^3$, or equivalently $| \mathcal{A}_S^f | = | \mathcal{A}_S^{f^U} |$. Thus, the continuous analogues of S in both digital spaces (R^3, f) and E^U are combinatorial surfaces.

Proposition 13. *Let S be a digital surface in a homogeneous (26, 6)-connected digital space (R^3, f). Then $f(S, \delta) = f^U(S, \delta)$ for each cell $\delta \in R^3$.*

Proof. If $\dim \delta = 3$ the equality $f(S, \delta) = f^U(S, \delta)$ is an immediate consequence of Remark 2(a). If $\dim \delta = 2$ (or $\dim \delta = 0$) we derive the result from the definition of f^U and Theorem 3 (or Theorem 1, respectively). In case $\delta = \langle \alpha_1, \alpha_2 \rangle$ is

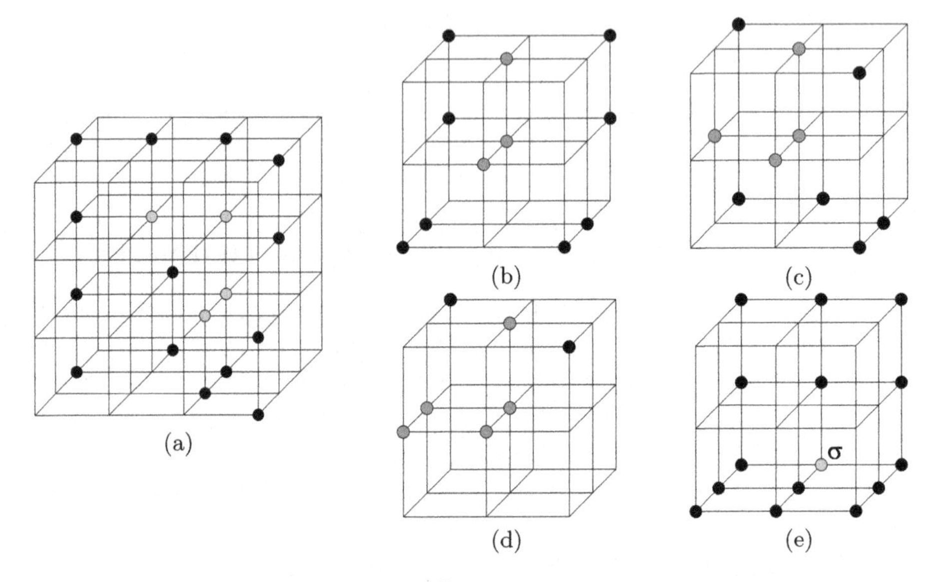

Fig. 5. Some surfaces in (R^3, f^U) which are not simplicity 26-surfaces

an edge and $\delta \notin \mathrm{supp}(S)$, then $f(S,\delta) = f^U(S,\delta) = 0$ by Axiom 2. Otherwise, if $\delta \in \mathrm{supp}(S)$, the triple $[S, f, \delta]$ satisfies one of the conditions (1)-(3) in Theorem 2 and the definition of f^U yields the equality $f^U(S,\delta) = f(S,\delta)$.

Remark 6. Notice that, while Proposition 13 is deduced from Theorems 1-3, these theorems can be conversely derived from Proposition 13, since one readily checks that digital surfaces in (R^3, f^U) satisfy Theorems 1-3.

By using the identification in Remark 1 between 3-cells in R^3 and points in $\mathbb{Z}^3$, we can compare the family of digital surfaces in the universal space (R^3, f^U) with other families of surfaces defined on $\mathbb{Z}^3$ by means of the $(26,6)$-adjacency. As far as we know, the space (R^3, f^U) provides a set of surfaces larger than any other definition in the literature. For example, we next show that the set of simplicity 26-surfaces [6] and, hence, both strong 26-surfaces [4] and Morgenthaler's $(26,6)$-surfaces [10], are strictly contained in this family.

Indeed, Theorem 1 in [5] shows that each simplicity 26-surface is a digital surface in the $(26,6)$-connected homogeneous digital space (R^3, f^{ss}), and hence a surface in (R^3, f^U) by Theorem 4. We recall that the lighting function f^{ss} is defined in [5] as follows. Given a digital object $O \subseteq \mathrm{cell}_3(R^3)$ and a cell $\delta \in R^3$, $f^{ss}(O,\delta) = 1$ if and only if: (a) $\dim\delta \geq 2$ and $\delta \in \mathrm{supp}(O)$; (b) $\dim\delta = 0$, $\delta \in \mathrm{supp}(O)$ and $\mathrm{patt}(O,\delta) \neq D_6^a$; and, $\dim\delta = 1$ and either $\mathrm{st}_3(\delta; R^3) \subseteq O$ or $\mathrm{st}_3(\delta; O) = \{\sigma, \tau\}$, with $\delta = \sigma \cap \tau$, and $f^{ss}(O,\alpha_1) = f^{ss}(O,\alpha_2)$ for the two vertices $\alpha_1, \alpha_2 < \delta$.

On the other hand, it is not difficult to check that each of the objects in Fig. 5(a)-(d) is (a piece of) a digital surface in (R^3, f^U). However, none of them is a simplicity 26-surface since they contain the patterns D_4^c, D_3^b, D_4^d and D_5^a (dots in grey colour), respectively, which are not allowed in this kind of surfaces (see Theorem 17 in [6]). Alternatively, it can also be checked that these objects are not surfaces in the space (R^3, f^{ss}). It is worth to point out that, in [8], Malandain et al. propose to consider the object depicted in Fig. 5(a) as a surface.

Finally, as mentioned in the Introduction, digital manifolds in suitable digital spaces (R^n, f) satisfy a digital version of the Jordan-Brouwer Theorem. More precisely, Theorem 5.3 in [3] shows that if the continuous analogue $|\mathcal{A}_{R^n}|$ is the Euclidean space $\mathbb{R}^n$, then the complement $\text{cell}_n(R^n) - M$ of any finite digital $(n-1)$-manifold M has two M-components, one of them finite. Moreover, if (R^n, f) is a solid space (see Proposition 12) a digital counterpart of the Index Theorem (see Theorem 3.16 in [2]) characterizes the finite M-component. Therefore, since (R^3, f^U) is a (26,6)-connected digital space, Proposition 12 yields the following

Theorem 5. *Each 26-connected digital surface S in (R^3, f^U) separates its complement* $\text{cell}_3(R^3) - S$ *into two 6-components.*

Despite of this result, several surfaces of the space (R^3, f^U) should be considered pathological examples since the deletion of one of their points might yield an object that still divides its complement into two 6-components, as it is the case of the 3-cell σ in Fig. 5(e). It can be checked that any non-pathological surface S is a *strongly separating* object (i.e., any 3-cell of S is 6-adjacent to both 6-components of $\text{cell}_3(R^3) - S$), and so we call S a *strongly separating surface.* Finally, we notice that the family of simplicity 26-surfaces is still strictly contained in the set of strongly separating surfaces, since the surfaces pictured in Fig. 5(a)-(d) belong to this class.

References

1. R. Ayala, E. Domínguez, A.R. Francés, A. Quintero. Digital Lighting Functions. *Lecture Notes in Computer Science.* **1347** (1997) 139–150.
2. R. Ayala, E. Domínguez, A.R. Francés, A. Quintero. A Digital Index Theorem. *Int. J. Patter Recog. Art. Intell.* **15**(7) (2001) 1–22.
3. R. Ayala, E. Domínguez, A.R. Francés, A. Quintero. Weak Lighting Functions and Strong 26-surfaces. *Theoretical Computer Science.* **283** (2002) 29–66.
4. G. Bertrand, R. Malgouyres. Some Topological Properties of Surfaces in $\mathbb{Z}^3$. *Jour. of Mathematical Imaging and Vision.* **11** (1999) 207–221.
5. J.C. Ciria, E. Domínguez, A.R. Francés. Separation Theorems for Simplicity 26-surfaces. *Lecture Notes in Computer Science.* **2301** (2002) 45–56.
6. M. Couprie, G. Bertrand. Simplicity Surfaces: a new definition of surfaces in $\mathbb{Z}^3$. *SPIE Vision Geometry V.* **3454** (1998) 40–51.
7. T.Y. Kong, A.W. Roscoe. Continuous Analogs of Axiomatized Digital Surfaces. *Comput. Vision Graph. Image Process.* **29** (1985) 60–86.

8. G. Malandain, G. Bertrand, N. Ayache. Topological Segmentation of Discrete Surfaces. *Int. Jour. of Computer Vision.* **10:2** (1993) 183–197.
9. Maunder, C. R. F., *Algebraic Topology.* Cambridge University Press 1980.
10. D.G. Morgenthaler, A. Rosenfeld. Surfaces in three–dimensional Digital Images. *Inform. Control.* **51** (1981) 227-247.
11. C.P. Rourke, and B.J. Sanderson. *Introduction to Piecewise-Linear Topology.* Ergebnisse der Math. **69**, Springer 1972.

Simple Points and Generic Axiomatized Digital Surface-Structures

Sébastien Fourey

GREYC Image – ENSICAEN, 6 bd maréchal Juin,
14050 Caen cedex – France
Sebastien.Fourey@greyc.ensicaen.fr

Abstract. We present a characterization of topology preservation within digital axiomatized digital surface structures (GADS), a generic theoretical framework for digital topology introduced in [2]. This characterization is based on the digital fundamental group that has been classically used for that purpose. More briefly, we define here simple points within GADS and give the meaning of the words: preserving the topology within GADS.

1 Introduction

In [2], a generic framework for digital topology has been introduced. This framework is in fact a whole axiomatic theory that allows us to prove results that become valid for any two dimensional digital space that satisfies the axioms of the theory. Some results such as a generic Jordan theorem has already been proved within this framework.

In this paper, we address a classical problem of digital topology: the characterization of topology preservation [15, 12, 18]. The main question being: When can we say that the deletion of one or several pixels (or voxels) from an image preserves the topology? In all cases, the answer to this question comes with the definition of *simple pixels/voxels*.

On the other hand, the digital fundamental group ([11]) has proved to be a convenient tool in order to characterize topology preservation in digital surfaces (see [16, 5]) as well as in the classical three dimensional digital space $\mathbb{Z}^3$ (see [6]). Here, we state in a very straightforward way a definition of the digital fundamental group of a GADS (Generic Axiomatized Digital Surface-Structure). Then, we present a characterization of topology preservation within a GADS, by removal of a simple point, based on the fundamental group.

2 Definition of GADS and pGADS

We recall here the basic notions and definitions from [2]. We should first summarize the motivation for the definition of a GADS. This starts with an observation: many results in digital topology come with a proof that *depends* on the digital

R. Klette and J. Žunić (Eds.): IWCIA 2004, LNCS 3322, pp. 307–317, 2004.

space that is considered. For example, a proof of a Jordan curve theorem exists for the space $\mathbb{Z}^2$ with the classical $(4,8)$ or $(8,4)$ pairs of adjacency relations. A similar result holds for the hexagonal grid, as well as $\mathbb{Z}^2$ with the Khalimsky adjacency relation. What is unsatisfactory here is that the proof of such a result has to be written for each of the considered spaces. The axiomatic definition of what actually is an admissible digital space is a response to this observation. The purpose of GADS as introduced in [2] is to define a generic framework that allows to state and prove results of digital topology which becomes valid for any admissible digital space. Thus, a single result would no longer need (sometimes similar) multiple proofs.

2.1 Basic Concepts and Notations

For any set P we denote by $P^{\{2\}}$ the set of all unordered pairs of distinct elements of P (equivalently, the set of all subsets of P with exactly two elements). Let P be any set and let $\rho \subseteq P^{\{2\}}$.[1] Two elements a and b of P [respectively, two subsets A and B of P] are said to be *ρ-adjacent* if $\{a,b\} \in \rho$ [respectively, if there exist $a \in A$ and $b \in B$ with $\{a,b\} \in \rho$]. If $x \in P$ we denote by $N_\rho(x)$ the set of elements of P which are ρ-adjacent to x; these elements are also called the *ρ-neighbors* of x. We call $N_\rho(x)$ the *punctured ρ-neighborhood* of x.

A *ρ-path* from $a \in P$ to $b \in P$ is a finite sequence $(x_0, \ldots, x_l)$ of one or more elements of P such that $x_0 = a$, $x_l = b$ and, for all $i \in \{0, \ldots, l-1\}$, $\{x_i, x_{i+1}\} \in \rho$. The nonnegative integer l is the *length* of the path. A ρ-path of length 0 is called a *one-point path.* For all integers m, n, $0 \leq m \leq n \leq l$, the subsequence $(x_m, \ldots, x_n)$ of $(x_0, \ldots, x_l)$ is called an *interval* or *segment* of the path. For all $i \in \{1, \ldots, l\}$ we say that the elements x_{i-1} and x_i are *consecutive* on the path, and also that x_{i-1} *precedes* x_i and x_i *follows* x_{i-1} on the path. Note that consecutive elements of a ρ-path can never be equal.

A ρ-path $(x_0, \ldots, x_l)$ is said to be *simple* if $x_i \neq x_j$ for all distinct i and j in $\{0, \ldots, l\}$. It is said to be *closed* if $x_0 = x_l$, so that x_0 follows x_{l-1}. It is called a *ρ-cycle* if it is closed and $x_i \neq x_j$ for all distinct i and j in $\{1, \ldots, l\}$. One-point paths are the simplest ρ-cycles. Two ρ-cycles $c_1 = (x_0, \ldots, x_l)$ and $c_2 = (y_0, \ldots, y_l)$ are said to be *equivalent* if there exists an integer k, $0 \leq k \leq l-1$, such that $x_i = y_{(i+k) \bmod l}$ for all $i \in \{0, \ldots, l\}$.

If $S \subseteq P$, two elements a and b of S are said to be *ρ-connected in S* if there exists a ρ-path from a to b that consists only of points in S. ρ-connectedness in S is an equivalence relation on S; its equivalence classes are called the *ρ-components of S*. The set S is said to be *ρ-connected* if there is just one ρ-component of S.

Given two sequences $c_1 = (x_0, \ldots, x_m)$ and $c_2 = (y_0, \ldots, y_n)$ such that $x_m = y_0$, we denote by $c_1.c_2$ the sequence $(x_0, \ldots, x_m, y_1, \ldots, y_n)$, which we call the *concatenation of c_1 and c_2*. Whenever we use the notation $c_1.c_2$, we are also implicitly saying that the last element of c_1 is the same as the first element

[1] ρ can be viewed as a binary, symmetric and irreflexive relation on P, and (P, ρ) as an undirected simple graph.

of c_2. It is clear that if c_1 and c_2 are ρ-paths of lengths l_1 and l_2, then $c_1.c_2$ is a ρ-path of length $l_1 + l_2$.

For any sequence $c = (x_0, \ldots, x_m)$, the *reverse of* c, denoted by c^{-1}, is the sequence $(y_0, \ldots, y_m)$ such that $y_k = x_{m-k}$ for all $k \in \{0, \ldots, m\}$. It is clear that if c is a ρ-path of length l then so is c^{-1}.

A *simple closed* ρ*-curve* is a nonempty finite ρ-connected set C such that each element of C has exactly two ρ-neighbors in C. (Note that a simple closed ρ-curve must have at least three elements.) A ρ-cycle c of length $|C|$ that contains every element of a simple closed ρ-curve C is called a ρ*-parameterization of* C. Note that if c and c' are ρ-parameterizations of a simple closed ρ-curve C, then c' is equivalent to c or to c^{-1}.

If x and y are ρ-adjacent elements of a simple closed ρ-curve C, then we may say that x and y are ρ*-consecutive* on C. If x and y are distinct elements of a simple closed ρ-curve C that are not ρ-consecutive on C, then each of the two ρ-components of $C \setminus \{x, y\}$ is called a ρ*-cut-interval* (of C) associated with x and y.

2.2 Definition of a GADS

Definition 1 (2D Digital Complex). *A* 2D digital complex *is an ordered triple* $(V, \pi, \mathcal{L})$*, where*

- V *is a set whose elements are called* vertices *or* spels,
- $\pi \subseteq V^{\{2\}}$*, and the pairs of vertices in* π *are called* proto-edges,
- $\mathcal{L}$ *is a set of simple closed* π*-curves whose members are called* loops,

and the following four conditions hold:

(i) V *is* π*-connected and contains more than one vertex.*
(ii) For any two distinct loops L_1 *and* L_2*,* $L_1 \cap L_2$ *is either empty, or consists of a single vertex, or is a proto-edge.*
(iii) No proto-edge is included in more than two loops.
(iv) Each vertex belongs to only a finite number of proto-edges.

When specifying a 2D digital complex whose vertex set is the set of points of a grid in $\mathbb{R}^n$, a positive integer k (such as 4, 8 or 6) may be used to denote the set of all unordered pairs of k-adjacent vertices. We write $\mathcal{L}_{2\times 2}$ to denote the set of all unit lattice squares in $\mathbb{Z}^2$. The triple $(\mathbb{Z}^2, 4, \mathcal{L}_{2\times 2})$ is a simple example of a 2D digital complex.

Definition 2 (GADS). *A* generic axiomatized digital surface-structure, *or* GADS, *is a pair* $\mathcal{G} = ((V, \pi, \mathcal{L}), (\kappa, \lambda))$ *where* $(V, \pi, \mathcal{L})$ *is a 2D digital complex (whose vertices, proto-edges and loops are also referred to as vertices, proto-edges and loops of* $\mathcal{G}$*) and where* κ *and* λ *are subsets of* $V^{\{2\}}$ *that satisfy Axioms 1, 2 and 3 below. The pairs of vertices in* κ *and* λ *are called* κ-edges *and* λ-edges, *respectively.* $(V, \pi, \mathcal{L})$ *is called the* underlying complex *of* $\mathcal{G}$.

Axiom 1. *Every proto-edge is both a κ-edge and a λ-edge:* $\pi \subseteq \kappa \cap \lambda$.

Axiom 2. *For all* $e \in (\kappa \cup \lambda) \setminus \pi$, *some loop contains both vertices of* e.

Axiom 3. *If* $x, y \in L \in \mathcal{L}$, *but* x *and* y *are not* π*-consecutive on* L, *then*

(a) $\{x, y\}$ *is a* λ*-edge if and only if* $L \setminus \{x, y\}$ *is not* κ*-connected.*
(b) $\{x, y\}$ *is a* κ*-edge if and only if* $L \setminus \{x, y\}$ *is not* λ*-connected.*

Regarding Axiom 2, note that if $e \in (\kappa \cup \lambda) \setminus \pi$ (i.e., e is a κ- or λ-edge that is not a proto-edge) then there can only be one loop that contains both vertices of e, by condition (ii) in the definition of a 2D digital complex.

As illustrations of Axiom 3, observe that both $((\mathbb{Z}^2, 4, \mathcal{L}_{2\times 2}), (4, 8))$ and $((\mathbb{Z}^2, 4, \mathcal{L}_{2\times 2}), (8, 4))$ satisfy Axiom 3, but $((\mathbb{Z}^2, 4, \mathcal{L}_{2\times 2}), (4, 4))$ violates the "if" parts of the axiom, while $((\mathbb{Z}^2, 4, \mathcal{L}_{2\times 2}), (8, 8))$ violates the "only if" parts of the axiom.

A GADS is said to be *finite* if it has finitely many vertices; otherwise it is said to be *infinite*. The set of all GADS can be ordered as follows:

Definition 3 ($\subseteq$ order, subGADS). *Let* $\mathcal{G} = ((V, \pi, \mathcal{L}), (\kappa, \lambda))$ *and* $\mathcal{G}' = ((V', \pi', \mathcal{L}'), (\kappa', \lambda'))$ *be* GADS *such that*

- $V \subseteq V'$, $\pi \subseteq \pi'$ *and* $\mathcal{L} \subseteq \mathcal{L}'$.
- *For all* $L \in \mathcal{L}$, $\kappa \cap L^{\{2\}} = \kappa' \cap L^{\{2\}}$ *and* $\lambda \cap L^{\{2\}} = \lambda' \cap L^{\{2\}}$.

Then we write $\mathcal{G} \subseteq \mathcal{G}'$ *and say that* $\mathcal{G}$ *is a* subGADS *of* $\mathcal{G}'$. *We write* $\mathcal{G} \subsetneq \mathcal{G}'$ *to mean* $\mathcal{G} \subseteq \mathcal{G}'$ *and* $\mathcal{G} \neq \mathcal{G}'$. *We write* $\mathcal{G} < \mathcal{G}'$ *to mean* $\mathcal{G} \subsetneq \mathcal{G}'$ *and* $\mathcal{L} \neq \mathcal{L}'$.

The following simple but important property of GADS is an immediate consequence of the symmetry of Axioms 1, 2 and 3 with respect to κ and λ:

Property 1. *If* $((V, \pi, \mathcal{L}), (\kappa, \lambda))$ *is a* GADS *then* $((V, \pi, \mathcal{L}), (\lambda, \kappa))$ *is also a* GADS. *So any statement which is true of every* GADS $((V, \pi, \mathcal{L}), (\kappa, \lambda))$ *remains true when* κ *is replaced by* λ *and* λ *by* κ.

2.3 Interior Vertices and pGADS

We are particularly interested in those GADS that model a surface without boundary. The next definition gives a name for any such GADS.

Definition 4 (pGADS). *A* pGADS *is a* GADS *in which every proto-edge is included in two loops. (The* p *in* pGADS *stands for* pseudomanifold.*)*

A finite pGADS models a closed surface. A pGADS that models the Euclidean plane must be infinite.

A vertex v of a GADS $\mathcal{G}$ is called an *interior vertex* of $\mathcal{G}$ if every proto-edge of $\mathcal{G}$ that contains v is included in two loops of $\mathcal{G}$. If follows that a GADS $\mathcal{G}$ is a pGADS if and only if every vertex of $\mathcal{G}$ is an interior vertex.

Below are pictures of some pGADS.

Example 1. $\mathbb{Z}^2$ *with the 4- and 8-adjacency relations*

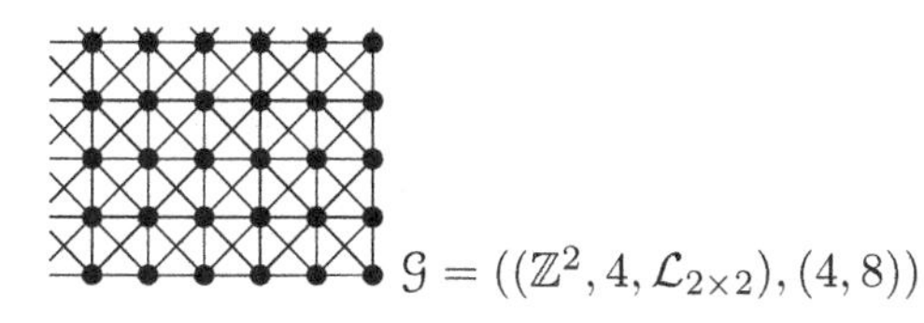

$\mathcal{G} = ((\mathbb{Z}^2, 4, \mathcal{L}_{2\times 2}), (4, 8))$

Example 2. $\mathbb{Z}^2$ *with Khalimsky's adjacency relation*

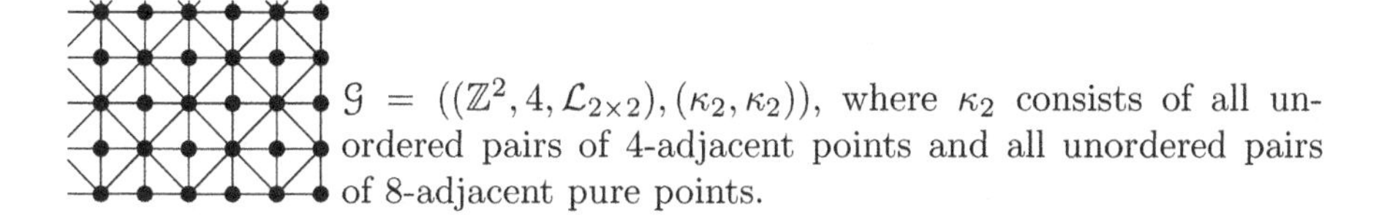

$\mathcal{G} = ((\mathbb{Z}^2, 4, \mathcal{L}_{2\times 2}), (\kappa_2, \kappa_2))$, where κ_2 consists of all unordered pairs of 4-adjacent points and all unordered pairs of 8-adjacent pure points.

Example 3. *A torus-like* pGADS

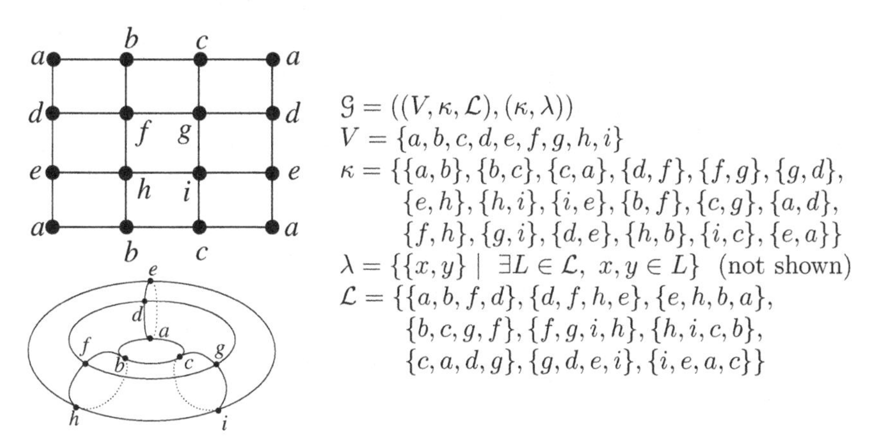

$\mathcal{G} = ((V, \kappa, \mathcal{L}), (\kappa, \lambda))$
$V = \{a, b, c, d, e, f, g, h, i\}$
$\kappa = \{\{a, b\}, \{b, c\}, \{c, a\}, \{d, f\}, \{f, g\}, \{g, d\}, \{e, h\}, \{h, i\}, \{i, e\}, \{b, f\}, \{c, g\}, \{a, d\}, \{f, h\}, \{g, i\}, \{d, e\}, \{h, b\}, \{i, c\}, \{e, a\}\}$
$\lambda = \{\{x, y\} \mid \exists L \in \mathcal{L},\ x, y \in L\}$ (not shown)
$\mathcal{L} = \{\{a, b, f, d\}, \{d, f, h, e\}, \{e, h, b, a\}, \{b, c, g, f\}, \{f, g, i, h\}, \{h, i, c, b\}, \{c, a, d, g\}, \{g, d, e, i\}, \{i, e, a, c\}\}$

In the sequel of this paper, $\mathcal{G} = ((V, \pi, \mathcal{L}), (\kappa, \lambda))$ is a GADS.

3 Simple Points in a GADS

In this section, we will define *simple points* in a GADS. In the classical meaning, simple points are points that can be deleted while preserving the topology of an image. By "preserving the topology" we mean preserving connectivity and holes. Proving that our definition is suitable in this sense will be the purpose of Section 5.

Several definitions and characterizations have been given for simple points in classical (2D or 3D) digital spaces. See for example [17, 1, 15, 19] for an overview. Our purpose here is to state a definition and a characterization within the generic framework of GADS, thus generalizing this classical notion for any admissible "surface like" digital space.

We can give in intuitive words a first definition of a simple point. Indeed, a point $x \in X \subset V$ is said to be κ-simple for X if and only if:

- X and $X \setminus \{x\}$ have the same number of κ-connected components;
- $\overline{X}$ and $\overline{X} \cup \{x\}$ have the same number of λ-connected components;
- X and $X \setminus \{x\}$ have the same holes.

We will now define a few notations that will allow us to state a formal definition of a simple point.

For any vertex v of $\mathcal{G}$, the *punctured loop neighborhood* of v in $\mathcal{G}$, denoted by $N_{\mathcal{L}}(v)$, is defined to be the union of all the loops of $\mathcal{G}$ which contain v, minus the vertex v itself.

Let $x \in V$. The axioms given in Section 2 somehow guarantee that loops are topological disks. However, $N_{\mathcal{L}}(x)$ needs not to be a topological disk (see the punctured loop neighborhood of any of the points in Example 3). Thus, we need to define a topology on $N_{\mathcal{L}}(x)$ under which it is a topological disk. Let y and y' be two points of $N_{\mathcal{L}}(x) \cup \{x\}$. We say that y and y' are κ_x-adjacent (respectively λ_x-adjacent) if $\{y, y'\} \in \kappa$ (respectively $\{y, y'\} \in \lambda$) and y and y' are both contained in a loop containing x. If $X \subset V$, we denote by $G_\kappa(x, X)$ (resp. $G_\lambda(x, X)$) the graph whose vertices are the points of $N_{\mathcal{L}}(x) \cap X$ and whose edges are pairs of κ_x-adjacent points (resp. λ_x-adjacent points) of $N_{\mathcal{L}}(x) \cap X$. Let $\rho = \kappa$ or $\rho = \lambda$. We denote by $\mathcal{C}_\rho^x(G_\rho(x, X))$ the set of connected components of $G_\rho(x, X)$ that are ρ-adjacent to x. Note that $\mathcal{C}_\rho^x(G_\rho(x, X))$ is a set of subsets of points of V and not a set of points.

Definition 5. *We call x a ρ-isolated point of X if $N_\rho(x) \cap X = \emptyset$ and a ρ-interior point if $N_\rho(x) \cap \overline{X} = \emptyset$.*

We can now state our definition of a simple point, which is also a local characterization.

Definition 6 (Simple Point). *A point x is said to be κ-simple in X if and only if the number $\mathtt{Card}(\mathcal{C}_\kappa^x(G_\kappa(x, X)))$ of connected components of $G_\kappa(x, X)$ which are κ-adjacent to x is equal to 1 and x is not interior to X.*

The following Lemma is a first step towards the justification of Definition 6.

Lemma 1. *Let $X \subset V$ and $x \in X$ be a κ-simple point of X. Then:*

- *X and $X \setminus \{x\}$ have the same number of κ-connected components;*
- *$\overline{X}$ and $\overline{X} \cup \{x\}$ have the same number of λ-connected components.*

We just give here the main argument for the proof of this Lemma. In order to prove that no κ-connected component is created, just consider two κ-connected points in X that are no longer connected in $X \setminus \{x\}$. If c is a shortest κ-path in X between the two points, then the point a that precedes x in c and the point b that follows x in c are both contained in C, the only connected component of $G_\kappa(x, X)$ which is κ-adjacent to x. Therefore, a and b are κ-connected in $X \setminus \{x\}$, like the two initial points.

It remains to be proved that the removal of a simple point preserves holes. This will be the purpose of Section 5.

4 Homotopic Paths and the Digital Fundamental Group of a GADS

In this section, ρ is a subset of $V^{\{2\}}$ such that $\rho \in \{\kappa, \lambda, \pi\}$, and X is a ρ-connected subset of V.

Loosely speaking, two ρ-paths in X with the same initial and the same final vertices are said to be *ρ-homotopic* within X in $\mathcal{G}$ if one of the paths can be transformed into the other by a sequence of small local deformations within X. The initial and final vertices of the path must remain fixed throughout the deformation process. The next two definitions make this notion precise.

Definition 7 (Elementary $\mathcal{G}$-Deformation). *Two finite vertex sequences c and c' of $\mathcal{G}$ with the same initial and the same final vertices are said to be* the same up to an elementary $\mathcal{G}$-deformation *if there exist vertex sequences c_1, c_2, γ and γ' such that $c = c_1.\gamma.c_2$, $c' = c_1.\gamma'.c_2$, and* either *there is a proto-edge $\{x, y\}$ for which one of γ and γ' is (x) and the other is (x, y, x),* or *there is a loop of $\mathcal{G}$ that contains all of the vertices in γ and γ'.*

Definition 8 (Homotopic ρ-Paths). *Two ρ-paths c and c' in X with the same initial and the same final vertices are* ρ-homotopic within X in $\mathcal{G}$ *if there exists a sequence of ρ-paths $c_0, \ldots, c_n$ in X such that $c_0 = c$, $c_n = c'$ and, for $0 \leq i \leq n-1$, c_i and c_{i+1} are the same up to an elementary $\mathcal{G}$-deformation. Two ρ-paths with the same initial and the same final vertices are said to be* ρ-homotopic in $\mathcal{G}$ *if they are ρ-homotopic within V in $\mathcal{G}$.*

The next proposition states a useful characterization of ρ-homotopy that is based on a more restrictive kind of local deformation than was considered above, which allows only the insertion or removal of either a "ρ-back-and-forth" or a cycle that parameterizes a simple closed ρ-curve in a loop of $\mathcal{G}$.

Definition 9 (Minimal ρ-Deformation). *Two ρ-paths c and c' with the same initial and the same final vertices are said to be* the same up to a minimal ρ-deformation in $\mathcal{G}$ *if there exist ρ-paths c_1, c_2 and γ such that one of c and c' is $c_1.\gamma.c_2$, the other of c and c' is $c_1.c_2$, and* either *$\gamma = (x, y, x)$ for some ρ-edge*

$\{x, y\}$ or γ *is a* ρ*-parameterization of a simple closed* ρ*-curve whose vertices are contained in a single loop of* $\mathcal{G}$.

This concept of deformation is particularly simple when $\rho = \pi$, because a simple closed π-curve whose vertices are contained in a single loop of $\mathcal{G}$ must in fact be a loop of $\mathcal{G}$, since a loop of $\mathcal{G}$ is a simple closed π-curve.

Proposition 1. *Two* ρ*-paths* c *and* c' *in* X *with the same initial and the same final vertices are* ρ*-homotopic within* X *in* $\mathcal{G}$ *if and only if there is a sequence of* ρ*-paths* $c_0, \ldots, c_n$ *in* X *such that* $c_0 = c$, $c_n = c'$ *and, for* $0 \leq i \leq n-1$, c_i *and* c_{i+1} *are the same up to a minimal* ρ*-deformation in* $\mathcal{G}$.

The proof of this proposition is not particularly difficult, and we leave it to the interested reader.

Now, let $b \in X$ be a point called the *base point.* We denote by $A_b^\rho(X)$ the set of all closed ρ-paths $c = (x_0, \ldots, x_l)$ which are contained in X such that $x_0 = x_l = b$. The ρ-homotopy relation is an equivalence relation on $A_b^\rho(X)$, and we denote by $\Pi_1^\rho(X, b)$ the set of equivalence classes of this equivalence relation. The concatenation of paths is compatible with the ρ-homotpy relation, hence it defines an operation on $\Pi_1^\rho(X, b)$ which to the class of c_1 and c_2 associates the class of $c_1.c_2$. This operation provides $\Pi_1^\rho(X, b)$ with a group structure. We call this group the *digital* ρ*-fundamental group of* X.

Now, we consider $Y \subset X \subset V$ and $b \in X$ a base point. Any closed ρ-path in Y is a particular case of a closed ρ-path in X. Furthermore, if two closed ρ-paths in Y are ρ-homotopic in Y, then they are also ρ-homotopic in X. These two properties enable us to define a canonical morphism $i_* : \Pi_1^\rho(Y, b) \longrightarrow \Pi_1^\rho(X, b)$ induced by the inclusion map $i : Y \longrightarrow X$. To the class of a closed ρ-path $c \in A_b^\rho(Y)$ in $\Pi_1^\rho(Y, b)$ the morphism i_* associates the class of the same ρ-path in $\Pi_1^\rho(X, b)$.

5 Simple Points and the Digital Fundamental Group

Here, we show that simple points have been properly defined. For this purpose, we use the formalism of the digital fundamental group. Indeed, it allows us to prove that "holes are preserved" when one removes a simple point from X, a subset of the set of points of a GADS.

In this section, ρ is either equal to κ or equal to λ, and X is a ρ-connected subset of V.

Lemma 2. *Let* $b \in X$ *and let* $x \in X$ *be a* ρ*-simple point distinct from* b. *Then any* ρ*-path of* $A_b^\rho(X)$ *is* ρ*-homotopic to a* ρ*-path contained in* $X \setminus \{x\}$.

Proof: Let $c = (x_0, \ldots, x_p)$ be a ρ-path in X such that $x_0 \neq x$ and $x_p \neq x$. We define a ρ-path $P(c)$ as follows: For any maximal sequence $\gamma = (x_k, \ldots, x_l)$ with $0 \leq k \leq l \leq p$ of points of c such that for all $i = k, \ldots, l$ we have $x_i \neq x$, we define $s(\gamma) = \gamma$. For any maximal sequence $\gamma = (x_k, \ldots, x_l)$ with $0 \leq k \leq l \leq p$

of points of c such that for $i = k, \ldots, l$ we have $x_i = x$, we define $s(\gamma)$ as equal to the shortest ρ-path from x_{l-1} to x_{k+1} in the single connected component of $G_\rho(x, X)$. Now $P(c)$ is the concatenation of all $s(\gamma)$ for all maximal sequences $\gamma = (x_k, \ldots, x_l)$ of points of c such that either for $i = k, \ldots, l$ we have $x_i \neq x$ or for $i = k, \ldots, l$ we have $x_i = x$. Now, it is readily seen that c is ρ-homotopic to $P(c)$ in X. □

Remark 1. *If $x \in X$ is a ρ-simple point in X and C is the single connected component of $G_\rho(x, X)$, then any two ρ-paths in C with the same extremities are ρ-homotopic in $X \setminus \{x\}$.*

Lemma 3. *Let $b \in X$ and let $x \in X$ be a ρ-simple point distinct from b. If two closed ρ-paths c_1 and c_2 of $A_b^\rho(X \setminus \{x\})$ are ρ-homotopic in X, then they are ρ-homotopic in $X \setminus \{x\}$.*

Proof: Let $P(c_1)$ and $P(c_2)$ be the two paths as defined in the proof of Lemma 2. Following Proposition 1, it is sufficient to prove that if c_1 and c_2 are the same up to a minimal ρ-deformation in X, then the two ρ-paths $P(c_1)$ and $P(c_2)$ are ρ-homotopic in $X \setminus \{x\}$. Thus, we suppose that $c_1 = c.\gamma.c'$ and $c_2 = c.c'$ where γ is either equal to (a, b, a) (with $\{a, b\}$ being a ρ-edge) or is a simple closed ρ-curve included in a loop L of $\mathcal{L}$. First, if we suppose that x does not belong to γ, then it is immediate that $P(c_1)$ and $P(c_2)$ are ρ-homotopic. Therefore, we may suppose without loss of generality that x belongs to γ. Furthermore, in order to clarify the proof, we suppose in the sequel that all paths are such that any two consecutive points are distincts. Let C be the only connected component of $G_\rho(x, X)$.

In a first case, we suppose that $\gamma = (a, b, a)$ whith $x = a$. We write $c_1 = \mu.(y, x, z).\mu'$ with $c = \mu.(y, x)$ and $c' = (x, z).\mu'$. Thus, we have $c_2 = \mu.(y, x, b, x, z).\mu'$. Now, y, z and b all belong to C. Let α be the shortest ρ-path from y to b in C and let β be the shortest ρ-path from b to z in C. Finally, let γ' be the shortest ρ-path from y to z in C. We have $P(c_1) = P(\mu).\gamma'.P(\mu')$ and $P(c_2) = P(\mu).\alpha.\beta.P(\mu')$. From the previous remark and since γ' and $\alpha.\beta$ are ρ-paths in C with the same extremities, we obtain that $P(c_1)$ and $P(c_2)$ are ρ-homotopic in $X \setminus \{x\}$.

The proof is similar in the case when γ is a closed ρ-path from the vertex x to x.

In the case when $\gamma = (a, b, a)$ with $x = b$ we obtain that $P(c_1) = P(c_2)$.

Remains the case when γ is a simple closed ρ-curve included in a loop and containing x (not being an extremity). But in this case, $P(\gamma)$ is included in C and therefore ρ-homotopic to the path reduced to its extremities. Since in this case $P(c_1) = P(c).P(\gamma).P(c')$ and $P(c_2) = P(c).P(c')$, we obtain that $P(c_1)$ and $P(c_2)$ are ρ-homotopic. □

Theorem 1. *Let $b \in X$ and let $x \in X$ be a ρ-simple point of X distinct from b. The morphism $i_* : \Pi_1^\rho(X \setminus \{x\}, b) \longrightarrow \Pi_1^\rho(X, b)$ induced by the inclusion map of $X \setminus \{x\}$ in X is a group isomorphism.*

Proof: Lemma 2 implies that i_* is onto and Lemma 3 implies that i_* is one to one. □

The latter lemma is the main result of this section. Indeed, it states that when one removes a simple point (following Definition 6) from a connected set X, then no hole is created nor removed. This, added to the fact that the removal of a simple point cannot create some new connected components nor remove any one (see Lemma 1), leads to the justification of the following affirmation:

"Removing a simple point preserves the topology."

This achieves the justification of the local characterization of simple points in a GADS given by Definition 6.

6 Concluding Remarks

We have introduced two new notions in the context of GADS: the digital fundamental group and the notion of simple point. In so doing, we illustrate the power of this axiomatic theory that allows us to prove very general results of digital topology. Indeed, the two previously mentionned notions are now valid for any two dimensional digital space that one can reasonably consider.

References

1. G. Bertrand. Simple points, topological numbers and geodesic neighborhoods in cubics grids. *Patterns Recognition Letters*, 15:1003–1011, 1994.
2. S. Fourey, T. Y. Kong, and G. T. Herman. Generic axiomatized digital surface-structures. *Discrete Applied Mathematics*, 139:65–93, April 2004.
3. S. Fourey and R. Malgouyres. Intersection number and topology preservation within digital surfaces. In *Proceedings of the Sixth International Workshop on Parallel Image Processing and Analysis (IWPIPA '99, Madras, India, January 1999)*, pages 138–158, 1999.
4. S. Fourey and R. Malgouyres. Intersection number of paths lying on a digital surface and a new Jordan theorem. In G. Bertrand, M. Couprie, and L. Perroton, editors, *Discrete Geometry for Computer Imagery: 8th International Conference (DGCI '99, Marne la Vallée, France, March 1999), Proceedings*, pages 104–117. Springer, 1999.
5. S. Fourey and R. Malgouyres. Intersection number and topology preservation within digital surfaces. *Theoretical Computer Science*, 283(1):109–150, June 2002.
6. S. Fourey and R. Malgouyres. A concise characterization of 3D simple points. *Discrete Applied Mathematics*, 125(1):59–80, January 2003.
7. G. T. Herman. Oriented surfaces in digital spaces. *Graphical Models and Image Processing*, 55:381–396, 1993.
8. G. T. Herman. *Geometry of digital spaces.* Birkhäuser, 1998.
9. G. T. Herman and J. K. Udupa. Display of 3D discrete surfaces. In *Proceeddings of SPIE*, volume 283, 1983.

10. E. D. Khalimsky, R. D. Kopperman, and P. R. Meyer. Computer graphics and connected topologies on finite ordered sets. *Topology and Its Applications*, 36:1–17, 1990.
11. T. Y. Kong. A digital fundamental group. *Computers and Graphics*, 13:159–166, 1989.
12. T. Y. Kong. On topology preservation in 2-d and 3-d thinning. *International Journal of Pattern Recognition and Artificial Intelligence*, 9(5):813–844, 1995.
13. T. Y. Kong and E. D. Khalimsky. Polyhedral analogs of locally finite topological spaces. In R. M. Shortt, editor, *General Topology and Applications: Proceedings of the 1988 Northeast Conference*, pages 153–164. Marcel Dekker, 1990.
14. T. Y. Kong, A. W. Roscoe, and A. Rosenfeld. Concepts of digital topology. *Topology and Its Applications*, 46:219–262, 1992.
15. T. Y. Kong and A. Rosenfeld. Digital topology : introduction and survey. *Computer Vision, Graphics and Image Processing*, 48:357–393, 1989.
16. R. Malgouyres and A. Lenoir. Topology preservation within digital surfaces. *Graphical Models (GMIP)*, 62:71–84, 2000.
17. A. Rosenfeld. Connectivity in digital pictures. *Journal of the Association for Computing Machinery*, 17:146–160, 1970.
18. A. Rosenfeld, T.Y. Kong, and A. Nakamura. Topology-preserving deformations of two-valued digital pictures. *Graphical Models and Image Processing*, 60(1):24–34, January 1998.
19. Azriel Rosenfeld, T. Yung Kong, and A. Nakamura. Topolgy-preserving deformations of two-valued digital pictures. *Graphical Models and Image Processing*, 60(1):24–34, January 1998.
20. J. K. Udupa. Multidimensional digital boundaries. *CVGIP: Graphical Models and Image Processing*, 56:311–323, 1994.

Minimal Non-simple Sets in 4-Dimensional Binary Images with (8,80)-Adjacency

T. Yung Kong[1] and Chyi-Jou Gau[2]

[1] Department of Computer Science, Queens College, City University of New York, 65-30 Kissena Boulevard, Flushing, NY 11367, U.S.A.
[2] Doctoral Program in Computer Science, Graduate School and University Center, City University of New York, 365 Fifth Avenue, New York, NY 10016, U.S.A.

Abstract. We first give a definition of simple sets of 1's in 4D binary images that is consistent with "(8,80)-adjacency"—i.e., the use of 8-adjacency to define connectedness of sets of 1's and 80-adjacency to define connectedness of sets of 0's. Using this definition, it is shown that in any 4D binary image every *minimal* non-simple set of 1's must be isometric to one of eight sets, the largest of which has just four elements. Our result provides the basis for a fairly general method of verifying that proposed 4D parallel thinning algorithms preserve topology in our "(8,80)" sense. This work complements the authors' earlier work on 4D minimal non-simple sets, which essentially used "(80,8)-adjacency"—80-adjacency on 1's and 8-adjacency on 0's.

1 Introduction

In this paper we use the term n-dimensional (or nD) *binary image* to mean a partition of $\mathbb{Z}^n$ into two subsets, one of which is a finite set of points called 1's, and the other of which is a (necessarily infinite) set of points called 0's. Our main goal is to establish the result stated as the Main Theorem below. In addition to its theoretical interest, this result provides the basis for a fairly general method of verifying that a proposed parallel thinning algorithm for 4D binary images "preserves topology" for all possible input images, in a sense that is consistent with the use of 8-adjacency to define connectedness within the set of 1's of the image (i.e., within the set that is thinned) and the use of 80-adjacency to define connectedness within the set of 0's.

Here 8- and 80-adjacency are symmetric binary relations on $\mathbb{Z}^4$ that are analogous to the familiar concepts of 4- and 8-adjacency on $\mathbb{Z}^2$ and the concepts of 6- and 26-adjacency on $\mathbb{Z}^3$: Two points $x, y \in \mathbb{Z}^4$ are *80-adjacent* if $x \neq y$ but each coordinate of x differs from the same coordinate of y by at most 1. Two points $x, y \in \mathbb{Z}^4$ are *8-adjacent* if they are 80-adjacent and they differ in just one of the four coordinates. For $\kappa = 80$ or 8, each point in $\mathbb{Z}^4$ is κ-adjacent to just κ points. Note that 8-adjacency on $\mathbb{Z}^4$ is a 4D analog of *4*-adjacency on $\mathbb{Z}^2$ (and not of 8-adjacency on $\mathbb{Z}^2$).

Let I be any subset of $\mathbb{Z}^4$. A point $p \in I$ is said to be (8,80)-*simple* in I if the deletion of $\{p\}$ from I preserves topology in the sense of Definition 1 below.

R. Klette and J. Žunić (Eds.): IWCIA 2004, LNCS 3322, pp. 318–333, 2004.

A subset S of I is said to be (8,80)-*simple* in I if S is finite and the points of S can be arranged in a sequence whose first point is (8,80)-simple in I, and each of whose subsequent points is (8,80)-simple in the set obtained from I by deleting all of the preceding points. An immediate consequence of this definition is that if $p \in I$ then the singleton set $\{p\}$ is (8,80)-simple in I if and only if the point p is (8,80)-simple in I. Another fundamental property of (8,80)-simple sets, which follows from Definition 1, is that if a set $D \subset I$ is (8,80)-simple in I, then the deletion of D from I preserves topology in the sense of Definition 1. We say that a subset S of I is *hereditarily* $(8,80)$-*simple* in I if every subset of S (including S itself) is $(8,80)$-simple in I.

Given sets $S \subseteq I \subseteq \mathbb{Z}^4$, we say that S is (8,80)-*minimal-non-simple* (or (8,80)-*MNS*) in I if S is not (8,80)-simple in I, but every proper subset of S is (8,80)-simple in I. Since (8,80)-simple sets are finite, so are (8,80)-MNS sets. Evidently, any finite subset of I that is not hereditarily (8,80)-simple in I must contain an (8,80)-MNS set of I. We say that a set $S \subset \mathbb{Z}^4$ *can be* $(8,80)$-*MNS* if there exists a set $I \subseteq \mathbb{Z}^4$ such that $S \subseteq I$ and S is (8,80)-MNS in I.

Our Main Theorem will tell us that there are, up to isometry, just eight different sets that can be (8,80)-MNS: In fact, it will say that a subset of $\mathbb{Z}^4$ can be (8,80)-MNS if and only if it is a set of one of the following eight kinds:

A. a singleton set
B. a set of two 8-adjacent points
C. a set of two diagonally opposite points of a 2×2 square
D. a set of two diametrically opposite points of a $2 \times 2 \times 2$ cube
E. a set of two diametrically opposite points of a $2 \times 2 \times 2 \times 2$ block
F. a set of three points in a $2 \times 2 \times 2$ cube that are the vertices of an equilateral triangle with a side-length of $\sqrt{2}$
G. a set of three points in a $2 \times 2 \times 2 \times 2$ block that are the vertices of an isosceles triangle whose side-lengths are $\sqrt{3}$, $\sqrt{3}$, and $\sqrt{2}$
H. a set of four points in a $2 \times 2 \times 2 \times 2$ block such that there is a fifth point in the block (but not in the set) that is 8-adjacent to each of the four

To see how the Main Theorem can be used to verify that a proposed 4D parallel thinning algorithm preserves topology (in the "(8,80)" sense of Definition 1) for all possible input binary images, suppose $\mathbf{O}$ is a 4D parallel local operator[1] that the thinning algorithm applies at some iteration or subiteration. How can we show that whenever the parallel operator $\mathbf{O}$ is applied to a 4D binary image, the set of 1's that are deleted by $\mathbf{O}$ is (8,80)-simple in the set of all 1's of that image?

The Main Theorem tells us that we can establish this by verifying that the parallel local operator $\mathbf{O}$ has the following property:

Property **A**: For every set $S \subset \mathbb{Z}^4$ of the eight kinds A – H, S is (8,80)-simple in I whenever I satisfies the following conditions:

[1] This is the 4D analog of the familiar concept of a parallel local 2D image processing operator (as defined, e.g., on p. 41 of [8]). Operators used for thinning binary images delete certain 1's (i.e., change those 1's to 0's) but leave 0's unchanged.

(i) $S \subseteq I \subseteq \mathbb{Z}^4$.
(ii) Every point of S is deleted when **O** is applied to the 4D binary image whose set of 1's is I.

It is easy to see that if Property **A** holds then, when **O** is applied to a 4D binary image whose set of 1's is J (say), the set of 1's that are deleted by **O** must actually be hereditarily (8,80)-simple in J. For otherwise the set of deleted 1's would contain some set S that is (8,80)-minimal-non-simple in J, and (by the Main Theorem) such an (8,80)-MNS set S must be a set of one of the eight kinds A – H, which would contradict Property **A** (since **A** says all such sets are (8,80)-simple in J and therefore are not (8,80)-MNS in J).

The concept of a minimal non-simple set, and the idea of proving that a parallel local deletion operator always preserves topology by verifying that it never deletes a minimal non-simple subset of the 1's, were introduced by Ronse [14] in the 1980's for 2D Cartesian grids. Hall [5] showed how Ronse's proof method was essentially equivalent to an older, path-based, proof method introduced by Rosenfeld in the 1970's [15]; ref. [5] also extended Ronse's proof method to a 2D hexagonal grid. In the 1990's Ma and Kong investigated minimal non-simple sets on a 3D Cartesian grid [10, 13], and the present authors studied minimal non-simple sets on a 3D face-centered cubic grid [3]. The concepts of minimal non-simple and hereditarily simple sets are related to Bertrand's concept of a *P-simple point* [1]: A subset P of a finite set $I \subseteq \mathbb{Z}^n$ is hereditarily simple in I if and only if every point of P is P-simple in I.

Exactly which sets of points can be minimal non-simple? This fundamental question was answered by the above-mentioned work of Ronse, Ma, and the authors in two and three dimensions, for definitions of topology-preserving deletion that are consistent with $(8,4)$- or $(4,8)$-adjacency on the 2D Cartesian grid, $(26,6)$-, $(6,26)$-, $(18,6)$-, or $(6,18)$-adjacency on the 3D Cartesian grid, and $(18,12)$-, $(12,18)$-, or $(12,12)$-adjacency on the 3D face-centered cubic grid.[2] Here *(κ, λ)-adjacency* refers to the use of κ-adjacency to define connectedness on the set from which grid points are to be deleted, and the use of λ-adjacency to define connectedness on the complement of that set. In a more recent paper [4], the authors answered the same question for the 4D Cartesian grid and $(80,8)$-adjacency. Our Main Theorem answers this question for the 4D Cartesian grid and $(8,80)$-adjacency.

2 The Xel $X(p)$, the Complementary Polyhedron $CP(S)$, and a Definition of $(2n, 3^n - 1)$-Topology Preservation

In this section n is an arbitrary positive integer, I denotes an arbitrary subset of $\mathbb{Z}^n$, and D denotes an arbitrary finite subset of I. We are going to define

[2] On the face-centered cubic grid, 12- and 18-adjacency relate each grid point to each of its 12 nearest neighbors and 18 nearest neighbors, respectively.

what we mean when we say that the deletion of D from I "preserves topology". (The statement and proof of our Main Theorem will actually depend only on the $n = 4$ case of this and other concepts defined in the present section and the next. But consideration of lower-dimensional cases may help the reader to understand the 4D case.) Our definition of topology-preserving deletion of D from I will be consistent with the use of $2n$-adjacency to define connectedness within I and the use of $(3^n - 1)$-adjacency to define connectedness within $\mathbb{Z}^n \setminus I$.

Here $2n$- and $(3^n - 1)$-adjacency are symmetric binary relations on $\mathbb{Z}^n$ that generalize the 4- and 8-adjacency relations on $\mathbb{Z}^2$: Writing p_i for the ith coordinate of a point p in $\mathbb{Z}^n$, x is *$2n$-adjacent to y*, and is a *$2n$-neighbor of y*, if and only if $\sum_{i=1}^{n} |x_i - y_i| = 1$; x is *$(3^n - 1)$-adjacent to y*, and is a *$(3^n - 1)$-neighbor of y*, if and only if $\max_{1 \le i \le n} |x_i - y_i| = 1$. For all $p \in \mathbb{Z}^n$, we write $N(p)$ to denote the set consisting of p and its $(3^n - 1)$-neighbors.

Now let $\alpha = 2n$ or $3^n - 1$, and let $S \subseteq \mathbb{Z}^n$. Then the restriction to S of the reflexive transitive closure of the α-adjacency relation is an equivalence relation on S; each of its equivalence classes is called an *α-component* of S. The set S is *α-disconnected* if it consists of more than one α-component; otherwise S is *α-connected*. An *α-path* is a nonempty sequence of points of $\mathbb{Z}^n$ such that each point of the sequence after the first is α-adjacent to its immediate predecessor; an α-path whose first and last points are respectively p and q is called an *α-path from p to q*.

We will use the term *n-xel in $\mathbb{R}^n$* to mean an "upright" closed n-dimensional unit hypercube in Euclidean n-space $\mathbb{R}^n$ that is centered on a point in $\mathbb{Z}^n$. Thus an n-xel in $\mathbb{R}^n$ is a Cartesian product $[i_1 - 0.5, i_1 + 0.5] \times \ldots \times [i_n - 0.5, i_n + 0.5]$ of n closed unit intervals whose centers $i_1, \ldots, i_n$ are all integers. For each point p in $\mathbb{Z}^n$, we write $X(p)$ to denote the n-xel in $\mathbb{R}^n$ that is centered at p.

Let S be any subset of $\mathbb{Z}^n$. Then the *open consolidation* of S, denoted by $\mathcal{O}(S)$, and the *complementary polyhedron*[3] of S, denoted by $\mathcal{CP}(S)$, are defined as follows (where **int** is the topological interior operator):

$$\mathcal{O}(S) = \textbf{int} \bigcup_{p \in S} X(p) \quad \text{and} \quad \mathcal{CP}(S) = \mathbb{R}^n \setminus \mathcal{O}(S) = \bigcup_{p \in \mathbb{Z}^n \setminus S} X(p)$$

It is readily confirmed that the connected components of $\mathcal{O}(S)$ and of $\mathcal{CP}(S)$ (with the usual, Euclidean, topology) are related to the $2n$-components of S and the $(3^n - 1)$-components of $\mathbb{Z}^n \setminus S$ as follows:

I. The set $\{E \cap \mathbb{Z}^n \mid E \text{ is a connected component of } \mathcal{O}(S)\}$ is exactly the set of all $2n$-components of S.
II. The set $\{C \cap \mathbb{Z}^n \mid C \text{ is a connected component of } \mathcal{CP}(S)\}$ is exactly the set of all $(3^n - 1)$-components of $\mathbb{Z}^n \setminus S$.

We now define topology-preserving deletion in terms of the complementary polyhedron:

[3] We use the term *polyhedron* to mean a set that is a union of a locally finite collection of simplexes. A polyhedron is a closed set, but need not be bounded.

Definition 1. *Let $I \subseteq \mathbb{Z}^n$ and let D be a finite subset of I. Then we say that the deletion of D from I is* $(2n, 3^n - 1)$**-topology-preserving** *if there is a deformation retraction of $\mathcal{CP}(I \setminus D)$ onto $\mathcal{CP}(I)$.*

This definition says that the deletion of D from I is $(2n, 3^n - 1)$-topology-preserving if and only if the complementary polyhedron of $I \setminus D$ can be continuously deformed over itself onto the complementary polyhedron of I *in such a way that all points that are originally in the latter set remain fixed throughout the deformation process.*[4] It can be shown that if such a continuous deformation exists then each connected component of $\mathcal{O}(I)$ contains just one connected component of $\mathcal{O}(I \setminus D)$, and each connected component of $\mathcal{CP}(I \setminus D)$ contains just one connected component of $\mathcal{CP}(I)$. It follows from this, and facts I and II above, that if the deletion of D from I is $(2n, 3^n - 1)$-topology-preserving then:

1. Each $2n$-component of I contains just one $2n$-component of $I \setminus D$.
2. Each (3^n-1)-component of $\mathbb{Z}^n \setminus (I \setminus D)$ contains just one (3^n-1)-component of $\mathbb{Z}^n \setminus I$.

Condition 1 says that no $2n$-component of I is split or completely eliminated as a result of the deletion of D. Condition 2 says that no two (3^n-1)-components of $\mathbb{Z}^n \setminus I$ are merged as a result of the deletion of D, and also that no new (3^n-1)-component of $\mathbb{Z}^n \setminus I$ is created.

We say that a point $p \in I$ is $(2n, 3^n - 1)$-*simple* in I if the deletion of $\{p\}$ from I is $(2n, 3^n - 1)$-topology-preserving. Thus p is $(2n, 3^n - 1)$-simple in I if and only if there is a deformation retraction of the polyhedron $\mathcal{CP}(I \setminus \{p\})$ onto the polyhedron $\mathcal{CP}(I)$.

While this definition of $(2n, 3^n - 1)$-simple points involves continuous deformation, Theorem 1 below gives essentially discrete necessary and sufficient local conditions for p to be $(2n, 3^n - 1)$-simple in I when $n \leq 4$. Our proof of the Main Theorem will be based on this discrete local characterization of $(2n, 3^n - 1)$-simple points.

In the case $n = 2$, $(2n, 3^n - 1) = (4, 8)$, it can be shown that the conditions 1 and 2 above are sufficient as well as necessary for the deletion of D from I to be $(4, 8)$-topology-preserving in the sense of Definition 1. These two conditions can arguably be regarded as the "standard" definition of (4,8)-topology preservation when a finite set D of 1's is deleted from the set I of all 1's of a 2D binary image—see, e.g., [12–p. 366] or [6–p. 156].

In the case $n = 3$, $(2n, 3^n - 1) = (6, 26)$, there is no "standard" definition of what it means for the deletion of D from I to be (6,26)-topology-preserving if D contains more than one point. However one can deduce from Theorem 1 below that the standard concept of a $(6, 26)$-simple point (see, e.g., [2–p. 117]) is equivalent to our concept of such a point.

[4] Deformation retraction is defined in, e.g., [7]. This concept is sometimes called *strong* deformation retraction (as in [16]).

3 A Discrete Local Characterization of $(2n, 3^n - 1)$-Simple Points, for $1 \leq n \leq 4$

In Section 2 we defined n-xels in $\mathbb{R}^n$. We now define the more general concept of a k-xel in $\mathbb{R}^n$, where k may be less than n. For any integer i, the singleton set $\{i + 0.5\}$ will be called an *elementary* 0-*cell*, and the closed unit interval $[i - 0.5, i + 0.5]$ of the real line will be called an *elementary* 1-*cell*. (Thus an elementary 1-cell is the same thing as a 1-xel in $\mathbb{R}$.)

A *xel* is a Cartesian product $E_1 \times \ldots \times E_m$, where m can be any positive integer and each of the E's is an elementary 1-cell or an elementary 0-cell. If k of the m E's are elementary 1-cells and the other $m - k$ E's are elementary 0-cells, then the xel is called a k-*xel*. Note that a k-xel $E_1 \times \ldots \times E_m$ is an upright closed k-dimensional unit (hyper)cube in $\mathbb{R}^m$, and its vertices are located at points each of whose coordinates differs from an integer by 0.5.

If a xel Y is a subset of a xel X then we say Y is a *face of* X. This is denoted by $Y \leq X$. If Y is a k-xel and $Y \leq X$, then we say Y is a k-*face of* X. It is easy to verify that, for all integers $0 \leq j \leq k$, every k-xel has just $2^{k-j}\binom{k}{j}$ different j-faces. If $Y \leq X$ and $Y \neq X$, then we say Y is a *proper* face of X; this is denoted by $Y < X$. A *xel-complex in* $\mathbb{R}^n$ is a finite collection $\mathbf{K}$ of xels in $\mathbb{R}^n$ with the property that $\mathbf{K}$ contains every proper face of each of its members (i.e., $Y \in \mathbf{K}$ whenever $X \in \mathbf{K}$ and $Y < X$).

The *Euler number* of a xel-complex $\mathbf{K}$ in $\mathbb{R}^n$, which is denoted by $\chi(\mathbf{K})$, is the integer $\sum_{i=0}^{n}(-1)^i c_i(\mathbf{K})$, where $c_i(\mathbf{K})$ is the number of i-xels in $\mathbf{K}$. If P is any finite union of xels in $\mathbb{R}^n$ then the *Euler number* of P, which is denoted by $\chi(P)$, is defined to be the number $\chi(\mathbf{K})$, where $\mathbf{K}$ is the xel-complex consisting of all the xels in $\mathbb{R}^n$ that are contained in the polyhedron P (i.e., $\mathbf{K}$ is the xel-complex such that $\bigcup \mathbf{K} = P$). It is not hard to show that if X is any xel, then $\chi(X) = 1$. Note also that $\chi(\emptyset) = 0$.

The *boundary complex* of a k-xel X in $\mathbb{R}^n$, denoted by $\mathbf{Boundary}(X)$, is the xel-complex in $\mathbb{R}^n$ consisting of all the proper faces of X.

Let $p \in I \subseteq \mathbb{Z}^n$. Recall that $X(p)$ denotes the n-xel in $\mathbb{R}^n$ that is centered at p. The *coattachment complex of* p *in* I, denoted by $\mathbf{Coattach}(p, I)$, is the xel-complex $\{Y \in \mathbf{Boundary}(X(p)) \mid \exists q \,.\, q \in \mathbb{Z}^n \setminus I \text{ and } Y < X(q)\}$ in $\mathbb{R}^n$. Thus $\mathbf{Coattach}(p, I)$ consists of those faces of $X(p)$ that also belong to the boundary complex of at least one n-xel in the complementary polyhedron of I. The set $\bigcup \mathbf{Coattach}(p, I) = X(p) \cap \mathcal{CP}(I)$ will be called the *coattachment set* of p in I; this is a polyhedron in $\bigcup \mathbf{Boundary}(X(p))$.

When $n = 1, 2, 3$, or 4, necessary and sufficient conditions for a point p to be $(2n, 3^n - 1)$-simple in a subset I of $\mathbb{Z}^n$ can be stated in terms of $\mathbf{Coattach}(p, I)$:

Theorem 1. *Let $I \subseteq \mathbb{Z}^n$, where $n = 1$, 2, 3, or 4. Let p be a point in I. Then p is $(2n, 3^n - 1)$-simple in I if and only if the following conditions all hold:*

1. *The polyhedron $\bigcup \mathbf{Coattach}(p, I)$ is connected.*
2. *The set $(\bigcup \mathbf{Boundary}(X(p))) \setminus \bigcup \mathbf{Coattach}(p, I)$ is connected.*
3. $\chi(\mathbf{Coattach}(p, I)) = 1$.

In the rest of this section we explain how Theorem 1 can be deduced from Theorem 2.10 in [10] and Theorem 2 in [4]. Let $p \in J \subseteq \mathbb{Z}^n$. Then we denote the xel-complex $\{Y \in \mathbf{Boundary}(X(p)) \mid \exists q \,.\, q \in J \setminus \{p\} \text{ and } Y < X(q)\}$ in $\mathbb{R}^n$ by $\mathbf{Attach}(p, J)$. We say that the point p is $(3^n - 1, 2n)$-*simple* in J if there is a deformation retraction of the polyhedron $\bigcup_{q \in J} X(q)$ onto the polyhedron $\bigcup_{q \in J\setminus\{p\}} X(q)$. (The reason for the name "$(3^n - 1, 2n)$-simple" is that deletion of such a point p from J may reasonably be said to "preserve topology" in a sense that is consistent with the use of $(3^n - 1)$-adjacency to define connectedness within J and the use of $2n$-adjacency to define connectedness within $\mathbb{Z}^n \setminus J$.) It is easily verified that if $p \in I \subseteq \mathbb{Z}^n$, and $J = (\mathbb{Z}^n \setminus I) \cup \{p\}$, then p is $(2n, 3^n - 1)$-simple in I if and only if p is $(3^n - 1, 2n)$-simple in J, and $\mathbf{Coattach}(p, I) = \mathbf{Attach}(p, J)$. It follows that Theorem 1 is equivalent to the following theorem:

Theorem 2. *Let $p \in J \subseteq \mathbb{Z}^n$, where $n =$ 1, 2, 3, or 4. Let p be a point in J. Then p is $(3^n - 1, 2n)$-simple in J if and only if the following conditions all hold:*

1. *The polyhedron $\bigcup \mathbf{Attach}(p, J)$ is connected.*
2. *The set $(\bigcup \mathbf{Boundary}(X(p))) \setminus \bigcup \mathbf{Attach}(p, J)$ is connected.*
3. $\chi(\mathbf{Attach}(p, J)) = 1$.

When $n = 1$, the truth of Theorem 2 can be verified by case-checking. When $n = 2$ or 3, Theorem 2 follows from Theorem 2.10 in [10] and well known topological facts. In the case $n = 4$, which is the case that our proof of the Main Theorem will depend on, Theorem 2 is essentially equivalent[5] to Theorem 2 in [4]; for an elementary proof of the "if" part, see [11]. This justifies Theorem 1.

4 The Schlegel Diagram of Boundary $(X(p))$, for $p \in Z^4$

When discussing the boundary complex of $X(p)$ for a point $p \in \mathbb{Z}^4$ whose coordinates are (p_1, p_2, p_3, p_4), a xel $E_1 \times E_2 \times E_3 \times E_4 \in \mathbf{Boundary}(X(p))$ may be denoted by a sequence of four symbols $a_1 a_2 a_3 a_4$, where:

- The symbol a_i is "+" if $E_i = \{p_i + 0.5\}$.
- The symbol a_i is "−" if $E_i = \{p_i - 0.5\}$.
- The symbol a_i is "±" if E_i is the closed interval $[p_i - 0.5, p_i + 0.5]$.

For example, $-{+}{+}-$ denotes the 0-xel $\{(p_1 - 0.5, p_2 + 0.5, p_3 + 0.5, p_4 - 0.5)\}$; $-{\pm}{-}+$ denotes the 1-xel $\{p_1 - 0.5\} \times [p_2 - 0.5, p_2 + 0.5] \times \{p_3 - 0.5\} \times \{p_4 + 0.5\}$; the 3-xel $[p_1 - 0.5, p_1 + 0.5] \times [p_2 - 0.5, p_2 + 0.5] \times [p_3 - 0.5, p_3 + 0.5] \times \{p_4 + 0.5\}$ is denoted by $\pm\pm\pm+$.

[5] Theorem 2 in [4] is directly applicable only if J is a finite set, but in proving Theorem 2 we may assume this. For if $N^k(p)$ denotes the set $\{x \in \mathbb{Z}^n \mid \|x - p\|_\infty \leq k\}$, then for any $k \geq 1$ conditions 1 – 3 of the theorem hold if and only if they hold when J is replaced by the finite set $J \cap N^k(p)$; and the definition of $(3^n - 1, 2n)$-simpleness implies that p is $(3^n - 1, 2n)$-simple in J if and only if p is $(3^n - 1, 2n)$-simple in $J \cap N^k(p)$ for some $k \geq 1$. (Theorem 2 implies that the latter is true even for $k = 1$, but we cannot use this fact in proving Theorem 2.)

For any $p \in \mathbb{Z}^4$, the xels in **Boundary**$(X(p))$ can be usefully represented by a *Schlegel diagram* in $\mathbb{R}^3 \cup \{\infty\}$ (where ∞ denotes a single "point at infinity" whose topological neighborhoods are the complements of bounded subsets of $\mathbb{R}^3$). This is shown in Figure 1. Each 0-xel in **Boundary**$(X(p))$ is represented by the vertex with the corresponding label in the figure. With one exception, each 1-, 2-, and 3-xel Y in **Boundary**$(X(p))$ is represented by the convex hull (in $\mathbb{R}^3$) of the vertices of the Schlegel diagram that represent the 0-faces of Y. The exception is that the 3-xel $\pm\pm\pm+$ is represented by the closure of the "outside" region of the diagram (i.e., the closure of the complement of the union of the parts of the diagram that represent the other xels).

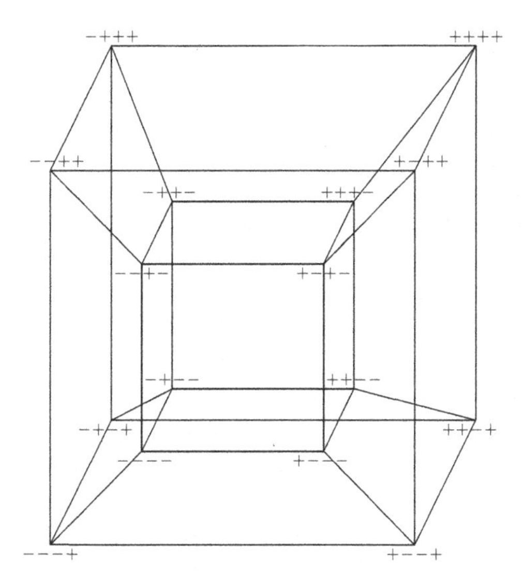

Fig. 1. A Schlegel diagram of **Boundary**$(X(p))$ in $\mathbb{R}^3 \cup \{\infty\}$ is shown here, for an arbitrary point $p \in \mathbb{Z}^4$. The large cube is subdivided into seven closed convex cells, one of which is the small cube in the center. Each of the seven closed convex cells represents one of the eight 3-xels in **Boundary**$(X(p))$. The closure of the unbounded outside region represents the other 3-xel in **Boundary**$(X(p))$.

If P is any union of xels in **Boundary**$(X(p))$, then let $\text{Schlegel}_p(P)$ denote the union of the parts of the Schlegel diagram of **Boundary**$(X(p))$ that represent the xels in P. It is not hard to show that the Schlegel diagram is a topologically faithful representation of **Boundary**$(X(p))$, in the sense that there is a homeomorphism of $\bigcup$**Boundary**$(X(p))$ onto $\mathbb{R}^3 \cup \{\infty\}$ that maps each xel $Y \in$ **Boundary**$(X(p))$ onto $\text{Schlegel}_p(Y)$. It follows that if P is any union of xels in **Boundary** $X(p)$, then:

(a) P is connected if and only if $\text{Schlegel}_p(P)$ is connected.
(b) $(\bigcup$**Boundary**$(X(p)))\setminus P$ is connected if and only if $(\mathbb{R}^3 \cup \{\infty\}) \setminus \text{Schlegel}_p(P)$ is connected.

The next two lemmas state properties of the xels of **Boundary**$(X(p))$ that are evident from (b) and inspection of Figure 1.

Lemma 1. *Let P be a union of xels in* **Boundary**$(X(p))$ *and let Y be a 1-xel in* **Boundary**$(X(p))$. *Then* $(\bigcup \textbf{Boundary}(X(p))) \setminus (P \cup Y)$ *is connected if* $(\bigcup \textbf{Boundary}(X(p))) \setminus P$ *is connected.* □

Lemma 2. *Let P be a union of xels in* **Boundary**$(X(p))$ *and let Y be a 2-xel in* **Boundary**$(X(p))$ *such that at most three of the four 1-faces of Y lie in P. Then* $(\bigcup \textbf{Boundary}(X(p))) \setminus (P \cup Y)$ *is connected if* $(\bigcup \textbf{Boundary}(X(p))) \setminus P$ *is connected.* □

5 Properties of (8, 80)-MNS Sets

In the Introduction we defined (8, 80)-MNS sets in $\mathbb{Z}^4$. In this section we present some important properties of such sets.

We begin with four basic facts. Let I be an arbitrary subset of $\mathbb{Z}^4$, let p be any point in I, and let D be any finite subset of $I \setminus \{p\}$. Then:

S0 $\emptyset$ is (8, 80)-simple in I.
S1 If D is (8, 80)-simple in I, and p is (8, 80)-simple in $I \setminus D$, then $D \cup \{p\}$ is (8, 80)-simple in I.
S2 If D is (8, 80)-simple in I, and $D \cup \{p\}$ is also (8, 80)-simple in I, then p is (8, 80)-simple in $I \setminus D$.
S3 p is (8, 80)-simple in I if and only if p is (8,80)-simple in $N(p) \cap I$.

Here **S0** and **S1** are immediate consequences of the definition of an (8, 80)-simple set, and **S3** is an easy consequence of Theorem 1. The truth of **S2** can be deduced from properties of deformation retraction.[6]

Our next theorem gives a convenient characterization (8,80)-MNS sets. This theorem is a fairly straightforward consequence of **S0**, **S1**, and **S2**, as is shown on p. 74 of [9].

Theorem 3. *Let $D \subseteq I \subseteq \mathbb{Z}^4$. Then D is (8, 80)-MNS in I if and only if D is nonempty and finite, and the following conditions hold for all $p \in D$:*

[6] Indeed, suppose D and $D \cup \{p\}$ are both (8, 80)-simple in I. Consider the inclusion maps $i_1 : \mathcal{CP}(I) \to \mathcal{CP}(I \setminus D)$ and $i_2 : \mathcal{CP}(I \setminus D) \to \mathcal{CP}((I \setminus D) \setminus \{p\})$. Since D is (8,80)-simple in I, there is a deformation retraction of $\mathcal{CP}(I \setminus D)$ onto $\mathcal{CP}(I)$. This implies that the map i_1 is a homotopy equivalence. Let $i_1' : \mathcal{CP}(I \setminus D) \to \mathcal{CP}(I)$ be a homotopy inverse of i_1, so that i_1' is a homotopy equivalence as well. Since $D \cup \{p\}$ is (8,80)-simple in I, there is a deformation retraction of $\mathcal{CP}((I \setminus D) \setminus \{p\})$ onto $\mathcal{CP}(I)$, and so the inclusion map $i_2 \circ i_1 : \mathcal{CP}(I) \to \mathcal{CP}((I \setminus D) \setminus \{p\})$ is a homotopy equivalence. As the inclusion map $i_2 : \mathcal{CP}(I \setminus D) \to \mathcal{CP}((I \setminus D) \setminus \{p\})$ is homotopic to $(i_2 \circ i_1) \circ i_1'$, it too must be a homotopy equivalence. By Cor. 1.4.10, Thm. 1.4.11, and Cor. 3.2.5 in [16], this implies there is a deformation retraction of $\mathcal{CP}((I \setminus D) \setminus \{p\})$ onto $\mathcal{CP}(I \setminus D)$, and so p is (8,80)-simple in $I \setminus D$, as **S2** asserts.

1. p is not $(8,80)$-simple in $I \setminus (D \setminus \{p\})$.
2. p is $(8,80)$-simple in $I \setminus T$ for every $T \subsetneq D \setminus \{p\}$. □

We now deduce three important but fairly easy corollaries from Theorem 3. Recall from the Introduction that we say a set $S \subset \mathbb{Z}^4$ *can be* $(8,80)$*-MNS* if there exists a set $I \subseteq \mathbb{Z}^4$ such that $S \subseteq I$ and S is (8,80)-MNS in I.

Corollary 1. *Let $X \subsetneq D \subseteq \mathbb{Z}^4$. Then if D can be (8,80)-MNS, so can $D \setminus X$.*

Proof. Suppose $D \subseteq I \subseteq \mathbb{Z}^4$ and D is $(8,80)$-MNS in I. Then Theorem 3 implies that $D \setminus X$ is $(8,80)$-MNS in $I \setminus X$. □

From **S3** and Theorem 3, we see that if $D \subseteq I \subseteq \mathbb{Z}^4$ and D is $(8,80)$-MNS in I, then $q \in N(p)$ for all pairs of points p and q in D. [Indeed, under these hypotheses Theorem 3 tells us that, for all distinct points p and q in D, p *is not* $(8,80)$-simple in $I \setminus (D \setminus \{p\})$, but p *is* $(8,80)$-simple in $I \setminus (D \setminus \{p,q\})$, which (together with **S3**) implies $N(p) \cap (I \setminus (D \setminus \{p\})) \neq N(p) \cap (I \setminus (D \setminus \{p,q\})$, so that $q \in N(p)$.] In other words:

Corollary 2. *D cannot be $(8,80)$-MNS if D is not a subset of a $2 \times 2 \times 2 \times 2$ block of points in $\mathbb{Z}^4$.* □

Corollary 3. *Let D be a subset of a $2 \times 2 \times 2 \times 2$ block in $\mathbb{Z}^4$ and suppose there exist distinct points $p, q, r \in D$ such that q lies on one of the shortest 8-paths (in $\mathbb{Z}^4$) from p to r. Then D cannot be $(8,80)$-MNS.*

Proof. Let I be any superset of D in $\mathbb{Z}^4$. Then $X(p) \cap X(r) \subseteq X(p) \cap X(q) \subseteq \bigcup \mathbf{Coattach}(p, I \setminus (D \setminus \{p,r\}))$. It follows that $\bigcup \mathbf{Coattach}(p, I \setminus (D \setminus \{p\})) = \bigcup \mathbf{Coattach}(p, I \setminus (D \setminus \{p,r\})) \cup (X(p) \cap X(r)) = \bigcup \mathbf{Coattach}(p, I \setminus (D \setminus \{p,r\}))$. Hence (by Theorem 1) if p is not (8,80)-simple in the set $I \setminus (D \setminus \{p\})$, then p is also not (8,80)-simple in $I \setminus (D \setminus \{p,r\})$. This and Theorem 3 imply that D is not $(8,80)$-MNS in I. □

6 The Main Theorem

In this section we state our Main Theorem, which identifies all sets that can be (8,80)-MNS. Note that we already know every such set is contained in some $2 \times 2 \times 2 \times 2$ block (by Corollary 2 of Theorem 3).

Let $p, q \in \mathbb{Z}^4$. The set $\{p,q\}$ will be called an *antipodean pair* (of a $2 \times 2 \times 2 \times 2$ block) if p is 80-adjacent to q and none of the four coordinates of p is equal to the same coordinate of q. The set $\{p,q\}$ will be called a $\sqrt{3}$*-pair* if p is 80-adjacent to q and just one of the four coordinates of p is equal to the same coordinate of q. The set $\{p,q\}$ will be called a $\sqrt{2}$*-pair* if p is 80-adjacent to q and just two of the four coordinates of p are equal to the same coordinate of q. Thus if p is 80-adjacent to q then p and q constitute an antipodean pair, a $\sqrt{3}$-pair, a $\sqrt{2}$-pair, or a pair of 8-adjacent points according to whether $X(p) \cap X(q)$ is a 0-xel, a 1-xel, a 2-xel, or a 3-xel in $\mathbb{R}^4$.

A set $\{p,q,r\}$ will be called a $\sqrt{2}$-*equilateral triple* if each of $\{p,q\}$, $\{q,r\}$, and $\{p,r\}$ is a $\sqrt{2}$-pair. We will call $\{p,q,r\}$ a $(\sqrt{3},\sqrt{3},\sqrt{2})$-*isosceles triple* if two of the three pairs $\{p,q\}$, $\{q,r\}$, $\{p,r\}$ are $\sqrt{3}$-pairs and the other is a $\sqrt{2}$-pair. (It is easily verified that every $(\sqrt{3},\sqrt{3},\sqrt{2})$-isosceles triple lies in a unique $2\times2\times2\times2$ block.) A set $\{p,q,r,s\}$ will be called a $\sqrt{2}$-*spanning quadruple* (of a $2\times2\times2\times2$ block) if there is a $2\times2\times2\times2$ block B and a point $x \in B$ such that p, q, r, and s are the four 8-neighbors of x in B. (Each 2-point subset of a $\sqrt{2}$-spanning quadruple is a $\sqrt{2}$-pair, but this property does not uniquely characterize $\sqrt{2}$-spanning quadruples: A $2\times2\times2\times1$ block contains 4-point sets in which every 2-point subset is a $\sqrt{2}$-pair, but these sets are not considered to be $\sqrt{2}$-spanning quadruples.)

We are now ready to state the Main Theorem:

Theorem 4 (Main Theorem). *A set $D \subset \mathbb{Z}^4$ can be $(8,80)$-MNS if and only if D is one of the following:*

1. *a singleton set*
2. *a pair of 8-adjacent points*
3. *a $\sqrt{2}$-pair*
4. *a $\sqrt{3}$-pair*
5. *an antipodean pair of a $2\times2\times2\times2$ block*
6. *a $\sqrt{2}$-equilateral triple*
7. *a $(\sqrt{3},\sqrt{3},\sqrt{2})$-isosceles triple*
8. *a $\sqrt{2}$-spanning quadruple of a $2\times2\times2\times2$ block*

7 Proof of the "If" Part of the Main Theorem

The eight kinds of set listed in the Main Theorem have the property that if D_1 and D_2 are any two sets of the same kind, then there is an isometry of $\mathbb{Z}^4$ onto itself that maps D_1 onto D_2. So if we can show that one set of each of these kinds can be $(8,80)$-MNS, then it will follow that all sets of these kinds can be $(8,80)$-MNS and that the "if" part of the Main Theorem holds.

In fact, it suffices to verify that one set of each of the following four kinds can be $(8,80)$-MNS:

- a pair of 8-adjacent points
- an antipodean pair of a $2\times2\times2\times2$ block
- a $(\sqrt{3},\sqrt{3},\sqrt{2})$-isosceles triple
- a $\sqrt{2}$-spanning quadruple of a $2\times2\times2\times2$ block

This is because each set of the other four kinds listed in the Main Theorem is a subset of a set of one of these four kinds, and Corollary 1 of Theorem 3 tells us that if a set can be (8,80)-MNS, then so can each of its nonempty subsets.

For brevity, when a,b,c, and d are single digit numbers we write $abcd$ to denote the point (a,b,c,d) in $\mathbb{Z}^4$. For example, 1203 denotes the point (1,2,0,3). With the aid of the Schlegel diagram of **Boundary**$(X(1111))$, we can see from Theorems 1 and 3 that the antipodean pair $\{1111,2222\}$ and the $\sqrt{2}$-spanning quadruple $\{1111, 2112, 2121, 2211\}$ are (8,80)-MNS in I if I is the $2 \times 2 \times 2 \times 2$ block $\{1,2\} \times \{1,2\} \times \{1,2\} \times \{1,2\}$. (For this I, **Coattach**$(1111, I)$ consists of all the xels in **Boundary**$(X(1111))$ that do *not* have ++++ as a 0-face.) We can similarly see that the $(\sqrt{3}, \sqrt{3}, \sqrt{2})$-isosceles triple $\{1111, 1222, 2221\}$ is (8,80)-MNS in the same set I, with the aid of the Schlegel diagrams of **Boundary**$(X(1111))$ and **Boundary**$(X(2221))$. Finally, with the aid of the Schlegel diagram of **Boundary**$(X(1111))$, we can see that the pair of 8-adjacent points $\{1111, 2111\}$ is (8,80)-MNS in the $2 \times 3 \times 1 \times 1$ block $\{1,2\} \times \{0,1,2\} \times \{1\} \times \{1\}$. It follows that the "if" part of the Main Theorem is true.

8 Proof of the "Only If" Part of the Main Theorem

We already know (from Corollary 2 of Theorem 3) that a set can only be (8,80)-MNS if it is a subset of a $2 \times 2 \times 2 \times 2$ block in $\mathbb{Z}^4$. Our next lemma classifies subsets of $2 \times 2 \times 2 \times 2$ blocks in a useful way.

If D and E are subsets of $\mathbb{Z}^4$ then we say that D is *E-like* if there is an isometry of $\mathbb{Z}^4$ onto itself that maps D onto E. We give two examples. Writing $abcd$ to denote the point (a, b, c, d) as before, a set is an antipodean pair if and only if it is $\{1111, 2222\}$-like, and is a $\sqrt{2}$-spanning quadruple if and only if it is $\{1111, 2112, 2121, 2211\}$-like.

Lemma 3. *Let D be a subset of a $2 \times 2 \times 2 \times 2$ block of points in $\mathbb{Z}^4$. Then D satisfies one of the following conditions:*

1. *D is a set of one of the eight kinds listed in the Main Theorem.*
2. *D* strictly *contains a pair of 8-adjacent points.*
3. *D* strictly *contains an antipodean pair.*
4. *D contains a 4-point set that is $\{1111, 1221, 2121, 2212\}$-like, $\{1111, 1222, 2121, 2212\}$-like, or $\{1111, 1221, 2121, 2211\}$-like.*

Proof. This lemma is easily verified by case-checking. (Note that it is enough to check that the lemma holds whenever D has five or fewer points. This is because condition 1 cannot hold if D has five points, and, for any set D_0, if one of the other three conditions holds when D is some 5-point subset of D_0, then it also holds when $D = D_0$.) We leave the details to the reader. □

Lemma 4. *Let D be a subset of a $2 \times 2 \times 2 \times 2$ block such that D strictly contains either a pair of 8-adjacent points or an antipodean pair. Then D cannot be (8,80)-MNS.*

Proof. This lemma follows from Corollary 3 of Theorem 3. Indeed, let a, b, and c be distinct points in a $2 \times 2 \times 2 \times 2$ block. If $\{a, b\}$ is an antipodean pair, then c lies on one of the shortest 8-paths from a to b. If a and b are 8-adjacent points that differ just in the ith coordinate, then c has the same ith coordinate as one of the two points a and b, and that point lies on one of the shortest 8-paths from the other of those two points to c. □

It follows from Lemmas 3 and 4 that we can now complete the proof of the "only if" part of the Main Theorem by showing that a set $D \subset \mathbb{Z}^4$ cannot be (8,80)-MNS if D satisfies condition 4 of Lemma 3. In view of Corollary 1 of Theorem 3, it is enough to show that none of the three sets $\{1111, 1221, 2121, 2212\}$, $\{1111, 1222, 2121, 2212\}$, and $\{1111, 1221, 2121, 2211\}$ can be (8,80)-MNS.

Consider the first set, $\{1111, 1221, 2121, 2212\}$. We claim that:

$$X(1111) \cap X(1221) \cap X(2121) \cap X(2212) = X(1111) \cap X(2121) \cap X(2212) \neq \emptyset \quad (1)$$

This is easily seen because

$$\begin{aligned} X(1111) \cap X(1221) &= \pm{+}{+}\pm \text{ in } \mathbf{Boundary}(X(1111)) \\ X(1111) \cap X(2121) &= {+}\pm{+}\pm \text{ in } \mathbf{Boundary}(X(1111)) \\ X(1111) \cap X(2212) &= {+}{+}\pm{+} \text{ in } \mathbf{Boundary}(X(1111)) \end{aligned}$$

which shows that the triple and the quadruple intersections in (1) are both equal to the 0-xel $++++$ in $\mathbf{Boundary}(X(1111))$. It is readily confirmed that the second set, $\{1111, 1222, 2121, 2212\}$, and the third set, $\{1111, 1221, 2121, 2211\}$, satisfy analogous conditions:

$$X(1111) \cap X(1222) \cap X(2121) \cap X(2212) = X(1111) \cap X(2121) \cap X(2212) \neq \emptyset$$
$$X(1111) \cap X(1221) \cap X(2121) \cap X(2211) = X(1111) \cap X(2121) \cap X(2211) \neq \emptyset$$

Theorem 5 below will tell us that this makes it impossible for these three sets to be (8,80)-MNS in a 4D binary image.

We will deduce Theorem 5 from the next two lemmas. These lemmas will be proved using the following identity, which holds when each of $P_1, P_2, \ldots, P_k$ is a finite union of xels in $\mathbb{R}^n$: $\chi(\bigcup_{i=1}^k P_i) = \sum_{T \subseteq \{1,2,\ldots,k\}, T \neq \emptyset} (-1)^{|T|-1} \chi(\bigcap_{i \in T} P_i)$. We call this the *Inclusion-Exclusion Principle for Euler numbers.* It can be deduced from the definition of $\chi(P)$ and the Inclusion-Exclusion Principle for cardinalities of finite sets.

Lemma 5. *Let $p \in \mathbb{Z}^4$, let Y be a xel in* $\mathbf{Boundary}(X(p))$*, and let P be a union of xels in* $\mathbf{Boundary}(X(p))$ *such that $\chi(P) = \chi(P \cup Y) = 1$. Then:*

1. *$\chi(P \cap Y) = 1$.*
2. *$P \cup Y$ is connected if P is connected.*
3. *If Y is a 1-xel or a 2-xel, then $(\bigcup \mathbf{Boundary}(X(p))) \setminus (P \cup Y)$ is connected if $(\bigcup \mathbf{Boundary}(X(p))) \setminus P$ is connected.*

Proof. It follows from the Inclusion-Exclusion Principle for Euler numbers that $\chi(P \cup Y) = \chi(P) + \chi(Y) - \chi(P \cap Y)$. This implies assertion 1, since $\chi(P) = \chi(P \cup Y) = 1$ and since $\chi(Y) = 1$ (as Y is a xel). Assertion 1 implies that $P \cap Y \neq \emptyset$. This implies assertion 2, since Y is connected. Assertion 3 is trivially valid if $Y \subseteq P$, so let us assume that $Y \not\subseteq P$. Now if Y is a 2-xel then assertion 1 implies that at most three of the four 1-faces of Y lie in P. Hence assertion 3 follows from Lemmas 1 and 2. □

Lemma 6. *Let $p \in \mathbb{Z}^4$, let X_1, X_2, and X_3 be xels in* **Boundary**$(X(p))$ *such that $X_1 \cap X_2 = X_1 \cap X_2 \cap X_3 \neq \emptyset$, and let P be a union of xels in* **Boundary**$(X(p))$ *such that $\chi(P) = \chi(P \cup X_1) = \chi(P \cup X_2) = \chi(P \cup X_3) = \chi(P \cup X_1 \cup X_2) = \chi(P \cup X_1 \cup X_3) = \chi(P \cup X_2 \cup X_3) = 1$. Then $\chi(P \cup X_1 \cup X_2 \cup X_3) = 1$.*

Proof. Any nonempty intersection of xels is a xel. So, since X_1, X_2, and X_3 are xels and $X_1 \cap X_2 \cap X_3 \neq \emptyset$, we have

$$\chi(X_i) = \chi(X_i \cap X_j) = \chi(X_1 \cap X_2 \cap X_3) = 1 \tag{2}$$

for all i and j in $\{1, 2, 3\}$. By assertion 1 of Lemma 5, we also have

$$\chi(P \cap X_1) = \chi(P \cap X_2) = \chi(P \cap X_3) = 1 \tag{3}$$

For all i and j in $\{1, 2, 3\}$, it follows from the Inclusion-Exclusion Principle for Euler numbers that

$$\begin{aligned} \chi(P \cup X_i \cup X_j) = \chi(P) + \chi(X_i) + \chi(X_j) \\ -\chi(P \cap X_i) - \chi(P \cap X_j) - \chi(X_i \cap X_j) \\ +\chi(P \cap X_i \cap X_j) \end{aligned} \tag{4}$$

Since $\chi(P) = \chi(P \cup X_i \cup X_j) = 1$, it follows from (2), (3), and (4) that

$$\chi(P \cap X_1 \cap X_2) = \chi(P \cap X_1 \cap X_3) = \chi(P \cap X_2 \cap X_3) = 1 \tag{5}$$

The Inclusion-Exclusion Principle for Euler numbers also implies that:

$$\begin{aligned} \chi(P \cup X_1 \cup X_2 \cup X_3) = \chi(P) + \chi(X_1) + \chi(X_2) + \chi(X_3) \\ - \sum_{1 \leq i < j \leq 3} \chi(X_i \cap X_j) - \sum_{i=1}^{3} \chi(P \cap X_i) \\ +\chi(X_1 \cap X_2 \cap X_3) + \sum_{1 \leq i < j \leq 3} \chi(P \cap X_i \cap X_j) \\ -\chi(P \cap X_1 \cap X_2 \cap X_3) \end{aligned} \tag{6}$$

Since $\chi(P) = 1$, and since $\chi(P \cap X_1 \cap X_2 \cap X_3) = \chi(P \cap X_1 \cap X_2) = 1$ (by (5) and the fact that $X_1 \cap X_2 \cap X_3 = X_1 \cap X_2$), it follows from (2), (3), (5), and (6) that $\chi(P \cup X_1 \cup X_2 \cup X_3) = 1$. □

Theorem 5. *Let D be a subset of a $2 \times 2 \times 2 \times 2$ block in $\mathbb{Z}^4$ and suppose there exist distinct points q_1, q_2, q_3, and p in D such that $X(p) \cap X(q_1) \cap X(q_2) = X(p) \cap X(q_1) \cap X(q_2) \cap X(q_3) \neq \emptyset$. Then D cannot be $(8, 80)$-MNS.*

Proof. By Corollary 1 of Theorem 3, it is enough to show that the four-point set $\{p, q_1, q_2, q_3\}$ cannot be (8,80)-MNS.

Suppose $\{p, q_1, q_2, q_3\} \subseteq I \subseteq \mathbb{Z}^4$ and $\{p, q_1, q_2, q_3\}$ is $(8, 80)$-MNS in I. Let $X_i = X(p) \cap X(q_i)$ for $i = 1$, 2, and 3, so that $X_1 \cap X_2 = X_1 \cap X_2 \cap X_3 \neq \emptyset$. It follows from Lemma 4 that each of X_1, X_2, and X_3 is a 1-xel or a 2-xel.

Let $P = \bigcup \mathbf{Coattach}(p, I)$. Then $\bigcup \mathbf{Coattach}(p, I \setminus \{q_i\}) = P \cup X_i$ and $\bigcup \mathbf{Coattach}(p, I \setminus \{q_i, q_j\}) = P \cup X_i \cup X_j$ for all $i, j \in \{1,2,3\}$; and we also have $\bigcup \mathbf{Coattach}(p, I \setminus \{q_1, q_2, q_3\}) = P \cup X_1 \cup X_2 \cup X_3$.

Since $\{p, q_1, q_2, q_3\}$ is $(8, 80)$-MNS in I, it follows from Theorem 3 that p is (8,80)-simple in I, in $I \setminus \{q_i\}$, and in $I \setminus \{q_i, q_j\}$, for all $i, j \in \{1,2,3\}$. Hence it follows from Theorem 1 that P, X_1, X_2, and X_3 satisfy the hypotheses of Lemma 6, and so we have:

$$\chi(\mathbf{Coattach}(p, I \setminus \{q_1, q_2, q_3\})) = \chi(P \cup X_1 \cup X_2 \cup X_3) = 1 \tag{7}$$

Since $\chi(P) = \chi(P \cup X_1) = \chi(P \cup X_1 \cup X_2) = \chi(P \cup X_1 \cup X_2 \cup X_3) = 1$, and since Theorems 1 and 3 imply that both $P = \bigcup \mathbf{Coattach}(p, I)$ and $(\bigcup \mathbf{Boundary}(X(p))) \setminus P$ are connected, we can deduce from Lemma 5 that $\bigcup \mathbf{Coattach}(p, I \setminus \{q_1, q_2, q_3\}) = P \cup X_1 \cup X_2 \cup X_3$ is connected, and that $(\bigcup \mathbf{Boundary}(X(p))) \setminus \mathbf{Coattach}(p, I \setminus \{q_1, q_2, q_3\}) = (\bigcup \mathbf{Boundary}(X(p))) \setminus (P \cup X_1 \cup X_2 \cup X_3)$ is also connected. But this, (7), and Theorem 1 imply that p is (8,80)-simple in $I \setminus \{q_1, q_2, q_3\}$, which contradicts Theorem 3 (since $\{p, q_1, q_2, q_3\}$ is $(8, 80)$-MNS in I). □

As explained above, Theorem 5 and Corollary 1 of Theorem 3 imply that a set $D \subset \mathbb{Z}^4$ cannot be (8,80)-MNS if D satisfies condition 4 of Lemma 3. It follows from this, and Lemmas 3 and 4, that the "only if" part of the Main Theorem is valid.

References

[1] G. Bertrand. On P-simple points. *C. R. Acad. Sci. Paris, Série I*, 321:1077–1084, 1995.
[2] G. Bertrand. A Boolean characterization of three-dimensional simple points. *Pattern Recogn. Lett.*, 17:115–124, 1996.
[3] C. J. Gau and T. Y. Kong. Minimal nonsimple sets of voxels in binary images on a face-centered cubic grid. *Int. J. Pattern Recogn. Artif. Intell.*, 13:485–502, 1999.
[4] C. J. Gau and T. Y. Kong. Minimal nonsimple sets in 4D binary images. *Graph. Models*, 65:112–130, 2003.
[5] R. W. Hall. Tests for connectivity preservation for parallel reduction operators. *Topology and Its Applications*, 46:199–217, 1992.
[6] R. W. Hall. Parallel connectivity-preserving thinning algorithms. In T. Y. Kong and A. Rosenfeld, editors, *Topological Algorithms for Digital Image Processing*, pages 145–179. Elsevier/North-Holland, 1996.
[7] L. C. Kinsey. *Topology of Surfaces*. Springer, 1993.
[8] R. Klette and P. Zamperoni. *Handbook of Image Processing Operators*. Wiley, 1996.
[9] T. Y. Kong. On the problem of determining whether a parallel reduction operator for n-dimensional binary images always preserves topology. In R. A. Melter and A. Y. Wu, editors, *Vision Geometry II (Boston, September 1993), Proceedings*, pages 69–77. Proc. SPIE 2060, 1993.
[10] T. Y. Kong. On topology preservation in 2D and 3D thinning. *Int. J. Pattern Recogn. Artif. Intell.*, 9:813–844, 1995.

[11] T. Y. Kong. Topology preserving deletion of 1's from 2-, 3- and 4-dimensional binary images. In E. Ahronovitz and C. Fiorio, editors, *Discrete Geometry for Computer Imagery: 7th International Workshop (DGCI '97, Montpellier, France, December 1997), Proceedings*, pages 3–18. Springer, 1997.
[12] T. Y. Kong and A. Rosenfeld. Digital topology: Introduction and survey. *Computer Vision, Graphics, and Image Processing*, 48:357–393, 1989.
[13] C. M. Ma. On topology preservation in 3D thinning. *CVGIP: Image Understanding*, 59:328–339, 1994.
[14] C. Ronse. Minimal test patterns for connectivity preservation in parallel thinning algorithms for binary digital images. *Discrete Appl. Math.*, 21:67–79, 1988.
[15] A. Rosenfeld. A characterization of parallel thinning algorithms. *Information and Control*, 29:286–291, 1975.
[16] E. H. Spanier. *Algebraic Topology.* Springer, 1989.

Jordan Surfaces in Discrete Antimatroid Topologies

Ralph Kopperman[1] and John L. Pfaltz[2]

[1] City University of New York, New York, NY
rdkcc@cunyvm.cuny.edu
[2] University of Virginia, Charlottesville, VA
jlp@virginia.edu

Abstract. In this paper we develop a discrete, T_0 topology in which (1) closed sets play a more prominent role than open sets, (2) atoms comprising the space have discrete dimension, which (3) is used to define boundary elements, and (4) configurations within the topology can have connectivity (or separation) of different degrees.

To justify this discrete, closure based topological approach we use it to establish an n-dimensional Jordan surface theorem of some interest. As surfaces in digital imagery are increasingly rendered by triangulated decompositions, this kind of discrete topology can replace the highly regular pixel approach as an abstract model of n-dimensional computational geometry.

1 Axiomatic Basis

Let $\mathbf{U}$ be a universe of arbitrary elements, or as we will call them, atoms. We let R denote a binary relation on $\mathbf{U}$. We denote the identity relation I on $\mathbf{U}$ by R^0. Relational composition is defined in the usual way, so $R^k = R \circ R^{k-1}$, and in particular, $R^1 \circ R^0 = R \circ I = R$. Notationally, we denote elements $(x, z) \in R^k$ by $x.R^k.z$.[1] Then, $x.R^k = \{z \mid x.R^k.z\}$ and $X.R^k = \{z \mid \exists x \in X, x.R^k.z\}$.

In addition to R, we assume an integer function $\delta : \mathbf{U} \to \mathbf{Z}$ that satisfies the following basic axiom

$$x.R.z \quad \text{implies} \quad \delta(x) > \delta(z). \tag{1}$$

An easy induction on k establishes that $x.R^k.z$ also implies $\delta(x) > \delta(z)$. Consequently,

Lemma 1. *If $x.R^m.z$ and $z.R^n.x$ then $n = m = 0$ and $x = z$.*

Proof. If $x.R^m.z$ then $\delta(x) > \delta(z)$, so if $z.R^n.x$ we have $\delta(z) > \delta(x)$, a contradiction unless $m = n = 0$ and $x = z$. □

[1] We employ a dot notation to clearly delineate the operator symbol from its argument(s).

R. Klette and J. Žunić (Eds.): IWCIA 2004, LNCS 3322, pp. 334–350, 2004.

Consequently, we can let

$$\delta(x) = min\{k \mid x.R^k \neq \emptyset, x.R^{k+1} = \emptyset\} \tag{2}$$

We should note that the implication of Lemma 1 coupled with the definition of δ in (2) together imply (1) and so these could be taken as the axiomatic basis instead.

Any relation R satisfying the functional constraint (1) is anti-symmetric. It is a pre-partial order; and its transitive closure $R^* = \bigcup_{k \geq 0} R^k$ is a partial order. Given such a relation R and function δ, we create a discrete topology τ by defining a closure operator φ. Since closure φ is just a relation on $\mathbf{U}$, as is R, we use a similar kind of notation. A topology $\tau = (\mathbf{U}, R, \delta, \varphi)$ is said to be **locally finite** if for all atoms $x \in \mathbf{U}$, their closure $x.\varphi$ is finite. We will always assume that τ is locally finite.

Depending on one's choice of δ and φ, there are many varieties of discrete topology τ capable of describing the structure that R imposes on $\mathbf{U}$. For the closure operator φ on $\mathbf{U}$ we choose, in this paper, to use the ideal, or "downset", operator with respect to R, that is X is closed in $\mathbf{U}$ (with respect to φ), if $x \in X$ implies $z \in X$ for all $z \in x.R^k, k \geq 0$, or equivalently $X.\varphi = X.R^*$. Ideal operators are antimatroid, in that they satisfy the following anti-exchange axiom

$$\text{if} \quad x, y \notin Z.\varphi, \quad \text{then} \quad y \in (x \cup Z).\varphi \quad \text{implies} \quad x \notin (y \cup Z).\varphi. \tag{3}$$

(We often elide the braces { ... } around singleton sets in expressions such as $(\{x\} \cup Z)$). We can visualize any topological space $\tau^{(n)}$ as R on an n-partite space such as Figure 1.

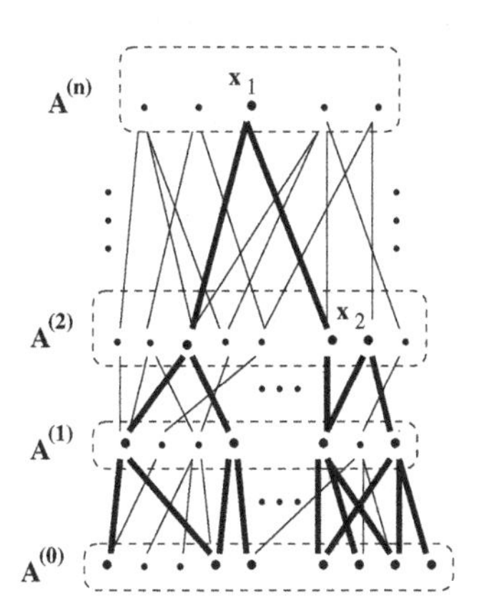

Fig. 1. An n-partite topological space, $\tau^{(n)}$

We may assume $\delta : \mathbf{U} \rightarrow [0, n]$ because

Lemma 2. *If $\tau = (\mathbf{U}, R, \delta, \varphi)$ is locally finite and φ is the ideal operator on R, then $\forall x \in \mathbf{U}, \exists k, x.R^k = \emptyset$.*

Proof. Suppose not. Since $R^* = \bigcup_{k \geq 0} R^k$, $|x.R^*| = |x.\varphi| > k$ contradicting local finiteness. □

By this definition, $\delta(\emptyset) = -1$, and those atoms x of $\mathbf{U}$ which are minimal with respect to R^* have $\delta(x) = 0$. If $\delta(x) = 1$, for every $y \in x.R$, $\delta(y) = 0$. More generally, if $\delta(x) = k$, then for every $y \in x.R$, $\delta(y) \leq k - 1$, and there exists at least one y, $\delta(y) = k - 1$. Clearly, $\delta : \mathbf{U} \rightarrow [0, n]$, so defined satisfies (1), and it is very reasonable in the context of the specific domain we will be examining later in this paper. As a notational convenience, we denote the collections of all atoms x, such that $\delta(x) = k$ by $A^{(k)}$. Thus the universe of all atoms $\mathbf{U} = \bigcup_{k=1,n} A^{(k)}$. An arbitrary collection Z of atoms we will call a **configuration** which we will denoted by $Z = [\, Z^{(0)}, Z^{(1)}, \ldots Z^{(n)} \,]$, where $Z^{(k)} = Z \cap A^{(k)}$. Sets and configurations we denote with uppercase letters; elements and atoms by lower case. A configuration Z is closed if $Z.\varphi = Z.R^* = Z$. In Figure 1, a closed configuration $Z = [\, \emptyset, \emptyset, \{x_2\}, \ldots, \{x_1\} \,].\varphi$ has been indicated by darker lines. We extend δ to configurations by letting $\delta(Z) = k$ where $k = max\{i \mid \delta(a) = i, a \in Z\}$. One can regard δ as a *dimension* concept.

2 Generators, Separation and Connectivity

A closure operator φ is a relation on $\mathbf{U}$ that is closed under intersection, that is $(X \cap Y).\varphi = X.\varphi \cap Y.\varphi$. Alternatively, it satisfies the standard closure axioms, *e.g.* it's monotone and idempotent, When the closure operator φ is defined by ideals in R, that is $\varphi = R^*$, these properties are evident. Moreover the anti-symmetry of R^* ensures it satisfies the anti-exchange property (3) and so φ is antimatroid. Equally important, this kind of ideal closure operator also has $(X \cup Y).\varphi = X.\varphi \cup Y.\varphi$, so it is also a "topological" closure operator.[2]

A topological space is T_0 if for any pair of points x and y, there exists at least one closed set containing one of them, but not the other [5]. Thus the reason for wanting the closure operator φ to be antimatroid is evident with the following theorem.

Theorem 1. *A discrete topology $\tau = (\mathbf{U}, R, \delta, \varphi)$ is T_0 if and only if its topological closure operator φ is antimatroid.*

Proof. Let φ be antimatroid. If we let $Z = \emptyset$ in (3), it immediately follows that τ is T_0.

Conversely, let $x, y \notin Z.\varphi$ and let $y \in (Z \cup x).\varphi$. We must show that $x \notin (Z \cup y).\varphi$. Since τ is T_0, there exists a closed set C containing precisely one of x or y, but not both. Suppose first that $x \in C$. Since $Z.\varphi \cup C$ is closed $(Z \cup x).\varphi \subseteq Z.\varphi \cup C$. But, now $y \notin Z.\varphi$ and $y \in (Z \cup x).\varphi$ imply that $y \in C$,

[2] The Kuratowski closure axioms [10] assume closure under union. This is not true for most closure operators.

contradicting choice of C. So, we must have $y \in C, \quad x \notin C$. Again, since $x.\varphi \cup C$ is closed, $(Z.\varphi \cup y).\varphi \subseteq Z.\varphi \cup C$. Then $x \notin Z.\varphi, \quad x \notin C$ imply $x \notin (Z \cup y).\varphi$ □

Theorem 2 will establish that the closed configurations of a discrete topology can be "shelled", one atom at a time.

Theorem 2. *Let $Z = [\, Z^{(0)}, Z^{(1)}, \ldots, Z^{(k)}, \ldots, Z^{(n)}\,]$ be a closed configuration of dimension k in $\tau^{(n)}$. For every atom, $z \in Z^{(k)}$, $Z-\{z\}$ is closed.*

Proof. Let $Z = [\, Z^{(0)}, Z^{(1)}, \ldots, Z^{(k)}, \ldots, Z^{(n)}\,]$ be any closed configuration of dimension k in $\mathcal{L}_\varphi$. Thus, $Z^{(k)} \neq \emptyset$, but for $\forall m > k, Z^{(m)} = \emptyset$. For any atom $z \in Z^{(k)}$, $Z-\{z\} = [\, Z^{(0)}, Z^{(1)}, \ \ldots, \ Z^{(k)}-z, \ \ldots, \ Z^{(n)}\,]$ and because $z \notin Z.R$, $(Z-z).R^* \subseteq Z-z$. Readily, $Z-z \subseteq (Z-z).R^*$, so $Z-z$ is closed. □

This is actually a well-known consequence of the antimatroid nature of φ in $\tau^{(n)}$. See [9, 11]. Alternately, Theorem 2 can be regarded as another proof that φ is antimatroid; one that is based solely on the definition of closure φ as an ideal R^*.

A set Y generates a closed set Z if $Y.\varphi = Z$. We say Y is a *generator* of Z, denoted $Z.\gamma$, if it is a minimal set that generates Z. The generator concept is fundamental in closure theory. For example, if closure is defined by a convex hull operator, then the generators of a convex polytope are its vertices. It is not hard to show that a closure operator is antimatroid if and only if every closed configuration has a unique minimal generator [11]. The set $X = \{x_1, x_2\}$ is the unique generator of the closed configuration of Figure 1. Many closure systems are not uniquely generated, therefore not antimatroid [4]. It is shown in [11] that

Theorem 3. *Let Z be closed in an antimatroid space $\tau^{(n)}$. Y is a maximal closed subset of Z if and only if $Z-Y = \{x_i\}$, where $x_i \in Z.\gamma$.*

This has been called the "Fundamental Covering Theorem" since it completely defines the covering relationships in the lattice of closed subspaces of $\tau^{(n)}$.

It is evident from the definition of $\delta(Z)$ and of generators that $\delta(Z) = \delta(Z.\gamma)$. If $Z.\gamma \subseteq A^{(k)}$, we say Z is homogeneously generated, or just **homogeneous**. Readily $\delta(Z) = k$. The entire space $\tau^{(n)}$ must be closed; it is homogeneous if $\tau^{(n)}.\gamma \subseteq A^{(n)}$. Although the entire space as illustrated in Figure 1 is homogeneous, the closed set generated by $X = \{x_1, x_2\}$ is not.

Let $\tau^{(n)} = R$ over $A^{(0)}, A^{(1)}, \ldots, A^{(n)}$ be an antimatroid topology. The restriction of R to $A^{(0)}, \ldots, A^{(k)}$, $k < n$, denoted $\tau_k^{(n)}$, is called the kth **subtopology** of $\tau^{(n)}$. If $\tau^{(n)}$ is the topology of Figure 1, then $\tau_2^{(n)}$ is just the lower tri-partite graph. Readily, if $\tau^{(n)}$ is homogeneous then $\tau_k^{(n)}$ is homogeneous.

A configuration Y is said to be **separable**, or **disconnected**, if there exist non-empty, disjoint, closed configurations Z_1, Z_2 such that $Y.\varphi = Z_1 \cup Z_2$.[3] A

[3] A more customary definition would have $Y \subseteq Z_1 \cup Z_2$, with $Y \cap Z_k \neq \emptyset$ [5]. But, since $Z_1 \cup Z_2$ is closed and $Y.\varphi$ is the smallest closed set containing Y, this definition is preferable.

configuration Y is **connected** if it is not separable. A configuration X connects Y_1, Y_2 if $Y_1.\varphi \cap Y_2.\varphi = X \neq \emptyset$. Readily, only closed configurations can connect closed configurations.

This is just the classical sense of separation and connectivity cast in terms of closure. But, in discrete systems, it is often useful to consider connectivity of different "strengths". We say that X is k-**separable** if there exist closed configurations Z_1, Z_2 such that $X.\varphi = Z_1 \cup Z_2$, $\delta(Z_1 \cap Z_2) = k \geq -1$.[4] When $\delta(Z_1 \cap Z_2) = k < \delta(Z_i)$ we will say that $Z_1 \cap Z_2$ k-separates Z_1 and Z_2.

X is *k-connected* if it is not (k-1)-separable. X is **disconnected** if it is (-1)-separable, that is $Z_1 \cap Z_2 = \emptyset$. X is 0-connected if $Z_1 \cap Z_2 \subseteq A^{(0)}$.

Theorem 4. *X is k-connected if and only if $X.\varphi$ is k-connected.*

Proof. Let Z_1, Z_2 be closed configurations such that $Z_1 \cap Z_2$ (k-1)-separates X. $X = Z_1 \cup Z_2$. Since $Z_1 \cup Z_2$ is closed, $X.\varphi = Z_1 \cup Z_2$, so $Z_1 \cap Z_2$ also (k-1)-separates $X.\varphi$.

Proof of the converse is similar. □

Thus closure cannot increase connectivity. In particular, disconnected configurations cannot become connected by closure.

Lemma 3. *In $\tau^{(n)}$, if $\delta(X) = k \leq n$ then X is at most (k-1)-connected.*

Proof. Let $X.\varphi = Z_1 \cup Z_2$ where $Z_i, i = 1, 2$, is non-empty and closed. Readily, $\delta(Z_i) = \delta(X) = k$.

Suppose $\delta(Z_1 \cap Z_2) = k$, that is there exists $x \in Z_1 \cap Z_2 \cap A^{(k)}$. Readily, $X.\varphi = (Z_1 - \{x\}) \cup Z_2$. By Theorem 2, $Z_1 - \{x\}$ is also closed. Use finite induction to remove all common atoms of dimension k until $\delta(Z_1 \cap Z_2) = k - 1$. □

If X is not (k-1)-separable, it cannot be (j-1)-separable, where $j < k$. So,

Lemma 4. *If X is k-connected, then X is j-connected for all $j \leq k$.*

From which it follows that

Lemma 5. *If X is i-connected in $\tau^{(n)}$, then X is i-connected in $\tau_k^{(n)}$, for all $0 < i \leq k$.*

Two atoms $x, z \in A^{(k)}$ are said to be **pathwise** i-**connected** if there exists a sequence $\rho_i = < y_0, \ldots, y_m >, m \geq 0$ such that $x = y_0, y_m = z$ and $y_j.\varphi \cap y_{j+1}.\varphi \cap A^{(i)} \neq \emptyset$. That is, y_j and y_{j+1} are at least i-connected. Pathwise connectivity can be regarded as a relation ρ_i on the atoms of the space with $(x, z) \in \rho_i$ if they are pathwise i-connected. Demonstrating that ρ_i is an equivalence relation is an easy exercise.

One would like to show that topological connectivity and pathwise connectivity are equivalent concepts, that is, a configuration X is topologically k-connected if and only if it is pathwise k-connected. Unfortunately, this is only partially true. To begin, it is easy to show that,

[4] Recall that in Section 1 we had defined $\delta(\emptyset) = -1$.

Theorem 5. *If a configuration X is pathwise k-connected it is topologically k-connected.*

Proof. Let X be pathwise k-connected and suppose there exists closed, non-empty Z_1, Z_2 such that $X = Z_1 \cup Z_2$ and $\delta(Z_1 \cap Z_2) \leq k-1$. By Lemma 3, $\delta(X) = \delta(Z_1 \cup Z_2) > k$. Let $x \in Z_1, z \in Z_2$ where $\delta(x) = \delta(z) = k+1$. Let $\rho_k =< y_0, \ldots, y_m >$ be a k connected chain of k+1 atoms between x and z which exists by hypothesis.

Since $y_0 \in Z_1, y_m \in Z_2$ there exists some pair of atoms $y_i \in Z_1, y_{i+1} \in Z_2$. But, $y_i.R \subseteq y_i.\varphi \subseteq Z_1$ and $y_{i+1}.R \subseteq y_{i+1}.\varphi \subseteq Z_2$. So $\delta(Z_1 \cap Z_2) = k > k-1$, a contradiction. □

To see that the converse need not be true, consider the simple counter example of Figure 2. Readily the entire space $\tau^{(n)} = [\,A^{(0)}, A^{(1)}, A^{(2)}\,] = [\,\{x_1, x_2, x_3, x_4\}, \{y_1, y_2\}, \{z\}\,]$ is topologically connected. But, the two 1-atoms y_1 and y_2 are not pathwise 0-connected.

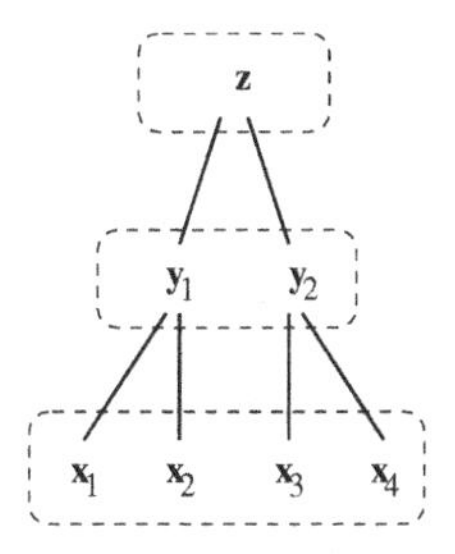

Fig. 2. A topologically connected configuration that is not pathwise 0-connected

A configuration X in $\tau^{(n)}$ is said to be **completely k-connected** if in each subtopology $\tau_i^{(n)}, i \leq k$, X is (i-1)-connected. If $X \in \tau^{(n)}$ is completely (n-1)-connected, we just say it is **completely connected**. Complete connectivity and pathwise connectivity are equivalent concepts because,

Theorem 6. *If a configuration Z is completely k-connected it is pathwise k-connected.*

Proof. This is most easily shown by the contrapositive. Suppose X is not pathwise k-connected. That is, there exists no sequence of (k+1) atoms between a pair of atoms $z_1, z_2 \in Z$. Consider the subtopology $\tau_{k+1}^{(n)}$. Let G_1 be the configuration of all (k+1)-atoms that can be reached by a k-path from z_1. Similarly, let G_2 be the configuration of all (k+1)-atoms reachable by a k-path from z_2.

Now, let $Z_1 = G_1.\varphi$ and $Z_2 = G_2.\varphi$. W.l.o.g we can assume $Z = Z_1 \cup Z_2$. (If not, we can form Z_3 in the same way from remaining (k+1)-atoms.) Readily, $\delta(Z_1 \cap Z_2) < k$ else there would be a k-path from z_1 to z_2. Z is not k-connected. □

We can use virtually the same proof to show that complete connectedness is inherited by subconfigurations.

Corollary 1. *If X is a k-connected subconfiguration of Z which is completely k-connected, then X is completely k-connected.*

Lemma 6. *Let $X \in \tau^{(n)}$. If X is completely k connected then $X.\gamma \subseteq A^{(k+1)} \cup \ldots A^{(n)}$.*

Proof. Suppose $x \in X.\gamma \cap A^{(i)}, i \leq k$. Since x is a generator, for all $j < i, x.R^j \cap X = \emptyset$, (*i.e.* x is maximal in the n-partite representation). But, then x can at most be (k-1)-connected to any other atom, contradiction assumption of k-connectivity. □

As suggested by this lemma, one would like to be able to somehow equate homogeneity and complete connectivity. But this need not be true. Figure 3(a) is homogeneous, but not 1-connected. Figure 3(b) is completely 1-connected (by default), but not homogeneous. However, when $\tau^{(n)}$ is completely (n-1)-connected we do have the corollary:

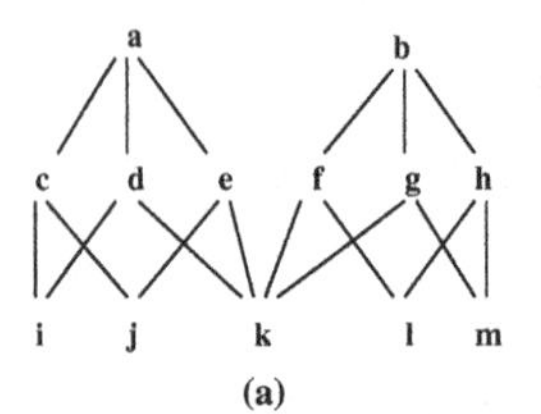

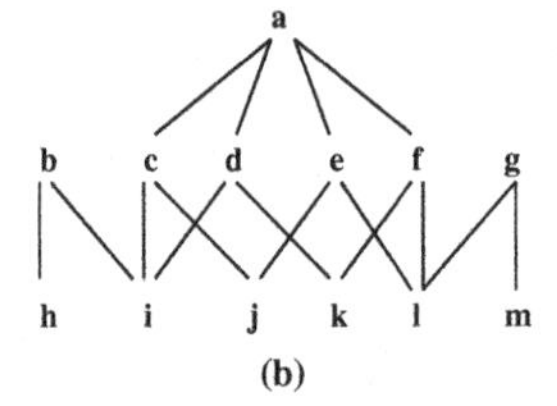

Fig. 3. Contrasting homogeneity and complete connection

Corollary 2. *If X is completely connected in $\tau^{(n)}$, then X is homogeneously generated.*

In contrast to complete connectivity which is a global property, we have weak connectivity which is local. A configuration X with $\delta(X) = m$ is **weakly connected at** p if there exist atoms $x, z \in X$,

(a) $\delta(x) = \delta(z) = m$,
(b) $x.\varphi \cap z.\varphi = p$,
(c) $\delta(p) = k < m - 1$, and
(d) $X \cap p.R^{-1}.\varphi$ is not (k+1)-connected.

In Figure 3(a), $Z = \{a, b\}.\varphi$ is weakly connected at k. Figure 3(b) is not weakly connected at i because $\delta(b) = \delta(i) + 1 = 1$ violating condition (c) above. The point of this condition is to prevent the strongest possible connectivity from being called “weak”.

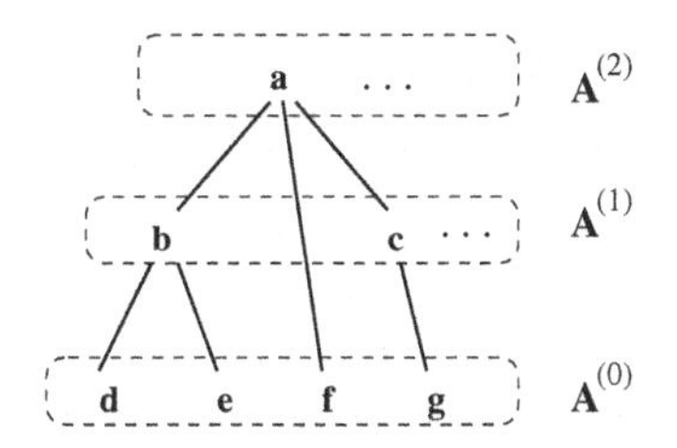

Fig. 4. A non-normal configuration

3 Normality and Boundaries

A space is said to be *normal* if for all $k > 0, x \in A^{(k)}$ implies $x.R \subseteq A^{(k-1)}$ and $|\, x.R \,| \geq 2$. The topology of Figure 4 is non-normal at both atoms c and f. Heretofore all our examples have been normal, even though none of the proofs have required it. From now on we will assume that topologies are normal, even though it is still unnecessary for many of the results of this section.

By the **boundary** of a closed configuration Z, denoted $Z.\beta$, we mean that configuration of atoms, $Z.\beta = \{x \in Z.\varphi \,|\, x.(R^{-1})^* \not\subseteq Z\}$. We say $y \in Z.\beta$ is a **face** of Z if y is a generator of $Z.\beta$, that is $y \in Z.\beta.\gamma$.

Lemma 7. *$Z.\beta$ is closed.*

Proof. $Z.\beta.\varphi = \{x \,|\, \exists y \in Z.\beta, x \in y.R^*\}$. Since $y \in Z.\varphi$, $z \in z.R^*$ for some $z \in Z$, and by transitivity, $x \in z.R^*$. Further, since $y.(R^{-1})^* \not\subseteq Z$,and $y.(R^{-1})^* \subseteq x.(R^{-1})^*, x.(R^{-1})^* \not\subseteq Z$. So $x \in Z.\beta$. □

Lemma 8. *If y is a face of $Z.\beta$ then $y.R^{-1} \not\subseteq Z$*

Proof. $y \in Z.\varphi$ implies $y.R^{-1} \cap Z.\varphi \neq \emptyset$. Since $y.(R^{-1})^* \not\subseteq Z$, there exists $y' \in y.R^{-1}$ such that $y' \in Z.\beta$. $y \in y'.\varphi$ implies y is not a minimal generator. □

We often think of $Z.\beta$ as separating Z from its complement in $\tau^{(n)}$.

Lemma 9. *Let Z be any configuration in $\tau^{(n)}$, and let $W = \tau^{(n)} - Z$. Then $Z.\beta = W.\beta$.*

Proof. By definition $y \in Z.\beta$ if there exists $z \in Z$ and $w \in W$ such that $y \in z.R^* \cap w.R^*$. By symmetry, $y \in W.\beta$, and conversely. □

Lemma 10. *If $\tau^{(n)}$ is homogeneously generated and $\delta(Z) < n$, then $Z.\beta = Z.\varphi$.*

Proof. Readily $Z.\beta \subseteq Z.\varphi$. Conversely, $\forall y \in z.\varphi$, $y.(R^{-1})^* \cap A^{(n)} \neq \emptyset$ (since $\tau^{(n)}$ is homogeneous) so $y.(R^{-1})^* \not\subseteq Z$ implying $Z.\varphi \subseteq Z.\beta$. □

The boundaries of homogeneous configurations are homogeneous.

Lemma 11. *Let Z be a configuration in a completely connected space $\tau^{(n)}$. If $Z.\gamma \subseteq A^{(n)}$, then $Z.\beta.\gamma \subseteq A^{(n-1)}$.*

Proof. Let $Z.\gamma \subseteq A^{(n)}$. Suppose $b \in Z.\beta.\gamma \cap A^{(k)}, k < n-1$. $b \in Z.\beta \subseteq Z.\varphi$ implies there exists $a \in Z.\gamma \subseteq A^{(n)}$ such that $b \in a.\varphi$. $b \in Z.\beta$ also implies there exists $a' \notin Z.\varphi$. Because by Corollary 2, $\tau^{(n)}$ is homogeneous, we may assume w.l.o.g. that $a' \in A^{(n)}$. Now $a \in b.(R^{-1})^*$ and $a' \in b.(R^{-1})^*$.

Since a and a' are k-connected (through b) and $\tau^{(n)}$ is completely connected, a and a' are (n-1)-connected, say through $c \in A^{(n-1)}$. By definition $c \in Z.\beta$, and by transitivity $b \in c.R^{-1}$, so b cannot be in $Z.\beta.\gamma$. □

The converse need not be true.

A generator x of a topology $\tau^{(n)}$ is said to be on the **border** of the space if there exists $y \in x.R$, such that $y.R^{-1} = \{x\}$. Let $\tau^{(n)}.B$ denote the collection of border generators. Note that $\tau^{(n)}.B$ may be empty. A configuration Z is said to be in **interior** position if $Z.\varphi \cap \tau^{(n)}.B.\varphi = \emptyset$.

Theorem 7. *Let Z be a completely connected configuration in interior position of $\tau^{(n)}$, $n \geq 2$, then $Z.\beta$ is completely (n-2)-connected.*

Proof. By induction on $|Z.\gamma|$. Readily, if $|Z.\gamma| = 1$, then $Z.\beta$ is (n-2)-connected because Z in interior position ensures that $Z.\beta \subseteq Z.\gamma.R^*$, and normality then ensures connectivity of $Z.\beta$.

To make the induction work, we must establish that when $|Z.\gamma| = n$ there exists a generator $x \in Z.\gamma$ which has a face in $Z.\beta$, whose removal will still leave Z (n-1)-connected.

The tricky induction step is when $|Z.\gamma| = 2$. Let $Z.\gamma = \{x_1, x_2\}$, where Z is completely connected. Because, Z is (n-1)-connected, Z is homogeneous (Corollary 2) with $x_1, x_2 \in A^{(n)}$, and $x_1.R \cap x_2.R = y \in A^{(n-1)}$; $y \notin Z.\beta$. But, since $\tau^{(n)}$ is normal, $x_2.R - \{y\} \neq \emptyset$, so x_2 has a face in $Z.\beta$. Readily, one can remove x_2 from Z so that $Z - \{x_2\}$ is still (n-1)-connected. Now, let $|Z.\gamma| = n$. Let $x \in Z.\gamma$ be any generator with a face in $Z.\beta$. If $Z - \{x\}$ is still (n-1)-connected, remove x. Otherwise, consider either of the two (n-2)-separated configurations Z_1 or Z_2, $Z_1 \cup Z_2 = Z$. $|Z_i.\gamma| < |Z.\gamma|$, so by induction there exists a generator $x \in Z_i.\gamma$ satisfying our requirements.

Remove x. $(Z - \{x\}).\beta$ is pathwise (n-2)-connected as is $\{x\}.\beta$ by induction. Since the faces common to $(Z - \{x\}).\beta$ and $\{x\}.\beta$ are each pathwise (n-2)-connected to the remaining faces, $(Z - \{x\} \cup \{x\}).\beta$ is pathwise (n-2)-connected. □

The converse need not be true. Even though a boundary is pathwise connected it may bound a weakly connected configuration. However we do know that:

Lemma 12. *If $Z.\beta$ is completely (n-2)-connected in $\tau^{(n)}$, then Z is at least (n-2)-connected.*

Proof. By Theorem 4, we may assume Z is closed. Since it is not (n-1)-connected, there exist two closed configurations Z_1 and Z_2 such that $Z = Z_1 \cup Z_2$, and $\delta(Z_1 \cap Z_2) \leq n-2$. But, $Z.\beta \subseteq Z.\varphi = Z$, so $\delta(Z_1 \cap Z_2) \geq n-2$. □

4 Geometric Spaces

In the rest of this paper we develop a specific discrete topology which is appropriate for digital images. It assumes an ideal closure and bounded dimension $\delta : \mathbf{U} \to [0, n]$. Its culmination will be another "Jordan Surface Theorem" which has attracted so much attention in the digital topology literature [3, 6–8].

Intuitively, a discrete n-dimensional, geometric space is formed by subdividing the space with (n-1)-dimensional constructs, whose intersections yield (n-2)-dimensional objects, *etc.* A 2-dimensional space is subdivided by lines which intersect in points, as in Figure 5(a). We will begin using geometric terms and call 0-atoms, "points"; 1-atoms, "lines". Instead of calling 2-atoms, regions, we prefer to use "tiles"; and instead of volumes, we will call 3-atoms "bricks". Computer applications, such as digital image processing, typically expect much more regular topologies such as Figure 5(b). In this field, "pixels" and "voxels" are a standard terminology. Atoms in $x.R$ and $x.R^{-1}$ are said to be **incident** to x. Thus, line 6 is incident to tile I and to the points b and h, but tiles I and II are not incident to each other. Terminology with a visual basis can help intuitive understanding; but fundamentally, any discrete topology can still be represented as an n-partite graph such as Figure 6.

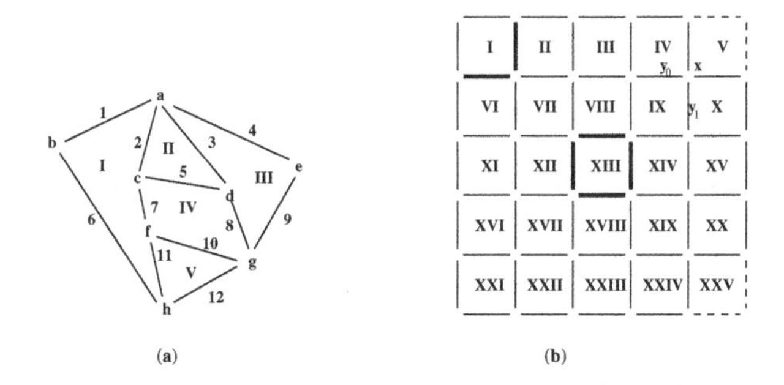

Fig. 5. Two geometric $\tau^{(2)}$ topologies

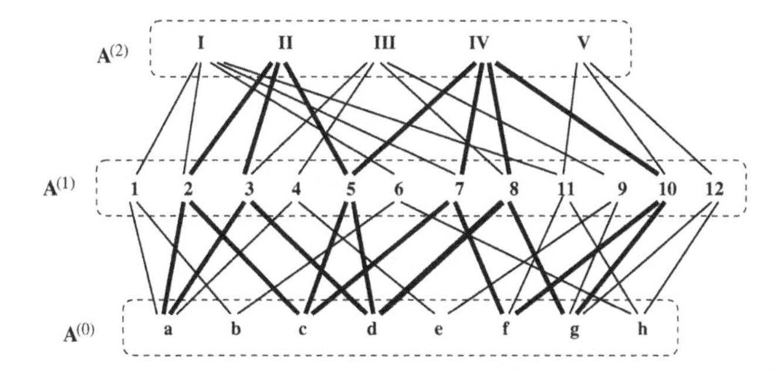

Fig. 6. The n-partite representation of Figure 5(a)

The notion of a boundary becomes more intuitive in geometric spaces such as Figures 5(a) and (b). Let $Z = \{II, IV\}.\varphi$ in Figure 5(a) corresponding to the darkened edges of Figure 6. Then $Z.\beta = [\,\{a, c, d, f, g\}, \{2, 3, 7, 8, 10\}, \emptyset\,]$. Observe that the line 5 is not in $Z.\beta$. Intuitively, an atom is in the boundary of Z only if it is incident to some atom not in Z. Readily, the generator of $Z.\beta$ is $Z.\beta.\gamma = \{2, 3, 7, 8, 10\}$. These are the faces of Z Only the generating tiles, two lines, and one point have been labelled in Figure 5(b). Here $\{XIII\}.\beta$ consists of the surrounding four bold lines; but $\{I\}.\beta$ consists of just the two bold lines. The remaining lines in $\{I\}.R$ are not incident to a tile "not in in Z". Tile I is a border tile of $\tau^{(n)}$. The atoms $\{1, 4, 6, 9, 12\} \subseteq A^{(1)}$ of Figure 6 are covered by singleton atoms. These singletons $\{I, III, V\} \in A^{(2)}$ constitute $\tau^{(n)}.B$.

A homogeneously generated n-dimensional topology $\tau^{(n)}$, will be called **geometric** if [5]

G1: $x \in y.R, y \in z.R$ implies there exists a unique $y' \neq y$ such that $x \in y'.R$ and $y' \in z.R$.
G2: for all $k > 0, x \in A^{(k)}$ implies $|\, x.R \,| \geq k + 1$.
G3: for all $k < n, x \in A^{(k)}$ implies $|\, x.R^{-1} \,| \geq n - k + 1$, and
G4: $y \in A^{(n-1)}$ implies $|\, y.R^{-1} \,| \leq 2$.

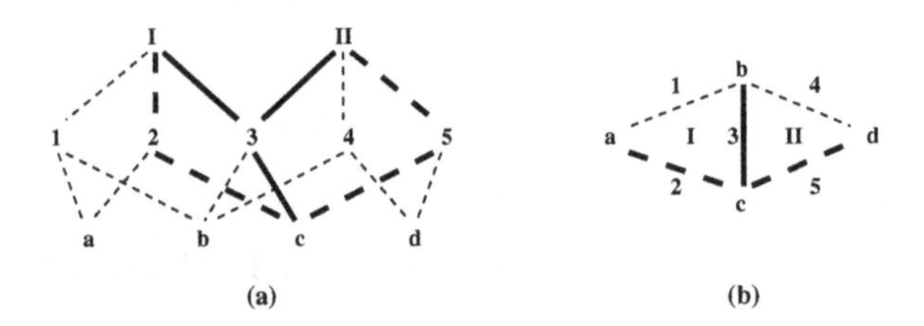

Fig. 7. Illustration of G1: (a) the n-partite representation of the geometry shown in (b)

Figure 7(a) illustrates the G1 property. Given the presence of line 3, with incident point c, that is incident to tiles I and II then the existence of two more lines, which we have labelled 2 and 5, that are also incident to tiles I and II and point c is forced. The other lightly dashed lines in Figures 7(a) and (b), the geometric equivalent of Figure 7(a), denote possible configurations; but other boundaries of the tiles I and II are quite possible. Only the two lines 2 and 5 are forced.

Property G2 further strengthens the usual normal constraint of $|\, x.R \,| \geq 2$ if $\delta(x) > 0$. It says that each line must have at least 2 end points, each tile must have at least 3 bounding lines, and each brick must have at least 4 bounding tiles. This corresponds to simplicial decomposition of physical space as we normally view it into triangles and tetrahedrons. Condition G3 implies that in

[5] That G1 thru G4 are properties of "geometric" topologies can be easily verified. Whether they are sufficient to characterize these topologies is unknown.

a 2-dimensional space, any point must be in the boundary (an endpoint) of at least 3 lines. Otherwise, the point is topologically redundant. Similarly, in a 3-dimensional space, each line segment must be incident to at least 3 tiles, because otherwise it too would be topologically redundant. Condition G4 which asserts that any (n-1)-atom can be the face of at most two atoms ensures that connected 1 and 2-dimensional topologies are strings and planar surfaces respectively. The closure structure of a geometric $\tau^{(1)}$ is the connected ordered topological space, or COTS, described in [6, 7, 8] and later illustrated in Figure 9. In 3, and higher dimensions, it asserts that a topological hyperplane separates exactly two regions.

Condition G3 says that (n-1)-atoms must separate at least two n-atoms, while condition G4 says they can separate no more than two n-atoms. Consequently, (n-1)-atoms must separate exactly 2 n-atoms, *except* possibly when the n-atom is at the *border*, $\tau^{(n)}.B$, of the entire space. Finite, discrete spaces often have borders where the expected properties of geometric spaces no longer hold. For example, in Figure 5(a), lines 1, 4, 6, 9 and 12 are incident to only one tile, and the points b and e are incident to only two lines. We must allow for these exceptions.

In Figure 5(a) and Figure 6, tiles I, III, V are border generators of $\tau^{(2)}.B$. There are no interior tiles in these two figures. In Figure 5(b), the 16 "outside" tiles constitute the border. All the remaining tiles satisfy the geometric constraints G1 through G4, but only tile XIII is an interior tile. Generators in interior position are well removed from the border of $\tau^{(n)}$, if there is one. In the discussions that follow we will assume all configurations have only generating atoms in interior position.

We now focus on the properties of the boundaries $Z.\beta$ of configurations Z in interior position in geometric topologies.

Lemma 13. *Let Z be a homogeneous n-dimensional configuration in interior position in a geometric topology $\tau^{(n)}$. Let y be a face of $Z.\beta$ and let $x \in y.R$. There exists a unique $y' \neq y$ such that*

> *(a) $x \in y'.R$, and*
> *(b) $y.R^{-1} \cap Z \subseteq y'.R^{-1} \cap Z$.*

Proof. Since $y \in Z.\beta$, y separates some $z \in Z$ from some $w \in \tau^{(n)} - Z$. Because $y \in A^{(n-1)}$, $y.R^{-1} = \{w, z\}$, so $y.R^{-1} \cap Z = z$. Since $y \in z.R, x \in y.R$ the existence of a unique y' with $x \in y'.R$ follows from G1. $y' \in z.R$ assures (b). □

Note that y' need not be an element of $Z.\beta$.

The following "continuation" theorem asserts that given a face $y_0 \in Z.\beta$, one can move in any "direction", *e.g.* across any face x of y_0, and find another face, possibly several, in $Z.\beta$. To see the import of this theorem, consider the two configurations of Figure 8. In $Z.\beta$ of Figure 8(a), line 2 in $A^{(1)}$ is 0-connected to lines 1 and 3 in $Z.\beta$. Since $Z.\beta$ is defined with respect to the n-atoms of Z and $\tau^{(n)} - Z$, we cannot ignore the topology of these configurations as we investigate $Z.\beta$. In Figure 8(a) tile V is pathwise 1-connected to III and V (itself) of

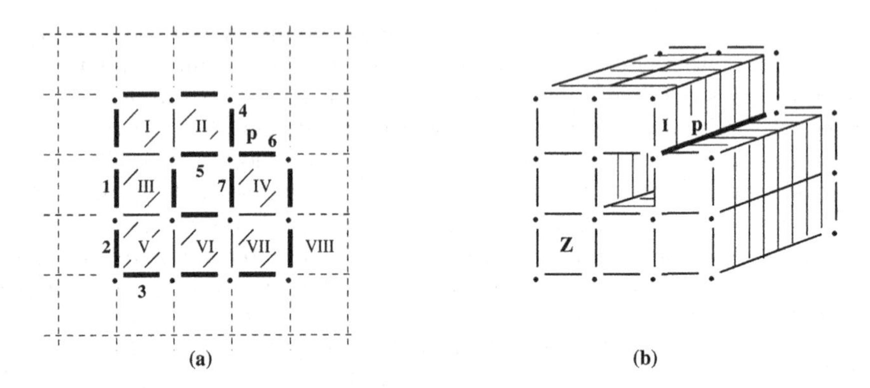

Fig. 8. Two configurations that are weakly connected at p

which 1 and 3 are faces. Similarly the corresponding tiles of the complement are pathwise connected. It is at places where Z is weakly (n-2) connected that possible complications can arise. Continuing face 4 of $Z.\beta$ through p could lead to faces 5, 6, or 7. Face 5 is pathwise connected *through* Z; and face 6 is pathwise *connected* through its complement. To make these ideas more formal we modify our path notation somewhat. Assuming the connectivity k is known, we now let $\rho_x(y_0, \ldots, y_n)$ denote a k-connected path such that $x \in y_i.\beta$ for $0 \leq i \leq n$. $\rho_x(y_0, \ldots, y_n)$ can be visualized as a path *around* x. In Figure 8(b) the tile labelled I in $Z.\beta$ is (n-2) path connected to three distinct tiles around p, two of which are hidden in the "tunnel".

Theorem 8. *Let Z be a homogeneous n-dimensional configuration in interior position in a geometric topology $\tau^{(n)}$, $n \geq 2$. Let y_0 be any face in $Z.\beta$ separating $z_0 \in Z$ from $w_0 \in W = \tau^{(n)} - Z$, and let $x \in y_0.R$. There exists a face y_n of $Z.\beta$ separating $z_n \in Z$ from $w_m \in W$, such that x separates y_0 from y_n and either*

(a) y_n is unique, in which case z_0, z_n are pathwise (n-1)-connected in Z and

w_0, w_m are pathwise (n-1)-connected in W,

or else

(b) Z (and W) is weakly (n-2)-connected at x.

Proof. Let $W = \tau^{(n)} - Z$ and let y_0 separate $z_0 \in Z$ from $w_0 \in W$. Application of Lemma 13 using x, y_0, z_0 ensures the existence of a unique y_1 such that $x \in y_1.R, y_1 \in z_0.R$. Since $y_1 \in A^{(n-1)}, y_1$ separates z_0 from some $a \in A^{(n)}$.

If $a \in W$, y_1 is a face of $Z.\beta$. Let $y_n = y_1$.

If $a \in Z$, let $z_1 = a$ and iterate the application of Lemma 13 using x, y_i, z_i. By local finiteness, this construction must terminate with a face y_n separating $z_n \in Z$ from $w' \in W$ such that x separates y_0 from y_n. Observe that we have created an (n-1)-connected path $\rho_x(z_0, z_n)$ such that $x \in z_i.R^2, 0 \leq i \leq n$.

We now repeat this construction using x, y_0, w_0 to first obtain y_1', and if necessary continue the construction to yield $y_m' \in Z.\beta$ separating $z' \in Z$

from $w_m \in W$. $\rho_x(w_0, w_m)$ is another (n-1)-connected path with $x \in w_j.R^2$, $0 \leq j \leq m$.

If $y_m = y_n$ then $z' = z_n$ and y_n is a unique face.

If $y_m \neq y_n$, we let $z'_0 = z'$, $y'_0 = y_m$ and repeat the construction using x, y'_0, z'_0.

Eventually, we obtain a face $y'_k \in Z.\beta$ separating z'_k from $w'' \in W$ where $\rho(z'_0, z'_k)$ is (n-1)-connected and $x \in z'_i.R^2, 0 \leq i \leq k$.

Since weak connectivity is a local property, we need only observe that $\{z_0, z_1, \ldots z_n\} \subset Z$ is weakly connected to $\{z'_0, z'_1, \ldots z'_m\} \subset Z$ at x. □

The role of "interior position" in this theorem can be visualized using Figure 5(b). Suppose tile V has been deleted from the space and that Z consists of the central 9 tiles. Suppose y_0 is the face of IX separating it from IV. Rotating "counterclockwise" around x one gets y_1 separating IX from tile $X \in W$. But tiles IV and X in W are not pathwise connected because V is missing.

5 Jordan Surface Theorem

A traditional statement of the Jordan Curve Theorem is:

Theorem 9 (Jordan Curve Theorem). *Let C be a simple closed curve in $\mathbf{R}^2$. Then $\mathbf{R}^2 - C$ consists of exactly two components A and B. Moreover, $C = \overline{A} - A = \overline{B} - B$.[2].*

Is the essence of the "Jordan curve property" that of subdividing the space into precisely two components, with the purpose of the theorem to show that any simple closed curve in R^2 has this property; or rather is the property a constraint on the curve, which one then shows separates the space. Examples of both interpretations can be found in the literature. Our approach is to define a **discrete Jordan surface** in a geometric topology $\tau^{(n)}$, to be a configuration S that separates $\tau^{(n)}$ into precisely two pathwise (n-1)-connected components, W and Z such that $S = W.\beta = Z.\beta$, where one is in interior position.[6] Neither W nor Z need be simply connected. For example, a discrete torus floating in $\tau^{(3)}$ could be a Jordan surface. If $n \geq 3$ either W or Z can be locally weakly connected. The boundary $X.\beta$ of Figure 8(b) is a Jordan surface; the boundary in Figure 8(a) is not.

We briefly review the considerable history associated with configurations such as Figure 8(a); which has become known as the "Rosenfeld Paradox". The common assumption has been that a "curve" in a pixel space is a "thin" connected sequence of pixels, no more than one pixel wide. If Z, the foreground configuration of Figure 8(a), is regarded as a closed, 0-connected "curve" then it does not separate the background into two components because the complement too is 0-connected at p. If it is not a closed curve, *i.e.* not everywhere 1-connected,

[6] Requiring one component to be in interior position, by convention Z, eliminates analogs of hyperplanes from being considered as Jordan surfaces.

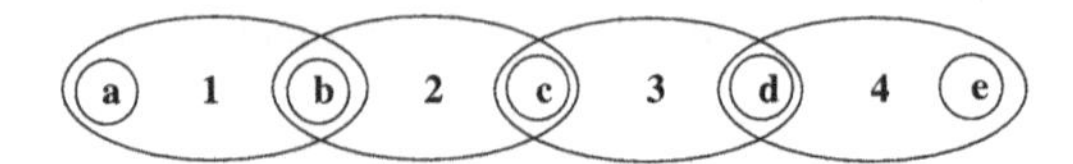

Fig. 9. A COTS, or Khalimsky topology

then it *does* separate the space.[7] A common means of resolving the paradox is to use 0-connectivity for the foreground (or background) and 1-connectivity for its complement[12, 13, 14]. The paradox partially arises from the assumption that pixels are the only elements of the space. This is a perfectly natural assumption if one is analyzing digital images. Then Figure 8(a) may, or may not, be regarded as a "thin, continuous, closed curve" which may, or may not, be a Jordan curve separating the space.

Several authors have resolved this paradox by introducing spatial elements of lower dimension much as we have. One of us has begun with a connected ordered topological space (COTS), or Khalimsky topology shown in Figure 9. It has two kinds of element. It is T_0; it is antimatroid. Its direct product is equivalent to the geometric $\tau^{(2)}$ of Figure 5(b) ith the "pure" direct product of lines corresponding to tiles; the "pure" direct product of points would be a point; and the "mixed" direct products of a line with s point would be the same as our line [7, 8].

Although he only considers pixel elements, Gabor Herman [3] effectively introduces (n-1)-dimensional elements by considering the "boundary" of Z, denoted $\partial(Z, \mathbf{U}-Z)$, to be the collection of pixel pairs whose first element is in Z, with the second in $\mathbf{U}-Z$. When the topology is a pixel plane, these pixel pairs are equivalent to line segments; when it's a 3-dimensional array of voxels, it is a collection of voxel faces. It corresponds to our interpretation of a Jordan surface S in $\tau^{(n)}$ as being a boundary where $S.\gamma \subseteq A^{(n-1)}$.

Theorem 10. *Let Z be a homogeneous n-dimensional configuration in interior position in a geometric topology $\tau^{(n)}$. The boundary $Z.\beta$ is a Jordan surface if*

(a) $Z.\beta$ is completely (n-2) connected, and

(b) if Z is weakly (n-2) connected at $X \subset Z.\beta$, then there exists neither a subset Z' of Z such that $X \subseteq Z'.\beta \subset Z.\beta$, nor a subset W' of the complement, $W = \mathbf{U}-Z$, such that $X \subseteq W'.\beta \subset Z.\beta$.

Proof. (Necessity) If $Z.\beta$ is a Jordan surface, then both Z and W are completely (n-1)-connected, so (a) follows from Theorem 7 and Lemma 12 assures us that all points where $Z.\beta$ is (n-2)-connected, Z must be at least (n-2)-connected as well. The complete connectivity of Z and W ensures that neither Z' nor W' can exist at these points.

(Sufficiency) Because $Z.\beta = W.\beta$ is (n-2)-connected, both Z and W are at least (n-2)-connected by Lemma 12. Moreover, because Z is in interior position,

[7] We should note that Rosenfeld used the terms 8-connected and 4-connected instead of 0-connected and 1-connected [12]. This designated the number of "connected" pixels in a rectangular pixel space; it is standard in image processing.

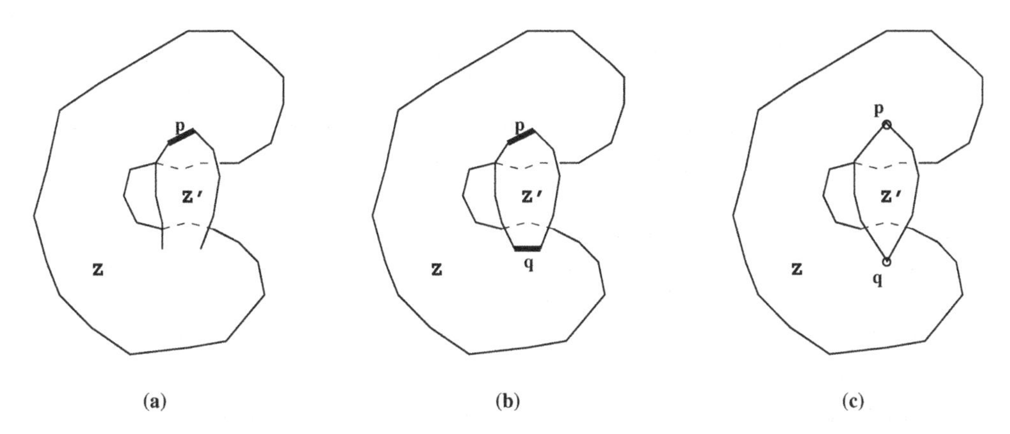

Fig. 10. Three configurations Z with different connectivities: (a) Z' is (n-1)-connected to Z, (b) Z' is (n-2)-connected to Z, (c) Z' is (n-3)-connected to Z

our definition of "interior" ensures that all generators of W that are border atoms belong to a single (n-1)-connected component. Now suppose that Z (or W) is not (n-1)-connected. Say, $Z = Z_1 \cup Z_2$, where Z_1 and Z_2 are only (n-2)-connected. Then $Z_1, \beta \subset Z.\beta$. But, this is explicitly ruled out by condition (b). Thus $Z.\beta$ must be a Jordan surface. □

Theorem 10 can be more easily visualized, where $n = 3$, by Figure 10. In Figure 10(a), $Z.\beta$ is 1-connected at p, but it is still a Jordan surface because no matter how the subconfiguration $Z' \subset Z$ is chosen, its boundary, $Z'.\beta \not\subset Z.\beta$. The portion at the base of Z' cannot be in the boundary of Z. In contrast, the boundary of Z in Figure 10(b) is not a Jordan surface because $Z'.\beta \subset Z.\beta$. Neither is the boundary of Z in Figure 10(c) a Jordan surface, but in this case it is because $Z.\beta$ is not (n-2)-connected.

Corollary 3. *Let Z be a homogeneous n-dimensional configuration in interior position in a geometric topology $\tau^{(n)}$. If Z is (n-1)-connected and nowhere weakly k-connected, where $k < n-1$, then $Z.\beta$ is a Jordan surface.*

Theorem 10 is of interest for two reasons. First, it completely characterizes all those surfaces which subdivide a geometric topology $\tau^{(n)}$ into precisely two completely (n-1)-connected components, one of which is in interior position. That component, which we have been denoting a Z, can be wildly contorted with many weak connectivities. But, we have shown that only at points of weak (n-2)-connectivity (where the argument of Theorem 8 shows the complement W must also be (n-2) connected) is further examination is required.

Second, it provides a theoretical basis for procedures to decide whether a specific surface of (n-1) atoms is a Jordan surface, thus reducing a potentially n-dimensional problem to one of (n-1) dimensions. Although provision of such an algorithm is beyond the scope of this paper, it is easy to envision one utilizing Theorem 8 which marks faces as examined and at each new face stacks all possible unmarked "adjacent" faces. For more sophisticated "face crawling" algorithms, the reader is refered to [3]. But, remember that those algorithms depend on a regular decomposition of the space. Our topologies need not be at all regular. They can easily arise from Voronoi decompositions [1] or polygonal mesh refinements [15].

References

1. Franz Aurenhammer. Voronoi diagrams — a survey of a fundamental geometric data structure. *ACM Computer Surveys*, 23(3):345–406, Sept. 1991.
2. Dick Wick Hall and Guilford L. Spencer. *Elementary Topology.* Wiley, New York, 1955.
3. Gabor T. Herman. *Geometry of Digital Spaces.* Birkäuser, Boston, 1998.
4. Robert E. Jamison and John L. Pfaltz. Closure Spaces that are not Uniquely Generated. In *Ordinal and Symbolic Data Analysis, OSDA 2000*, Brussels, Belgium, July 2000.
5. John L. Kelley. *General Topology.* Van Nostrand, 1955.
6. E. D. Khalimsky, Ralph Kopperman, and Paul R. Meyer. Computer Graphics and Connected Topologies on Finite Ordered Sets. *Topology and its Applications*, 36:1–17, Jan. 1990.
7. T. Yung Kong, Ralph Kopperman, and Paul R. Meyer. A Topological Approach to Digital Topology. *Am. Math. Monthly*, 98(10):901–917, Dec. 1991.
8. Ralph Kopperman, Paul R. Meyer, and R. G. Wilson. A Jordan Surface Theorem for Three-dimensional Digital Spaces. *Discrete and Computational Geometry*, 6:155–161, 1991.
9. Bernhard Korte, László Lovász, and Rainer Schrader. *Greedoids.* Springer-Verlag, Berlin, 1991.
10. Kazimierz Kuratowski. *Introduction to Set Theory and Topology.* Pergamon Press, 1972.
11. John L. Pfaltz. Closure Lattices. *Discrete Mathematics*, 154:217–236, 1996.
12. Azriel Rosenfeld. Picture processing by computer. *ACM Computer Surveys*, 1(3), Sept. 1969.
13. Azriel Rosenfeld and Avinash C. Kak. *Digital Picture Processing.* Academic, New York, 1982.
14. Azriel Rosenfeld and John L. Pfaltz. Distance Functions on Digital Pictures. *Pattern Recog.*, 1(1):33–61, Sept. 1968.
15. Colin Smith, Przemyslaw Prusinkiewicz, and Faramarz Samavati. Local Specification of Surface Subdivision Algorithms. *Second Int'l. Workshop AGTIVE 2003*, Lecture Notes in Computer Science, # 3062:313–328, Sept. 2004.

How to Find a Khalimsky-Continuous Approximation of a Real-Valued Function

Erik Melin

Uppsala University, Department of Mathematics,
Box 480, SE-751 06 Uppsala, Sweden
melin@math.uu.se

Abstract. Given a real-valued continuous function defined on n-dimensional Euclidean space, we construct a Khalimsky-continuous integer-valued approximation. From a geometrical point of view, this digitization takes a hypersurface that is the graph of a function and produces a digital hypersurface—the graph of the digitized function.

Keywords: Khalimsky topology, digitization, discretization, digital surface.

1 Introduction and Background

The increasing use of multi-dimensional images in applications of computer imagery calls for a development of digital geometry in three dimensions and higher. In particular, digital curves, surfaces, planes and other digital counterparts of Euclidean geometrical objects have been extensively studied. Several different approaches have been used in this study. Historically, the first attempts to define digital objects were algorithmic; a digital object was defined to be the result of a given algorithm. We may here mention the works by Bresenham (1965) and Kaufman (1987). One major drawback of this approach is that it may be hard to determine the geometrical properties of the defined objects.

To get precise definitions, a more recent approach is to let digital objects be defined by their local properties. This point of view is generally graph-theoretic but is often called topological, although no topology is actually involved. Keywords sometimes used by the advocates of this direction are topological and geometrical consistency; however, from time to time it is a little unclear what these concepts really mean. Much has been written in this field and we have no ambition to provide all significant references. A survey of the field with many references has been written by Kong and Rosenfeld (1989). Concerning digital surfaces a pioneering work is Morgenthaler & Rosenfeld's (1981) paper. This study was continued by Kim (1984), Rosenfeld et al. (1991), Cohen-Or and Kaufman (1995) and by Chen et al. (1999). A slightly different approach was suggested by Herman (1998). Here, the space consists of voxels and surface elements (surfels) where the surfels alternatively can be thought of as adjacency

R. Klette and J. Žunić (Eds.): IWCIA 2004, LNCS 3322, pp. 351–365, 2004.

relations between voxels. A different method of describing linear digital objects is through Diophantine inequalities (Reveillès 1991). This sort of description is often called arithmetic or analytic.

An important aspect of the subject is the problem of finding a digital representation of, say, a surface in $\mathbb{R}^3$. This process is sometimes called *discretization* and sometimes *digitization*. There are many ways to perform digitization but a general goal is that the digitized object should preserve characteristic properties of the original object. Straight lines in the plane are naturally of fundamental importance and Rosenfeld (1974) clarified the properties of the grid intersection digitization of straight lines. A new digitization of straight lines in the plane was suggested by Melin (2003*a*), where the digital lines respect the Khalimsky topology. The present paper treats a generalization of this digitization to higher dimensions. Some advantages of working in a topological space compared to the purely graph-theoretical approach has been discussed by for example Kong, Kopperman & Meyer (1991) and in Kong (2003) it is shown that the Khalimsky topology is in a sense the natural choice.

Another digitization scheme that should be mentioned is the supercover. Ronse and Tajine (2000, 2002) showed that the supercover is a Hausdorff discretization. A generalization of the supercover to a partition of $\mathbb{R}^n$, where the quotient topology is homeomorphic to Khalimsky n-space, has been considered by Couprie, Bertrand & Kenmochi (2003). However, since the space in that paper is viewed as partition av $\mathbb{R}^n$, the properties of this digitization is quite different from our digitization.

2 Mathematical Background

In this section we present a mathematical background for this paper. The first subsection contains some general topology for digital spaces. After that, the two following subsections give an introduction to the Khalimsky topology and to Khalimsky-continuous functions. Some properties of such functions are proved. The mathematical background is concluded by a result on continuous extension which is needed.

2.1 Topology and Smallest-Neighborhood Spaces

In any topological space, a finite intersection of open sets is open, whereas the stronger requirement that an arbitrary intersection of open sets be open is not satisfied in general. Alexandrov (1937) considers topological spaces that fulfill the stronger requirement, where arbitrary intersections of open sets are open. Following Kiselman (2002), we will call such spaces *smallest-neighborhood spaces*. Another name that has been used is *Alexandrov spaces*.

Let B be a subset of a topological space X. The *closure* of B is a very well known notion, and the closure of B usually denoted by $\overline{B}$. In this paper, we will instead denote the closure of B in X by $C_X(B)$. This allows us to specify in what space we consider the closure and is also a notation dual to N_X defined below. The closure of a set is defined to be the intersection of all closed sets containing it.

Dually, we define $N_X(B)$ to be the intersection of all open sets containing B. In general $N_X(B)$ is not an open set, but in a smallest-neighborhood space it is. Clearly $N_X(B)$ is the smallest neighborhood of the set B. If there is no danger of ambiguity, we will just write $N(B)$ and $C(B)$ instead of $N_X(B)$ and $C_X(B)$. If x is a point in X, we define $N(x) = N(\{x\})$ and $C(x) = C(\{x\})$. Note that $y \in N(x)$ if and only if $x \in C(y)$.

We have already remarked that $N(x)$ is the smallest neighborhood of x. Conversely, the existence of a smallest neighborhood around every point implies that an arbitrary intersection of open sets is open; hence this existence could have been used as an alternative definition of a smallest-neighborhood space. A topological space X is called *connected* if the only sets which are both closed and open are the empty set and X itself. A point x is called *open* if the set $\{x\}$ is open, and is called *closed* if $\{x\}$ is closed. If a point x is either open or closed it is called *pure*, otherwise it is called *mixed*.

Two distinct points x and y in X are called *adjacent* if the subspace $\{x, y\}$ is connected. It is easy to check that x and y are adjacent if and only $y \in N(x)$ or $x \in N(y)$. Another equivalent condition is $y \in N(x) \cup C(x)$. The *adjacency set* in X of a point x, denoted $A_X(x)$, is the set of points adjacent to x. Thus we have $A_X(x) = (N_X(x) \cup C_X(x)) \smallsetminus \{x\}$. Often, we just write $A(x)$. A point adjacent to x is called a *neighbor* of x. This terminology, however, is somewhat dangerous since a neighbor of x need not be in the smallest neighborhood of x.

Kolmogorov's separation axiom, also called the T_0 axiom, states that given two distinct points x and y, there is an open set containing one of them but not the other. An equivalent formulation is that $N(x) = N(y)$ implies $x = y$ for every x and y. The $T_{1/2}$ axiom states that all points are pure. Clearly any $T_{1/2}$ space is also T_0. Smallest-neighborhood spaces satisfying the T_1 axiom must have the discrete topology and are therefore not so interesting. A useful observation is that if X and Y are topological spaces and $x \in X$ and $y \in Y$, then $N_{X\times Y}(x, y) = N_X(x) \times N_Y(y)$ and similarly for the closure.

2.2 The Khalimsky Topology

We will construct a topology on the digital line, $\mathbb{Z}$, originally introduced by Efim Khalimsky (see Khalimsky, Kopperman & Meyer (1990) and references there). Let us identify with each even integer m the closed, real interval $[m-1/2, m+1/2]$ and with each odd integer n the open interval $]n - 1/2, n + 1/2[$. These intervals form a partition of the Euclidean line $\mathbb{R}$ and we may therefore consider the quotient space. Identifying each interval with the corresponding integer gives us the *Khalimsky topology* on $\mathbb{Z}$. Since $\mathbb{R}$ is connected, the Khalimsky line is connected. It follows readily that an even point is closed and that an odd point is open. In terms of smallest neighborhoods, we have $N(m) = \{m\}$ if m is odd and $N(n) = \{n \pm 1, n\}$ if n is even.

Let a and b, $a \leqslant b$, be integers. A *Khalimsky interval* is an interval $[a, b] \cap \mathbb{Z}$ of integers with the topology induced from the Khalimsky line. We will denote such an interval by $[a, b]_{\mathbb{Z}}$ and call a and b its endpoints. A *Khalimsky arc* in a topological space X is a subspace that is homeomorphic to a Khalimsky interval. If any two points in X are the endpoints of a Khalimsky arc, we say that X is *Khalimsky arc-connected.*

Theorem 1. *A T_0 smallest-neighborhood space is connected if and only if it is Khalimsky arc-connected.*

Proof. See for example Theorem 11 of Melin (2004). Slightly weaker is Khalimsky, Kopperman & Meyer's (1990, Theorem 3.2c) result. □

On the digital plane, $\mathbb{Z}^2$, the Khalimsky topology is given by the product topology. Points with both coordinates odd are open and points with both coordinates even are closed. These are the pure points in the plane. Points with one odd and one even coordinate are mixed. Let us call a point q such that $\|q - p\|_\infty = 1$ an *l^∞-neighbor* of p. We may note that a mixed point p is connected only to its four pure l^∞-neighbors $(p_1 \pm 1, p_2)$ and $(p_1, p_2 \pm 1)$, whereas a pure point p is connected to all eight l^∞-neighbors.

$$(p_1 \pm 1, p_2),\ (p_1, p_2 \pm 1),\ (p_1 + 1, p_2 \pm 1) \text{ and } (p_1 - 1, p_2 \pm 1)$$

More generally, Khalimsky n-space is $\mathbb{Z}^n$ equipped with product topology. Here, points with all coordinates odd are open and points with all coordinates even are closed. Let $\mathbb{P}_n$ denote the set of pure points in $\mathbb{Z}^n$. Note that $\mathbb{P}_n$ is not a product space: $\mathbb{P}_n \neq \mathbb{P}_1^n = \mathbb{Z}^n$.

Let $\mathrm{ON}(p) = \{x \in A(p);\ x \text{ is open}\}$ be the set of open neighbors of a point p in $\mathbb{Z}^n$ and similarly $\mathrm{CN}(p) = \{x \in A(p);\ x \text{ is closed}\}$ be the set of closed neighbors. The cardinality of a set X is denoted by $\mathrm{card}(X)$. If c is the number of even coordinates in p and d is the number of odd coordinates, then $\mathrm{card}(\mathrm{ON}(p)) = 2^c$ and $\mathrm{card}(\mathrm{CN}(p)) = 2^d$. Define also $\mathrm{PN}(p) = \mathrm{CN}(p) \cup \mathrm{ON}(p)$ to be the set of all pure neighbors of a point p. A pure point in $\mathbb{Z}^n$ has always 2^n pure neighbors. For mixed points, however, the situation is different. In $\mathbb{Z}^2$ every mixed point has 4 pure neighbors. In $\mathbb{Z}^3$ a mixed point has $2^1 + 2^2 = 6$ pure neighbors. But in $\mathbb{Z}^4$ a mixed point may have $2^1 + 2^3 = 10$ or $2^2 + 2^2 = 8$ pure neighbors. Obviously, the number of possibilities increases even more in higher dimension. These different types of points have different topological properties and may cause the digitization process to become more complex in higher dimension, cf. Remark 20.

2.3 Continuous Functions

Unless otherwise stated we shall assume that $\mathbb{Z}$ is equipped with the Khalimsky topology from now on. This makes it meaningful to consider (for example) continuous functions $f\colon \mathbb{Z} \to \mathbb{Z}$. Continuous integer-valued functions will sometimes be called Khalimsky-continuous, if we want to stress that a function is not real continuous. We will discuss some properties of such functions; more details can be found in e.g. (Kiselman 2002). Suppose that f is continuous. Since $M = \{m, m+1\}$ is connected it follows that $f(M)$ is connected, but this is the case only if $|f(m) - f(m+1)| \leqslant 1$. Hence f is Lipschitz with Lipschitz constant 1; we say f is Lip-1. Lip-1, however, is not sufficient for continuity. If $y = f(x)$ is odd, then $U = f^{-1}(\{y\})$ must be open, so if x is even then $x \pm 1 \in U$. This means that $f(x \pm 1) = f(x)$. In a more general setting, this generalizes to the following:

Proposition 2. *Let X be a topological space and $f\colon X \to \mathbb{Z}$ be a continuous mapping. Suppose that $x_0 \in X$. If $f(x_0)$ is odd, then f is constant on $N(x_0)$ and $|f(x) - f(x_0)| \leqslant 1$ for all $x \in C(x_0)$. If $f(x)$ is even, then f is constant on $C(x_0)$ and $|f(x) - f(x_0)| \leqslant 1$ for all $x \in N(x_0)$.*

Proof. Let $y_0 = f(x_0)$ be odd. Then $\{y_0\}$ is an open set. Hence $f^{-1}(\{y_0\})$ is open and therefore $N(x_0) \subset f^{-1}(\{y_0\})$, that is, $f(N(x_0)) = \{y_0\}$. Moreover, the set $A = \{y_0, y_0 \pm 1\}$ is closed. But then the set $f^{-1}(A)$ is closed also and this implies that $f(x) \in A$ for all $x \in C(x_0)$. The even case is dual. □

Let us define the *length* of a Khalimsky arc A to be the number of points in A minus one, $L(A) = \operatorname{card} A - 1$. Theorem 1 allows us to define the *arc metric* on a T_0 connected smallest-neighborhood space X to be the length of the shortest arc connecting x and y in X:

$$\rho_X(x, y) = \min(L(A);\ A \subset X \text{ is a Khalimsky arc containing } x \text{ and } y).$$

Proposition 3. *Let X be a connected, T_0 smallest-neighborhood space. If f is a continuous mapping $X \to \mathbb{Z}$, then f is Lip-1 for the arc metric.*

Proof. This is Proposition 15 of Melin (2004). □

Theorem 4. *Let X be a connected, T_0 smallest-neighborhood space and consider a family of continuous mappings $f_j\colon X \to \mathbb{Z}$, $j \in J$. If the set $\{f_j(a);\ j \in J\}$ is bounded for some $a \in X$, then the mappings $f_\star(x) = \inf_{j\in J} f_j(x)$ and $f^\star(x) = \sup_{j\in J} f_j(a)$ are continuous.*

Proof. We first prove that $f_\star$ and $f^\star$ are finite everywhere. Assume the contrary and suppose that there is a sequence of indices $j_k \in J$, $k = 1, 2, \ldots$ and a point $x \in X$ such that $f_{j_k}(x) \to +\infty$ (for example) as $k \to \infty$. Then also $f_{j_k}(a) \to +\infty$, contrary to our assumption, since each f_j is continuous and therefore, by Proposition 3, Lip-1 for the arc metric.

To demonstrate continuity, we have to prove that for each $x \in X$ we have $N_X(x) \subset f^{-1}(N_{\mathbb{Z}}(f(x)))$. Therefore we fix $x \in X$ and suppose $y \in N_X(x)$. Let $f_\star(x) = m$ and let $j \in J$ be an index such that $f_j(x) = m$.

We consider two cases separately. The relations that come from Proposition 2 are marked with a †. If m is even, then $f_\star(y) \leqslant f_j(y) \leqslant^\dagger m + 1$. We may conclude that for any index $i \in J$, we have $f_i(y) \geqslant^\dagger m - 1$ since $x \in C(y)$ and $f_i(x) \geqslant m$. But then $f(y) \in N_{\mathbb{Z}}(m) = \{m, m \pm 1\}$ as required. If m is odd, we still have $f_i(y) \geqslant m-1$ for the same reason as above. But $m-1$ is even, so if $f_i(y) = m-1$, then also $f_i(x) =^\dagger m-1 < f_\star(x)$, which is a contradiction. Finally, note that $f_\star(y) \leqslant f_j(y) =^\dagger f_j(x)$. Hence, $f_\star(y) = m \in N_{\mathbb{Z}}(m)$ and therefore $f_\star$ is continuous. Continuity of $f^\star$ is proved in the same way. □

Given a mapping $f\colon X \to Y$ between any two sets, the graph of f is defined by $G_f = \{(x, f(x));\ x \in X\} \subset X \times Y$. Suppose now that X and Y are topological spaces. It is a general topological fact that if $f\colon X \to Y$ is continuous, then G_f is homeomorphic to X. This means that the graph of a Khalimsky-continuous map $f\colon \mathbb{Z}^n \to \mathbb{Z}$ is homeomorphic to $\mathbb{Z}^n$. An intuitive interpretation of this is that the graph has no redundant connectedness and in particular is bubble-free, cf. Andrès (2003).

2.4 Continuous Extension

We will find use for a result on continuous extension in Khalimsky spaces, which can be found in Melin (2003*b*, Theorem 12). To formulate it, we need first the following definition:

Definition 5. Let $A \subset \mathbb{Z}^n$ and let $f\colon A \to \mathbb{Z}$ be a function. Let x and y be two distinct points in A.

If one of the following conditions are fulfilled for some $i = 1, 2, \dots, n$,

1. $|f(x) - f(y)| < |x_i - y_i|$ or
2. $|f(x) - f(y)| = |x_i - y_i|$ and $x_i \equiv f(x) \pmod 2$,

then we say that the function is **strongly Lip-1 with respect to (the points) x and y.** If the function is strongly Lip-1 with respect to every pair of distinct points in A then we simply say that f is **strongly Lip-1**.

Theorem 6. *Let $A \subset \mathbb{Z}^n$, and let $f\colon A \to \mathbb{Z}$ be any function. Then f can be extended to a continuous function on all of $\mathbb{Z}^n$ if and only if f is strongly Lip-1.*

3 Khalimsky-Continuous Digitization

Let X be a set and Z an arbitrary subset of X. A *digitization* of X is a mapping $\mathfrak{D}\colon \mathrm{P}(X) \to \mathrm{P}(Z)$. Given a subset $A \subset X$, we think of $\mathfrak{D}(A)$ as a digital representation of A. In this paper, we will mainly be interested in the case when X is the Euclidean space $\mathbb{R}^n$ and Z is $\mathbb{Z}^n$ equipped the Khalimsky topology.

Given a digitization $\mathfrak{D}\colon \mathrm{P}(\mathbb{R}^n) \to \mathrm{P}(\mathbb{Z}^n)$ and a function $f\colon \mathbb{R}^{n-1} \to \mathbb{R}$, we can define a set-valued mapping $(\mathfrak{D}^s f)\colon \mathbb{Z}^{n-1} \to \mathrm{P}(\mathbb{Z})$ via the graph of f.

$$(\mathfrak{D}^s f)(p) = \{m \in \mathbb{Z};\ (p, m) \in \mathfrak{D}(G_f)\}$$

We would like to construct a integer-valued mapping from this set-valued mapping. Obviously it may happen that $(\mathfrak{D}^s f)(x) = \emptyset$ for some x, The set of points where this does not occur is of interest, hence we define the *digitized domain* as the set

$$\mathrm{Dom}(\mathfrak{D} f) = \{p \in \mathbb{Z}^{n-1};\ (\mathfrak{D}^s f)(x) \neq \emptyset\} \tag{1}$$

We shall also assume from now on that the function and the digitization are such that $(\mathfrak{D}^s f)(x)$ is a finite set for each x. Under this assumption we may define two integer-valued mappings, namely an *upper digitization* of f

$$\mathfrak{D}^\diamond f\colon \operatorname{Dom}(\mathfrak{D}f) \to \mathbb{Z},\ z \mapsto \max(m \in L;\ (z,m) \in \mathfrak{D}(G_f))$$

and similarly a *lower digitization*

$$\mathfrak{D}_\diamond f\colon \operatorname{Dom}(\mathfrak{D}f) \to L,\ z \mapsto \min(m \in L;\ (z,m) \in \mathfrak{D}(G_f)).$$

If it happens that $\mathfrak{D}^\diamond f = \mathfrak{D}_\diamond f$, we can define the *restricted digitization* of f at a point as the common value, $\mathfrak{D}f = \mathfrak{D}^\diamond f = \mathfrak{D}_\diamond f$. The digitized domain is in general not equal to all of $\mathbb{Z}^{n-1}$ and therefore the restricted digitization need to be extended in some way.

The next task is to define the a digitization ($\mathfrak{P}$) that will form the foundation for the remaining parts of this paper. Then the main goal of this section will be to prove Theorem 14 which states that the following algorithm will result in the desired approximation.

Algorithm 7. Khalimsky-Continuous Digitization

1. Apply the pure digitization, $\mathfrak{P}$, to the graph of f (see (3)) to obtain the pure points in the digitization of the graph.
2. Extend the obtained function to be defined on all pure points in the domain. This is a local operation, which depends only on $\mathfrak{P}f$ in a small neighborhood of each pure point. See Definition 10.
3. Extend the digital function to all of $\mathbb{Z}^n$ using the formulas of Theorem 14. This is again a local operation. □

The first goal is to define a digitization of $\mathbb{R}^n$ into $\mathbb{P}_n$, the set of pure points in $\mathbb{Z}^n$. Let

$$U_n = \{x \in \mathbb{R}^n;\ |x_i| = 1/2 \text{ for } i = 1, 2, \ldots, n-1 \text{ and } x_n = 1/2\}$$

and define

$$C_n = \bigcup_{x \in U_n} \{tx;\ t \in \left]-1, 1\right]\}.$$

Thus C_n is a cross with 2^n arms. Let

$$H_n(0) = C_n \cup \{x \in \mathbb{R}^n;\ \|x\|_\infty < 1/2\}, \tag{2}$$

be the union of this cross and an open cube in $\mathbb{R}^n$. This definition is illustrated in Fig. 1. Note that $H_n(0)$ is in fact the open cube together with finitely many (2^{n-1}) points added to half of the vertices, i.e.,

$$H_n(0) = U_n \cup \{x \in \mathbb{R}^n;\ \|x\|_\infty < 1/2\}$$

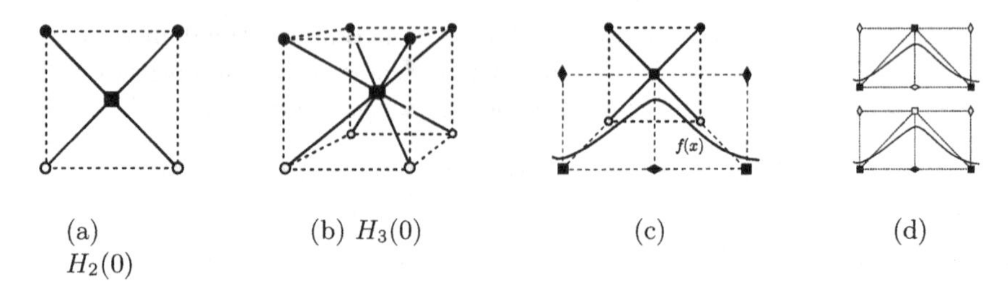

(a) $H_2(0)$ (b) $H_3(0)$ (c) (d)

Fig. 1. (a) and (b) The sets H_2 and H_3. (c) A curve, $f(x)$, that does not intersect U_2 but intersects H_2. (d) The upper image shows the result of the digitization with H_2, the lower shows what the result would be if only U_2 were used (see Definintion 10 on page 359). Thus the inclusion of the open cube improves the approximation

The reason for us to use (2) as the definition is that from the topological point of view the important fact is that H_n contain all the diagonal arms of C_n. The cube only improves the metric approximation as illustrated in Fig. 1 (c) and (d) and as discussed in Sect. 4.

For each $p \in \mathbb{P}_n$ let $H_n(p) = H_n(0) + p$ be $H_n(0)$ translated by the vector p. Note that $H_n(p) \cap H_n(q)$ is empty if $p \neq q$ and that $\bigcup\{H_n(p);\ p \in \mathbb{P}_n\}$ contains every diagonal grid line of the type $\{tx + p; t \in \mathbb{R}\}$ where $p \in \mathbb{P}_n$ and x is a vector in U_n. Note also that if $x, y \in H_n(p)$, then $-1/2 < x_n - p_n \leqslant 1/2$ and in particular $|x_n - y_n| < 1$. Since most of the time, we will consider a fixed dimension n, we shall just write $H(p)$ instead of $H_n(p)$ to simplify notation. Using the set $H(p)$, we define the *pure digitization* of a subset $A \subset \mathbb{R}^n$ as:

$$\mathfrak{P}(A) = \{p \in \mathbb{P}_n;\ H(p) \cap A \neq \emptyset\}. \tag{3}$$

Lemma 8. *Suppose That the Mapping* $f\colon \mathbb{R}^n \to \mathbb{R}$ *is Lip-1 for the* l^∞*-metric. Then* $\mathfrak{P}^\diamond f = \mathfrak{P}_\diamond f$ *so that* $\mathfrak{P} f = \mathfrak{P}^\diamond f = \mathfrak{P}_\diamond f$ *can be defined. Furthermore,* $\mathfrak{P} f$ *is also Lip-1 for the* l^∞*-metric in* $\mathbb{Z}^n$*.*

Proof. Let $p \in \mathrm{Dom}(\mathfrak{P} f) \subset \mathbb{Z}^n$ and suppose that $i, j \in \mathbb{Z}$, $i \neq j$, are integers such that $(p, i) \in \mathfrak{P}(G_f)$ and $(p, j) \in \mathfrak{P}(G_f)$. Then there are $x, y \in \mathbb{R}^n$ such that $(x, f(x)) \in H(p, i)$ and $(y, f(y)) \in H(p, j)$. Clearly this implies that $\|x - y\|_\infty \leqslant 1$. Since (p, i) and (p, j) are pure points, i and j have the same parity and therefore $|i - j| \geqslant 2$. But then it follows that $|f(x) - f(y)| > 1$ and this contradicts the fact that f is Lip-1. Hence $\mathfrak{P} f$ can be defined.

For the second part, let $p, q \in \mathrm{Dom}(\mathfrak{P} f)$ where $p \neq q$. Define $d = \|p - q\|_\infty$, $i = (\mathfrak{P} f)(p)$ and $j = (\mathfrak{P} f)(q)$. Again there are points $x, y \in \mathbb{R}^{n-1}$ such that $(x, f(x)) \in H(p, i)$ and $(y, f(y)) \in H(q, j)$. We must show that $|i - j| \leqslant d$. Since $\|x - y\|_\infty \leqslant d + 1$, the Lip-1 assumption gives that $|f(x) - f(y)| \leqslant d + 1$. Suppose that $|i - j| > d$. Since (p, i) and (q, j) are pure points it follows that $d \equiv |i - j| \pmod 2$, and therefore $|i - j| \geqslant d + 2$. But then

$$|f(x) - f(y)| > |i - j| - 1 \geqslant d + 2 - 1 = d + 1 \geqslant |f(x) - f(y)|,$$

which is contradictory. □

Clearly, $\mathrm{Dom}(\mathfrak{P}f)$ need not be equal to all of $\mathbb{P}_n$. If $f = 0$ is the zero function, then $\mathrm{Dom}(\mathfrak{P}f)$ consists of precisely the closed points in $\mathbb{P}_n$. Note that if p is an open point in $\mathbb{P}_n$, then all its neighbors are closed and therefore in $\mathrm{Dom}(\mathfrak{P}f)$. The following lemma shows that this is not a coincidence. Note that only the diagonal gridlines of C_n are used in the proof.

Lemma 9. *Suppose That $f\colon \mathbb{R}^n \to \mathbb{R}$ is Lip-1 for the l^∞-metric, and let $\mathfrak{P}f$ be its pure digitization. If $p \in \mathbb{P}_n$ does not belong to $\mathrm{Dom}(\mathfrak{P}f)$, then every pure neighbor of p belongs to $\mathrm{Dom}(\mathfrak{P}f)$. Furthermore, there is an integer r such that $(\mathfrak{P}f)(q) = r$ for every pure neighbor q of p.*

Proof. Let us say, for definiteness, that p is an open point. Since $(p, k) \notin \mathfrak{P}(G_f)$ for any (odd) $k \in \mathbb{Z}$, there must be an even integer r such that $|f(p) - r| < 1$. Let q be a pure neighbor of p. This means that $q = p + (a_1, a_2, \ldots, a_n)$ where $|a_i| = 1$ and that q is closed. The point $(q, r) \in \mathbb{P}_{n+1}$ is closed and we will show that it belongs to the digitized graph.

If $f(q) = r$, then clearly $(q, r) \in \mathfrak{P}(G_f)$. Suppose $f(q) < r$ (the case $f(q) > r$ is similar). Let $\psi\colon [0,1] \to \mathbb{R}^n$ be the parameterization of the real line segment $[p, q]$ given by $\psi(t) = q + t(p - q)$. In particular we have $\psi(0) = q$ and $\psi(1) = p$. Now, we define a mapping $g\colon [0,1] \to \mathbb{R}$ by $g(t) = f(\psi(t)) - (r - t)$. Note that $g(0) = f(q) - r < 0$ and that $g(1) = f(p) - r + 1 > 0$. Hence there is a ξ such that $g(\xi) = 0$. Define $x = \psi(\xi)$. By construction, the point $(x, f(x))$ is on the diagonal grid line between the pure points (q, r) and $(p, r - 1)$, i.e., the line segment $\{(\psi(t), r - t);\ t \in [0, 1]\}$. Therefore, either (q, r) or $(p, r - 1)$ belongs to the digitized graph, but $(p, r - 1)$ does not by assumption. Hence $q \in \mathrm{Dom}(\mathfrak{P}f)$ and $(\mathfrak{P}f)(q) = r$. □

Using this result, we will extend $\mathfrak{P}f$ to a mapping defined on all pure points. Let $p \in \mathbb{P}_n \smallsetminus \mathrm{Dom}(\mathfrak{P}f)$. It is easy to see that in fact, with r as in the lemma above, $|f(x) - r| \leqslant 1/2$ for every x with $\|x - p\|_\infty \leqslant 1/2$. Therefore, it is reasonable to let the extension take the value r at p. Since f takes the value r at all neighbors of p, by the lemma, it is also clear that this choice results in a function that is strongly Lip-1 in $\mathbb{Z}^n$. Note that it is easy to find this r given a function f and a $p \in \mathbb{P}_n$; let $r = r(f, p)$ be the integer r such that (p, r) is mixed and $|f(p) - r| \leqslant 1/2$.

Definition 10. Suppose that $f\colon \mathbb{R}^n \to \mathbb{R}$ is Lip-1 for the l^∞-metric. Then the **pure Khalimsky digitization**, $\mathfrak{K}_p f\colon \mathbb{P}_n \to \mathbb{Z}$ is defined by:

$$(\mathfrak{K}_p f)(p) = \begin{cases} (\mathfrak{P}f)(p) & \text{if } p \in \mathrm{Dom}(\mathfrak{P}f) \\ r(f, p) & \text{otherwise} \end{cases} \tag{4}$$

Example 11. The Khalimsky line has no mixed points. Therefore, given a Lip-1 function $f\colon \mathbb{R} \to \mathbb{R}$, the pure Khalimsky-continuous digitization is a mapping $\mathfrak{K}_p f\colon \mathbb{Z} \to \mathbb{Z}$. This digitization is illustrated in Fig. 2. If $f(x) = kx + m$ where $|k| \leqslant 1$, then this digitization agrees with the Khalimsky-continuous lines treated in Melin (2003*a*).

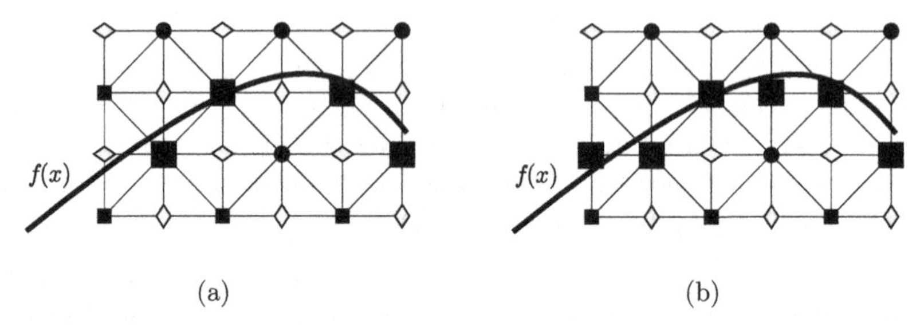

Fig. 2. Khalimsky-continuous digitization in two dimensions. In (a) the pure points in the graph are defined (large squares) and (b) shows the extension of this mapping to all integers

Note that the pure digitization of a set is a set *of* pure points and the pure digitization of a function is a function defined *on* pure points. The digitization on the mixed points of $\mathbb{Z}^{n-1}$ remains to be defined. The values of a continuous function on the pure points do not determine the whole function uniquely, as the following example in $\mathbb{Z}^2$ demonstrates.

Example 12. Let $f\colon \mathbb{P}_2 \to \{0,1\} \subset \mathbb{Z}$ be defined by $f(p_1,p_2) \equiv p_1 \pmod 2$. For each mixed point we can extend f continuously by defining it to be either 0 or 1. Since each point can be treated independently, there are uncountable many different extensions. Confer Remark 20.

Definition 13. Let $f\colon \mathbb{R}^n \to \mathbb{R}$ be Lip-1 for the l^∞-metric. Then the **lower Khalimsky digitization**, $\mathfrak{K}_\star f\colon \mathbb{Z}^n \to \mathbb{Z}$, is the infimum of all continuous extensions of $\mathfrak{K}_p f$. Similarly, the **upper Khalimsky digitization**, $\mathfrak{K}^\star f$, is the supremum of all continuous extensions of $\mathfrak{K}$.

By Theorem 4, the lower and upper Khalimsky digitizations are continuous. We will now give an explicit way to calculate them. Suppose $f\colon \mathbb{R}^n \to \mathbb{R}$ is Lip-1 for the l^∞-metric, and let q be a mixed point. Define the set $E(q)$ by:

$$E(q) = \{m \in \mathbb{Z};\ (\mathfrak{K}_p f) \cup \{(q,m)\} \text{ is strongly Lip-1}\}.$$

The notation $(\mathfrak{K}_p f) \cup \{(q,m)\}$ is set theoretic and means the extension of $(\mathfrak{K}_p f)$ at point q with the value m. Since $\mathfrak{K}_p$ is strongly Lip-1, Theorem 6 guarantees that $E(q)$ is never the empty set. It is easy to see that it is only necessary to check the strongly Lip-1 condition with respect to the points in the pure neighborhood of q, $\mathrm{PN}(q)$. And there are not so many possibilities. Let $(\mathfrak{K}_p f)(\mathrm{PN}(q)) \subset \mathbb{Z}$ denote the image of this neighborhood. Since the l^∞-distance between two point in a neighborhood is at most two and $\mathfrak{K}_p f$ is Lip-1, it follows that

$$\max(\mathfrak{K}_p f)(\mathrm{PN}(q)) - \min(\mathfrak{K}_p f)(\mathrm{PN}(q)) \leqslant 2. \tag{5}$$

Suppose that the difference in (5) equals two. Then the only possible value for a strongly Lip-1 extension of $\mathfrak{K}_p f$ at q is the mean of these extreme values, since this extension must necessarily be Lip-1.

Next, suppose that the difference in (5) equals zero. Since q is not pure, it has at least one closed and one open neighbor. One of these neighbors will fail to match in parity with the value in $(\mathfrak{K}_p f)(\mathrm{PN}(p))$. Therefore again, the only possible extension at the point q is this value.

Finally, suppose the difference in (5) equals one. Then $\mathfrak{K}f$ takes both an even and an odd value in the neighborhood of q. If $(\mathfrak{K}_p f)$ is even for all points in $\mathrm{CN}(q)$ and odd for all points in $\mathrm{ON}(q)$, then we have a choice; $E(q)$ consists of these two values. If on the other hand, $(\mathfrak{K}_p f)(p)$ is odd for some point $p \in \mathrm{CN}(q)$, then $E(q)$ can contain only this value—and similarly if $(\mathfrak{K}_p f)(p)$ is even for some point $p \in \mathrm{ON}(q)$. Since we know that the extension exists, it cannot happen that $(\mathfrak{K}_p f)$ is odd for some point in $\mathrm{CN}(q)$ and even for some point in $\mathrm{ON}(p)$.

By Theorem 6, every function $\mathfrak{K}_p f$ extended at a mixed point q with a value in $E(q)$ can be extended to a continuous function defined on all of $\mathbb{Z}^n$. To sum up, we get the following result:

Theorem 14. *The lower and upper Khalimsky digitizations of a Lip-*1 *function* $f \colon \mathbb{R}^n \to \mathbb{R}$ *can be calculated by the following formulas:*

$$(\mathfrak{K}_\star f)(x) = \begin{cases} (\mathfrak{K}_p f)(x) & \text{if } x \in \mathbb{P}_n \\ \min E(x) & \text{otherwise.} \end{cases} \tag{6}$$

$$(\mathfrak{K}^\star f)(x) = \begin{cases} (\mathfrak{K}_p f)(x) & \text{if } x \in \mathbb{P}_n \\ \max E(x) & \text{otherwise.} \end{cases} \tag{7}$$

Figure 3 shows the result of a Khalimsky-continuous digitization of a function of two variables. We remark that the Khalimsky-continuous digitization is increasing in the following sense: If f and g are Lip-1 mappings $f, g \colon \mathbb{R}^n \to \mathbb{R}$ and $f \leqslant g$, then $\mathfrak{K}^\star f \leqslant \mathfrak{K}^\star g$ and $\mathfrak{K}_\star f \leqslant \mathfrak{K}_\star g$. This is straightforward to prove from the definitions.

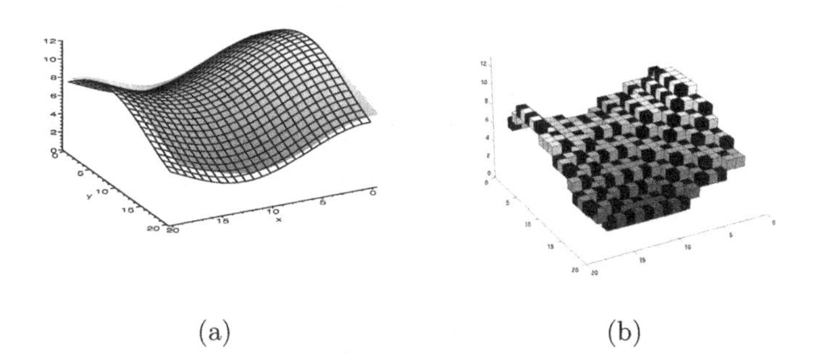

(a) (b)

Fig. 3. The function $f(x, y) = 7 + \frac{5}{2} \sin \frac{x}{5} + 2 \cos \frac{y}{4}$. A continuous picture is showed in (a) and the lower Khalimsky continuous digitization is showed in (b). The black voxels are the pure points in the digitization

4 Approximation Properties

By just using rounding, it is immediate that there is a integer-valued approximation F of a real-valued function f such that $|F(x) - f(x)| \leqslant 1/2$ for all x in the domain. When we in addition require that the approximation be Khalimsky-continuous, it is reasonable to expect that the approximation property deteriorates.

Theorem 15. *Let $f\colon \mathbb{R}^n \to \mathbb{R}$ be Lipschitz for the l^∞-metric with Lipschitz constant $\alpha \leqslant 1$ and let $F\colon \mathbb{Z}^n \to \mathbb{Z}$ be either the upper or the lower Khalimsky continuous digitization of f. Then $|f(p) - F(p)| \leqslant (1 + 3\alpha)/2$ for each $p \in \mathbb{Z}^n$.*

Proof. We have to treat different types of points in the digitization separately. First, suppose that $p \in \mathbb{Z}^n$ is pure and let $r = F(p)$. If $(p, r) \in \mathbb{Z}^{n+1}$ is pure, then the graph of f must intersect $H(p, r)$ and it follows that $|f(p) - F(p)| \leqslant 1/2+\alpha/2$. If instead (p, r) is mixed, then for all $x \in \mathbb{R}^n$ such that $\|x - p\|_\infty \leqslant 1/2$ the inequality $|f(x) - F(x)| \leqslant 1/2$ must hold and of course, in particular, this is true for $x = p$. Now, let q be a mixed point in $\mathbb{Z}^n$. Suppose first that a pure neighbor, p, of q is mapped to a mixed point in the graph, i.e., $(q, F(q))$ is mixed. This implies $F(q) = F(p)$. Let $x = \frac{1}{2}(p + q)$ be the point halfway between p and q. Then $F(q) = F(p)$ and $|f(x) - F(p)| \leqslant 1/2$ by the argument above, so that

$$|f(q) - F(q)| \leqslant |F(q) - F(x)| + |f(q) - f(x)| \leqslant \frac{1}{2} + \alpha\|x - q\|_\infty \leqslant \frac{1 + \alpha}{2}.$$

Next, we consider the case where all pure neighbors of q are mapped to pure points in the graph. There are two sub-cases to consider. Suppose first that the difference in (5) is two so that there are points $p_1, p_2 \in \mathrm{PN}(q)$ such $F(p_1) = r$ and $F(p_2) = r + 2$. Starting at p_1 and using an estimation similar to the one above, we obtain $f(q) \leqslant r + 1/2 + 3\alpha/2$. If we instead start at p_2, we obtain $f(q) \geqslant r+3/2-3\alpha/2$. By definition, $F(q) = r+1$ so we can estimate $|F(q) - f(q)|$ from above and below:

$$f(q) - (r + 1) \leqslant -1/2 + 3\alpha/2 = (-1 + \alpha)/2 + \alpha \leqslant \alpha$$

and

$$f(q) - (r + 1) \geqslant 1/2 - 3\alpha/2 = (1 - \alpha)/2 - \alpha \geqslant -\alpha$$

so $|f(q) - F(q)| \leqslant \alpha$. Note that this case can occur only if $\alpha > 1/3$. Finally, we consider the case when the difference in (5) is one; say that $F(p_1) = r$ and $F(p_2) = r + 1$. Here, there is a difference depending on whether we use the lower or the upper digitization; we may have $F(q) = r$ or $F(q) = r + 1$. Let us consider the lower digitization, i.e., $F(q) = r$. Then

$$f(q) - F(q) \leqslant f(p_1) + \alpha - r \leqslant (1 + 3\alpha)/2, \tag{8}$$

and from below we have

$$f(q) - F(q) \geqslant f(p_2) - \alpha - r \geqslant (1/2 - \alpha/2) - \alpha = (1 - 3\alpha)/2 \geqslant -\alpha,$$

so that $|f(q) - F(q)| \leqslant (1 + 3\alpha)/2$ and the proposition is proved. □

Since α is bounded by 1, we obtain $|f(q) - F(q)| \leqslant 2$ for any mapping where the Khalimsky digitization is defined. The following example shows that the bound in the theorem is sharp.

Example 16. Let $f\colon \mathbb{R}^2 \to \mathbb{R}$ be defined by $f(x,y) = \min(x+1, 3-x)$. It is easy to check that $(\mathfrak{K}_p f)(0,0) = (\mathfrak{K}_p f)(2,0) = 0$ and that $(\mathfrak{K}_p f)(1,1) = (\mathfrak{K}_p f)(1,-1) = 1$. Thus $(\mathfrak{K}_\star f)(1,0) = 0$, while $f(1,0) = 2$. More generally, for $0 < \alpha \leqslant 1$, define a Lip-α mapping as

$$f_\alpha(x,y) = \min\left(\alpha(x + \tfrac{1}{2}) + \tfrac{1}{2}, \alpha(\tfrac{5}{2} - x) + \tfrac{1}{2}\right)$$

We have $(\mathfrak{K}_p f_\alpha)(0,0) = (\mathfrak{K}_p f_\alpha)(2,0) = 0$ and $(\mathfrak{K}_p f)(1,\pm 1) = 1$ as before. Therefore, we again get $(\mathfrak{K}_\star f_\alpha)(1,0) = 0$, while $f_\alpha(1,0) = (3\alpha + 1)/2$.

If one checks the proof of Theorem 15, one sees that it is only in one case that we get the bound $(1+3\alpha)/2$. It is in (8). In the proof, the lower digitization is considered and there we have the bad bound above. If instead one considers the upper digitization, then the bad bound is from below. In all other cases, the bound is α or $(1+\alpha)/2$, so in the one-dimensional case we obtain the following corollary, since there are no mixed points in the domain.

Corollary 17. *Let F be the Khalimsky-continuous digitization of a mapping $f\colon \mathbb{R} \to \mathbb{R}$, which is Lip-1. Then $|f(p) - F(p)| \leqslant (1+\alpha)/2$ for each $p \in \mathbb{Z}$.*

In two dimensions, it is also possible to improve the approximation. Since a mixed point in the Khalimsky plane never has a mixed neighbor, we can define the *optimal Khalimsky-continuous digitization* as follows:

$$(\mathfrak{K} f)(x) = \begin{cases} (\mathfrak{K}^\star f)(x) \text{ if } |(\mathfrak{K}^\star f)(x) - f(x)| < |(\mathfrak{K}_\star f)(x) - f(x)| \\ (\mathfrak{K}_\star f)(x) \text{ otherwise.} \end{cases} \tag{9}$$

Corollary 18. *Let F be the optimal Khalimsky-continuous digitization of a mapping $f\colon \mathbb{R}^2 \to \mathbb{R}$ which is Lipschitz for the l^∞-metric with Lipschitz constant $\alpha \leqslant 1$. Then $|f(p) - F(p)| \leqslant (1+\alpha)/2$ for each $p \in \mathbb{Z}^2$.*

The following example shows that in general, the bound $(\alpha+1)/2$ cannot be improved. It is stated in one dimension, but can clearly be extended to higher dimensions.

Example 19. Let $0 \leqslant \alpha \leqslant 1$ and define $f\colon \mathbb{R} \to \mathbb{R}$, $f_\alpha(x) = \alpha x + (1-3\alpha)/2$. Suppose that F_α is a Khalimsky-continuous approximation of f_α. Since $f_\alpha(1) = 1 - (1+\alpha)/2$, it is necessary that $F_\alpha(1) = 0$, if F is to approximate f_α better than $\mathfrak{K} f_\alpha$. By continuity, it follows then that $F(2) = 0$, while $f_\alpha(2) = (\alpha+1)/2$.

Remark 20. The definition of the optimal Khalimsky digitization is utterly dependent on the fact that mixed points in the plane are not connected, and therefore can be treated one by one. In three and more dimensions, this is no longer true. If, for example, we have $f\colon \mathbb{Z}^3 \to \mathbb{Z}$ and define $f(1,0,0) = 1$ then necessarily $f(1,1,0) = 1$ if f is to be continuous. One way out of this is to define an order among the mixed points. We can decide to first define the extension of $\mathfrak{K}_p$ on the points with precisely two odd coordinates (which are independent) and then on the remaining mixed points.

5 Conclusions

We have shown that it is possible to find a reasonable Khalimsky-continuous approximation of Lip-1, real-valued function of real variables. The Lip-1 condition is of course a restriction, although it is clearly impossible to avoid. For a sufficiently nice function defined on, say, a bounded set, it is of course always possible to rescale the function to satisfy this condition. In a forthcoming paper we will investigate more carefully the properties of the Khalimsky-continuous digitization when applied to hyperplanes. A future task is to generalize these results and definitions to more general surfaces than the graphs of functions.

Acknowledgement

I am grateful to Christer Kiselman for comments on earlier versions of this manuscript and to Ola Weistrand for letting me use his routines for voxel visualization.

References

Alexandrov, Paul. 1937. "Diskrete Räume." *Mat. Sb.* 2(44):501–519.

Andrès, Eric. 2003. "Discrete linear objects in dimension n: the standard model." *Graphical Models* 1–3(65):92–111.

Bresenham, Jack E. 1965. "Algorithm for computer control of a digital plotter." *IBM Systems Journal* 4(1):25–30.

Chen, Li, Donald H. Cooley & Jianping Zhang. 1999. "The equivalence between two definitions of digital surfaces." *Inform. Sci.* 115(1-4):201–220.

Cohen-Or, Daniel & Arie Kaufman. 1995. "Fundamentals of surface voxelization." *Graphical Models and Image Processing* 57(6):453–461.

Couprie, Michel, Gilles Bertrand & Yukiko Kenmochi. 2003. "Discretization in 2D and 3D orders." *Graphical Models* 65(1–3):77–91.

Herman, Gabor T. 1998. *Geometry of digital spaces.* Applied and Numerical Harmonic Analysis Boston, MA: Birkhäuser Boston Inc.

Kaufman, Arie. 1987. "Efficient algorithms for 3D scan-conversion of parametric curves, surfaces and volumes." *Computer Graphics* 21(4):171–179.

Khalimsky, Efim, Ralph Kopperman & Paul R. Meyer. 1990. "Computer graphics and connected topologies on finite ordered sets." *Topology Appl.* 36(1):1–17.

Kim, Chul E. 1984. "Three-dimensional digital planes." *IEEE Trans. Pattern Anal. Machine Intell.* PAMI-6(5):639–645.

Kiselman, Christer O. 2002. *Digital geometry and mathematical morphology.* Lecture notes. Uppsala University. Available at `www.math.uu.se/~kiselman`.

Kong, T. Yung. 2003. "The Khalimsky topologies are precisely those simply connected topologies on $\mathbb{Z}^n$ whose connected sets include all $2n$-connected sets but no $(3^n - 1)$-disconnected sets." *Theoret. Comput. Sci.* 305(1-3):221–235.

Kong, T. Yung & Azriel Rosenfeld. 1989. "Digital topology: Introduction and survey." *Comput. Vision Graph. Image Process.* 48(3):357–393.

Kong, T. Yung, Ralph Kopperman & Paul R. Meyer. 1991. "A topological approach to digital topology." *Amer. Math. Monthly* 98(10):901–917.

Melin, Erik. 2003*a*. "Digital straight lines in the Khalimsky plane." (To appear in *Mathematica Scandinavica*).

Melin, Erik. 2003*b*. Extension of continuous functions in digital spaces with the Khalimsky topology. U.U.D.M. Report 2003:29 Uppsala University. Available at `www.math.uu.se/~melin`.

Melin, Erik. 2004. Continuous extension in topological digital spaces. U.U.D.M. Report 2004:2 Uppsala University.

Morgenthaler, David G. & Azriel Rosenfeld. 1981. "Surfaces in three-dimensional digital images." *Inform. and Control* 51(3):227–247.

Reveillès, Jean-Pierre. 1991. Géométrie discrète, calcul en nombres entiers et algorithmique. Ph.D. thesis. Université Louis Pasteur, Strasbourg.

Ronse, Christian & Mohamed Tajine. 2000. "Discretization in Hausdorff space." *J. Math. Imaging Vision* 12(3):219–242.

Rosenfeld, Azriel. 1974. "Digital straight line segments." *IEEE Trans. Computers* C-23(12):1264–1269.

Rosenfeld, Azriel, T. Yung Kong & Angela Y. Wu. 1991. "Digital surfaces." *CVGIP: Graph. Models Image Process.* 53(4):305–312.

Tajine, Mohamed & Christian Ronse. 2002. "Topological properties of Hausdorff discretization, and comparison to other discretization schemes." *Theoret. Comput. Sci.* 283(1):243–268.

Algorithms in Digital Geometry Based on Cellular Topology

V. Kovalevsky

University of Applied Sciences, Berlin
kovalev@tfh-berlin.de
www.kovalevsky.de

Abstract. The paper presents some algorithms in digital geometry based on the topology of cell complexes. The paper contains an axiomatic justification of the necessity of using cell complexes in digital geometry. Algorithms for solving the following problems are presented: tracing of curves and surfaces, recognition of digital straight line segments (DSS), segmentation of digital curves into longest DSS, recognition of digital plane segments, computing the curvature of digital curves, filling of interiors of n-dimensional regions (n=2,3,4), labeling of components (n=2,3), computing of skeletons (n=2, 3).

1 Introduction

First of all let us discuss the question, why should one use cell complexes in digital geometry. The author supposes that there are three categories of researchers in this field. Those of the first category would prefer a mathematical proof of the assertion that cell complexes belong to the class of locally finite topological spaces that are in agreement with the axioms and with most definitions of the classical topology.

Researcher of the second category would perhaps ask, whether classical axioms are really important for applications. They may believe that it is possible to find another set of axioms and deduce the properties of the topological space best suitable for applications in image analysis and in computer graphics from these new axioms.

Researcher of the third category pay no attention at all to axiomatic theories. They are only interested in methods enabling one to develop efficient solutions of geometric and topological problems in image analysis and in computer graphics.

The author hopes to satisfy with the present paper the desires of researchers of all three categories.

Let us start with the second category. First of all it is necessary to mention that we need not considering spaces in which each neighborhood of a point contains an infinite number of points, since such a space cannot be explicitly represented in a computer. A space in which each element has a neighborhood containing a finite number of elements is called a *locally finite space* (LFS).

The author believes that everybody will agree that the features of a space, most important for applications are those of connectivity and of boundary. Therefore we suggest the following set of axioms concerned with these notions. We denote the elements of the space not as "points" since, as we shall see soon, there are in an LFS elements with different neighborhoods. They have different topological properties and thus must have different notations.

R. Klette and J. Žunić (Eds.): IWCIA 2004, LNCS 3322, pp. 366–393, 2004.

Axiom 1: For each space element e there are certain subsets containing e, which are neighborhoods of e. Since the space is locally finite there exists the smallest neighborhood of e.

We shall denote the smallest neighborhood of e by SON(e). The exact definition of the neighborhood will be derived from the whole set of the Axioms below.

Axiom 2: There are space elements whose SON consists of more than one element. If a and b are space elements and $b \in \text{SON}(a)$ then the set $\{a, b\}$ is connected. Also a set $\{a\}$ consisting of a single space element is connected.

We shall say that a and b are *directly connected* or *incident* to each other. This is a binary relation *Inc*. Since $\{a, b\}$ and $\{b, a\}$ denote one and the same set, the incidence relation *Inc* is symmetric. It is reflexive since according to Axiom 2 the set $\{a\}$ is connected.

Axiom 3: The connectivity relation is the transitive hull of the incidence relation. It is symmetric, reflexive and transitive. Therefore it is an equivalence relation.

Let us now formulate the axioms related to the notion of a boundary. The classical definition of a boundary (exactly speaking, of the topological boundary or of the frontier) is as follows:

Definition BD: The topological boundary of a subset T of the space S is the set of all space elements whose *each* neighborhood intersects both T and its complement $S–T$.

In the case of a locally finite space, it is obviously possible to replace "each neighborhood" by "smallest neighborhood". We shall denote the topological boundary i.e. the frontier of the subset T of the space S by Fr(T,S). Now we introduce the notion of a *thin* boundary:

Definition TB: The boundary Fr(T,S) of a subset T of an n-dimensional space S is called *thin* if it contains no n-dimensional cube of 2^n mutually incident space elements.

Axiom 4: The topological boundary of any subset T is thin and is the same as the topological boundary of the complement $S–T$.

Let us remain the reader the classical axioms of the topology. The topology of a space S is defined if a collection of subsets of S is declared to be the collection of the *open subsets*. These subsets must satisfy the following Axioms:

Axiom C1: The whole set S and the empty subset $\varnothing$ are open.
Axiom C2: The union of any number of open subsets is open.
Axiom C3: The intersection of a finite number of open subsets is open.
Axiom C4: The space has the separation property.

There are (at least) three versions of the separation property and therefore three versions of Axiom C4 (Fig. 1.1) :

Axiom T_0: For any two distinct points x and y there is an open subset containing exactly one of the points.

Axiom T_1: For any two distinct points x and y there is an open subset containing x but not y and another open subset, containing y but not x.

Axiom T_2: For any two distinct points x and y there are two non-intersecting open subsets containing exactly one of the points.

A space with the separation property T_2 is called *Hausdorff space*. The well-known Euclidean space is a Hausdorff space.

Fig. 1.1. A symbolic illustration to the separation axioms

It is easily seen that if a single point is not an open subset, then only the Axiom T_0 may be applied to a locally finite space (LFS). Really, Axioms T_1 and T_2 demand that the open subsets under consideration contain infinitely many points, no matter how small they are. Such subsets do not exits in a LFS and cannot be explicitly represented in a computer. Therefore only Axiom T_0 is relevant for an LFS.

The author has proved [11] that the above set of Axioms C1, C2, C3 and T_0 is equivalent to the set of our suggested set of Axioms 1 to 4. This means that Axioms C1 to C3 and T_0 may be deduced from Axioms 1 to 4 as theorems. This proves that the classical notion of open subsets is important for providing a space with the features of connectivity and boundary that satisfy the "obviously true" demands formulated as our Axioms 1 to 4.

Unfortunately, it is impossible to repeat here the proof because of lack of space. We shall only show that the classical Axiom T_0 follows from the demands that the boundaries be thin and that the boundary of a subset T be the same as the boundary of its complement $S–T$.

Theorem NT. If the neighborhood of a space element e in the Definition BD contains besides e itself all elements that stay in a *symmetric* binary relation with e then there exist subsets whose boundary is not thin.

Proof: We presume that two elements whose one coordinate differs by 1 while all other coordinates are equal satisfy the symmetric relation mentioned in Theorem NT.

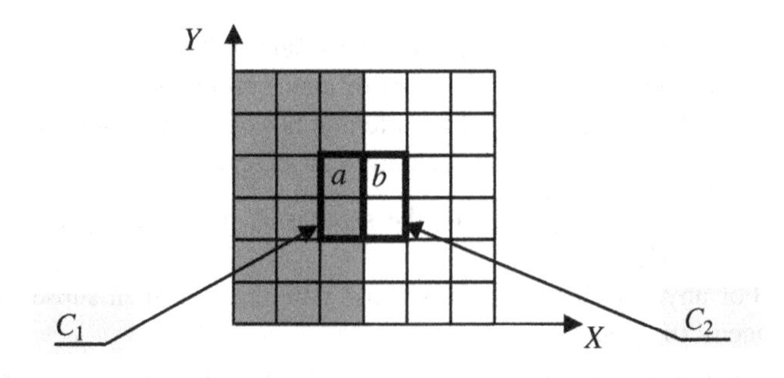

Fig. 1.2. An illustration to the proof of Theorem NT for n=2

Let T be a subset of the space S that contains an (n−1)-dimensional cube $C_1=\{x_1=m\}\times\{x_2, x_2+1\}\times...\times\{x_n, x_n+1\}$ while all elements of the "adjacent" cube $C_2=\{x_1=m+1\}\times\{x_2, x_2+1\}\times...\times\{x_n, x_n+1\}$ belong to the complement $S-T$ (Fig. 1.2). The neighborhood of any element $a\in C_1$ contains the element $b\in C_2$ whose coordinate x_1 is equal to m+1 and all other coordinates are equal to the corresponding coordinates of a. Thus $b\in S-T$ and a belongs to the boundary of T. Since the relation defining the neighborhoods is symmetric, a belongs to the neighborhood of b, that intersects T at a and $S-T$ at b. Therefore b belongs to the boundary of T. This is true for all elements of the cube C_2. Thus both cubes C_1 and C_2 are in the boundary of T. Their union is an n-dimensional cube and hence the boundary is not thing.

There are two possibilities to achieve that the boundary be thin for any subset:

1. To change the Definition BN of the boundary so that only elements of T may belong to the boundary of T.
2. To define the neighborhood by means of an *antisymmetric relation*, which means that if b is in the smallest neighborhood of a then a is not in the smallest neighborhood of b.

The first possibility leads to different boundaries of T and of its complement $S-T$, which may be considered as a topological paradox and contradicts our Axiom 4. The remaining possibility 2 demands that the smallest neighborhoods satisfy the classical Axiom T_0. This is exactly what we wanted to demonstrate.

In the next chapters we will describe some algorithms in digital geometry that are based on the theory of abstract cell complex (AC complexes). In an AC complex the neighborhoods are defined by means of an *antisymmetric* bounding relation. Therefore the boundaries in AC complexes are thin and they satisfy the classical axioms. The author hopes that this will satisfy the desire of a reader of the first category mentioned at the beginning of Introduction. As to a reader of the second category, we have demonstrated that an attempt to "invent" a new "obviously true" set of axioms leads to no new concept of a topological space: the set of the new axioms has turned out to be equivalent to that of classical axioms. We suppose that this will be the case of all other sets of axioms that are in accordance with our "healthy" understanding of topology.

As to the readers of the third category, we hope that they will be convinced and satisfied after having read Part II "The Algorithms".

Part I – Theoretical Foundations

2 Short Remarks on Cell Complexes

We presume that the reader is acquainted with the theory of AC complexes. To make the reading of the paper easier we have summarized the most important definitions in the Appendix. For more details we refer to [2, 5, 8].

3 Data Structures

In this Section we shall describe some data structures used in algorithms for solving topological and geometrical problems when using AC complexes.

3.1 The Standard Raster

Two- and three-dimensional images are usually stored in a computer in arrays of the corresponding dimension. Each element of the array contains either a gray value, or a color, or a density. This data structure is called the *standard raster*. It is not designed for topological calculations, nevertheless, it is possible to perform topological calculations without changing the data structure. For example, it is possible to trace and encode the frontier of a region in a two-dimensional image in spite of the apparent difficulty that the frontier according to Definition FR (see Appendix) consists of 0- and 1-cells, however, the raster contains only pixels which *must* be interpreted as 2-cells. The reason is that a pixel is a carrier of an optical feature that is proportional to certain elementary area. Thus pixels, which are the 2-cells, must correspond to elementary areas rather than 0- or 1-cells whose area is zero. On the same reason voxels must correspond to 3-cells.

When considering an image as an AC complex one must admit that a 2-dimensional image contains besides the pixels also cells of lower dimension 0 and 1. This often arises objections since we can see on a computer display only the pixels, i.e. the cells of the highest dimension. However, this fact is not important: really, one should think about the rendering of a 3-dimensional object that is a set of voxels. What we see are not the 3-dimensional voxels but rather their 2-dimensional facets and 1-dimensional edges. This fact does not prevent us from working with voxels which we do not see on the display. All these are peculiarities of the human viewing system rather than that of the objects. By the way, it is no problem to make, if desired, the lower dimensional cells visible on a display [2].

The tracing in the standard raster is possible because the concept of an AC complex is the *way of thinking* about topological properties of digitized images rather than a way of encoding them. Let us explain this idea for the case of tracing frontiers.

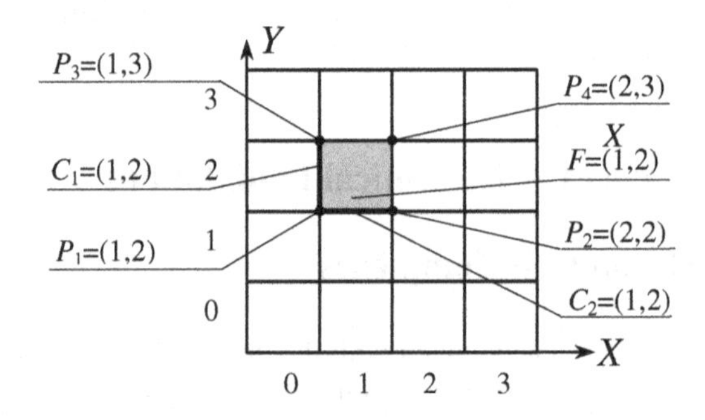

Fig. 3.1. Non-topological coordinates of cells of lower dimensions

We think of a two-dimensional (2D) image as of a 2D Cartesian complex (Appendix) containing cells of dimensions form 0 to 2. The 2-cells (pixels) have in the standard raster coordinates which, unlike to topological coordinates of a pixel being always odd (Appendix and Fig. A.1), may take in the standard raster *any integer values*, both odd and even. Pixels are explicitly represented in the raster, however, the 0- and 1-cells are present implicitly.

Coordinate Assignment Rule: Each pixel F of a 2D image gets one 0-cell assigned to it as its "own" cell. This is the 0-cell lying in the corner of F which is the nearest to the origin of the coordinates (P_1 in Fig. 3.1). Also two 1-cells incident to F and to P are declared to be own cells of F (C_1 and C_2 in Fig. 3.1). Thus each pixel gets three own cells of lower dimensions. *All own cells of F get the same coordinates as F.* They can be distinguished by their type.

In the three-dimensional case each voxel gets seven own cells of lower dimensions which are arranged similarly. These seven cells get the same coordinates as the corresponding voxel.

Unfortunately, some cells in the boundary of the raster remain without an "owner". In most applications this is of no importance. Otherwise the raster must be correspondingly enlarged.

According to the above rule, it is not difficult to calculate the coordinates of all pixels incident to a given point. This coordinates are used to get the gray values of the pixels from the array containing the image. Depending on these gray values the *virtual* tracing point P moves for one position to the next. The movement is represented by the changing coordinates of the virtual point. The details of the tracing algorithm are described in Section 4.

The majority of low level topological problems in image processing may be solved in a similar way, i.e. without representing cells of lower dimension as elements of some multidimensional array. A typical exception is the problem of filling the interior of a region defined by its frontier (Section 8). The solution is simpler when using two array elements per pixel: one for the pixel itself and one more for its own vertical crack (1-cell). The solution consists in reading the description of the frontier (e.g. its crack code), labeling all vertical cracks of the frontier in the array and in counting the labeled cracks in each row (starting with 0) and filling the pixels between the crack with an even count $2 \cdot i$ and the next crack (with the count $2 \cdot i+1$). The details of this algorithm are described in Section 8.

Even more complicated topological problems may be solved by means of the standard raster. For example, when tracing surfaces (Section 7) or producing skeletons (Section 11) the so called simple pixels must be recognized. It is easier to correctly recognize all simple pixels if *cells of all dimensions* of the region under consideration are labeled. To perform this in a standard raster, it is possible to assign a bit of a raster element representing the pixel F to each own cell of F. For example, suppose that one byte of a two-dimensional array is assigned to each pixel of the image shown in Fig. 3.1. Consider the pixel F with coordinates (1, 2) and the byte assigned to it. The bit 0 of the byte may be assigned to the 0-cell P_1, the bit 1 to the 1-cell C_1, the bit 2 to the 1-cell C_2. The remaining bits may be assigned to F itself. Similar assignments are also possible in the 3D case.

As we see, there is no necessity to allocate memory space for each cell of a complex, which would demand four times more memory space than that needed for pixels only, or eight times more than that needed for the voxels in the 3D case.

3.2 The Topological Raster

To explicitly represent an image as a Cartesian AC complex the so called *topological raster* [8] must be used. It is again a 2- or 3-dimensional array in which a memory element is reserved for each cell of each dimension.

In a topological raster *each cell has different coordinates* which simultaneously are the indices of the array. Each coordinate axis is a topological line (Definition TL, Appendix). The 0-cells of the axis have even coordinates, the 1-cells have odd coordinates (Fig. A1, Appendix). The dimension and the orientation (if defined) of any cell may be calculated from its topological coordinates. The dimension of any cell is the *number* of its odd coordinates, the orientation is specified by indicating *which* of the coordinates are odd. For example, the cell C_1 in Fig. A.1 has *one* odd coordinate and this is its X-coordinate. Thus it is a *one*-dimensional cell oriented along the X-axis. The 2-cell F has two and the 0-cell P has no odd coordinates. In a three-dimensional complex the orientation of the 2-cells may be specified in a similar way: if the ith coordinate of a 2-cell F is the only even one then the normal to F is parallel to the ith coordinate axis.

The advantages of the topological raster are shaded by its drawback that it demands 2^n times more memory space than the standard raster. Also the time needed to process it is correspondingly greater. Therefore it is reasonable to use the topological raster to save images only for not too large images or for research purposes where the processing time is not so important.

The best way to use the topological coordinates consists in the following. When thinking about the problem to be solved one should use the topological coordinates. Also a program which must solve a topological or geometrical problem in which notions like the connectivity, boundary, incidence etc. are involved should work with topological coordinates. Only the storage of values assigned to the ground cells, i.e. to pixels or voxels, should be performed in the standard raster to save memory space. This is possible since the cells of lower dimension do not carry values like color, gray value or density. They only carry indices of subsets to which they belong, i.e. foreground or background. These indices may be specified by means of predefined rules called membership rules [2].

For example, to compute the connected components of a binary 2D image the rule may be formulated which says that each cell of dimension 0 or 1, which is incident to a foreground pixel, belongs also to the foreground. There are relatively seldom problems the solution of which demands that certain values must be assigned to cells of lower dimensions and then evaluated during the computation. For example, when tracing the boundaries of all connected components of a binary 2D image it is necessary to label the cells of the boundaries which are already processed. Otherwise one and the same component may be found many times. This, however, does not mean that one must label all cells of a boundary. If the process of finding a new component looks for vertical 1-cells of a boundary while comparing the values of two adjacent pixels in a row then it is sufficient to label *only the vertical cracks* of the boundary being processed. It is sufficient to have one bit of memory to save such a label. This bit may be located in the byte of the standard raster, which byte is assigned to the pixel lying in the same row as the crack to be labeled just to the right from the crack. In this case 7 bits remain for the gray values of the pixels which is sufficient in the most cases.

It is also possible to have another array for the cracks along that for the pixels. In this case one needs two times more memory space than for the pixels only. However, it is still more economical than using a topological raster which needs 4 times more memory space. We shall show more examples of using the standard raster while working with cells of lower dimensions in the algorithms of Part II.

Part II – The Algorithms

We describe here some algorithms for computing topological features of subsets in 2D and 3D digitized images. Since the programming languages of the C-family are now more popular than that of the PASCAL-family, we use here a pseudo-code that resembles the C-language.

4 Boundary Tracing in 2D Images

Boundary tracing becomes extremely simple when thinking of a 2D image as of a 2D complex. The main idea of the algorithm consists in the following: at each boundary point (0-cell) *P* (Fig. 4.1) find the next boundary crack *C* incident to *P* and make a step along the crack to the next boundary point. Repeat this procedure until the starting point is reached again. Starting points of all boundary components must be found during an exhaustive search through the whole image. The following subroutine `Trace()` is called each time when a not yet visited boundary point of a region is found.

To avoid calling `Trace()` more than once for one and the same foreground component vertical cracks must be labeled (e.g. in a bit of `Image[]`) as "already visited". `Trace()` follows the boundary of one foreground region while starting and stopping at the given point `(x,y)`. Points and cracks are present only implicitly (see Section 3). `Trace()` starts always in the direction of the positive Y-axis.

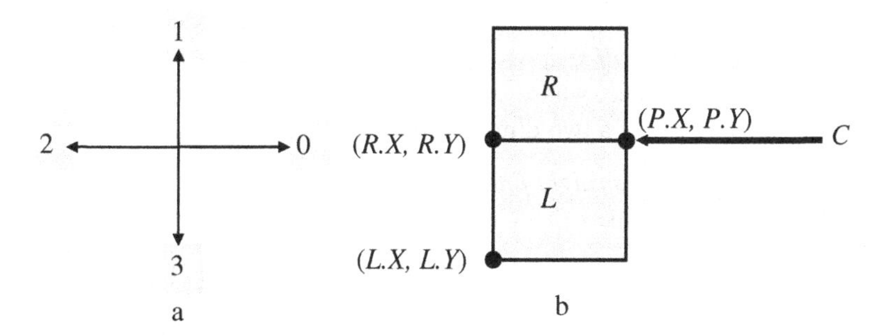

Fig. 4.1. The four directions (a) and the right and left pixels (b)

After each move along a boundary crack C the values of *only two pixels* R and L of SON(P) of the end point P of C must be tested since the values of the other two pixels of SON(P) have been already tested during the previous move. Note that the cracks in a 2D Cartesian complex have only 4 different directions. Therefore the variable "direction" takes the values from 0 to 3 (Fig. 4.1a).

The point corresponding to the corner of a pixel, which is the nearest to the coordinate origin, has the same coordinates as the pixel. For a detailed description of this algorithm see [4].

The Pseudo-Code of `Trace()`: `Image[NX, NY]` is a 2D array (standard raster) whose elements contain gray values or colors. The variables `P`, `R`, `L` and the elements of the arrays `right[4]`, `left[4]` and `step[4]` are structures each representing a 2D vector with integer coordinates, e.g. `P.X` and `P.Y`. The operation "+" stands for vector addition. Text after // is a comment.

```
void Trace(int x, int y, char image[])
{ P.X=x; P.Y=y; direction=1;
  do
  { R=P+right[direction];   // R is the "right" pixel
    L=P+left[direction];     // L is the "left" pixel
    if (image[R]==foreground)
      direction=(direction+1) MOD 4; // right turn
    else
      if (image[L]==background)
        direction=(direction+3) MOD 4; // left turn
    P=P+step[direction]; //a move in the new direction
  } while( P.X!=x || P.Y!=y);
} // end Trace
```

This simple algorithm is used in any program for the analysis of boundaries in 2D images, e.g. in segmenting the boundary into digital straight segment, in the polygonal approximation of boundaries, in calculating the curvature etc.

5 Segmentation of Digital Curves into Longest DSSs

5.1 Theoretical Preliminaries

Definition HS: A *digital half-space* is a region (Definition RG, Appendix) containing all ground cells of the space, whose coordinates satisfy a linear inequality. A *digital half-plane* is a half-space of a two-dimensional space.

Definition DSS: A *digital straight line segment* (DSS) is any connected subset of the frontier of a digital half-plane.

Fig. 5.1 shows an example of the half-plane defined by the inequality $2 \cdot x - 3 \cdot y + 3 \geq 0$. All pixels of the half-plane are represented by shaded squares.
We suggest to consider two kinds of digital curves: *visual curves* as sequences of pixels and *boundary curves* as sequences of alternating points and cracks. We

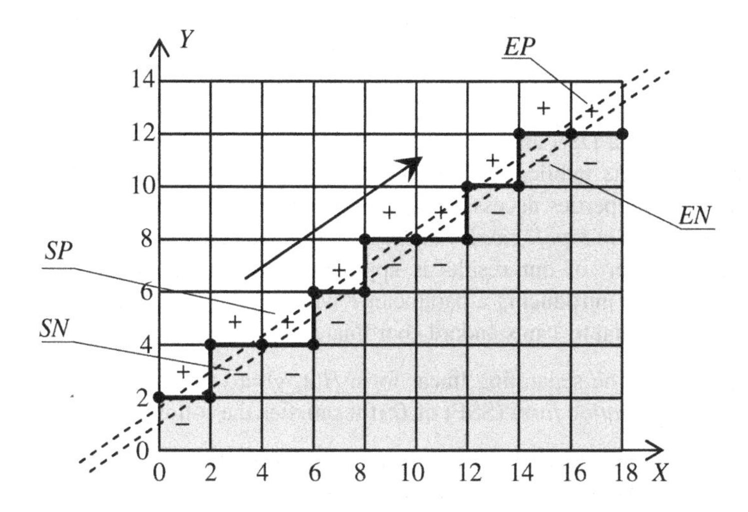

Fig. 5.1. Examples of a half-plane and of a DSS in topological coordinates

consider DSSs as *boundary curves* while in the most publication (e.g. in [14]) they are considered as visual curves, i.e. as sequences of pixels.

Since in practice DSSs are mostly used in image analysis rather than in computer graphics, considering them as boundary curves is more adequate to the demands of applications as this will be shown below in Section 5.3.

5.2 Most Important Properties of a DSS

To investigate the properties of DSSs it is easier and more comprehensible to consider at first the 2-cells (pixels) incident to the cells of a DSS rather than the cells of the DSS themselves. Consider a digital curve K in a two-dimensional space. It is possible to assign an orientation to K and thus to the 1-cells of K. Suppose that K does not intersect the boundary of the space. Then each 1-cell C of K is incident to exactly two pixels. One of them lies to the positive side and the other to the negative side of the ordered pair of the end points of C. (The cell c lies to the positive side of the ordered pair (a, b) if the rotation from b to c about a is in the mathematically positive direction, as e.g. the rotation from the positive X-axis to the positive Y-axis).

Definition PP: The pixel P which is incident with an oriented crack C of the curve K and lies to the positive side of the ordered pair of the end points of C is called the *positive pixel* of K. Similarly, the incident pixel lying to the negative side of the ordered pair of the end points of C is called the *negative pixel* of K.

In Fig. 5.1 the positive pixels are labeled by "+" and the negative ones by "–".

The set of all positive pixels of K will be called the *positive pixel set* of K and denoted by $SP(K)$. The set of all negative pixels of K will be called the *negative pixel set* of K and denoted by $SN(K)$.

If K is a DSS then $SP(K)$ lies in a half-plane while $SN(K)$ lies in its complement. Therefore there is a linear form $H(x, y)$ such that $H(x, y) \geq 0$ for all positive pixels of K

and $H(x, y)<0$ for the negative ones. We shall call $H(x, y)$ the *separating linear form* of the DSS.

The properties of a DSS which we are interested in and which are important for the recognition of a DSS are known from the literature [Kim 90] during many years. The majority of the publications consider *visual lines in standard coordinates*. We repeat here the properties necessary for the recognition of DSS and reformulate them for *boundary lines in topological coordinates*. The necessary proofs are to be found in [12]. The most part of our results is applicable for both standard and topological coordinates due to introducing a parameter e which is the minimum distance between two pixels: e is equal to 1 in standard coordinates and to 2 in topological ones.

Definition SSF: The separating linear form $H(x, y)=a \cdot x+b \cdot y+r$ of a DSS D is called the *standard separating form* (SSF) of D if it satisfies the following conditions:

$H(x, y) \geq 0$ for all positive pixels of D, $H(x, y)<0$ for all negative ones;
The coefficients of $H(x, y)$ are integers while a and b are mutually prime;
There are either at least two positive pixels P_1 and P_2 of D at which $H(x, y)$ takes its minimum value *with respect to* all positive pixels of D or at least two negative pixels N_1 and N_2 of D at which $H(x, y)$ takes its maximum value *with respect to* all negative pixels of D.

Definition BS: The set of positive pixels of a DSS D at which the SSF of D takes its minimum value with respect to all positive pixels of D is called the *positive base* of D. Similarly, the set of negative pixels of a DSS D at which the SSF of D takes its maximum value with respect to all negative pixels of D is called the *negative base* of D.

The pixel of the base B, which is the nearest to the starting point (respectively, end point) of the oriented DSS is called the *starting pixel* (respectively, the *end pixel*) of B. In Fig. 5.1 the starting pixel of the positive base is denoted by SP, its end pixel is denoted by EP. Similarly, the starting and the end pixel of the negative base are denoted by SN and EN.

It is well-known and proved in [12] that a DSS contains cracks of at most two different directions and that the values of $H(.)$ satisfy the inequalities:

$$0 \leq H(x, y) \leq e \cdot (\max(|a|,|b|)-1) \quad \text{for positive pixels;}$$
$$-e \cdot \max(|a|,|b|) \leq H(x, y) \leq -e \quad \text{for negative ones.}$$

Let S and E be the starting and the end pixel of those base of the DSS, which contains more than one pixel. Then the vector parallel to $E-S$ whose components are mutually prime is called the *base vector* of the DSS. If both bases contain a single pixel, which is only the case, when the DSS consists of a single crack, then the base vector is a unit vector parallel to that crack.

The problem most important for applications is that of segmenting a given digital curve into as long as possible DSSs. The author has developed an algorithm [3] which starts with the first two cracks of the curve (two adjacent cracks with the point between them compose always a DSS) and then tests the following cracks one crack after another whether the sequence of cracks is still a DSS. Section 5.3 describes a slightly modified more comprehensible version of the algorithm. The idea is as follows.

The first two cracks uniquely specify the bases and the SSF $H(V)$ of the DSS. The algorithm checks whether the direction of the next crack C is allowed. If not then C does not belong to the actual DSS which ends at the starting point of C. Otherwise the values $H(P)$ and $H(N)$ of the SSF for the two pixels P and N incident to C must be checked. If both $H(P)$ and $H(N)$ are in the allowed intervals then C belongs to the actual DSS. If one of these values is at the boundary of the allowed interval then the corresponding pixel P or N belongs to the corresponding base which must be prolonged until P or N. If one of the values $H(P)$ and $H(N)$ deviates from the boundary of the allowed interval by the value of e (e=1 for standard coordinates and e=2 for topological coordinates) then the coefficients of $H(V)$ must be redefined: the sequence of cracks is a DSS slightly different from the actual DSS. If one of the values $H(P)$ and $H(N)$ deviates from the boundary of the allowed interval by a value greater than e then there is no DSS containing C and all previous cracks. The actual DSS ends at the starting point of C.

The author has proved the correctness of the algorithm [12]. The proofs of all necessary lemmas and theorems take about 10 pages and cannot be repeated here.

5.3 The Algorithm "Subdivide In DSS"

In this section we describe an algorithm which traces a given digital curve *CV* in a 2D image, e.g. the boundary of a region, and subdivides the curve into the longest possible DSSs. The output of the algorithm is a list of the coordinates of the endpoints of the DSSs.

Tracing the Curve: Call the subroutine *Ini* (see below). Choose an arbitrary crack *SC* of *CV*, choose its orientation while specifying its direction *dir* as one of the numbers 0 to 3 as shown in Fig. 4.1a. *SC* is the starting crack. Save the coordinates of the starting point of *SC* as the starting point of the first DSS. Set the running crack *C*=*SC* and start the tracing loop. The loop runs through all cracks of *CV*. For each new location of the running crack *C* call the subroutine *Reco*(*C*, *dir*) (see below) where *dir* is the direction of the oriented crack *C*. If the return value of *Reco* is zero go over to the next crack. Otherwise save the starting point of the running crack *C* as the end point of the current DSS and start the recognition of the next DSS while calling *Ini* and *Reco*(*C*, *dir*) again. Stop when the starting crack *SC* is reached again. The curve *CV* is subdivided into as long as possible DSSs by the saved points.

Subroutine *Ini*: Set the crack counter *CC* to zero and the two prohibited directions *Proh*1 and *Proh*2 to −1 as "unknown".

Subroutine *Reco*: It gets as parameters the coordinates of the running crack *C* and its direction *dir*. The subroutine performs as follows.

As the first step it calculates the coordinates of the positive *P* and the negative *N* pixels incident to *C* while considering the value of *dir*.

If the counter *CC*=0 then set the first prohibited direction *Proh*1 as opposite to *dir*. Set the starting pixel *StartP* and the end pixel *EndP* of the positive base both equal to *P*. Similarly set *StartN* and *EndN* equal to *N*. Set the base vector $(b, -a)$ equal to the unit vector with the direction *dir*. Increment *CC*; return 0.

If $CC\neq0$ then, independently upon the value of CC, test whether *dir* is equal to one of the prohibited directions. If this is the case then return 1: the actual crack *C* does not belong to the running DSS. If *dir* is not prohibited and *Proh*2 is still unknown then set *Proh*2 as opposite to *dir*.

If $CC=1$ then
BEGIN

Set *EndP*=*P* and *EndN*=*N* and increment *CC*. If *Proh*2 is still unknown return 0: the first two cracks have the same direction, the parameters *a* and *b* remain unchanged. Otherwise calculate the coefficients *a*, *b*, *r* of the separating linear form $H(x, y)=a\cdot(x-StartP.x)+b\cdot(y-StartP.y)$ so that $H(x, y)=0$ for the positive pixels and $H(x, y)=-e$ for the negative ones. Concretely: if the endpoints of the positive base coincide then

$$a=-(EndN.y-StartN.y)/e;$$
$$b=(EndN.x-StartN.x)/e;$$

otherwise (5.3.1)

$$a=-(EndP.y-StartP.y)/e;$$
$$b=(EndP.x-StartP.x)/e;$$

Return 0.
END IF

For all subsequent cracks ($CC>1$) calculate the values *HP* and *HN* of the separating form for *P* and for *N*: $HP=H(P)$ and $HN=H(N)$.

The following steps of the Algorithm are the decisive ones:

If $HP<-e$ or $HN>0$ then return 1: the actual crack *C* does not belong to the running DSS.
If $HP=0$ then set *EndP*=*P*: *P* lies on the positive base which must be prolonged.
Otherwise, if $HP=-e$ then redefine both bases while setting *EndP*=*P*; *StartN*=*EndN* and redefine the parameters *a* and *b* of $H(x, y)$ according to the redefined bases.
If $HN=-e$ then set *EndN*=*N*: *N* lies on the negative base which must be prolonged.
Otherwise, if $HN=0$ then redefine both bases while setting *EndN*=*N*; *StartP*=*EndP* and redefine the parameters *a* and *b* of $H(x, y)$ according to the redefined bases.
return 0.
End of Subroutine *Reco*.

The Pseudo-Code of Reco:

```
int CRecoDSS::Reco(CPoint Crack, int dir)
{ P=Crack+ToPos[dir];     // The Array "ToPos[4]" contains vectors pointing
                          // from a crack to its positive pixel P
  N=Crack-ToPos[dir];     // N is the negative pixel of Crack
  if (CC==0)              // "CC" is the number of tested cracks
  { Prohibit1=Opposite(dir);  Prohibit2=-1; CC=1;
    StartP=EndP=P;    StartN=EndN=N;
    a=-Param[dir].y;  b=Param[dir].x; // a unit vector along "dir"
    return 0;
  }
```

```
  if (dir==Prohibit1 || dir==Prohibit2) return 1;
  if (dir!=Opposite(Prohibit1) && Prohibit2==-1)
            Prohibit2=Opposite(dir);
if (CC==1)
  { EndP=P; EndN=N; CC=2;
    if (Prohibit2==-1) return 0; // only one direction
    if (EndP==StartP)
    { a=-(EndN.y-StartN.y)/e;  b=(EndN.x-StartN.x)/e;
    }
    else
    { a=-(EndP.y-StartP.y)/e;   b=(EndP.x-StartP.x)/e;
    }
    return 0;       // any two allowed cracks compose a DSS
  }
  int HP=GetH(P);  int HN=GetH(N); //the values of the SSF "H"
  if (HP<-e || HN>0) return HP;     // not a DSS
  if (HP==0)  EndP=P;
  else
    if (HP==-e)
    { EndP=P; StartN=EndN;
      a=-(EndP.y-StartP.y)/e; b=(EndP.x-StartP.x)/e;
    }
  if (HN==-e) EndN=N;
  else
    if (HN==0)
    { EndN=N; StartP=EndP;
      a=-(EndN.y-StartN.y)/e; b=(EndN.x-StartN.x)/e;
    }
  return 0;
} //************************** end Reco ********************************
```

The recognition of DSS is one of fastest and most economical methods of encoding geometrical objects. For example, it is possible to encode the boundary of a region as a sequence of DSSs while saving the coordinates of a single point and four integer parameters for each DSS [6]. The parameter *exactly* specify the location of the DSS in the image thus enabling the *exact* reconstruction of the region. The author has developed an economical code which needs on the average 2.3 bytes per one DSS. When encoding a quantified gray value image a compression rate of 3.1 was reached [6]. The method is very fast: for example, it encodes a binary image of 640×480 pixels containing about 30 disk-shaped objects in 20 ms on a PC with a processor of 700 MHz. In that time also the recognition of the disks and estimating their locations and diameters was performed.

6 Recognition of Digital Plane Patches (DPP)

The notion of a digital plane patch (or segment) is well known from the literature. [1, Rev 95]. The DPPs were mostly considered as sets of voxels, i.e. as three-dimensional objects ("thick DPPs"). However, it is more appropriate to consider a

DPP as a subset of a surface, i.e. as a two-dimensional object. Our approach based on cell complexes makes it possible.

Definition DPP: A connected subset of the frontier of a three-dimensional half-space (Definition HS, Section 5.1) is called a *digital plane patch* (DPP).

A DPP contains no voxels (as any frontier in a 3D space does). It contains only 2-cells called facets, 1-cells called cracks and 0-cells called points.

6.1 The Problem of the Segmentation of Surfaces into DPPs

This problem is of great practical importance since its solution promises an economical and precise encoding of 3D scenes.

The Problem Statement:

Given: a surface in a 3D space, e.g. the boundary of a connected subset.

Find: the minimum number of DPPs representing the surface in such a way that it is possible to reconstruct the surface, at least with a predefined precision, from the code of the DPPs.

The problem consists of two partial problems:

1. *Recognition:* Given a set of facets decide whether it is a DPP or not.
2. *Choice*: Given a surface S and a subset of facets, which is known to be a DPP, decide which facet of S should be appended to the subset to achieve that the number of the DPPs in the segmentation of S be minimal.

6.2 The Partial Problem of the Recognition of a DPP

Similarly as in the case of the DSS (Section 5), a surface S specifies two sets of voxels incident to the facets of S: a positive and a negative set. The voxels of the positive set lie outside of the body whose boundary is S, that of the negative set lie inside. If a subset T^2 of facets of S is a DPP then it lies in the frontier of a half-space whose voxels satisfy a linear inequality. The inequality separates the positive voxels of T from its negative voxels.

To decide whether T^2 is a DPP it is sufficient to solve the following system of $2 \cdot N$ linear inequalities in the components H_k, k=1, 2, 3; of a 3D vector $\boldsymbol{H}$ and a scalar value C where N is the number of facets in T^2.

$$\begin{aligned} &\Sigma_k H_k \cdot V_k^+(F_i) - C \geq 0; \\ &\qquad\qquad \text{with } F_i \in T^2;\ i=1...N; \\ &\Sigma_k H_k \cdot V_k^-(F_i) - C < 0; \end{aligned} \qquad \text{(UV)}$$

$V_k^+(F_i)$ stays for the kth topological coordinate of the positive voxel incident to the face F_i. Similarly, $V_k^-(F_i)$ is the kth topological coordinate of the negative voxel incident to F_i. The vector $\boldsymbol{H}$ is the normal to a plane separating all positive voxels from the negative ones, while C specifies the distance of the plane from the coordinate origin.

It is possible to solve the problem by a fast method similar to that of recognizing a DSS (Section 5). The corresponding algorithm and the related theory are, however, rather complicated. Their presentation here is impossible because of the page limit. We describe a rather simple algorithm [13] whose only drawback is its low speed: it is an $O(N^2)$ algorithm. Nevertheless, the algorithm is well suited for research purposes.

The algorithm solves the following problem:

Given are two sets M^+ and M^- of points in an n-dimensional space.
Find a $(n-1)$-dimensional hyperplane HP separating the sets.

The Solution: HP is specified by two vectors A^+ and A^- as the middle perpendicular to the line segment (A^+, A^-). Let $Dist(P)$ be the signed distance of the point P to HP.

The Algorithm:

1. Set A^+ equal to an arbitrary point from M^+ and A^- equal to an arbitrary point from M^-. Carry out a sequence of the following iterations.
2. During each iteration test all points P from M^+ and M^- as follows:
   If the point $P \in M^+$ lies on the wrong side of HP which means $Dist(P)<0$, then set:
   A^+:= *Foot of the perpendicular from* A^- *to the segment* (A^+, P).
   If $P \in M^-$ and $Dist(P)>0$, then set:
   A^-:= *Foot of the perpendicular from* A^+ *to the segment* (A^-, P).
3. If there is no point on the wrong side of HP stop the Algorithm. The separating hyperplane is the middle perpendicular to the line segment (A^+, A^-).
4. If the distance between A^+ and A^- is less than a predefined threshold then there exists no separating hyperplane; the convex hulls of M^+ and M^- intersect.

Fig. 6.1 shows an example of separating two point sets in a 2D space. The point sets are represented by "+" and "–" signs, the vectors A^+ and A^- by encircled signs; the old and the new separation plane by a dashed and a solid line. The point P lies on the wrong side of the old plane. It is connected with the old vector A^+ and a perpendicular has been dropped from A^- onto the segment (A^+, P). The foot of the perpendicular is the new vector A^+.

We have applied this method to recognize DPPs while using the set of the positive voxels V^+ as M^+ and that of the negative voxels V^- as M^-.

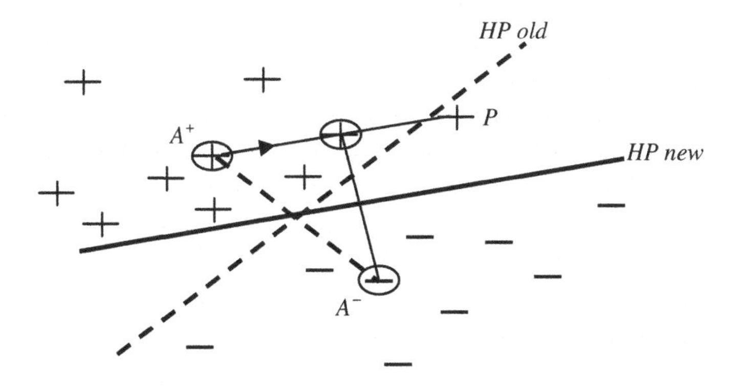

Fig. 6.1. An example of the correction of the separating plane

6.3 The Partial Problem of the "Choice"

There is no efficient method for the solution of the partial problem "Choice" known until now. It should be mentioned that this problem is much more difficult in the case of DPPs as in the case of DSSs. Really, in the latter case there are exactly two possibilities to continue a partially recognized DSS: forward or backward along the digital curve. But in the case of a DPP there are as many possibilities to continue as the number of facets adjacent to a partially recognized DPP. There is no known criterion to decide which of them should be preferred. When choosing the next facet arbitrarily then the found DPPs look chaotic even for surfaces of regular polyhedrons (Fig. 6.2).

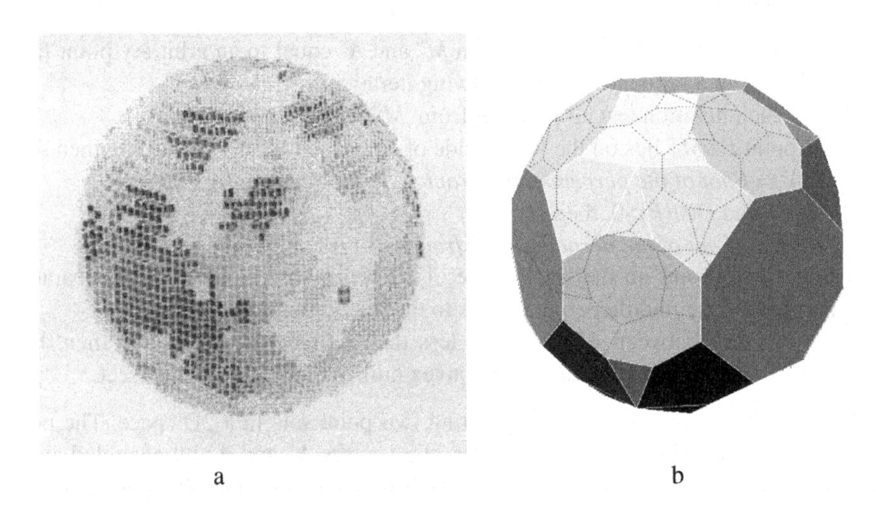

Fig. 6.2. Examples of segmenting the boundary of a digital ball into DPPs by arbitrarily choosing the next facet (a) and by computing its convex hull (b)

7 Tracing Surfaces in 3D

The algorithm [7] presented here traces a surface *S* of a three-dimensional body in a way similar to that of peeling a potato: the facets of *S* are visited one after another composing a continuous sequence. Each facet is encoded by one byte. If the genus of *S* is zero then each facet is contained in the encoded sequence only once. The code of a surface of a greater genus contains a small number of facets many times which makes the code a little longer. In any case it is possible to reconstruct the surface and the body from the code exactly.

To explain the algorithm we need the following notions from the theory of AC complexes:

Definition OF: The *open frontier* Of(*L*, *S*) of a subcomplex *L* of a complex *S relative to S* is the subcomplex of *S* containing all cells *C* of *S* whose closure Cl(*C*, *S*) contains both cells of *L* as well as cells of the complement *S*–*L*.

Definition SI: A facet F of S is called *simple relative to the subcomplex L* if $F \notin L$, the intersection $\mathrm{Cl}(F, S) \cap L$ is connected and $\mathrm{Cl}(F, S) \cap (S-L) \neq \varnothing$.

The algorithm works as follows: It chooses an arbitrary facet of the surface S as the starting one and labels its closure. Then it traces the open frontier $\mathrm{Of}(L, S)$ of the set L of labeled cells, encodes the facets of $\mathrm{Of}(L, S)$ (1 byte per facet), and labels the closures of simple facets.

This ensures that L remains homeomorphic to a closed 2-ball (a disk).

The author has proved that if the surface S is homeomorphic to a sphere then the traced sequence is a Hamilton path: each facet is visited exactly once. Otherwise there remain a few non-simple facets which are visited at least twice. Their code elements are attached to the end of the sequence of simple facets. Thus the code sequence is always connected. A verbal description of the algorithm follows. A detailed description and the related proofs may be found in [7].

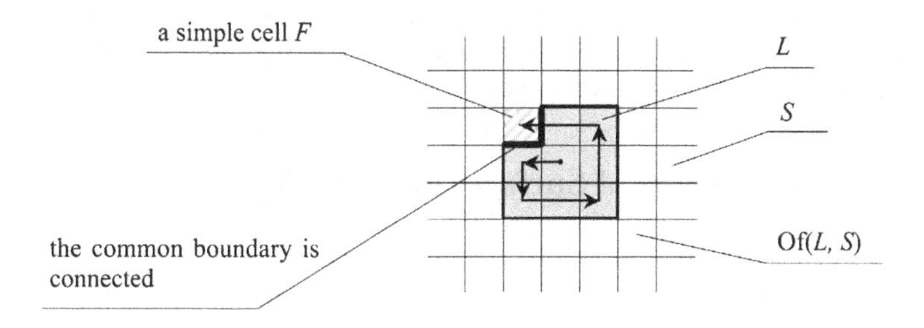

Fig. 7.1. The moves at the beginning of the tracing

The Algorithm:

Notations: S is the surface to be traced. $L \subset S$ is the subset of labeled cells; it is homeomorphic to a closed 2-ball. The "rest sequence" is the set of non-simple facets at the stage when all simple facets of S are already labeled. The rest sequence consists of a single facet if the genus of S is zero.

1. Take any facet of S as the starting facet F_0, label its closure and save its coordinates as the starting coordinates of the code. This is the seed of L, i.e. $L=\{F_0\}$. Denote any one crack of the boundary $\mathrm{Fr}(F_0,S)$ as C_{old} and find the facet F of S which is incident to C_{old} and adjacent to F_0. Set F_{old} equal to F_0 and the logical variable *REST* to FALSE. *REST* indicates that the tracing of the rest sequence is running.
2. (Start of the main loop) Find the crack C_{new} as the first unlabeled crack of $\mathrm{Fr}(F,S)$ encountered during the scanning of $\mathrm{Fr}(F,S)$ clockwise while starting with the end point of C_{old}, which is in $\mathrm{Fr}(L,S)$. If there is no such crack and F is labeled stop the Algorithm: the encoding of S is finished.
3. If F is simple label its closure.

4. Put the direction of the movement from F_{old} to C_{old} and that of the movement from C_{old} to F into the next byte of the code. If the facet F is non-simple set the corresponding bit in the code (to recognize codes of non simple facets in the ultimate sequence).
5. If *REST* is TRUE check, whether F is equal to F_{stop} and C_{new} is equal to C_{stop}. (These variables were defined in item 6 of the previous loop). If this is the case stop the Algorithm and analyze the rest sequence to specify the genus of S as explained in [7]. Delete multiple occurrences of facets from the rest sequence.
6. If F is simple set *REST* equal to FALSE; else set F_{stop} equal to F, C_{stop} equal to C_{new} and *REST* equal to TRUE.
7. Set F_{old} equal to F. Find the facet F_{new} of S incident to C_{new} and adjacent to F. Set F equal to F_{new} and C_{old} equal to C_{new}. Go to item 2.

End of the Algorithm

The algorithm was successfully tested by the author for surfaces of genus up to 5. The bodies were exactly reconstructed. A graduate student of the University of Applied Sciences Berlin [16] has also programmed the algorithm and has made a lot of successful experiments with very complicated bodies of high genus.

8 Filling the Interiors of Surfaces in Multi-dimensional Images

To test whether an n-cell P of an n-space lies in the interior of a given closed hyper-surface S it is sufficient to count the intersections of S with a ray from P to any point outside the space: iff the count is odd then P is within S. However, it is difficult to distinguish between intersection and tangency (Fig. 8.1 a and b).

The solution becomes easy if the surface is given as one or many $(n-1)$-dimensional manifolds in an n-dimensional Cartesian AC complex and the "ray" is a sequence of alternating n- and $(n-1)$-cells all lying in one row of the raster (Fig. 8.1c).

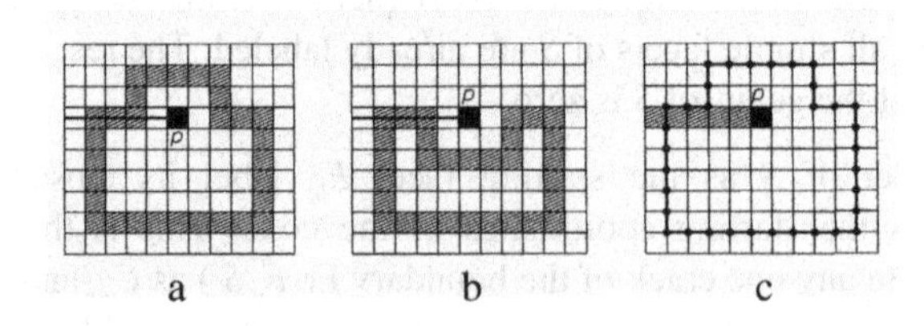

Fig. 8.1. Intersection (a) and tangency (b) are difficult to distinguish in "thick" boundaries; this is easy at boundaries in complexes (c)

In a 2D image the "surface" must be a closed sequence of cracks and points (Fig. 8.1c). Then intersections are only possible at vertical cracks and the problem of distinguishing between intersections and points of tangency does not occur. If one knows which n-cells are within S then one can fill the interior of S by labeling these cells. The method has been successfully implemented for dimensions n=2, 3, 4.

The Pseudo-Code:

Denote by `F` the current *n*-cell of the *n*-dimensional standard raster. Choose a coordinate axis *A* of the Cartesian space (e.g. *A*=*X* in the 2D case). Denote by `C(F)` the (*n*−1)-cell incident to `F`, whose normal is parallel to *A* (e.g. the vertical crack incident to `F` in the 2D case). Label all (*n*−1)-cells of *S* whose normal is parallel to *A*. In the 2D case when *A*=*X* these are the vertical cracks of *S*.

```
for each row R parallel to A do
{ BOOLEAN fill=FALSE;
  for each n-cell F in the row R do
  { if C(F) is labeled then fill=NOT fill;   // inverting fill
    if fill is TRUE then F=foreground;
    else                 F=background;
  }
}
```

9 Component Labeling in an n-Dimensional Space

We consider here the simplest case of a 2D binary image in a standard raster while the algorithm is applicable also to multi-valued and multi-dimensional images in a topological raster. In a standard raster a function must be given which specifies which raster elements are adjacent to each other and thus are connected if they have the same color. In our simple 2D example we use the well-known "(8,4)-adjacency". In the general case the adjacency of the *n*-cells of an *n*-dimensional complex must be specified by rules specifying the membership of cells of lower dimensions [2] since an (*a*, *b*)-adjacency is not applicable for multi-valued images [5].

In a topological raster the connectivity of two cells is defined by their incidence which in turn is defined by their topological coordinates (Section 3.2).

The Algorithm:

It is expedient to consider a multi-dimensional image as a one-dimensional array *Image*[*N*]. For example, in the 2D case the pixel with coordinates (*x*, *y*) may be accessed as *Image*[*y*·*NX*+*x*] where *NX* is the number of pixels in a row. The value *y*·*NX*+*x* is called the *index* of the pixel (*x*, *y*).

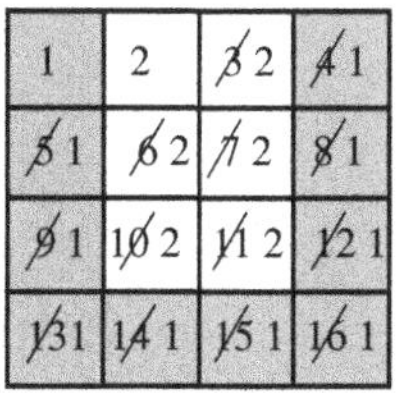

1	2	~~3~~ 2	~~4~~ 1
~~5~~ 1	~~6~~ 2	~~7~~ 2	~~8~~ 1
~~9~~ 1	~~10~~ 2	~~11~~ 2	~~12~~ 1
~~13~~ 1	~~14~~ 1	~~15~~ 1	~~16~~ 1

first run

1	2	2	1
1	2	2	1
1	2	2	1
1	1	1	1

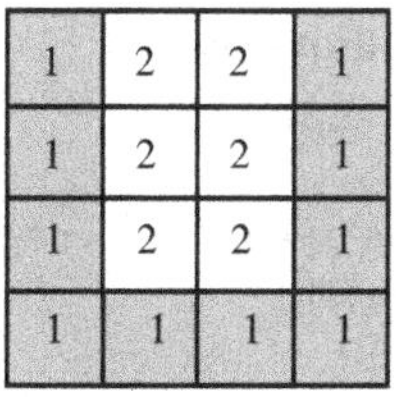

second run

Fig. 9.1. Illustration to the algorithm of component labeling

Given is a binary array `Image[]` of `N` elements and two functions `NumberNeighb(color)` and `Neighb(i,k)`. The first function returns the number of adjacent pixels depending on the color of a given pixel; the second one returns the index of the *k*th neighbor of the *i*th pixel. As the result of the labeling each pixel gets additionally (in another array `Label[]`) the label of the connected component which it belongs to.

The Pseudo-Code:

Allocate the array `Label[N]` of the same size as `Image[N]`. Each element of `Label` is initialized by its own index:

```
for (i=1; i<N; i++) Label[i]=i; //first loop
for (i=1; i < N; i++)
{ color=Image[i];
  for (j=0; j<NumberNeighb(color); j++)
  { k=Neighb(i, j); //the index of the jth neighbor of i
    if (Image[k]==color) SetEquivalent(i,k,Label);
  }
} // end of the first run
SecondRun(Label,N); // end of the Algorithm
```

Each element of `Label[N]` must have at least $\log_2 N$ bits N being the number of elements in `Image[N]`. The subroutine `SetEquivalent()` makes the preparation for labeling the pixels having the indices `i` and `k` as belonging to one and the same component. For this purpose the subroutine find the "roots" of both pixels `i` and `k`, and the greater root gets the smaller one as its label. The function `Root()` (see below) returns the last value in the sequence of indices where the first index `k` is that of the given pixel, the next one is the value of `Label[k]` etc. until `Label[k]` becomes equal to `k`. The subroutine `SecondRun()` replaces the value of `Label[k]` by the value of a component counter or by the root of `k` depending on whether `Label[k]` is equal to `k` or not.

Pseudo-Codes of the Subroutines:

```
subroutine SetEquivalent(i,k,Label)
{ if (Root(i,Label)<Root(k,Label))
       Label[Root(k,Label)]=Root(i,Label);
  else Label[Root(i,Label)]=Root(k,Label);
} // end of SetEquivalent
int Root(k, Label)
{ do
  { if (Label[k]==k) return k;
    k=Label[k];
  } while(1);
} // end of Root
subroutine SecondRun(Label,N)
{ count=1;
  for (i=0; i<N; i++)
```

```
  { value=Label[i];
    if (value==i)
    { Label[i]=count;  count=count+1;
    }
    else Label[i]=Label[value];
  }
} // end of SecondRun
```

10 Computing the Curvature of Digital Curves

The most of the algorithms for computing the curvature suggested during the last decades have a common drawback: they have a very low precision. The reason is that the precision of calculating the curvature of a curve depends dramatically on the precision of estimating the coordinates of the points of the curve. Coordinates of points in digital images are specified with a precision of about ±0.7 pixel. The author has demonstrated [9] that estimating the curvature with a precision of for example 10% is impossible for digital curves with the curvature radius less than 270 pixels because of not sufficient precision of the coordinates.

We suggest a method of using the gray values in a digital image to essentially increase the precision of estimating the coordinates. The method is applicable for gray value images of objects having an almost constant brightness against a homogeneous background.

In this case the gray values at the boundary of the object contain information about the subpixel position of the boundary.

The precision of estimating the coordinates is about 1/100 of a pixel. The curvature may be calculated as the inverse of the radius of a circle through three subpixel points. The optimal distance between the points must be calculated as a function of a coarse estimate of the expected curvature [9].

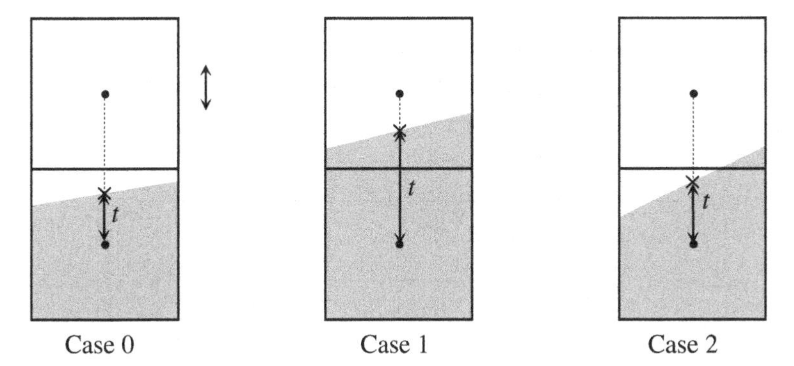

Fig. 10.1. Dependence of the portion of a pixel covered by the object on the subpixel location of its boundary (marked by a cross)

The precision of estimating the curvature is essentially higher than that of known methods. Fig. 10.2 and 10.3 show an example.

The curvature radius of the left and right outer boundary of the object shown in Fig. 10.2 is about 25 pixels. The curvature was estimated with a relative error of about 4% (see Fig. 10.3). When estimating the curvature starting with pixel coordinates without using the gray values the relative error would be according to equation (13) of [9] at least 33%.

Fig. 10.2. The gray value image of a link of a bicycle chain

Fig. 10.3 shows the curvature of the outer boundary of the object of Fig. 10.2 calculated by our method.

The relative error is of about 4%. We hope that Fig. 10.3 gives a general impression of the accuracy of the method as applied to real objects.

11 Skeleton of a Subset in 2D and 3D Images

Definition SK: The skeleton of a given set T of an n-dimensional (n=2,3) image I is a subset $S \subset T$ with the following properties:

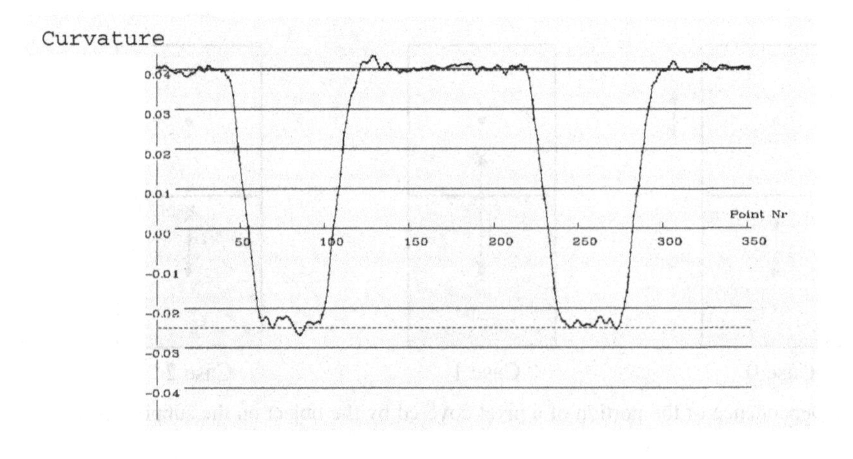

Fig. 10.3. The curvature of the outer boundary of the object of Fig. 10.2; the dashed lines show the true values of the minimum and maximum curvature

a) S has the same number of connected components as T;
b) The number of connected components of $I-S$ is the same as that of $I-T$;
c) Certain singularities of T are retained in S;
d) S is the smallest subset of T having the properties a) to c).

Singularities may be defined e.g. as the "end points" in a 2D image or "borders of layers" in a 3D image etc.

When considering an n-dimensional image (n=2,3) as an AC complex I the problem of the skeletonization consists in finding the condition under which a cell of the foreground T may be reassigned to the background $B=I-T$ without changing the number of the components of both T and B. To derive the condition we need the following notion:

Definition IS: The complex IS(C, I)=SON(C, I)∪Cl(C, I)−{C} is called the *incidence structure* of the cell C relative to the space I. It is the complex consisting of all cell incident to the cell C excluding C itself [10].

The author has proved that the membership of a cell C of any dimension may be changed between T and B without changing the number of the components of both T and B iff each of the intersections IS(C, I)∩T and IS(C, I)∩B are not empty and connected. We shall call such cells *IS-simple* relative to the set T. In the rest of this Section we call them "simple".

Thus to calculate the skeleton of a set T one must remove all IS-simple cells of T while reassigning them to B and regarding the singularities.

A well-known difficulty in calculating skeletons is that it is impossible to remove all simple pixels simultaneously without violating the skeleton conditions. However, representing an image as a complex C makes it possible to calculate the skeleton by a procedure which may be either sequential or parallel. It is based on the notion of the *open frontier* (s. Section 7 above). The procedure consist in removing IS-simple non-singular cells of T alternatively from the frontier Fr(T, C) and from the open frontier Of(T, C). We present below a simple version of the algorithm for the 2D topological raster.

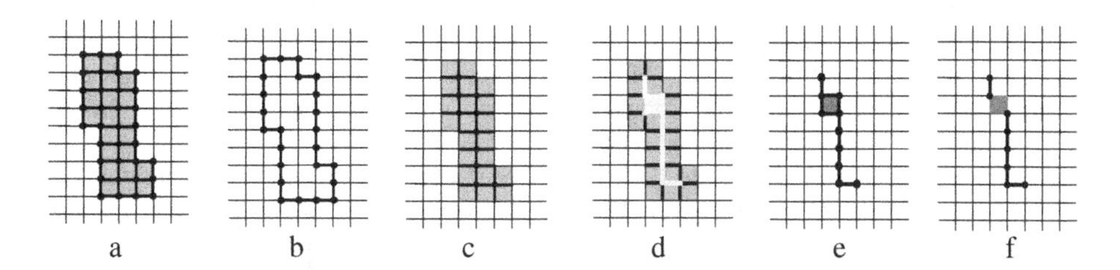

Fig. 11.1. a) a given 2D subcomplex T; b) its frontier Fr; c) the set $T-Fr$: the simple cells of the frontier deleted; d) the open frontier Of of the set $T-Fr$; e) the set $T-Fr-Of$: the simple cells of the open frontier deleted; f) the skeleton

The Algorithm:

Let $I[NX, NY]$ be a 2D array with topological coordinates. The subset T is given by labeling cells of all dimensions of T: $I[x, y]$ >0 iff the cell $(x, y)\in T$. To delete a cell means to set its label $I[x, y]$ to zero. A cell C is *singular* iff it is incident to exactly one labeled cell other than C.

To calculate the skeleton of T run the following loop:

do { Scan I and delete all simple and non-singular cells of $T \cap \mathrm{Fr}(T, I)$;
CountClose = number of cells deleted during this scan;
Scan I and delete all simple and non-singular cells of $T \cap \mathrm{Of}(T, I)$;
CountOpen = number of cells deleted during this scan;
} while (*CountClose*+*CountOpen* > 0);
// end of Algorithm

Fig. 11.1 above shows an example. The result may be, if desired, easily transformed either to a sequence of pixels or to a one-dimensional complex containing only points and cracks.

In the case of 3D images the algorithm can produce either 1D or 2D skeletons depending upon the choice of the kind of the singularities.

References

(most of the publications by the author are available at "www.kovalevsky.de")

[1] Andres, E.: Le Plan Discret, Colloque "Geometrie Discrete en Imagerie", University of Strasbourg, pp. 45-61, 1993.

[2] Kovalevsky, V.: Finite Topology as Applied to Image Analysis. Computer Vision, Graphics and Image Processing 45 (1989) 141-161

[3] New Definition and Fast Recognition of Digital Straight Segments and Arcs, 10th International Conference on Pattern Recognition, Atlantic City, June 17-21, IEEE Press, **vol. II**, pp. 31-34, 1990.

[4] Kovalevsky, V.: Finite Topology and Image Analysis. In: Hawkes, P. (ed.): Advances in Electronics and Electron Physics, Vol. 84. Academic Press (1992) 197-259

[5] Kovalevsky, V.: Digital Geometry Based on the Topology of Abstract Cell Complexes", Proceedings of the Third International Colloquium "Discrete Geometry for Computer Imagery", University of Strasbourg, September 20-21, pp. 259-284, 1993.

[6] Kovalevsky, V.: Applications of Digital Straight Segments to Economical Image Encoding, In: Ahronovitz, E., Fiorio, Ch. (eds), Discrete Geometry for Computer Imagery, Lecture Notes in Computer Science, Vol. 1347, Springer-Verlag, Berlin Heidelberg New York (1997), pp. 51-62

[7] Kovalevsky, V.: A Topological Method of Surface Representation. In: Bertrand, G., Couprie, M., Perroton, L. (eds.): Discrete Geometry for Computer Imagery. Lecture Notes in Computer Science, Vol. 1568. Springer-Verlag, Berlin Heidelberg New York (1999), 118-135

[8] Kovalevsky, V,: Algorithms and Data Structures for Computer Topology. In: Bertrand, G. et all (eds.), Lecture Notes in Computer Science, Vol. 2243: Special issue on Digital and Image Geometry, Springer-Verlag, Berlin Heidelberg New York (2001), pp. 37-58.

[9] Kovalevsky, V,: Curvature in Digital 2D Images, International Journal of Pattern Recognition and Artificial Intelligence, Vol. 15, No. 7, (2001) pp. 1183 - 1200

[10] Kovalevsky, V,: Multidimensional Cell Lists for Investigating 3-Manifolds. Discrete Applied Mathematics, Vol. 125, Issue 1, (2002) pp. 25-43.

[11] Kovalevsky, V.: Axiomatic Digital Topology, to be published, 2004.

[12] Kovalevsky, V.: Recognition of Digital Straight Segments in Cell Complexes, to be published, 2004.
[13] Kozinets, B.N.: An Iteration-Algorithm for Separating the Convex Hulls of two Sets. In Vapnik, V.N. (ed): Lerning Algorithms for Pattern Recognition (in Russian language), Publishing house "Sovetskoe Radio", Moscow, 1973.
[14] Reveillès, J.P.: Structure des Droit Discretes. Journée mathématique et informatique, (1989).
[15] Reveillès, J.P.: Combinatorial Pieces in Digital Lines and Planes. In: Vision geometry III. Proceedings of SPIE, Vol. 2573 (1995) pp. 23-34.
[16] C. Urbanek: Computer Graphics Tutorials - Visualizing of the Kovalevsky Algorithm for Surface Presentation, Graduation Thesis, University of Applied Sciences, Berlin, 2003.

Appendix: The Topology of Abstract Cell Complexes

An abstract cell complex (AC complex) is a locally finite topological space (LFS). Elements of this space are called *cells*. As we have seen in Introduction, in a space satisfying the Axioms 1 to 4 which are equivalent to the classical Axioms C1 to C4, neighborhoods must be defined by means of an *antisymmetric* binary relation. It is usual to use in the topology of cell complexes the so called *bounding relation* which is antisymmetric, irreflexive and transitive. Thus it is a partial order.

It is possible either to consider the cells as subsets of an Euclidean space or as some abstract objects having certain properties and relations to each other. In the first case the complex is called Euclidean complex, in the second case it is an abstract cell complex, or AC complex. The author is successfully working with AC complexes and is convinced that considering besides the complex itself also the Euclidean space in which the complex is embedded brings no advantages. The concept of AC complexes opens the exiting possibility to develop digital topology and digital geometry independently of the general topology and of Euclidean geometry.

The most important notions are surely acquainted to the reader from earlier publications (e.g. [2, 5, 8]. Therefore we shall repeat here only the definitions which are necessary to follow the presentation. Other definitions, necessary for certain algorithms are given at the corresponding place.

Definition AC: An *abstract cell complex* (AC complex) $C=(E, B, dim)$ is a set E of abstract elements (cells) provided with an antisymmetric, irreflexive, and transitive binary relation $B \subset E \times E$ called the *bounding relation*, and with a dimension function $dim: E \rightarrow I$ from E into the set I of non-negative integers such that $dim(e') < dim(e'')$ for all pairs $(e',e'') \in B$.

The maximum dimension of the cells of an AC complex is called its dimension. We shall mainly consider complexes of dimensions 2 and 3. Their cells with dimension 0 (0-cells) are called *points*, cells of dimension 1 (1-cells) are called *cracks* (edges), cells of dimension 2 (2-cells) are called *pixels* (or facets) and that of dimension 3 are the *voxels*.

If $(e', e'') \in B$ then it is usual to write $e'<e''$ or to say that the cell e' *bounds* the cell e''. Two cells e' and e'' of an AC complex C are called *incident to each other in C* iff either $e'=e''$, or e' bounds e'', or e'' bounds e'. In AC complexes no cell is a subset of

another cell, as it is the case in simplicial and Euclidean complexes. Exactly this property of AC complexes makes it possible to define a topology on the set of abstract cells independently from any Hausdorff space.

The topology of AC complexes with applications to computer imagery has been described in [2]. We recall now a few most important definitions. In what follows we say "complex" for "AC complex".

Definition SC: A *subcomplex* $S = (E', B', dim')$ of a given complex $C = (E, B, dim)$ is a complex whose set E' is a subset of E and the relation B' is an intersection of B with $E' \times E'$. The dimension dim' is equal to dim for all cells of E'.

Since a subcomplex is uniquely defined by the subset E' it is possible to apply set operations as union, intersection and complement to complexes. We will often say "subset" while meaning "subcomplex".

The *connectivity* in complexes is the *transitive hull of the incidence relation.* It can be shown that the connectivity thus defined corresponds to classical connectivity.

Definition OP: A subset OS of cells of a subcomplex S of a complex C is called *open in S* if it contains all cells of S bounded by cells of OS. An n-cell c^n of an n-dimensional complex C^n is an open subset of C^n since c^n bounds no cells of C^n.

Definition SON: The smallest subset of a set S which contains a given cell c of S and is open in S is called the *smallest (open) neighborhood* of c relative to S and is denoted by SON(c, S).

The word "open" in "smallest open neighborhood" may be dropped since the smallest neighborhood is always open, however, we prefer to retain the notation "SON" since it has been used in many publications by the author.

Definition CL: The smallest subset of a set S which contains a given cell c of S and is closed in S is called the *closure* of c relative to S and is denoted by Cl(c, S).

Definition FR: The *frontier* Fr(S, C) of a subcomplex S of a complex C *relative to C* is the subcomplex of C containing all cells c of C whose SON(c, C) contains both cells of S as well as cells of the complement C–S.

Illustrations to AC complexes, SONs and closures of cells of different dimensions may be found in [2, 4, 10].

Definition OF: The *open frontier* Of(S, C) of a subcomplex S of a complex C *relative to C* is the subcomplex of C containing all cells c of C whose closure Cl(c, C) contains both cells of S as well as cells of the complement C–S.

Definition BD: The (combinatorial) boundary ∂S of an n-dimensional subcomplex S of a complex C is the union of the closures of all $(n-1)$-cells of C each of which bounds exactly one n-cell of S.

Definition HN: An n-dimensional subcomplex S of a complex C is called *homogeneously n-dimensional* iff each cell of S bounds at least one n-cell of S.

Definition RG: An n-dimensional subcomplex S of a complex C is called a *region* of C iff it is homogeneously n-dimensional and C–S–∂C is also homogeneously n-dimensional.

Definition TL: A connected one-dimensional complex whose each cell, except two of them, is incident to exactly two other cells, is called a *topological line*.

It is easily seen that it is possible to assign integer numbers to the cells of a topological line in such a way that a cell incident to the cell having the number k has the number $k-1$ or $k+1$. These numbers are called the *topological coordinates* of the cells [8].

Definition CR: A Cartesian (direct) product C^n of n topological lines is called an n-dimensional *Cartesian* complex [8].

The set of cells of C^n is the Cartesian product of n sets of cells of the topological lines which are the *coordinate axes* of the n-dimensional space C^n. They are denoted by A_i, i=1,2,...,n. A cell of C^n is an n-tuple $(a_1, a_2,..., a_n)$ of cells a_i of the corresponding axes: $a_i \in A_i$. The bounding relation of C^n is defined as follows: the n-tuple $(a_1, a_2,..., a_n)$ is bounding another distinct n-tuple $(b_1, b_2,..., b_n)$ iff for all i=1,2,...n the cell a_i is incident to b_i in A_i and $dim(a_i) \le dim(b_i)$ in A_i.

The dimension of a product cell is defined as the sum of dimensions of the factor cells in their one-dimensional spaces. Topological coordinates of a product cell are defined by the vector whose components are the coordinates of the factor cells in their axes.

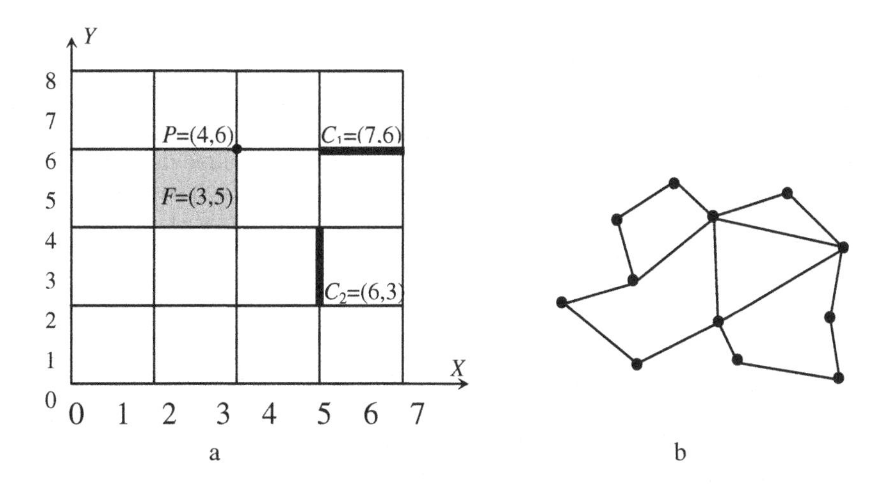

Fig. A.1. Example of a two-dimensional Cartesian (a) and non-Cartesian (b) complexes

Fig. A.1a shows four cells in a two-dimensional Cartesian complex: P is a 0-cell (point), C_1 and C_2 are 1-cells (a horizontal and a vertical crack), F is a 2-cell (pixel).

If we assign even numbers to the 0-cells and odd ones to the 1-cells of the axes then the dimension of a cell in a Cartesian complex is equal to the number of its odd coordinates.

An Efficient Euclidean Distance Transform

Donald G Bailey

Institute of Information Sciences and Technology,
Massey University, Palmerston North, New Zealand
D.G.Bailey@massey.ac.nz

Abstract. Within image analysis the distance transform has many applications. The distance transform measures the distance of each object point from the nearest boundary. For ease of computation, a commonly used approximate algorithm is the chamfer distance transform. This paper presents an efficient linear-time algorithm for calculating the true Euclidean distance-squared of each point from the nearest boundary. It works by performing a 1D distance transform on each row of the image, and then combines the results in each column. It is shown that the Euclidean distance squared transform requires fewer computations than the commonly used 5x5 chamfer transform.

1 Introduction

Many image analysis applications require the measurement of objects, the components of objects or the relationship between objects. One technique that may be used in a wide variety of applications is the distance transform or Euclidean distance map [1,2]. Let the pixels within a two-dimensional digital image $I(x,y)$ be divided into two classes – object pixels and background pixels.

$$I(x,y) \in \{Ob, Bg\} \tag{1}$$

The distance transform of this image, $I_d(x,y)$ then labels each object pixel of this binary image with the distance between that pixel and the nearest background pixel. Mathematically,

$$I_d(x,y) = \begin{cases} 0 & I(x,y) \in \{Bg\} \\ \min\left(\left\|x - x_0, y - y_0\right\|, \forall I(x_0,y_0) \in Bg\right) & I(x,y) \in \{Ob\} \end{cases} \tag{2}$$

where $\|x,y\|$ is some two-dimensional distance metric. Different distance metrics result in different distance transformations. From a measurement perspective, the Euclidean distance is the most useful because it corresponds to the way objects are measured in the real world. The Euclidean distance metric uses the L_2 norm and is defined as

$$\|x,y\|_{L_2} = \sqrt{x^2 + y^2} \tag{3}$$

R. Klette and J. Žunić (Eds.): IWCIA 2004, LNCS 3322, pp. 394–408, 2004.

This metric is isotropic in that distances measured are independent of object orientation, subject of course to the limitation that the object boundary is digital, and therefore in discrete locations. The major limitation of the Euclidean metric, however is that it is not easy to calculate efficiently for complex shapes. Therefore several approximations have been developed that are simpler to calculate for two-dimensional digital images using a rectangular coordinate system. The first of these is the city block, or Manhattan metric, which uses the L_1 norm

$$\|x, y\|_{L_1} = |x| + |y| \tag{4}$$

where the distance is measured by the number of horizontal and vertical steps required to traverse (x,y). If each pixel is considered a node on a graph with each node connected to its 4 nearest neighbours, the city block metric therefore measures the distance as the minimum number of 4-connected nodes that must be passed through. Diagonal distances are over-estimated by this metric because a diagonal connection counts as 2 steps, rather than $\sqrt{2}$.

Another measure commonly used is the chessboard metric, using the L_∞ norm

$$\|x, y\|_{L_\infty} = \max(|x|, |y|) \tag{5}$$

which measures the number of steps required by a king on a chess board to traverse (x,y). The chessboard metric considers each pixel to be connected to its 8 nearest neighbours, and measures the distance as the minimum number of 8-connected nodes that must be passed through. Diagonal distances are under-estimated by this metric as a diagonal connection counts as only 1 step.

A wide range of other metrics have been proposed that aim to approximate the Euclidean distance while retaining the simplicity of calculation of the city block and chessboard metrics. Perhaps the simplest of these is to simply average the city block and chessboard distance maps:

$$\begin{aligned} \|x, y\|_{Hybrid} &= \tfrac{1}{2}\left(|x| + |y| + \max(|x|, |y|)\right) \\ &= \max(|x|, |y|) + \tfrac{1}{2}\min(|x|, |y|) \end{aligned} \tag{6}$$

Fig. 1 graphically compares these different metrics in measuring the distance from a point in the centre of an image. The anisotropy of the non-Euclidean distance measures is clearly visible.

1.1 Morphological Approach – Grassfire Transform

Calculation of the distance transform directly using Eq. (2) is impractical because it involves measuring the distance between every object pixel and every background pixel. The time required would be proportional to the number of pixels in the image squared. Therefore more efficient algorithms have been developed to reduce the computational complexity.

Intuitively, the simplest approach to calculate the distance transform is to iteratively label each pixel starting from the edges of the object. The so-called grassfire transform imagines that a fire is started at each of the edge pixels which burns with constant velocity. An object pixel's distance from the boundary is therefore given by the time it takes the fire to reach that pixel. The grassfire transform is initialised by labelling all of the background pixels as 0. In iteration i, each unlabelled object pixel that is adjacent to (using 4-connections for the city block metric or 8-connections for the chessboard metric) a pixel labelled i-1 is labelled i. The iterations continue until all of the pixels have been labelled.

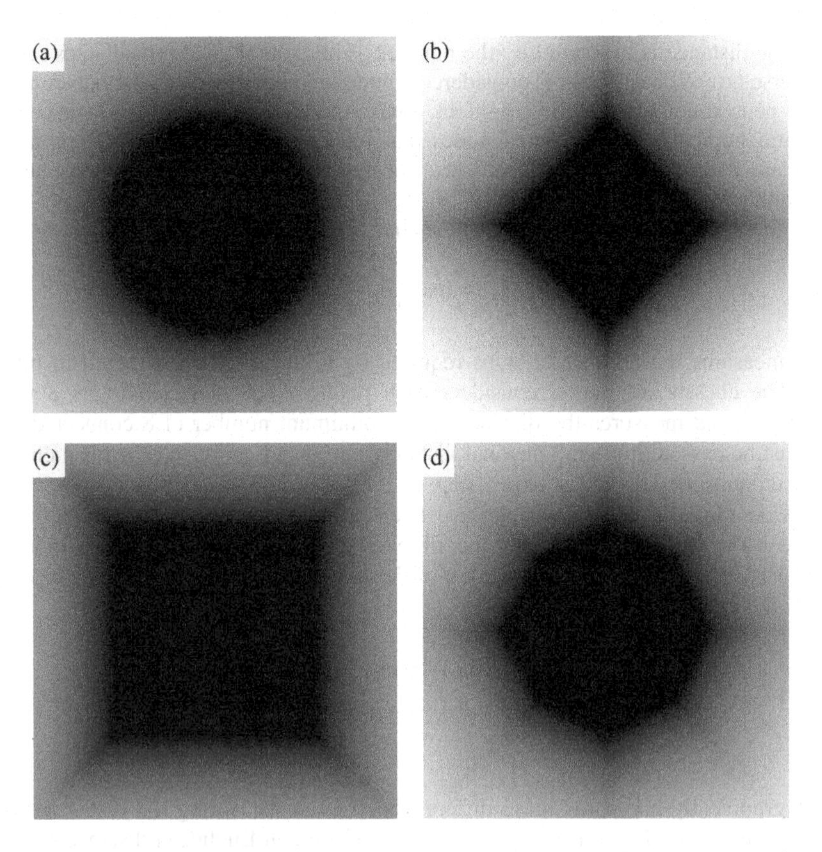

Fig. 1. Four commonly used distance metrics – measuring the distance from the centre of the image: (a) Euclidean metric, Eq. (3); (b) city block metric, Eq. (4); (c) chessboard metric, Eq. (5); (d) a hybrid metric, Eq. (6)

This iterative approach is like peeling the layers of an onion. This may be achieved by using a morphological filter to erode the object by one layer at each iteration. The shape of the structuring element (see Fig. 2) determines which distance metric is being applied. Each pixel is then labelled by the iteration number at which it was eroded from the original image. The hybrid distance metric of Eq. (6) may be achieved by alternating the cross and square structure elements at successive iterations [2].

The major limitation of using such small structuring elements is that many iterations are required to label large objects. Also, the hybrid metric provides only a crude approximation of the Euclidean distance. Both of these limitations may be overcome by greyscale morphology with a conical structuring element. In general, larger structuring elements require fewer iterations and the final result more closely approximates the Euclidean distance. However, the cost is that larger structuring elements are more computationally intensive at each iteration. For this reason, much research has gone into ways of decomposing the conical structuring element to reduce the computational burden (see for example [3-5]).

Fig. 2. Structure elements for: (a) city block erosion; and (b) chessboard erosion

1.2 Two Pass Algorithms – Chamfer Distance Transform

The iterations required by successive use of morphological filters may be removed by making the observation that successive layers will be adjacent. Therefore the distance may be calculated by propagating the distances from adjacent pixels. This approach requires only two passes through the image, one from the top left corner to the bottom right corner and the second from the bottom right back through the image to the top left corner. These two passes propagate the distances from the top and left edges of the object, and from the bottom and right edges of the object respectively. Each pass uses only values that have already been calculated.

If using a 3x3 window, the first pass propagates the distance from the 3 pixels above, and the one pixel to the left of the current pixel, adding an increment that depends on whether the pixel is 4- or 8-connected. Background pixels are assigned a distance of 0.

$$I_d(x,y) = \min\begin{pmatrix} I_d(x-1,y-1)+b, & I_d(x,y-1)+a, & I_d(x+1,y-1)+b, \\ I_d(x-1,y)+a & & \end{pmatrix} \tag{7}$$

The second pass propagates the distance from the 3 pixels below and the one pixel to the right of the current pixel. The second pass only replaces the distance calculated in the first pass if it is smaller, which will be the case if the pixel is closer to the bottom or right edges of the object.

$$I_d(x,y) = \min\begin{pmatrix} & I_d(x,y), & I_d(x+1,y)+a, \\ I_d(x-1,y+1)+b, & I_d(x,y+1)+a, & I_d(x+1,y+1)+b \end{pmatrix} \tag{8}$$

Different increments, a and b, will result in different distance metrics. The city block distance is given with a=1 and b=2; the chessboard distance with a=b=1; and the hybrid distance of Eq. (6) is given with a=1 and b=1.5, or equivalently with a=2 and b=3 (to maintain integer arithmetic), and dividing the result by 2. A better approximation to the Euclidean distance may be obtained by using the integer weights a=3 and b=4, and dividing the result by 4, although this still results in an octagonal pattern similar to that seen in (d).

A more accurate distance measure may be obtained by optimising the increments, or by using a larger window size [6]. A larger window size compares more terms (4 for a 3x3 window, 8 for a 5x5 window, and 16 for a 7x7 window - see Fig. 3), and provides a more accurate estimate of distances that are off-diagonal. A 5x5 window provides a reasonable compromise between computational complexity and approximation accuracy, and is commonly used when a closer approximation to the true Euclidean distance is required. As the number of operations is fixed for each pixel, the time required to execute the chamfer distance algorithms is proportional to the number of pixels in the image.

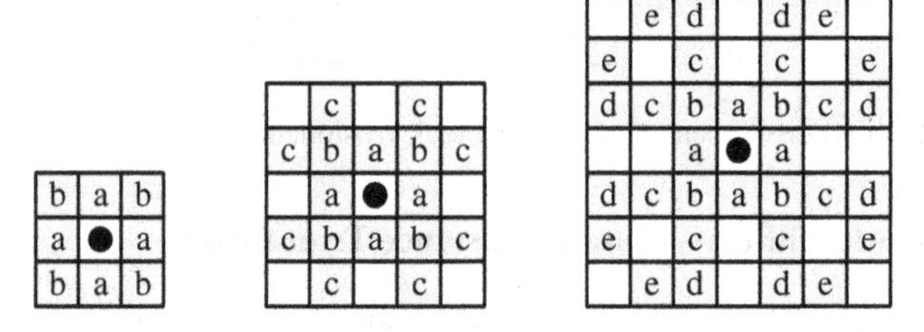

Fig. 3. The location of increments within 3x3, 5x5 and 7x7 windows. The blank spaces do not need to be tested because they are multiples of smaller increments

1.3 Vector Propagation

The two-pass chamfer distance algorithm may be adapted to measure the Euclidean distance by propagating vectors instead of the scalar distance [7-9]. The basic approach remains the same – as the window is scanned through the image, the distance is calculated by minimising an incremental distance from its neighbours. Measuring the Euclidean distance requires a square-root operation. However, if the minimum distance is selected, then the distance squared will also be minimised. This reduces the number of expensive square root operations that are actually needed. In many applications, the distance squared transform is suitable, avoiding square roots altogether.

Whereas the chamfer distance only requires an image of scalars, measurement of the Euclidean distance requires an intermediate image of vectors with x and y offsets. Background pixels are assigned a vector of (0,0). The minimum distance is calculated by propagating the vector components of each of the neighbours that have already been calculated. A similar operation is performed for the second pass, from the bottom right back up the image to the top left.

Consider an isolated background pixel (a single pixel hole in an "infinite" object). The first pass will propagate the correct distances downwards in the image as illustrated in Fig. 4. The pixels in the lower right quadrant have 3 redundant paths from adjacent pixels. The redundancy is less in the lower left octant because the pixel immediately to the right of the current pixel has not been processed yet, so has an unknown distance from the background.

The second pass is more revealing. The top left quadrant is fully redundant. In the top right quadrant, there are no direct right propagations because the pixel immediately to the left of the current pixel has not yet been processed. The bottom left octant has no redundancy. The propagation path to any pixels in this region follows the bottom left diagonal in the first pass, and then left from that in the second pass.

This lack of redundancy means that every pixel on the propagation path must be closer to the original background point than any other background pixel. If not, for example if any of the diagonal pixels is closer to another background pixel, then the pixels within this region will have the incorrect minimum distance. Fig. 5 shows a construction where this will be the case, and there is an error in the derived distance.

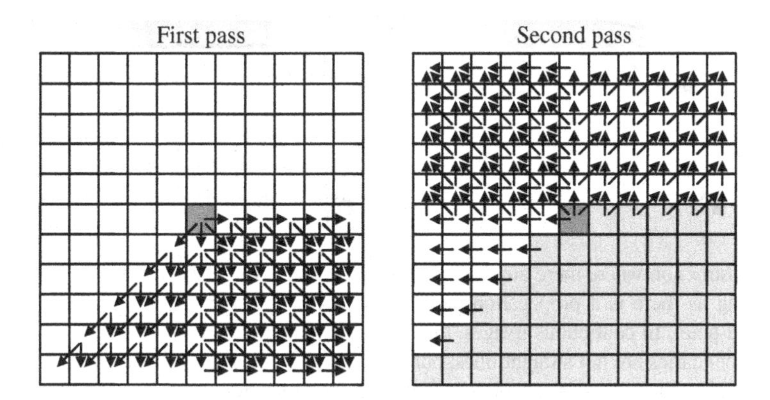

Fig. 4. Propagation of distances from a single background pixel within the two passes of the algorithm. Arrows show where the minimum comes from. Where there are multiple arrows entering a pixel, all of the paths equal the minimum

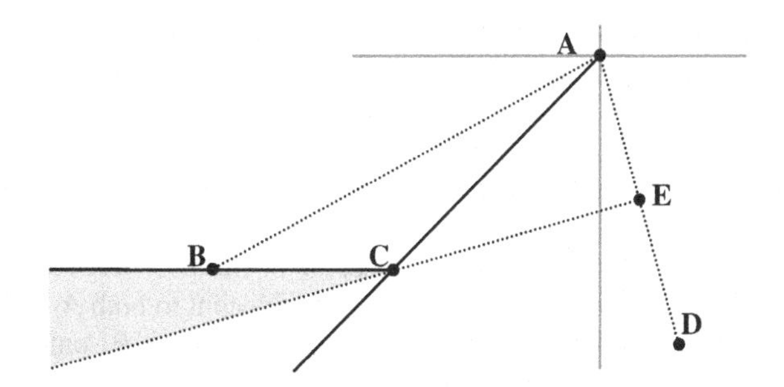

Fig. 5. Construction illustrating regions which will have an incorrect distance calculated. **A** and **D** are background pixels. Line **CE** is the perpendicular bisector of line **AD**, and consists of the points which are equidistant from **A** and **D**; points above this line are closer to **A** and points below this line are closer to **D**. The propagation of minimum distance from **A** to **B** follows the lower-left diagonal from **A** to **C**, then left to **B** (see Fig. 4). Pixels below **BC** that are above **CE** (shaded region) will be using diagonal pixels that have been labeled as being closer to **D** and so will have incorrect distances

These errors may be corrected by allowing a more direct path between point **A** and **B**. This requires a third pass through the image to provide the missing diagonal connections in this octant. It is also necessary to include the right to left propagation to correct any errors resulting from the propagation of incorrect distances in this direction. To accommodate both these propagations, it is necessary for the third pass to proceed from the top right corner to the bottom left corner, traversing right to left along each row of pixels. The redundancy added by this pass will enable the correct distances to be obtained when there are two background points. (This source of errors is overlooked in Shih and Fu [9]).

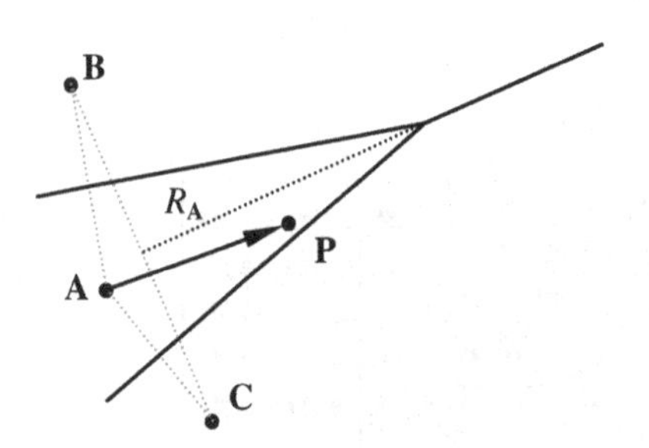

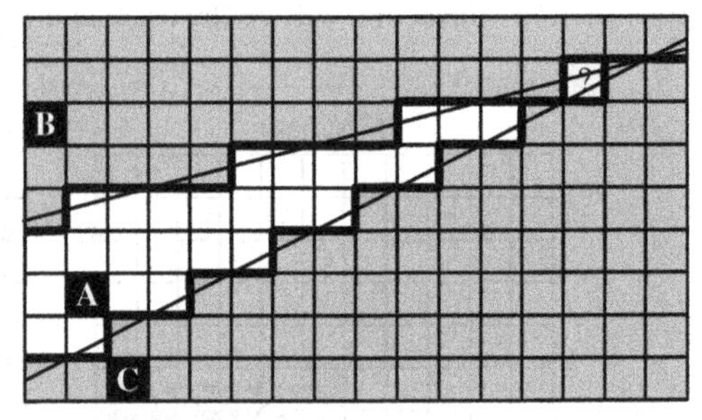

Fig. 6. Construction where there are three background pixels. The correct distances will be measured as long as there is a propagation path completely within the region associated with each background pixel. In continuous images, this will always be the case. With digital images, however, the boundaries are not straight lines, but jagged digital lines. This is illustrated in the example on the right, where there is an isolated pixel failing this criterion

Now consider the case where there are three background points, as illustrated in Fig. 6. The perpendicular bisectors between each pair of points govern the boundaries between the regions made up of points closest to each of the three background pixels. Consider pixel **A**, the central point of the three, and its associated region, $R_{\mathbf{A}}$. If there is a connected propagation path completely within the region associated with that point, then each pixel will have the correct distance. This is because each pixel along the path will be propagating the correct distance.

In the general case, when there are many background pixels within an image, the region $R_{\mathbf{A}}$ consisting of all of the point closest to a given background pixel, **A**, may be constructed as follows. The perpendicular bisector of the line between **A** and another background pixel **B** consists of all points that are equidistant to both **A** and **B**. All points on the **A** side of the bisector are closer to **A**. A point is in region $R_{\mathbf{A}}$ only if it is on the **A** side of all such bisectors. Therefore, $R_{\mathbf{A}}$ consists of the intersection of all such regions:

$$R_{\mathbf{A}} = \bigcap \left\{ \mathbf{P} \,\middle|\, \|\mathbf{P}-\mathbf{A}\| \le \|\mathbf{P}-\mathbf{B}\|, \forall \mathbf{P} \in Ob, \mathbf{B} \in Bg, \mathbf{A} \neq \mathbf{B} \right\} \tag{9}$$

The division of an image into regions in this manner is called the Voronoi diagram. The Voronoi diagram effectively associates each point within an image with the nearest feature (or background) point. Therefore obtaining the distance transform from the Voronoi diagram is a relatively simple matter [10,11].

From a vector propagation standpoint, since the Voronoi region R_A is convex, the line segment between **A** and any point within $R_{\mathbf{A}}$ will lie completely within $R_{\mathbf{A}}$. As a result, provided distances may propagate along this line segment, the correct distance will be obtained for every point in $R_{\mathbf{A}}$ and by generalization, any object point.

For continuous images, this will always be the case. However, for digital images, the boundaries of R_A are not continuous lines, but are digital lines, and are distorted by the pixel grid. When two digital bisectors approach at an acute angle, as shown in the example in Fig. 6, there may be an isolated pixel, or short string of pixels that are not 8-connected with the rest of the region [12]. Consequently, there will not be a continuous 8-connected path between such groups and the nearest background pixel for the distances to propagate along. These groups will therefore not have the correct minimum distances assigned to them. It can be shown that using a small local window cannot prevent such errors [12].

1.4 Boundary Propagation

Another class of techniques combines the idea of the grassfire transform with the propagation approach described in the previous section. These methods maintain a list of boundary pixels, and propagate these in a non-raster fashion [12-14]. Redundant comparisons may be avoided by only testing based on the direction of the nearest boundary pixel [14]. Errors such as that shown in Fig. 6 may be avoided by propagating vectors past the maximum until the difference exceeds 1 pixel [13]. While this extended propagation overcomes these errors, if care is not taken these additional propagations can result in large numbers of unnecessary comparisons [12].

1.5 Independent Scanning of *x* and *y*

The definition of Euclidean distance in Eq. (3) leads to a different class of algorithms. From Pythagoras' theorem, the distance squared to a background pixel can be determined by considering the x and y components separately. Therefore it is possible to independently consider the rows and columns. The first step looks along each row to determine the distance of each object point from the nearest boundary point on that row. This requires two scans, from left to right and right to left to measure the distances from the left and right edges of the object respectively. The second step then considers each column, and for each pixel in that column determines the closest background point by examining only the row distances in that column:

$$I_d^2(x, y) = \min_n \left(I(x, y_n)^2 + (y - y_n)^2 \right) \tag{10}$$

Thus the search has been reduced from two dimensions in Eq. (2) to one dimension. The search can be accomplished with a scan down and up the column propagating the row distances and selecting the global minima at each pixel [15]. Unfortunately, as applied, this algorithm requires that multiple row points be propagated simultaneously. The effect is that in the worst case the algorithm as described is not linear in the number of pixels (as are the chamfer and vector propagation algorithms).

2 Linear Time Independent Scanning

The key to making an independent scanning algorithm operate in linear time is to determine in advance exactly which pixels in a column that a particular row will influence. This information may be obtained by constructing a partial Voronoi diagram for each column.

2.1 Row Scanning

The first step operates on each row independently. It consists of two passes – from left to right and then right to left. The left to right pass determines the distance to the left boundary of an object

$$I(x,y)=\begin{cases} I(x-1,y)+1 & I(x,y)\in Ob \\ 0 & I(x,y)\in Bg \end{cases} \tag{11}$$

If the pixel on the edge of the image is an object pixel, its distance is set to ∞. The right to left pass replaces this with the distance to the right boundary if it is shorter:

$$I(x,y)=\begin{cases} \min(I(x,y),I(x+1,y)+1) & I(x,y)\in Ob \\ 0 & I(x,y)\in Bg \end{cases} \tag{12}$$

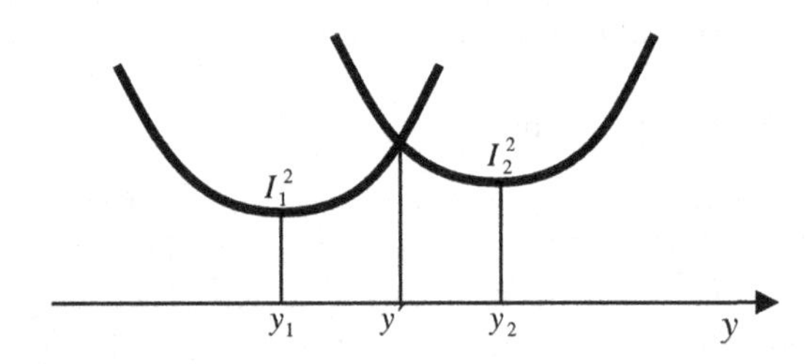

Fig. 7. The distance squared along a column, showing the regions of influence of two background points

2.2 Column Scanning

Consider an image with two background pixels at $I(x_1,y_1)$ and $I(x_2,y_2)$, with $y_1<y_2$. Let I_1 and I_2 be the corresponding minimum row distances in column x. The distance squared function in column x is illustrated in Fig. 7. The column is split into two with part of the column coming under the influence of (x_1,y_1) and part coming under the influence of (x_2,y_2). The boundary between the two regions is given from the intersection of the two parabola:

$$I_1^2+(y'-y_1)^2=I_2^2+(y'-y_2)^2 \tag{13}$$

Solving this for the position of the intersection gives:

$$y'=y_2+\frac{I_2^2-I_1^2-(y_2-y_1)^2}{2(y_2-y_1)} \tag{14}$$

Note that there will always be exactly 1 intersection point, corresponding to where the perpendicular bisector between $I(x_1,y_1)$ and $I(x_2,y_2)$ intersects column x, although the bisector may not necessarily be between y_1 and y_2. As the distance is only evaluated for integer values of y, it is not necessary to know the precise location of the intersection, only which two pixels it falls between. This means that integer division may be used, and the remainder or fractional part discarded. If the numerator is positive, the number calculated is the last pixel under the influence of y_1. If negative, it is the first pixel under the influence of y_2.

Assume that the image is being scanned in the increasing y direction. Now consider adding a third background point $I(x_3,y_3)$, where $y_2<y_3$ with intersection between parabolas 2 and 3 at y''. If $y' < y''$ then there are three regions of influence, corresponding to the sets of points nearest to each of the background pixel. However if $y'' < y'$ then background point 2 has no influence in column x because its parabola will be greater than the minimum of parabolas 1 and 3 at every point. The boundary between parabolas 1 and 3 may then be found from Eq. (14).

Extending this search to N points would require N^2 tests in the worst case. However, by making use of the fact that the points are ordered, and scanning in only one direction at a time, the number of tests may be reduced to N.

The basic data structure used to maintain the information is a stack. Each stack item of contains a pair of values (y,y^I) representing respectively a row number, y, and the maximum row which that row influences, y^I. The stack is initialized as $(0,N)$. This is saying that in the absence of further information, the first row will influence the whole image.

For each successive row, Eq. (14) is evaluated with y_1 as the row number from the top of stack, and y_2 the new row. There are three cases of interest:

1. $y' > N$. The boundary of influence between y_1 and y_2 is past the end of the image, so the new row will have no influence.
2. $y' > y_0^I$, where y_0^I is the influence from the previous stack entry, and corresponds to the start of the influence of row y_1. In this case row y_1 has a range of influence, and y_1^I is set to y'. The new row, y_2 is added to the stack, with y_2^I set to N.
3. $y' \leq y_0^I$. In this case, row y_1 has no influence on the distance transform in this column. Row y_1 is therefore popped off the top of the stack, and Eq. (14) is re-evaluated with the new top of stack. This process is repeated until either the stack is empty (the new row will influence all previous rows) or case 2 is met (the start of the influence of the new row has been found).

After processing all of the rows, the boundary points between each of the key influencing rows is known. Since the row that will provide the minimum distance for each row is known, it is simply a matter of using the stack as a queue for a second pass down the column to evaluate the distances.

Since Eq. (14) may be evaluated multiple times for each row, it is necessary to demonstrate that this algorithm actually executes in linear time. Observe that in cases 1 and 2, Eq. (14) is evaluated once as the new row is added (or discarded). If case 3 is selected, one existing row will always be eliminated from the stack for each additional time Eq. (14) is evaluated. These subsequent evaluations may therefore be may be associated with the row being eliminated rather than the row being added. As a row may only be eliminated once at most, the total number of times that Eq. (14) is evaluated will be between N and $2N$. Therefore the total number of operations is proportional to N and the above algorithm executes with time proportional to the number of pixels in the image.

3 Efficient Implementation

First note that both Eq. (10) and (14) involve squaring operations. Rather than calculate this each time using multiplications, a lookup table can be precalculated and used. The maximum size of this lookup table is the maximum of the number of rows or columns in the image. Rather than use multiplications to populate the lookup table, it may be filled as follows:

$$x^2 = \begin{cases} 0 & x = 0 \\ (x-1)^2 + 2x - 1 & x > 0 \end{cases} \tag{15}$$

3.1 Row Scan

The minimum operation of Eq. (12) may be eliminated if the width of the object on row y is known. So as the row is scanned, the distance from the left edge of the object is determined, as in Eq. (11). However, when the next background pixel is encountered, the width of the object is known from the distance of the last pixel filled. Therefore as the line is filled back, it only needs to be filled back half of the width. This right-to-left fill is performed immediately rather than waiting for a second pass since the position of the right edge is now known.

Rather than store the distance, storing the distance squared is more useful since it needs to be squared for in Eq. (14).

3.2 Column Scan – Pass 1

The most expensive operation within the column scanning is the division in Eq. (14). Therefore the speed may be increased by reducing the number of times Eq. (14) is evaluated. Since, in general, many of the rows are eliminated, if those rows may be eliminated beforehand this can save time. Separating the scan into two passes, first down the column and then up the column, and propagating the distances while scanning can achieve this.

Referring to Eq. (14), observe that if $I_2 \leq I_1$ then $y' < y_2$. This implies that if the image is being scanned in the positive direction, the intersection point has already been passed, and as far as the rest of the scan is concerned, y_1 may be eliminated. For a typical image, this implies that approximately half of the initial scans in the first pass may be eliminated by a simple comparison.

Secondly, in assigning the distances during the first pass, if the distance on any row is decreased, that row will have no influence in the second pass. This is because any background pixel that causes such a reduction must be closer to that object pixel (for the reduction to occur) and also be in a row above it (to have influence in the first pass). In the second pass, back up the column, if Eq. (14) was applied to those two rows, the boundary would be below the row that was modified. This implies that it will have no influence in the upward pass. Therefore all such rows may be ignored in the second pass. This may be accomplished by setting the minimum row distance of that pixel to ∞.

Taking these into account, the first column pass may be implemented as follows:

1. Skip over background pixels – they will have zero distance. Reset the stack and push the row number of the last background pixel onto the stack. To avoid scanning through these pixels in the second pass, the location of the first background pixel may be recorded in a list.
2. If $I^2(x, y)$ is infinite (there are no background pixels in this row), skip to step 10 to update the distance map.
3. If the stack is empty, skip to step 5. Otherwise calculate the new distance that would be propagated to the current row from the bottom of the stack, y_c:

$$I_{new}^2(x, y) = I(x, y_c)^2 + (y - y_c)^2 \tag{16}$$

4. If $I_{new}^2(x, y) < I^2(x, y)$ then the previous rows have no influence over the current row. Therefore the complete stack is reset, and the current row number is pushed onto the stack. Proceed with processing the next pixel (step 11).

Steps 5 to 9 consist of a loop that updates the stack.

5. If the stack is empty, push the current row onto the stack, and go to step 10.
6. If the current distance is less than that on the row pointed to by the top of stack, $(I^2(x, y) < I_{tos}^2(x, y))$ then the current top of stack will no longer have any influence. Pop the entry from the top of the stack, and loop back to step 5.
7. Calculate the influence boundary between the top of stack and the current row using Eq. (14). If this boundary is past the end of the image, the current row will have no influence. Set $I^2(x, y)$ to ∞ and skip to step 10.
8. If the boundary is greater than that of the previous stack entry (top-of-stack – 1, if it exists) then adjust the boundary on the top of stack to the value just calculated. Push the current row onto the stack and skip to step 10.
9. Otherwise the current top of stack has no influence, so pop the top entry from the stack and loop back to step 5.
10. If the new value was not calculated in step 3, then calculate it now (if the stack is empty, skip to step 11). This value is written to the output image, $I_d^2(x, y)$. If $I_{new}^2(x, y) < I^2(x, y)$ then set $I^2(x, y)$ to ∞ because this row will not have any influence on the second pass. If the boundary of influence of the entry on the bottom of the stack ends at the current row, then the entry may be pulled from the bottom of the stack (that entry will have no further influence on the rest of the column).
11. Move to the next pixel in the current row, and repeat.

At this stage, all of the distances that need to be propagated down the image will have been propagated. Most of the rows that are unlikely to influence the propagation back up the image have also been eliminated.

3.3 Column Scan – Pass 2

The second column scan, from the bottom of the image to the top proceeds in the much the same manner as the first scan. The exceptions are:

Step 1: Rather than scanning through the background pixels a second time, use the previously recorded top of the run.

Step 4: Also check if $I_{new}^2(x, y) > I_d^2(x, y)$. In this case, the distance being propagated up will no longer have any influence (the pixels have already been set with a lower distance). Therefore clear the stack, and continue scanning with the next pixel (step 11).

Steps 7 and 10: $I^2(x, y)$ does not need to updated, as this is not used any more.

3.4 Analysis of Complexity

Scanning through the image requires 1 increment and 1 comparison for every pixel visited. During the row pass, the whole image is scanned once in the left to right direction. Half of the object pixels are scanned a second time from right to left to update the distance from the right edge. Testing to see if a pixel is object or background requires 1 comparison. While the object pixels are being updated, a separate counter is maintained to keep track of the distance, requiring 1 addition, and a squaring operation (via table lookup).

For the column scanning, the exact complexity of the algorithm is made more difficult to calculate by the loop in steps 5 to 9 of the column pass. However, it was argued that Eq. (14) would be evaluated somewhere between 1 and 2 times per object pixel on average. The worst case is actually be less than 2 because that would imply that no row had any influence! The average gains made by splitting the column analysis into two passes will not necessarily result in gains in the worst case.

The whole image is scanned during the first pass of column scanning This results in 1 increment and 1 comparison per pixel, plus a test for a background pixel at each pixel. In the second pass, only the object pixels are processed.

The tests in steps 2-4 require 1 comparison each, and are executed during both passes through the object rows. Eq. (16) is evaluated either on step 3 or 10, and requires 2 additions, 1 squaring operation and 1 stack access to obtain the row to be propagated. It will be evaluated at most twice per object pixel (once in each pass). The test of step 4 ensures that the loop (steps 5-9) will only be entered in only one of the passes. Therefore the operations in the loop may be executed up to 2 times per object pixel. Accessing the top of stack (an array lookup) is performed in steps 6 and 8 (with a subtraction in 8 to access the previous entry). Evaluation of Eq. (14) requires 3 additions, one squaring, one division, and one stack access. The tests in steps 5-8 require 1 comparison each. As a result of the tests, a value is either pushed onto the stack (an addition to adjust the stack pointer, and a stack access) or popped off the stack (adjusting the stack pointer only). As these are also associated with the looping, they will be executed once each per object pixel in the worst case. Finally, in step 10, there are 2 comparisons, a stack access, and an addition to adjust the stack if the bottom entry has no further influence.

These results are sumarised in Table 1, and compared with the number of operations required to implement 3x3 and 5x5 chamfer distance transforms. It should be emphasized that the results for the new algorithm are worst case, and for more typical data, many of the comparisons made in steps 2-4 would result in the loop (steps 5-9) being bypassed, reducing the average number of operations per object pixel to ~45.

The number of operations per pixel is the same as that for the chamfer algorithms because only two full passes are made through the image. Although the independent scanning algorithm makes two passes along both rows and columns, after the first pass the object boundaries are knows so the second pass only needs to scan the object pixels. While the algorithmic complexity of independent scanning is considerably greater than that of the simpler chamfer algorithms, the worst case computational complexity is similar to that of the 5x5 chamfer transform. For a more typical image, the computation complexity is expected to be between that of the 3x3 and 5x5 chamfer transforms. In many applications the distance-squared transform produced by this algorithm is suitable, although if necessary a square root operation may be applied during the second column pass.

Table 1. Summary of the number operations required to implemnent Euclidean distance transformation in the worst case. Key: + additions or subtractions; < comparisons; [ ] array indexing, including accessing the image, the stack, and the squaring lookup table; / divisions. Scanning includes checking for background pixels. The total is the total only per object pixel, assuming all operations are of equal complexity. It is acknowledged that division will take longer than the other operations. For comparison, the totals from the 3x3 and 5x5 chamfer algorithms are also given

	Per image pixel			Per object pixel				
	+	<	[ ]	+	<	[ ]	/	Total
Row scanning	1	2	1	½	½			
Distance calculation				1½		1½		
Column Scanning	1	2	1	1	1	1		
Steps 2-4				4	6	4		
Steps 5-9				10	8	9	2	
Step 10				2	4	3		
TOTAL	2	4	2	19	19½	18½	2	59
3x3 Chamfer	2	4	2	15	7	11		33
5x5 Chamfer	2	4	2	28	15	19		62

4 Summary

This paper has demonstrated that a linear-time Euclidean distance-squared transform may be implemented efficiently in terms of computation using only integer arithmetic. If the actual distance map is required, then a square root will be necessary. It is shown that in the worst case, the computational complexity of the proposed distance transform is similar to that of the commonly used 5x5 chamfer distance unless a square root is required. On more typical images, the complexity is expected to be between the 3x3 and 5x5 chamfer distance transforms, while providing exact results.

The algorithm is implemented by first forming a distance map along each of the rows, and then combining these distances in the columns. Since each row and column are operated on independently, such an implementation may be efficiently parallelised. This approach is also readily extended to higher dimensions or anisotropic sampling, where the different axes may have different sample spacing. The independent scanning approach inherently avoids the distance errors that are associated with the simpler vector propagation algorithms (using either raster or contour propagation).

The implementation described is also efficient in terms of its memory utilisation. If the transformation is performed in place (the same image array is used for both input and output) then modest additional scratch memory is required. A lookup table is used for performing squaring operations – this needs to be the larger of the number of rows or columns in the image. Memory is also required for the stack. It can be shown that the maximum number of stack entries is half of the height of each column. Temporary storage is also required to hold the results of the first column pass. This also needs to be the height of the image. While this is not as good as the chamfer algorithms (which need no additional storage), it is a significant savings over the vector propagation approaches which require a scratch image of vectors.

References

1. A. Rosenfeld and J. Pfaltz, "Sequential Operations in Digital Picture Processing", Journal of the ACM, 13:4, pp 471-494 (1966).
2. J.C. Russ, "Image Processing Handbook", 2nd edition, CRC Press, Boca Raton, Florida (1995).
3. C.T. Huang and O.R. Mitchell, "A Euclidean Distance Transform Using Grayscale Morphology Decomposition", IEEE Transactions on Pattern Analysis and Machine Intelligence, 16:4, pp 443-448 (1994).
4. F.M. Waltz and H.H. Garnaoui, "Fast Computation of the Grassfire Transform Using SKIPSM", SPIE Conf on Machine Vision Applications, Architectures and System Integration III, Vol 2347, pp 396-407 (1994).
5. R. Creutzburg and J. Takala, "Optimising Euclidean Distance Transform Values by Number Theoretic Methods", IEEE Nordic Signal Processing Symposium, pp 199-203 (2000).
6. M.A. Butt and P. Maragos, "Optimal Design of Chamfer Distance Transforms", IEEE Transactions on Image Processing, 7:10, pp 1477-1484 (1998).
7. P.E. Danielsson, "Euclidean Distance Mapping", Computer Graphics and Image Processing 14, pp 227-248 (1980).
8. I. Rangelmam, "The Euclidean Distance Transformation in Arbitrary Dimensions", Pattern Recognition Letters, 14, pp 883-888 (1993).
9. F.Y. Shih and Y.T. Wu, "Fast Euclidean Distance Transformation in 2 Scans Using a 3x3 Neighborhood", Computer Vision and Image Understanding, 93, pp 109-205 (2004).
10. H. Breu, J. Gil, D. Kirkatrick, and M. Werman, "Linear Time Euclidean Distance Transform Algorithm", IEEE Transactions on Pattern Analysis and Machine Intelligence, 17:5 pp 529-533 (1995).
11. W. Guan and S. Ma, "A List-Processing Approach to Compute Voronoi Diagrams and the Euclidean Distance Transform", IEEE Transactions on Pattern Analysis and Machine Intelligence, 20:7: pp 757-761 (1998).
12. O. Cuisenaire and B. Macq, "Fast Euclidean Distance Transformation by Propagation using Multiple Neighbourhoods", Computer Vision and Image Understanding, 76, pp 163-172 (1999).
13. L. Vincent, "Exact Euclidean distance function by chain propagations", IEEE Computer Society Conference on Computer Vision and Pattern Recognition, pp 520-525 (1991).
14. H. Eggers, "Two Fast Euclidean Distance Transformations in Z^2 Based on Sufficient Propagation", Computer Vision and Image Understanding, 69, pp 106-116 (1998).
15. T. Saito and J.I. Toriwaki, "New Algorithms for Euclidean Distance Transformations of an N-dimensional Digitised Picture with Applications", Pattern Recognition, 27, pp 1551-1565 (1994).

Two-Dimensional Discrete Morphing*

Isameddine Boukhriss, Serge Miguet, and Laure Tougne

Université Lyon 2, Laboratoire LIRIS,
Bâtiment C, 5 av. Pierre Mendès-France,
69 676 Bron Cedex, France
{iboukhri, smiguet, ltougne}@liris.univ-lyon2.fr
http://liris.cnrs.fr

Abstract. In this article we present an algorithm for discrete object deformation. This algorithm is a first step for computing an average shape between two discrete objects and may be used for building a statistical atlas of shapes. The method we develop is based on discrete operators and works only on digital data. We do not compute continuous approximations of objects so that we have neither approximations nor interpolation errors. The first step of our method performs a rigid transformation that aligns the shapes as best as possible and decreases geometrical differences between them. The next step consists in searching the progressive transformations of one object toward the other one, that iteratively adds or suppresses pixels. These operations are based on geodesic distance transformation and lead to an optimal (linear) algorithm.

1 Introduction

Many medical images are produced every day and their interpretation is a very challenging task. 3D atlases can be of great interest since they allow to help this interpretation by very precise models. Most of the time, these atlases are built manually and represent a considerable amount of work for specialists of the domain. Moreover, they only contain static information corresponding to a single patient or potientially an average shape corresponding to a small set of patients. It would be very useful to compute these atlases in an automated way from a set of images: it would allow to compute not only an average shape between all input data but also statistical measures indicating the interindividual variability of these shapes. This is the basic idea of the statistical atlas [FLD02].

In this paper, our goal is to study the progressive deformation from one object to another one which is the first step for the computation of an average object. For sake of simplicity, we focus on 2D binary images but the proposed approach could easily be generalized to 3D. Our method is decomposed into two steps: the first one consists in making a rigid registration of the two objects and the second one in computing the deformation.

* This work was supported by the RAGTIME project of the Rhône-Alpes region.

R. Klette and J. Žunić (Eds.): IWCIA 2004, LNCS 3322, pp. 409–420, 2004.

2 State of the Art

Concerning the rigid registration, many algorithms exist in the literature. Some of them are based on intensity and they use similarity measures such as correlation coefficients, correlation ratios [RMP+98] and mutual information [WV96]. These algorithms do not need segmentation of the images and are based on the statistic dependences between the images to be registered. Other algorithms are based on geometrical properties such as surfaces, curvatures [MMF98] and skeletons [LB98]. These methods are generally faster than the previous ones, nevertheless less precise. As a matter of fact, the extraction of the surface and the computation of the curvature are noise sensitive and they may induce imprecision in the registration. A discrete method has been proposed by Borgefors [Bor88] which uses distance transforms. In this method, for one of the images the associated distance card is computed. The object of interest in the second image is approximated by a polygon which is superimposed on the previous distance card. Then, the squared values of the pixels of the distance card in which the polygon is superimposed are averaged to obtain a contour distance. The author searches for the rigid transformation of the polygon that minimizes this distance. However, it is difficult to estimate the number of necessary iterations.

We can also find lots of methods in the literature that make the morphing of two shapes. An important part of them is based on the interpolation of the positions and/or the colours of the pixels in the two images [Iwa02]. In our case, we consider the morphing as generation of intermediate images.

A recent study [BTSG00] computes the average shape between two continuous shapes. First, it makes the registration of the two images and it computes the skeleton of the difference between the two shapes. Using an elimination process, it only keeps the points of the skeleton that are equidistant of two borders of two different objects. This method preserves the topology and the initial shape of the objects. However, it only allows to generate the median shape and not a continuous deformation of one shape to the other one. The method we present in this paper is a generalized discrete version of the previous one. The generalization we propose allows to compute not only a median shape but also the different intermediate shapes.

In the following section we recall some notions necessary for the comprehension of the remainder of the text. Section 4 is the heart of the article and describes the proposed method. In section 5, we present some examples in which we have applied our method. Finally, we conclude and present some future works.

3 Preliminaries

Let us give the formal context of our study and recall some basic notions concerning the inertia moments.

3.1 Neighborhood, Connectivity and Distance

We consider 2D shapes in the $\mathbb{Z}^2$ space. The pixels of the shape have the value 1 and the pixels that belong to the background have the value 0. The object is considered as 8-connected and background is considered as 4-connected. We work in 3×3 neighborhood. Let a and b denote two binary shapes, we denote the symmetric difference by $a \Delta b = \{a \cup b\} \backslash \{a \cap b\}$.

In the following we will use the distance transform. This is a classical tool [RP68] which associates to each point of the object the distance of the nearest pixel of the background. In our study, we will use the chamfer distance 3-4 which is a good approximation of the Euclidean distance. The figure 3.1 gives an example of a distance transform obtained in this way.

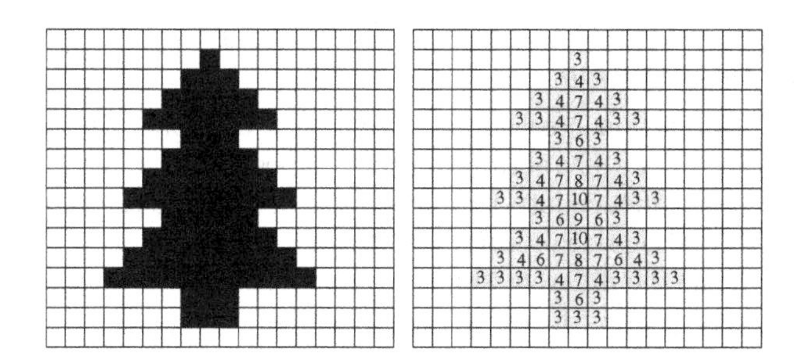

Fig. 1. An example of distance transform using the chamfer 3-4 distance

3.2 Inertia Moments

In order to make the registration, we will use the moments associated to the shapes. Such descriptors are especially interesting in order to determine the position, the orientation and scale of an object. Moreover, one of their main advantages is their small sensitivity to noise.

Let us consider a two-dimensional image I. The general form of the discrete moments is:

$$M_{pq} = \sum_x \sum_y (x - X_c)^p (y - Y_c)^q I(x, y)$$

with $0 \leq p, q \leq 2$.

X_c and Y_c are the coordinates of the barycentre of the shape :

$$X_c = \frac{1}{N} \sum_i X_i \qquad Y_c = \frac{1}{N} \sum_i Y_i$$

From these moments we can compute the principal inertia axis of the shape. Such a computation is made with the help of the inertia matrix:

$$MI = \begin{pmatrix} M_{20} & -M_{11} \\ -M_{11} & M_{02} \end{pmatrix}$$

This matrix is normalized and diagonalized in order to obtain the eigenvectors V_1 and V_2 and the associated eigenvalues λ_1 and λ_2. Let us suppose $\lambda_1 > \lambda_2$. V_1 represents the maximal elongation axis of the object.

In the following section, we describe the proposed method to obtain all the intermediary images between two given images.

4 Methodology

The method is based on an operation that consists in aligning the two figures. This operation is described in subsection 4.1. When the two shapes are superimposed, in the same referential, we can start the morphing step. The morphing operation is described in subsection 4.2.

4.1 Rigid Registration

The rigid transformation consists in a sequence of global operations applied on the shapes in order to superimpose the shapes in the same lattice. We can decompose such a sequence into two parts: the first one is the rigid registration itself and the second one deals with scaling and re-sampling.

In order to show the different steps, we have chosen two shapes, presented in the figure 4.1, representing fishes we want to align.

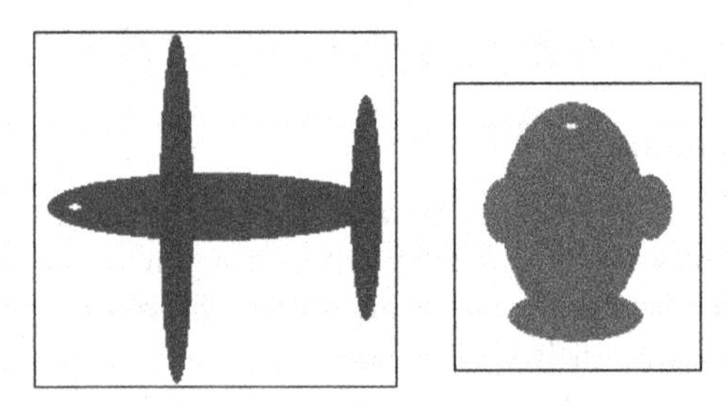

Fig. 2. Two shapes we want to align

The first step is directly deduced from the computation of the inertia moments presented in subsection 3.2 and consists in a translation and a rotation. Simply, it consists in computing the inertia moments of order one and two for each shape. It is then followed by the application of Backward Mapping [Wol90] to make the transformation. Figure 3 gives the resulting image.

The scaling and re-sampling operations are very important because we cannot compare two shapes that are not digitalized on the same grid. Our goal is to make homothetic scaling in order to preserve the global aspect of the shapes. An intuitive method would be to decrease the size of the biggest shape or to increase the size of the smallest one. In fact, it would be equivalent in a continuous domain. But, in the discrete space the objects can have different samplings and if we reduce the size we may lose important information; on the contrary, if

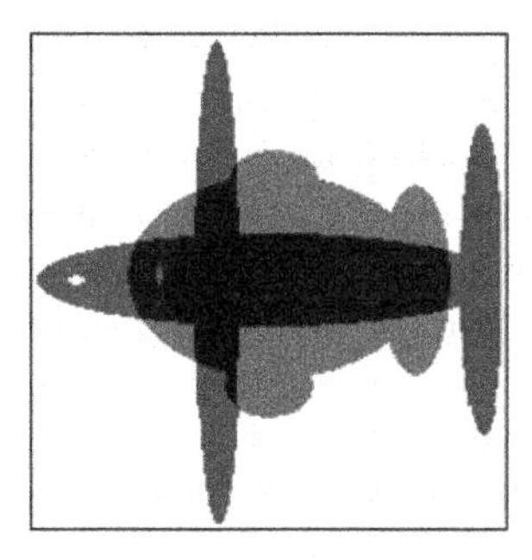

Fig. 3. Rigid registration: translation and rotation

we zoom a digital object we may generate aliasing artefacts. We have chosen a compromise which aligns the principal vectors on a new system of coordinates and which chooses an intermediate scale between the two shapes (it reduces the biggest and increases the smallest). Moreover, as we have mentioned previously, the two shapes may not have the same sampling resolution: the pixel size along the x axis may not be the same as along the y axis and, they also may be different between the two images. This is an other reason why we do not re-sample only one of the shapes using the grid of the other one but prefer to use a third lattice that is good adapted to the re-sampling of the two shapes. Figure 4 shows the resulting image after the scaling and re-sampling operations in our example.

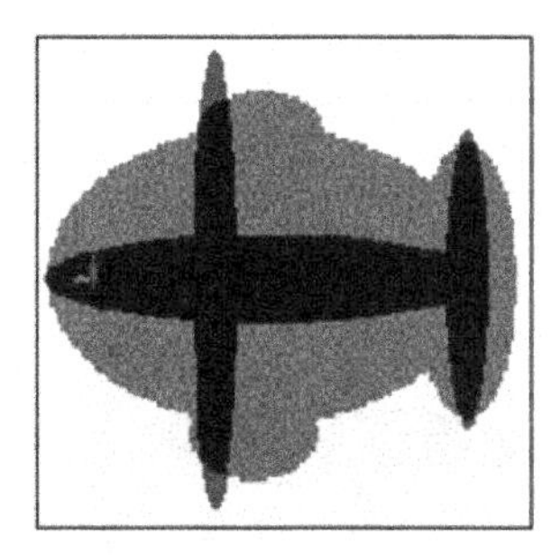

Fig. 4. The two shapes after registration, scaling and re-sampling

Note however that we have shown here the different steps in order the make it more easily understandable, but in fact all the corresponding transformation matrices are combined together to obtain only one transformation matrix T that is applied at once. For classical reasons linked to discrete transformations using different lattices, we have chosen to use the backward mapping technique to apply the transformation. Using bounding boxes of the input shapes, we determine the limits of the output space and for each pixel of the output space we compute its predecessor using T^{-1}. Even if this technique may generate aliasing we will see in the final results that it will be negligible for the conservation of the global aspect of shapes.

The principal axis is a direction of the maximal elongation of a shape. The alignment of the principal axis we have described in this section might be subject to orientation errors. Among the two possible orientations we have for direct transformations, we select the one that maximizes the Hamming distance between the two shapes.

4.2 Discrete Morphing

The two shapes are now discretized in the same lattice. We want to define a transformation that progressively deforms one of them into the second. Just remark here that the transformation we propose is symmetrical: we can choose as first image indifferently one of the two images. In order to obtain the intermediate shapes, we iteratively add pixels to shape a (those belonging to $b \backslash a$) and delete other pixels from shape a (those belonging to $a \backslash b$) until the object converges to shape b. The decision concerning the adding or deleting is determined by the distance information. As a matter of fact, the first step consists in assigning to the symmetric difference of the two shapes a distance information. In a second step, we make a dilation or an erosion of the difference according to this distance information. However the propagation speed cannot be the same in all directions: the longer the contour of a has to progress toward the contour of b, the faster has to be the propagation speed. We have thus to compute the different 4-connected components of $a \Delta b$ (see figure 5) and to set up the propagation speed in each component proportional to the largest distance information in this component (see details below).

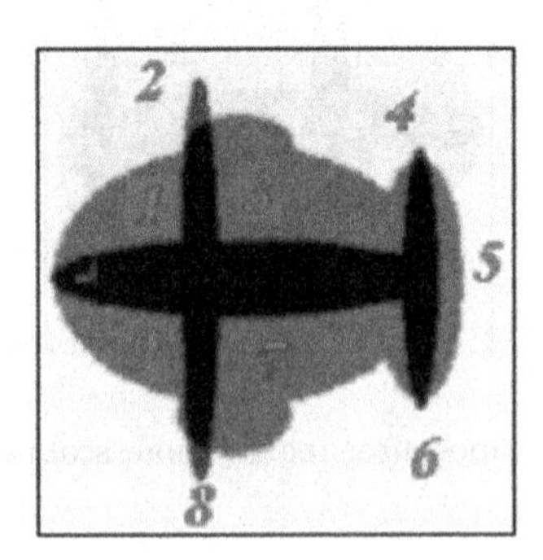

Fig. 5. Connected components labelling

Geodesic Propagation. Using the distance cards described in subsection 3.1, we construct two kinds of geodesic waves: the first one is initialized by the intersection of the two shapes and propagates in the difference of the two shapes toward the background. The second one is initialized by the background and propagates in the difference toward the intersection. We denote by d_1 the 3-4 distance from

$a \cap b$ to $a \Delta b$, and d_2 the 3-4 distance from the complement of $a \cup b$ to $a \Delta b$. The figure 6 shows the two geodesic waves in our example.

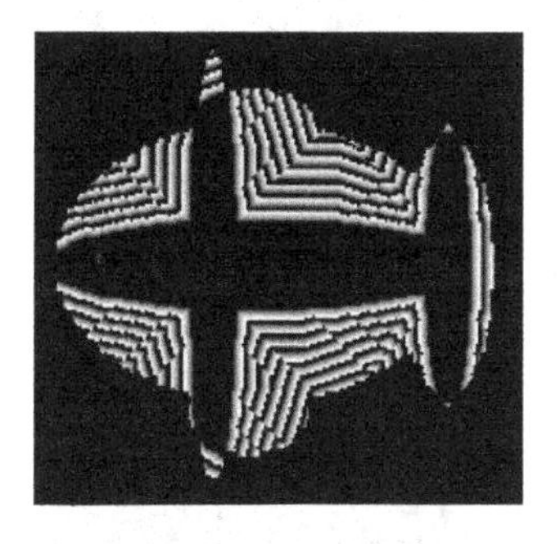
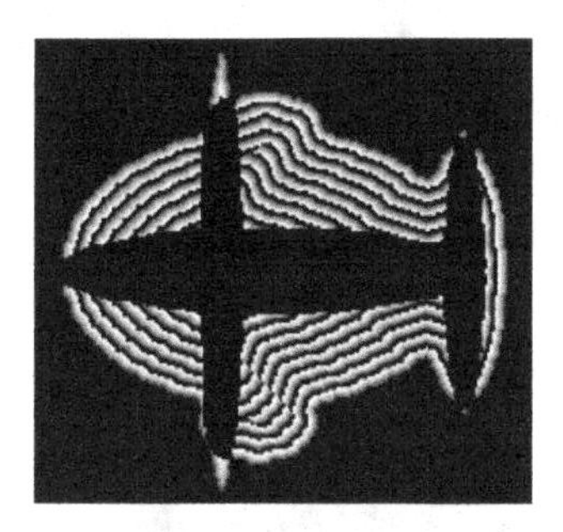

Fig. 6. The two geodesic waves: d1 (left) and d2(right)

Consequently, to each pixel of the difference we associate two distances d_1 and d_2 that are the basis of the morphing.

Erosion and Dilation. Let us denote by β a parameter, varying between 0 to 1, that gives the degree of progression of the morphing. The value $\beta = 0$ gives shape a, $\beta = 1$ gives shape b and any other value between 0 and 1 gives an intermediate shape between a and b. To obtain exactly the median image, the parameter β must be equal to $\frac{1}{2}$.

The erosion and dilation process is applied to each connected component of the difference. Some parts will grow and other will thin. In order to make proportional growing and thinning on all the connected components, we compute for the connected component labelled i its maximal distance d_{1i} and d_{2i}.

Then the decision of erosion or dilation is taken as follows. We start with a shape c initialized with pixels of shape a. A pixel of $a \backslash b$ in the ith component of $a \Delta b$ is removed from c if it is labeled with distances d_1 and d_2 verifying:

$$d_1 \geq d_{1i}(1 - \beta) \text{ or } d_2 \leq d_{2i}\beta$$

A pixel of $b \backslash a$ in the ith component of $a \Delta b$ is added to c if it is labeled with distances d_1 and d_2 verifying:

$$d_1 \leq d_{1i}\beta \text{ and } d_2 \geq d_{2i}(1 - \beta)$$

Figure 7 shows the different intermediate images for our example corresponding to β equals to $\frac{1}{2}$, $\frac{1}{3}$, $\frac{1}{4}$ and $\frac{1}{5}$.

The dilation and erosion step is the final step to generate the average shape. It can also be used to generate arbitrary intermediate shapes between any two shapes.

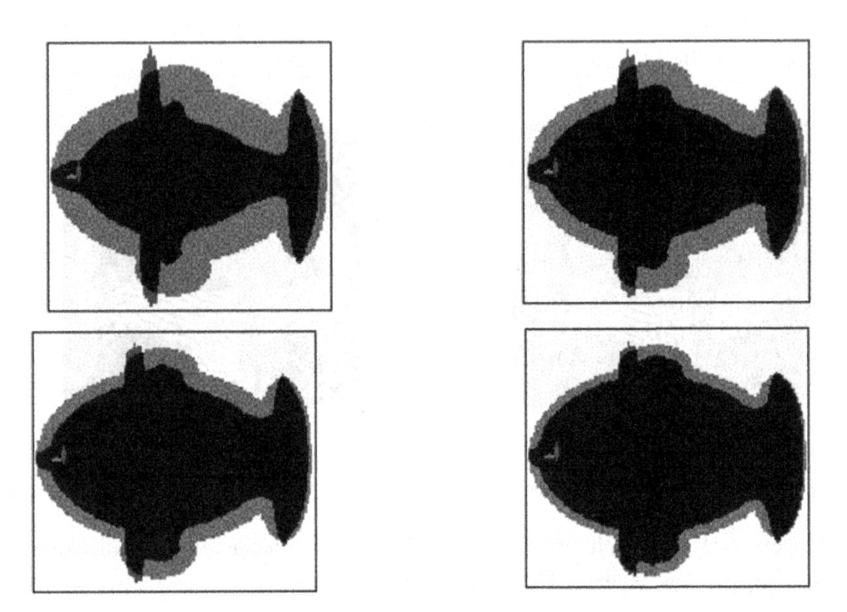

Fig. 7. Examples of intermediate images with $\beta = \frac{1}{2}$, $\beta = \frac{1}{3}$, $\beta = \frac{1}{4}$ and $\beta = \frac{1}{5}$

Table 1. Pseudo-code of the morphing algorithm

```
a=Data_file(1)
b=Data_File(2)
a and b are the two input objects
(a,b)=Registration(a,b)
Component_labeling(aΔb)
/* each 4-component of aΔb has its own label i */
Geodesic_distance_transform(aΔb, background)
Geodesic_distance_transform(aΔb, a ∩ b)
β=morphing_parameter
c = a
For each connected component i of (aΔb)
  d1i=compute_maximum(d1 in i)
  d2i=compute_maximum(d2 in i)
  for each pixel p in component i
    if(p ∈ a\b) /* Erosion */
      if(d1[p] ≥ d1i(1 − β) or d2[p] ≤ d2iβ)
        c = c\{p}
    else /* (p ∈ b\a): Dilation */
      if(d1[p] ≤ d1iβ and d2[p] ≥ d2i(1 − β))
        c = c ∪ {p}
return c
```

4.3 Pseudo-Code and Complexity

Table 1 is a pseudo-code Summarizing all the steps needed to deform one discrete object to another one: the complexity of our method is $O(n)$ where n is the number of pixels of the images (assumed to be of the same order of magnitude in the input and the output images). The distance information we manipulate cannot be computed with the traditional distance transform algorithms [Blu67] that are not adapted for geodesic distance transformations since we work on non-convex domains. We use an adapted version of Dijkstra Graphsearch algorithm using a priority queue indexed by the distance. Since the maximal possible distance value is at worst proportional to the size of the input, we can use a bucket data structure [CLRS01] with all points at distance i stored in a chained list of index i.

Thus, each new pixel will be stored in a chained list of index i witch is its distance value. As we can update pixels with other distance values due to new waves of geodesic propagation, it is logical to follow a rule:

- If the new distance value is bigger than the original one: we do nothing
- If it is smaller, we update the pixel distance value by deleting it from his original list and adding it in the appropriate chained list, indexed with the new distance value, at the last position.
- If for a given distance value there are no elements, we move to the next chained list indexed with the next distance value.

So, at each step we are not obliged to sort values (find the pixels with smallest values), we can obtain them by selecting elements from first chained list associated to the smallest distance value.

This structure allows us to insert a new element and to select the smallest element in constant time. A given pixel can be inserted in the bucket a constant number of times only, leading to a total coast proportional to the number of pixels of the images. In the following section we give some examples that illustrates the flexibility of our approach.

5 Results

In order to validate our method, we have made tests between discrete objects of different natures. This indicates that the approach can be very adaptable to various shapes. In figure 8 we try to find an average shape between a circular object and a fish. Another example is in figure 9 and computes an average shape between a rectangular object and a fish.

It can be noted that it is equivalent to transform the object a into object b using $\beta = \beta_0$ than to transform object b into object a using $\beta = 1 - \beta_0$. The extension to 3D is straightforward and does not need any additional operation. We use the 3x3x3 neighborhood and the 3-4-5 chamfer distance.

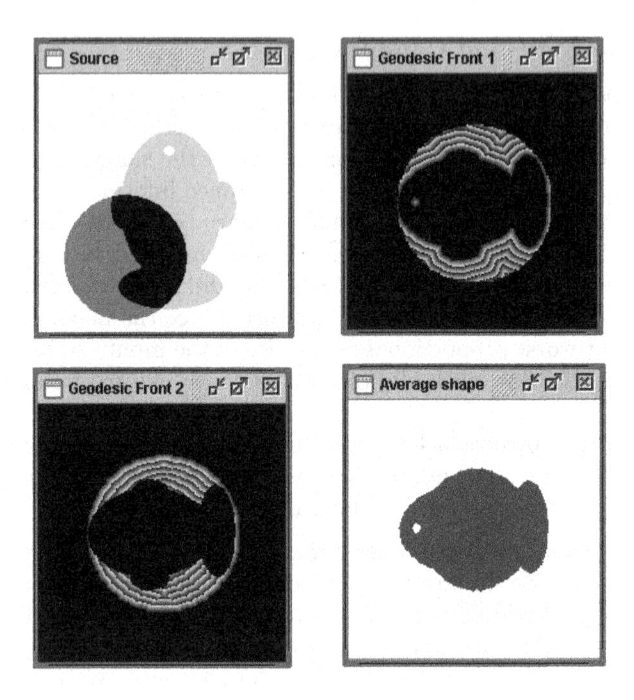

Fig. 8. Deformation of various shapes with β=1/2

6 Conclusion and Future Works

We have presented a linear algorithm for computing an intermediate shape between two arbitrary input binary shapes. A progression parameter β ranging from 0 to 1 allows to control the influence of each input shape. A generalization of this problem to n shapes could be achieved by recursively computing an average between the first $(n-1)$ ones then computing an intermediate between this new shape and the nth one, using $\beta = \frac{1}{n}$. It would be interesting to study the influence, on the final shape, of the order in which these shapes are processed.

In this paper we have used the chamfer distance as a good integer approximation of Euclidian distance. An improvement in precision could be the use of the Euclidian distance itself which can also be computed in linear time as shown in [CMT04] at least for 2D domains. To our knowledge, there exist no linear algorithm for computing Euclidian geodesic distance transformation in 3D. The chamfer distance is thus probably a good choice in 3D if performance is important.

In all the examples we have shown in this paper, the topology of intermediate shapes is preserved if the two input objects have the same topology. In some cases, however, if the objects are too different it can happen that the topology of intermediate objects is not preserved (holes or disconnected components might temporarily appear). If the two objects have the same topology and their intersection after the registration step also has the same topology, we could expect to

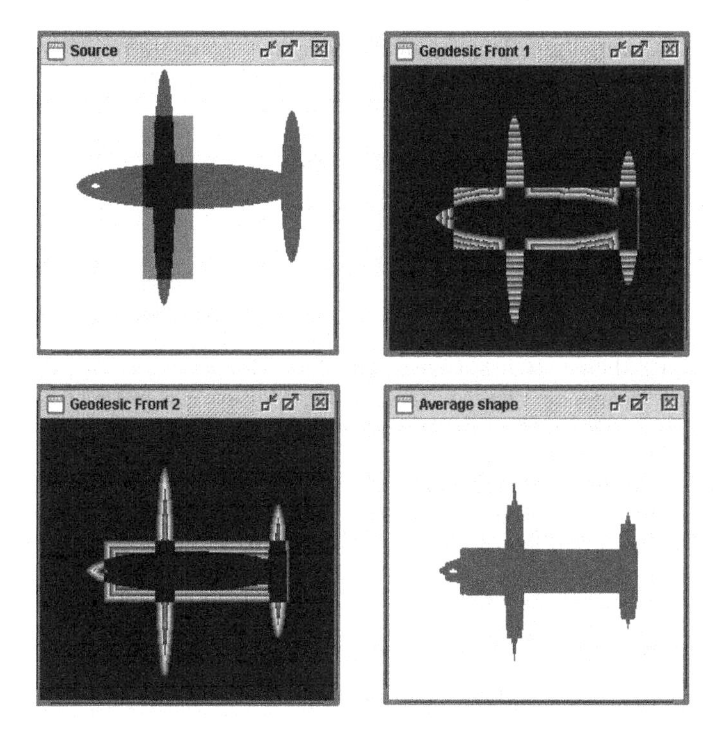

Fig. 9. Deformation of various shapes with β=1/2

find a transformation whose intermediate shapes also have the same properties. A heuristic for reaching this goal could be to use thinning or thickening operators, allowing to ensure that the removal or adding of a point [Ber96] does not change the topology of intermediate objects. These topics should be the subject of further studies.

References

[Ber96] G. Bertrand. A Boolean Characterization of Three-dimensional Simple Points. *Pattern Recognition Letters*, 17:115–124, 1996.

[Blu67] Blum. A Transformation for Extracting new Descriptors of Shape. *In models for perception of speech and visual form*, pages 362–380, 1967.

[Bor88] G. Borgefors. Hierarchical Chamfer Matching: a Parametric Edge Matching Algorithm . *IEEE transactions on pattern analysis and machine intelligence*, 1988.

[BTSG00] R. Blanding, G. Turkiyyah, D. Storti, and M. Ganter. Skeleton-based Three Dimensional Geometric Morphing. *Computational Geometry*, 15:129–148, 2000.

[CLRS01] Thomas H. Cormen, Charles H. Leiserson, Ronald L. Rivest, and C. Stein. *Inroduction à l'Algorithmique*. xDunod, Paris, 2001.

[CMT04] D. Coeurjolly, S. Miguet, and L. Tougne. 2D and 3D Visibility in Discrete Geometry: An Application to Discrete Geodesic Paths. *Pattern Recognition Letters*, 25:561–570, 2004.

[FLD02] M. Fleute, S. Lavallé, and L. Desbat. Integrated approach for matching statistical shape models with intra-operative 2D and 3D data. In *MICCAI 2002*, volume part II, pages 364–372, Springer 2002.

[Iwa02] M. Iwanowski. Image Morphing Based on Morphological Interpolation Combined with Linear Filtering, 2002. Available at: `http://wscg.zcu.cz/wscg2002/Papers_2002/A23.pdf`.

[LB98] A. Liu and E. Bullitt. 3D/2D Registration via Skeletal near Projective Invariance in Tubular Objectives. In *MICCAI*, pages 952–963, 1998.

[MMF98] C.R. Maurer, R.J. Maciunas, and J.M. Fitzpatrick. Registration of Head CT Images to Physical Space Using Multiple Geometrical Features. In *Proc. SPIE Medical Imaging 98*, volume 3338, pages 72–80, San diego CA, February 1998.

[RMP+98] A. Roche, G. Malandain, X. Pennec, P. Cathier, and N. Ayache. Multimodal Image Registration by Maximisation of the Correlation Ratio. Technical Report 3378, INRIA, 1998. Available at : `http://www.inria.fr/rrrt/rr-3378.html`.

[RP68] A. Rosenfeld and J.L. Pfaltz. Distance functions on digital pictures. *Pattern Recognition*, 1:33–61, 1968.

[Wol90] G. Wolberg. *Digital Image Warping*. IEEE Computer Society Press, Los Alamitos, CA, 1990.

[WV96] A. Wells and P. Viola. Multimodal Volume Registration by Maximization of Mutual Information. *Medical Image Analysis*, 1:32–52, 1996.

A Comparison of Property Estimators in Stereology and Digital Geometry

Yuman Huang and Reinhard Klette

CITR, University of Auckland, Tamaki Campus, Building 731,
Auckland, New Zealand

Abstract. We consider selected geometric properties of 2D or 3D sets, given in form of binary digital pictures, and discuss their estimation. The properties examined are perimeter and area in 2D, and surface area and volume in 3D. We evaluate common estimators in stereology and digital geometry according to their multiprobe or multigrid convergence properties, and precision and efficiency of estimations.

1 Introduction

Both stereology as well as digital geometry are mainly oriented towards property estimations. Stereologists estimate geometric properties based on stochastic geometry and probability theory [8]. Key intentions are to ensure isotropic, uniform and random (IUR) object-probe interactions to ensure the unbias of estimations. The statistical behavior of property estimators is also a subject in digital geometry. But it seems that issues of algorithmic efficiency and multigrid convergence became more dominant in digital geometry.

Both disciplines attempt to solve the same problem, and sometimes they follow the same principles, and in other cases they apply totally different methods. In this paper, a few property estimators of stereology and digital geometry are comparatively evaluated, especially according to their multiprobe or multigrid convergence behavior, precision and efficiency of estimations.

From the theoretical point of view, one opportunity for comparison is to study how testing probe and resolution affect the accuracy of estimations. We define multiprobe convergence in stereology analogously to multigrid convergence in digital geometry. Both definitions can be generalized to cover not only the estimation of a single property such as length, but also of arbitrary geometric properties, including area in 2D, and surface area and volume in 3D. Measurements follow stereological formulas, which connect the measurements obtained using different probes with the sought properties. We define multiprobe convergence in stereology analogously to multigrid convergence in digital geometry.

Definition 1. *Let Q be the object of interest for estimation, and assume we have a way of obtaining discrete testing probes about Q. Consider an estimation method E which takes testing probes as input and estimates a geometric property $X(Q)$ of Q. Let $X_E(T_n)$ be the value estimated by E with input T_n, where T_n*

R. Klette and J. Žunić (Eds.): IWCIA 2004, LNCS 3322, pp. 421–431, 2004.

contains exactly $n > 0$ *testing probes. The method* E *is said to be* multiprobe convergent *iff it converges as the number* n *of testing probes tends to infinity (i.e.,* $\lim_{n\to\infty} X_E(T_n) = c,\ c \in \mathbb{R}^2$*), and it converges to the true value (i.e.,* $c = X(Q)$*).*

Testing probes are measured at, along or within points, lines, planes, disectors and so forth. The geometric property $X(Q)$ might be the length (e.g., perimeter), area, surface area, or volume, and we write L, A, S, or V for these, respectively. (Note that the true value $X(Q)$ is typically unknown in applications.) The method E can be defined by one of the stereological formulas.

The grid resolution h of a digital picture is an integer specifying the number of grid points per unit. There are different models of digitizing sets Q for analyzing them based on digital pictures. Grid squares in 2D or grid cubes in 3D have grid points as their centers. The Gauss digitization $G(Q)$ is the union of the grid squares (or grid cubes) whose center points are in Q. The study of multigrid convergence is a common approach in digital geometry [5]. However, we recall the definition:

Definition 2. *Let* Q *be the object of interest for estimation, and assume we have a way of digitizing* Q *into a 2D or 3D picture* P_h *using grid resolution* h*. Consider an estimation method* E *which takes digital pictures as input and estimates a geometric property* $X(Q)$ *of* Q*. Let* $X_E(P_h)$ *be the value estimated by* E *with input* P_h*. The method* E *is said to be* multigrid convergent *iff it converges as the number* h *of grid resolution tends to infinity (i.e.,* $\lim_{n\to\infty} X_E(P_h) = c,\ c \in \mathbb{R}^2$*), and it converges to the true value (i.e.,* $c = X(Q)$*).*

For simplicity we assume that our 2D digital pictures P_h are of size $h \times h$, digitizing always the same unit square of the real plane. Analogously, we have $h \times h \times h$ in the 3D case. An estimation method in digital geometry is typically defined by calculating approximative geometric elements (e.g. straight segments, surface patches, surface normals) which can be used for subsequent property calculations.

2 Estimators

Perimeter and surface area estimators in stereology and in digital geometry are totally different by applied method. In stereology, for example a line testing probe (a set of parallel straight lines in 2D or 3D space) is used, and the perimeter or surface area of an object is approximated by applying stereological formulas with the count of object-line intersections.

In digital geometry, estimators for measuring perimeter or surface area can be classified as being either local or global (see, e.g., [1, 4, 5]). [10] suggested using a fuzzy approach for perimeter and area estimations in gray level pictures. The chosen perimeter estimator for our evaluation (for binary pictures) is global and based on border approximations by subsequent digital straight segments (DSS) of maximum length.

The surface area can be estimated for "reasonable small" 3D objects by using time-efficient local polyhedrizations such as weighted local configurations [7]. For higher picture resolutions it is recommended to apply multigrid convergent techniques. For example, [5] illustrates that local polyhedrizations such as based on marching cubes are not multigrid convergent. Global polyhedrization techniques (e.g., using digital planar segments) or surface-normal based methods can be applied to ensure multigrid convergence. Because of the space constraint, none of these surface area estimators are covered in this paper. The 2D area and volume estimators of stereology and those of digital geometry follow the same point-count principle. In stereology, a point probe (a set of systematic or random points) is placed in 2D or 3D space, the points within the object are counted, and the stereology formulas are applied to estimate the area or volume of an object of interest.

In digital geometry, a common estimator of area is pixel counting, whereas that of volume is voxel counting. These two estimators are also a special case of point counting in stereology where a set of pixel or voxel centers is used as the point probe. Because the resolution directly relates to the number of pixels or voxels in the digital picture, it will consequently affect the precision of estimations if the stereological formulas are applied.

2.1 Perimeter and Surface Area

Line intersection count method (LICM) is a common stereology estimator for 2D perimeter and 3D surface area. This estimator involves four steps: generate a line probe, find object border in a digital picture, count object-probe intersections, and apply stereological formulas to obtain an estimation of the property.

The line probes in 2D or 3D may be a set of straight or circular lines (see, e.g., [3, 8, 9]), or made up of cycloid arcs. We use a set of straight lines as the testing probe of both perimeter and surface area estimations.

There are many possible ways of defining a digital line (see, e.g., [5]). In this paper, a 2D or 3D straight line in digital pictures is represented as a sequence of pixels or voxels whose coordinates are the closest integer values to the real ones. There exist different options for defining borders of objects in 2D or 3D pictures. In this paper, in 2D the 8-border of Q is used to observe intersections between Q and the line probe, whereas in 3D the 26-border of Q is used (for definitions of 8-border or 26-border see, for example, [5]). Isotropic, uniform, and random (IUR) object-probe intersections are required for obtaining unbiased estimation results. Note that any digital line (which is an 8-curve) intersects an 8-border (which is a 4-curve) if it "passes through".

Definition 3. *A pixel (voxel) visited along a digital line is an* intersection *iff it is an 8-border pixel (26-border voxel) of the object Q and successor (along the line) is a non-border pixel (voxel).*

Since the objects of interest are general (not specified as a certain type), they may not be IUR in 2D or 3D space. To ensure IUR object-probe intersections, either the object or the probe must be IUR, or the combination of both must

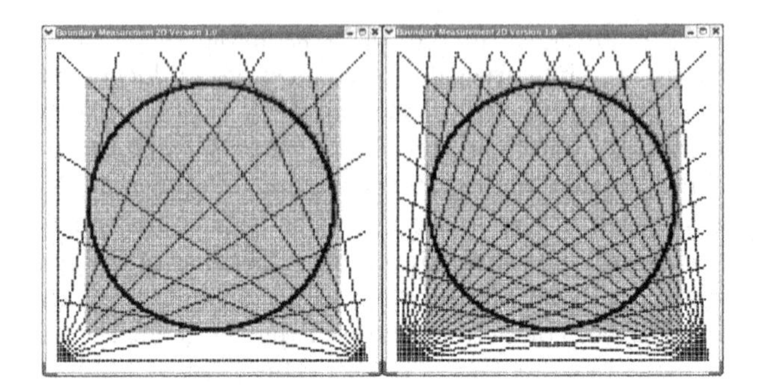

Fig. 1. Screen shots of pivot lines of the LICM estimator using line probes of 16 (left) and 32 (right) directions

be isotropic [8]. In the experiments we attempt to generate an IUR straight line probe for the stereology estimator (LICM) to produce unbiased results for perimeter and surface area measurements.

Perimeter. The perimeter estimators examined are the stereology estimator (LICM) and the digital geometry estimator based on maximum length digital straight segments (DSS).

The stereology estimator corresponds to the stereological formula

$$L_A = \frac{\pi}{2} \cdot P_L = \frac{\pi}{2} \cdot \frac{P}{L_T}$$

which is basically the calculation of perimeter density (length per unit area) of a 2D object. P is the number of intersections between line probe and (border of the) object Q. L_T is the total length of the testing line probe, and P_L is the point count density (intersections per unit length). As a corollary of this, the perimeter of an object can be estimated by multiplying L_A by the total area A_T that the line probe occupies:

$$\widehat{L} = L_A \cdot A_L = \frac{\pi}{2} \cdot \frac{P}{L_T} \cdot A_T$$

At the beginning of our experiments we consider how the estimation precision is influenced by an increase in the number of directions n of line probes. The direction of line probes are selected by dividing 180° equally by n. For instance, if there are 4 directions required, then lines of slopes 0°, 45°, 90°, and 135° are generated. Figure 1 illustrates the lines for 16 or 32 directions. Note that only one "pivot line" per direction is shown in this figure.

To avoid errors or a bias caused by direction or position of line probes, every pivot line of one direction is shifted along the x and y axes by just one pixel. Assume that the pivot line which is incident with $(0,0)$ (i.e., the left lower corner of a picture) intersects the frontier of the unit square $[0,1] \times [0,1]$ again at point p, with an Euclidean distance L between $(0,0)$ and p. The total length of a line probe in direction 0°, 45° or 90° is hL, and equal to

$$\frac{3h - h\tan\alpha}{2}L$$

where α is the smaller angle between 0° and 45° defined by the pivot line in our unit square.

However, a stereological bias [8] can not be avoided in this case due to the preselection of start position and direction of line probes for perimeter estimation of non-IUR objects. Therefore, we also include an estimator which uses lines at random directions, generated using a system function *rand()* (which is supposed to generate uniformly distributed numbers in a given interval). Although a random number generator is used, the generated line probes are not necessarily isotropic in 2D space. (An improved IUR direction generator is left for future research.) For the digital geometry estimator DSS, we start at the clockwise lower-leftmost object pixel and segment a path of pixels into subsequent DSSs of maximum length. Debled-Rennesson and Reveillès suggested an algorithm in [2] for 8-curves, and earlier Kovalevsky suggested one in [6] for 4-curves. In our evaluation, both methods have been used, but we only report on the use of the second algorithm (as implemented for the experiments reported in [4]) in this paper.

The DSS estimator is multigrid convergent, whereas the multiprobe convergent behavior of the LICM estimator depends on the used line probe. The digital geometry estimator DSS is time-efficient because it traces only borders of objects, and used one of the linear on-line DSS algorithms.

The time-efficiency of the stereology estimator LICM depends on the number of line pixels involved, since it checks every pixel in a line probe to see whether there is an intersection. In both implementations of LICM (i.e., n directions by equally dividing 180°, and random directions generated by the system function *rand()*), every pivot line into one direction is translated along x- and y-axes at pixel distance (in horizontal or vertical direction we have a total of h lines; in any other direction we have a total of $2h - 1$ lines), which results into multiple tests (intersection?) at all pixels just by considering them along different lines. In case of pictures of "simple objects" we can improve the efficiency by checking along borders only instead of along all lines. However, normally we can not assume that for pictures in applications.

Surface Area. We tested speed and multiprobe convergence of the stereology estimator LICM for surface area measurements, using the stereological formula

$$S_V = 2 \cdot I_L = 2 \cdot \frac{P_L}{L_T}$$

where I_L is the density of intersections of objects with the line probe, and this is equal to the result of the line intersection count P_L divided by the total length L_T of the line probe.

Consequently, the surface area of Q can be estimated by multiplying its surface density (obtained from the previous relationship) by the total volume of the testing space V_T (i.e., h^3, which is the occupied volume of a 3D picture), $\widehat{S} = S_V \cdot V_T$.

Similar to the stereology estimator of perimeter measurement, a way of creating an IUR straight line probe in 3D is required for unbiased surface area estimations. The LICM estimator may not be multiprobe convergent because of the used line probes. The efficiency of the estimator depends on the total number of line voxels.

2.2 Area and Volume

The stereology estimator for both 2D area (3D volume) measurements counts 2D (3D) points which are within the object of interest. When the point probes used are the centers of *all* pixels (voxels) of a given picture, then this estimator coincides with the method used in digital geometry.

Basic stereological formulas (see, e.g., [3]) such as

$$\widehat{A} = A_P \cdot P = \frac{A_T}{P_T} \cdot P = \Delta x \cdot \Delta y \cdot P$$

are applied for 2D area estimation. The area of an object A can be estimated by multiplying the number of incident points P by the area A_P per point. In other words, the area occupied by all incident points approximates the area of the measured object. The area A_P per point is the total area A_T of the picture (i.e., width Δx times height Δy) divided by the total number P_T of points of probe T.

The principle of Bonaventura Cavalieri (1598-1647) is suggested for estimating the volume of an object Q in 3D space in stereology books such as [3, 8, 9]. [8] combines the Cavalieri principle with the point count method. A series of parallel 2D planes is used as test probe, where a set of points is placed into each plane to obtain the intersected area of the plane with the object. If there are m planes used in the process, the volume of the object V can be estimated by multiplying the distance θ between the planes by the sum of the intersected areas $A_1, A_2, A_3, \ldots, A_m$ of all planes, $\widehat{V} = \theta \cdot (A_1 + A_2 + A_3 + \cdots A_m)$.

Russ and Dehoff use a set of 3D points as the test probe in [9] to estimate the volume fraction V_V from calculating a point fraction P_P, the ratio of incident points P (with the object) to the population of points P_T, $V_V = P_P = \frac{P}{P_T}$. As a corollary of this, the volume V of the object is equal to the product of the volume fraction and the total volume occupied by the testing probe V_T (i.e., h^3 in an $h \times h \times h$ picture), which can be calculated by multiplying the number of incident points by the volume per point V_P,

$$\widehat{V} = V_V \cdot V_T = \frac{P}{P_T} \cdot h^3 = P \cdot V_P$$

The results of both methods are identical when all the 2D planes are coplanar at equal distances, the set of points chosen in every plane is uniform, and the interspacing between these 2D points equals to the distance between planes.

For area and volume estimations, since the estimators in both fields follow the same point count principles, the choice of point probes will definitely influence the performance of the algorithm. If a point probe is randomly picked from

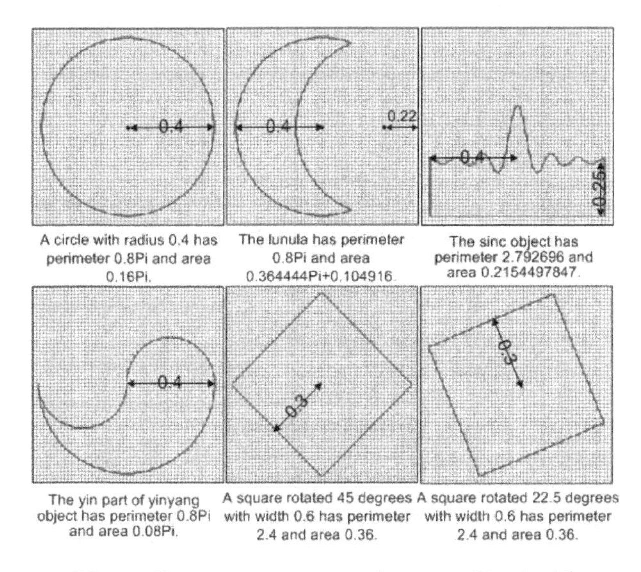

Fig. 2. Six pictures suggested as test data in [4]

regular 2D or 3D grid points, and the chosen number of points is less than the total number of pixels or voxels in the picture, then the method is (trivially) more time-efficient than the digital geometry approach.

Area and volume estimators are "very precise", suggesting a fast multiprobe convergence (see experiments in Section 3); they have been widely used in research and commercial fields. The digital geometry estimators (i.e., considering all pixels or voxels) are known [5] to be multigrid convergent (e.g., for particular types of convex sets).

3 Evaluation

We tested 2D perimeter and area estimators by using six objects shown in Figure 2. For the area of the lunula we used the formula $b \cdot r - a(r-h)$ for the area of the "removed" segment of the disk, with arc length $b = 2\pi r \cdot \alpha/360$ (using integral part $\alpha = 139$ of the estimate $\alpha = 139.0253698\ldots$), segment height $h = 0.26$, and $a = 2 \cdot \sqrt{0.1404}$. In case of 3D objects, we used a cylinder, sphere, cube, and an ellipsoid for volume estimation, and the first three objects are also used for surface area estimation.

The relative error E_r of experiments is a percentage, it is equal to the absolute value of the estimated value V_e minus the true value V_t, divided by the true value, then multiplied by 100, $E_r = \frac{|V_e - V_t|}{V_t} \cdot 100$.

Perimeter and Surface Area. We compare results of the DSS estimator, the LICM estimator with four preselected directions (which are 0°, 45°, 90°, and 135°), and the LICM estimator with four random directions (called LICM_R; generated using the system function *rand()*), see Figure 3.

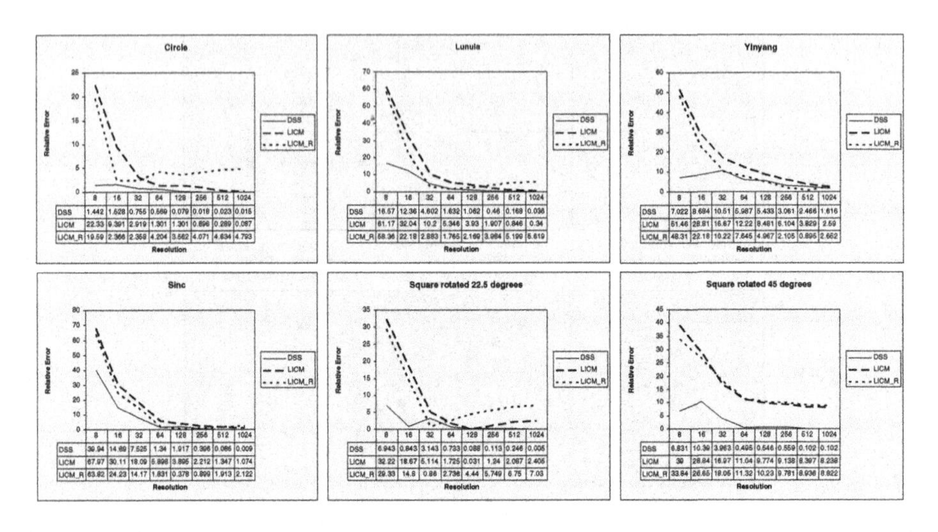

Fig. 3. Comparison of DSS, LICM (4 directions) and LICM_R (LICM with 4 random directions) on six test objects

The obtained results show that the precision of the DSS estimator is the best of these three estimators, whereas the precision of the LICM estimator with four preselected directions is better than that of LICM_R. Multiprobe or multigrid convergence is apparent in most of the diagrams. Obviously, four random numbers are not "able" to define an IUR direction generator. (The figure shows results for directions 151.23°, 70.99°, 140.96° and 143.72°; LICM_R appears to be not multiprobe convergent on the circle, lunula, and square rotated 22.5° for these values.) Figure 4 (left) illustrates the relative errors averaged over all six test objects. The obtained LICM_R errors are slightly increasing between resolution 256 and 1024.

We also tested the multiprobe behavior for an increase in numbers of directions (up to 128 different directions in experiments). Surprisingly, results did not

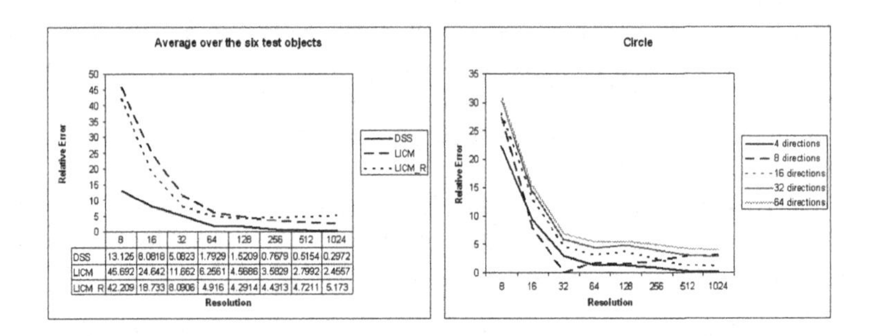

Fig. 4. Left: comparison of average relative errors over all the six test objects, for DSS, LICM and LICM_R. Right: LICM-estimation of the perimeter for the disk using different numbers of directions

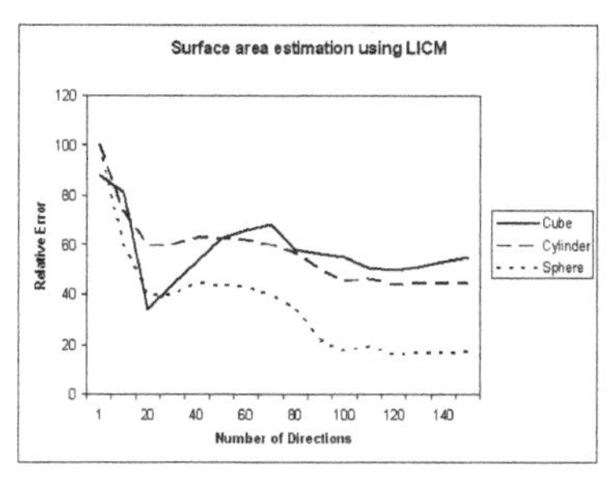

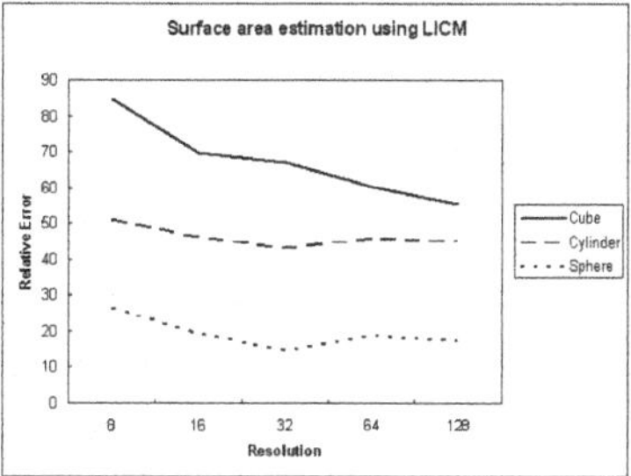

Fig. 5. LICM_R-estimation of surface area for the cube, cylinder, and sphere. Left: using a 128×128 picture with different number of directions. Right: using 100 lines at different picture resolutions

steadily improve by increasing the number of directions for LICM or LICM_R, and in some cases the error even increased for larger numbers of directions. See Figure 4 (right) for the example of a disk, where the error is smallest in general for just 4 directions! A possible explanation is that more directions increase the number of lines which do not intersect the circle at all.

The surface area of 3D objects is estimated using the LICM_R estimator with digital rays in 3D space starting at random positions, and into random directions (generated by the system function *rand()*). (We tested up to a picture resolution of 128. Rays are only traced within the space of the picture.) Now, in 3D space we generated every ray individually (i.e., not shifting pivot lines or rays anymore as in 2D). So, the number of rays is now reduced to be equal to the number of directions!

Figure 5 (left) shows results for a constant resolution ($h = 128$) and increases in numbers of directions (i.e., numbers of rays). Figure 5 (right) shows results for a constant number of directions (100), and an increase in picture resolution. The results indicate relatively large errors. (Obviously, this is certainly related to the smaller number of rays compared to the number of lines in 2D.) However, it can be seen that the results for the sphere are better than those for the cube and cylinder in both Figures.

Area and Volume. The results shown on the left of Figure 6 are using the pixel count estimator which checks all 2D grid points of the picture. They are very precise on all six test objects, always with less than 0.1 percent error from the true area when the picture resolution is 1024.

Trends for circle and yinyang are similar due to the yinyang shape being formed by circular curves. The estimation for the square rotated 22.5° converges fastest, with an error of a bit more than 0.0001 percent at resolution 1024.

We also estimated volumes of 3D objects using the voxel count method with regular (grid-point) point probes, which is equivalent to applying the Cavalieri principle for a very special case (see discussion above).

The results on the right of Figure 6 indicate reasonably small errors, which are all below 0.1 percent at picture resolution 1024. They all reflect multigrid

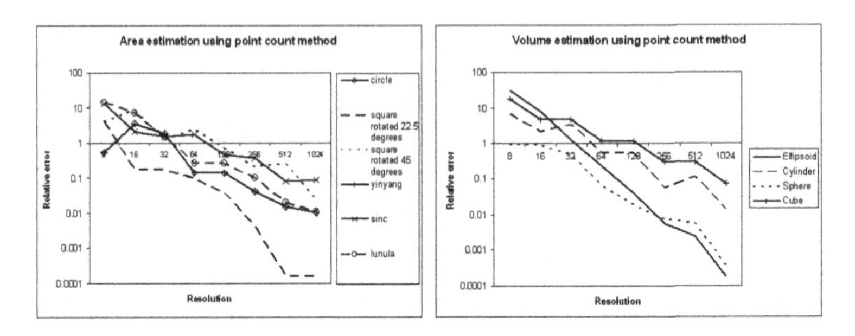

Fig. 6. Left: area estimation for the six test objects using the point count method with regular (grid-point) point probes. Right: volume estimation for sphere, cube, ellipsoid, and cylinder using the Cavalieri principle in a special form (probes at all grid points in the 3D picture)

and multiprobe convergent behavior (both are theoretically known) of the voxel count estimator on ellipsoid, cylinder, sphere and cube.

4 Conclusions

The point count estimators in stereology and digital geometry are different by motivation. Because a point is zero-dimensional, estimations using randomly chosen 2D or 3D points, or all pixels or voxels of the picture are unbiased and precise. Because estimators which use all pixel or voxel centers are very precise, there is no need to apply a random point generator for area and volume estimations.

If using the LICM for estimating perimeter or surface area, the IUR object-probe interaction must be guaranteed in order to make an unbiased observation of the object structure. Position or direction of the line probe cannot be preselected in this case to avoid bias. However, in our experiments a bias could not be totally removed even if lines of random directions are used (using a uniform number generator). It might be worth to spend more efforts on building an IUR line generator for the unbiased estimation of perimeter or surface area. Without such an ideal IUR line generator, the DSS-estimator appears to be the more time-efficient and faster converging method for perimeter estimations instead of the stereology estimator LICM.

The pixel and voxel count estimators for 2D area and volume are theoretically known to be multiprobe or multigrid convergent. Results for the DSS-estimator for length also corresponded to its known multigrid convergence, whereas the multiprobe convergence of the LICM estimator for length depends on the chosen line probes.

In practice, when the segmented objects are “complicated and irregular by shape”, such as biological tissues and material microstructure, the stereology estimator LICM may be more efficient than the digital geometry estimator DSS as we do not need to trace all borders in the picture. In future experiments, more shapes generated randomly in size and position should be used and the average over all results of shapes should be considered.

References

1. D. Coeurjolly and R. Klette. *A comparative evaluation of length estimators.* In *Proc. ICPR*, **IV**: 330–334, 2002.
2. I. Debled-Rennesson and J. Reveillès. A linear algorithm for segmentation of digital curves. *Int. J. Pattern Recognition and Artificial Intelligence*, **9**: 635-662, 1995.
3. C.V. Howard and M.G. Reed. *Unbiased Stereology: Three-Dimensional Measurement in Microscopy.* BIOS Scientific Publishers, Oxford, 1998.
4. R. Klette, V. Kovalevsky, and B. Yip. On the length estimation of digital curves. Vision Geometry VIII, in *Proc. of SPIE*: 117-128, July 1999.
5. R. Klette and A. Rosenfeld. *Digital Geometry: Geometric Methods for Digital Picture Analysis.* Morgan Kaufmann, San Francisco, 2004.
6. V.A. Kovalevsky. New definition and fast recognition of digital straight segments and arcs. In *Proc. 10th Intl. Conf. on Pattern Recognition*, pages 31-34, 1990.
7. J. Lindblad. Surface area estimation of digitized planes using weighted local configurations. *DGCI 2003*, LNCS,**2886**: 348-357, 2003.
8. P.R. Mouton. *Principles and Practices of Unbiased Stereology: An Introduction for Bioscientists.* The Johns Hopkins University Press, Baltimore, 2002.
9. J.C. Russ and R.T. Dehoff. *Practical Stereology*, 2nd edition. Plenum, New York, 2000.
10. N. Sladoje, I. Nyström and P.K. Saha. Measuring perimeter and area in low resolution images using a fuzzy approach. *SCIA 2003*, LNCS, **2749**: 853-860, 2003.

Thinning by Curvature Flow

Atusihi Imiya[1], Masahiko Saito[2], and Kiwamu Nakamura[2]

[1] IMIT, Chiba University
[2] School of Science and Technology, Chiba University,
Yayoi-cho 1-33, Inage-ku, Chiba 263-8522, Japan
imiya@faculty.chiba-u.jp

Abstract. In this paper, we define digital curvature flow for spatial digital objects. We define the principal normal vectors for points on the digital boundary of a binary spatial object. We apply the discrete curvature flow for the skeletonisation of binary objects in a space, and develop a transform which yields the curve-skeletons of binary objects in a space.

1 Introduction

In this paper, we introduce a new transform for the discrete binary set, which we call "digital curvature flow." Digital curvature flow is discrete curvature flow for the boundary of a digital object, that is, digital curvature flow describes the motion of a boundary which is controlled by the curvature of the boundary of binary digital images in a space. As applications of digital curvature flow, we develop thinning. The thinning algorithm preserves the geometry of junctions.

The skeletons of a binary object in a plane and in a space are a tree-form curve [1–6] and the collection of part of the curved surface [7–11], respectively. These skeletons are called the medial axis and surface, or more generally, the medial set of an object [7–9]. The skeleton of an object is a fundamental geometric feature for image and shape analysis. Therefore, skeletonisation has been studied in the field of pattern recognition and computer vision for a long time.

Classical thinning for planar objects based on the discrete transform usually transforms T- and V-shape junctions to Y-shape junctions. These changes of junctions affect the final form of thinning process yielding unexpected needles and branches which are not in the original forms. Hilditch's thinning is an algorithm that does not produce unexpected needles and branches [12]. The method contains processes based on the configurations in the neighbourhood of each point.

Our method based on curvature flow in a space is defined using configurations of vertices on an isotetic polyhedron derived from the 6-connected boundary. This is an advantage of our algorithm compared to distance-transform-based thinning, since the evolution of shape for thinning in each step is based on the configurations of vertices.

A unified treatment of shape deformation is required for intelligent editing of image contents for multimedia technology. Curvature flow and the diffusion

R. Klette and J. Žunić (Eds.): IWCIA 2004, LNCS 3322, pp. 432–442, 2004.

process on surfaces provide mathematical foundations for a unified treatment of the deformation of surfaces [13]. These deformation operations for boundaries are discussed in the framework of the free boundary problem in the theory of partial differential equations. For the construction of solutions of the partial differential equation representing deformed surfaces, the numerical computation is achieved using an appropriate discretisation scheme. Bruckstein *et al.* derived a discrete version of this problem for planar shapes [14]. Furthermore, Imiya and Eckhardt [15] proposed a spatial version of Bruckstein *et al.*'s discrete treatment of curvature flow. Then, in this paper, we apply the discrete curvature flow for the skeletonisation of binary objects in a space, and develop a transform which yields curve-skeletons of binary objects in a space.

2 Connectivity and Neighbourhood

Setting $\mathbf{R}^2$ and $\mathbf{R}^3$ to be two- and three-dimensional Euclidean spaces, we express vectors in $\mathbf{R}^2$ and $\mathbf{R}^3$ as $\boldsymbol{x} = (x, y)^\top$ and $\boldsymbol{x} = (x, y, z)^\top$, respectively, where $\top$ is the transpose of the vector. Setting $\mathbf{Z}$ to be the set of all integers, the two- and three-dimensional discrete spaces $\mathbf{Z}^2$ and $\mathbf{Z}^3$ are sets of points such that both x and y are integers and all x, y, and z are integers, respectively. For $n = 2, 3$ we define the dual set for the set lattice points $\mathbf{Z}^n$ as

$$\overline{\mathbf{Z}^n} = \{\boldsymbol{x} + \frac{1}{2}\boldsymbol{e} | \boldsymbol{x} \in \mathbf{Z}^n\}, \tag{1}$$

where $\boldsymbol{e} = (1, 1)^\top$ and $\boldsymbol{e} = (1, 1, 1)^\top$ for $n = 2$ and $n = 3$, respectively. We call $\mathbf{Z}^n$ and $\overline{\mathbf{Z}^n}$ the lattice and the dual lattice, respectively.

On $\mathbf{Z}^2$ and in $\mathbf{Z}^3$,

$$\mathbf{N}^4((m, n)^\top) = \{(m \pm 1, n)^\top, (m, n \pm 1)^\top\} \tag{2}$$

and

$$\mathbf{N}^6((k, m, n)^\top) = \{(k \pm 1, m, n)^\top, (k, m \pm 1, n)^\top, (k, m, n \pm 1)^\top\} \tag{3}$$

are the planar 4-neighbourhood of point $(m, n)^\top$ and the spatial 6-neighbourhood of point $(k, m, n)^\top$, respectively. In this paper, we assume the 4-connectivity on $\mathbf{Z}^2$ and the 6-connectivity in $\mathbf{Z}^3$.

For integers k, m, and n, the collection of integer-triplets (k', m', n') which satisfies the equation

$$(k - k')^2 + (m - m')^2 + (n - n')^2 = 1 \tag{4}$$

define points in the 6-neighbourhood of point $(k, m, n)^\top$. If we substitute $k = k'$, $m = m'$, and $n = n'$ to eq. (4), we obtain the equations,

$$(m - m')^2 + (n - n')^2 = 1, \tag{5}$$

$$(k - k')^2 + (n - n')^2 = 1, \tag{6}$$

and

$$(m - m')^2 + (n - n')^2 = 1, \tag{7}$$

respectively. These equations define points in the planar 4-neighbourhoods. Therefore, setting one of x, y, and z to be a fixed integer, we obtain two-dimensional sets of lattice points such that

$$\mathbf{Z}_1^2((k,m,n)^\top) = \{(k,m,n)^\top | \exists k, \forall m, \forall n \in \mathbf{Z}\}, \tag{8}$$

$$\mathbf{Z}_2^2((k,m,n)^\top) = \{(k,m,n)^\top | \forall k, \exists m, \forall n \in \mathbf{Z}\}, \tag{9}$$

and

$$\mathbf{Z}_3^2((k,m,n)^\top) = \{(k,m,n)^\top | \forall k, \forall m, \exists n \in \mathbf{Z}\}. \tag{10}$$

These two dimensional discrete spaces are mutually orthogonal. Denoting

$$\mathbf{N}_1^4((k,m,n)^\top) = \{(k,m\pm 1,n)^\top, (k,m,n\pm 1)^\top\}, \tag{11}$$

$$\mathbf{N}_2^4((k,m,n)^\top) = \{(k\pm 1,m,n)^\top, (k,m,n\pm 1)^\top\}, \tag{12}$$

and

$$\mathbf{N}_3^4((k,m,n)^\top) = \{(k\pm 1,m,n)^\top, (k,m\pm 1,n)^\top\}, \tag{13}$$

the relationship

$$\mathbf{N}^6((k,m,n)^\top) = \mathbf{N}_1^4((k,m,n)^\top) \cup \mathbf{N}_2^4((k,m,n)^\top) \cup \mathbf{N}_3^4((k,m,n)^\top) \tag{14}$$

holds, since $\mathbf{N}_i^4((k,m,n)^\top)$ is the 4-neighbourhood on plane $\mathbf{Z}_i^2((k,m,n)^\top)$ for $i = 1,2,3$ [16]. Equation (14) implies that the 6-neighbourhood is decomposed into three mutually orthogonal 4-neighbourhoods.

A pair of points $(k,m,n)^\top$ and $\boldsymbol{x} \in \mathbf{N}^6((k,m,n)^\top)$ is a unit line segment in $\mathbf{Z}^3$; furthermore, four 6-connected points which form a circle define a unit plane segment in $\mathbf{Z}^3$ with respect to the 6-connectivity. Therefore, we assume that our object is a complex of $2\times 2\times 2$ cubes which share at least one face with each other [18]. Thus, the surface of an object is a collection of unit squares which are parallel to planes $x = 0$, $y = 0$, and $z = 0$.

Assuming the 4-connectivity and 6-connectivity for planar objects and spatial objects, respectively, an object is a complex of discrete simplexes.

Definition 1. *A discrete object is a collection of n-simplexes which are connected by $(n-1)$-simplexes in $\mathbf{Z}^n$ for $n = 2,3$.*

This definition agrees with that of Kovalevsky [19] for planar objects. Since in $\mathbf{R}^n$ there exist k-simplexes for $k = 1,2,\cdots,n$, we define the collection of these simplexes. Since an object is a complex, n-simplexes of an object share $(n-1)$-simplexes. However, some $(n-1)$-simplices are not shared by a pair of simplexes.

Definition 2. *The boundary of an object is a collection of $(n-1)$-simplexes which are not shared by a pair of n-simplexes of an object.*

In Figure 1, we show simplexes, objects and the boundaries of objects in $\mathbf{Z}^n$ for $n = 2, 3$. On the top column of Figure 2, we show a 1-simplex (a), a 2-simplex (b), an object (c), and the boundary of this object on a plane. Furthermore, in the bottom column of the same figure, we show a 2-simplex (a), a 3-simplex (b), an object (c), and the boundary of this object (d) in a space. In the following, we deal with the thinning procedures for 6-connected discrete objects.

Since we are concerned with a binary discrete object, we affix values of 0 and 1 to points in the background and in objects, respectively. On $\mathbf{Z}^2$, three types of point configurations are illustrated in Figure 2, exist in the neighbourhood of a point $\times$ on the boundary. In Figure 2, $\bullet$, and $\circ$, are points on the boundary and in the background, respectively. Setting $f_i \in \{0, 1\}$ to be the value of point $\boldsymbol{x}_i$ such that

$$\begin{array}{ccc} & \boldsymbol{x}_3 = (m, n+1)^\top, & \\ \boldsymbol{x}_5 = (m-1, n)^\top, & \boldsymbol{x}_0 = (m, n)^\top, & \boldsymbol{x}_1 = (m+1, n)^\top, \\ & \boldsymbol{x}_7 = (m, n-1)^\top, & \end{array} \tag{15}$$

the curvature of point $\boldsymbol{x}_0$ is defined by

$$r(\boldsymbol{x}_0) = 1 - \frac{1}{2}\sum_{k \in N} f_k + \frac{1}{4}\sum_{k \in N} f_k f_{k+1} f_{k+2}, \tag{16}$$

where $N = \{1, 3, 5, 7\}$ and $k + 8 = k$. The curvature indices of configurations (a), (b) and (c) are positive, zero, and negative, respectively. Therefore, we call these configurations of (a), (b), and (c) convex, flat, and concave, respectively, and affix the indices $+$, 0, and $-$, respectively.

Using combinations of planar curvature indices on three mutually orthogonal planes which pass through a point $\boldsymbol{x}_0$, we define the curvature index of a point $\boldsymbol{x}_0$ in $\mathbf{Z}^3$ since the 6-neighbourhood is decomposed into three 4-neighbourhoods. Setting α_i to be the planar curvature index on plane $\mathbf{Z}^2_i(\boldsymbol{x}_0)$ for $i = 1, 2, 3$, the curvature index of a point in $\mathbf{Z}^3$ is a triplet of two-dimensional curvature indices $(\alpha_1, \alpha_2, \alpha_3)$ such that $\alpha_i \in \{+, -, 0, \emptyset\}$. Here, if $\alpha_i = \emptyset$, the curvature index of a point on plane $\mathbf{Z}^2_i(\boldsymbol{x}_0)$ is not defined. Therefore, for the boundary points, seven configurations

$$\begin{array}{c} (+,+,+), (+,+,-), (+,0,0), \\ (0,0,\emptyset), \\ (-,-,-), (+,-,-), (-,0,0) \end{array} \tag{17}$$

and their permutations are possible.

Since a triplet of mutually orthogonal planes separates a space into eight parts, we call one eighth of the space an octspace. The number of octspaces determines the configurations of points in a $3 \times 3 \times 3$ cube. There exist nine configurations in the $3 \times 3 \times 3$ neighbourhood of a point on the boundary since these configurations separate $\mathbf{Z}^3$ into two parts which do not share any common points. These configurations have also been the same things introduced by Françon [20] for the analysis of discrete planes. The curvature analysis of discrete surfaces also yields these configurations. In Figure 4, we show 9 configurations on the 6-connected boundary.

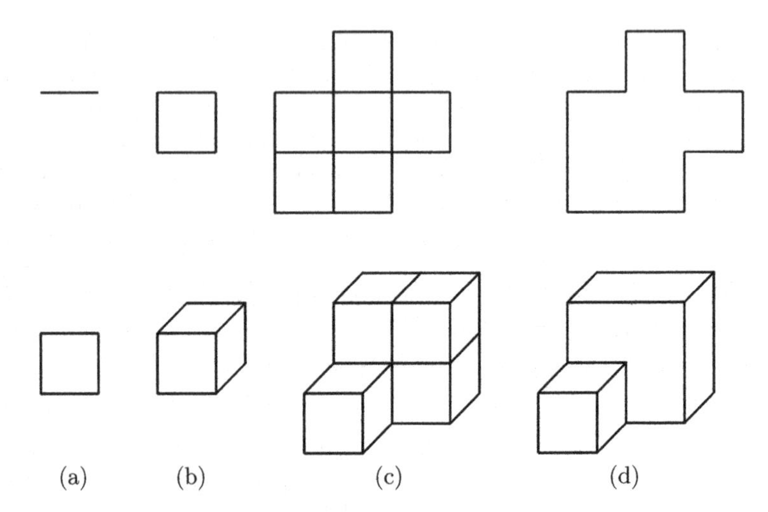

Fig. 1. In the top column, the 1-simplex (a), the 2-simplex (b), an object (c), and the boundary of this object (d) on a plane, and in the bottom column, the 2-simplex (a), the 3-simplex (b), an object (c), and the boundary of this object in a space

3 Thinning by Curvature Flow

Using the two-dimensional curvature code $\gamma(\boldsymbol{x}_j)$ of each point $\boldsymbol{x}_j$, we classify points on slices of boundary $\mathbf{C}$ into types $\mathbf{N}_+$ and $\mathbf{N}_-$, such that

$$\mathbf{C} = \mathbf{N}_+ \bigcup \mathbf{N}, \ \mathbf{N}_+ \bigcap \mathbf{N}_- = \emptyset, \tag{18}$$

where

$$\mathbf{N}_- = \bigcup_j \mathbf{N}_-(j), \ \mathbf{N}_+ = \mathbf{C} \setminus \mathbf{N}_- \tag{19}$$

for

$$\mathbf{N}_-(j) = \{\boldsymbol{x}_\beta | \gamma(\boldsymbol{x}_\beta) = 0, \, j < \beta < j+m, \gamma(\boldsymbol{x}_j) = -, \ \gamma(\boldsymbol{x}_{j+m}) = -\}. \tag{20}$$

Each $\mathbf{N}_-(j)$ is a sequence of flat points whose two endpoints are concave points. Furthermore, $\mathbf{N}_-$ is the union of these sequences on the boundary. In Figure 4, we show the motions of boundary edges of this rule.

Using the outward normal vector for each point on the boundary, we define a transform from point set $\mathbf{C}$ on $\mathbf{Z}^2$ to point set $\overline{\mathbf{C}}$ on $\overline{\mathbf{Z}^2}$. Setting $\mathbf{C} = \mathbf{N}_+$, we derive the rule which moves all edges in the inward direction.

$$\overline{\boldsymbol{x}}_j = \begin{cases} \boldsymbol{x}_j - \boldsymbol{n}_j, & \text{if } \gamma(\boldsymbol{x}_j) \neq 0, \\ \boldsymbol{x}_j - \boldsymbol{n}_j \pm \frac{1}{2}\boldsymbol{n}_j, & \text{if } \gamma(\boldsymbol{x}_j) = 0, \ \boldsymbol{x}_j \in \mathbf{N}_+. \end{cases} \tag{21}$$

Although this transformation transforms a point on a corner to a point, a flat point is transformed to a pair of points. Therefore, for a corner, transformation

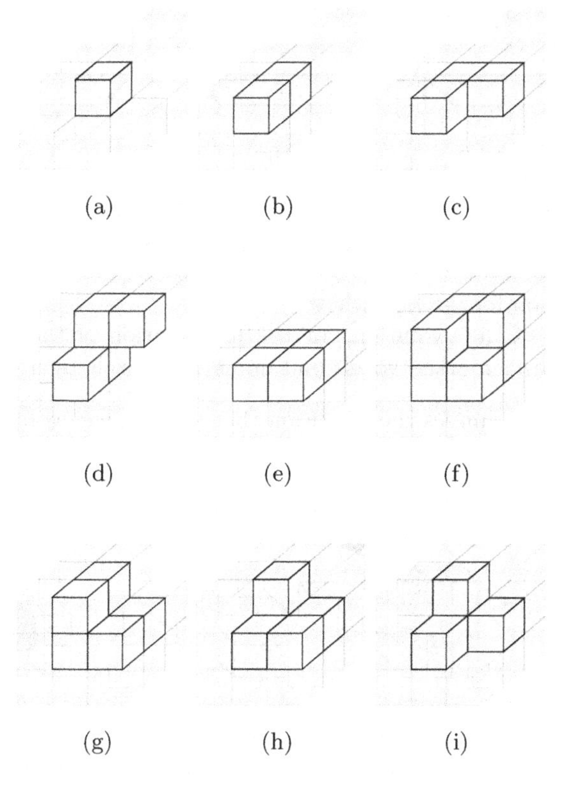

Fig. 2. Angles and configurations on the boundary of three-dimensional 6-connected objects

acts as curvature flow, though for flat points this transformation acts as diffusion on the boundary curve.

Using the curvature on slices we define the motion of boundary points as

$$\overline{\boldsymbol{x}_j} = \boldsymbol{x}_j + \frac{1}{2}(\varepsilon_1 \boldsymbol{n}_1 + \varepsilon_2 \boldsymbol{n}_2 + \varepsilon_3 \boldsymbol{n}_3) \tag{22}$$

from the point $\boldsymbol{x}$ to the point $\boldsymbol{y}$ on the dual lattice, where

$$\varepsilon_\alpha = \begin{cases} +1, \; r(\boldsymbol{x}) \leq 0, \\ -1, \; r(\boldsymbol{x}) \geq 0. \end{cases} \tag{23}$$

This is the three-dimensional version of the two-dimensional motion on slices of the boundary.

For the construction of the one-voxel-thick curve, we add the following rules for the evolution of flow.

1. During the evolution of the thinning by curvature flow, once the thickness of parts becomes one, stop the evolution for these parts and mark these voxels.

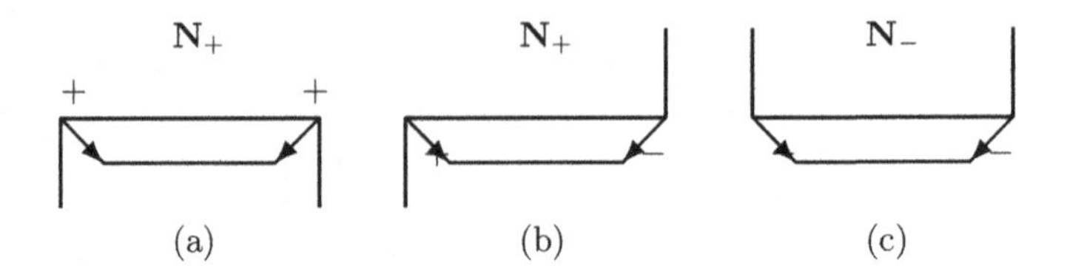

Fig. 3. Motions of the positive and negative lines on the boundary

2. Apply evolution for voxels without marks.
3. At each step of the evolution, for the preservation of the topology of the skeleton, connect marked voxels and unmarked voxels by a voxel.

The first rule guarantees the condition that the surface evolution by eq. (22) stops after yielding a one-voxel-thick object from the original object.

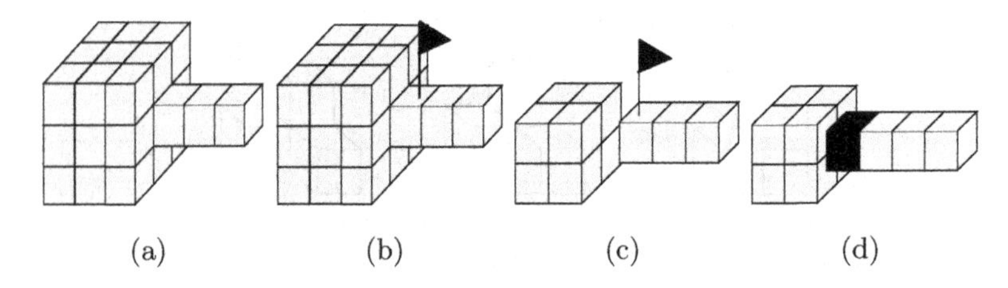

Fig. 4. Evolution of thinning process with marking to a voxel. Once a part of an object becomes one-voxel thick as shown in (a), the algorithm affixes the label on the voxel on the end of a one-thick part which connects ton non-one-thick parts as shown in (b). Then, the algorithm applies thinning process to non-one-thick parts and does not apply thinning process to the one-thick part as shown in (c). Furthermore, the algorithm connects the non-thick part with the results of thinning as shown in (d). The algorithm iterates this process for the thinning

Figure 5 expresses this process. In Figure 5, marks are illustrated as flags on voxels. The process shows that once a part of an object becomes one-voxel thick as shown in (a), the algorithm affixes the label on the voxel on the end of a one-thick part which connects ton non-one-thick parts as shown in (b). Then, the algorithm applies thinning process to non-one-thick parts and does not apply thinning process to the one-thick part as shown in (c). Furthermore, the algorithm connects the non-thick part with the results of thinning as shown in (d). The algorithm iterates this process for the thinning.

We call the motion of boundary points caused by the successive application of this transformation digital curvature flow. Therefore, odd and even steps of digital curvature flow transform points on the lattice to points on the dual lattice and points on the dual lattice to points on the lattice, respectively.

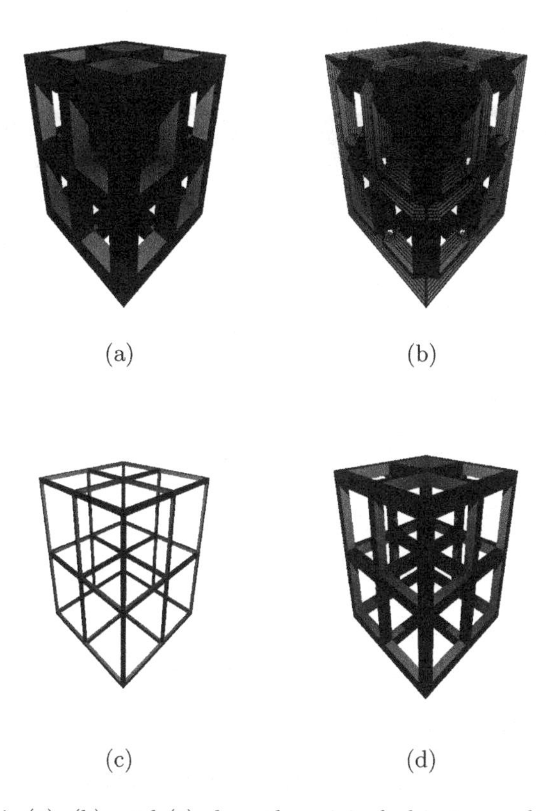

Fig. 5. Results 1: (a), (b), and (c) show the original objects, results of the distance-transform, and the results of the new thinning method, respectively. (d) shows intermediate shape extracted by our method

4 Examples

In Figure 6, (a), (b), and (c) show the original objects, results of the distance-transform, and the results of the new thinning method, respectively. Furthermore, in Figure 6, (d) shows an intermediate shape for skeletonisationb extracted by our method. This result shows that it is possible to control the thickness of the curve skeleton using our method preserving topology of the result in each step of iterations. In Figure 7, the curve-skeletons of (c) and (d) are extracted from a pair of objects which are mutually congruent as shown in Figures 7 (a) and (b), respectively. These figures show that the marking process of the algorithm is effective for the preservation of the topology.

In Figure 9, for comparison, we show the results of the distance-transform and our algorithm for a planar object. These results show that our algorithm yields one-voxel-thickness skeletons and that the algorithm preserves topology of the original objects.

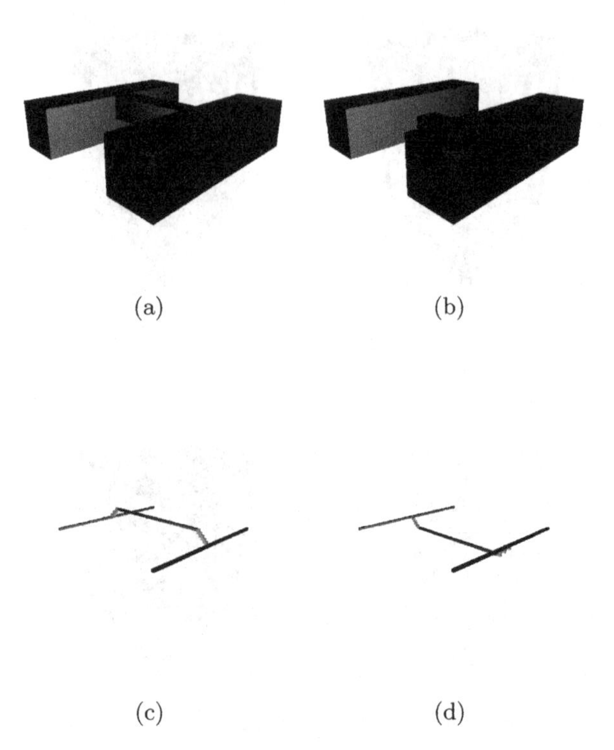

Fig. 6. Results 3: The curve-skeletons of (c) and (d) are extracted from a pair of objects which are mutually congruent in (a) and (c). These figures show that the marking process of the algorithm is effective for the preservation of the topology

These results show that our method preserves the topology and the geometry of junctions, although the distance-transform-based thinning does not preserve these properties. Furthermore, our method yields curve-skeletons of objects.

5 Connection Conversion

In reference [17], we introduced Gauss-Bonnet theorem for 8-connectivity boundary. Using this theory, the 8-connected boundary is converted to 4-connected boundary preserving topology of the boundary. After conversion of the connectivity, the boundary curve is deformed for the extraction of the skeleton.

In reference [16], we have introduced the rules for the conversion of 18- and-26 connected local configurations to 6-connected ones for 3-dimensional boundary elements. These rules eliminate edges whose lengths are $\sqrt{2}$ and $\sqrt{3}$, and convert parts with these edges to collections of edges whose lengths are 1. After converting the connectivity, it is possible to apply our thinning algorithm for the skeletonisation of discrete 18- and 26- connected objects. The conversion rules often yield needles and walls, which are 1- and 2-dimensional simplexes, respectively.

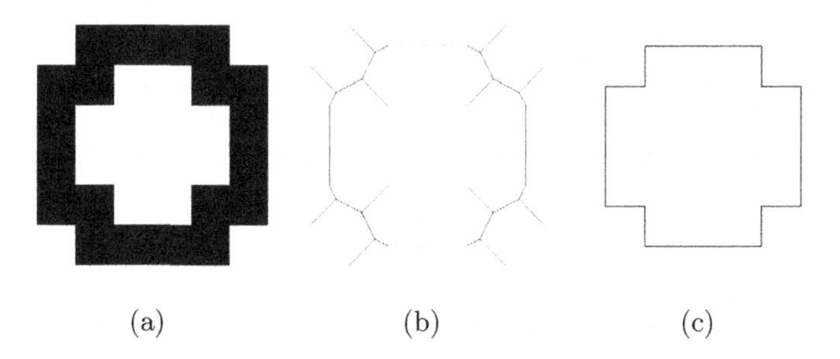

(a) (b) (c)

Fig. 7. Results of two-dimensional skeletonisation: (a), (b), and (c) show the original objects, results of the distance-transform, and the results of the new thinning method, respectively

A point on a unit-height wall is a point on a unit-length needle in each slice plane. In each slice, if we consider the four connectivity, there exist three configurations in Figure 1 and unit-length needles for the boundary of the quasi object yielded from a 18- and 26-connected object. For the concave, flat, convex, and needle point, the number of points in the four neighbourhood is 0, 1, 2, and 3, respectively. Therefore, after rewriting the boundary configurations from 18- and 26-connectivity to 6-connectivity, eliminating points which satisfy the relation $|\boldsymbol{N}_\alpha(\boldsymbol{x})| = 3$ for

$$\boldsymbol{N}_\alpha(\boldsymbol{x}) = \{\boldsymbol{y}|\boldsymbol{y} \in \mathbf{N}^4_\alpha(\boldsymbol{x}), \boldsymbol{y} \in \mathbf{F}^c\}, \tag{24}$$

where $\mathbf{F}^c$ is the compliment of point set $\mathbf{F}$, and $|\mathbf{A}|$ is the number of elements in point set $\mathbf{A}$.

6 Conclusions

In this paper, we introduced a new transform for the binary digital set, which we call digital curvature flow. Digital curvature flow is a digital version of curvature flow. As applications of discrete curvature flow, we developed thinning. The thinning algorithm preserves the geometry of junctions. Our method based on curvature flow is defined using configurations of vertices of an isotetic polyhedron derived from the 6-connected boundary.

In the previous paper [16], we introduced the rewriting rules for the connection conversion from 18- and 26- connectivities to 6-connectivities. After converting the connectivity, it is possible to apply our thinning algorithm for the skeletonisation of discrete 18- and 26- connected objects.

References

1. Blum, H., Biological shape and visual science, J. Theoretical Biology, **38**, 205-285, 1963.

2. Rosenfeld, A., Axial representations of shapes, CVGIP, **33**, 156-173, 1986.
3. Bookstein, F. L., The line-skeleton, CVGIP, **11**, 1233-137, 1979.
4. Amenta, N., Bern, M., Eppstein, D., The crust and the β-skeleton: Combinatorial curve reconstruction, Graphical Models and Image Processing, **60**, 125-135, 1998.
5. Attali, D. and Montanvert, A., Computing and simplifying 2D and 3D continuous skeletons, CVIU, **67**, 261-273, 1997.
6. Giblin, P. J. and Kimia, B. B., On the local form and transitions of symmetry sets and medial axes, and shocks in 2D, Proceedings of ICCV, 385-391, 1999.
7. Nystrom, I., Sanniti di Baja, G., Svensson, S., Curve skeletonization by junction detection Lecture Notes in Computer Science, **2059**, 229-238, 2001.
8. Svensson, S., Nystrom, I., Sanniti di Baja, G., Curve skeletonization of surface-like objects in 3D images guided by voxel classification, Pattern Recognition Letters, **23**, 1419-1426, 2002.
9. Sanniti di Baja, G., Svensson, S., Surface skeletons detected on the D6 distance transform. Lecture Notes in Computer Science **1876**, 387-396, 2000.
10. Svensson, S., Borgefors, G., Nystrom, I., On reversible skeletonization using anchor-points from distance transforms Journal on Visual Communication and Image Representation, **10**, 379-397, 1999.
11. Svensson, S., Sanniti di Baja, G., Using distance transforms to decompose 3D discrete objects, Image and Vision Computing, **20**, 529-540, 2002.
12. Hilditch, J. C., Linear skeletons from square cupboards, in Meltzer, B., and Michie, D. eds., *Machine Intelligence 4*, 403-422, Edinburgh University Press: Edinburgh, 1969.
13. Sethian, J. A.,*Level Set Methods: Evolving Interfaces in Geometry Fluid Mechanics, Computer Vision, and Material Science.* Cambridge University Press, Cambridge, 1996.
14. Bruckstein, A.M., Shapiro, G., Shaked, D., Evolution of planar polygons, Journal of Pattern Recognition and Artificial Intelligence, **9**, 991-1014, 1995.
15. Imiya, A. and Eckhardt, U., Discrete curvature flow, Lecture Notes in Computer Science, **1682**, 477-483, 1999.
16. Imiya, A. and Eckhardt, U., The Euler characteristics of discrete objects and discrete quasi-objects, CVIU, **75**, 307-318, 1999.
17. Imiya, A., Saito., M., Tatara, K., Nakamura, K., Digital curvature flow and its application to skeletonization, Journal of Mathematical Imaging and Vision, **18**, 55-68, 2003.
18. Imiya, A. and Eckhardt, U., Discrete curvature flow, Lecture Notes in Computer Science, **1682**, 477-483, 1999.
19. Kovalevsky, V.A., Finite topology as applied to image analysis, Computer Vision, Graphics and Image Processing, **46**, 141-161, 1989.
20. Françon, J., Sur la topologie d'un plan arithmétique, Theoretical Computer Sciences, **156**, 159-176, 1996.
21. Toriwak, J.-I., *Digital Image Processing for Computer Vision, Vols. 1 and 2*, Sokodo: Tokyo, 1988.

Convex Functions on Discrete Sets

Christer O. Kiselman

Uppsala University, P. O. Box 480, SE-751 06 Uppsala, Sweden
kiselman@math.uu.se
http://www.math.uu.se/~kiselman

Abstract. We propose definitions of digital convex sets and digital convex functions and relate them to a refined definition of digital hyperplanes.

Keywords: Digital convex set, digital convex function, naive digital line, digital hyperplane, Fenchel transformation.

1 Introduction

Digital geometry is a branch of geometry which is inspired by the use of computers in creating images and of importance for the proper understanding and creation of many algorithms in image processing. In Euclidean geometry convex sets play an important role, and convex functions of real variables are of importance in several branches of mathematics, especially in optimization.

All this forms the background of the present paper, where we will propose definitions of convex sets (Definition 3.1) and convex functions (Definition 4.1) in a digital setting, definitions that have many desirable properties. They are in fact very simple—some may call them naive—but it seems to be necessary to investigate them first before one can go on to more sophisticated definitions. We shall show that functions which are both convex and concave have interesting relations to a refined definition of digital hyperplanes.

The notion of a digital straight line received a satisfying definition in Rosenfeld's seminal paper (1974), where he explained how to digitize a real straight line segment. Since then, variants of this digitization have been introduced, among them digitizations which respect the Khalimsky topology; see Melin (2003). Here we shall not consider the Khalimsky topology, however. Instead, we shall look at definitions of digital hyperplanes, in particular that of Reveillès (1991), and compare them with the notion of digitally convex and concave functions.

We cannot mention here all the work done on convex sets and digital planes; we refer the reader to the surveys by Eckhardt (2001) and Rosenfeld & Klette (2001) and the many papers referred to there. Eckhardt studies no less than five different notions of convexity; one of them he calls H-convexity (2001:218)—this is the notion studied in the present paper.

We write $\mathbf{Z}$ for the ring of integers and $\mathbf{R}$ for the field of real numbers. When defining functions with integer values, we shall often use the *floor* and *ceiling functions* $\mathbf{R} \ni t \mapsto \lfloor t \rfloor, \lceil t \rceil \in \mathbf{Z}$. They are uniquely determined by the

R. Klette and J. Žunić (Eds.): IWCIA 2004, LNCS 3322, pp. 443–457, 2004.

requirement that $\lfloor t \rfloor$ and $\lceil t \rceil$ be integers for every real number t and by the inequalities

$$t-1 < \lfloor t \rfloor \leqslant t < \lfloor t \rfloor + 1; \qquad \lceil t \rceil - 1 < t \leqslant \lceil t \rceil < t+1, \qquad t \in \mathbf{R}. \qquad (1.1)$$

2 The Real Case

Let E be a vector space over the field of real numbers. A subset A of E is said to be *convex* if the segment $[a,b] = \{(1-t)a + tb; 0 \leqslant t \leqslant 1\}$ is contained in A for every choice of $a, b \in A$; in other words if $\{a,b\} \subset A$ implies $[a,b] \subset A$. And convex functions are most conveniently defined in terms of convex sets: a function $u\colon E \to [-\infty, +\infty] = \mathbf{R} \cup \{+\infty, -\infty\}$ is said to be *convex* if its *epigraph*

$$\operatorname{epi} u = \{(x,t) \in E \times \mathbf{R}; u(x) \leqslant t\}$$

is a convex set in $E \times \mathbf{R}$. For functions $f\colon P \to [-\infty, +\infty]_{\mathbf{Z}} = \mathbf{Z} \cup \{+\infty, -\infty\}$, where P is a subset of E, we define the epigraph as a subset of $P \times \mathbf{Z}$:

$$\operatorname{epi} f = \{(p,q) \in P \times \mathbf{Z}; f(p) \leqslant q\}.$$

We shall also need the *strict epigraph* of u, which is the set

$$\operatorname{epi_s} u = \{(x,t) \in E \times \mathbf{R}; u(x) < t\}.$$

It is convex if and only if u is convex.

Related to these notions are the *graph* and *hypograph* of a function, defined by

$$\operatorname{graph} u = \{(x,t) \in E \times \mathbf{R}; u(x) = t\} \text{ and } \operatorname{hypo} u = \{(x,t) \in E \times \mathbf{R}; u(x) \geqslant t\},$$

respectively.

It is also possible to go in the other direction and define convex sets in terms of convex functions: a set A in E is convex if and only if its indicator function i_A is convex, where we define $i_A(x) = 0$ if $x \in A$ and $i_A(x) = +\infty$ otherwise. Naturally we would like to keep these equivalences in the digital case.

Important properties of the family of convex sets in a vector space are the following.

Proposition 2.1. *If C_j, $j \in J$, are convex sets, then the intersection $\bigcap C_j$ is convex. If the index set J is ordered and filtering to the right, and if $(C_j)_{j \in J}$ is an increasing family of convex sets, then its union $\bigcup C_j$ is convex.*

Because of this result, the intersection

$$\operatorname{cvx} A = \bigcap \big(C \in \mathscr{P}(E); C \text{ is convex and } C \supset A\big), \qquad A \in \mathscr{P}(E),$$

of all convex sets containing a given subset A of E is itself convex; it is called the *convex hull of A*.

Proposition 2.2. *If u_j, $j \in J$, are convex functions on a vector space, then $\sup u_j$ is convex. If the index set J is ordered and filtering to the right, and if*

$(u_j)_{j\in J}$ is a decreasing family of convex functions, then its infimum $\inf u_j$ *is convex.*

To a given function $u\colon E \to [-\infty, +\infty]$ we associate two convex functions, viz. the supremum v of all convex minorants of u and the supremum w of all affine minorants of u. According to Proposition 2.2 these functions are themselves convex, and of course $w \leqslant v \leqslant u$. We shall denote v by $\operatorname{cvx} u$, and w by $\tilde{\tilde{u}}$, a notation which will become clear when we have introduced the Fenchel transformation below. The function $v = \operatorname{cvx} u$ will be called the *convex hull of* u. For functions $f\colon P \to [-\infty, +\infty]$, P being a subset of E, we shall use the same notation. Such a function can be extended to a function u defined in all of E simply by taking $u = +\infty$ in the complement of P (then u and f have the same epigraph), and we define $\operatorname{cvx} f = \operatorname{cvx} u$.

In many cases, but not always, $\tilde{\tilde{u}}$ is equal to $\operatorname{cvx} u$ (see Examples 2.3 and 2.4 below). To understand this, we note that $\tilde{\tilde{u}}$ has two extra properties in addition to being convex, properties that are not always shared by $\operatorname{cvx} u$. The first is that $\tilde{\tilde{u}}$ is lower semicontinuous for any topology for which the affine functions are continuous. The second is that if u takes the value $-\infty$ at a point, then $\tilde{\tilde{u}}$ must be identically equal to $-\infty$ (there are no affine minorants), whereas $\operatorname{cvx} u$ may take also finite values or $+\infty$.

Example 2.3. Let $P \subset \mathbf{R}^2$ be the set of all points $(p_1, 1/(1+p_1^2))$, $p_1 \in \mathbf{Z}$. Define a function $f\colon P \to [0, +\infty]$ by $f(p) = 0$ for $p \in P$. Then $\operatorname{cvx} f$ takes the value 0 when $0 < x_2 < 1$ or $x = (0, 1)$. On the other hand $\tilde{\tilde{f}}$ takes the value 0 in the closed strip $0 \leqslant x_2 \leqslant 1$ and $+\infty$ elsewhere, so the two functions differ when $x_2 = 0$ or $x_2 = 1$, $x_1 \neq 0$.

Example 2.4. Let α be an irrational number and define $f\colon \mathbf{Z}^2 \to [0, +\infty]$ by $f(p) = 0$ if $p_2 = \lceil \alpha p_1 \rceil$, $p_1 \neq 0$; $f(0) = 3$; and $f(p) = +\infty$ otherwise. Then $(\operatorname{cvx} f)(x) = 0$ if x is in the open strip $\alpha x_1 < x_2 < \alpha x_1 + 1$; $(\operatorname{cvx} f)(0) = 3$; and $(\operatorname{cvx} f)(x) = +\infty$ elsewhere. On the other hand $\tilde{\tilde{f}}(x) = 0$ when $\alpha x_1 \leqslant x_2 \leqslant \alpha x_1 + 1$ and $+\infty$ outside this closed strip.

We thus have

$$w = \tilde{\tilde{u}} \leqslant v = \operatorname{cvx} u \leqslant u. \tag{2.1}$$

However, in our research it will not be enough to study these functions: it is necessary to look at their epigraphs.

The epigraph $\operatorname{epi} u$ of u is a subset of $E \times \mathbf{R}$ and its convex hull $C = \operatorname{cvx}(\operatorname{epi} u)$ is easily seen to have the property

$$(x, s) \in C, s \leqslant t \text{ implies } (x, t) \in C. \tag{2.2}$$

The function $V_C(x) = \inf\big(t; (x, t) \in C\big)$ satisfies

$$\operatorname{epi}_{\mathrm{s}} V_C \subset \operatorname{cvx}(\operatorname{epi} u) \subset \operatorname{epi} V_C. \tag{2.3}$$

It is clear that V_C is convex and equal to the largest convex minorant $v = \operatorname{cvx} u$ of u already introduced. Thus $\operatorname{cvx} u$ can be retrieved from $\operatorname{cvx}(\operatorname{epi} u)$ but not conversely. The inequality (2.1) and the inclusion relation (2.3) can be combined to

$$\operatorname{epi}_{\mathrm{s}} u \subset \operatorname{epi}_{\mathrm{s}}(\operatorname{cvx} u) \subset \operatorname{epi}_{\mathrm{s}}(\operatorname{cvx} u) \cup \operatorname{epi} u \subset \operatorname{cvx}(\operatorname{epi} u) \subset \operatorname{epi}(\operatorname{cvx} u) \subset \operatorname{epi} \tilde{\tilde{u}}, \tag{2.4}$$

and in general we cannot claim that $\operatorname{cvx}(\operatorname{epi} u)$ is an epigraph (see Examples 4.2 and 4.3).

In this paper, convex sets which are squeezed in between the epigraph and the strict epigraph of a function will play an important role. Such sets C satisfy $\operatorname{epi}_{\mathrm{s}} u \subset C \subset \operatorname{epi} u$ for some function u. This means that C is obtained from the strict epigraph by adding some points in the graph:

$$C = \operatorname{epi}_{\mathrm{s}} u \cup \{(x, u(x)); x \in A\} \subset \operatorname{epi}_{\mathrm{s}} u \cup \operatorname{graph} u = \operatorname{epi} u.$$

Extreme examples are the following. If u is strictly convex, like $u(x) = \|x\|_2^p$, $x \in \mathbf{R}^n$, with $1 < p < +\infty$, then any such set is convex, even though A may be very irregular. If on the other hand $u = 0$, then such a set is convex if and only if A itself is convex.

Definition 2.5. *Let E be a real vector space and denote by $E^\star$ its algebraic dual (the set of all real-valued linear forms on E). For any function $u \colon E \to [-\infty, +\infty]$ we define its* Fenchel transform $\tilde{u}$ *by*

$$\tilde{u}(\xi) = \sup_{x \in E} \big(\xi(x) - u(x)\big), \qquad \xi \in E^\star.$$

For any function $v \colon F \to [-\infty, +\infty]$ defined on a vector subspace F of $E^\star$ we define its Fenchel transform *by*

$$\tilde{v}(x) = \sup_{\xi \in F} \big(\xi(x) - v(\xi)\big), \qquad x \in E.$$

The second Fenchel transform $\tilde{\tilde{u}}$ of u is well-defined if we fix a subspace F of $E^\star$. This subspace can be anything between $\{0\}$ and all of $E^\star$, in particular we can take F as the topological dual E' of E if E is equipped with a vector space topology.

The restriction $\tilde{u}|_F$ of the Fenchel transform to a subspace F of $E^\star$ describes all affine minorants of u with linear part in F: a pair $(\xi, \beta) \in F \times \mathbf{R}$ belongs to $\operatorname{epi} \tilde{u}$ if and only if $x \mapsto \xi(x) - \beta$ is a minorant of u. This implies that $\tilde{\tilde{u}}$ is the supremum of all affine minorants of u with linear part in F. This function is a convex minorant of u, but it has the additional properties that it cannot take the value $-\infty$ unless it is the constant $-\infty$, and it is lower semicontinuous with respect to the topology $\sigma(E, F)$, the weakest topology on E for which all linear forms in F are continuous. One can prove that $\tilde{\tilde{u}}$ is the largest convex minorant of u with these properties. General references for the Fenchel transformation are Hörmander (1994), Singer (1997) and Hiriart-Urruty & Lemaréchal (2001).

3 Convex Sets

Definition 3.1. *Let E be a real vector space and fix a subset P of E. A subset A of P is said to be* P-convex *if there exists a convex set C in E such that $A = C \cap P$.*

We are mostly interested in the case $E = \mathbf{R}^n$, $P = \mathbf{Z}^n$.

For digitizations of convex sets the mapping $C \mapsto C \cap \mathbf{Z}^n$ is not always satisfactory, because it yields the empty set for some long and narrow convex sets C. One might then want to replace it by a mapping like $C \mapsto (C + B) \cap \mathbf{Z}^n$, where B is some fixed set which guarantees that the image is nonempty when C is nonempty, e.g., $B = B_{\leqslant}(0, r)$, where $r = 1/2$ if we use the l^∞ norm in $\mathbf{R}^n$, $r = \sqrt{n}/2$ if we use the l^2 norm, or $r = n/2$ if we use the l^1 norm. However, for our purpose, when we apply this operation to the epigraph of a function, this phenomenon will not appear: the epigraph of a function with finite values always intersects $\mathbf{Z}^n \times \mathbf{Z}$ in a nonempty set.

Lemma 3.2. *Given a vector space E and a subset P of E, the following properties are equivalent for any subset A of P.*

1. *A is P-convex;*
2. *$A = (\operatorname{cvx} A) \cap P$;*
3. *$A \supset (\operatorname{cvx} A) \cap P$.*
4. *For all n, all $a_0, \dots, a_n \in A$, and for all nonnegative numbers $\lambda_0, \dots, \lambda_n$ with $\sum_0^n \lambda_j = 1$, if $\sum_0^n \lambda_j a_j \in P$, then $\sum_0^n \lambda_j a_j \in A$.*

Proof. This is easy. As far as property 4 is concerned, we can, in view of Carathéodory's theorem, let n be the dimension of E if the space is finite dimensional; otherwise we must use all n.

Definition 3.3. *Fix two subsets P and Q of a vector space E and define an operator $\gamma = \gamma_{P,Q} \colon \mathscr{P}(E) \to \mathscr{P}(P)$ by $\gamma(A) = \operatorname{cvx}(A \cap Q) \cap P$.*

We can think of $E = \mathbf{R}^n$, $P = m\mathbf{Z}^n$, $m = 1, 2, \ldots$, and $Q = \mathbf{Z}^n$. We note that $\gamma(C)$ is P-convex if C is convex in $\mathbf{R}^n$.

Lemma 3.4. *The mapping γ is increasing; it satisfies $\gamma(\gamma(A)) \subset \gamma(A)$; and it satisfies $A \subset \gamma(A)$ if $A \subset P \cap Q$. Thus $\gamma|_{\mathscr{P}(P)}$ is a closure operator in $\mathscr{P}(P)$ if $Q \supset P$.*

Proof. The mapping $\gamma = j_P \circ \operatorname{cvx} \circ j_Q$ is a composition of three increasing mappings, viz. j_Q (intersection with Q), cvx (taking the convex hull), and j_P (intersection with P), and as such itself increasing. The composition $\gamma \circ \gamma$ is equal to $j_P \circ \operatorname{cvx} \circ j_Q \circ j_P \circ \operatorname{cvx} \circ j_Q$, which is smaller than $j_P \circ \operatorname{cvx} \circ \operatorname{cvx} \circ j_Q = j_P \circ \operatorname{cvx} \circ j_Q = \gamma$. Finally, it is clear that $\gamma(A)$ contains A if A is contained in $P \cap Q$. If $Q \supset P$, then γ is increasing, idempotent and extensive, thus a closure operator in $\mathscr{P}(P)$.

Proposition 3.5. *Let E be a real vector space and P any subset of E. Then A is P-convex iff $A = \gamma(A)$ for all $Q \supset P$ iff $A = \gamma(A)$ for some $Q \supset P$.*

Proof. If A is P-convex, $A = C \cap P$, then $\gamma(A) = \gamma(C \cap P) = \operatorname{cvx}(C \cap P \cap Q) \cap P = C \cap P = A$ for all $Q \supset P$.

If $A = \gamma(A)$ for some choice of $Q \supset P$, then $A = \operatorname{cvx}(A \cap Q) \cap P = C \cap P$ if we define $C = \operatorname{cvx}(A \cap Q)$, so that A is P-convex.

Corollary 3.6. *If $A = C \cap P$, then $C \supset \gamma(A)$ for any Q.*

Thus in the definition of P-convex sets we may always take $C = \gamma(A) = \operatorname{cvx} A$ provided $Q \supset P$.

It is now easy to prove the following result.

Proposition 3.7. *Let E be a vector space and P any subset of E. If A_j, $j \in J$, are P-convex sets, then the intersection $\bigcap A_j$ is P-convex. If the index set J is ordered and filtering to the right, and if $(A_j)_{j \in J}$ is an increasing family of P-convex sets, then its union $\bigcup A_j$ is also P-convex.*

Proof. For each A_j we have $A_j = C_j \cap P$, where $C_j = \operatorname{cvx} A_j$ is a convex set in E. Then $\bigcap A_j = \bigcap (C_j \cap P) = \left(\bigcap C_j\right) \cap P$. The last set is P-convex in view of Proposition 2.1.

For the union we have $\bigcup A_j = \bigcup (C_j \cap P) = \left(\bigcup C_j\right) \cap P$, so Proposition 2.1 gives also the second statement—the family $(C_j) = (\operatorname{cvx} A_j)$ is increasing since (A_j) is.

While the intersection of two P-convex epigraphs gives a reasonable result, the intersection of an epigraph and a hypograph may consist of two points quite far from each other:

Example 3.8. Let $A = \{p \in \mathbf{Z}^2; p_2 \geqslant p_1/m\}$ and $B = \{p \in \mathbf{Z}^2; p_2 \leqslant p_1/m\}$, where $m \in \mathbf{N} \smallsetminus \{0\}$. Then A and B are $\mathbf{Z}^2$-convex and their intersection consists of all points (mp_2, p_2), $p_2 \in \mathbf{Z}$. We can easily modify the example so that the intersection consists of exactly two points, $(0,0)$ and $(m,1)$, where m is as large as we please.

4 Convex Functions

Definition 4.1. *Let E be a vector space and P any of its subsets. A function $f\colon P \to [-\infty, +\infty]_{\mathbf{Z}}$ is said to be $(P \times \mathbf{Z})$-convex if its epigraph*

$$\operatorname{epi} f = \{(p,t) \in P \times \mathbf{Z}; f(p) \leqslant t\}$$

is a $(P \times \mathbf{Z})$-convex subset of $E \times \mathbf{R}$.

We have mainly the case $E = \mathbf{R}^n$ and $P = \mathbf{Z}^n$ in mind.

If $f\colon P \to [-\infty, +\infty]_{\mathbf{Z}}$ is a P-convex function, then there is a convex set C in $E \times \mathbf{R}$ such that $C \cap (P \times \mathbf{Z}) = \operatorname{epi} f$. In view of Corollary 3.6, the smallest such set C is the convex hull of epi f. However, a set C such that $C \cap (P \times \mathbf{Z}) = \operatorname{epi} f$ does not necessarily have the property (2.2), so we introduce

$$C^+ = \{(x,t) \in E \times \mathbf{R}; \exists s \leqslant t \text{ with } (x,s) \in C\}.$$

There is a function $V_{C^+}: E \to [-\infty, +\infty]$ such that

$$\operatorname{epi_s} V_{C^+} \subset C^+ \subset \operatorname{epi} V_{C^+}.$$

It would perhaps seem natural to require that C^+ be closed or open so that one could always take either the epigraph or the strict epigraph of V_{C^+}, but simple examples (see below) show that this is not possible. We note that when we take $C = \operatorname{cvx}(\operatorname{epi} f)$, then $C^+ = C$.

Some care is needed, because even if epi f is closed, its convex hull need not be closed:

Example 4.2. Let $f_0(p) = \lceil \alpha p \rceil$, $p \in \mathbf{Z}$, where α is irrational. We also define $f_1(p) = f_0(p)$ for $p \in \mathbf{Z} \smallsetminus \{0\}$ and $f_1(0) = 1$. These functions are easily seen to be $(\mathbf{Z} \times \mathbf{Z})$-convex. Indeed, $\operatorname{cvx}(\operatorname{epi} f_1)$ is the open half plane $C_1 = \{(x,t); t > \alpha x\}$, a strict epigraph, and $\operatorname{cvx}(\operatorname{epi} f_0)$ is the convex set $C_0 = C_1 \cup \{(0,0)\}$, which is neither an epigraph nor a strict epigraph. (However, also the closed half plane $\{(x,t); t \geqslant \alpha x\}$ intersects $\mathbf{Z}^2$ in epi f_0.) We finally note that the functions $-f_0$ and $-f_1$ are $(\mathbf{Z} \times \mathbf{Z})$-convex as well.

A convex function need not be determined by its restriction to the complement of a point: in Example 4.2 above, f_0 and f_1 agree on $\mathbf{Z} \smallsetminus \{0\}$. This kind of ambiguity is, however, something we have to live with if we want results like Proposition 2.2 to hold. In the example f_0 is a supremum of convex functions without this ambiguity, and f_1 is the limit of a decreasing sequence of convex functions without the ambiguity. To make this precise, define $g_s(p) = \lceil \alpha p + s \rceil$, $p \in \mathbf{Z}$, where s is a real parameter. Then $g_s \to f_1$ as s tends to zero through positive values, and $g_s \to f_0$ as s tends to zero through negative values. We have $g_s(0) = 1$ for $0 < s < 1$, and we claim that $g_s(0)$ is determined by the restriction of g_s to the complement of the origin when $0 < s < 1$. Indeed, let an extension take the value $c \in \mathbf{Z}$ at the origin. Then $c \leqslant 0$ is impossible for $s > 0$, and $c \geqslant 2$ is impossible for all s such that $0 < s < 1$. Similarly g_s with s negative is determined from its restriction to $\mathbf{Z} \smallsetminus \{0\}$. So the functions g_s with s small and nonzero do not have this kind of ambiguity, whereas their limits as $s \to 0\pm$ do.

Example 4.3. Let a set A of even integers be given and define $g_A(p) = \lceil \frac{1}{2}p \rceil = \frac{1}{2}p + \frac{1}{2}$, $p \in \mathbf{Z}$, p odd, and $g_A(p) = \frac{1}{2}p$ when p is even, $p \in A$, and $g_A(p) = \frac{1}{2}p + 1$ when p is even and $p \notin A$. This function is $(\mathbf{Z} \times \mathbf{Z})$-convex if and only if A is an interval in $2\mathbf{Z}$. To see this, we note that $\operatorname{cvx}(\operatorname{epi} g_A)$ is the convex set

$$C_I = \{(x,t) \in \mathbf{R} \times \mathbf{R}; t > \tfrac{1}{2}x\} \cup \{(x, \tfrac{1}{2}x) \in I \times \mathbf{R}\},$$

where I is the convex hull of A. Then $C_I \cap (\mathbf{Z} \times \mathbf{Z})$ is equal to epi g_A if and only if A is an interval of even integers. We thus easily get examples of functions which are $(\mathbf{Z} \times \mathbf{Z})$-convex as well as examples of functions which are not. The set C_I is in general neither an epigraph nor a strict epigraph.

Proposition 4.4. *Let $u\colon E \to [-\infty, +\infty]$ be a convex function on a vector space E. Let P be a subset of E. Then the restrictions $\lfloor u \rfloor \big|_P$ and $\lceil u \rceil \big|_P$ are $(P \times \mathbf{Z})$-convex. In particular $\lceil \operatorname{cvx} g \rceil \big|_P$ and $\left\lceil \widetilde{\widetilde{g}} \right\rceil \Big|_P$ are $(P \times \mathbf{Z})$-convex for any function $g\colon P \to [-\infty, +\infty]_{\mathbf{Z}}$.*

Proof. Writing $f = \lfloor u \rfloor \big|_P$ and $g = \lceil u \rceil \big|_P$ we have (cf. (1.1))

$$u - 1 < f \leqslant u \text{ and } u \leqslant g < u + 1 \text{ in } P,$$

which implies that $\operatorname{epi}_{\mathrm{s}}(u-1) \cap (P \times \mathbf{Z}) = \operatorname{epi} f$ and $\operatorname{epi} u \cap (P \times \mathbf{Z}) = \operatorname{epi} g$. Hence the functions f and g are $(P \times \mathbf{Z})$-convex.

Theorem 4.5. *Let E be a vector space and P one of its subsets. For any $(P \times \mathbf{Z})$-convex function $f\colon P \to \mathbf{Z}$ we have $\operatorname{cvx} f \leqslant \lceil \operatorname{cvx} f \rceil \leqslant f \leqslant \operatorname{cvx} f + 1$ in P.*

Proof. The inequality $\operatorname{cvx} f \leqslant f$ holds for any function. Hence $\operatorname{cvx} f \leqslant \lceil \operatorname{cvx} f \rceil \leqslant \lceil f \rceil = f$.

For the last inequality we argue as follows. Let $C = \operatorname{cvx}(\operatorname{epi} f)$ and $v = \operatorname{cvx} f$. Then $C \cap (P \times \mathbf{Z}) = \operatorname{epi} f$ and $\operatorname{epi}_{\mathrm{s}} v \subset C \subset \operatorname{epi} v$. If $v(p) < q$, then $(p, q) \in C$, which implies that $(p, q) \in \operatorname{epi} f$, i.e., $f(p) \leqslant q$. Take now $q = \lceil v(p) + \varepsilon \rceil$, where $\varepsilon > 0$. Then $v(p) < q$, so that $f(p) \leqslant \lceil v(p) + \varepsilon \rceil < v(p) + 1 + \varepsilon$. Letting ε tend to zero we see that $f(p) \leqslant v(p) + 1 = (\operatorname{cvx} f)(p) + 1$. This completes the proof of the theorem.

We define

$$P^j = \{p \in P; f(p) = \lceil (\operatorname{cvx} f)(p) \rceil + j\}, \qquad j = 0, 1.$$

In view of the last theorem we have $P = P^0 \cup P^1$. We also define

$$A^j = \{p \in P; f(p) = (\operatorname{cvx} f)(p) + j\}, \qquad j = 0, 1.$$

Corollary 4.6. *With f as in the theorem, P can be divided into three disjoint sets: $P^0 \smallsetminus A^0$, A^0, and $A^1 = P^1$. The first set is precisely the set of points p such that $(\operatorname{cvx} f)(p)$ is not an integer.*

Proof. It is clear that the three sets $P^0 \smallsetminus A^0$, A^0 and P_1 are pairwise disjoint. It is also easy to see that $p \in A^0 \cup A^1$ if and only if $(\operatorname{cvx} f)(p)$ is an integer. It follows that $A^j \subset P^j$. Finally, we shall prove that $P^1 \subset A^1$. If $p \in P^1$, then $\lceil (\operatorname{cvx} f)(p) \rceil$ is equal to $f(p) - 1$. But we always have $(\operatorname{cvx} f)(p) \geqslant f(p) - 1$, so that $(\operatorname{cvx} f)(p) = \lceil (\operatorname{cvx} f)(p) \rceil$ and p belongs to A^1.

Let us say that a function $u\colon \mathbf{R}^n \to [-\infty, +\infty]$ is *of fast growth* if for any constant c the set $\{x \in \mathbf{R}^n; u(x) \leqslant c\|x\|_2\}$ is bounded. The same terminology applies to a function defined in a subset P of $\mathbf{R}^n$; we understand that it takes the value $+\infty$ outside P. In particular, if f is equal to plus infinity outside a bounded set, it is of fast growth.

Theorem 4.7. *Let P be a discrete subset of $\mathbf{R}^n$ and let $f\colon P \to [-\infty, +\infty]_{\mathbf{Z}}$ be a function of fast growth. Then f is $(P \times \mathbf{Z})$-convex if and only if $f = \lceil \operatorname{cvx} f \rceil$, in other words the set P^1 is empty, and we have*

$$(\operatorname{cvx} f)(p) \leqslant f(p) < (\operatorname{cvx} f)(p) + 1, \qquad p \in P. \tag{4.1}$$

It is equivalent to say that there exists a convex function u such that $f = \lceil u \rceil$.

Proof. We already know from Proposition 4.4 that the condition is sufficient.

To prove necessity, assume that f is $(P \times \mathbf{Z})$-convex. Then $C = \operatorname{cvx}(\operatorname{epi} f)$ is a convex subset of $\mathbf{R}^n \times \mathbf{R}$ such that $C \cap (P \times \mathbf{Z}) = \operatorname{epi} f$. However, we now know that C is closed, so that actually $C = \operatorname{epi} \operatorname{cvx} f$. We also know from the previous theorem that $\operatorname{cvx} f \leqslant f \leqslant \operatorname{cvx} f + 1$ in P. A point $(p, f(p) - 1)$ does not belong to $\operatorname{epi} f$ and hence not to C. Since C is closed and its boundary is defined by $\operatorname{cvx} f$, we must have $f(p) - 1 < (\operatorname{cvx} f)(p)$, which was to be proved.

We can now take a look again at Examples 4.2 and 4.3.

Example 4.8. In Example 4.2 we find that

$$(\operatorname{cvx} f_0)(x) = \tilde{\tilde{f}}_0(x) = (\operatorname{cvx} f_1)(x) = \tilde{\tilde{f}}_1(x) = \alpha x, \qquad x \in \mathbf{R}.$$

Thus $\operatorname{cvx} f_0 \leqslant f_0 < \operatorname{cvx} f_0 + 1$, but $(\operatorname{cvx} f_0)(0) = 0 \neq 1 = f_1(0)$, so that $f_1(0) = (\operatorname{cvx} f_1)(0) + 1$. This shows that some condition is necessary in the theorem.

Example 4.9. In Example 4.3 we also have that $\tilde{\tilde{g}}_A = \operatorname{cvx} g_A$. We find that $(\operatorname{cvx} g_A)(x) = \frac{1}{2}x$ for all $x \in \mathbf{R}$ when A is nonempty. Therefore $\lceil \operatorname{cvx} g_A \rceil \neq g_A$ if A is nonempty but not equal to all of $2\mathbf{Z}$. In fact, we then have $\lceil (\operatorname{cvx} g_A)(p) \rceil = \frac{1}{2}p < g_A(p) = \frac{1}{2}p + 1$ when $p \in 2\mathbf{Z} \smallsetminus A$. Still g_A is convex if A is an interval of $2\mathbf{Z}$. When A is empty we have $(\operatorname{cvx} g_\varnothing)(x) = \frac{1}{2}x + \frac{1}{2}$.

Given $u\colon \mathbf{R}^n \to [-\infty, +\infty]$ we define $u_r(x) = u(x)$ if $\|x\|_2 \leqslant r$ and $u_r(x) = +\infty$ otherwise. We also define $u_{[r]}(x) = \max(u(x), \|x\|_2^2 - r)$. Then u_r and $u_{[r]}$ are of fast growth, and we note that u is convex if and only if all the u_r are convex, or, equivalently, all the $u_{[r]}$ are convex. The functions u_r and $u_{[r]}$ decrease to u as r tends to plus infinity. The same applies to functions $f\colon P \to [-\infty, +\infty]$ or $f\colon P \to [-\infty, +\infty]_{\mathbf{Z}}$.

Corollary 4.10. *Let P be a discrete subset of $\mathbf{R}^n$ and let a function $f\colon P \to [-\infty, +\infty]_{\mathbf{Z}}$ be given. Then f is $(\mathbf{Z}^n \times \mathbf{Z})$-convex if and only if $f_r = \lceil \operatorname{cvx} f_r \rceil$ for all $r \in]0, +\infty[$, equivalently if and only if $f_{[r]} = \lceil \operatorname{cvx} f_{[r]} \rceil$ for all $r \in \mathbf{N}$, where f_r and $f_{[r]}$ are defined as before the statement of the corollary.*

Proposition 4.11. *Let E be a vector space and P any of its subsets. If f_j, $j \in J$, are $(P \times \mathbf{Z})$-convex functions, then $\sup f_j$ is $(P \times \mathbf{Z})$-convex. If the index* *set J is ordered and filtering to the right, and if $(f_j)_{j\in J}$ is a decreasing family of $(P \times \mathbf{Z})$-convex functions, then its infimum* $\inf f_j$ *is $(P \times \mathbf{Z})$-convex as well.*

Proof. We note that $\operatorname{epi}(\sup_j f_j) = \bigcap_j \operatorname{epi} f_j$. The latter set is $(P \times \mathbf{Z})$-convex according to Proposition 3.7. Hence $\sup_j f_j$ is $(P \times \mathbf{Z})$-convex.

For the second part we note that $\operatorname{epi}_\mathrm{s}(\inf_j f_j) = \bigcup_j \operatorname{epi}_\mathrm{s} f_j$. Now $\operatorname{epi}_\mathrm{s} u$ and $\operatorname{epi} u$ are convex at the same time, so it follows from Proposition 3.7 that the latter set is $(P \times \mathbf{Z})$-convex. Hence $\operatorname{epi}_\mathrm{s}(\inf_j f_j)$ is $(P \times \mathbf{Z})$-convex.

5 Functions Which Are Both Convex and Concave

A function u such that $-u$ is convex is called *concave*. A real-valued function on $\mathbf{R}^n$ which is both convex and concave is necessarily *affine*, i.e., of the form $u(x) = \alpha \cdot x + \beta$ for some $\alpha \in \mathbf{R}^n$ and $\beta \in \mathbf{R}$. In this section we shall investigate such functions in the discrete case.

Proposition 5.1. *Let P be a nonempty subset of a vector space E and $f\colon P \to \mathbf{R}$ a real-valued function. Given a linear form $\alpha \in E^\star$ and a real number β we let $h_{\alpha,\beta}$ be the smallest constant $h \in [0, +\infty]$ such that*

$$0 \leqslant \alpha(p) + \beta \leqslant f(p) \leqslant \alpha(p) + \beta + h, \qquad p \in P. \tag{5.1}$$

We let $h_\alpha = \inf_{\beta\in\mathbf{R}} h_{\alpha,\beta}$ be the smallest constant h such that (5.1) holds for some $\beta \in \mathbf{R}$. Then $h_\alpha = \widetilde{f}(\alpha) + \widetilde{g}(-\alpha)$, where for ease in notation we have written g for $-f$. Moreover, $h_\alpha = h_{\alpha,\beta}$ for a unique β, viz. $\beta = -\widetilde{f}(\alpha)$.

Proof. The inequality $\alpha(p) + \beta \leqslant f(p)$ for all $p \in P$ is equivalent to $\widetilde{f}(\alpha) \leqslant -\beta$, and the inequality $f(p) = -g(p) \leqslant \alpha(p) + \beta + h$ for all $p \in P$ is equivalent to $\widetilde{g}(-\alpha) \leqslant \beta + h$. Therefore (5.1) implies that $\widetilde{f}(\alpha) + \widetilde{g}(-\alpha) \leqslant -\beta + (\beta + h) = h$.

Conversely, if h is a real number and $\widetilde{f}(\alpha) + \widetilde{g}(-\alpha) \leqslant h$, then $\widetilde{f}(\alpha)$ is a real number: $\widetilde{f}(\alpha) = -\infty$ would imply that f is identically equal to $+\infty$, which is excluded by hypothesis, and $\widetilde{g}(-\alpha) = -\infty$ would imply that f is identically $-\infty$, which is also excluded by hypothesis; finally, the inequality excludes that $\widetilde{f}(\alpha)$ is equal to $+\infty$. Therefore $\beta = -\widetilde{f}(\alpha)$ (obviously the best choice of β) yields $\widetilde{f}(\alpha) \leqslant -\beta$ and $\widetilde{g}(-\alpha) \leqslant \beta + h$, which, as already noted, is equivalent to (5.1). The infimum of all real h satisfying (5.1) is equal to the infimum of all real h satisfying $\widetilde{f}(\alpha) + \widetilde{g}(-\alpha) \leqslant h$, which completes the proof.

Proposition 5.2. *Let E be a vector space and P a subset such that $\operatorname{cvx} P = E$. Let a real-valued function $f\colon P \to \mathbf{R}$ be given, and let $h_* = \inf_{\alpha\in E^\star} h_\alpha$ be the smallest constant such that (5.1) holds for some $\alpha \in E^\star$ and some $\beta \in \mathbf{R}$. Assume that h_* is finite. Then $-\operatorname{cvx} f - \operatorname{cvx}(-f)$ is constant and equal to h_*.*

Proof. Let h be a number such that $\alpha+\beta \leqslant f \leqslant \alpha+\beta+h$ in P for some $\alpha \in E^\star$ and some $\beta \in \mathbf{R}$. Then

$$\alpha+\beta \leqslant u \leqslant f \leqslant -v \leqslant \alpha+\beta+h \text{ in } P,$$

where $u = \operatorname{cvx} f$ and $v = \operatorname{cvx}(-f)$. Adding v to all members we obtain

$$\alpha+\beta+v \leqslant u+v \leqslant f+v \leqslant 0 \leqslant \alpha+\beta+h+v \text{ in } P. \tag{5.2}$$

We see that $u+v$ is a convex function which is nonpositive in all of P, thus also in $\operatorname{cvx} P$, which by hypothesis is equal to E. But such a function must be constant; let us define $\omega = -(u+v) \geqslant 0$. By the same argument, $v+\alpha$ is a constant γ. We now have $\gamma+\beta \leqslant -\omega \leqslant 0 \leqslant \gamma+\beta+h$, which shows that $h \geqslant \omega$, and, by taking the infimum over all such h, that $h_* \geqslant \omega$.

Conversely, we note that $-\omega \leqslant f+\gamma-\alpha \leqslant 0$, thus $\alpha-\gamma-\omega \leqslant f \leqslant \alpha-\gamma$, which shows that $\omega \geqslant h_\alpha \geqslant h_*$. We conclude that $\omega = h_*$.

Theorem 5.3. *Let E be a vector space and P a subset of E such that $\operatorname{cvx} P = E$. If both functions $f\colon P \to \mathbf{Z}$ and $-f$ are $(P \times \mathbf{Z})$-convex, then f deviates at most by $\frac{1}{2}$ from an affine function: there exist a linear form $\alpha \in E^\star$ and constants $\beta, \omega \in \mathbf{R}$ such that*

$$0 \leqslant f(p) - \alpha(p) - \beta \leqslant \omega \leqslant 1, \qquad p \in P. \tag{5.3}$$

The best constant ω is equal to the constant $-\operatorname{cvx} f - \operatorname{cvx}(-f)$. Also $(\operatorname{cvx} f)(x) = \alpha(x)+\beta$ and $\operatorname{cvx}(-f)(x) = -\alpha(x)-\beta-\omega$ if ω is chosen as the smallest possible constant.

Proof. We know from Theorem 4.5 that the two convex functions $u = \operatorname{cvx} f$ and $v = \operatorname{cvx}(-f)$ satisfy

$$u \leqslant f \leqslant u+1 \text{ and } v \leqslant -f \leqslant v+1 \text{ in } P. \tag{5.4}$$

The functions u and v are real-valued convex functions and possess affine minorants. This implies that f satisfies (5.1) with some finite h. From Proposition 5.2 and (5.2) we know that $u+v$ is a constant $-\omega$ and that $h_* = \omega$ is the best constant in (5.1) when we are allowed to vary both α and β.

It remains to be seen that $\omega \leqslant 1$. The first inequality in (5.4) can be rewritten in the notation of the previous proof as

$$-\omega-\gamma \leqslant f-\alpha \leqslant -\omega-\gamma+1,$$

which shows that $h = 1$ is an admissible choice; thus the infimum h_* of all such h cannot exceed 1.

The last statement follows from the inequality

$$\alpha(p)+\beta \leqslant u \leqslant -v \leqslant \alpha(p)+\beta+\omega, \qquad p \in P,$$

where we now know that $u+v = -\omega$, so that

$$\alpha(p) + \beta + v \leqslant -\omega \leqslant 0 \leqslant \alpha(p) + \beta + \omega + v,$$

which forces $\alpha(p) + \beta + \omega + v$ to be equal to 0. The proof is complete.

We rewrite the theorem in the most common situation:

Corollary 5.4. *If both $f\colon \mathbf{Z}^n \to \mathbf{Z}$ and $-f$ are $(\mathbf{Z}^n \times \mathbf{Z})$-convex, then there exist $\alpha \in \mathbf{R}^n$ and $\beta \in \mathbf{R}$ such that*

$$0 \leqslant f(p) - \alpha \cdot p - \beta \leqslant \omega, \qquad p \in \mathbf{Z}^n, \tag{5.5}$$

where ω is the constant $-\operatorname{cvx} f - \operatorname{cvx}(-f) \leqslant 1$.

Is it possible to take one of the inequalities in (5.5) strict, like in (4.1)? We shall see that this is not always so.

Example 5.5. In Example 4.2 we see that $\alpha p \leqslant f_0(p) < \alpha p + 1$, whereas $\alpha p < f_1(p) \leqslant \alpha p + 1$. In each case we have one strict inequality. Both inequalities are optimal ($\omega = 1$).

Example 5.6. In Example 4.3 we see that

$$\tfrac{1}{2}p \leqslant g_A(p) \leqslant \tfrac{1}{2}p + 1, \qquad p \in \mathbf{Z}. \tag{5.6}$$

If A is empty this can be improved to $\frac{1}{2}p + \frac{1}{2} \leqslant g_{\varnothing}(p) \leqslant \frac{1}{2}p + 1$. If A is equal to all of $2\mathbf{Z}$, then we have $\frac{1}{2}p \leqslant g_{2\mathbf{Z}}(p) \leqslant \frac{1}{2}p + \frac{1}{2}$. Thus in these two cases the graph of g_A is contained in a strip of height $\omega = \frac{1}{2}$. In all other cases we see that none of the inequalities in (5.6) can be replaced by a strict inequality. We already remarked above that g_A is $(\mathbf{Z} \times \mathbf{Z})$-convex if and only if A is an interval of even numbers. We note that both g_A and $-g_A$ are $(\mathbf{Z} \times \mathbf{Z})$-convex if and only if $A = \varnothing$ or $A = 2\mathbf{Z}$ or A is a semi-infinite interval.

The example shows that there is a choice between the intervals $[0, \omega[$ and $]0, \omega]$ in the inequality (5.3) for different values of p. This choice is made precise in the following result.

Theorem 5.7. *Let $f\colon \mathbf{Z}^n \to \mathbf{Z}$ and $-f$ be $(\mathbf{Z}^n \times \mathbf{Z})$-convex and let $\alpha \in \mathbf{R}^n$ and $\beta \in \mathbf{R}$ be such that (5.3) holds with $\omega = h_*$, i.e., with the smallest h possible. Define*

$$D^j = \{(p, f(p)) \in \mathbf{Z}^n \times \mathbf{Z}; f(p) = \alpha \cdot p + \beta + j\omega\}, \qquad j = 0, 1,$$

and

$$A^j = \pi_{n+1}(D^j) = \{p \in \mathbf{Z}^n; f(p) = \alpha \cdot p + \beta + j\omega\}, \qquad j = 0, 1,$$

where $\pi_{n+1}\colon \mathbf{Z}^n \times \mathbf{Z} \to \mathbf{Z}^n$ denotes the projection which forgets the last coordinate. Assume that $\omega > 0$. Then A^0 and A^1 are disjoint, and D^0 and D^1 are $(\mathbf{Z}^n \times \mathbf{Z})$-convex.

Proof. That A^0 and A^1 are disjoint follows from the fact that f takes different values in them: $f(p) = \alpha \cdot p + \beta$ when $p \in A^0$ while $f(p) = \alpha \cdot p + \beta + \omega$ if $p \in A^1$.

The sets $T^j = \{(p, q) \in \mathbf{Z}^n \times \mathbf{Z}; q = \alpha \cdot p + \beta + j\omega\}$ are $(\mathbf{Z}^n \times \mathbf{Z})$-convex, as are $\operatorname{epi} f$ and $\operatorname{hypo} f$. Therefore so are the intersections $D^0 = T^0 \cap \operatorname{epi} f$ and $D^1 = T^1 \cap \operatorname{hypo} f$.

6 Digital Hyperplanes

The concept of *naive discrete line* was introduced by Reveillès (1991:48). Such a line is defined to be the set of all integer points $p \in \mathbf{Z}^2$ such that $0 \leqslant \alpha_1 p_1 + \alpha_2 p_2 < \max(|\alpha_1|, |\alpha_2|)$, where α_1 and α_2 are relatively prime integers. Generalizing this slightly, we define a *naive digital hyperplane* as the set of all points $p \in \mathbf{Z}^n$ which satisfy the double inequality

$$0 \leqslant \alpha \cdot p + \beta < h,$$

for some $\alpha \in \mathbf{R}^n \smallsetminus \{0\}$ and some $\beta \in \mathbf{R}$, where $h = \|\alpha\|_\infty$. We remark that one can always interchange the strict and the non-strict inequalities: the set just defined can equally well be defined by

$$0 < (-\alpha) \cdot p - \beta - \omega \leqslant h.$$

The precise size of h is important for the representation of the hyperplane as the graph of a function of $n - 1$ variables as shown by the following result.

Theorem 6.1. *Define*

$$T = \{p \in \mathbf{Z}^n; 0 \leqslant \alpha \cdot p + \beta \leqslant h\} \text{ and } T_s = \{p \in \mathbf{Z}^n; 0 < \alpha \cdot p + \beta < h\}, \quad (6.1)$$

where $\alpha \in \mathbf{R}^n \smallsetminus \{0\}$, $\beta \in \mathbf{R}$ *and* $h > 0$, *and let*

$$T^j = \{p \in \mathbf{Z}^n; \alpha \cdot p + \beta = jh\}, \qquad j = 0, 1. \quad (6.2)$$

Let D be a subset of $\mathbf{Z}^n$ which is contained in T and contains T_s and define $D_s = D \cap T_s$ and $D^j = D \cap T^j$. Fix an integer $k = 1, \dots, n$ and let $\pi_k \colon \mathbf{Z}^n \to \mathbf{Z}^{n-1}$ be the projection which forgets the kth coordinate. Then $\pi_k\big|_D$ is injective if $h < |\alpha_k|$, and $\pi_k\big|_D$ is surjective if $h > |\alpha_k|$. If $h = |\alpha_k|$, then $\pi_k\big|_D$ is injective if and only if $\pi_k(D^0)$ and $\pi_k(D^1)$ are disjoint, and $\pi_k\big|_D$ is surjective if and only if $\pi_k(D^0 \cup D^1) = \pi_k(T^0 \cup T^1)$.

Proof. For ease in notation we let $k = n$ and write $p' = (p_1, \dots, p_{n-1})$ and similarly for α. Then p belongs to T if and only if

$$-\alpha' \cdot p' - \beta \leqslant \alpha_n p_n \leqslant -\alpha' \cdot p' - \beta + h, \quad (6.3)$$

and p belongs to T_s if and only if

$$-\alpha' \cdot p' - \beta < \alpha_n p_n < -\alpha' \cdot p' - \beta + h. \quad (6.4)$$

Clearly for every p' there is at most one p_n which satisfies the inequalities if $h < |\alpha_n|$ or if $h = |\alpha_n|$ and $(\alpha' \cdot p' + \beta)/h$ is not an integer. Also there is at least one p_n if $h > |\alpha_n|$ or if $h = |\alpha_n|$ and $(\alpha' \cdot p' + \beta)/h$ is not an integer. Here it does not matter whether we use (6.3) or (6.4), so the conclusion holds also for D.

The case when $h = |\alpha_n|$ and $(\alpha' \cdot p' + \beta)/h$ is an integer remains to be considered. Then we see that there are two values of p_n which satisfy (6.3) and none that satisfies (6.4). Hence there is at most one p_n such that (p', p_n) belongs to $D = D^0 \cup D_{\mathrm{s}} \cup D^1$ if and only if $\pi_k(D^0)$ and $\pi_k(D^1)$ are disjoint. There is at least one p_n such that (p', p_n) belongs to D if and only if $\pi_k(D^0 \cup D^1)$ contains every point in the projection of $T^0 \cup T^1$. This completes the proof.

We do not suppose here that $h = \|\alpha\|_\infty$. However, this is the most natural case: we then know that $\pi_k\big|_D$ is a bijection for any k such that $|\alpha_k| = \|\alpha\|_\infty$ and the conditions on the D^j are satisfied, and that $\pi_j\big|_D$ is surjective for all j such that $|\alpha_j| < \|\alpha\|_\infty$.

In view of Theorems 5.7 and 6.1 it seems reasonable to propose the following definition.

Definition 6.2. *A* refined digital hyperplane *is a* $\mathbf{Z}^n$*-convex subset* D *of* $\mathbf{Z}^n$ *which is contained in* T *and contains* T_{s}*, where* T *and* T_{s} *are the slabs defined by (6.1) for some* $\alpha \in \mathbf{R}^n \smallsetminus \{0\}$*,* $\beta \in \mathbf{R}$*, and* $h > 0$*; and in addition is such that, for at least one* k *such that* $|\alpha_k| = h$*, the sets* $D^j = D \cap T^j$ *have disjoint projections* $\pi_k(D^j)$*, and* $\pi_k(D^0 \cup D^1) = \pi_k(T^0 \cup T^1)$*.*

The naive hyperplanes now appear as a special case, viz. when $D^0 = T^0$, and D^1 is empty, or conversely, and $|\alpha_k| = \|\alpha\|_\infty$.

Example 6.3. Define $D = (D^0 \times \{0\}) \cup (D^1 \times \{1\})$, where D^j, $j = 0, 1$, are two subsets of $\mathbf{Z}^{n-1}$ such that $D^1 = \mathbf{Z}^{n-1} \smallsetminus D^0$. Then D is a refined digital hyperplane if and only if both D^0 and D^1 are $\mathbf{Z}^{n-1}$-convex.

Example 6.4. Define $D = \{(p_1, p_1) \in \mathbf{Z}^2; p_1 \leqslant 0\} \cup \{(p_1, p_1 + 1) \in \mathbf{Z}^2; p_1 \geqslant 0\}$. This is a refined digital hyperplane with $|\alpha_1| = |\alpha_2| = \|\alpha\|_\infty = 1$. The projection π_1 satisfies the requirements in the definition, but π_2 does not.

The following result motivates the definition just given and relates it to the digitally convex functions we have introduced.

Theorem 6.5. *A subset* D *of* $\mathbf{Z}^n$ *is a refined digital hyperplane if and only if it is the graph of a function* $f\colon \mathbf{Z}^{n-1} \to \mathbf{Z}$ *such that both* f *and* $-f$ *are* $(\mathbf{Z}^{n-1} \times \mathbf{Z})$*-convex.*

Proof. Let f be a $(\mathbf{Z}^{n-1} \times \mathbf{Z})$-convex function such that also $-f$ is $(\mathbf{Z}^{n-1} \times \mathbf{Z})$-convex. Then $D = \operatorname{graph} f$ is a refined digital hyperplane according to Theorem 5.7.

Conversely, if D is a refined digital hyperplane and $h = |\alpha_n|$, then the projection $\pi_n\big|_D$ is bijective, and this allows us to define a function $f\colon \mathbf{Z}^{n-1} \to \mathbf{Z}$, $f(p') = -\alpha' \cdot p' - \beta + jh$ as in the proof of Theorem 6.1 with $j = 0$ or 1 being uniquely determined by the requirements on the D^j. This function as well as its negative are $(\mathbf{Z}^{n-1} \times \mathbf{Z})$-convex, since both its epigraph and its hypograph are $\mathbf{Z}^n$-convex. To wit, assuming α_n to be positive, its epigraph is equal to $D + (\{0\} \times \mathbf{N})$, and its hypograph is equal to $D + (\{0\} \times (-\mathbf{N}))$.

7 Conclusions

In this paper we have studied a simple definition of convex sets in $\mathbf{Z}^n$ and of convex functions defined on $\mathbf{Z}^n$ and having integer values. The definitions are actually given not only for functions defined on $\mathbf{Z}^n$ but for other subsets of $\mathbf{R}^n$ as well.

We have shown that the functions so defined share important properties of convex functions defined on vector spaces, viz. concerning the relation between convex sets and convex functions, and suprema and infima of families of functions. We have also clarified how much a convex digital function can deviate from a convex function of real variables.

From several points of view the definitions seem to be satisfying. They are extremely simple and easy to grasp; nevertheless, there are nontrivial difficulties in checking whether a given function is convex.

A kind of ambiguity in the values of a convex function is shown to be inevitable: in general a convex function is not determined by its restriction to the complement of a point.

Functions that are both convex and concave are of interest as candidates for defining digital hyperplanes; in fact we have shown that they define sets which are precisely the sets satisfying a refined definition of digital hyperplanes.

Acknowledgment. I am grateful to Erik Melin for comments on earlier versions of the manuscript, to Jean-Pierre Reveillès for clarifying the history of the notion of a naive digital line, and to an anonymous referee for help with references.

References

Eckhardt, Ulrich 2001. Digital lines and digital convexity. *Lecture Notes in Computer Science* **2243**, pp. 209—228.

Hiriart-Urruty, Jean-Baptiste; Lemaréchal, Claude 2001. *Fundamentals of Convex Analysis.* Springer-Verlag. X + 259 pp.

Hörmander, Lars 1994. *Notions of Convexity.* Boston: Birkhäuser. viii + 414 pp.

Melin, Erik 2003. Digital straight lines in the Khalimsky plane. Uppsala University, Department of Mathematics, Report 2003:30. Accepted for publication in *Mathematica Scandinavica.*

Reveillès, J[ean]-P[ierre] 1991. *Géométrie discrète, calcul en nombres entiers et algorithmique.* Strasbourg: Université Louis Pasteur. Thèse d'État, 251 pp.

Rosenfeld, Azriel 1974. Digital straight line segments. *IEEE Transactions on Computers.* **c-32**, No. 12, 1264—1269.

Rosenfeld, Azriel; Klette, Reinhard 2001. Digital straightness. *Electronic Notes in Theoretical Computer Science* **46**, 32 pp. `http://www.elsevier.nl/locate/entcs/volume46.html`

Singer, Ivan 1997. *Abstract Convex Analysis.* John Wiley & Sons, Inc. xix + 491 pp.

Discrete Surfaces Segmentation into Discrete Planes

Isabelle Sivignon[1], Florent Dupont[2], and Jean-Marc Chassery[1]

[1] Laboratoire LIS - Institut National Polytechnique de Grenoble,
961, rue de la Houille Blanche - BP46,
38402 St Martin d'Hères Cedex
{sivignon, chassery}@lis.inpg.fr
[2] Laboratoire LIRIS - Université Claude Bernard Lyon 1,
Bâtiment Nautibus - 8, boulevard Niels Bohr,
69622 Villeurbanne Cedex
fdupont@liris.cnrs.fr

Abstract. This paper is composed of two parts. In the first one, we present an analysis of existing discrete surface segmentation algorithms. We show that two classes of algorithms can actually be defined according to discrete surface and plane definitions. In the second part, we prove the link between the two classes presented. To do so, we propose a new labelling of the surface elements which leads to a segmentation algorithm of the first class that can be easily transformed into a segmentation algorithm of the second class.

1 Introduction

The segmentation of a discrete object surface into pieces of discrete planes is the first step of more global processes like polyhedrization of the surface [1] or surface area estimation [2]. This problem can be stated as follows: given a discrete object, a pair of connectivities used for the object and the background, and a definition of the discrete surface, label all the surface elements of the object such that the elements with the same label belong to the same piece of discrete plane. Each set of surface elements with the same label is called a discrete face.

This statement raises the problem that both discrete surfaces and discrete planes have to be defined before the segmentation process. In a general way, there exist two families of discrete surfaces definition: either the voxels or the surfels are the surface elements. In the same way, two types of discrete planes are mainly used: the naive planes and the standard planes.

In the literature, those discrete surface and discrete plane definitions are all used in segmentation algorithms, but most of the time, few arguments are given to justify the use of one definition instead of another. Some authors even propose several algorithms using different planes and surfaces definitions, but did not explore the relation between the different algorithms [3]. Indeed, by now, no work has been accomplished in order to study the link between, first, the discrete

R. Klette and J. Žunić (Eds.): IWCIA 2004, LNCS 3322, pp. 458–473, 2004.

surface and discrete plane definitions, and second, the segmentation processes that use different kinds of discrete planes.

In Section 2, we define those discrete surfaces and point out the links between the surfaces and the different types of discrete planes. Then, we propose a short overview of existing segmentation algorithms, in relation with the surface and plane definitions given previously. Next, in Section 3, we recall a segmentation algorithm we proposed in a previous work. This algorithm uses naive planes over the surface voxels and is quite different from existing algorithms since the voxels' faces are labelled instead of the voxels themselves. We moreover show how the 6-connectivity between the object voxels can be handled in this algorithm. Finally, in Section 4, we propose a transformation from this segmentation algorithm with naive planes into a segmentation algorithm with standard planes. This transformation shows that the two segmentation processes are in fact equivalent. Some image results are given before a few words of conclusion.

2 Different Discrete Planes for Different Kind of Surfaces

2.1 Two Types of Surfaces

The two existing approaches differ by the surface elements considered: the surface elements are either voxels (dimension 3) or voxels' faces (dimension 2).

In his PhD thesis, Perroton [4] calls those approaches homogeneous and heterogeneous respectively, since in the first one the surface elements are space elements, whereas in the second one the dimension of the surface elements is lower than the dimension of the discrete space.

Homogeneous Approach. In 1982, a definition of discrete surface in dimension n is proposed by Udupa et al. [5], and a surface tracking algorithm is described. Nevertheless, no Jordan's theorem proving that this surface separates the object voxels from the background voxels has been shown with this definition. The year before, Morgenthaler et al. [6] gave a local definition of a discrete surface, and a global Jordan's theorem based on this definition is shown. Malgouyres also proposed [7] a local definition for 26-connected discrete objects with a 6-connected background. A Jordan's theorem is also shown and the link between Morgenthaler's surfaces is stated [8, 7]. Another surface definition involving n-dimensional surface elements has been recently proposed in [9], where separating pairs of connectivity are studied according to this definition.

Nevertheless, this approach causes some problems [4]: first, the surface of an object and the surface of the background are different, and second, for some objects containing a cavity, the two surfaces expected (two surfaces should separate the object from the two background components: the cavity, and the infinite background) may have voxels in common.

Heterogeneous Approach. This approach is based on the abstract cellular complex introduced by Kovalevsky [10] for discrete images topology. In dimension 3, the following cells are defined (see Figure 1(a)):

- *voxel*: cell of dimension 3
- *surfel*: cell of dimension 2
- *linel*: cell of dimension 1
- *pointel*: cell of dimension 0.

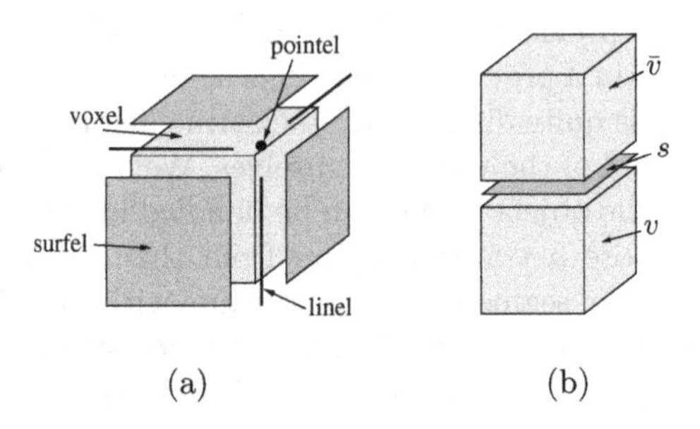

Fig. 1. (a) Cellular decomposition of a voxel. (b) A surfel defined by two 6-connected voxels

Surfels can also be defined as pairs of voxels:

Definition 1. *A surfel is the common face of a couple of* 6*-connected voxels. We denote* $s = \{v, \bar{v}\}$ *the surfel defined by the two* 6*-connected voxels* v *and* $\bar{v}$*.*

The surface of an object is directly derived from this definition:

Definition 2. *Let* $\mathcal{O}$ *be a 3D discrete object. The surface of* $\mathcal{O}$ *is defined as the set of surfels* $s = \{v, \bar{v}\}$ *such that* $v \in \mathcal{O}$ *and* $\bar{v} \notin \mathcal{O}$*.*

In 1992, Herman [11] generalizes this definition to any dimension. This definition has moreover been used to write surface tracking algorithms: first, Artzy et al. [12] proposed a tracking algorithm for 18-connected objects; then, Gordon and Udupa [13] extend this algorithm for 6-connected objects improving it; in 1994, Udupa [14] adds a connectivity definition over surfels, and proposes a very efficient tracking algorithm; in [15], Perroton describes a parallel surface tracking algorithm for 26-connected objects.

In the following, we will see how this definition may be used in two different ways in the context of surface segmentation.

2.2 Which Planes for Which Surfaces?

In a segmentation process, the discrete points of the object surface are labelled according to the discrete face they belong to. In the literature, two main approaches have been developed: the discrete points are either the voxels or the pointels of the object.

The definition of a discrete surface given in Definition 2 states that the surface is composed of all the surfels $s = \{v, \bar{v}\}$ such that $v \in \mathcal{O}$ and $\bar{v} \notin \mathcal{O}$. From this definition, we derive that the segmentation is done over the pointels of the object surface. Nevertheless, another point of view is given by the definition of the inner boundary of an object:

Definition 3. *Let $\mathcal{O}$ be a 3D discrete object. The inner boundary of $\mathcal{O}$ is defined as:*

$$II(\mathcal{O}) = \{v \mid \{v, \bar{v}\} \text{ is a surfel of } \mathcal{O}\text{'s surface}\}$$

The voxels of $II(\mathcal{O})$ are the underlying voxels of the surface surfels. When a segmentation is done over the object voxels, the voxels of $II(\mathcal{O})$ are usually labelled.

Let us recall the analytical definition of a discrete plane [16, 17]:

Definition 4. *A discrete plane of parameters (a, b, c, μ) (a, b, c and μ belong to $\mathbb{Z}$ and a, b and c are relatively prime) and thickness ω is the set of voxels (x, y, z) verifying the following inequalities:*

$$0 \leq ax + by + cz + \mu < \omega$$

The thickness parameter ω defines many plane classes, but two types of discrete planes are mainly used in the literature. When $\omega = \max(|a|, |b|, |c|)$, the discrete plane is called *naive*, and when $\omega = |a| + |b| + |c|$, the plane is called *standard*. Naive planes are the thinnest planes without 6-connected holes, and standard planes are the thinnest planes without tunnels [17]. For further information about digital planarity, see for instance [18].

Table 1 presents the links between discrete planes and discrete surfaces. On one hand, if the discrete points are the voxels of the inner boundary, then naive planes are used for the segmentation. Indeed, the voxels that are 18-connected but not 6-connected with the background are needed for a segmentation with standard planes, but do not belong to the inner boundary of the object. On the first line of the table, a naive plane recognition has been done over the inner boundary of the object: the voxels of $II(\mathcal{O})$ are labelled.

On the other hand, the pointels belonging to the object's surface are 6-connected, which suits well to a segmentation with standard planes. On the second line of the table, a standard plane recognition has been done over the pointels of the surfel boundary of the object: a surfel is colored if and only if its four adjacent pointels belong to the discrete face.

2.3 Short Overview of Existing Algorithms

The first segmentation algorithm was proposed by Debled-Rennesson [19]. She applies the naive plane recognition algorithm she proposed on the inner boundary of a discrete object. However, her algorithm works only for objects having known symmetries. With a totally different idea, Vittone proposes in [20] a segmentation algorithm based on the fact that any naive plane can be decomposed into basic structures [21]. But only a few preliminary results are given and this method has not been further extended by now. The following year, Vittone [22] wrote another segmentation algorithm using the naive plane recognition algorithm she proposed. This recognition algorithm is based on a dual transformation, which we also use in our work. This will be detailed in the next section. In [3], Papier proposes another segmentation algorithm based on Debled-Rennesson naive

Table 1. Links between the discrete surfaces and the discrete planes used for the segmentation process

Surface definition	Discrete points	Discrete planes	Example
Inner boundary	Voxels	Naive planes	
Set of surfels	Pointels	Standard planes	

plane recognition algorithm. However, his algorithm does not give satisfactory results since many small discrete faces (with a few voxels) are recognized.

This observation led him to propose a segmentation algorithm of the surfel boundary with standard planes [1]. The discrete faces recognized are forced to be connected and homeomorphic to disks in order to be used in a polyhedrization process. A Fourier-Motskin elimination algorithm is used to recognize the standard planes, which makes the overall algorithm complexity high. Moreover, many small discrete faces are still recognized by this algorithm, and some tricks like discrete face size limitations are proposed to better the results. Finally, Klette and Sun proposed in [2] a segmentation algorithm close to the one of Papier in order to estimate the object surface area. In this case, no particular constraint over the shape of the discrete faces is required, except the connectivity, which is implicitly obtained since a breadth-first tracking of the surfels is done.

For the algorithms of the first paragraph, a segmentation of the inner boundary is done, and the voxels are labelled. The algorithms of the second paragraph use a standard plane segmentation of the object pointels, and a surface surfel is labelled by a discrete face when its four adjacent pointels belong to the face.

3 Surfel Labelling Using Naive Planes

In this section, we propose a segmentation algorithm of the inner boundary of an object using naive planes. But instead of labelling the voxels of the inner boundary, we label only some of the boundary's surfels. Indeed, all the existing algorithms using naive planes label the voxels of the object inner boundary. Nevertheless, the inner boundary definition is derived from the surfel surface definition. Thus the surface elements are the surfels, and it seems more appropriate to label the surfels. In Section 4, we show how this segmentation of the inner boundary can be easily transformed into a segmentation into standard planes, which link the two segmentation approaches existing by now.

Since the final segmentation uses standard planes, we consider a 6-connectivity relationship for the voxels of the object. Indeed, 18- and 26- connectivities do not enable the description of the object surfel surface as a 2-dimensional combinatorial manifold, which is inconsistent with the use of standard planes [23].

3.1 Algorithm Description

Part of this algorithm is presented in [24] and we first recall here the main points of this method.

The general idea of this algorithm is to get the following labelling of the surface surfels: if a surfel $s = \{v, \bar{v}\}$ of the surface (v belongs to the object and $\bar{v}$ does not) is labelled with the discrete face f, then there exists a Euclidean plane p crossing the segment $[v\bar{v}[$. In other words, the discretization of the plane p with the OBQ discretization [25] scheme contains the voxels of f.

In the following, we denote by l one direction of the set $\{(1,0,0),(-1,0,0),(0,1,0),(0,-1,0),(0,0,1),(0,0,-1)\}$. Then for each voxel v of the inner boundary of an object, there exist at least one direction l such that the voxel $v + l$ does not belong to the object. The naive plane recognition used is a directional recognition algorithm:

Definition 5. *Let V be a set of voxels. The directional recognition in direction l computes the set of Euclidean planes crossing all the segments $[v\bar{v}[$ such that $v \in V$ and $\bar{v} = v + l$.*

This recognition is achieved using Vittone's naive plane recognition algorithm [22]. This algorithm is based on a dual transformation. Consider for instance the direction $l = (0, 0, 1)$. Then the Euclidean planes of parameters $(\alpha_0, \beta_0, \gamma_0)$ which cross the segment $[v, v+l[$ where $v = (x, y, z)$ are those fulfilling the double inequality $0 \leq \alpha_0 x + \beta_0 y + z + \gamma_0 < 1$. Then, in the parameter space (α, β, γ), the set of Euclidean planes computed by the directional algorithm on a set V is a convex polyhedron defined as the intersection of all the double inequalities related to the voxels of V.

But this recognition algorithm enables to label only the surfels $s = \{v, v'\}$ where $v' = v+l$ if l is the direction of recognition (see Figure 2(a)). A second step of the algorithm is to label the inner surfels of the plane, which are the surfels sharing two edges with surfels already labelled by the directional algorithm (see Figure 2(b) and (c)). It is easy to prove that the Euclidean planes that cut the segments related to the two surfels defining an inner surfel also cut the segment related to this inner surfel.

After this short presentation of the recognition process used, let us see how this algorithm is applied on a surface. Since the quality of the segmentation is not the main point of this paper, and for the sake of clarity of the rest of this

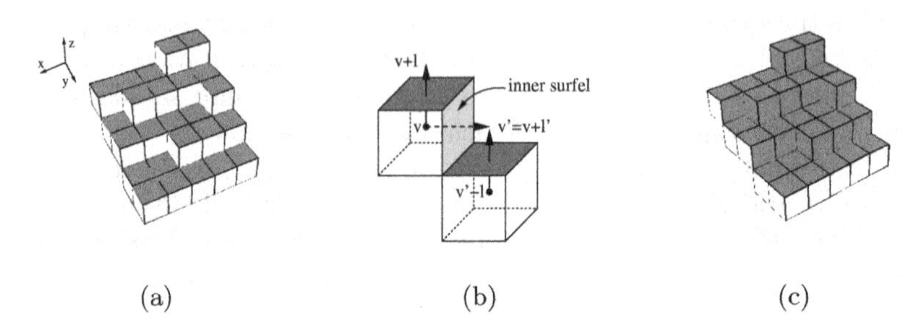

Fig. 2. Labelling of the surfels of a piece of plane of parameters $(1, 3, -5, 0)$ using the directional algorithm in direction $(0, 0, 1)$ (from [24]): (a) just after the directional algorithm; (b) illustration of an inner surfel; (c) after the labelling of inner surfels

work, we do not present the algorithm described in [24], but quickly explain a simple method. A seed voxel v_0 is chosen for the discrete face to recognize in direction l. Neighbour voxels are added one by one if they belong to the same naive plane as the voxels of f and if they fulfill the condition presented in the next paragraph. This process starts over with a new seed voxel until all the surfels are labelled.

3.2 6-Connected Objects

Since 6-connectivity is considered for the discrete object, a discrete face recognized on the object surface should not be split over two different 6-connected components. Since naive planes are 18-connected, we have to define the cases where two voxels are 18-connected in a 6-connected object.

Definition 6. *Let v and v' be two voxels of a 6-connected object $\mathcal{O}$. Then v and v' are 18-neighbours in $\mathcal{O}$ if and only if one of the following two conditions is fulfilled (see Figure 3):*

- *v and v' are 6-connected (they share a surfel)*
- *v and v' share a linel, and one out of the other two voxels sharing this linel belongs to $\mathcal{O}$.*

Now consider a directional recognition in direction l.

Definition 7. *Two voxels v and v' are l-neighbours in a 6-connected object $\mathcal{O}$ if only if they fulfill the following three conditions:*

- *v and v' belong to the inner boundary of $\mathcal{O}$ and the voxels $v + l$ and $v' + l$ do not belong to $\mathcal{O}$*
- *the projection of v and v' along direction l are 4-connected*
- *v and v' are 18-neighbours in $\mathcal{O}$ (see Definition 6).*

Figure 4 presents some examples of voxels that are or are not l-neighbours. Finally, during the segmentation process, a new voxel can be added to a discrete face f recognized in direction l if and only if it is l-neighbour with a voxel of f.

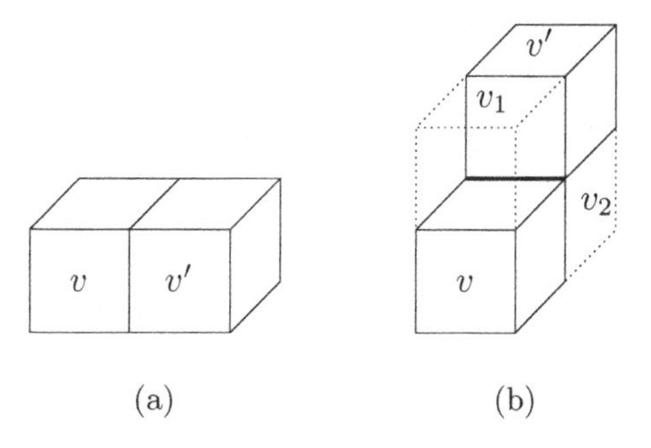

Fig. 3. The voxels v and v' are 18-connected in a 6-connected object if and only if they are in configuration (a) or if v_1 or v_2 belongs to the object in configuration (b)

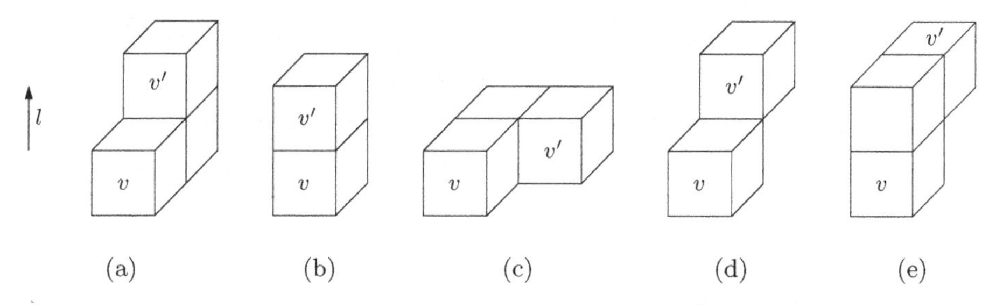

Fig. 4. The two voxels v and v' are l-neighbours (see l on the left) only in configuration (a) : in (b), the first condition of Definition 7 is not fulfilled since the surfel $\{v, v+l\}$ does not belong to the surface; in (c), the second condition is not fulfilled since the projections of v and v' along direction l are not 4-connected; in (d), the two voxels dot not fulfill the conditions of Definition 6; in (e), the two last conditions are fulfilled, but like in (b), the first one is not

Examples of segmentation results using 6-connectivity are proposed in Figure 5. The recognition in direction $(0,0,1)$ only has been computed on those examples. In (a), one unique plane is recognized; in (b), two planes are recognized since the object is composed of two 6-connected components; in (c), three planes are recognized, one for each 6-connected component.

3.3 Results

Some results of the algorithm proposed in [24] with the 6-connectivity constraint are presented in Figure 6. The results obtained with the simple algorithm presented in this paper simply contain more small faces. The first image (a) is an ellipsoid of parameters $(20, 16, 10)$ which is composed of 4190 surface surfels. 131 discrete faces are recognized on this surface, and 15 among them are composed of less than 5 surfels. The image in (b) is a microscopic scale view of part of a bone. Its surface is composed of 5298 surfels, and 201 discrete faces are recognized by the algorithm. 65 discrete faces are composed of less than 5 surfels.

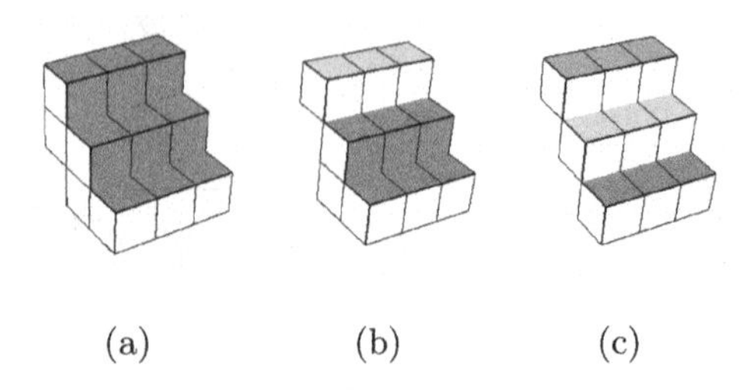

(a) (b) (c)

Fig. 5. Influence of the 6-connectivity on the segmentation computed

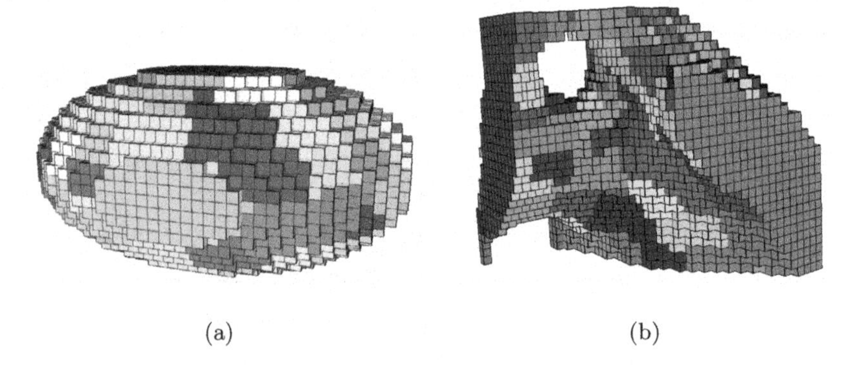

(a) (b)

Fig. 6. Results of the segmentation algorithm of the inner boundary with naive planes: (a) an ellipsoid of parameters $(20, 16, 12)$ and (b) a piece of bone

4 Segmentation Using Standard Planes

In this section, we show that the segmentation using naive planes over the inner boundary voxels presented in the previous section can be very easily transformed into a segmentation of the surface pointels into standard planes.

4.1 Transformation

Let us consider an infinite 3D discrete object defined as a half-space bounded by a naive plane of parameters (a, b, c, μ). Let us moreover suppose that the normal vector of this plane is directed towards the discrete object itself. Then we study the surfel surface of this object (see Figure 7(a)).

We use here the definition of standard plane proposed in [26] and directly derived from the standard digitization scheme:

Definition 8. *Let P be a Euclidean plane defined by the equation $ax+by+cz+\mu = 0$, with a,b,c and μ integers. Then the standard digitization of P is the set of voxels (x, y, z) such that:*

$$-\frac{|a|+|b|+|c|}{2} \leq ax+by+cz+\mu < \frac{|a|+|b|+|c|}{2}$$

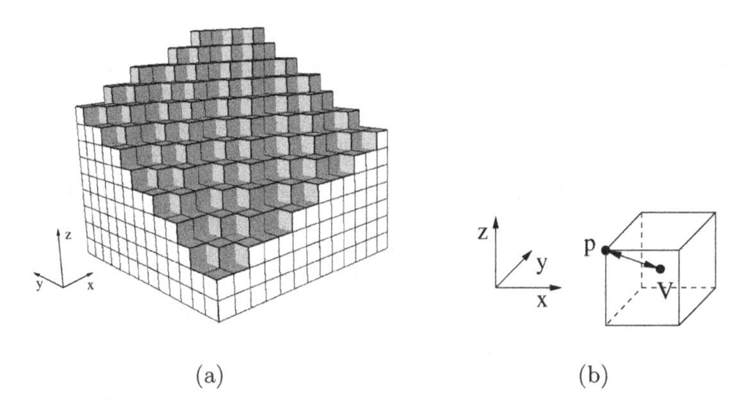

Fig. 7. (a) Discrete half-space bounded by the naive plane of parameters $(2, 3, -5)$. The surfels of the surface are in grey. (b) Illustration of the bijective transformation between a voxel V and a pointel p for given plane parameters

if $(a > 0)$ or $(a = 0$ and $b > 0$ or $(a = 0$ and $b = 0$ and $c > 0)$ (called standard orientation). Otherwise, the left inequality is strict whereas the right one is not.

Note that the possible values of $ax + by + cz + \mu$ are centered around zero. The following theorem makes the link between the naive plane parameters and the pointels of the surface.

Theorem 1. *Let P be a naive plane of parameters (a, b, c, μ) and $\mathcal{S}$ be the surfel surface described as above. Then the pointels belonging to $\mathcal{S}$ belong to the standard plane of parameters (a, b, c, μ') in the discrete grid translated by $t = (\frac{1}{2}, \frac{1}{2}, \frac{1}{2})$, with:*

$$\mu' = \tau - \frac{|a| + |b| + |c|}{2} - v.(a, b, c)$$

where

- $\tau = \begin{cases} \mu \text{ if } (a > 0) \text{ or } (a = 0 \text{ and } b > 0) \text{ or } (a = 0 \text{ and } b = 0 \text{ and } c > 0) \\ \mu + 1 \text{ otherwise} \end{cases}$
- $v = -\frac{1}{2}(sgn(a), sgn(b), sgn(c)) + (\frac{1}{2}, \frac{1}{2}, \frac{1}{2})$

($sgn(x)$ is equal to 1 if $x > 0$, -1 if $x < 0$, and 1 or -1 if $x = 0$).

Proof. This proof works by construction. Figure 8 illustrates the different steps of this construction from a naive 2D segment to a standard 2D segment. This example will be detailed in the next paragraph.

Since we transform a naive plane into a standard plane, the first thing is to thicken the naive plane P. This is equivalent to adding some voxels that fulfill larger inequalities (the thickness determines the bounds of the inequalities in a discrete plane definition). This is done in three steps:

- Since the inequalities defining a standard plane depend on the signs of the parameters, a preprocessing step is required when ($a < 0$) or ($a = 0$ and $b < 0$) or ($a = 0$ and $b = 0$ and $c < 0$). We use the fact that the two double-inequalities $0 \leq ax + by + cz + \mu < \omega$ and $0 < ax + by + cz + \mu + 1 \leq \omega$ are equivalent when integer values are considered to do the parameters translation $(a, b, c, \mu) \to (a, b, c, \mu + 1)$ in the case of a non standard orientation. This operation simply swap the large and strict inequalities in the discrete plane definition. In the following, we suppose that the parameters are in a standard orientation since the two cases are similar.
- The thickness of a standard plane is $|a| + |b| + |c|$. Since we supposed that the normal vector is directed towards the object, the voxels $V = (x, y, z)$ fulfilling the condition $\max(|a|, |b|, |c|) \leq ax + by + cz + \mu < |a| + |b| + |c|$ must be added to thicken the plane inside the discrete object.
- At this point, our plane is defined by the double inequality $0 \leq ax + by + cz + \mu < |a| + |b| + |c|$. To be consistent with Definition 8, a last translation of the parameters of vector $(0, 0, 0, -\frac{|a|+|b|+|c|}{2})$ must be done.

All the voxels of the discrete object that are 18-connected or 26-connected with the background are added by those transformations, and the translation $(a, b, c, \mu) \to (a, b, c, \mu(+1) - \frac{|a|+|b|+|c|}{2})$ has been done. The next part of the transformation consists in moving the discrete grid such that the discrete points are now the pointels.

A one-to-one and onto transformation is defined between the voxels of the plane P after thickening and the pointels of the surface: the pointel $p = (x - \frac{1}{2}\text{sgn}(a), y - \frac{1}{2}\text{sgn}(b), z - \frac{1}{2}\text{sgn}(c))$ is associated to the voxel $V = (x, y, z)$. This transformation is of course bijective but we must prove that if V belongs to P, then p belongs to the surface of $\mathcal{O}$. Let us consider a plane of parameters (a, b, c) with $a > 0$, $b > 0$ and $c < 0$. The other cases are symmetrical. Then the transformation maps the point V to the pointel $p = (x - \frac{1}{2}, y - \frac{1}{2}, z + \frac{1}{2})$ (see Figure 7(b)). Only three types of voxels may appear in the thickened discrete plane P:

- : 6-connected with the background
- : 18-connected with the background 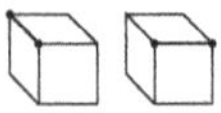
- : 26-connected with the background 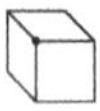

For each case, the pointels that belong to the surface are underlined. We note that for each voxel (x, y, z) of the thickened plane P, the pointel $(x - \frac{1}{2}, y - \frac{1}{2}, z + \frac{1}{2})$ always belongs to $\mathcal{O}$'s surface. For other parameters' signs a similar construction can be done, and finally, the transformation always maps a voxel of P into a pointel of the surface. If one of the parameters a, b or c is equal to zero, then two pointels of each voxel of P are always on $\mathcal{O}$'s surface. Then two bijective transformations are possible and correspond to the two possible values of $sgn(x)$ when $x = 0$. When two out of the three parameters are equal to zero, then 4 pointels of each voxel belong to $\mathcal{O}$'s surface, and then four transformations are possible.

This transformation defines the first part of the translation vector v of the theorem, and a simple translation of the grid of the vector $t = (\frac{1}{2}, \frac{1}{2}, \frac{1}{2})$ leads to the final result. □

The following Corollary shows how to transform a segmentation with naive planes into a segmentation with standard planes:

Corollary 1. *Let f be a discrete face recognized on the inner boundary of a discrete object $\mathcal{O}$ with the algorithm presented in Section 3. If the OBQ discretization of the Euclidean plane of parameters (a, b, c, μ) contains the voxels of f, then the standard discretization of the plane (a, b, c, μ') (where μ' is defined as in Theorem 1) contains all the pointels belonging to the labelled surfels of f, in a discrete grid translated by $t = (\frac{1}{2}, \frac{1}{2}, \frac{1}{2})$.*

The proof of this corollary is straightforward from Theorem 1.

4.2 Example

Figure 8 is an example of the transformation steps of Theorem 1 for a discrete segment. The first column represents the discrete segment, the second one represents the set of Euclidean solution lines in the parameter space, and the third one the parameters of the Euclidean lines related to the vertices of the convex polygon of the parameter space.

On the first line, a segment of the naive line of parameters $(2, -3, 3)$ is represented. On the second line, the naive segment has been thickened in order to get a standard segment. The parameters translation corresponding to this thickening is done. The third line represents the standard pointel segment derived from the standard voxel segment of the second line. A second translation of the Euclidean planes parameters is accomplished.

4.3 Results

Since the transformation presented above only concerns the parameters of the Euclidean planes recognized, visual results are hard to show. Nevertheless, Figure 9 presents a result of this transformation for a catenoid [1]. In (a), the surfel labelling using a naive plane segmentation over the inner boundary is represented: all the voxels containing surfels of the same color belong to the same discrete naive plane. In (b), the pointel labelling is represented after the transformation defined in Theorem 1: all the pointels with the same color belong to the same standard discrete plane. The same colors are used for the two images, so that the comparison is easier. Note that since each pointel belongs to several discrete faces, its color is the one of the first recognized face it belongs to.

[1] This image has been generated using the volgen program available on `http://www.cb.uu.se/tc18/code_data_set/Code/Volgen/`

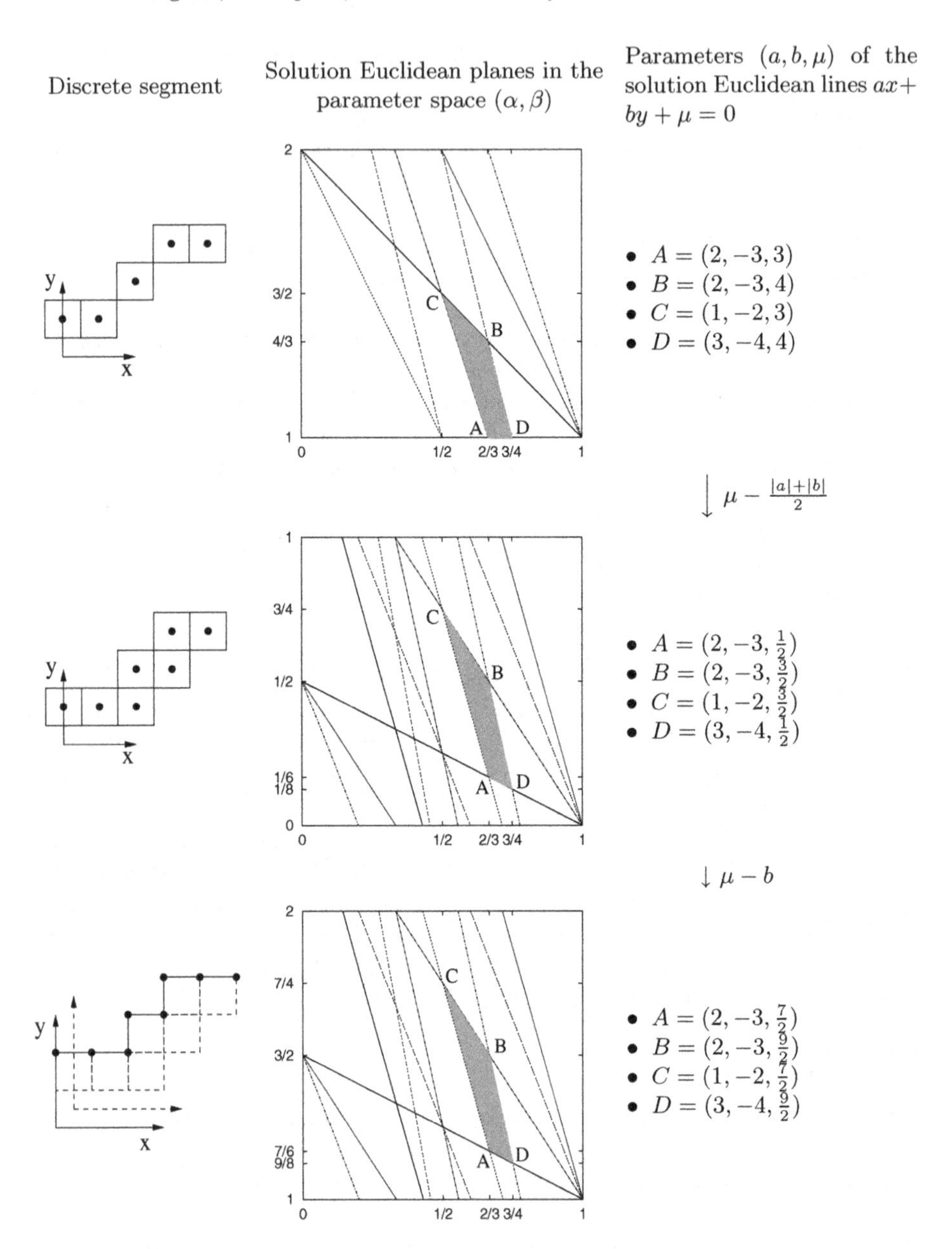

Fig. 8. All the steps from a naive voxel segment to a standard pointel segment

5 Conclusion

In this paper, we showed that two classes of discrete surface segmentation algorithms can actually be considered: the first one consists in looking for naive planes over the surface inner boundary, and the second one uses standard planes

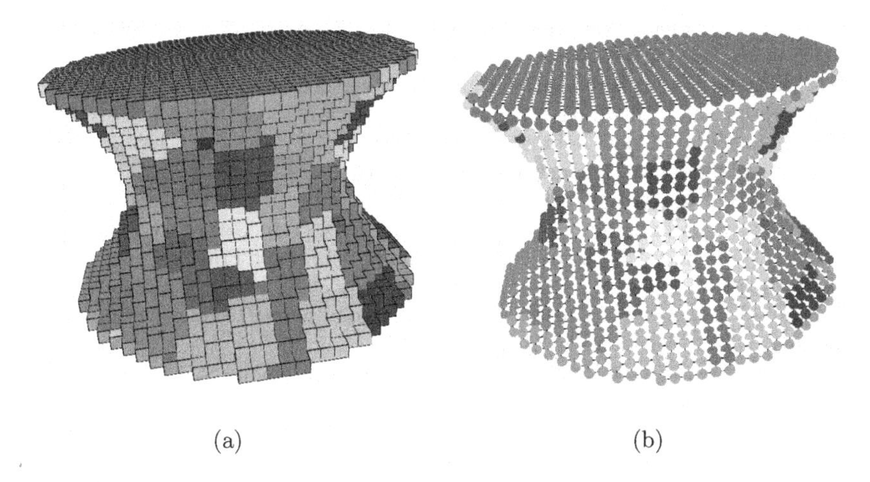

Fig. 9. Segmentation of a discrete catenoid surface: (a) surfel labelling; (b) pointel labelling

over the surface surfels. We recalled the different existing segmentation algorithms highlighting the fact that they all belong to one of those two classes.

The second part of this work was to show that those two types of segmentation are equivalent, and we expressed the transformation that changes from one to the other. To do this study, we proposed a segmentation algorithm using naive planes and generating a surfel labelling for 6-connected discrete objects.

Since the two segmentations are equivalent, an interesting future work would be to propose a complete study on the quality of segmentation algorithms, pointing out the influence of the choice of the first voxels of the discrete faces for instance. Such a study is very useful since the quality of the surface segmentation induces part of the quality of the polyhedrization or surface area estimation based on it.

References

1. Françon, J., Papier, L.: Polyhedrization of the boundary of a voxel object. In Bertrand, G., Couprie, M., Perroton, L., eds.: Discrete Geometry for Computer Imagery. Volume 1568 of Lect. Notes on Comp. Sci., Marne-la-Vallée, France, Springer-Verlag (1999) 425–434
2. Klette, R., Sun, H.J.: Digital planar segment based polyhedrization for surface area estimation. In Arcelli, C., Cordella, L.P., Sanniti di Baja, G., eds.: International Workshop on Visual Form. Volume 2059 of Lect. Notes on Comp. Sci., Capri, Italie, Springer-Verlag (2001) 356–366
3. Papier, L.: Polyédrisation et visualisation d'objets discrets tridimensionnels. PhD thesis, Université Louis Pasteur, Strasbourg, France (1999)
4. Perroton, L.: Segmentation parallèle d'images volumiques. PhD thesis, Ecole Normale Supérieure de Lyon, Lyon, France (1994)

5. Udupa, J.K., Srihari, S.N., Herman, G.T.: Boundary detection in multidimension. IEEE Trans. on Pattern Anal. and Mach. Intell. **4** (1982) 41–50
6. Morgenthaler, D.G., Rosenfeld, A.: Surfaces in three-dimensional digital images. Information and Control **51** (1981) 227–247
7. Malgouyres, R.: A new definition of surfaces of $\mathbb{Z}^3$. A new 3D Jordan theorem. Theoretical Computer Science **186** (1997) 1–41
8. Bertrand, G., Malgouyres, R.: Some topological properties of discrete surfaces. In Miguet, S., Montanvert, A., Ubéda, S., eds.: Discrete Geometry for Computer Imagery. Volume 1176 of Lect. Notes on Comp. Sci., Lyon, France, Springer-Verlag (1996) 325–336
9. Brimkov, V., Klette, R.: Curves, surfaces and good pairs. Technical Report CITR-TR-144, CITR, Auckland (NZ) (2004)
10. Kovalevsky, V.: Finite topology as applied to image analysis. Computer Vision, Graphics and Image Processing **46** (1989) 141–161
11. Herman, G.T.: Discrete multidimenional jordan surfaces. Computer Vision, Graphics and Image Processing **54** (1992) 507–515
12. Artzy, E., Frieder, G., Hreman, G.T.: The theory, design, implementation and evaluation of a three-dimensional surface detection algorithm. Computer Graphics and Image Processing **15** (1981) 1–24
13. Gordon, D., Udupa, J.K.: Fast surface tracking in three-dimensional binary images. Computer Vision, Graphics and Image Processing **45** (1989) 196–214
14. Udupa, J.K.: Multidimensional digital boundaries. Computer Vision, Graphics and Image Processing **56** (1994) 311–323
15. Perroton, L.: A new 26-connected objects surface tracking algorithm and its related pram version. Journal of Pattern Recognition and Artificial Intelligence **9** (1995) 719–734
16. Réveillès, J.P.: Géométrie discrète, calcul en nombres entiers et algorithmique. Thèse d'etat, Université Louis Pasteur, Strasbourg, France (1991)
17. Andrès, E., Acharya, R., Sibata, C.: Discrete analytical hyperplanes. Graphical Models and Image Processing **59** (1997) 302–309
18. Brimkov, V., Coeurjolly, D., Klette, R.: Digital Planarity - A Review. Technical Report CITR-TR-142, CITR, Auckland (NZ) (2004)
19. Debled-Rennesson, I.: Etude et reconnaissance des droites et plans discrets. PhD thesis, Université Louis Pasteur, Strasbourg, France (1995)
20. Vittone, J.: Caractérisation et reconnaissance de droites et de plans en géométrie discrète. PhD thesis, Université Joseph Fourier, Grenoble, France (1999)
21. Vittone, J., Chassery, J.M.: $(n-m)$-cubes and farey nets for naive planes understanding. In Bertrand, G., Couprie, M., Perroton, L., eds.: Discrete Geometry for Computer Imagery. Volume 1568 of Lect. Notes on Comp. Sci., Marne-la-Vallée, France, Springer-Verlag (1999) 76–87
22. Vittone, J., Chassery, J.M.: Recognition of digital naive planes and polyhedrization. In: Discrete Geometry for Computer Imagery. Volume 1953 of Lect. Notes on Comp. Sci., Springer-Verlag (2000) 296–307
23. Françon, J.: Sur la topologie d'un plan arithmétique. Theoretical Computer Science **156** (1996) 159–176
24. Coeurjolly, D., Guillaume, A., Sivignon, I.: Reversible discrete volume polyhedrization using marching cubes simplification. In Latecki, L.J., Mount, D.M., Wu, A.Y., eds.: Vision Geometry XII. Volume 5300 of Proceedings of SPIE., San Jose (2004) 1–11
25. Groen, F.C.A., Verbeek, P.W.: Freeman-coe probabilities of object boundary quantized contours. Computer Graphics and Image Processing **7** (1978) 391–402

26. Andrès, E.: Defining discrete objects for polygonlization : the standard model. In Braquelaire, A., Lachaud, J.O., Vialard, A., eds.: Discrete Geometry for Computer Imagery. Volume 2301 of Lect. Notes on Comp. Sci., Bordeaux, France, Springer-Verlag (2002) 313–325

Sketch-Based Shape Retrieval Using Length and Curvature of 2D Digital Contours

Abdolah Chalechale, Golshah Naghdy, and Prashan Premaratne

School of Electrical, Computer & Telecommunications Engineering,
University of Wollongong, NSW 2522, Australia
{ac82, golshah, prashan}@uow.edu.au

Abstract. This paper presents a novel effective method for line segment extraction using chain code differentiation. The resulting line segments are employed for shape feature extraction. Length distribution of the extracted segments along with distribution of the angle between adjacent segments are exploited to extract compact hybrid features. The extracted features are used for sketch-based shape retrieval. Comparative results obtained from six other well known methods within the literature have been discussed. Using MPEG-7 contour shape database (CE-1) as the test bed, the new proposed method shows significant improvement in retrieval performance for sketch-based shape retrieval. The Average Normalized Modified Retrieval Rank (ANMRR) is used as the performance indicator. Although the retrieval performance has been improved using the proposed method, its computational intensity and subsequently, its feature extraction time are slightly higher than some other methods.

1 Introduction

Humans can easily recognize objects from their shapes. Many applications including computer vision, object recognition, image retrieval, and indexing are likely to use shape features. Shape feature extraction has received a great deal of attention over the last decades [1–3]. In content-based image retrieval (CBIR), shape is exploited as one of the primary image features for retrieval purposes [4, 5]. Shape representation techniques fall into three main categories: feature vector approach (most popular technique), transformation approach, and relational approach. The choice of a particular representation is usually driven by application needs. These three categories are briefly explained in the following.

A shape is represented as a numerical vector in the feature vector approach. The difference between two shapes is evaluated based on a suitable distance. The Euclidean distance is the most widely used distance metric. However, other distances such as Hausdorff distance can also be employed. In the transformation methods [6, 7], shapes are distinguished by measuring the effort needed to transform one shape to another. Similarity is measured as a transformation distance. The methods in this category perform run-time evaluation of shape differences and do not support indexing, their retrieval performances are inefficient. In the relational approach [5, 8], complex shapes (or the scene) are broken down into a set of salient component parts. These are individually

R. Klette and J. Žunić (Eds.): IWCIA 2004, LNCS 3322, pp. 474–487, 2004.

described through suitable feature vectors. The overall description includes both the description of the individual parts and the relation between them. This approach is not commonly used in shape-based retrieval but, instead, is widely employed for the recognition of complex shapes or scenes.

Techniques in the feature vector approach are generally divided into two different classes: contour-based and region-based [9]. The former exploits shape boundary information to generate feature vectors while the latter uses the whole area of the shape. Contour-based techniques have gained more popularity based on the simplicity of shape contour feature acquisition and the sufficiency of the contour to represent shape in many applications. The Fourier descriptor (FD) and autoregressive methods, two representative techniques in this class, are compared in [10]. The curvature scale space (CSS) technique is adopted by MPEG-7 for extraction of contour-based descriptor [11]. It is computationally expensive and highly dependent on the contour continuity.

In region-based techniques, all the pixels within a shape region are taken into account to obtain the shape representation. Moment descriptors are commonly employed in region-based methods to describe shape. Zernike moments and angular radial transform (ART) methods are exploited in MPEG-7 to extract region-based shape descriptors [12, 11]. Recently Hoynck and Ohm [13] proposed a modification in the ART descriptor extraction procedure for considering partial occlusions.

The 2D Fourier transform in polar coordinates is employed for shape description in [14]. Its superiority over 1D Fourier descriptors, curvature scale space descriptor and Zernike moments has been shown in [15]. The polar Fourier descriptor (PFD) is extracted from frequency domain by exploiting two-dimensional Fourier transform on polar raster sampled image. It is able to capture image features in both radial and spiral directions.

Although much work has been done in the area of image retrieval using shape queries, very few have considered hand-drawn shapes as the input query. Matusiak *et al.* [16] and Horace *et al.* [17] have previously reported using rough and simple hand-drawn shape as input query for sketch-based shape retrieval. The approach in [16] is based on the curvature scale space technique that is computationally expensive and has been shown to be less efficient than the Fourier descriptors and Zernike moments techniques[15]. In [17] several dominant points are extracted for each contour using information derived from the convex hull and the contour curvature.

The angular partitioning (AP) method has been used for sketch-based image retrieval [18]. In this approach an image consisting of a non-homogeneous background and several objects is analyzed for the retrieval task. Scale and rotation invariant features are extracted using (a) spatial distribution of edge pixels in an abstract image, and (b) the magnitude of coefficients of a 1D Fourier transform. The abstract image is generated from original image by a statistical approach and from sketched query by a morphological thinning. An enhanced version of the approach which uses a more precise partitioning scheme has been recently proposed [19]. Here, the abstract image is overlayed by an angular radial partitioning (ARP) mask. The invariance features are extracted using several 1D Fourier transforms applied to the concentric partitions. The essence of an image including one isolated shape is different from an image containing multiple objects. The former possesses an image plane which is more scarce than the image plane of the latter which has more pixels therein. Hence, the philosophy behind angular or radial partitioning, used in the AP and ARP methods, is not adequate for shape description even though it is reasonably sufficient for multi-component images.

In this paper, we propose a new method for contour polygonization using chain code representation of the boundary shapes. The chain coded curve is filtered first for noise elimination. Next, it is submitted to a shifting operation to reduce wraparound effect followed by applying a Gaussian filter to smooth the curve. Finally, curvature/break points are determined using curve derivation. The result is a line segment set which is exploited for shape feature extraction using the existing length and curvature distributions. The hybrid features are integrated to generate a distance measure for measuring the similarity between the hand-drawn sketches and the original shapes.

The outline of the paper is as follows: the new proposed method is detailed in the next section followed by comparative results and discussion in Section 3. Section 4 concludes the chapter.

2 Contour Polygonization Using Chain Code Differentiation (CPCD)

In this section the details of the chain-based line segment extraction method are discussed. The input of the method is a digitized curve C derived by any contour extraction technique on a planar shape. In addition, any thinned sketched contour can be used as the input C. First, the starting point of an 8-connectivity chain code is determined using raster scanning the curve plane I [20]. The macro chain $A_i = \{a_1, a_2 \dots a_{n_i}\}, i = 1, 2, \dots, m$, where m is the number of chains in I and n_i is the chain length, is obtained and put in a chain set $\{A_i\}$.

Note that for a simple shape from the database, there usually exists only one closed contour C but for a corresponding sketched query, there are sometimes more than one curve per shape. This is due to occasional disconnectivities resulting from free-hand-drawing, scan resolution, and associated noise. The proposed method can cope with multiple-contour shape as well as one-contour shape since it considers several chains to be in the set $\{A_i\}$. For each A_i in $\{A_i\}$ we apply the following steps (see Fig. 1):

1. Eliminating chain noise: noisy points which make the chain over oscillating are eliminated by median filtering. Applying a third order one-dimensional median filter on the vector A_i reduces the effect of such points adequately. Figs. 1-b and c show the effect of reducing the number of chain points by median filtering.
2. Shifting operation: the standard chain code representation has the wraparound drawback [21]. For example, a line along $-22.5°$ direction (in the forth quarter of the trigonometric circle) is coded as $\{707070\dots\}$ using standard chain code representation. To eliminate or reduce such wraparound, a new modified code $B_i = \{b_1, b_2 \dots b_{n_i}\}$ for each $A_i = \{a_1, a_2 \dots a_{n_i}\}$ can be extracted by a shifting operation defined recursively as:

$$\begin{cases} b_1 = a_1 \\ b_k = g_k, g_k \in Z \mid (g_k - a_k) \bmod 8{=}0 \text{ and} \\ \qquad |g_k - b_{k-1}| \text{ is minimized for } k = 2, 3, \dots n_i \end{cases} \quad (1)$$

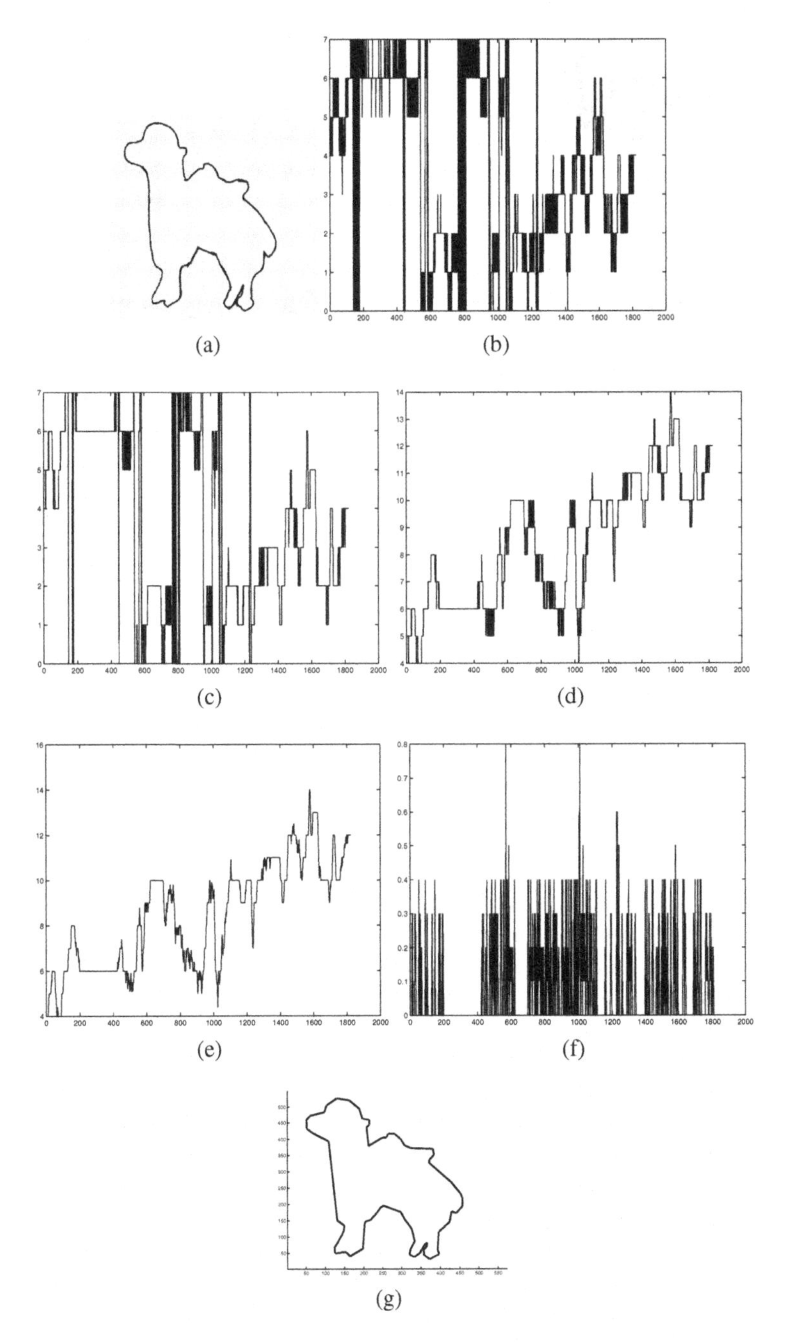

Fig. 1. (a) an example thinned hand-drawn sketch, (b) the corresponding chain code (c) median filtered code, (d) shifted code, (e) smoothed representation, (f) derivative, and (g) extracted line segments

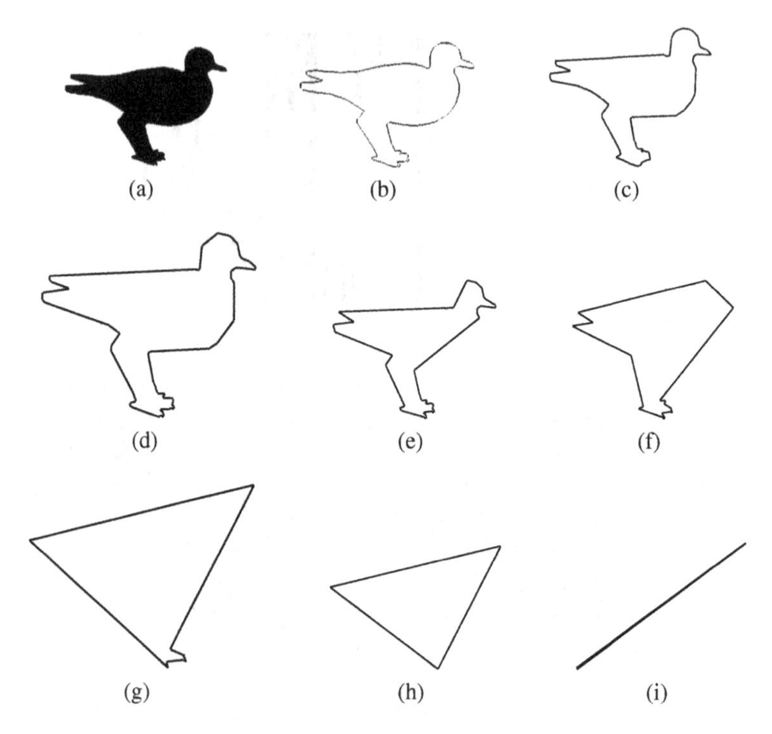

Fig. 2. The effect of variable τ : (a) an original database image, (b) the contour (c) polygonized contour with $\tau = 0.1$, NoS=112, (d) $\tau = 0.3$, NoS=64, (e) $\tau = 0.4$, NoS=41, (f) $\tau = 0.5$, NoS=18, (g) $\tau = 0.7$, NoS=8, (h) $\tau = 0.9$, NoS=3, (i) $\tau = 1$, NoS=2

The line along $-22.5°$ direction is now coded as $\{787878\ldots\}$. Comparison of Figs. 1-c and d shows the wraparound effect.

3. Smoothing operation: the shifted chain code B_i is then smoothed by a five-point Gaussian filter $\{0.1, 0.2, 0.4, 0.2, 0.1\}$ [21]. $\Gamma(\theta)$ is the shifted and smoothed waveform where θ is the traversing variable (Fig. 1-e).
4. First derivative and break points extraction: $d\Gamma/d\theta$ determines the rate of change of $\Gamma(\theta)$ with respect to θ. The extreme points of this derivative is considered as break points (ζ_k), if they are greater than a threshold τ. The τ value determines the degree of coarse-to-fine approximation of the input curve with a polygon. Figs. 1-f and g depict the resulting derivative and the corresponding extracted line segments to rebuild the shape.

The selection of threshold τ has a great influence on the rebuilding of the shape. Fig. 2 shows the effect of the variable τ and the resulting number of segments (NoS) for an example shape. As can be seen, smaller values of τ make the resulting polygon to resemble the contour curve more closely (with more line segments). For shape retrieval, we are not interested in very fine polygons because the extracted features need to represent only the overall structure of the shape and the details are not important. Therefore,

$\tau = 0.3$ is chosen in our experiments. It is interesting to note that higher τ's can be used for extracting a major axis of the shape as shown in Fig. 2-i.

The line segment l_k which connects ζ_k to ζ_{k+1} is considered as the lineal approximation of the micro chain lying between the two points. This segment is employed to construct the approximating polygon. Although a finer approximation for the micro chain from ζ_k to ζ_{k+1} can be employed to obtain more line segments and consequently more precisely fitted polygons, experiments on many test data showed that there is no significant improvement in the retrieval performance by applying such extra computations. This arises from the fact that overall structure of the shape can be well captured by a moderate number of segments. Therefore the set

$$L_i = \{l_k\} \tag{2}$$

where l_k is the straight line segment connecting ζ_k to ζ_{k+1} will be used as the line segment set of chain A_i. Finally, the union of all L_i sets, say L, for $i = 1, 2, \ldots, m$, is obtained:

$$L = \{L_1\} \cup \{L_2\} \cup \ldots \{L_m\} \tag{3}$$

L is the line segment set of the underlying shape which is used as polygonal approximation of the boundary curve C.

2.1 A Hybrid Shape Similarity Measure

The length of extracted line segments ($l_k \in L$) are rotation and translation invariant. Normalizing the length by the maximum length makes it scale invariant as well. In addition, the angle between successive segments in L is scale, translation and rotation invariant.

For the purpose of shape recognition and retrieval, we employ the aforementioned principles to extract two discriminating and affine transforms invariant vectors Ψ and Φ using the following procedure:

1. Initialize Ψ (30 entry) and Φ (18 entry).
2. For each segment in L:
   - Compute the length.
   - Normalize the length with the maximum length.
   - Uniformly quantize the normalized length to 30 equal parts and add one to the corresponding entry of Ψ.
3. For each segment pair in L which are adjacent:
   - Compute the angle between the segments (corner angle).
   - Uniformly quantize the corner angle to 18 equal parts (to make steps of $10°$) and add one to the corresponding entry of Φ.

Next, we combine these two feature vectors to make hybrid shape features used for measuring the similarity. For this, the distance between two different shapes is computed using the combination of Euclidean distances obtained individually from corresponding Ψ and Φ vectors. One of the difficulties involved in integrating different distance measures is the difference in the range of associated distance values. In order to overcome the problem, the two distance values are normalized first to be within the same range of [0,1] and then are integrated with a weighting scheme. More precisely, let Q be a query image and P be a database image. Let $D^n_\Psi(Q,P)$ denotes the normalized Euclidean distance between Q and P on the basis of geometric segment's length and $D^n_\Phi(Q,P)$ denotes their normalized Euclidean distance on the basis of corner angle. The normalization is accomplished as follows:

$$\begin{aligned} D^n_\Psi(Q,P) &= [D_\Psi(Q,P) - mindist_\Psi] \,/\, [maxdist_\Psi - mindist_\Psi] \\ D^n_\Phi(Q,P) &= [D_\Phi(Q,P) - mindist_\Phi] \,/\, [maxdist_\Phi - mindist_\Phi] \end{aligned} \tag{4}$$

where $D_\Psi(Q,P)$ and $D_\Phi(Q,P)$ are the Euclidean distances between Q and P based on geometric length and based on corner angle respectively. $mindist$ and $maxdist$ are the minimum and the maximum distance values of the query image Q to the database images according to the corresponding distance used (i.e. D_Ψ or D_Φ).

Finally, an integrated and hybrid distance D between Q and P is defined as:

$$D(P,Q) = \frac{w_1 \times D^n_\Psi(P,Q) + w_2 \times D^n_\Phi(P,Q)}{w_1 + w_2} \tag{5}$$

where w_1 and w_2 are the weights assigned to to the length-based distance and the angle-based distance, respectively. In current implementation we have used $w_1 = 1$ and $w_2 = 1.5$. This is to put more emphasis on the curvature features than on the length features.

3 Experimental Results

To evaluate the retrieval performance of the proposed method, in comparison with other well-known methods, seven different approaches were implemented. The Fourier descriptor (FD) [9], PFD method [14], Zernike moment invariants (ZMI) [12], and the ART [4, 11], AP [18], ARP [19] and the proposed CPCD methods (Section 2) were employed to extract features for sketch-based shape retrieval. The comparative results are presented in this section.

The MPEG-7 contour shape database CE-1, set A1 and A2 [9], was used as the common test bed for all methods. The database consists of pre-segmented shapes, defined by single closed contours acquired from real world objects. It takes into consideration the common shape distortions and the inaccurate nature of shape boundaries in segmented shapes. Set A1 is for the test of scale invariance and contains 420 shapes of 70 classes (6 in each class). An image of each class was scaled by the following factors: 200%, 30%, 25%, 20%, and 10%. In a similar manner, set A2 contains 420 images created by rotation of the original 70 shapes by the following angles: 9°, 36°, 45°, 90°, and 150°. Consequently, appending sets A1 and A2 forms a database of 840 images in 70 groups, each with 12 similar images. Fig. 3 depicts one image from each class in the database and Fig. 4 shows examples of variation within a class.

Since the database images are all binary, we obtained the boundary contour of each images as the set of foreground pixels which have at least one neighboring background pixel. We also collected 105 different hand-drawn sketches similar to randomly selected shapes (Fig. 5 shows some examples). They were morphologically thinned to represent

Fig. 3. Shapes from 70 different classes existing in the MPEG-7, CE-1 database

Fig. 4. Examples of variations within the deer class

shape's boundary contour. To be able to evaluate scale and rotation invariance properties, the database and the sketched images were chosen to have varying sizes and directions.

The ANMRR criterion is used as the retrieval performance measure. This measure considers not only the recall and precision information but also the rank information among the retrieved images. It is defined in the MPEG-7 standard [22] as follows:

$$AVR(q) = \sum_{\kappa=1}^{NG(q)} \frac{Rank(\kappa)}{NG(q)}$$

$$MRR(q) = AVR(q) - 0.5 - \frac{NG(q)}{2}$$

$$NMRR(q) = \frac{MRR(q)}{K+0.5-0.5*NG(q)}$$

$$ANMRR = \frac{1}{N} \sum_{q=1}^{N} NMRR(q)$$

where $NG(q)$ is the number of ground truth images for a query q. $K = \min(4 * NG(q), 2 * GTM)$ where GTM is max $\{NG(q)\}$ for all q's of a data set. $Rank(k)$ is the rank of the found ground truth images, counting the rank of the first retrieved image as one. A $Rank$ of $K + 1$ is assigned to each of the ground truth images which are not in the first K retrievals. For example, suppose a given query q_i has 10 similar images in an image database ($NG = 10$). If an algorithm finds 6 of them in the top 20 retrievals ($K = 20$) in the ranks of 1,5,8,13,14 and 18, then the $AVR(q_i) = 14.3$, $MRR(q_i) = 8.8$ and $NMRR(q_i) = 0.5677$.

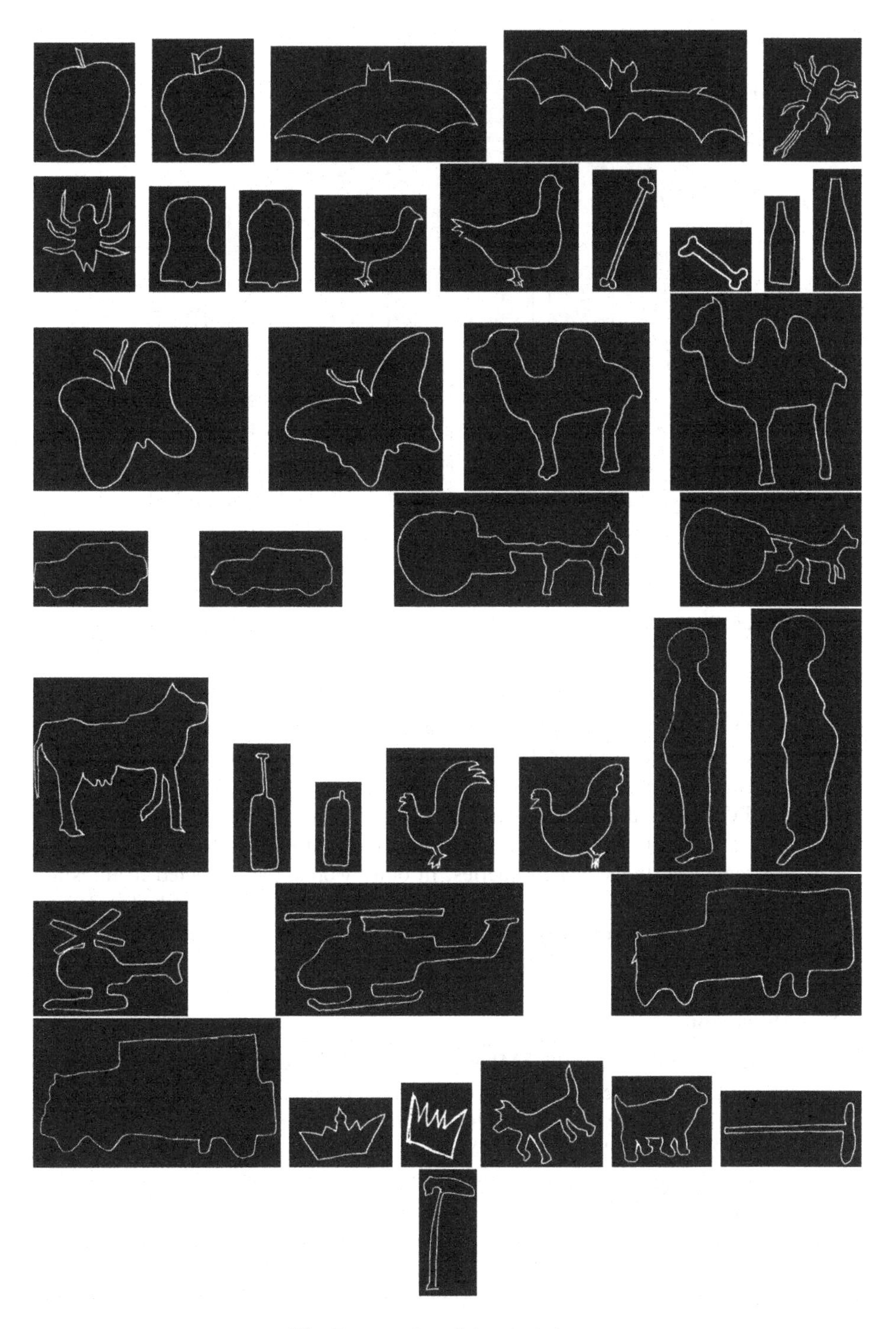

Fig. 5. Examples of sketched shapes

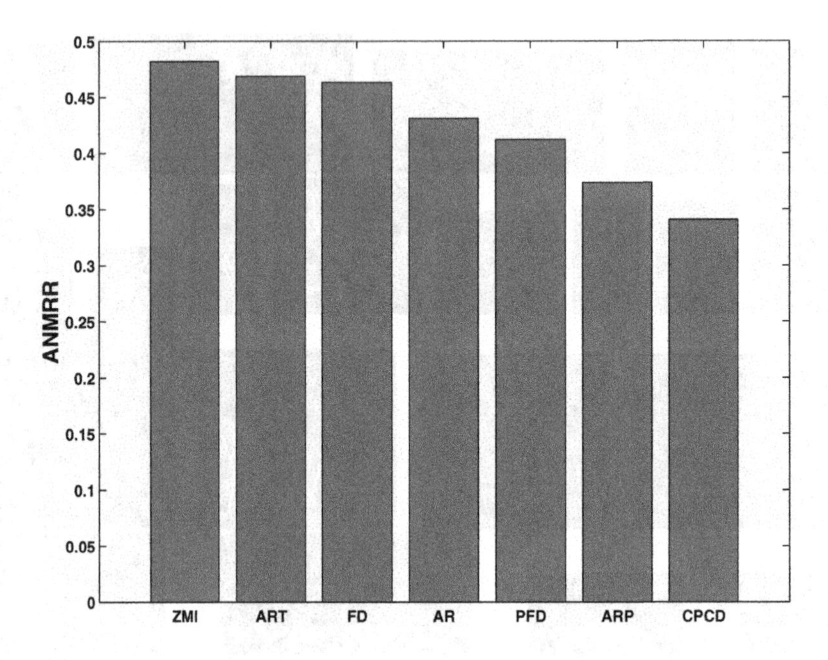

Fig. 6. Retrieval performance of different methods expressed by the ANMRR

Note that the NMRR and its average (ANMRR) will always be in the range of $[0, 1]$. Based on the definition of ANMRR, the smaller the ANMRR, the better the retrieval performance. In our experiments the $NG(q) = 12$ for all q's, $K = 24$ and $N = 105$.

Fig. 6 exhibits the resulting ANMRR for different methods. As can be seen, the proposed CPCD method shows the best retrieval performance (i.e., ANMRR=0.3403). This is due to the ability to capture global and local structural similarities between database images and the sketched queries. In other words, the extracted features are capturing the overall structural properties of the shape using the length distribution of predominant sides. In addition, the local properties of the shape are also exploited utilizing angle size distribution of the corner points.

3.1 Discussion

It is worthwhile to note that as the ZMI and ART methods are region-based approaches, their retrieval performances are the lowest in the current application, i.e. 0.4819 and 0.4687, respectively. The ARP method shows more effective performance (0.3739) than the FD (0.4632), AP (0.4312), and PFD (0.4125) methods, respectively. The varying degree of performance in these methods arises from different algorithms they employ for feature extraction.

The FD and PFD methods are based on the Fourier transform in the Cartesian and polar coordinates, respectively. The basis functions used in these methods can effectively capture the similarity between different contour shapes as they exploit magnitude

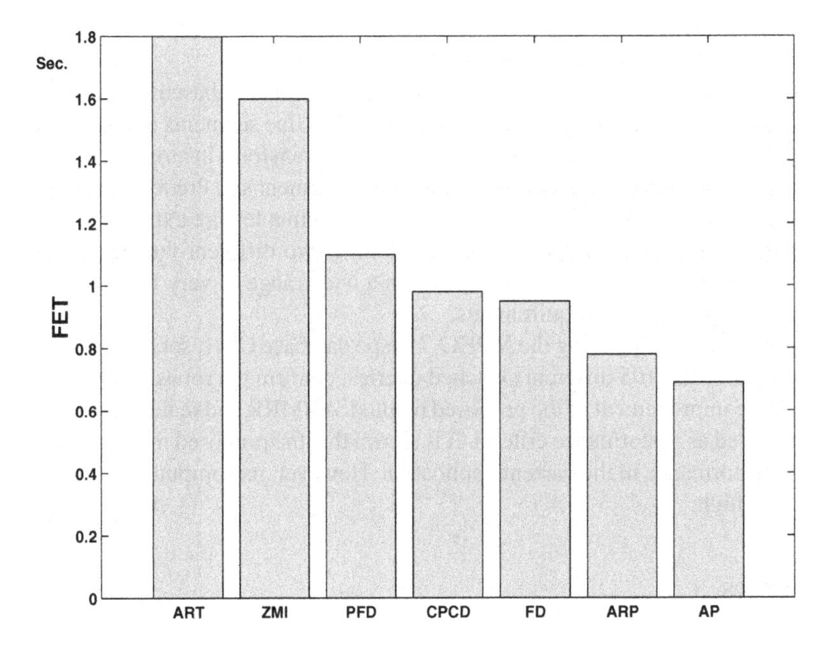

Fig. 7. Average feature extraction time for different methods using both database and sketched shapes on a Pentium-III, 1000 MHz machine

and phase information of the Fourier coefficients. However, they cannot tolerate disconnectivities. Moreover, this kind of transform is more effective for contour-based shape retrieval, that is when the outline of the shape is used for feature extraction. The AP and ARP methods exhibit good performances where there is enough spatial information in the relevant slices or sectors. In the current application (sketch-based shape retrieval), that uses only the contour data, there is no sufficient information in terms of the number of pixels in such areas. This shortcoming can be partially overcome by refining the partitioning scheme used in the AP and ARP approaches while adversely increases the feature extraction time (FET).

Fig. 7 shows average feature extraction times for the aforementioned algorithms using a Pentium-III, 1000 MHz machine on a collection of both the database and the sketched shapes. The AP and ARP methods posses the shortest feature extraction times as they are performed in the pixel domain and simply collect data from a few associated regions. The AP extraction time is shorter as its partitioning scheme is less complicated than the ARP's scheme. The FD, CPCD, and PFD methods need a moderate time for feature extraction and the ZMI and ART methods are the slowest. Longer extraction times for the transform-based methods (i.e., FD, PFD, ZMI, and ART) arise from higher computational cost of handling complex basis functions. For the CPCD method, the convolution and differentiation operations involved in the smoothing and the break point extraction are the most time consuming steps in the algorithm.

4 Conclusion

A novel and effective feature extraction approach for sketch-based shape retrieval is proposed. The approach is based on the distribution of line segments extracted by chain code differentiation. The extracted features are affine transform invariant. The boundary contour is approximated by a polygon using a line segment set. Predominant characteristics of the polygon, sides and corners, are employed in a feature extraction algorithm. Extracted features have a hybrid nature combining two different feature vectors. The approximating polygon can be adjusted within a wide range of very fine to very coarse based on the application's requirements.

Experimental results using the MPEG-7 shape database CE-1, set A1 and A2 including 840 shapes, and 105 different sketched queries, confirm the robustness and retrieval performance improvement of the proposed method. ANMRR and feature extraction time (FET) are used as performance criteria. It is shown that the proposed method has the best retrieval performance in the current application. However, its computational intensity is moderately high.

References

1. Pavlidis, T.: Survey: A review of algorithms for shape analysis. Computer Graphics and Image Processing **7** (1978) 243–258
2. Loncaric, S.: A survey of shape analysis techniques. Patt. Recog. **31** (1998) 983–1998
3. Jain, A.J., Vailaya, A.: Shape-based retrieval: a case study with trademark image databases. Patt. Recog. **31** (1998) 1369–1390
4. Bober, M.: MPEG-7 visual shape descriptors. IEEE Trans. Circ. and Syst. for Video Tech. **11** (2001) 716–719
5. Bimbo, A.D.: Visual Inform. retrieval. Morgan Kaufmann Publishers (1999)
6. Widrow, B.: The "rubber-mask" technique-ii. pattern storage and recognition. Patt. Recog. **5** (1973) 199–211
7. Kass, M., Witkin, A., Terzopoulos, D.: Snakes: active contour models. Int. Journal of Computer Vision **3** (1988) 321–331
8. Smith, J.R., Chang, S.F.: VisualSEEk: a fully automated content-based image query system. In: Proc. ACM Multimedia 96., USA (1996) 87–98
9. Zhang, D., Lu, G.: Evaluation of MPEG-7 shape descriptor against other shape descritors. multimedia systems **9** (2003) 15–30
10. Kauppinen, H., Seppanen, T., Pietikainen, M.: An experimental comparison of autoregressive and Fourier-based descriptors in 2D shape classification. IEEE Trans. Patt. Anal. and Mach. Intel. **17** (1995) 201–207
11. ISO/IEC JTC1/SC29/WG11/N4358: Text of ISO/IEC 15938-3/FDIS information technology – multimedia content description interface – part 3 visual, Sydney (2001)
12. ISO/IEC JTC1/SC29/WG11/N3321: MPEG-7 visual part of experimentation model version 5, Nordwijkerhout (2000)
13. Hoynck, M., Ohm, J.R.: Shape retrieval with robustness against partial occlusion. In: IEEE Int. Conf. Acoustics, Speech, and Signal Processing. Volume 3. (2003) 593–596
14. Zhang, D., Lu, G.: Generic Fourier descriptor for shape-based image retrieval. In: Proc. IEEE Int. Conf. Multimedia and Expo. Volume 1. (2002) 425–428
15. Zhang, D., Lu, G.: Shape-based image retrieval using generic Fourier descriptor. Signal Processing: Image Commun. **17** (2002) 825–848

16. Matusiak, S., Daoudi, M., Blu, T., Avaro, O.: Sketch-based images database retrieval. In: Proc. 4th Int. Workshop Advances in Multimedia Inform. Syst. MIS'98. (1998)
17. Ip, H.H.S., Cheng, A.K.Y., Wong, W.Y.F., Feng, J.: Affine-invariant sketch-based retrieval of images. In: Proc. IEEE Int. Conf. Comput. Graphics. (2001) 55–61
18. Chalechale, A., Naghdy, G., Mertins, A.: Sketch-based image matching using angular partitioning. IEEE Trans. Systems, Man, Cybernetics - Part A: Systems and Humans (2004)
19. Chalechale, A., Naghdy, G., Premaratne, P.: Image database retrieval using sketched queries. In: Proc. IEEE Int. Conf. Image Processing (ICIP'04), Singapore (2004)
20. Gonzalez, R.C., Woods, R.E.: Digital Image Processing. Addison-Wesley (1992)
21. Li, H., Manjunath, B.S., Mitra, S.K.: A contour-based approach to multisensor image registration. IEEE Trans. Image Processing **4** (1995) 320–334
22. ISO/IEC JTC1/SC29/WG11-MPEG2000/M5984: Core experiments on MPEG-7 edge histogram descriptors, Geneva (2000)

Surface Smoothing for Enhancement of 3D Data Using Curvature-Based Adaptive Regularization*

Hyunjong Ki, Jeongho Shin, Junghoon Jung, Seongwon Lee, and Joonki Paik

Image Processing and Intelligent Systems Lab,
Department of Image Engineering,
Graduate School of Advanced Imaging Science, Multimedia, and Film,
Chung-Ang University,
221 Huksuk-Dong, Tongjak-Ku, Seoul 156-756, Korea
paikj@wm.cau.ac.kr
http://ipis.cau.ac.kr

Abstract. This paper presents both standard and adaptive versions of regularized surface smoothing algorithms for 3D image enhancement. We incorporated both area decreasing flow and the median constraint as multiple regularization functionals. The corresponding regularization parameters adaptively changes according to the local curvature value. The combination of area decreasing flow and the median constraint can efficiently remove various types of noise, such as Gaussian, impulsive, or mixed types. The adaptive version of the proposed regularized smoothing algorithm changes regularization parameters based on local curvature for preserving local edges and creases that reflects important surface information in 3D data. In addition to the theoretical expansion, experimental results show that the proposed algorithms can significantly enhance 3D data acquired by both laser range sensors and disparity maps from stereo images.

1 Introduction

As the 3D modeling technique covers wider applications in the computer vision area, 3D image processing attracts more attention. Recently, various modes of signals such as intensity and range images are widely used in 3D reconstruction, but observed data are corrupted by many different types of noise and often need to be enhanced before further applications. Among various 3D image processing techniques, surface smoothing is one of the most basic operations for processing the surface.

* This work was supported in part by Korean Ministry of Science and Technology under the National Research Lab. Project, in part by Korean Ministry of Education under Brain Korea 21 Project, and in part by grant No.R08-2004-000-10626-0 from the Basic Research Program of the Korea Science & Engineering Foundation.

R. Klette and J. Žunić (Eds.): IWCIA 2004, LNCS 3322, pp. 488–501, 2004.

In order to reconstruct a surface, many regularization-based algorithms have been studied in the field of early vision. Some of them adopted a smoothing inhibition on edge pixels by means of a Boolean-valued line process[1-3]. In [4], Gokmen used continuously varying regularization parameters for edge detection and surface reconstruction, which is an extended version of the 2D processing [5].

These methods, however, did not consider impulsive noise that frequently arises in both range data and disparity maps. In general, range data includes heavy noise in the mixed form of Gaussian and impulsive noise. Although the existing regularized noise smoothing algorithms can easily remove Gaussian noise, there has not been many researches to deal with impulsive noise included in observed range data. A few approaches to smoothing impulsive noise have been proposed in [6, 7], but they consist of two independent steps, and could not provide a structured way to get the optimal solution.

This paper presents a regularized noise smoothing algorithm for 3D image using multiple regularization functionals for smoothing both Gaussian and impulsive noise. For removing Gaussian noise we adopt area decreasing flow as the first smoothness constraint. On the other hand, for removing impulsive noise, we incorporate the second smoothness constraint that minimizes difference between the median filtered data and the original data. Existing adaptive regularization methods used the 2D edge information to adaptively change the weight of the regularization parameter to preserve edges. The use of 2D edge map in range data, however, is not effective because surface curvature is not considered. Therefore we adaptively change regularization parameter according to curvature information of the input range data.

The rest of this paper is organized as follows: In Section 2 we summarize the regularizied surface smoothing with area decreasing flow and the median constraint. In Section 3 we present adaptive edge preservation based on the curvature estimation in conjunction with the proposed regularization. In Section 4 we summarized the proposed regularization algorithm. Experimental results and conclusions are given in Sections 5 and 6, respectively.

2 Regularizied Surface Smoothing with Area Decreasing Flow and Median Constraint

The 3D image smoothing problem corresponds to a constrained optimization problem. Regularization is the most widely used method to solve practical optimization problems with one or more constraints. In Section 2.1, we present the regularized-viewpoint invariant reconstruction of a surface from 3D data. In Section 2.2, the area decreasing flow is applied to 3D data smoothing. Median constraints for removing impulsive noise is presented in Section 2.3.

2.1 Viewpoint Invariant Model

The mathematical model for surface estimation can be written as

$$g = f + \eta, \tag{1}$$

where g represents an observed surface corrupted by noise η, f and denotes the original surface. In order to solve (1), we have to solve the following optimization problem

$$\hat{f} = \arg\min_{f} O(f), \tag{2}$$

where $\hat{f}$ represents the optimally estimated surface and

$$O(f) = \|g - f\|^2 + \lambda\|Cf\|^2. \tag{3}$$

In (3) C is a matrix which represents a high-pass filter, λ and represents the regularization parameter that controls the fidelity to the original surface, $\|g - f\|^2$, and smoothness of the restored surface, $\|Cf\|^2$.

In the viewpoint-invariant case, the cost function can be written

$$\sum_{i,j} [f_{i,j} - g_{i,j}]^2 \cos^2\phi + \lambda[Cf_{i,j}]^2 = \sum_{i,j} [f_{i,j} - g_{i,j}]^2 \frac{1}{1 + z_x^2 + z_y^2} + \lambda[Cf_{i,j}]^2, \tag{4}$$

where z_x and z_y represent the first order derivative at the $(i,j)^{th}$ measurement in the horizontal and vertical directions, respectively. The cost function given in (3) can be minimized by solving the linear equation.

$$(I + \lambda C^T C)f = g. \tag{5}$$

2.2 Area Decreasing Constraints

Traditionally, raw images are preprocessed using some filtering method. With the regularization method, we can control the tradeoff between smoothness and data compatibility by using an appropriate stabilizing functional related to the 3D surface properties.

Traditionally, the surface is considered as a graph $z(x, y)$, which can be expressed as $z_{i,j}$ over a rectangular grid. Letting $c_{i,j}$ represent the observed data, the cost function can be written as

$$O(z) = \sum_{i,j} (z_{i,j} - c_{i,j})^2 cos^2\phi + \lambda F_s. \tag{6}$$

In practice, $cos\phi$ represents approximated surface slant with respect to the incident direction of measurement. The larger the angel ϕ between the surface normal and the direction of measurement, the smaller the confidence is. Because $(z_{i,j} - c_{i,j})cos^2\phi$ represents the perpendicular distance between the estimated and real surfaces, it is viewpoint-invariant. The stabilizing function F_s can take various forms. For example, first order regularization is used in [8] while a second order model is investigated in [9].

Estimating the elevation $z_{i,j}$ is feasible for sparse data. But in dense range images from a range scanner with a spherical coordinate system, $z(x, y)$ is no longer a graph and but estimating the elevation may result in overlap in range measurement. Therefore, we estimate the range $r_{i,j}$ instead of $z_{i,j}$ so that all refinement takes place a long the line of measurement.

For the Perceptron range scanner, the range value of each pixel $R_{i,j}$ is converted to $(x_{i,j}, y_{i,j}, z_{i,j})$ in the Cartesian coordinate. We adopt the calibration model, described in [8], as

$$\begin{cases} x_{i,j} = dx + rsin\alpha \\ y_{i,j} = dy + rcos\alpha sin\beta \\ z_{i,j} = dz - rcos\alpha cos\beta, \end{cases}$$

$$\begin{cases} \alpha = \alpha_0 + H_0(col/2 - j)/N_0 \\ \beta = \beta_0 + V_0(row/2 - i)/M_0, \end{cases}$$

$$\begin{cases} dx = (p_2 + dy)tan\alpha \\ dy = dztan(\theta + 0.5\beta) \\ dz = -p_1(1.0 - cos\alpha)/tan\gamma, \end{cases} \tag{7}$$

and

$$\begin{cases} r_1 = (dz - p_2)/\delta \\ r_2 = \sqrt{dx^2 + (p + dy)^2}/\delta \\ r = (R_{i,j} + r_0 - r_1 - r_2)/\delta \end{cases} \tag{8}$$

where $p_1, p_2, \gamma, \theta, \alpha_0, \beta_0, H_0, V_0, r_0, \delta$ represents the set of calibration parameters of the scanner, and (M_0, N_0) refers to the image size. For estimating r, we can use the following parameterization.

$$X(\alpha, \beta) = (rsin\alpha, rcos\alpha sin\beta, -rcos\alpha cos\beta), \tag{9}$$

and small values denoted by dx, dy, and dz are ignored in the analysis process.

The coefficients of the first fundamental form, which will be used shortly in the computation of the surface area, are given, in the basis of X_α, X_β, as

$$\begin{cases} E = X_\alpha \cdot X_\alpha = r^2 + r_\alpha^2 \\ F = X_\alpha \cdot X_\beta = r_\alpha r_\beta \\ G = X_\beta \cdot X_\beta = r^2 cos^2\alpha + r_\beta^2, \end{cases} \tag{10}$$

where

$$X_\alpha = \frac{\partial X}{\partial \alpha}, X_\beta = \frac{\partial X}{\partial \beta}, r_\alpha = \frac{\partial r}{\partial \alpha}, r_\beta = \frac{\partial r}{\partial \beta}, \tag{11}$$

and C represents the observed value of r. Range data smoothing can then be performed by minimizing the following cost function,

$$f = \sum_{i,j} (r_{i,j} - c_{i,j})^2/\sigma_{i,j}^2 + F_s. \tag{12}$$

We let the stabilizing function h be the surface area, which can be calculated as

$$F_s = h_{area} = \int\int_{\bar{D}} \sqrt{EG - F^2} d\alpha d\beta$$
$$= \sum_{i,j} \lambda_{i,j} (r_{i,j}^4 cos^2\alpha + r_{i,j}^2 r_\beta^2 cos^2\alpha), \quad (13)$$

where D represents the domain of (α, β).

2.3 Median Constraints for Removing Impulsive Noise

In general, range data is corrupted by the mixture of Gaussian and impulsive noises. Although the existing regularized surface smoothing algorithm can easily remove Gaussian noise, impulsive noise, caused by random fluctuation of the sensor acquisition or by incorrect disparity measurement, is not easy to be removed from observed 3D image data. It is also difficult to remove noise near edge when the existing 2D edge-based adaptive regularization is used. Since it is well-known that median filter is very effective in removing impulsive noise from images. The second smoothness constraint is additionally incorporated into the existing regularization algorithm. The proposed method minimizes the difference between the median filtered data and the estimated data. The corresponding regularization term can be formulated as

$$h_{median} = \lambda_M \sum_{i,j} (r_{i,j} - M)^2, \quad (14)$$

where M represents the median filtered data of a noisy data.

3 Adaptive Regularization Using Curvature Estimation

Incorporation of the regularizing term into the cost function given in (12) tends to suppress local activity in the range image. Although the smoothing function is good for suppressing undesired noise, it also degrades important features such as edges, corners, and segmentation boundaries. Existing adaptive regularization methods used the 2D edge detection operation to adaptively change the weight of the regularization parameter so that edges are preserved during the regularization. But the use of 2D edge map in smoothing 3D data is not effective because surface curvature is not considered. Therefore we utilize the result of curvature analysis to adaptively weight the regularization parameter λ so that real 3D edges are preserved during the regularization process.

Curvature estimation is an important task in 3D object description and recognition. Differential geometry [9] provides several measures of curvature including Gaussian and mean curvatures. The combination of these curvature values enables to categorize the local surface types. Especially, Gaussian curvature gives significant information of surfaces. Its sign determines the convexity of the surface. If $K > 0$ $K = 0$ or $K < 0$ at a surface x, the corresponding surface is called elliptic, parabolic, or hyperbolic, respectively.

The magnitude of the Gaussian curvature represents the area of the changing normals in relation to the area of the underlying surface. Thus, in surfaces which contain shape curve or spicular patches, the magnitude of the curvature should be relatively high, whereas in areas such as a plane, curvature value should become low. Fig. 1(b) shows Gaussian curvature image of the lobster range image shown in Fig. 1(a). As shown in Fig. 1(b), the curvature is where there is no bendings or bumps, such as in the middle of the lobster. Curvature is high in areas in 3D edges or creases. This observation, together with the theoretical definition, justifies the use of Gaussian curvature as a measure of adaptive controlling the regularization parameters.

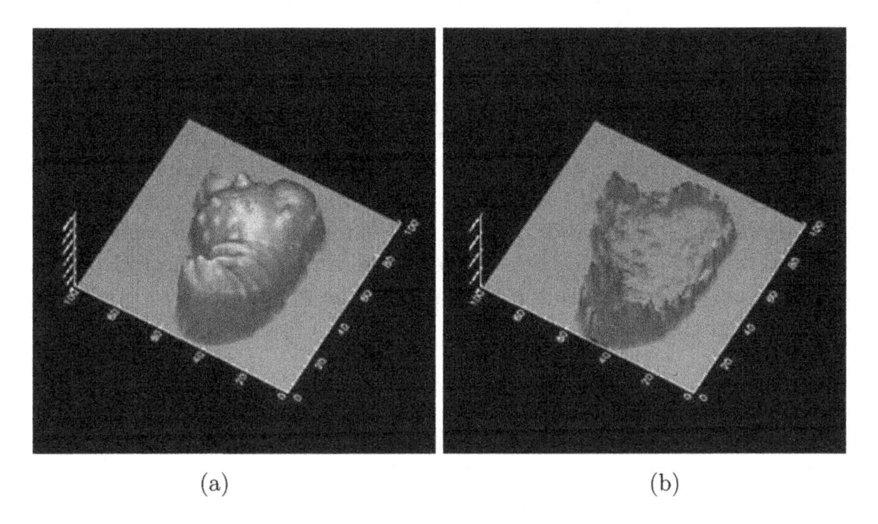

(a) (b)

Fig. 1. (a) Original range data "lobster" and (b) The magnitude of its Gaussian curvature

A formula for Gaussian curvature is written as

$$K = \frac{r_{xx} r_{yy} - r_{xy}^2}{(1 + r_x^2 + r_y^2)^2}, \tag{15}$$

where r represents a range image, and its subscripts indicate partial differentiations in the corresponding direction. the regularization term in (13) can adaptively be weighted by using

$$\lambda_{i,j} = \frac{\rho}{1 + \theta K^2(i,j)}, \tag{16}$$

where θ, $0 < \theta < 1$, represents a parameter that determines sensitivity of 3D edge strength, and ρ is a scaling parameter. The selection of ρ generally depends on the desired data compatibility as well as the level of noise reduction.

4 The Proposed Regularization Algorithm

From (12), (13) and (14), we formulate the proposed regularized energy function as

$$\begin{aligned} f_{i,j}(r) &= \sum_{i,j}(r_{i,j}-c_{i,j})^2 cos^2\phi + \lambda_M h_{median} \\ &= \sum_{i,j}(r_{i,j}-c_{i,j})^2 cos^2\phi \\ &+ \lambda_A \sum_{i,j}(r_{i,j}^4 cos^2\alpha + r_{i,j}^4 cos^2\alpha + r_{i,j}^2 r_\beta^2 + r_\alpha^2 r_{i,j}^2 cos^2\alpha) \\ &+ \lambda_M \sum_{i,j}(r_{i,j}-M)^2. \end{aligned} \tag{17}$$

The proposed regularization algorithm can be implemented in two different versions, such as standard and adaptive versions. In the adaptive version of the regularization algorithm, parameter λ_A in (17) adaptively changes according to curvature estimation by (16), while it is fixed in the standard version.

Among various optimization methods, the simple gradient descent method is adopted to minimize (17) because convergence can easily be controlled with an adaptively varying regularization parameter. The estimation $r_{i,j}^{'}$ of each measurement $r_{i,j}$ is given as

$$r_{i,j}^{n+1} = r_{i,j}^n - w\frac{\partial f_{i,j}}{\partial r_{i,j}} \tag{18}$$

where w represents the iteration step size, and

$$\begin{aligned} \frac{\partial f}{\partial r_{i,j}} &= 2(r_{i,j}-c_{i,j})/\sigma_{i,j}^2 + \lambda_{i,j}\{4r_{i,j}^3 cos^2\alpha \\ &+ [2r_{i,j}(r_{i+1,j}-r_{i,j})^2 - 2r_{i,j}^2(r_{i+1,j}-r_{i,j}) + 2r_{i,j}^2(r_{i,j}-r_{i-1,j})](\frac{1}{\partial\beta})^2 \\ &+ [2r_{i,j}(r_{i,j+1}-r_{i,j})^2 - 2r_{i,j}^2(r_{i+1,j}-r_{i,j}) + 2r_{i,j-1}^2(r_{i,j}-r_{i,j-1})](\frac{cos\alpha}{\partial\alpha})^2\}. \end{aligned} \tag{19}$$

In calculating the derivative of f in (17), the following forward difference approximations were used

$$r_\alpha = \frac{r_{i,j+1}-r_{i,j}}{d\alpha} \quad and \quad r_\beta = \frac{r_{i+1,j}-r_{i,j}}{d\beta} \tag{20}$$

Alternatively, the central difference approximation can also be used for r_α and r_β, such as

$$r_\alpha = \frac{r_{i,j+1}-r_{i,j-1}}{2d\alpha} \quad and \quad r_\beta = \frac{r_{i+1,j}-r_{i-1,j}}{2d\beta} \tag{21}$$

with this central difference approximation the derivative of f is then obtained as

$$\frac{\partial f}{\partial r_{i,j}} = 2(r_{i,j} - c_{i,j})/\sigma_{i,j}^2 + \lambda_{i,j}\{4r_{i,j}^3 cos^2\alpha + [2r_{i,j}(r_{i+1,j} - r_{i-1,j})^2$$
$$- 2r_{i+1,j}^2(r_{i+2,j} - r_{i,j}) + 2r_{i-1,j}^2(r_{i,j} - r_{i-2,j})](\frac{1}{2\partial\beta})^2 + [2r_{i,j}(r_{i,j+1} - r_{i,j-1})^2$$
$$- 2r_{i,j}^2(r_{i+2,j} - r_{i,j}) + 2r_{i,j-1}^2(r_{i,j} - r_{i,j-2})](\frac{cos\alpha}{2\partial\alpha})^2\}. \quad (22)$$

Algorithm 1 Surface Smoothing with Multiple Regularization Functionals

1. Set an initial guess as r^0, and $n \leftarrow 0$.
2. Define regularization energy function $f_{i,j}(r)$
3. Obtain experimentally optimal regularization functionals λ_A and λ_M in $f_{i,j}(r)$.
4. Evaluate $\nabla_r f_{i,j}(r)|_{r=r^n}$ by differentiating objective function $f_{i,j}(r)$.
5. Choose iteration step size W.
6. Update the measurement point by $r^{n+1} \leftarrow r^n + w - \nabla f_{i,j}(r^n)$.
7. If $|r^{n+1} - r^n|$ is smaller than δ, a pre-specified error limit, then stop the algorithm, and the current measurement point will be the estimate of the solution. Otherwise, $k \leftarrow k + 1$, and go to step2. ($\delta = 0.0001$)

Algorithm 2 Adaptive Regularizied Smoothing

1. Set an initial guess as r^0, and $n \leftarrow 0$.
2. Define regularization energy function $f_{i,j}(r)$
3. Calculate $\lambda_{A(i,j)} = \frac{\rho}{1+\theta K^2(i,j)}$ in $f_{i,j}(r)$, where $K = \frac{r_{xx}r_{yy} - r_{xy}^2}{(1+r_x^2+r_y^2)^2}$
4. Obtain experimentally optimal regularization functionals λ_M in $f_{i,j}(r)$.
5. Evaluate $\nabla_r f_{i,j}(r)|_{r=r^n}$ by differentiating objective function $f_{i,j}(r)$.
6. Choose iteration step size W.
7. Update the measurement point by $r^{n+1} \leftarrow r^n + w - \nabla f_{i,j}(r^n)$.
8. If $|r^{n+1} - r^n|$ is smaller than δ, a pre-specified error limit, then stop the algorithm, and the current measurement point will be the estimate of the solution. Otherwise, $k \leftarrow k + 1$, and go to step2. ($\delta = 0.0001$)

5 Experiment Results

To demonstrate the performance of the proposed algorithm, we experimented with two types of 3D data, such as range data using a laser range sensor and disparity data from a pair of stereo images.

5.1 Experiment with Range Data

In this subsection, we present experimental results of range data acquired by a laser range sensor. In the first experiment, Lobster range data of size 100×100, shown in Fig. 2(a), is used. Fig. 2(b) shows the corrupted version of the range

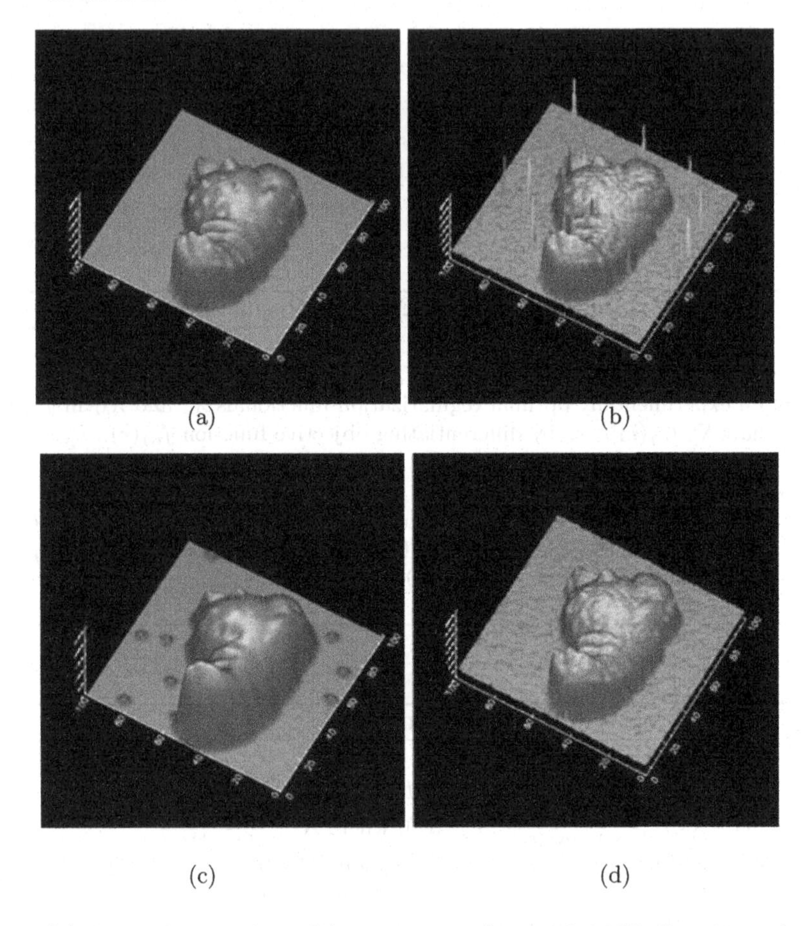

Fig. 2. (a) Original range data, (b) noisy range data with 30dB Gaussian noise and impulsive noise, (c) smoothed range data using Gaussian smoothing filter (σ=1.5), and (d) smoothed range data using a median filter(4×4)

data by 30dB Gaussian noise and 0.5% impulsive noise. Figs. 2(c) and 2(d) respectively show the reconstructed range data using a simple Gaussian smoothing filter with σ=1.5 and a 4×4 median filter. As shown in Fig. 2(c), shape of the lobster is over-smoothed and impulsive noise is still remaining. In Fig. 2(d), impulsive noise is removed to some degree but Gaussian noise is still remaining. The standard and adaptive versions of the proposed regularization results are respectively shown in Figs. 3(a) and 3(b). In case of the standard version, which uses only area decreasing constraint, we choose $w = 10^{-7}$ and $\lambda_{area} = 5 \times 10^{-3}$. In case of the adaptive version, λ_{area} changes according to estimated curvature value. The standard regularization algorithm results in blurred edges in most areas. On the other hand, as shown in Fig. 3(b), the smoothed range data using the proposed adaptive algorithm keeps meaningful edge clear. For

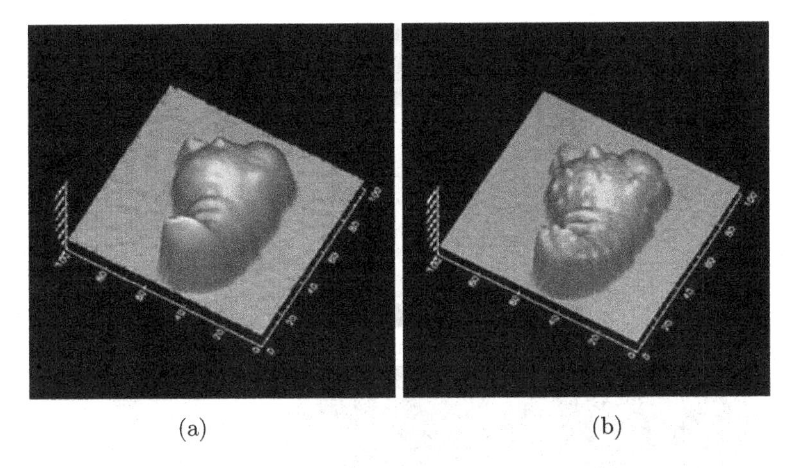

(a) (b)

Fig. 3. (a) Smoothed range data using the proposed non-adaptive regularization and (b) smoothed range data using the proposed adaptive regularization

the proposed adaptive regularization, we used $w = 10^{-8}$, $\rho = 0.2$, $\theta = 0.5$, and $\lambda_{median} = 2 \times 10^6$. As shown Fig. 3(b), both Gaussian and impulsive noises are completely removed with preserving meaningful edges. Table 1 shows objective performance of various smoothing algorithms. Both standard and adaptiveversions of the proposed regularization algorithms outperforms other existing methods in the sense of PSNR.

Table 1. The objective comparison of various smoothing methods

Each smoothing methods for noisy range data	SNR(dB)
Noisy range data	27.8971
Averaging filter	29.7065
Gaussian filter	29.7619
The proposed non-adaptive regularization method	30.5328
The proposed adaptive regularization methods	35.1605

5.2 Disparity Data Using Stereo Images

In this section, we deal with experimental result of disparity data which is calculated by using stereo image. A pair of stereo images of size 256×256 are used for the experiment as shown in Figs. 5(a) and 5(b). Fig. 5(c) shows disparity data computed by the correlation technique from a pair of stereo images. Fig. 6(a) shows the original disparity data in the range format. We can notice that the original disparity data contains significant amount of noise component due to correlation measurement error. The noise distribution in the disparity data is similar to that of the range data shown in Figs. 2 and 4. Figs 6(b)

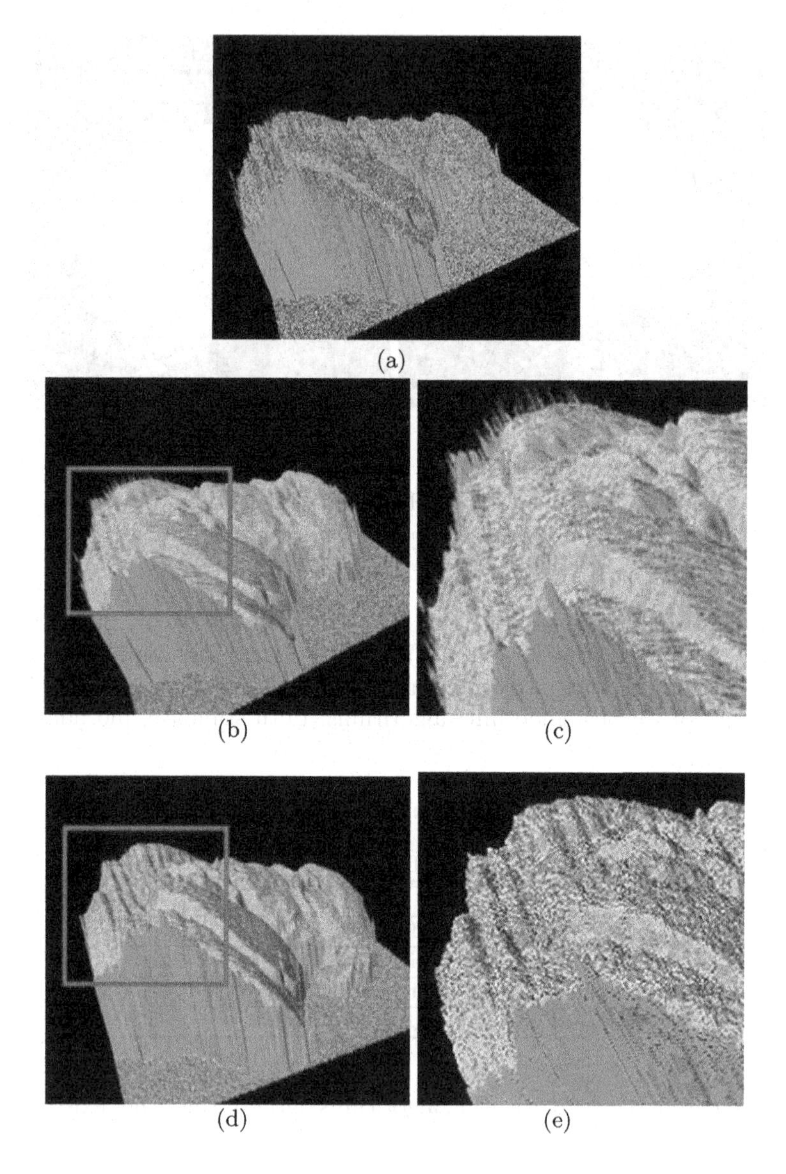

Fig. 4. (a) Real noisy range data, (b) smoothed range data using the proposed non-adaptive regularization, (c) the magnified version of (b), (d) smoothed range data using the proposed adaptive algorithm, and (e) the magnified version of (d)

and 6(c) respectively show the reconstructed range data using the Gaussian smoothing filter with $\sigma = 1.5$ and a 4×4 median filter. Figs. 6(d) and 6(e) show the smoothing results using the standard and adaptive versions of the proposed regularization algorithms, respectively. The same regularization parameter is used as given in subsection 5.1. As shown in the experiment result, we can find that the propose smoothing algorithm also effectively remove noise of the disparity data near edge when the 3D curvature-based adaptive regularization is used.

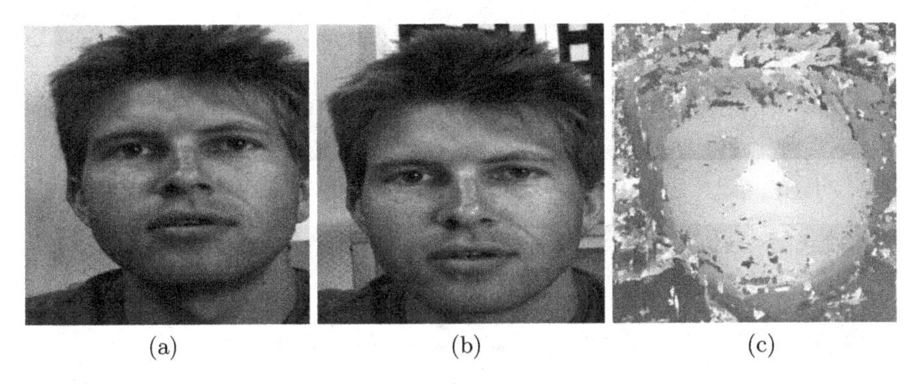

(a) (b) (c)

Fig. 5. (a) The first stereo image, (b) The second stereo image, and (c) the corresponding disparity data

6 Conclusions and Future Works

We presents both standard and adaptive versions of regularized surface smoothing algorithm for 3D image enhancement. The basic idea of this work is to incorporate multiple regularization functionals for suppressing various types of noise and to adaptively change the weight of regularization based on curvature analysis. For this end, we incorporated both area decreasing flow and the median constraint as multiple regularization functionals. For removing Gaussian noise we adopted area decreasing flow as the first smoothness constraint. On the other hand, for removing impulsive noise, we incorporated the second smoothness constraint that minimizes difference between the median filtered data and the original data. At the same time, an adaptive version of regularization
algorithm changed adaptively regularization parameters based on local curvature for preserving local edges and creases that reflects important surface information in 3D data. To demonstrate the performance of the proposed algorithm, we experimented with two types of 3D data, such as range data using laser sensor and disparity data using a pair of stereo images. As a result, the proposed algorithms can significantly enhance 3D data acquired by both laser range sensors and disparity maps from stereo images while preserving meaningful edge.

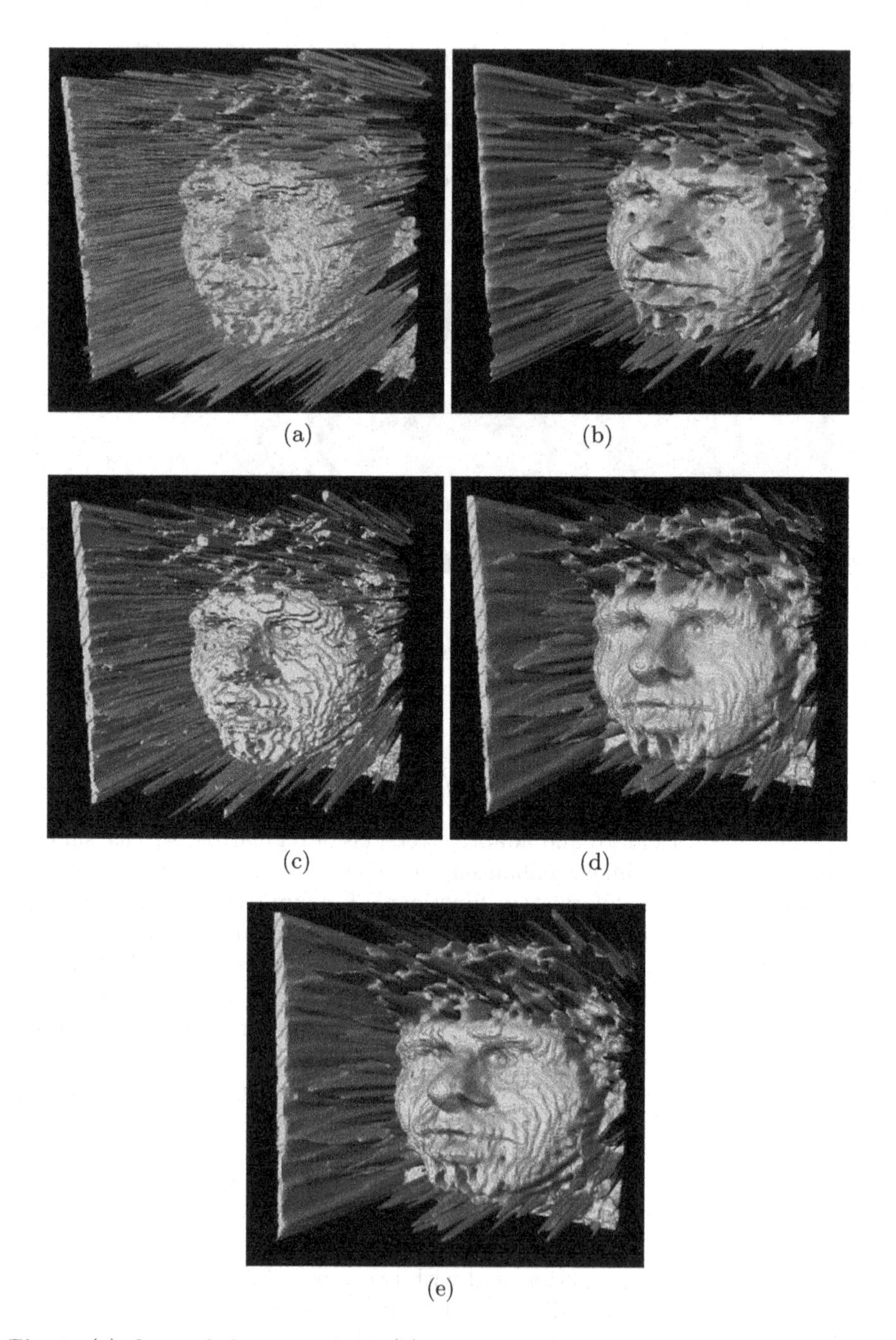

Fig. 6. (a) Original disparity data, (b) smoothed disparity data using a Gaussian smoothing filter($\sigma = 1.5$), (c) smoothed disparity data using a median filter ($4{\times}4$) (d) smoothed disparity data using the proposed non-adaptive regularization, and (e) smoothed disparity data using the proposed adaptive regularization

References

1. Blake, A., Zisserman, A.: Visual Reconstruction, MIT Press (1987)
2. Geman, S., Geman, D.: Stochastic relaxation, Gibbs distributions and the Bayesian restoration of images. IEEE Trans. Pattern Analysis, Machine Intelligence **6** (1984) 721-741
3. Terzopoulos, D.: The computation of visual surface representations, IEEE Trans. Pattern Analysis, Machine Intelligence **10** (1988) 417-438
4. Gokmen, M., Li, C.-C.: Edge detection and surface reconstruction using refined regularization, IEEE Trans. Pattern Analysis, Machine Intelligence **15** (1993) 492-499
5. Katsaggelos, A.K., Biemond, J., Schafer, R.W.: Mersereau, R. M.: A regularized iterative image restoration algorithms, IEEE Trans. Signal Processing **39** (1991) 914-929
6. Sinha, S.S., Schunck, B.G.: A two-stage algorithm for discontinuity-preserving surface reconstruction, IEEE Trans. Pattern Analysis, Machine Intelligence **14** (1992) 36-55
7. Umasuthan, M., Wallace, A.M.: Outlier removal and discontinuity preserving smoothing of range data, IEE Proc.-Vis. Image Signal Process **143** (1996) 191-200
8. Hoover, A.: The Space Envelope Representation for 3D Scenes, PhD thesis, Department of Computer Science and Engineering, University of South Florida (1996.)
9. Manfredo Do Carmo.: Differential geometry of curves and surfaces, Prentice Hall (1976)
10. June, H.Yi, David, M.Chelberg.: Discontinuity-preserving and viewpoint invariant reconstruction of visible surface using a first order regularization, IEEE Trans. Pattern Analysis, Machine Intelligence **17** (1995)
11. Stevenson, R.L., Delp, E.J.: Viewpoint invariant recovery of visual surface from sparse data, IEEE Trans. Pattern Analysis, Machine Intelligence **14** (1992) 897-909
12. Trucco, E., Verri, A.: Introductory Techniques for 3-D Computer Vision, Prentice-Hall (1998)
13. Shin, J.H., Sun, Y., Joung, W.C., Paik, J.K., Abidi, M.A.: Regularized noise smoothing of dense range image using directional Laplacian operators, Proc. SPIE Three-Dimensional Image Capture and Applications IV **4298** (2001) 119-126
14. Sun, Y., Paik, J.K., Price, J.R., Abidi, M.A.: Dense range image smoothing using adaptive regularization, Proc. 2000 Int. Conf. Image Processing **2** (2000) 10-13
15. Ki, H., Shin, J., Paik, J.: Regularized surface smoothing for enhancement of range data, Proc. IEEK **26** (2003) 1903-1906, in Korean

Minimum-Length Polygon of a Simple Cube-Curve in 3D Space

Fajie Li and Reinhard Klette

CITR, University of Auckland, Tamaki Campus, Building 731,
Auckland, New Zealand

Abstract. We consider simple cube-curves in the orthogonal 3D grid of cells. The union of all cells contained in such a curve (also called the tube of this curve) is a polyhedrally bounded set. The curve's length is defined to be that of the minimum-length polygonal curve (MLP) fully contained and complete in the tube of the curve. So far, only a "rubber-band algorithm" is known to compute such a curve approximately. We provide an alternative iterative algorithm for the approximative calculation of the MLP for curves contained in a special class of simple cube-curves (for which we prove the correctness of our alternative algorithm), and the obtained results coincide with those calculated by the rubber-band algorithm.

1 Introduction

The analysis of cube-curves is related to 3D image data analysis. A cube-curve is, for example, the result of a digitization process which maps a curve-like object into a union S of face-connected closed cubes. The computation of the length of a cube-curve was the subject in [3], and the suggested local method has its limitations if studied with respect to multigrid convergence. [1] presents a rubber-band algorithm for an approximative calculation of a minimum-length polygonal curve (MLP) in S. So far it was still an open problem to prove whether results of the rubber-band algorithm always converge to the exact MLP or not. In this paper we provide a non-trivial example where the rubber-band algorithm is converging against the MLP. So far, MLPs could only be tested manually for "simple" examples. This paper also presents an algorithm for the computation of approximate MLPs for a special class of simple cube-curves. (The example for the rubber-band algorithm is from this class.)

Following [1], a grid point $(i, j, k) \in \mathbb{Z}^3$ is assumed to be the center point of a *grid cube* with *faces* parallel to the coordinate planes, with *edges* of length 1, and *vertices* as its corners. *Cells* are either cubes, faces, edges, or vertices. The intersection of two cells is either empty or a joint *side* of both cells. A *cube-curve* is an alternating sequence $g = (f_0, c_0, f_1, c_1, \ldots, f_n, c_n)$ of faces f_i and cubes c_i, for $0 \leq i \leq n$, such that faces f_i and f_{i+1} are sides of cube c_i, for $0 \leq i \leq n$ and $f_{n+1} = f_0$. It is *simple* iff $n \geq 4$ and for any two cubes $c_i, c_k \in g$ with $|i-k| \geq 2 \pmod{n+1}$, if $c_i \bigcap c_k \neq \phi$ then either $|i-k| \geq 2 \pmod{n+1}$ and $c_i \bigcap c_k$ is an edge, or $|i-k| \geq 3 \pmod{n+1}$ and $c_i \bigcap c_k$ is is a vertex.

R. Klette and J. Žunić (Eds.): IWCIA 2004, LNCS 3322, pp. 502–511, 2004.

A *tube* **g** is the union of all cubes contained in a cube-curve g. A tube is a compact set in $\mathbb{R}^3$, its frontier defines a polyhedron, and it is homeomorphic with a torus in case of a simple cube-curve. A curve in $\mathbb{R}^3$ is *complete* in **g** iff it has a nonempty intersection with every cube contained in g. Following [4, 5], we define that a *minimum-length polygon* (MLP) of a simple cube-curve g is a shortest simple curve P which is contained and complete in tube **g**. The length of a simple cube-curve g is defined to be the length $l(P)$ of an MLP of g.

It turns out that such a shortest simple curve P is always a polygonal curve, and it is uniquely defined if the cube-curve is not only contained in a single layer of cubes of the 3D grid (see [4, 5]). If contained in one layer, then the MLP is uniquely defined up to a translation orthogonal to that layer. We speak about *the* MLP of a simple cube-curve.

A *critical edge* of a cube-curve g is such a grid edge which is incident with exactly three different cubes contained in g. Figure 1 shows all the critical edges of a simple cube-curve. If e is a critical edge of g and l is a straight line such that $e \subset l$, then l is called a *critical line of* e *in* g or *critical line* for short.

Assume a simple cube-curve g and a triple of consecutive critical edges e_1, e_2, and e_3 such that $e_i \perp e_j$, for all $i, j = 1, 2, 3$ with $i \neq j$. If the x-coordinates (y-coordinates, or z-coordinates) of two vertices (i.e., end points) of e_1 and e_3 are equal when e_2 is parallel to the x-axis (y-axis, or z-axis), we say that e_1, e_2 and e_3 form an *end angle*, and g *has an end angle*, denoted by $\angle(e_1, e_2, e_3)$; otherwise we say that e_1, e_2 and e_3 form a *middle angle*, and g *has a middle angle*. Figure 1 shows a simple cube-curve which has 5 end angles $\angle(e_{21}, e_0, e_1)$, $\angle(e_4, e_5, e_6)$, $\angle(e_6, e_7, e_8)$, $\angle(e_{14}, e_{15}, e_{16}))$, $\angle(e_{16}, e_{17}, e_{18})$, and many middle angles (e.g., $\angle(e_0, e_1, e_2)$, $\angle(e_1, e_2, e_3)$, and $\angle(e_2, e_3, e_4)$).

A simple cube-curve g is called *first class* iff each critical edge of g contains exactly one vertex of the MLP of g. This paper focuses on first-class simple cube-curves which have at least one end angle (as the one in Figure 1).

Let $S \subseteq \mathbb{R}^3$. The set $\{(x, y, 0) : \exists z(z \in \mathbb{R} \wedge (x, y, z) \in S)\}$ is the *xy-projection* of S, or *projection* of S for short. Analogously we define the *yz-* or *xz-projection* of S. The paper is organized as follows: Section 2 describes theoretical fundamen-

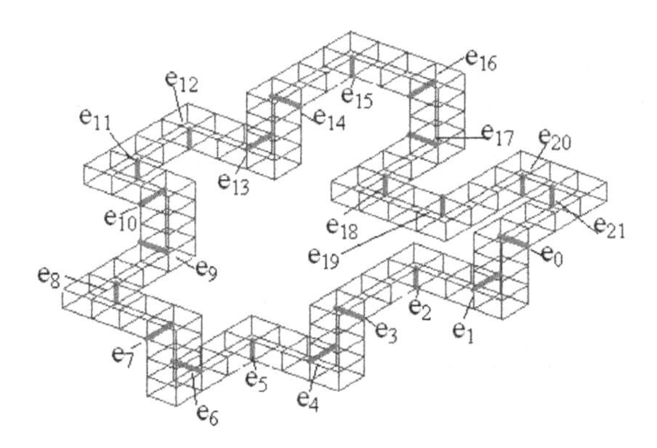

Fig. 1. Example of a first-class simple cube-curve which has middle and end angles

tals for the length calculation of first-class simple cube-curves. Section 3 presents our algorithm for length computation. Section 4 gives experimental results of an example and a discussion of results obtained by the rubber-band algorithm for this particular input. Section 5 gives the conclusions.

2 Basics

We provide mathematical fundamentals used in our algorithm for computing the *MLP* of a first-class simple cube-curve. We start with citing a basic theorem from [1]:

Theorem 1. *Let g be a simple cube-curve. Critical edges are the only possible locations of vertices of the MLP of g.*

This theorem is of fundamental importance for both the rubber-band algorithm and our algorithm (to be defined later in this paper). Let $d_e(p,q)$ be the Euclidean distance between points p and q.

Let e_1, e_2, and e_3 be three (not necessarily consecutive) critical edges in a simple cube-curve, and let l_1, l_2, and l_3 be the corresponding three critical lines. We express a point $p_2(t_2) = (x_2 + k_{x_2}t_2, y_2 + k_{y_2}t_2, z_2 + k_{z_2}t_2)$ on l_2 in general form, with $t_2 \in \mathbb{R}$. Analogously, let $p_1(t_1)$, $p_3(t_3)$ be points on l_1, l_3, respectively.

Lemma 1. *Let $d_2(t_1, t_2, t_3) = d_e(p_1, p_2) + d_e(p_2, p_3)$. It follows that $\frac{\partial^2 d_2}{\partial {t_2}^2} > 0$.*

Space in this short conference note does not allow to provide proofs.[1]

Let l_i be a critical line, $p_i \in l_i$, where $i = 0, 1, 2, \ldots, n$. Let $d(t_0, t_1, \ldots, t_n) = \sum_{i=0}^{n-1} d_e(p_i, p_{i+1})$. Assume $n+1$ reals t_{i_0} $(i = 0, 1, \ldots, n)$ which define a minimum $d(t_{0_0}, t_{1_0}, \ldots, t_{n_0})$ of function $d(t_0, t_1, \ldots, t_n)$. By Lemma 1 we immediately obtain

Lemma 2. *For any two reals t_{i_1} and t_{i_2}, we have*

$$d(t_{0_0}, \ldots, t_{i_0}, \ldots, t_{n_0}) < d(t_{0_0}, \ldots, t_{i_1}, \ldots, t_{n_0}) < d(t_{0_0}, \ldots, t_{i_2}, \ldots, t_{n_0})$$

if $t_{i_0} < t_{i_1} < t_{i_2}$, and

$$d(t_{0_0}, \ldots, t_{i_1}, \ldots, t_{n_0}) > d(t_{0_0}, \ldots, t_{i_2}, \ldots, t_{n_0}) > d(t_{0_0}, \ldots, t_{i_0}, \ldots, t_{n_0})$$

if $t_{i_1} < t_{i_2} < t_{i_0}$.

Let e_1, e_2, and e_3 be three critical edges, and let l_1, l_2, and l_3 be their critical lines, respectively. Let p_1, p_2, and p_3 be three points such that p_i belongs to l_i, where $i = 1, 2, 3$. Let the coordinates of p_2 be $(x_2 + k_{x_2}t_2, y_2 + k_{y_2}t_2, z_2 + k_{z_2}t_2)$. Let $d_2 = d_e(p_1, p_2) + d_e(p_2, p_3)$.

[1] See our online CITR-TR-147 for a longer version, also containing proofs.

Lemma 3. *The function* $f(t_2) = \frac{\partial d_2}{\partial t_2}$ *has a unique real root.*

Let l_i be a critical line, $p_i \in l_i$, the coordinates of p_i be $(x_i + k_{x_i}t_i, y_2 + k_{y_i}t_i, z_i + k_{z_i}t_i)$, where $i = 1, 2, \ldots, n$. Let $d(t_0, t_1, \ldots, t_n) = \sum_{i=0}^{n-1} d_e(p_i, p_{i+1})$.

Theorem 2. *There is a unique* $(n+1)$*-tuple of reals* t_{i_0} $(i = 0, 1, \ldots, n)$ *defining the minimum* $d(t_{0_0}, t_{1_0}, \ldots, t_{n_0})$ *of* $d(t_0, t_1, \ldots, t_n)$, *with* $\frac{\partial d}{\partial t_i}(t_{0_0}, t_{1_0}, \ldots, t_{n_0}) = 0$, *for* $i = 0, 1, \ldots, n$.

Let e_1, e_2 and e_3 be three consecutive critical edges of a simple cube-curve g. Let $D(e_1, e_2, e_3)$ be the dimension of the linear space generated by e_1, e_2 and e_3. Let l_{13} be a line segment with its two end points at e_1 and e_3. Let $d_{e_ie_j}$ be Euclidean distance between e_i and e_j (i.e., the minimum distance between points p on e_i and q on e_j), where $i, j = 1, 2, 3$.

Lemma 4. *The line segment* l_{13} *is not completely contained in* **g** *if* $D(e_1, e_2, e_3) = 3$, $\min\{d_{e_1e_2}, d_{e_2e_3}\} \geq 1$ *and* $\max\{d_{e_1e_2}, d_{e_2e_3}\} \geq 2$, *or if* $D(e_1, e_2, e_3) \leq 2$ *and* $\min\{d_{e_1e_2}, d_{e_2e_3}\} \geq 2$.

Let g be a simple cube-curve such that any three consecutive critical edges e_1, e_2 and e_3 do satisfy that either $D(e_1, e_2, e_3) = 3$, $\min\{d_{e_1e_2}, d_{e_2e_3}\} \geq 1$ and $\max\{d_{e_1e_2}, d_{e_2e_3}\} \geq 2$, or $D(e_1, e_2, e_3) \leq 2$ and $\min\{d_{e_1e_2}, d_{e_2e_3}\} \geq 2$. By Lemma 4, we immediately obtain

Lemma 5. *Every critical edge of* g *contains at least one vertex of* g*'s MLP.*

Let g be a simple cube-curve, and assume that every critical edge of g contains at least one vertex of the MLP. Then we also have the following:

Lemma 6. *Every critical edge of* g *contains at most one vertex of* g*'s MLP.*

Proof. Assume that there exists a critical edge e such that e contains at least two vertices v and w of the MLP P of g. Without loss of generality, we can assume that v and w are the first (in the order on P) two vertices which are on e. Let u be a vertex of P, which is on the previous critical edge of P. Then line segments uv and uw are completely contained in **g**. By replacing $\{uv, uw\}$ by uw we obtain a polygon of length shorter than P, which is in contradiction to the fact that P is an MLP of g. □

Let g be a simple cube-curve such that any three consecutive critical edges e_1, e_2, and e_3 do satisfy that either $D(e_1, e_2, e_3) = 3$, $\min\{d_{e_1e_2}, d_{e_2e_3}\} \geq 1$ and $\max\{d_{e_1e_2}, d_{e_2e_3}\} \geq 2$, or $D(e_1, e_2, e_3) \leq 2$ and $\min\{d_{e_1e_2}, d_{e_2e_3}\} \geq 2$. By Lemma 5 and Lemma 6, we immediately obtain

Theorem 3. *The specified simple cube-curve* g *is first class.*

Let e_1, e_2, and e_3 be three consecutive critical edges of a simple cube-curve g. Let p_1, p_2, and p_3 be three points such that $p_i \in e_i$, for $i = 1, 2, 3$. Let the coordinates of p_i be $(x_i + k_{x_i}t_i, y_2 + k_{y_i}t_i, z_i + k_{z_i}t_i)$, where $k_{x_i}, k_{y_i}, k_{z_i}$ are either 0 or 1, and $0 \leq t_i \leq 1$, for $i = 1, 2, 3$. Let $d_2 = d_e(p_1, p_2) + d_e(p_2, p_3)$.

Theorem 4. $\frac{\partial d_2}{\partial t_2} = 0$ *implies that we have one of the following representations for* t_3*: we can have*

$$t_3 = \frac{-c_2 t_1 + (c_1 + c_2) t_2}{c_1} \tag{1}$$

if $c_1 > 0$*; we can also have*

$$t_3 = 1 - \sqrt{\frac{c_1^2 (t_2 - a_2)^2}{(t_2 - t_1)^2} - c_2^2} \quad or \tag{2}$$

$$t_3 = \sqrt{\frac{c_1^2 (t_2 - a_2)^2}{(t_2 - t_1)^2} - c_2^2} \tag{3}$$

if a_2 *is either 0 or 1, and* c_1 *and* c_2 *are positive; and we can also have*

$$t_3 = 1 - \sqrt{\frac{(t_2 - a_2)^2 [(t_1 - a_1)^2 + c_1^2]}{(t_2 - b_1)^2} - c_2^2} \quad or \tag{4}$$

$$t_3 = \sqrt{\frac{(t_2 - a_2)^2 [(t_1 - a_1)^2 + c_1^2]}{(t_2 - b_1)^2} - c_2^2} \tag{5}$$

if a_1*,* a_2*, and* b_1 *are either 0 or 1, and* c_1 *and* c_2 *are positive reals.*

The proof of Case 3 of Theorem 4 and Lemma 3 show the following:

Lemma 7. *Let* g *be a first class simple cube-curve. If* e_1, e_2 *and* e_3 *form a middle angle of* g *then the vertex of the* MLP *of* g *on* e_2 *can not be an endpoint (i.e., a grid point) on* e_2*.*

Lemma 8. *Let* $f(x)$ *be a continuous function defined on interval* $[a, b]$*, and assume* $f(\xi) = 0$ *for some* $\xi \in (a, b)$*. Then, for every* $\varepsilon > 0$*, there exist* a' *and* b' *such that for every* $x \in [a', b']$ *we have* $|f(x)| < \varepsilon$*.*

Lemma 9. *Let* $f(x)$ *be a continuous function on an interval* $[a, b]$*, with* $f(\xi) = 0$ *at* $\xi \in (a, b)$*. Then for every* $\varepsilon > 0$*, there are two integers* $n > 0$ *and* $k > 0$ *such that for every* $x \in [\frac{(k-1)(b-a)}{n}, \frac{k(b-a)}{n}]$*, we have* $|f(x)| < \varepsilon$*.*

3 Algorithm

This section contains main ideas and steps of our algorithm for computing the MLP of a first class simple cube-curve which has at least one end angle.

3.1 Basic Ideas

Let p_i be a point on e_i, where $i = 0, 1, 2, \ldots, n$. Let the coordinates of p_i be $(x_i + k_{x_i} t_i, y_2 + k_{y_i} t_i, z_i + k_{z_i} t_i)$, where $i = 0, 1, \ldots,$ and n. Then the length of the polygon $p_0p_1 \ldots p_n$ is $d = d(t_0, t_1, \ldots, t_n) = \sum_{i=0}^{n} d_e(p_i, p_{i+1})$. If the polygon $p_0p_1 \ldots p_n$ is the MLP of g, then (by Theorem 2) we have $\frac{\partial d}{\partial t_i} = 0$, where $i = 0, 1, \ldots, n$.

Assume that e_i, e_{i+1}, and e_{i+2} form an end angle, and also e_j, e_{j+1}, and e_{j+2}, and no other three consecutive critical edges between e_{i+2} and e_j form an end angle, where $i \leq j$ and $i, j = 0, 1, 2, \ldots, n$. By Theorem 4 we have $t_{i+3} = f_{i+3}(t_{i+1}, t_{i+2}), t_{i+4} = f_{i+4}(t_{i+2}, t_{i+3}), t_{i+5} = f_{i+5}(t_{i+3}, t_{i+4}), \ldots, t_j$, and $t_{j+1} = f_{j+1}(t_{j-1}, t_j)$. This shows that $t_{i+3}, t_{i+4}, t_{i+5}, \ldots, t_j$, and t_{j+1} can be represented by t_{i+1}, and t_{i+2}. In particular, we obtain an equation $t_{j+1} = f(t_{i+1}, t_{i+2})$, or

$$g(t_{j+1}, t_{i+1}, t_{i+2}) = 0, \tag{6}$$

where t_{j+1}, and t_{i+1} are already known, or

$$g_1(t_{i+2}) = 0. \tag{7}$$

Since e_i, e_{i+1}, and e_{i+2} form an end angle it follows that $e_{i+1} \perp e_{i+2}$. By Theorem 4 we can express $\frac{\partial d_2}{\partial t_{i+2}}$ either in the form

$$\frac{t_{i+2} - t_{i+1} - a_1}{\sqrt{(t_{i+2} - t_{i+1} - a_1)^2 + b_1^2}} + \frac{t_{i+2} - a_2}{\sqrt{(t_{i+2} - a_2)^2 + (t_{i+3} - b_2)^2 + c_2^2}} \tag{8}$$

or in the form

$$\frac{t_{i+2} - b_1}{\sqrt{(t_{i+1} - a_1)^2 + (t_{i+2} - b_1)^2 + c_1^2}} + \frac{t_{i+2} - a_2}{\sqrt{(t_{i+2} - a_2)^2 + (t_{i+3} - b_2)^2 + c_2^2}} \tag{9}$$

If t_{i+2} satisfies Equation (8), then $\frac{\partial d_2}{\partial t_{i+2}}(a_1') < 0$, and $\frac{\partial d_2}{\partial t_{i+2}}(a_2') > 0$, where $a_1' = \min\{t_{i+1}+a_1, a_2\}$, and $a_2' = \max\{t_{i+1}+a_1, a_2\}$. It follows that Equation (8) has a unique real root between a_1' and a_2'. If t_{i+2} satisfies Equation (9), then Equation (9) has a unique real root between a_2 and b_1. In summary, there are two real numbers a and b such that Equation (9) has a unique root in between a and b. If $g_1(a)g_1(b) < 0$, then we can use the bisection method (see [2–page 49]) to find an approximate root of Equation (9). Otherwise, by Lemma 9, we can also find an approximate root of Equation (9). Therefore we can find an approximate root for $\frac{\partial d}{\partial t_k} = 0$, where $k = i+2, i+3, \ldots$, and j, and an exact root for $\frac{\partial d}{\partial t_k} = 0$, where $k = i+1$ and $j+1$. In this way we will find an approximate or exact root t_{k_0} for $\frac{\partial d}{\partial t_k} = 0$, where $k =$ 1,2, $\ldots$, and n. Let $t'_{k_0} = 0$ if $t_{k_0} < 0$ and $t'_{k_0} = 1$ if $t_{k_0} > 1$, where $k = 1, 2, \ldots, n$. Then (by Theorem 2) we obtain an approximation of the MLP (its length is $d(t'_{1_0}, t'_{2_0}, \ldots, t'_{i_0}, \ldots, t'_{n_0})$) of the given first class simple cube-curve.

3.2 Main Steps of the Algorithm

The input is a first class simple cube-curve g with at least one end angle. The output is an approximation of the MLP and a calculated length value.

Step 1. Represent g by the coordinates of the endpoints of its critical edges e_i, where $i = 0, 1, 2, \ldots, n$. Let p_i be a point on e_i, where $i = 0, 1, 2, \ldots, n$. Then the coordinates of p_i should be $(x_i + k_{x_i} t_i, y_2 + k_{y_i} t_i, z_i + k_{z_i} t_i)$, where only one of the parameters k_{x_i}, k_{y_i} and k_{z_i} can be 1, and the other two are equal to 0, for $i = 0, 1, \ldots, n$.

Step 2. Find all end angles $\angle(e_j, e_{j+1}, e_{j+2}), \angle(e_k, e_{k+1}, e_{k+2}), \ldots$ of g. For every $i \in \{0, 1, 2, \ldots, n\}$, let $d_{i+1} = d_e(p_i, p_{i+1}) + d_e(p_{i+1}, p_{i+2})$. By Lemma 3, we can find a unique root $t_{(i+1)_0}$ of equation $\frac{\partial d_{i+1}}{\partial t_{i+1}} = 0$ if e_i, e_{i+1} and e_{i+2} form an end angle.

Step 3. For every pair of two consecutive end angles $\angle(e_i, e_{i+1}, e_{i+2})$ and $\angle(e_j, e_{j+1}, e_{j+2})$ of g, apply the ideas as described in Section 3.1 to find the root of equation $\frac{\partial d_k}{\partial t_k} = 0$, where $k = i+1, i+2, \ldots,$ and $j+1$.

Step 4. Repeat Step 3 until we find an approximate or exact root t_{k_0} for $\frac{\partial d}{\partial t_k} = 0$, where $d = d(t_0, t_1, \ldots, t_n) = \sum_{i=1}^{n-1} d_i$, for $k = 0, 1, 2, \ldots, n$. Let $t'_{k_0} = 0$ if $t_{k_0} < 0$, and $t'_{k_0} = 1$ if $t_{k_0} > 1$, for $k = 0, 1, 2, \ldots, n$.

Step 5. The output is a polygonal curve $p_0(t'_{1_0})p_1(t'_{2_0}) \cdots p_n(t'_{n_0})$ of total length $d(t'_{1_0}, t'_{2_0}, \ldots, t'_{i_0}, \ldots, t'_{n_0})$, and this curve approximates the MLP of g.

We give an estimate of the time complexity of our algorithm in dependency of the number of end angles m and the accuracy (tolerance ε) of approximation.

Let the accuracy of approximation be $\frac{1}{2^k}$. By [2–page 49], the bisection method needs to know the initial end points a and b of the search interval $[a, b]$. In the best case, if we can set $a = 0$ and $b = 1$ to solve all the forms of Equation (7) by the bisection method, then the algorithm completes each run in $O(mk^2)$ time. In the worst case, if we have to find out the values of a and b for every of the forms of Equation (7) by the bisection method, then by Lemma 9, and let us assume that we need 2^{k_0} steps to find out the values of a and b, the algorithm completes each run in $O(mk^2 2^{k_0})$ time.

4 Experiments

We provide one example where we compare results obtained with our algorithm with those of the rubber-band algorithm as described in [1].

We approximate the MLP of the first-class simple cube-curve of Figure 1.

Step 1. We identify all coordinates of the critical edges $e_0, e_1, \ldots, e_{21}$ of g. Let p_i be a point on the critical line of e_i, where $i = 0, 1, \ldots, 21$.

Step 2. We calculate the coordinates of p_i, where $i = 0, 1, \ldots 21$, as follows: $(1 + t_0, 4, 7)$, $(2, 4 + t_1, 5)$, $(4, 5, 4 + t_2)$, $(4 + t_3, 7, 4)$, $(5, 7 + t_4, 2)$, $(7, 8, 1 + t_5)$ $\ldots$ $(2, 2, 7 + t_{21})$.

Step 3. Now let $d = d(t_0, t_1, \ldots, t_{21}) = \sum_{i=0}^{21} d_e(p_i, p_{i+1(\mathrm{mod}\ 22)})$. Then we obtain

$$\frac{\partial d}{\partial t_0} = \frac{t_0 - 1}{\sqrt{(t_0 - 1)^2 + t_{21}^2 + 4}} + \frac{t_0 - 1}{\sqrt{(t_0 - 1)^2 + t_1^2 + 4}} \tag{10}$$

$$\frac{\partial d}{\partial t_1} = \frac{t_1}{\sqrt{(t_0 - 1)^2 + t_1^2 + 4}} + \frac{t_1 - 1}{\sqrt{(t_1 - 1)^2 + (t_2 - 1)^2 + 4}} \tag{11}$$

$$\frac{\partial d}{\partial t_2} = \frac{t_2 - 1}{\sqrt{(t_1 - 1)^2 + (t_2 - 1)^2 + 4}} + \frac{t_2}{\sqrt{t_2^2 + t_3^2 + 4}} \tag{12}$$

$$\frac{\partial d}{\partial t_3} = \frac{t_3}{\sqrt{t_2^2 + t_3^2 + 4}} + \frac{t_3 - 1}{\sqrt{(t_3 - 1)^2 + t_4^2 + 4}} \tag{13}$$

$$\frac{\partial d}{\partial t_4} = \frac{t_4}{\sqrt{(t_3 - 1)^2 + t_4^2 + 4}} + \frac{t_4 - 1}{\sqrt{(t_4 - 1)^2 + (t_5 - 1)^2 + 4}} \tag{14}$$

and

$$\frac{\partial d}{\partial t_5} = \frac{t_5 - 1}{\sqrt{(t_4 - 1)^2 + (t_5 - 1)^2 + 4}} + \frac{t_5 - 1}{\sqrt{(t_5 - 1)^2 + t_6^2 + 4}} \tag{15}$$

By Equations (10) and (15) we obtain $t_0 = t_5 = 1$.

Similarly, we have $t_7 = t_{15} = 0$, and $t_{16} = 1$. Therefore we find all end angles as follows: $\angle(e_{21}, e_0, e_1)$, $\angle(e_4, e_5, e_6)$, $\angle(e_6, e_7, e_8)$, $\angle(e_{14}, e_{15}, e_{16})$, and $\angle(e_{15}, e_{16}, e_{17})$.

By Theorem 4 and Equations (11), (12), (13) it follows that

$$t_2 = 1 - \sqrt{\frac{(t_1 - 1)^2[(t_0 - 1)^2 + 4]}{t_1^2} - 4} \tag{16}$$

$$t_3 = \sqrt{\frac{t_2^2[(t_1 - 1)^2 + 4]}{(t_2 - 1)^2} - 4} \tag{17}$$

and

$$t_4 = \sqrt{\frac{(t_3 - 1)^2[t_2^2 + 4]}{t_3^2} - 4} \tag{18}$$

By Equation (14) we have

$$t_4^2[(t_5 - 1)^2 + 4] = (t_4 - 1)^2[(t_3 - 1)^2 + 4]$$

Let

$$g_1(t_1) = t_4^2[(t_5 - 1)^2 + 4] - (t_4 - 1)^2[(t_3 - 1)^2 + 4] \tag{19}$$

By Equation (16) we have $t_1 \in (0, 0.5)$, $g_1(0.4924) = 3.72978 > 0$, and also $g_1(0.4999) = -51.2303 < 0$. By Theorem 2 and the Bisection Method we obtain the following unique roots of Equations (19), (16), (17), and (18):

$t_1 = 0.492416$, $t_2 = 0.499769$, $t_3 = 0.499769$, and $t_4 = 0.507584$,

with error $g_1(t_1) = 4.59444 \times 10^{-9}$. These roots correspond to the two consecutive end angles $\angle(e_{21}, e_0, e_1)$ and $\angle(e_4, e_5, e_6)$ of g.

Step 4. Similarly, we find the unique roots of equation $\frac{\partial d}{\partial t_i} = 0$, where $i = 6, 7, \ldots, 21$. At first we have $t_6 = 0.5$, which corresponds to the two consecutive end angles $\angle(e_4, e_5, e_6)$ and $\angle(e_6, e_7, e_8)$; then we also obtain

$t_8 = 0.492582, t_9 = 0.494543, t_{10} = 0.331074, t_{11} = 0.205970, t_{12} = 0.597034$, $t_{13} = 0.502831, t_{14} = 0.492339$, which correspond to the two consecutive end

Table 1. Comparison of results of both algorithms

Critical points	t_{i_0} (our algorithm)	t_{i_0} (Rubber-Band Algorithm)
p_0	1	1
p_1	0.492416	0.4924
p_2	0.499769	0.4998
p_3	0.499769	0.4998
p_4	0.507584	0.5076
p_5	1	1
p_6	0.5	0.5
p_7	0	0
p_8	0.492582	0.4926
p_9	0.494543	0.4945
p_{10}	0.331074	0.3311
p_{11}	0.205970	0.2060
p_{12}	0.597034	0.5970
p_{13}	0.502831	0.5028
p_{14}	0.492339	0.4923
p_{15}	0	0
p_{16}	1	1
p_{17}	0.501527	0.5015
p_{18}	0.77824	0.7789
p_{19}	0.56314	0.5641
p_{20}	0.32265	0.3235
p_{21}	0.2151	0.2157

angles $\angle(e_6, e_7, e_8)$ and $\angle(e_{14}, e_{15}, e_{16})$; followed by $t_{15} = 0, t_{16} = 1$, which correspond to the two consecutive end angles $\angle(e_{14}, e_{15}, e_{16})$ and $\angle(e_{15}, e_{16}, e_{17})$; and finally $t_{17} = 0.501527, t_{18} = 0.77824, t_{19} = 0.56314, t_{20} = 0.32265$, and $t_{21} = 0.2151$, which correspond to the two consecutive end angles $\angle(e_{15}, e_{16}, e_{17})$ and $\angle(e_{21}, e_0, e_1)$.

Step 5. In summary, we obtain the values shown in the first two columns of Table 1. The calculated approximation of the MLP of g is $p_0(t'_{1_0})p_1(t'_{2_0})\cdots p_n(t'_{n_0})$, and its length is $d(t'_{1_0}, t'_{2_0}, \ldots, t'_{i_0}, \ldots, t'_{n_0}) = 43.767726$, where $t'_{i_0} = t_{i_0}$ for i limited to the set $\{0, 1, 2, \ldots, 21\}$.

The Rubber-Band Algorithm [1] calculated the roots of Equations (10) through (15) as shown in the third column of Table 1. Note that there is only a finite number of iterations until the algorithm terminates. No threshold needs to be specified for the chosen input curve.

From Table 1 we can see that both algorithms converge to the same values.

5 Conclusions

We designed an algorithm for the approximative calculation of an MLP for a special class of simple cube-curves (first-class simple cube-curves with at least one end angle). Mathematically, the problem is equivalent to solving equations with one variable each. Applying methods of numerical analysis, we can compute their roots with sufficient accuracy. We illustrated by one non-trivial example that the Rubber-Band Algorithm also converges to the correct solution (as calculated by our algorithm).

Acknowledgements. The authors thank Dr Thomas Bülow for providing the source code of and correct results of the Rubber-Band Algorithm. The IWCIA reviewers' comments have been very helpful for revising an earlier version of this paper.

References

1. T. Bülow and R. Klette. Digital curves in 3D space and a linear-time length estimation algorithm. *IEEE Trans. Pattern Analysis Machine Intelligence*, **24**:962–970, 2002.
2. R.L. Burden and J.D. Faires. BF*Numerical Analysis*. 7th edition, Brooks Cole, Pacific Grove, 2000.
3. A. Jonas and N. Kiryati. Length estimation in 3-D using cube quantization. *J. Mathematical Imaging Vision*, **8**:215–238, 1998.
4. F. Sloboda, B. Zaťko, and R. Klette. On the topology of grid continua. SPIE *Vision Geometry VII,* **3454**:52–63, 1998.
5. F. Sloboda, B. Zaťko, and J. Stoer. On approximation of planar one-dimensional continua. In R. Klette, A. Rosenfeld, and F. Sloboda, editors, *Advances in Digital and Computational Geometry*, pages 113–160. Springer, Singapore, 1998.

Corner Detection and Curve Partitioning Using Arc-Chord Distance

Majed Marji[1], Reinhard Klette[2], and Pepe Siy[1]

[1] Daimler Chrysler Corporation, 800 Chrysler Drive, Auburn Hills, MI 48326-2757, USA
[2] The University of Auckland, CITR, Tamaki campus, Bldg. 731, Morrin Road, Glen Innes, Auckland 1005, New Zealand

Dedicated to Professor Azriel Rosenfeld

Abstract. There are several algorithms for curve partitioning using the arc-chord distance formulation, where a chord whose associated arc spans k pixels is moved along the curve and the distance from each border pixel to the chord is computed. The scale of the corners detected by these algorithms depends on the choice of integer k. Without a priori knowledge about the curve, it is difficult to choose a k that yields good results. This paper presents a modified method of this type that can tolerate the effects of an improper choice of k to an acceptable degree.

1 Introduction

Partitioning of digital planar curves is often based on using the arc-chord distance formulation. A polygonal approximation of the digital curve is then generated by connecting the attained partitioning points - also referred to as *corners* - with straight lines. In the arc-chord distance method, a *chord* whose associated arc spans k pixels (i.e., a k*-point arc* of the curve) is moved along the curve and the distance from each pixel on the k-point arc to the chord is computed. A *significance* (also called *cornerity*) *measure* is formulated using these distances (e.g., maximum, distance-accumulation etc.) and processed in order to define corners of the curve.

Ramer's algorithm [1] recursively partitions an arc at the point whose distance from the chord is a maximum. Rutkowski [2] computes the maximum distance of each point p on the curve from any chord having a given arc length and having p on its arc, and partitions the curve at local maxima of this distance.

The algorithm of Fischler and Bolls [3] labels each point on a curve as belonging to one of three categories: 1) a point in a smooth interval, 2) a critical point, or 3) a point in a noisy interval. To make this choice, the algorithm analyzes the deviations of the curve from a chord or "stick" that is iteratively advanced along the curve. If the curve stays close to the chord, points in the interval spanned by the chord will be labeled as belonging to a smooth section. If the curve makes a single excursion from the chord, the point in the interval that is farthest from the chord will be labeled a critical point (actually, for each placement of the

R. Klette and J. Žunić (Eds.): IWCIA 2004, LNCS 3322, pp. 512–521, 2004.

chord, an accumulator associated with the farthest point will be incremented by the distance between the point and the chord). If the curve makes two or more excursions, points in the interval will be labeled as noise points.

Phillips and Rosenfeld [4] presented a modified version of the algorithm presented in [2]. They also suggested an approach to choosing good values of k in a given part of the curve. To find a good value of k, they determined the best fit straight line for each k-point arc of the curve, and computed the RMS error corresponding to this fit. This process is repeated for a sequence of arc lengths, producing a sequence of fit measures for each border point. In a given part of the border "good" values of k are taken as those which produce local minima in the fit measure.

Han [5] proposed a method similar to that of Fischler and Bolls [3] but used the signed distance to the chord. The algorithm keeps two separate accumulators for the positive and negative arc-chord distances to distinguish between concave and convex corners. For a given chord-length L, a line is drawn from point p_i to point p_{i+L} on the curve. The signed distances from all points $p_{i+1}, \ldots, p_{i+L-1}$ to this line are calculated. The point with positive maximum distance is defined as p_+, and the point with the negative minimum distance is defined as p_- . If the absolute value of the maximum (minimum) distance exceeds a given threshold $D_{\min}$, the counter $(h(p_+)/h(p_-))$, associated with the point that corresponds to the maximum (minimum) is incremented (decremented). The line is advanced by one pixel and the process is repeated until the entire curve is scanned. In other words, this algorithm counts how many times each border point happened to be the farthest point from the line $p_i p_{i+L}$. At the end of the calculation, the points whose accumulator value exceeds a given threshold $H_{\min}$ are marked as concave (convex) points.

Lin, Dou and Wang [6] proposed a new shape description method termed the *arc height function* and used it to detect the corners of the border. A chord that joins border points p_i and p_{i+k} is advanced along the curve, one pixel at a time. A straight-line perpendicular to the chord passing through its center p_c intersects the border at point p_x. The distance between p_c and p_x is the arc height, which, when computed for all positions of the chord gives the arc height function. The corner points of the border correspond to the local maxima of the arc height function.

The algorithm of Aoyama and Kawagoe [7] starts by finding all occurrences of digital straight-line patterns and marking their endpoints as candidate vertices. The best approximating straight-lines are determined by considering the ratio between the height H and the length L of the chord, which is termed the *pseudo curvature* G. Two modifications were made to the calculation of H and G. First, the pseudo curvature calculation was modified in such a way to prevent a long straight-line from being approximated as an inclined line. Second, the distance calculation was modified to take into account cases where the perpendicular line does not intersect the line segment.

Wu and Wang [8] combined corner detection and polygonal approximation. For a given parameter k, a significance measure is assigned to each border point

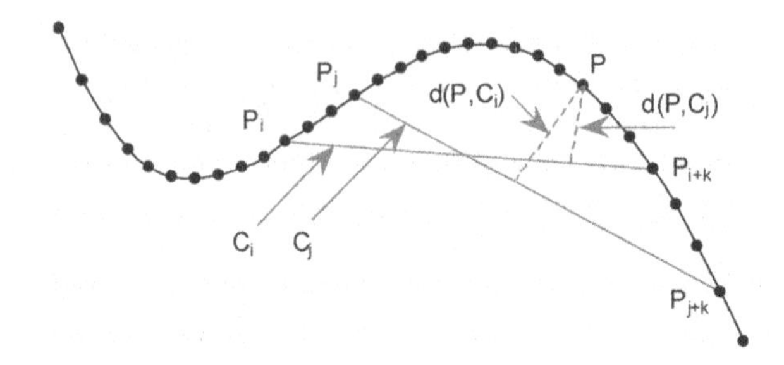

Fig. 1. Illustration of the Phillips and Rosenfeld algorithm

p_i as the ratio $\frac{d_i}{L_i}$, where d_i is the distance between point p_i and the chord (p_{i-k}, p_{i+k}), and L_i is the length of the chord. Local maxima points whose significance is greater than a threshold were taken as potential corners and used as the starting points for polygonal approximation. The border points within each segment (between two corners) are sorted according to their significance (most significant first). The sorted points are tested sequentially by calculating their distance to the chord that joins the end points of the segment; if this distance exceeds a given threshold, the corresponding point is marked as a corner.

The work of Fischler and Wolf [9] extends the technique of [3]. An important contribution of their work over [3] is a major revision of the approach to filtering the critical points, based on comparisons at a given scale as well as across different scales (i.e., different values of the input parameter k). In addition, the sign of the computed saliency measure is taken into consideration.

Han and Poston [10] proposed an enhanced version of the algorithm presented in [5]. Here, instead of incrementing a counter when the distance exceeds a threshold as in [5], the actual signed Euclidean distance is accumulated.

In this work, we propose a new algorithm based on the work of Phillips and Rosenfeld [4] that can tolerate the effect of an improper choice of k to an acceptable degree.

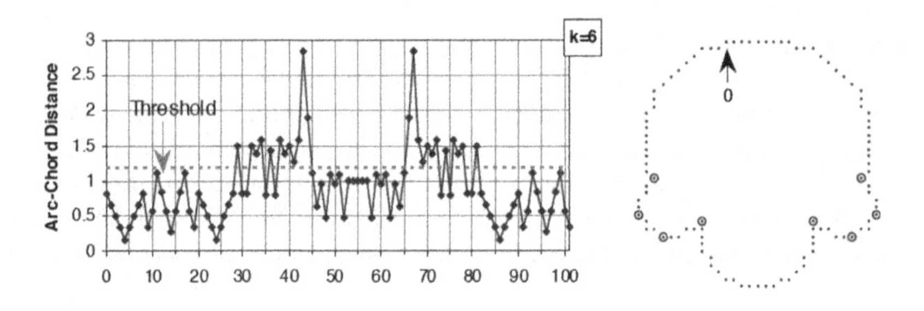

Fig. 2. Arc-chord distance measure and the corners detected by the Phillips-Rosenfeld algorithm using $k = 6$

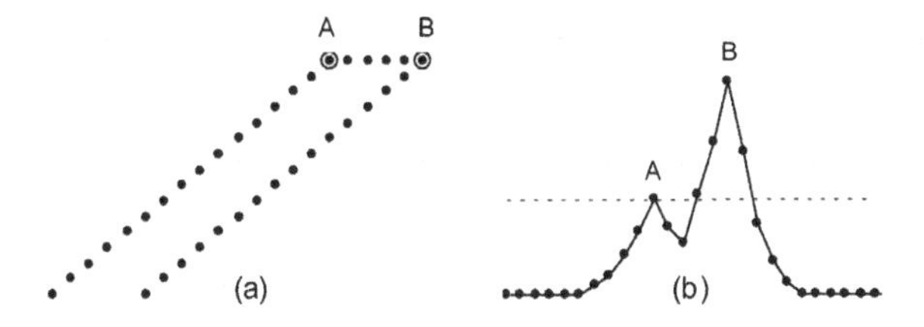

Fig. 3. An isolated corner model (a) and its associated (b) arc-chord distance using $k = 6$

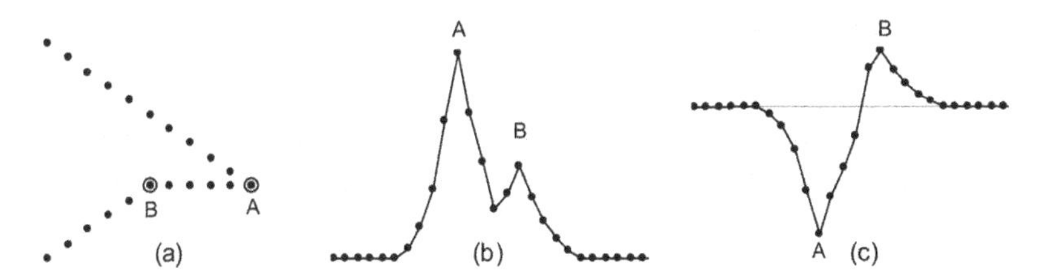

Fig. 4. An isolated corner model (a) and its associated unsigned (b) and signed (c) arc-chord distance measure using $k = 6$

2 The Method of Phillips and Rosenfeld

The algorithm is illustrated with the aid of Figure 1. Let p be a point on the curve and let k be the chosen arc length. For each chord C whose arc has length k and has p in its interior, let $d(p, C)$ be the perpendicular distance from p to C. Let $M(p, C)$ be the maximum of these distances for all such chords. Point p is called a *partition point* if the value of $M(p, C)$ is a local maximum (for the given k) and also exceeds a threshold $t = k/5 \cong (k/2)\cos(135°/2)$, which is the altitude of an isosceles triangle whose vertex angle is 135° and whose equal sides have lengths $k/2$. Point p is considered a local maximum point if the following condition is satisfied:

$$M(p, C) \geq M(p_x, C), \text{ for all } p_x \in \{p_{i-(k/2)}, \dots, p_{i+(k/2)}\}$$

To demonstrate the effect of thresholding, the Semicircles shape [11] and its associated arc-chord distance for $k = 6$ are shown in Figure 2. It is clear from this example that we cannot expect the suggested threshold of $k/5$ to work in all cases. Although lowering the threshold value will enable us to detect the missed corners in this example, it may result in many spurious corners for other test shapes.

A potential problem with the local maximum determination scheme is illustrated in Figure 3, which shows an isolated corner model[1] and its arc-chord distance using $k = 6$. In this example, peak A will be suppressed by some points

[1] A synthetic curve segment with a single corner. Thus, there are no near by corners that may affect the resulting arc-chord distance measure.

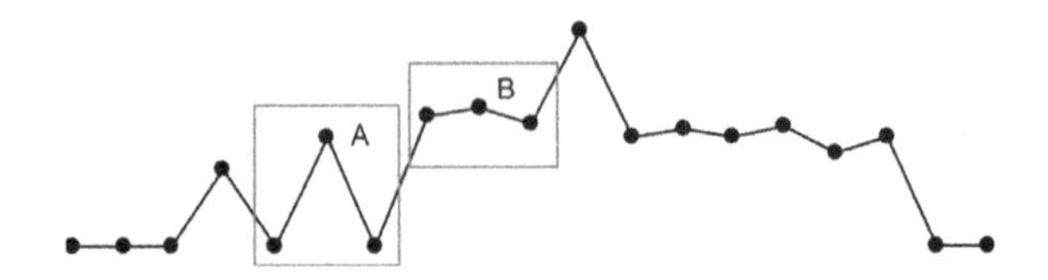

Fig. 5. The height of a peak is not indicative of its prominence

in its neighborhood with higher significance although non-of these points satisfy the local maximum criterion.

Figure 4 demonstrates that the inclusion of the sign information in the definition of the arc-chord distance can prevent some peaks from being masked by other neighboring peaks of opposite concavity (convexity). This figure shows an isolated corner model and its associated signed and unsigned arc-chord distance measures using $k = 6$. While peak B in the unsigned measure of Figure 4(b) will be discarded by the non-maximum suppression scheme, it has a better chance of being detected if the signed measure of Figure 4(c) is used instead. In addition, the inclusion of the sign information provides valuable evidence about the concavity and convexity of the curve without introducing any overhead on the subsequent calculations.

Figure 5 demonstrates that the height of a peak is not sufficient by itself to quantify the peak. In this example, although peak A is more visually prominent, it may be suppressed by peak B whose height is larger than that of peak A. This suggests that other criteria should be considered to quantify the strength of the peaks.

The final issue that was not explicitly discussed in [4] is that of plateaus. It is possible to have adjacent curve points with equal arc-chord distances, and trying to resolve these ties arbitrarily may result in detecting false corners. Although this may not cause noticeable problems for real borders, a properly designed algorithm should be able to handle these cases at least systematically.

3 The New Algorithm

The steps of the new algorithm will be described below and will be illustrated with the aid of the Semicircles shape shown in Figure 6.

1. Compute the arc-chord distance for each border point using the method of Phillips and Rosenfeld. Here however, we use the signed distance from the point to the chord instead of the absolute distance value. For the Semicircles shape, this measure is shown in Figure 7 using $k = 6$.
2. Separate the signed arc-chord distance function $d(p)$ into two functions $d_+(p)$ and $d_-(p)$ as follows

$$d_+(p) = \begin{cases} d(p), \text{ if } d(p) \geq 0 \\ 0 \quad\quad \text{otherwise} \end{cases} \quad \text{and} \quad d_-(p) = \begin{cases} \mid d(p) \mid, \text{ if } d(p) < 0 \\ 0 \quad\quad\quad \text{otherwise} \end{cases}$$

   Figure 8 shows these two functions for the measure of Figure 7.

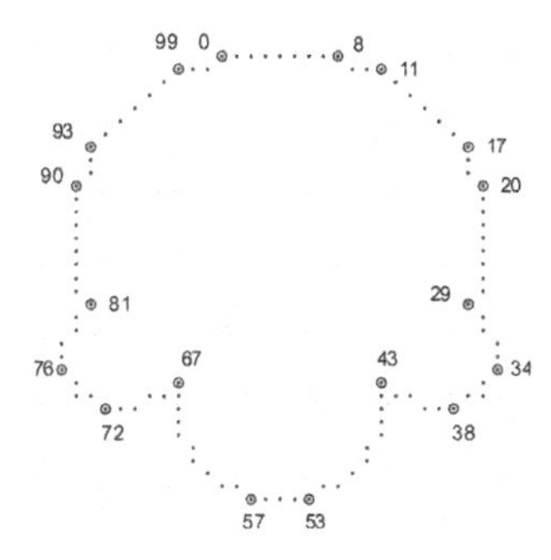

Fig. 6. Semicircles shape used to illustrate the new algorithm. Border points have been numbered for convenience

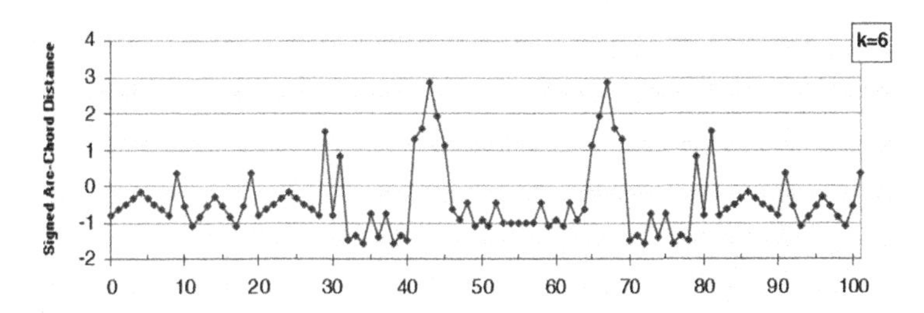

Fig. 7. Signed arc-chord distance for the Semicircles shape using $k = 6$

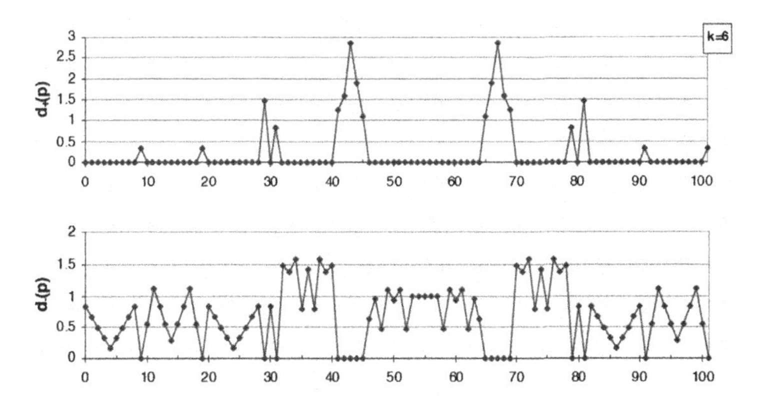

Fig. 8. The functions $d_+(p)$ and $d_-(p)$ for the Semicircles shape using $k = 6$

3. The signals $d_+(p)$ and $d_-(p)$ are processed separately where a search procedure is applied to detect the local maximum points. For each point p_i, we attempt to find the largest possible window that contains p_i such that the significance of all of the points in that window to both the left and right of p_i is strictly decreasing. If such a window exists, then p_i is considered a local maximum point, and the leftmost $P_L(p_i)$ and rightmost $P_R(p_i)$ points

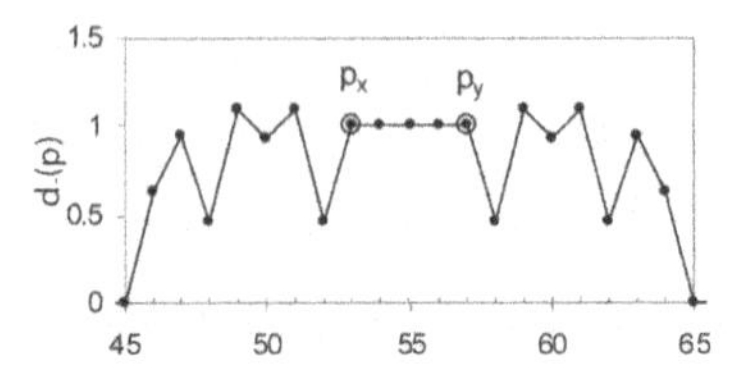

Fig. 9. Handling plateaus

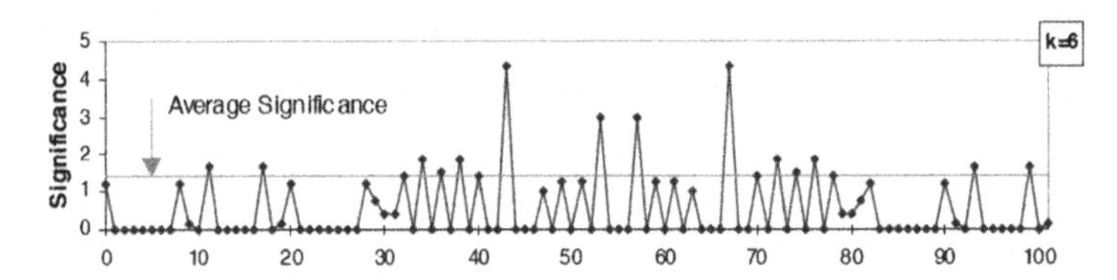

Fig. 10. Significance measure for the Semicircles shape using $k = 6$

of that window are recorded. For example, in Figure 8 ($d_-(p)$), point 28 is a local maximum point with $P_L(p_i) = 24$ and $P_R(p_i) = 29$.

The two endpoints of *valid* plateaus are handled differently. A plateau whose leftmost and rightmost end points are, respectively, p_x and p_y, is considered to be valid if $d(p_x) > d(p_x - 1)$ and $d(p_y) > d(p_y + 1)$. In this case, we set $P_R(p_x) = P_R(p_y)$ and $P_L(p_y) = P_L(p_x)$. This is illustrated in Figure 9, which represents a segment of the function of Figure 8. In this example, $P_L(p_y) = P_L(p_x) = 52$ and $P_R(p_x) = P_R(p_y) = 58$.

4. The significance of each local maximum point p_i found in the previous step is evaluated as the area of the polygon whose vertices are the points $(p_x, d_\pm(p_x))$ where $p_x \in [P_L(p_x), P_R(p_x)]$. This is shown in Figure 10 for the Semicircles shape.
5. The *mean significance* μ is calculated for all the local maximum points. In the above example (for instance) we have 40 local maximum points whose mean significance evaluates to 1.39 (see Figure 10).
6. All local maximum points whose significance value is greater than or equal to the average μ are marked as candidate corners. For the Semicircles test shape, this results in the following 18 points: 11, 17, 32, 34, 36, 38, 40, 43, 53, 57, 67, 70, 72, 74, 76, 78, 93, and 99.

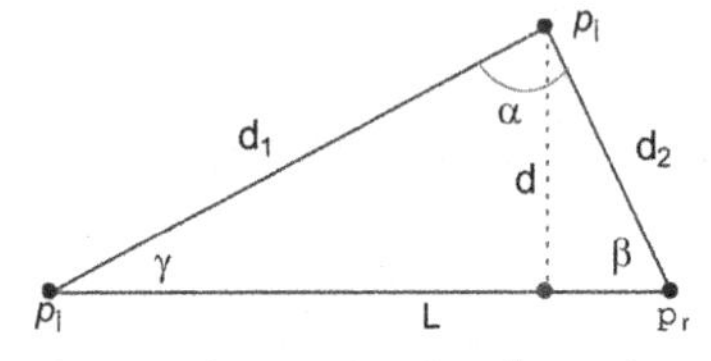

Fig. 11. Conditions for testing local maximum point p_i

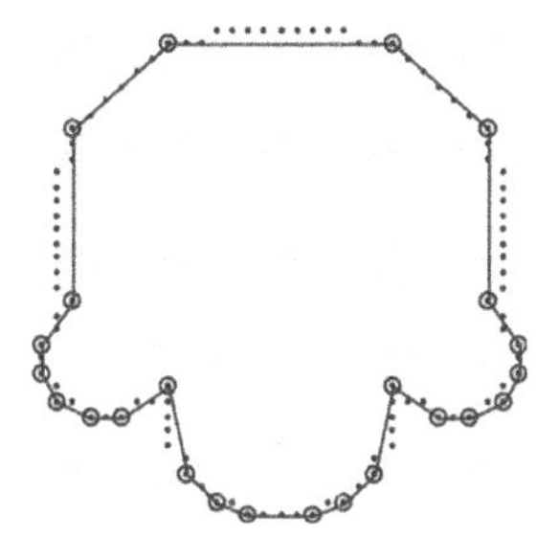

Fig. 12. Candidate corners after processing all local maxima points

7. The remaining local maxima points are sorted according to their significance in descending order (most significant first) and processed sequentially. For every local maximum point p_i, we consider the two candidate corners that proceed and succeed p_i; denote these two points by p_l and p_r, respectively, as shown in Figure 11.

   Then p_i is considered a candidate corner if

$$d \geq 1 \quad \text{and} \quad \frac{d}{L} \geq \frac{d_1 d_2}{L^2} \sin \alpha$$

   where α is set to 155°. The first condition is based on the fact that a slanted straight line is quantized into a set of horizontal and vertical line segments separated by one-pixel steps. In addition, we assume that the "border noise" is no more than one pixel, and if the noise level is known a priori, this threshold can be adjusted accordingly. The second condition allows us to detect the vertex of a triangle whose vertex angle is less than α. In order to preserve the symmetry of the shape, all local maxima points with equal significance level are processed in the same iteration.

   For the Semicircles shape, the first iteration examines points 49, 51, 59 and 61 (since all of them have the same significance); all these points satisfy the two conditions and are hence marked as candidate dominant points. The second iteration examines points 8, 20, 28, 82, and 90; none of these points satisfy the two conditions. The process continues until all local maxima points are examined. The output of this step is shown in Figure 12.
8. Because in step 6 we added all the peaks with "above-average" significance without paying attention to their proximity (in terms of border pixels), it is reasonable to believe that some of the marked candidate corners do not correspond to true corners of the curve. The purpose of the current step is to suppress the false corners (if any). First, we calculate the ratio d/L (see Figure 11) for all candidate corners and sort these corners in ascending order (lowest first). Candidate corners with $d < 1$ are considered insignificant and marked for deletion. Here also, we process all points with identical d/L value in the same iteration. The final result after this step is shown in Figure 13. The figure also shows the corners detected by the Phillips-Rosenfeld algorithm.

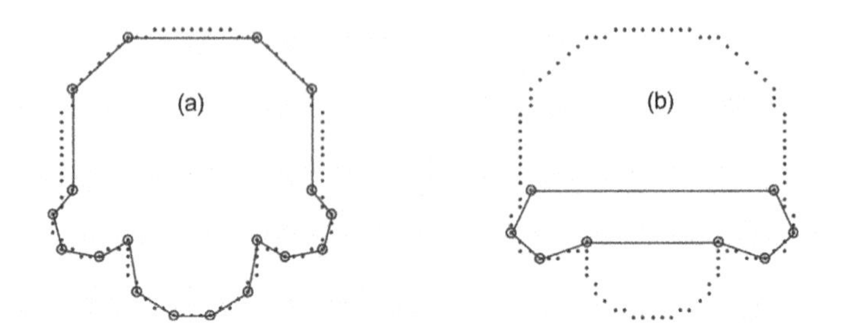

Fig. 13. Corners of the Semicircles shape using $k = 6$ produced by the current algorithm (a) and the Phillips-Rosenfeld algorithm (b)

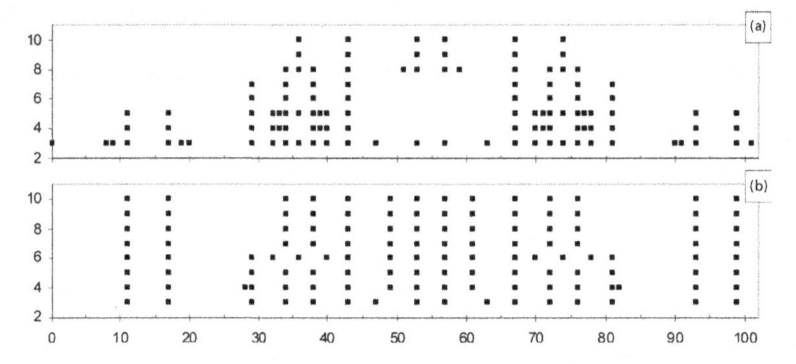

Fig. 14. Scale-space map for the Semicircles shape: (a) Phillips-Rosenfeld algorithm and (b) new algorithm

4 Experimental Results and Conclusions

To see the difference between the new algorithm and the Phillips-Rosenfeld algorithm, the scale space map for the Semicircles shape is shown in Figure 14 using k values in the range $[3, 10(= N/10)]$. Note that the Phillips-Rosenfeld algorithm did not detect any corners for the curve segment $[43, 67]$ for several scales whereas the results of the new algorithm were consistent to within a tolerance of $1 - 2$ pixels. In fact, the polygon generated by the new algorithm did provide a visually pleasing approximation of the shape for all the considered values of k.

We have described a new algorithm for curve partitioning using the arc-chord distance formulation. The algorithm can tolerate the effect of the scale parameter k to an acceptable degree.

References

1. U. Ramer. An iterative procedure for the polygonal approximation of plane closed curves. *Computer Graphics Image Processing*, **1**:244–256, 1972.

2. W.S. Rutkowski and A. Rosenfeld. A comparison of corner-detection techniques for chain-coded curves. Technical Report TR-623, Computer Science Center, University of Maryland, 1978.
3. M. Fischler and R. Bolles. Perceptual organization and curve partitioning, *IEEE Trans. Pattern Analysis Machine Intelligence*, **8**:100–105, 1986.
4. T.Y. Phillips and A. Rosenfeld. A method for curve partitioning using arc-chord distance. *Pattern Recognition Letters*, **5**: 285–288, 1987.
5. J.H. Han. Detection of convex and concave discontinuous points in a plane curve. In *Proc. 3rd Int. Conf. Computer Vision*, December 4-7, Osaka, Japan, pages 71–74, 1990.
6. Y. Lin, J. Dou, and H. Wang. Contour shape description based on an arc height function. *Pattern Recognition*, **25**: 17–23, 1992.
7. H. Aoyama and M. Kawagoe. A piecewise linear approximation method for preserving visual feature points of original figures. *CVGIP: Graphical Models and Image Processing*, **53**: 435–446, 1991.
8. S.-Y. Wu and M.-J. Wang. Detecting the dominant points by the curvature-based polygonal approximation. *CVGIP: Graphical Models and Image Processing*, **5**: 79–88, 1993.
9. M.A. Fischler and H.C. Wolf. Locating perceptually salient points on planar curves. *IEEE Trans. Pattern Analysis Machine Intelligence*, **16**:113–129, 1994.
10. J.H. Han and T. Poston. Chord-to-point distance accumulation and planar curvature: a new approach to discrete curvature. *Pattern Recognition Letters*, **22**:1133–1144, 2001.
11. C.-H. The and R.T. Chin. On the detection of dominant points on digital curve. *IEEE Trans. Pattern Analysis Machine Intelligence*, **11**: 859–872, 1989.

Shape Preserving Sampling and Reconstruction of Grayscale Images

Peer Stelldinger

Cognitive Systems Group, University of Hamburg,
Vogt-Kln-Str. 30, D-22527 Hamburg, Germany

Abstract. The expressiveness of a lot of image analysis algorithms depends on the question whether shape information is preserved during digitization. Most existing approaches to answer this are restricted to binary images and only consider nearest neighbor reconstruction. This paper generalizes this to grayscale images and to several reconstruction methods. It is shown that a certain class of images can be sampled with regular and even irregular grids and reconstructed with different interpolation methods without any change in the topology of the level sets of interest.

1 Introduction

Much of the information in an analog image may get lost under digitization. An image analysis algorithm can only be successful, if the needed information is preserved during the digitization process. Since a lot of image analysis algorithms are based on level sets, isosurface contours, and their shapes, it is important to know how to guarantee that the shapes of level sets are preserved. Up to now the problem of shape preserving digitization has mostly been dealt with for binary images.

It is well known that so-called r-regular binary images (see definition 1) can be digitized with square or hexagonal grids of a certain density without changing the shape in a topological sense [2, 11, 12]. Recently Kthe and the author were able to show that this is true for *any* grid of a certain density and that this still holds if the image is blurred by a disc shaped point spread function [7, 13]. In case of square grids, this is also proved for square shaped point spread functions [8, 9]. But all this work is not only restricted to binary images but also to nearest neighbor reconstruction in combination with thresholding. The only exception is the work of Florncio and Schafer [2], which allows other morphological reconstruction methods, too, but then only guarantees a bounded Hausdorff error instead of topological equivalence. In general, reconstruction means extending the domain of a discrete image function from the set of sampling points to the whole plane $\mathbb{R}^2$. In another paper [3] Florncio and Schafer show that even certain grayscale images can be sampled and reconstructed with a bounded Hausdorff error, when using a regular grid and some morphological reconstruction method.

R. Klette and J. Žunić (Eds.): IWCIA 2004, LNCS 3322, pp. 522–533, 2004.

All the mentioned approaches use several different ways to compare an image with its reconstructed digital counterpart. The strongest mentioned similarity criterion is strong r-similarity [13], which subsumes the others and which is used in this paper. The prior results are generalized to grayscale images and to a broad class of important reconstruction methods.

2 Regular Images and 2D Monotony

In this section some basic concepts are defined, which are necessary for the following work. Namely a definition of *r-regular* graylevel images and a generalization of *monotony* to 2D is given. Additionally some connections between these two ideas are shown, which are used in the proofs of the following sections.

At first some basic notations are given: The Complement of a set A will be noted as A^c. The boundary ∂A is the set of all common accumulation points of A and A^c. The interior A^0 of A is defined as $A \setminus \partial A$ and the closure $\overline{A}$ is the union of A and ∂A. A set A is open, if $A = A^0$ and it is closed if $A = \overline{A}$. $\mathcal{B}_r(c) := \{x \in \mathbb{R} | (x - c)^2 \leq r^2\}$ denotes the closed disc and $\mathcal{B}_r^0(c) := (\mathcal{B}_r(c))^0$ denotes the open disc of radius r and center c. The ε-dilation of a set A is defined as the set of all points having a distance of at most ε to some point in A. $L_t(f)$ shall be the level set with threshold value t of an image function $f : \mathbb{R}^2 \to \mathbb{R}$: $L_t(f) := \{\boldsymbol{x} \in \mathbb{R}^2 | f(x) \geq t\}$. A set $A \subset \mathbb{R}^2$ is called *simple 2D (simple 1D)* if it is homeomorphic to the unit disc $\mathcal{B}_1(0)$ (to the unit interval). Obviously compact subsets of the plane are simple 2D iff their boundary is a Jordan curve.

Definition 1. *A compact set $A \subset \mathbb{R}^2$ is called* r-regular *if for each boundary point of A it is possible to find two osculating open discs of radius r, one lying entirely in A and the other lying entirely in A^c. A grayscale* image function *$f : \mathbb{R}^2 \to \mathbb{R}$ is r-regular, if each level set is r-regular.*

Note, that an r-regular grayscale image does not contain isolated extrema or saddle points, but plateaus. Each local extremum is a plateau with r-regular shape. The property that extrema become plateaus is similar to the concept of one-dimensional local monotonic functions, as defined in [1]. These functions, which are monotonic in any interval of some restricted size, do not change under median filtering. Additionally they are invariant under morphological opening and closing, which is also true for r-regular binary images as already stated by Serra [12]. This suggests to ask for the relationship between the concepts of monotony and r-regularity. Therefore a suitable generalization of monotony to 2D is needed.

Our approach is to understand local monotony as a topological criterion of the neighborhood: When applying an arbitrary threshold function to a 1D locally monotonic function, the resulting binary set can have at most one component in each interval of some restricted size. This can easily be generalized to higher dimensions:

Definition 2. *Let $A \subset \mathbb{R}^2$ be a simple 2D set. A closed set $B \subset \mathbb{R}^2$ is called* monotonic in A, *if both $B \cap A$ and $\overline{B^c} \cap A$ are simple 2D, empty or one-point-*

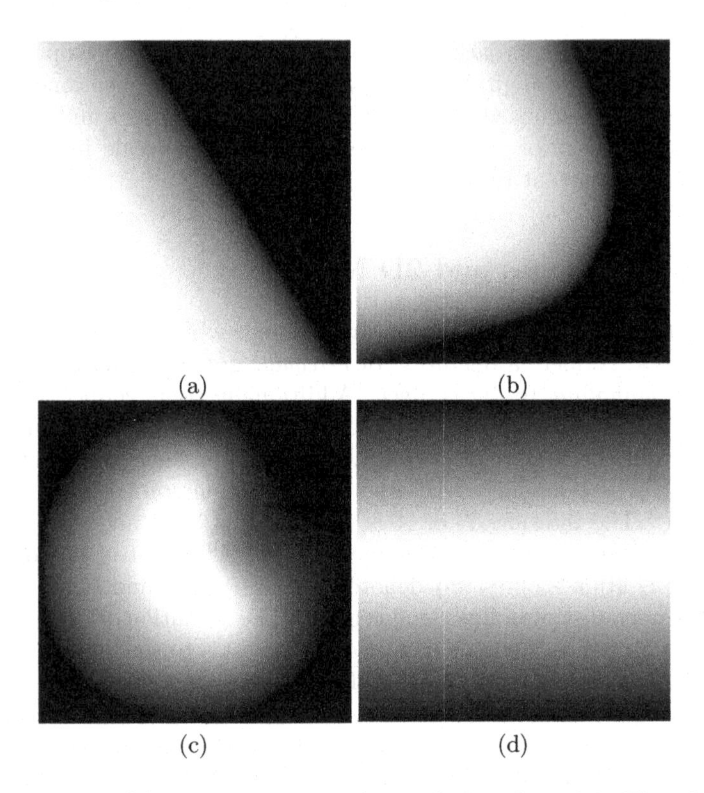

Fig. 1. The image in (a) is monotonic on each straight line through it. Thus obviously it can be called monotonic in 2D. (b) shows an image, which is homeomorphic to (a), and thus monotonic. There is no local maximum or minimum in the shown area, except of exactly one maximum and one minimum at the boundary. (c) shows a local maximum and thus is not monotonic in the shown area. The same is true for image (d), since there are two minimal regions

sets. B is called constant in A *if $B \cap \partial A = \emptyset$. An image function $f : \mathbb{R}^2 \to \mathbb{R}$ is called* monotonic (constant) in A, *if $L_t(f)$ is monotonic (constant) in A for each threshold value $t \in \mathbb{R}$.*

This definition of monotony is a generalization of monotony on paths, since a function is monotonic on some path if the level sets and their complements are simple 1D, empty or one-point-sets. If you have an image function being monotonic in a simple 2D set A, there exists from each point in A a monotonic decreasing path to any minimal boundary point of A and a monotonic increasing path to any maximal boundary point. Figure 1 illustrates, what monotony in 2D means: If an image is monotonic in some area, then this part of the image is homeomorphic to an image which is (in the classical sense) monotonic on each straight line though it.

Definition 3. *Let $f : \mathbb{R}^2 \to \mathbb{R}$ be an image function and let $A \subset \mathbb{R}^2$ be a simple 2D set. Further let $A_v = A + \boldsymbol{v}$ be the result of the translation of A by a* *vector* $\boldsymbol{v} \in \mathbb{R}^2$. *Then* f *is called* locally monotonic w.r.t. A, *if* f *is monotonic in* A_v *for* any $\boldsymbol{v}$.

Lemma 1. *An image function* f *is locally monotonic w.r.t. a simple 2D set* A *iff each level set* $L_t(f)$ *with* $t \in \mathbb{R}$ *is locally monotonic w.r.t.* A.

Proof. The lemma follows directly from the definition. □

The next lemmas show the connection between local monotony and r-regularity. Since 2D monotony is a fundamental property of several reconstruction methods (see section 3), the lemmas explain why r-regular images have been used for nearly all shape preserving sampling theorems:

Lemma 2. *Let* A *be a disc with radius smaller than some* $r \in \mathbb{R}$ *or let* A *be an intersection of a finite number of such discs. Then every* r*-regular image* f *is locally monotonic w.r.t.* A.

Proof. For each threshold value $t \in \mathbb{R}$, $L_t(f)$ is an r-regular set. Due to r-regularity, no three boundary points of $L_t(f)$ lie on a common circle of some radius smaller than r. Since no three boundary points of A lie on a common circle with a radius of at least r, A and $L_t(f)$ can have at most two boundary points in common. If $L_t(f) \cap A$ or $\overline{(L_t(f))^c} \cap A$ is empty, $L_t(f)$ is monotonic in A. Otherwise one or two of such boundary points exist, because no component of an r-regular image can lie completely in A. If there is only one such boundary point, $L_t(f) \cap A$ or $\overline{(L_t(f))^c} \cap A$ has to be a one-point-set, which implies monotony. Finally if there are two such boundary points, the boundary part $\partial L_t(f) \cap A$ cuts A into two simple 2D parts $L_t(f) \cap A$ and $\overline{(L_t(f))^c} \cap A$, which implies monotony. Obviously this is also true for any translated version of A and thus f is locally monotonic w.r.t. A. □

Thus you can take for example any disc shaped area of some r-regular image and it will be monotonic, if the disc has some radius smaller than r. Or you can take a finite intersection of such areas, e.g. Reuleaux triangles (see [13] for a definition).

Lemma 3. *Let* A *be an* r*-regular set and* $B, C \subset \mathbb{R}^2$ *be two simple 2D sets such that*

- B *is a subset of some* r'*-disc and* C *is a subset of some* r''*-disc with* $r', r'' < r$,
- A *is monotonic in both* B *and* C,
- ∂B *crosses* ∂C *in exactly two points* p_1, p_2,
- ∂A *crosses* ∂B *in exactly two points* b_1, b_2 *both different from* p_1 *and* p_2, *and*
- ∂A *crosses* ∂C *in exactly two points* c_1, c_2 *both different from* p_1 *and* p_2.

Then A *is monotonic either in* $\overline{B \setminus C}$ *or in* $\overline{C \setminus B}$. *Furthermore* A *is monotonic in either* $\partial B \cap C$ *or* $\partial C \cap B$ *(see Fig. 2).*

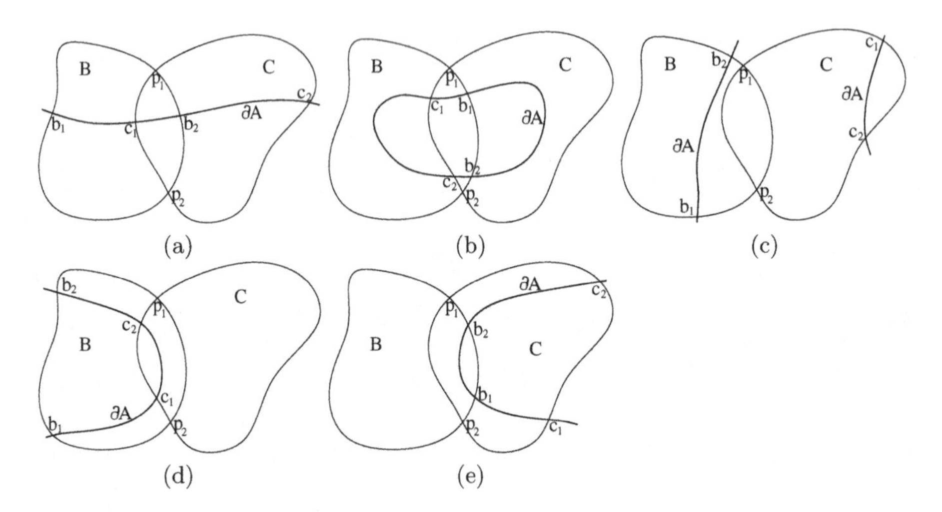

Fig. 2. There are only 5 possibilities, how ∂A can go through B and C, such that their boundaries are each only crossed twice but not in p_1, and p_2

Proof. Figure 2 shows the different possibilities of ∂A going through B and C. Both $\partial A \cap B$ and $\partial A \cap C$ consist of exactly one component, since they cannot be empty and since A is monotonic in B and C. Thus $\partial A \cap (B \cup C)$ consists of two components if $\partial A \cap (B \cap C)$ is empty, and of one component otherwise. If $\partial A \cap (B \cap C)$ is empty, $A \cap B \cap C$ is empty or equal to $B \cap C$ (see Fig. 2c) then A is monotonic in both $\overline{B \setminus C}$ and $\overline{C \setminus B}$ and A is even constant in both $\partial B \cap C$ and $\partial C \cap B$.

Otherwise $\partial A \cap (B \cup C)$ must be connected. If the intersection of A and $\overline{B \setminus C}$ is empty or equal to $\overline{B \setminus C}$ (see Fig. 2e) or if the intersection of A and $\overline{C \setminus B}$ is empty or equal to $\overline{C \setminus B}$ (see Fig. 2d), A is monotonic in $\overline{B \setminus C}$ (then A is constant and thus monotonic in $\partial C \cap B$) or in $\overline{C \setminus B}$ (then A is constant in $\partial B \cap C$), respectively.

Else $\partial A \cap (B \cup C)$ must go through $\overline{B \setminus C}$, $\overline{C \setminus B}$ and $B \cap C$ without meeting p_1 or p_2. There are only two remaining possibilities: First, ∂A enters $\overline{B \setminus C}$ at some point b_1 on $\partial B \setminus C$, next enters $B \cap C$ at some point c_1 on $\partial B \cap C^0$ and leaves it at some point b_2 on $\partial C \cap B^0$, before finally leaving $\overline{C \setminus B}$ through some point c_2 on $\partial C \setminus B$ (see Fig. 2a). Second, ∂A does not intersect $\partial(B \cup C)$. Starting in $B \setminus C$, it goes through $(B \cap C)^0$ into $C \setminus B$ and on another path back through $(B \cap C)^0$ into $B \setminus C$ to the starting point (see Fig. 2b). In both cases both $\overline{B \setminus C}$ and $\overline{C \setminus B}$ are cut by ∂A into two simple 2D parts, which implies monotony.

In the first case ∂A intersects both $\partial B \cap C$ and $\partial C \cap B$ in only one point and thus A is monotonic on these paths. The second case is in contradiction to the r-regularity of A, since $B \cup C$ is subset of the union of two discs of radii smaller than r and no such union can cover an r-regular set. □

Lemma 4. *Let A, B be simple 2D sets, such that A is monotonic in B and let $S = (s_0, s_1, \ldots, s_n = s_0)$ be a clockwise ordered cyclic list of points lying* *on ∂B, such that there exist no four points s_a, s_b, s_c, s_d with $a < b < c < d$ such that $s_a, s_c \in A$ and $s_b, s_d \in A^c$ or vice versa. Further let $P = \{p_{0,1}, p_{1,2}, \ldots, p_{n-1,n}\}$,$p_{i,j} \subset B$, be a set of non-intersecting (except of their endpoints) simple paths in B between neighboring points of S, such that A is monotonic in each path (the existence of such paths is shown in the proof of Theorem 1). Then the area enclosed by the paths is simple 2D and A is monotonic in it.*

Proof. The paths define a closed curve. Since they do not intersect (except of their endpoints), this curve is a jordan curve. The closed set C, which is circumscribed by this curve, is a simple 2D set and a subset of B. Each path p_i cuts $\partial A \cap B$ in at most two parts, where only one part can intersect C. It follows by induction that $\partial A \cap C$ consists of at most one component, hitting ∂C in two points. This component separates C into two simple 2D sets (bounded by jordan curves) $A \cap C$ and $\overline{A^c} \cap C$. Thus A is monotonic in C. □

3 Sampling and Reconstruction

In order to compare analog with digital images, a definition of the processes of sampling and reconstruction is needed. The most obvious approach for sampling is to restrict the domain of the image function to a set of sampling points. This set is called a sampling grid. In most approaches only special grids like square or hexagonal ones are taken into account [2, 3, 4, 5, 8, 9, 11, 12]. A more general approach only needs that a grid is a countable subset of $\mathbb{R}^2$, with the sampling points being not too sparse or too dense anywhere [6, 7, 13]. There the pixel shapes are introduced as Voronoi regions.

Definition 4. *A countable set $S \subset \mathbb{R}^2$ of* sampling points, *where the Euclidean distance from each point $\boldsymbol{x} \in \mathbb{R}^2$ to the next sampling point is at most $r \in \mathbb{R}$, is called an r*-grid *if for each bounded set $A \in \mathbb{R}^2$ the subset $S \cap A$ is finite. The* Voronoi region *of a sampling point is the set of all points lying at least as near to this point than to any other sampling point. A maximal set of sampling points s_i, whose Voronoi regions have a common corner point p, is called* Delaunay tuple *and its convex hull is the* Delaunay cell. *Obviously, the sampling points s_i lie on a common circle with center p (see Fig. 3c). p is called the* center point *of the Delaunay cell. An r-grid is named* degenerated *if at least one of the associated Delaunay cells (not necessarily triangles) does not contain its center point. A Delaunay tuple is called* well composed *regarding to an image function f, if the clockwise ordered cyclic list of sampling points $s_1, s_2, \ldots s_{n-1}, s_n = s_0$ has only one local maximum and one local minimum (plateaus are allowed) in f, or in other words that there exist no four points s_a, s_b, s_c, s_d with $a < b < c < d$ such that*

$f(s_a) > f(s_b)$, $f(s_b) < f(s_c)$, $f(s_c) > f(s_d)$ and $f(s_d) < f(s_a)$, or
$f(s_a) < f(s_b)$, $f(s_b) > f(s_c)$, $f(s_c) < f(s_d)$ and $f(s_d) > f(s_a)$.

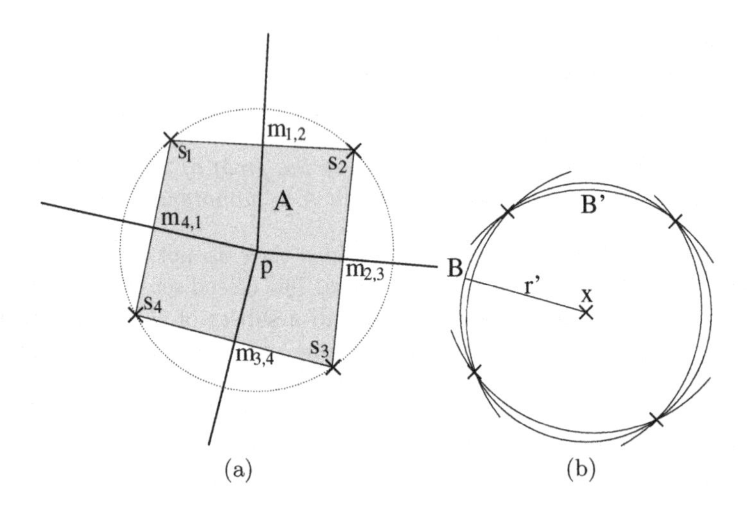

Fig. 3. (a): Definition of linear reconstruction (definition 5). (b): construction of monotonic covering of a Delaunay cell (Theorem 1)

Note, that for any well composed set in the sense of Latecki [9], each Delaunay tuple is well composed. The definition even matches the definition of extended well composed sets by Wang and Bhattacharya [14].

Since reconstruction is a local process where the value of a point is influenced by the surrounding sampling points, several reconstruction methods are defined piecewise. One obvious idea is to define the reconstruction for each Delaunay cell of the sampling grid separately. Then the basic idea is that an image, which is locally monotonic, should have a locally monotonic reconstruction. Therefore a class of such reconstruction functions is defined, which includes generalizations of bilinear and nearest neighbor reconstruction to arbitrary grids.

Definition 5. *Let* Img *be the class of all grayscale image functions* $f : \mathbb{R}^2 \to \mathbb{R}$ *and let* $S \subset \mathbb{R}^2$ *be an* r*-grid for some* r*. A function* $\text{rec}_S : \text{Img} \to \text{Img}$ *is called a* reconstruction function *if* $\forall f \in \text{Img}, \forall s \in S : (\text{rec}_S(f))(s) = f(s)$ *and* $\forall f, g \in \text{Img} : f|_S = g|_S \Rightarrow \text{rec}_S(f) = \text{rec}_S(g)$*. A reconstruction function is called* Delaunay-monotonic *if for each well composed Delaunay tupel the reconstruction is monotonic in the corresponding Delaunay cell, and – restricted to this cell – each nonempty level set of the reconstruction and its complement contains at least one sampling point, respectively.*

Now let S *be a non-degenerated grid and* f *be an image function.*

The linear reconstruction function rec_S *is defined as follows. Let* A *be a Delaunay cell of the grid and* p *be its center point (see Fig. 3a). Then the Delaunay cell can be divided by the median lines from* p *to the boundary edges of the Delaunay cell into quadrangles. Now the reconstruction function is defined by a bilinear interpolation in each of the quadrangles, where the values for the four vertices are the following: At the sampling points* s_i, $\text{rec}_S(f)$ *is equal to* f*; at* p, $\text{rec}_S(f)$ *is the mean of the values of* f *at the sampling points (*p *has equal* *distance to all of these sampling points); at the median points* $m_{i,j}$, $\text{rec}_S(f)$ *is the mean value of the two corresponding sampling points* s_i, s_j.

The nearest neighbor reconstruction function rec_S *is defined by giving each point the value of the nearest of the sampling points which correspond to the actual Delaunay cell. If there is no unique nearest sampling point, the one with the highest value is taken.*

In a Delaunay-monotonic reconstruction, the image is monotonic in each Delaunay cell, where this is possible. Of course it cannot be monotonic in a Delaunay cell where the clockwise ordered cyclic list of sampling points has not only one local maximum and one local minimum in the image function.

In case of square grids, these definitions are equivalent to the standard bilinear interpolation and the nearest neighbor interpolation. Any linear or nearest neighbor reconstruction function is Delaunay-monotonic, since no overshooting can occur. Note that even more complex reconstruction methods like biquadratic interpolation only need slight modifications (cutting off the overshootings) in order to be Delaunay-monotonic. In case of the nearest neighbor reconstruction, the value of each sampling point is simply set to the whole Voronoi region, which is equal to the reconstructions used in [2, 3, 7, 8, 9, 11, 12, 13]. Even the marching squares algorithm, a two-dimensional simplification of the well-known marching cubes algorithm [10], defines a Delaunay-monotonic reconstruction function.

4 Shape Preserving Sampling Theorems

In this section at first two minor sampling theorems are proved. The first theorem is only for binary images, where no sampling point lies on the boundary of any foreground component. The second theorem extends this to any binary image. After that the third and main result finally generalizes these theorems to grayscale images.

In the following, the well-defined similarity criterion *strong* r*-similarity* (see [13]) is used to compare shapes before and after digitization. Two shapes $A, B \subset \mathbb{R}^2$ being strongly r-similar means that there exists a homeomorphic deformation f of the plane $\mathbb{R}^2$, with $f(A) = B$ and where the movement of each point is bounded by r: $\forall \boldsymbol{x} \in A : |\boldsymbol{x} - f(\boldsymbol{x})| \leq r$. Such a restricted homeomorphism is called r*-homeomorphism*. This criterion is stricter than both the preservation of topology used by Pavlidis and Latecki et al. [8, 9, 11] and the isomorphy of homotopy trees used by Serra [12]. It additionally sets a bound for the Hausdorff-distance of the original and the reconstructed set and of their boundaries. Strong r-similarity is originally a criterion for binary images like shapes. If we have a grayscale image, we can investigate the shapes given by the level sets of the image. So two grayscale images are called strongly r-similar if this is true for all of their level sets.

Theorem 1. *Let* A *be an* r*-regular set and* S *be a non-degenerated* r'*-grid with* $r' < r$*, such that no sampling point* $x \in S$ *lies on* ∂A*. Further let* rec_S *be a Delaunay-monotonic reconstruction function. Then there exists a* $2r'$*-homeo-*

morphism from A to the reconstruction $L_v(\text{rec}_{\mathcal{S}}(A))$ for each threshold value $v \in (0,1)$.

Proof. In the following such a homeomorphism is defined by partitioning the original image and the reconstruction into homeomorphic parts corresponding to the Delaunay cells.

Let p be the center point of some Delaunay cell of the grid. Then there exists a disc with radius of at most r' and center in p, such that each element of the Delaunay tuple lies on its boundary. The set X of all so defined discs (one for each corner point) covers the whole plane and A is monotonic in each disc.

Now we replace each disc B by the intersection B' of new discs of slightly bigger radius $r'' < r$, which cover all the sampling points lying on ∂B and which each have exactly two neighboring sampling points on their boundary (see Fig. 3b). Doing this with an appropriate radius r'' we can guarantee that $\partial A \cap B'$ is no one-point-set. The so constructed set X' still covers the plane, while A being monotonic in each element of X' according to Lemma 2.

Due to the construction of X', the boundaries of each two elements of X' intersect in two or zero points. Now let B_1, B_2 be two elements of X', such that their boundaries intersect in two points p_1 and p_2. Since the interior of each element of X' does not contain any sampling point, the path $\partial B_1 \cap B_2$, going from p_1 to p_2 does not hit any sampling point except of the possible sampling points p_1 and p_2. If p_1 and p_2 are both sampling points, we can choose one such path P between them due to Lemma 3, with A being monotonic in P. By doing this for every pair of neighboring sampling points we can map each edge of the Delaunay graph to such a path. Each of these paths is covered by both discs of X, which intersect with the endpoints of the path. Now we modify this set of paths Y, such that no two paths intersect in non-sampling points:

If two paths intersect in a common subpath, which is not an isolated point, we can displace one path in this area by a small distance, such that the two new paths intersect only in the endpoints of the formerly common subpath and in common intersection points with ∂A. This is possible without changing the monotony of A on the paths since ∂A intersects the paths in at most one isolated point. The resulting paths can be guaranteed to be covered by the corresponding elements of X, due to the construction of X'. Thus the resulting paths only intersect in isolated points. Now let P_1 and P_2 be two paths, crossing each other in two points p_1 and p_2. Then by swapping the parts of the paths between p_1 and p_2 we get two paths, which intersect in p_1 and p_2, but do not cross each other in these points. If two paths cross each other in only one point there are two cases: First, if they intersect also in an endpoint we can use the same swapping technique. Second, if they intersect in no other point, one of the endpoints of one path must be enclosed by the circular set of paths covered by one of the corresponding discs of X, which is impossible since no sampling point lies in the interior of such a disc.

Since the swapped path parts are covered by the same discs of X, the covering properties do not change under this swap operation. By induction we get a set

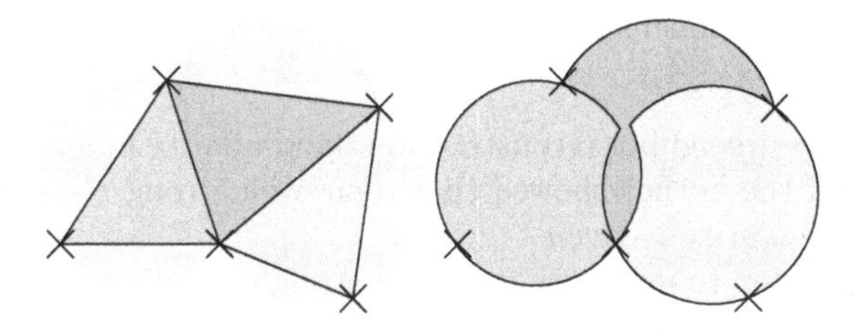

Fig. 4. The reconstructed image function is monotonic in each Delaunay cell (colored regions in the left figure). One can find corresponding regions in the original image, where the image function is monotonic, too (colored regions in the right figure)

of paths Y' where no two paths cross each other. The paths can only intersect at some single points without crossing. At these points we can displace one path by a small distance such that the paths do not intersect anymore.

All these path modifications do not change the monotony of A on them. Thus the resulting set of paths defines an embedding of the Delaunay graph into the plane, such that each region can be covered by an r'-disc. Figure 4 shows an example of neighboring Delaunay cells and their corresponding embeddings in the reconstructed image. With Lemma 4 the Delaunay cells and their corresponding regions due to the above construction are each homeomorphic. Since two such corresponding regions can be covered by a common r'-disc, each homeomorphism is a $2r'$-homeomorphism. Obviously we can choose homeomorphisms which are identical on the constructed paths, which implies strong r'-similarity of A and $L_v(\text{rec}_{\mathcal{S}}(A))$ for each theshold value $v \in (0,1)$. □

This means that if a shape is r-regular, it can be sampled with any sampling grid of a certain density and any Delaunay-monotonic reconstruction method, such that the resulting reconstruction has exactly the same topological properties and only a bounded Hausdorff distance to the original shape. The only restriction is that the sampling points are not allowed to lie on the shape boundary. The next theorem shows that this restriction is not really existent.

Theorem 2. *Let A be an r-regular set and S be a non-degenerated r'-grid with $r' < r$. Further let* rec *be a Delaunay-monotonic reconstruction function. Then there exists for any $\varepsilon > \mathbb{R}$ a $(2r' + \varepsilon)$-homeomorphism from A to the reconstruction $L_v(\text{rec}_{\mathcal{S}}(A))$ for each threshold value $v \in (0,1)$.*

Proof. Let d be the minimal distance between A and any sampling point not lying in A – this is uniquely defined since there is only a finite number of sampling points having a distance of at most r due to the compactness of A. Now let ε be any number with $0 < \varepsilon < \min(d, r - r')$. Then the ε-dilation A' of A is ε-homeomorphic to A and has exactly the same reconstruction, since the values at the sampling points did not change. Due to Theorem 1 the $(r - \varepsilon)$-regular set A' is $(2r')$-homeomorphic to the reconstruction and in case of the nearest neighbor reconstruction even r'-homeomorphic to it. The concatenation of the two homeomorphisms defines a $(2r' + \varepsilon)$-homeomorphism from A to the reconstruction $L_v(\text{rec}_{\mathcal{S}}(A))$. □

In case of nearest neighbor reconstruction the similarity bound is even stronger. In [13] Kthe and the author showed that then even strong r'-similarity instead of strong $(2r')$-similarity is given.

Now the final step to grayscale images is straightforward, since each level set of an r-regular grayscale image is r-regular, too.

Corollary 1. *Let f be an r-regular image function and S be a non-degenerated r'-grid with $r' < r$. Further let* rec *be a Delaunay-monotonic reconstruction function. Then the reconstruction is strongly $(2r)$-similar to f for any theshold value, which is not equal to the image value at some sampling point.*

If the threshold value is equal to the image value at some sampling point, the corresponding level set of the reconstructed image function is not necessarily simple 2D, but can contain one dimensional parts and thus the topology changes. But one can show that this does not happen, if the original image is $2r$-regular, because then each plateau is reconstructed topologically correctly.

5 Conclusions

It was proved that any grayscale image, which has only r-regular level sets, can be sampled by arbitrary sampling grids of sufficient density and reconstructed by a non-overshooting interpolation method, and still remains strongly r'-similar (for some bounded r') for any threshold value, which is not an image value at some sampling point. This implies that most level sets do not change topology under digitization and thus you can say the topology of the image is preserved. Each maximum or minimum plateau of the original image can be found in the digitization having the same height.

References

1. Acton, S.T.: *A PDE technique for generating locally monotonic images.* Proceedings of the IEEE International Conference on Image Processing, 1998.
2. Florncio, D.A.F., Schafer, R.W.: *Homotopy and Critical Morphological Sampling.* Proceedings of SPIE Visual Communications and Image Processing, pp. 97-109, 1995.
3. Florncio, D.A.F., Schafer, R.W.: *Critical Morphological Sampling and its Applications to Image Coding.* In J. Serra, P. Soille (Eds.): Mathematical Morphology and Its Applications to Image Processing, Computational Imaging and Vision, vol. 2, pp. 109-116, Kluwer Academic Publishers, Dordrecht, 1994.
4. Haralick, R.M., Zhuang, X., Lim, C., Lee, J.S.J.: *The Digital Morphological Sampling Theorem.* IEEE Transactions on Acoustics, Speech and Signal Processing **37**, pp. 2067-2090, 1989.
5. Heijmans, H.J.A.M., Toet, A.: *Morphological Sampling.* Computer Vision, Graphics and Image Processing: Image Understanding **54**, pp. 384-400, 1991.

6. Herman, G.T.: *Geometry of Digital Spaces.* Birkhuser Boston, 1998.
7. Köthe, U., Stelldinger, P.: *Shape Preserving Digitization of Ideal and Blurred Binary Shapes.* In: I. Nystrm et al. (Eds.): DGCI 2003, LNCS **2886**, pp. 82-91, Springer, 2003.
8. Latecki, L.J., Conrad, C., Gross, A.: *Preserving Topology by a Digitization Process.* Journal of Mathematical Imaging and Vision **8**, pp. 131–159, 1998.
9. Latecki, L.J.: *Discrete Representation of Spatial Objects in Computer Vision.* Kluwer, 1998.
10. Lorensen, W.E., Cline, H.E.: *Marching Cubes: A High Resolution 3D Surface Construction Algorithm.* Computer Graphics **21**, no. 4, 1987.
11. Pavlidis, T.: *Algorithms for Graphics and Image Processing.* Computer Science Press, 1982.
12. Serra, J.: *Image Analysis and Mathematical Morphology.* Academic Press, 1982.
13. Stelldinger, P., Köthe U.: *Shape Preservation During Digitization: Tight Bounds Based on the Morphing Distance.* In: B. Michaelis, G. Krel (Eds.): Pattern Recognition, LNCS **2781**, pp. 108-115, Springer, 2003.
14. Wang, Y., Bhattacharya, P.: *Digital Connectivity and Extended Well-Composed Sets for Gray Images.* Computer Vision and Image Understanding **3**, pp. 330–345,1997.

Comparison of Nonparametric Transformations and Bit Vector Matching for Stereo Correlation

Bogusław Cyganek

AGH - University of Science and Technology,
Al. Mickiewicza 30, 30-059 Kraków, Poland
cyganek@uci.agh.edu.pl

Abstract. The paper describes and compares stereo matching methods based on nonparametric image transformations. The new nonparametric measures for local neighborhoods of pixels are proposed as well. These are extensions to the well known Census transformation, successively used in many computer vision tasks. The resulting bit-fields are matched with the binary vectors comparison measures: Hamming, Tanimoto and Dixon-Koehler. The presented algorithms require only integer arithmetic what makes them very useful for real-time applications and hardware implementations. Many experiments with the presented techniques, employed to the stereovision, showed their robustness and competing execution times.

1 Introduction

The paper concerns stereo matching methods with special stress on development of reliable algorithms suitable for hardware implementations. For such platforms the bit or integer arithmetic is preferable. In this paper we compare different nonparametric image transformations with respect to the diverse bit vector matching measures applied to the task of stereo correlation.

The key concept behind the nonparametric correlation measures lies in changing a given value from a statistical sample by its rank among all the other values in that sample. This way, the resulting list of new values has the *uniform* distribution function. Therefore the correlating statistics deals only with uniform sets of integers. The two examples of such correlation measures are Spearman's ρ or Kendall's τ [13].

The nonparametric measures were introduced to image society by Zabih and Woodfill [15], as well as by Bhat and Nayar [2]. Their *Census* and *Rank* measures are built in local neighborhoods of pixels based on relations among neighboring pixels. This way transformed images are then compared to find out the correlation measure. They have proved great usefulness in stereo image matching [1][4], and also in image processing with neural networks [3][5].

The more complex versions of the nonparametric *Census* transformation are proposed here. The new versions of this transformation have been developed to convey more information than the original *Census*. In result, each pixel is transformed to a bit-stream. Comparison of such data is a little different than comparison of the luminance signals. Usually the Hamming distance is employed to find a correlation

R. Klette and J. Žunić (Eds.): IWCIA 2004, LNCS 3322, pp. 534–547, 2004.

between bit-streams. In this paper we compare other useful metrics such as the Tanimoto distance [14][9] or its modification proposed by Dixon and Koehler [8]. The last two measures have found great popularity in the molecular biology and chemistry but can be also used by the computer vision community.

2 Nonparametric Measures in Pixel Neighborhoods

The *Rank* and *Census* transforms were proposed by Zabih and Woodfill [15] for computation of correspondences by means of the local nonparametric transformation applied to the images before matching process. Both transformations start from the image intensity signals and are computed in a certain region around a central pixel. Size and shape of this region can be set arbitrarily. Usually it is a square and such square regions are also assumed in this paper, although this assumption can be relaxed.

For a given central pixel and its closest neighborhood the *Rank* transform is defined as the number of pixels in that region for which the intensity signal is greater or equal than that of the central pixel. For the same setup the *Census* transform returns an ordered stream of bits where a bit at a given position is set if and only if intensity of the central pixel is less or equal to intensity of a pixel from the neighborhood around this central pixel. Fig. 1a explains the ideas behind the *Rank* and *Census* transformations. Fig. 1b depicts assumed pixel orders for computation of the *Census* values for two 3×3 and 5×5 neighborhoods. An interesting observation for *Census* is that a value of the central pixel is taken only as a reference and does not go into the output bit stream. Therefore for 3×3 and 5×5 regions we obtain computer efficient representations of eight and twenty four bits (i.e. one and three bytes), respectively.

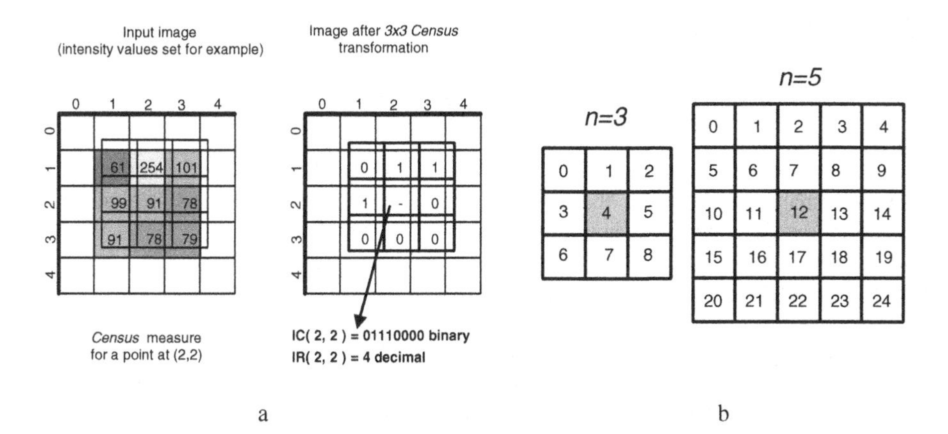

Fig. 1. Explanation of the *Census IC* and *Rank IR* measures for an exemplary pixel at position (2,2) (a). Assumed pixel numbering in the square 3×3 and 5×5 neighborhoods of pixels when computing *Census* (b)

For a given central pixel at *(i,j)* and its closest square neighborhood containing *n* pixels, the corresponding *Census* measure *IC(i,j)* can be expressed as a series of bits:

$$IC(i,j) = b_{n^2-1} \ldots b_k \ldots b_3 b_2 b_1 b_0 ,$$
$$\text{where } k \in \left[0,\ldots,n^2-1\right] \; / \; \left\{ \; \left\lfloor \tfrac{n^2}{2} \right\rfloor \; \right\}. \tag{1}$$

Bit-pixel numbering in a neighborhood depicts Fig. 1b. The b_k parameter can be expressed as follows:

$$b_k = \begin{cases} 1, & when \;\; I\left(i - \left\lfloor \frac{n}{2} \right\rfloor + \left\lfloor \frac{k}{n} \right\rfloor, j - \left\lfloor \frac{n}{2} \right\rfloor + k \bmod n \right) > I(i,j) \\ 0 & otherwise \end{cases}, \tag{2}$$

where $I(i,j)$ denotes the intensity value for an image at a point at *(i,j)* in image coordinates, $\lfloor k/n \rfloor$ means an integer division of *k* by *n*, k mod n is a modulo *n* division. In (2) we additionally assume that all *(i,j)* locations denote valid image indexes. If not then we assign 0 for that location.

Both transformations are resistive to the radiometric gain and bias imbalances among images to be matched [15]. However, the *Rank* transform loses information of spatial distribution of pixels since all of the pixels surrounding a central one are encoded into a single value. The only advantage of this limiting process comes from the reduction of memory necessary for representation of transformed images. For example in a case of eight bits per pixel in an original image, after the *Rank* transform only four bits per pixel are required (or even three with some additional assumptions). Contrary to the *Rank,* the *Census* transformation preserves the spatial distribution of pixels by encoding them in a bit stream. Therefore *Census* is much more robust for block matching.

Usually the block matching for *Rank* representation of images is performed using the L_1 norm. For *Census* often the Hamming distance between bit streams is computed.

In practice, these measures become very useful for strongly noised or not equally lightened images. Moreover, the two transforms showed to be very attractive if considered for custom hardware implementations or robotics [1].

2.1 Detailed Neighborhood Relation

Let us enhance the concept of the *Census* measure to convey more detailed information on pixel relations. This can be done by allocation of *more than one bit* for the relation between two pixels, which relation can now be defined as follows:

$$R(k) = I(c) - I(k), \tag{3}$$

where *c* is an index of the central pixel, *k* is a free index for all neighboring pixels.

Table 1 presents a proposition of a nonparametric measure in a form of the several fuzzy-like [7][16] relations *S(k)* between pixels with the corresponding bit-encoding *E(k)*. This relation among neighboring pixels will be called *Detail 3×3*, since it is computed in the 3×3 pixel neighborhood.

Table 1. Rules for the relation between pixels based on their relative intensity values

	Relation type *S(k)*		Bit encoding *E(k)*
1	strongly	smaller	011
2		smaller	010
3	little	smaller	001
4		equal	000
5	little	greater	101
6		greater	110
7	strongly	greater	111

The sigmoidal and hyperbolic fuzzy membership functions for the relation *S(k)* were proposed in [7]. Alternatively, the rules in Table 1 can be approximated by a piecewise-linear function and implemented by providing six threshold values on *R(k)*. However, this can be a little cumbersome in practice. Therefore the bit encoding proposed in Table 1 has been chosen to form information in accordance with the scheme presented in Table 2.

Table 2. Bit encoding scheme of the proposed *Detail* nonparametric measure

Bit number	2	1	0
Bit meaning	The sign of difference	Difference category, bit 1	Difference category, bit 0

For computer implementation the bit encoding method was devised that relies only on bit shifting and an increment by one. In each step the positive remainder *R* is shifted right by two bits and if the result is still different than zero, the encoding value is incremented by 1. Such a bit procedure leads to the following thresholds on *R*:

- If $|R| \in [0,3]$ then *S(k)* = "equal".
- If $|R| \in (3, 15]$ then *S(k)* = "little smaller/greater",
- If $|R| \in (15, 63]$ then *S(k)* = "smaller/greater",
- If $|R| \geq 64$ then we classify as *S(k)* = "strongly smaller/greater".

The same thresholds hold for negative *R*. The proposed algorithm can be easily implemented in assembly or in C++. For example see the following C++ code fragment:

```
int E = 0x00; // init the encoding
if( R < 0 ) {
  R = -R;      // make R positive
```

```
    E = 0x04;     // set a sign
}
for( int i = 0; i < 3; ++ i )
  if( ( R >>= 2 ) != 0 )
    ++ E;         // shift R, and if not 0 then increment E
```

The code above is very simple and can be implemented in any assembly language or in hardware (FPGA).

3 Correlation of the Bit-Stream Image Representations

For comparison of two bit-vectors of equal size the Hamming distance can be used:

$$D_H(\mathbf{a},\mathbf{b}) = \frac{1}{N}\sum_{i=1}^{N} a_i \otimes b_i \,, \tag{4}$$

where a, b are the compared *binary* vectors of the same length N (henceforth, we assume a column representation of vectors), ⊗ denotes the bit XOR operation. Based on (4) it is evident that D_H only accounts for bits that do not match.

The other measure, know also as the Tanimoto distance [14][9], is defined as follows:

$$D_T(\mathbf{a},\mathbf{b}) = \begin{cases} 1 & if \quad \mathbf{a}=\mathbf{b}=\mathbf{0} \\ 1 - \dfrac{\mathbf{a}^{\mathrm{T}}\mathbf{b}}{\mathbf{a}^{\mathrm{T}}\mathbf{a}+\mathbf{b}^{\mathrm{T}}\mathbf{b}-\mathbf{a}^{\mathrm{T}}\mathbf{b}} & otherwise \end{cases}, \tag{5}$$

where a, b are the binary vectors of the same length N. D_T favors situations when a match is done on bits with value '1' rather than '0'.

The modification made by Dixon and Koehler [8] is a composition of (4) and (5):

$$D_{DK}(\mathbf{a},\mathbf{b}) = D_H(\mathbf{a},\mathbf{b})D_T(\mathbf{a},\mathbf{b}). \tag{6}$$

D_{DK} tends to equalize the opposing size effects of the Hamming and Tanimoto coefficients.

Table 3. Examples of the measures D_H, D_T and D_{DK} for different vectors. A value 0 means a perfect match, 1 means no match at all. The lower the measure value, the better match

Ex.	**a**	**b**	D_H	D_T	D_{DK}
1	011001	011001	0	0	0
2	011001	111111	0.5	0.5	0.25
3	011001	000000	0.5	1	0.5
4	101010	010110	0.66	0.8	0.53
5	101011	010111	0.66	0.66	0.44
6	101011	010100	1	1	1

Table 3 presents the three measures (4), (5) and (6) for exemplary vectors *a* and *b* which length *N* is always 6. It can be noticed that all the measures produce *0* for the perfect match (i.e. all bits are the same) and *1* for a total mismatch (i.e. all bits are different). However, observing examples in the rows 2 and 3 of Table 3 we notice that D_T assigns *1* also in a case of comparison with a zero vector. The same effect can be observed in examples at rows 4 and 5 from Table 3 where the two vectors *a* and *b* match only on the last two bits. However, in the example 4 one of these bits is '0' and therefore D_T gives worse match score (i.e. its value is higher). This favor of '1s' can be used in image matching since for the *Census* and the transformation from Table 1, '0' is assigned for pixels with the same intensity. Such pixels, i.e. with the same intensity values, cannot be used for further matching (e.g. for stereo or motion detection) since in this case every comparison metric does not exhibit significant extreme values and therefore does not lead to a reliable match. Indeed, for normed vector spaces a norm value taken from the vector difference is 0 if the vectors are the same.

It is also interesting to analyze behavior of the D_{DK} which is a product of the D_H and D_T. Comparing the examples 2 and 3 we see that in both cases there is a match of three bits. However, in the former these bits are '1s', in the latter do match '0s'. D_{DK} reflects this situation assigning "better" matching score to the match of '1s'. This is an effect of the D_T factor. At the other hand, the total mismatch indicated by the D_T for the example in row 3 seems to be to restrictive in many applications and therefore D_{DK} seems to be more appropriate.

4 Experimental Results

All of the presented algorithms were implemented in C++ and built into the image matching software framework [6]. The platform run-time parameters are as follows: IBM PC with Pentium 2GHz, 512 RAM.

For experimental comparisons ten different stereo-pairs were selected; two of them are presented in Fig. 2. "Tsukuba" is an artificial image of size 384×288; "Parkmeter" is a real image of size 256×240.

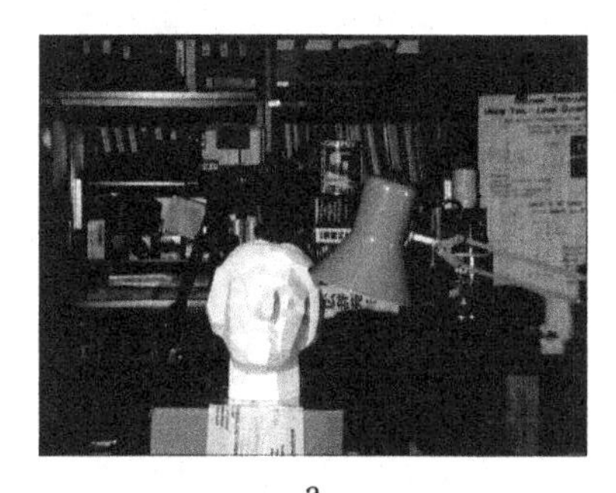

a

b

Fig. 2. Left images of the stereo pairs: "Tsukuba" 384×288 (a), "Parkmeter" 256×240 (b)

We performed comparisons of the three local pixel neighborhoods:

1. *Census* 3×3: 8 bits per pixel (formula (1)).
2. *Census* 5×5: 24 bits per pixel (formula (1)).
3. *Detail* 3×3: 24 bits per pixel (8 neighbors, 3 bits per neighbor, coding from Table 1).

Comparisons were done in respect to the following bit-vectors matching measures:

1. Hamming, formula (4).
2. Tanimoto, formula (5).
3. Dixon-Koehler, formula (6).

The chosen stereo method is an area-based version with a matching window 3×3 pixels. The area matching window should not be confused here with the neighborhood window for computation of *Census* or *Detail* measures. Stereo correlation is done in a canonical setup. For reduction of false matches the disparity map cross checking method and median 3×3 filtering were employed [17][10][5][6]. The three comparison breakdowns are presented in Fig. 3, Fig. 5 and Fig. 7.

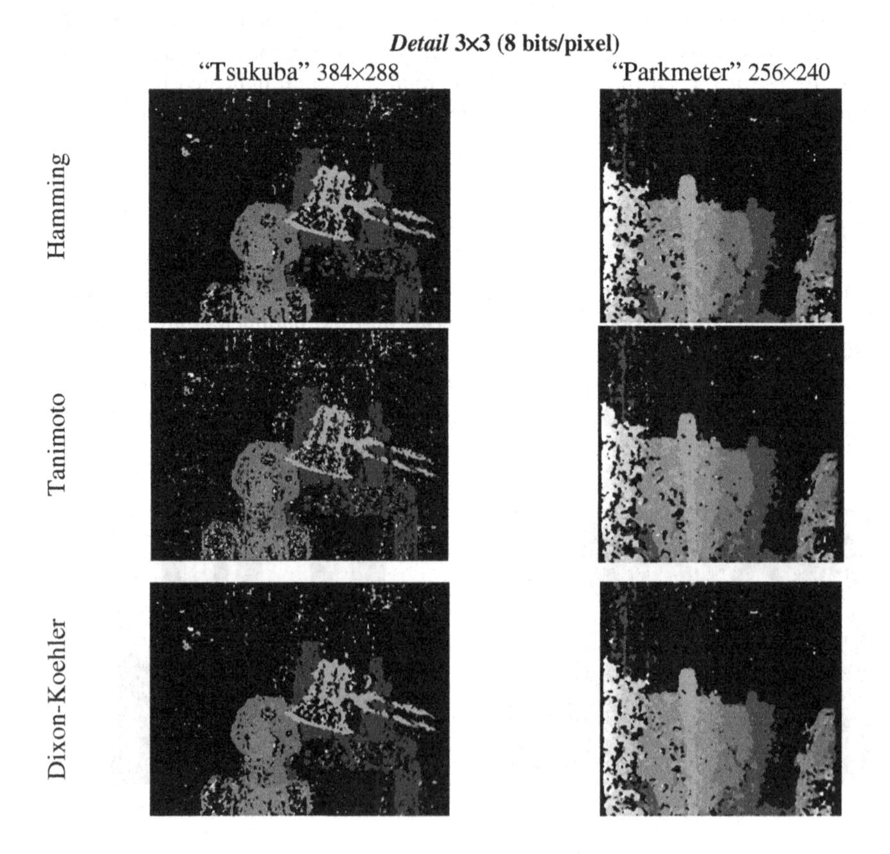

Fig. 3. Disparity maps of the two test stereo-pairs "Tsukuba" and "Parkmeter". The pixel neighborhood coding *Detail* 3×3. Bit-fields compared with measures: Hamming, Tanimoto and Dixon-Koehler (from top to bottom). Block matching window size 3×3 pixels

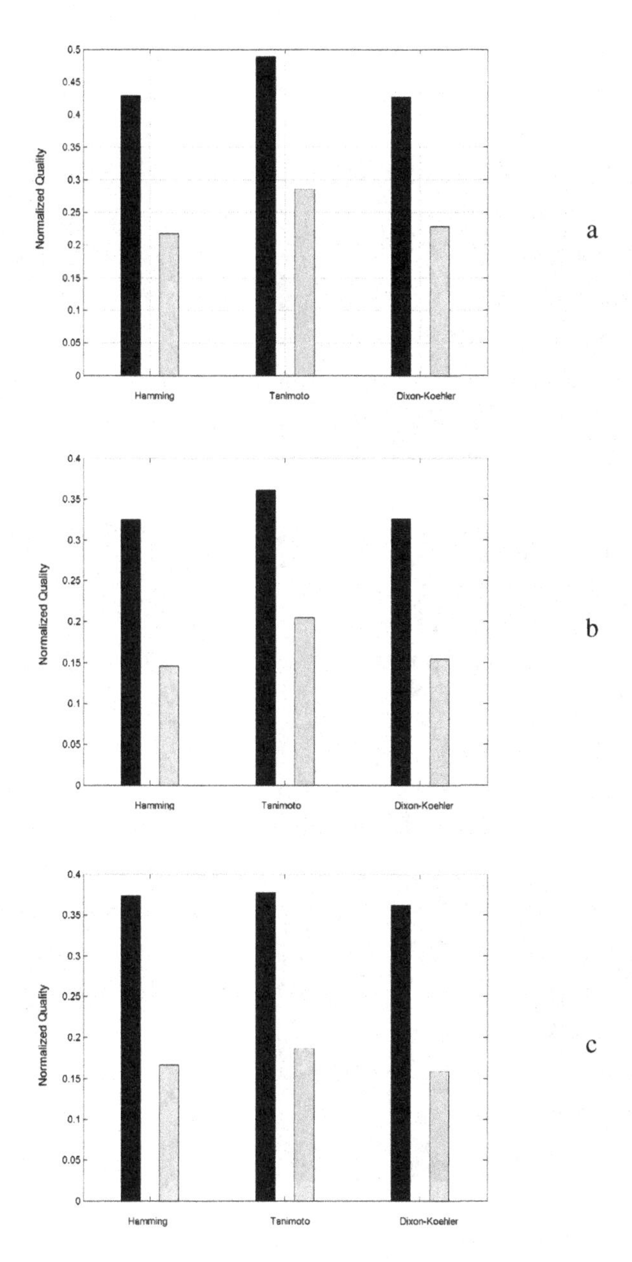

Fig. 4. Comparison of the quality of disparity maps measured as a normalized amount of false matches (the vertical axes). Each plot compares two disparity maps (left bar - "Tsukuba" and right - "Parkmeter") in respect to the three binary comparison measures (Hamming, Tanimoto, Dixon-Koehler) and different pixel neighborhood coding: *Census* 3×3 (a), *Census* 5×5 (b), *Detail* 3×3 (c)

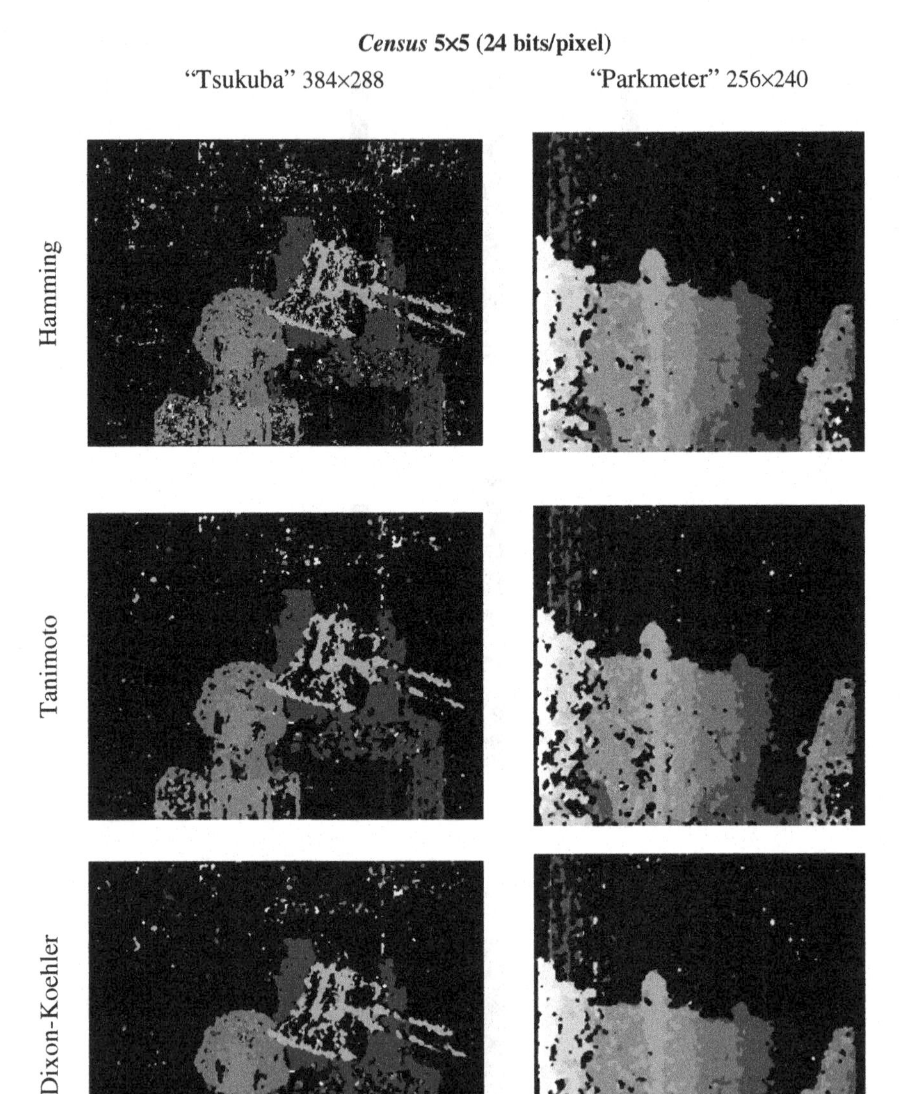

Fig. 5. Disparity maps of the two test stereo-pairs "Tsukuba" and "Parkmeter". The pixel neighborhood coding *Census* 5×5. Bit-fields compared with measures: Hamming, Tanimoto and Dixon-Koehler (from top to bottom). Block matching window size 3×3

Comparison of the quality of disparity maps measured as amount of false matches detected by the cross-checking is presented in Fig. 4. Each plot in Fig. 4 compares two disparity maps of "Tsukuba" and "Parkmeter" in respect to the three binary comparison measures: Hamming, Tanimoto, Dixon-Koehler, as well as

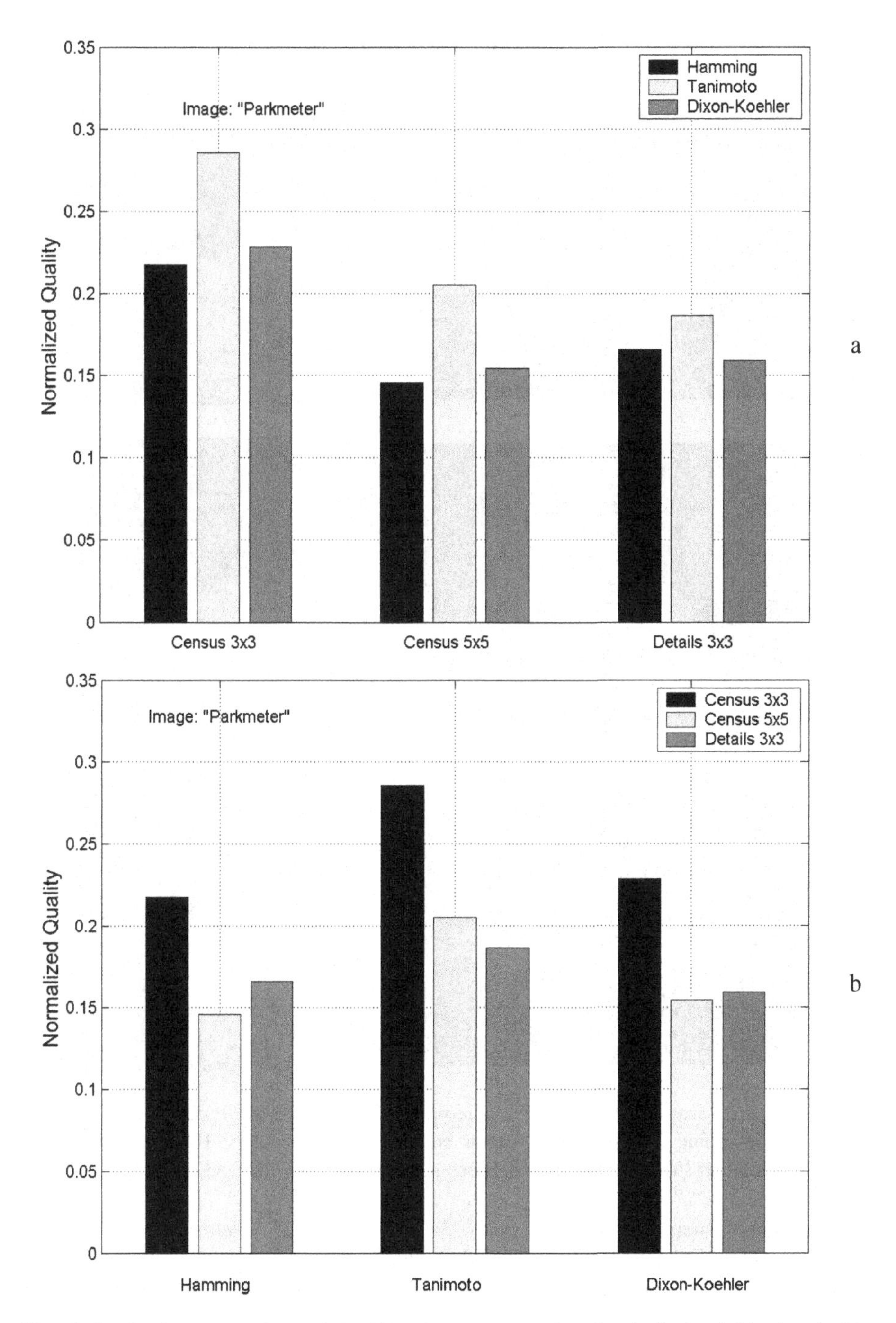

Fig. 6. Qualitative comparison of the binary measures against local pixel neighborhoods (a), and vice versa (b). The vertical axis denotes number of false matches detected by the cross-checking, divided by a total number of pixels. The lower bar, the better quality

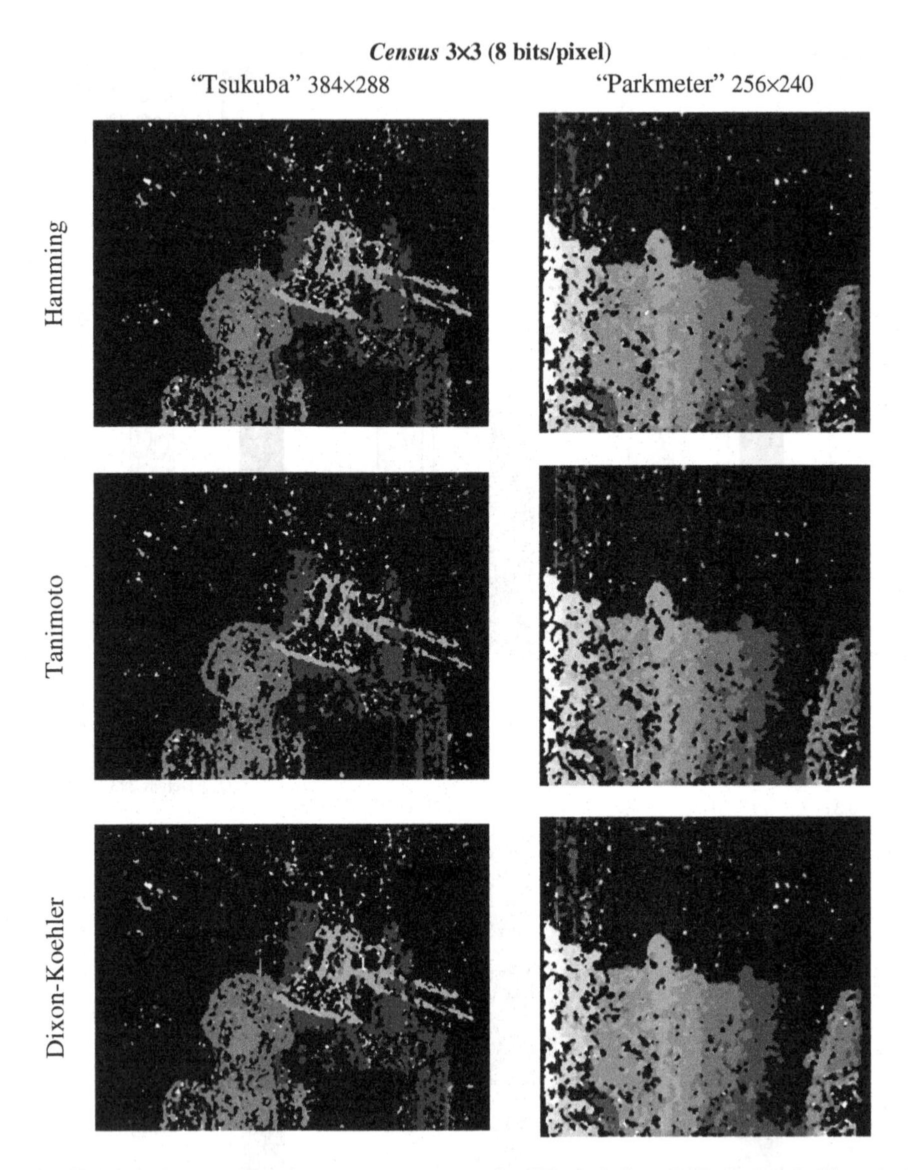

Fig. 7. Disparity maps of the two test stereo-pairs "Tsukuba" and "Parkmeter". The pixel neighborhood coding *Census* 3×3. Bit-fields compared with measures: Hamming, Tanimoto and Dixon-Koehler (from top to bottom). Block matching window size 3×3

different pixel neighborhoods: *Census* 3×3, *Census* 5×5, *Detail* 3×3. Based on additional experimental results with other stereo-pairs we can state that the best quality results are obtained for the *Census* 5×5 neighborhood with Hamming and Detail neighborhood 3×3 with Dixon-Koehler measures. Both setups use 24 bits per single pixel. However, the size of neighborhoods is quite different. This result is also affirmed in Fig. 6.

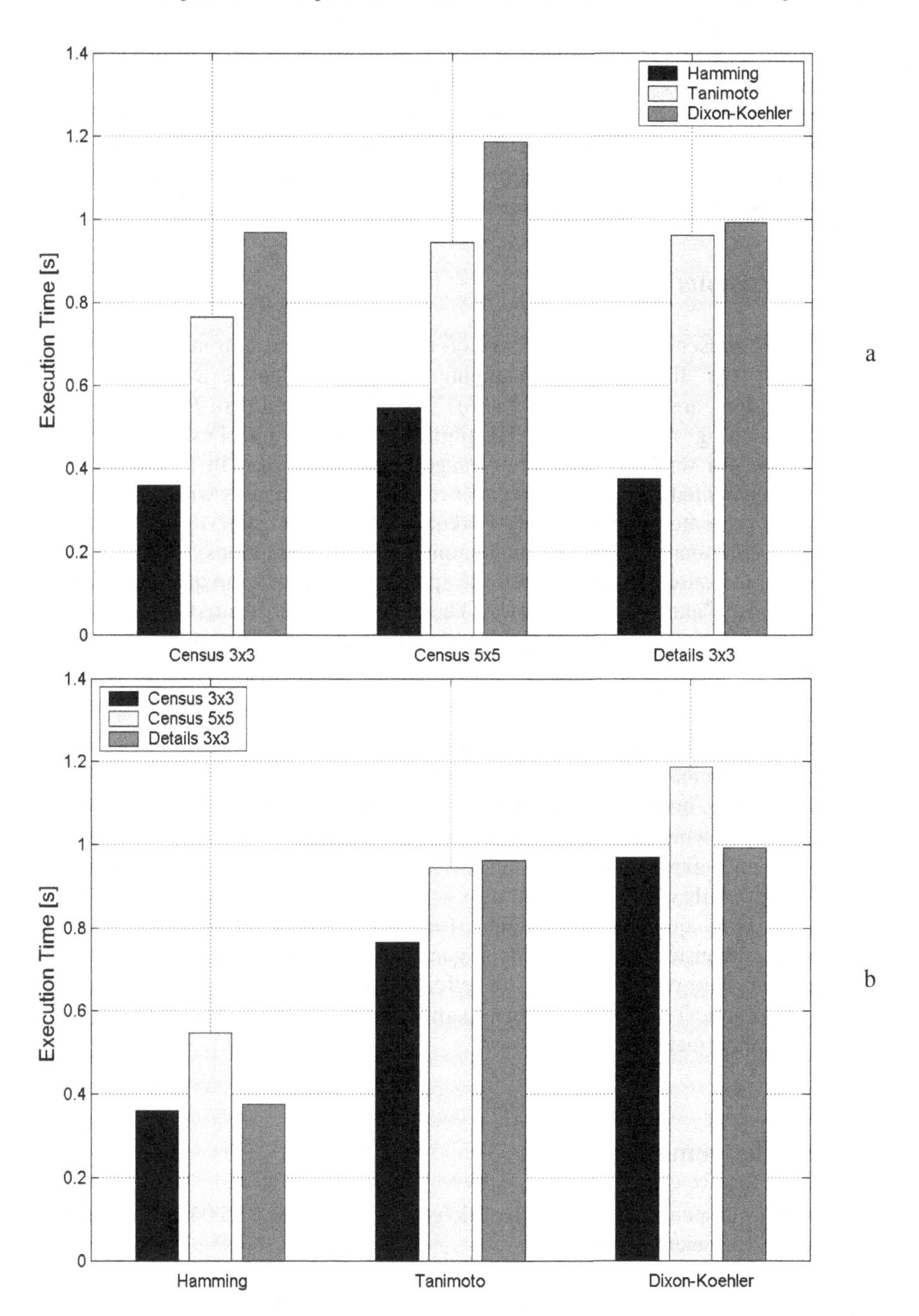

Fig. 8. Averaged execution times in seconds of the stereo-matching algorithms against local pixel neighborhoods (a), and comparison measures (b). The lower bar, the lower execution time

Further experiments revealed that when *Detail* 3×3 is extended from 3×3 neighborhoods to 5×5 then the increase of quality is about 15-25%.

The situation is different when one compares execution times presented in Fig. 8. It is evident that the Hamming measure outperforms the Tanimoto and Dixon-Koehler in respect to the computation complexity. Certainly, execution times shown in Fig. 8 can be significantly shortened in time optimized implementations.

5 Conclusions

This paper presents the comparison of the nonparametric transformations with respect to the three binary matching measures. The tested nonparametric transformations are: *Census 3×3*, *Census 5×5*, and devised here *Detailed 3×3*. The binary matching measures are: Hamming, Tanimoto and Dixon-Koehler. The experiments for ten different stereo images were performed in the software test-bench implemented in C++. In this paper results of experiments with the two stereo-pairs were presented: the artificial "Tsukuba" and the real stereo-pair "Parkmeter". Different combinations of the nonparametric transformations, stereo-pairs and correlation measures were tested. Of the special interest was the quality of resulting disparity maps and execution times. The former was measured in terms of the normalized number of false matches rejected during cross-checking. The smaller the outliers, the better quality of a given method. After experiments we can draw the following conclusions:

1. The use of nonparametric transformations increases resistance to the noise and distortions encountered in the input images.
2. Regarding quality, the nonparametric measures allowed for much smaller matching windows compared to such measures as SAD, SSD, etc. [4]. In the presented experiments we matched square windows of only 3×3 pixels thus small details were not filtered out.
3. The best quality is obtained for larger pixel neighborhoods in the nonparametric transformations at considerably small matching windows (thus allowing for preservation of image features).
4. The practical trade off between quality and speed offers *Census 5×5* with the Hamming measure.

Acknowledgement

This work was sponsored by the scientific grant no. KBN 3T11C 045 26 of the Polish Committee for Scientific Research.

References

1. Banks J., Bennamoun M., Corke P.: Non-Parametric Techniques for Fast and Robust Stereo Matching. CSIRO Manufacturing Science and Technology, Australia (1997)

2. Bhat D.N., Nayar S.K.: Ordinal Measures for Image Correspondence. IEEE Transaction on Pattern Analysis and Machine Intelligence Vol. 20 No. 4 (1998)
3. Cyganek B.: Neural Networks Application to The Correlation-Based Stereo-Images Matching, Engineering Applications of Neural Networks, Proceedings of the 5th International Conference EANN '99, Warsaw, Poland (1999) pp. 17-22
4. Cyganek, B., Borgosz, J.: A Comparative Study of Performance and Implementation of Some Area-Based Stereo Algorithms, LNCS 2124 (2001) 709-716
5. Cyganek, B.: Three Dimensional Image Processing, (in Polish) EXIT Warsaw (2002)
6. Cyganek, B., Borgosz, J.: An Object-Oriented Software Platform for Examination of Algorithms for Image Processing and Compression, LNCS 2658 (2003) 713-720
7. Cyganek B., Borgosz J.: Fuzzy Nonparametric Measures for Image Matching, Springer Lecture Notes in Artificial Intelligence 3070 (Subseries of LNCS), L. Rutkowski et.al. (Eds.), Proceedings of the 7th International Conference on Artificial Intelligence and Soft Computing – ICAISC 2004, Zakopane, Poland (2004) 712-717
8. Dixon, S.L., Koehler, R.T. J. Med Chem. 42 (1999) 2887–2900
9. Duda, R.O., Hart, P.E., Stork, D.G.: Pattern Classification. Wiley (2001)
10. Fua P.: A Parallel Stereo Algorithm that Produces Dense Depth Maps and Preserves Image Features, INRIA Technical Report No 1369 (1991)
11. Fusiello, A. et.al..: Efficient stereo with multiple windowing. CVPR 858–863 (1997)
12. Hartley, R.I., Zisserman A.: Multiple View Geometry in Computer Vision. CUP (2000)
13. Press W.H., Teukolsky S.A., Vetterling W.T., Flannery B.P.: Numerical Recipes in C. The Art of Scientific Computing. Second Edition. Cambridge University Press (1999)
14. Sloan Jr., K. R., Tanimoto, S. L.: Progressive Refinement of Raster Images, IEEE Transactions on Computers, Vol. 28, No. 11 (1979) 871-874
15. Zabih, R., Woodfill, J.: Non-Parametric Local Transforms for Computing Visual Correspondence. Proc. Third European Conf. Computer Vision (1994) 150-158
16. Zadeh, L.A.: Fuzzy sets. Information and Control, 8 (1965) 338-353
17. Zhengping, J.: On the Mutli-Scale Iconic Representation for Low-Level Computer Vision. Ph.D. Thesis. The Turing Institute and University of Strathclyde (1988) 114-118

Exact Optimization of Discrete Constrained Total Variation Minimization Problems

Jérôme Darbon[1,2] and Marc Sigelle[2]

[1] EPITA Research and Development Laboratory (LRDE),
14-16 rue Voltaire, F-94276 Le Kremlin-Bicêtre, France
jerome.darbon@{lrde.epita.fr, enst.fr}
[2] ENST TSI / CNRS LTCI UMR 5141,
46 rue Barrault, F-75013 Paris, France
marc.sigelle@enst.fr

Abstract. This paper deals with the total variation minimization problem when the fidelity is either the L^2-norm or the L^1-norm. We propose an algorithm which computes the exact solution of these two problems after discretization. Our method relies on the decomposition of an image into its level sets. It maps the original problems into independent binary Markov Random Field optimization problems associated with each level set. Exact solutions of these binary problems are found thanks to minimum-cut techniques. We prove that these binary solutions are increasing and thus allow to reconstruct the solution of the original problems.

1 Introduction

Image reconstruction and deconvolution methods are often based on the minimization of the constrained total variation [1, 2] of an image. These problems have minimizers in the space of functions of bounded variation [3] which allows for discontinuities and thus preserve edges and sharp boundaries. Suppose u is defined on a rectangle Ω of $\mathbb{R}^2$. Then the total variation (TV) of u is $TV(u) = \int_\Omega |\nabla u|$, where the gradient of u is taken in the distributional sense. A classical way to minimize the TV is achieved by a gradient descent which yields the following evolution equation: $\frac{\partial u}{\partial t} = \text{div}\left(\frac{\nabla u}{|\nabla u|}\right)$. The last term corresponds to the curvature of u. In order to avoid division by zero, a classical approximation is to replace $|\nabla u|$ by $\sqrt{|\nabla u|^2 + \epsilon}$. However, this scheme tends to smooth discontinuities and although it converges towards the solution when ϵ tends to 0, it does not provide an exact solution. Other formulation of TV minimization using duality is presented in [4]. A fast algorithm which converges towards the solution can be derived from this formulation. In [5], a fast approximation minimization algorithm for Markov Random Field (MRF) is presented. It relies on minimum cost cut and the result is a local minimum.

In [6], a fast algorithm to compute the exact solution in 1D for the TV minimization problem subject to the L^2 constraint is presented. However, the

R. Klette and J. Žunić (Eds.): IWCIA 2004, LNCS 3322, pp. 548–557, 2004.

algorithm does not scale to higher dimensions. In 1D, one can find an exact solution using dynamic programming [7], provided that the label state is discrete. The complexity of such a method is $\Theta(N^2|\Omega|)$, where N and $|\Omega|$ are the cardinality of the label state and the number of pixels in the discrete domain Ω, respectively. In [8], Ishikawa presents an algorithm to find the exact solution for MRF with convex priors in a polynomial time.

In this paper, we focus on TV minimization with L^1 or L^2 fidelity. Thus, we are interested in minimizing the following functionals:

$$E_\alpha(u,\beta) = \int_\Omega |u(x) - v(x)|^\alpha \, dx + \beta \int_\Omega |\nabla u| \ ,$$

where $\alpha \in \{1,2\}$ and $\beta \geq 0$. The use of the L^1 fidelity has already been studied in [9–11]. Our main contribution is an exact optimization of a discretization of the two functionals $E_\alpha(.,\beta)$. It relies on reformulating the original problem into several independent binary problems which are expressed through the MRF framework. It is based on the decomposition of a function into its level sets.

The rest of this paper is as follows. The decomposition of the considered problems into independent binary problems is described in section 2. In section 3, reconstruction of the solution from solutions of the binary problems is shown. Minimization algorithm and results are presented in section 4. Finally we draw some conclusions in section 5.

2 Formulation Through Level Sets

In this section, we show that minimization of the TV minimization problem with L^1 or L^2 fidelity can be decomposed into the minimization of independent binary problems. For each level $\lambda \in [0, N-1]$, we consider the thresholded images u^λ of an image: $u^\lambda = 1\!\!1_{u \leq \lambda}$. Note that this decomposition is sufficient to reconstruct the gray-level image: $u(x) = \min\{\lambda, u^\lambda(x) = 1\}$.

2.1 Coarea Formula

For any function u which belongs to the space of bounded variation, the Coarea formula [3] gives $TV(u) = \int_{\mathbb{R}} P(u^\lambda) d\lambda$,for almost all λ and where $P(u^\lambda)$ is the perimeter of u^λ. In the discrete lattice version we define for each site s its grey level u_s and $u_s^\lambda = u^\lambda(s) = 1\!\!1_{u_s \leq \lambda}$. We estimate the perimeter using pairs of neighboring pixels: $TV(u) = \sum_{\lambda=0}^{N-1} \sum_{s \sim t} R_{s,t}(u_s, v_s, \lambda)$, where $s \sim t$ denotes neighboring pixels and $R_{s,t}(u_s, v_s, \lambda) = w_{s,t}|u_s^\lambda - u_t^\lambda|$ ($w_{s,t}$ is some coefficient). For our experiments we use two different contour length estimators. The first one consists in considering only the four-connected neighborhood and setting $w_{s,t}$ to 1. The second one, as proposed in [12], sets $w_{s,t}$ to 0.26 and 0.19 for the four and eight connected neighborhood respectively. Note that the latter estimation is not accurate for small regions.

2.2 Expressing L^1 and L^2 Through Level Sets

We reformulate L^1 fidelity into level sets. We decompose the domain into the following two disjoint sets $\{s : u_s < v_s\}$ and $\{s : u_s > v_s\}$. This yields

$$\begin{aligned}\sum_{s\in\Omega} |u_s - v_s| &= \sum_{u_s<v_s} |v_s - u_s| + \sum_{u_s>v_s} |u_s - v_s| = \sum_{s\in\Omega}\sum_{\lambda=0}^{N-1} (\mathbb{1}_{u_s\le\lambda<v_s} + \mathbb{1}_{v_s\le\lambda<u_s}) \\ &= \sum_{\lambda=0}^{N-1}\sum_{s\in\Omega} \mathbb{1}_{u_s\le\lambda}\ \mathbb{1}_{\lambda<v_s} + \mathbb{1}_{v_s\le\lambda}\ \mathbb{1}_{\lambda<u_s} = \sum_{\lambda=0}^{N-1}\sum_{s\in\Omega} u_s^\lambda\ (1-v_s^\lambda) + (1-u_s^\lambda)\ v_s^\lambda \\ &= \sum_{\lambda=0}^{N-1}\sum_{s\in\Omega} |u_s^\lambda\ - v_s^\lambda| = \sum_{\lambda=0}^{N-1}\sum_{s\in\Omega} D_1(u_s, v_s, \lambda) \qquad (1)\end{aligned}$$

where $D_1(x,y,\lambda) = |x^\lambda - y^\lambda|$, and where we used the property: $|a-b| = a+b-2ab$ for binary variables a, b. Note that this formulation shows that the L^1-norm treats level sets of the image u independently of their associated gray-levels. This can be seen as adopting a geometrical point of view.

The same approach is used for the decomposition of L^2 into level sets. However, contrary to the L^1 norm, the decomposition cannot be independent of its gray-levels. We begin with separating the sum according previous disjoint sets and using the formula $\sum_{k=1}^{M}(2k-1) = M^2$:

$$\sum_{s\in\Omega} (u_s - v_s)^2 = \sum_{u_s<v_s}\sum_{k=1}^{v_s-u_s} (2k-1) + \sum_{u_s>v_s}\sum_{l=1}^{u_s-v_s} (2l-1).$$

Then for the first sum we make the following change of variable $k \leftarrow v_s - \lambda$, while we do $l \leftarrow \lambda - v_s + 1$ for the second one. It leads to:

$$\begin{aligned}\sum_{s\in\Omega} (u_s - v_s)^2 &= \sum_{u_s<v_s}\sum_{\lambda=u_s}^{v_s-1} (2(v_s-\lambda)-1) + \sum_{u_s>v_s}\sum_{\lambda=v_s}^{u_s-1} (2(\lambda-v_s)+1) \\ &= \sum_{s\in\Omega}\sum_{\lambda=0}^{N-1} \mathbb{1}_{u_s\le\lambda<v_s}\ (2(v_s-\lambda)-1) + \sum_{\lambda=0}^{N-1} \mathbb{1}_{v_s\le\lambda<u_s}\ (2(\lambda-v_s)+1) \\ &= \sum_{s\in\Omega}\sum_{\lambda=0}^{N-1} (\mathbb{1}_{u_s\le\lambda}\ \mathbb{1}_{\lambda<v_s}\ - \mathbb{1}_{v_s\le\lambda}\ \mathbb{1}_{\lambda<u_s}\ )\ (2(v_s-\lambda)-1) \\ &= \sum_{\lambda=0}^{N-1}\sum_{s\in\Omega} (u_s^\lambda\ (1-v_s^\lambda)\ - (1-u_s^\lambda)\ v_s^\lambda\ )\ (2(v_s-\lambda)-1) \\ &= \sum_{\lambda=0}^{N-1}\sum_{s\in\Omega} (u_s^\lambda\ - v_s^\lambda)\ (2(v_s-\lambda)-1) = \sum_{\lambda=0}^{N-1}\sum_{s\in\Omega} D_2(u_s, v_s, \lambda) \qquad (2)\end{aligned}$$

where $D_2(x,y,\lambda) = (x^\lambda\ - y^\lambda)\ (2(y-\lambda)-1)$. This formulation shows that L^2 can be decomposed into level sets where their associated gray-levels are taken into account.

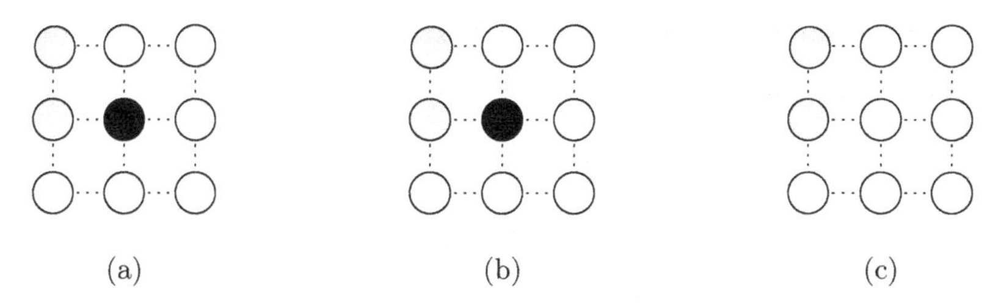

Fig. 1. Since $E_1(., \beta)$ is not strictly convex, minimizers can be non-unique. The original image is depicted in (a) where 4-connectivity is considered. Black and white circles refer to sites whose value is 0 and 1, respectively. If $\beta = 0.25$ then there are two minimizers depicted in (b) and (c), whose associated energy is 1

2.3 Independent Optimizations

Finally, both energies can be re-written as follows:

$$E_\alpha(u, \beta) = \sum_{\lambda=0}^{N-1} \left(\sum_{s \in \Omega} D_\alpha(u_s, v_s, \lambda) + \beta \sum_{s \sim t} w_{s,t} |u_s^\lambda - u_t^\lambda| \right) = \sum_{\lambda=0}^{N-1} E_\alpha^\lambda(u^\lambda, \beta) \ ,$$

where $E_\alpha^\lambda(u^\lambda, \beta) = \sum_{s \in \Omega} D_\alpha(u_s, v_s, \lambda) + \beta \sum_{s \sim t} w_{s,t} |u_s^\lambda - u_t^\lambda|$. Note that each term $E_\alpha^\lambda(u^\lambda, \beta)$ is a 2D MRF which only involves binary variables and pairwise interactions. Pairwise interactions only deal with the same gray-level component (λ) of two neighboring pixels (u_s and u_t). The data fidelity term can use different gray-level components of the observed image v, such as the L^2-norm case for instance. This is possible provided that the data fidelity energy can be linearly decomposed with respect to each component u_s^λ. The prior is an Ising model [13].

Now suppose that for each λ, we independently find the best binary configuration $\hat{u}^\lambda$ which minimizes the energy of the MRF. Clearly, the summation will be minimized. Thus we will find a minimizer for $E_\alpha(., \beta)$ provided that the following property of monotony holds for binary minimizers:

$$\hat{u}^\lambda \leq \hat{u}^\mu \ \forall \lambda < \mu \ . \tag{3}$$

Indeed, if this property holds, then the minimizer $\hat{u}$ of $E_\alpha(., \beta)$ is given [14] by $\hat{u}_s = \min\{\lambda, \hat{u}_s^\lambda = 1\} \ \forall s$. The monotone property is proved in the next section.

3 Reconstruction of the Solution

In this section, we prove the monotone property defined by (3). However, since $E_1(., \beta)$ is not strictly convex, it leads to non-unique minimizers in general. Such a situation is depicted in Figure 1. The monotone property can be violated in that case. However the following Lemma will be useful.

3.1 A Lemma Based on Coupled Markov Chains

Lemma. *If the local conditional posterior energy at each site s can be written up to a constant, as:*

$$E_\alpha(u_s \mid \{u_t\},\ v_s) = \sum_{\lambda=0}^{N-1} \phi_s(\lambda)\ u_s^\lambda \tag{4}$$

where $\phi_s(\lambda)$ is a non-increasing function of λ, then one can exhibit a "coupled" stochastic algorithm minimizing each total posterior energy $E_\alpha^\lambda(u^\lambda, \beta)$ while preserving the monotone condition: $\forall s$, $u_s^\lambda \nearrow$ with λ .

In other words, given a binary solution $u^\star$ to the problem E_α^k, there exists at least one solution $\hat{u}$ to the problem E_α^l such that $u^\star \leq \hat{u}\ \forall k \leq l$. The proof of the Lemma relies on coupled Markov chains [15].

Proof: We endow the space of binary configurations by the following order : $u \leq v$ iff $u_s \leq v_s\ \forall s \in \Omega$. From the decomposition (4) the local conditional posterior energy at level value λ is $\phi_s(\lambda)\ u_s^\lambda$. Thus the related Gibbs local conditional posterior probability is

$$P(u_s^\lambda = 1 \mid \{u_t^\lambda\},\ v_s^\lambda) = \frac{\exp -\phi_s(\lambda)}{1 + \exp -\phi_s(\lambda)} = \frac{1}{1 + \exp \phi_s(\lambda)}\ . \tag{5}$$

With the conditions of the Lemma, this latter expression is clearly a monotone non-decreasing function of λ.

Let us now design a "coupled" Gibbs sampler for the N binary images in the following sense: first consider a visiting order of the sites (tour). When a site s is visited, pick up a *single* random number ρ uniformly distributed in $[0, 1]$. Then, for each value of λ, assign: $u_s^\lambda = 1$ if $\rho \leq P(u_s^\lambda = 1 \mid \{u_t^\lambda\},\ v_s^\lambda)$ or else $u_s^\lambda = 0$.[1] From the non-decreasing monotony of (5) it is seen that the set of assigned binary values at site s satisfies $u_s^\lambda = 1 \Rightarrow u_s^\mu = 1\ \forall \mu > \lambda$. The monotone property $u^\lambda \leq u^\mu\ \forall\ \lambda < \mu$ is thus preserved. Clearly, this property also extends to a series of N coupled Gibbs samplers having *the same* positive temperature T when visiting a given site s: it suffices to replace $\phi_s(\lambda)$ by $\phi_s(\lambda)\ /\ T$ in (5). Hence, this property also holds for a series of N coupled Simulated Annealing algorithms [16] where a *single* temperature T boils down to 0 (either after each visited site s or at the beginning of each tour [13] .) □

Several points should be emphasized here:

- The coupled monotony-preserving Gibbs samplers described in [15] relate to the *same* MRF but for various initial conditions, while here, our N coupled Gibbs samplers relate to N *different* posterior MRF's (one for each level λ).
- It must also be noticed that our Lemma gives a *sufficient* condition for the simultaneous, "level-by-level independent" minimization of posterior energies while preserving the monotone property.

[1] This is the usual way to draw a binary value according to its probability, except that we use here the same random number for all the N binary images.

3.2 The L^1 and L^2 Cases

Let us show that both L^1 regularization and attachment to data energies feature property (4); so will do their sum and thus the total posterior energy. Using previous property for binary variables a, b: $|a - b| = a + b - 2ab$, this yields:

$$\sum_{t\sim s} |u_s - u_t| = \sum_{\lambda=0}^{N-1} \sum_{t\sim s} |u_s^\lambda - u_t^\lambda| = \sum_{\lambda=0}^{N-1} \sum_{t\sim s} (1 - 2u_t^\lambda)\, u_s^\lambda \; + C$$

where $C = \sum_{\lambda=0}^{N-1} \sum_{t\sim s} u_t^\lambda$ is a "constant" since it only depends on the $\{u_t^\lambda\}$. Thus $\phi_s(\lambda) = \sum_{t\sim s} (1 - 2u_t^\lambda)$, which is by essence a non-increasing function of λ.

Similarly starting from (1) for the L^1 attachment to data term:

$$|u_s - v_s| = \sum_{\lambda=0}^{N-1} |u_s^\lambda - v_s^\lambda| = \sum_{\lambda=0}^{N-1} (1 - 2v_s^\lambda)\, u_s^\lambda \; + C' \quad , \quad C' = \sum_{\lambda=0}^{N-1} v_s^\lambda$$

The approach for the L^2 relies on the same method. From (2) one can write:

$$(u_s - v_s)^2 = \sum_{\lambda=0}^{N-1} \phi_s(\lambda)\, u_s^\lambda + C'' \; ,$$

where $\phi_s(\lambda) = 2(v_s - \lambda) - 1$ clearly fulfills our requirement.

Thus in both cases, $\phi_s(\lambda)$ is a non-increasing function, so that TV regularization with either L^1 and L^2-fidelity both follow the conditions of our Lemma. Although we have proved the monotone property, it does not provide an algorithm to compute the solution. Indeed, using a Simulated Annealing process, one knows it has no stopping criteria. We propose an algorithm in the next section.

4 Computations and Experiments

In this section, we describe our algorithm and present some experiments.

Greig et al. [17] were the first ones to propose an exact optimization for binary MRF. It is based on constructing a graph such that its minimum cut (MC) gives an optimal labelling. Since this seminal work, other graph constructions were proposed to solve some non-binary problems exactly [8, 18]. In [19], the authors propose a necessary condition for binary functions to be minimized via MCs along with a graph construction. Our Ising model fulfills the condition.

For each level we construct the graph as proposed in [19] and compute a MC. However, since uniqueness cannot be assured with L^1 fidelity, the algorithm returns one of the optimal configurations. Since these minimizations are independently performed, the monotone property can be violated. To reconstruct the solution, one flips every pixel where this property is violated. This flipping process also gives an optimal labelling since the energy does not change.

Table 1. Time results (in seconds) with L^1 fidelity for the image "hand"

Size	4-Connectivity	8-Connectivity
151x121	4.97	7.58
343x243	21.02	30.56

(a)

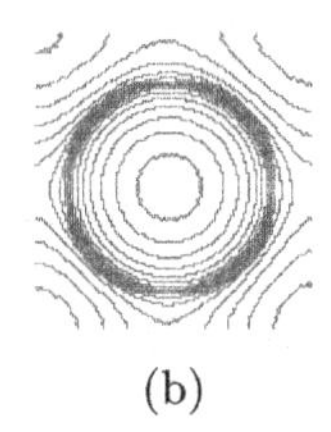

(b)

(c)

Fig. 2. Minimizers of TV under L^2 constraints ($\beta = 1$). The original image is depicted in (a). The level lines resulting from the gradient descent algorithm are presented in (b). The level lines of the exact solution, using our algorithm, are depicted in (c)

To compute the MC, we used the algorithm described in [20]. For our binary problems, this algorithm gives near-linear performance with respect to the number of pixels $|\Omega|$. Since we compute N cuts, the complexity of our algorithm is near-linear both with respect to N and $|\Omega|$. Time results (on a 1.6GHz Pentium IV) for our method are presented in table 1 for L^1 fidelity. This is in contrast with the near-quadratic behavior of [8] with respect to N.

For our experiments, we always use the 8-connectivity. In [21] the authors give exact and analytic solutions for TV minimization with L^2 attachment for radial symmetric functions. For instance, if the observed image is a circle then the solution is a circle with the same radius : only its gray-levels change. Figure 2 depicts the level-lines of the solutions for our algorithm and the gradient descent algorithm. For the latter, we approximate $TV(u)$ by $\sqrt{|\nabla u|^2 + \epsilon}$ with $\epsilon = 1$. Note how many level lines are created by the gradient descent algorithm.

TV minimization is well-known for its high performance in image restoration. Figure 3 depicts a cartoon image and its noisy version corrupted by an additive Gaussian noise ($\sigma = 30$). It also presents the results of the restoration using the gradient descent method and our algorithm. Although the results visually look the same, the exact solution provides a much better result in terms of level lines. Note how corners of the squares are smoothed. This is predicted by the theory [22] which states that a square cannot arise as a solution. Results of the regularization using L^1-fidelity are depicted in figure 4. The higher the coefficient β, the more fine structures are removed while the contrast remains preserved.

5 Conclusion

In this paper we have presented an algorithm to compute the exact solution of the discrete TV-based restoration problem when fidelity is the L^1 or L^2 norm.

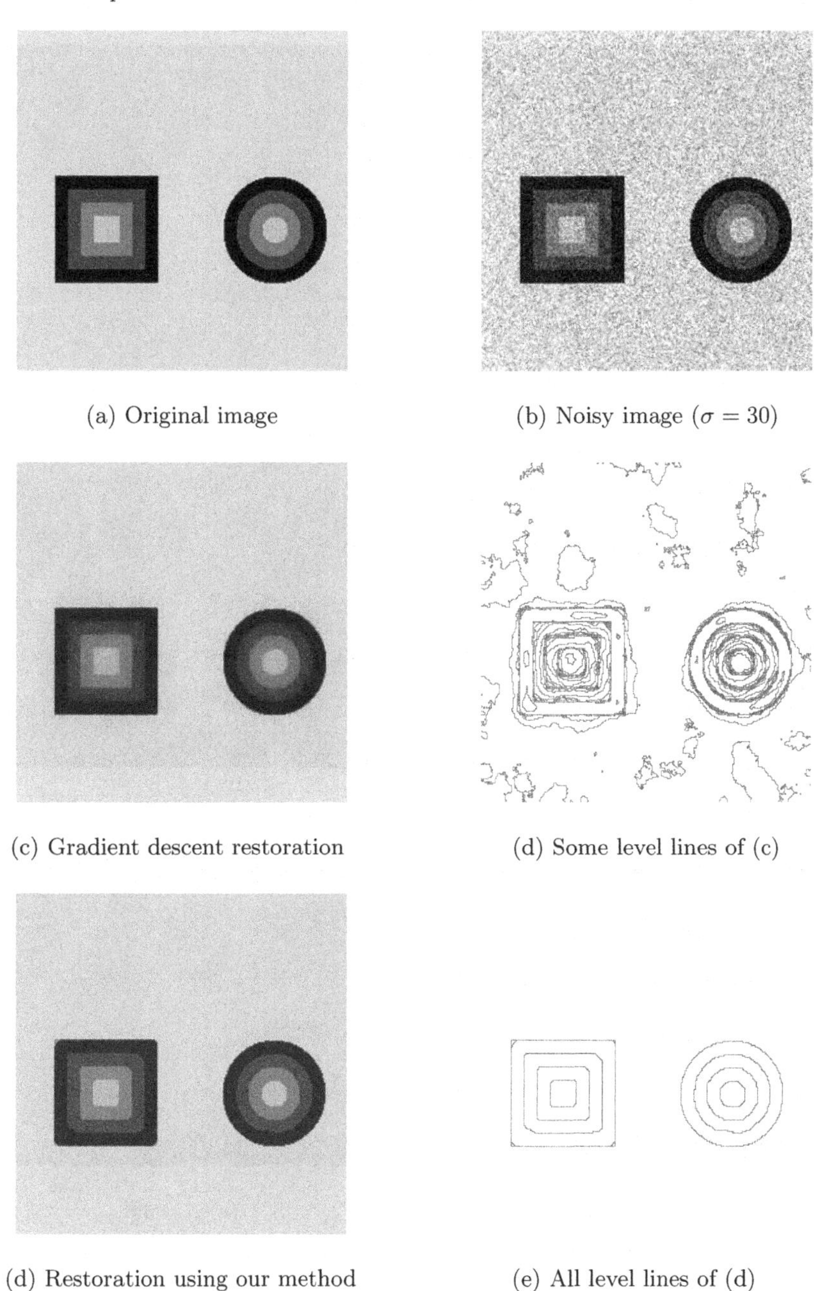

(a) Original image

(b) Noisy image ($\sigma = 30$)

(c) Gradient descent restoration

(d) Some level lines of (c)

(d) Restoration using our method

(e) All level lines of (d)

Fig. 3. Restoration of a blocky image corrupted by a Gaussian noise. Results of TV minimization with L^2 fidelity for the gradient descent algorithm and our method. Only level lines multiples of 5 are displayed on (d)

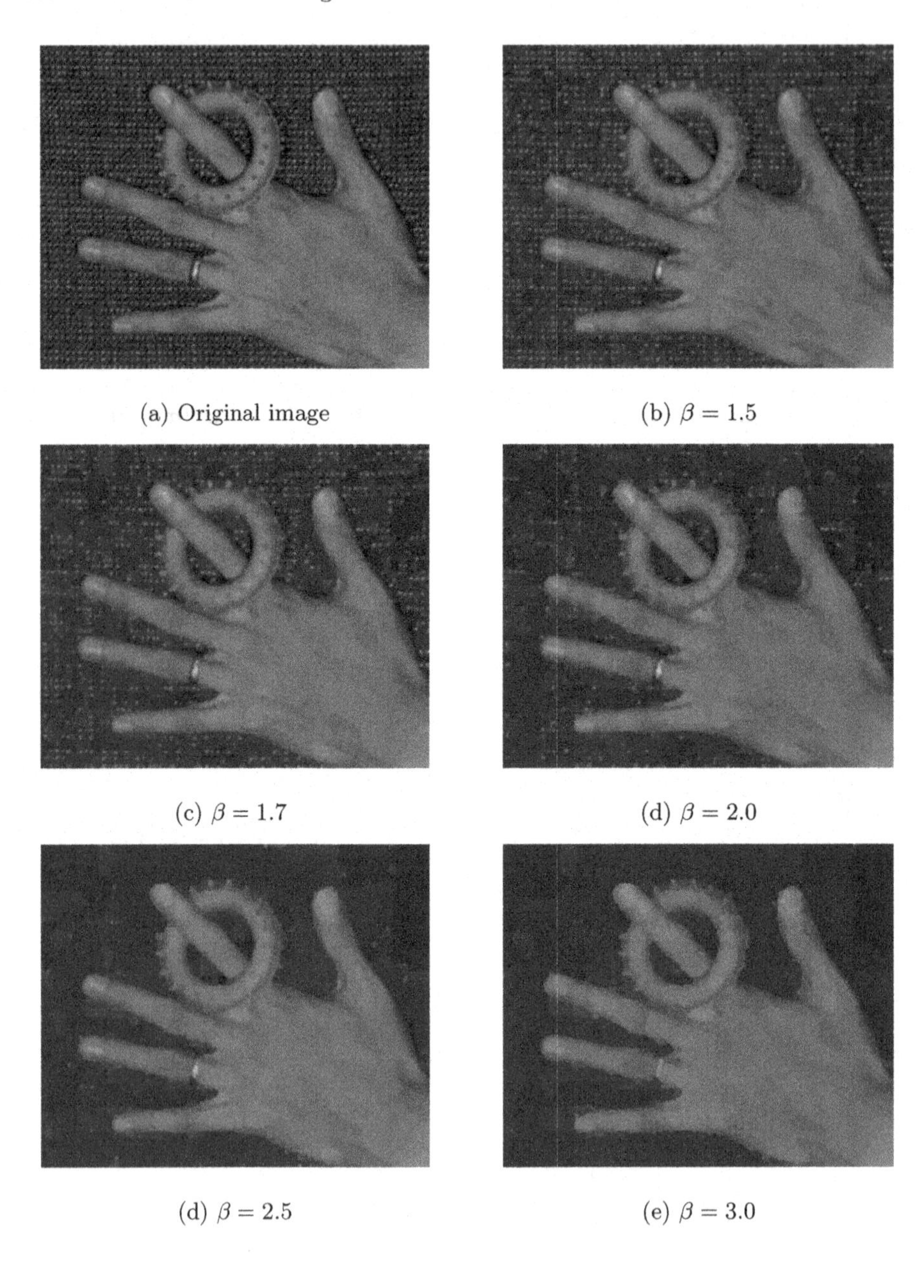

(a) Original image

(b) $\beta = 1.5$

(c) $\beta = 1.7$

(d) $\beta = 2.0$

(d) $\beta = 2.5$

(e) $\beta = 3.0$

Fig. 4. Minimizers of TV with L^1 fidelity

It relies on the decomposition of the problem into binary ones thanks to a levelset formulation. It allows for an algorithm whose complexity is near-linear both with respect to the image size and the number of labels.

Extension of this method to other types of fidelity is in progress. We will show that the condition stated by our Lemma is equivalent to the fact that each local conditional posterior energy is a convex function. Finally a faster minimization algorithm which takes into account the monotone property is under study. Comparisons with other exact minimization algorithms must also be made.

References

1. Rudin, L., Osher, S., Fatemi, E.: Nonlinear total variation based noise removal algorithms. Physica D. **60** (1992) 259–268
2. Sauer, K., Bouman, C.: Bayesian estimation of transmission tomograms using segmentation based optimization. IEEE Nuclear Science **39** (1992) 1144–1152
3. Evans, L., Gariepy, R.: Measure Theory and Fine Properties of Functions. CRC Press (1992)
4. Chambolle, A.: An algorithm for total variation minimization and applications. Journal of Mathematical Imaging and Vision **20** (2004) 89–97
5. Boykov, Y., Veksler, O., Zabih, R.: Fast approximate energy minimization via graph cuts. IEEE PAMI **23** (2001) 1222–1239
6. Pollak, I., Willsky, A., Huang, Y.: Nonlinear evolution equations as fast and exact solvers of estimation problems. to appear in IEEE Signal Processing (2004)
7. Amini, A., Weymouth, T., Jain, R.: Using dynamic programming for solving variational problems in vision. IEEE PAMI **12** (1990) 855–867
8. Ishikawa, H.: Exact optimization for Markov random fields with priors. IEEE PAMI **25** (2003) 1333–1336
9. Alliney, S.: A property of the minimum vectors of a regularizing functional defined by means of the absolute norm. IEEE Signal Processing **45** (1997) 913–917
10. Chan, T., Esedoḡlu, S.: Aspect of total variation regularized l^1 function approximation. Technical Report 7, UCLA (2004)
11. Nikolova, M.: Minimizers of cost-functions involving nonsmooth data-fidelity terms. SIAM J. Num. Anal. **40** (2002) 965–994
12. Nguyen, H., Worring, M., van den Boomgaard, R.: Watersnakes: Energy-driven watershed segmentation. IEEE PAMI **23** (2003) 330–342
13. Winkler, G.: Image Analysis, Random Fields and Dynamic Monte Carlo Methods. Applications of mathematics. Springer-Verlag (2003)
14. Guichard, F., Morel, J.: Mathematical morphology "almost everywhere". In: Proceedings of ISMM, Csiro Publishing (2002) 293–303
15. Propp, J.G., Wilson, D.B.: Exact sampling with coupled Markov chains and statistical mechanics. Random Structures and Algorithms **9** (1996) 223–252
16. Geman, S., Geman, D.: Stochastic relaxation, gibbs distributions, and the bayesian restoration of images. IEEE PAMI **6** (1984) 721–741
17. Greig, D., Porteous, B., Seheult, A.: Exact maximum a posteriori estimation for binary images. Journal of the Royal Statistics Society **51** (1989) 271–279
18. Roy, S.: Stereo without epipolar lines: A maximum-flow formulation. International Journal of Computer Vision **34** (1999) 147–162
19. Kolmogorov, V., Zabih, R.: What energy can be minimized via graph cuts? IEEE PAMI **26** (2004) 147–159
20. Boykov, Y., Kolmogorov, V.: An experimental comparison of min-cut/max-flow algorithms for energy minimization in vision. IEEE PAMI **26** (2004) 1124–1137
21. Strong, D., Chan, T.: Edge preserving and scale-dependent properties of total variation regularization. Inverse Problem **19** (2003) 165–187
22. Meyer, Y.: Oscillating patterns in image processing and nonlinear evolution equations. University Lecture Series **22** (2001)

Tensor Algebra: A Combinatorial Approach to the Projective Geometry of Figures

David N.R. McKinnon and Brian C. Lovell

Intelligent Real-Time Imaging and Sensing (IRIS) Group,
School of Information Technology and Electrical Engineering,
University of Queensland, St. Lucia QLD 4066, Australia
mckinnon@itee.uq.edu.au

Abstract. This paper explores the combinatorial aspects of symmetric and anti-symmetric forms represented in tensor algebra. The development of geometric perspective gained from tensor algebra has resulted in the discovery of a novel projection operator for the Chow form of a curve in $\mathbb{P}^3$ with applications to computer vision.

1 Introduction

The notation used in this paper is adapted from [16] with the novel addition of the symmetric operators which we will use to derive the representations for curves, surfaces and other non-linear algebraic maps. Also, we will maintain the use of vectorizations of the (anti)symmetric tensor forms, that offer an equivalent expression of the algebra as a coefficient ring defined over the field of real numbers ($\mathbb{R}$). This in turn can be expressed as the elements of a vector space on a computer. This approach to geometry is analytically equivalent to the approach taken by other authors studying geometry of linear objects [17], invariants [15, 8], multiple view geometry [16, 10] and also in the theory symmetric functions [14]. The major contribution of this paper is the projection operator for the Chow form of a curve in space. This projection operator has allowed for a new class of curve based approaches to multiple view geometry.

2 Tensor Basics

Tensors are a generalization of the concept of vectors and matrices. In fact vectors and matrices are 1 and 2-dimensional instances of a tensor. Tensors make it easier to understand the interaction of algebraic expressions that involve some type of multilinearity in their coefficients. The following sections will briefly introduce some of the algebra underlying the types of tensors we are most interested in.

2.1 Vector Spaces

Tensors are composed entirely from vector spaces. Vector spaces can be combined using a range of standard operators resulting in differently structured tensors. We will limit

R. Klette and J. Žunić (Eds.): IWCIA 2004, LNCS 3322, pp. 558–567, 2004.

our study of the geometry herein to projective vector space $\mathbb{P}^n$. An element of an n-dimensional projective vector space in the tensor notation is denoted as $\mathbf{x}^{{}_mA_i^s} \in \mathbb{P}^n$. The symbol ${}_mA_i^s$ is called an indeterminant and identifies several important properties of the vector space. Firstly in order to better understand the notation we must rewrite $\mathbf{x}^A$ in the standard vector form. This is achieved by listing the elements of the vector space using the indeterminant as the variables of the expression. In this manor the symbol that adjoins the indeterminant is merely cosmetic, for example the equivalent vector space is, $\mathbf{x}^{{}_mA_i^s} \equiv [{}_mA_0^s, {}_mA_1^s, \ldots, {}_mA_n^s]^\top$, where m identifies the multilinearity of the indeterminant, s depicts the degree (or step) of the indeterminant. We show in the next section that there are several different types of degree that we will be concerned with and that these are used to denote a vectorization of the tensor form. The last element specifying the indeterminant is i, this a *choice* of the positioning of the elements in the vector, we most commonly refer to i as the index of the indeterminant. The standard indexing is $i \in \{0 \ldots n\}$ for an n-dimensional projective vector space.

Indeterminants of a regular vector (vertical) space ($\mathbb{P}^n$) are called *contravariant* and indeterminants of a dual (horizontal) vector space in ${}^*\mathbb{P}^n$ (covector) are called *covariant*. The notation for a dual vector (covector) space is analogous to that for a regular vector space, $\mathbf{x}_{{}^*_mA_i^s} \equiv [{}^*_mA_0^s, {}^*_mA_1^s, \ldots, {}^*_mA_n^s]$, the only difference being that the vector is transposed. In the interests of compactness and clarity often we will abandon the entire set of labels for an indeterminant via an initial set of assignments. If this is the case assume that i is any arbitrary scalar between 0 and n and $s, m = 1$. If an indeterminant is used in a covariant expression then the $*$ may also be omitted.

2.2 Tensors and Contraction

Tensor contraction is the process of eliminating vector spaces from a given tensor. Tensor contraction is achieved via a dot product of elements from a regular and dual vector space, resulting in a cancellation of both indeterminants.

$$\mathbf{x}_{{}^*_mA_i^s}\mathbf{x}^{{}_mA_i^s} \equiv [{}^*_mA_0^s, {}^*_mA_1^s, \ldots, {}^*_mA_n^s] \begin{bmatrix} {}_mA_0^s \\ {}_mA_1^s \\ \vdots \\ {}_mA_n^s \end{bmatrix} = \alpha \tag{1}$$

Since our vector spaces are projective the contraction results in the scalar α. Most often we will be concerned with algebraically exact contractions that result in the scalar 0, such a contraction is referred to as an incidence relation. Geometrically this usually corresponds to an exact point-hyperplane pair. The rules for tensor contraction are as follows;

- Contraction may only occur between common indeterminants, $\mathbf{x}_A\mathbf{y}^{AB} = \mathbf{z}^B$. Whereas is the case of $\mathbf{x}_A\mathbf{y}^{CB}$ no contraction can occur.
- Contraction occurs independent of the ordering of the indeterminants. For example $\mathbf{x}_{AB}\mathbf{y}^A = \mathbf{z}_B$ is equivalent to $\mathbf{y}^A\mathbf{x}_{BA} = \mathbf{z}_B$, this is called the Einstein summation notation.

3 Tensor Products

The basic tools used to construct the algebraic/geometric entities in the tensor notation are called operators. There are three different types of operators that we will use in this paper and for each operator we will maintain two differing representations, that of a tensor form and its equivalent vector form (Table 1). In Table 1 the symbols $\nu_n^d = \binom{d+n}{d} - 1$, $\eta_n^k = \binom{n+1}{k} - 1$ and $\pi_n^d = \prod_{i=1}^d n_i$.

The two different forms of the tensor are representative of the fact that we can always rewrite any tensor expression as an ordered vector of its coefficients. We call this alternative to the tensor form the vector form. Writing the tensor as a vector of coefficients abandons any symmetry present in the tensor this results in a less fruitful representation since it limits the way in which an equation can be contracted for symbolic derivations but in turn reduces the redundancy associated with the symmetry resulting in a more efficient representation for mappings between vector spaces.

Table 1. Tensor Operators

Operator	Symbol	Tensor Form	Vector Form
Segre	-	$\mathbf{x}^{A_i \dots B_j}$	$\mathbf{x}^{\alpha^d} \in \mathbb{P}^{\pi_n^d}$ where $\mathbf{x}^{A_i} \in \mathbb{P}^{n_A}$
Antisymmetric (Step-k)	$[\dots]$	$\mathbf{x}^{[A_i \dots B_j]}$	$\mathbf{x}^{\alpha^{[k]}} \in \mathbb{P}^{\eta_n^k}$
Symmetric (Degree-d)	$(\dots)$	$\mathbf{x}^{(A_i \dots A_j)}$	$\mathbf{x}^{\alpha^{(d)}} \in \mathbb{P}^{\nu_n^d}$

3.1 Segre Operator

The Segre or *tensor product* operator is the most familiar of operators as it is generalization of the outer product rule for multiplication from linear algebra. A simple example of the Segre product operator is in the outer product multiplication of two vectors $\mathbf{x}^A \in \mathbb{P}^n$ and $\mathbf{y}^B \in \mathbb{P}^m$ the resulting tensor form is a purely contravariant matrix.

$$\mathbf{x}^A \mathbf{y}^B \equiv \begin{bmatrix} A_0B_0 & A_0B_1 & \cdots & A_0B_m \\ A_1B_0 & A_1B_1 & \cdots & A_1B_m \\ \vdots & \vdots & \ddots & \vdots \\ A_nB_0 & A_nB_1 & \cdots & A_nB_m \end{bmatrix} \equiv \mathbf{z}^{\alpha^2} \in \mathbb{P}^{nm} \tag{2}$$

If the multiplication had occurred in the order $\mathbf{y}^B \mathbf{x}^A$ the then resulting matrix $\mathbf{z}^{BA}$ will simply be $\mathbf{z}^{AB}$ transposed. There is no analogue of the transpose from linear algebra in the tensor notation. Instead the equivalent of a transpose operation is just a shuffling of the indeterminants in the symbolic expression.

The equivalent vector form of the Segre product example given above is found by listing the elements of the resulting matrix $\mathbf{z}^{AB}$ in a vector ordered by the first indeterminant A, $\mathbf{x}^A \mathbf{y}^B \equiv \mathbf{z}^{\alpha^2} \in \mathbb{P}^{nm} = \begin{bmatrix} A_0B_0 & A_0B_1 & \cdots & A_nB_m \end{bmatrix}^\top$.

3.2 Antisymmetric Operator

The next operator of interest is the anti-symmetric operator ($[\dots]$), the number of times the operator is applied to the vector space (k) is referred to as the step. Antisymmetrization in tensor algebra can be summarized as the process of multiplying tensor spaces according to following rules;

- Antisymmetrization may only occur between projective vector spaces of equivalent dimension, ie. for $\mathbf{x}^{[A}\mathbf{y}^{B]}$ to be admissible then given $\mathbf{x}^A \in \mathbb{P}^n$ and $\mathbf{y}^B \in \mathbb{P}^m$ then $m = n$.
- Labeling the result of $\mathbf{x}^{[A}\mathbf{y}^{B]}$ as $\mathbf{z}^{[AB]}$ where $\mathbf{x}^A, \mathbf{y}^B \in \mathbb{P}^n$. Denoting the indeterminants of $\mathbb{P}^n$ as belonging to the set $\alpha_i \in \{0 \ldots n\}$ then the following rules apply to the indeterminants $A_{\alpha_i} B_{\alpha_j}$,
  1. if $\alpha_i = \alpha_j$ then $[A_{\alpha_i} B_{\alpha_j}] = 0$
  2. if $\alpha_i \neq \alpha_j$ then $[A_{\alpha_i} B_{\alpha_j}] = \frac{1}{p!}\text{sign}(\beta\alpha_i\alpha_j)A_{\alpha_i}B_{\alpha_j}$ where $\beta = \alpha/\{\alpha_i, \alpha_j\}$, which is the entire set of indeterminants modulo the ones contained in the expression (α_i, α_j), also p is the number of different ways you can reorder $A_{\alpha_i} B_{\alpha_j}$.
- The antisymmetrization of step $n+1$ or greater of elements from $\mathbb{P}^n$ will be 0 for projective vector spaces ($\mathbf{x}^{[A_0 \cdots}\mathbf{y}^{A_n \cdots}\mathbf{z}^{B_j]} = 0$).

Now we apply these rules to a simple example,

$$\mathbf{x}^{[A}\mathbf{y}^{B]} \equiv \begin{bmatrix} 0 & [A_0B_1] & [A_0B_2] \\ [A_1B_0] & 0 & [A_1B_2] \\ [A_2B_0] & [A_2B_1] & 0 \end{bmatrix} = \begin{bmatrix} 0 & \frac{A_0B_1}{2} & -\frac{A_0B_2}{2} \\ -\frac{A_1B_0}{2} & 0 & \frac{A_1B_2}{2} \\ \frac{A_2B_0}{2} & -\frac{A_2B_1}{2} & 0 \end{bmatrix} \equiv \mathbf{z}^{[AB]} \in \mathbb{P}^{\eta_2^2} \tag{3}$$

Analyzing the sign of the elements in the tensor in a little more detail we see that $[A_0B_1] = A_0B_1$ since (201) is an even permutation of (012) resulting in positive sign. Also $[A_0B_2] = -A_0B_2$ since (102) is an odd permutation of (012). Another point to note about the tensor given in the example above is that assuming the field for the tensor operations is commutative their are only 3 unique elements involved in its construction. These elements are repeated in an antisymmetric fashion across the main diagonal.

This is due to the fact that k anitsymmetrizations of a n-dimensional projective vector space contains $\eta_n^k + 1$ indeterminants (from Table 1). This is precisely equivalent to the number of ways that we can reorder the members of α in a *strictly increasing* manor. For example the unique elements and associated vector spaces of $\mathbb{P}^{\eta_2^2}$ and $\mathbb{P}^{\eta_3^3}$ are; $[+01, -02, +12]$ and $[-012, +013, -023, +123]$ Here we have ordered the elements of these sets in a lexicographic order.

3.3 Symmetric Operator

The next operator of interest is the symmetrization operator. The symmetrization operator allows us to create symmetric expansions of the vector space at hand, the number of symmetrizations applied is referred to the degree d of operator. The vectorized version of the symmetrization operator is known in the literature as the Veronese embedding [9]. Symmetrization may also be summarized according to the following set of rules;

- Symmetrization may only occur between projective vector spaces of equivalent dimension, ie. for $\mathbf{x}^{(A}\mathbf{y}^{B)}$ to be admisable then given $\mathbf{x}^A \in \mathbb{P}^n$ and $\mathbf{y}^B \in \mathbb{P}^m$ then $m = n$. As a matter of convention we will usually denote a symmetrization with the same indeterminant repeated ie. $\mathbf{x}^{(A}\mathbf{y}^{A)}$.

- Labeling the result of $\mathbf{x}^{(A} \cdots \mathbf{y}^{A)}$ as $\mathbf{z}^{(A \cdots A)}$ where $\mathbf{x}^A \in \mathbb{P}^n$. Denoting the indeterminants of $\mathbb{P}^n$ as belonging to the set $\alpha = \{0 \dots n\}$ then the following rule applies to the indeterminants $A_{\alpha_i} \cdots A_{\alpha_j}$,
  - for all $\alpha_i \dots \alpha_j$, $(A_{\alpha_i} \cdots A_{\alpha_j}) = \mathrm{perms}(A_{\alpha_i} \cdots A_{\alpha_j})$
- $(A_{\alpha_i} B_{\alpha_j}) = \frac{1}{p!} A_{\alpha_i} B_{\alpha_j}$ where p is the number of unique permutations possible by reordering $A_{\alpha_i} B_{\alpha_j}$.

These rules state that the symmetrization produces indeterminants that are equal for every possible reordering of the indexes, thus enabling the symmetry. Applying this to a simple example we have,

$$\mathbf{x}^{(A}\mathbf{y}^{A)} \equiv \begin{bmatrix} (A_0A_0) & (A_0A_1) & (A_0A_2) \\ (A_1A_0) & (A_1A_1) & (A_1B_2) \\ (A_2A_0) & (A_2B_1) & (A_2A_2) \end{bmatrix} = \begin{bmatrix} A_0A_0 & \frac{A_0A_1}{2} & \frac{A_0A_2}{2} \\ \frac{A_0A_1}{2} & A_1A_1 & \frac{A_1A_2}{2} \\ \frac{A_0A_2}{2} & \frac{A_1A_2}{2} & A_2A_2 \end{bmatrix} \equiv \mathbf{z}^{(AB)} \in \mathbb{P}^{\nu_2^2} \tag{4}$$

Again assuming a commutative field we have only 6 ($\nu_2^2 + 1 = 6$, Table 1) unique combinations of indeterminants in the tensor given above. This is precisely equivalent to the number of ways that we can reorder the members of α in a *strictly non-decreasing* manor. For example the unique elements of ν_2^2 and ν_1^3 are; $[00, 01, 11, 02, 12, 22]$ and $[000, 001, 011, 111]$. Again we have ordered the elements of these sets in a lexicographic order.

4 Linear Features in P^2 and P^3

The application of the tensor operators given in Table 1 to vector spaces gives us a means represent the geometry of various features we encounter in computer vision as the embedding of the (codimension 1) coefficient ring of the feature into a vector space.

4.1 Geometry of the Antisymmetrization Operator

Linear features are defined as any feature the that can be expressed in terms of strictly linear coefficients, this is to say the highest degree of the monomials composing the geometric form is 1. In order to construct the total set of linear features in a projective vector space $\mathbb{P}^n$ we use a geometric interpretation of the antisymmetric operator defined in the previous section.

This amounts to viewing the application of the antisymmetrization operator to a set of contravariant vectors as the join ($\wedge$) of two vectors (ie. $\mathbf{x}^{[A}\mathbf{y}^{B]} \equiv \mathbf{x}^A \wedge \mathbf{y}^B$). Likewise the application of the antisymmetrization operator to two or more covariant vectors is the meet ($\vee$) (ie. $\mathbf{x}_{[A}\mathbf{y}_{B]} \equiv \mathbf{x}_A \vee \mathbf{y}_B$). This equivalent interpretation of the antisymmetrization operator originates from the Grassmann-Cayley algebra [5, 17].

The Degrees Of Freedom (DOF) of contravariant hyperplanes is the the same as the size of the space they are embedded in, whereas covariant hyperplanes will always have 1 DOF. Generically contravariant vectors are points in a projective space and covariant vectors are lines, planes for $\mathbb{P}^2$, $\mathbb{P}^3$ respectively.

The interpretation of the join ($\wedge$) and meet ($\vee$) operators is quite literal as they denote the joining of two or more contravariant vectors (points) and the intersection of two or more covariant vectors (lines, planes, etc.).

In $\mathbb{P}^2$ or the projective plane the only linear features not including the plane itself are the point and line. Table 2 summarizes the representation and the DOF for linear features in the projective plane ($[A_0, A_1, A_2] \in \mathbb{P}^2$). Similarly, Table 3 summarizes the representation and the DOF for linear features in projective space ($[a_0, a_1, a_2, a_3] \in \mathbb{P}^3$).

Table 2. Linear features and their duals in $\mathbb{P}^2$

Feature	$\mathbb{P}^2$	$^*\mathbb{P}^2$	$\mathbf{DOF}_i$	Embedding
Points	$\mathbf{x}^{A_0}$	$\mathcal{A}_* : \mathbf{x}^{A_0} \rightarrow \epsilon_{A_0A_1A_2}\mathbf{x}^{A_0} = \mathbf{x}_{[A_1A_2]}$	2	$\mathbb{P}^2$
Lines	$\mathbf{x}^{[A_0A_1]}$	$\mathcal{A}_* : \mathbf{x}^{[A_0A_1]} \rightarrow \epsilon_{A_0A_1A_2}\mathbf{x}^{A_0A_1} = \mathbf{x}_{A_2}$	1	$\mathbb{P}^2$

Table 3. Linear features and their duals in $\mathbb{P}^3$

Feature	$\mathbb{P}^3$	$^*\mathbb{P}^3$	$\mathbf{DOF}_s$	Embedding
Points	$\mathbf{x}^{a_0}$	$\mathcal{A}_* : \mathbf{x}^{a_0} \rightarrow \epsilon_{a_0a_1a_2a_3}\mathbf{x}^{a_0} = x_{[a_1a_2a_3]}$	3	$\mathbb{P}^3$
Lines	$\mathbf{x}^{[a_0a_1]}$	$\mathcal{A}_* : \mathbf{x}^{[a_0a_1]} \rightarrow \epsilon_{a_0a_1a_2a_3}\mathbf{x}^{a_0a_1} = x_{[a_2a_3]}$	4	$\mathbb{P}^5$
Planes	$\mathbf{x}^{[a_0a_1a_2]}$	$\mathcal{A}_* : \mathbf{x}^{[a_0a_1a_2]} \rightarrow \epsilon_{a_0a_1a_2a_3}\mathbf{x}^{a_0a_1a_2} = x_{a_3}$	1	$\mathbb{P}^3$

Tables 2 and 3 also demonstrate the process of dualization for linear feature types via the dualization mapping ($\mathcal{A}_*$) [7]. Dual representations in the antisymmetric (Grassmann-Cayley) algebra are equivalent covariant forms of the same geometric object interchanging the position and structure indeterminants using an alternating contraction ($\epsilon_{\beta_0\ldots\beta_n}$). In addition to this the dualization function ($\mathcal{A}_*$) in the antisymmetric algebra is commutative.

4.2 Lines in P^3

The only feature included in the prior discussion that cannot be classified as a hyperplane is the variety of the line embedded in $\mathbb{P}^3$. From Table 3 we denote the line joining two points as $\mathbf{x}^{[a_0}\mathbf{y}^{a_1]} \simeq \mathbf{l}^{[a_0a_1]} \equiv \mathbf{l}^\omega \in \mathbb{P}^5$ or dually as the intersection of two planes $\mathbf{x}_{[a_0}\mathbf{y}_{a_1]} \simeq \mathbf{l}_{[a_0a_1]} \equiv \mathbf{l}_\omega \in {}^*\mathbb{P}^5$. Therefore, analyzing the 6 coefficients of the line,

$$\mathbf{l}^{[a_ia_j]} \equiv \mathbf{l}^\omega \in \{[a_0a_1], [a_0a_2], [a_1a_2], [a_0a_3], [a_1a_3], [a_2a_3]\} \tag{5}$$

we see that they are degree two monomials formed by taking the determinant of rows i and j (where $i < j$) of the the $[4\times2]$ matrix formed by the two contravariant points on the line. Similarly the dual representation of the line is formed by taking the determinant of columns i and j from the $[2\times4]$ matrix formed by the two covariant vectors depicting the planes that meet to form the line. This representation of the line is called the Plucker line.

The coefficients of the Plucker line have only 4 DOF (instead of 6). This is due to the loss of one DOF in the projective scaling and another due to a special relationship between the coefficients of the line called the quadratic Plucker relation. Writing the equations of the line in the skew-symmetric tensor form and taking the determinant of $\mathbf{l}^{[a_ia_j]}$ we find, $([a_0a_1][a_2a_3] - [a_0a_2][a_1a_3] + [a_0a_3][a_1a_2])$ which is the quadratic Plucker relation for the line. If the quadratic Plucker relation does not equal zero then the coefficient vector in $\mathbb{P}^5$ does not correspond to a line in $\mathbb{P}^3$.

5 Degree-d Features in P^2 and P^3

The next group of features we are interested in expressing in tensor algebra are curves and surfaces in $\mathbb{P}^2$ and $\mathbb{P}^3$. These features are essentially non-linear as they are composed of monomials which are of degree ≥ 2.

5.1 Hypersurfaces in P^2 and P^3

Curves in $\mathbb{P}^2$ and surfaces in $\mathbb{P}^3$ are both instances of codimension 1 hypersurfaces, which we will construct from the symmetric operator $(\ldots)$ as demonstrated in Table 4. In the Algebraic-Geometry literature hypersurfaces are referred to as the coefficient ring corresponding to the degree-d Veronese embedding of a complex vector space $\mathbf{x}^\alpha \in \mathbb{CP}^n$ [9]. This means that hypersurfaces are generically points in a $\mathbb{CP}^{\nu_n^d}$ dimensional space, where $\nu_n^d = \binom{n+d}{n} - 1$, thus they have $\nu_n^d - 1$ DOF. This results in a degree-d hypersurface $\mathbf{H}_{\alpha^{(d)}}$ (where $\mathbf{x}^\alpha \in \mathbb{P}^n$) satisfying the equation $\mathbf{H}_{\alpha^{(d)}} \mathbf{x}^{\alpha^{(d)}} = 0$, which is the incidence relation for hypersurfaces. Hypersurfaces allow us to categorize implicit curves and surfaces into different classes according to their total degree (d) of the embedding.

Table 4. Degree-d hypersurfaces and their duals in $\mathbb{P}^2$ & $\mathbb{P}^3$

Hypersurface	Regular	Dual	DOF	Embedding
$\mathbb{CP}^2$	$\mathbf{x}_{(A\ldots A)}$	$\mathcal{S}_* : \mathbf{x}_{(A\ldots A)} \to \mathbf{x}^{(A\ldots A)}$	ν_2^d	$\mathbb{CP}^{\nu_2^d}$
$\mathbb{CP}^3$	$\mathbf{x}_{(a\ldots a)}$	$\mathcal{S}_* : \mathbf{x}_{(a\ldots a)} \to \mathbf{x}^{(a\ldots a)}$	ν_3^d	$\mathbb{CP}^{\nu_3^d}$

Curves in P^2. A common example of an implicit hypersurface in $\mathbb{P}^2$ is the quadratic hypersurface (conic) We can write this as a vector of coefficients and use the incidence relation of hypersurfaces to get the typical equation for a point $\mathbf{x}^A$ on a degree 2 curve (conic).

$$\mathbf{c}_{(AA)} \mathbf{x}^A \mathbf{x}^A \equiv \mathbf{c}_{A^{(2)}} \mathbf{x}^{A^{(2)}} = 0 \tag{6}$$

All quadratic curves in the plane can be given in terms of this incidence relation by varying the coefficient vector $\mathbf{c}_{A^{(2)}}$ of the hypersurface. This concept can be used to define curves of any degree in $\mathbb{P}^2$ and likewise surfaces of any degree in $\mathbb{P}^3$. The dualization operator for symmetrically embedded hypersurfaces ($\mathcal{S}_*$) has a much more complicated action on the coefficient ring (see [6] for more details), however it simplifies in the case of degree 2 hypersurfaces to being simply the adjoint of the original symmetric matrix of coefficients (ie. $\mathcal{S}_*(\mathbf{c}_{(AA)}) \equiv \mathrm{adj}(\mathbf{c}_{(AA)}) = \mathbf{c}^{(AA)}$). Also in the degree 2 case the tangent cone at a single point $\mathbf{p}^\alpha$ to the surface or curve is identical since degree 2 hypersurfaces embedded in $\mathbb{P}^n$ (where $n \geq 2$) have no shape characteristics.

5.2 Hypersurfaces in P^5 : Chow Forms

For practical purposes we wish to have a single equation (codimension 1 hypersurface) to define the locus of a curve in space. Arthur Cayley in the first of two papers [1, 2] on the topic describes the problem like this; '*The ordinary analytical representation of a curve in space by the equations of two surfaces passing through the curve is, even in the case where the curve is the complete intersection of the two surfaces, inappropriate as involving the consideration of surfaces which are extraneous to the curve'.*

The use of the *extraneous* surfaces in the representation of the curve can be abandoned if insted we consider the curve as the intersection of a cone of Plucker lines having as its apex a variable point in $\mathbf{p}^a \in \mathbb{P}^3$. In this manor the equation of the curve depicts the intersection of every cone of Plucker lines with apex $\mathbf{p}^a$ not incident with the curve. So for this purpose we represent the equation of a curve embedded in $\mathbb{P}^3$ as the degree-d embedding of the Plucker line $\mathbf{x}^\omega \in \mathbb{P}^5$ (Table 3).

This results in a degree-d curve satisfying the equation $\mathbf{C}_{\omega^{(d)}}\mathbf{x}^{\omega^{(d)}} = 0$ this type of equation is referred to as the *Chow Form* of the curve (see [6] 1-cycle) after its more contemporary definition by Chow and van der Waerden [3]. There exists $\nu_5^d = \binom{d+5}{d}$ coefficients in the ring of the Chow Form of a degree-d curve in $\mathbb{P}^3$. Due to the redundancy of the Plucker relation for a line, the $\mathbf{DOF}$ of the Chow Form are [9, 12], $\mathbf{DOF}_{\text{cf}} = \xi_5^d = \binom{d+5}{d} - \binom{d-2+5}{d-2} - 1$.

It is important to be able to characterise the Plucker relations that lead to the ancillary constraints on the Chow Form. Cayley [1, 2] was able to show that this condition for curves in space is equivalent to the following equation,

$$\frac{\partial^2 \mathbf{C}_{\omega^{(d)}}}{\partial\omega_0\partial\omega_5} + \frac{\partial^2 \mathbf{C}_{\omega^{(d)}}}{\partial\omega_1\partial\omega_4} + \frac{\partial^2 \mathbf{C}_{\omega^{(d)}}}{\partial\omega_2\partial\omega_3} = 0 \tag{7}$$

which we can manufacture by symbolic differentiation of the Chow polynomial. This condition does suffice for $d \leq 3$ however for $d \geq 4$ further constraints constructed from higher order derivatives of the Chow form must be used.

A key property of the Chow form of the curve is its invariance to projective space transforms, this property is inherited from the underlying Plucker line representation. This makes the Chow form of the curve ideal for use in solving problems based in projective geometry (eg. those relating to projective observations) since the shape properties for all curves are invariant.

6 Projection Operators

The projection operator is an in injective projective transformation of vector space. Projection (in a geometric context) is the process projecting a feature to a lower dimensional embedding. The abstract definition of a projection is the mapping,

$$\pi_p : \mathbb{P}^n - \{p\} \rightarrow \mathbb{P}^{n-1} \tag{8}$$

which is projection from $\mathbb{P}^n$ to $\mathbb{P}^{n-1}$ from a point p where $p \in \mathbb{P}^n$ but doesn't intersect $\mathbf{X}$ where $\mathbf{X}$ is the projective variety (feature) that is being projected. Primarily we will be interested in projections of features in $\mathbb{P}^3$ to $\mathbb{P}^2$.

6.1 Embedded Projection Operators

Projection Operators for Surfaces. First we will develop the projection operator for surfaces in $\mathbb{P}^3$. In this setting the projection of a surface into the image is achievable by a projection of the dual tangent cone to the surface with the camera center as its apex ($\mathbf{S}^{a^{(d)}}$)

to the dual plane curve in the image ($\mathbf{c}^{A^{(d)}}$). This projection utilizes the vectorization of the degree d symmetric embedding of the Point-to-Point projection operator. However, since the degree of the dual of an algebraic surface is $d(d-1)^2$ in a practical setting only the projection of a degree 2 surface is reasonably attainable. Furthermore its seems to be an open question as to whether or not there exists a closed form manor of deriving the dual of surface with degree > 2. This operator has been noted and used by several authors [4, 11, 13] in the degree 2 case.

$$\mathbf{c}^{A^{(d)}} \simeq \mathbf{P}_{a^{(d)}}^{A^{(d)}} \mathbf{S}^{a^{(d)}} \tag{9}$$

Projection Operators for Curves in P^3. We now develop a novel projection operator for the Chow form of the curve in $\mathbb{P}^3$. We saw in section 5.2 that a curve in $\mathbb{P}^3$ is represented as the coefficient ring $\mathbf{c}_{\omega^{(d)}}$ where $\mathbf{x}^\omega \in \mathbb{P}^5$ is the coefficient ring of a Plucker line. Using the transpose of the Line-to-Point projection operator $\mathbf{P}_A^{[a_0 a_1]}$ [13] and rewriting it as $\mathbf{P}_A^{[a_0 a_1]} \equiv \mathbf{P}_A^\omega$, where $\mathbf{x}^\omega \in \mathbb{P}^5$ is the Plucker embedding of the line. We now have the basic linear mapping between a Plucker line passing through the camera center ($\mathbf{e}^a$) and the point at which it intersects the image plane. The vectorization of the symmetric degree d embedding of the transpose of the Line-to-Point operator gives us the projection operator,

$$\mathbf{x}_{A^{(d)}} \simeq \mathbf{P}_{A^{(d)}}^{\omega^{(d)}} \mathbf{c}_{\omega^{(d)}} \tag{10}$$

which projects the coefficients of the Chow form of the curve from $\mathbb{P}^3$ to its image a hypersurface (or curve) in $\mathbb{P}^2$. This projection operation is invariant to projective transforms of $\mathbb{P}^3$ since the underlying object in the embedding is the Plucker line which itself is invariant to projective transforms of $\mathbb{P}^3$.

7 Conclusions and Future Work

In this paper we presented the use of the symmetric and anti-symmetric tensor algebras for exemplifying the geometry of linear and non-linear figures. We showed that the anti-symmetric algebra naturally encompasses the range of linear objects in $\mathbb{P}^2$ and $\mathbb{P}^3$, also providing us with a means producing projections between these spaces.

We broadened the application of the symmetric tensor algebra to include the representation of degree d hypersurfaces, as far as the authors are aware this is first application of symmetric tensor algebra as a geometrically constructive operator in the computer vision literature. This understanding has allowed the generalization of earlier results for the projection of surfaces and also a novel operator for the projection of the Chow form of a curve in $\mathbb{P}^3$. These discoveries have resulted in a number of practical algorithms to compute location of curves and surfaces in space, as well as solving for the location of the cameras viewing known curves or surfaces, this will be a feature of future work.

References

1. A. Cayley. On a new analytical representation of curves in space. *Quart. J. of Pure and Appl. Math.*, 3, 1860.

2. A. Cayley. On a new analytical representation of curves in space ii. *Quart. J. Math.*, 5, 1862.
3. W. Chow and B. van der Waerden. Zur algebraische geometrie ix. *Math. Ann.*, 113:692–704, 1937.
4. G. Cross. *Surface Reconstruction from Image Sequences : Texture and Apparent Contour Constraints*. PhD thesis, University of Oxford, Trinity College, 2000.
5. O. D. Faugeras and B. Mourrain. On the geometry and algebra of the point and line correspondences between N images. *Int. Conf. on Computer Vision*, pages 951–956, 1995.
6. I. M. Gelfand, M. M. Kapranov, and A.V. Zelvinsky. *Discriminants, Resultants and Multidimensional Determinants*. Birkhauser, 1994.
7. W. Greub. *Multilinear Algebra*. Springer-Verlag, 1978.
8. Frank D. Grosshans. *Invariant theory and superalgebras*. Regional conference series in mathematics, no. 69. America Mathematical Society, 1987.
9. Joe Harris. *Algebraic Geometry, A first course*. Springer Verlag, first edition, 1992.
10. R. I. Hartley and A. Zisserman. *Multiple View Geometry in Computer Vision*. Cambridge University Press, 2000.
11. F. Kahl and A. Heyden. Using conic correspondences in two images to estimate the epipolar geometry. *Int. Conf. on Computer Vision*, 1998.
12. J.Y. Kaminski, M. Fryers, A. Shashua, and M. Teicher. Multiple view geometry of non-planar algebraic curves. *Int. Conf. on Computer Vision*, 2001.
13. David N. McKinnon and Brian C. Lovell. Towards closed form solutions to the multiview constraints of curves and surfaces. *DICTA03*, pages 519–528, 2003.
14. Bruce E. Sagan. *The Symmetric Group : Representations, Combinatorial Algorithms and Symmetric Functions*. Graduate Texts in Mathematics, Vol. 203. Springer, 2001.
15. B. Sturmfels. *Algorithms in Invariant Theory*. Springer-Verlag, 1993.
16. B. Triggs. The geometry of projective reconstruction i: Matching constraints and the joint image. *Int. Conf. on Computer Vision*, pages 338–343, 1995.
17. Neil White. Geometric applications of the grassmann-cayley algebra. *Handbook of Discrete and Computational Geometry*, pages 881–892, 1997.

Junction and Corner Detection Through the Extraction and Analysis of Line Segments

Christian Perwass*

Institut für Informatik und Praktische Mathematik,
Christian-Albrechts-Universität zu Kiel,
Christian-Albrechts-Platz 4, 24118 Kiel, Germany
chp@ks.informatik.uni-kiel.de

Abstract. An algorithm is presented that analyzes the edge structure in images locally, using a geometric approach. A local edge structure that can be interpreted as a corner or a junction is assumed to be representable by a set of line segments. In a first step a segmentation of the local edge structure into line segments is evaluated. This leads to a graph model of the local edge structure, which can be analyzed further using a combinatorial method. The result is a classification as corner or junction together with the absolute orientation and internal structure, like the opening angle of a corner, or the angles between the legs of a junction. Results on synthetic and real data are given.

1 Introduction

In many areas of Computer Vision the detection of feature points in an image plays an important role. For example, in the field of object recognition the use of "key features" for an object shows some promising results [12, 9]. Also tracking and 3D-reconstruction algorithms use feature points in images. These feature points are quite often corners, or more specifically, intrinsically two dimensional (i2D) structures. The advantage of i2D structures over intrinsically one dimensional (i1D) structures, i.e. edges, is that they can be identified with a specific position in an image, whereas i1D structures only allow for a localization along one direction. Note that the intrinsic dimensionality of a structure is synonymous with its co-dimension.

Many algorithms have been developed to detect corners and edges, see e.g. [7, 10, 8, 13, 6]. At a signal level, edge detectors basically locate places in an image where the image gradient is high. Different types of edges may also be distinguished by evaluating the local phase of the image signal (cf. [5]). In order to detect i2D structures, usually the image gradients within an image patch are combined in some way. One method often used is the summation of the structure tensor over an image patch. The rank of the resultant matrix then indicates

* This work has been supported by DFG Graduiertenkolleg No. 357 and by EC Grant IST-2001-3422 (VISATEC).

R. Klette and J. Žunić (Eds.): IWCIA 2004, LNCS 3322, pp. 568–582, 2004.

whether the image gradient vectors in the respective image patch span a 1D-space or a 2D-space. In the first case an edge is present in the image patch, since all gradients point (approximately) in the same direction. In the second case a corner or junction must be present, since the gradients point in two or more different directions. For example, the Förstner operator [7] and the corner detector by Harris and Stevens [10] are based on this principle.

Methods using the structure tensor in this way can only distinguish between i1D and i2D structures. A further distinction between i2D signals is not possible, since the structure tensor can at most encode two directions. Nevertheless, it would be advantageous to distinguish between i2D image structures like corners, line crossings, Y-junctions and T-junctions and also to measure the orientations of their different parts. Consequently, there has been some effort to analyze i2D structures further, see e.g. [1, 4, 11].

In this paper we propose a method to extract the type of i2D image structures and to evaluate their parameters, like the opening angle of a corner, for example. Instead of analyzing the image gradients directly, we use a two step approach. In a first step the image gradients are used to find edges. At this step an appropriate scale and smoothing has to be selected for the given image. The result of this first step is an image containing only the edges of the initial image.

In a second step the local geometry of the edges is analyzed. This analysis is again split into a number of steps. First we observe that the line structures we are interested in can be represented by a pair of conics in a useful way. Note that we do not use the form of the fitted conics directly to analyze the image structure, as for example in [15], but consider their intersections instead. From this a weighted graph representing the local edge structure can be constructed. Which particular structure is present may then be deduced from a combinatorial analysis of the graph.

The remainder of this paper is structured as follows. First we discuss the fitting of conic pairs to edge data and how this can be used to extract a graph representing the local edge geometry. The next part is dedicated to the combinatorial analysis of the extracted graph. This is followed by the presentation of experimental results and some concluding remarks.

2 Fitting Conic Pairs

As mentioned in the introduction, an image is reduced to a set of edge pixels in an initial preprocessing step. The assumption we then make is that within a local area the edge pixels can be segmented into a set of line segments. A corner, for example, consists of two line segments that meet in the local area. Even though we are looking for line segments, it is not obvious how to fit lines to the data, since it is in general not known how many line segments are present. A Y-junction, for example, has three line segments, while a corner or a T-junction only have two.

The basic idea we follow here is to perform an eigenvector analysis of the edge data in a local area. However, instead of using the data directly we first transform it to some other space, where the eigenvectors represent conics. From a geometric point of view, we try to find the conics that best fit the data. How this can be used to segment the data into line segments will be described later. First the embedding used and the eigenvector analysis are discussed.

As mentioned before, the edge geometry is evaluated locally. That is, given an edge image, we move a window of comparatively small size over it and try to analyze the local edge geometry within the window for all window positions. For each window position the pixel coordinates of the edge points are transformed to coordinates relative to the center of the window, such that the top left corner of the local area is at position $(-1, 1)$ and the bottom right corner at the position $(1, -1)$. The main reason for this transformation is to improve the numerical stability of the algorithm. Let the position vector of the ith edge point in the local area in transformed coordinates be denoted by the column vector $\mathsf{w}_i = (u_i, v_i)^\mathsf{T}$. These position vectors are embedded in a 6D-vector space of symmetric matrices, which allows us to fit conics to the set of edge points. The details of this embedding are as follows.

2.1 The Vector Space of Conics

It is well known that given a symmetric 3×3 matrix A, the set of vectors $\mathsf{x} = (x, y, 1)^\mathsf{T}$ that satisfy $\mathsf{x}^\mathsf{T}\,\mathsf{A}\,\mathsf{x} = 0$, lie on a conic. This can also be written using the scalar product of matrices, denoted here by $\cdot$, as $\left(\mathsf{x}\,\mathsf{x}^\mathsf{T}\right) \cdot \mathsf{A} = 0$. It makes therefore sense to define a vector space of symmetric matrices in the following way. If a_{ij} denotes the component of matrix A at row i and column j, we can define the transformation $\mathcal{T}$ that maps elements of $\mathbb{R}^{3\times3}$ to $\mathbb{R}^6$ as

$$\mathcal{T} : \mathsf{A} \in \mathbb{R}^{3\times3} \mapsto \left(a_{13},\, a_{23},\, \tfrac{1}{\sqrt{2}}\, a_{33},\, \tfrac{1}{\sqrt{2}}\, a_{11},\, \tfrac{1}{\sqrt{2}}\, a_{22},\, a_{12}\right)^\mathsf{T} \in \mathbb{R}^6. \tag{1}$$

A vector $\mathsf{x} \in \mathbb{R}^3$ may now be embedded in the same six dimensional space via

$$\mathbf{x} := \mathcal{T}\left(\mathsf{x}\,\mathsf{x}^\mathsf{T}\right) = \left(x,\, y,\, \tfrac{1}{\sqrt{2}},\, \tfrac{1}{\sqrt{2}}\, x^2,\, \tfrac{1}{\sqrt{2}}\, y^2,\, x\,y\right)^\mathsf{T} \in \mathbb{R}^6. \tag{2}$$

If we define $\mathbf{a} := \mathcal{T}(\mathsf{A})$, then $\mathsf{x}^\mathsf{T}\,\mathsf{A}\,\mathsf{x} = 0$ can be written as the scalar product $\mathbf{x}^\mathsf{T}\,\mathbf{a} = 0$. Finding the vector $\mathbf{a}$ that best satisfies this equation for a set of points $\{\mathbf{x}_i\}$ is usually called the algebraic estimation of a conic [2].

In the following we will denote the 6D-vector space in which 2D-conics may be represented by $\mathbb{D}^2 \equiv \mathbb{R}^6$. A 2D-vector $(x, y) \in \mathbb{R}^2$ is transformed to $\mathbb{D}^2$ by the function

$$\mathcal{D} : \quad (x, y) \in \mathbb{R}^2 \;\mapsto\; (x,\, y,\, \tfrac{1}{\sqrt{2}},\, \tfrac{1}{\sqrt{2}}\, x^2,\, \tfrac{1}{\sqrt{2}}\, y^2,\, xy) \in \mathbb{D}^2. \tag{3}$$

2.2 The Eigenvector Analysis

In order to analyze the edge data, we embed the data vectors $\{\mathsf{w}_i\}$ in the vector space of symmetric matrices as described above, i.e. $\mathbf{w}_i := \mathcal{D}(\mathsf{w}_i)$. Note that we use a different font to distinguish image vectors $\mathsf{w}_i \in \mathbb{R}^2$ and their embedding $\mathbf{w}_i \in \mathbb{D}^2$. Denote by $\mathbf{W}$ the matrix constructed from the $\{\mathbf{w}_i\}$ as

$\mathbf{W} = (\mathbf{w}_1, \ldots, \mathbf{w}_N)^\top$, where N is the number of data vectors. A conic $\mathbf{a} = \mathcal{T}(\mathsf{A})$ that minimizes $\|\mathbf{W}\,\mathbf{a}\|^2$ is then a best fit to the data in an algebraic sense. The key to our algorithm is not just to look at the best fit but at the *two* eigenvectors of $\mathbf{W}$ with the smallest eigenvalues.

We evaluate the eigenvectors and eigenvalues of $\mathbf{W}$ by performing a singular value decomposition (SVD) on $\mathbf{W}^\top \mathbf{W}$, which is symmetric. The singular vectors are then simply the eigenvectors and the square root of the singular values gives the eigenvalues of $\mathbf{W}$.

If $\mathbf{W}$ has two small eigenvalues, this means that the whole subspace spanned by the corresponding eigenvectors is a good fit to the data. In mathematical terms this can be written as follows. Throughout this text we will use $\mathbf{c}_1, \mathbf{c}_2$ to denote the two eigenvectors with smallest eigenvalues of $\mathbf{W}$. For any $\alpha, \beta \in \mathbb{R}$, $\mathbf{c} = \alpha\,\mathbf{c}_1 + \beta\,\mathbf{c}_2$ is a good fit to the data. This may also be termed a pencil of conics. The base points that define this pencil of conics are those that lie on all conics in this pencil. These points are simply the intersection points of the conics $\mathbf{c}_1$ and $\mathbf{c}_2$. It therefore seems sensible that the intersection points of $\mathbf{c}_1$ and $\mathbf{c}_2$ also contain important information about the structure of the data from which $\mathbf{W}$ was constructed. An example that this is indeed the case can be seen in figure 1. The dots in this figure represent the data points and the two hyperbolas are the conics represented by the two eigenvectors of the corresponding $\mathbf{W}$ matrix with the smallest eigenvalues. It can immediately be seen that each conic by itself does not represent the data distribution. However, their intersection points lie exactly in the clusters formed by the data.

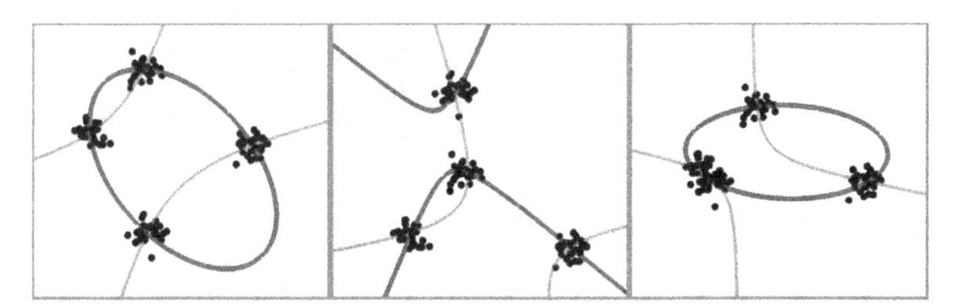

Fig. 1. Examples of four and three clusters of data points and the conics represented by the two eigenvectors with the smallest eigenvalues

By intersecting conics $\mathbf{c}_1$ and $\mathbf{c}_2$, we basically try to represent the data in terms of up to four points. In effect, this is not much different from a principal component analysis (PCA). The space of intersections of eigenvectors of $\mathbf{W}$ may actually be expressed as the vector space of bivectors in a Clifford algebra over the vector space of symmetric matrices. In this sense, the space of intersections may be regarded as a kind of "second order" null space of $\mathbf{W}$. See [14] for more details.

2.3 Analyzing Image Data

The type of data that we want to analyze with the above described method, are sets of a few line segments, like those shown in figure 2. Images 1 and 2 show Y-junctions. Note that the fitting of conics to the data also works for gray scale edge images and thick lines as shown in image 2. Image 3 shows a T-junction, images 4 and 5 different types of corners, image 6 a line crossing and images 7 and 8 show lines. A standard PCA approach on 2D position vectors will not be of any use in this case, since this would not allow us to distinguish between differently oriented line segments in the same data subspace. Instead we observe that the intersections of conics $\mathbf{c}_1$ and $\mathbf{c}_2$, represent the data in a very useful way, which can be seen in figure 2, where the two conics are drawn on top of the image structures. The images show that the intersection points of $\mathbf{c}_1$ and $\mathbf{c}_2$ lie on the line segments, and that the line segments always lie approximately between two intersection points. Unfortunately, we cannot give an analytic proof that this always has to be the case. However, we can give a motivation for this behavior.

First of all consider the case where only two line segments are present as in corners, crossings and T-junctions. By fitting projective conics to the data, line pairs can be represented well, since they are simply degenerate conics. Hence, the best fitting projective conic will approximate the actual structure in the image. The next best fitting conic is orthogonal to the first and also has to pass somehow through the data, since it is still a fairly good fit. Therefore, the two conics have to intersect on the line segments.

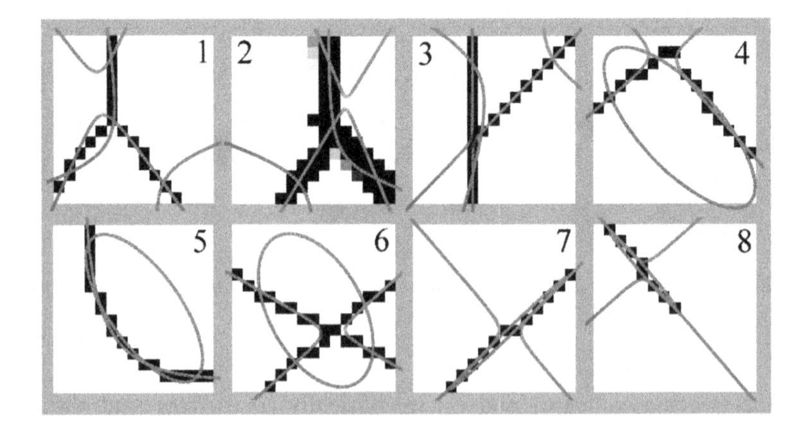

Fig. 2. Examples of typical image structures

The more complex case is the one where three different line segments meet in a single point, as is the case for Y-junctions. In this case one pair of line segments can be represented by one branch of one conic, and the last line segment by one branch of the other conic. Hence, the two conics again have to meet on or near the line segments.

In the following we will denote the set of intersection points of $\mathbf{c}_1$ and $\mathbf{c}_2$ in $\mathbb{R}^2$ as $\mathbb{S}_E \subset \mathbb{R}^2$. We use the subscript E for $\mathbb{S}$, since $\mathbb{S}_E$ contains the intersection points in Euclidean space $\mathbb{R}^2$. If $|\mathbb{S}_E| = 4$, that is, $\mathbf{c}_1$ and $\mathbf{c}_2$ intersect in four points, then there are six unique point pairs between which lines could occur. Typically, only a few of these lines are actually present in the image, though. Therefore, we are not finished once we have found $\mathbb{S}_E$. We also have to check which of the possible six lines have support in the image. Once we have identified such a subset, the last step will be to analyze the extracted line segments and to decide which type of structure, if any, is currently present.

2.4 Intersection of Conics

Finding the intersection points of two 2D conics is not trivial. In general one has to solve a polynomial equation of degree at most four. The method we use is described in detail in [14]. In short, given two conics we find a linear combination of them that represents a degenerate conic, for example a line pair. This degenerate conic then also passes through the intersection points of the two initial conics. This allows us to evaluate the intersection points of the two conics by evaluating the intersection of the degenerate conic with one of the initial conics. This is much simpler than solving a polynomial of degree four, since it results in two polynomial equations of degree two. The only numerically sensitive operation we have to use is the evaluation of eigenvectors and eigenvalues, for which many stable numerical algorithms exist.

2.5 Finding Support for Lines

Given the set of intersection points $\mathbb{S}_E$ of two conics, the question now is which of the $\binom{|\mathbb{S}_E|}{2}$ lines, is actually present in the data, if any. The basic idea is as follows: the number of data points along a line segment should be at least as high as the separation between the two corresponding intersection points measured in pixels. Since the data points give the coordinates of edge pixels, this condition basically says that there is a closed line of pixels between two intersection points. In order to weaken this condition somewhat, we use the following mathematical approach to implement the idea.

Denote by $\mathbb{W} \subset \mathbb{R}^2$ the set of data points, i.e. the set of edge pixels in a local area. Furthermore, let $N = |\mathbb{W}|$ be the number of data points. We take as distance between a data point and a line segment the orthogonal separation of the point from the line segment, if the data point projects onto the line segment. If it does not, then the distance is taken as infinity. The latter condition implements the idea that a data point that does not project onto a line segment should not count at all towards the support of a line segment.

The support of the jth line segment is then given by

$$q_j^{sup} = \sum_{i=1}^{N} \exp\left(-\frac{1}{2}\left(\frac{d_{ij}}{\lambda\, d_{pix}}\right)^2 \right), \tag{4}$$

where $d_{ij} \in \mathbb{R}$ is the distance measure between data point i and line segment j, $d_{pix} \in \mathbb{R}$ gives the width of a pixel, and $\lambda \in \mathbb{R}$ is a scale factor. When $d_{ij} = 0$, then a data point lies directly on the line segment in question. This will then add unity to the support measure q_j^{sup}. The factor λ sets the support data points off the line segment add towards q_j^{sup}. If $d_{ij} \to \infty$, then this will add nothing to q_j^{sup}, i.e. the corresponding data point adds no support to the respective line segment.

In order to decide whether an evaluated support measure q_j^{sup} represents good or bad support for a line segment, we have to evaluate the support that could ideally be expected for the line segment. Ideal support for a line segment means, that the maximum number of pixels possible along the line segment were present. If this is the case, the value of q_j^{sup} will be just this number of pixels. Since we only count those data points that appear between the end points of the line segment, the value q_j^{exp} we should expect for q_j^{sup} can be evaluated as

$$q_j^{exp} := \frac{1}{d_{pix}} \max\left\{ |r_j^1|, |r_j^2| \right\} - 1, \tag{5}$$

where $\mathsf{r}_j := (r_j^1, r_j^2)$ is the direction vector of the jth line segment.

If $q_j^{sup} \geq q_j^{exp}$ we can be sure that the jth line segment has good support in the image. If, however, $q_j^{sup} < q_j^{exp}$ we should give the respective line segment a lower confidence value. The final quality measure for a line segment is therefore evaluated as

$$q_j := \begin{cases} \exp\left(-\frac{1}{2} \left(\frac{q_j^{sup} - q_j^{exp}}{\tau\, q_j^{exp}} \right)^2 \right) & : \ q_j^{sup} < q_j^{exp} \\ 1 & : \ q_j^{sup} \geq q_j^{exp} \end{cases}, \tag{6}$$

where $\tau \in \mathbb{R}$ gives a measure of how close q_j^{sup} has to be to q_j^{exp} in order for it to give a high q_j value.

Every $q_j \in [0, 1]$ gives a measure of support for a line segment. The closer the value of q_j to unity, the more likely it is that the respective line segment is also present in the local image area under inspection. Which particular structure is present in the local image area depends on the combination of line segments with good support. It is therefore useful to collect the separate support measures in a support matrix. Let us denote the support value of the line segment between intersection points $\mathsf{s}_i \in \mathbb{S}_E$ and $\mathsf{s}_j \in \mathbb{S}_E$ by $q_{i,j} = q_{j,i}$. The support matrix Q is now defined as

$$\mathsf{Q} := \begin{pmatrix} 0 & q_{1,2} & q_{1,3} & q_{1,4} \\ q_{2,1} & 0 & q_{2,3} & q_{2,4} \\ q_{3,1} & q_{3,2} & 0 & q_{3,4} \\ q_{4,1} & q_{4,2} & q_{4,3} & 0 \end{pmatrix} \tag{7}$$

We can also regard Q as a weight matrix, giving the weights of the edges of a fully connected graph with four vertices. Note that if less than four conic intersection points are found, Q is reduced accordingly.

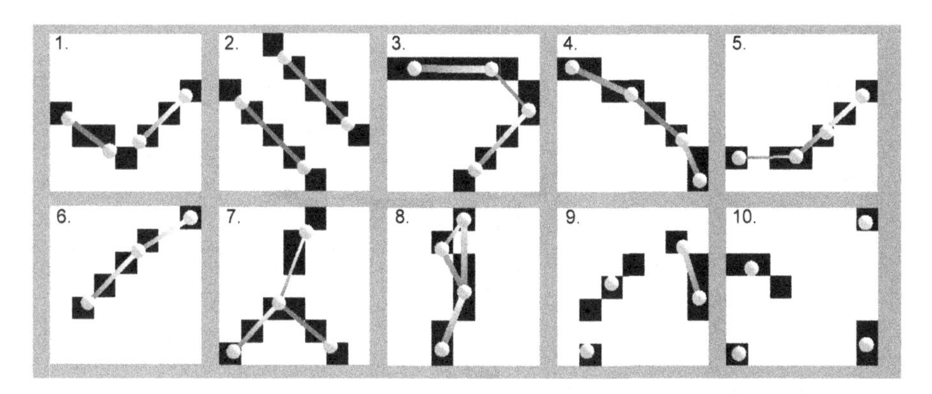

Fig. 3. Examples of analyzed image structures

3 Analyzing the Line Segments

After the support for the set of possible line segments has been evaluated, we still have to analyze the set of lines and decide on the type of image structure that is present. Figure 3 shows a set of typical structures that are encountered. In this collection of images the round points represent the conic intersection points found and the lines drawn show those lines for which sufficient support was found. The thicker a line, the higher its support value as evaluated in equation (6).

Images 1 and 2 of figure 3 show line pairs. The structures in images 3, 4 and 5 will be called 4-chains. The structure in image 6 is called a 3-chain and image 7 shows a star. The remaining images show spurious structures. An example not shown here is that of a line. When a line is the only element in the local area that is analyzed, the four conic intersection points also lie almost on that line. In the following we will neglect such structures and concentrate on the detection of corners and junctions. The structures we will interpret are thus a line pair, a 4-chain, a 3-chain and a star.

Given a set of intersection points and line segments, the next step is to test the line segment structure for one of the different patterns described above. We will describe the method used with the help of an example. Figure 4 shows the intersection points and the line segments with their respective weights found for an image structure. Let Q denote the support quality matrix for this structure as defined in equation (7). In this case, the values $q_{1,2}$, $q_{1,3}$ and $q_{1,4}$ are close to unity and the values $q_{2,3}$, $q_{3,4}$ and $q_{4,2}$ are close to zero. We can therefore evaluate a measure of confidence that the present structure is a star as

$$C = \big(q_{1,2}\, q_{1,3}\, q_{1,4}\big) \big(1 - q_{2,3}\, q_{3,4}\, q_{4,2}\big) \tag{8}$$

That is, we have to test for a positive and a negative pattern. Since the numbering of the intersection points is arbitrary, the above measure will in general have to be evaluated for all permutations of $\{1, 2, 3, 4\}$. In order to formulate this mathematically, let us denote by $\mathbf{i}$ an index vector defined as $\mathbf{i} := (i_1, i_2, i_3, i_4)$.

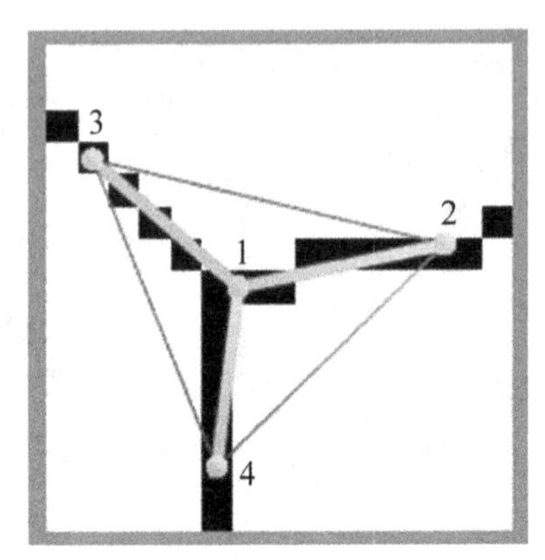

Fig. 4. Example of a junction structure

We can then define a positive (p^+) and a negative (p^-) pattern that we expect for a particular structure. In the case of the star structure, these patterns are

$$\begin{aligned} p^+(\mathbf{i}) &= \Big((i_1, i_2),\, (i_1, i_3),\, (i_1, i_4)\Big), \\ p^-(\mathbf{i}) &= \Big((i_2, i_3),\, (i_3, i_4),\, (i_2, i_4)\Big). \end{aligned} \tag{9}$$

In the following let p_k^+ denote the kth index pair of p^+, and analogously for p^-. In order to improve the readability of the following formulas, we will also write $\mathsf{Q}[i_1,\, i_2]$ to denote the element q_{i_1,i_2}. The confidence value for the star pattern for a particular $\mathbf{i}$ may then be written as

$$C\Big(p^+(\mathbf{i}),\, p^-(\mathbf{i})\Big) = \Big(\prod_k \mathsf{Q}[p_k^+(\mathbf{i})]\Big)\Big(1 - \prod_l \mathsf{Q}[p_l^-(\mathbf{i})]\Big) \tag{10}$$

The permutation of $\mathbf{i}$ that gives the largest value of $C(p^+(\mathbf{i}),\, p^-(\mathbf{i}))$ then allows us to evaluate the central point of the star (i_1) and the three end points ($i_2,\, i_3,\, i_4$). We will denote this value of $\mathbf{i}$ as $\hat{\mathbf{i}}$, with

$$\hat{\mathbf{i}} = \arg \max_{\mathbf{i} \in \text{perm}\{(1,2,3,4)\}} C\Big(p^+(\mathbf{i}),\, p^-(\mathbf{i})\Big), \tag{11}$$

where $\text{perm}\{(1,2,3,4)\}$ denotes the set of index vectors of permutations of $(1,2,3,4)$.

For each structure that we would like to test for, we can define a positive (p^+) and a negative (p^-) pattern and then evaluate the confidence value $C(p^+(\hat{\mathbf{i}}),\, p^-(\hat{\mathbf{i}}))$ on a given Q matrix. In our implementation of the algorithm we test for the star, the 4-chain, the 3-chain, the 3-chain with a disjoint point and the line pair. Typical examples of these structures are shown in figure 5. Note that image 5 shows a 3-chain with a disjoint point and images 2 and 3 both show 4-chains. The latter two structures should be interpreted in different ways. While

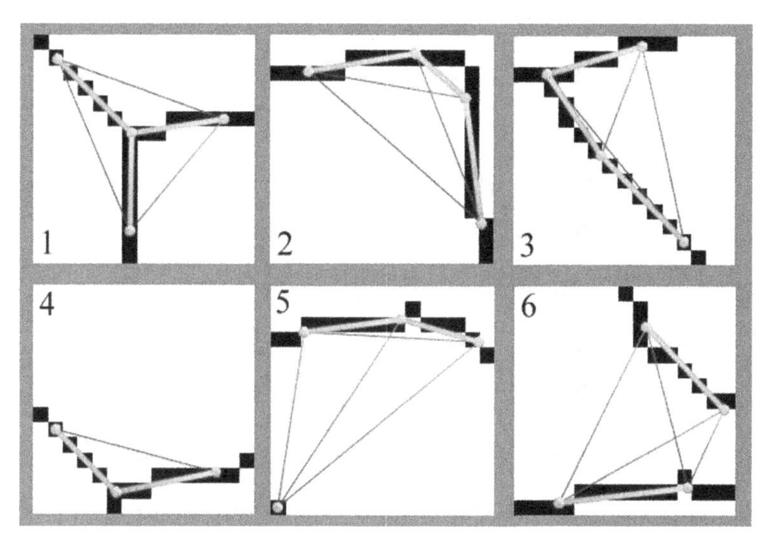

Fig. 5. Examples of structures tested for. 1. Star, 2. & 3. 4-Chain, 4. & 5. 3-Chain, 6. Line Pair

image 2 can be interpreted as a double corner, image 3 should be interpreted as a single corner. This shows that by finding the best matching structure to a local image area, we still cannot make a final decision on what the structure represents.

3.1 Analyzing Line Structures

For each of the structures we test for, we obtain a confidence value $C(p^+(\hat{\mathbf{i}}), p^-(\hat{\mathbf{i}}))$. The structure with the highest confidence value is then analyzed further to decide whether it represents a corner, a double corner or a junction. One could also test for a curve, a line or a line pair, but in this text we are mainly interested in finding corners and junctions.

The Star. The star can be interpreted immediately. Since we have $\hat{\mathbf{i}}$ we know which of the intersection points is the central point and which are the three edge points. From this the position of the junction in the image and the angles between the legs can be readily evaluated. If the angle between two legs is nearly 180 degrees one may also call the junction a T-junction and otherwise a Y-junction.

The 4-Chain. A 4-chain can represent a number of different entities: a corner, a double corner, a curve and a line. The difference between these entities cannot be defined strictly in general. Which structure is present depends on the angles between the legs of the 4-chain. Here thresholds have to be set that are most appropriate for the current application. In this text we will concentrate on distinguishing between corners and junctions. Therefore, a curve will also be interpreted as a corner with large opening angles. See figure 6 for examples of these structures.

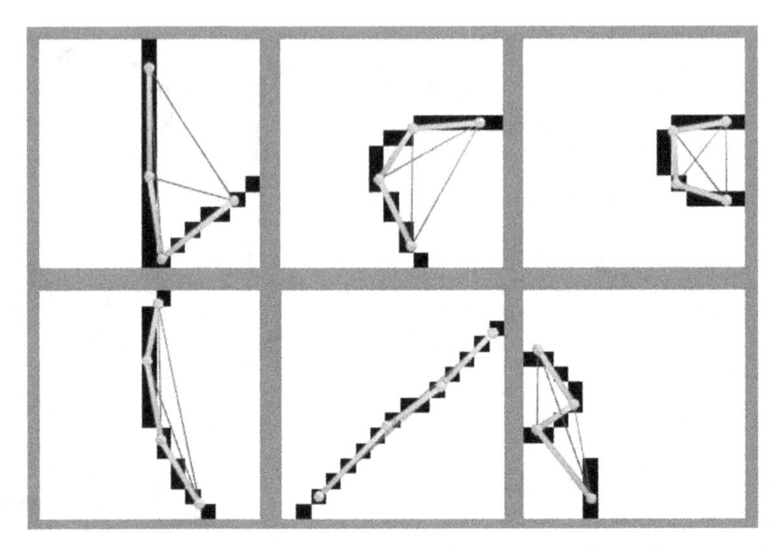

Fig. 6. Examples of entities that can be described by a 4-chain. From top-left to bottom-right: corner, double corner, double corner, curve, line, "snake"

Two angles (α_1, α_2) can be evaluated between the three legs of a 4-chain. Since we always take the smaller angle between two line segments, we have to make sure that the present 4-chain does not have a form as in the bottom-right image of figure 6, which we will call a "snake". This can be checked by evaluating the cross products of the directions of the line segment pairs from which the angles are evaluated. If the resultant vectors point in opposite directions, then the 4-chain describes a snake.

The other structures are distinguished using α_1 and α_2 as follows.

- **Line**, $\alpha_1 > 170°$ and $\alpha_2 > 170°$.
- **Double Corner**, $\alpha_1 < 150°$ and $\alpha_2 < 150°$.
- **Corner**, in all other cases. The corner is given by the two line segments with the smaller angle between them.

The 3-Chain. A 3-chain either describes a corner or a line. It usually appears if one of the intersection points of the conics lies outside the local image area under investigation and is thus neglected. A 3-chain can also appear if two intersection points are so close to each other that they are combined into a single point.

The Line Pair. A line pair either describes two disjunct lines which we will not interpret further, or a crossing of two lines. These two cases can be distinguished quite easily by evaluating the intersection point of the two lines given by the extension of the line segments. If the intersection point lies on both line segments, then we found a crossing.

3.2 Translation Invariance of Structure Analysis

Using the analysis described in the previous sections, we can obtain for each local area in an image an interpretation of the area's structure, where we distinguish between corners and junctions. For every corner we obtain its location, its opening angle and its orientation. Junctions may be separated into Y-junctions, T-junctions and crossings. For each of these we also obtain their location, orientation and angles.

The same structure found in one local area is also likely to be found in neighboring areas, whereby each time the structure has the same absolute position. This follows, since the method described here is translation and rotation invariant for one data set. In a real image, however, translating a local area will remove some edge points from the local area and others will appear. For some examples of translation invariance see [14].

Nevertheless, typically a particular corner or junction may not only be found at one particular test position. Instead, strong structures are likely to appear for a set of neighboring test positions. This offers the possibility of applying a clustering procedure to the corners and junctions found, in order to stabilize the output of the algorithm. However, this has so far not been implemented.

4 Experiments

Before the structure analysis algorithm can be applied, the edges of an image have to be extracted. This was done using the Canny edge detector [3]. The initial image and the result of the edge detection can be seen in figure 7. The algorithm was applied to this edge image, whereby a test window of 15×15 pixels was moved over the image in steps of two pixels. The factor λ from equation (4) was set to 0.1 and the factor τ from equation (6) to 0.2.

Recall that equation (11) gives a confidence value for a structure. This confidence can be used to measure the confidence we can have in a corner or junction found. The junctions found are shown in figure 8. Here the left image shows all junctions and the right image only those junctions with a confidence value of 0.90 or higher. The images in figure 9 show those corners with a confidence value of 0.99 or higher and an opening angle between 0 and 150 degrees, and 0 and 110 degrees, for the left and right image, respectively.

From the images shown here it can be seen that the algorithm finds all important corners and also gives a good measure of their opening angle. Furthermore, almost all junctions were found. Junctions that were not detected have fairly large gaps in their contour with respect to the size of the test window. Three spurious junction were found. These false positives occurred at places where the gap between two separate structures became so small that they appear locally as one structure with some missing pixels. The problem that manifests itself here is, that within a small test window, structures can only be interpreted locally. Global relationships are not taken into account which leads to false positives and false negatives.

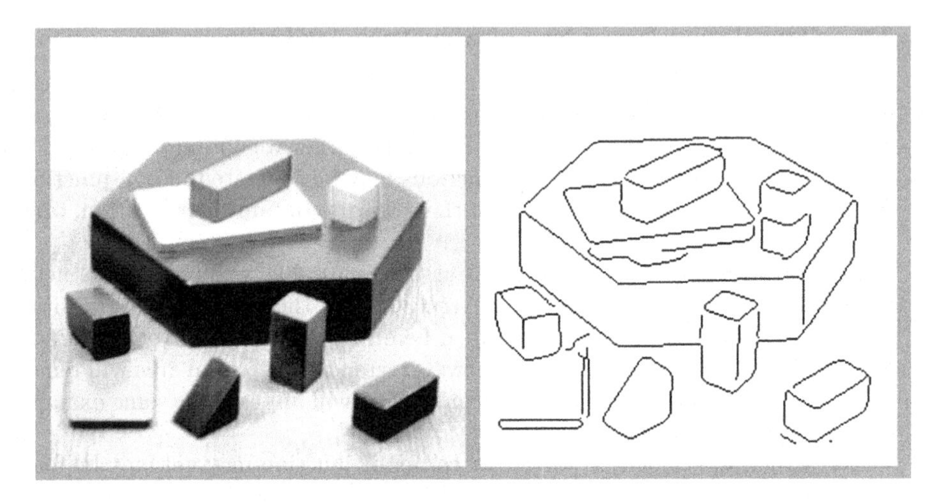

Fig. 7. Example image "blox" (left), and the extracted edges (right)

The two main problems the algorithm faces are the following:

- Corners and junctions only become apparent at a particular scale. If the scale is chosen too small, many spurious corners may be found. If it chosen too large, too much structure may be present in a test window such that the algorithm fails.
- Edges may be incomplete. If there are only small gaps in the edges, the algorithm may still give good results. However, if the gaps become too large with respect to the test window, structures will not be detected correctly. Here

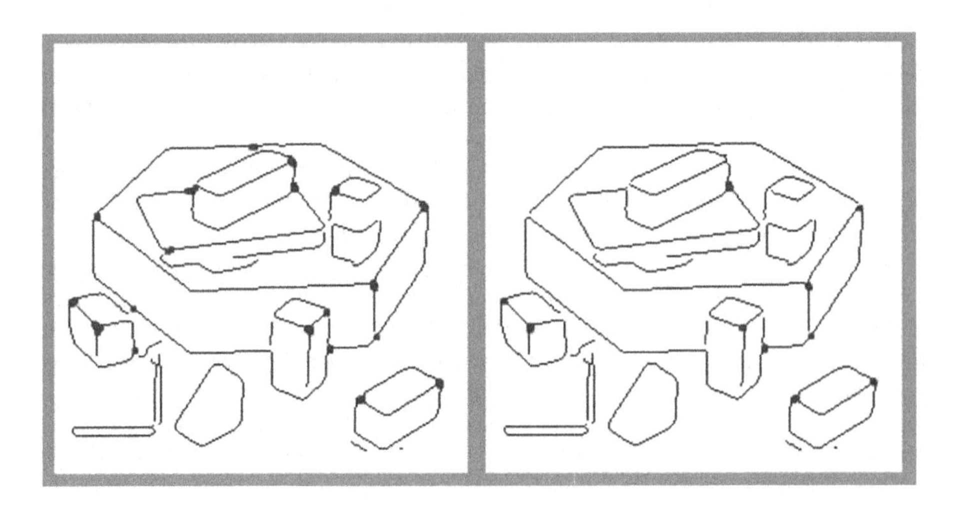

Fig. 8. Detected junctions in example image "blox", with confidence value ≥ 0.50 (left) and ≥ 0.90 (right)

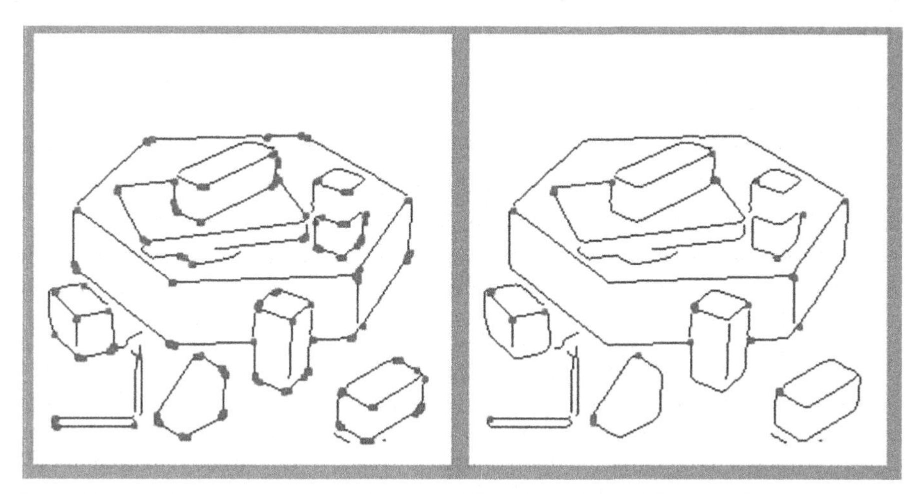

Fig. 9. Detected corners in example image "blox", with confidence value ≥ 0.90 and opening angles between 0 and 150 degrees (left), and 0 and 110 degrees (right)

the balance has to be found between bridging gaps and detecting corners and junction where there are none.

Compared with algorithms that use the gradient field directly for the corner and junction detection, the algorithm presented here does not work directly on the image data. Instead, an initial edge detection abstracts somewhat from it. This means that the edge detection algorithm has to deal with most of the noise present in an image. The type of noise the analysis algorithm then has to be able to cope with are incomplete edges. Clearly, the better the edge detection, the better the results the analysis algorithm generates. Note that more experimental results can be found in [14].

5 Conclusions

We have presented an algorithm that uses conics to analyze local image structure. The main idea is to fit intersections of conics to the data. It was found that these intersections can represent local image structures in a very useful way: the line segments that make up the local image structure lie between intersection point pairs. This basically reduces the search space of possible line segments to at most six specific ones. It was then shown that through a combinatorial analysis of the resultant graph it is possible to extract corners and junctions from an image and to evaluate their parameters.

A potential advantage of the presented algorithm over standard corner detectors is that it can distinguish between different types of i2D structures, like corners and junctions. Furthermore, it can be used to extract parameters of the local image structures, like the opening angle of a corner.

References

1. S. Baker, S. K. Nayar, and H. Murase. Parametric feature detection. *IJCV*, 27(1):27–50, 1998.
2. F.L. Bookstein. Fitting conic sections to scattered data. *Comp. Graph. Image Proc.*, 9:56–71, 1979.
3. J. Canny. A computational approach to edge detection. *IEEE Transactions on Pattern Analysis and Machine Intelligence*, 8(6), November 1986.
4. M.A. Cazorla, F. Escolano, R. Rizo, and D. Gallardo. Bayesian models for finding and grouping junctions. In *Second International Workshop on Energy Minimization Methods in Computer Vision and Pattern Recognition*, 1999. York.
5. M. Felsberg and G. Sommer. The monogenic signal. *IEEE Transactions on Signal Processing*, 49(12):3136–3144, December 2001.
6. M. Felsberg and G. Sommer. Image features based on a new approach to 2D rotation invariant quadrature filters. In A. Heyden, G. Sparr, M. Nielsen, and P. Johansen, editors, *Computer Vision, ECCV02, Kopenhagen, 2002*, volume 2350 of *LNCS*, pages 369–383. Springer, 2002.
7. W. Förstner. A feature based correspondence algorithm for image matching. *Intl. Arch. of Photogrammetry and Remote Sensing*, 26:150–166, 1986.
8. W. Förstner. A framework for low level feature extraction. In J. O. Eklundh, editor, *Computer Vision - ECCV'94*, volume 2 of *LNCS 801*, pages 383–394. Springer-Verlag, 1994.
9. Gösta H. Granlund and Anders Moe. Unrestricted recognition of 3-D objects using multi-level triplet invariants. In *Proceedings of the Cognitive Vision Workshop*, Zürich, Switzerland, September 2002. URL: http://www.vision.ethz.ch/cogvis02/.
10. C. G. Harris and M. J. Stevens. A combined corner and edge detector. In *Proc. of 4th Alvey Vision Conference*, 1988.
11. U. Köthe. Edge and junction detection with an improved structure tensor. In B. Michaelis and G. Krell, editors, *Pattern Recognition*, LNCS 2781, pages 25–32. Springer-Verlag, 2003.
12. D. G. Lowe. Local feature view clustering for 3d object recognition. In *IEEE Conference on Computer Vision and Pattern Recognition*, pages 682–688, 2001.
13. F. Mokhtarian and R. Suomela. Curvature scale space for robust image corner detection. In *Proc. International Conference on Pattern Recognition*, pages 1819–1821, 1998.
14. C. Perwass. Analysis of local image structure using intersections of conics. Technical Report Number 0403, Christian-Albrechts-Universität zu Kiel, Institut für Informatik und Praktische Mathematik, July 2004.
15. M. Shpitalni and H. Lipson. Classification of sketch strokes and corner detection using conic sections and adaptive clustering. *Trans. of ASME J. of Mechanical Design*, 119(2):131–135, 1997.

Geometric Algebra for Pose Estimation and Surface Morphing in Human Motion Estimation

Bodo Rosenhahn and Reinhard Klette

University of Auckland (CITR),
Computer Science Department,
Private Bag 92019 Auckland, New Zealand
{bros028, r.klette}@cs.auckland.ac.nz

Abstract. We exploit properties of geometric algebras (GAs) to model the 2D-3D pose estimation problem for free-form surfaces which are coupled with kinematic chains. We further describe local and global surface morphing approaches with GA and combine them with the 2D-3D pose estimation problem. As an application of the presented approach, human motion estimation is considered. The estimated joint angles are used to deform surface patches to gain more realistic human models and therefore more accurate pose estimation results.

1 Introduction

A geometric algebra is a Clifford algebra with a specific geometric interpretation. The term geometric algebra was introduced by D. Hestenes, who applied Clifford Algebras on classical geometry and mechanics in the early 1960's [13]. Due to its properties, geometric algebra unifies mathematical systems which are of interest for computer graphics and computer vision. Examples of such systems are quaternions, dual-quaternions, Lie algebras, Lie groups, screw geometry in Euclidean, affine, projective or conformal geometry. In this contribution we show the applicability of conformal geometric algebra (CGA) [12, 15, 18] for solving the 2D-3D pose estimation problem. We use the example of human motion modeling and estimation [1, 4, 6, 9, 10] to show how it is possible to apply a unified approach to extend a basic scenario to a complex application by exploiting properties of CGA.

Pose estimation is a common task in computer vision. For a definition of the pose problem, we quote [11]: *By pose we mean the transformation needed to map an object model from its inherent coordinate system into agreement with the sensory data.* We deal with the *2D-3D pose estimation problem*: we assume an image of an object captured by a calibrated camera. Additionally to these 2D sensory data we also assume that a 3D representation of an object model is given. 2D-3D pose estimation means to specify a rigid motion (containing both 3D rotation and 3D translation) which fits the object model data with the 2D sensory data. The problem of 2D-3D pose estimation can be tackled from different points of view such as geometric or numerical perspectives. In the literature,

R. Klette and J. Žunić (Eds.): IWCIA 2004, LNCS 3322, pp. 583–596, 2004.

Euclidean, projective and conformal approaches can be found in combination with Kalman-filter, SVD, Newton-Raphson or gradient descent approaches. It is further crucial how objects are represented. The literature deals with point and line based representations, kinematic chains, higher order curves/surfaces, up to free-form contours or free-form surfaces. See [19] for an overview.

Geometric algebras can handle different object representations due to their multi-vector concepts and they allow to transform entities (rotation, translation, screw motion, reflection, refraction, etc.) with the help of the geometric product. It is therefore a useful tool to model geometric and numerical aspects in a unified language. For an introduction to geometric algebras, the reader is referred to [7, 8, 13, 18, 19, 21]. A brief list of homepages and research projects on Clifford algebras (with further links) can be found in [14].

For complex tasks, such as human motion estimation, a representation of the human body as a simple joint and skeleton model can be inadequate. Indeed the coupling of joints within a 2-parametric surface model gives a good initial guess about human motion (see Figure 1), but the human anatomy allows for much more degrees of freedom than, for example, three revolute joints for the shoulder, two for the elbow and two for the wrist. The human being is able to move the shoulder backward and forward, is able to raise and lower the shoulders to certain degrees, and if such additional degrees of freedom are not modeled, the pose results can become inaccurate or worthless. There is a need for an interaction of computer vision and computer graphics: a realistic model is needed to achieve accurate pose estimations, and, vice versa, accurate pose estimations help to refine a realistic model. In this contribution we show how to use geometric algebra for adding surface morphing and joint deformation approaches into surface modeling of a human being. We further show how to use this more complex, but more realistic model in the theory of CGA for pose estimation. In this contribution we are not dealing with the problem of recognizing or identifying

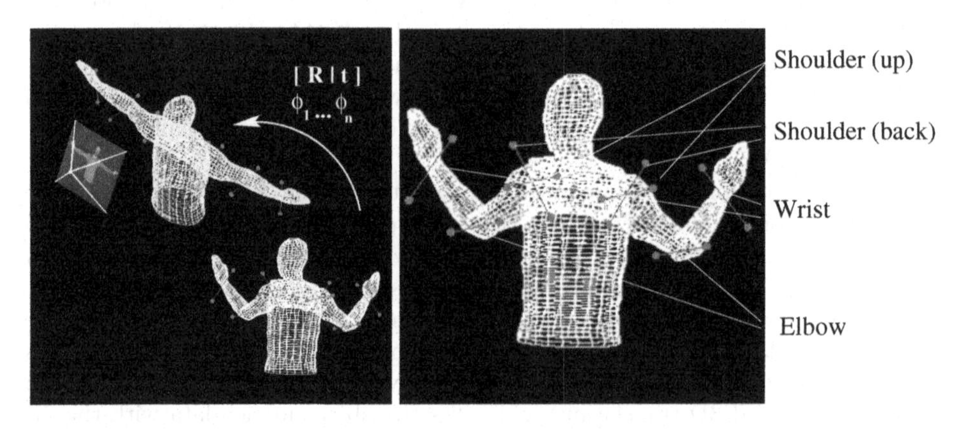

Fig. 1. Left: The pose scenario: the aim is to estimate the pose $\boldsymbol{R}$, $\boldsymbol{t}$ and the joint angles ϕ_i. Right: The names of the used joints

a moving human. Instead we are dealing with the estimation of object-specific parameters, like the pose and the joint angles.

We start this contribution with an introduction to silhouette based 2D-3D pose estimation of free-form contours and free-form surfaces. We want to quote Besl [3] for a definition: *A free-form surface has a well defined surface that is continuous almost everywhere except at vertices, edges and cusps.* Since we already model the pose problem and surface representation in CGA, we will introduce surface morphing in CGA in Section 3, so that we can directly use morphing concepts in the pose scenario without changing our main algorithms. Section 4 presents some experimental results and Section 5 ends with a brief discussion.

2 Preliminary Work

Clifford or geometric algebras [21] can be used to deal with geometric aspects of the pose problem. We only list a few properties which are important for our studies. The elements in geometric algebras are called multivectors which can be multiplied by using a geometric product. It allows a coordinate-free and symbolic representation. We use conformal geometric algebra (CGA) for modeling the pose problem. The CGA is build up on a conformal model which is coupled with a homogeneous model to deal with kinematics and projective geometry simultaneously. In conclusion, we deal with the Euclidean, kinematic and projective space in a uniform framework and can therefore cope with the pose problem within one theory. In the equations we will use the inner product $\cdot$, the outer product $\wedge$, the commutator $\underline{\times}$, and anticommutator $\overline{\times}$ product, which can be derived from the geometric product. Though we will also present equations formulated in conformal geometric algebra, we only explain these symbolically and want to refer to [19] for more detailed information.

2.1 Point Based Pose Estimation

For 2D-3D point based pose estimation we use constraint equations which compare 2D image points with 3D object points. Assume an image point $\boldsymbol{x}$ and the optical center $\boldsymbol{O}$. These define a 3D projection ray $\underline{\boldsymbol{L}}_x = \mathbf{e} \wedge (\boldsymbol{O} \wedge \boldsymbol{x})$ as a Plücker line [17]. The motor $\boldsymbol{M}$ is defined as exponential of a twist Ψ, $\boldsymbol{M} = \exp(-\frac{\theta}{2}\Psi)$, and formalizes the unknown rigid motion as a screw motion [17]. The motor $\boldsymbol{M}$ is applied on an object point $\underline{\boldsymbol{X}}$ as versor product, $\underline{\boldsymbol{X}}' = \boldsymbol{M}\underline{\boldsymbol{X}}\widetilde{\boldsymbol{M}}$, where $\widetilde{\boldsymbol{M}}$ represents the reverse of $\boldsymbol{M}$. Then the rigidly transformed object point $\underline{\boldsymbol{X}}'$ is compared with the reconstructed line $\underline{\boldsymbol{L}}_x$ by computing the error vector between the point and the line. This specifies a constraint equation in geometric algebra:

$$(\boldsymbol{M}\underline{\boldsymbol{X}}\widetilde{\boldsymbol{M}}) \underline{\times} (\mathbf{e} \wedge (\boldsymbol{O} \wedge \boldsymbol{x})) = 0$$

Note that we deal with a 3D formalization of the pose problem. The constraint equations can be solved by linearization (i.e., solving the equations for the twist-parameters which generate the screw motion) and by applying the Rodrigues formula for a reconstruction of the group action [17]. Iteration leads to a gradient descent method in 3D space. This is presented in [19] in more detail where similar equations have been introduced to compare 3D points with 2D lines (3D planes), and 3D lines with 2D lines (3D planes).

Joints along a kinematic chain can be modeled as special screws with no pitch. In [19] we have shown that the twist then corresponds to a scaled Plücker line $\Psi = \theta \underline{\boldsymbol{L}}$ in 3D space, which gives the location of the general rotation. Because of this relation it is simple to move joints in space, and they can be transformed by a motor $\boldsymbol{M}$ in a similar way $\Psi' = \boldsymbol{M}\Psi\widetilde{\boldsymbol{M}}$ such as plain points.

2.2 Contour-Based Pose Estimation

We now model free-form contours and discuss their role for solving the pose problem. The pose estimation algorithm for surface models (as introduced in this paper) relies onto a contour based method. Therefore, a brief recapitulation of [19] on contour based pose estimation is of importance. The main idea is to interpret a 1-parametric 3D closed curve as three separate 1D signals which represent the projections of the curve along the x, y and z axis, respectively. Since the curve is assumed to be closed, the signals are periodic and can be analyzed by applying a 1D discrete Fourier transform (1D-DFT). The inverse discrete Fourier transform (1D-IDFT) enables us to reconstruct low-pass approximations of each signal. Subject to the sampling theorem, this leads to the representation of the 1-parametric 3D curve $C(\phi)$ as follows:

$$C(\phi) = \sum_{m=1}^{3} \sum_{k=-N}^{N} \boldsymbol{p}_k^m \exp\left(\frac{2\pi k \phi}{2N+1} \boldsymbol{l}_m\right)$$

The parameter m represents each dimension and the vectors $\boldsymbol{p}_k^m$ are phase vectors obtained from the 1D-DFT acting on dimension m. In this equation we have replaced the imaginary unit $i = \sqrt{-1}$ by three different rotation planes, represented by the bivectors $\boldsymbol{l}_i$, with $\boldsymbol{l_i}^2 = -1$. Using only a low-index subset of the Fourier coefficients results in a low-pass approximation of the object model which is used to regularize the pose estimation algorithm. For pose estimation, this model is then combined with a version of an ICP-algorithm [23].

2.3 Silhouette-Based Pose Estimation of Free-Form Surfaces

We assume a two-parametric surface [5] of the form

$$F(\phi_1, \phi_2) = \sum_{i=1}^{3} f^i(\phi_1, \phi_2) \mathbf{e}_i$$

with three 2D functions $f^i(\phi_1, \phi_2) : \mathbb{R}^2 \to \mathbb{R}$ acting on the different Euclidean base vectors $\mathbf{e}_i$ $(i = 1, \ldots, 3)$. A two-parametric surface allows that two independent parameters ϕ_1 and ϕ_2 are used for sampling a 2D surface in 3D space. For a discrete number of sampled points, $f^i_{n_1,n_2}$, $(n_1 \in [-N_1, N_1]; n_2 \in [-N_2, N_2];$ $N_1, N_2 \in \mathbb{N}$, $i = 1, \ldots, 3)$ on the surface, we can then interpolate the surface by using a 2D discrete Fourier transform (2D-DFT), and we apply an inverse 2D discrete Fourier transform (2D-IDFT) for each base vector separately. Subject to the sampling theorem, the surface can be written as a Fourier representation

$$F(\phi_1, \phi_2) = \sum_{i=1}^{3} \sum_{k_1=-N_1}^{N_1} \sum_{k_2=-N_2}^{N_2} \boldsymbol{p}^i_{k_1,k_2} \exp\left(\frac{2\pi k_1 \phi_1}{2N_1+1} \boldsymbol{l}_i\right) \exp\left(\frac{2\pi k_2 \phi_2}{2N_2+1} \boldsymbol{l}_i\right)$$

The complex Fourier coefficients are contained in the vectors $\boldsymbol{p}^i_{k_1,k_2}$ that lie in the plane spanned by $\boldsymbol{l}_i$. We will also call them phase vectors. These vectors can be obtained by a 2D-DFT of the sample points $f^i_{n_1,n_2}$ on the surface. We now continue with the algorithm for silhouette-based pose estimation of surface models.

We assume a properly extracted image contour of our object (i.e., in a frame of the sequence). To compare points on the image silhouette with the 3D surface model, we consider rim points on the surface (i.e., which are on an occluding boundary of the object). This means we work with the 3D silhouette of the surface model with respect to the camera. To ensure this, we project the 3D surface on a virtual image. Then the contour is calculated and from the image contour the 3D silhouette of the surface model is reconstructed. The contour model is then applied within the contour-based pose estimation algorithm. Since aspects of the surface model are changing during ICP-cycles, a new silhouette will be estimated after each cycle to deal with occlusions within the surface model. The algorithm for pose estimation of surface models is summarized in Figure 2, and it is discussed in [20] in more detail.

2.4 Human Motion Estimation

We continue with our way how to couple kinematic chains within a surface model. Then we present a pose estimation algorithm which estimates the pose and angle configurations simultaneously.

A surface is given in terms of three 2-parametric functions with respect to the parameters ϕ_1 and ϕ_2. Furthermore, we assume a set of joints J_i. By using an extra function $\mathcal{J}(\phi_1, \phi_2) \rightarrow [J_i | J_i : i\text{th joint}]$, we are able to give every node a joint list along the kinematic chain. Note that we use $[,]$ and not $\{,\}$, since

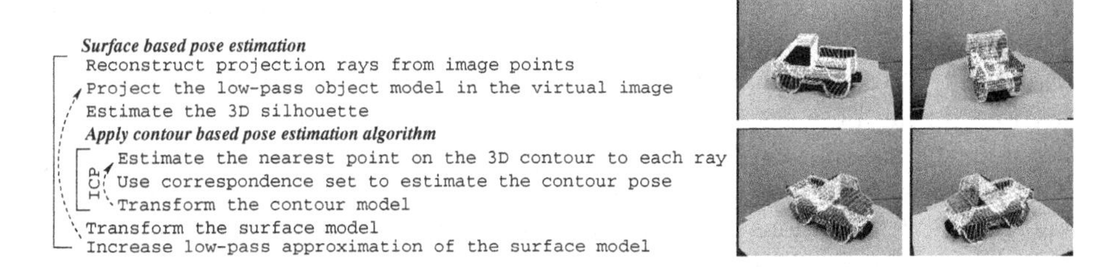

Fig. 2. Left: The algorithm for pose estimation of surface models. Right: A few example images of a tracked car model on a turn-table

the joints are given as an ordered sequence along the kinematic chain. Since the arms contain two kinematic chains (i.e., for the left and right arm, separately), we introduce a further index to separate the joints on the left arm from the ones on the right arm. The joints themselves are represented as objects in an extra field in form of a look-up table, and their parameters can be accessed immediately from the joint index numbers. Furthermore, it is possible to transform the location of the joints in space (as clarified in Section 2.1). For pose estimation of a point $\underline{\boldsymbol{X}}_{n,i_n}$ attached to the nth joint along the kinematic chain, we generate constraint equations of the form

$$(\boldsymbol{M}(\boldsymbol{M}_1 \ldots \boldsymbol{M}_n \underline{\boldsymbol{X}}_{n,i_n} \widetilde{\boldsymbol{M}_n} \ldots \widetilde{\boldsymbol{M}_1})\widetilde{\boldsymbol{M}}) \underline{\times}\, \mathbf{e} \wedge (\boldsymbol{O} \wedge \boldsymbol{x}_{n,i_n}) = 0$$

To solve a set of such constraint equations we linearize the motor $\boldsymbol{M}$ with respect to the unknown twist Ψ, and the motors $\boldsymbol{M}_i$ with respect to the unknown angles θ_i. The twists Ψ_i are known a priori.

The basic pose estimation algorithm is visualized in Figure 3. We start with simple image processing steps to gain the silhouette information of the person by using a color threshold and a Laplace operator. Then we project the surface mesh in a virtual image and estimate its 3D contour. Each point on the 3D contour carries a given joint index. Then we estimate the correspondences by using an ICP-algorithm, generate the system of equations, solve them, transform the object and its joints, and iterate this procedure. During iteration we start with a low-pass object representation and refine it by using higher frequencies. This helps to avoid local minima during iteration.

First results of the algorithm are shown on the left of Figure 4. The figure contains two pose results; on each quadrant it shows the original image and overlaid the projected 3D pose. The other two images show the estimated joint angles in a virtual environment to visualize the error between the ground truth and the estimated pose. The tracked image sequence contains 200 images. In this sequence we use just three joints on each arm and neglect the shoulder (back) joint. The right diagram of Figure 4 shows the estimated angles of the joints during the image sequence. The angles can easily be identified with the sequence. Since the movement of the body is continuous, the estimated curves are also "relatively smooth".

3 More Realistic Human Models

We are interested in using skinning approaches to model "more realistic" human motions during pose estimation. The hypothesis at this stage is that the more accurate the human being is modeled, the more accurate the pose result will be. This requires two things in a first step, namely, to model joint transformations and surface deformations during joint motions. So far, skin and muscle deformations are not yet modeled. Take, for example, the shoulder joint: If the shoulder is moving, muscles are tensed and the skin is morphing. The task is to model such deformations dependent on the joint angle. To achieve surface morphing,

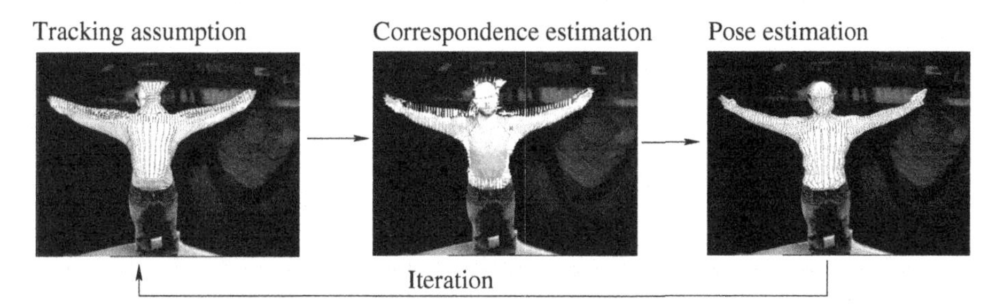

Fig. 3. The basic algorithm: Iterative correspondence and pose estimation

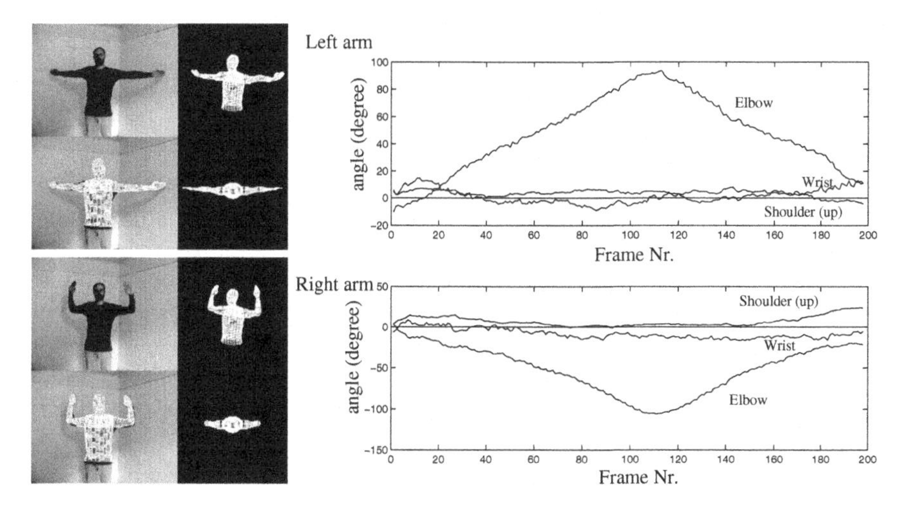

Fig. 4. Left: First pose results with a 6 DOF kinematic chain. Right: Angles of the left and right arm during the tracked image sequence

we will express two known approaches for surface morphing in CGA. These are a global approach and a local approach; the latter one uses radial basis functions.

3.1 Joint Motions

Joints along the kinematic chain can be modeled as screws with no pitch. We already have shown that its twist then corresponds to a scaled Plücker line $\Psi = \theta \underline{\boldsymbol{L}}$ in space, which gives the location of the general rotation. Because of this relation it is simple to move joints in space, and they can be transformed by a motor $\boldsymbol{M}$ in a similar manner $\Psi' = \boldsymbol{M}\Psi\widetilde{\boldsymbol{M}}$ as 3D points. To interpolate between two given joint locations Ψ and Ψ', we can use a motor $\boldsymbol{M} = \exp(-\frac{\rho}{2}\Xi)$, with the property that for $\rho = 2\pi$ it holds $\boldsymbol{M}\Psi\widetilde{\boldsymbol{M}} = \Psi'$. Then $\boldsymbol{M}_t = \exp(-\frac{t2\pi}{2}\Xi)$ for $t \in [0\ldots1]$ leads to a motor which interpolates the joint location Ψ via $\boldsymbol{M}_t\Psi\widetilde{\boldsymbol{M}}_t$ towards Ψ'.

3.2 Global Surface Interpolation

We assume two free-form surfaces given as two-parametric functions in CGA as follows:

$$F_1(\phi_1,\phi_2) = \sum_{i=1}^{3} f_1^i(\phi_1,\phi_2)\mathbf{e}_i \text{ and } F_2(\phi_1,\phi_2) = \sum_{i=1}^{3} f_2^i(\phi_1,\phi_2)\mathbf{e}_i$$

For a given parameter $t \in [0\ldots1]$, the surfaces can be linearly interpolated by evaluating

$$F_t(\phi_1,\phi_2) = \left(\sum_{i=1}^{3} f_1^i(\phi_1,\phi_2)\mathbf{e}_i\right) t + \left(\sum_{i=1}^{3} f_2^i(\phi_1,\phi_2)\mathbf{e}_i\right)(1-t)$$

We perform a linear interpolation along the nodes, and this results in the following:

$$F_t(\phi_1,\phi_2) = \begin{cases} \sum_{i=1}^{3} f_1^i(\phi_1,\phi_2)\mathbf{e}_i = F_1(\phi_1,\phi_2) \text{ , for } t=1 \\ \sum_{i=1}^{3} f_2^i(\phi_1,\phi_2)\mathbf{e}_i = F_2(\phi_1,\phi_2) \text{ , for } t=0 \end{cases}$$

Figure 5 shows examples of morphing a male into a female torso. Note that we are only morphing surfaces with known and predefined topology. This means that we have knowledge about the correspondences between the surfaces, and morphing is realized by interpolating the corresponding nodes on the mesh.

The linear interpolation can be generalized by using an arbitrary function $\omega(t)$ with the property

$$\omega(t) = \begin{cases} 0 \text{ , for } t=1 \\ 1 \text{ , for } t=0. \end{cases}$$

Then, an interpolation is still possible by using

$$F_t(\phi_1,\phi_2) = \left(\sum_{i=1}^{3} f_1^i(\phi_1,\phi_2)\mathbf{e}_i\right)\omega(t) + \left(\sum_{i=1}^{3} f_2^i(\phi_1,\phi_2)\mathbf{e}_i\right)(1-\omega(t))$$

Figure 6 shows different possible functions which result in different interpolation dynamics. and Figure 7 illustrates these different dynamics: using the square root function for weighting leads to a faster morphing at the beginning, which slows down at the end, whereas squared weighting leads to a slower start and a faster ending. Therefore, we can use non-linear weighting functions to gain a natural morphing behavior dependent on the joint dynamics.

Figure 8 shows a comparison of the non-modified model (right) with a morphed joint-transformed model (left). As it can be seen, the shoulder joint is moving down and in-wards during motion, and, simultaneously, does the surface of the shoulder part morph. The amount of morphing and joint transformation is steered through the angle of the shoulder (up) joint (left and right, respectively). As can be seen, the left motion appears more natural than the right one.

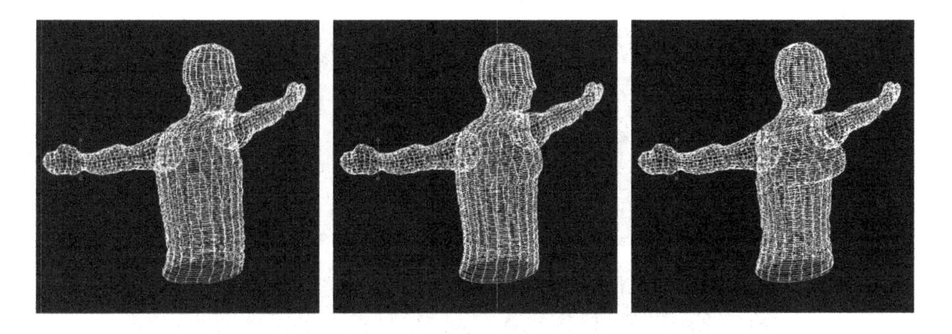

Fig. 5. Morphing of a male into a female torso

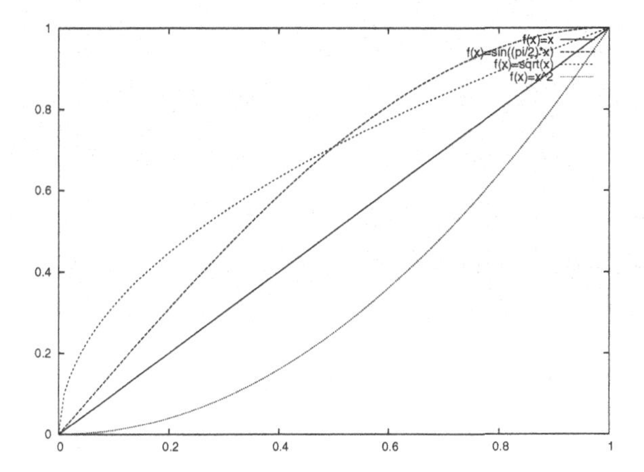

Fig. 6. Different weighting functions during interpolation

3.3 Local Surface Morphing

The use of radial basis functions for local morphing is common practice for modeling facial expressions.

The basic idea is as follows: we move a node on the surface mesh, and we move the neighboring nodes in a similar manner, but decreasingly with increasing distance to the initiating node. It is further possible to deform the radial basis function to allow a realistic morphing in the presence of bones or ligaments. The classic equation for a radial basis function is

$$r(x,y) = \exp\left(-\frac{(x-c_x)^2}{r_x}\right)\exp\left(-\frac{(y-c_y)^2}{r_y}\right)$$

with the centre (c_x, c_y) and the radius (r_x, r_y). The values $(c_x, c_y) = (0,0)$ and $(r_x, r_y) = (1,1)$ lead to the classic Gaussian form as shown on the left of Figure 9.

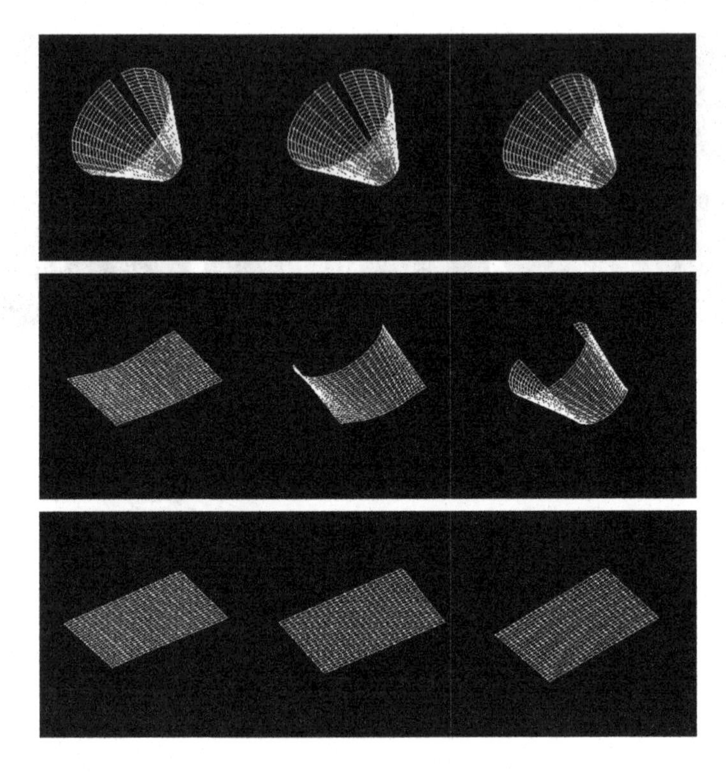

Fig. 7. Different interpolations. Left: square root, middle: linear and right: square interpolation

The coupling of a radial basis function with the surface mesh leads to

$$F_R(\phi_1, \phi_2) = \boldsymbol{T}_r \left(\sum_{i=1}^{3} f^i(\phi_1, \phi_2) \mathbf{e}_i \right) \widetilde{\boldsymbol{T}_r}$$

with

$$\boldsymbol{T}_r = 1 + \frac{\mathbf{e}\boldsymbol{t}}{2} \quad \text{for} \quad \boldsymbol{t} = \mathbf{e}_3 r(\phi_1, \phi_2)$$

$\boldsymbol{T}$ is a translator which translates a node along an orientation, and the amount of translation is steered through the value of the radial basis function. Note that $\boldsymbol{T}$ is dependent on (ϕ_1, ϕ_2) and different for each node. In this case we model a deformation along the $\mathbf{e}_3$-axes, but it can be any orientation, and the Gaussian function can be arbitrarily scaled. For our human model we additionally steer the amount of morphing through the joint angle θ_1 of the shoulder (back) joint. This means, if the shoulder is not moving forwards or backwards, we will not have any morphing, but the more the arm is moving, the larger will be the amount of morphing. With a scaling parameter λ, the morphing translator is completely given as

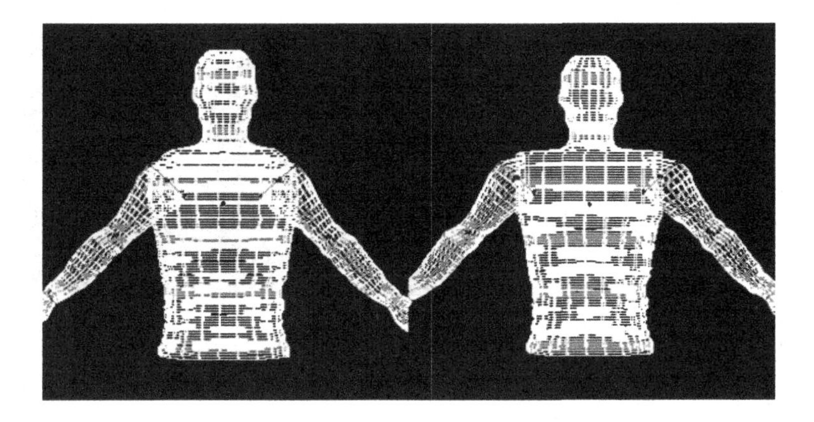

Fig. 8. Different arm positions of the morphed joint-transformed model (left) and non-modified model (right)

$$\boldsymbol{T}_r = 1 + \frac{\mathbf{e}\boldsymbol{t}}{2} \quad \text{with} \quad \boldsymbol{t} = \lambda \theta_1 r(\phi_1, \phi_2) \mathbf{e}_3$$

In contrast to global morphing, local approaches have the advantage that they can more easily be used in the context of multiple morphing patches. For example, simultaneous shoulder morphing up/down and forwards/backwards is hardly possible with a global approach, but simple with a local one.

Figure 9 shows on the left a typical radial basis function to realize local surface morphing. The images on the right show a double morphing on the shoulder: moving the arms up or down and forwards or backwards leads to a suited deformation of the shoulder patch and a similar motion of the joint locations.

4 Experiments

This section presents a few experiments using global and combined (i.e., global and local) morphing methods. The implementation is done in C++ on a standard Linux PC (2.4 GHz) and we need 100ms for each frame, including image processing, pose estimation and surface morphing.

The morphing effect during an image sequence can be seen in Figure 10. A person is moving his arms down, and as it can be seen in the left images, the shoulder is moving downwards, too. The pose result for the morphed/joint transformed model is shown in the middle image, and the result for the non-modified model is shown in the right image. As shown, the matching result of the morphed/joint transformed object model is much better.

Figure 11 shows example images for a double-morphing in the shoulder area. The shoulder is morphing downwards when the arms are moving down, and forwards when the arms are moving forward. Both motions can occur simultaneously, leading to a more realistic motion behavior then using rigid joints. Since

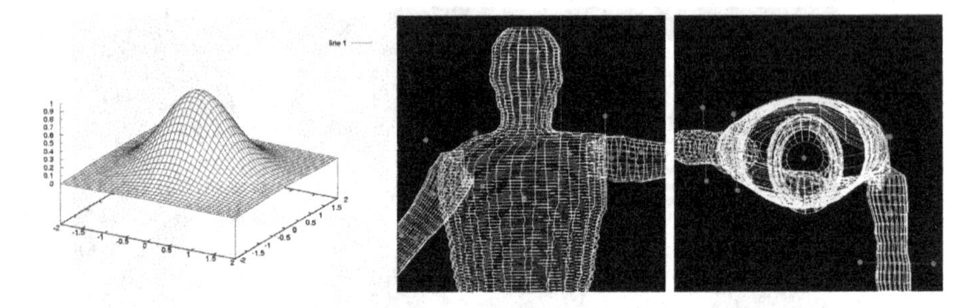

Fig. 9. Left: A 2D-radial basis function. Right: Double surface morphing on the shoulder

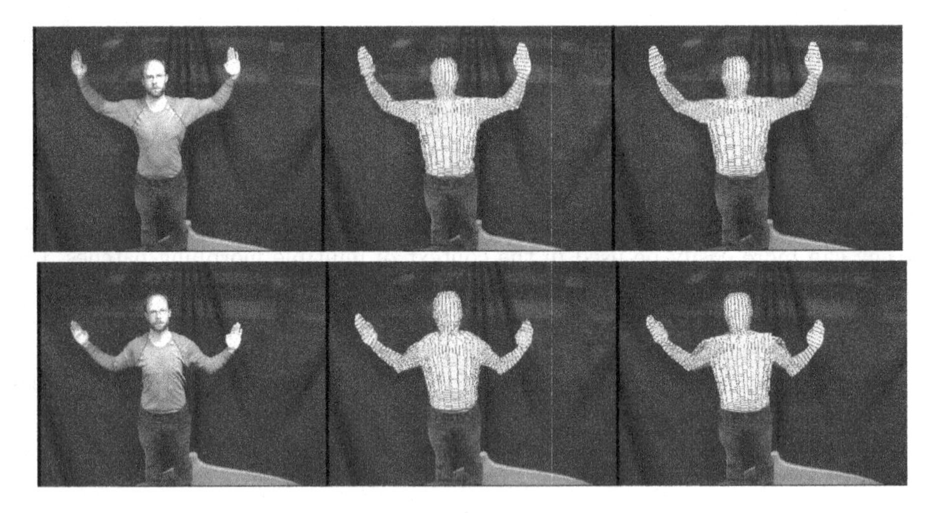

Fig. 10. Comparison of pose results for the morphed/joint transformed model and the non-modified model

Fig. 11. Pose results using two shoulder morphing operators

the morphing effect can be seen more easily during the whole sequence (in contrast to a few snap shots), the reader is invited to see the sequence at `http://www.citr.auckland.ac.nz/~bodo/DMorph.mpg`.

5 Discussion

This contribution presents an embedding of global and local morphing techniques in CGA. The motivation for this paper was to show the applicability of geometric algebras to model complex geometric problems. At first we recalled the 2D-3D pose estimation problem for free-form surface models. Then we extended a surface model by joints and used the human motion estimation problem as an example scenario for discussing CGA. Due to the complexity of human motions, we introduced local and global morphing approaches in CGA to gain a more realistic human model. The amount of deformation is steered through a related joint angle. It is further possible to deform (e.g., the shoulder patch) even with non-linear weighting functions or as coupled local and global deformation. The experiments showed the usefulness of this approach to obtain more accurate tracking results for human motions.

Acknowledgments

This work has been supported by the EC Grant IST-2001-3422 (VISATEC) and by the DFG grants RO 2497/1-1 and RO 2497/1-2.

References

1. Allen B., Curless B. and Popovic Z. Articulated body deformation from range scan data. In *Proceedings 29th Annual Conf. Computer Graphics and Interactive Techniques*, San Antonio, Texas, pp. 612 - 619, 2002.
2. Arbter K. and Burkhardt H. Ein Fourier-Verfahren zur Bestimmung von Merkmalen und Schätzung der Lageparameter ebener Raumkurven. *Informationstechnik*, Vol. 33, No. 1, pp. 19-26, 1991.
3. Besl P.J. The free-form surface matching problem. *Machine Vision for Three-Dimensional Scenes, Freemann H. (Ed.)*, pp. 25-71, Academic Press, 1990.
4. Bregler C. and Malik J. Tracking people with twists and exponential maps. *Conf. on Computer Vision and Pattern Recognition*, Santa Barbara, California, pp. 8-15, 1998.
5. Campbell R.J. and Flynn P.J. A survey of free-form object representation and recognition techniques. *Computer Vision and Image Understanding*, Vol. 81, pp. 166-210, 2001.
6. Chadwick J.E., Haumann D.R. and Parent R.E. Layered construction for deformable animated characters *Computer Graphics*, Vol. 23, No. 3, pp. 243-252, 1989.
7. Dorst L. The inner products of geometric algebra. In *Applied Geometric Algebras for Computer Science and Engineering, Dorst L., Doran C. and Lasenby J. (Eds.)*, Birkhäuser Verlag, pp. 35-46, 2001.
8. Dorst L. Honing geometric algebra for its use in the computer sciences. In [21], pp. 127-152, 2001.
9. Fua P., Plänkers R., and Thalmann D. Tracking and modeling people in video sequences. *Computer Vision and Image Understanding*, Vol. 81, No. 3, pp.285-302, March 2001.

10. Gavrilla D.M. The visual analysis of human movement: A survey. *Computer Vision and Image Understanding*, Vol. 73 No. 1, pp. 82-92, 1999.
11. Grimson W. E. L. *Object Recognition by Computer.* The MIT Press, Cambridge, MA, 1990.
12. Hestenes D., Li H. and Rockwood A. New algebraic tools for classical geometry. In [21], pp. 3-23, 2001.
13. Hestenes D. and Sobczyk G. *Clifford Algebra to Geometric Calculus.* D. Reidel Publ. Comp., Dordrecht, 1984.
14. Homepages Clifford (geometric) algebra
`http://www.ks.informatik.uni-kiel.de`
`http://www.clifford.org/`
`http://modelingnts.la.asu.edu/GC_R&D.html`
`http://www.mrao.cam.ac.uk/~clifford/`
`http://www.science.uva.nl/ga/`
`http://clifford.physik.uni-konstanz.de/~fauser/P_cl_people.shtml`
`http://www-groups.dcs.st-and.ac.uk/~history/Mathematicians/Clifford.html`
15. Li H., Hestenes D. and Rockwood A. Generalized homogeneous coordinates for computational geometry. In [21], pp. 27-52, 2001.
16. Mikic I., Trivedi M, Hunter E, and Cosman P. Human body model acquisition and tracking using voxel data *Computer Vision* , Vol. 53, Nr. 3, pp. 199–223, 2003.
17. Murray R.M., Li Z. and Sastry S.S. *A Mathematical Introduction to Robotic Manipulation.* CRC Press, 1994.
18. Perwass C. and Hildenbrand D. Aspects of Geometric Algebra in Euclidean, Projective and Conformal Space. An Introductory Tutorial. *Technical Report 0310, Christian-Albrechts-Universität zu Kiel, Institut für Informatik und Praktische Mathematik*, 2003.
19. Rosenhahn B. Pose Estimation Revisited. (PhD-Thesis) *Technical Report 0308, Christian-Albrechts-Universität zu Kiel, Institut für Informatik und Praktische Mathematik*, 2003. Available at `www.ks.informatik.uni-kiel.de`
20. Rosenhahn B., Perwass C. and Sommer G. Pose estimation of free-form surface models. In *Pattern Recognition, 25th DAGM Symposium*, B. Michaelis and G. Krell (Eds.), Springer, Berling, LNCS 2781, pp. 574-581, 2003.
21. Sommer G. (Ed.). *Geometric Computing with Clifford Algebra.* Springer, Berlin, 2001.
22. Theobalt C., Carranza J., Magnor A. and Seidel H-P. A parallel framework for silhouette based human motion capture *Proc. Vision, Modeling, Visualization 2003*, Munich, Nov. 19-21, pp. 207-214, 2003.
23. Zang Z. Iterative point matching for registration of free-form curves and surfaces. *Computer Vision*, Vol. 13, No. 2, pp. 119-152, 1999.

A Study on Supervised Classification of Remote Sensing Satellite Image by Bayesian Algorithm Using Average Fuzzy Intracluster Distance

Young-Joon Jeon, Jae-Gark Choi, and Jin-Il Kim

Department of Computer Engineering, Dongeui University,
San24, Gaya-dong, Busanjin-gu, Busan, 614-714, Korea
j4017@chol.com, {cjg, jikim}@deu.ac.kr

Abstract. This paper proposes a more effective supervised classification algorithm of remote sensing satellite image that uses the average fuzzy intracluster distance within the Bayesian algorithm. The suggested algorithm establishes the initial cluster centers by selecting training samples from each category. It executes the extended fuzzy c-means which calculates the average fuzzy intracluster distance for each cluster. The membership value is updated by the average intracluster distance and all the pixels are classified. The average intracluster distance is the average value of the distance from each data to its corresponding cluster center, and is proportional to the size and density of the cluster. The Bayesian classification algorithm is performed after obtaining the prior probability calculated by using the information of average intracluster distance of each category. While the data from the interior of the average intracluster distance is classified by fuzzy algorithm, the data from the exterior of intracluster is classified by Bayesian classification algorithm. The testing of the proposed algorithm by applying it to the multispectral remote sensing satellite image resulted in showing more accurate classification than that of the conventional maximum likelihood classification algorithm.

1 Introduction

Remote sensing is the science of gaining information about the earth's surface by analyzing date acquired from a distance. Remote sensing has proved a powerful technology for monitoring the earth's surface and atmosphere at a global, regional, and even local scale [1]. Since the first resource satellite was launched in 1972, the remote sensing community has witnessed impressive progress in remotely sensed imagery, in both quality and quantity. The greatest progress has been in the improvement of the spectral and spatial resolutions. High spectral-resolution images have hundreds of bands to monitor the earth's surface down to the molecular level. High spatial-resolution images can be used essentially in a manner similar to large-scale aerial photos. Up to the present, various methods have been developed for extracting land use/cover information from remote sensing data by performing multi-spectral classification on the electromagnetic spectrum of geometric registered remote sensing data [2][3].

R. Klette and J. Žunić (Eds.): IWCIA 2004, LNCS 3322, pp. 597–606, 2004.

There are many methodologies to apply remote sensing image classification, such as fuzzy logic, neural network theory, etc. compared with the previous statistical procedures. For instance, neural network can accomplish efficiently the classification of the non-normal distributional category (which is hard to classify under pre-existing statistical method) [4][5]. Zadeh introduced the concept of fuzzy sets in which imprecise knowledge can be used to define an event [6]. Fuzzy logic provides useful concepts and tools to deal with the uncertainty inherent in the remote sensing spectral data. In fuzzy logic methods, a membership grade matrix is applied for each pixel and proportions of component land use/cover classes in a pixel can be estimated from the membership grades. The fuzzy c-means (FCM) algorithm has been extensively used as an unsupervised clustering technique. Most analytic fuzzy clustering approaches are derived from Bezdek's fuzzy c-means (FCM) algorithm [7][8]. The FCM method creates a membership grade to belong to the cluster using a fuzzy coefficient and it also allocates pixels to the cluster. Statistical pattern recognition is the most widely used approach for image classification. Statistical classifiers are based on some quite sophisticated statistics and probability theories [1]. Bayesian classification is one of the most extensively used techniques for the classification of such data [9][10][11].

This research proposes a supervised classification algorithm integrated by fuzzy and Bayesian algorithms. The supervised classification algorithm of remote sensing satellite image uses the average intracluster distance within the fuzzy and Bayesian algorithm. Fuzzy Gustafson-Kessel (GK) algorithm [12] was used in the form extended for the FCM algorithm. By using the training data, the supervised classification algorithm establishes the initial cluster center and executes fuzzy clustering algorithm without repeating. Then the algorithm calculates the average intracluster distance. Bayesian classification method is performed after obtaining the prior probability calculated by using the information of intraclusters. Then the category is decided by comparing fuzzy membership value with likelihood rate. The proposed method is applied by Landsat TM satellite image for the verifying test.

2 Bayesian Classification Algorithm

This section is an introduction of the concept of Bayesian maximum likelihood classification algorithm. The maximum likelihood classification (MLC) algorithm is the common decision rule for supervised classification. This decision rule is based on the probability that a pixel belongs to a particular class [9][10][11]. It assumes that these probabilities are equal for all classes, and that the input bands have normal distributions. The maximum likelihood classification is performed according to the following decision rule:

$$x \in w_i, \text{ if } P(w_i \mid x) > P(w_j \mid x) \text{ for all } j \neq i \tag{1}$$

where:

x : the position vector, a column vector of brightness values for the pixel;

w_i : the spectral classes for an image, $w_i = 1, 2, \cdots, c$, where c is the total number of classes

$P(w_i \mid x)$: the conditional probabilities, which gives the likelihood that the correct class is w_i for a pixel at position x .

The pixel at x belongs to class(category) w_i , if $P(w_i \mid x)$ is the largest.

To obtain $P(w_i \mid x)$ in (1), a probability distribution $P(x \mid w_i)$ can be estimated from training data for each ground cover type. The desired $P(w_i \mid x)$ in (1) and available $P(x \mid w_i)$ are related by Bayes rule:

$$P(w_i \mid x) = \frac{P(x \mid w_i)P(w_i)}{P(x)} \tag{2}$$

where $P(x \mid w_i)$ is the class conditional probability that a feature vector x occurs in a given class w_i, $P(w_i)$ is the prior probability. The probability for any pixel to belong to the class w_i irrespective to its feature, and $P(x)$ is the unconditional probability that the pixel x occurs in the image.

Supervised classification estimates these prior probability using training samples from the data. The maximum likelihood decision rule assumes that the probability distributions for the classes are of the form of multivariate normal models, therefore,

$$P(x \mid w_i) = (2\pi)^{-\frac{N}{2}} \left|\sum{}_i\right|^{-\frac{1}{2}} \exp\left\{-\frac{1}{2}(x-m_i)^T \sum{}_i^{-1}(x-m_i)\right\} \tag{3}$$

where m_i and $\sum_i$ are the mean vector and covariance matrix of the data in class w_i.

Resulting from applying natural logarithm to (3) and mathematical simplifications the discriminant function for the Bayesian maximum likelihood classifier is :

$$D_i(x) = \ln P(w_i) - \frac{1}{2}\ln\left(\left|\sum{}_i\right|\right) - \frac{1}{2}(x-m_i)^T \sum{}_i^{-1}(x-m_i) \tag{4}$$

where:

$P(w_i)$: the probability that class w_i occurs in the image (equal for all classes, or is entered from a priori knowledge);

$\left|\sum_i\right|$: determinant of $\sum$;

$\sum_i^{-1}$: inverse of $\sum$;

ln : natural logarithm function;

T : transposition function.

The pixel is assigned to the class w_i , for which discriminant $D_i(x)$ is the largest.

3 Proposed Algorithm

This paper proposes supervised classification algorithm of remote sensing satellite image by using the average intracluster distance within the fuzzy GK and Bayesian algorithm. The suggested algorithm uses the fuzzy GK algorithm in the form extended for the FCM. Different cluster distributions and sizes usually lead to sub optimal results with FCM. In order to adopt to different structures in data, GK algorithm used the covariance matrix to capture ellipsoidal properties of cluster. It makes classification of the remote sensing satellite image with multidimensional data possible. Fuzzy algorithm generally iterates the execution until there is almost no change in membership value. However, this research does not iterate the execution due to the usage of training data as initial and central value. This algorithm classifies the input image by performing fuzzy GK algorithm once from the training data that was selected from each category by the analyst. Giving consideration to the membership value of each category, the fuzzy GK algorithm is again performed and the average intracluster distance is calculated for each category. The membership value is updated by the average intracluster distance and all the pixels are classified. The prior probability is obtained by using the information of pixels placed in the intracluster of each category, and Bayesian classification algorithm is executed. The final classification category is determined by comparing Bayesian and fuzzy algorithms.

The suggested classification method is where the analyst first decides the category for classification and then selects the training area according to each category from the input image. Training areas are selected from the image such that pixels within that geographic area are representative of the spectral response pattern of a particular information category(class). For the classification to be accurate, the training data for each information class must capture the variability within objects in the class. When a classification is run, the computer uses the training data to form decision boundaries, and then assigns unknown pixels into information classes based on these established boundaries. We establish the initial cluster center (v_i^0) of fuzzy GK algorithm by calculating average value from each selected training sample. Using the selected training sample from each category, covariance matrix is calculated.

$$F_i = \frac{\sum_{j=1}^{M} \mu_{ij}^m (X_{ijb} - V_{ib})(X_{ijb} - V_{ib})^T}{\sum_{j=1}^{N} \mu_{ij}^m} = \frac{\sum_{j=1}^{M} \mu_{ij}^m \left[\begin{pmatrix} x_{ij1} \\ x_{ij2} \\ \vdots \\ x_{ijN} \end{pmatrix} - \begin{pmatrix} v_{i1} \\ v_{i2} \\ \vdots \\ v_{iN} \end{pmatrix} \right] \bullet \left[\begin{pmatrix} x_{ij1} \\ x_{ij2} \\ \vdots \\ x_{ijN} \end{pmatrix} - \begin{pmatrix} v_{i1} \\ v_{i2} \\ \vdots \\ v_{iN} \end{pmatrix} \right]^T}{\sum_{j=1}^{N} \mu_{ij}^m} \quad (5)$$

where F_i is covariance matrix of classification category w_i, b is the number of bands, $i = \{1, \cdots, c\}$ is the category number of category(class) w_i and $V = (v_1, \cdots, v_c)$ is a vector of cluster centers. μ_{ij} establishes every pixel as 1.

The prototypes are typically selected to be idealized geometric forms such as linear varieties or points. When point prototypes are used, the general form of the distance measure is given by

$$d_{ij}^2 = (x_j - v_i)^T A_i (x_j - v_i) \tag{6}$$

where the norm matrix A_i is a positive definite symmetric matrix. The FCM algorithm uses Euclidian distance measure, while the fuzzy GK algorithm uses the Mahalanobis distance. Covariance matrix(F_i) is used when calculating Maharanobis distance. For every pixel of the input image, d_{ij}^2 is calculated.

$$d_{ij}^2 = (X_{ijb} - V_{ib})^T \left[|F_i|^{\frac{1}{b}} \cdot F_i^{-1} \right] (X_{ijb} - V_{ib}) \tag{7}$$

where d_{ij}^2 represents Maharanobis distance from the center of category w_i to pixel j. F_i^{-1} is inverse matrix of F_i. For every pixel in the input image, membership value(μ_{ij}) of each category is calculated using d_{ij}^2. Membership value of each category is updated using

$$u_{ij} = \left[\sum_{k=1}^{c} \left(\frac{d_{ij}^2}{d_{jk}^2} \right)^{\frac{2}{m-1}} \right]^{-1}, \ 1 \le k \le c; \ 1 \le j \le N \tag{8}$$

The class(category) centers is updated using

$$v_i = \frac{\sum_{j}^{N} \mu_{ij}^m \cdot x_j}{\sum_{j=1}^{N} \mu_{ij}^m} \tag{9}$$

The process of fuzzy GK algorithm is again performed from formula (5) to (9) using updated v_i and μ_{ij}. In this case, F_i uses the classified result of each category by the first process of fuzzy GK algorithm. F_i is obtained giving consideration to its position, and then Maharanobis distance d_{ij}^2 from center v_i and jth pixel of updated category is calculated. Membership values and class(category) centers is newly updated from the result of the fuzzy GK algorithm that is again performed. And then all pixels of the input image are classified by each category through the membership value μ_{ij}. Average intracluster distance is obtained for each category according to the result of classification. The average intracluster distance is the average value of all the distance from each data to its corresponding cluster center, and is proportional to the size and density of the cluster. The average fuzzy intracluster distance(η_i) is calculated using a formula (10) from PCM(possibilistic clustering algorithms) suggested by Krishanapuram [13]. Typically K is chosen to be 1.

$$\eta_i = K \frac{\sum_{j=1}^{n} \mu_{ij}^m d_{ij}^2}{\sum_{j=1}^{n} \mu_{ij}^m}, \quad K=1 \tag{10}$$

The memberships of PCM are updated as follows

$$\mu_{ij} = \frac{1}{1+\left(\frac{d_{ik}}{\eta_i}\right)^{\frac{1}{m-1}}} \tag{11}$$

The value of η_i determines the distance at which the membership value of pixel in a cluster becomes 0.5. Every pixel of the image is classified according to membership value. The group of data belonging within the interior of average intracluster distance is called intracluster. Intracluster is referred to the pixels of $\mu_{ij} \geq 0.5$.

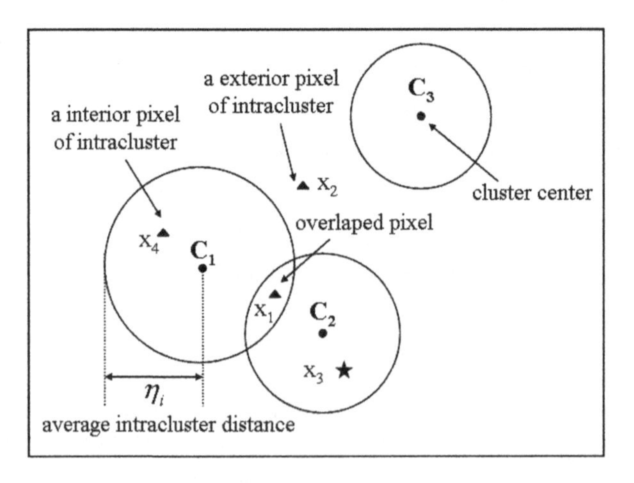

Fig. 1. The average intracluster distance of each category and the distance relationship between intracluster and pixel

Fig.1 shows the average intracluster distance of each category and the distance relationship between intracluster and pixel. X_1 is a pixel located in the interior of average intracluster distance while being overlapped with clusters C_1 and C_2. X_2 is a pixel located in the exterior of average intracluster distance of each category. Pixels X_3 and X_4 are located in the interior of average intracluster distance and are not overlapped within two clusters. X_3 and X_4 are classified into C_1 and C_2 according to the fuzzy algorithm. The categories of pixels X_1 and X_2 are determined after comparing membership value of fuzzy algorithm with the likelihood rate of Bayesian classification. That is, the pixels belonging to the overlapped area of clusters and the pixels from $\mu_{ij} < 0.5$ are determined into final classified category after comparing the result of Bayesian classification and fuzzy classification. The classification categories for X_1 and X_2 pixels are decided and allocated to the corresponding category when the results of classification by fuzzy algorithm and Bayesian algorithm belong to the same item. If the distribution result of two algorithms belongs to the different item, the data in the interior of intracluster is allocated by the distance to the center of cluster and the data in the exterior of intracluster by Bayesian algorithm.

Supervised Bayesian classification algorithm is performed after obtaining the prior probability calculated by using the information of intraclusters.

$$D_i(x) = \ln P(w_i) - \frac{n}{2}\ln 2\pi - \frac{1}{2}\ln|F_i| - \frac{1}{2}(X_{ijb} - V_{ib})^T \left[|F_i|^{\frac{1}{b}} \cdot F^{-1} \right] (X_{ijb} - V_{ib}) \quad (12)$$

$$P(w_i) = \frac{\text{number of intracluster pixels from category } w_i}{\text{total number of all intracluster pixels}}$$

F_i is covariance matrix of the pixels belonging to the intracluster of the category w_i. X_{ij} is pixels to belong to the intracluster of the category w_i. $P(w_i)$ is the a prior probability. The probability for any pixels to belong to the intracluster of the category w_i. That is, $P(w_i)$ is a ratio of intracluster pixels from each category to the intracluster pixels of all categories. Because the data of intracluster shows a normal distribution, intracluster can be used in calculating covariance matrix and maharanobis distance, both of which are then applied to Bayesian algorithm. The pixel is assigned to the category w_i, for which $D_i(x)$ is largest. The data in intracluster executes the classification by fuzzy algorithm. The data outside the average intracluster distance determines the classification category after performing Bayesian classification and comparing it with its result. Bayesian classification uses the intracluster information as prior probability.

4 Experiments and Results

The proposed classifier is applied to both areas in Seoul, Korea, from the satellite Landsat TM image and in Busan, Korea, from the satellite IKONOS image.

Landsat TM satellite image records 7 bands of multispectral data at 30m resolution. The selected bands were two, four, and seven. The four classification categories were forest, water, crop and urban areas. As for the training data, four areas were chosen from each of four classification categories from input image. This research tested if the area selected by training data of each category from the satellite image is categorized exactly into the same category after completing the classification when classifying using proposed algorithm. This experiment produced a result that was averaged after repeatedly testing for 40 times. As a result, the overall accuracy of the conventional maximum likelihood algorithm, FCM algorithm, and proposed algorithm were 92.50%, 90.22%, and 93.92% respectively. The proposed algorithm from Landsat TM satellite image showed better results than conventional maximum likelihood classification and FCM algorithm. Table 1 shows the classification results by proposed algorithm, conventional maximum likelihood classification algorithm and FCM algorithm from Landsat TM satellite image. Fig. 2 shows classification result images of maximum likelihood algorithm and proposed algorithm using Landsat TM satellite image.

Table 1. The classification results by proposed algorithm and conventional maximum likelihood and FCM algorithm from Landsat TM satellite image

Category		Forest	Water	Crop	Urban	Overall accuracy
Number of training pixels		1024	1024	1024	1024	4096
Classification methods	Maximum likelihood	90.43%	98.05%	92.48%	89.06%	92.50%
	FCM	85.15%	97.46%	85.83%	92.45%	90.22%
	Proposed method	91.25%	98.53%	93.45%	92.45%	93.92%

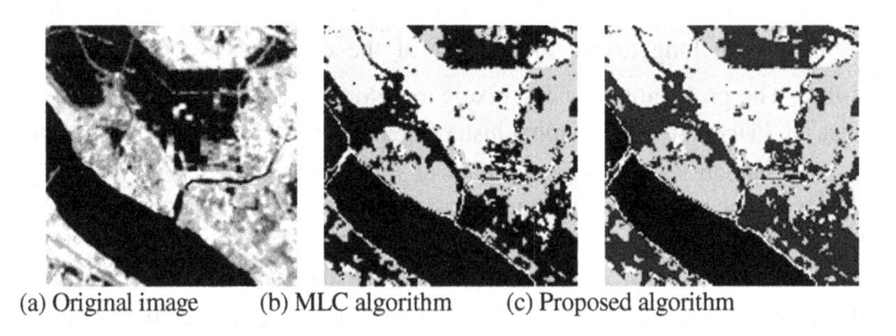

(a) Original image (b) MLC algorithm (c) Proposed algorithm

Fig. 2. Classification result images of maximum likelihood algorithm (MLC) and proposed algorithm using Landsat TM satellite image

The following is the same experiment as before, using IKONOS image satellite to confirm if it is possible to apply proposed algorithm to the high resolution satellite image. IKONOS satellite image is high resolution satellite image that records 4 bands of multispectral data at 4m resolution and one panchromatic band with one meter resolution. The image size used for the experiment is 1000 by 1000 pixels and it is consisted of 4 bands. Forest, water, soil, urban areas were selected as four classification categories. In the input image, two to four areas were chosen as training data from each classification category. Table 2 show the classification results by proposed algorithm, FCM and conventional maximum likelihood classification method from IKONOS high resolution satellite image. IKONOS high resolution satellite image also shows a better classification result than conventional maximum likelihood classification method. This result showed that it is possible to apply proposed algorithm effectively to the classification of high resolution satellite image. Fig. 3 show classification result images of maximum likelihood algorithm and proposed algorithm using IKONOS satellite image.

Table 2. The classification results by proposed algorithm, conventional maximum likelihood and FCM algorithm from IKONOS high resolution satellite image

Category		Forest	Water	Soil	Urban	Overall accuracy
Number of training pixels		9249	11323	13785	19426	53783
Classification method	Maximum likelihood	9110	11285	12753	18080	95.94%
	FCM	8820	10951	12896	17094	93.42%
	Proposed method	9235	11291	13127	18586	97.62%

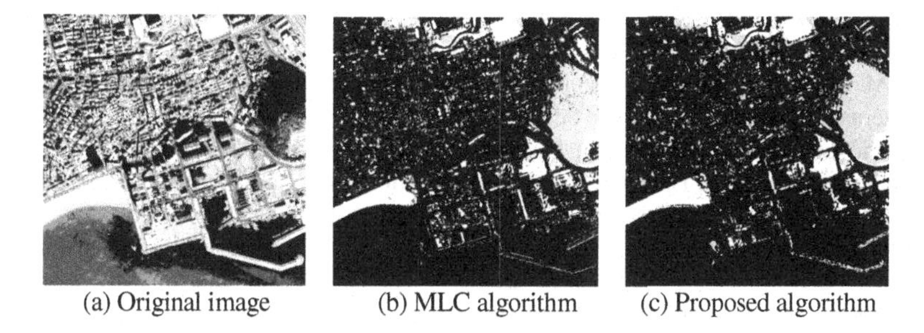
(a) Original image (b) MLC algorithm (c) Proposed algorithm

Fig. 3. Classification result images of maximum likelihood (MLC) and proposed algorithm from IKONOS satellite image

As a result of the experiment using Landsat TM satellite image and IKONOS high resolution satellite image, it shows that the proposed method could improve performance of classification rather than the conventional maximum likelihood classification method. This result improved the classification result of the proposed algorithm with giving a positive effect to the classification of Bayesian algorithm that used a prior probability of the information obtained from the execution of fuzzy GK algorithm in which the information of the pixels within the calculated average intracluster distance had a normal distribution.

5 Conclusions

In this research, a supervised classification algorithm of remote sensing satellite image was proposed by using the average intracluster distance within the fuzzy and Bayesian algorithms. Fuzzy GK algorithm was used in the form extended for the FCM algorithm. By using the training data, it establishes the initial cluster center and executes fuzzy clustering algorithm. The prior probability is obtained by using the information of pixels placed in the average intracluster distance of each category, and then the Bayesian classification algorithm is executed.

The proposed algorithm was tested by applying it to the Landsat TM and IKONOS remote sensing satellite images. As a result, the overall accuracy showed a better outcome than the FCM algorithm or the conventional maximum likelihood classification algorithm. Average intracluster distance is the average value of the distance from each data to the cluster center and is in ratio to the size and density of cluster. The information of the pixels within the average intracluster distance has a positive normal distribution, and the whole classification result was improved by giving a positive effect to the classification of Bayesian algorithm that used the information as prior probability. The proposed algorithm proved to be practical and effective applied to the high resolution image satellite as shown in the classification result of IKONOS image satellite. The next research task will be the solution of pixels for urban and shadow portions in classification.

References

1. John A. Richards : Remote Sensing Digital Image Analysis : An Introduction, Second, Revised and Enlarged Edition, Springer-Verlag (1994) 229-262,
2. J. M. Cloutis : Hyperspectral geological remote sensing: Evaluation of analytical techniques, International Journal of Remote Sensing, 17(12) (1996) 2215~2242,.
3. David Landgrebe : Information Extraction Principles and Methods for Multispectral and Hyperspectral Image Data, Chapter 1 of Information Processing for Remote Sensing, edited by C. H. Chen, published by the World Scientific Publishing Co., Inc (1999) 1-30.
4. Mehmet I Saglam, Bingul Yazgan, Okan K Ersoy : Classification of Satellite Images by using Self-organizing map and Linear Support Vector Machine Decision tree, GISdevelopment Conference Proceedings of Map Asia (2003).
5. Zhaocong Wu : Research on remote sensing image classification using neural network based on rough sets, Info-tech and Info-net, 2001. Proceedings. ICII 2001-Beijing. 2001 International Conferences on, Vol. 1 (2001) 279-28429.
6. L.A.Zadeh : Fuzzy sets as a basis for theory of possibility, Fuzzy sets and Systems, Vol. 35 (1978) 3-28.
7. N.R. Pal and J.C. Bezdek : On cluster validity for the fuzzy c-means model, IEEE Transactions on Fuzzy Systems, Vol. 3, No. 3 (1995) 370-379.
8. Melgani, F., Hashemy B.A.R. and Taha S.M.R. : An explicit fuzzy supervised classification method for multispectral remote sensing images, Geoscience and Remote Sensing, IEEE Transactions on, Vol. 38, Issue 1 Part 1 (2000) 287-295.
9. B.Gorte and A. Stein : Bayesian classification and class area estimation of satellite images using stratification, IEEE Trans. On Geoscience and Remote Sensing, Vol.36, No.3 (1998) 803-812.
10. Amal S. Perera, Masum H. Serazi, William Perrizo : Performance Improvement for Bayesian Classification on Spatial Data with P-Trees, 15th International Conference on Computer Applications in Industry and Engineering (2002).
11. Qilian Liang, : MPEG VBR video traffic classification using Bayesian and nearest neighbor classifiers, Circuits and Systems, 2002. ISCAS 2002. IEEE International Symposium on, Vol. 2 (2002) II-77-II-80.
12. D.Gustafson and W.Kessel : Fuzzy clustering with a fuzzy covariance matrix, In Proc. IEEE CDC, San Diego, USA, (1979) 761-766.
13. R. Krishnapuram and J. M. Keller : A possibilistic approach to clustering, IEEE Trans. on Fuzzy Systems, Vol. 1, No. 2 (1993) 98-110.

Tree Species Recognition with Fuzzy Texture Parameters

Ralf Reulke[1] and Norbert Haala[2]

[1] Institut für Informatik, Computer Vision, Humboldt-Universität zu Berlin, 10099 Berlin, Deutschland
reulke@informatik.hu-berlin.de
http://www.informatik.hu-berlin.de/reulke
[2] Institut für Photogrammetrie (ifp), Universität Stuttgart, 70174 Stuttgart, Deutschland
norbert.haala@ifp.uni.stuttgart.de
http://www.ifp.uni-stuttgart.de/institut/staff/haala.htm

Abstract. The management and planning of forests presumes the availability of up-to-date information on their current state. The relevant parameters like tree species, diameter of the bowl in defined heights, tree heights and positions are usually represented by a forest inventory. In order to allow the collection of these inventory parameters, an approach aiming at the integration of a terrestrial laser scanner and a high resolution panoramic camera has been developed. The integration of these sensors provides geometric information from distance measurement and high resolution texture information from the panoramic images. In order to enable a combined evaluation, in the first processing step a co-registration of both data sets is required. Afterwards geometric quantities like position and diameter of trees can be derived from the LIDAR data, whereas texture parameters are derived from the high resolution panoramic imagery. A fuzzy approach was used to detect trees and differentiate tree species.

1 Introduction

The development of tools allowing for an automatic collection of information required to build up conventional forest inventories is one of the main objectives of the NATSCAN project (see [1]). This project provides the framework for the approaches presented in this paper. Forest inventory parameters of interest are tree species, tree height and diameter at breast height (DBH) or crown projection area. Additionally, information about the numbers and types of wood quality features and the length of the branch free bowl have to be provided, since these parameters determine the value of the timber products. Due to the recent development of tools for automatic data collection, traditional manual approaches can be replaced and thus the individual influence of human measurement can be eliminated. Within the NATSCAN project, the automatic feature collection is

R. Klette and J. Žunić (Eds.): IWCIA 2004, LNCS 3322, pp. 607–620, 2004.

carried out by the application of LIDAR measurements from airborne and terrestrial platforms, respectively ([3],[10]). This paper describes an approach aiming at the further improvement of the LIDAR based data collection from a terrestrial platform by the application of a panoramic camera. This camera additionally captures high resolution color images, which can be used as a complementary source of information and thus support the analysis of the range images from LIDAR measurement. In the following section the sensor system, i.e. the terrestrial laser scanner and the panoramic camera, is described. The coregistration of the collected data sets, which is a prerequisite for further processing, is presented in section 3. In section 4 approaches aiming at the collection of the required inventory parameters based on a combined data interpretation are shown. Finally, the tree recognition with fuzzy texture parameters is discussed in section 5.

2 Data Collection

Within the project, the IMAGER 5003 system from Zoller+Fröhlich was used for the collection of the terrestrial laser scanning data. This sensor utilizes phase difference measurements to capture the required distance information, which limits the maximum range of the system to 53.5 m. For longer distances an ambiguity problem for phase analysis exists. The main advantages of the system are the high speed of data collection and the large field of view, which allows the realization of 360° horizontal and 310° vertical turns during measurements. For the IMAGER 5003, the divergence of the laser beam of 0.3 mrad results in a spot size of the laser footprint of 1.9 cm at a distance of 50 m.

Fig. 1. Range image captured from LIDAR scanner

Fig. 1 presents the range image of the test area close to the city of Freiburg (Germany), which was collected by the LIDAR sensor. In this example, the pixel coordinates of the image are defined by the horizontal and vertical direction of the laser beam, respectively. The size of the image is 5000 pixels in height and 8400 pixels in width. Fig. 2 depicts a reflectance image of the same measurement. This image can be additionally captured based on the intensity measurement of the respective laser foot-prints. The spectral information as provided from this reflectivity image is limited to the wavelength of the laser beam, which is

Fig. 2. Reflectivity image captured from LIDAR scanner

in the near infrared spectrum. The spacing is exactly the same for range and intensity image, since they relate to the same point measurements. Thus, the main advantage of the intensity measurement is the exact coregistration of both data sets. This is demonstrated in Fig. 3, which shows a 3D view of a tree trunk generated from the range and reflectance image as provided from the laser scanner.

Fig. 3. 3D view of a tree surface. The texture is from the reflectivity image of LIDAR measurement

One general problem of LIDAR measurements is the limited spatial resolution as compared to digital images of good quality. This resolution is sufficient for the collection of geometric parameters like tree position or diameter at breast height. Nevertheless, the discrimination of different tree types, which i.e. requires an analysis of the tree bark's structure, is only feasible if an increased spatial resolution is available. Thus, for our investigations, a panoramic camera is additionally applied in order to collect high resolution color imagery.

Fig. 4. EYESCAN camera

Fig. 4 depicts the panoramic camera EYESCAN, which was used in our project. The camera was developed by the KST GmbH in a cooperation with the German Aerospace Center (DLR). Image collection is realized with a CCD line, which is mounted parallel to the rotation axis of a turntable [7]. Thus, the height of the panoramic image is determined by the number of detector elements of the CCD line, resulting in an image height of 10,200 pixels. A 360° turn of the camera, which is performed to collect a full panorama, results in an image width of 43868 columns. Since the CCD is a RGB triplet, true color images are captured. The radiometric resolution of each color channel is 14 bit, the focal length of the camera is 60 mm and the pixel size is 7 μm. This, for example, results in a sampling distance of 6 mm at a distance of 50 m for the collected imagery.

3 Data Coregistration

As a first processing step, range data and high resolution panoramic imagery have to be coregistrated to allow for the subsequent combined processing of both data sets. This coregistration can be achieved similar to a standard orthoimage generation by mapping the panoramic image to the surface from the LIDAR measurement. In order to perform this process, information on the full geometric model of the camera, which is described by [8], is required. In addition to the interior orientation, the exterior orientation of the camera has to be determined i.e. by a spatial resection based on control point measurements. In our configuration, both data sets were collected from the same position. Additionally, since laser scanner and panoramic camera are rotating on turntables, both data sets can be represented in a cylindrical coordinate system. For these reasons, the coregistration process could be simplified considerably by mapping the panoramic image to the range and reflectivity images from LIDAR measurement using a second order polynomial.

Fig. 5. Measurement of corresponding points in panoramic (top) and reflectance image (bottom)

The parameters of this polynomial can be determined easily from corresponding points in the respective data sets. Since no signalized points were available, natural points were selected for this purpose. Measurement and coregistration was realized within the software ERDAS, which is normally used for remote sensing applications. As it is depicted in fig. 5, for point measurement the reflectivity image of the LIDAR data was applied. By these means, the manual identification of corresponding points could be simplified since the reflectance data is more similar to the panoramic image than the corresponding range measurements. Still, the identification and exact measurement of points in a forest environment is relatively difficult. Problems can, for example, occur due to wind movement of the tree branches. For our data sets additional problems resulted from the relatively large time difference (4 months) between the collection of panoramic image and LIDAR data. These problems resulted in remaining differences between the corresponding points after mapping in the order of 10 pixels (RMSE).

Nevertheless, for our applications these differences were acceptable, as is exemplarily demonstrated in Fig. 6. The 3D view of this tree trunk was already presented in Fig. 3. Whereas in this figure the reflectance image of the laser scanner is used for texture mapping, in Fig. 6 the high resolution color image is used for the same purpose. Since the panoramic image was collected at a different epoch, the signalized point, which is depicted in Fig. 3, is no longer available.

Fig. 6. 3D view of tree surface, texture from high resolution panoramic image

For our investigations, the range data and the high resolution panoramic image were captured by different instruments. If, similar to airborne applications, both sensors are integrated into a single system, data processing can be simplified considerably. This integration is realized in airborne systems e.g. for the simultaneous acquisition of range and image data. In this configuration, the differences in position and orientation between both sensors are exactly calibrated. Thus, the different data sets can be exactly coregistrated without any additional tie-point measurement.

In addition to a potential increase of the mapping accuracy by a more rigorous transformation using the full geometric model of the camera, the availability of a fully integrated system would optimize the process of data coregistration. If an exact calibration of the differences in position and orientation between laser scanner and panoramic camera is available, tie-point measurement is no longer required. Thus, a much simpler and faster data collection would become possible.

4 Data Interpretation

After the coregistration of the range data and the high resolution panoramic image, a combined evaluation can be performed in order to collect the required forest inventory parameters. In this context, the LIDAR measurement is very well suited for the collection of geometric properties like the automatic localization of trees and the computation of tree diameters. This can, for example, be realized by the application of a Hough-Transform, which detects circular structures in the 3D point clouds derived from range measurement [9].

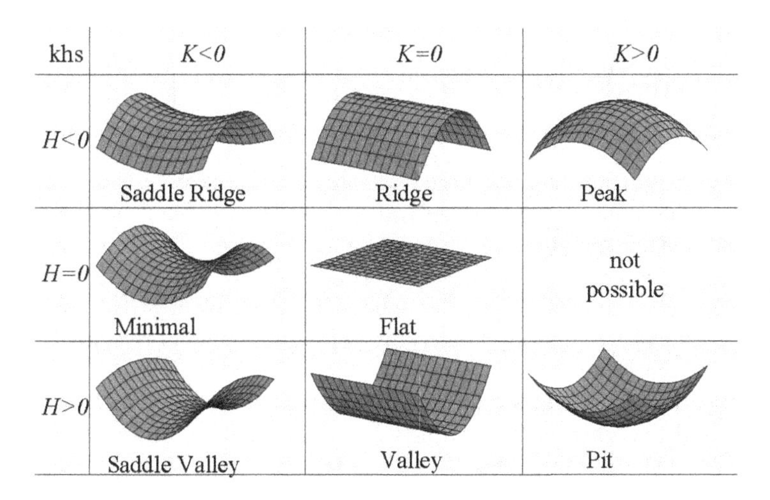

Fig. 7. Fundamental surface types by mean and Gaussian curvature signs

A similar type of information can be extracted from range image analysis by the application of curvature based segmentation. Fig. 7 depicts different surface types which can be derived from range data based on the sign of the mean curvature H and Gaussian curvature K [2].

The translation and rotation invariant values are calculated in a local coordinate system based on the first and second fundamental form of a parameterized surface

$$K = \frac{eg - f^2}{EG - F^2} \qquad H = \frac{1}{2} \cdot \frac{eG - 2fF + gE}{EG - F^2} \tag{1}$$

Since the surface measured by the laser scanner can be described by the function $z = f(x, y)$ a simple parameterization is feasible by

$$\begin{pmatrix} u \\ v \\ f(u, v) \end{pmatrix}$$

The partial derivatives of this surface are defined by

$$x_u = \begin{pmatrix} 1 \\ 0 \\ f_u \end{pmatrix} \quad \text{and} \quad x_v = \begin{pmatrix} 0 \\ 1 \\ f_v \end{pmatrix}$$

In order to obtain the coefficients $E = 1 + f_u^2$, $F = f_u f_v$, $G = 1 + f_v^2$ and $e = \frac{f_{uu}}{s}$, $f = \frac{f_{uv}}{s}$, $g = \frac{f_{vv}}{s}$ with $s := \sqrt{f_u^2 + f_v^2 + 1}$ the derivatives of the surface have to be calculated. This can be achieved by a convolution of the surface image with the discrete masks

$$D_u = \frac{1}{8}\begin{pmatrix} -1 & 0 & 1 \\ -2 & 0 & 2 \\ -1 & 0 & 1 \end{pmatrix}, \quad D_{uu} = \frac{1}{4}\begin{pmatrix} 1 & -2 & 1 \\ 2 & 4 & 2 \\ 1 & -2 & 1 \end{pmatrix}, \quad D_{uv} = \frac{1}{8}\begin{pmatrix} -1 & 0 & 1 \\ 0 & 0 & 0 \\ 1 & 0 & -1 \end{pmatrix}, \quad \text{etc.}$$

Alternative approaches are based on the least-squares estimation of local bivariate polynomials, where the derivatives used are defined by the estimated polynomial coefficients [4].

If the mean and Gaussian curvature H and K are calculated, according to Fig. 7 the respective surface type is defined for each pixel of the range image. Thus, regions can be generated easily by combining neighbored pixels of corresponding surface type.

Fig. 10 gives an example of this surface type analysis, which was used to classify cylindrical objects (type Ridge $K = 0$, $H < 0$) in order to localize tree trunks.

Fig. 8. Result of curvature based segmentation

Complementary to the analysis of the range data for the collection of geometric features, the high resolution panoramic images can be applied for a texture based classification. By these means, the structure of the respective tree bark is analyzed in order to discriminate between different types of trees in a subsequent step.

5 Tree Recognition with Fuzzy Texture Parameters

The main goal when processing the tree bark textures is to find suitable parameters that describe each texture to a certain degree and allow for separating it from others. For analysis, 6 different trees $(a \ldots f)$ and 12 texture samples (400x400 pixels size) per tree have been chosen. The tree categories are marked with symbols $(+, *, \triangle, \square, \times$ and $bg)$. For analysis of these samples there are many approaches of texture evaluation available, which are often either statistical or structural approaches [6].

Statistical methods for feature extraction are based on the investigation of the gray level values for each single pixel. In first order statistics, only the pixel itself is measured. If also neighboring pixels (in a certain distance and direction) are taken into consideration the approach is said to be of second order. Working examples for the latter are color co-occurrence approaches where second order statistics on the color histogram reveal information on a texture [11]. Structural texture analysis extracts elements and determines their shapes, striving to find placement rules to describe how these elements are occurring, e.g. their density, spatial distance and regularity of layout [6].

Fig. 9. Tree bark texture samples (12 patches each)

Fuzzy set theory provides another way of seeing images and textures. When dealing with images as arrays of gray values, often information losses are accepted just in order to get "crisp" results. Fuzzy approaches try to avoid the loss of information by giving uncertainty a place. In the first step of "fuzzyfying" an image, it is converted from an array of gray values to an array of "membership". A priori knowledge of the context could be brought into the process by choosing and parameterization of the membership function. The conversion from gray value images to the fuzzyfied version can be as simple as directly mapping gray values to the range $[0, 1]$ thus expressing the property of brightness of each pixel. Depending on the application, one can parameterize the membership function to enhance particular features in the image.

For our demands, we simply fuzzified the bark texture images by using Zadeh's S-Function [12] to get brightness membership values

$$\mu(x) = \begin{cases} 0 & x \leq a, \\ 2\left(\frac{x-a}{c-a}\right)^2 & a < x \leq b, \\ 1 - 2\left(\frac{x-c}{c-a}\right)^2 & b < x \leq c, \\ 1 & x > c. \end{cases} \tag{2}$$

which is 0 until the gray values reach a, then rising in an S-form to 1, which it reaches right when approaching c. $b = (a + c)/2$ lies right in the middle between a and c and is the point of inflection, where the functions curvature changes sign. Furthermore, when reaching b, the membership value $\mu(b) = 0.5$.

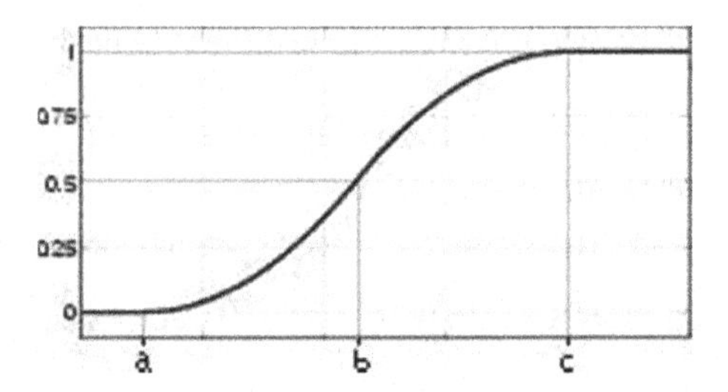

Fig. 10. Plot of Zadeh's S-function

The limit of its first derivative is 0 when approaching the lower and the top end of the gray scale (a and c respectively), producing little change in "brightness-membership" there. The maximum of change is right at its inflection point b.

The conversion from gray values to brightness membership doesn't directly lead to the discrimination of tree species. There exist several fuzzy measures though, that can now be calculated for the image, namely area, perimeter, compactness and logarithmic entropy [5].

$$a(X) = \sum_m \sum_n \mu_{mn} = \sum_l \mu(l)h(l) \tag{3}$$

The area, as a measure of how many pixels belong to what extent to the set of "bright" pixels, or how much "brightness membership" can be found in the image, simply sums up all membership values. An image half black, half of it white can not be distinguished from an all gray alternative by only using the area-parameter. It's value can grow arbitrarily for larger images.

$$p(X) = \sum_{m=1}^{M} \sum_{n=1}^{N-1} |\mu_{mn} - \mu_{m,n+1}| + \sum_{n=1}^{N} \sum_{m=1}^{M-1} |\mu_{mn} - \mu_{m+1,n}| \tag{4}$$

The perimeter is the sum-up of all membership changes between neighbors in 4-neighborhoods, i.e. pixels that are horizontally or vertically connected. By summing up membership differences to the adjacent pixels to the right and bottom sides, the whole neighborhood is accounted for. Like the area parameter, the perimeter's limit is given only by the size of the image. We cannot directly compare perimeters between image samples of different sizes. It is a measure for the rate of change between neighboring memberships, therefor a measure of "contrast" in the image. Remarkably, images of only one constant brightness, regardless of value, would always yield a 0 perimeter.

$$comp(X) = \frac{a(X)}{p^2(X)} \tag{5}$$

The compactness, aggregating area and perimeter, provides us with a measure that doesn't change for smaller or bigger images, i.e. while the area and perimeter of a texture grow boundless for larger images, the compactness just wouldn't change.

To calculate the entropy, a function is needed for the evaluation of membership values in terms of "vagueness" or "uncertainty". The function $T_e(l)$ assigns a vagueness value according to its distance to the "certain" bounds 0 and 1. Membership in the middle of the interval $[0, 1]$ yield the highest vagueness value, which is $ln(2)$ for that function.

$$T_e(l) = -\mu(l) ln\mu(l) - (1 - \mu(l)) ln(1 - \mu(l)) \tag{6}$$

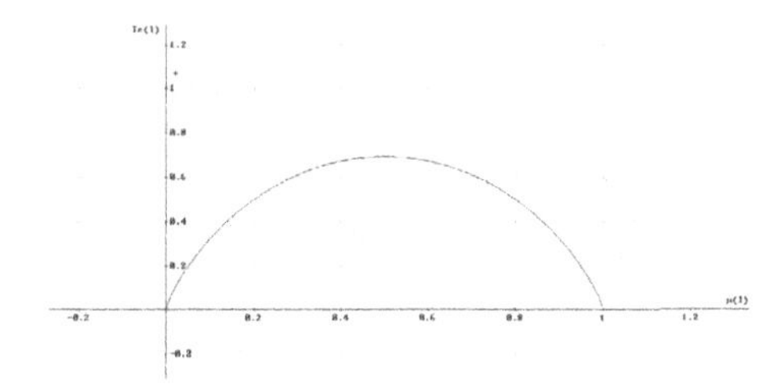

Fig. 11. The "vagueness" or "uncertainty" of membership in the range [0,1]

Fig. 11 shows the plot: the more the memberships are displaced from the 0 and 1 margins, the more their vagueness increase.

$$H(X) = \frac{1}{MN\ ln2} \sum_{l} T_e(l) h(l) \tag{7}$$

The sum of all vagueness values in the image, normalized to the size of the image and the maximum of T_e, is called its entropy. It serves as a measure for the overall uncertainty found in the image.

Now the fuzzy parameters of all tree bark textures can be calculated and compared for different trees.

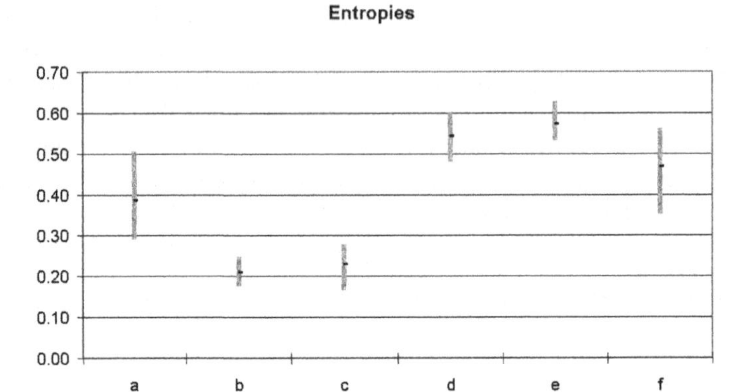

Fig. 12. Entropy values calculated for all texture categories

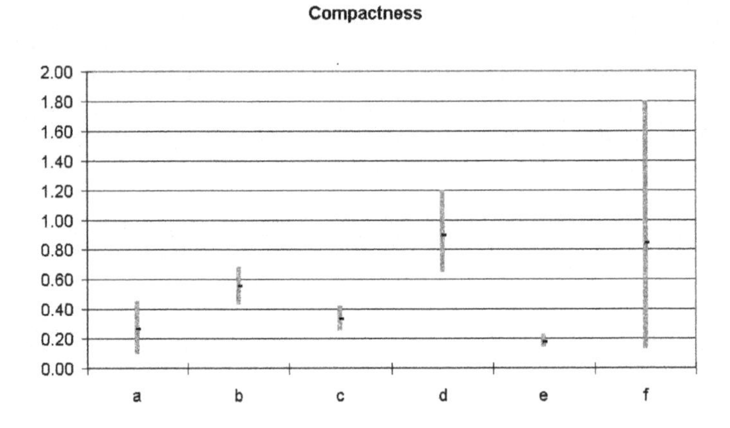

Fig. 13. Compactness values calculated for all texture categories

The entropy-plot in Fig. 12 shows that this feature alone allows distinguishing between some but not all samples, e.g. it would not be possible to tell textures from trees *d* and *e* apart, since they are much too close in terms of entropy values. As mentioned at the beginning of this section, the selection of suitable parameters is most important to find a way of classifying patterns. The solution in this approach is to use a two-dimensional feature space incorporating entropy and one (well-chosen) other parameter. The area parameter is out of the question, because it can be computed from entropy. In Fig. 13 it can be seen that tree texture parameters cluster differently (e.g. trees *d* and *e* are easily discriminated in that form).

When taking the compactness as the second feature classifier and bringing both into a two-dimensional plot, one can see that entropy and compactness of each texture make it possible to tell different trees apart. The plotted symbols are referring to the texture samples in Fig. 9.

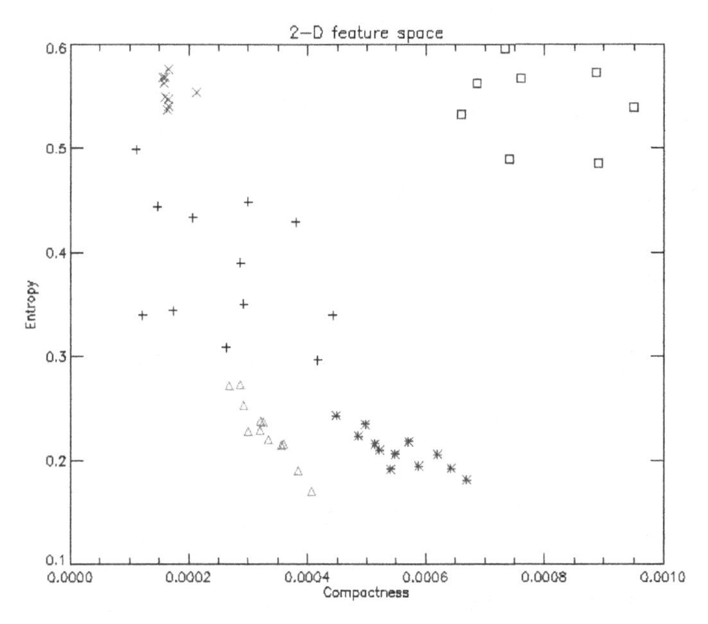

Fig. 14. Texture parameter 2-D feature space

The next step would be to find a suitable decision boundary to bring the model to practice and decide to which category an unknown texture sample most likely belongs. It can be clearly seen that all tree textures are clustering very well and possibly allow further processing of the data to obtain reliable classification decisions.

6 Conclusion

This paper described the combined evaluation of high resolution imagery from a panoramic camera and LIDAR data from a terrestrial laser scanner. Data collection based on these types of sensors has already been demonstrated for applications like virtual city modelling or heritage documentation. In contrast, the presented application on the automated collection and determination of tree parameters required for forest inventories is a novel approach and thus demonstrates the potential of this type of integrated sensor configuration and combined data evaluation also for natural environments. The LIDAR system is suited very well to detect tree trunks when using the shown curvature segmentation approach. Since that method is not sensitive to leaves occluding the trunk, it often should be preferred to camera approaches. Finally, fuzzification of the bark textures and selection of appropriate fuzzy parameters make it possible to distinguish tree species from each other.

Acknowledgement

We thank the Institute for Forest Growth, Universität Freiburg for the provision of the LIDAR data used for our investigations. The authors would also like to thank Frederik Meysel for his valuable assistance with the analysis and the formulation.

References

[1] Web: http://www.natscan.de/ger/welcome.php.
[2] P.J. Besl. Segmentation through variable order surface fitting. *IEEE Transactions on Pattern Analysis and Machine Intelligence*, 10(2):167–192, 1988.
[3] Koch B. Friedlaender, H. First experience in the application of laserscanner data for the assessment of vertical and horizontal forest structures. *IAPRS*, XXXIII, Part B7:693–700, 2000.
[4] R.M. Haralick. Digital step edges from zero-crossings of second directional derivatives. *IEEE Transactions on Pattern Analysis and Machine Intelligence*, PAMI-6(1):58–68, 1984.
[5] Pal S.K. Kundu, M.K. Automatic selection of object enhancement operator with quantitative justification based on fuzzy set theoretic measures. *Pattern Recognition Letters*, 11:811–829, 1990.
[6] M. K. (Ed.) Pietikainen. *Texture Analysis in Machine Vision.* World Scientific Publishing Company, 2000.
[7] Korsitzky H.-Reulke R. Scheele M. Solbrig M. Scheibe, K. EYESCAN - a high resolution digital panoramic camera. *RobVis 2001*, pages 77–83, 2001.
[8] Maas H.-G. Schneider, D. Geometric modelling and calibration of a high resolution panoramic camera. *Optical 3-D Measurement Techniques VI*, II:122–129, 2003.
[9] Aschoff T.-Spiecker H. Thies M. Simonse, M. Automatic determination of forest inventory parameters using terrestrial laserscanning. *Proceedings of the ScandLaser Scientific Workshop on Airborne Laser Scanning of Forests*, pages 251–257, 2003.
[10] Aschoff T.-Spiecker H. Thies, M. Terrestrische laserscanner im forst - für forstliche inventur und wissenschaftliche datenerfassung. *AFZ/Der Wald 58*, 22:1126–1129, 2003.
[11] G. Gimel'farb Yu, L. Image retrieval using colour co-occurrence histograms. *Image and Vision Computing New Zealand 2003, Palmerston North, New Zealand*, pages 42–47, 2003.
[12] L. A. Zadeh. Fuzzy sets as a basis for a theory of possibility. *Fuzzy Sets and Systems*, 1:3–28, 1978.

Fast Segmentation of High-Resolution Satellite Images Using Watershed Transform Combined with an Efficient Region Merging Approach

Qiuxiao Chen[1,2], Chenghu Zhou[1], Jiancheng Luo[1], and Dongping Ming[1]

[1] The State Key Lab of Resources & Environmental Information System, Chinese Academy of Sciences, Datun Road, Anwai, Beijing, 100101, China
{chenqx, zhouch, luojc, mingdp}@lreis.ac.cn

[2] Dept. of Regional and Urban Planning, Yuquan Campus, Zhejiang University, 38 Zheda Road, Hangzhou, 310027, China
chen_qiuxiao@zju.edu.cn

Abstract. High-resolution satellite images like Quickbird images have been applied into many fields. However, researches on segmenting such kind of images are rather insufficient partly due to the complexity and large size of such images. In this study, a fast and accurate segmentation approach was proposed. First, a homogeneity gradient image was produced. Then, an efficient watershed transform was employed to gain the initial segments. Finally, an improved region merging approach was proposed to merge the initial segments by taking a strategy to minimize the overall heterogeneity increased within segments at each merging step, and the final segments were obtained. Compared with the segmentation approach of a commercial software eCognition, the proposed one was a bit faster and a bit more accurate when applied to the Quickbird images.

1 Introduction

Although being a key research field in various domains, image segmentation is still on its research stage and is a bit far from wide application. It is particularly a truth in the remote sensing field. Being an increasing important information source, remote sensing images are playing an increasing role in many fields like environment monitoring, resource investigation, precision agriculture, urban planning and management, etc. In order to improve remote sensing images processing ability and extend and expand further their application fields, researches on segmenting this kind of image are necessary and urgent.

Among recent limited studies on remote sensing images segmentation, much attention has been paid to segment SAR (Synthetic Aperture Radar) images [6][8][9][13][14] and the medium resolution remote sensing images, say TM or SPOT images [1][5][11][15]. The relevant researches [16][17] on high-resolution remote sensing images, say IKONOS or QuickBird images, are rather insufficient.

Using region based segmentation approaches to partition images is a very natural thing since the target of a segmentation task is to obtain meaningful regions. Region based approaches like region growing and watershed transform have been employed to

R. Klette and J. Žunić (Eds.): IWCIA 2004, LNCS 3322, pp. 621–630, 2004.

segment the remote sensing images for years. However they are mainly applied to medium or coarse resolution remote sensing images, the counterpart studies on high resolution images are seldom conducted. Due to the large application potential of high-resolution satellite images, such studies will probably attract more attention in the near future. In consideration of the difficulty to judge whether two neighboring pixels belong to the same region and when to stop the region growing process if we utilize region growing approach to segment images, we will focus only on the watershed transform approach in this study.

The premise of watershed transform is to construct a suitable gradient image. To high-resolution satellite images like QuickBird images, the optimum gradient should incorporate both spectral and texture features. A simply way to produce a gradient feature named H for multiband images was proposed in [12]. The higher/lower H is, the more likely the relevant pixel locates at regions boundary/center. Therefore, it is more suitable to use watershed approach than region growing method [12] to segment H feature image.

We should keep in mind some drawback of the watershed transform approach. It always produces over-segmentation results, its implementation is not as fast as we have expected when applied to remote sensing images. In order to overcome the above shortcoming, both the pro- and post-processing are needed, and some relevant modifications are also necessary. Another thing we should pay attention to is the fake basins produced by watershed transform probably due to images noise. In order to remove these fake basins, two thresholds are pre-defined. To reduce the segments number further, a region merging process is utilized as a post-processing operation. Generally, the conventional region merging approach is relatively too slow. During its implementation, much time has been wasted on searching, removing and sorting. By utilizing another data structure proposed by us, the computing cost of region merging is reduced greatly, and the implementation speed of region merging is raised substantially.

The content of this paper is organized as follows: in section one some background is introduced and the reasons why we study the method proposed are addressed, the methodology is presented in section two, the sample image is introduced in section three, segmentation results and some discussion are included in section four, and conclusions and future research advice are made and proposed in the last section.

2 Methodology

The methodology of the current work can be described as follows, first a homogeneity gradient image (H-image) was produced; then watershed transform was applied to the gradient image, the initial segmentation results were obtained. Finally, an improved region growing method was utilized to obtain the final segments. To validate the accuracy and efficiency of the proposed approach, a comparison with the segmentation approach of a commercial software named eCognition was also conducted.

2.1 Local Homogeneity Gradient Extraction

So far there are many texture models developed. However, the majority of them are not suitable for performing segmentation of remote sensing images because of their relatively poor implementation efficiency, which is, for a large extent, due to the complexity and large size of remote sensing images On the other hand, most of these texture models just aim at one band image. In [12], a rather simply homogeneous index – H index for multiband images was proposed. In the current study, we will utilize it directly.

By calculating each pixel's H value, an H-image can be obtained. The dark and bright areas in the H-image actually represent the region centers and region boundaries respectively. In fact, H is a gradient feature, so watershed approach is more suitable to segment H-image when compared with region growing approach utilized in [7][12].

2.2 Watershed Transform

In mathematical morphology, an image is usually considered as a topographical surface where the grey level of each pixel corresponds to an altitude, the grey value can be gradient, spectral intensity or other features, and generally it refers to the former. To search for contours in images, it is quite straightforward to estimate the crest lines on the gradient image. Watershed transformation is just a way of extracting these lines from the modulus of the gradient image [13]. In general, region boundaries correspond to watershed lines (with high gradient), and region interior (with low gradient) being the catchment basin.

So far, there are several watershed transform algorithms developed. Among them, the floating point based rainfalling watershed algorithm [19] is perhaps the most efficient one. In the current study, we will use it to obtain the initial segmentation results.

Though being easy to implement, the rainfalling approach also produces over-segmentation (too many segments), and to conduct subsequent image processing or analyses based on a large quantity of segments becomes very difficult and inefficient, which will obviously prevent its further application. Therefore, some form of pre- and post-processing are required in order to produce a segmentation that better reflects the arrangement of objects within the image. An easy way to prevent producing too many catchment basins is to set a threshold -- h-threshold for the local minimum, that is, to set all the pixels with a gradient value lower than this threshold as a local minimum. We can also set another threshold -- size-threshold to remove small basins probably caused by image noise. As to post-processing approaches, the commonly used one is region merging, which will be discussed in the next subsection.

2.3 Region Merging

RAG is the commonly used data structure for representing partitions [3][21]. However, it is rather slow to implement the region merging process based on such data structure [10]. In order to improve the implementation efficiency, we made some modifications on the RAG. The modified RAG (MRAG) consists of three parts—V, E, N. Though being the nodes set like that of the general RAG, here V utilizes a different data structure – a two-dimensional dynamic array to record the region information. Because each node (region) probably has the different number of adjacency regions and such number of a certain node may increase or decrease after each merging step, we should allocate sufficient memory for each node in order to prevent the overflow if we use a static array, which means a considerable waste of memory. Thus, we utilize a two-dimensional dynamic data structure, through which, it is very convenient to add or remove adjacency regions, and memory is saved as well. E has the same meaning as that of the general RAG, only records the pseudo-address of each edge. N is a cost matrix recording the merging cost of each pair of regions, the cost of non-adjacency region pairs is set to an infinite value. For each region merged into other regions, its region label will be changed and all relevant edges are set to be dead edges (their cost in N is set to an infinite value).

Based on MRAG, the region merging process (Fig. 2.) can be described as follows,

- Search in E for the miniCostEdge (the edge with the minimum merging costs) and remove those dead edges in E according to the cost matrix N, set miniCostEdge to be dead edge in N and remove it from the E;
- Assume the relevant nodes (regions) of the miniCostEdge are p and q, let q be a region merged into p, then execute the following step: (1) read the neighbor information of q from V; (2) set q to be dead edge; (3) update the region information of p in V – features needed for calculating merging cost and add p's new neighbors (from q) in V; (4) update the cost of edges relevant to p in N; (5) insert new edges of p into E.

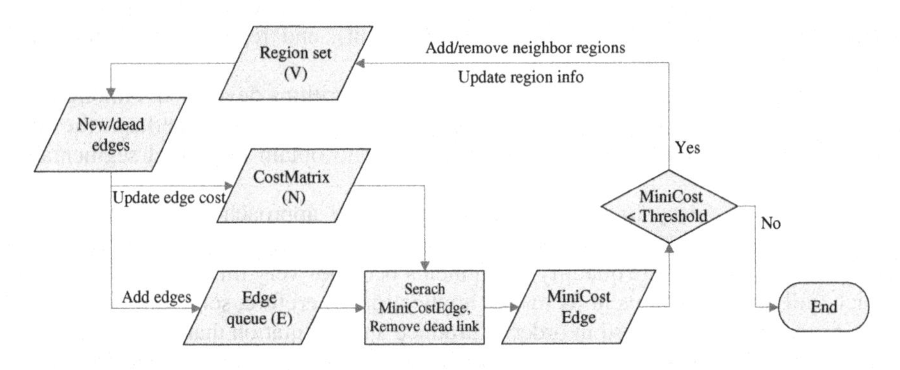

Fig. 1. The flow chart of region merging process

By using MRAG, the information and adjacency regions of a certain region can be obtained quickly. Except for searching the miniCostEdge, no other search will be needed. Since edges with a merging cost larger than merging threshold will never happen, such edges can be skipped.

At each merging step, we wish that the most similar adjacency region pair is merged. There are several approaches to calculate the similarity of two segments, to compute the Euclidean Distance of their feature vector is probably the simplest one. Generally, we wish that a segmentation routine maximizes the overall homogeneity within segments while maximizing the overall heterogeneity among segments. To maximize the overall homogeneity within segments is the same thing as to minimize the overall heterogeneity. Generally, feature variation can be regarded as the heterogeneity, so we can fulfill the above target if we minimize the variation increased at each merging step. Assume there are two adjacency regions A and B, and C being the new region if A and B is merged, the cost to merge A and B can be calculated as follows,

$$\cos t = \sqrt{\sum_{j=1}^{k} S_A \cdot A_{.j}^2 + \sum_{j=1}^{k} S_b \cdot B_{.j}^2 - \sum_{j=1}^{k} (S_A + S_B) \cdot C_{.j}^2}\,, \tag{1}$$

here, S_A and S_B are the size of the segment A and B respectively, k denotes the feature dimension, $A_{.j}$, $B_{.j}$, and $C_{.j}$ is the mean of the feature value in the jth dimension of A, B and C respectively. Since C is more heterogeneous than A or B, with the merging cost increasing, the homogeneity of the new formed region will not be ensured. So, we set a threshold of merging cost in advance. If the minimum merging cost exceeds the threshold, the merging process will be stopped. Through this mechanism, we can ensure the homogeneity of each segment and the overall homogeneity of all image segments as well.

2.4 Segmentation Approach of eCognition

eCognition is a software for satellite image classification developed by Definiens Imaging GmbH in Germany. At least in remote sensing community, it is the first commercially available product for object-oriented image analysis. Due to its unique contribution, Definiens Imaging GmbH was awarded the 2002 European 1st Prize, which was organized by the European Council of Applied Sciences and Engineering (Euro-CASE). So far, this product has been utilized by several government agencies, universities and corporations for various application purposes like forest management, land cover/use mapping, agriculture observation and mapping, etc.

In eCognition, the patented multi-resolution segmentation technique is perhaps one of the most critical components since it aims at object-oriented image analysis. It can be used for many kinds of earth observation data (either optical or radar). Utilizing it, image objects can be founded even in textured data like SAR images, high-resolution satellite data or airborne data (for more detail, please refer to [2]). Since we aim at segmenting high-resolution satellite images, it is natural to take multi-resolution segmentation technique for comparison. In the current study, the eCognition 4.0 trial version (with the same functions as eCognition 4.0 Professional except for image size limitation) is used.

3 Experimental Data

To evaluate the performance of the proposed segmentation approach, a QuickBird image was utilized in the present work. Such image data has five bands -- a pan band with a spatial resolution 0.7 meters and four multispectral bands with a spatial resolution 2.8 meters. The satellite image used here was acquired on Oct. 10, 2003 covering the part of Taizhong County in Taiwan Province, China. Since the eCognition trial version can only handle the image with a size not larger than 1024*1024, the sample image cut down from the whole image was 1000 rows by 1000 columns, and only multispectral bands will be used. To facilitate the visual inspection of segmentation results, each multispectral band was fused with the pan band using Erdas Imagine 8.4, thus the final spatial resolution of the test image was 70cm by 70cm.

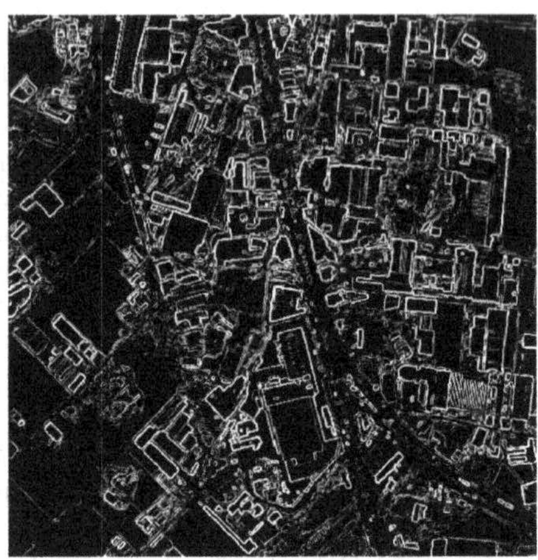

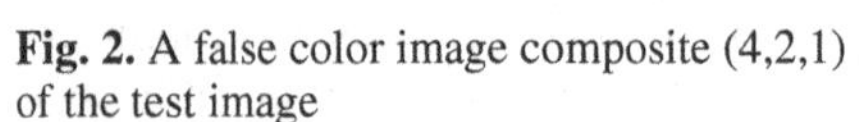

Fig. 2. A false color image composite (4,2,1) of the test image

Fig. 3. The homogeneity gradient image -- H-image

4 Results and Discussions

Utilizing the H index proposed in [12], an H-image was produced after normalization (Fig. 3.). Due to the large size of the test image and its complexity, there will be too many segments produced if we employ watershed transform without other operations. In order to alleviate the computing cost, an h-threshold was adopted, pixels with its H value below the h-threshold were considered from valleys, and basins were to be formed through these pixels. Because of unavoidable noise in the image, there existed a considerable amount of small fake basins. A basin with its size smaller than a preset value--size-threshold was considered to be such fake basin and was to be merged into the neighboring big basins. Based on these two thresholds, segments obtained after watershed transform were reduced greatly. In this study, we took h-threshold and size-threshold as 39 and 30 respectively. An initial segmentation based on watershed transform was shown in Fig.4., 3619 segments were obtained.

In remote sensing images, a segment with a large size means a real geo-object, we should be very careful about the merging between these segments. Otherwise, the merging process will be out of control, and the final segmentation result will be far from satisfactory. In this study, a segment with a size larger than 600 pixels was considered as a big segment, a merging between big segments occurs only when the corresponding merging cost was rather small when compared with the merging cost threshold (the former being smaller than 5% of the latter). By utilizing the proposed merging process, the final segmentation results with 499 segments were obtained (Fig. 5.), and the merging cost threshold used was 2000.0.

Because of its high resolution character of the test image, we can check the segmentation results by visual inspection easily. From Fig. 5., we can see that the agricultural patches with relatively uniform color features (mainly distributed on the left part and the upper right corner of the image, generally being red, grey or a bit bright when naked, and being dark blue when immersing with the water) were

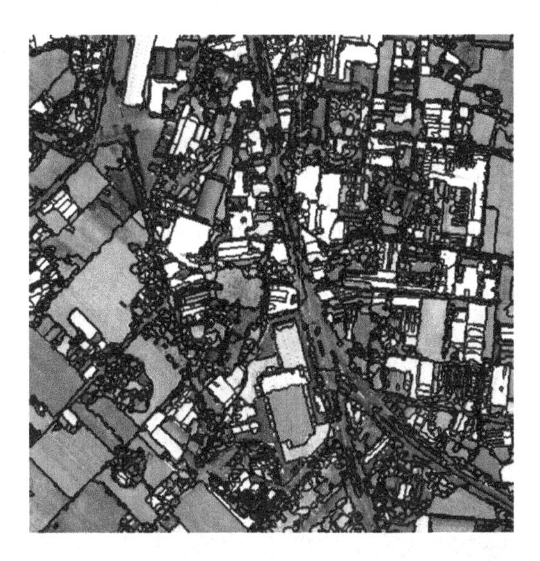

Fig. 4. Initial results (3619 segments) with h-threshold being 39 and size-threshold being 30

segmented successfully even in the condition when existing textures or spectral value variation in them. Man-made constructions with a relative high radiation (being light blue or white) were also segmented quite well. The segmentation result of green areas (being red with texture features) was also quite good.

In all, both the segmentation approaches obtained quite good results. However, with a more detail comparison, we can learn that boundaries of segments in Fig. 5. were a bit more accurate than those in Fig. 6. and Fig. 7. In Fig. 5., geo-objects with a

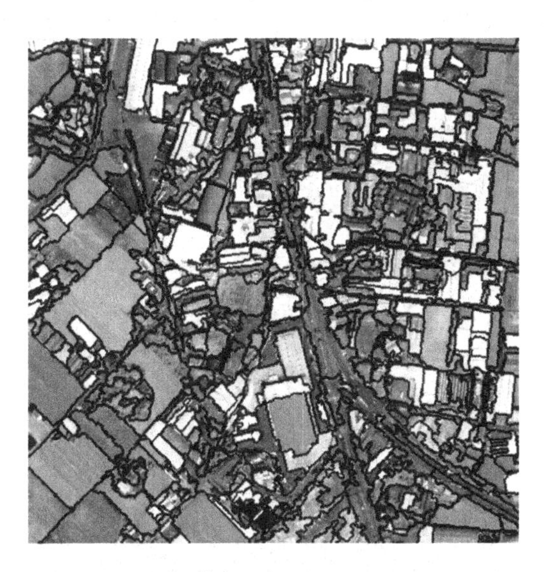

Fig. 5. Final results (499 segments) with merging threshold being 2000

large size were segmented as well as those with a relative small size. Using a notebook with an Intel P4-M processor with 1.8 GHz and a 384 MB memory, it only took 25 seconds for the proposed approach to segment the sample image in the Microsoft Windows XP operating system environment, about 12 seconds for watershed transform and about 13 seconds for region merging. While using eCognition, to obtain segmentation results shown in Fig. 6. or Fig. 7., 35 seconds were spent. It seemed that the improved region merging approach proposed in this study was very efficient.

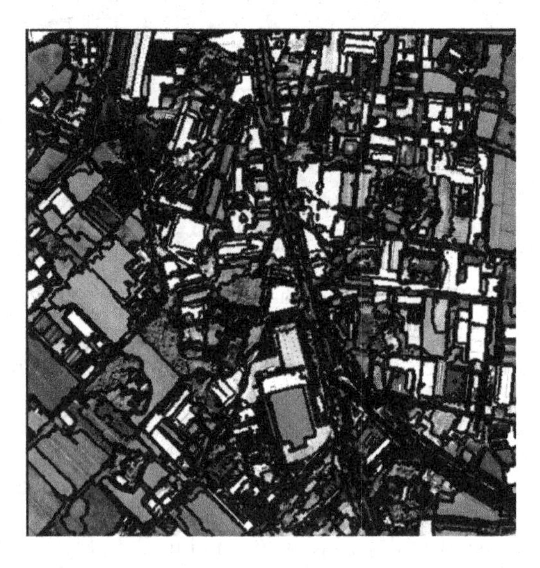

Fig. 6. Segmentation results using eCognition (510 segments) with scale parameter being 100

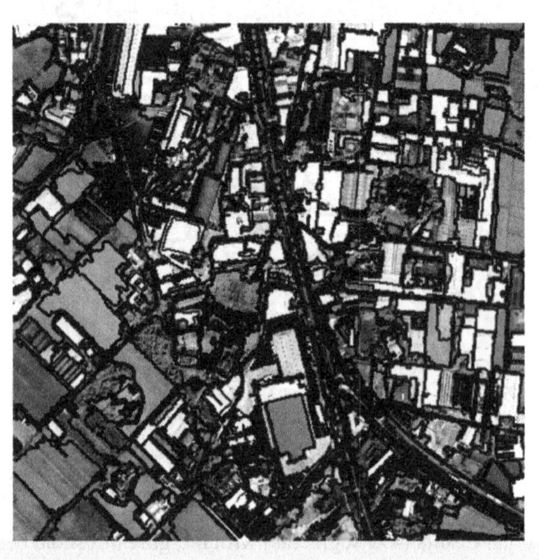

Fig. 7. Segmentation results using eCognition (276 segments) with scale parameter being 150

However, it should be pointed out that road segments extracted were not so satisfying, and segmentation results were a bit sensitive to the watershed transform parameters, say, h-threshold, size-threshold and the merging cost threshold.

5 Conclusions

In all, the segmentation approach was well adapted to the unique character of high resolution satellite images. On the one hand, the quick implementation was realized through the easy extraction of multiband gradient feature and the efficient rainfalling watershed transform algo-rithm. By utilizing h-threshold and size-threshold, small basins were removed which accelerated the merging process. Utilizing a two dimensional dynamic array assisted by the cost matrix, a modified RAG was proposed, through which it was very convenient to locate each region, obtain/update its information and neighbor regions, and remove unused edges (dead edges). Experiments showed that the proposed approach had a quick implementation, and the segmentation result was rather promising when compared with the segmentation approach of a commercial software.

In order to improve segmentation accuracy further, we will focus our future work on the following aspects: (1) to incorporate the shape index in segmentation routine to extract objects with regular shape (e.g. road, building); (2) to find an automatic way to set a suitable value for parameters like merging cost threshold, h-threshold and size-threshold.

Acknowledgements

This study is supported partially by the National Natural Science Foundation of China under Grant No. 40301030 and by High Tech Research and Development (863) Program under Grant No. 2002AA716101.

References

1. M. Acharyya, R.K. De and M.K. Kundu. Segmentation of Remotely Sensed Images Using Wavelet Features and Their Evaluation in Soft Computing Framework. IEEE Transactions on Geoscience and Remote Sensing, 41(12):2900-2905, 2003.
2. M. Baatz and A. Schäpe. Multiresolution Segmentation – an optimization approach for high quality multi-scale image segmentation. In: Strobl, J. et al. (eds.): Angewandte Geographische Infor-mationsverarbeitung XII. Wichmann, Heidelberg, 12-23, 2000.
3. D. Ballard and C. Brown, Computer Vision. Englewood Cliffs, NJ: Prentice-Hall, 1982.
4. S. Beucher and F. Meyer. The Morphological Approach to Segmentation: the Watershed Transformation. In Mathematical Morphology and its Applications to Image Processing, E.R. Dougherty, Ed. New York: Marcel Dekker, 433–481, 1993.
5. J. Bosworth, T. Koshimizu and S.T. Acton. Multi-resolution Segmentation of Soil Moisture Imagery by Watershed Pyramids with Region Merging. Int. J. Remote Sensing, 24(4):741–760, 2003.

6. P.B.G. Dammert; J.I.H. Askne and S. Kuhlmann. Unsupervised Segmentation of Multitemporal Interferometric SAR Images. IEEE Transactions on Geoscience and Remote Sensing, 37(5): 2259 – 2271, 1999.
7. Y. Deng, B.S. Manjunath and H. Shin. Color Image Segmentation. Proc. of IEEE Computer Society Conf. on Computer Vision and Pattern Recognition, CVPR'99, 2:446-451, 1999.
8. Y. Dong, B.C. Forester, and A.K. Milne. Segmentation of Radar Imagery Using the Gaussian Markov Random Field Model. Int. J. Remote Sensing, 120(8):1617-1639, 1999.
9. Y. Dong, B.C. Forster and A.K. Milne. Comparison of Radar Image Segmentation by Gaussian- and Gamma-Markov Random Field Models. Int. J. Remote Sensing, 24(4):711–722, 2003.
10. K. Haris, S. Efstratiadis, N. Maglaveras and A. Katsaggelos. Hybrid Image Segmentation Using Watersheds and Fast Region Merging. IEEE Trans. Image Process., 7(12):1684-1699, 1998.
11. R.A. Hill. Image Segmentation for Humid Tropical Forest Classification in Landsat TM Data. Int. J. Remote Sensing, 20(5):1039-1044, 1999.
12. F. Jing, M.J. Li, H.J. Zhang and B. Zhang. Unsupervised Image Segmentation Using Local Homogeneity Analysis. Proc. IEEE International Symposium on Circuits and Systems, 2003.
13. W. Li, G.B. B•ni• and D.C. He, et al. Watershed-based Hierarchical SAR Image Segmentation. Int. J. Remote Sensing, 20(17): 3377-3390, 1999.
14. J. Lira and L. Frulla. An Automated Region Growing Algorithm for Segmentation of Texture Regions in SAR Images. Int. J. Remote Sensing, 19(18):3595-3606, 1998.
15. S.K. Pal, A. Ghosh and B.U. Shankar. Segmentation of Remotely Sensed Images with Fuzzy Thresholding, and Quantitative Evaluation. Int. J. Remote sensing, 21(11):2269–2300, 2000.
16. M. Pesaresi and J.A. Benediktsson. A New Approach for the Morphological Segmentation of High-resolution Satellite Imagery. IEEE Transactions on Geoscience and Remote Sensing, 39(2): 309-320, 2001.
17. A. Pekkarinen. A Method for the Segmentation of Very High Spatial Resolution Images of Forested Landscapes. Int. J. Remote Sensing, 23(14):2817-2836, 2002.
18. D. Raucoules and K.P.B. Thomson. Adaptation of the Hierarchical Stepwise Segmentation Algorithm for Automatic Segmentation of a SAR Mosaic. Int. J. Remote Sensing, 20(10):2111-2116, 1999.
19. P.D. Smet and R.L. Pires. Implementation and analysis of an optimized rainfalling watershed algorithm. Proc. SPIE, 3974:759-766, Image and Video Communications and Processing, 2000
20. L. Vincent and P. Soille. Watershed in Digital Spaces: an Efficient Algorithm Based on Immersion Simulation. IEEE T ransactions on Pattern Analysis and Machine Intelligence, 13:583 - 598, 1991.
21. X. Wu. Adaptive Split-and-merge Segmentation Based on Piecewise Least-square Approximation. IEEE Trans. Pattern Anal. Machine Intell., 15:808-815, 1993.

Joint Non-rigid Motion Estimation and Segmentation

Boris Flach[1] and Radim Sara[2]

[1] Dresden University of Technology
[2] Czech Technical University Prague

Abstract. Usually object segmentation and motion estimation are considered (and modelled) as different tasks. For motion estimation this leads to problems arising especially at the boundary of an object moving in front of another if e.g. prior assumptions about continuity of the motion field are made. Thus we expect that a good segmentation will improve the motion estimation and vice versa. To demonstrate this we consider the simple task of joint segmentation and motion estimation of an arbitrary (non-rigid) object moving in front of a still background. We propose a statistical model which represents the moving object as a triangular (hexagonal) mesh of pairs of corresponding points and introduce an provably correct iterative scheme, which simultaneously finds the optimal segmentation and corresponding motion field.

1 Introduction

Even though motion estimation is a thoroughly investigated problem of image processing, which had attracted attention for decades, we should admit, that at least a handful crucial open problems remain on the agenda. One of them are object boundaries and occlusions (imagine e.g. an object moving in front of a background: If motion estimation is considered as a separate task, then usually some continuity or smoothness prior assumptions for the motion field are modelled [1, 2], which regularise and thus improve the result almost everywhere but not at boundaries and partial occlusions. Thus we can expect that a good segmentation will improve the motion estimation and vice versa. In case of strict a-priori knowledge – e.g. rigid body motion – this segmentation can be modelled directly in terms of admissible motion fields [3]. In case of more general situations like non-rigid motion this is not possible. To overcome this problem, we propose a joint motion estimation and segmentation model. To start with, we consider this task for the simple situation of an arbitrary (non-rigid) object moving in front of a still background.

We propose a statistical model, which represents both, the moving foreground object and the background in terms of labelling vertices of a triangular /hexagonal) mesh. The state (label) of each vertex is a segment label and a displacement vector. This allows to incorporate prior assumptions for the objects shape and the motion field by either hard or statistical restrictions. For instance, to avoid

R. Klette and J. Žunić (Eds.): IWCIA 2004, LNCS 3322, pp. 631–638, 2004.

motion fields which do not preserve the topology of the object (given by the segmentation), we require coherent orientations for the displacement states of elementary triangles of the lattice in the foreground segment. Image similarity as well as consistency with some colour models for the segments are modelled in a statistical way. Consequently, we obtain a statistical model for the (hidden) state field and the images, which is a Gibbs probability distribution of higher order in our case. This allows to pose the joint segmentation and motion estimation as a Bayes task with respect to a certain loss function.

2 The Model

Consider non-rigidly moving object against stationary background. The object can have holes but must not self-occlude during the motion. There are two images and the task is to segment the moving object and estimate the motion field.

Let R be a set of vertices associated with some subset of image pixels chosen in a regular way. We consider a hexagonal lattice on these vertices (see Fig. 1). Its edges are denoted by $e \in E$, where E_1 denotes the subset of edges forming the rectangular sub-lattice. The elementary triangles of the hexagonal lattice are denoted by $t \in T$. Each vertex $r \in R$ has a compound label $\big(x(r), v(r)\big)$, where $x(r) \in \{0,1\}$ is a segment label and $v(r) \in V$ is an integer-valued displacement vector. A complete labelling is thus a pair of mappings $x\colon R \to \{0,1\}$, $v\colon R \to V$ and defines a simultaneous segmentation and motion field.

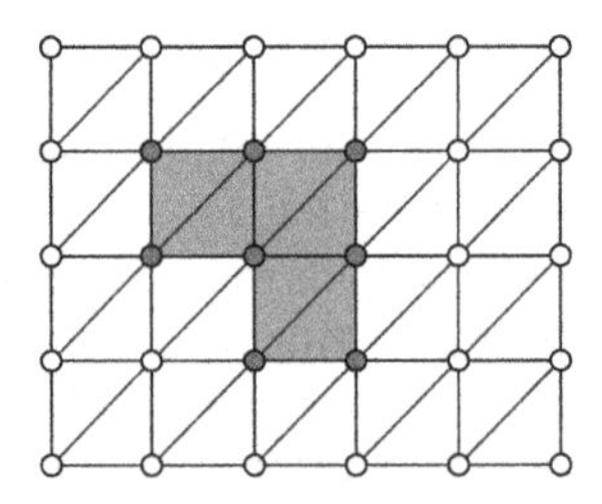

Fig. 1. Hexagonal lattice and segmentation

We consider a prior statistical model for such labellings which favours compact foreground segments and continuous displacement fields and preserves the topology of the foreground segment:

$$p(x,v) = \frac{1}{Z} \exp\Big[-\beta \sum_{(r,r')\in E_1} \|x(r)-x(r')\|^2 - \sum_{t\in T} H_t\big(x(t),v(t)\big) - \sum_{e\in E_1} H_c\big(x(e),v(e)\big) - \sum_{r\in R} H_b\big(x(r),v(r)\big)\Big] \quad (1)$$

The first sum is over all edges in E_1 and represents a Potts model for segmentation. The second sum is over all elementary triangles of the hexagonal lattice.

The function H_t is infinity if all three vertices $r \in t$ are labelled as foreground and their displacements reverse the orientation of that triangle. It is zero in all other cases. Hence, this term zeroes the probability for displacement fields which do not preserve the topology of the foreground segment. The third sum is over all edges in E_1. The function H_c is infinity if both vertices are labelled by foreground and their displacement vectors differ more than a predefined value. It is zero in all other cases. Hence, this term zeroes the probability of non-continuous displacement fields for the foreground segment. The last sum is over all vertices and H_b is infinity in case a vertex marked background has an nonzero displacement vector and is zero otherwise. Hence, this term reflects our assumption of a still background. Combining all together, we get a third order Gibbs probability distribution aka Markov Random Field.

Our measurement model is as follows. Let $y, \tilde{y}$ be two feature fields taken from two consecutive images of the scene i.e. $y, \tilde{y}\colon R \to F$, where F is the set of feature values (which might be a set of colours in the simplest case). To express the conditional probability of obtaining y and $\tilde{y}$ given segmentation x and displacement field v, we introduce the following subsets of vertices. Let $S(x) \subseteq R$ denote the subset of vertices labelled as foreground, $S(x) = \{r \in R \mid x(r) = 1\}$. All vertices labelled as background are divided into two sets: $O(x)$ represents those which are occluded by foreground in the second image and $B(x)$ are those which are visible in both images:

$$O(x) = \big\{r \in R \mid x(r) = 0,\, \exists\, r'\colon x(r') = 1,\, r = r' + v(r')\big\}, \tag{2}$$

$$B(x) = R \setminus \big(S(x) \cup O(x)\big). \tag{3}$$

Using these sets, the conditional probability is as simple as

$$p(y, \tilde{y} \mid x, v) = \exp\Bigg[\sum_{r \in S(x)} q_f\Big(y(r),\, \tilde{y}\big(r + v(r)\big)\Big) + \\ + \sum_{r \in B(x)} q_b\big(y(r),\, \tilde{y}(r)\big) + \sum_{r \in O(x)} \bar{q}_b\big(y(r)\big)\Bigg], \tag{4}$$

where $q_f(f,\, f')$ and $q_b(f,\, f')$ are the log-likelihoods to obtain feature values f and f' for corresponding image points in foreground and background, respectively and where $\bar{q}_b$ is the log-likelihood to obtain the feature value f for a background image point. These probabilities can be easily estimated from foreground and background feature distributions if a simple independent noise model is assumed for the camera. It is worth noting that the second and third sum in (4) are non-local: occlusion of a vertex r depends on the states of all vertices which might occlude r.

Having a complete statistical model for quadruples $(y,\, \tilde{y},\, x,\, v)$, we formulate the recognition task as Bayes decision with respect to the following loss function

$$C(y,\, \tilde{y},\, x,\, v) = \sum_r \Big[\mu_1\, \mathbf{1}\{x(r) \neq x'(r)\} + \mu_2\, \|v(r) - v'(r)\|^2\Big], \tag{5}$$

which is locally additive. Each local term penalises wrong decisions with respect to segmentation and displacement vectors, respectively. Minimising the average loss (i.e. the risk) gives the following Bayes decision [6]

$$x^*(r) = \arg\max_k p_r(x(r) = k \mid y,\, \tilde{y}), \tag{6}$$

$$v^*(r) = \sum_v v \cdot p_r(v(r) = v \mid y,\, \tilde{y}), \tag{7}$$

where $p_r(x(r) \mid y,\, \tilde{y})$ and $p_r(v(r) \mid y,\, \tilde{y})$ are the marginal posterior probabilities for segment label and displacement vector, respectively:

$$p_r(x(r) = k \mid y,\, \tilde{y}) = \sum_{x:\, x(r)=k} \sum_v p(x,\, v \mid y,\, \tilde{y}) \tag{8}$$

and similarly for $p_r(v(r) = v \mid y,\, \tilde{y})$. Hence, we need this probability for the Bayes decision. Note that the (7) gives non-integer decisions for the displacement vectors, though the states are considered as integer-valued vectors.

We do not know how to perform the sums in (7) and (8) over all state fields explicitly and in polynomial time. Nevertheless, it is possible to estimate the needed probabilities using a Gibbs sampler [5]. In one sampling step, one chooses a vertex r, fixes the states in all other vertices and randomly generates a new state according to its posterior conditional probability $p(x(r), v(r) \mid x(R \setminus \{r\}),\, v(R \setminus \{r\}),\, y,\, \tilde{y})$ given the fixed states $x(R \setminus \{r\})$, $v(R \setminus \{r\})$ in all other vertices. According to [5] the relative state frequencies observed during the sampling process converge to the needed marginal probabilities.

3 Experiments

In this section we show results on three image pairs: `rect`, `hand` and `ball`, see Figs. 2, 3 and 4, respectively. The `rect` are prepared artificially and have size 100×100 The size of the `hand` images is 580×500 pixels, and the `ball` images 253×256 pixels. Both pairs are JPEG images shot by a standard compact digital camera.

The motion in the `rect` pair is a combination of a translation and a projective transform. In this dataset, the segmentaion itself is not as simple, but being combined with motion estimation, the result is fairly good.

The motion in the `hand` pair is almost uniform in the direction towards the lower-left image corner. It is rather non-uniform in the `ball` pair, although the dominant motion is translation, again towards the lower left corner of the image, the additional components are in-plane counter-clockwise rotation due to wrist rotation and out-of plane rotation due to closing the hand towards the forearm. The ball moves rigidly but the wrist does not.

In the first natural dataset, the segmentation task itself is relatively easy, based on the foreground-background colour difference but the motion is hard to estimate due to the lack of sufficiently strong features in the skin region in sub-quantised low-resolution images. In the second dataset, however, the motion estimation task is more easy based on rich colour structure (up to highlights and uniform-colour patches) but the segmentation would be more difficult on the colour alone.

The `hand` images are re-quantised to 32 levels per colour channel and the `ball` images to 16 levels. The image features f were the re-quantised RGB triples. The log-likelihoods q_f, q_b and $\bar{q}_b$ are estimated from the re-quantised images based on rough manual pre-segmentation (although their automatic estimation from data is possible, see e.g. [4], it is not the focus of the present paper).

The spacing of the regular lattice was four pixels in the `hand` pair and three pixels in the `ball` pair. In the `hand` pair, the expected motion range was -24 ± 12 pixels in the horizontal direction and ± 12 pixels in the vertical direction; in the `ball` pair, the corresponding ranges were -18 ± 6 pixels and 11 ± 4 pixels,

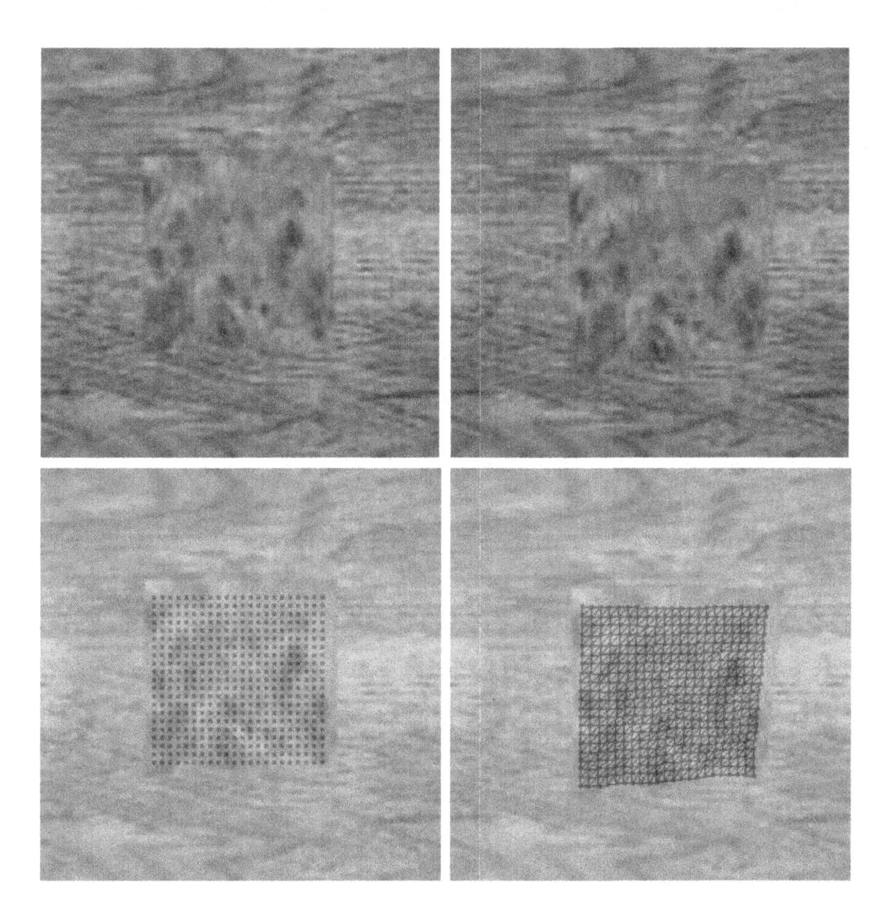

Fig. 2. Results on the `rect` pair. Top row: input images. Bottom row: segmentation in the first frame and the deformed triangular mesh in the second frame, both overlaid over the corresponding images

respectively. The initial estimate for the field x were based on local (vertex-wise) decisions using the log-likelihoods q_f and q_b. The initial estimate for the motion field v was zero.

Results for the `hand` pair are shown in Fig. 3. The bottom left overlay shows those vertices of the hexagonal lattice that are labelled as foreground (in red). The bottom right overlay shows the lattice after displacement by estimated motion field. The moving boundary at the lower edge of the wrist is captured, although it is fuzzy because of body hair seen against the background.

Results for the `ball` pair are shown in Fig. 4. We used a stronger β for the Potts model compared to the `hand` pair, due to the more complex colour structure. Again, the bottom left overlay shows in red those vertices of the hexagonal lattice that are labelled as foreground. The bottom right overlay shows the vertices displaced by the motion field as blue dots and the residual motion field after subtracting the mean motion vector $\bar{v} = (-20.5, 11.0)$ to see the other modes of the motion.

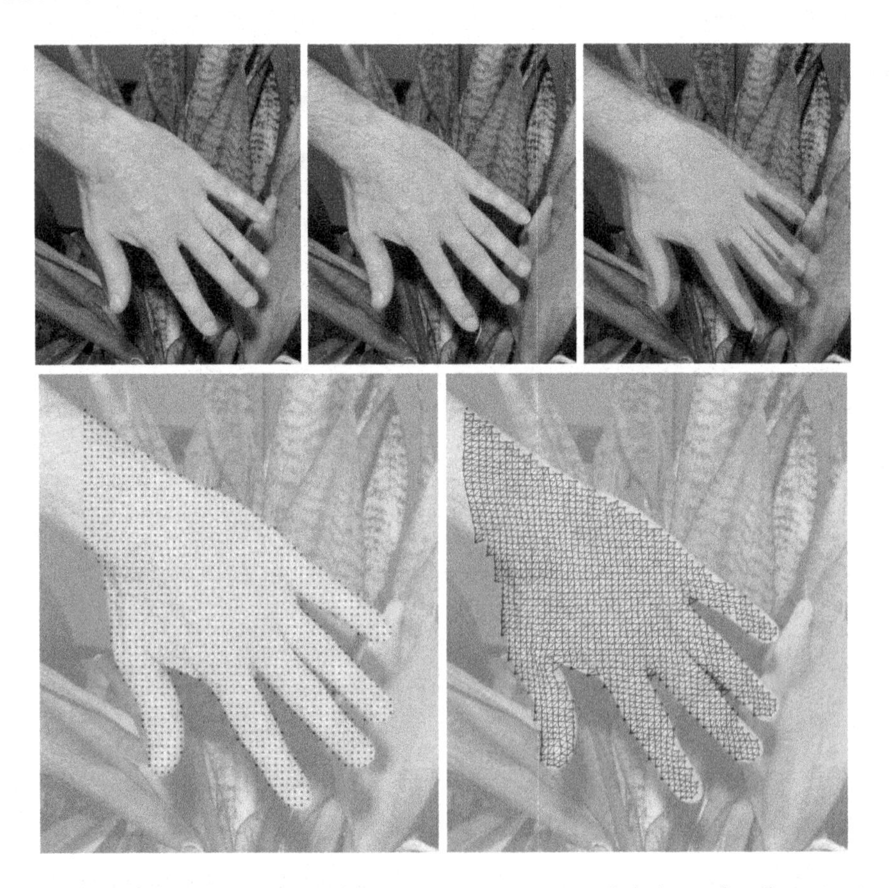

Fig. 3. Results on the `hand` pair. Top row: input images and their overlay. Bottom row: segmentation in the first frame and the deformed triangular mesh in the second frame, both overlaid over the corresponding images

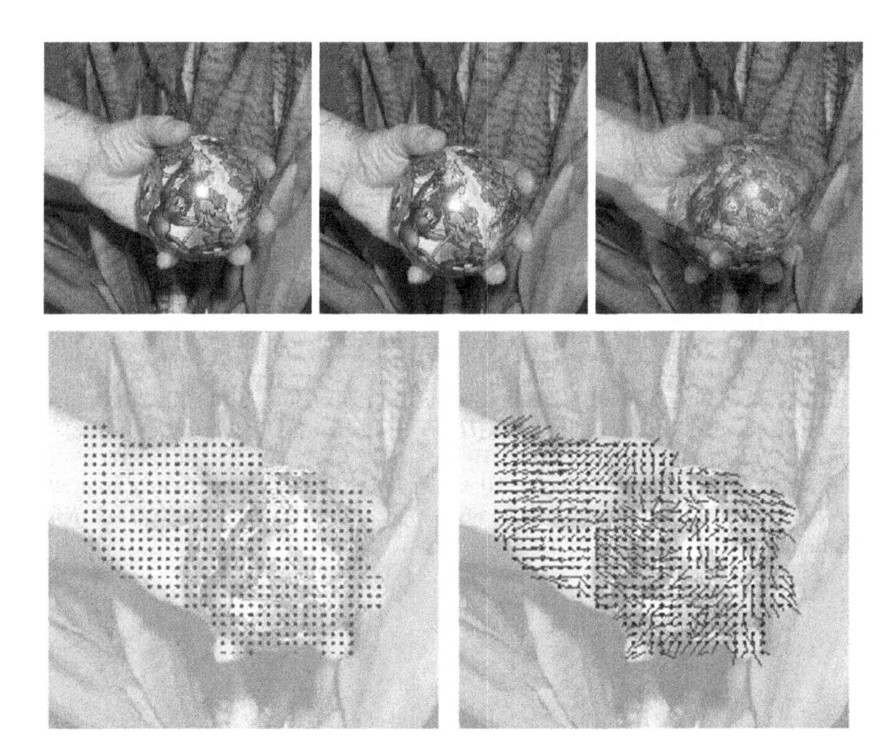

Fig. 4. Results on the `ball` pair. Top row: input images and their overlay. Bottom row: segmentation in the first frame and the residual motion field after subtracting mean motion vector

4 Conclusions

In this paper we present our preliminary results on a simplified version for joint segmentation and motion estimation. Though these results seem to promising, there are some open problems:

1. The segmentation model is too simple.
2. the model is not symmetric with respect to time reversal.
3. The topology preservation condition does not enforce the boundary to be a Jordan-curve.
4. Self occlusions of the foreground are not allowed.

We believe these can be addressed in our future work.

References

1. Y. Boykov, O. Veksler, and R. Zabih. Fast approximate energy minimization via graph cuts. In *Proc. of the 7. Intl. Conf. on Computer Vision*, volume 1, pages 377–384, 1999.

2. Thomes Brox, Andres Bruhn, Nils Papenberg, and Joachim Weickert. High accuracy optical flow estimation based on a theory for warping. In T. Pajdla and J. Matas, editors, *Computer Vision*, volume 3024 of *Lecture Notes in Computer Science*, pages 25–36. Springer, 2004. Proc. of ECCV2004.
3. Daniel Cremers and Christoph Schnörr. Motion competition: Variational integration of motion segmentation and shape recognition. In Luc van Gool, editor, *Pattern Recognition*, volume 2449 of *Lecture Notes in Computer Science*, pages 472–480. Springer, 2002. Proc. of DAGM2002.
4. Boris Flach, Eeri Kask, Dmitrij Schlesinger, and Andriy Skulish. Unifying registration and segmentation for multi-sensor images. In Luc Van Gool, editor, *Pattern Recognition*, volume 2449 of *Lecture Notes in Computer Science*, pages 190–197. Springer Verlag, 2002.
5. Stuart Geman and Donald Geman. Stochastic relaxation, Gibbs distributions and the bayesian restoration of images. *IEEE Transactions on Pattern Analysis and Machine Intelligence*, 6(6):721–741, 1984.
6. Michail I. Schlesinger and Vaclav Hlavač. *Ten Lectures on Statistical and Structural Pattern Recognition*, volume 24 of *Computational Imaging and Vision*. Kluwer Academic Press, 2002.

Sequential Probabilistic Grass Field Segmentation of Soccer Video Images

Kaveh Kangarloo[1,2] and Ehsanollah Kabir[3]

[1] Dept. of Electrical Eng., Azad University, Central Tehran Branch, Tehran, Iran
[2] Dept. of Electrical Eng., Azad University, Science and Research unit, Tehran, Iran
kangarloo@iauctb.org
[3] Dept. of Electrical Eng., Tarbiat Modarres University, Tehran, Iran
kabir@modares.ac.ir

Abstract. In this paper, we present a method for segmentation of grass field of soccer video images. Since the grass field is observed as a green and nearly soft region, the hue and a feature representing the color dispersion in horizontal and vertical directions are used to model the grass field as a mixture of Gaussian components. At first, the grass field is roughly segmented. On the base of grass field model, the probability density function of non-grass field is estimated. Finally using the Bayes theory, in a recurrent process the grass field is finally segmented.

Keywords: Football, Grass-Field, Color, Texture, Gaussian Mixture Model, Bayes theory, Segmentation.

1 Introduction

Segmentation is the process of partitioning an image to homogeneous regions. By this manner similar regions must be categorized into the same group. Mainly the similarity is defined based on color, shape or texture [1,2].

In surveillance, tracking or traffic control systems, image segmentation is based on motion [3]. Differential methods and background modeling are the most important techniques used to estimate the motion vectors. On the other hand in applications such as face recognition, image segmentation is based on color similarity. Color space clustering is an efficient method applied in this respect [4]. K-means, FCM, Bayesian approaches and neural network based methods as SOM are known as the most important clustering techniques [5].

In Graph based segmentation, key points in the feature space are considered as distinct nodes. The similarity of two nodes is specified by an edge, which connects them as an arm. The goal is to partition the set of vertices into disjoint sets, based on a predefined similarity measure. By this manner the main image is segmented into homogenous regions [6].

Grass field Segmentation plays a fundamental role in detection and tracking of players, deep compression and content-based football image retrieval. Seo *et al.* [7] proposed a thresholding motion and color histogram technique for tracking the

R. Klette and J. Žunić (Eds.): IWCIA 2004, LNCS 3322, pp. 639–645, 2004.

players. In other words, static and green pixels are labeled as grass. In another paper, at first by thresholding color histogram, the green pixels are segmented. Considering motion and edge density in the resulting region, players are detected [8].

Similar methods are proposed to analyze the baseball or tennis video. At first, applying a threshold value on color histogram, the land is segmented. Then considering the camera movement, tennis serve or baseball pitch is detected [9,10].

In another paper, the predominant color in random selection of frames is selected as grass color. By thresholding color histogram, the grass field is segmented [11].

In this paper, we propose a probabilistic method in which the grass field could be segmented in a highly accurate manner. For this purpose, at first based on color and its dispersion, the grass field is roughly segmented. In the second step, using a recurrent algorithm based on the Bayes theory, grass field is finally segmented. This paper is organized as follow. Section 2 introduces the applied features for grass field modeling. In section 3, based on the clusters related to grass and non-grass samples, the classification method is selected. Section 4, provides the experimental results and draws the conclusion.

2 Grass Field Modeling

Manily the grass field is observed as a solid, green and soft region. In this regard, color and texture features, are suitable for modeling and segmentation. Tracking, virtual reality, guarding system and image retrieval are samples of machine vision applications in which color segmentation plays the main role. Cameras, produce RGB signal that is sensitive to the scene illumination and gamma coefficient changes.

Mainly in dynamic scenes different color spaces such as normalized *RGB*, *CIE-Luv*, *CIE-Lab*, *YUV* and *HSI* are utilized [12,13]. Based on the application, it was decided to use the *HSI* color space. H, S and I stand for hue, saturation and intensity respectively. By this manner, the 'green color' is in the range of $0.1<hue<0.35$.

In this research, the texture analysis is based on color dispersion estimation [14]. For this purpose, a window by the size of N is utilized to estimate the color dispersion in horizontal and vertical directions (Eq.1-4). The obtained feature is used to represent the image texture (Fig. 1).

$$M_1 = \sum_i \sum_j A \times \left| (h_{(i,j)} - Mean_W) \right| \qquad i, j \in W \tag{1}$$

$$M_1 = \sum_i \sum_j A \times \left| (h_{(i,j)} - Mean_W) \right| \qquad i, j \in W \tag{2}$$

$$Mean_W = 1/N^2 \sum_i \sum_j h_{(i,j)} \tag{3}$$

$$\phi = \begin{cases} \dfrac{|M_1 - M_2|}{M_1 + M_2} & M1 + M2 \neq 0 \\ 0 & M1 + M2 = 0 \end{cases} \tag{4}$$

Where A and B are masks used to estimate the color dispersion in horizontal and vertical directions (table 1). *Meanw* is the mean of hue inside the window. M_1 and M_2 are the color dispersions in horizontal and vertical directions respectively.

Table 1. The 7*7 sample masks applied to estimate the color dispersions in horizontal and vertical directions (A,B)

0	0	1	1	1	0	0
0	0	1	1	1	0	0
0	0	0	0	0	0	0
0	0	0	0	0	0	0
0	0	0	0	0	0	0
0	0	1	1	1	0	0
0	0	1	1	1	0	0

0	0	0	0	0	0	0
0	0	0	0	0	0	0
1	1	0	0	0	1	1
1	1	0	0	0	1	1
1	1	0	0	0	1	1
0	0	0	0	0	0	0
0	0	0	0	0	0	0

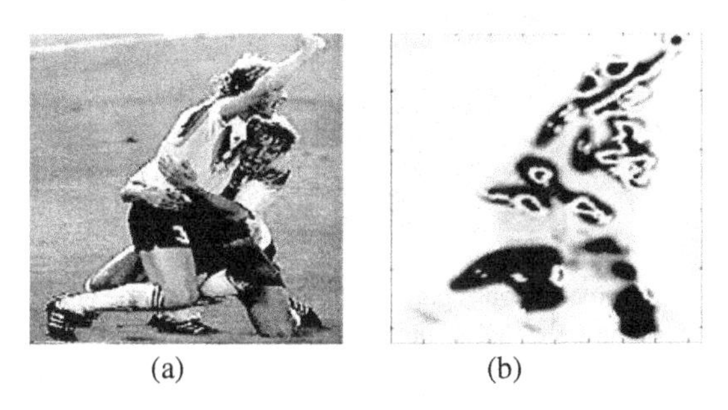

(a) (b)

Fig. 1. (a) Main image, (b) Color dispersion (Φ)

As a result, the two features, *hue* and Φ, should be small values in the grass field. Considering two-dimensional color-texture histogram, the grass field appears as a Gaussian dispersion form (Fig. 3).

3 Grass Field Segmentation

In this paper grass field segmentation is implemented in two phases. First, using a multi-layer perceptron (MLP) classifier, the grass field is roughly segmented. In the next stage based on the probability density function of non-grass field, the grass field is segmented in a highly accurate manner.

For learning the applied 2-3-5-1 MLP classifier, 700 samples have been selected. The network learns and would be able to classify samples as grass or non-grass sets. Table 2, shows the classification results according to the different sizes of windows. Considering the results, we decided to apply the 7*7 masks for color dispersion analysis.

In order to partition an image to grass and non-grass field, thresholding of the MLP output should be made. But selecting the optimal threshold value is an important factor that influences the segmentation accuracy (Fig. 2).

Table 2. The classification results according to the mask size

Window Size		5*5	7*7	9*9	11*11
Recognition Rate	Non-Grass	%96	%98	%93	%83
	Grass	%97	%99	%95	%84

Fig. 2. The grass field segmentation regarding to a threshold value applied to the MLP output. First row: main image and the MLP output (The probability of belonging to grass-field). Second row, left to right: grass field segmentation based on applying the threshold values of 0.7 and 0.5 respectively

To select the optimal threshold value, we have applied a Bayesian approach [15]. Regarding to the grass and non-grass field models, the probability of belonging to grass field is:

$$P(G|x) = \frac{P(x\mid G).P(G)}{P(x\mid G).P(G) + P(x\mid nG).P(nG)} \quad (5) \quad (5)$$

Where P(G) and P(nG) are the probability of grass and non-grass field observation(P(G)+P(nG)=1). P(x|G) and P(x|nG) are the probability density function of grass and non-grass field respectively.

Since the grass field appears as a Gaussian dispersion form, on the 2D hue-Φ histogram, it is modeled as a mixture of Gaussian components (Fig 3). The specifications of these components can be calculated applying grass field samples selected from football images. On this basis, the probability density function of grass field is:

$$P(x\mid G) = \sum_{i=1}^{n} \alpha_i P_i(x\mid G_i) \quad (6) \quad (6)$$

$$P(x\mid G_i) = \frac{1}{2\pi(|\Sigma_i|)^{1/2}} e^{-1/2(x-\mu_i)^T \Sigma_i^{-1}(x-\mu_i)} \quad (7) \quad (7)$$

Where μ_i and $\sum i$ are the corresponding parameters of the i_{th} component and α_i is the conditional probability that observation x belongs to i_{th} components. If the number of desired components that define the grass field model is known, the unknown parameters could be calculated based on the well known, EM[1] algorithm [16].

In this paper the grass field is modeled based on three Gaussian components. For this purpose, 350 grass samples of soccer images have been selected. The said two features, *hue* and Φ, have been calculated and applied for grass field modeling.

Fig. 3. Two-dimensional color-texture histogram, Left column): main images, Right column): two-dimensioned histograms, Horizontal and vertical axes are bases on hue and Φ

The probability density function of non-grass field, $P(x|nG)$, and the probability of grass field observation are not known. In order to estimate the probability density function of non-grass field, it is recommended that region covered with grass, be roughly segmented. The color-texture histogram of remaining portion of image can be used as an estimation of probability density function of non-grass field.

For this purpose the output of the MLP classifier is compared with a threshold value of +0.7. The zones exceeding this threshold value are labeled as grass. Then this region is cut and the color-texture histogram of the remaining portion of image is used to approximate the probability density function of non-grass field. In order to estimate the other unknown variable, $P(G)$, it is assumed that the probability of grass and non-grass field observation are proportional to their size in the image.

By this manner, the grass field is segmented and the sizes of obtained regions are estimated. If the calculated grass-field region size is different grossly from that as presupposed, the process is repeated. The new calculated value, $P(G)$, is used as observation probability and the findings are calculated again. The process is repeated until the difference between the two successive grass-field region size (observation probability) be less than a predetermined value of %1 (Fig. 4).

[1] Expectation Maximization

4 Conclusion

This paper presents a method by which the grass field in football images could be segmented. As mentioned, the hue and color dispersion in horizontal and vertical directions used as features for grass field modeling. By this manner, the grass field is segmented in two phases. At first utilizing an MLP classifier the grass field is roughly segmented and then based on the Bayes theory a fine segmentation is performed.

It goes without say that if we are able to segment the grass field, we would be able to send a large portion of football images only by a flag as grass. By this manner, we could be able to compress the football images to great amount. For this purpose, based on the proposed algorithm, grass field is segmented. The main image is divided into 8*8 blocks, those located in grass field are sent with just a flag and the non-grass field as foreground, is encoded in MPEG [17]. The proposed algorithm was applied to some related sequences. Considering the size of background and final encoded sequences, an average compression ratio, 99/5 percent was obtained.

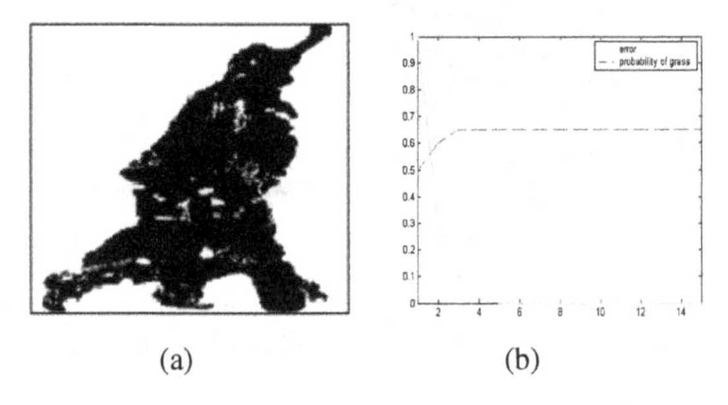

Fig. 4. (a) The grass field segmentation result after 15 iteration. (b) The error and grass field observation probability during 15 iteration

References

1. Pal, N.R, Pal, S.K: A Review on Image Segmentation Techniques. Pattern Recognition Letters, Vol. 26, no. 9, (1993) 1277-1294.
2. Southall, B., Buxton, B., Marchant, J. and Hague, T.: On the Performance Character- ization of Image Segmentation Algorithms: A Case Study, ECCV00, Vol. 2. (2000) 351-365
3. Stiller, C. and Konrad, J: Estimating motion in image sequences. IEEE Signal Processing Magazine, vol. 16. (1999) 70-91
4. Cheng, H. D., Jiang, X. H, Sun, Y. and Wang, J.: Color image segmentation: advances and prospects. Pattern Recognition, vol.34. (2001) 2259-2281
5. Fraley, C. and Raftery, A. E.: How many clusters? Which clustering method? Answers via model-based cluster analysis. Technical Report 329, Department of Statistic University of Washington, (1998)
6. Jianbo, S and Malik, J.: Normalized cuts and image segmentation. IEEE Transaction on Pattern analysis and Machine Intelligence, Vol. 22, (2000) 888-905
7. Seo, Y., Choi, S., Kim, H. and Hong, K.S.: Where are the ball and players?: Soccer Game Analysis with Color-based Tracking and Image Mosaic. Proceedings of Int. Conference on Image Analysis and Processing (ICIAP), (1997) 196-203

8. Utsumi, O., Miura, K., Ide, I., Sakai, S. and Tanaka, H.: An Object Detection Method for Describing Soccer Games from Video. Proceedings of IEEE Int. Conference on Multimedia and Expo. (ICME), vol. 1, (2002) 45-48
9. Sudhir, G., Lee, J. C. M., Jain, A. K.: Automatic Classification of Tennis Video for High-level Content-based Retrieval. International Workshop on Content-Based Access of Image and Video Databases (CAIVD), (1998) 81-90
10. Hua, W., Han, M. and Gong, Y.: Baseball Scene Classification Using Multimedia Features. Proceedings of IEEE International Conference on Multimedia and Expo., Vol. 1, (2002) 821-824
11. Xu, P., Xie, L., Chang, S.F., Divakaran, A., Vetro, A. and Sun, H.: Algorithms and System for Segmentation and Structure Analysis in Soccer Video. Proceedings of IEEE Int. Conference on Multimedia and Expo.(ICME), (2001) 928-931
12. Raja, Y., McKenna, S. and Gong, S.: Color model selection and adaptation in dynamic scenes. In 5th European Conference on Computer Vision, (1998) 460–474
13. Liu, J. G. and Moore, J. M.: Hue image RGB color composition. A simple technique to suppress shadow and enhance spectral signature. International Journal of Remote Sensing, vol. 1, (1990) 1521–1530
14. Materka, A. and Strzelecki, M.: Texture analysis methods – A review. Technical Report, University of Lodz, Institute of Electronics, COST B11 Technical Report, Brussels 1998.
15. Williams, C.K.I. and Barber, D.: Bayesian classification with Gaussian processes, IEEE Transactions on Pattern Analysis and Machine Intelligence, Vol.20, (1998) 1342-1352
16. McLachlan and Krishnan, T.: The EM algorithm and extensions. Wiley, (1997)
17. ISO-13818-2: Generic Coding of moving pictures and associated audio (MPEG-2).

Adaptive Local Binarization Method for Recognition of Vehicle License Plates

Byeong Rae Lee[1], Kyungsoo Park[2], Hyunchul Kang[2], Haksoo Kim[3], and Chungkyue Kim[4]

[1] Dept. of Computer Science, Korea National Open University, 169 Dongsung-Dong, Chongro-Ku, Seoul, 110-791, Korea
brlee@knou.ac.kr

[2] Dept. of Information and Telecommunication Engineering, University of Incheon, 177 Dowha-Dong, Nam-Gu, Incheon, 402-749, Korea
{kspark, hckang}@incheon.ac.kr

[3] Dept. of Information and Telecommunication Engineering, Sungkonghoe University, 1-1 Hang-Dong, Kuro-Ku, Seoul, 152-716, Korea
hskim@mail.skhu.ac.kr

[4] Dept. of Computer Science and Engineering, University of Incheon, 177 Dowha-Dong, Nam-Gu, Incheon, 402-749, Korea
ckkim@incheon.ac.kr

Abstract. A vehicle license-plate recognition system is commonly composed of three essential parts: detecting license-plate region in the acquired images, extracting individual characters, and recognizing the extracted characters. But in the process, the problems like damage of license-plate and unequal light effect make it difficult to detect accurate vehicle license-plate region and to extract letters in that region. In this paper, to extract characters accurately in the license-plate region, a local adaptive binarization method which is robust under non-uniform lighting environment is proposed. To get better binary images, region-based threshold correction based on a prior knowledge of character arrangement in the license-plate is applied. With the proposed binarization method, 96% of 650 sample vehicle license-plates images are correctly recognized. Compared to existing local threshold selection methods, about 5% of improvement in recognition rate is obtained with the same recognition module based on LVQ.

1 Introduction

Recently the traffic density has been rapidly increased. On this account, the research regarding the automatic vehicle identification system on the road for managing the roads, inspecting the vehicles automatically and investigating the abnormal vehicles has been actively advance. This automatic number recognition system is composed largely with image capturing, number plate and character extracting and recognizing departments. A vehicle license-plate recognition system is commonly composed of three essential parts: detecting license-plate region in the acquired images, extracting individual characters, and recognizing the extracted characters.

R. Klette and J. Žunić (Eds.): IWCIA 2004, LNCS 3322, pp. 646–655, 2004.

Several methods for detecting license-plate region have been proposed.

First, a template matching method describes a collection of standard template of the license-plate region previously and then extracts the license-plate region from images by comparing the images with the predefined set of prototypes. This method shows high recognition rate when the license-plate is similar to a template in the predefined collection. This method, however, has some problems. It needs large number of templates to accept diverse types of vehicles and accordingly requires much recognition time.

Second, in the method using Hough transform, edges in the input images are detected first. Then Hough transform is applied to detect the license-plate regions. This method, however, has difficulty in extracting license-plate region when the boundary of the license-plate is not clear or the images contain lots of vertical and horizontal edges around the radiator grilles, and takes long processing time.

For the last, there is a method which uses brightness change in the image. This method is based on the property that the brightness change in the license-plate region is more remarkable and more frequent than the other region. Since this method does not depend on the edge of license-plate boundary, this method can be applied to the image with unclear license-plate boundary and can be implemented simple and fast algorithm. The correct license-plate extraction rate is relatively high compared to the former methods.

To recognize the numbers and letters in the license-plate region, the individual characters should be extracted. Some of the images taken from the roads, however, contain damaged license-plates that cause incorrect segmentation. Since this can lead to incorrect recognition, a robust segmentation algorithm is vital for vehicle license-plate recognition systems.

Several studies have been done to binarize vehicle license-plate images, including global threshold selection methods and local binarization methods which use local threshold selection techniques.

In this paper, we suggest a robust binarization method which operates individual character level, and refines the result based on a priori knowledge about approximate position and size of each character regions.

This paper consists of four chapters: the recognition process of the vehicle license-plate is discussed in chapter 2, the existing methods and the proposed method to binarize license-plate images and to extract individual characters are described in chapter 3, experimental results are provided in chapter 4, and finally conclusions are in chapter 5.

2 Recognition of the Vehicle License Plate

The system for recognizing the vehicle license plate consists of four parts, those are taking the images with cameras, segmenting the license-plate region, extracting the individual characters, and recognizing the extracted characters.

2.1 Extraction of the Vehicle License Plate

To recognize the vehicle license from its front images taken on the roads, the first thing to do is to extract the license-plate regions in the images. A 256-level gray-scale image is captured as input to the recognition system. The components that have characteristic features of license-plate in the image are detected and the similarity measure such as size of the component, relation between components, and concentration of the components is examined. The characteristic attributes of the license-plate used in this study is as follows.

1. More than 4 numeric characters are included in the license-plate region.
2. These numeric characters have relatively little damage. The difference of brightness between the numeric characters and background is relatively large.
3. The regions of the numeric characters are separated each other and have sizes within a certain range.
4. The numeric-character regions are close to each other.
5. The border of the license-plate contains horizontal and vertical edge components, and exists within a certain distance from the numeric character region.

These features are considerably robust. Damages can make the license-plate region not be extracted. But partial failure of detecting one or two character regions can be restored by inferring the lost region based on the regions of the neighboring characters.

To make this method effective, it is needed to minimize the influence of brightness variation. To do this, we perform license-plate extraction with an image that is obtained by removing low frequency components from the original image. The entire flow of the algorithm extracting the license plate is shown in Figure 1.

After sub-sampling as pre-processing, candidate regions for license plate are extracted from the car images. The low frequency images are obtained by applying a 7×7 smoothing filter on the original images.

These images are binarized with multiple thresholds to cope with uneven brightness distribution. The license-plates regions are searched in these binary images. Then the candidate regions for license-plate are sent to a recognizer. The recognizer generates license string by extracting the individual characters and identifying each characters.

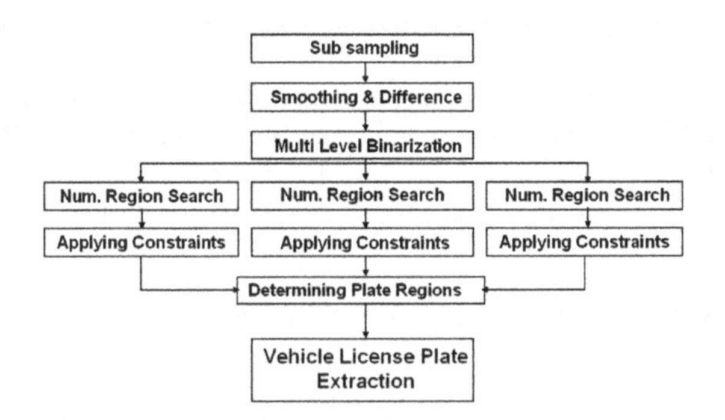

Fig. 1. Extracting license plate region

2.2 Recognition of the License Plate

The license plate image extracted from the input image is sent to character recognition module. Figure.2 shows a brief flow of the recognition process. A binary image is obtained from the extracted license plate image. Then individual characters are segmented and recognized.

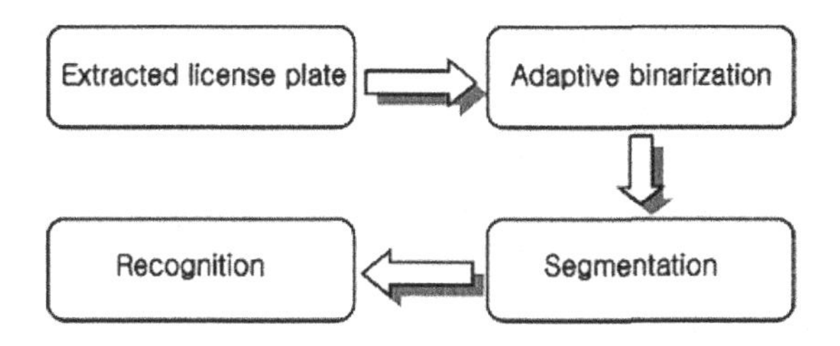

Fig. 2. Character recognition in license plate

2.3 The Recognition Module

To obtain individual character region from extracted license plate images, character and number regions in plate are segmented through adaptive binarization. The segmented characters are delivered to the recognition module that is implemented using a neural network based on LVQ (Learning Vector Quantizer).

The structure of LVQ neural network is shown in Figure 3. Each of the neurons in the competitive layer forms a reference vector representing a subset of training samples during the learning process.

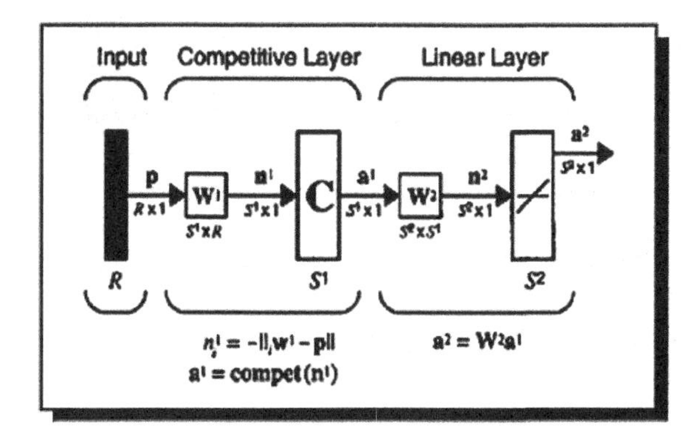

Fig. 3. Structure of LVQ neural network

3 Recognition of the Vehicle License Plate

It is necessary to binarize the images to distinguish between the character and background of the vehicle license plate as a pre-processing. To binarize the images of license plate, the binarization within individual character region is proposed in this paper.

3.1 The Established Binarization Methods

3.1.1 Global Binarization

Global binarization is a method that has single threshold *T* for entire image. The pixels with gray level greater than *T* are labeled as '1', while the others are labeled as '0'. Therefore the result of global binarization depends on single threshold *T*.

In the existing methods, the threshold is selected according to histogram analysis, complexity of images, and edges. In some extreme cases, users select thresholds manually. All of those methods, however, have problems when the brightness distribution in the image is not uniform. In that case, objects cannot be separated from the background with single threshold.

If an image is taken in a uniform lighting environment and the license plate is in good condition, global binarization can be an efficient method. In many cases, however, one threshold cannot produce proper result because of physical damages or uneven lighting conditions. Figure 4 shows an example with an uneven lighting environment. No single threshold can clearly discriminate all the characters from the background.

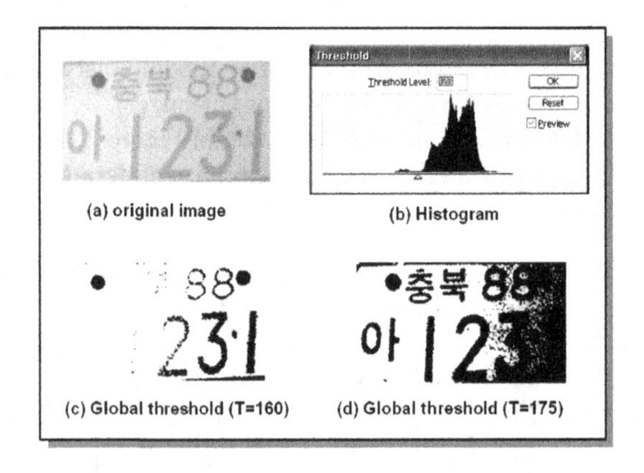

Fig. 4. License plate image with uneven lighting environment

3.1.2 Local Binarization

The brightness distribution of various positions in a license plate image may vary because of the condition of the plate and the effect of lighting environment. Since the binarization with one global threshold cannot produce useful results in such cases, local binarization methods are widely used. In many local binarization methods, an image is divided into m×n blocks and a threshold is chosen for each block. Figure.5 shows an example obtained using a local threshold method with block size 64×16.

The block level local binarization method can make better result than global binarization methods. This method, however, has problem caused by the discontinuity between adjacent blocks. As shown in Figure 6, one character can be placed on more than 2 adjacent blocks. If the difference in the thresholds between two adjacent blocks is too large, the binary shape of the character can be deformed and has considerable difference compared to training patterns. Consequently it is possible to decrease recognition rate.

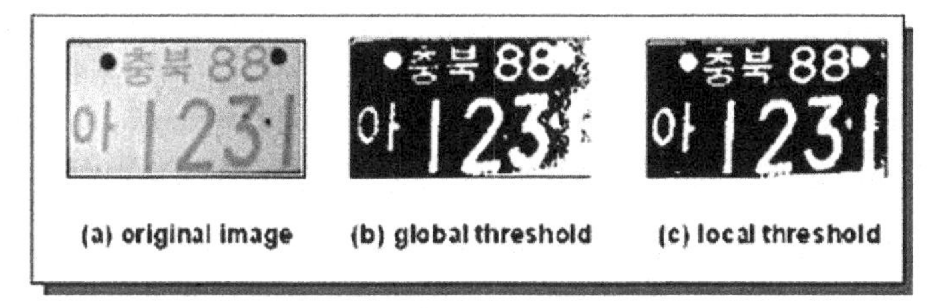

Fig. 5. Images of global and local binarization

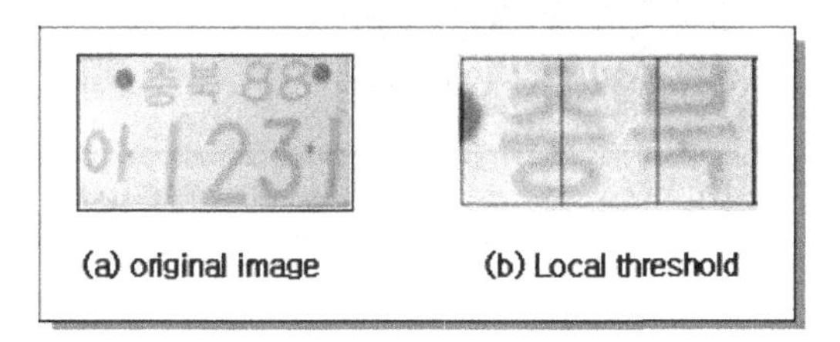

Fig. 6. Problem of local threshold

3.2 Proposed Binarization Method

Even though local binarization method can improve the problems in global methods, it still has problems in applying each threshold among blocks. To minimize the effect of discontinuity between adjacent blocks, a revised local binarization method which determines a threshold for each character region is proposed.

3.2.1 Adaptive Local Processing

As stated before, local binarazation methods which select thresholds for a given image divided into same size blocks can cause blocking effects, especially when a character is placed in more than 2 blocks. In this case, defining a block as each character region can be a effective method to handle this problem. Figure 7 shows a comparison of the shapes of blocks in existing local threshold binarization method and proposing adaptive local binarization method.

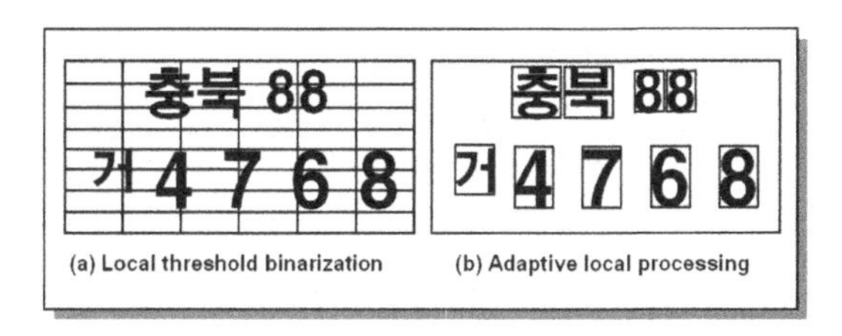

Fig. 7. Comparison of block-shapes in local binarization methods

In the proposed method, positions and sizes of rectangles that contain characters in a license plate must be determined first. For this purpose, histograms of pixel numbers corresponding to the characters in horizontal and vertical direction are examined.

To minimize the influence of noise, binarized edge image is used in horizontal and vertical projection. Figure 8 shows the pixel cumulation in horizontal direction to separate the rows of strings. The histogram is obtained by counting the pixels whose values are '1' along each horizontal line. The boundary region is defined as the point where the pixel cumulation is minimum within 1/4 to 3/4 of the vertical range.

Fig. 8. Region separation

The character region is searched to separate the characters in each row. For some Korean characters, two separate regions form one character region. In this case the two adjacent regions are merged into one region. This operation is based on the information such as average character width and type of character (Korean character or numeric character) at that location. Figure 9 is region segmentation result for the image in Figure 8. The character regions are segmented for upper and lower row respectively.

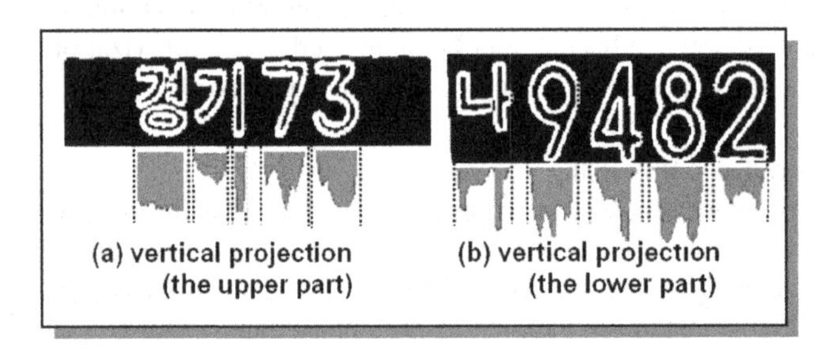

Fig. 9. Character region segmentation

Thresholds are determined for the regions respectively. Then each region is binarized with its own threshold value. Sometimes small meaningless regions can be detected. These regions are regarded as background if the sizes of the regions are small compared to normal characters.

3.2.2 Adaptive Region Split

Even in a region for one character, the brightness can vary due to physical bent license plate. Non-uniform illumination environment can cause similar effect, too. In majority of those cases, brightness change occurs mainly in vertical direction. The result of this problem is usually given in the form of partially missing region or broken region. To cope with such a problem, horizontal pixel cumulation histogram in a character region is checked to find out if there is split or missing in character region. If one character is divided into two parts or part of it is missing, the region is split into two regions and thresholds are determined for these regions respectively.

Figure 10 shows the results of binarization of a license plate image whose lower part is slightly bent. In Figure 10(a), four numeric characters in the second row are partially removed because the brightness of lower part is darker than upper part due to reduced light. Figure 10(b) is the result of proposed method. It can be seen that the missing parts of the characters are restored.

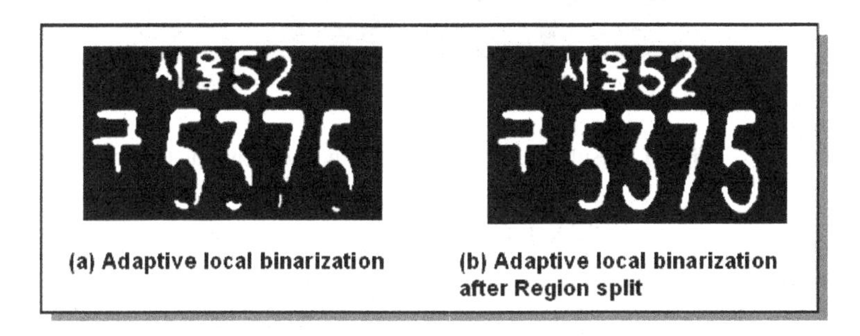

Fig. 10. Result of binarization after region split

4 Experiment and Effect

The experiment has been performed to evaluate the efficiency of modified local binarization which is presented in this paper with car images taken in tollgates. The size of the images is 1300×1024 and the algorithm is implemented using Visual C++6.0 on P-IV 2.4GHz, Windows-XP. Global binarization method, local binarization method and modified local binarization method proposed in this paper is applied to license plates images respectively to obtain binary images.

Figure 11 is an example showing the improvement of proposed method compared to other methods. Figure 11(b) shows the result of global binarization, and the right part of the binary image is not obtained properly. Figure 11(c) is the result of local binarization using uniform grid that divide the image. It can be noticed that part of the characters in the upper row are missing. Improved result obtained using proposed method is shown in Figure 11(d).

650 license plates images are used to evaluate the performance improvement. In Table 1, correct recognition rates obtained using the binary images obtained using local binarization method and proposed binarization method are shown. 96.76% of the license plates are successfully recognized with proposed method. More than 5% of correct recognition rate is obtained with the method.

There are some recognition failures in the experiment. The old version license plates with thin font are sensitive to noise. In spite of improvement in proposed method, significant physical damage on license plates make binarization and character extraction extremely difficult and cause recognition failure.

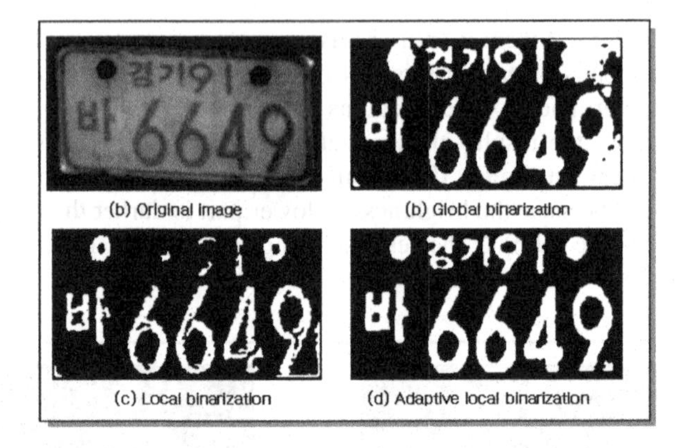

Fig. 11. Results of binarization with noisy license plate image

Table 1. Comparison of recognition rate

	Established Binarization Processing	Proposed method	Improvement
No. of total license plates	650		5%
Recognized license plates	596	629	
Success rate	91.69%	96.76	

5 Conclusion

In this paper, modified region-based local binarization method for license plate recognition system is proposed. Thresholds are chosen for character region in a license plate image.

Using the proposed method, improved binarization result can be obtained. As a result, characters in license plates can be accurately extracted, and recognition rate can be raised. Since the proposed method is based on detecting character region, the process needed to remove noise and border line of license plates is simplified. 96% of 650 license plate images are correctly recognized, and more than 5% of improvement is achieved compared to existing binarization methods.

Even though the recognition performance is improved, more effort is needed to extract license plate regions accurately in the images. Recognition algorithm should be improved too so that it is more robust in noisy environment.

References

1. D. Gao and J. Zhou.: Car License Plate Detection from Complex Scene. In Proceedings of International Conference on Signal Processing (2000) 1409-1414

2. J. Rosa and T. Pavlidis.: A Shape Analysis Model with Applications to Character Recognition System. IEEE Trans. Pattern Anal. Mach. Intell., Vol. PAMI-16 (1994) 393-404
3. N. A. Khan et. Al.: Synthetic Patttern Recognizer for Vehicle License Plates. IEEE Transaction on Vehicular Technology, Vol. 4, No. 4 (1995) 790-799
4. H. A. Hrgt, et. Al.: A high Performance License Plate Recognition System. Proc. IEEE intl. Conf. On System, Man and Cybernetics, Vol. 5 (1998) 4357-4362
5. Rafael C. Gonzales, Richard E. Woods.: Digital Image Procesing, 2nd Ed., Prentice Hall (1992)
6. Martin T. Hagan, Howard B. Demuth, Mark Beal.: Neural Network Design, Chapman & Hall (1996) 14_16-14_23
7. J. Ohya, A. Shio and S. Akamatsu.: Recognizing characters in scene images, IEEE Trans., Vol. PAMI-16 (1994) 214-220.
8. N. Otsu.: A Threshold Selection Method from Gray-scale histogram, IEEE Trans. On System, Man, and Cyberetics, SMC-8 (1978) 62-66.

Blur Identification and Image Restoration Based on Evolutionary Multiple Object Segmentation for Digital Auto-focusing*

Jeongho Shin, Sunghyun Hwang, Kiman Kim, Jinyoung Kang, Seongwon Lee, and Joonki Paik

Image Processing and Intelligent Systems Lab,
Department of Image Engineering,
Graduate School of Advanced Imaging Science, Multimedia, and Film,
Chung-Ang University,
221 Huksuk-Dong, Tongjak-Ku, Seoul 156-756, Korea
paikj@cau.ac.kr
http://ipis.cau.ac.kr

Abstract. This paper presents a digital auto-focusing algorithm based on evolutionary multiple object segmentation method. Robust object segmentation can be conducted by the evolutionary algorithm on an image that has several differently out-of-focused objects. After segmentation is completed, point spread functions (PSFs) are estimated at differently out-of-focused objects and spatially adaptive image restorations are applied according to the estimated PSFs. Experimental results show that the proposed auto-focusing algorithm can efficiently remove the space-variant out-of-focus blur from the image with multiple, blurred objects.

1 Introduction

A demand for digital multi-focusing techniques is rapidly increasing in many visual applications, such as camcorders, digital cameras, and video surveillance systems. Multi-focusing refers to a digital image processing technique that restores multiple, differently out-of-focused objects in an image. Conventional focusing techniques, such as manual focusing, infra-red auto-focusing (IRAF), through the lens auto-focusing (TTLAF), and semi-digital auto-focusing (SDAF), cannot inherently deal with multi-focusing function. Multi-focusing can be realized with fully digital auto-focusing (FDAF) based on PSF estimation and restoration. However, it additionally requires robust segmentation of multiple objects to estimate their individual PSFs.

* This work was supported in part by Korean Ministry of Science and Technology under the National Research Lab. Project, in part by Korean Ministry of Education under Brain Korea 21 Project, and in part by grant No.R08-2004-000-10626-0 from the Basic Research Program of the Korea Science & Engineering Foundation.

R. Klette and J. Žunić (Eds.): IWCIA 2004, LNCS 3322, pp. 656–668, 2004.

Given multiple objects at different distances from a camera, target objects are well-focused and the others are out-of-focus. The other objects are blurred in different degrees depending on their distances from the camera [1]. The amount of out-of-focus blur also depends on camera parameters such as lens position with respect to the image detector, focal length of the lens, and diameter of the camera aperture [1, 2]. Thus it is impractical to make all objects to be well-focused, when cameras have fixed focal lens or shallow depth of field.

In this paper, we propose a novel digital auto-focusing algorithm based on evolutionary multiple objects segmentation, which can restore all differently out-of-focused objects, simultaneously. The robust segmentation method of out-of-focused objects is one of the essential parts of the digital auto-focusing algorithm because the objects are distorted by out-of-focus blur. The existing algorithms such as Snake [3] or modified active contour models [6] do not have enough robustness with real image because they rely on the detection of the local gradient and can be easily distracted by noise or edges on the background. We adopt an evolutionary algorithm for segmenting differently out-of-focused objects.

Evolutionary algorithms [5, 6] solely depend on the survival of the fittest individuals which will replace their parents. Once the segmentation is completed, the restoration of the objects is performed by the PSF estimated with the segmented objects. In this paper, the PSF estimation technique is the modified version of Kim's algorithm [7], which has the improved accuracy of the estimated PSF using an isotropic PSF model and least-squares optimization. A block diagram of the proposed digital auto-focusing based on evolutionary multiple object segmentation algorithm is shown in Fig. 1.

This paper is organized as follows: In section 2, we explain the technique for segmenting multiple objects using evolutionary algorithm. In section 3, the corresponding isotropic PSFs in the image formation system are respectively identified according to each segmented region or objects, which can be considered as the estimation process in blind image restoration [7, 8, 9, 10]. Section 4 presents the spatially adaptive image restoration scheme that combines object segmentation and PSF estimation for restoring multiple objects. Experimental results and conclusions are given in section 5 and 6, respectively.

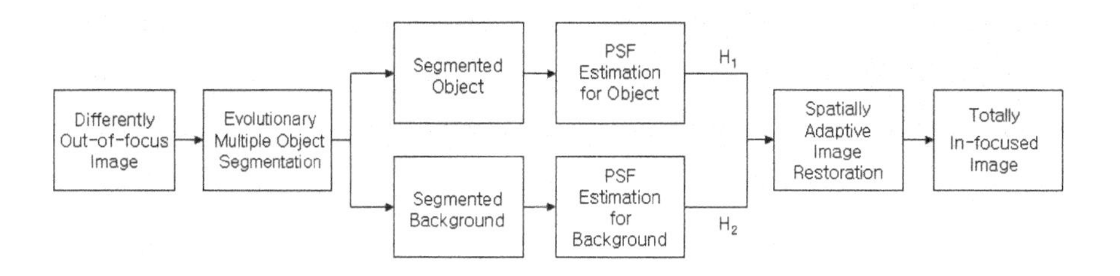

Fig. 1. A block diagram of the proposed digital auto-focusing based on evolutionary multiple object segmentation algorithm

2 Object Segmentation Using Evolutionary Algorithm

In this section, the object segmentation using the evolutionary algorithm is presented. To extract the precise boundaries of objects, the extracted initial boundaries can be used as a parent in the following evolutionary algorithm-based object segmentation.

The evolutionary algorithms are stochastic search methods, which incorporate aspects of natural selection or survival of the fittest. In other words, an evolutionary algorithm maintains a population of structures (i.e., it is usually randomly generated initially) that evolves according to rules of selection, crossover, mutation and survival, referred to as genetic operators. A shared 'environment', determines the fittest or performance of each individual in the population. The fittest individuals are more likely to be selected for reproduction (i.e., retention or duplication), while crossover and mutation modify those individuals, yielding potentially superior ones.

From the extracted initial boundary, the precise boundary can be obtained by the proposed evolutionary segmentation algorithm to be presented in this section. Starting from an initial population of feasible boundaries, the object boundary evolves throughout crossover and mutation operations that induce gradually better solution. The selection is performed by evaluating the fitness of the solutions. For example, Fig. 2 shows an example of the evolutionary algorithm using a synthetic noisy and bumpy image. The latter image was simulated using 10dB zero-mean Gaussian noise. For the experiment 1,000 initial populations were generated and 100 iterations were performed. The more complicated a shape becomes, the more number of initial populations and iterations are needed.

We will show the main components of the evolutionary algorithm for object segmentation in the remainder of this section. The proposed object segmentation based on evolutionary algorithm is summarized in Algorithm 1.

2.1 Initial Population

In order to start the evolutionary algorithm, we first generate an initial population of boundaries at random. The initial population can be generated under

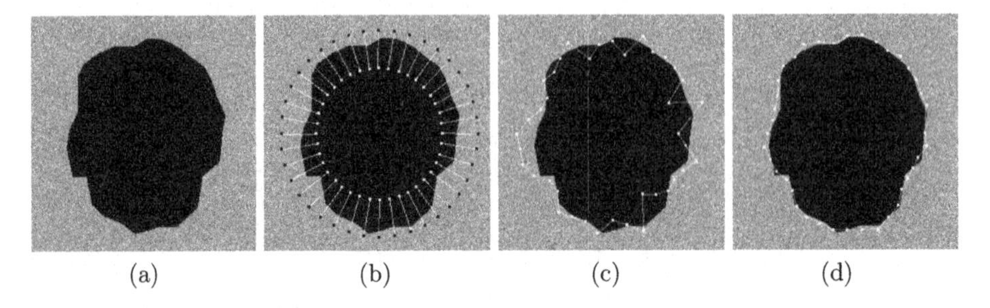

Fig. 2. Example of boundary extraction by evolutionary algorithm using a bumpy shape: (a) a bumpy shape with 10dB Gaussian noise, (b) magnified and reduced contours for initial population, (c) the initial contour, and (d) the detected contour of (a)

Algorithm 1 Evolutionary Object Segmentation.

1. Given an initial boundary, denoted as x, generate both magnified and reduced versions of x by 20%, denoted as x^R and x^M, respectively.
2. Take a random boundary in the range bounded by x^R and x^M. Repeat this step until the initial population size becomes N.
3. Crossover: Randomly choose a pair of boundaries, and choose one location and swap edge points on the corresponding location. Repeat this step N times.
4. Mutation: Choose an edge point from the first boundary, and perform mutation on it. Repeat this step for all N shapes.
5. Among $3N$ shapes, produced by steps 3, 4, and 5, choose N best shapes based on a fitting criterion.
6. Repeat steps 3 to 6 until the best fitted shape converges to the desired one.

assumption that any feasible boundary should be near the boundary of target object. The initial population represents a coarse boundary which can be approximated by some vertices. The vertex of the boundary is denoted as a two dimensional vector whose elements represent the position of the vertex.

In our algorithm the initial population can be obtained by generating random shapes between magnified and reduced versions of the initial boundary. As shown in Fig. 2b, a vertex can move along the line connecting the vertex of the inner contour and the corresponding vertex of the outer contour. The jth vertex of the ith contour can be generated by

$$v_i^j(s) = v_{inner}^j(s) + Rand(s)\{v_{outer}^j(s) - v_{inner}^j(s)\}, \tag{1}$$

where $v_i^j(s)$, $v_{outer}^j(s)$, and $v_{inner}^j(s)$ respectively denote the jth vertex of the ith contour, and the jth vertices of outer and inner contours. In Eq. (1), $Rand(s)$ generates a uniformly-distributed random number in the range $(0 < x < 1)$, with regard to the position of s. As a result, the initial population can be created, which usually consists of a large number of shapes. The appropriate number of the initial population depends on a specific application. The better the initial population, the easier and faster the search for the optimal boundary.

2.2 Crossover and Mutation

Crossover in biological terms refers to the blending of chromosomes from the parents to produce new chromosomes for the offspring. The analogy carries over to crossover in evolutionary algorithms. The evolutionary algorithm selects two shapes at random from the mating pool. Whether the selected boundaries are different or identical does not matter. We randomly choose one location and produce new boundaries by swapping all edge points on the corresponding location.

Mutation randomly changes an edge point of the boundaries of the population from time to time. The main reason for mutation is the fact that some local configurations of edge points of the boundaries of the population can be totally lost as a result of reproduction and crossover operations. Mutation protects evolutionary algorithms against such irrecoverable loss of good solution features.

2.3 Fitness and Selection

An evolutionary algorithm performs a selection process in which the "most fit" members of the population survive, and the "least fit" members are eliminated. As a result, the selected shapes are allowed to mate at random, creating offspring which is subject to crossover and mutation with constant probabilities. The evaluation of the degree of fitness depends on the value of the objective function.

In the proposed object segmentation algorithm, the selection criterion describes how close it is from the optimal boundary. The fitness of each population as a contour can be evaluated as

$$f(c_i) = \sum_{j=0}^{N} \left[\sum_{k=-3}^{3} \sum_{l=-3}^{3} |\nabla I(x_i^j + k, y_i^j + l)|^2 - |v_i^{j-1} - 2v_i^j + v_i^{j+1}|^2 \right], \quad (2)$$

where the vertex $v_i^j = [x_i^j, y_i^j]^T$, and $I(m,n)$ and c_i respectively represent the intensity at the position of (m,n) and ith contour. In Eq. (2), the first term of the ri ght side makes the contour be placed at the features such as edge and line, and the other term retains continuity between vertices.

3 Isotropic Blur Identification

In this section we present an isotropic PSF model and estimation algorithm without any recursive procedure or iterative optimization [8–10]. By considering a two dimensional (2D) isotropic PSF as a discrete model with a finite number of free coefficients, which determine a set of concentric circles, we can systematically estimate the PSF by solving a linear equation. The number of equations, that is constraints, must be equal to or larger than that of different coefficients to avoid underdetermined condition. It is highly desirable to have more measurements than the number of unknown coefficients to obtain reliable estimation results under noise-prone condition.

3.1 2D Isotropic PSF Model

The discrete approximation of an isotropic PSF is shown in Fig. 3. As shown in Fig. 3, many pixels are located off concentric circles within the region defined as

$$S_R = \{(m,n) | \sqrt{m^2 + n^2}\}, \quad (3)$$

where R is the radius of the PSF. Each pixel within the support is located either on the concentric circles or not. The pixels on a concentric circle are straightforwardly represented as the PSF coefficients. On the other hand, pixels

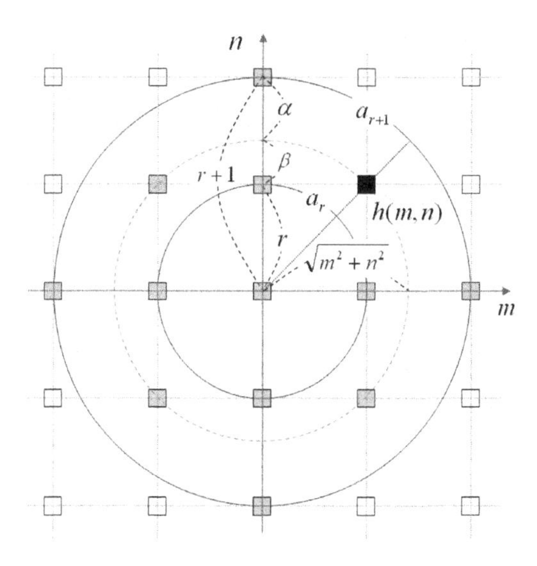

Fig. 3. The geometrical representation of a 2D isotropic discrete PSF: The gray rectangles represent inner pixels of the PSF and the empty ones are outer pixels of the PSF

off a concentric circle are not described by those. So these pixels should be interpolated by using adjacent pixel on the concentric circle as

$$\tilde{h}(m,n) = \begin{cases} \alpha a_r + \beta a_{r+1}, & \text{if}(m,n) \in S_R, \\ 0, & \text{elsewhere}, \end{cases} \tag{4}$$

where a_r and a_{r+1} respectively represent the rth and the $r+1$st entries of the PSF coefficient vector. In Eq. (4), index r is determined as

$$r = \left\lfloor \sqrt{m^2+n^2} \right\rfloor, \tag{5}$$

where $\lfloor \cdot \rfloor$ is the truncation operator to integer. Based on Fig. 3, α and β are determined as

$$\alpha = r - 1 + \sqrt{m^2+n^2}, \text{ and } \beta = 1 - \alpha. \tag{6}$$

This approximation of 2D discrete PSF is available to the popular isotropic blurs, such as Gaussian out-of-focus blur, uniform out-of-focus blur, and x-ray scattering.

3.2 Estimation of the PSF Coefficients

In order to estimate all PSF coefficients, we call the fundamental relationship between 2D PSF and 1D step response [1]. Now, let assume that an original image is same as simple pattern image $f_P(k,l)$, shown in Fig. 4a, can be represented as

$$f_P(k,l) = \begin{cases} i_L, & \text{if } 0 \le k < N, \text{ and } 0 \le l < t, \\ i_H, & \text{if } 0 \le k < N, \text{ and } t \le l < N, \end{cases} \tag{7}$$

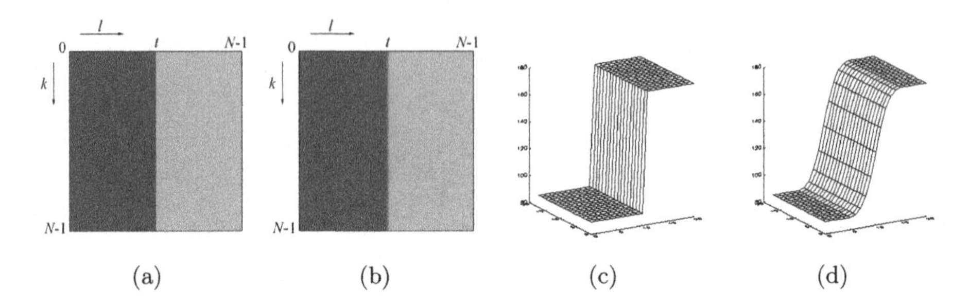

Fig. 4. The pattern images and its edges: (a) original simple pattern image, (b) blurred pattern image, (c) edges of (a), and (d) edges of (b)

where the constant t is a boundary of left and right of the pattern image. The blurred pattern image $g_P(k,l)$, shown in 4b, is obtained by convolving Eq. (7) and Eq. (4) as

$$g_P(k,l) = \begin{cases} i_L, & \text{if } 0 \leq k < N, \text{ and } 0 \leq l < t-R, \\ s(i), & \text{if } 0 \leq k < N, \text{ and } t-R \leq l \leq t+R, \\ i_H, & \text{if } 0 \leq k < N, \text{ and } t+R < l < N, \end{cases} \tag{8}$$

where R is the radius of the PSF, and the sequence $s(i)$ represents the 1D step response and $s(i) = \{s_0,\ s_1,\ \ldots,\ s_{2R}\}$. Each observed 1D step response corresponds to partial summation of the PSF as

$$s_i = i_H \left[\sum_{m=-R}^{R} \sum_{n=-R}^{-R+i} \tilde{h}(m,n) \right] + i_l, \quad 0 \leq i < 2R. \tag{9}$$

Unpacking the summation in Eq. (9) yields the following linear simultaneous equations:

$$\begin{aligned} s_0 &= i_H \{\tilde{h}(0,-R) + 2[\tilde{h}(1,-R) + \cdots + \tilde{h}(R,-R)]\} + i_l, \\ s_1 &= i_H \{\tilde{h}(0,-R+1) + 2[\tilde{h}(1,-R+1) + \cdots + \tilde{h}(R,-R+1)]\} + s_0, \\ &\vdots \\ s_{2R} &= i_H \{\tilde{h}(0,R) + 2[\tilde{h}(1,R) + \cdots + \tilde{h}(R,R)]\} + s_{2R-1}. \end{aligned} \tag{10}$$

Furthermore, using the discrete approximation, we can substitute all PSF elements into linear combination of free coefficients of the PSF as derived in Eq. (4). For examples, first and last equations in Eq. (10) always become

$$s_0 = i_H a_R + i_L, \quad \text{and} \quad s_{2R} = i_h a_R + s_{2R-1}. \tag{11}$$

Eq. (10) can be rewritten in a compact matrix-vector form as

$$\mathbf{s} = \mathbf{D}\mathbf{a}, \tag{12}$$

where $\mathbf{s}$ is a $R+1$ dimensional vector defined as

$$\mathbf{s} \equiv \frac{1}{i_H} \begin{bmatrix} s_0 - i_L & s_1 - i_L & \cdots & s_{2R} - i_L \end{bmatrix}^T, \tag{13}$$

and $\mathbf{a}$ is a $R+1$ dimensional PSF coefficients vector. The matrix $\mathbf{D}$ is not specified in a closed form, but has a form of

$$\mathbf{D} = \begin{pmatrix} \times & 0 & \cdots & 0 \\ \times & \times & \ddots & 0 \\ \vdots & \vdots & \ddots & 0 \\ \times & \times & \times & \times \\ \vdots & \vdots & \vdots & \vdots \\ \times & \times & \times & \times \end{pmatrix}, \tag{14}$$

where $\times$'s denote arbitrary nonzero entries. In equation uncorrupted data should be contaminated by the noise which was added to the degradation process. If s contains measurement error, then the corrupted version can be represented as

$$\hat{\mathbf{s}} = \mathbf{Da}. \tag{15}$$

Because the measurement error $\mathbf{e}$ is unknown, the best we can then do is to choose an estimator $\hat{\mathbf{a}}$ that minimizes the effect of the errors in some sense. For mathematical convenience, a natural choice is to consider the least-squares criterion,

$$\varepsilon_{LS} = \frac{1}{2}\|\mathbf{e}\|^2 = \frac{1}{2}(\hat{\mathbf{s}} - \mathbf{Da})^T(\hat{\mathbf{s}} - \mathbf{Da}) \tag{16}$$

Minimization of the least-squares error in Eq. (16) with respect to the unknown coefficients $\mathbf{a}$ leads to so-called *normal equations* [11]

$$(\mathbf{D}^T\mathbf{D}) = \hat{\mathbf{a}}_{LS} = \mathbf{D}^T\hat{\mathbf{s}}, \tag{17}$$

which determines the least-squares estimate of $\mathbf{a}$. Note that the shape of the observation matrix $\mathbf{D}$ guarantees its columns to be independent. Thus, the $(R+1)\times(R+1)$ *Grammian matrix* $\mathbf{D}^T\mathbf{D}$ is *positive-definite* [11] and we can explicitly solve the normal equations by rewriting Eq. (17) as

$$\hat{\mathbf{a}}_{LS} = (\mathbf{D}^T\mathbf{D})^{-1}\mathbf{D}^T\hat{\mathbf{s}} = \mathbf{D}^+\hat{\mathbf{s}}, \tag{18}$$

where $\mathbf{D}^+ = (\mathbf{D}^T\mathbf{D})^{-1}\mathbf{D}^T$ is the *pseudoinverse* of $\mathbf{D}$. The optimal coefficients are used in constructing PSF which was modeled in Eq. (4).

4 Spatially Adaptive Image Restoration

Iterative image restoration is the most suitable for multi-focusing because: (i) There is no needto determine or implement the inverse of an operator, (ii) knowledge about the solution can be incorporated into the restoration process, (iii) the solution process can be monitored as it progresses, and (iv) constraints can be used to control the effects of noise [12].

The image degradation model is given as

$$\mathbf{y} = \mathbf{H}\mathbf{x}, \tag{19}$$

where $\mathbf{y}$, $\mathbf{H}$, and $\mathbf{x}$ respectively represent the observed image, the degradation operator, and the original image. A general image restoration process based on the constrained optimization approach is to find $\hat{\mathbf{x}}$ which minimizes

$$\|\mathbf{C}\hat{\mathbf{x}}\|^2, \tag{20}$$

subject to

$$\|\mathbf{y} - \mathbf{H}\hat{\mathbf{x}}\|^2 \leq \epsilon^2, \tag{21}$$

where $\hat{\mathbf{x}}$, $\mathbf{C}$, and ϵ^2 respectively represent the restored image, a high-pass filter for incorporating a priori smoothness constraint, and an upper bound of error residual of the restored image. The constrained optimization problem, described in Eq. (20) and Eq. (21), can be solved by minimizing the following functional

$$\hat{\mathbf{x}} = \overset{\arg\min}{\mathbf{x}} \; f(\mathbf{x}), \tag{22}$$

where

$$f(\mathbf{x}) = \|\mathbf{y} - \mathbf{H}\mathbf{x}\|^2 + \|\mathbf{C}\mathbf{x}\|^2. \tag{23}$$

In Eq. (23), $\mathbf{C}$ represents a high-pass filter, and $\|\mathbf{C}\mathbf{x}\|$ represents a stabilizing functional whose minimization suppresses high frequency components due to noise amplification. The regularization parameter λ controls the fidelity to the original image and smoothness of the restored image. The argument $\mathbf{x}$ that minimizes the functional $f(\mathbf{x})$ is given by the following expression [13]

$$(\mathbf{H}^T\mathbf{H} + \lambda\mathbf{C}^T\mathbf{C})\mathbf{x} = \mathbf{H}^T\mathbf{y}. \tag{24}$$

The operator $\left(\mathbf{H}^T\mathbf{H} + \lambda\mathbf{C}^T\mathbf{C}\right)$ has a continuous inverse [13] or it is a better conditional matrix than $\mathbf{H}$ or $\mathbf{H}^T\mathbf{H}$, assuming that the matrix $\mathbf{C}$ has been chosen properly. The regularized solution given by Eq. (24) can be successively approximated by means of the iteration

$$\begin{aligned} \mathbf{x}_0 &= 0, \\ \mathbf{x}_{k+1} &= \mathbf{x}_k + \beta\left[\mathbf{H}^T\mathbf{y} - \left(\mathbf{H}^T\mathbf{H} + \lambda\mathbf{C}^T\mathbf{C}\right)\mathbf{x}_k\right], \end{aligned} \tag{25}$$

where β is a multiplier that can be constant or it can depend on the iteration index.

In order to remove the space-variant out-of-focus blur, Eq. (25) should be modified, for each pixel $x(p)$ which is included in the segmented region denoted by s, as

$$x^{k+1}(p) = x^k(p) + \beta\left(\mathbf{e}_p^T\mathbf{H}_s\mathbf{y} - \mathbf{e}_p^T\mathbf{T}_s\mathbf{x}^k(p)\right), \text{ for } p = (i-1)M + j, \; 0 < s \leq L-1, \tag{26}$$

where $\mathbf{H}_s$ represents the degradation matrix for the segment containing $x(p)$, $\mathbf{e}_p$ the pth unit vector, L the number of different segmented regions, and finally $\mathbf{T}_s = \mathbf{H}^T\mathbf{H} + \lambda\mathbf{C}^T\mathbf{C}$.

5 Experimental Results

In order to demonstrate the performance of the proposed algorithm, we used a set of real images with one or more differently out-of-focused objects from background. We also present the results of both the evolutionary segmentation algorithm and snake [3] for comparing the performance of contour extraction.

5.1 Simultaneously Digital Auto-focusing for an Object and Background

In the first experiment, we used a real image captured by using a Nikon-D100 digital still camera. The image has an object differently out-of-focused from background, as shown in Fig. 5a. The initial populations are randomly produced within the user defined region which is determined as the inner and outer circles with a priori knowledge. In this experiment, 3,000 initial populations are produced for boundary extraction. The user defined region and an extracted contour are depicted in Fig. 5b and Fig. 5d, respectively. The contour is approximated by combining 30 points by nature. For the comparison purpose, snake is performed for the same image, and the result is shown in Fig. 5c with of $\alpha = 0.2$ and $\beta = 1.2$ for internal force and $\gamma = 3$ for image force [3]. It takes a little long time to accomplish boundary extraction with evolutionary algorithm, but this algorithm gives more acceptable result than the gradient-based contour extraction techniques. Then the object can be segmented by applying the morphological processing of the region filling [12] to the contour. The seed point is determined as the mean position of all contour points. The segmented result is shown in Fig. 5e. Next, blurs are estimated in the object and background, separately. The method proposed in section 3 were used for estimating each blur. The restored version of the original blurred image is obtained by using the method proposed in section 4. The restored image is shown in Fig. 5f.

5.2 Digital Auto-focusing for Two or More Objects

In this experiment, we used an image that has two differently out-of-focused objects. The image is also captured by the same digital camera used in previous subsection and is shown in Fig. 6a.

The evolutionary contour extraction algorithm adjusts to the image containing two objects. We assume that the location of each object is given a priori without loss of practical applicability. The contours extracted by the evolutionary algorithm are shown in Fig. 6b. Two objects are segmented by the morphological region filling with each seed chosen by the same method in the previous subsection. The result of this segmentation is shown in Fig. 6c where different gray values distinguish two objects. The restored image is shown in Fig. 6d.

6 Conclusions

We proposed a fully digital auto-focusing algorithm for restoring the image with multiple, differently out-of-focused objects, which is based on evolutionary mul-

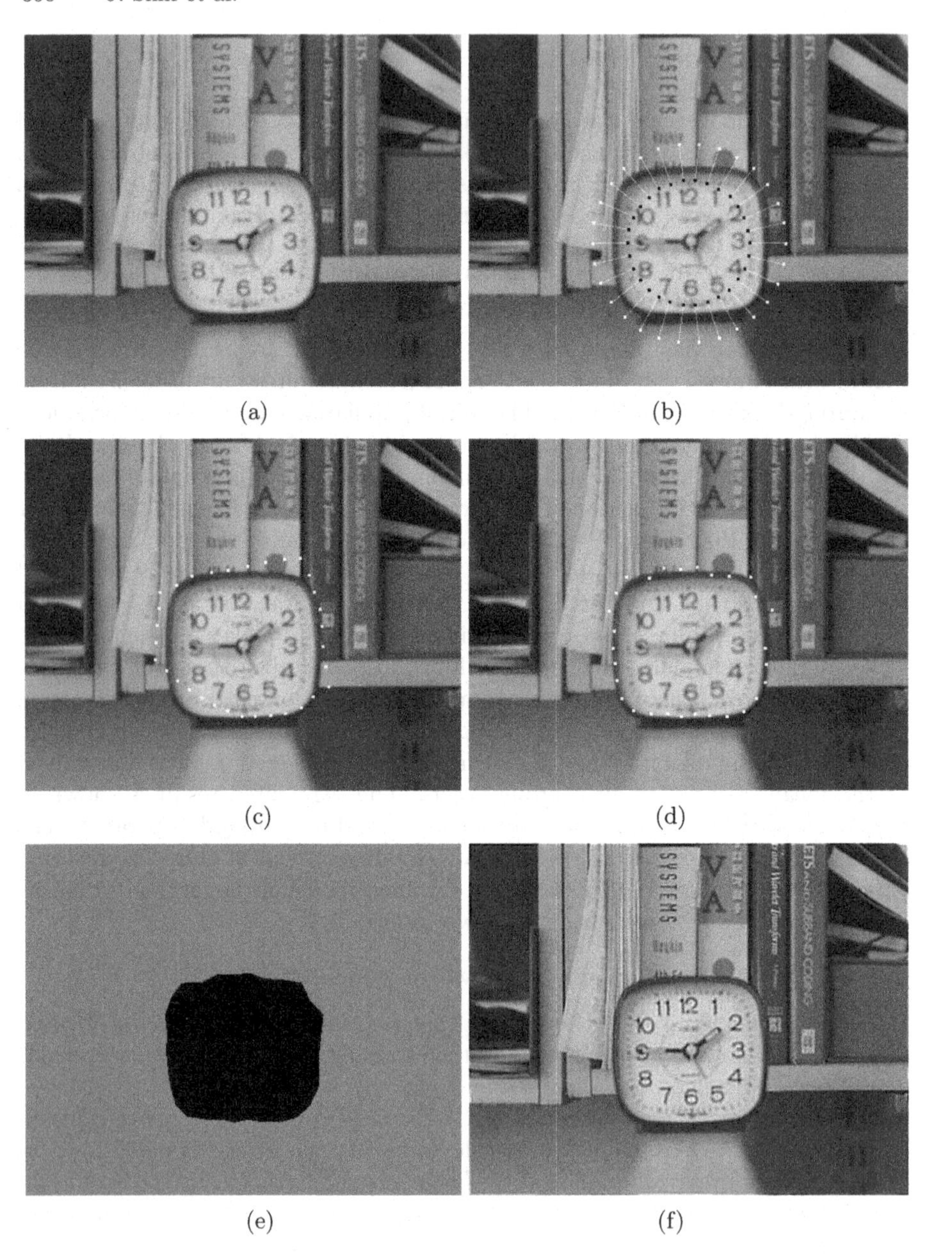

Fig. 5. The results of contour extraction and restoration for simultaneously digital auto-focusing with a real captured image: (a) Original clock image in which a clock is differently out-of-focus from background, (b) the user defined region, (c) the extracted contour by using snake, (d) the extracted contour using the evolutionary algorithm, (e) the segmented object from (d), and (f) the in-focused image by using the proposed restoration algorithm

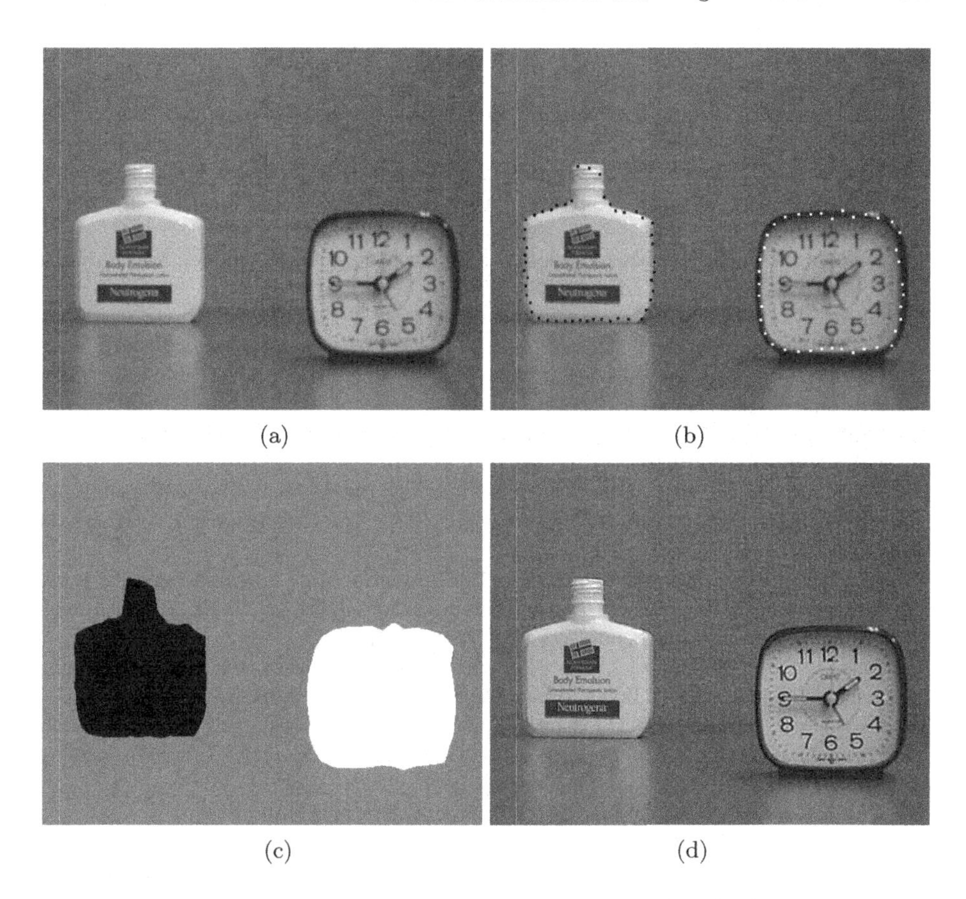

Fig. 6. The results of contour extraction and restoration for simultaneously digital auto-focusing with a real captured image involving two differently blurred objects: (a) Original image with two differently blurred objects, (b) the extracted contours using the evolutionary algorithm, (c) the segmented objects, and (d) the in-focused image by using the proposed restoration algorithm

tiple object segmentation. The proposed algorithm can be developed by object based image segmentation and an adaptive image restoration algorithm. For space-variant image restoration, the input image should be segmented into differently out-of-focused objects or background. In order to obtain robustly segmented objects, evolutionary algorithms are applied to the proposed object segmentation algorithm. In the experimental results, we showed that the proposed restoration algorithm can efficiently remove the space-variant out-of-focus blur from the image with multiple blurred objects. In addition, it is possible to apply the proposed digital auto-focusing algorithms to the surveillance video for human tracking if the methods of deformable contour extraction, such as active shape model, can be combined efficiently.

References

1. Andrews, H.C., Hunt, B.R.: Digital Image Restoration. Prentice-Hall New Jersey (1977)
2. Subbarao, M., Tyan, J.K.: Selecting the optimal focus measure for autofocusing and depth-from-focus. IEEE Trans. Pattern Analysis and Machine Intelligence **20** (1998) 864–870
3. Kass, M., Witzkin, A., Terzopoulos, D.: Snake: Active contour model. International Journal of Computer Vision (1988) 321–331
4. Blake, A., Isard, M.: Active Contours. Springer-Verlag, Berlin Heidelberg New York (1998)
5. Goldberg, D.: Genetic Algorithm in Search, Optimization and Machine Learning. Addision-Wesley (1989)
6. Sonka, M., Hlavac, V., Boyle, R.: Image Processing, Analysis, and Machine Vision. PWS Publishing (1999)
7. Kim, S.K., Park, S.R., Paik, J.K.: Simultaneous out-of-focus blur estimation and restoration for digital auto-focusing system. IEEE Trans. Consumer Electronics **34** (1998) 1071–1075
8. Lagendijk, R.L., Biemond, J., Boekee, D.E.: Identification and restoration of noisy blurred image using the expectation-maximization algorithm. IEEE Trans. Acoustic, Speech and Signal Processing **38** (1990) 1180–1191
9. Reeves, S.J., Mersereau, M.R.: Blue identification by the method of generalized cross-validation. IEEE Trans. Image Processing **1** (1992) 301–311
10. Lun, D.P.K., Chan, T.C.L., Hsung, T.C., Feng, D.D., Chan, Y.H.: Efficient blind restoration using discrete periodic radon transform. IEEE Trans. Image Processing **13** (2004) 188–200
11. Noble, B., Daniel, J.: Applied Linear Algebra. Prentice Hall (1988)
12. Katsaggelos, A.K.: Iterative image restoration algorithms. Optical Engineering **287** (1989) 735-748
13. Miller, K.: Least-squares method for ill-posed problems with a prescribed bound. SIAM J. Math. Anal. **1** (1970) 52-57
14. Pratt, W. K., Digital Image Processing. 2nd Ed. John Wiley (1991)

Performance Evaluation of Binarizations of Scanned Insect Footprints

Young W. Woo

Division of Computer & Visual Engineering,
College of Engineering, Dongeui University,
San 24, Gaya-Dong, Pusanjin-Gu, Pusan, 614-714, Korea

Abstract. The paper compares six conventional binarization methods for the special purpose of subsequent analysis of scanned insect footprints. We introduce a new performance criterion for performance evaluation. The six different binarization methods are selected from different methodologically categories, and the proposed performance criterion is related to the specific characteristics of insect footprints having a very small percentage of object areas. The results indicate that a higher-order entropy binarization algorithm, such as proposed by Abutaleb, offers best results for further pattern recognition application steps for the analysis of scanned insect footprints.

Keywords: binarization, insect footprints, pattern recognition.

1 Introduction

In order to obtain accurate pattern recognition results using binarized images, it is obviously important to choose a suitable binarization algorithm. There exists a wide variety of binarization algorithms [1], and there is no "generally best" binarization algorithm for all kinds of gray level images. Some binarization algorithms are good for certain types of grey images but bad for other types of grey images. The project of interest (at CITR Auckland) is the classification of insects based on scanned tracks. Insects walk across a preinked card (as produced by Connovation Ltd., New Zealand), and leave a track on a white card which will be scanned on a standard flatbed scanner. The generated pictures are of very large scale due to the required resolution. Applications are related to, for example, pest control, environment protection, or monitoring of insect numbers.

We compared six representative binarization algorithms for binarizing scanned insect footprints, which have a relatively small object area compared to the non-object area, and selected the best binarization algorithm for pattern recognition for this particular application using a new performance criterion. The proposed binarization performance criterion is based on characteristics of insect footprints. In Section 2, we detail methods and formulas of six binarization algorithms. In Section 3, the proposed binarization performance criterion is presented. In Section 4, test images for experiments are shown and experimental results are presented. Finally we provide conclusions in Section 5.

R. Klette and J. Žunić (Eds.): IWCIA 2004, LNCS 3322, pp. 669–678, 2004.

2 Six Binarization Algorithms

Research on binarization of gray images dates back for more than thirty years. The paper [1] compares 40 binarization algorithms by two kinds of test data set. The first test data set is the set of 40 NDT (Non Destructive Testing) images, and the second test data set is the set of 40 document images. Binarization evaluation ranking tables in [1] show that the performance of each algorithm is different when different types of test data sets are applied. Only the binarization algorithm by Kittler and Illingworth [4] is best for both kinds of test data sets. So, as a first conclusion we know that conventional binarization algorithms' performance highly depends on the kind of gray images used in a particular application. In our case we have to binarize scanned insect footprints. We selected six different binarization algorithms(three global binarization algorithms and three dynamic binarization algorithms) that behaved "relatively good" for the two kinds of test data set in [1]. We chose three global binarization algorithms because they have advantage in processing time. If a certain global binarization algorithm has enough good performance on scanned insect footprints, the recognition system using the global binarization algorithm will have a merit in computing time. The selected six algorithms are as follows:

- Rosenfeld's convex hull binarization algorithm [2],
- Otsu's clustering binarization algorithm [3],
- Kittler and Illingworth's minimum error binarization algorithm [4],
- Kapur's entropic binarization algorithm [5],
- Abutaleb's higher-order entropy binarization algorithm [6], and
- Bernsen's local contrast binarization algorithm [7].

For a brief explanation of each algorithm, we use the following notation. The histogram and the probability mass function (PMF) of the image are indicated, respectively, by $h(g)$ and $p(g)$, for $g = 0 \ldots G_{max}$, where G_{max} is the maximum gray level in the image (which is typically 255). If the gray level range is not explicitly limited to a subinterval $[g_{\min}, g_{\max}]$, it will be assumed to be from 0 to G_{max}. The cumulative probability function is defined as

$$P(g) = \sum_{i=0}^{g} p(i) \tag{1}$$

The object (or *foreground*) and non-object (or *background*) PMFs are expressed as $P_f(g)$, for $0 \leq g \leq T$, and $P_b(g)$, for $T+1 \leq g \leq G$, respectively, where T is the threshold value. The object and non-object area probabilities are calculated as follows:

$$P_f(T) = P_f = \sum_{g=0}^{T} p(g) \quad \text{and} \quad P_b(T) = P_b = \sum_{g=T+1}^{G} p(g) \tag{2}$$

The Shannon entropy, parametrically dependent on the threshold value T for foreground and background, is formulated as follows:

$$H_f(T) = -\sum_{g=0}^{T} p_f(g)\log p_f(g) \quad \text{and} \quad H_b(T) = -\sum_{g=T+1}^{G} p_b(g)\log p_b(g) \tag{3}$$

The mean and variance of the foreground and background as functions of the thresholding level T are denoted as follows:

$$m_f(T) = \sum_{g=0}^{T} g \cdot p(g) \quad \text{and} \quad \sigma_f^2(T) = \sum_{g=0}^{T} [g - m_f(T)]^2 p(g)$$

$$m_b(T) = \sum_{g=T+1}^{G} g \cdot p(g) \quad \text{and} \quad \sigma_b^2(T) = \sum_{g=T+1}^{G} [g - m_b(T)]^2 p(g) \tag{4}$$

2.1 Rosenfeld's Convex Hull Binarization Algorithm

This algorithm is based on analyzing the concavities of the histogram $h(g)$ defined by its convex hull, $H(g)$; that is the set-theoretic difference $|H(g) - p(g)|$. When the convex hull of the histogram is calculated, the "deepest" concavity points become candidates for a threshold. In case of competing concavities, some object attribute feedback, such as low busyness of the edges of the thresholded image, can be used to select one of them. In this algorithm, the following equation is used for finding an optimal threshold value:

$$T_{opt} = \arg\max\{[p(g) - H(g)]\} \tag{5}$$

2.2 Otsu's Clustering Binarization Algorithm

This algorithm is to minimize the weighted sum of within-class variances of the foreground and background pixels to establish an optimum threshold. Recall that minimization of within-class variances is tantamount to the maximization of between-class scatter. This method gives satisfactory results when the numbers of pixels in each class are close to each other. The Otsu's algorithm is today one of the most referenced binarization algorithms. In this algorithm, the following equation is used for finding optimal threshold value:

$$T_{opt} = \arg\max\{\frac{P(T)[1 - P(T)][m_f(T) - m_b(T)]^2}{P(T)\sigma_f^2(T) + [1 - P(T)]\sigma_b^2(T)}\} \tag{6}$$

2.3 Kittler and Illingworth's Minimum Error Binarization Algorithm

This algorithm assumes that the image can be characterized by a mixture distribution of foreground and background pixels: $p(g) = P(T) \cdot p_f(g) + [1 - P(T)] \cdot p_b(g)$.

Kittler and Illingworth's algorithm does not need the assumption that the foreground and background distribution function is an equal variance Gaussian density function and, in essence, addresses a minimum error Gaussian density-fitting problem. In this algorithm, the following equation is used for finding optimal threshold value:

$$\begin{aligned} T_{opt} = \arg\min\{&P(T)\log\sigma_f(T) + [1-P(T)]\log\sigma_b(T) \\ &-P(T)\log P(T) - [1-P(T)]\log[1-P(T)]\} \end{aligned} \tag{7}$$

where $\{\sigma_f(T),\ \sigma_b(T)\}$ are foreground and background standard deviations.

2.4 Kapur's Entropic Binarization Algorithm

This algorithm assumes the image foreground and background as two different signal sources, so that when the sum of the two class entropies reaches its maximum, the image is said to be optimally binarized. In this algorithm, the following equation is used for finding an optimal threshold value:

$$T_{opt} = \arg\max[H_f(T) + H_b(T)] \text{ with} \tag{8}$$

$$H_f(T) = -\sum_{g=0}^{T} \frac{p(g)}{P(T)} \log\frac{p(g)}{P(T)} \quad \text{and} \quad H_b(T) = -\sum_{g=T+1}^{G} \frac{p(g)}{P(T)} \log\frac{p(g)}{P(T)}$$

2.5 Abutaleb's Higher-Order Entropy Binarization Algorithm

This algorithm assumes the joint entropy of two related random variables, namely, the image gray level g at a pixel, and the average gray level $\bar{g}$ of a neighborhood centered at that pixel. Using the 2-D histogram $p(g,\ \bar{g})$, for any threshold pair $(T,\bar{T})$, one can calculate the cumulative distribution $P(T,\ \bar{T})$, and then define the foreground entropy as

$$H_f = -\sum_{i=1}^{T}\sum_{j=1}^{\bar{T}} \frac{p(g,\bar{g})}{P(T,\bar{T})} \log\frac{p(g,\bar{g})}{P(T,\bar{T})} \tag{9}$$

Similarly, one can define the background region's second order entropy. Under the assumption that the off-diagonal terms (i.e., the quadrants $[(0,\ T),\ (\bar{T}\ ,G)]$ and $[(T,\ G),\ (0,\ \bar{T})]$) are negligible and contain elements only due to image edges and noise, the optimal pair $(T,\ \bar{T})$ can be found as the minimizing value of the 2-D entropy functional. In this algorithm, the following equation is used for finding an optimal threshold value:

$$(T_{opt}, \bar{T}_{opt}) = \arg\min\{\log[P(T,\bar{T})[1-P(T,\bar{T})]] + H_f/P(T,\bar{T}) + H_b/[1-P(T,\bar{T})]\}$$

where $H_f = -\sum_{i=1}^{T}\sum_{j=1}^{\bar{T}} \frac{p(g,\bar{g})}{P(T,\bar{T})} \log\frac{p(g,\bar{g})}{P(T,\bar{T})}$ and

$$H_b = -\sum_{i=T+1}^{G}\sum_{j=\bar{T}+1}^{G} \frac{p(g,\bar{g})}{1-P(T,\bar{T})} \log\frac{p(g,\bar{g})}{1-P(T,\bar{T})} \tag{10}$$

2.6 Bernsen's Local Contrast Binarization Algorithm

In the local binarization algorithm of Bernsen, the gray value of a pixel is compared with the average of the gray values in some neighborhood(50×50 window suggested) about the pixel, chosen by footprint size. Then, the threshold is set at the midrange value, which is the mean of the minimum $I_{low}(i,j)$ and maximum $I_{high}(i,j)$ gray values in the local window of suggested size w =50. However, if the contrast $C(i,j) = I_{high}(i,j) - I_{low}(i,j)$ is below a certain threshold (this contrast threshold was 50), then that neighborhood is said to consist only of background. In this algorithm, the following equation is used for finding an optimal threshold value:

$$T(i,j) = 0.5\{\max_{w}[I(i+m,j+n)] + \min_{w}[I(i+m.j+n)]\} \quad (11)$$

where $w = 50$, provided contrast $C(i,j) = I_{high}(i,j) - I_{low}(i,j) \geq 50$.

In order to find an appropriate binarization algorithm for scanned insect footprints, the above 6 binarization algorithms have been implemented and their performances are evaluated using the proposed binarization performance criterion discussed in the next section.

3 The Proposed Binarization Performance Criterion

There are various conventional performance criteria for evaluation of binarization algorithms. In [1], five performance criteria are used for evaluating conventional binarization algorithms. In these criteria, only NU(region NUniformity) needs no ground-truth images for comparison. In case of scanned insect footprints, it is almost impossible to acquire ground-truth images because nobody can decide easily which area is foreground or background. Because of this difficulty, we decided to use the NU criterion. The criterion NU measure is defined as follows:

$$NU = \frac{|F_T|}{|F_T + B_T|}\frac{\sigma_f^2}{\sigma^2} \quad (12)$$

where σ^2 represents the variance of the whole image, σ_f^2 represents the foreground variance, F_T, and B_T denote the areas of foreground and background

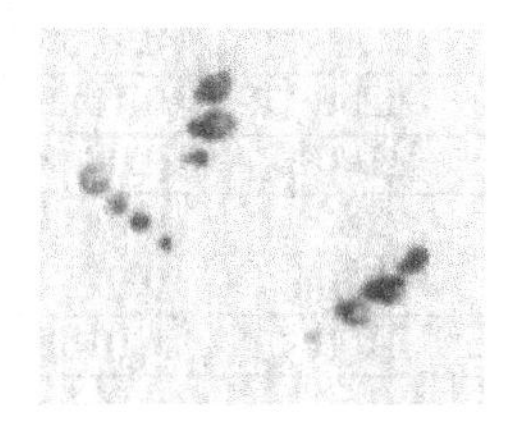

Fig. 1. A sample image

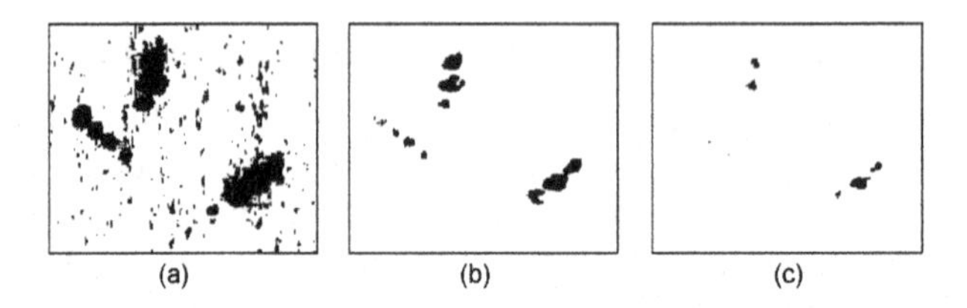

Fig. 2. The binarized images of Fig. 1 for threshold value = (a)230, (b)180, (c)130

pixels in the test image, and $|.|$ is the cardinality of the set. It is expected that a well-segmented image will have a nonuniformity measure close to 0, while the worst case of NU=1 corresponds to an image for which background and foreground are indistinguishable up to second order moments [1]. However we found that there is a problem when this criterion is applied to scanned insect footprints' binarizations. The problem is that the value of this criterion goes nearer and nearer to 0 when a threshold value goes smaller and smaller. To show this result we have chosen some area from a randomly selected insect footprint. The chosen area is shown as Figure 1. The binarized images at several threshold values are shown in Figure 2.

We evaluated the NU values using the sample image of Figure 1, varying threshold values from 130 to 230. The result is shown in Table 1. In this table, we can see that the threshold value goes smaller, the NU value becomes smaller, or, in other words, "better". But when we closely look at the above three binarized images, Figure 2(b) is better than Figure 2(c). So the performance criterion of NU could not give an appropriate result on the scanned insect footprints.

We propose a new binarization performance criterion named *Minimum Number of Foreground Segments* (MNFS). In Figure 2(a), it is decided that too many background noise area pixels are converted to 0 (i.e., foreground spots). In contrast, it is decided that too many foreground spots' area pixels are converted

Table 1. NU values for the sample image of Figure 1

Threshold	NU
230	0.521458
220	0.227144
210	0.153669
200	0.103933
190	0.068894
180	0.044197
170	0.025973
160	0.013889
150	0.006888
140	0.003250
130	0.001024

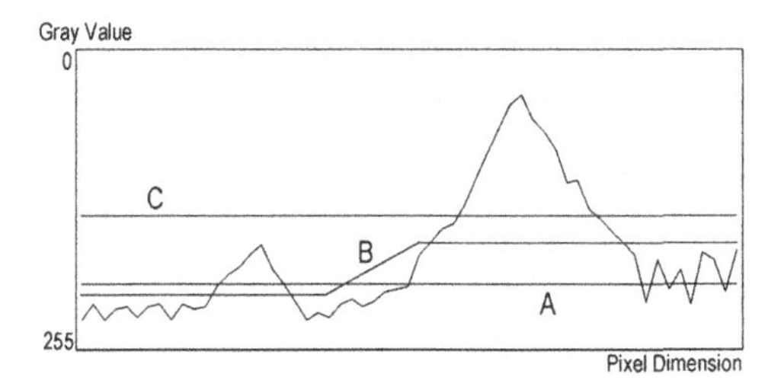

Fig. 3. Three cases of threshold line for binarization

to 255 (background area) in Figure 2(c). In these results, we can assume that "good" binarized images have a small number of disconnected regions but a large number of foreground pixels to be converted to 0 (foreground) in case of insect footprints. We illustrate the concept of this idea in Figure 3.

In Figure 3, B is the assumed optimal threshold line that can produce the best binarized image because the only two "real spots" (one is a "clear spot" but another is a "dim spot") are converted to foreground objects. A is the threshold line that converts lots of background noise area to foreground objects because the threshold value is too high. C is the threshold line that misses the dim spot because the threshold value is too low. In case of the sample image of Figure 1, A means Figure 2(a), B is similiar with Figure 2(b) (because B means, in fact, locally adaptive binarization method) and C means Figure 2(c).

How can we choose a threshold method like B? The key idea for the choice of an optimal threshold value or locally adaptive binarization method is to compare the number of disconnected segments (NDS) with the number of foreground pixels ($|F_T|$) and to compare the variance of background area. If the ratio of the two numbers ($NDS/|F_T|$) is smaller (i.e., threshold value becomes smaller than A) and the normalized variance of background area ($\sigma_b^2(T)/\sigma^2(T)$) is smaller (i.e., threshold value becomes lager than C), the threshold value can be considered to be "better". So we can consider the binarization as the best for binarization of scanned insect footprints if we achieve a minimum value of the product of ratio and normalized variance. The proposed binarization performance criterion is defined as follows:

$$MNFS = \frac{NDS}{|F_T|} \cdot \frac{\sigma_b^2(T)}{\sigma^2(T)} \tag{13}$$

where NDS indicates the number of disconnected foreground segments. For example, let us evaluate the MNFS values using the sample image of Figure 1, varying threshold values from 130 to 230. The result is shown in Table 2. We knew that the MNFS values are not linearly increasing or decreasing as in the criterion NU. In this sample image, the global threshold value of 218 has the minimum MNFS value when only global threshold values are applied.

Table 2. The MNFS values for the sample image of Figure 1

Threshold	MNFS
230	0.00289038
220	0.000359181
218*	0.000342245
210	0.000554109
200	0.000790112
190	0.00146493
180	0.00299014
170	0.00541835
160	0.00717066
150	0.0113167
140	0.018377
130	0.0265409

4 Test Images and Experimental Results

Our test images consisted of a variety of 16 images of American Cockroach, 30 images of Black Cockroach, and 25 images of Native Bush Cockroach. All images are scanned by 1200 DPI in 8-bit gray image format. Several test images are shown in Figure 4. The two images on the left are American Cockroaches, the two images in the middle are Black Cockroaches, and the two images on the right are Native Bush Cockroaches.

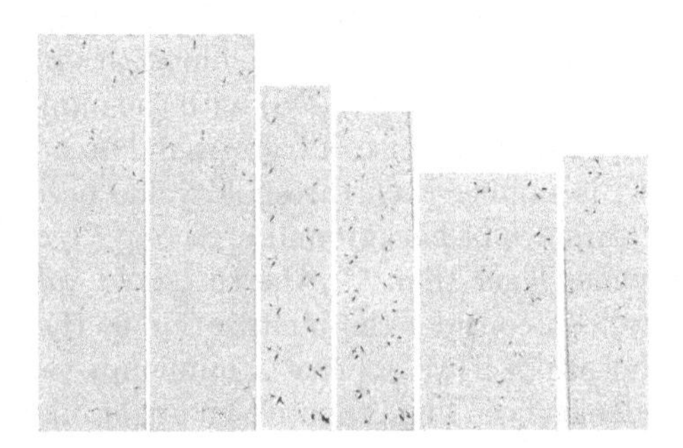

Fig. 4. Six sample images for testing

Results are shown in Figures 5 and 6 for the sample image of Figure 1. A number in parenthesis in the caption is the calculated global threshold value for the algorithm. Rosenfeld's, Abutaleb's, and Bernsen's algorithms do not have global threshold values because they are locally adaptive algorithms.

Fig. 5. Binarized images using Rosenfeld's, Otsu's[198] and Kittler's[222] algorithm

Fig. 6. Binarized images using Kapur's[214], Abutaleb's and Bernsen's algorithm

Table 3. Average MNFS values

Algorithm	Rosenfeld	Otsu	Kittler	Kapur	Abutaleb	Bernsen
MNFS	0.005607	0.004225	0.005145	0.004982	0.004034	0.007406

The average MNFS values are given in Table 3 using 71 test images. In the experiment, we found that no global threshold value could get better MNFS values than the MNFS values of the Abutaleb's algorithm. In several test images, Otsu's and Kittler's algorithms result in incorrect threshold values (often too large threshold values). "Incorrect threshold value" means the binarized image by the value is not considered as a normal insect foorprints when a human expert looks into the image. Otsu's algorithm produced incorrect threshold values on 6 test images, Kittler's algorithm produced incorrect threshold values on 13 test images. The MNFS values shown in Table 3 correspond to these incorrectly binarized images. We concluded that Otsu's and Kittler's algorithms are inadequate regardless of the MNFS value, and Abutaleb's algorithm is (in general) best for binarization of scanned insect footprints. A binarized test image using Abutaleb's algorithm is shown in Figure 7. An incorrectly binarized test image using Kittler's algorithm is shown in Figure 8.

Fig. 7. Binarized test image using Abutaleb's algorithm

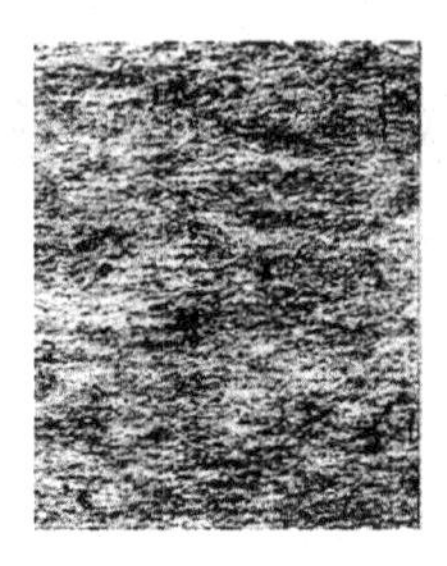

Fig. 8. Incorrectly binarized test image using Kittler's algorithm

5 Conclusions

We compared six different binarization algorithms and proposed a new binarization performance criterion to analyze the best performance for scanned insect footprints. The experimental results showed that Abutaleb's binarization method based on higher-order entropy produced (in general) the best binarized images. Binarized footprints have been further used in projects at CITR for property calculation, geometric modelling, and towards insect recognition. Results will be reported in forthcoming reports.

Acknowledgment. The reported binarization evaluation is part of an insect track recognition project at CITR, initiated and supported by Connovation Ltd., Auckland. The support of CITR, especially by Reinhard Klette, and of Connovation Ltd., especially by Warren Agnew, is very much appreciated by the author during his sabbatical stay at CITR.

References

1. M. Sezgin and B. Sankur. Survey over image thresholding techniques and quantitative performance evaluation. *Journal of Electronic Imaging* **13**: 146-165, Jan, 2004.
2. A. Rosenfeld and P. De la Torre. Histogram concavity analysis as an aid in threshold selection. *IEEE Trans. on System, Man and Cybernetics* **SMC-13**: 231–235, 1983.
3. N. Otsu. A threshold selection method from gray level histograms. *IEEE Trans. on System, Man and Cybernetics* **SMC-9**: 62–66, 1979.
4. J. Kittler and J. Illingworth. Minimum error thresholding. *Pattern Recognition*, **19**: 41–47, 1986.
5. J.N. Kapur, P.K. Sahoo, and A.K.C. Wong. A new method for gray-level picture thresholding using the entropy of the histogram. *Graphics Models Image Processing*, **29**: 273–285, 1985.
6. A.S. Abutaleb. Automatic thresholding of gray-level pictures using two-dimensional entropy. *Computer Vision Graphics Image Processing*, **47**: 22–32, 1989.
7. J. Bernsen. Dynamic thresholding of gray level images. In Proc. of ICPR'86, pages 1251–1255, 1986.

2D Shape Recognition Using Discrete Wavelet Descriptor Under Similitude Transform

Kimcheng Kith and El-hadi Zahzah

Université de La Rochelle,
Laboratoire d'Informatique, Images, Interactions,
Avenue M Crepeau La Rochelle 17042 , France
(kkimchen, ezahzah)@univ-lr.fr

Abstract. The aim of this paper is to propose a 2D shape recognition method under similitude transforms using the *Discrete Dyadic Wavelet Transform with decimation* (DWT). We propose four technics to fix the starting point necessary to obtain the same object representation in the framework of objects retrieval system. These representations are obtained by applying the DWT on the contour signature. We also propose a method to select a decomposition level of the DWT. These different solutions are assessed and compared using 1400 $2D$ objects extracted from the MPEG7 database.

1 Introduction

Object recognition is a main problem in computer vision. Several works were led on this field. These works belong to two main categories : contour- based and region-based methods. The criterion used to classify a method to one category or another is to see if the descriptors are calculated on the contour or on the region. For a good overview of the various representations, description and recognition techniques see Zhang and Lu. 2004. In this paper only 2D shape are considered in the framework of image retrieval using a large database. The technic used, build up an object representation based on the application of DWT on the contour signature. The DWT choice rather than *Fourier Transform* for instance, is motivated by the differences existing between Fourier analysis and wavelets. Fourier basis functions are localized in frequency but not in time. Small frequency changes in the Fourier transform will produce changes everywhere in the time domain. Wavelets are local in both frequency/scale (via dilatation) and in time (via translation). This localization is an advantage in many cases. Another advantage is that many classes of functions can be represented by wavelets in a more compact way. For example, functions with discontinuities and functions with sharp spikes usually take substantially fewer wavelet basis function than sine-cosine basis function to achieve a comparable approximation. In addition, the epithet "Fast" for Fourier transform can, in most cases be substituted by "Faster" for the wavelets. It is well known that the computational complexity of the FFT is $O(N\log_2(N))$, for the DWT the computational complexity goes

R. Klette and J. Žunić (Eds.): IWCIA 2004, LNCS 3322, pp. 679–689, 2004.

down to $O(N)$. Finally the DWT enables to change the basis functions, and users are free to create their own basis functions adapted to their application.

The implementation of the DWT proposed by Mallat in Mallat 1989, suffers of its non time-invariance to translation contrary to the continuous wavelet transform, and this is why the DWT is not frequently used specially in computer vision. Yang et al., in Hee Soo Yang and Lee 1998, proposed a solution to resolve this problem, but the implementation is relatively complex.

Khalil and Bayoumi 2002 Tieng and Boles 1997 proposed another solution used it in pattern recognition which is the non decimated version of the DWT called the SWT for *(Stationary Wavelet Transform)* in Misiti *et al.* 2003 which has been proposed at the first time by Mallat in Mallat 1991. The use of the SWT needs always a post-processing step consisting in an algorithm to match the query and the model descriptor for object retrieval. This step is a big time consuming, furthermore the complexity of the SWT is of $O(N \log_2(N))$ greater than the DWT complexity. In this paper, we propose four simple solutions to resolve the problem of the non-invariance to translation of the DWT consisting in fixing the starting point. The knowledge of the starting point enables to avoid the post-processing described above. We have applied and compared these methods in the framework of the 2D shape recognition. These methods can individually be used or combined to improve the efficiency and performances. We also propose a technic to select the decomposition level before using the DWT.

2 Shape Descriptor

Generally, the shape signature must represent in a unique way the shape by a one dimensional function derived from contour points. Many shape signatures exist based on centroidal profile, complex coordinates, centroid distance, tangent angle, cumulative angle, curvature, area, and chord length. In this section, we show the different steps to built up the signature and the shape descriptor. The contour associated to a close curve is re-parameterized by the normalized arc-length, and then it is re-sampled uniformly into 2^k points. The four proposed solutions to resolve the non-invariance to translation of the DWT are presented, and finally the shape descriptor is deduced by using the DWT on the contour signature.

2.1 Arc Length Parameterization and Normalization

Let assume that the contour is defined by a closed curve $\Gamma = (x_i, y_i)$ with $1 \leq i \leq n$ where n is the number of points on the contour. To parameterize the contour by the arc length one must compute $l(i)$ which is the length of the segment on the contour between the starting point (x_1, y_1) and a given point (x_i, y_i):

$$l(i) = \begin{cases} 0 & \text{for } i = 1 \\ \sum_{k=1}^{i-1} \sqrt{(x_i - x_{i-1})^2 + (y_i - y_{i-1})^2} & \text{for } 2 \leq i \leq n \end{cases}$$

$l(n)$ is the total length of the contour. For the scale invariance, the total length of contour $l(n)$ is normalize to a unity i.e the coordinates (x_i, y_i) and the parameter $l(i)$ are substituted respectively by $(\frac{x_i}{l(n)}, \frac{y_i}{l(n)})$ and $\frac{l(i)}{l(n)}$ for each i $1 \leq i \leq n$. Before applying the DWT on the contour signature, the contour $(\frac{x_i}{l(n)}, \frac{y_i}{l(n)})$ is re-sampled uniformly into $N = 2^k$ points. For simplification reasons, the new contour is also denoted by (x_i, y_i) $1 \leq i \leq N$ and verifies the property that the distance between two successive points is equal to a constant $\frac{1}{N}$. Let's note that after this processing, the contour is always normalized according to scale change. In order to normalize the contour according to translation and rotation, we define $s(i)_{1 \leq i \leq N}$ which represents the signature of the contour by

$$s(i) = \sqrt{(x_i - \dot{x})^2 + (y_i - \dot{y})^2}$$
$$\dot{x} = \tfrac{1}{N} \textstyle\sum_{i=1}^{N} x_i, \text{ and } \dot{y} = \tfrac{1}{N} \textstyle\sum_{i=1}^{N} y_i$$

2.2 Fixing the Starting Point

As recalled in the introduction section, the well-known problem with DWT is that it is not invariant to translation. To avoid this drawback, and in order to compare the descriptors (obtained by the DWT application on the contour signature), the same starting point must be used for the signatures representing the same contour (object). In the following, four solutions are proposed to fix this specific starting point on the contour before applying the DWT.

1. **Furthest Distance**
   The first easiest solution is to consider the starting point at the contour point such that its distance to the centroid is maximal. Formally this is equivalent to consider the point $(x_{i_{max}}, y_{i_{max}})$ as a starting point with $i_{max} = \underset{1 \leq i \leq N}{arg\ max}\ s(i)$. The choice of a point if one finds many points verifying this condition, is discussed later.
2. **Maximum Curvature**
   Another solution is to choose a point which has the maximum curvature. Many equivalent definitions of curvature exist, in this paper the curvature k_i at a given point (x_i, y_i) is given by :

$$k_i = \frac{\dot{x}_i \ddot{y}_i - \ddot{x}_i \dot{y}_i}{(\dot{x}_i^2 + \dot{y}_i^2)^{3/2}} \tag{1}$$

   where $(\dot{x}_i, \dot{y}_i)$ et $(\ddot{x}_i, \ddot{y}_i)$ represent respectively the first and the second derivative at (x_i, y_i). Many ways can be used to calculate the curvature k_i of equation (1). The approach we adopt here is to substitute the first and the second derivative at the point (x_i, y_i) by the first and the second derivative of the convolved version at the same point with the gaussian function $g_\sigma(i)$ as in S. Abbasi and Kittler 2000Mokhtarian and Abbasi 2002, hence the equation (1) becomes:

$$k_i = \frac{\dot{x}(i,\sigma)\ddot{y}(i,\sigma) - \ddot{x}(i,\sigma)\dot{y}(i,\sigma)}{[\dot{x}^2(i,\sigma) + \dot{y}^2(i,\sigma)]^{3/2}}$$

with $\dot{x}(i,\sigma) = (x * \dot{g}_\sigma)(i)$ and $\ddot{x}(i,\sigma) = (x * \ddot{g}_\sigma)(i)$ the same formula are used for $\dot{y}(i,\sigma)$ and $\ddot{y}(i,\sigma)$.

3. **Principal Axis**
   One can also consider the starting point as the intersection point between the principal axis and the contour. The principal axis is deduced from the well known principal Component Analysis (PCA). The eigen vectors and values are calculated from the covariance matrix C given by:

$$C = \sum_{i=1}^{N} (x_i - \dot{x}, y_i - \dot{y})^t \, (x_i - \dot{x}, y_i - \dot{y})$$

   If e_1 is the eigen vector with magnitude equal to one corresponding to the greatest eigen value of the covariance matrix C, the two intersection points (x_{s_1}, y_{s_1}) and (x_{s_2}, y_{s_2}) between the contour and the principal axis (e_1) are deduced calculating the subscripts P_1 and P_2:

$$P_1 = \underset{i}{argmin} \|(x_i - \dot{x}, y_i - \dot{y}) - e_1 s_i\|$$

$$P_2 = \underset{i}{argmin} \|(x_i - \dot{x}, y_i - \dot{y}) + e_1 s_i\|$$

   The intersection between the principal axis and the contour gives two point, the selection of a unique point is discussed later.

4. **The Natural Axis**
   Another possibility to fix the starting point is to extract from the so called *natural axis*, the point which coordinates $(x_{i_{natural}}, y_{i_{natural}})$ such as the subscript $i_{natural}$ is defined by

$$i_{natural} = \lfloor \frac{\Theta_{natureal}}{2\pi/N} \rfloor + 1$$

$$\text{with} \quad \Theta_{natural} = angle(\sum_{m=1}^{N} s_m e^{2\pi i(m-1)/N})$$

   with $angle(z)$ returns the angle between the interval $[0, 2\pi[$ of the complex number z and $\lfloor x \rfloor$ returns the integer number i such as $i \leq x < i+1$. $\Theta_{natural}$ correspond to the natural orientation of the signature $(s_i)_{1 \leq i \leq N}$.

   The value of s_m of the above equation can be substituted by the value s_m^p with $p > 0$ and instead of considering $\Theta_{naturel}$ as the angle of the first Fourier coefficient of the signature s_m^p, one can substituted it by the first non zero Fourier coefficient of the signature s_m^p, it means that if the first coefficient is zero, one takes the second if it is non zero and the third if not, and so on.

Let's note that

- to choose the starting point, the strategy using the furthest distance, the maximum curvature and the principal axis can be changed with respectively the minimum distance, minimum curvature and secondary axis.

- in case of multiple points verifying the furthest distance and maximum curvature, one can select randomly a given point among this choice to fix the starting point, another way is to select the two or the three first as the starting point and taking the one which verifies the minimum distance. The same processing can be used for the intersection points P_1 and P_2 between the principal axis and the contour.
- the methods proposed to choose the starting point could be combined in a way that if an ambiguity appears for example when using the furthest distance, a second technic as the maximum curvature or the one based on the natural axis could be used.
- to reinforce the robustness of the method, one can consider the nearest point to the starting point as another starting point and finally, choose the one which verify the criteria of the minimum distance.

2.3 The Choice Decomposition Level

After the stage of the contour parameterization and re-sampling, and after the starting point determination using one of the four methods described above (2.2), each contour object can be now represented by a characteristic vector obtained by the application of the DWT on the contour signature. Before applying the DWT, one needs to choose which level of decomposition is to be useful to represent the contour?. Let's note that there is k possible levels of decomposition ($k = \log_2 N$ given at the beginning when re-sampling uniformly the contour). Here, we explain why and how to choose the decomposition level using the signature energy.

Let's $S^j = (s_i^j)_{1 \leq i \leq N}$ is the jth contour signature of the database. As the energy of S^j represents information contained in S^j, and more the energy is high, better the signal S^j is represented. The figure (1) illustrates this idea. In this figure, one can notice that more the decomposition level is high, more the signal approximation energy decreases and hence the approximation quality goes to be badly. We keep in mind that the signal energy gives global information, i.e that if two signals are close, the energy difference of these signals is low. note that the contrary is not always true. i.e, if two signals have the same energy, these signals are not necessarily close. In our application, to select a level L which represents at best the global signature of the database models, we use the average energy of the database on k decomposition levels. The figure (2) and the table (1) give the average values of energy of the whole 1400 contour signatures of the database. According to these table and graph, the best and the reasonable choice is to take $L = 3$ or $L = 4$ because the energy difference in term of percentage between level 3 and the other lower levels is small, and the difference between the level 4 and the the upper levels is high enough. According to our experimentation, $L = 3$ gives the best result. We recall that the goal here is to use a minimum DWT coefficients which represent at best the signal. The contours S^j can be now represented by the approximation coefficients $A_L(S^j)$ at level $L = 3$.

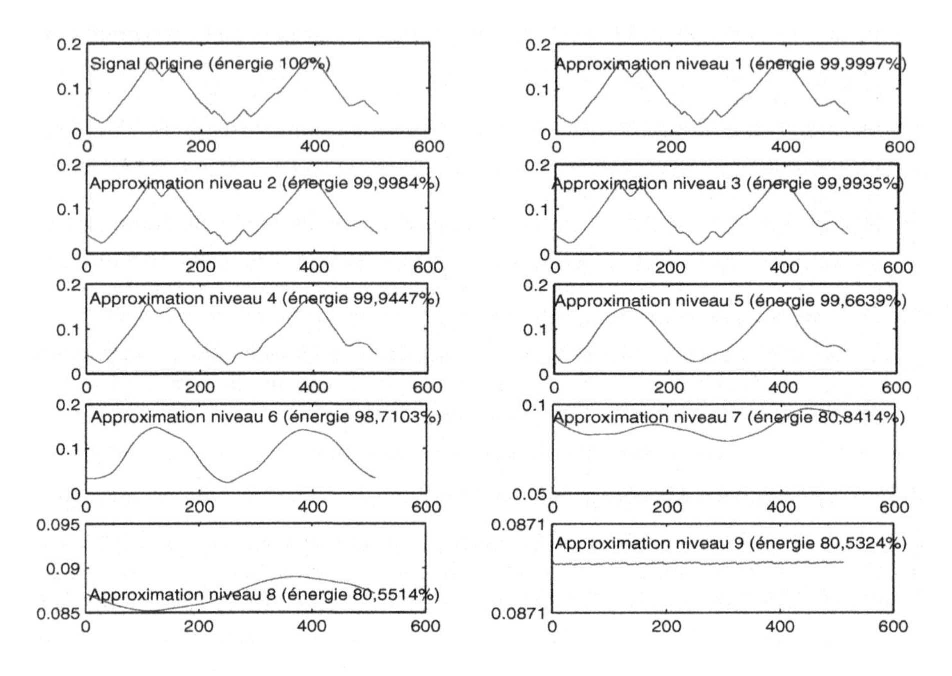

Fig. 1. Signal energy and its approximations on 7 decomposition levels

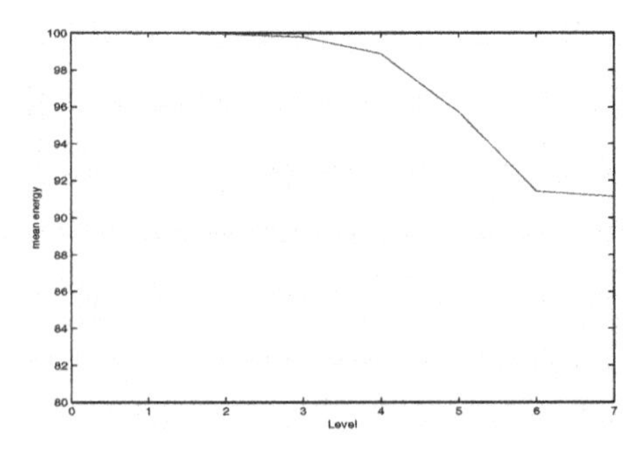

Fig. 2. Representation of the mean energy of signature decomposition by the *DWT* on 7 decomposition levels

Table 1. The percentage in average of signature energy of the database on the 7 decomposition levels

Level	0	1	2	3	4	5	6	7
Mean energy	100	99.994	99.971	99.788	98.893	95.736	91.432	91.154

2.4 Matching and Complexity

A detailed study can be found in Zhang 2003 on the the effect of using different similarity measurement as *Histogram Intersection, Mahalanobis Distance, Minkowski-form distance, cosine,* χ^2... for image retrieval. The results have shown that *city block* distance (which is a particular case of the Minkowski distance) and χ^2 *statistic measure* outperform other distance measure in terms of both retrieval accuracy and retrieval efficiency. As the city bloc distance is cheaper and easy to calculate comparing to χ^2, then to measure the dissimilarity degree between signatures S^i and S^j, the city-block distance between two vectors $A_L(S^i)$ and $A_L(S^j)$ has been used. In order to take into account the symmetric contour, the curve is scanned in both directions clockwise and counter-clockwise.

For the computational complexity, one can easily show that the parameterization and the starting point localization is calculated in $O(N)$ operations. The complexity of the *DWT* is linear according to input signal dimension ($O(N)$).

Finally, the global computational complexity is $O(N)$, without taking into account the time used to retrieve similar contours, because this stage, depends on the number of contours existing in the database and the strategy used to structure data for indexing and comparing the characteristic vectors.

3 Experimentation

For our experimentation, we use the CE-1 MPEG7 database which contains 3 sets of contours A, B, C described in details in Latecki *et al.* 2000.

- The set A is dedicated to test the invariance according to rotation and scaling.
- The set B is dedicated to test the recognition according to the similitude transforms. This set contains 1400 contours of 70 classes and each class contains 20 different contours of the same object.
- The set C is dedicated to test the recognition according to the affine transform.

In this paper, we limit our experience to the set B. The value of N is fixed to 128 and the value of decomposition level L is 3. For the curvature of equation (1), σ is set to 1.4. All the database contours are closed, hence, all the signatures are periodic and each one is represented by a descriptor vector of dimension 16 which represent the approximation coefficients at level 3.

To assess the performance of our method, we use the most common measure well known by *precision* (P) and *recall* (R) of retrieval used by Bimbo 1999, and defined respectively by:

$$P = \frac{r}{n} = \frac{\textit{number of relevant retrieved contour}}{\textit{number of retrieved image in the database}}$$

$$R = \frac{r}{m} = \frac{\textit{number of relevant retrieved contour}}{\textit{number of relevant image in the database}}$$

Precision P is defined as the ratio of the number of retrieved relevant shapes r to the total number of retrieved shapes n. Precision P measures the accuracy of the retrieval and the speed of the recall. Recall R is defined as the ratio of the number of retrieved relevant images r to the total number m of relevant shapes in the whole database. Recall measures the robustness of the retrieval performance used also by Zhang and Lu 2002.

Three experimentations are presented here. The first one concerns the evaluation using different wavelet basis functions, in the second experimentation the four solutions presented in section (2.2) are compared, and the last one is the comparison between our method and the one using the Fourier descriptor. For each experimentation the values of precision P and recall R are calculated.

For each query, the precision values are calculated corresponding to the recall values between 10%, and 100%. The precision values of our system are actually the average values of precision of the whole 1400 queries.

Wavelet Basis Functions Comparison

At the introduction section, we have recalled that one advantage of using the DWT according to the others well known transforms is that the users are free to use or to built up their own basis functions. In the following several well known wavelet basis functions are used and compared with the method we propose, as Daubechies, Coiflets, Symlets, Discrete Meyer, Discrete Bi-orthogonal and the inverse bi-orthogonal. The time consuming of the DWT depend principally on the filter length which is set to 12 in our experimentations except the Discrete Meyer length filter which is set to 62. Using the furthest distance to the the centroid to fix the starting point, the average performance of retrieval using different basis wavelets functions are given in figure (3) (a-c). The table (2) gathers all the graph data of figure (3) (a-c).

According to this table and to the graphs of figure (3) (a-c), we can see that the performance is similar for the five basis functions db6, sym6, coif2, biorl.5 and rbiol.5, the graphs corresponding to basis functions db6, sym6 and coif2 are overlapped in figure (3) (a). One can also see that the two basis functions bior3.5 and rbio5.5 have the same performance greater than the others ones. On the contrary, the performance of the basis function of dmey is the lowest compared to all the others basis functions used in our experimentations. The performance of the basis functions bior5.5 and rbio3.5 is also small but greater than of dmey.

Comparison of the Four Starting Point Fixation Technics

In this paragraph, the performance evaluation is done by comparing the four solutions to fix the starting point proposed in section (2.2). The wavelet basis function used for this comparison is rbior5.5.

The figure (4)(a) and the table (3) show the graphic and the data corresponding to this evaluation. From this table one can read that the technic to fix the starting point using the furthest distance is the best comparing to the three others technics.

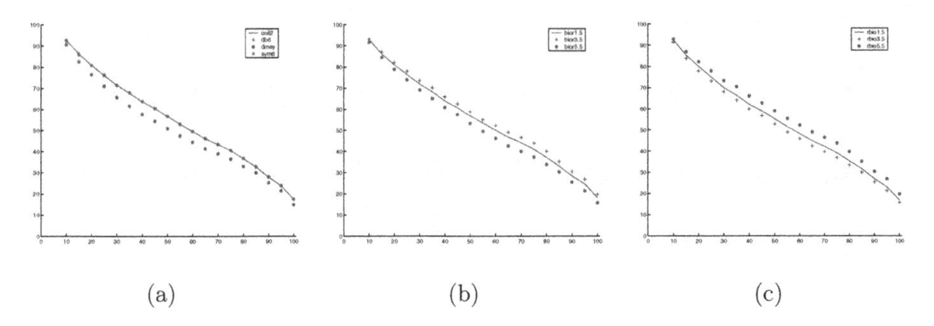

(a) (b) (c)

Fig. 3. Performance of the retrieval in average using different wavelet basis functions

Table 2. The Precision table corresponding to different recall values and different wavelet basis functions

Wavelet \ Recall	10	20	30	40	50	60	70	80	90	100	Average
db6	92.6	80.8	71.3	63.7	56.8	49.5	43.4	36.7	28.1	17.4	54.0
sym6	92.5	80.8	71.4	63.8	56.6	49.5	43.3	36.7	28.0	17.4	54.0
coif2	92.7	80.8	71.5	63.6	56.7	49.4	43.2	36.6	27.8	17.45	54.0
bior1.5	92.9	81.0	71.9	63.8	56.8	49.9	44.1	37.1	28.4	18.0	54.4
bior3.5	93.1	82.1	73.6	65.8	58.8	52.1	46.5	40.0	30.4	19.6	56.2
bior5.5	91.6	78.8	69.1	60.8	53.1	46.1	40.0	33.8	25.5	15.7	51.4
rbio1.5	92.4	80.0	70.0	62.1	55.1	48.2	42.1	35.3	27.0	16.9	52.9
rbio3.5	91.5	77.9	68.1	59.9	52.7	45.7	39.6	33.3	25.3	15.8	51.0
rbio5.5	92.9	82.2	73.3	66.1	59.0	52.2	46.4	39.6	30.3	19.7	56.2
dmey	90.5	76.5	65.7	57.7	50.9	44.4	38.9	33.0	25.2	14.9	49.8

Comparison with Fourier Transform

The literature is very affluent on Fourier Descriptors. On the contour based processing, the difference of these descriptors is mainly on the contour representation. The contour can be represented for instance by signature, the cartesian coordinates, complex numbers etc. Zhang Zhang 2002 shows that the most robust descriptor is the one obtained from the signature.

To be objective in the comparison, our method using the furthest distance to fix the starting point is compared with the Fourier descriptor obtained from the contour signature using 15 coefficients (see Zhang 2002 for the details of calculating the Fourier Descriptor). The figure (4) (b) and the table (4) show the comparison results. On can see the the performance of both methods are very similar. This result is very encouraging, because we believe that it can be improved by selecting for example an adapted basis wavelet function or taking into account not only the approximation coefficients but also the details coefficients. The performance may also be improved by applying the DWT with other types

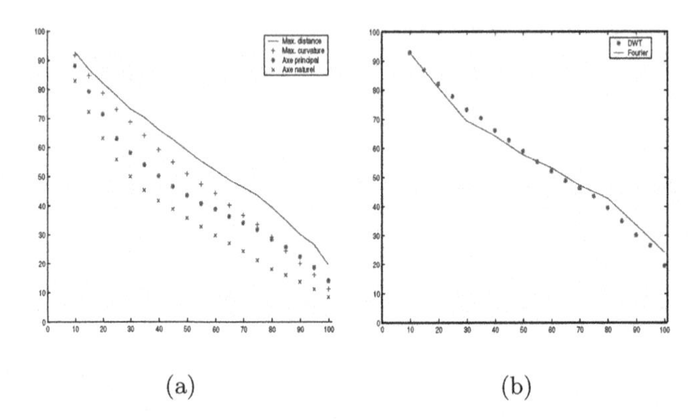

Fig. 4. Performance Retrieval Comparison (a) Between the four solutions to fix the starting point (b) Between the *DWT* using the furthest distance to the centroid and the Fourier transform

Table 3. Average Performance Comparison Table using the four proposed technics

Starting Point \ Recall	10	20	30	40	50	60	70	80	90	100	Average
Maximum Distance	92.9	82.2	73.3	66.1	59.0	52.2	46.4	39.6	30.3	19.7	56.2
Maximum Curvature	91.7	78.7	68.7	59.3	50.9	44.3	36.8	29.4	20.3	11.4	49.2
Principal Axis	88.1	71.5	58.3	50.3	43.6	38.9	34.3	28.6	22.6	14.3	45.0
Naturel Axis	83.0	63.2	50.2	41.8	35.9	30.0	24.6	18.3	13.9	8.71	36.9

Table 4. The average comparison Table of performance of the proposed and Fourier method

Method \ Recall	10	20	30	40	50	60	70	80	90	100	Average
Fourier	93	81	69.5	64.3	57.8	53.4	47.4	42.9	33.8	24.5	56.76
DWT	92.9	82.2	73.3	66.1	59.0	52.2	46.4	39.6	30.3	19.7	56.2

of representation, keeping in mind that the *DWT* is faster the the *FFT* in term of computational complexity.

4 Conclusion

In this paper we propose a 2D shape recognition method based on the Discrete Wavelet Transform. To compare objects regardless of their geometric properties in translation, rotation and scale using the *DWT* coefficients, one must fix the starting point. To do that, we proposed four easy solutions based on the distance of the contour to the centroid, the object curvature, the principal axis, and the natural axis. Furthermore to represent at best the signature contour by the *DWT* coefficients, we proposed a solution to select automatically the decomposition level, based on the average energy contained in the whole database. The experimentations done show that the results are very encouraging and quite similar to those obtained by Zhang 2002. In the future works we project to improve the results by using for example other contour representation or other invariant obtained from both approximation and detail coefficients of the *DWT*.

References

A. Del Bimbo. visual information retrieval. *morgan kauffman, san francisco usa*, pages 56–57, 1999.

Sang Uk Lee Hee Soo Yang and Kyoung Mu Lee. Recognition of 2d object contours using starting-point-independent wavelet coefficient matching. *Journal of Visual Communication and Image Representation*, 9(2):171–191, 1998.

M. Khalil and M. Bayoumi. Affine invariant for objet recognition using the wavelet transform. *Pattern Recognition Letters*, 23:57–72, 2002.

L.J Latecki, R Lakamper, and U Eckhardt. Shape descriptors for non-rigid shapes with a single closed contour. *IEEE conf on Computer Vision and Pattern Recognition (CVPR)*, pages 4124–429, 2000.

Stephan Mallat. A theory for multiresolution signal decomposition : the wavelet representation. *IEEE Transaction on Pattern Analysis and Machine Intelligence*, 11:674–693, 1989.

Stphan Mallat. Zero-crossings of a wavelet transform. *IEEE Transactions on Information Theory*, 37:1019–1033, 1991.

Michel Misiti, Yves Misiti, Georges Oppenheim, and Jean-Michel Poggi. *Les ondelettes et leurs applications.* hermes Sciences, 2003.

Farzin Mokhtarian and Sadegh Abbasi. Shape similarity retrieval under affine transforms. *Pattern Recognition*, 35:31–41, 2002.

F. Mokhtarian S. Abbasi and J. Kittler. Enhancing css-based shape retrieval for objects with shallow concavities. *Image and Vision Computing*, 18:199–211, 2000.

Q. M. Tieng and W. W. Boles. Wavelet-based affine invariant representation: A tool for recognizing planar objects in 3d space. *IEEE Transaction on Pattern Analysis and Machine Intelligence*, 19(8):846–857, 1997.

Dengsheng Zhang and Guojun Lu. *A comparative study of curvature scale space and fourier desriptors for shape-based image retrieval.* PhD thesis, Gippsland school of computing and Info Tech Monach Univer, 2002.

D. S. Zhang and G. Lu. Review of shape representation and description techniques. *Pattern Recognition*, 37(1):1–19, 2004.

Dengsheng Zhang. *Image Retrieval Based on Shape.* PhD thesis, Faculty of Information Technologie, March 2002.

Dengsheng Zhang. *Evaluation of similarity measurement for image retrieval.* PhD thesis, Gippsland school of computing and Info Tech Monach Univer, March 2003.

Which Stereo Matching Algorithm for Accurate 3D Face Creation ?

Ph. Leclercq, J. Liu, A. Woodward, and P. Delmas

CITR,
Department of Computer Science,
The University of Auckland,
New Zealand
patrice@cs.auckland.ac.nz

Abstract. This paper compares the efficiency of several stereo matching algorithms in reconstructing 3D faces from both real and synthetic stereo pairs. The stereo image acquisition system setup and the creation of a face disparity map benchmark image are detailed. Ground truth is build by visual matching of corresponding nodes of a dense colour grid projected onto the faces. This experiment was also performed on a human face model created using OpenGL with mapped texture to create as perfect as possible a set for evaluation, instead of real human faces like our previous experiments. Performance of the algorithms is measured by deviations of the reconstructed surfaces from a ground truth prototype. This experiment shows that contrary to expectations, there is seemingly very little difference between the currently most known stereo algorithms in the case of the human face reconstruction. It is shown that by combining the most efficient but slow graph-cut algorithm with fast dynamic programming, more accurate reconstruction results can be obtained.

1 Introduction

Reliable human-computer interface systems are today under active research. As their essential component, realistic 3D face generation recently received much attention [1]. Potential applications vary from facial animation and modelling for augmented reality and head tracking to face authentication for security purposes. Since the nineteen eighties [2], researchers have been trying to use planar images of faces for any of the above applications. Lately, more advanced methods such as Active Appearance Models [3] have been able to retrieve 3D pose and geometry of the face in real-time using both 2D image information and a statistical model of a 3D face. Still, results are highly influenced by viewpoint and illumination variations and usually provide a very rough description of face features such as eyes, mouth, eyebrows. At the same time, 3D face acquisition techniques have been found less dependent to the viewpoint and illumination; and for humans, to posture and expression variations. Furthermore, such techniques are expected to be reliable, compact, accurate and low-cost. Although many approaches (such as

R. Klette and J. Žunić (Eds.): IWCIA 2004, LNCS 3322, pp. 690–704, 2004.

laser range scanner devices, stripe pattern generator-based tools, marker-based techniques) may be used to generate 3D faces, imaging techniques involving only off-the-shelf low cost digital still cameras have been the most widely researched as they are cheap, require no special hardware equipment or complicated system setup and may be applicable in the future to a wide range of situations. Here we propose to assess the performance of the most widely known stereo correspondence algorithms for 3D face generation. First, the theory beneath the nine different stereo algorithms studied will be described. Next, two different ways to generate human face stereo pairs using off-the-shelf cameras will be presented. Performances of the stereo algorithms on the generated ground truths are then exhibited and commented.

2 Algorithms

Notations. $I_L(P)$, respectively $I_R(P)$, is the information (intensity, luminance or any other colour channel) at point P in the left, respectively right, image. Correlation functions in each image are evaluated over a square 'window' $w(P,r)$ of $(2 \cdot r+1) \times (2 \cdot r+1)$ neighbouring pixels, with P, its centre and r its radius.

Corr1: Normalised Square of Differences Correlation (C_1 in Faugeras *et al.* [4]). The first correlation cost, $C(P)$, is the normalised intensity difference:

$$C_1(P) = \sqrt{1} \frac{\sum_{P' \in w(P,\beta)} (I_L(P') - I_R(P'))^2}{\sqrt{\sum_{P' \in w(P,\beta)} I_L(P')^2} \cdot \sqrt{\sum_{P' \in w(P,\beta)} I_R(P')^2}} \quad (1)$$

Corr2: Normalised Multiplicative Correlation (C_2 in Faugeras *et al.* [4]). The second cost is a normalised multiplicative correlation function:

$$C_2(P) = \frac{\sum_{P' \in w(P,\beta)} I_L(P') \cdot I_R(P')}{\sqrt{\sum_{P' \in w(P,\beta)} I_L(P')^2} \cdot \sqrt{\sum_{P' \in w(P,\beta)} I_R(P')^2}} \quad (2)$$

Sum of Absolute Differences (*SAD*). The third cost simply sums absolute differences without normalisation:

$$SAD(P) = \sum_{P' \in w(P,\beta)} |I_R(P') - I_L(P')| \quad (3)$$

A variant of the SAD cost is the Sum of Squared Differences (SSD):

$$SSD(P) = \sum_{P' \in w(P,\beta)} (I_R(P') - I_L(P'))^2$$

More details regarding SAD and possible parallel implementation can be found in [5]

Census. Zabih *et al.* use intensities in a different way [6]: bit-vectors V are created by concatenating the results of the *Census transform* - $\xi(P, P')$ - first in $w(P, \beta)$, then over a correlating window $w(O, \alpha)$:

$$V(O, \alpha) = \bigotimes_{P \in w(O,\alpha)} \Big(\bigotimes_{P' \in w(P,\beta)} \xi(P, P') \Big), \quad \xi(P, P') = \begin{cases} 1 \text{ if } I(P) < I(P') \\ 0 \text{ otherwise.} \end{cases}$$

The correlation cost is obtained using the Hamming distance between $V(O_L, \alpha)$ and $V(O_R, \alpha)$ bit vectors at point O in left and right image.

$$C_{Census} = V(O_L, \alpha) \ominus V(O_R, \alpha)$$

where $\ominus$ denotes the Hamming distance.

Chen and Medioni. The Chen-Medioni algorithm [7] takes a volumetric approach to calculate a disparity map of a 3D surface. A correct disparity map corresponds to a maximal surface in the u-v-d volume, where u and v are pixel co-ordinates, d is the disparity. The disparity surface is extracted by first locating seed points in the volume (correspondences with a high likelihood of being correct matches), and propagating outward from these positions, thus tracing out a surface in the volume. An upper threshold is used to discriminate these seed points. Propagation utilizes the disparity gradient limit [8] which states that the rate of change in disparity is less than the rate of change in pixel position. Therefore only directly adjacent neighbours along a scan line are analyzed. A lower threshold is be used to reject poor matches at the propagation stage. Normalized cross-correlation is used as the similarity measure. The initial step for the algorithm requires the location of seed points. A multi-resolution approach is utilized where a mipmap of the original stereo pair is created. At the lowest resolution, the left image is divided into buckets and pixels inside are randomly selected for seed points suitability. Next, as a result of surface tracing, a low resolution disparity map is obtained. The disparities from this map are inherited as seed points for each higher resolution, until the full scale disparity map is constructed. By propagating from initial seed points a level of local coherence is obtained.

Symmetric Dynamic Programming Stereo: SDPS. Gimel'farb proposes a dynamic programming approach [9] to the stereo problem based on a Markov chain modelling of possible transitions to form a *continuous* graph of profile variants. Then, the reconstructed profile has to maximise the likelihood ratio - chosen as the log-likelihood ratio for instance - with respect to a purely random one.

This more traditional algorithm sequentially estimates a collection of continuous 1D epipolar profiles of a 2D disparity map of a surface by maximising the likelihood of each individual profile. Regularisation with respect to partial occlusions is based on the Markov chain models of epipolar profiles and image signals along each scanline. The algorithm accounts for non-uniform photometric distortions of stereo images by mutually adapting the corresponding signals. The adaptation is based on sequential estimation of the signals for each binocularly visible point along the profile variant by averaging the corresponding signals in the stereo images. The estimated signals are then adapted to each image by changing the corresponding increments to within an allowable range.

Roy and Cox Maximum Flow Formulation. Roy and Cox introduce the *local-coherence* constraint [10]: the algorithm[1] assesses the full 2D problem instead of the usual 1D epipolar constraint. The stereo correspondence problem is posed as a maximum flow problem and minimal cut is used to extract the disparity map.

Graph Cut Variants. This approach performs a statistically optimal estimation of a disparity map for a 3D human face from a given stereo pair of images. The estimation is based on general local 2D coherence constraints [10], instead of the previous traditional 1D ordering ones. To perform the 2D optimization, the matching is formulated as a maximum-flow graph problem. The images and maps are described with a simple Markov Random Field (MRF) model taking into account the differences between grey values in each pair of corresponding points and the spatial relationships between the x-disparities. The goal is to find a piecewise smooth x-disparity map consistent with the observed data which minimises the total energy based on the Gibbs potentials of pairwise interactions between the disparities and corresponding signals. Epipolar surface profiles are stacked together to form a 3D matching cube. The disparity map of a desired surface is obtained by approximately solving the maximum-flow/minimum-cut problem on a graph linking discrete 3D points in the cube [11]. Here, three graph cut algorithm variants[2] have been evaluated: BVZ in [12], KZ1 in [13] and KZ2 in [14].

3 Experimental Setups and Benchmark Definitions

This chapter introduces two different methods to generate dense ground truth for the assessment of stereo matching algorithms performance on stereo set of human faces. The first method uses standard off the shelf digital camera while the latter integrates data acquired with these cameras and an OpenGL 3D face model to create a synthetic stereo pair with corresponding disparity map.

3.1 Stereo Bench Setup

Two Cannon PowerShot A60 cameras are manually placed as close as possible to the epipolar geometry configuration as illustrated in Fig. 1. The cameras focal length and synchronized frame capturing (1600×1200 pixels uncompressed) are controlled via an embedded Cannon software for image acquisition. The

[1] `www2.iro.umontreal.ca/~roys/publi/iccv98/code.html`
[2] `www.cs.cornell.edu/People/vnk/software.html\#MAXFLOW`

Fig. 1. Stereo bench setup. The 2 cameras are placed in vertical epipolar position

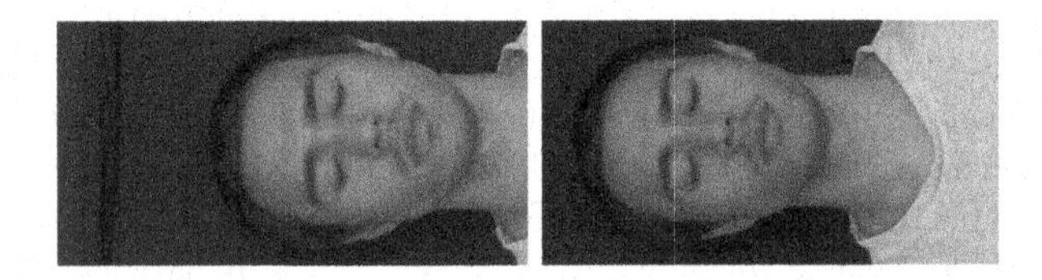

Fig. 2. Stereopairs of Alex placed in vertical epipolar position

baseline length is $\sim 11\ cm$ and the subject-camera distance is around $110\ cm$. Further studies on optimal subject-camera distances for best accuracy as well as full camera calibration are not detailed here but can be found in [15]. Most stereo-matching algorithms scan epipolar stereo images line per line, to find corresponding, pixels for sake of simplicity and speed. Usually stereo cameras are placed horizontally with left camera lines matched to their respective right camera rows. However, for faces, as suggested in [7] and confirmed by our experiments, cameras should be placed one on top of another to avoid symmetric face features, such as eyebrows, eyes, nostrils and mouth to appear line wise on the stereo image set. Furthermore, this setup minimizes the creation of outliers around the nose region as both cameras see the face from a symmetric vertical viewpoint.

Typical stereo pairs of faces (as in Fig. 2) occupy a 650×450 pixels rectangle in the image and have a disparity range of 20 to 30 pixels

In many cases, Stereo-matching algorithms have been compared to artificial set of images corrupted with added Gaussian noise [16]. The created stereo set is then processed by each algorithms and the obtained disparity map compared to the perturbed synthetic disparity. For human faces, it is very unlikely that artificially created stereopairs will exhibit realistic surface and texture. Therefore, the ground truth has to be generated from acquired face data.

To initiate our data acquisition a coloured (red, green, and blue, see Fig. 3) grid was projected onto a human test subject. A stereo pair of images with grid projections, and a third without (for texture mapping purposes), was taken. Correspondence between pixels in the left and the right images was carried out

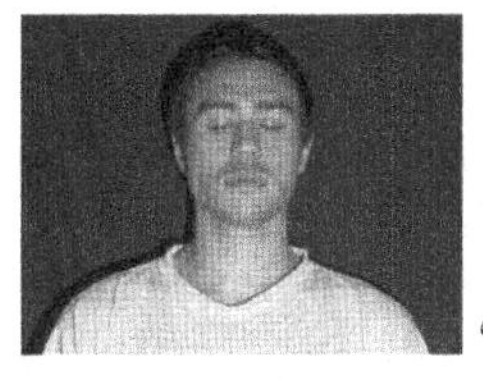
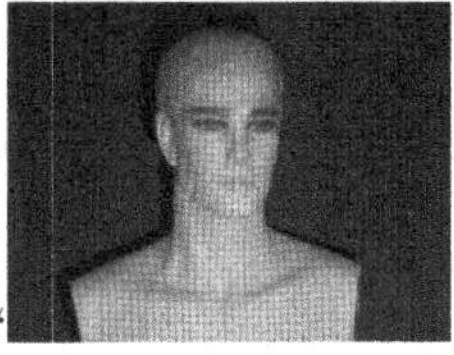

Fig. 3. Projection of a coloured grid on stereo images of test subjects for creating the reference disparity maps

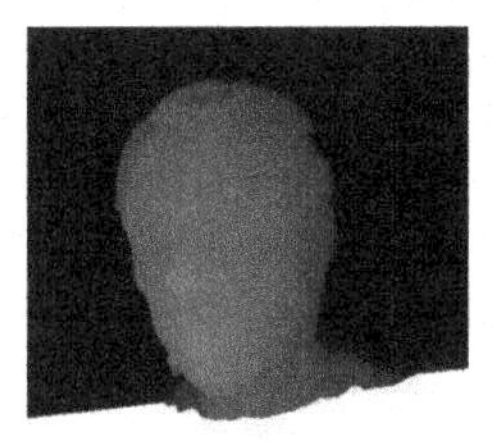

Fig. 4. Depth map, with and without texture, of a test subject

through semi-automatic registration using the points located at the intersections of the grid lines. Once acquired, this sparse dataset, composed of approximately 350 points evenly distributed over the whole facial surface, was interpolated, through cubic spline interpolation, to obtain dense ground truth disparities (see Fig. 4).

Note that as stereo pairs were acquired in real conditions, left and right images might exhibit different illuminations as well as slight epipolar errors. For the latter, manually generating the sparse maps showed epipolar irregularities of about $1 \sim 2$ pixels, eventhough cameras were carefully adjusted beforehands.

3.2 3D-Modelled Stereo Pairs

In order to develop a test environment for accurate comparison of stereo reconstruction algorithms, the provision of ground truth disparity data is necessary. A methodology for synthetic ground truth was developed involving three steps:

- manual acquisition of a sparse set of data,
- generation of a dense set and creation of a synthetic 3D model and
- virtual camera setup allowing sub-pixel ground truth disparities.

Obtaining an Initial Dataset. After acquiring the data set, as described in section 3.1, it is tessellated to form a surface; the relatively small facet size means the resultant surface has smoothness at an ample scale. Since each pixel in the acquired image corresponds to a surface's vertex, generating texture mapping coordinates is straight forward. This composition provides an accurate synthetic representation of a true human face.

Virtual Test Bench Setup. Rendering of our face model utilised the *OpenGL API* which contains number of features for quick camera locations and parameters setup. To mirror the existing *Canon Powershot A60* cameras setup used in our lab: the baseline was set to $100mm$ and the focal length was derived from a 1/2.7" (*i.e.* $5.27mm \times 3.96mm$) CCD sensor in simple geometric terms:

$$FocalLength = \frac{SensorHeight}{2 \cdot \tan(FieldOfView/2)} = 5.4mm$$

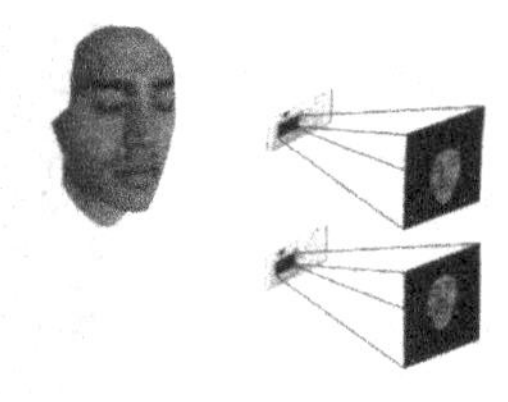

Fig. 5. The camera configuration used for this experiment consisted of a vertical arrangement, coplanar images planes, and equivalent parameter settings between both cameras. Generated images were taken at 800×600 resolution while cameras were placed on their sides to allow the human face to fully exist within the frame. This also followed the convention of standard epipolar searches restricted to scan lines

Extraction of Ground Truth and Synthetic Stereo Pair. To extract ground truth disparities the scene is first rendered from the lower camera, this will result in the top image becoming our reference frame. The OpenGL depth buffer is then read from, providing us with a window-space z-value at each pixel position, *i.e.* the pseudo-depth value expressed in the default range of $[0.0, 1.0]$. A value of 0.0 corresponds to a point lying on the near clipping plane, and 1.0, the far one. The clipping planes exist axis-aligned along the optical axis in front of the virtual camera. Once this information is obtained we can then un-project each pixel location back into the scene to find out its existence in the world-space frame. This transform uses the contextual knowledge of the current OpenGL *modelview* and *projection* matrices along with the current *viewport*. The scene is now setup and rendered from the upper camera, obtaining a new *modelview* matrix which can be utilised to project our known 3D points defined in world-space, into window-space co-ordinates (defined at sub-pixel accuracy). All points that exist on the far clipping plane (after un-projection) can be rejected as they are not of interest. Due to the epipolar setup, sub-pixel accurate disparities are found simply through the parallax across scan-lines as: $Disparity = X_{top} - X_{bottom}$.

Note: since we can find a disparity for every pixel within the virtual camera frame, this technique provides disparities at occluded areas within the stereo pair. The benefit of a virtual setup allows a variety of stereo pairs based on different camera configurations to be quickly and easily generated. It provides a relevant test bench for the special case of stereo reconstruction for human faces.

4 Results

4.1 Experimental face stereopairs

"Alex" and "Mannequin" stereopairs were processed using the previously described algorithms, the computed disparity maps were compared to the benchmark by point-to-point subtraction. Both visual appearances of the disparity maps obtained - see Fig. 10 and 9 - and their statistical error behaviour - see Table 2 and 1 - provide useful insights. Clearly, The Chen-Medioni (C-M) algorithm shows a very large error range and seems unsuitable for faces lacking texture such as "Philippe" (Fig. 6) while retaining acceptable results on "Mark" (Fig. 6). This is easily understandable as the algorithm first looks for seeds, defined as pixels with high likelihood of good matches, to propagate its search for corresponding pixels through the face. Generally all algorithms have better results on "Alex", because of its richer texture compared to "Mannequin". As expected, SDPS (Fig. 7) performs well (Table 2) offering the best trend between statistical behaviour (best average, best standard deviation) on "Mannequin" while taking only 27 seconds to process. Although correlation approaches are outperformed, *Census* and C_2 show decent matching and would probably make use of bigger correlation windows. Due to its bit-pattern, Census is penalised on standard PC architecture but could be a good candidate for hardware implementation. C_2 (described by Faugeras as the best performing correlation cost) exhibits overall better results due to its robustness to lighting conditions change. Slight shift from perfect epipolar geometry as well as illumination variation may be responsible for the, otherwise performing, Minimal-Cut algorithms bad results. However, Minimal-Cut variants and *SDPS* show very tight distribution and would show much higher correct matches percentages if the constraint was to be relaxed to ± 1 *pixel*.

4.2 3D Modelled Stereopair

Using ground truth benchmark derived from real stereopairs of human faces showed that most stereo algorithms are highly influenced by slight shift in epipolar geometry configuration as well as variation of illumination within the stereo pair images. In this experiment we intend to use the synthetic stereo pair created in Section 3.2 to annihilate the effect of most noise sources.

Table 3 shows that in the case of the human face, correlation algorithms are not as clearly outperformed by dynamic programming as in other situations - see [16] - except in the case of the error range where dynamic programming algorithms provides sensibly tighter distributions. Results confirm the trend observed with the previous experiment where Graph-Cut based algorithms do not show a good percentage of correct matching points while retaining a low mean and standard deviation error range. On error-free ground truth Correlations methods do provide good correct matching points ratio as well as a low mean but do have a somehow larger standard deviations. Although the Roy-Cox algorithm provides the lowest error range mean, standard deviation and second best correct matching point value it takes up to 20 minutes to process a stereo pair. It can be noted that SAD combined with a $Median[5,5]$ filter outscores Roy-Cox, with 87

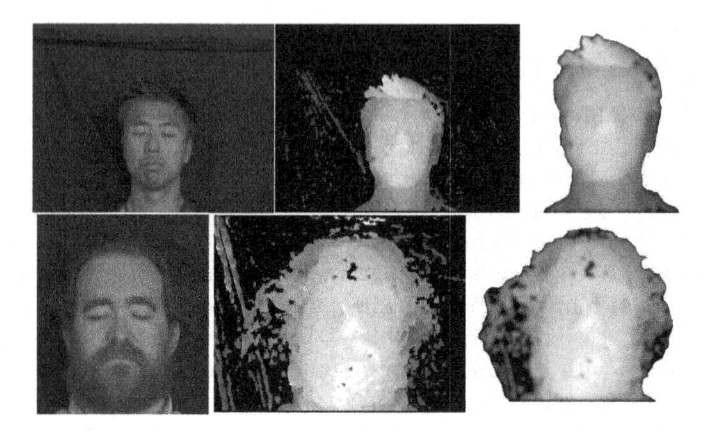

Fig. 6. Disparity map "Mark" (top left) and "Philippe" (bottom left) obtained with the Chen-Medioni algorithm with (right) and without (middle) median filtering

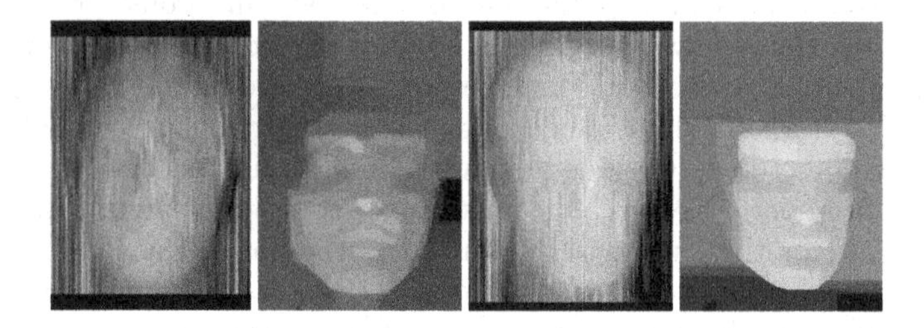

Fig. 7. Disparity map obtained on alex (first 2 images) and the mannequin (last 2 images) using SDPS and MCS

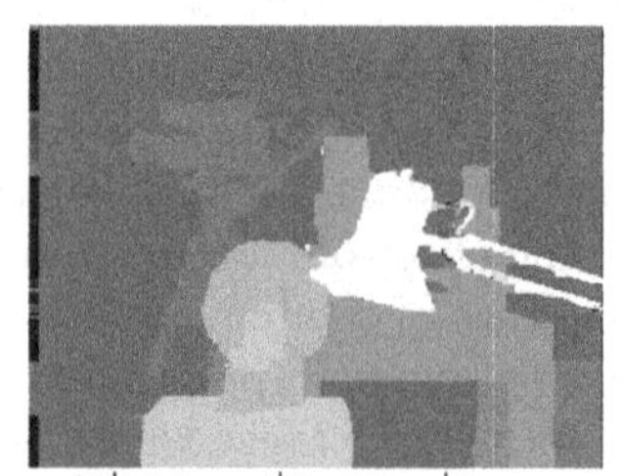

Algorithm	Correct %	$\|Mean\|$	StdDev	Error range	Timing (s)
BVZ	81.3	0.72	1.9	$-14 \sim 14$	23.7
KZ1	85.0	0.67	1.9	$-14 \sim 14$	40.1
KZ2	85.6	0.59	1.7	$-14 \sim 14$	65.3

Fig. 8. KZ2 disparity map (top), Our results (bottom) - after integrating Kolmogorov's code to our software - on the "Tsukuba" stereopair for comparisons with [12, 13, 14], with post-precessing applied

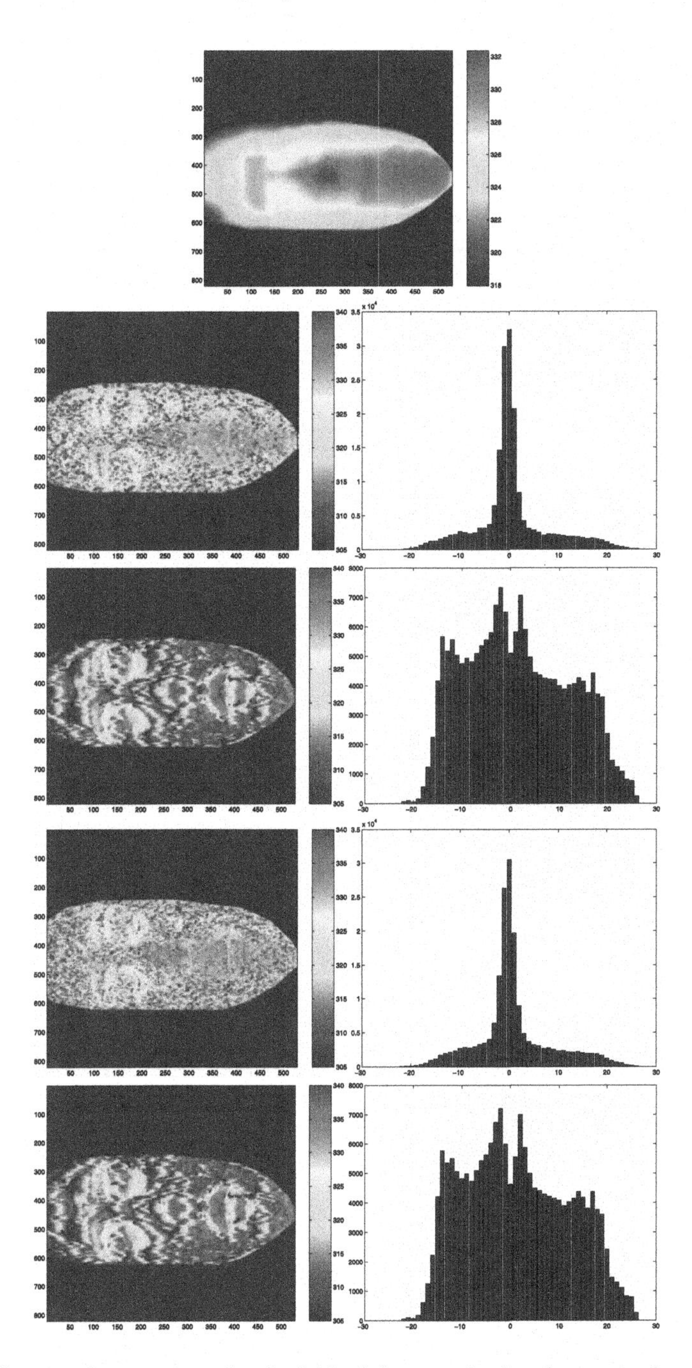

Fig. 9. All examples are given for the "Alex" Stereopair. On the top is the grountruth disparity map. On the left is the computed disparity map and on the right the corresponding error histogram in pixels. The face has 184920 pixels. From top to bottom, $Census(r = 2)$, $C_1(r = 2)$, $C_2(r = 2)$ and $SAD(r = 2)$

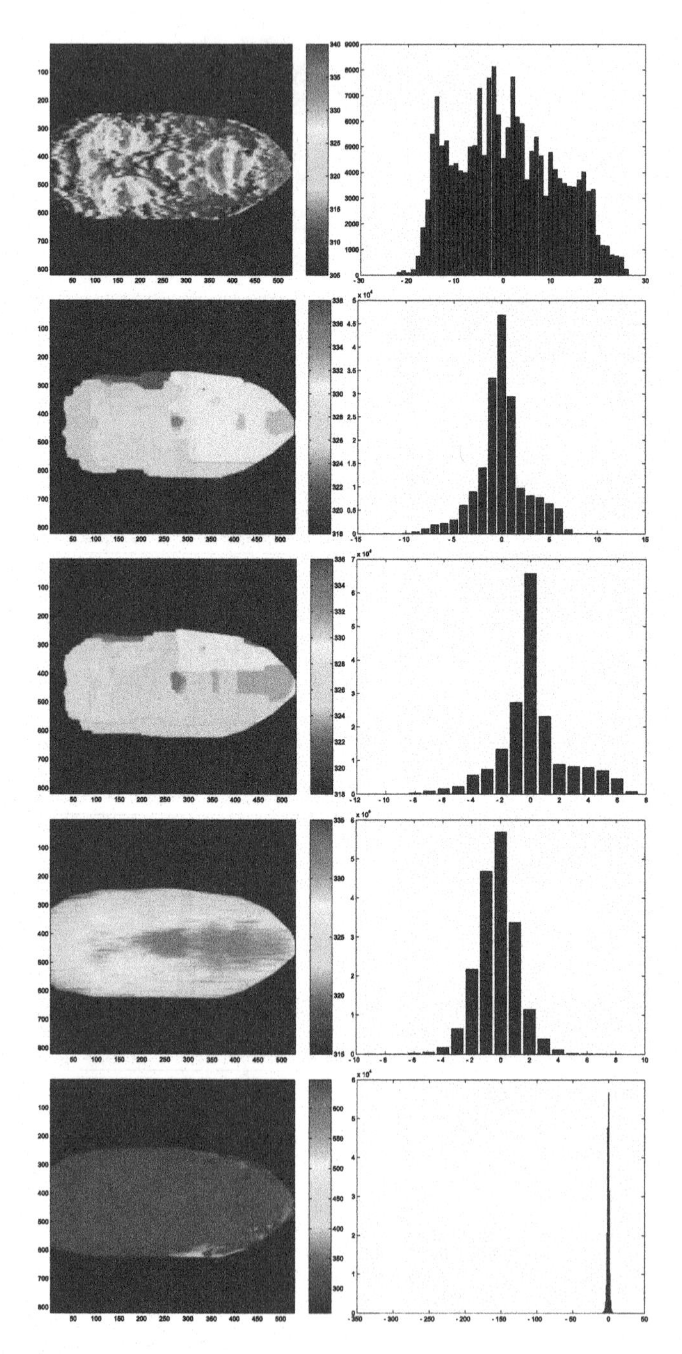

Fig. 10. All examples are given for the "Alex" stereopair. On the left is the computed disparity map and on the right the corresponding error histogram in pixels. The face has 184920 pixels. From top to bottom, BVZ, $KZ1$, $KZ2$, $SDPS$ and $Chen-Medioni$

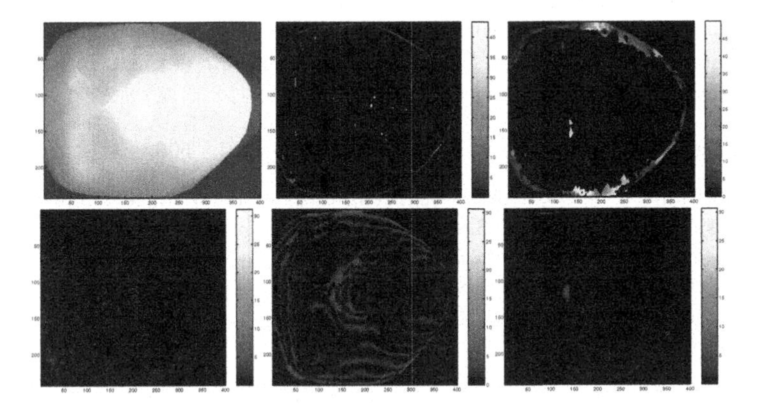

Fig. 11. From top to bottom and left to right:
generated ground truth, SAD, Chen and Medioni, SDPS, BVZ and Roy and Cox error distributions

Table 1. Results of several algorithms on the "Alex" stereopair: the percentage of good matches, *i.e.* $|error| < 0.5$ *pixels*, mean of absolute errors, error distribution standard deviation and error range

* denotes algorithm working on a smaller window, *i.e.* 820×529.

Algorithm	Correct %	$\|Mean\|$	StdDev	Error range	Timing (s)
$Census(\alpha = 2, \beta = 2)$	17.5	4.61	7.2	$-22 \sim 27$	167*
$Census(\alpha = 4, \beta = 4)$	29.2	2.23	4.4	$-22 \sim 26$	1532*
$Corr1(r = 4)$	7.3	8.03	10.0	$-22 \sim 27$	49*
$Corr2(r = 4)$	27.9	2.38	4.6	$-22 \sim 26$	44*
$SAD(r = 4)$	3.2	8.31	10.2	$-22 \sim 27$	34*
BVZ	9.5	7.65	9.8	$-22 \sim 26$	2036
$KZ1$	25.3	1.83	2.6	$-12 \sim 11$	4447
$KZ2$	35.5	1.59	2.4	$-10 \sim 7$	2214
$SDPS$	30.7	1.06	1.4	$-9 \sim 8$	27
$Chen - Medioni$	30.7	1.23	3.6	$-318 \sim 13$	89

Conclusion

Nine different stereo-matching algorithms have been applied to a set of real and synthetic referenced stereo pairs of faces. Simple correlation-based (C_1, C_2 and SAD) algorithms result in a high noise level of their face disparity map. The Chen-Medioni algorithm encounters difficulty to process textureless area of faces which results in entire zone of the face (often on the forehead) being left with no disparity information. Otherwise described as one of the best stereo algorithms in terms of accuracy [16], Minimal-Cut algorithms, have problems dealing with the specific geometry of faces as well as exhibiting penalising processing time. As expected, SDPS exhibits good time-accuracy behaviour. Overall, this ex-

Table 2. Results of several algorithms on the "Mannequin" stereopair: the percentage of good matches, *i.e.* $|error| < 0.5\ pixels$, mean of absolute errors, error distribution standard deviation and error range

* denotes algorithm working on a smaller window, *i.e.* 590×500.

Algorithm	Correct %	$\|Mean\|$	StdDev	Error range	Timing (s)
$Census(\alpha = 2, \beta = 2)$	8.8	6.12	7.8	$-20 \sim 24$	130*
$Census(\alpha = 4, \beta = 4)$	12.7	5.31	7.2	$-19 \sim 24$	369*
$Corr1(r = 4)$	6.8	8.38	10.1	$-20 \sim 24$	39*
$Corr2(r = 4)$	12.9	4.69	6.6	$-20 \sim 24$	35*
$SAD(r = 4)$	6.2	8.54	10.3	$-20 \sim 24$	27*
BVZ	9.5	8.92	10.7	$-19 \sim 24$	2320
$KZ1$	9.7	7.57	9.1	$-18 \sim 21$	3301
$KZ2$	11.1	7.44	8.9	$-18 \sim 20$	1916
$SDPS$	22.6	1.60	2.2	$-11 \sim 14$	22
$Chen - Medioni$	15.2	15.34	59.7	$-393 \sim 341$	175

Table 3. Main results for the studied algorithms: their percentage of correct matches *i.e.* $|error| < 0.5 pixels$, the mean of absolute errors, the standard deviation, the error range and the running time. Tests executed on a Pentium III 1266MHz with 5BG of RAM with no parallel implementation

* denotes time compensation for correlation algorithms, *i.e.* these are ran on a small window around the face and have been normalised for $800 \times 600 = 480000\ pixels$ instead of 87584.

Algorithm	Correct %	$\|Mean\|$	Stdev.	Error Range	Time(s)
$Census(\alpha = 2, \beta = 2)$	79.6	0.4	2.2	$-35 \sim 42$	315.4*
$C_1(r = 2)$	82.3	0.3	2.0	$-42 \sim 44$	47.2*
$C_2(r = 2)$	75.4	0.7	2.9	$-32 \sim 45$	41.3*
$SAD(r = 2)$	85.9	0.3	1.8	$-32 \sim 44$	25.3*
$Chen - Medioni(R_{thr.} = 0.6)$	67.7	1.6	6.5	$-39 \sim 41$	18.8
$SDPS(P_B = 0.9, P_M = 0.1)$	71.0	0.3	1.2	$-31 \sim 17$	10.4
$GraphCut(KZ1)$	36.0	1.0	2.0	$-32 \sim 36$	182.3
$GraphCut(KZ2)$	35.5	1.0	1.8	$-33 \sim 34$	603.4
$GraphCut(BVZ)$	40.0	0.8	1.6	$-33 \sim 19$	87.5
$Roy - Cox(d_{smooth} = 10)$	80.5	0.2	1.1	$-31 \sim 17$	1329.9

periment illustrated that most of background or recent stereovision algorithms provide close results in the case of the human face. Even though Graph-Cut variants produce smaller percentages of correct matches, they have a low standard deviation. Also, our stringent criterion might explain the difference with other studies - [16] - using a $\pm 1.5\ pixel$ criterion leads to correct percentages of $\simeq 85\%$. It seems that Energy minimisation methods tend to have an overall optimisation criterion instead of a local, which explains why it has a lower percentage of correct matches with our stringent criterion but still delivers a tight error distribution. Further studies involve assessing the accuracy of stereo algorithms on face features (rather than on the overall face), which are especially important as they carry most of the audiovisual and biometric information expressed, as well as introducing a new stereo algorithm adapted to the specific geometry profile of the face.

References

1. Q. Wang, H. Zhang, T. Riegeland, E. Hundt, G. Xu, and Z. Zhu. Creating animatable MPEG4 face. In *International Conference on Augmented Virtual Environments and Three Dimensional Imaging*, Mykonos, Greece, 2001.
2. E.D. Petajan, B. Bischoff, D. Bodoff, and N.M. Brooke. An improved automatic lipreading system to enhance speech recognition. *CHI88*, pages 19–25, 1988.
3. J. Xiao, S. Baker, I. Matthews, and T. Kanade. Real–time combined 2D+3D active appearance models. *IEEE International Conference on Computer Vision and Pattern Recognition*, june 2004.
4. O. Faugeras, B. Hotz, H. Mathieu, T. Viéville, Z. Zhang, P. Fua, E. Théron, L. Moll, G. Berry, J. Vuillemin, P. Bertin, and C. Proy. Real time correlation-based stereo: algorithm, implementatinos and applications. Technical Report 2013, Institut National De Recherche en Informatique et en Automatique (INRIA), 06902 Sophia Antipolis, France, 1993.
5. Philippe Leclercq and John Morris. Robustness to noise of stereo matching. In IEEE Computer Society, editor, *Proceedings of the International Conference on Image Analysis and Processing (ICIAP'03)*, pages 606–611, Mantova, Italy, September 2003.
6. Ramin Zabih and John Woodfill. Non-parametric local transforms for computing visual correspondence. *Lecture Notes in Computer Science (LNCS)*, 800:151–158, 1994.
7. Q. Chen and G. Medioni. Building 3-D human face models from two photographs. In *Journal of VSLI Signal Processing*, pages 127–140, 2001.
8. S.T. Barnard and M.A. Fischler. Computational stereo. In *Surveys*, volume 14, pages 553–572, 1982.
9. Georgy Gimel'farb. Binocular stereo by maximising the likelihood ratio relative to a random terrain. In Reinhard Klette, Shmuel Peleg, and Gerald Sommer, editors, *International Workshop Robot Vision '01*, number 1998 in Lecture Notes in Computer Science (LNCS), pages 201–208, Auckland, New-Zealand, February 2001.
10. Sébastien Roy and Ingemar J. Cox. A maximum-flow formulation of the n-camera stereo correspondence problem. In *IEEE Proceedings of International Conference on Computer Vision (ICCV'98)*, Bombay, 1998.
11. Yu. Boykov, O. Veksler, and R. Zabih. Fast approximate energy minimization via graph cuts. In *Proc. 7th Int. Conf. Computer Vision (ICCV 1999)*, pages 377–384, Kerkyra, Corfu, Greece, September 20–25 1999. IEEE Computer Society.
12. Yuri Boykov, Olga Veksler, and Ramin Zabih. Markov random fields with efficient approximations. In *IEEE Computer Vision and Pattern Recognition Conference (CVPR'98)*, June 1998.
13. Vladimir Kolmogorov and Ramin Zabih. Computing visual correspondence with occlusions via graph cuts. In *International Conference on Computer Vision (ICCV'01)*, July 2001.

14. Vladimir Kolmogorov and Ramin Zabih. Multi-camera scene reconstruction via graph cuts. In A. Heyden et al., editor, *In European Conference on Computer Vision (ECCV'02)*, number 2352 in Lecture Notes in Computer Science (LNCS), pages 82–96, 2002.
15. M. Chan, C.F. Chen, G. Barton, P. Delmas, G. Gimel'farb, P. Leclercq, and T. Fisher. A strategy for 3D face analysis and synthesis. *Image and Vision Computing New Zealand*, pages 384–389, 2003.
16. Daniel Scharstein and Richard Szeliski. A taxonomy and evaluation of dense two-frame stereo correspondence algorithms. *International Journal of Computer Vision*, 47:7–42, April–June 2002.

Video Cataloging System for Real-Time Scene Change Detection of News Video

Wanjoo Lee[1], Hyoki Kim[2], Hyunchul Kang[2], Jinsung Lee[3], Yongkyu Kim[4], and Seokhee Jeon[5]

[1] Division of Computer & Information, YongIn University, Samga-Dong 470, Yong-In, Gyunggi-Do, 449-714, Korea
wjlee@yongin.ac.kr
[2] Dept. of Information & Telecommunication Engineering, University of Inchoen, Dowha-Dong, Nam-Gu, Inchoen, 402-749, Korea
{coolbrain, hckang}@incheon.ac.kr
[3] Hee-Jung B/D 4F, Banpo-4Dong 49-11, Seocho-gu, Seoul, 137-044, Korea
jinsemi@neomedia.co.kr
[4] Dept. of Information Communication Engineering, Sungkyul University, Anyang 8-Dong, Manan-Gu, Anyang, 430-742, Gyunggi-Do
ykkim@sungkyul.edu
[5] Dept. of Electronics Engineering, University of Incheon, Dowha-2Dong, Nam-Gu, Incheon, 402-749, Korea

Abstract. It is necessary for various multimedia database applications to develop efficient and fast storage, indexing, browsing, and retrieval of video. We make a cataloging system, which has important information of video data. In this paper, we proposed a method to process huge amount of broadcasting data. It can detect scene-change quickly and effectively for MPEG stream. To extract the DC-image, minimal decoding is employed form MPEG-1 compressed domain. Scene-change is detected by the modified histogram comparative method that combines luminance and color component. A neural network is used to judge whether it is scene-change with an anchorperson and to raise precision of an anchorperson scene extraction.

1 Introduction

In the past, information retrieval was dependent on key words that are extracted from document(sentence). It is necessary for more relevant search to use contents based search which is based not only on character but also on image and audio related the character[2]. In particular, about 73% of many billions website are including image data[2]. It will be important element to use image and video information for searching. There is video data for 6 million hours in all around world. Every year, video is increasing by 10%[2]. In addition, video that has 1 hour requires about 10 hours to add an indexing to the video [2]. Thus, it is impossible and inefficient to make literal explanation to image and video, because that is subjective and has huge amount of information. Therefore, as image and video have become more important subjects for information retrieval, a cataloging system is needed for more efficient search of the media.

R. Klette and J. Žunić (Eds.): IWCIA 2004, LNCS 3322, pp. 705–715, 2004.

In this paper, we intend to achieve real-time processing to deal with huge amount of broadcast data. We also want to detect correct scene-change with fast method. For example of news video, every accident report consists of anchorperson part and reporter part repeatedly. We wish to analyze the structure information from video stream. In structure information of the news video, anchorperson scene is the best for analysis. We create a model for anchorperson by neural network, and we detect anchorperson scene. This structural information as well as time data play an important role in retrieval. As the result of parsing, it is possibility to provide video summary and video plot.

2 Video Cataloging System and Characteristic of News Video

2.1 Video Cataloging System

The Video Cataloging System extract semantic information such as key frame, letter and figure from inputted video data, and effectively stores the information for later search. This system functionally consists of video data formation, browsing system, automatic index, and content-based retrieval system. Figure 1 shows an example of the video cataloging system. In order to build in this video cataloging system, video summary system, which searches scene-change, extracts key-frame, and removes similar frame, must be developed.

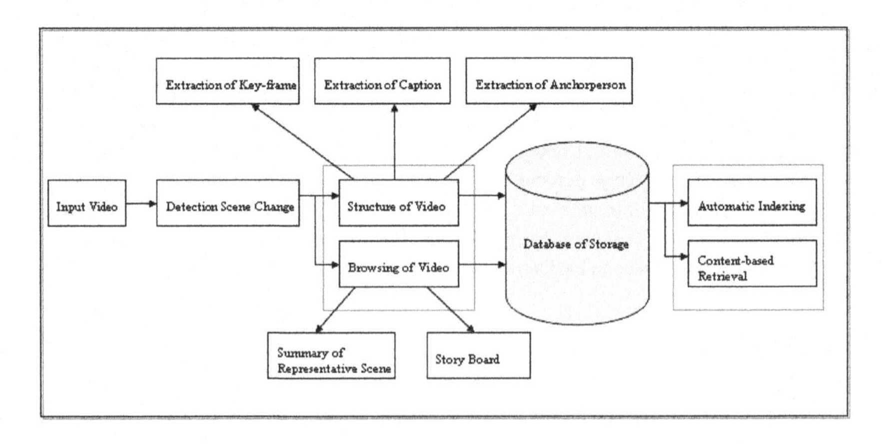

Fig. 1. Video Cataloging System

This paper proposed the video processing system for video cataloging system. The video processing system extract video structure information from input video stream, and it consist of scene-change detection process, an anchorperson screen extraction process and a video summary creation process. Also the extracted representative screen is stored with the time code, and the compression of this input stream follows a MPEG standard form.

The current content-based retrieval cannot do automatic meaning analysis of contents. In order to overcome this limit, many researches about news video data, which have the character with fixed type, are become accomplished mainly in the inside and outside of the country for content-based retrieval.

2.2 Characteristic of News Video

According to each genre, content-information with high level will be different. With the genre of news, many content-information with high level may exist such as anchorperson scene, and graphic scene, but obviously anchorperson scene is the most important. We will be able to summarize the full story by anchorperson-scene. And we will be able to separate different articles as it is placed at the beginning of each article video data.

News video data helps scene segmentation, composition of index, extraction of representative content by using the formal knowledge of time and space. Temporal composition of news video is explained in figure 2. The beginning scenes tell that news has been started. Each news program has different beginning scene, and furthermore each broadcasting system has different beginning scene.

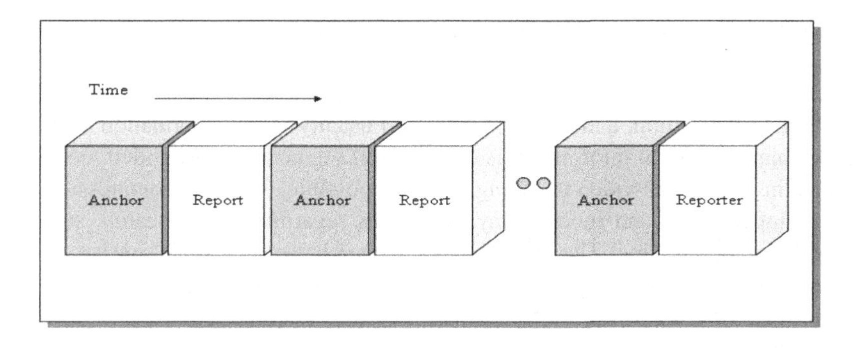

Fig. 2. Temporal Composition of News Video

Anchorperson scene after beginning scene tells the news story, and one or more anchorperson may show up. The place where the news contents which briefly shown in anchorperson scene is reporter scene. The reporter scene may consist of data, interview, artificial pictures and tables, which are connected the content of the news.

Spatial composition of news video, consists of anchorperson scene and reporter scene as shown in figure 3. The anchorperson scene consists of news icon, news caption, anchorperson part, anchorperson's name, news program title, and name of broadcasting company. Usually the location and the lasting-time of each components are uniform.

News icon, the name of a broadcasting company and the caption for one frame is only a portion of the whole video stream, but they give important information regarding the news.

Fig. 3. Spatial Composition of News Video

3 Scene Change Detection

Content-based retrieval can be categorized by which is used among visual information, literal information, audio information, and usually visual information is used for video summary. Visual information is usually used segmentation for video scenes, and through this, structural video browsing becomes possible. As video compression skills are developed more and more, many researches regarding video search, summary, browsing are being done. The algorithm for scene change generally consists of compressed domain and non-compressed domain. In this paper, for the purpose of real time process, the algorithm of compressed domain is applied. We do an automatic scene change detection and anchorperson scene detection, which are basis for the index of content-based. Video data stored in MPEG-1 video stream.

3.1 Cut-Comparison for Scene Change Detection

As Figure 4 shows, a D-picture is generated from the DC-values of DCT coefficients in each 8x8 blocks. For example, a original image with 352x240 has a D-picture with 44x30. This D-picture is used for cut-scene change detection as a source frame. Figure 4 shows cut-scene change detection using D-picture[8].

There are various methods, which the MPEG video frame can be compared with the created D-Picture. But, some problem would be happened in case the P,B frame uses, because the frame refer the previous or the after. Therefore, these comparing methods with only I-frames lead to faster processing time[9]. In this paper, we use no P and B frames that require much time to decode a part of frame because of real time processing.

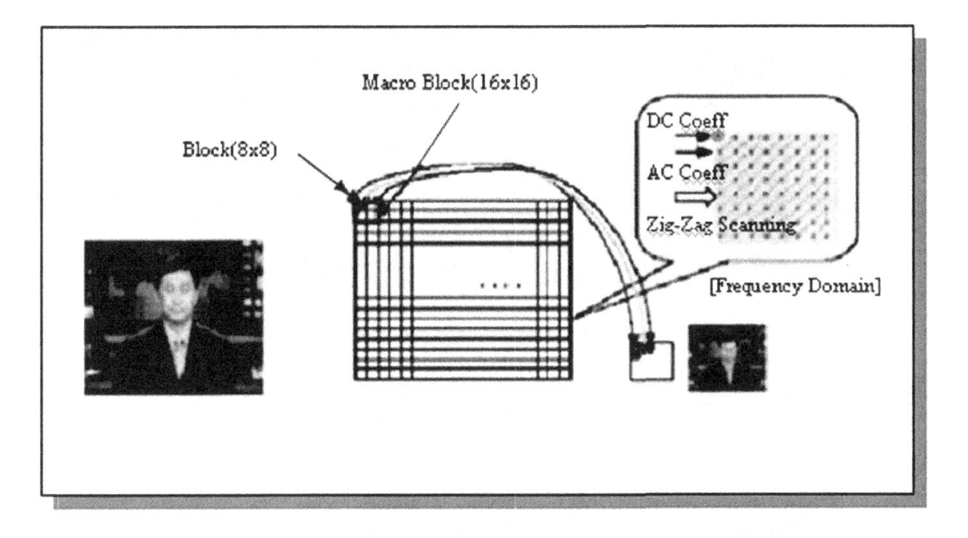

Fig. 4. Original Image and DC Image

3.2 Proposed Method for Scene Change Detection

The Figure 5 shows a feature in news video through the histogram difference of frames. In the figure, We can found out characteristic of the news video data that an anchor introduce to the summary at the induction before a reporter says in detail. The whole flow of news video is very static and the frequency of the scene-change is relatively less than the one of advertisements or music videos.

In general, the generation sequence in the video is about 5 sec such as TV program. The NTSC MPEG with 29.97 frame per second usually includes under 15 pictures on a GOP, this GOP show I-picture one time at least. Therefore, it will be right to detect a cut even if I-picture only uses, because I-picture is a representative frame in GOP.

As we decode DC coefficients that represent pixel values in 8x8 blocks, the shape of histogram shows a sharp Gaussian curve. This shape would not have much difference, in case of detected feature vector through quantization with luminance component, because the most quantization has the similar value.

For calculating the distance of the compressed domain, it usually would use the YUV of the DC Image. The Figure 6 shows which compared with the difference of the decoding DC Image average value, variance value, luminance component(Y component) and the suggested methods in this paper.

In this paper, to emphasize the distance of scenes, we use the variation of a method that gets the difference of cumulative histogram between two frames [4,5,6]. Luminance represents a whole image, but dose not contain any color information. So the similar distribution of luminance in images results an errors in the detection. To emphasize the distance of scenes, we use not only luminance but also chrominance.

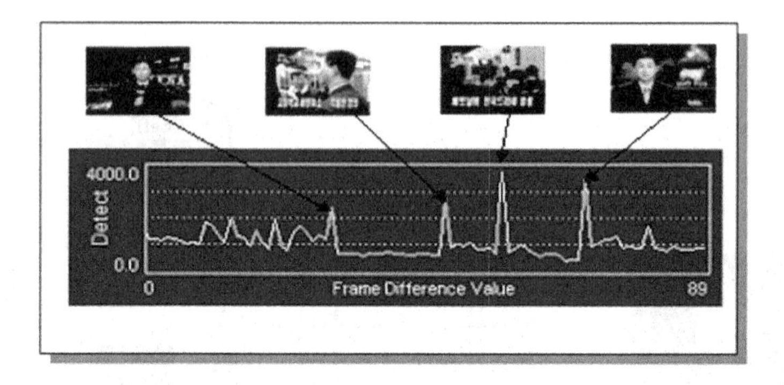

Fig. 5. Histogram Difference of News Video

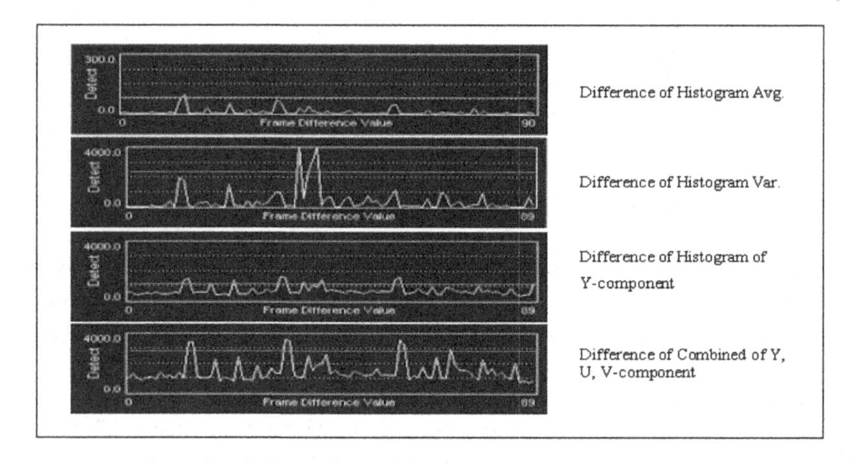

Fig. 6. Feature Value of Frame Comparison

Equations to compute mean (D_1), distribution (D_2), cumulative histogram value (D_3), and the value (D_4) from the proposed method are.

$$D_1(i,i+1)mean = \left| \frac{1}{n}\sum_{j=1}^{n}\overline{Y}_i - \frac{1}{n}\sum_{j=1}^{n}\overline{Y_{i+1}} \right| \quad \text{---(1)}$$

$$D_2(i,i+1)\mathrm{var} = \left| \frac{1}{n}\sqrt{(\sum_{j=1}^{n}\overline{Y}_i - \mu_i)^2} - \frac{1}{n}\sqrt{(\sum_{j=1}^{n}\overline{Y_{i+1}} - \mu_{i+1})^2} \right| \quad \text{---(2)}$$

$$D_3(i,i+1)histo = \left| \sum_{j=1}^{n} YH_i - \sum_{j=1}^{n} YH_{i+1} \right| \quad \text{---(3)}$$

$$D_4(i,i+1) = \left| \sum_{j=1}^{n} YH_i - \sum_{j=1}^{n} YH_{i+1} \right| + w \times \left(\left| \sum_{j=1}^{n} UH_i - \sum_{j=1}^{n} UH_{i+1} \right| + \left| \sum_{j=1}^{n} VH_i - \sum_{j=1}^{n} VH_{i+1} \right| \right) \quad \text{---(4)}$$

Where n is the number of entire pixels, Y is the luminance, U and V are the chrominance and YH , UH , and VH is the histogram values of Y , U , and V in each ith frame respectively. In case of (D_4), we make the decision of threshold easy by adding [w x the cumulative histogram of chrominance] to the cumulative histogram of luminance, where w represent weight that we set as 2 in our experiment. As Figure 6 shows, the difference of distance between scenes through the proposed method than the others is much lager.

With a I-picture in MPEG-1 video sequence, we generate a DC image from the least decoding and compute the distance of the images, shortly, the cumulative histogram, between the frames. This method is robust against noise or light.

We use the YCbCr color model and first compute the cumulative histograms for Y, Cb, and Cr. If the distance, using the proposed method in this paper, between images is lager than Th(threshold), then we regard a scene change as detected.

4 Anchorperson Shot Detection

For the efficient retrieval of news videos, it is required to extract the exact terms of anchorpersons. There are many different data in news videos, and anchorpersons, graphics, interviews, reports, and so on would be high-level data. Especially, the scenes of anchorpersons are useful to divide the items of news.

4.1 Characteristic of Anchorperson Scene

Figure 7 shows an example of the anchorperson-frame of spatial composition. News icon may not exist, and exist to anchorperson-frame. But however, we can see that anchorperson's face exists without correlation with existence availability of news icon. For an anchorperson-frame extraction, we analyze the structural feature of the frame where the anchorperson exists. Then we search the feature of anchorperson's face region, and recognize anchorperson's face using this feature. Recognized result is used information of key-frame.

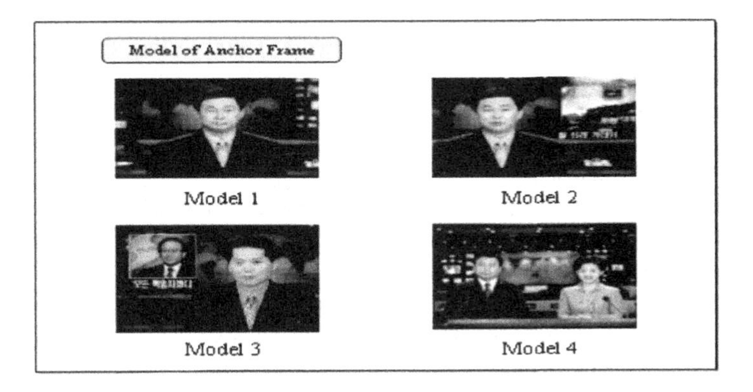

Fig. 7. Anchorperson Model

In this paper, we wish to recognize the anchor face with anchor facial existence area without considering existence availability of news icon.

In this paper, we define anchorperson by the following features:

1. Anchorperson's face always appear.
2. Anchorperson's faces have uniform size.
3. Anchorperson's face is little motion.
4. Anchorperson's background is similar.

4.2 Proposed Extraction Algorithm of Anchorperson

Generally, there are many methods that use the color information or the organizational feature to detect the scenes of an anchorperson, or anchorpersons. But, in this paper, we use the ADALINE(Adaptive Linear Neuron) neural network, unlike other existing methods. Because, whether the scene contains an anchorperson (or anchorpersons) is regarded as pattern recognition. Neural network is good for deciding where a class belongs in, according to feature vectors, defined as input.

Figure 8 shows an example that forms a DC image into the input pattern of neural network. Binary coded procedure is performed for 40x30 DC image, not full-size image. Threshold in this procedure becomes the mean of intensities of a DC image.

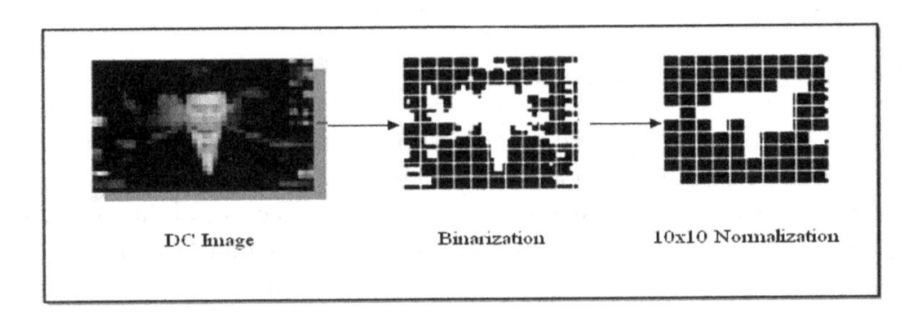

Fig. 8. ADALINE Input Pattern

By the property of the ADLINE neural network, input vectors should be normalized into 10x10 input values. On a normalized image, input pattern becomes -1 for 0 pixel value and +1 for 255. The reason of using luminance as input vector is that in most news videos the scenes of anchorperson occur indoors and in this case a human face becomes the brightest areas.

In respect of contents-base retrieval news icons play an important role. So, we divided these models into pure anchor and anchor with news icon while the purpose in this experiment is to search for the term of anchorperson. But, comparing the anchorperson with the reporter, input patterns for these two scenes are nearly same. Therefore, we should separate these two scenes. In this paper, we created an additional input model for the non-anchorperson.

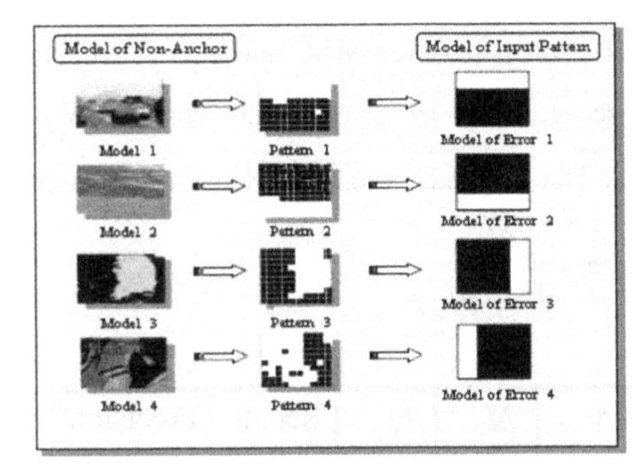

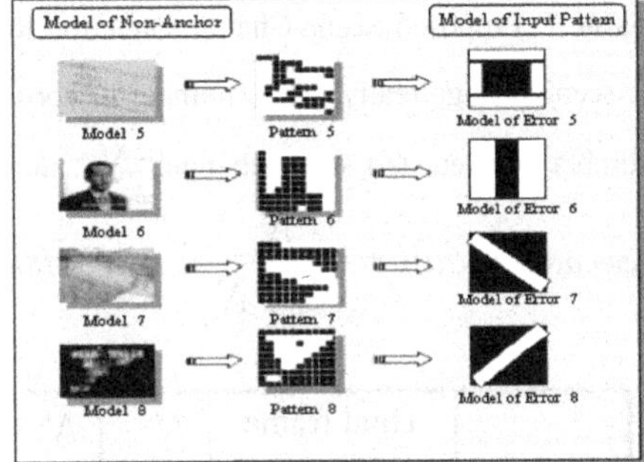

Fig. 9. Non-Anchorperson Pattern Modeling

5 Experiment Result and Analysis

For our experiment, we used a Pentium IV 1.7GHz with Widows XP-OS and Visual C++ 6.0 for programming language. To verify our algorithm proposed in this paper, we used news videos of TV Program. New video is decoded by MPEG-1 standard.

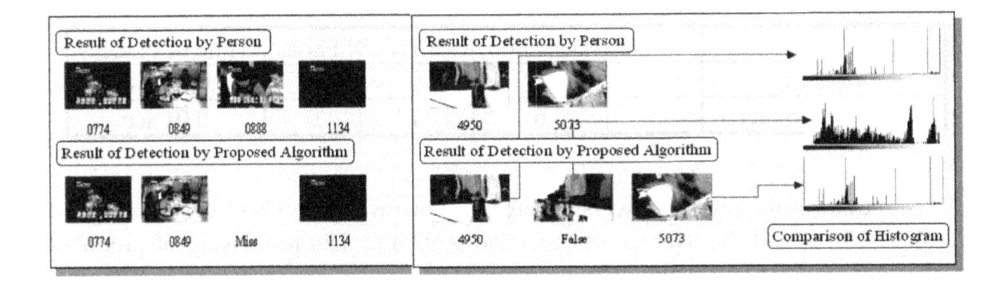

Fig. 10. Results of Missed Detection & False Detection

Figure 10 shows a result of the missed detection. Missed detection occurs in the case which has the distribution where the histogram is similar. Figure 10 shows a result of the false detection. False detection dose not have similar histogram. False detection occurs in the case which has the effect of panning. Figure 11 shows a result of the Anchorperson detection.

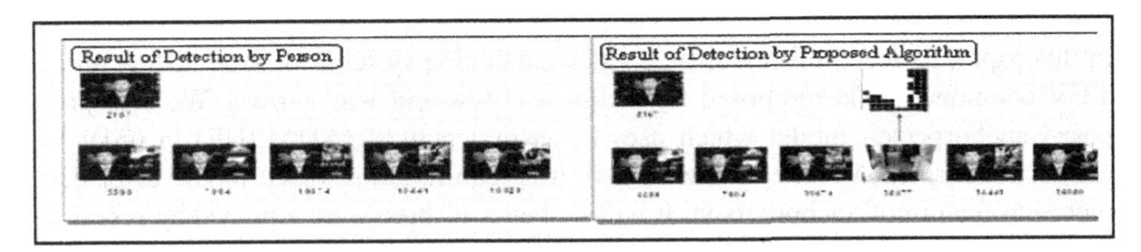

Fig. 11. Result of Anchorperson Detection

Table 1. Proposed Scene Change Dtetction Result. This paper used MBC news. N_r : number of scene-change really; N_c : number of correct scene-change in detected scene-change; N_d : number of detected scene-change; N_t : number of false detection; N_m : number of missed detection; $recall = \frac{N_c}{N_c + N_f}$; $percision = \frac{N_c}{N_c + N_m}$;

	Total frame Time	N_r	N_c	N_d	N_t	N_m	Recall	Precision
News1	16063	121	110	114	4	7	96.5%	94.0%
News2	15284	106	95	100	5	6	95.0%	94.1%
News3	27452	221	196	207	11	4	94.7%	93.3%
Average							95.1%	93.8%
Time of process	0.016 sec (Per frame processing time)							

Table 2. Proposed Anchorperson Detection Result

Class	Total	Detection	Miss	False	Time of Process
Anchor	1	1	0	0	0.01 sec
Anchor+News Icon	5	6	1	0	0.01 sec

The result of the proposed scene-change is show in Figure 12-14. The average of recall is 95.1% and the average of precision is 93.8%. The processing of proposed scene-change is 0.016 second. The proposed detection of anchorperson is 0.01 second. These algorithms will be able to accomplish with a real-time.

However, in distribution of similar histogram and the panning of the camera, proposed algorithms occur miss and false. And anchorperson detection preceding learning is necessary. When the learning which is sufficient is not accomplished, it will not be able to use in anchorperson detection. Also, training many non-anchorperson model is necessary, and the error of judge of anchorperson is demand the 2nd decision.

6 Conclusion

In this paper, to accomplish real-time video cataloging system, we used histogram of YUV component. The proposed algorithm was fast and was correct. We also proposed anchorperson model which uses the neural network(ADALINE) in order to extract anchorperson scene in news video. It is required to research input vector that represent feature of anchorperson. It will be better if there were a preceding research for speaker identification system.

References

1. T. Sikira. MPEG Digital Video-Coding. In *IEEE Signal Processing Magazine*, Vol. 14, 82-100. Sep. 1997.
2. Young-Min Kim, Song-Ha Choe, Bon-Woo Hwang and Yoon-Mo Yang. Content-Based Retrieval of Video. In *Korea Information Science Society*, Vol. 16, No. 8, 39-47. Aug. 1998.
3. Shin You. A Study on the Video Segmentation for an Education VoD Service System. In *University of Incheon, Korea*, Dec. 1999.
4. B. L. Yeo and B. Liu. On the extraction of DC sequence from MPEG compressed video. In *IEEE Proc ICIP*, Vol. 12, 260-263. Oct. 1995.
5. P. Arman and H. Zhang. Content-Based Representation and Retrieval of Visual Media:A State-of-the-Art Review. In *Multimedia Tools and Application*, Vol. 3, 179-202, Nov. 1996.
6. J. Meng, Y. Juang and S.F. Chang. Scene change detection in a MPEG compressed video sequene. In Proc SPIE, Vol.2419, 14-25, Feb. 1995.
7. Jun-Il Hong. A Study on the Extraction of Caption and Recognition for Knowledge-Based Retrieval. In *University of Incheon, Korea,* Dec. 2001.
8. Boon-Lock Yeo and Bede Liu. Rapid Scene Analysis on Compressed Video. In *IEEE Trans. On Circuit and Systems for Video Technology*, Vol. 5, No. 6, 533-544, Dec. 1995.
9. Nilesh V.Patel and Ishwar K.Sethi. Compressed Video Processing for cut Detection. In *IEEE Proc. Video, Image and Signal Processing*, Vol. 143, No. 5, 315-522, Oct. 1996.
10. H. Luo and Q. Haung. Automatic Model-based Anchorperson Detection. In *Processings of SPIE on Storage and Retrieval for Media Database*, Vol. 4315, 536-544, Oct. 2001.
11. D. Chai and K.N. Ngan. Face Segmentation Using Skin-Color Map in Videophone Applications. In *IEEE Trans. On Circuit and Systems for Video Technlogy*, Vol. 9, No. 4, 551-564, June. 1999.
12. A. Hanjalic, R.L. Lagensijk and J. Biemond. Template-based detection of anchorperson shots in news programs. In *IEEE Proc ICIP*, Vol. 3, 148-152. Oct. 1998

Automatic Face Recognition by Support Vector Machines*

Huaqing Li, Shaoyu Wang, and Feihu Qi

Department of Computer Science and Engineering,
Shanghai Jiao Tong University, Shanghai 20030, P.R. China
{waking_lee, wang, fhqi}@sjtu.edu.cn

Abstract. Automatic face recognition, though being a hard problem, has a wide variety of applications. Support vector machine (SVM), to which model selection plays a key role, is a powerful technique for pattern recognition problems. Recently lots of researches have been done on face recognition by SVMs and satisfying results have been reported. However, as SVMs model selection details were not given, those results might have been overestimated. In this paper, we propose a general framework for investigating automatic face recognition by SVMs, with which different model selection algorithms as well as other important issues can be explored. Preliminary experimental results on the ORL face database show that, with the proposed hybrid model selection algorithm, appropriate SVMs models can be obtained with satisfying recognition performance.

1 Introduction

Support vector machine (SVM) is considered to be one of the most effective algorithms for pattern recognition (classification) problems. Generally, it works as follows for binary problems [11]: First the training examples are mapped, through a mapping function ϕ, into a high (even infinite) dimensional feature space $\mathcal{H}$. Then the optimal separating hyperplane in $\mathcal{H}$ is searched for to separate examples of different classes as possible, while maximizing the distance from either class to the hyperplane. In implementation, the use of kernel functions avoids the explicit use of mapping functions and makes SVM a practical tool. However, as different kernel functions lead to different SVMs with probably quite different performance, it is very important to select appropriate types and parameters of the kernel function for a given problem.

Automatic face recognition (AFR), being a special case of pattern recognition, has a wide variety of applications including access control, personal identification, human-computer interaction, law enforcement, etc. In recent yeas, AFR has attracted lots of research efforts and a large amount of algorithms have been proposed in literature [3]. Not surprisingly, exploiting SVMs for face recognition also has been widely carried out [4–8]. Though excellent performance was

* This work is supported by the National Natural Science Foundation of China (No. 60072029).

R. Klette and J. Žunić (Eds.): IWCIA 2004, LNCS 3322, pp. 716–725, 2004.

reported, no details about how the authors have selected the SVMs parameters for their problems were given. Hence the reported results might have been overestimated. On the other hand, if we have to select a kernel function and its parameters manually for every given problem, AFR by SVMs becomes semiautomatic. These, together with the inherent difficulty of SVMs model selection, explain why no commercial face recognition software is based on SVMs [1, 2].

In this paper, we propose a framework for investigating AFR by SVMs, which is shown in Figure 1. It has the following work flows: In the training phase, images are preprocessed and features are extracted. Then SVMs hyperparameters are automatically selected. Finally SVMs models are trained on the training set. In the testing phase, the same image preprocessing and feature extraction operations are performed on the test images, and those trained SVMs models are employed for class prediction. The "user settings" module in the training phase involves four user-dependant choices: type of kernel function, model selection algorithm, decomposing strategy for multi-class problems and SVMs outputs combining strategy. Note that, this module is for investigating purpose and there is no need to change the choices from problem to problem in practical use.

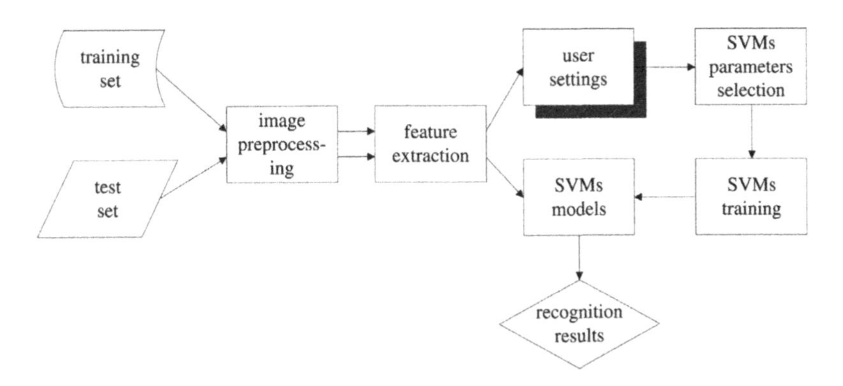

Fig. 1. Framework of automatic face recognition by SVMs

This paper focuses on SVMs with the radius basis function (RBF) kernel. Three model selection algorithms are explored: the cross-validation algorithm, a gradient-descent based algorithm with the radius/margin bound and a hybrid algorithm. Their performance is evaluated in terms of the quality of the chosen SVMs models. Preliminary experimental results on the ORL face database show that, with the proposed hybrid algorithm, appropriate SVMs models can be chosen to achieve satisfying recognition performance. The rest of the paper is organized as follows: In Section 2, we briefly review the basic theory of SVMs. Model selection algorithms are described in Section 3. Section 4 introduces schemes of using SVMs for multi-class classification problems. Experimental results and analysis are given in Section 5. Finally, Section 6 concludes the paper.

2 SVMs for Pattern Recognition

Given a set of linearly separable training examples $\{x_i, y_i\}$, $i = 1, 2, \ldots, l$, where $x_i \in R^n$ is the i-th training vector and $y_i \in \{-1, 1\}$ is the corresponding target label. An SVM searches for the optimal separating hyperplane which separates the largest possible fraction of examples of the same class on the same side. This can be formulated as follows:

$$\min \quad \frac{1}{2}||w||^2 \ , \tag{1}$$

$$\text{s.t.} \quad y_i(w \bullet x_i + b) - 1 \geq 0 \quad \forall i \ ,$$

where w is normal to the hyperplane, b is the threshold, $||\cdot||$ is the Euclidean norm, $\bullet$ stands for dot product. Introducing Lagrangian multipliers α_i, we obtain

$$\min \quad L_P = \frac{1}{2}||w||^2 - \sum_{i=1}^{l} \alpha_i y_i (x_i \bullet w + b) + \sum_{i=1}^{l} \alpha_i \ . \tag{2}$$

(2) is a convex quadratic programming problem, we can equally solve the Wolfe dual

$$\max \quad L_D = \sum_{i=1}^{l} \alpha_i - \frac{1}{2} \sum_{i=1}^{l} \sum_{j=1}^{l} \alpha_i \alpha_j y_i y_j (x_i \bullet x_j) \ , \tag{3}$$

$$\text{s.t.} \quad \sum_{i=1}^{l} \alpha_i y_i = 0 \ , \quad \alpha_i \geq 0 \quad \forall i \ ,$$

with the solution

$$w = \sum_{i=1}^{l} \alpha_i y_i x_i \ . \tag{4}$$

For a test example x, the classification is $f(x) = sign(w \bullet x + b)$.

The above linearly separable case results in an SVM with hard margin, i.e. no training errors occur. If the training set is nonlinearly separable, we can first map, through function ϕ, the original inputs into a feature space $\mathcal{H}$ wherein the mapped examples are linearly separable. Then we search for the optimal separating hyperplane in $\mathcal{H}$. The corresponding formulas of (3) and (4) are

$$\max \quad L_D = \sum_{i=1}^{l} \alpha_i - \frac{1}{2} \sum_{i=1}^{l} \sum_{j=1}^{l} \alpha_i \alpha_j y_i y_j (\phi(x_i) \bullet \phi(x_j)) \ . \tag{5}$$

$$w = \sum_{i=1}^{l} \alpha_i y_i \phi(x_i) \ . \tag{6}$$

The corresponding classification rule is

$$\begin{aligned} f(x) &= sign(w \bullet \phi(x) + b) \\ &= sign(\textstyle\sum_{i=1}^{l}(\phi(x_i) \bullet \phi(x)) + b) \ . \end{aligned} \tag{7}$$

As the only operation involved is dot product, kernel functions can be used to avoid the use of the mapping function ϕ via $K(x_i, x_j) = \phi(x_i) \bullet \phi(x_j)$. One of the most popular kernels is the RBF kernel

$$K(x_i, x_j) = \exp(-\sigma||x_i - x_j||^2) .$$

If the training set is nonseparable, slack variables ξ_i have to be introduced. Then the constraints of (1) are modified as

$$y_i(w \bullet x_i + b) - 1 + \xi_i \geq 0 \quad \forall i . \tag{8}$$

Two objectives exist under such cases. One is the L1 soft margin formula:

$$\min \quad \frac{1}{2}||w||^2 + C\sum_{i=1}^{l} \xi_i , \tag{9}$$

where C is the penalty parameter. The other is the L2 soft margin formula:

$$\min \quad \frac{1}{2}||w||^2 + \frac{C}{2}\sum_{i=1}^{l} \xi_i{}^2 . \tag{10}$$

(9) and (10) differ in that (10) can be treated as hard margin cases through some transformation while (9) can not. Since some theoretical bound (e.g. the radius/margin bound) based model selection algorithms are applicable only to SVMs with hard margin, we employ (10) in this paper. The kernel employed is the RBF kernel. Thereby, there are two parameters to be tuned by model selection algorithms, the penalty parameter C and the kernel parameter σ.

3 SVMs Model Selection Algorithms

There are mainly two categories of algorithms for SVMs model selection. Algorithms from the first category estimate the prediction error by testing error on a data set which has not been used for training, while those from the second category estimate the prediction error by theoretical bounds.

3.1 The Cross Validation Algorithm

At present, the cross validation (CV) algorithm, which falls into the first category, is one of the most popular and robust algorithms [13, 14]. The CV algorithm first divides the original training set into several subsets of nearly the same size. Then each subset is sequentially used as the validation set while the others are used as the training set. Finally SVMs performance on all validation sets is summed to form the cross validation rate.

Generally the CV algorithm employs an exhaustive grid-search strategy in some predefined parameters ranges. In [12], Chung et al. pointed out that trying exponentially growing sequences of C and σ is a practical method to identify good parameters for SVMs with the RBF kernel. However, a standard grid-search is very computational expensive when dealing with even moderate problems.

In [14], Staelin proposed a coarse-to-fine search strategy for the CV algorithm based on ideas from design of experiments. Experimental results showed that it is robust and worked effectively and efficiently. The strategy can be briefly described as follows: Start the search with a very coarse grid covering the whole search space and iteratively refine both the grid resolution and search boundaries, keeping the number of samples roughly constant at each iteration. In this paper, a similar search strategy like this is employed for the CV algorithm.

3.2 The Algorithm with the Radius/Margin Bound

The gradient-descent based algorithm with the radius/margin bound is another popular algorithm for SVMs model selection [16–18]. The radius/margin bound is proposed by Vapnik and only suitable for SVMs with no training errors [19]. It is an upper bound on the number of errors of the leave-one-out procedure and can be expressed as

$$T = \frac{1}{l}\frac{R^2}{\gamma^2} = \frac{1}{l}R^2||w||^2 , \tag{11}$$

where R is the radius of the minimal sphere which encloses all mapped examples in $\mathcal{H}$, $\gamma = \frac{1}{||w||}$ is the margin of the optimal separating hyperplane obtained by an SVM. Note that R can be easily computed [15].

Chapelle et al. argued that the radius/margin bound is differentiable with respect to the SVMs parameters and gave generalized formulas for computing the derivatives [16]. Based on their work, the following gradient of SVMs with the BRF kernel can be obtained [17]:

$$\frac{\partial T}{\partial C} = \frac{1}{l}\left[\frac{\partial ||w||^2}{\partial C}R^2 + ||w||^2\frac{\partial R^2}{\partial C}\right] , \tag{12}$$

$$\frac{\partial T}{\partial \sigma^2} = \frac{1}{l}\left[\frac{\partial ||w||^2}{\partial \sigma^2}R^2 + ||w||^2\frac{\partial R^2}{\partial \sigma^2}\right] . \tag{13}$$

The derivatives of $||w||^2$ are given by

$$\frac{\partial ||w||^2}{\partial C} = \frac{1}{C^2}\sum_{i=1}^{l}{\alpha_i}^2 , \tag{14}$$

$$\frac{\partial ||w||^2}{\partial \sigma^2} = -\sum_{i=1}^{l}\sum_{j=1}^{l}\alpha_i\alpha_j y_i y_j\frac{\partial K(x_i,x_j)}{\partial \sigma^2} . \tag{15}$$

The derivatives of R^2 are given by

$$\frac{\partial R^2}{\partial C} = -\frac{1}{C^2}\sum_{i=1}^{l}\lambda_i(1-\lambda_i) , \tag{16}$$

$$\frac{\partial R^2}{\partial \sigma^2} = -\sum_{i=1}^{l}\sum_{j=1}^{l} \lambda_i \lambda_j \frac{\partial K(x_i, x_j)}{\partial \sigma^2} \ . \tag{17}$$

and

$$\frac{\partial K(x_i, x_j)}{\partial \sigma^2} = -K(x_i, x_j)||x_i - x_j||^2 \ . \tag{18}$$

Hence the gradient of T can be easily obtained once T has been computed. Here (18) is different from (11) in [17] due to the difference in RBF kernel formulas.

With this good property of the radius/margin bound, gradient-descent based algorithms can be employed to obtain good SVMs parameters. In this paper, we consider the BFGS quasi-Newton method described in [12]. The gradient-descent based algorithm is very fast and can theoretically obtain the best parameters. However its practical performance depends on the initialization a lot. An inappropriate initialization may leads to very poor parameters.

3.3 The Hybrid Algorithm

To overcome the shortcoming of the gradient-descent based algorithm, we propose a hybrid algorithm which works as follows: First the CV algorithm is employed to select the SVMs parameters. Then with these parameters as initials, the gradient-descent based algorithm is employed to obtain even better parameters. Since the course-to-fine search strategy described in Section 3.1 is employed, the first step is not very expensive. Such a scheme makes the gradient-descent based algorithm more stable and can always obtain good parameters. Note that, this is also mentioned in [14]. However no experiments were performed there.

4 SVMs for Multi-class Classification Problems

As SVM is dedicated to binary classification problems, two popular strategies have been proposed to apply it to multi-class problems. Suppose we are dealing with a K-class problem. One scheme can be used is the *one-against-rest* method [10], which trains one SVM for each class to distinguish it from all the other classes. Thus K binary SVMs need be trained. The other scheme is the *one-against-one* method [9], which trains $K(K-1)/2$ binary SVMs, each of which discriminate two of the K classes. In this paper, the latter strategy is employed.

Another important issue is to combine the outputs of all binary SVMs to form the final decision. The most simple method is Max-Voting, which assigns a test example to the class with the most winning two-class decisions [20]. More sophisticated methods are available, given that each binary SVM outputs probabilities. Unfortunately, as shown in Section 2, SVMs can not produce such probabilistic outputs. In [21], Platt suggested to map original SVM outputs to probabilities by fitting a sigmoid after the SVM:

$$P(y=1|x) = \frac{1}{1+\exp(Af(x)+B)} \ . \tag{19}$$

Parameters A and B are found by minimizing the negative log likelihood of the training data:

$$\min \quad -\sum_{i=1}^{l} t_i \log(p_i) + (1-t_i)\log(1-p_i) \ , \tag{20}$$

where

$$p_i = \frac{1}{1+\exp(Af(x_i)+B)} \ , \quad t_i = \frac{y_i+1}{2} \ . \tag{21}$$

Actually Platt argued more sophisticated method to assign the target probabilities t_i. However, empirically we find that when the training set is small, (21) is more suitable. Care should also be taken when generating the sigmoid training set $(f(x_i), y_i)$.

With probabilistic outputs $r_{ij} = P(i|i \ or \ j, x) = \frac{1}{1+\exp(Af_{ij}(x)+B)}$ of SVMs, some pairwise coupling algorithms can be employed. In [22], Wu et al. proposed an approach to combine the *pairwise probabilities*, which are the probabilistic outputs of all binary SVMs, into a common set of posterior probabilities P_i. They found the optimal P through solving the following optimization problem:

$$\min \quad \sum_{i=1}^{K} \sum_{j:j \neq i} (r_{ji}P_i - r_{ij}P_j)^2 \ , \tag{22}$$

$$\text{s.t.} \quad \sum_{i=1}^{K} P_i = 1 \ , \quad P_i \geq 0 \ \forall i \ .$$

Note that (22) can be reformulated as

$$\min \quad 2P^T Q P \quad \equiv \quad \min \frac{1}{2} P^T Q P \ , \tag{23}$$

where

$$Q_{ij} = \begin{cases} \sum_{s:s \neq i} r_{si}^2 & \text{if } i = j \ , \\ -r_{ji}r_{ij} & \text{if } i \neq j \ . \end{cases} \tag{24}$$

Then P can be obtained by solving the following linear system:

$$\begin{bmatrix} Q & \mathbf{e} \\ \mathbf{e}^T & 0 \end{bmatrix} \begin{bmatrix} P \\ b \end{bmatrix} = \begin{bmatrix} \mathbf{0} \\ 1 \end{bmatrix} \ . \tag{25}$$

5 Experimental Results

Experiments are performed on the ORL face database. The database contains 10 different images of 40 distinct subjects. For some of the subjects, the images were taken at different times, varying lighting slightly, facial expressions (open/closed eyes, smiling/non-smiling) and facial details (glasses/no-glasses). All the images were taken against a dark homogeneous background and the subjects are in

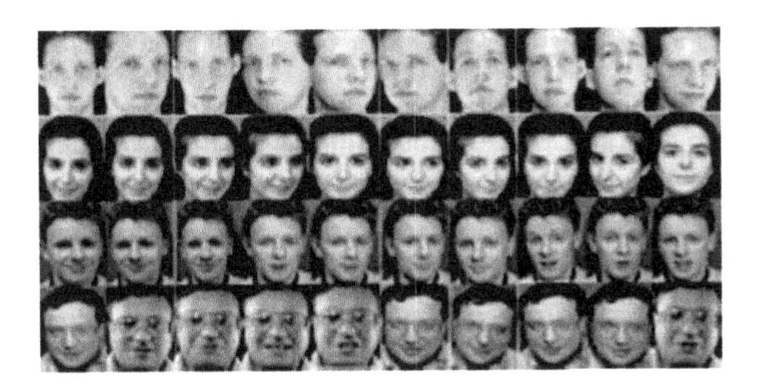

Fig. 2. Example faces of four subjects from ORL, each row corresponds to one subject

up-right, frontal position (with tolerance for some side movement). The original resolution of the images was 92×112, 8-bit grey levels. Some example faces are show in Figure 2.

In our experiments, the ORL database is randomly divided into two sets with equal size. Five images of each subject are used for SVMs training, the rest five are used for SVMs testing. Feature extraction is done with the Eigenface algorithm.

LIBSVM [13] is employed for SVMs training. Parameters selection is performed in the $\log_2$-space with ranges of $\log_2 C \in \{-5, -4, \ldots, 15\}$ and $\log_2 \sigma \in \{-15, -14, \ldots, 3\}$. For the five-fold CV algorithm, the search is done with a resolution of five for both parameters, totally five iterations are performed. For the gradient-descent based algorithm, the initialization is $\log_2 C = 0$ and $\log_2 \sigma = 0$. Note that, the fitting of sigmoid functions has no effect on the gradient-descent based algorithm. But it makes the CV algorithm more complex.

Experimental results are shown in Table 1, where *t-ratio* stands for the recognition accuracy achieved on the test set by a trained SVM. The *"best" t-ratio* is also obtained through a coarse-to-fine search process but with a finer resolution of ten for both parameters, wherein recognition accuracies on the test set are directly used to assess the quality of parameters. Totally ten iterations are performed, which result in a much more exhaustive search than that of the CV algorithm.

From Table 1, we can see that the hybrid algorithm can select SVMs models with the best recognition ratio, which is very close to the "best" t-ratio, in all cases. While the other two perform comparatively in most cases. The performance of the CV algorithm is not surprising, as it is known to be an accurate method for choosing parameters of any learning algorithm. When 20 eigenfaces are used for feature extraction, the gradient-descent based algorithm obtains the poorest SVMs models, which, from another point of view, verifies the importance of the initialization of the algorithm. As expected, the hybrid algorithm is more robust and can always obtain good SVMs models. It is interesting to note that the selected parameters by different algorithms may differ widely even when the resulted SVMs have the same recognition accuracy (e.g. when 40 features are used for recognition).

Table 1. SVMs model selection results and corresponding face recognition accuracy

#eigen faces	the CV algorithm			the gradient algorithm			the hybrid algorithm			"best"
	$\log_2 C$	$\log_2 \sigma$	t-ratio	$\log_2 C$	$\log_2 \sigma$	t-ratio	$\log_2 C$	$\log_2 \sigma$	t-ratio	t-ratio
20	3.27	-0.38	94.5%	5.34	-0.72	94%	0.63	-1.24	**95.5%**	*96.5%*
40	4.38	-0.47	**95%**	9.23	-1.47	**95%**	6.31	-0.53	**95%**	*95.5%*
60	3.45	-1.37	95%	3.46	-1.73	94.5%	2.94	-2.12	**96%**	*96%*
80	2.44	-0.87	94%	1.28	-0.03	**95.5%**	4.25	-2.74	**95.5%**	*96%*
100	3.75	-0.09	**95.5%**	0.76	-1.12	95%	3.72	0.92	**95.5%**	*95.5%*

It should be pointed out that the obtained SVMs performance is slightly inferior to those reported in many literatures [6, 8]. This may due to two issues: the difference in dividing the ORL database and the strategy employed to combine binary SVMs outputs. In [6], a similar result of 95.13% on the ORL database was reported, which is about 2-3% lower than those obtained with more sophisticated combining strategies, which are currently under our investigation.

6 Conclusion

In this paper, we propose a framework for investigating automatic face recognition by support vector machines. Three model selection algorithms are explored: the cross-validation algorithm, a gradient-descent based algorithm with the radius/margin bound and a hybrid algorithm. Their performance is evaluated in terms of the quality of chosen SVMs models. Preliminary experimental results on the ORL face database show that, with the proposed hybrid algorithm, nearly "best" recognition ratio can be achieved.

References

1. http://www.viisage.com/ww/en/pub/viisage___products/viisage___products___facetools.htm
2. `http://www.identix.com/products/pro_sdks_id.html`
3. Zhao, W., Chellappa, R., Phillips, P. J., Rosenfeld, A.: Face recognition: A Literature Survey. ACM Computing Surveys. **35** (2003) 399-458
4. Dai, G., Zhou, C.: Face Recognition Using Support Vector Machines with the Robust Feature. In: Proc. of IEEE workshop on Robot and Human Interactive Communication. (2003) 49-53
5. Déniz, O., Castrillón, M., Hernández, M.: Face Recognition Using Independent Component Analysis and Support Vector Machines. Pattern Recognition Letters. **24** (2003) 2153-2157
6. Li, Z., Tang, S.: Face Recognition Using Improved Pairwise Coupling Support Vector Machines. In: Proc. of Intl. Conf. on Neural Information Processing. **2** (2002) 876-880

7. Jonsson, K., Matas, J., Kittler, J., Li, Y.P.: Learning Support Vector Machines for Face Verification and Recognition. In: Proc. of IEEE Intl. Conf. on Automatic Face and Gesture Recognition. (2000) 208-213
8. Guo, G., Li S., Kapluk, C.: Face Recognition by Support Vector Machines. In: Proc. of IEEE Intl. Conf. on Automatic Face and Gesture Recognition. (2000) 196-201
9. Kressel, U.: Pairwise Classification and Support Vector Machines. In: Schölkopf, B., Burges, C., Smola, A. (eds.): Advances in Kernel Methods: Support Vector Learning. MIT Press (1999) 255-268
10. Hsu, C.-W, Lin, C.-J.: A Comparison of Methods for Multi-Class Support Vector Machines. IEEE Trans. on Neural Networks. **13** (2002) 415-425
11. Burges, C.J.: A Tutorial on Support Vector Machines for Pattern Recognition. Data Mining and Knowledge Discovery. **2** (1998) 121–267
12. Chung, K.-M., Kao, W.-C., Sun, T., Wang, L.-L., Lin, C.-J.: Radius Margin Bounds for Support Vector Machines with the RBF Kernel. Neural Computation. **11** (2003) 2643–2681
13. Chang, C.-C., Lin, C.-J.: LIBSVM: A Library for Support Vector Machines. (2002) Online at `http://www.csie.ntu.edu.tw/~cjlin/papers/libsvm.pdf`
14. Staelin, C.: Parameter Selection for Support Vector Machines. (2003) Online at `http://www.hpl.hp.com/techreports/2002/HPL-2002-354R1.pdf`
15. Li, H.-Q., Wang, S.-Y., Qi, F.-H: Minimal Enclosing Sphere Estimation and Its Application to SVMs Model Selection. In: Proc. of IEEE Intl. Sympo. on Neural Networks. (2004) 487-493
16. Chapelle, O., Vapnik, V., Bousquet, O., Mukherjee, S: Choosing Multiple Parameters for Support Vector Machines. Machine Learning. **46** (2002) 131-159
17. Keerthi, S.S.: Efficient Tuning of SVM Hyperparameters Using Radius/Margin Bound and Iterative Algorithms. IEEE Trans. on Neural Networks. **13** (2002) 1225-1229
18. Chung, K.-M., Kao, W.-C., Sun, T., Wang, L.-L., Lin, C.-J.: Radius Margin Bound for Support Vector Machines with the RBF Kernel. In: Wang, L., Rajapakse, J.C., Fukushima, K., Lee, S.Y., Yao, X. (eds.): Proc. of the 9th Intl. Conf. on Neural Information Processing. **2** (2002) 893-897
19. Vapnik, V.: Statistical Learning Theory. Wiley (1998)
20. Friedman, J.: Another approach to polychotomous classification. Technical report, Stanford University, (1996)
21. Platt, J.: Probabilistic Outputs for Support Vector Machines and Comparison to Regularized Likelihood Methods. In: Smola, A.J., Bartlett, P.L., Schölkopf, B., Schuurmans, D. (eds.): Advances in Large Margin Classifiers. MIT Press (2000) 61-74
22. Wu, T.-F., Lin, C.-J., Weng, R.C.: Probability Estimates for Multi-Class Classification by Pairwise Coupling. Journal of Machine Learning Research. **5** (2004) 975-1005
23. Goh, K., Chang, E., Cheng, T.: Support Vector Machine Pairwise Classifiers with Error Reduction for Image Classification. In: Proc. of ACM Intl. Conf. on Multimedia (MIR Workshop). (2001) 32-37

Practical Region-Based Matching for Stereo Vision

Brian McKinnon and Jacky Baltes

Department of Computer Science,
University of Manitoba,
Winnipeg, MB,
Canada R3T 2N2
jacky@cs.umanitoba.ca
http://avocet.cs.umanitoba.ca

Abstract. Using stereo vision in the field of mapping and localization is an intuitive idea, as demonstrated by the number of animals that have developed the ability. Though it seems logical to use vision, the problem is a very difficult one to solve. It requires the ability to identify objects in the field of view, and classify their relationship to the observer. A procedure for extracting and matching object data using a stereo vision system is introduced, and initial results are provided to demonstrate the potential of this system.

1 Introduction

This paper describes our research into stereo vision for simultaneous localization and mapping (SLAM). In this paper we present a novel algorithm for matching stereo images based on regions extracted from a stereo pair. Another notable feature of our approach is that it is implemented using commodity hardware.

The overall goal of the project is to develop urban search and rescue (USAR) robots that can generate maps of an unstructured environment given a sequence of stereo images and find victims within this environment. In our research, we focus on the use of vision as sole sensor. The use of vision as the sole sensor on a mobile robot is considered by some to be a radical idea. Today, most work in SLAM uses LADAR (laser scanners), which provide highly accurate point clouds in 3D space. The advantage of vision is that cheap commodity hardware can be used. Furthermore, developing methods for extracting and representing the necessary information (e.g., ground plane, walls, doors, other objects) is a wide open problem. The USAR domain adds to this complexity because the real-time constraints imposed on the mobile robot.

To test our approach, we take part in the NIST sponsored USAR competitions. The NIST domain (see Fig. 1) is a challenging domain for today's robots (especially ones using vision) since it includes uneven lighting, glass, mirrors, and debris.

In this paper, the first stage of the vision-based mapping and localization system is described. This involves image processing to extract and represent

R. Klette and J. Žunić (Eds.): IWCIA 2004, LNCS 3322, pp. 726–738, 2004.

Fig. 1. NIST Reference Test Arenas for Autonomous Mobile Robots at the AAAI 2000 in Austin, Texas, USA

objects in the captured image. Included is a description of the methods currently being used by other researchers. Additionally, the problem of localization will be discussed and a variety of systems will be examined, including vision-based and more traditional distance sensor-based solutions.

2 Related Work

Both region extraction and localization are active areas of research. There are a wide variety of solutions currently being investigated, many of these yielding promising results.

The two most important steps in region extraction are identification and representation of features in the image. Research is still active in this area, since current systems encounter environments that cause failure rates to become unmanageable. Examples of systems currently being studied include [6] [3] and [4].

2.1 Scale Invariant Feature Extraction

In [6] an object recognition system is introduced that has become known as Scale Invariant Feature Extraction (SIFT). It uses a feature representation that is invariant to scaling, translation, rotation, and partially invariant to changes in illumination. The output of this system is a set containing the orientation, position, relative location, and colour gradient of key features within an image. Scale invariance is achieved through the use of the Gaussian kernel as described in [5]. For rotational invariance and efficiency, key locations are selected at the maxima and minima from the difference of the Gaussian function applied in scale space. A threshold is applied to the gradient magnitude for robustness. This is useful since illumination changes may greatly affect the gradient magnitude, however it should have little impact on the direction. Once a set of keys are defined for a given object, live images are scanned and objects are selected using a best-bin-first search method. Bins containing at least three entries for an object are matched to known objects using a least square regression. Experimental results show that the system is effective at detecting known objects, even in the presence of occlusion, since only three keys are necessary for a match to occur. This can be seen in figure 2. Using this method, a localization system has been implemented in [8].

Fig. 2. Using SIFT the sample objects(above) are searched from in the image(left). The keys generated are used to match the samples to the image(right). Images are from [6]

2.2 Blobworld Representation

The Blobworld representation is introduced in [3] as a means of performing image retrieval. Pixels in an image are assigned to a vector containing their colour, texture, and position. Colours are smoothed to prevent incorrect segmentation due to textures, and are stored using the L*a*b* colour format. L*a*b* representation shown in figure 3 contains three colour channels, L for luminosity, a for the colour value between red and green, and b for the colour value between yellow and blue. Texture features that are used include contrast, anisotropy, and polarity. Contrast is the difference between light and dark area, anistropy indicates the direction of the texture, and polarity measures how uniformly the texture is oriented. Regions are grouped spatially if they belong to the same colour/texture cluster. A gradient is generated in the x and y direction, containing the histogram value of pixels in that region. For matching, the user starts by selecting blobs from the image that will be used for comparison against the database. Regions are matched to the database by the quadratic distance between their histograms' x and y values. The Euclidean distance for the contrast and anisotropy texture are also used in the comparison. The best image is selected based on the similarity values for the selected blobs as shown in figure 4. This method was used as a basis for an optimal region match in [2], however it is unclear how robustly the method handles translation of the blobs. This system is not directly usable for localization, since blobs must be selected manually.

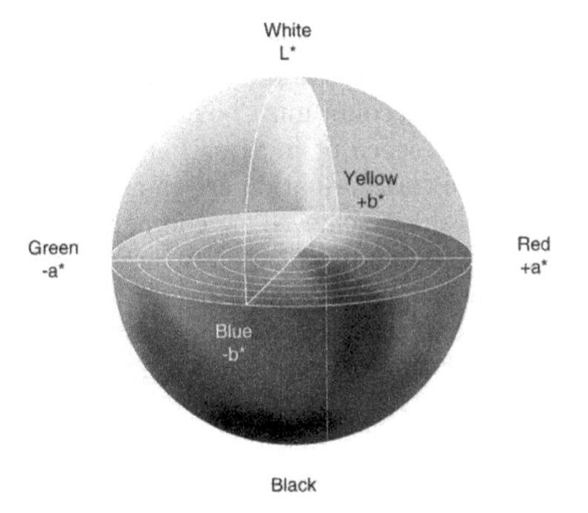

Fig. 3. The L*a*b* colour wheel. Image is from http://www.colorspan.com/support/tutorials/cmpl/lab.asp

2.3 Wavelet-Based Region Fragmentation

A recent approach to region extraction is known as Wavelet-based Indexing of Images using Region Fragmentation (WINDSURF) [1]. In this approach the wavelet transform is employed to extract colour and texture information from an image. The wavelet transform is similar in principle to a fast Fourier transform where data is represented using a wave form. A clustering algorithm is used on the output coefficients from the wavelet transform, producing regions that contain similar colour and texture waves. By using only the coefficients, regions are clustered without considering spatial information. This means that images

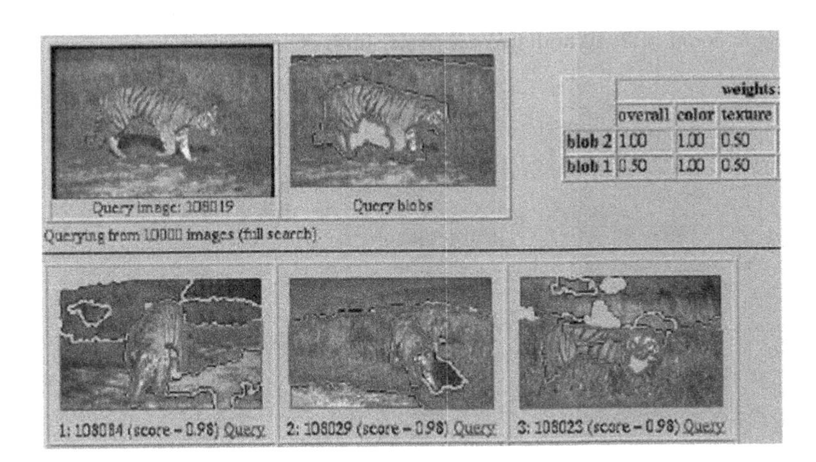

Fig. 4. This sample output from [3] shows the blobs generated from the sample, and images containing similar blobs that were retrieved

cannot be compared based on the location of the regions, however it allows matching that is invariant to position and orientation differences. One limitation in this system is that the region count must be defined, so clustering of dissimilar regions can occur in the presence of images that contain more features than expected. The results of the transformation are shown in figure 5 using different region counts. Image retrieval is achieved by comparing the size, centroid, and features of regions from the sample image to those in the database images. Experimental results show that the system is capable of retrieving images which are semantically correlated to the sample. In the case of figure 6 a match could consist of images containing similar sized and textured blue and gray regions.

3 Implementation

In order to perform stereo matches we must first process the raw captured images. The process implemented in this paper consists of 5 steps:

1. Colour Correction
2. Image Blur
3. Edge Detection

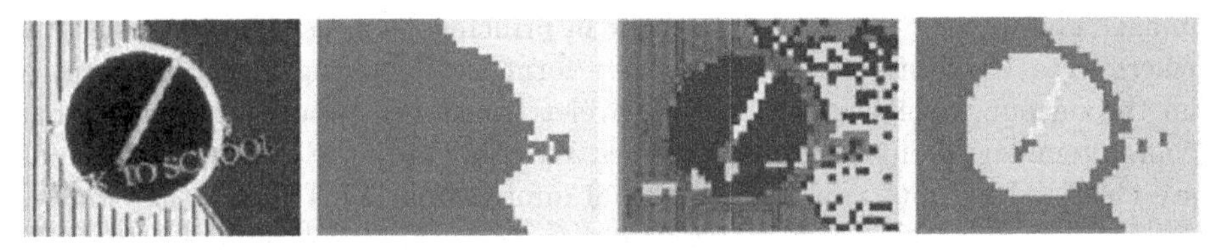

Fig. 5. This image shows the results of applying the wavelet transform to sample image, using region count of 2, 10 and 4 respectively. Images are from [1]

Fig. 6. This shows the results of a sample wavelet query using the dome of St. Peter's in Rome. Images are from [1]

4. Region Extraction
5. Stereo Matching

Each processing stage has an important role in transforming the image. The result must be an accurate representation of the objects contained in the image.

3.1 Colour Correction

Normalization of the colour channels is necessary to correct imbalances in the colour response of the two cameras. This is useful in situations where images are too bright or dark, or if a colour channel is over saturated. This allows more accurate matches between two separate images. The method used in this system relies on the mean and standard deviation of the colour channels.

Calculating the mean and standard deviation of an image usually requires at least two passes through the image. In the first pass, the mean is calculated and in the second the differences to the mean are summed up. Applying the colour correction would result in a third pass. It is impossible to perform this much pre-processing for each image and maintain a reasonable frame rate.

To speed up the computation of colour normalization, we use a pipelined approach. The key idea here is that mean and standard deviation of the image sequence are relatively constant over time. Therefore, we use the mean and standard deviation of previous images. At time t, the mean of the image at time $t-2$ and the standard deviation at time $t-1$ is being used. This means that colour correction has a negligible impact on the runtime.

The colour of each channel is bounded by using the mean as the centre value and setting the range to a distance of two standard deviations plus and minus from the mean. This range is then stretched to cover the entire possible range of the colour channel. Using a distance of two standard deviations allows outliers to be discarded, which gives a more accurate representation of the true colour range of the image.

3.2 Image Blur

The blur that is applied to the image is important since raw images are prone to noise. This noise can be generated by lighting problems, textured surfaces, or a low quality capture device. The goal is to smooth the image so small inconsistencies can be ignored by the edge detection. There are many methods of applying a blur to an image. The simplest method, which was implemented in the original version of the application, transformed the centre pixel to contain the average colour of the surrounding neighbourhood. The primary advantage of this method was simplicity, but it also operated very quickly on an image.

The Gaussian blur is an improved method that provides smoothness and circular symmetry [9]. One feature that a Gaussian blur provides over simple blurring is the ability to repeat small areas of blur to generate large areas. The values used in a Gaussian blur are generally based on the binomial coefficients, or Pascal's triangle. To apply a blur of $\mathrm{N} = \mathrm{k}$, coefficients are selected such that $\mathrm{i} + \mathrm{j} = \mathrm{k}$. For example, to apply a 3x4 filter, $\mathrm{k} = 5$ is selected, and then the

Fig. 7. The normalization process helps balance the colours in the image by stretching the colour range of pixels

coefficients at a depth of 2 and 3 are used. Depth 2 corresponds to the set {1 2 1}, and 3 corresponds to {1 3 3 1}. One set is applied to the image horizontally, and the second is applied vertically to the result of the first. The result stored in each pixel is normalized, with a division by the sum of the two coefficients.

3.3 Edge Detection

Edge detection is an essential component in the object extraction process. It allows boundaries to form, which provides assistance in the region growing process. For this, a Sobel edge detection is implemented, since it is simple, robust, and versatile. It involves applying convolution masks across the entire image. The Sobel edge masks used in the horizontal and vertical direction respectively are:

$$\begin{pmatrix} -1 & 0 & 1 \\ -2 & 0 & 2 \\ -1 & 0 & 1 \end{pmatrix} \; and \; \begin{pmatrix} -1 & -2 & -1 \\ 0 & 0 & 0 \\ 1 & 2 & 1 \end{pmatrix}$$

The results of each mask are normalized with a division by four. The value from these transformations are tested against a threshold, that once exceeded indicates the presence of an edge. The edge pixels are stored within an edge map used in the region extraction.

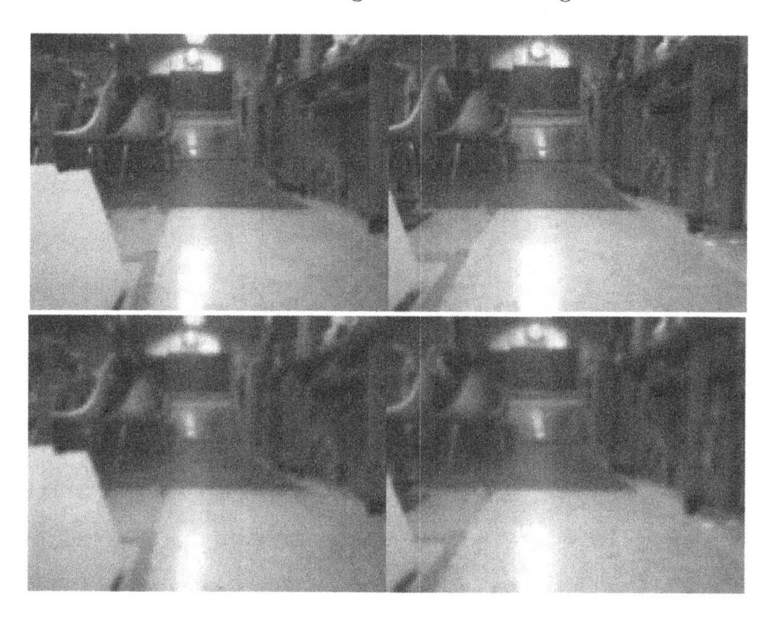

Fig. 8. The product of applying a Gaussian blur on the raw image

3.4 Region Extraction

The region growing method that has been explored uses a stack-based growth that identifies matching pixels based on the colour channels match with the previously accepted pixels. The image is scanned linearly for pixels that have not been detected as edges and have not been previously examined. When a pixel is found, it is set as the start point for the region growing, and is pushed on to the stack to become the examined pixel. The neighbouring pixels to the top, bottom, left, and right are considered for addition to the region provided they have not been identified as edge pixels. Colour match is calculated by using the sum of squared error across all the colour channels (RGB). If the resulting value falls below a defined threshold then the match is considered to be acceptable. Initially, the neighbouring pixels are added to the region based on an acceptable colour match with only the examined pixel. Once a threshold size has been reached, neighbours are added based on the colour match with the mean colour value of all pixels in that region. Each time a pixel is accepted, it is pushed onto the stack and the process is repeated. Termination occurs once no pixels are acceptable matches with the current region. If the region has reached a threshold size then it is added to the list of identified regions. Regions below the threshold size are rejected, and each pixel is marked as previously examined to prevent additional small regions from being generated at a starting pointing in this area.

The threshold value for the colour match is very small, and as a result a single object in the image could be separated into several small regions. There is no restriction on a new regions ability to expand into another region, and this

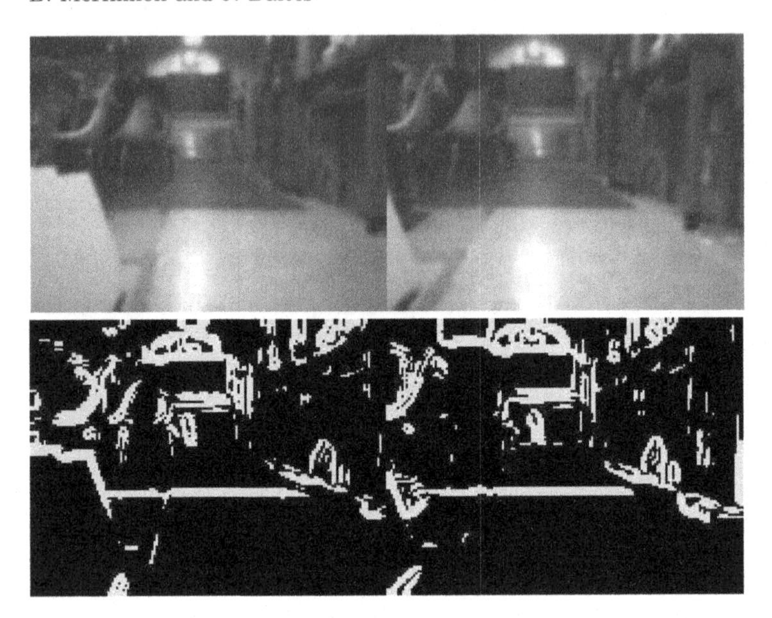

Fig. 9. The result of applying a Sobel edge detection to the blurred image

overlap is used as the basis for merging regions into objects. Neighbouring regions are combined if the overlap exceeds a threshold of either a percentage of the pixels in the region (generally used to merge smaller regions), or a threshold number of pixels (used for matching larger regions). This method of overlap merging the objects, allows either shadow or glare to be joined properly without setting the colour match threshold to an excessively high value.

Once completed, the objects are defined by:

1. Size (in pixels),
2. Mean colour value,
3. Centroid x and y,
4. Bounding box,and
5. Region pixel map (i.e., mapping from pixels to associated regions)

These objects are then used as the feature points in the stereo matching process.

3.5 Stereo Matching

Once an image has been segmented, the shape of the object is used to identify matches between the stereo images. Objects from the two images are superimposed onto each other at the centroid, and the size of the union between the two is calculated. The union size is calculated by identifying pixels that are contained in both images. If a minimum percentage of pixels overlap, then the two objects are identified as a stereo pair. Though it is possible for an object to produce

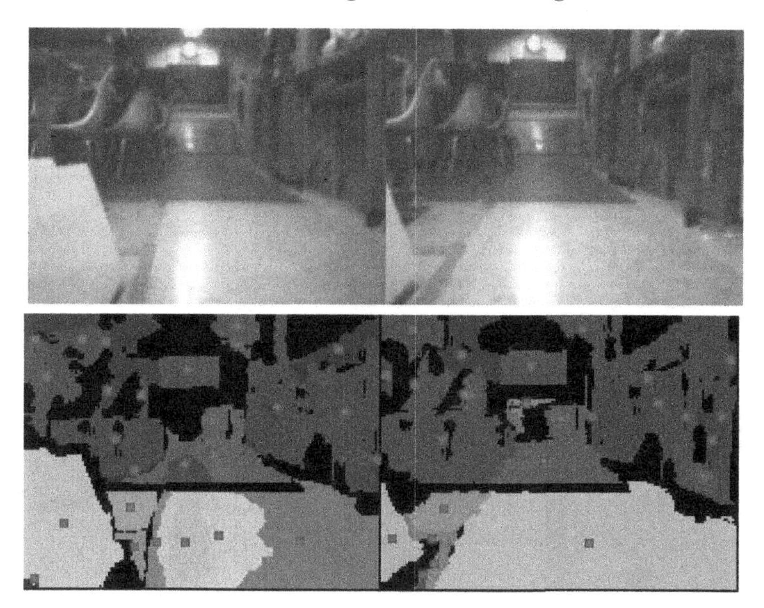

Fig. 10. Shown here is the resulting set of regions. Red dots indicate the centroid, and the colour represents the average colour of the region

more than a single stereo pair, only the strongest match is considered. The displacement between the objects is calculated, and will later be used to determine the distance to the object. Before a match of stereo regions is considered, it must appear over a series of images. This is useful when attempting to ignore noisy and incorrect matches.

Currently, the matching is done uncalibrated, so no consideration is given to the intrinsic or extrinsic properties of the camera. Therefore, matches can occur even in the presence of unreasonable displacement in the vertical direction. This does increase the number of incorrect matches, however, a substantial number of the matches are still correct. Since there are still many correct matches, it may be possible to calibrate stereo cameras with no human interaction even in the presence of a significant alignment difference between the two cameras. Once calibrated the accuracy of the matching should increase dramatically.

4 Experiments

This system has been implemented and tested using an unmodified area in the Computer Science department at the University of Manitoba. The robot used in this project, named Spike, is a one-sixth scale model of a Pt Cruiser with rear wheel drive. The radio controller has been modified to allow the vehicle to be controlled through the parallel port on any standard PC. The micro-controller is a C3 Eden VIA Mini-ITX running at 533MHz, with a 256Mb flash card. We developed our own mini version of the Debian Linux distribution refined for use

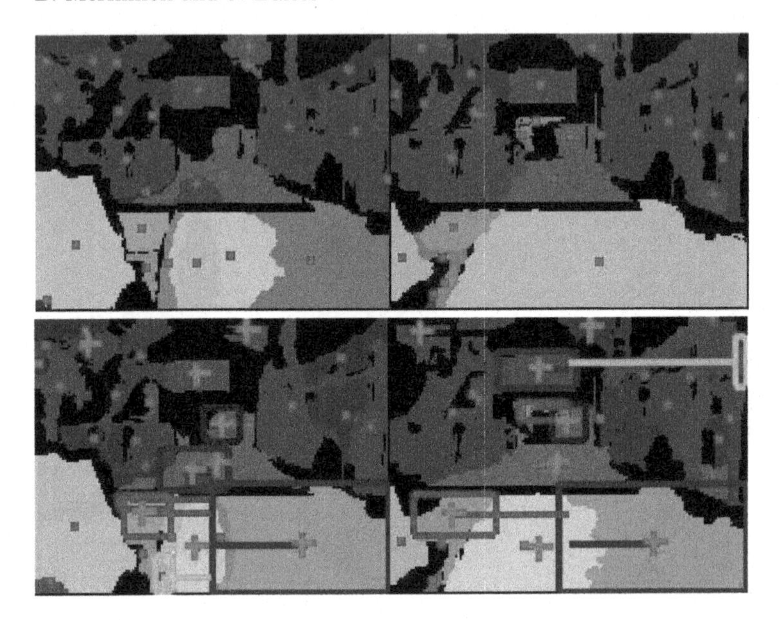

Fig. 11. Matching regions are bounded by a coloured box, with a line going from the centroid to the centre of the image. Pink crosses indicate the presence of a strong match

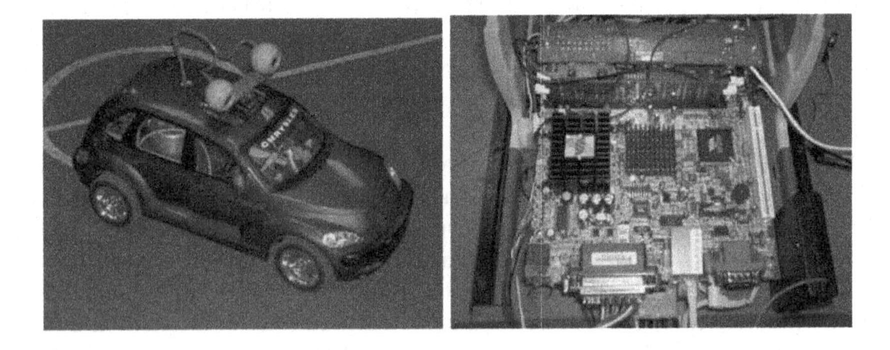

Fig. 12. Spike, the mobile robot used for this experiment

on systems with reduced hard drive space. The vision hardware consists of two USB cameras capable of capturing 320 by 240 pixel images. The cameras are mounted on a servo that allows the stereo rig to pan in a range of plus or minus forty-five degrees.

In figure 13 we demonstrate the matching ability of this system. The raw image has been segmented into 21 objects for the right image, and 25 objects for the left image. In this sample, seven stereo pairs have been generated, as indicated by the coloured boxs surrounding the matched pairs. Over the compete runtime in this sample, nine pairs were selected as strong stereo pairs, as indicated by

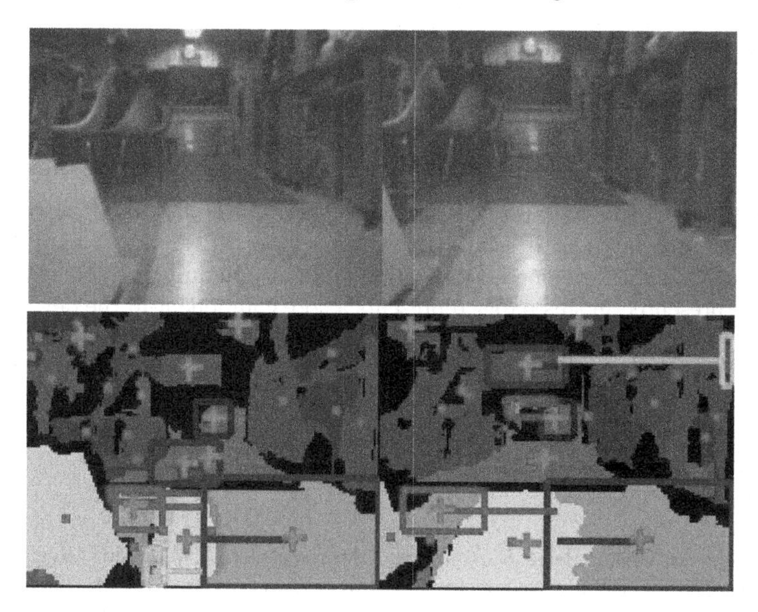

Fig. 13. Demonstration of matching in an unknown environment

the pink crosses on the resulting image. Five of these are both correct, and the regions appropriately represent the object in the image. Two of the matches are spatially correct, however these two should be represented as one. The system still has some difficulty when bright glares are reflected from smooth surfaces. The remaining two objects have been matched incorrectly.

Currently, the complete system runs at 0.5 frames per second. We are currently investigating methods for speeding up this process. In particular, the matching of regions is computationally expensive because the stereo system is uncalibrated at the moment. Therefore, regions must be compared against all other regions. By adding at least a rough calibration, most of these comparisons can be avoided, since for example, matching regions must have similar Y coordinates.

5 Conclusion

This paper presented the initial steps taken in the development of a vision-based autonomous robot. With the region-based object extraction and stereo matching implemented the ground work is laid for the development of the remaining components. These include 3D scene interpretation, mapping, localization, and autonomous control.

With the ability to extract objects from an image, and generate stereo pairs from a set of images, the next step will involve the development of a camera calibration system. The goal is to design a self-calibrating system that can produce the Fundamental Matrix without human interaction [7]. The Fundamental Matrix allows the object matching search to be constrained to a single line, rather than the entire image. This will improve the run-time and accuracy of the stereo pair matching process.

Once a set of stereo pairs can be generated, the next step is to calculate the distance to the objects. This is done by measuring the disparity, or horizontal offset, of the object as observed in each image. Once the set of objects have a distance associated with them, they can be used to generate a map. Once a mapping system is developed, localization and path planning can be added. The research presented in this paper represents a core component in the development of a fully autonomous robot, that is able to view its environment, interpret the images into a 3D model, and given this information is able to create a map of its surroundings and localize itself within this environment.

References

1. Stefania Ardizzoni, Ilaria Bartolini, and Marco Patella. Windsurf: Region-based image retrieval using wavelets. In *DEXA Workshop*, pages 167–173, 1999.
2. Ilaria Bartolini, Paolo Ciaccia, and Marco Patella. A sound algorithm for region-based image retrieval using an index. In *DEXA Workshop*, pages 930–934, 2000.
3. Chad Carson, Megan Thomas, Serge Belongie, Joseph M. Hellerstein, and Jitendra Malik. Blobworld: A system for region-based image indexing and retrieval. In *Third International Conference on Visual Information Systems*. Springer, 1999.
4. Hiroshi Ishikawa and Ian H. Jermyn. Region extraction from multiple images. In *Eigth IEEE International Conference on Computer Vision*, July 2001.
5. Tony Lindeberg. Scale-space: A framework for handling image structures at multiple scales. In Egmond aan Zee, editor, *Proc. CERN School of Computing,*, September 1996.
6. David G. Lowe. Object recognition from local scale-invariant features. In *Proc. of the International Conference on Computer Vision ICCV, Corfu*, pages 1150–1157, 1999.
7. Quang-Tuan Luong and Olivier Faugeras. The fundamental matrix: theory, algorithms, and stability analysis. *The International Journal of Computer Vision*, 17(1):43–76, 1996.
8. Stephen Se, David Lowe, and Jim Little. Mobile robot localization and mapping with uncertainty using scale-invariant visual landmarks. *I. J. Robotic Res*, 21:735–760, 2002.
9. F. Waltz and J. Miller. An effecient algorithm for gaussian blur using finite-state machines. In *SPIE Conf. on Machine Vision Systems for Inspection and Metrology VII*, 1998.

Video Program Clustering Indexing Based on Face Recognition Hybrid Model of Hidden Markov Model and Support Vector Machine*

Yuehua Wan[1], Shiming Ji[1], Yi Xie[2], Xian Zhang[1], and Peijun Xie[1]

[1] Zhejiang University of Technology, 310032, Hangzhou, P.R.China
(Wanyuehua, jishiming)@zjut.edu.cn
[2] Hangzhou University of Commerce, 310035, Hangzhou, P.R.China
xieyi@mail.hzic.edu.cn

Abstract. Human face is a very important semantic cue in video program. Therefore, this paper presents to implement video program content indexing based on Gaussian clustering after face recognition through Support Vector Machine (SVM) and Hidden Markov Model (HMM) hybrid model. The task consists of following steps: first, SVM and HMM hybrid model is used to recognize human face by Independent basis feature of face apparatus; then, the recognized faces are clustered for video content indexing by Mixture Gaussian. From the experiments, the precision of the mixed model for face recognition is 97.8 percent, and the recall is 95.2, which is higher than the complexion model. And the precision of the face clustering indexing is 94.6 percent of the mixed model for compere new program. The indexing result of clustering is famous.

1 Introduction

The information is expressed in the style of multimedia in digital libraries. And the readers acquire knowledge from the text, figure, image, video and audio. How to find the needed information from the great deal data of multimedia has become a hot research topic in the field of multimedia indexing based on content [1-4]. In the multimedia indexing technology based on content of digital library, the video flow information must be accurate sorted and indexed firstly, i.e. structured [5]. The main idea is to achieve the aim of sorting, identifying and indexing the multimedia information by using vision characteristics or hearing characteristics or fusing those two characteristics [6].

To sorting and indexing the video flow may based on different semantics levels. One is the advanced semantics level, which is the high abstract result of some video and audio multimedia event occurring in different time and palace. Two is intermediate semantics level, which is the description of single event, not deal with

* This work is supported by Zhejiang Provincial Natural Science Foundation of China; Grant Number: M503099.

R. Klette and J. Žunić (Eds.): IWCIA 2004, LNCS 3322, pp. 739–749, 2004.

some events' intersection. The last is elementary semantics level, which is to sort multimedia data basically using seeing or hearing information. To indexing multimedia data using semantic realizes the courses of multimedia data from no structure to be structured, and organizes data flow efficiently. It makes indexing facility.

The traditional method of sorting video only uses the basic characteristic of seeing and hearing, not considers the vision object composed of those characteristics, i.e. face. In fact, face is the familiar object of vision program, and it contains a lot of semantics information. The indexing aim of video content is achieved by picking-up the important characters of face first, then identifying the characters, processing the clustering indexing of the identify result lastly.

Because Support Vector Machine (SVM) has the better sorting ability and the divert probability of Hidden Markov Model (HMM)'s every state can restrict the topology of face's every apparatus, this paper brings forward a new method that SVM and HMM may be used to identify the face's apparatus, and cluster the recognized faces for video content indexing.

2 The Video Face Identifying Based on SVM and HMM

The essence of face identifying is the matching problem of conversion 3D mold object to 2D projection. Its difficult consists of the face pattern diversity and ambiguity in the process of acquiring image. This requests the right rate high in the process of identifying, though the face sample is a few.

This paper divides a face into five parts: forehand, eyes, nose, mouth and chin, then seeks the independent basis feature of those apparatus and constructs the SVM to identify those apparatus. In order to use the topology restriction, this paper uses HMM containing five states to restrict the identify result, and forms the mixed model of video face identifying based on SVM and HMM (Fig.1).

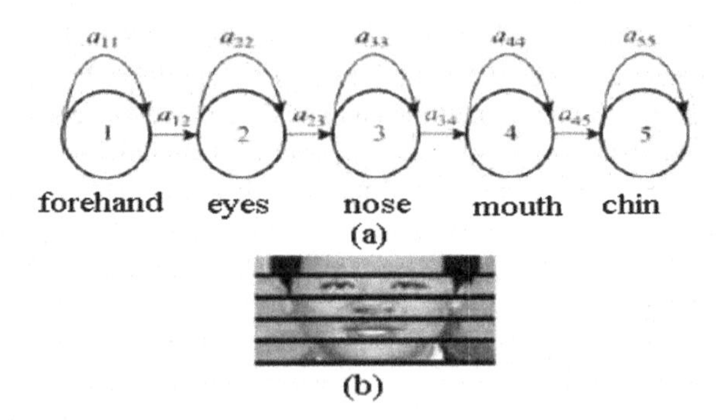

Fig. 1. The mixed model of video face identifying based on SVM and HMM

In Fig.1, the located square area which a face may be in is transformed into a 50*70 gray image (shown in the top right of Fig.1b), then the gray image is scanned from below to top using 10*70 image and every apparatus of face can be found (eyes area found by scan is shown in the middle of Fig.1b). Moreover, every apparatus is transformed into a 25*30 gray image (the transformed image of eyes area is shown in the below of Fig.1b). The training face samples may be transacted also. This gains the transformed images of every apparatus. Then the independent basis feature can be sought. Five SVM are created through training those independent basis features. Gaussian model used in HMM is replaced by the created SVM. And HMM is reconstructed which transfers from left to right and has five states. Those five states in HMM are corresponding to five face apparatus. The five states and transfer probability between those states forms the develop restriction to face every apparatus. The issues of the mixed model construction, face identifying based on the mixed model and face-clustering indexing are discussed below.

2.1 The Mixed Identifying Model of Face Based on SVM and HMM

SVM brought forward by Vapink is a statistic learning theory based on Structural Risk Minimization (SRM) [7-8]. It can be used to analyze the sort and regression problems. The decision-making hyperplane is formed based on SVM theory, which makes the most sort margin between every data. For linearity impartibility data, SVM mapped the data of lower dimension space to multi - dimension space according the Cover theory [9]. The nucleus function of SVM has three kinds mostly: the polynomial function, the radial basis function and multilayer sigmoidal NN function. In this paper, the sigmoidal NN function is used. In the case of lower dimension impartibility data, for each unknown demanded distinguish sample u, the best hyperplane sort function f is calculated:

$$\mathrm{f(u)} = \sum_{\mathrm{i=1}}^{\mathrm{S}} \mathrm{a_i} y_i k(u, x_i) + b \tag{1}$$

Where, $\mathrm{a_i}$ is the Lagrange operator corresponding to support vector $\mathrm{x_i}$, $\mathrm{y_i}$ is the sort label of support vector $\mathrm{x_i}$ (the value is 1or -1). S is sum of support vector; K is the nucleus function, and the sign of $\mathrm{f(x)}$ shows the different sorts.

Because HMM has stability math foundation and can simulate dynamic variety of data better, it is used in the field of face identifying proverbially [10-11]. In order to combine SVM and HMM, the method is to replace the Gaussian of HMM by SVM. So the output value of SVM must be a probability. But the standard SVM doesn't create probability (the output data of SVM is a distance, not probability). The problem was solved by Platt [12]· He trained SVM $\mathrm{c_i}$ first, then using the sigmoid nucleus function, he calculated the probability of the unknown data u (the unknown data is independent basis feature of apparatus) relative to $\mathrm{c_i}$:

$$\mathrm{p(c_i \mid u)} = \frac{1}{1 + \exp(\mathrm{Ax} + \mathrm{B})} \tag{2}$$

Where, A and B is the parameter of sigmoid function.

So, the mixed model λ of SVM and HMM may be expressed by the five elements array $\lambda = (Q, M, a, b, \pi)$ [13]. Where, Q equal 5, expressing the state total number of λ (denoting every apparatus of the face); M equal 1, which means for each state there is one SVR corresponding to the face apparatus basis feature which had been adopted; a is a N*N matrix, expressing the state diverting probability in λ(the probability from one face apparatus shift to another). There is only the diverting probability which state shifts from the left to right, other one is zero. The effect of the diverting probability is to restrict the develop distributing of face apparatus; b is a Q*M matrix, which means the probability of the observed event turns to some state. $b = \{b_j(k)\}$, where $b_j(k) = P[o_t = v_k \mid q_t = j]$ ($1 \le k \le M$), expresses the probability that the observed event o_t equal v_k, when the current state is j. The value is calculated by formular2; $\pi = \{\pi_i\}$, where $\pi_i = P[q_1 = i]$ ($1 \le i \le Q$), expresses in one random process serial, the probability that the first state is state (the probability is promised equal in the experiment).

2.2 The Face Identifying Based on Mixed Model

Sid face training samples that needed to be identified are collected. To each sample, the basis features are picked-up and trained. Then the SVR, which is used to identify those apparatus, may be acquired (when training, the face which doesn't belong to some body is a counter example). Through Baum-Welch expect arithmetic, the corresponding diverting probability parameters are calculated in HMM, and the mixed model Sid_i ($1 \le i \le Sid$), which is token some face is created, Sid is the faces' sort. For face block, which is located through random complexion model, the forward operator is used to calculate the appearance probability $P(Sid_i \mid block)$ that the face block relative to the mixed model Sid_i.

From Bayesian, $P(Sid_i \mid block) = \dfrac{P(block \mid Sid_i)P(Sid_i)}{\sum_{1 \le i \le Sid} P(block \mid Sid_i)P(Sid_i)}$, where $P(block \mid Sid_i)$ is the probability of block appearing for each Sid_i; $P(Sid_i)$ is the prior probability of each model appearing. Over here, the probability of each model appearing is thought to be equal; also because to each $P(block \mid Sid_i)$, $\sum_i P(block \mid Sid_i)P(Sid_i)$ is equal. Therefore

$P(Sid_i | block) \sim P(block | Sid_i)$, and then $P(block | Sid_i)$ is calculated through forward operator. Seeing about the size relationship of $P(Sid_i | block)$, through select the Sid_i making the $P(block | Sid_i)$ most, so the $P(Sid_i | block)$ is the most also, then the block is identified to be face Sid_i.

2.3 The Clustering Indexing of Face Characteristic

For the face video frame identified by the mixed model, the video contents they contain must be indexed. The identified similarity faces are mustered together by clustering operator. So the significative video scene is formed.

When face clustering, we use all the identified face apparatus to buildup an integer eigenvector, then put up the Gaussian clustering by three Gaussian model. The parameter of Gaussian mixed model is acquires by EM operator [14]. Supposed the identified face total number is $nFace$, and the parameter in the mixed model $\theta = \{\theta_i\} (1 \leq i \leq 3)$. Eigenvector $x = nFace * d$ consists of all the basis frame of the identified face's forehand, eyes, nose, mouth and chin. Three row vectors selected random from x is the clustering center. The new parameter θ_i of each clustering subset is calculated by k average operator, such as the clustering center, prior probability, mean, covariance and posterior probability.

Actually the EM operator consists of parameter estimation and maximum likelihood operator. When the step $k+1$ iterativeness is calculated, the estimation of mixed model parameter θ_i is counted by the fellow formula:

$$P(x_j | \theta_i)^{[k+1]} = \frac{\left|\sum_i^k\right|^{-1/2} \exp\left\{-\frac{(x_j - u_i^{[k]})^T \sum_i^{[k]-1} (x_j - u_i^{[k]})}{2}\right\}}{\sum_{l=1}^3 \left|\sum_l^{[k]}\right|^{-1/2} \exp - \frac{(x_j - u_l^{[k]})^T \sum_l^{[k]-1} (x_j - u_l^{[k]})}{2}} \quad (3)$$

$$u_i^{k+1} = \frac{\sum_{j=1}^{nFace} P(x_j | \theta_i)^{[k]} x_j}{\sum_{j=1}^{nFace} P(x_j | \theta_i)^{[k]}} \quad (4)$$

$$\sum_i^{[k+1]} = \frac{\sum_{j=1}^{nFace} P(x_j | \theta_i)^{[k]} (x_j - u_i^{[k]})(x_j - u_i^{[k]})^T}{\sum_{j=1}^{nFace} P(x_j | \theta_i)^{[k]}} \quad (5)$$

$$P(\theta_i)^{[k+1]} = \frac{\sum_{j=1}^{nFace} P(x_j | \theta_i)^{[k]}}{nFace} \quad (6)$$

$$Center_i^{[k+1]} = \sum_{j=1}^{nFace} P(x_j | \theta_i)^{[k]} x_j \quad (7)$$

The hereinbefore formula express the posterior probability. $P(x_j | \theta_i)^{[k+1]}$, mean $u_i^{[k+1]}$, covariance $\sum_i^{[k+1]}$, prior probability $P(\theta_i)^{[k+1]}$ and clustering center $Center_i^{[k+1]}$ which is corresponding to each Gaussian model, when the estimation is in step $k+1$ on condition that the step k parameters are acquired.

In step $k+1$, the parameter of maximum likelihood method may be predigested as follow inequality:

$$\left| P(x | \theta^{[k]}) - P(x | \theta^{[k+1]}) \right| < \varepsilon \quad (8)$$

If formula (8) is right, the iterativeness will stop. Then for each row of x (they are corresponding to each identified face), $P(x_j | \theta_i)$ is calculated. Finally each x_j is classified to Gaussian distributing θ_i, which is corresponding to the most $P(x_j | \theta_i)$. In this way, all the identified faces are clustering to three types, and the clustering result of face is used to indexing video content.

3 The Experiments and Data Analyses

3.1 The Face Basis Characters' Picked-Up

In the experiments, VC and Matlab are used to process simulating. The face database which is used in the face identification is the Olivetti Research Laboratory database of

AT&T laboratory in Cambridge College. The face database has 400 gray face images came from 40 men in differ complexion, race and sex (each men has 10 photographs), and the image is photographed in different time and different illumination. The pose and expression aren't ilk, and some face is little deflexion.

While identifying the face, we find that it is difficult to distinguish the chin and mouth. So to the 50*70 face block, the block of chin (mouth), nose, eyes and forehand is picked-up separately from underside to top according to width of 14 pixels, 8 pixels, 10 pixels and 18 pixels. Upon that the corresponding mixed model has four states only. Then those blocks are transformed to 25*30 images and the characteristic ICA is picked-up. In the experiment, to each body six faces are selected to form the training database, the other is testing face. Eight face bases ICA and each apparatus basis ICA are shown in Fig.2 that is acquired from 60 face-training samples.

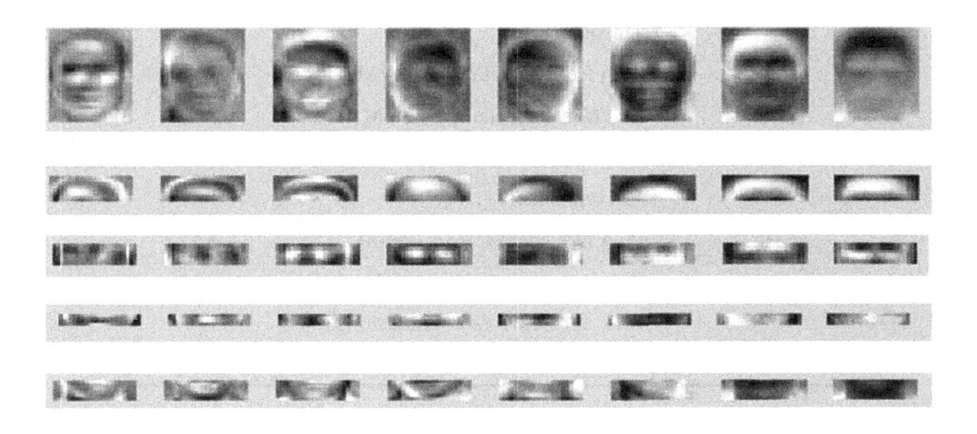

Fig. 2. The basis ICA space (from top the downside it is face, forehand, eyes, nose and mouth (chin))

If all the training and testing samples are looked upon the face space, the eight face basis in the first line of Fig.2 are used to express the every face in training and testing samples. Also the eight forehand bases in the second line of Fig.2 are used to express the every forehand of the forehand space that is formed by all the samples. The other face apparatus is expressed by the same method. The denote method is coefficient ICA which is formed by those bases.

3.2 The Face Identifying of Mixed Model

In order to carry through the contrast experiment, ICA frames are picked-up from each body's six training samples in training database and the mixed model is trained. Furthermore, PCA and ICA frame are picked-up also to train SVM sorter. The identifying efficiency result of those two models is shown in table 1. Table 1is the average right ratio of the 40 identified faces (for each body the testing faces number is four). When training, the opposite samples of each identified template aren't belong to this body's sample.

From table 1, we can see the identified result of mixed model is famous in condition that the ICA dimension is proper. The reason for identifying error is that eight percent face in the face database isn't the full frontal face.

Table 1. The different basis frame and identifying ratio of sorting model

The dimension of frame	The face characteristic + identifying method	The identifying ratio (%)
8	PCA face characteristic +SVM	23.3
	ICA face characteristic +SVM	26.8
	ICA apparetus face characteristic +mixed model	25
20	PCA face characteristic +SVM	75.6
	ICA face characteristic +SVM	83.7
	ICA apparetus face characteristic +mixed model	85.4
36	PCA face characteristic +SVM	82.7
	ICA face characteristic +SVM	91.1
	ICA apparetus face characteristic +mixed model	93

Nefian[15] presented the inline HMM to identify face. Each face apparatus is identified by the inline HMM in which every state is corresponding to Gaussian distributing. And the restriction is realized by a super inline HMM. But the training of inline HMM will take long time, and the repeat training will lead of model complex. The identifying ratio will fall also. Its identifying ratio is 86 percent only, under this paper's ratio. In the textual mixed model, the least frame risk characteristic of SVR and the restriction of inline HMM to face apparatus make the identifying ratio high in small training samples.

3.3 The Face Clustering Semantic Indexing

From ATT face database 2 samples are selected of each body, 20 faces is acquired. The mixed model for face identifying is created by picking-up and training the ICA frame of face.

We first choose the compere new program, conversation program and other video program. Each program lasts 2minutes and the sampling frequency is 5 frames per second. Then we apply the trained mixed model to identify the possible face block which is located by complexion model. 952 faces are gained. Finally to the ICA frame of those apparatus, the Gaussian clustering based on three mixed model is progressed.

According to the clustering result, the indexing is:(1) If the face in some video flow belong to one great kind mostly, the segment is to be indexed as "the compere new program"; (2) If the mainly face belong to other side great kind mostly and subsection belong to one small kind, the segment is to be indexed as "the conversation program program"; (3) The rest is indexed as the other.

(a) the compere program

(b) the conversation program

(c) the other program

Fig. 3. The semantic indexing based on Gaussian clustering of mixed face model

Table 2. The face block locating, face identifying and clustering result

(a) The locating result of face block and face identifying result

	Precision	Recall
The identified result of complexion model	92.3%	88.7%
The identified result of mixed model	97.8%	95.2%

(b) The clustering indexing result of face

	Precision	Recall
the compere new program	94.6%	89%
conversation program	91%	83%
Other program	83%	70%

The indexing result is shown in Fig. 3. The compere faces are clustered to one great sort and indexed as the compere new program. To the conversation program, because the cover face and other face belong to different sort, the program is indexed as conversation program. The rest program is indexed as other program.

The result of complexion locating, identifying based on mixed model and clustering indexing is shown in table 2(a). From table 2, we can see, though the precision of impossible face block locating isn't high, the veracity of identifying is famous. Besides, even some faces block isn't be located, the clustering result isn't too affected considering the space-time relativity of video flow. From the clustering indexing of 952 faces, we can see the precision is high.

4 Conclusions

Because the face delegates some semantic, the indexing of video content is realized by clustering different face. The precision of using SVR and HMM to index face is high. Then we realized the video content indexing by clustering the identified data.

The content for face recognition is extensive. At present, the method usually aims for some types of problems, and the theory and relevance technology aren't maturity. There are lots of problems to be researched.

1. The face's direction affects the identifying precision mostly. How to acquire the higher precision to the face that is decline or covered half is always challenge, almost in video programs the gesture and direction is variability.
2. How to acquire much more content indexing according the identifying result is another problem to be resolved.
3. The face is only a medium containing semantics in multimedia data flow. How to combinative the face and other medium containing semantics, especially the audio information turning up with the face together, is the ultimate aim.
4. At present, the method for face recognition in complex background aims for obverse and modesty face's recognition mostly. There are lots of difficult to identify the flank face. Researches in this field are emphasis in the future.
5. Due to the complication of face recognition, that there is an available method isn't realistic. So resolving the issue for face recognition in the condition of particular restriction or some practical background is the main topic in face recognition field in future.

References

1. He Limin, Wan Yuehua. Key techniques of content-based image retrieval in digital library. The Journal of the Library Science in China,2002,28(6),26-36
2. He Limin, Wan Yuehua. Key techniques of content-based video retrieval in digital library. The Journal of the Library Science in China,2003,29(2):52-56
3. Petkovic, M. et al. Content-based video indexing for the support of digital library search. In: Proceedings 18th International Conference on Data Engineering,2002,494-495
4. Lin, Wei-Hao. Hauptmann, Alexander G. A wearable digital library of personal conversations. In:Proceedings of the ACM International Conference on Digital Libraries, 2002, 277-278
5. Lyu, Michael R; Yau, Edward; Sze, Sam. A multilingual, multimodal digital video library system. In: Proceedings of the ACM International Conference on Digital Libraries, 2002, 145-153
6. Y. Wang, Z. Liu and J. Huang. Multimedia content analysis using audio and visual information. IEEE Signal Processing Magazine,2000,17(6):12-36
7. Boser BE, Guyon IM. et al. A training algorithm for optimal margin classifiers. In: Proceedings of the fifth annual workshop on Computational learning theory, 1992, 144-152

8. Guo, Guodong, Li, Stan Z. Content-based audio classification and retrieval by support vector machines. IEEE Transactions on Neural Networks, 2003, 14(1):209-215
9. T. M. Cover. Geometrical and statistical properties of systems and linear inequalities with applications in pattern recognition. IEEE Trans. on Electronic Computers, 1965,19:326-334
10. Wallhoff, F. et al. A comparison of discrete and continuous output modeling techniques for a Pseudo-2D Hidden Markov Model face recognition system. In: IEEE International Conference on Image Processing, 2001, 2:685-688
11. Jin Hui, Gao Wen. Analysis And Recognition Of Facial Expression Image Sequences Based On Hmm, Acta Automatica Sinica,2002,28(4):646-650
12. Platt,J.C., Probabilistic Outputs for Support Vector Machines for Pattern Recognition, U.Fayyad, Editor, 1999, Kluwer Academic Publishers: Boston
13. L.R.Rabiner, A Tutorial on Hidden Markov Models and Selected Applications in Speech Recognition. Proceedings of the IEEE,1989,77(2):257-286
14. Dempster, A., Laird, N., Rubin, D. Maximum likelihood from incomplete data via the EM algorithm. Journal of the Royal Statistical Society, 1977,39(Series B):1-38
15. A.V. Nefian, and M.H.Hayes. Hidden Markov Models for Face Recognition. In: IEEE International Conference on Acoustics, Speech and Signal Processing, 1998, 5:2721-2724

Texture Feature Extraction and Selection for Classification of Images in a Sequence

Khin Win[1], Sung Baik[1], Ran Baik[2], Sung Ahn[3], Sang Kim[1], and Yung Jo[1]

[1] College of Electronics and Information Engineering,
Sejong University, Seoul, Korea
{kkwin, sbaik, sue9868, joyungki}@sejong.ac.kr
[2] Dept. of Computer Engineering,
Honam University, Gwangju, Korea
baik@honam.ac.kr
[3] School of Management Information System,
Kookmin University, Seoul, Korea
sahn@kookmin.ac.kr

Abstract. This paper presents texture feature extraction and selection methods for on-line pattern classification evaluation. Feature selection for texture analysis plays a vital role in the field of image recognition. Despite many approaches done previously, this research is entirely different from them since it comes from the fundamental ideas of feature selection for image retrieval.. The proposed approach is capable of selecting the best features without recourse to classification and segmentation procedures. In this approach, probability density function estimation and a modified Bhattacharyya distance method are applied for clustering texture features of images in sequences and for comparing multi-distributed clusters with one another, respectively.

1 Introduction

Beyond the thriving of digital images, new challenges have arisen as difficult issues to get the robust characterization and efficient extraction of features from various natures of texture images for classification and segmentation. Ongoing approaches of image recognition rely on color, texture, shape, and object spatial relations. However, the difficult issue of image analysis is an investigation of very similar pattern discrimination techniques. It is difficult to discriminate those images digitally by a machine vision. In general, the five major categories to identify the image are statistical, geometrical, structural, model-based and signal processing [1]. The most recently used methods in signal processing are the Fourier, Gabor and Wavelet transform. Unlike the Fourier transform, wavelets have both scale aspect and time aspects. The major advantage of the Wavelet transform is the ability to perform local analysis whereas a serious drawback of the Fourier transform is a loss of time information in the frequency domain, that is, it has no spatial extent. It is trivial if the signal properties do not change much over time but in the real world the nature of signals can be found such as complex, sometimes self-similarity and fractal characteristics. And so,

R. Klette and J. Žunić (Eds.): IWCIA 2004, LNCS 3322, pp. 750–757, 2004.

triviality becomes nontrivial and the Fourier analysis could not address such cases. Wavelet transform overcomes these problems with its local extent property. The use of wavelets has developed in many fields for analyzing, synthesizing, de-noising, and compressing signals and images. The discrete wavelet transform (DWT) is a simple and intuitive method to discriminate the similar images.

In this paper, we describe feature extraction methods to characterize the very similar patterns with the support of the DWT, Gabor and Laws' filters. A feature selection method has been applied to a sequence of images. Since feature selection is a complex problem, we need to form the criteria to measure and compare the class separability of feature sets. Then, we have to choose the robust feature values from each feature set by using the selection criteria, and build up the most reliable feature vector based on these selected feature values. So we made many comparisons and used several statistical parameters to obtain the best feature values from each feature set. Experiments have conducted with texture images in a sequence (See Fig. 1) used in the previous research for the adaptive texture recognition [2]. In each image, four different texture areas in a scene are divided into two groups (areas A & D and B & C). Each group consists of two classes of similar texture. The similarities of texture areas within the same group are considered more difficult for classification.

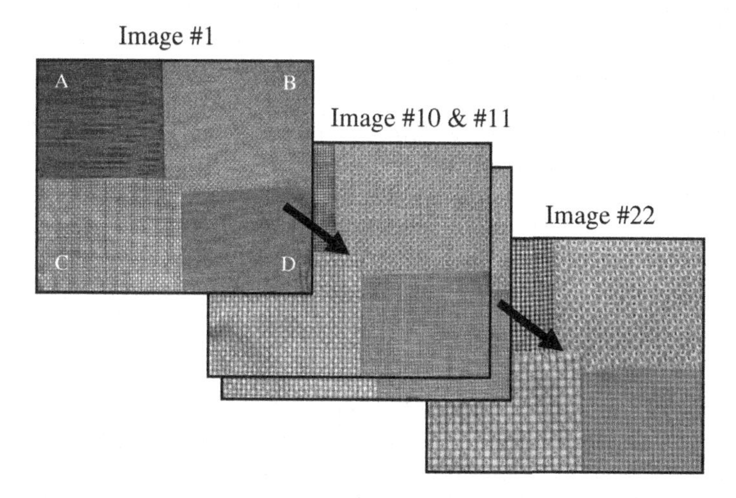

Fig. 1. A sequence of 22 texture images with four classes (In the first image, label A, B, C, and D represent different texture class areas)

2 Texture Feature Extractions

Each image is processed to extract texture features by using the most popular texture feature extraction methods, which are 1) Gabor spectral filtering [3], 2) Laws' energy filtering [4,5], and 3) Wavelet transform. Those methods have been widely used by researchers and perform very well for various classification and image segmentation tasks. The Wavelet transform has rich information such as scale, spatial, and frequency to represent texture features and is a robust and an efficient method, rich in functions and abilities for image processing. A pyramidal wavelet transform is a simple but reliable method to get the better performance. The Mother wavelet is formed by the Daubechies wavelet (db1, db2, db3) and the biorthogonal wavelet (bior1.3, bior2.4, bior3.7, bior4.4, bior5.5). An input image is decomposed into four sub-images at one, two and three scale.

In a pyramidal wavelet transform, an image is decomposed into four sub-images namely approximation (A), horizontal (H), vertical (V) and diagonal (D) at initial scale (one scale), and for the next scale decompositions, each approximation component is divided into four sub-images continually and goes down to desired scales (levels). All of these wavelet sub-images are called wavelet coefficients. Channel variances can be computed from each sub-image, and are used as texture features [6]. We can obtain 3n+1 sub-images for 'n' levels decompositions. Previous paper states that it can be computed as the subsequent procedures [7].

The most significant information for texture images appears only in middle frequency regions (horizontal and vertical frequency regions) [8]. For this reason, current work selects the horizontal and vertical sub-images for texture images and discards the unimportant frequency channels like diagonal and approximation. In addition, the performance of the wavelet transform procedures on input images at different level decompositions with several experimental learning and testing shows that strong discriminative power can be obtained from the horizontal and vertical sub-images. For this reason, the horizontal and vertical components at level one are selectable to discriminate the texture feature in a current scheme. This implies that the low frequency region may not necessarily contain significant information. The noticeable point to use the Daubechies wavelet is that the greater number of wavelet function (e.g. 'db16') can cause instability due to the nature of the wavelet translational invariant. Thus, the decomposition level (scale) and selection of sub-images (channels) are determined by the needs and nature of original texture images.

3 Texture Features Selection

Choosing the best feature set is one of the fundamental problems in texture classification and segmentation. The basic goals of feature selection are to identify which features are important in discriminating among texture classes and to choose the best subset of features for classification and segmentation.

Basically, the two quantitative approaches to the evaluation (selection) of image features are prototype performance and figure-of-merit [9]. In the prototype approach, a prototype image is classified by a classification procedure using various image features to be evaluated. The classification is then measured for each feature set. The best feature set provides the results in the least classification error. The problem of this approach is that the performance depends not only on the quality of features but also on the classification or segmentation ability of the classifier. The figure-of-merit approach uses some functional distance measurements between sets of image features. For example, a large distance implies a low classification error, and vice versa. The Bhattacharyya distance is a scalar function of the probability densities of features of a pair of classes.

In this work, the fundamental idea to select the best performance feature set depends on two main factors; 1) probability density function estimation and 2) Bhattacharyya distance, by the concept of which new criteria for feature selection are invented. The following are the detailed observations and discussions of selection and comparisons between different feature sets.

3.1 Probability Density Function Estimation (PDF)

A popular method used to observe the mixture distributions is probability density function estimation. The basic idea is that the density functions are used as the model of mixture distributions and class conditional density functions are used as the learning weights after applying the winning mechanism. The parameters of PDF are applicable to feature selection rules to discriminate the mixture distributions for each feature image.

The probability density function estimation yields significant information such as mean vectors, covariance matrices, mixing weights and the number of nodes for each mixture. The mixing weights less than 2% are discarded in principle [10]. Consequently, we can form a new feature set selection criterion by using these parameters and Bhattacharyya distance together..

3.2 Bhattacharyya Distance

Bhattacharyya distance is to measure the similarity between two distributions (groups) [7]. It can be expressed as:

$$\mu = \tfrac{1}{8}(M_2 - M_1)^T \left[\frac{\Sigma_1 - \Sigma_2}{2}\right]^{-1} (M_2 - M_1) + \tfrac{1}{2}\ln\frac{\left|\frac{\Sigma_1 + \Sigma_2}{2}\right|}{\sqrt{|\Sigma_1|}\sqrt{|\Sigma_2|}} \tag{2}$$

where μ = Bhattacharyya distance, M = mean vectors, Σ = covariance matrices. i = node number (Note: there are four nodes in each class)
M_{iA} = mean vectors of class A,
Σ_{iA} = variance matrices of class A,
M_{iB} = mean vectors of class B,
Σ_{iB} = variance matrices of class B.

We have to calculate the Bhattacharyya distances for two kinds of distributions; 1) normal distribution and 2) mixture distributions. In normal distribution, the Bhattacharyya distance is proportional to the Mahalanobis distance between the means of classes [11], while histogram representation contains only one mean and its related variance matrix. So far, the Bhattacharyya distance is well-suited for normal distribution with one mean and one variance. In mixture distributions, we modified the Bhattacharyya distance in order to measure the similarity between two classes with several distributions. The modified Bhattacharyya distance (See Fig. 2) is called Average Bhattacharyya Distance.

We can write out the mathematical expression for our proposed distance between two class distributions (e.g., class A and class B) by the following steps. First, we measure the individual distance between two nodes of class A and class B as follows:

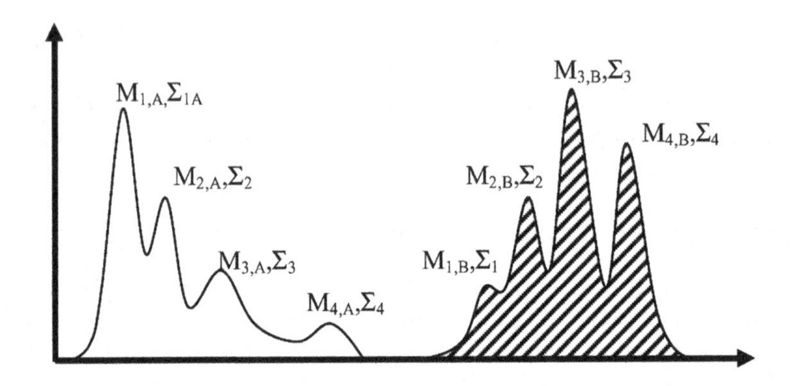

Fig. 2. Mixture distributions between class A and class B

$$d_{i,j}^{A,B} = \tfrac{1}{8}(M_{j,B} - M_{i,A})^T \left[\frac{\Sigma_{i,A} - \Sigma_{j,B}}{2}\right]^{-1} (M_{j,B} - M_{i,A}) + \tfrac{1}{2}\ln \frac{\left|\frac{\Sigma_{i,A} + \Sigma_{j,B}}{2}\right|}{\sqrt{|\Sigma_{i,A}|}\sqrt{|\Sigma_{j,B}|}} \tag{3}$$

where
$d_{i,j}^{A,B}$ = Bhattacharyya distance between class A at node i and class B at node j,
i, j = node index in class A and class B, respectively,
$M_{j,B}$, $\Sigma_{j,B}$ = mean and covariance of class B at node j,
$M_{i,A}$, $\Sigma_{i,A}$ = mean and covariance of class A at node i.

Second, we calculate the influence of each node,

$$d_{i,j}^{W} = d_{i,j}^{A,B}(w_i w_j) \tag{4}$$

where
$d_{i,j}^{W}$ = Weighted Bhattacharyya distance of class A and class B,
w_i, w_j = mixing weight at node i and j respectively,
$w_i, w_j \geq th1(e.g.,= 0.02)$ for each node.

Hence, the Average Bhattacharyya distance of two class distributions can be expressed as:

$$d_{avg}^{A,B} = \frac{1}{nm}\sum_{i=1}^{n}\sum_{j=1}^{m} d_{i,j}^{W} \tag{5}$$

where n, m = number of nodes in class A and class B, respectively.

3.3 Selection Criteria

Criteria for feature selection are considered in the following cases to discriminate several classes within each individual image, and to deal with a sequence of images with regard to the given class.

(1) Class Sensitivity: It concerns the difference of the same class between two consecutive images. The difference shows changes occurred due to different resolution or perceptual condition between these two images.
(2) Class Range: It indicates the similarity of the two mean values between different classes in each image and their class variances.
(3) Class Separability: It determines the separability of different classes according to the values of the Average Bhattacharyya distances. The high value of Average Bhattacharyya distances refers to strong separability.

The class measurement for these cases is as follows:

$$\left|difM_{i,i+1}^{Cl}\right| \leq th2(e.g.,=5) \tag{6}$$

$$\left|BD_{Cl_i,Cl_{i+1}}\right| \leq th3(e.g.,=0.3) \tag{7}$$

where

$N_i^{Cl} \leq \frac{(N_{\max}^{Cl} + N_{\min}^{Cl})_i}{2}$ with mixing weight $\geq th4(e.g.,=0.02)$,

$\left|difM_{i,i+1}^{Cl}\right|$= the difference of average mean values for same classes between two consecutive images (each mean value corresponds to each distribution of a class),

$\left|BD_{Cl_i,Cl_{i+1}}\right|$= Average Bhattacharyya distance between same classes between two consecutive images,

$\left|BD_{Cl_i,Cl'_{i+1}}\right|$= Average Bhattacharyya distance between different classes between two consecutive images,

N_i^{Cl}= number of nodes for each class, and

Cl, *Cl'* = class types (e.g. *Cl*= class A, *Cl'*= Class B).

4 Experimentation

Twenty two texture images in a sequence presented in Fig. 1 are used for experiments, in which we evaluate the performance of different feature extraction methods: 1) Gabor filter bank, 2) Laws' energy filter bank, and 3) Wavelet transform filter bank. The Wavelet bank has 24 wavelet filters. The Gabor and Laws' banks consist of 16 Gabor filters (8 orientations and 2 scales in the frequency domain) and 25 filters, respectively. The first step of each experiment is to locally select the best eight filters from each individual filter bank. The second step is to compare three feature sets obtained by those filters with each other and to finally select the best eight features for the efficient classification of images in a sequence.

According to feature selection rules, we tried to select the best feature set 1) to separate all classes immediately within each image, and 2) to strongly separate between some classes of images in a sequence by observing the graphical representation of mean value variations of each class for images in a sequence.

Table 1 presents the description of the eight features finally selected through several experiments. In particular, it is indisputable that the Wavelet transform is a preferable method to clearly discriminate the class A and D in each image whereas the Laws' energy filtering is very good to separate class B and C. Also, the Gabor filtering has strong discrimination on individual class separability in general and is also good in the class sensitivity. Therefore, the feature selection makes very good effects on the classification of images in a sequence when these eight features are well combined.

Table 1. Eight selected features and their specifications

Feature Set	Specification		
Gabor	Frequency, Angle		$2\sqrt{2}$, 45°
	Frequency, Angle		$2\sqrt{2}$, 135°
	Frequency, Angle		$4\sqrt{2}$, 0°
Laws'	E5S5	E5 = [-1 -2 0 2 1]	S5 = [-1 0 2 0 -1]
	S5E5	S5 = [-1 0 2 0 -1]	E5 = [-1 -2 0 2 1]
	L5E5	L5 = [1 4 6 4 1]	E5 = [-1 -2 0 2 1]
Wavelet	Wavelet Function		bior[1] 1.3
	Decomposition Level		3
	Wavelet Function		db[2] 3
	Decomposition Level		3

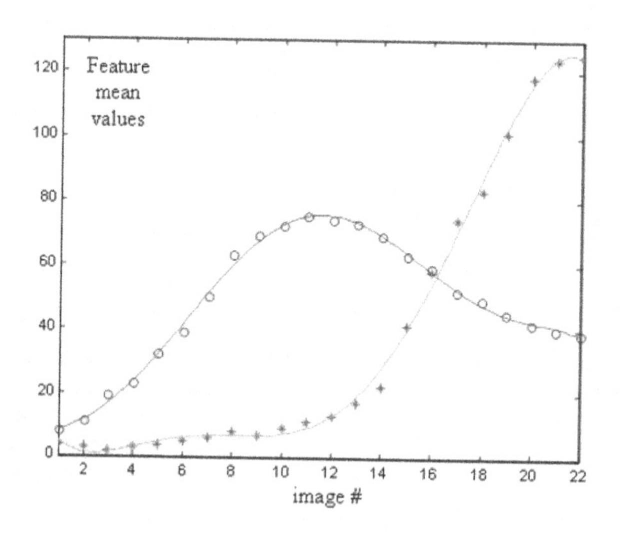

Fig. 3. The change of mean values of features obtained by one of Gabor filters over a sequence of images ('*' and 'o' marks indicates class A and C, respectively.)

According to the results of several experiments, the performance of each feature extraction is summarized as follows:

---

[1] Biorthogonal.
[2] Daubechies.

- Most of Wavelet filters are suitable to characterize the class A and D region in each image.
- E5S5, E5E5 and L5E5 of Laws' filters are well-suited to separate individual class discrimination for class B and class C.
- Some of Gabor filters provide smooth changes (with some patterns) of mean values of all classes over a sequence of images. In particular, class A and C are well separate along all images except image 16 as presented in Fig. 3.

References

1. T. Randen, Filtering for Texture Classification: A Comparative Study, IEEE Trans. Pattern Analysis and Machine Intelligence, Vol.. 21, No.. 4, April, 1999
2. S. W. Baik and P. Pachowicz, On-Line Model Modification Methodology for Adaptive Texture Recognition, IEEE Transactions on Systems, Man, and Cybernetics, Vol. 32, Issue. 7, 2002
3. M. Farrokhnia and A. Jain, A multi-channel filtering approach to texture segmentation, Proceedings of IEEE Computer Vision and Pattern Recognition Conference, pp. 346-370, 1990
4. M. Chantler, The effect of variation in illuminant direction on texture classification, Ph D Thesis, Dept. Computing and Electrical Engineering, Heriot-Watt University, 1994
5. K. Laws, Textured image segmentation, Ph.D. Thesis. Dept. of Electrical Engineering, University of Southern California, Los Angeles, 1980
6. C.. C. Chen, Filtering Methods for Texture Discrimination, Pattern Recognition Letters, Vol.. 20, pp. 783-790, 1999
7. O. Pichler, A. Teuner and B. J. Hosticka, A Comparison of Texture Feature Extraction using Adaptive Gabor Filtering, Pyramidal and Tree Structured Wavelet Transforms, Elsevier Science on Pattern Recognition, Vol. 29, No.. 5, pp.. 733-742, 1996
8. T. Chang, Texture Analysis and Classification with Tree-Structured Wavelet Transform, IEEE Trans. on Image Processing, Vol.. 2, No.. 4, pp. 429-441, October, 1993
9. K. P. William, Digital Image Processing, John Wiley & Sons, Inc, New York, NY 10158-0012, 2001
10. H. Yin and N. M. Allinson, Self-Organizing Mixture Networks for Probability Density Modelling, Neural Networks Proceedings, The IEEE World Congress on Computational Intelligence, Vol. 3, pp.. 2277-2281, May, 1998
11. B. Huet and E. R. Hancock, Cartographic Indexing into a Database of Remotely Sensed Images, Depart of Computer Science, University of York, Y01 5DD, UK

Author Index

Lecture Notes in Computer Science

For information about Vols. 1–3222

please contact your bookseller or Springer

Vol. 3337: J.M. Barreiro, F. Martin-Sanchez, V. Maojo, F. Sanz (Eds.), Biological and Medical Data Analysis. XI, 508 pages. 2004.

Vol. 3333: K. Aizawa, Y. Nakamura, S. Satoh (Eds.), Advances in Multimedia Information Processing - PCM 2004, Part III. XXXV, 785 pages. 2004.

Vol. 3332: K. Aizawa, Y. Nakamura, S. Satoh (Eds.), Advances in Multimedia Information Processing - PCM 2004, Part II. XXXVI, 1051 pages. 2004.

Vol. 3331: K. Aizawa, Y. Nakamura, S. Satoh (Eds.), Advances in Multimedia Information Processing - PCM 2004, Part I. XXXVI, 667 pages. 2004.

Vol. 3323: G. Antoniou, H. Boley (Eds.), Rules and Rule Markup Languages for the Semantic Web. X, 215 pages. 2004.

Vol. 3322: R. Klette, J. Žunić (Eds.), Combinatorial Image Analysis. XII, 760 pages. 2004.

Vol. 3316: N.R. Pal, N. Kasabov, R.K. Mudi, S. Pal, S.K. Parui (Eds.), Neural Information Processing. XXX, 1368 pages. 2004.

Vol. 3315: C. Lemaître, C.A. Reyes, J.A. González (Eds.), Advances in Artificial Intelligence – IBERAMIA 2004. XX, 987 pages. 2004. (Subseries LNAI).

Vol. 3312: A.J. Hu, A.K. Martin (Eds.), Formal Methods in Computer-Aided Design. XI, 445 pages. 2004.

Vol. 3311: V. Roca, F. Rousseau (Eds.), Interactive Multimedia and Next Generation Networks. XIII, 287 pages. 2004.

Vol. 3309: C.-H. Chi, K.-Y. Lam (Eds.), Content Computing. XII, 510 pages. 2004.

Vol. 3308: J. Davies, W. Schulte, M. Barnett (Eds.), Formal Methods and Software Engineering. XIII, 500 pages. 2004.

Vol. 3307: C. Bussler, S.-k. Hong, W. Jun, R. Kaschek, Kinshuk, S. Krishnaswamy, S.W. Loke, D. Oberle, D. Richards, A. Sharma, Y. Sure, B. Thalheim (Eds.), Web Information Systems – WISE 2004 Workshops. XV, 277 pages. 2004.

Vol. 3306: X. Zhou, S. Su, M.P. Papazoglou, M.E. Orlowska, K.G. Jeffery (Eds.), Web Information Systems – WISE 2004. XVII, 745 pages. 2004.

Vol. 3305: P.M.A. Sloot, B. Chopard, A.G. Hoekstra (Eds.), Cellular Automata. XV, 883 pages. 2004.

Vol. 3303: J.A. López, E. Benfenati, W. Dubitzky (Eds.), Knowledge Exploration in Life Science Informatics. X, 249 pages. 2004. (Subseries LNAI).

Vol. 3302: W.-N. Chin (Ed.), Programming Languages and Systems. XIII, 453 pages. 2004.

Vol. 3299: F. Wang (Ed.), Automated Technology for Verification and Analysis. XII, 506 pages. 2004.

Vol. 3298: S.A. McIlraith, D. Plexousakis, F. van Harmelen (Eds.), The Semantic Web – ISWC 2004. XXI, 841 pages. 2004.

Vol. 3295: P. Markopoulos, B. Eggen, E. Aarts, J.L. Crowley (Eds.), Ambient Intelligence. XIII, 388 pages. 2004.

Vol. 3294: C.N. Dean, R.T. Boute (Eds.), Teaching Formal Methods. X, 249 pages. 2004.

Vol. 3293: C.-H. Chi, M. van Steen, C. Wills (Eds.), Web Content Caching and Distribution. IX, 283 pages. 2004.

Vol. 3292: R. Meersman, Z. Tari, A. Corsaro (Eds.), On the Move to Meaningful Internet Systems 2004: OTM 2004 Workshops. XXIII, 885 pages. 2004.

Vol. 3291: R. Meersman, Z. Tari (Eds.), On the Move to Meaningful Internet Systems 2004: CoopIS, DOA, and ODBASE, Part II. XXV, 824 pages. 2004.

Vol. 3290: R. Meersman, Z. Tari (Eds.), On the Move to Meaningful Internet Systems 2004: CoopIS, DOA, and ODBASE, Part I. XXV, 823 pages. 2004.

Vol. 3289: S. Wang, K. Tanaka, S. Zhou, T.W. Ling, J. Guan, D. Yang, F. Grandi, E. Mangina, I.-Y. Song, H.C. Mayr (Eds.), Conceptual Modeling for Advanced Application Domains. XXII, 692 pages. 2004.

Vol. 3288: P. Atzeni, W. Chu, H. Lu, S. Zhou, T.W. Ling (Eds.), Conceptual Modeling – ER 2004. XXI, 869 pages. 2004.

Vol. 3287: A. Sanfeliu, J.F. Martínez Trinidad, J.A. Carrasco Ochoa (Eds.), Progress in Pattern Recognition, Image Analysis and Applications. XVII, 703 pages. 2004.

Vol. 3286: G. Karsai, E. Visser (Eds.), Generative Programming and Component Engineering. XIII, 491 pages. 2004.

Vol. 3285: S. Manandhar, J. Austin, U. Desai, Y. Oyanagi, A. Talukder (Eds.), Applied Computing. XII, 334 pages. 2004.

Vol. 3284: A. Karmouch, L. Korba, E.R.M. Madeira (Eds.), Mobility Aware Technologies and Applications. XII, 382 pages. 2004.

Vol. 3283: F.A. Aagesen, C. Anutariya, V. Wuwongse (Eds.), Intelligence in Communication Systems. XIII, 327 pages. 2004.

Vol. 3281: T. Dingsøyr (Ed.), Software Process Improvement. X, 207 pages. 2004.

Vol. 3280: C. Aykanat, T. Dayar, İ. Körpeoğlu (Eds.), Computer and Information Sciences - ISCIS 2004. XVIII, 1009 pages. 2004.

Vol. 3278: A. Sahai, F. Wu (Eds.), Utility Computing. XI, 272 pages. 2004.

Vol. 3274: R. Guerraoui (Ed.), Distributed Computing. XIII, 465 pages. 2004.

Vol. 3273: T. Baar, A. Strohmeier, A. Moreira, S.J. Mellor (Eds.), <<UML>> 2004 - The Unified Modelling Language. XIII, 454 pages. 2004.

Vol. 3271: J. Vicente, D. Hutchison (Eds.), Management of Multimedia Networks and Services. XIII, 335 pages. 2004.

Vol. 3270: M. Jeckle, R. Kowalczyk, P. Braun (Eds.), Grid Services Engineering and Management. X, 165 pages. 2004.

Vol. 3269: J. Lopez, S. Qing, E. Okamoto (Eds.), Information and Communications Security. XI, 564 pages. 2004.

Vol. 3266: J. Solé-Pareta, M. Smirnov, P.V. Mieghem, J. Domingo-Pascual, E. Monteiro, P. Reichl, B. Stiller, R.J. Gibbens (Eds.), Quality of Service in the Emerging Networking Panorama. XVI, 390 pages. 2004.

Vol. 3265: R.E. Frederking, K.B. Taylor (Eds.), Machine Translation: From Real Users to Research. XI, 392 pages. 2004. (Subseries LNAI).

Vol. 3264: G. Paliouras, Y. Sakakibara (Eds.), Grammatical Inference: Algorithms and Applications. XI, 291 pages. 2004. (Subseries LNAI).

Vol. 3263: M. Weske, P. Liggesmeyer (Eds.), Object-Oriented and Internet-Based Technologies. XII, 239 pages. 2004.

Vol. 3262: M.M. Freire, P. Chemouil, P. Lorenz, A. Gravey (Eds.), Universal Multiservice Networks. XIII, 556 pages. 2004.

Vol. 3261: T. Yakhno (Ed.), Advances in Information Systems. XIV, 617 pages. 2004.

Vol. 3260: I.G.M.M. Niemegeers, S.H. de Groot (Eds.), Personal Wireless Communications. XIV, 478 pages. 2004.

Vol. 3258: M. Wallace (Ed.), Principles and Practice of Constraint Programming – CP 2004. XVII, 822 pages. 2004.

Vol. 3257: E. Motta, N.R. Shadbolt, A. Stutt, N. Gibbins (Eds.), Engineering Knowledge in the Age of the Semantic Web. XVII, 517 pages. 2004. (Subseries LNAI).

Vol. 3256: H. Ehrig, G. Engels, F. Parisi-Presicce, G. Rozenberg (Eds.), Graph Transformations. XII, 451 pages. 2004.

Vol. 3255: A. Benczúr, J. Demetrovics, G. Gottlob (Eds.), Advances in Databases and Information Systems. XI, 423 pages. 2004.

Vol. 3254: E. Macii, V. Paliouras, O. Koufopavlou (Eds.), Integrated Circuit and System Design. XVI, 910 pages. 2004.

Vol. 3253: Y. Lakhnech, S. Yovine (Eds.), Formal Techniques, Modelling and Analysis of Timed and Fault-Tolerant Systems. X, 397 pages. 2004.

Vol. 3252: H. Jin, Y. Pan, N. Xiao, J. Sun (Eds.), Grid and Cooperative Computing - GCC 2004 Workshops. XVIII, 785 pages. 2004.

Vol. 3251: H. Jin, Y. Pan, N. Xiao, J. Sun (Eds.), Grid and Cooperative Computing - GCC 2004. XXII, 1025 pages. 2004.

Vol. 3250: L.-J. (LJ) Zhang, M. Jeckle (Eds.), Web Services. X, 301 pages. 2004.

Vol. 3249: B. Buchberger, J.A. Campbell (Eds.), Artificial Intelligence and Symbolic Computation. X, 285 pages. 2004. (Subseries LNAI).

Vol. 3246: A. Apostolico, M. Melucci (Eds.), String Processing and Information Retrieval. XIV, 332 pages. 2004.

Vol. 3245: E. Suzuki, S. Arikawa (Eds.), Discovery Science. XIV, 430 pages. 2004. (Subseries LNAI).

Vol. 3244: S. Ben-David, J. Case, A. Maruoka (Eds.), Algorithmic Learning Theory. XIV, 505 pages. 2004. (Subseries LNAI).

Vol. 3243: S. Leonardi (Ed.), Algorithms and Models for the Web-Graph. VIII, 189 pages. 2004.

Vol. 3242: X. Yao, E. Burke, J.A. Lozano, J. Smith, J.J. Merelo-Guervós, J.A. Bullinaria, J. Rowe, P. Tiňo, A. Kabán, H.-P. Schwefel (Eds.), Parallel Problem Solving from Nature - PPSN VIII. XX, 1185 pages. 2004.

Vol. 3241: D. Kranzlmüller, P. Kacsuk, J.J. Dongarra (Eds.), Recent Advances in Parallel Virtual Machine and Message Passing Interface. XIII, 452 pages. 2004.

Vol. 3240: I. Jonassen, J. Kim (Eds.), Algorithms in Bioinformatics. IX, 476 pages. 2004. (Subseries LNBI).

Vol. 3239: G. Nicosia, V. Cutello, P.J. Bentley, J. Timmis (Eds.), Artificial Immune Systems. XII, 444 pages. 2004.

Vol. 3238: S. Biundo, T. Frühwirth, G. Palm (Eds.), KI 2004: Advances in Artificial Intelligence. XI, 467 pages. 2004. (Subseries LNAI).

Vol. 3236: M. Núñez, Z. Maamar, F.L. Pelayo, K. Pousttchi, F. Rubio (Eds.), Applying Formal Methods: Testing, Performance, and M/E-Commerce. XI, 381 pages. 2004.

Vol. 3235: D. de Frutos-Escrig, M. Nunez (Eds.), Formal Techniques for Networked and Distributed Systems – FORTE 2004. X, 377 pages. 2004.

Vol. 3234: M.J. Egenhofer, C. Freksa, H.J. Miller (Eds.), Geographic Information Science. VIII, 345 pages. 2004.

Vol. 3233: K. Futatsugi, F. Mizoguchi, N. Yonezaki (Eds.), Software Security - Theories and Systems. X, 345 pages. 2004.

Vol. 3232: R. Heery, L. Lyon (Eds.), Research and Advanced Technology for Digital Libraries. XV, 528 pages. 2004.

Vol. 3231: H.-A. Jacobsen (Ed.), Middleware 2004. XV, 514 pages. 2004.

Vol. 3230: J.L. Vicedo, P. Martínez-Barco, R. Muñoz, M. Saiz Noeda (Eds.), Advances in Natural Language Processing. XII, 488 pages. 2004. (Subseries LNAI).

Vol. 3229: J.J. Alferes, J. Leite (Eds.), Logics in Artificial Intelligence. XIV, 744 pages. 2004. (Subseries LNAI).

Vol. 3226: M. Bouzeghoub, C. Goble, V. Kashyap, S. Spaccapietra (Eds.), Semantics of a Networked World. XIII, 326 pages. 2004.

Vol. 3225: K. Zhang, Y. Zheng (Eds.), Information Security. XII, 442 pages. 2004.

Vol. 3224: E. Jonsson, A. Valdes, M. Almgren (Eds.), Recent Advances in Intrusion Detection. XII, 315 pages. 2004.

Vol. 3223: K. Slind, A. Bunker, G. Gopalakrishnan (Eds.), Theorem Proving in Higher Order Logics. VIII, 337 pages. 2004.